i CLASSZONE.COM

Looking for ways to integrate the Web into your curriculum?

ClassZone, McDougal Littell's textbook-companion Web site, is the solution! Online teaching support for you and engaging, interactive content for your students!

ClassZone is your online guide to *Geometry, Concepts and Skills*

- Additional worked-out examples to supplement lessons
- Help getting started with homework exercises
- Career profiles and more about real-world applications
- Interactive review games
- Extra challenge problems
- Internet research tutorial
- Teacher Center for classroom planning

Log on to ClassZone at
www.classzone.com

With the purchase of *Geometry, Concepts and Skills* you have immediate access to ClassZone.

Teacher Access Code

MCD2WZXLQ24HY

Use this code to create your own user name and password. Then access both teacher only and student resources.

Geometry
Concepts and Skills

Teacher's Edition

Ron Larson

Laurie Boswell

Lee Stiff

McDougal Littell
A HOUGHTON MIFFLIN COMPANY

Evanston, Illinois • Boston • Dallas

About Geometry: Concepts and Skills

This book has been written so that all students can understand geometry. The course focuses on the key topics that provide a strong foundation in the essentials of geometry. Lesson concepts are presented in a clear, straightforward manner, supported by frequent worked-out examples. The page format makes it easy for students to follow the flow of a lesson, and the vocabulary and visual tips in the margins help students learn how to read the text and diagrams. Checkpoint questions within lessons give students a way to check their understanding as they go along. The exercises for each lesson provide many opportunities to practice and maintain skills, as well as to apply concepts to real-world problems.

ISBN: 0-618-14051-4   3456789–DWO–05 04 03

Internet Web Site: http://www.mcdougallittell.com

Geometry
Concepts and Skills

Teacher's Edition

Contents

About the Authors

RON LARSON is a professor of mathematics at Penn State University at Erie, where he has taught since receiving his Ph.D. in mathematics from the University of Colorado in 1970. He is the author of a broad range of instructional materials for middle school, high school, and college. Dr. Larson has been an innovative writer of multimedia approaches to mathematics, and his Calculus and Precalculus texts are both available in interactive form on the Internet.

LAURIE BOSWELL is a mathematics teacher at Profile Junior-Senior High School in Bethlehem, New Hampshire. A recipient of the 1986 Presidential Award for Excellence in Mathematics Teaching, she is also the 1992 Tandy Technology Scholar and the 1991 recipient of the Richard Balomenos Mathematics Education Service Award from the New Hampshire Association of Teachers of Mathematics. She has had leadership positions in state and NCTM committees.

LEE STIFF is a professor of mathematics education in the College of Education of North Carolina State University at Raleigh and has taught mathematics at the high school and middle school levels. He has served on the NCTM Board of Directors and was elected President of NCTM for the years 2000–2002. He is the 1992 recipient of the W. W. Rankin Award for Excellence in Mathematics Education presented by the North Carolina Council of Teachers of Mathematics.

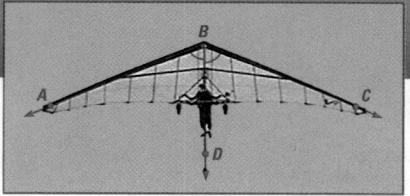 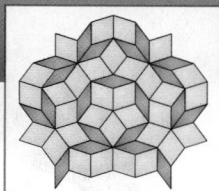

Chapter 1 Basics of Geometry

Contents **v**

Chapter 2 Segments and Angles

Chapter 3

Parallel and Perpendicular Lines

Chapter 4 Triangle Relationships

Chapter 5 Congruent Triangles

Chapter 6 Quadrilaterals

x **Contents**

Chapter 7 Similarity

STUDENT HELP
....

Visualize It! *356, 359, 373, 381, 384, 394*
Study Tip *357, 367, 374, 380, 394, 395*
Vocabulary Tip *365*
Skills Review *358*
Test Tip *405*

APPLICATION HIGHLIGHTS
....

Batting Average *357*
Map Scale *360*
Mural *370*
Aspect Ratio *371*
Hockey *374*
A-Frame *383*
Fractals *391*
Flashlight Image *397*

INTERNET
....

360, 363, 366, 370, 376, 377, 383, 387, 391, 397

Chapter 8 Polygons and Area

xii Contents

Chapter 9 Surface Area and Volume

Chapter 10

Right Triangles and Trigonometry

Chapter 11 ▸ Circles

Contents of
Student Resources

Program Overview

Build your students' **confidence** with the most **accessible** program available.

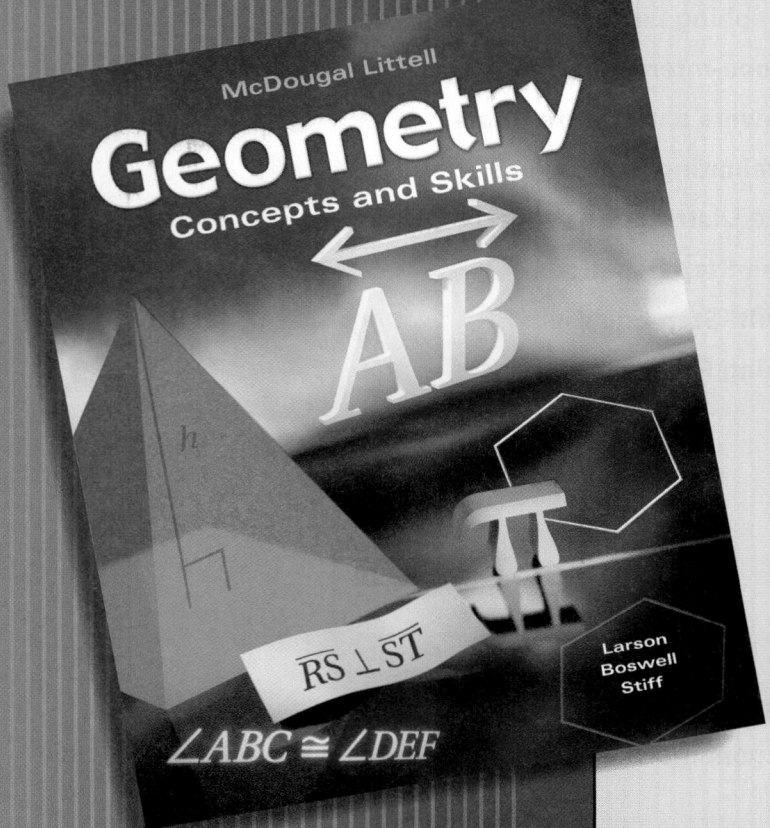

McDougal Littell

Geometry
Concepts and Skills

Because every student absolutely can understand geometry

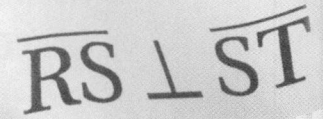

- ✓ Give your students the opportunity to read, see, and practice geometry skills with **manageable pacing**.

- ✓ Create a lasting impression with **visual strategies** that help students remember key ideas.

- ✓ Help students understand concepts with **built-in learning support**.

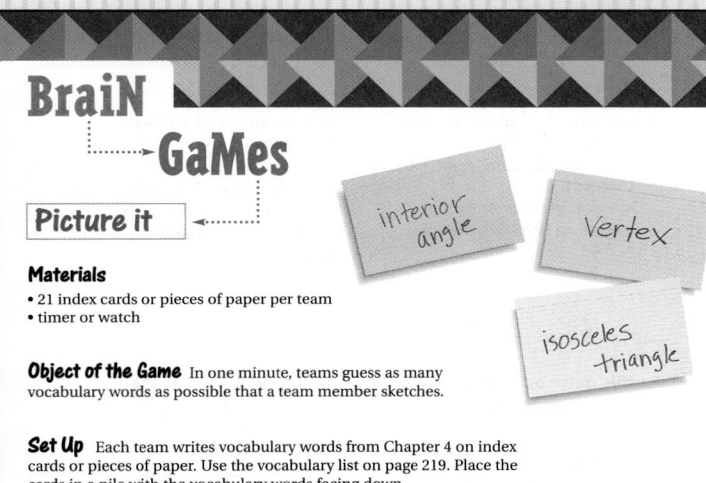

BraiN →GaMes

Picture it

interior angle

Vertex

isosceles triangle

Materials
- 21 index cards or pieces of paper per team
- timer or watch

Object of the Game In one minute, teams guess as many vocabulary words as possible that a team member sketches.

Set Up Each team writes vocabulary words from Chapter 4 on index cards or pieces of paper. Use the vocabulary list on page 219. Place the cards in a pile with the vocabulary words facing down.

How to Play

Step 1 ▶ Each team chooses a player to sketch.

Step 2 ▶ The person who is sketching, looks at the word on the top card, without anyone else seeing it. A one minute timer is set. The sketcher draws a picture of the word on the other side of the card.

Step 3 ▶ When a team member guesses correctly, the sketcher goes to the next card.

Step 4 ▶ At the end of the minute, each card is checked to see if the word was drawn correctly. Record a point for each correct card.

Step 5 ▶ Play continues until each team member has had a turn to sketch. The team with the most points wins.

Another Way to Play Extend each turn to three minutes. Instead of just saying the vocabulary word, team members also have to write the definition on the card.

EXAMPLE 1 Use the Base Angles Theorem

Find the measure of ∠L.

Solution

Angle L is a base angle of an isosceles triangle. From the Base Angles Theorem, ∠L and ∠N have the same measure.

ANSWER ▶ The measure of ∠L is 52°.

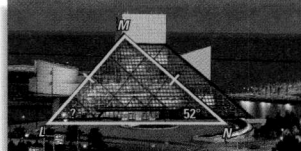

Rock and Roll Hall of Fame, Cleveland, Ohio

THEOREM 4.4

Converse of the Base Angles Theorem

Words If two angles of a triangle are congruent, then the sides opposite them are congruent.

Symbols If ∠B ≅ ∠C, then $\overline{AC} \cong \overline{AB}$.

Visualize It!

Base angles don't have to be on the bottom of an isosceles triangle.

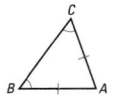

EXAMPLE 2 Use the Converse of the Base Angles Theorem

Find the value of x.

Solution

By the Converse of the Base Angles Theorem, the legs have the same length.

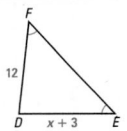

$DE = DF$	Converse of the Base Angles Theorem
$x + 3 = 12$	Substitute $x + 3$ for DE and 12 for DF.
$x = 9$	Subtract 3 from each side.

ANSWER ▶ The value of x is 9.

Checkpoint ✓ Use Isosceles Triangle Theorems

Find the value of y.

1. 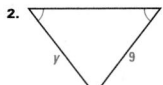 2. 3.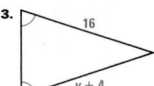

Geometry
Concepts and Skills
An understandable solution

The proof is in

Preview It!

A **Preview** provides a chapter overview that helps students make connections between what they know and what they are going to learn.

Prepare for It!

The **Chapter Readiness Quiz** helps teachers assess prerequisite skills and prepare students for the chapter.

Visualize It!

The **Visual Strategy** feature offers students a visual study tip that helps them see and remember concepts.

Chapter  4 Study Guide

PREVIEW What's the chapter about?

- Classifying triangles and finding their angle measures
- Using the Distance Formula, the Pythagorean Theorem, and its converse
- Showing relationships between a triangle's sides and angles

Key Words

- equilateral, isosceles, scalene triangles, *p. 173*
- equiangular, acute, right, obtuse triangles, *p. 174*
- interior, exterior angles, *p. 181*
- legs of an isosceles triangle, *p. 185*

- base angles of an isosceles triangle, *p. 185*
- hypotenuse, *p. 192*
- Pythagorean Theorem, *p. 192*
- Distance Formula, *p. 194*
- median of a triangle, *p. 207*
- centroid, *p. 208*

PREPARE Chapter Readiness Quiz

Take this quick quiz. If you are unsure of an answer, look at the reference pages for help.

Vocabulary Check *(refer to p. 61)*

1. In the figure shown, $\overrightarrow{BD}$ is the angle bisector of $\angle ABC$. What is the value of x?

    Ⓐ 10   **Ⓑ** 15   **Ⓒ** 20   **Ⓓ** 30

Skill Check *(refer to pp. 30, 55)*

2. What is the distance between $P(2, 3)$ and $Q(7, 3)$?

    Ⓕ 3   **Ⓖ** 4   **Ⓗ** 5   **Ⓙ** 7

3. What is the midpoint of a segment with endpoints $A(0, 2)$ and $B(6, 4)$?

    Ⓐ (3, 2)   **Ⓑ** (3, 3)   **Ⓒ** (4, 2)   **Ⓓ** (0, 3)

VISUAL STRATEGY Drawing Triangles

Visualize It! → When you sketch a triangle, try to make the angles roughly the correct size.

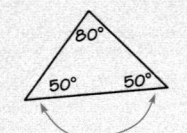
These angles are the same in an isosceles triangle.

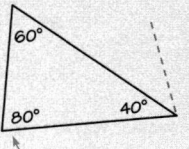
This 80° angle is twice as big as the 40° angle.

the pages

4.3 Isosceles and Equilateral Triangles

Goal
Use properties of isosceles and equilateral triangles.

Key Words
• legs of an isosceles triangle
• base of an isosceles triangle
• base angles

Geo-Activity — Properties of Isosceles Triangles

① Fold a sheet of paper in half. Use a straightedge to draw a line from the fold to the bottom edge. Cut along the line.

② Unfold and label the angles as shown. Use a protractor to measure ∠H and ∠K. What do you notice?

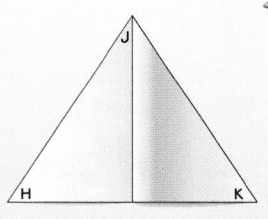

③ Repeat Steps 1 and 2 for different isosceles triangles. What can you say about ∠H and ∠K in the different triangles?

Try It!

Geo-Activity gives students the opportunity to perform concrete activities that demonstrate the concepts they are learning

Student Help

The Geo-Activity shows that two angles of an isosceles triangle are always congruent. These angles are opposite the congruent sides.

The congruent sides of an isosceles triangle are called **legs** .

The other side is called the **base** .

The two angles at the base of the triangle are called the **base angles** .

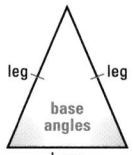

Isosceles Triangle

Read It!

Vocabulary Tip makes the language of geometry accessible to all students.

THEOREM 4.3

Base Angles Theorem

Words If two sides of a triangle are congruent, then the angles opposite them are congruent.

Symbols If $\overline{AB} \cong \overline{AC}$, then $\angle C \cong \angle B$.

Understand It!

Theorems are presented verbally, visually, and symbolically so students can see the concepts, read the concepts, and know how to write the concepts.

Getting Ready
A Guide to Student Help

Each chapter begins with a Study Guide

CHAPTER PREVIEW .... gives an overview of what you will be learning.

KEY WORDS .... lists important new words in the chapter.

READINESS QUIZ .... checks your understanding of words and skills that you will use in the chapter, and tells you where to go for review.

VISUAL STRATEGY .... suggests a visual way to make your learning easier.

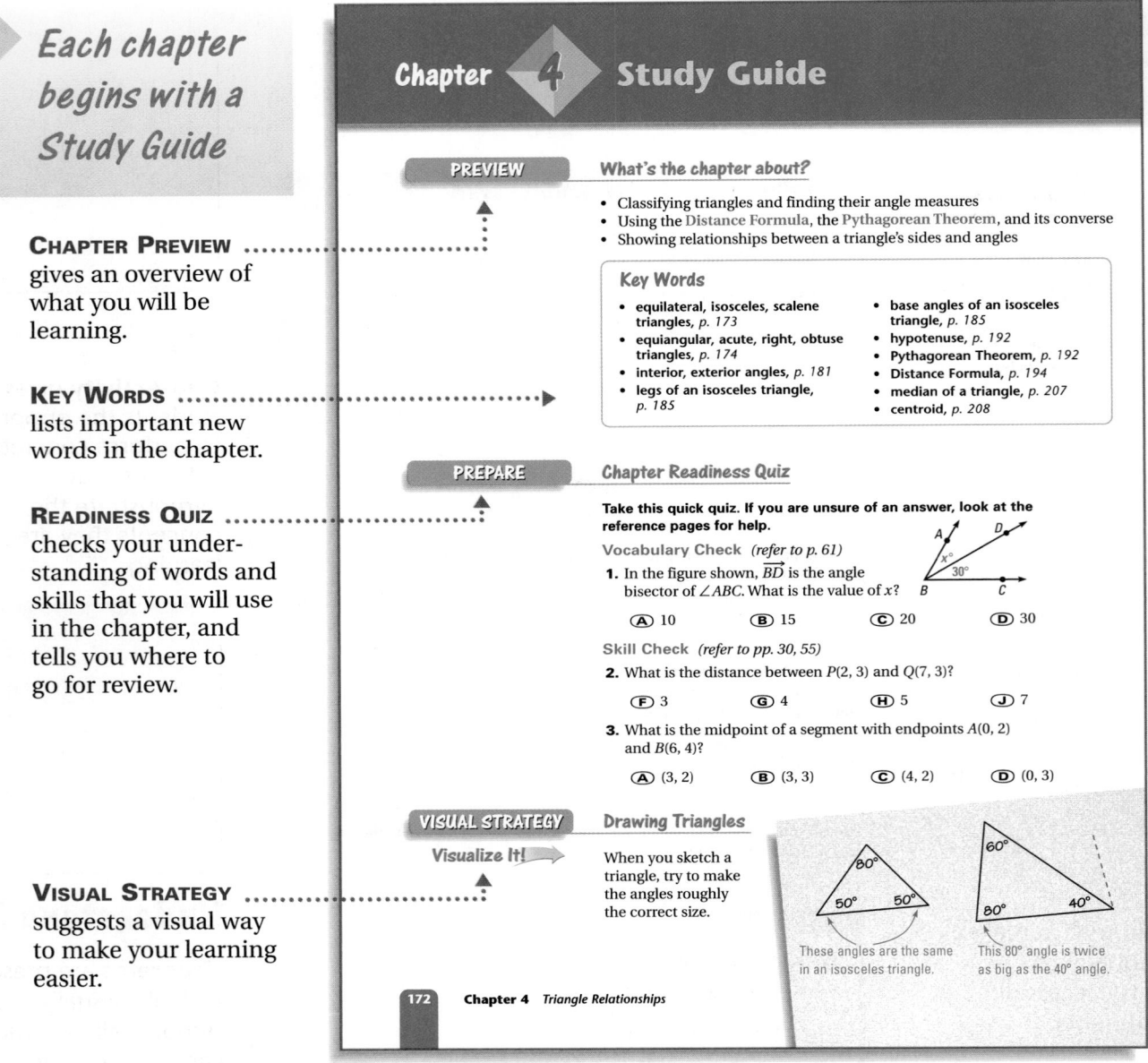

Chapter **4** Study Guide

PREVIEW — What's the chapter about?

- Classifying triangles and finding their angle measures
- Using the Distance Formula, the Pythagorean Theorem, and its converse
- Showing relationships between a triangle's sides and angles

Key Words

- equilateral, isosceles, scalene triangles, *p. 173*
- equiangular, acute, right, obtuse triangles, *p. 174*
- interior, exterior angles, *p. 181*
- legs of an isosceles triangle, *p. 185*
- base angles of an isosceles triangle, *p. 185*
- hypotenuse, *p. 192*
- Pythagorean Theorem, *p. 192*
- Distance Formula, *p. 194*
- median of a triangle, *p. 207*
- centroid, *p. 208*

PREPARE — Chapter Readiness Quiz

Take this quick quiz. If you are unsure of an answer, look at the reference pages for help.

Vocabulary Check *(refer to p. 61)*

1. In the figure shown, $\overrightarrow{BD}$ is the angle bisector of $\angle ABC$. What is the value of x?

     A 10   **B** 15   **C** 20   **D** 30

Skill Check *(refer to pp. 30, 55)*

2. What is the distance between $P(2, 3)$ and $Q(7, 3)$?

     F 3   **G** 4   **H** 5   **J** 7

3. What is the midpoint of a segment with endpoints $A(0, 2)$ and $B(6, 4)$?

     A (3, 2)   **B** (3, 3)   **C** (4, 2)   **D** (0, 3)

VISUAL STRATEGY — Drawing Triangles

Visualize It! When you sketch a triangle, try to make the angles roughly the correct size.

These angles are the same in an isosceles triangle.

This 80° angle is twice as big as the 40° angle.

172   **Chapter 4**   *Triangle Relationships*

Student Help Notes

········ **VOCABULARY TIPS**
explain the meaning
and origin of words.

STUDY TIPS help you ········
understand concepts
and avoid errors.

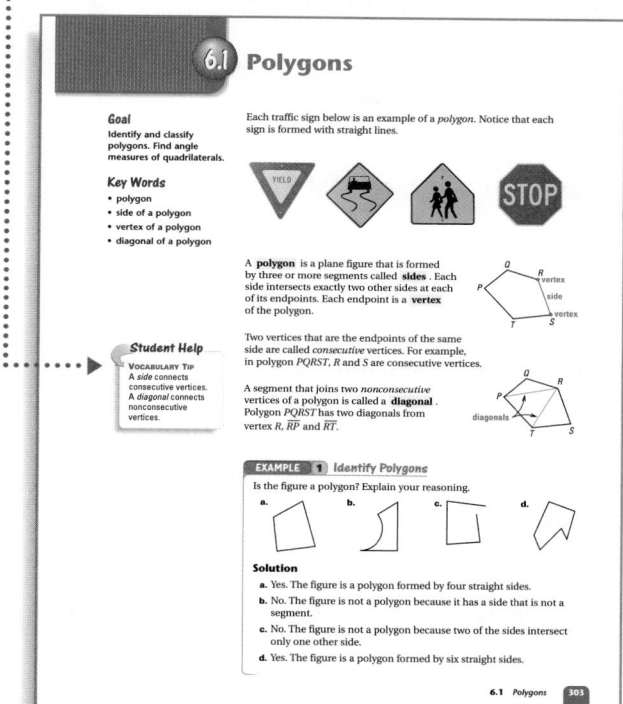

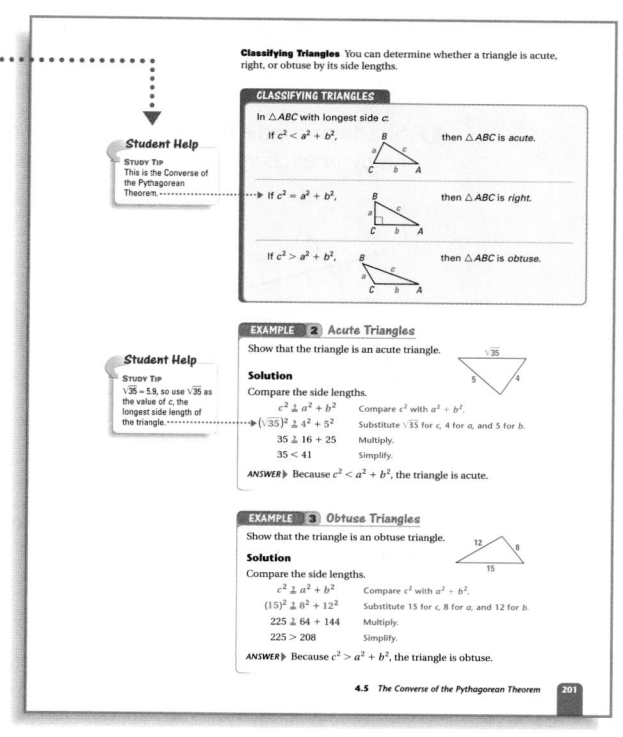

Other Notes Included Are ...

- **READING TIPS**   • **SKILLS REVIEW**
- **LOOK BACK**   • **EXTRA PRACTICE**

Internet References

MORE EXAMPLES
are at classzone.com.

HOMEWORK HELP
for some exercises is
at classzone.com.

Student Help
CLASSZONE.COM

MORE EXAMPLES
More examples at
classzone.com

Homework Help

HOMEWORK HELP chart
shows which examples
will help you with which
exercises.

Homework Help

Example 1: Exs. 11–16
Example 2: Exs. 17–41
Example 3: Exs. 42–47

Visualizing Geometry

VISUAL PRESENTATIONS of lessons use photographs and diagrams to develop concepts. Color is used in text and diagrams to assist learning.

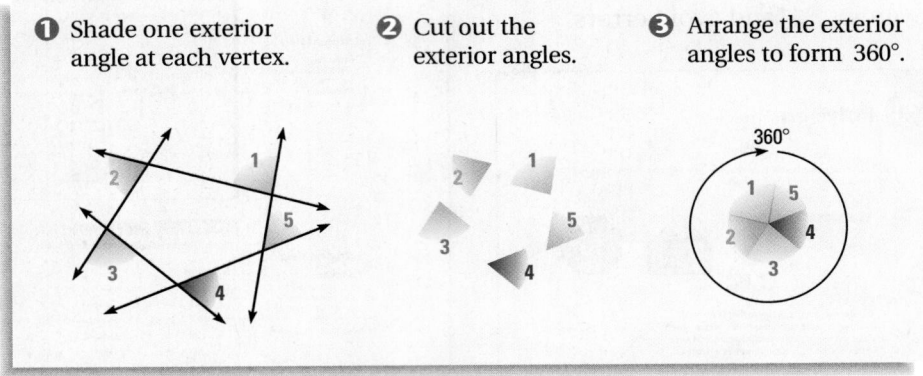

❶ Shade one exterior angle at each vertex.

❷ Cut out the exterior angles.

❸ Arrange the exterior angles to form 360°.

VISUALIZE IT! NOTES illustrate concepts and provide helpful hints for remembering ideas.

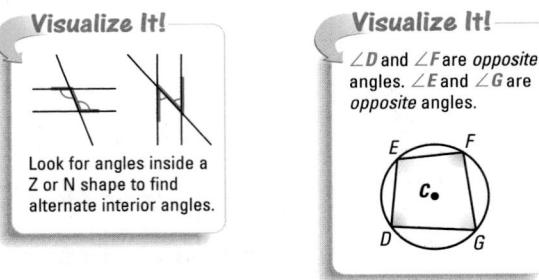

Visualize It!

Look for angles inside a Z or N shape to find alternate interior angles.

Visualize It!

∠D and ∠F are *opposite* angles. ∠E and ∠G are *opposite* angles.

VISUALIZE IT! EXERCISES ask you to make sketches that match descriptions, analyze diagrams, and draw conclusions from them.

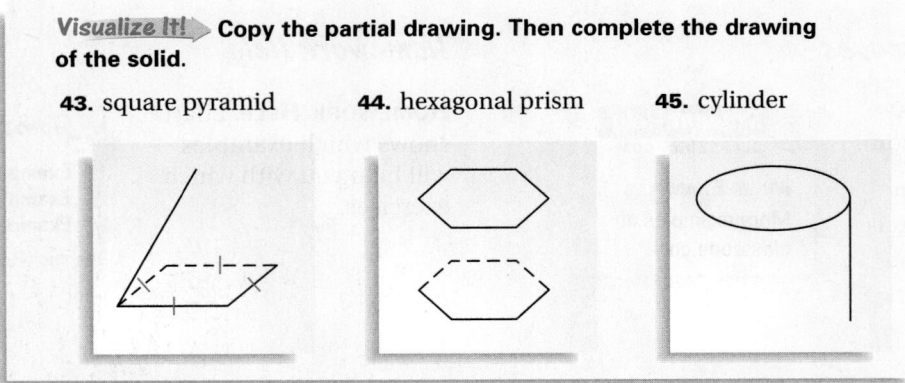

Visualize It! Copy the partial drawing. Then complete the drawing of the solid.

43. square pyramid   **44.** hexagonal prism   **45.** cylinder

TEACHER'S RESOURCE PACKAGE

This package is conveniently organized and includes a variety of materials to help you adapt the program to your teaching style and to the specific needs of your students!

Package Includes:

- Chapter Resource Books (one for each chapter, organized by lesson)
- Practice Workbook with Examples (TE)
- Warm-Up Transparencies and Daily Homework Quiz
- Worked-Out Solution Key

CHAPTER RESOURCE BOOKS

Chapter Resource Books allow you to carry the resources you have for a chapter in one manageable book. The materials in each Chapter Resource Book are organized by lesson so that you can easily see everything you have available.

Chapter Resource Books Include:

- **Tips for New Teachers**
- **Parent Guide for Student Success**
- **Strategies for Reading Mathematics**
- **Lesson Plans (Regular and Block Schedule)**
- **Warm-Up Exercises and Daily Homework Quizzes**
- **Technology Activities with Keystrokes**
- **Practice (Levels A and B)**
- **Reteaching with Practice**
- **Quick Catch-Up for Absent Students**
- **Learning Activities**
- **Real-Life Applications**
- **Quizzes**
- **Brain Games Support**
- **Chapter Review Games and Activities**
- **Chapter Tests (Levels A and B)**
- **SAT/ACT Chapter Test**
- **Alternative Assessment with Rubric and Math Journal**
- **Project with Rubric**
- **Cumulative Review**
- **Resource Book Answers**

TRANSPARENCY PACKAGES

The transparency packages give you many easy-to-use options for reviewing homework, starting class, and teaching visual strategies.

ANSWER TRANSPARENCIES FOR CHECKING HOMEWORK

WARM-UP TRANSPARENCIES AND DAILY HOMEWORK QUIZ

- **Warm-Up Exercises**
- **Daily Homework Quizzes**

VISUALIZE IT!

- **Promote visual teaching**
- **Reinforce key concepts**

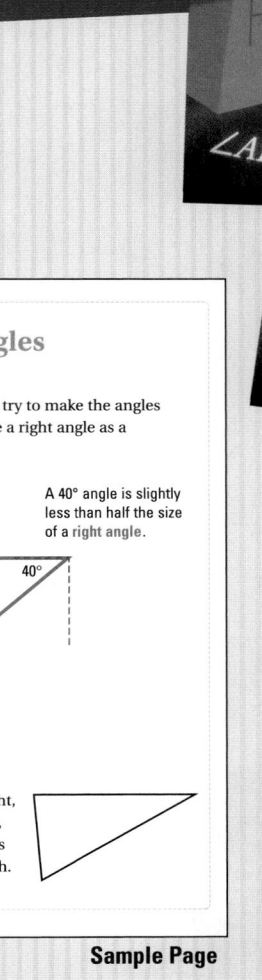

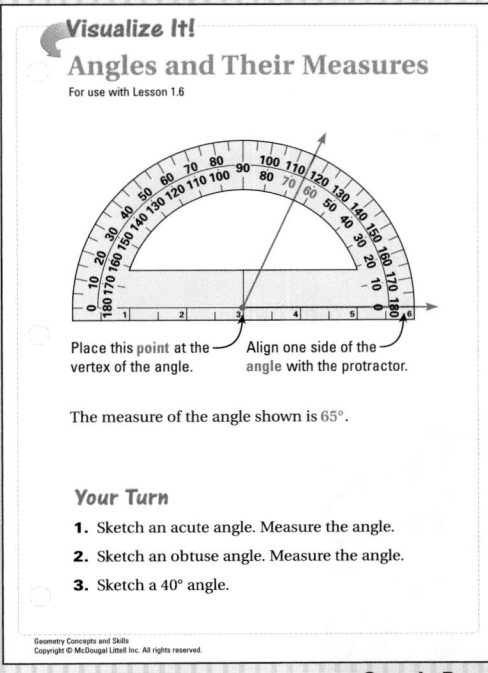

Visualize It!

Angles and Their Measures

For use with Lesson 1.6

Place this **point** at the vertex of the angle. Align one side of the **angle** with the protractor.

The measure of the angle shown is 65°.

Your Turn

1. Sketch an acute angle. Measure the angle.
2. Sketch an obtuse angle. Measure the angle.
3. Sketch a 40° angle.

Sample Page

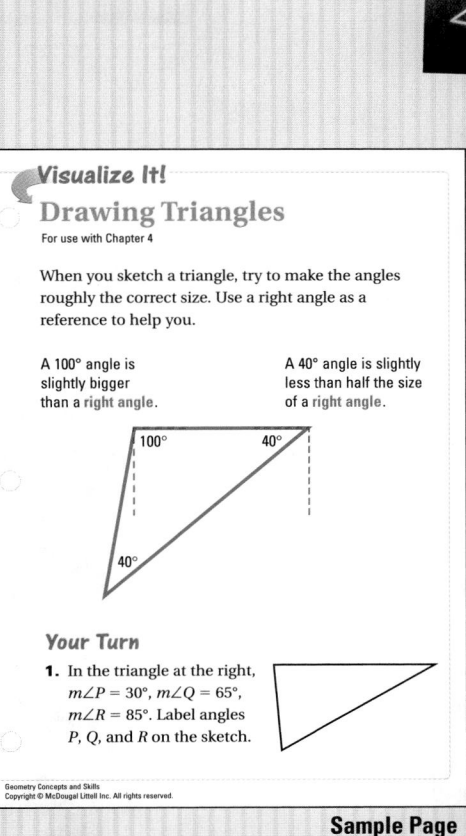

Visualize It!

Drawing Triangles

For use with Chapter 4

When you sketch a triangle, try to make the angles roughly the correct size. Use a right angle as a reference to help you.

A 100° angle is slightly bigger than a right angle.

A 40° angle is slightly less than half the size of a right angle.

100° 40°

40°

Your Turn

1. In the triangle at the right, $m\angle P = 30°$, $m\angle Q = 65°$, $m\angle R = 85°$. Label angles P, Q, and R on the sketch.

Sample Page

PRACTICE WORKBOOK

- Worked-out Examples
- Practice Exercises

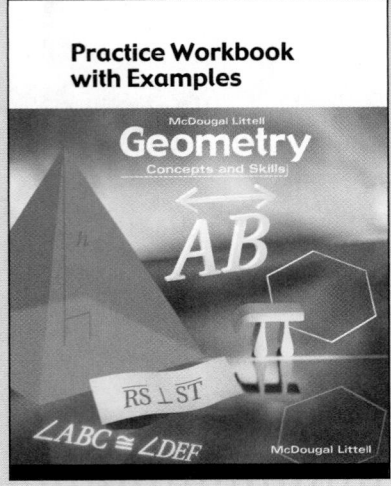

SPANISH RESOURCES

Resources in Spanish Include:

- Reteaching with Practice
- Quizzes
- Chapter Tests
- SAT/ACT Chapter Tests
- Alternative Assessments with Rubrics
- Math Journal

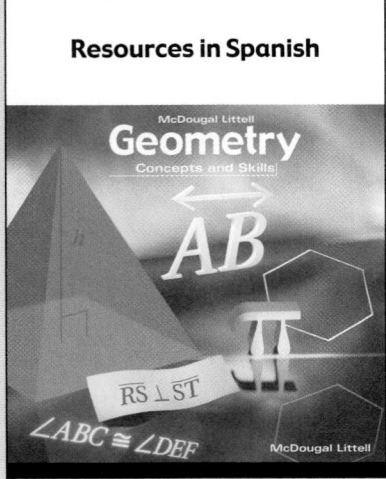

TECHNOLOGY RESOURCES

McDougal Littell technology resources help you and your students meaningfully use technology to enhance lessons and build understanding.

Technology for
Planning and Teaching

ONLINE LESSON PLANNER

- Create customized lesson plans
- Adjust to schedule changes
- Adapt the program to local and state objectives

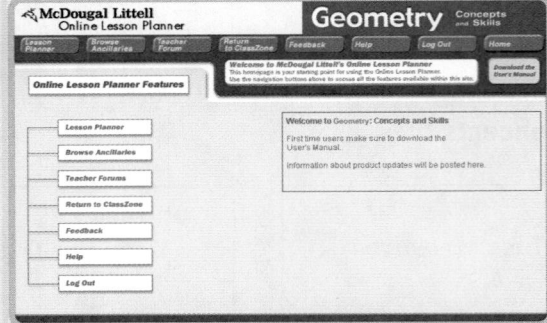

ELECTRONIC TEACHER TOOLS

This handy tool provides all your teaching resources on one CD-ROM.

Student Support

Review and Assessment

CLASSZONE

ClassZone is the companion Web site to McDougal Littell *Geometry, Concepts and Skills* that includes Student Help, career links, and more. To access ClassZone, go to *www.classzone.com.*

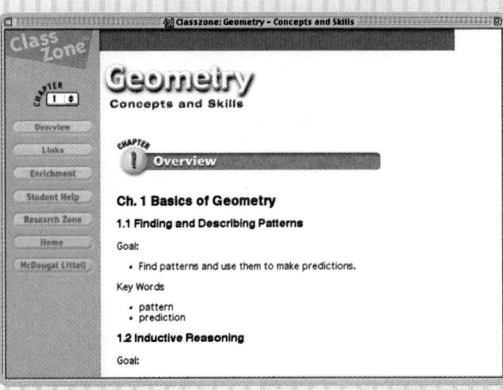

TIME-SAVING TEST AND PRACTICE GENERATOR

- Develop tests and practice sheets
- Instantly create answer keys

GEOMETRY IN MOTION VIDEO

The Geometry in Motion video brings geometry to life through animations that help students visualize and understand geometric concepts. Now you can quickly and easily help students see geometric concepts develop.

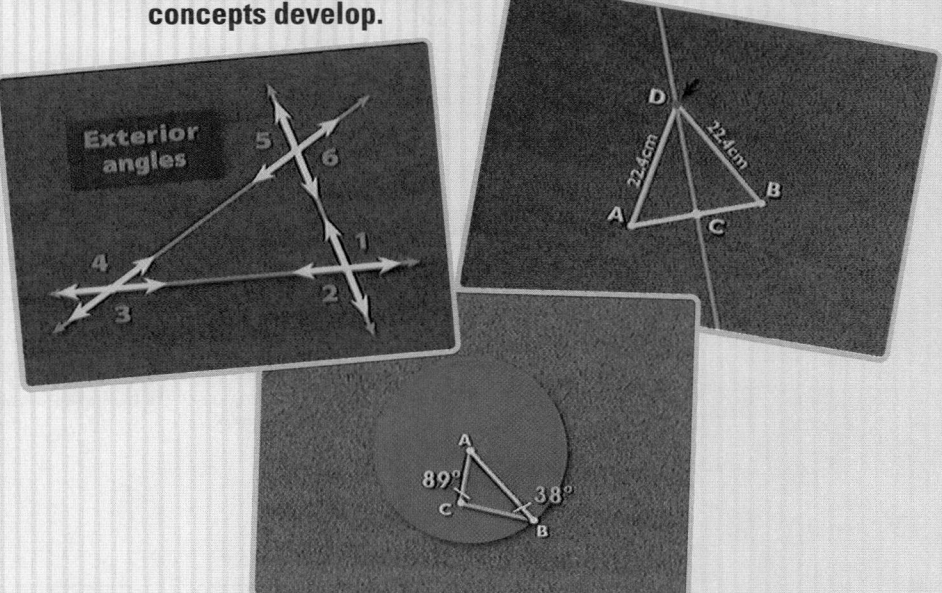

PLANNING THE CHAPTER

Complete planning support preceding each chapter

- Regular and block schedules for pacing the course

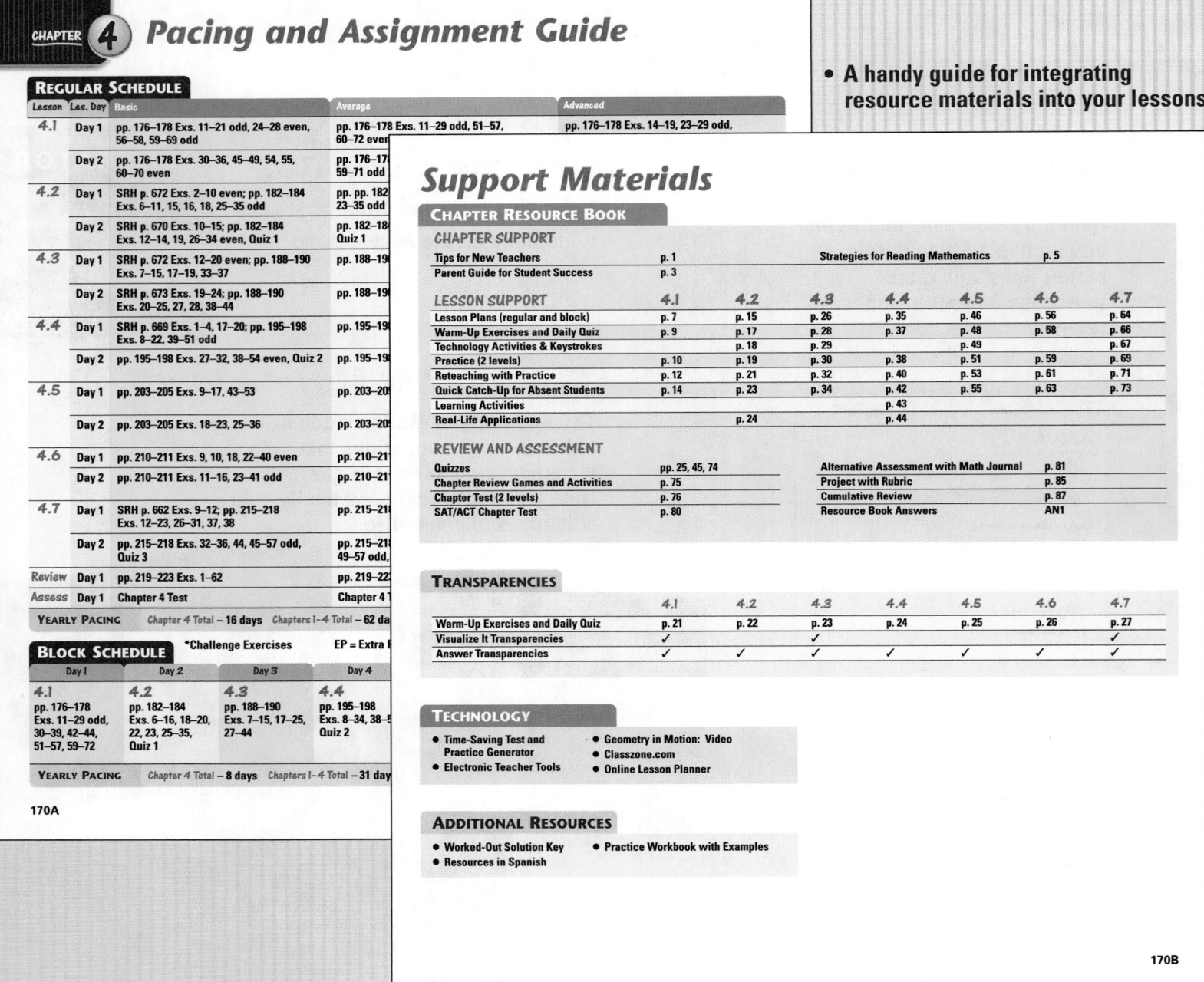

CHAPTER 4 *Pacing and Assignment Guide*

- A handy guide for integrating resource materials into your lessons

REGULAR SCHEDULE

Lesson	Les. Day	Basic	Average	Advanced
4.1	Day 1	pp. 176–178 Exs. 11–21 odd, 24–28 even, 56–58, 59–69 odd	pp. 176–178 Exs. 11–29 odd, 51–57, 60–72 even	pp. 176–178 Exs. 14–19, 23–29 odd,
	Day 2	pp. 176–178 Exs. 30–36, 45–49, 54, 55, 60–70 even	pp. 176–17 59–71 odd	
4.2	Day 1	SRH p. 672 Exs. 2–10 even; pp. 182–184 Exs. 6–11, 15, 16, 18, 25–35 odd	pp. pp. 182 23–35 odd	
	Day 2	SRH p. 670 Exs. 10–15; pp. 182–184 Exs. 12–14, 19, 26–34 even, Quiz 1	pp. 182–18 Quiz 1	
4.3	Day 1	SRH p. 672 Exs. 12–20 even; pp. 188–190 Exs. 7–15, 17–19, 33–37	pp. 188–19	
	Day 2	SRH p. 673 Exs. 19–24; pp. 188–190 Exs. 20–25, 27, 28, 38–44	pp. 188–19	
4.4	Day 1	SRH p. 669 Exs. 1–4, 17–20; pp. 195–198 Exs. 8–22, 39–51 odd	pp. 195–19	
	Day 2	pp. 195–198 Exs. 27–32, 38–54 even, Quiz 2	pp. 195–19	
4.5	Day 1	pp. 203–205 Exs. 9–17, 43–53	pp. 203–20	
	Day 2	pp. 203–205 Exs. 18–23, 25–36	pp. 203–20	
4.6	Day 1	pp. 210–211 Exs. 9, 10, 18, 22–40 even	pp. 210–211	
	Day 2	pp. 210–211 Exs. 11–16, 23–41 odd	pp. 210–211	
4.7	Day 1	SRH p. 662 Exs. 9–12; pp. 215–218 Exs. 12–23, 26–31, 37, 38	pp. 215–21	
	Day 2	pp. 215–218 Exs. 32–36, 44, 45–57 odd, Quiz 3	pp. 215–21 49–57 odd,	
Review	Day 1	pp. 219–223 Exs. 1–62	pp. 219–22	
Assess	Day 1	Chapter 4 Test	Chapter 4 T	

YEARLY PACING *Chapter 4 Total* – **16 days** *Chapters 1–4 Total* – **62 da**

BLOCK SCHEDULE   *Challenge Exercises   EP = Extra*

Day 1	Day 2	Day 3	Day 4
4.1 pp. 176–178 Exs. 11–29 odd, 30–39, 42–44, 51–57, 59–72	4.2 pp. 182–184 Exs. 6–16, 18–20, 22, 23, 25–35, Quiz 1	4.3 pp. 188–190 Exs. 7–15, 17–25, 27–44	4.4 pp. 195–198 Exs. 8–34, 38–5 Quiz 2

YEARLY PACING *Chapter 4 Total* – **8 days** *Chapters 1–4 Total* – **31 day**

170A

Support Materials

CHAPTER RESOURCE BOOK

CHAPTER SUPPORT

Tips for New Teachers	p. 1	Strategies for Reading Mathematics	p. 5
Parent Guide for Student Success	p. 3		

LESSON SUPPORT

	4.1	4.2	4.3	4.4	4.5	4.6	4.7
Lesson Plans (regular and block)	p. 7	p. 15	p. 26	p. 35	p. 46	p. 56	p. 64
Warm-Up Exercises and Daily Quiz	p. 9	p. 17	p. 28	p. 37	p. 48	p. 58	p. 66
Technology Activities & Keystrokes		p. 18	p. 29		p. 49		p. 67
Practice (2 levels)	p. 10	p. 19	p. 30	p. 38	p. 51	p. 59	p. 69
Reteaching with Practice	p. 12	p. 21	p. 32	p. 40	p. 53	p. 61	p. 71
Quick Catch-Up for Absent Students	p. 14	p. 23	p. 34	p. 42	p. 55	p. 63	p. 73
Learning Activities					p. 43		
Real-Life Applications		p. 24		p. 44			

REVIEW AND ASSESSMENT

Quizzes	pp. 25, 45, 74	Alternative Assessment with Math Journal	p. 81
Chapter Review Games and Activities	p. 75	Project with Rubric	p. 85
Chapter Test (2 levels)	p. 76	Cumulative Review	p. 87
SAT/ACT Chapter Test	p. 80	Resource Book Answers	AN1

TRANSPARENCIES

	4.1	4.2	4.3	4.4	4.5	4.6	4.7
Warm-Up Exercises and Daily Quiz	p. 21	p. 22	p. 23	p. 24	p. 25	p. 26	p. 27
Visualize It Transparencies	✓		✓				✓
Answer Transparencies	✓	✓	✓	✓	✓	✓	✓

TECHNOLOGY

- Time-Saving Test and Practice Generator
- Electronic Teacher Tools
- Geometry in Motion: Video
- Classzone.com
- Online Lesson Planner

ADDITIONAL RESOURCES

- Worked-Out Solution Key
- Resources in Spanish
- Practice Workbook with Examples

170B

- **Strategies for enabling all students to learn mathematics**

CHAPTER 4 *Providing Universal Access*

Strategies for Strategic Learners

USE A VISUAL MODEL
Students learn about triangle relationships in this chapter. They begin to identify various kinds of triangles and to use the properties of triangles to solve problems involving an unknown angle measure. Help students become familiar with triangles and the new terminology by having them draw a triangle using a ruler or straightedge. Instruct them to label the three vertices A, B, and C. Ask them to identify the three sides $\overline{AB}$, $\overline{BC}$, $\overline{AC}$. Have students measure the lengths of the sides with a ruler and find the measures of the angles using a protractor. You can use this activity to compare a variety of triangles. Since each student will draw his or her own triangle, the triangles will be of different sizes and shapes.

You might also consider asking that individual students draw a specific type of triangle, from isosceles or acute to right scalene or obtuse isosceles.

THINKING OUT LOUD
Sometimes mathematical procedures become so automatic for adults that we do not even realize all the steps we have taken. For example, in the following problem from page 192, there are several steps that some students might need to learn explicitly. Teachers can talk through their thinking while solving a problem. After students become familiar with their teacher's thinking, they can express their own thinking out loud. This gives the teacher an excellent opportunity to see where students are confused.

Problem: Find the unknown side length. Tell whether the side lengths form a Pythagorean triple.

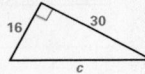

Solution: (The teacher should recite the following out loud to the class.) "Think to yourself. How do I find the length? What do I already know? I know I have a triangle. It has a right angle, so it must be a right triangle. I know the lengths of the other two sides. What is special about right triangles? The Pythagorean Theorem applies to right triangles. What does the Pythagorean Theorem say? It says that $(\text{hypotenuse})^2 = (\text{leg})^2 + (\text{leg})^2$."

"Which side is the hypotenuse? I remember that the hypotenuse is always opposite the right angle so the length c is the hypotenuse. This is also the side for which I'm trying to find the length. So I substitute the other two lengths into the equation:"

$$(\text{hypotenuse})^2 = (\text{leg})^2 + (\text{leg})^2$$
$$= (16)^2 + (30)^2$$
$$= 256 + 900$$
$$= 1156$$

"To find the length of the hypotenuse, I must find the square root of each side. I use my calculator to find the square root of

170C

- **Content and teaching strategies provide suggestions to help you present each lesson and to increase student understanding**

CHAPTER 4 *Content and Teaching Strategies*

This chapter focuses on triangles and their properties. In the first lesson, students are introduced to the classification of triangles by the lengths of their sides and by the measures of their angles. Lesson 4.2 presents the Triangle Sum Theorem, as well as two implications of this fundamental and important theorem. Lesson 4.3 begins the study of special triangles, investigating isosceles triangles and equilateral triangles. This study continues with right triangles in Lesson 4.4, where the Pythagorean Theorem is introduced. The converse of this theorem is presented in Lesson 4.5, where along with two related inequalities it is used to classify triangles by relating the squares of the lengths of the sides of a triangle. The concept of the median of a triangle is introduced in Lesson 4.6, as well as the related notion of the centroid of a triangle. Lesson 4.7 presents the basic inequalities involving the sides and angles of a general triangle, leading to the statement of the Triangle Inequality Theorem.

Lesson 4.1

CLASSIFYING TRIANGLES The classification of triangles by sides and angles can be confusing for students because there is a good deal of overlap in the categories. For example, as students may suspect, every equilateral triangle is also equiangular and every equiangular triangle is also equilateral. (These facts will be stated formally in Lesson 4.3.) Also, a right triangle may be isosceles, and in fact this is always the case for the two triangles formed when the diagonal of a square is drawn. Point out that any of the triangles in the Classification of Triangles by Sides summary box may be isosceles, and that an equiangular triangle is always isosceles. Stress that all but the equiangular triangle may be scalene.

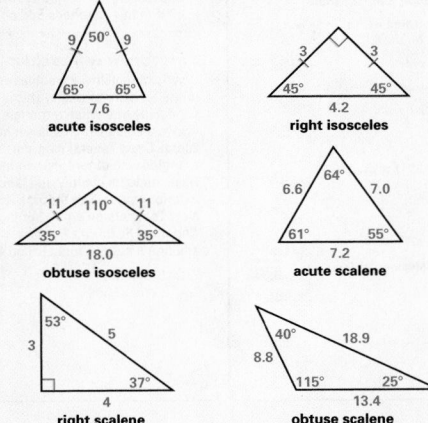

acute isosceles

right isosceles

obtuse isosceles

acute scalene

right scalene

obtuse scalene

170E

Lesson 4.2

QUICK TIP
When first discussing exterior angles, students may become confused by the fact that two different exterior angles can be drawn at each vertex of a triangle. Point out the pairs of exterior angles at each vertex of the first triangle at the top of page 181 and have students note that the angle pairs are vertical angles. Remind them that vertical angles have the same measure, leading to the fact that either exterior angle at vertex C could have been shown in the figure given with the Exterior Angle Theorem.

MODELING This lesson contains one of the most important theorems in geometry, the Triangle Sum Theorem, which states that the sum of the angle measures of any triangle is 180°. You can verify the truth of this theorem by cutting a triangle out of paper or cardboard (or asking students to do this) and tearing off the three corners, which can always be arranged, in pie fashion, to form a straight angle. The theorem is also remarkable in that it states that all triangles in a plane have the same angle sum.

You can demonstrate the truth of the Exterior Angle Theorem by noting that the sum of the measures of the three angles of the triangle is 180°, as is the sum of the measures of an exterior angle and its adjacent interior angle. If these two sums are set equal to one another and the measure of the adjacent interior angle is subtracted from both sides, the result is the symbolic form of the theorem.

USING ALGEBRA Both the Corollary to the Triangle Sum Theorem and the Exterior Angle Theorem are very useful in solving algebraic problems involving angle measures. As Example 2 illustrates, the corollary must be translated into an equation before it can be useful in solving numerical problems. Note that the corollary always enables you to find the measures of all the angles of a right triangle, given the measure of one of the acute angles.

Lesson 4.3

VISUALIZING DIAGRAMS As with parallel lines, students sometimes have difficulty seeing the relationships in an isosceles triangle if the triangle is oriented on the page so that its base is not horizontal. As before, a useful hint is to rotate the book or paper on which the diagram appears until the base becomes horizontal. Another suggestion for problem-solving that involves angle measures is to label as many angles of the triangle with their measures as possible. For example, if the measure of one base angle of an isosceles triangle is given, students should immediately label the other base angle with the same measure. They can then find the measure of the third angle of the triangle using the Triangle Sum Theorem. In dealing with the lengths of sides of a triangle, students should first look to see whether the two congruent sides of an isosceles triangle are labeled with algebraic expressions. If they are, an equation can immediately be written by setting the two expressions equal to each other.

PLANNING THE LESSON
Handy, point-of-use support for each lesson

Plan
- Pacing summary
- Teaching resources

4.2 Angle Measures of Triangles

Goal
Find angle measures in triangles.

Key Words
- corollary
- interior angles
- exterior angles

The diagram below shows that when you tear off the corners of any triangle, you can place the angles together to form a straight angle.

THEOREM 4.1

Triangle Sum Theorem

Words The sum of the measures of the angles of a triangle is 180°.

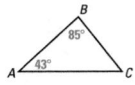

Symbols In $\triangle ABC$, $m\angle A + m\angle B + m\angle C = 180°$.

Student Help

READING TIP
Triangles are named by their vertices. $\triangle ABC$ is read "triangle ABC."

EXAMPLE 1 Find an Angle Measure

Given $m\angle A = 43°$ and $m\angle B = 85°$, find $m\angle C$.

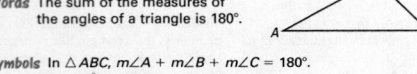

Solution

$m\angle A + m\angle B + m\angle C = 180°$	Triangle Sum Theorem
$43° + 85° + m\angle C = 180°$	Substitute 43° for $m\angle A$ and 85° for $m\angle B$.
$128° + m\angle C = 180°$	Simplify.
$128° + m\angle C - 128° = 180° - 128°$	Subtract 128° from each side.
$m\angle C = 52°$	Simplify.

ANSWER $\angle C$ has a measure of 52°.

CHECK ✓ Check your solution by substituting 52° for $m\angle C$.
$43° + 85° + 52° = 180°$

4.2 Angle Measures of Triangles **179**

179

1 Plan

Pacing
Suggested Number of Days

Basic: 2 days
Average: 2 days
Advanced: 2 days
Block Schedule: 1 block

Teaching Resources

📄 *Blacklines*
(See page 170B.)

📖 *Transparencies*
- Warm-Up with Quiz
- Answers

💻 *Technology*
- Electronic Teacher Tools
- Test and Practice Generator
- Online Lesson Planner
- Internet Support

📹 *Video*
- Geometry in Motion

2 Teach

Content and Teaching Strategies
For background information on geometric concepts and teaching strategies related to this lesson, see pages 170E and 170F in this Teacher's Edition.

Tips for New Teachers
Some students have trouble recognizing an exterior angle. This is especially true when a triangle is drawn with a base that is not horizontal. Draw several diagrams of triangles with different orientations. Have students identify and label the exterior angles. See the Tips for New Teachers on pp. 1–2 of the *Chapter 4 Resource Book* for additional notes about Lesson 4.2.

Extra Example 1
See next page.

...other angles are formed.
...angles.

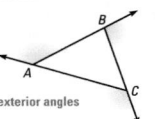
exterior angles

...or angles are the
...ly *one* exterior angle at

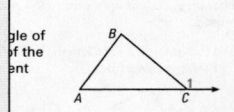

...ge of
...of the
...ent

...sure
...$m\angle 1$.

...le Theorem
...8° for $m\angle A$
...$m\angle C$.

Geometric Reasoning
Point out that an exterior angle forms a linear pair with the interior angle of the triangle that shares its vertex. Make sure students do not mistakenly think that any angle on the outside of a triangle is an exterior angle.

Extra Example 3
Given $m\angle D = 45°$ and $m\angle DFG = 100°$, find $m\angle E$. 55°

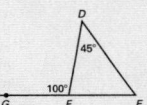

✓ **Concept Check**
How are the measures of interior angles and exterior angles of a triangle related? The measure of an exterior angle of a triangle is equal to the sum of the measures of the two nonadjacent interior angles. Also, the sum of the measures of an exterior angle and its adjacent interior angle is 180°.

🔲 **Daily Puzzler**
The ratio of the measures of the angles in a triangle is 3:2:1. Find the angle measures. 90°, 60°, 30°

Teach
- **Extra Example for each example in the book**
- **Geometric Reasoning notes to build conceptual understanding**
- **Concept Check question at the end of each lesson**
- **Teaching Tips**
- **... AND MORE**

181

Assess

- Mini-Quiz for each lesson
- Assessment resources

Apply

- Assignment Guide
- Extra practice references
- Homework Check exercises

➌ Apply

Assignment Guide

BASIC
Day 1: SRH p. 672 Exs. 2–10 even; pp. 182–184 Exs. 6–11, 15, 16, 18, 25–35 odd
Day 2: SRH p. 670 Exs. 10–15; pp. 182–184 Exs. 12–14, 19, 26–34 even, Quiz 1

AVERAGE
Day 1: pp. 182–184 Exs. 6–11, 15, 16, 18, 20, 23–35 odd
Day 2: pp. 182–184 Exs. 12–14, 19, 22, 26–34 even, Quiz 1

ADVANCED
Day 1: pp. 182–184 Exs. 9–11, 15–18, 20, 21, 23–25, 27–35 odd
Day 2: pp. 182–184 Exs. 12–14, 19, 22, 26–34 even, Quiz 1; EC: TE p. 170D*, classzone.com

BLOCK SCHEDULE
pp. 182–184 Exs. 6–16, 18–20, 22, 23, 25–35, Quiz 1

Extra Practice
- Student Edition, p. 681
- Chapter 4 Resource Book, pp. 19–20

Homework Check
To quickly check student understanding of key concepts, go over the following exercises:

Basic: 7, 8, 13, 18, 19
Average: 8, 9, 13, 18, 19
Advanced: 9, 10, 14, 19, 21

✗ Common Error
In Exercises 6–14, watch for students who try to measure the angles using a protractor, rather than using the Triangle Sum Theorem and the Exterior Angle Theorem to calculate the angle measures.

1. *Sample answer:*

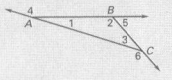

182

4.2 Exercises

Guided Practice

Vocabulary Check

Skill Check

Practice and Appl...

Extra Practice
See p. 679.

Homework Help
Example 1: Exs. 6–11, 15–21, 23, 24
Example 2: Exs. 6–11, 15–21
Example 3: Exs. 12–14, 18–22

➍ Assess

Assessment Resources
The Mini-Quiz below is also available on blackline (*Chapter 4 Resource Book*, p. 28) and on transparency. For more assessment resources, see:
- Chapter 4 Resource Book
- Standardized Test Practice
- Test and Practice Generator

Mini-Quiz
Find the measure of ∠1.

1.
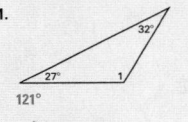
121°

2.
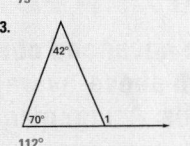
75°

3.
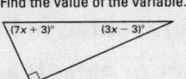
112°

4. Find the value of the variable. 9
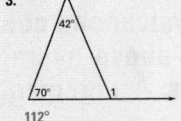

5. Raymundo is designing a logo that is formed by using two congruent right triangles as shown. The smaller acute angle of each triangle has a measure of 27°. Find the measure of each larger acute angle. 63°

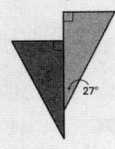

184

24. **Using Algebra** In △PQR, the measure of ∠P is 36°. The measure of ∠Q is five times the measure of ∠R. Find m∠Q and m∠R. m∠Q = 120°; m∠R = 24°

Standardized Test Practice

25. **Multiple Choice** Find the value of x. C
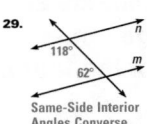

Ⓐ 8 Ⓑ 13
Ⓒ 16 Ⓓ 29

26. **Multiple Choice** Suppose a triangle has interior angle measures of 50°, 60°, and 70°. Which of the following is *not* an exterior angle measure? F

Ⓕ 100° Ⓖ 110° Ⓗ 120° Ⓙ 130°

Mixed Review

Showing Lines are Parallel Explain how you would show that *m* ∥ *n*. State any theorems or postulates that you would use. *(Lesson 3.5)*

27.
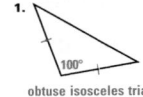
Corresponding Angles Converse

28.
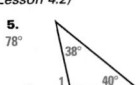
Alternate Exterior Angles Converse

29.
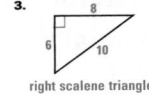
Same-Side Interior Angles Converse

Algebra Skills

Comparing Numbers Compare the two numbers. Write the answer using <, >, or =. *(Skills Review, p. 662)*

30. 1015 and 1051
1015 < 1051

31. 3.5 and 3.06
3.5 > 3.06

32. 8.09 and 8.1
8.09 < 8.1

33. 1.75 and 1.57
1.75 > 1.57

34. 0 and 0.5
0 < 0.5

35. 2.055 and 2.1
2.055 < 2.1

Quiz 1

Classify the triangle by its angles and by its sides. *(Lesson 4.1)*

1.
obtuse isosceles triangle

2.
acute scalene triangle

3.
right scalene triangle

Find the measure of ∠1. *(Lesson 4.2)*

4.
30°

5.
78°

6.
65°

PACING THE COURSE

The Pacing Chart below shows the number of days allotted for each chapter.
The Regular Schedule requires 160 days. The Block Schedule requires 80 days.
These time frames include days for review and assessment: 2 days per chapter
for the Regular Schedule and 1 day per chapter for the Block Schedule. Semester
and trimester divisions are indicated by red and blue rules, respectively.

	Semester 1							Semester 2			
Chapter	1	2	3	4	5	6	7	8	9	10	11
Regular Schedule	14	14	18	16	18	14	14	12	12	14	14
Block Schedule	7	7	9	8	9	7	7	6	6	7	7
	Trimester 1				Trimester 2				Trimester 3		

Assignments are provided with each lesson for a basic course, an
average course, an advanced course, and a block-schedule course.
Each of the four courses covers all eleven chapters.

BASIC COURSE

The basic course is intended for students who enter with
below-average mathematical and problem-solving skills.
Assignments include:

- spiral review of pre-course and on-level topics through
  Skills Review Handbook and Extra Practice references

- substantial work with the skills and concepts
  presented in the lesson

- straightforward applications of these skills and concepts

- test preparation and mixed review exercises

AVERAGE COURSE

The average course is intended for students who enter
with typical mathematical and problem-solving skills.
Assignments include:

- substantial work with the skills and concepts
  presented in the lesson

- applications of these skills and concepts

- test preparation and mixed review exercises

ADVANCED COURSE

The advanced course is intended for students who enter
with above-average mathematical and problem-solving
skills. Assignments include:

- substantial work with the skills and concepts
  presented in the lesson

- more complex applications and challenge exercises

- test preparation and mixed review exercises

- optional extra challenge exercises provided in the
  Teacher's Edition and on the ClassZone web site

BLOCK-SCHEDULE COURSE

The block-schedule course is intended for schools that
use a block schedule. The exercises assigned are
comparable to the exercises for the average course.

The Pacing and Assignment Guide for each chapter is located on the interleaved pages preceding the chapter. Part of the Pacing Chart for Chapter 4 is shown here.

The *Regular-Schedule Chart* provides pacing for the basic, average, and advanced courses.

REGULAR SCHEDULE

Lesson	Les. Day	Basic	Average	Advanced
4.1	Day 1	pp. 176–178 Exs. 11–21 odd, 24–28 even, 56–58, 59–69 odd	pp. 176–178 Exs. 11–29 odd, 51–57, 60–72 even	pp. 176–178 Exs. 14–19, 23–29 odd, 48–66 even
	Day 2	pp. 176–178 Exs. 30–36, 45–49, 54, 55, 60–70 even	pp. 176–178 Exs. 30–39, 42–44, 48–50, 59–71 odd	pp. 176–178 Exs. 30–41, 43–45, 55–71 odd; EC: TE p. 170D*, classzone.com
4.2	Day 1	SRH p. 672 Exs. 10–15; pp. 182–184 Exs. 6–11, 15, 16, 18, 25–35 odd	pp. 182–184 Exs. 6–11, 15, 16, 18, 20, 23–35 odd	pp. 182–184 Exs. 9–11, 15–18, 20, 21, 23–25, 27–35 odd
Review	Day 1	pp. 219–223 Exs. 1–62	pp. 219–223 Exs. 1–62	pp. 219–223 Exs. 1–62
Assess	Day 1	Chapter 4 Test	Chapter 4 Test	Chapter 4 Test

YEARLY PACING Chapter 4 Total – **16 days** Chapters 1–4 Total – **62 days** Remaining

* Challenge Exercises EP = Extra Practice SRH = Skills Review Handbook EC = Extra Challenge

The *Block-Schedule Chart* provides pacing for the block-schedule course

BLOCK SCHEDULE

Day 1	Day 2	Day 3	Day 4	Day 5
4.1 pp. 176–178 Exs. 11–29 odd, 30–39, 42–44, 51–57, 59–72	4.2 pp. 182–184 Exs. 6–16, 18–20, 22, 23, 25–35, Quiz 1	4.3 pp. 188–190 Exs. 7–15, 17–25, 27–44	4.4 pp. 195–198 Exs. 8–34, 38–55, Quiz 2	4.5 pp. 203–205 Exs. 8–23, 25–39, 42–53

YEARLY PACING Chapter 4 Total – **8 days** Chapters 1–4 Total – **31 days** Remaining

An *Assignment Guide* for each lesson is provided at the beginning of the exercise set. Assignments are given for basic, average, advanced, and block-schedule courses.

Assignment Guide

BASIC
Day 1: pp. 210–211 Exs. 11–16, 23–41 odd

AVERAGE
Day 1: pp. 210–211 Exs. 11–17, 19–23, 31–41 odd

ADVANCED
Day 1: pp. 210–211 Exs. 11–17, 19–23, 31–41 odd; EC: TE p. 170D*, classzone.com

BLOCK SCHEDULE
pp. 210–211 Exs. 9–41

Providing Universal Access

By Catherine Barkett

With careful planning, teachers can help *all* students reach a level of mathematical competence needed to continue their education in mathematics.

Introduction

In most classrooms, students present a variety of achievement levels, skills, and needs. The goal for all students is the same: We want them to develop sufficient computational, procedural, and problem solving skills to provide a solid foundation for further study in mathematics. However, all students do not arrive at these competencies at the same time or in the same way. In this article we suggest research-based strategies teachers can use to modify curriculum and instruction for special needs students. The basic instructional plan in *Geometry: Concepts and Skills* is designed for students who are achieving at near grade level; but just prior to each chapter we include specific suggestions for students who are achieving above and below grade level, and for students who are not fluent in English (see also the article titled "Adapting Curriculum and Instruction for English Learners").

Teachers may find it helpful to view students as members of four basic groups. (English learners can be found in all four groups.) Teachers do not need to place students in these groups; the categories are suggested so teachers can plan ahead to meet different needs of students. Note the use of the term *strategic learners* to emphasize the kind of instruction needed by students achieving below grade level. This term is not synomous with special education. It may include some special education pupils but includes many more students whose low achievement levels are the result of inadequate prior schooling or attendance, high mobility rates, or a host of other reasons that have nothing to do with their abilities or disabilities. The term *strategic learner* was selected because it is a positive term that emphasizes what needs to happen in order for these students to be successful and implies what we believe: students achieving below grade level *can* be successful in mathematics given carefully designed instruction.

Setting the Right Tone

There are three key strategies recommended for teachers as they adapt any program to students' needs:

- Use frequent assessment as a way to determine what each student does or does not know, and use that assessment as the basis for planning.

FOUR BASIC STUDENT GROUPS

ADVANCED GROUP	GRADE LEVEL GROUP	STRATEGIC GROUP	INTENSIVE NEEDS GROUP
Advanced students have already completed some of the grade-level material. They make rapid progress and become bored with repetition. They may or may not have been formally identified as gifted or talented in the area of mathematics.	Students achieving at grade level may have minor, occasional difficulties but they can be assisted to maintain their progress with extra practice and individual or group assistance on an ad hoc basis.	Strategic learners are not achieving at expected grade level but can, with a carefully designed program that provides targeted assistance. Systematic differentiation such as preteaching, reteaching, and additional instructional time should be planned for these students, as suggested in each chapter.	Intensive needs students are those whose performance is two or more standard deviations below the mean on standardized measures. These students will probably already be eligible for special education services. This is a very small percentage of the general population.

ADVANCED GROUP

SUGGESTED PLAN
1. Assess what these students already know.
2. Allow these students to "test out" of chapters or assignments.
3. Substitute challenging assignments for easier ones.
4. Modify instruction so that it is more complex or more in-depth.

GRADE LEVEL GROUP

SUGGESTED PLAN
1. Assess what these students already know.
2. Progress through *Geometry: Concepts and Skills* at the recommended pace and sequence.
3. On an ad hoc basis, review or provide additional practice as needed.

STRATEGIC GROUP

SUGGESTED PLAN
1. Assess what these students already know.
2. Provide additional scaffolding and the instructional variations suggested in this book.
3. Focus on the key concepts and present material systematically.
4. Vary the kinds of instruction so that students have several opportunities to understand.
5. Provide additional practice homework.

INTENSIVE NEEDS GROUP

SUGGESTED PLAN
1. Assess what these students already know.
2. Determine if these students have an IEP.
3. Refer students for special education testing or child study team discussion; enlist the help of specialists.
4. Carefully consider each student's most appropriate placement in mathematics.
5. Use the specific suggestions for strategic learners.

- Plan modifications of curriculum and instruction ahead of time so that you are ready to differentiate as the need arises.

- Use a variety of grouping strategies to facilitate learning. A combination of whole class instruction and temporary groupings of students, with groups organized around students' needs, will facilitate management of the variety of achievement levels and learning needs in the classroom.

Assessment, planning, and flexible grouping are essential to ensuring that your students have the optimal chance for success. In addition to these three key strategies, general guidelines for establishing a classroom designed to meet students' needs are:

1. Establish an atmosphere where students feel comfortable asking questions and are rewarded for asking about things they don't understand.

2. Maintain the same goals for all students. Allow additional time and practice for students who need it, and provide challenging alternatives for those who are ready to move more quickly.

3. Clearly identify the skill, concept, or standards you are working on and measure progress towards those ends.

4. Have students show their work. It is much easier for teachers to understand where a student gets confused if they have evidence of the student's thought process.

5. Try small modifications in curriculum and instruction before more drastic ones.

6. Don't persist with a strategy that is not working. Try something else.

7. Encourage effort and persistence, and celebrate successes with your students.

Varying Curriculum and Instruction

1. **Time** Most students whose achievement is below grade level will need more time. Students who are not fluent in English will need more time. The contents of this book might be offered over a two-year period, or two periods a day. Perhaps the day can be extended through study hall, regular homework assignments, tutoring, or Saturday, summer, or "off track" catch-up sessions. Advanced students might "test out" of portions of the book and complete the material in half a year, or they may compact two courses into one.

2. **Presentation** Instructing in a variety of ways and taking a single concept and explaining it verbally and visually with concrete and abstract examples provide students multiple opportunities for understanding. Area, for example, is a key concept in Geometry. In earlier grades students have found areas of simple objects, both by using tiles to cover an area, and by applying algorithms. Later, students practice with unusual shapes, dividing them into simpler ones. Finally, students tackle sophisticated examples like finding the shape with the maximum area given a fixed perimeter, using both trial and error and more abstract methods.

3. **Task parameters** Multi-step problems can be especially difficult for students. These types of problems are just combinations of simpler problems and can be broken down into those simpler steps, with additional help and practice at each step. Confusing elements can be minimized and extraneous material can be eliminated. For advanced students, simpler problems can be eliminated and more challenging ones (as suggested in each chapter) may be substituted.

4. **Methods of assessment** Students learning English may be able to demonstrate on paper what they cannot yet verbalize. Students with physical challenges may be unable to draw a graph but may be able to select the right graph from a series of options or verbally describe the graph so that someone else can draw it. Allow students to demonstrate their knowledge in a variety of ways while helping all students to master the skills and knowledge necessary to exhibit their understanding in standard ways.

Geometry: Concepts and Skills is organized so that much of the differentiation for special needs students is built into the design of the program. Note that simpler concepts are introduced before more complex ones. Ample practice is provided. Challenge exercises are included throughout the pupil text. Activities and Geo-Activities provide students with models for conceptual understanding of the mathematical reasoning behind each key concept. Mathematical reasoning is stressed throughout each chapter. Vocabulary words, examples, and Guided Practice exercises are standard features of each chapter. Each lesson includes Mixed Review exercises so that students recall and use skills and understandings from previous chapters. These features were designed to help you meet the needs of the students in your class.

For Students who have Trouble Paying Attention

Some students in your classroom may be formally identified as having Attention Deficit Hyperactivity Disorder (ADHD) or Attention Deficit Disorder (ADD). Others may exhibit the same learning challenges but may not be formally identified. Whether formally identified or not, students who have trouble paying attention generally share the following characteristics:

- Trouble paying attention is not just occasional. It occurs most or all of the time, across content areas, and is inappropriate for the age of the child.

- Forgetfulness, memory problems, losing things, disorganization

- Restlessness, fidgeting

- Socially inappropriate behavior such as excessive talking, interrupting others, and difficulty waiting their turn

These students may be very bright and capable in mathematics but have a hard time staying focused for long periods of time. They need to be taught strategies for organizing their work and keeping track of where they are. In general, students with attention problems need to be helped to develop coping strategies. The teacher should approach the student in a problem solving mode: "Let's find ways to help you concentrate and organize your work," rather than using one of the following strategies in a punitive way.

1. **Present the work in smaller chunks over smaller time periods** and then gradually increase expectations. If students have trouble completing long tests, for example, break the material into smaller quizzes and increase the length of the quizzes as the year progresses.

2. **Use cumulative review and practice.** Have students periodically review what they learned in previous chapters and provide additional practice if they have forgotten.

3. **Make it more obvious what the student should focus on.** For example, use the test generator to put only four problems on each page; use a large font; or use an index card or piece of cardboard with a hole cut out of the middle to place on the page so that the student can focus on one problem at a time. A pencil, finger, highlighter, or sticky paper can also be used by the student to keep track of which problem he or she is working on.

4. **Have students race against the clock.** For some students, racing against the clock to see how many problems can be completed accurately within a five-minute time period is more motivating than doing the same number of problems at their leisure. The time period can be extended gradually.

5. **Help students develop simple strategies for bringing work to and from class.** A two-pocket folder, where homework goes home in the left pocket and comes back in the right, is a simple way to keep track of assignments.

6. **Allow movement and schedule breaks.**

7. **Minimize distractions by seating students that are easily distracted near the teacher** and away from hallway noise. Tables with several students at a table are more distracting than rows of desks. When students are to work quietly, offer headphones to block out noise. Headphones can be set to play quiet music, "white noise," or can be used just as earplugs to help block out noise.

8. **Graphic organizers** such as Venn diagrams, tree diagrams, lists, outlines, tables, and charts can all provide structures for organizing and remembering information. Mental images, choral responses, or even hand signals can help students remember. Highlighters can be used to make sure that decimal points are lined up. Graph paper is excellent for keeping homework problems neat, even when a graph is not required.

9. **Keep instructions simple and clear, especially at the beginning of the year.** Establish routines (*e.g.*, the week's homework is always due on Thursday; assignments are written in a specific place on the board; the last ten minutes of class is used to make sure everyone understands what homework is expected and how to do it). Students who know the routine find it easier to work independently.

For Students who have Trouble Understanding the Concepts

Success in mathematics, as in music, sports, or other areas, comes for most students only with hard work and persistent effort. Concepts may seem difficult at first, but with repeated teaching and practice virtually all students can master the mathematics they need to graduate from high school, access a variety of jobs, and lay the foundation for further study in mathematics or a related field.

Several strategies can help students make steady progress in mathematics. These include:

1. Focus on key mathematical concepts.

2. Review key concepts and skills from earlier lessons, chapters, or years.

3. Preteach key concepts and vocabulary.

4. Anticipate problem areas.

5. Provide scaffolding (guided practice) for students who need extra help.

6. Think out loud to show hidden steps.

7. Provide a sample problem to which students can return when they get stuck.

8. Break problems into simpler components.

9. Explicitly teach students a variety of problem solving strategies and help them select one that fits the situation.

10. Present concepts in a variety of ways: visually, verbally, concretely, abstractly, etc.

11. Encourage students to draw a picture or use a visual aid such as a number line, graph, or diagram.

12. Provide sufficient practice.

Finally, good teachers are perpetual students themselves. They are always looking for ways to deepen their understanding of mathematics and for good ways to explain and teach mathematics to others.

For Advanced Students

Occasionally students can demonstrate mastery of all the mathematics expected to be learned in a given grade level. Repeating previously learned material for a year is deadly to these students. It can make them dislike mathematics. For these students, moving them up a grade level for math is a simple and cost-effective solution.

Most advanced students, however, are advanced in some areas but not in others. They tend to learn quickly and need more instructional material, as well as more difficult material. The pupil edition, teacher's edition, and Web site, to this program provide challenging exercises that can be used when students have demonstrated competence in a particular area. These challenge exercises should be substituted for the easier exercises in a homework assignment or lesson. When they have the time and interest, all students should be encouraged to work the challenge exercises.

General strategies for differentiating the curriculum for advanced learners include:

1. **Vary the pacing.** Allow advanced students some flexibility in how they progress through the course. Students who can demonstrate mastery of the objectives for a given lesson or chapter can be working on challenge exercises. Advanced students may become fascinated with a particular aspect of mathematics and want to spend *more* time on it.

2. **Differentiate in terms of depth.** Encourage advanced students to delve more in depth into mathematics. Looking at the details and the patterns; studying the language of the discipline; and looking at trends, themes, properties, theorems, proofs, and unanswered questions can enrich the curriculum for advanced students.

3. **Differentiate in terms of complexity.** Advanced students may be ready to connect ideas across disciplines in ways characteristic of older students or adults. Encourage them to investigate relationships between mathematics and art, history, science, and music, and to look at the development of mathematics over time.

Using Grouping to Benefit All Students

Grouping advanced learners together for investigations of challenge problems can provide you with time to work more closely with a group of students who need help in a particular area. Alternatively, while students who need more help are working on additional reinforcement activities or practice, you can work with a group of advanced students on a challenge project. Groups can be organized and revised daily, weekly, or by lesson according to how proficient students are with the concepts and skills targeted for that day, week, or lesson. At times you may have only one student who is ready for a challenge problem; at other times the whole class may be ready. Flexible grouping is the key to ensuring that students do not become "tracked." Asking advanced students to report to the whole class on their progress on challenge problems can provide the opportunity for the whole class to engage in more abstract and theoretical thinking.

Adapting Curriculum and Instruction for English Learners

By Olga Bautista and Catherine Barkett

Introduction

English learners come to the classroom with all the variety of English speakers in regard to mathematics achievement. They may be at, behind, or ahead of grade expectations in mathematics. They may be gifted or eligible for special education services. They may have been born in the United States, or they may have arrived in this country very recently. They may speak one or more languages, and they may be literate in one or more languages other than English. They may be nearly fluent in English or have beginning or intermediate levels of understanding and production.

They have in common one characteristic: They are all learning English.

With careful planning, teachers can maximize success for English learners in the mathematics classroom. Assessing each student's competencies in mathematics and English will form a basis for program planning.

Getting to know your students

Before school starts, check the cumulative folder on each student in your class to determine which ones are learning English. See if there is recent testing. Two types of testing are most useful: mathematics

ENGLISH LEARNERS

	LOW MATHEMATICS ACHIEVEMENT	HIGH MATHEMATICS ACHIEVEMENT
LOW READING ACHIEVEMENT	**Who is this student?** Student may be new to the class, school, or country. Student may have had inadequate schooling. Student may have moved a lot. Student may be unmotivated or have test anxiety. Low reading achievement may be depressing mathematics scores. Student may have gaps and holes in knowledge. Student may need special education assistance. **What to do?** Examine cumulative folder for other testing, notes, etc. Delay any testing for a week or two. Help student feel comfortable in the class during that period of time. Administer mathematics achievement test and reading test, preferably in an individual setting. Plan to assess this student at weekly intervals and closely monitor classroom work to determine if progress is being made. Look at the English Learners suggestions in each chapter.	**Who is this student?** Student has had good prior mathematics instruction. Mathematics is an area where this student can excel. Math achievement level may actually be higher than scores indicate. (Inability to read affects mathematics achievement as well.) Word problems will be especially difficult. **What to do?** Mathematics instruction should proceed at normal or near normal pace. Student should be involved in a systematic English language development program and intensive reading program outside of mathematics class. Spend part of each class period on mathematics vocabulary study. Provide a bilingual dictionary and grade-level mathematics text in the home language for home use. Look at the English Learners suggestions in each chapter for those that are most useful.
HIGH READING ACHIEVEMENT	**Who is this student?** Student may have been designated as an English learner because oral fluency is not at grade level. Student may be able to use reading ability to improve math scores. Student may not test well in mathematics. Most students can make rapid progress in mathematics; a few may have learning difficulties that require the help of a specialist. **What to do?** Assess mathematics achievement in a variety of ways. Concentrate on developing oral fluency. Focus on vocabulary specific to mathematics.	**Who is this student?** May be a student who is ready for re-designation as a fluent English speaker. May need extra study in academic vocabulary, *i.e.*, the specialized vocabulary of mathematics. Given systematic instruction, this student should be able to achieve at or above grade level. **What to do?** Scan all of the suggestions for English learners in each chapter and progress through the ones the student needs as quickly as possible. Monitor carefully to make sure this student continues to progress at a reasonable pace.

achievement and reading achievement levels. Use the chart on the facing page as a guide to understanding student assessment data:

Suggestions for Mathematics Teachers of English Learners

1. Allocate additional time for mathematics. Many students will be translating from English to their primary language and back again. The meaning of many words will not be immediately clear. When you ask questions, allow extra time for students to respond. Reading mathematics textbooks and understanding what is asked for in a word problem will be slower.

2. Use student's background knowledge. Some English learners will have developed substantial background in mathematics; others will have very little. Find out what students know and then build on that knowledge.

3. Reduce the amount and sophistication of the English language used. This may be done by reordering the lessons in each chapter to begin with key vocabulary, followed by problems with a minimum of written English, followed by at least one word problem each day. Monitor and simplify the speech you use. Speak more slowly, avoid idioms and slang, be precise and concise, and use short sentences and simple vocabulary. Using hand gestures and pictures as well as words aids communication.

4. Introduce one concept per day. Keeping the focus of each day simple will aid students in understanding the point of the lesson. Focus on key concepts, and use mathematics instructional time well.

5. Use a variety of different methods for getting a point across. Presenting concepts verbally and visually, with concrete examples and in abstract mathematical symbols, and using pictures, graphs, diagrams, and charts will enhance the chance that students will understand at least one of the presentations. As you introduce a new word, rule, or theorem, write it down.

6. Provide opportunities for English learners to interact with their English-speaking peers. Students who are learning a language need to hear native speakers using the language, and they need opportunities to use their new mathematics vocabulary in their speech and in their writing.

7. Provide opportunities for English learners to discuss their understandings with each other, confirm the homework assignments, or ask questions of each other in whatever language they may have in common.

8. Allow English learners to demonstrate what they know in a variety of ways. When students first learn a language, they are usually shy about speaking. They generally understand spoken language before they can reproduce it. Students who have recently arrived from another country with good prior schooling may be able to read in English but not speak it. Allow students to point, nod, gesture, draw a picture, or work math problems without words as they learn English.

9. Extend mathematics instructional time through homework, an extra class period, summer school, or tutoring. Many of the language-related suggestions for English learners in this series can be carried out by the language arts teacher.

10. Keep on hand picture dictionaries, foreign language dictionaries, and drawing materials.

Specific Suggestions

Prior to each chapter you will find suggestions to help you modify curriculum and instruction so that the content is accessible to English learners. Many of the suggestions are well suited for discussion in a language arts class, English as a second language class, or in a tutorial. In these sections we will provide you suggestions and activities designed to (1) teach the vocabulary commonly used in mathematics; (2) explain mathematical concepts in a variety of ways; and (3) dissect the structure of word problems. Much of the vocabulary study in this book may be review for your students, and in that case, you should feel free to work as quickly as possible through the activities. For those students who need more systematic study, progressing through the activities as indicated will ensure that students have refreshed their understanding of basic terms prior to statewide testing that generally occurs toward the end of each school year. We recommend that if you have English learners in your classroom, you skim all the chapter suggestions for English learners so that you may use them as you need them.

DECIMALS AND FRACTIONS

Skills Review
pp. 655–659

Evaluate.

1. $5.125 + 0.78$ 5.905   **2.** $130.5 - 1.09$ 129.41   **3.** 3.9×2.4 9.36   **4.** $9.6 \div 0.02$ 480

Evaluate. Write the answer in simplest form.

5. $\frac{1}{8} + \frac{5}{8}$ $\frac{3}{4}$ **6.** $\frac{3}{4} - \frac{1}{2}$ $\frac{1}{4}$ **7.** $\frac{4}{5} \times \frac{3}{10}$ $\frac{6}{25}$ **8.** $\frac{7}{8} \div \frac{1}{2}$ $\frac{7}{4}$

Write the fraction as a decimal. For repeating decimals, also round to the nearest hundredth for an approximation.

9. $\frac{2}{5}$ 0.4   **10.** $\frac{5}{8}$ 0.625   **11.** $\frac{1}{3}$ $0.\overline{3}$; 0.3   **12.** $\frac{1}{6}$ $0.1\overline{6}$; 0.17

Write the decimal as a fraction in simplest form.

13. 0.25 $\frac{1}{4}$ **14.** 0.375 $\frac{3}{8}$ **15.** 0.51 $\frac{51}{100}$ **16.** $0.\overline{6}$ $\frac{2}{3}$

RATIO AND PROPORTION

Skills Review
pp. 660–661

There are 22 students in a geometry class, 12 girls and 10 boys. Write each ratio in simplest form.

17. boys : girls $\frac{5}{6}$ **18.** girls : boys $\frac{6}{5}$

19. boys : students $\frac{5}{11}$ **20.** girls : students $\frac{6}{11}$

Solve the proportion.

21. $\frac{x}{5} = \frac{16}{20}$ 4   **22.** $\frac{9}{x} = \frac{12}{21}$ $15\frac{3}{4}$ **23.** $\frac{3}{4} = \frac{x}{30}$ $22\frac{1}{2}$ **24.** $\frac{5}{2} = \frac{4}{x}$ $1\frac{3}{5}$

INEQUALITIES AND ABSOLUTE VALUE

Skills Review
p. 662

Compare the two numbers. Write the answer using <, >, or =.

25. -4 and -8
$-4 > -8$ **26.** 6 and -6
$6 > -6$ **27.** -1.5 and -1.9
$-1.5 > -1.9$ **28.** 3.08 and 3.17
$3.08 < 3.17$

Evaluate.

29. $|-5|$ 5   **30.** $|-17|$ 17   **31.** $|3|$ 3   **32.** $|9.9|$ 9.9

INTEGERS

Skills Review p. 663

Evaluate.

33. $-3 + 16$ 13

34. $-8 - 4$ −12

35. $(-5)(-9)$ 45

36. $100 \div (-4)$ −25

THE COORDINATE PLANE; SLOPE OF A LINE

Skills Review pp. 664–665

Plot the points in a coordinate plane. 37–40. See margin.

37. $A(3, 1)$

38. $B(2, -4)$

39. $C(-1, 5)$

40. $D(0, -2)$

Find the slope of the line that passes through the points.

41. $(-3, -2)$ and $(0, 0)$ $\frac{2}{3}$

42. $(-2, 1)$ and $(5, 1)$ 0

43. $(0, 4)$ and $(6, -2)$ −1

POWERS AND SQUARE ROOTS

Skills Review pp. 668–669

Simplify.

44. $(-3)^3$ −27

45. 2^4 16

46. $\sqrt{81}$ 9

47. $\sqrt{64}$ 8

48. $\sqrt{50}$ $5\sqrt{2}$

49. $\sqrt{72}$ $6\sqrt{2}$

50. $\frac{4}{\sqrt{2}}$ $2\sqrt{2}$

51. $\sqrt{\frac{1}{9}}$ $\frac{1}{3}$

EVALUATING EXPRESSIONS; THE DISTRIBUTIVE PROPERTY

Skills Review pp. 670–671

Evaluate the expression when $n = -1$.

52. $2n^2 - 5$ −3

53. $n + 90$ 89

54. $n(n - 4)$ 5

55. $(10 - n)^2$ 121

Use the distributive property to rewrite the expression without parentheses.

56. $-8(x + 3)$ $-8x - 24$

57. $a(a - 7)$ $a^2 - 7a$

58. $(y - 6)(5)$ $5y - 30$

59. $(12 + z)z$ $12z + z^2$

SOLVING EQUATIONS

Skills Review pp. 672–673

Solve the equation.

60. $x + 11 = 4$ −7

61. $17 = m - 16$ 33

62. $12 - x = 15$ −3

63. $-8y = -2$ $\frac{1}{4}$

64. $\frac{x}{2} = -9$ −18

65. $3a - 1 = 8$ 3

66. $24 = \frac{5}{8}x + 4$ 32

67. $16z - 9 = 7z$ 1

37–40.

Pre-Course Practice

DECIMALS AND FRACTIONS

Skills Review
pp. 655–659

Evaluate.

1. $0.67 + 1.045$ 1.715   **2.** $8.5 + 1.52$ 10.02   **3.** $13.7 + 0.03$ 13.73   **4.** $\$4.19 + \$10.50$ $14.69

5. $0.45 - 0.08$ 0.37   **6.** $120 - 58.5$ 61.5   **7.** $16.8 - 3.72$ 13.08   **8.** $\$20 - \$1.99 - \$5$ $13.01

9. 3.4×6.1 20.74   **10.** 0.02×10 0.2   **11.** 8.4×1.05 8.82   **12.** $\$15 \times 0.06$ $.90

13. $33.5 \div 0.01$ 3350   **14.** $0.418 \div 2$ 0.209   **15.** $15 \div 1.5$ 10   **16.** $56.44 \div 8.3$ 6.8

Write two fractions equivalent to the given fraction.   17–21. Sample answers are given.

17. $\frac{2}{3}$ $\frac{4}{6}, \frac{20}{30}$ **18.** $\frac{1}{5}$ $\frac{2}{10}, \frac{3}{15}$ **19.** $\frac{10}{20}$ $\frac{1}{2}, \frac{5}{10}$ **20.** $\frac{9}{12}$ $\frac{3}{4}, \frac{18}{24}$ **21.** $\frac{4}{7}$ $\frac{8}{14}, \frac{20}{35}$

Find the reciprocal of the number.

22. $\frac{3}{8}$ $\frac{8}{3}$ **23.** 9 $\frac{1}{9}$ **24.** $\frac{1}{4}$ 4   **25.** $\frac{5}{2}$ $\frac{2}{5}$ **26.** 1 1

Add or subtract. Write the answer in simplest form.

27. $\frac{1}{4} + \frac{3}{4}$ 1   **28.** $\frac{5}{12} + \frac{5}{12}$ $\frac{5}{6}$ **29.** $\frac{5}{9} - \frac{2}{9}$ $\frac{1}{3}$ **30.** $\frac{6}{7} - \frac{1}{7}$ $\frac{5}{7}$

31. $\frac{2}{5} + \frac{1}{15}$ $\frac{7}{15}$ **32.** $\frac{5}{6} - \frac{2}{9}$ $\frac{11}{18}$ **33.** $\frac{7}{8} - \frac{3}{4}$ $\frac{1}{8}$ **34.** $\frac{2}{3} - \frac{1}{2}$ $\frac{1}{6}$

Multiply or divide. Write the answer in simplest form.

35. $\frac{3}{4} \times \frac{1}{4}$ $\frac{3}{16}$ **36.** $\frac{1}{2} \times \frac{4}{5}$ $\frac{2}{5}$ **37.** $15 \times \frac{3}{8}$ $5\frac{5}{8}$ **38.** $\frac{5}{8} \times \frac{2}{3}$ $\frac{5}{12}$

39. $\frac{3}{10} \div \frac{3}{5}$ $\frac{1}{2}$ **40.** $\frac{7}{12} \div \frac{1}{3}$ $1\frac{3}{4}$ **41.** $20 \div \frac{2}{3}$ 30   **42.** $\frac{7}{10} \div 8$ $\frac{7}{80}$

Write the fraction as a decimal. For repeating decimals, also round to the nearest hundredth for an approximation.

43. $\frac{1}{2}$ 0.5   **44.** $\frac{2}{3}$ $0.\overline{6}$; 0.67   **45.** $\frac{4}{9}$ $0.\overline{4}$; 0.44   **46.** $\frac{3}{8}$ 0.375   **47.** $\frac{3}{4}$ 0.75

Write the decimal as a fraction in simplest form.

48. 0.13 $\frac{13}{100}$ **49.** 0.05 $\frac{1}{20}$ **50.** 0.625 $\frac{5}{8}$ **51.** 0.3 $\frac{3}{10}$ **52.** $0.\overline{8}$ $\frac{8}{9}$

RATIO AND PROPORTION

Skills Review
pp. 660–661

There are 27 students in a geometry class, 15 boys and 12 girls. Write each ratio in simplest form.

1. boys : girls $\frac{5}{4}$

2. girls : boys $\frac{4}{5}$

3. girls : students $\frac{4}{9}$

4. boys : students $\frac{5}{9}$

Simplify the ratio.

5. $\frac{1\ \text{ft}}{4\ \text{in.}}$ $\frac{3}{1}$

6. $\frac{20\ \text{ft}}{5\ \text{yd}}$ $\frac{4}{3}$

7. $\frac{85\ \text{cm}}{1\ \text{m}}$ $\frac{17}{20}$

8. $\frac{2\ \text{kg}}{500\ \text{g}}$ $\frac{4}{1}$

Solve the proportion.

9. $\frac{x}{3} = \frac{20}{21}$ $2\frac{6}{7}$

10. $\frac{18}{x} = \frac{9}{4}$ 8

11. $\frac{5}{8} = \frac{x}{2}$ $1\frac{1}{4}$

12. $\frac{9}{14} = \frac{9}{x}$ 14

13. $\frac{5}{x} = \frac{2}{10}$ 25

14. $\frac{9}{10} = \frac{x}{4}$ $3\frac{3}{5}$

15. $\frac{x}{10} = \frac{9}{50}$ $1\frac{4}{5}$

16. $\frac{24}{36} = \frac{6}{x}$ 9

INEQUALITIES AND ABSOLUTE VALUE

Skills Review
p. 662

Compare the two numbers. Write the answer using <, >, or =.

1. -8 and -10 $-8 > -10$
2. 934 and 943 $934 < 943$
3. 0 and -5 $0 > -5$
4. -8 and 8 $-8 < 8$

5. $\frac{1}{4}$ and 0.25 $\frac{1}{4} = 0.25$
6. $\frac{1}{9}$ and $\frac{1}{7}$ $\frac{1}{9} < \frac{1}{7}$
7. 8.65 and 8.74 $8.65 < 8.74$
8. -0.5 and -0.55 $-0.5 > -0.55$

Write the numbers in order from least to greatest.

9. 4, -6, -9, 0, 3 $-9, -6, 0, 3, 4$

10. 3.06, 3.16, 3.6, 3.1, 3.61 3.06, 3.1, 3.16, 3.6, 3.61

11. 8652, 8562, 8256, 8265, 8526
8256, 8265, 8526, 8562, 8652

12. $-2.4, -1.6, -0.8, -1.9, -2.0$
$-2.4, -2.0, -1.9, -1.6, -0.8$

Evaluate.

13. $|9|$ 9
14. $|-16|$ 16
15. $|-1|$ 1
16. $|0|$ 0
17. $|2.5|$ 2.5
18. $|-4|$ 4

INTEGERS

Skills Review
p. 663

Evaluate.

1. $-7 + (-5)$ -12
2. $8 + (-2)$ 6
3. $-10 + 6$ -4
4. $20 + (-5) + 2$ 17

5. $-6 - 5$ -11
6. $19 - 24$ -5
7. $-1 - (-6)$ 5
8. $12 - (-9)$ 21

9. $8(-7)$ -56
10. $-1(20)$ -20
11. $-8(-4)$ 32
12. $-5(5)(6)$ -150

13. $48 \div (-4)$ -12
14. $-8 \div (-1)$ 8
15. $-75 \div 25$ -3
16. $-81 \div (-27)$ 3

Pre-Course Practice xxiii

T45

10–15.

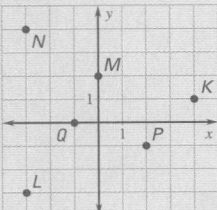

THE COORDINATE PLANE

Skills Review p. 664

Give the coordinates of each point.

1. A (1, 3)   **2.** B (−2, 1)   **3.** C (2, 0)

4. D (−2, −4)   **5.** E (−4, 2)   **6.** F (4, −2)

7. G (3, 3)   **8.** H (0, −3)   **9.** J (−3, −1)

Plot the points in a coordinate plane. 10–15. See margin.

10. K(4, 1)   **11.** L(−3, −3)   **12.** M(0, 2)

13. N(−3, 4)   **14.** P(2, −1)   **15.** Q(−1, 0)

SLOPE OF A LINE

Skills Review p. 665

Find the slope of the line that passes through the points.

1. (−3, −1) and (0, 0) $\frac{1}{3}$ **2.** (4, 2) and (1, 7) $-\frac{5}{3}$ **3.** (−3, 2) and (6, 2) 0

4. (−1, 2) and (1, −2) −2   **5.** (−2, −5) and (2, −3) $\frac{1}{2}$ **6.** (2, 6) and (4, 4) −1

Plot the points and draw a line that passes through them. Determine whether the slope is *positive, negative, zero,* or *undefined.* 7–12. Check graphs.

7. (4, −1) and (−2, −1) zero   **8.** (0, 0) and (3, −2) negative   **9.** (−1, 4) and (−1, 3) undefined

10. (0, −2) and (0, 5) undefined   **11.** (2, 1) and (−5, 1) zero   **12.** (2, 2) and (4, 5) positive

POWERS AND SQUARE ROOTS

Skills Review pp. 668–669

Evaluate.

1. 5^3 125   **2.** $(-2)^3$ −8   **3.** $(-4)^2$ 16   **4.** 100^2 10,000   **5.** $(-1)^4$ 1

Find all square roots of the number or write *no real square roots.*

6. 81 9, −9   **7.** 16 4, −4   **8.** −4 no real square roots   **9.** −100 no real square roots   **10.** 36 6, −6

Evaluate. Give the exact value if possible. Otherwise, approximate to the nearest tenth.

11. $\sqrt{25}$ 5   **12.** $\sqrt{9}$ 3   **13.** $\sqrt{10}$ 3.2   **14.** $\sqrt{21}$ 4.6   **15.** $\sqrt{49}$ 7

Simplify.

16. $\sqrt{20}$ $2\sqrt{5}$ **17.** $\sqrt{600}$ $10\sqrt{6}$ **18.** $\sqrt{75}$ $5\sqrt{3}$ **19.** $\sqrt{63}$ $3\sqrt{7}$ **20.** $\sqrt{52}$ $2\sqrt{13}$

21. $\sqrt{\frac{4}{81}}$ $\frac{2}{9}$ **22.** $\sqrt{\frac{1}{4}}$ $\frac{1}{2}$ **23.** $\frac{1}{\sqrt{2}}$ $\frac{\sqrt{2}}{2}$ **24.** $\frac{5}{\sqrt{3}}$ $\frac{5\sqrt{3}}{3}$ **25.** $\frac{10}{\sqrt{5}}$ $2\sqrt{5}$

EVALUATING EXPRESSIONS

Skills Review
p. 670

Evaluate the expression.

1. $12 - 8 \cdot 2 + 4$ 0   **2.** $20 - 3^2 - 5$ 6   **3.** $7^2 \cdot 2 + 2$ 100   **4.** $(8 + 19) \div (4 - 1)$ 9

5. $19 + 16 \div 4$ 23   **6.** $3(8 - 2)^2$ 108   **7.** $(-7)(5 - 2 \cdot 3)$ 7   **8.** $8 \cdot 4 - 5^2$ 7

Evaluate the expression when $n = -4$.

9. $2n^2$ 32   **10.** $(-5n)^2$ 400   **11.** $n(n + 6)$ -8 **12.** $90 - n$ 94

13. $-8n + 10$ 42   **14.** $11 + 6n$ -13 **15.** $(-n)^3$ 64   **16.** $n^2 - 100$ -84

THE DISTRIBUTIVE PROPERTY

Skills Review
p. 671

Use the distributive property to rewrite the expression.

1. $3(y - 5)$ $3y - 15$ **2.** $(4a - 1)6$ $24a - 6$ **3.** $-2(3x + 3)$ $-6x - 6$ **4.** $(10 + z)z$ $10z + z^2$

5. $2(x + y)$ $2x + 2y$ **6.** $(c + 8)(-5)$ $-5c - 40$ **7.** $x(x - 4)$ $x^2 - 4x$ **8.** $6x(x - 1)$ $6x^2 - 6x$

Simplify the expression.

9. $5x + 8 - 9x$
$-4x + 8$
13. $2(x - 5) + 13$
$2x + 3$

10. $-12y - y + 2$
$-13y + 2$
14. $4y - (y + 6)$
$3y - 6$

11. $4 + z + z - 5$
$2z - 1$
15. $18 - (9 - x)$
$9 + x$

12. $x^2 + 2x - 6x - 12$
$x^2 - 4x - 12$
16. $a(a + 2) - 5(a - 3)$
$a^2 - 3a + 15$

SOLVING EQUATIONS

Skills Review
pp. 672–673

Solve the equation.

1. $x - 4 = 18$ 22   **2.** $10 = m + 7$ 3   **3.** $x + 2 = -5$ -7 **4.** $y - 8 = -6$ 2

5. $4a = -24$ -6 **6.** $-x = 9$ -9 **7.** $6m - 5 = 1$ 1   **8.** $2n + 3 = 11$ 4

9. $5a + 7 = 4a$ -7 **10.** $8 - 7x = x$ 1   **11.** $10 - 2x = 2x$ $\frac{5}{2}$ **12.** $2n - 1 = 5n + 8$ -3

13. $\frac{n}{5} = 7$ 35   **14.** $-\frac{2}{3}z = 22$ -33 **15.** $\frac{x}{8} - 2 = -13$ -88 **16.** $8 = \frac{1}{2}x + 6$ 4

USING FORMULAS

Skills Review
p. 674

Use the formulas for a rectangle with length ℓ and width w.

1. A rectangle is 8 cm long and 6 cm wide. Find the perimeter. 28 cm

Area $= \ell w$
Perimeter $= 2\ell + 2w$

2. Find the area of a rectangle with length 14 ft and width 5 ft. 70 ft^2

3. Find the length of a rectangle with area 56 m^2 and width 8 m. 7 m

4. A rectangle has perimeter 20 in. and length 3 in. Find the width. 7 in.

Pre-Course Practice   **xxv**

T47

Pacing and Assignment Guide

REGULAR SCHEDULE

Lesson	Les. Day	Basic	Average	Advanced
1.1	Day 1	SRH p. 663 Exs. 1–8; pp. 5–7 Exs. 11–14, 17–20, 36, 41–53 odd	pp. 5–7 Exs. 13–24, 34–36, 39–53 odd	pp. 5–7 Exs. 15–25, 33–36, 39–53 odd
	Day 2	pp. 5–7 Exs. 26–28, 32, 34, 37, 38–52 even	pp. 5–7 Exs. 26–29, 32, 33, 37, 38–52 even	pp. 5–7 Exs. 26–32, 37, 38–52 even; EC: TE p. 1D*, classzone.com
1.2	Day 1	pp. 11–13 Exs. 8, 9, 13–15, 23, 24–40 even	pp. 11–13 Exs. 8–10, 12–15, 23, 24–40 even	pp. 11–13 Exs. 8–16, 23–41 odd
	Day 2	pp. 11–13 Exs. 17–20, 22, 25–41 odd	pp. 11–13 Exs. 17–20, 22, 25–41 odd	pp. 11–13 Exs. 17–20, 21*, 22–40 even; EC: classzone.com
1.3	Day 1	pp. 17–20 Exs. 15–18, 19–33 odd, 36, 38–41, 50– 55, 71–81 odd	pp. 17–20 Exs. 15–26, 32–48 even, 54, 55, 70–82 even	pp. 17–20 Exs. 15–26, 31–51 odd, 55, 57, 70–82 even
	Day 2	pp. 17–20 Exs. 20–34 even, 35, 42–49, 58–60, 63, 70, Quiz 1	pp. 17–20 Exs. 31–51 odd, 56, 58–61, 63, 64, 66, 68, Quiz 1	pp. 17–20 Exs. 34–48 even, 58–69, Quiz 1; EC: TE p. 1D*, classzone.com
1.4	Day 1	pp. 25–27 Exs. 7–12, 16–26 even, 35, 36, 38–40	pp. 25–27 Exs. 7–19, 23–27 odd, 36–40	pp. 25–27 Exs. 7–15 odd, 19–21, 23–26, 35–40
	Day 2	pp. 25–27 Exs. 28–31, 41–52	pp. 25–27 Exs. 28–32, 34, 41–53 odd	pp. 25–27 Exs. 28–32, 33*, 34, 42–54 even; EC: classzone.com
1.5	Day 1	pp. 31–33 Exs. 7–15 odd, 18–24 even, 35, 39–45 odd	pp. 31–33 Exs. 10–16, 17–23 odd, 35–45 odd	pp. 31–33 Exs. 10–14, 19–23, 31, 37–45 odd
	Day 2	SRH p. 664 Exs. 13–20; pp. 31–33 Exs. 23–31 odd, 32–44 even	pp. 31–33 Exs. 25–27, 30–44 even	pp. 31–33 Exs. 24–32 even, 33*, 34–36, 38–44 even; EC: classzone.com
1.6	Day 1	pp. 38–41 Exs. 15–23, 34, 37, 40, 41, 46–50, 53–59 odd	pp. 38–41 Exs. 15–23, 34–38 even, 40–42, 46–50, 53–59 odd	pp. 38–41 Exs. 15–23 odd, 35–39 odd, 42–44, 46–48, 51, 52–60 even
	Day 2	pp. 38–41 Exs. 24–31, 54–60 even, Quiz 2	pp. 38–41 Exs. 24–32, 44, 45, 54–60 even, Quiz 2	pp. 38–41 Exs. 24–29, 32, 33, 45, Quiz 2; EC: TE p. 1D*, classzone.com
Review	Day 1	pp. 42–45 Exs. 1–39	pp. 42–45 Exs. 1–39	pp. 42–45 Exs. 1–39
Assess	Day 1	Chapter 1 Test	Chapter 1 Test	Chapter 1 Test

YEARLY PACING   Chapter 1 Total – **14 days**   Chapter 1 Total – **14 days**   Remaining – **146 days**

*Challenge Exercises   EP = Extra Practice   SRH = Skills Review Handbook   EC = Extra Challenge

BLOCK SCHEDULE

Day 1	Day 2	Day 3	Day 4	Day 5	Day 6	Day 7
1.1 pp. 5–7 Exs. 13–24, 26–29, 32–39, 41–53 odd	**1.2** pp. 11–13 Exs. 8–15, 17–20, 22–25, 26–40 even	**1.3** pp. 17–20 Exs. 15–26, 28–52 even, 54–61, 63–66, 68, 70–82 even, Quiz 1	**1.4** pp. 25–27 Exs. 7–19, 23–27 odd, 28–32, 34, 36–40, 41–53 odd	**1.5** pp. 31–33 Exs. 10–16, 17–23 odd, 25–27, 30–34 even, 35–45	**1.6** pp. 38–41 Exs. 15–32, 34–38 even, 40–42, 45–50, 53–60, Quiz 2	**Review** pp. 42–45 Exs. 1–39 **Assess** Chapter 1 Test

YEARLY PACING   Chapter 1 Total – **7 days**   Chapter 1 Total – **7 days**   Remaining – **73 days**

Support Materials

CHAPTER RESOURCE BOOK

CHAPTER SUPPORT

Tips for New Teachers	p. 1	Strategies for Reading Mathematics	p. 5
Parent Guide for Student Success	p. 3		

LESSON SUPPORT

	1.1	1.2	1.3	1.4	1.5	1.6
Lesson Plans (regular and block)	p. 7	p. 16	p. 24	p. 35	p. 43	p. 53
Warm-Up Exercises and Daily Quiz	p. 9	p. 18	p. 26	p. 37	p. 45	p. 55
Technology Activities & Keystrokes			p. 27		p. 46	p. 56
Practice (2 levels)	p. 10	p. 19	p. 29	p. 38	p. 47	p. 59
Reteaching with Practice	p. 12	p. 21	p. 31	p. 40	p. 49	p. 61
Quick Catch-Up for Absent Students	p. 14	p. 23	p. 33	p. 42	p. 51	p. 63
Learning Activities						
Real-Life Applications	p. 15				p. 52	

REVIEW AND ASSESSMENT

Quizzes	pp. 34, 64	Alternative Assessment with Math Journal	p. 71
Chapter Review Games and Activities	p. 65	Project with Rubric	p. 73
Chapter Test (2 levels)	p. 66	Cumulative Review	p. 75
SAT/ACT Chapter Test	p. 70	Resource Book Answers	AN1

TRANSPARENCIES

	1.1	1.2	1.3	1.4	1.5	1.6
Warm-Up Exercises and Daily Quiz	p. 2	p. 3	p. 4	p. 5	p. 6	p. 7
Visualize It Transparencies			✓		✓	✓
Answer Transparencies	✓	✓	✓	✓	✓	✓

TECHNOLOGY

- Time-Saving Test and Practice Generator
- Electronic Teacher Tools
- Geometry in Motion: Video
- Classzone.com
- Online Lesson Planner

ADDITIONAL RESOURCES

- Worked-Out Solution Key
- Resources in Spanish
- Practice Workbook with Examples

Providing Universal Access

Prior to each chapter, we will outline modifications of curriculum and instruction designed to address the unique needs of English Learners, Advanced Learners, and Strategic Learners. English Learners are those who are not yet fluent in English; Advanced Learners are those whose mathematics achievement is above grade level; and Strategic Learners are those whose mathematics achievement is below grade level (in other words, those who need strategic and sustained assistance in order to be successful in mathematics). Each class of students is different, and you may find your whole class benefits from some of the suggestions. Most of the activities for English Learners and Strategic Learners would best be done in a second class period, as homework, or in a tutorial, as both groups of students need increased instructional time in mathematics. Some of the activities, particularly those involving vocabulary development, would fit nicely into a language arts period. The activities for Advanced Learners in these pages and throughout the text are meant as substitutes for the easier problems in the text.

Strategies for Strategic Learners

ASSESS STUDENT ACHIEVEMENT

This chapter includes a review of concepts that students should have already learned, such as making predictions based on patterns. It assumes that students have basic familiarity with the coordinate plane.

Teachers are encouraged to review the math test results available on each student from the previous year. End-of-year tests that came with the mathematics book the student used, as well as any standardized testing that might be available, can be utilized. Teachers may want to administer the Pre-Course Test provided on pages T42 and T43. Tests should be reviewed to see if there are common areas of weakness among students, and, if so, the prior year's textbook and ancillaries would provide ample teaching suggestions and practice exercises. Many students, especially those not in year-round schools, will simply have forgotten over the summer and will require just enough explanation and exercises to refresh their memories. Some students, however, will have gaps in their knowledge base. For these students, the teacher can select relevant portions of the prior year's textbook or problems in the ancillaries for review. Ideally, students who need extra help should be scheduled for an extra period of mathematics or tutorial help.

REVIEW PREVIOUSLY LEARNED CONCEPTS

COORDINATE PLANES A review of the basic components of a coordinate plane and how to plot points in the plane will be beneficial for many students. Discussing the vocabulary associated with coordinate planes, such as *x-axis*, *y-axis*, *origin*, and *quadrant*, as well as the terms *x-coordinate*, *y-coordinate*, and *ordered pair* encountered when plotting points will help students reacquire skills from their algebra course that may have eroded since the previous school year. Students will encounter examples and exercises in the lessons of Chapter 1 that involve points and segments shown in coordinate planes.

Strategies for English Learners

VOCABULARY DEVELOPMENT

Common English words and specialized mathematical terms are vital to understanding and expressing mathematical concepts. By high school, most native English speakers have developed a full vocabulary of common English words, including mathematical terms and concepts. Words that express temporal relationships, such as *first*, *second*, *third*, *before*, *after*, *yesterday*, *today*, and *tomorrow*; words that describe shapes, such as *circle*, *triangle*, and *square*; and words that describe the relationships of objects in space such as *over*, *under*, and *between*, are commonly understood by fluent English speakers in high school. But for the English learner, these words may represent large stumbling blocks as these students try to understand mathematical concepts, decide what is asked for in a word problem, and express their thinking in mathematical terms. Because of the particular difficulty that word problems present, you may want to start each chapter with the Key Words and Chapter Readiness Quiz on the Study Guide page before turning back to the Application presented on the chapter opener pages that begin the chapter.

MATHEMATICS VOCABULARY Plan to spend some time each day developing vocabulary related to mathematics. You might start with your English learners by reading the 500 most common words in the English language. Several such lists are readily available. These common words can form the basis of vocabulary study done outside math class, in language arts or in a tutorial setting.

WORDS AND SYMBOLS Mathematical notation can be confusing at this stage. Review with students the symbols related to the concepts of absolute value, squares, and square roots. Also review the order of operations, including powers and grouping symbols.

Strategies for Advanced Learners

DIFFERENTIATE INSTRUCTION IN TERMS OF COMPLEXITY

Advanced students may find it relatively easy to discern the pattern in many of the visual pattern problems posed in Lesson 1.1. Ask these students to create some patterns of their own that involve having more than one characteristic of the figures changing. For example, the pattern could involve a sequence of figures whose number of sides increase by one and which rotate around their center in relationship to the previous figure. The students' visual patterns can be displayed so that all students have an opportunity to determine the pattern and sketch the next figure.

DIFFERENTIATE INSTRUCTION IN TERMS OF DEPTH

Since advanced students tend to move more quickly through the instructional material than other students, the challenge problems in this book and the Teacher's Edition can be substituted for easier problems. Also, peruse the classzone.com web site with these students to look for additional problems and activities that can supplement their geometry course. Point out the challenge problems in each chapter. Encourage these students throughout the year to delve more deeply into mathematics content that interests them and to find connections between the mathematics in this course and the content of other courses, such as science, language arts, music, art, and history.

Challenge Problem for use with Lesson 1.1:

The arrangement of numbers below, called *Pascal's Triangle* after a French mathematician and philosopher of the seventeenth century, can be continued indefinitely.

```
          1
        1   1
      1   2   1
    1   3   3   1
  1   4   6   4   1
          ⋮
```

a. Each row starts and ends with 1. Do you see a relationship between the other entries in each row and those in the row directly above it? (*Yes; each of the other entries in a row is the sum of the two numbers to either side of it in the row above.*)

b. Write the next three rows of Pascal's Triangle.
Solution:
```
        1   5   10   10   5   1
      1   6   15   20   15   6   1
    1   7   21   35   35   21   7   1
```

c. If you take the second number from the left in each row, starting with the second row, what pattern do you get? Suppose you take the third number of each row, starting with the third row. What pattern do you get? (*The list of second numbers is a list of consecutive whole numbers beginning with 1. The list of third numbers has the form 1, 1 + 2, 1 + 2 + 3, 1 + 2 + 3 + 4, and so on (called the triangular numbers).*)

Challenge Problem for use with Lesson 1.3:

Can you arrange 10 points in a plane in five lines with four points in each line?
Solution:

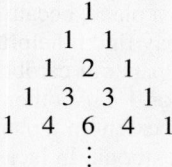

Challenge Problem for use with Lesson 1.6:

In the diagram, $\angle AXB$ and $\angle CXD$ are straight angles. $\angle 1$, $\angle 2$, and $\angle 3$ are congruent to each other, and $\angle 4$, $\angle 5$, and $\angle 6$ are congruent to each other. $\angle CXE$ is a right angle. Find $m\angle 6$. (*50°*)

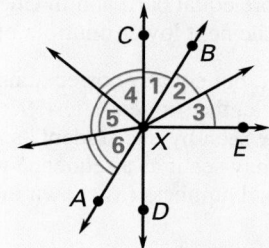

As a prelude to studying geometric figures, the first two lessons of the chapter introduce the related ideas of pattern recognition and inductive reasoning. In Lesson 1.3, the basic geometric concepts of point, line, and plane are introduced, along with the basic relationships among them. Lesson 1.4 offers instructions on sketching geometric objects, which will aid students in making drawings of situations that are described verbally and give them a better understanding of geometric relationships. The concept of length is introduced in Lesson 1.5. The notion of betweenness leads to a statement of the Segment Addition Postulate, which is then used to find the lengths of segments. Lesson 1.6 discusses angles, another basic concept of geometry. The system of measuring angles in degrees is introduced, and the idea of the interior of an angle leads to a statement of the Angle Addition Postulate.

Lesson 1.1

Stress that when examining a given sequence of figures or numbers, students should compare and contrast the entries to discern the pattern. Once they have determined the pattern, students can use it to predict the next figure or number.

QUICK TIP
But what if students are unable to discern the pattern? As a starting point, urge students to find one characteristic that the entries all have in common. For example, in Checkpoint Exercise 1 on page 3 each figure is a circle, and in Checkpoint Exercise 4 each number is a multiple of 5. Then instruct students to identify how each successive figure or number differs from the one that precedes it. In Checkpoint Exercise 1, each successive circle is divided into one more equal part; and in Checkpoint Exercise 4, each successive number is the next lower multiple of 5.

On occasion, you should expect different, but equally correct, answers to pattern recognition problems. For example, the sequence 1, 4, 9, 16, 25, 36, … (Example 2 of Lesson 1.1) may be seen by one student as a string of square numbers, but another student may see it as a sequence whose successive terms differ by consecutive odd numbers (as shown in Example 2 of Lesson 1.2).

Lesson 1.2

Emphasize that conjectures are only educated guesses and that they must be verified as true for all possible cases.

DISPROVING A CONJECTURE It sometimes happens, however, that conjectures are not true. Emphasize that one *counterexample* is enough to invalidate a conjecture. After refreshing your students' memories about prime numbers, present the following conjecture.

The expression $n^2 - n + 11$ produces a prime number whenever n is a positive integer.

Have students check the validity of this conjecture for $n = 1, 2, 3, …$. Students should discover that the conjecture is true for $n = 1, 2, 3, …, 10$, but that it fails for $n = 11$. Therefore, the conjecture is false.

Numbers can often describe geometric patterns. Several examples are mentioned in the Practice and Applications sections of Lessons 1.1 and 1.2. For a classroom project, have students try to divide a square into as many separate regions as possible for a given number of straight lines. Using one line, a square can be divided into 2 regions. Using two lines, the maximum number of regions that can be created is 4. The results for the first four positive integers are given in the following table. (*Note*: It becomes increasingly difficult to draw the figure as the number of lines increases. Students may need assistance drawing the fourth figure.)

Number of Lines	1	2	3	4
Number of Regions	2	4	7	11

Ask students to write a conjecture about the maximum number of regions that can be formed in a square by an increasing number of lines. Refer students back to the Geo-Activity on page 8 if they need help seeing the numerical pattern.

Lesson 1.3

Students may be confused by the phrase *undefined terms* used in this lesson. It is important to stress that the terms *point*, *line*, and *plane* are undefined because these are the basic terms used to define other terms used in geometry. Point out that students probably already have a good understanding of these terms based on the objects they see every day.

VISUALIZING DIAGRAMS Students sometimes have difficulty visualizing drawings that illustrate a line and a plane, because it is not clear when the line is *on* (or *in*) the plane. You may find it helpful to model this situation using a pencil and a piece of stiff paper or cardboard. Sketching a line in a plane is a topic discussed in Lesson 1.4. Another source of difficulty for students is visualizing the infinite extent of a plane when planes are represented by parallelograms in the textbook. In fact, three-dimensional visualization of any kind is often difficult for students. A good plan is to try to illustrate new geometric statements with models of some kind (boxes for solids, sheets of paper for planes, pencils for lines) whenever possible.

SYMBOLS/NOTATION Make sure students understand, and distinguish in their own writing, the notations for a line, a segment, and a ray. Emphasize the importance of the order of the points used to identify a ray, with the endpoint always being listed first. Doing this successfully indicates that students understand the geometric distinctions between these objects.

Lesson 1.4

QUICK TIP

Initially, sketching geometric figures is difficult for many students. If you have a rectangular desk, begin by having students draw the top of the desk from their perspective. It may help them to close one eye in order to suppress their depth perception. Most students should draw the desktop in the shape of a parallelogram (or a trapezoid if they are sitting directly in front of your desk). If some students draw rectangles, persuade them to see that this view can only be observed from a position directly above the desk. After students have drawn parallelograms representing the desktop, have them add a vertical plane to their drawings, which you could model by holding a large sheet of stiff paper or cardboard perpendicular to the desktop.

You can model a line intersecting a plane by taking a sharpened pencil and poking it through a sheet of stiff paper or cardboard. Hold the model so students can see both ends of the pencil and ask them to sketch their view of the model. Instruct them to show the visible ends of the pencil as solid segments in their figures and to show the part of the pencil they cannot see as a dashed segment. Stress that parts of a figure which are not visible are always shown using dashed segments. To show the point where the pencil passes through the paper (modeling the point where the line intersects the plane), instruct students to draw a dot at the location on their figure where the solid and dashed segments meet inside the parallelogram representing the plane. You can use the index card model from Activity 1.4 to show students while asking them to sketch the intersection of two planes.

Lesson 1.5

QUICK TIP

When measuring segments using a ruler, stress that students should always be sure to place the zero mark of the scale on one of the endpoints so they can simply read the ruler measure at the other endpoint. If your students use a ruler that has a customary scale on one side and a metric scale on the opposite side, remind students to be certain they are using the scale indicated in the exercise.

The concept of *betweenness* in mathematics can be confusing for students because in mathematics the word *between* has a precise meaning, whereas in everyday life its use is often less precise. You can model this distinction by placing 3 objects on a table or desk in such a way that one object lies between the other two in the everyday sense (not all in a line, but one item slightly out of alignment with the other two), but not *between* them in the mathematical sense. Stress that the word *between* will always be used in its strict mathematical sense in this course. Use the figure in Postulate 5 on page 29 to emphasize this concept.

SYMBOLS/NOTATION As students are beginning to discover, the study of geometry uses a "language" all its own. With the introduction of the term *congruent segments*, students encounter the use of tick marks in figures for the first time. Take time to emphasize the meaning behind their appearance in a figure. Be sure to relate them to the use of the congruence symbol, which is also new to students. Stress that *segments are congruent*, while their *lengths are equal*. Students need to make the distinction between the symbolisms $\overline{AB} \cong \overline{CD}$ and $AB = CD$. Watch for students who incorrectly write $\overline{AB} = \overline{CD}$ or $AB \cong CD$. Now is a good time to point out that congruence has no meaning for lines and rays.

Lesson 1.6

An important point for students to understand in their introduction to angles is that an angle is a way of indicating *rotation*. That is, one side of an angle can be thought of as having been rotated from an imaginary starting position that coincides with the other side. This idea can be modeled using the hands of a clock, but it may be more apparent when students use a protractor to measure an angle as seen in Example 2. While they are using protractors, look for students who are positioning their protractors incorrectly, especially if the center and zero marks are not along the bottom edge of their protractor. Point out the two scales on a protractor and caution students to choose the correct one when reading an angle measure. Being able to identify the angle as acute or obtuse before measuring it will lead students to the correct scale.

When discussing the classification of angles by their measure (as *acute*, *right*, *obtuse*, or *straight*), caution them against assuming that an angle which looks to be a right angle really is one. Stress that if an angle is a right angle, this fact will be stated in words or be indicated by the symbol for a right angle presented on page 36. A similar caution should be made regarding straight angles.

Finally, have students note the similarity between the Angle Addition Postulate and the Segment Addition Postulate.

SYMBOLS/NOTATION As with congruent segments, work with students to be sure they understand how to correctly write the congruence of angles. Stress that *angles are congruent*, while their *measures are equal*. Emphasize the distinction between the symbolisms $\angle ABC \cong \angle DEF$ and $m\angle ABC = m\angle DEF$. Watch for students who write them incorrectly as $m\angle ABC \cong m\angle DEF$ and $\angle ABC = \angle DEF$.

Chapter **1**

Basics of Geometry

How are runways named?

Runways are named based on the angles they form with due north, measured in a clockwise direction. These angles are called *bearings*. To determine the runway number, the bearing is divided by 10.

The plane in the diagram has a bearing of 50°.

50 ÷ 10 = 5

It is on runway 5.

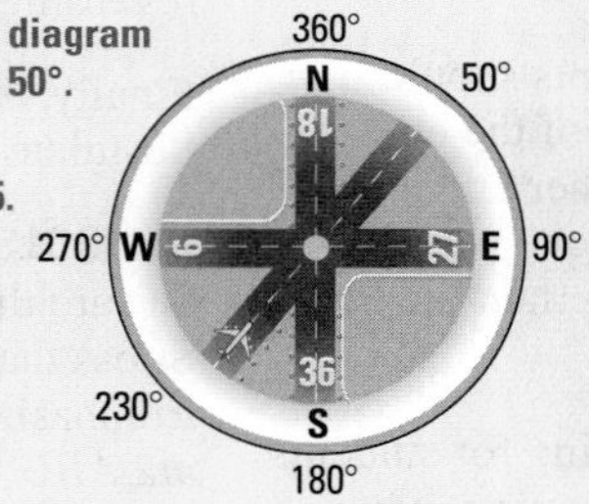

360°
N   50°
270° W   E 90°
230°
S
180°

Learn More About It

You will learn more about airport runways in Exercises 40–44 on p. 40.

Who uses Basics of Geometry?

SURVEYOR
Surveyors use angles to determine boundaries for roads, cell phone towers, and buildings. They also help determine safe speed limits on roads. (p. 39)

SCIENTIST
Scientists make observations, look for patterns, and use previous experiences to develop conjectures that can be tested. (p. 9)

How will you use these ideas?

- Understand how the Braille alphabet works. (p. 7)
- Predict when a full moon will occur. (p. 12)
- Understand why all tires of a three-wheeled car always touch the ground. (p. 19)
- Sketch intersecting lines in a perspective drawing. (p. 27)
- Use angle measures to locate cities by longitude. (p. 40)

Hands-On Activities

Activities (more than 20 minutes)
1.4 Exploring Intersections, p. 21
1.6 Kinds of Angles, p. 34

Geo-Activities (less than 20 minutes)
1.2 Making a Conjecture, p. 8

Projects

A project covering Chapters 1–2 appears on pages 102–103 of the Student Edition. An additional project for Chapter 1 is available in the *Chapter 1 Resource Book*, pp. 73–74.

⊞ Technology

- Electronic Teacher Tools
- Test and Practice Generator
- Online Lesson Planner

Internet Support
CLASSZONE.COM

- Application and Career Links
  6, 7, 12, 39, 40
- Student Help
  4, 10, 12, 18, 24, 32

Diagnostic Tools

The **Chapter Readiness Quiz** can help you diagnose whether students have the following skills needed in Chapter 1:
- Add and subtract decimals and integers.
- Evaluate absolute values.
- Add squares of numbers.

Reteaching Material

This resource is available for students who need additional help with the skills on the Chapter Readiness Quiz:

📖 **Chapter 1 Resource Book**
- Reteaching with Practice (Lessons 1.1–1.6)

Additional Resources

The following resources are provided to help you prepare for the upcoming chapter and customize review materials:

📖 **Chapter 1 Resource Book**
- *Tips for New Teachers*, pp. 1–2
- *Parent Guide*, pp. 3–4
- *Lesson Plans*, pp. 7, 16, 24, 35, 43, 53
- *Lesson Plans for Block Scheduling*, pp. 8, 17, 25, 36, 44, 54

💻 *Technology*
- Electronic Teaching Tools
- Online Lesson Planner
- Test and Practice Generator

Visualize It!

Additional suggestions for helping students visualize geometry are on pp. 15, 19, 23, 24, 26, 29, 30, 31, 36, 37, and 39.

PREVIEW **What's the chapter about?**

- Making predictions and **conjectures** based on patterns
- Sketching **points**, **lines**, **planes**, and their intersections
- Measuring **segments** and **angles**

Key Words

- **conjecture**, *p. 8*
- **inductive reasoning**, *p. 8*
- **counterexample**, *p. 10*
- **point, line, plane**, *p. 14*
- **postulate**, *p. 14*
- **segment**, *p. 16*
- **ray**, *p. 16*
- **distance, length**, *p. 28*

- **congruent segments**, *p. 30*
- **sides of an angle**, *p. 35*
- **vertex of an angle**, *p. 35*
- **measure of an angle**, *p. 36*
- **congruent angles**, *p. 36*
- **acute, right, obtuse, and straight angle**, *p. 36*

PREPARE **Chapter Readiness Quiz**

Take this quick quiz. If you are unsure of an answer, look at the reference pages for help.

Skill Check *(refer to pp. 655, 662, 663, 668)*

1. Evaluate $102.9 - 34.7$. **A**

    Ⓐ 68.2   **Ⓑ** 72.2   **Ⓒ** 78.2   **Ⓓ** 137.6

2. Evaluate $|-4|$. **J**

    Ⓕ -4 **Ⓖ** -2 **Ⓗ** 2   **Ⓙ** 4

3. Evaluate $-6 + (-5)$. **A**

    Ⓐ -11 **Ⓑ** -1 **Ⓒ** 1   **Ⓓ** 11

4. Evaluate $5^2 + (-1)^2$. **J**

    Ⓕ 8   **Ⓖ** 16   **Ⓗ** 24   **Ⓙ** 26

VISUAL STRATEGY **Learning Vocabulary**

Visualize It! ➡ Important words in this book are in **bold and highlighted** type. Keep a section in your notebook for definitions. Draw sketches near them to help you.

> *Vocabulary*
>
> *acute angle:*
> *an angle with*
> *a measure*
> *between 0° and 90°* 35°

1.1 Finding and Describing Patterns

Goal
Find patterns and use them to make predictions.

Key Words
- pattern
- prediction

In the bracelet shown at the right, the beads and knots follow a pattern. You can repeat the pattern to make the bracelet longer.

EXAMPLE 1 Describe a Visual Pattern

Describe a pattern in the figures.

  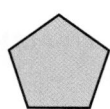 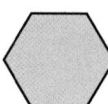

Solution

The figures have 3 sides, 4 sides, 5 sides, and 6 sides.

ANSWER ▶ The number of sides increases by one.

EXAMPLE 2 Describe a Number Pattern

Describe a pattern in the numbers.

a. 3, 6, 9, 12, 15, 18, . . .

b. 1, 4, 9, 16, 25, 36, . . .

Solution

a. Each number after the first is 3 more than the previous one.

b. The numbers are squares of consecutive numbers.

1	4	9	16	25	36
1^2	2^2	3^2	4^2	5^2	6^2

1. The number of equal parts into which the circle is divided increases by one.

2. The shaded triangle moves one space in a counterclockwise direction.

 Describe a Visual or Number Pattern

Describe a pattern.

1.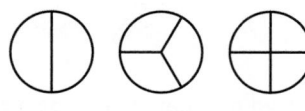

2.

3. Each number after the first is 4 more than the previous number.

4. Each number after the first is 5 less than the previous number.

3. 4, 8, 12, 16, 20, 24, . . .

4. 35, 30, 25, 20, 15, 10, . . .

Extra Example 1

Describe a pattern in the figures.

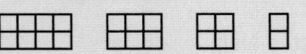

The number of squares decreases by two.

Extra Example 2

Describe a pattern in the numbers.

a. 6, 11, 21, 41, 81, 161, ... Each number is 1 less than twice the previous number.

b. 311, 297, 283, 269, 255, ... Each number is 14 less than the previous number.

Extra Example 3

Sketch the next figure you expect in the pattern.

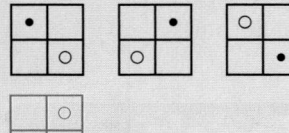

Extra Example 4

Sketch the next figure you expect in the pattern.

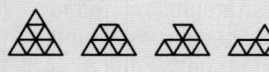
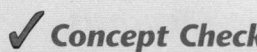

✓ Concept Check

Why is finding a pattern in a list of numbers or figures useful? You can use the pattern to make a prediction about other numbers or figures in the list.

🧩 Daily Puzzler

To get the next number in a pattern, you multiply the previous number by 2 and subtract 1. If the fourth number is 17, what is the first number in the pattern? 3

5, 6. See Additional Answers beginning on page AA1.

4

Student Help
CLASSZONE.COM

MORE EXAMPLES
More examples at classzone.com

EXAMPLE 3 Make a Prediction

Sketch the next figure you expect in the pattern.

Solution

The arrow's color changes back and forth between green and red. The arrow makes a quarter turn each time.

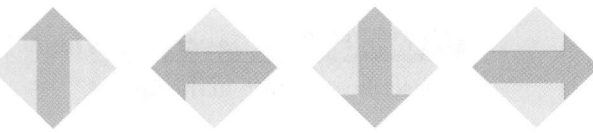

ANSWER ▶ The fourth figure is red. Its arrow points to the right.

EXAMPLE 4 Make a Prediction

Sketch the next figure you expect in the pattern.

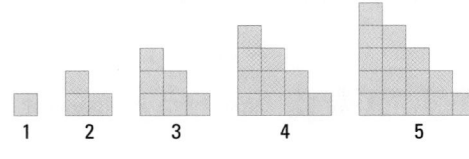

Solution

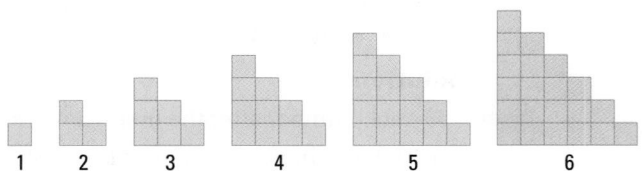

ANSWER ▶ The sixth figure has six squares in the bottom row.

Checkpoint ✓ Make a Prediction

Sketch the next two figures you expect in the pattern. 5–6. See margin.

5. **6.**

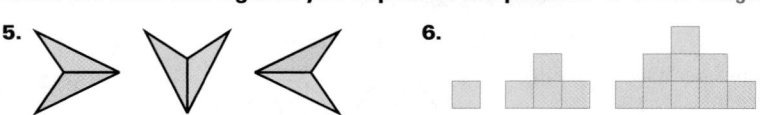

Write the next two numbers you expect in the pattern.

7. −2, −5, −8, −11, . . . −14; −17   **8.** 4, 10, 16, 22, . . . 28; 34

Guided Practice

Skill Check

5. Each number is 8 more than the previous number; 35; 43.

6. Each number is 3 times the previous number; 162; 486.

7. Each number is 0.5 more than the previous number; 9.0; 9.5.

8. Each number is 6 less than the previous number; −11; −17.

9. Each number is $\frac{1}{4}$ the previous number; 1; $\frac{1}{4}$.

10. The numbers in the odd-numbered positions alternate between 3 and −3; the numbers in the even-numbered positions are 0; −3; 0.

Sketch the next figure you expect in the pattern. 1–4. See margin.

1.

2.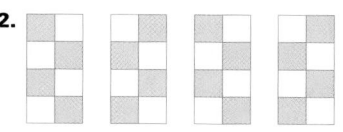

3.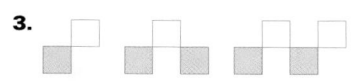

4.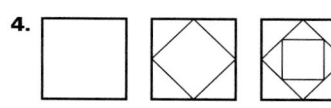

Describe a pattern in the numbers. Write the next two numbers you expect in the pattern.

5. 3, 11, 19, 27, . . .   **6.** 2, 6, 18, 54, . . .   **7.** 7.0, 7.5, 8.0, 8.5, . . .

8. 13, 7, 1, −5, . . .   **9.** 256, 64, 16, 4, . . .   **10.** 3, 0, −3, 0, 3, 0, . . .

Practice and Applications

Extra Practice

See p. 675.

17. Each number is 5 more than the previous number; 24.

18. Each number is 9 less than the previous number; −34.

19. Each number is $\frac{1}{4}$ the previous number; $\frac{5}{4}$.

20. Each number is twice the previous number; 40.

21. Begin with 1 and add 2, then 3, then 4, and so on; 21.

22. Begin with 5 and add 2, then 4, then 6, then 8, and so on; 35.

Describing Visual Patterns Sketch the next figure you expect in the pattern. 11–16. See margin.

11.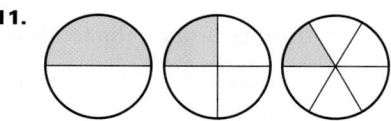

12.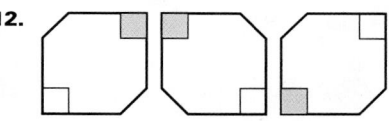

13.

14.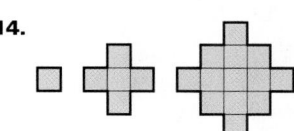

15.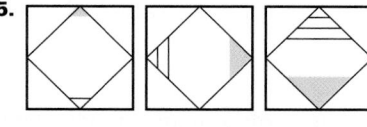

16.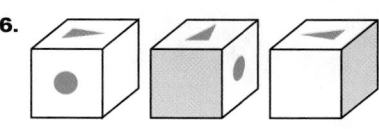

Describing Number Patterns Describe a pattern in the numbers. Write the next number you expect in the pattern.

17. 4, 9, 14, 19, . . .   **18.** 2, −7, −16, −25, . . .

19. 320, 80, 20, 5, . . .   **20.** 2.5, 5, 10, 20, . . .

21. 1, 3, 6, 10, 15, . . .   **22.** 5, 7, 11, 17, 25, . . .

Homework Help

Example 1: Exs. 11–16
Example 2: Exs. 17–25
Example 3: Exs. 26–32
Example 4: Exs. 26–32

1.1 Finding and Describing Patterns **5**

Teaching Tip

In Exercise 35, make sure students notice that there are hidden blocks in the second and third figures of the pattern. Students can easily forget to count blocks that are not readily visible.

31.
```
F—F—F—F—F        F—F—F—F—F—F
|   |   |   |   |        |   |   |   |   |   |
F—C—C—C—C—F     F—C—C—C—C—C—F
|   |   |   |   |        |   |   |   |   |   |
F—F—F—F—F        F—F—F—F—F—F
```

33. Each of the second ten letters has the same pattern of dots as the letter above it, with an additional dot inserted in the first column of the third row.

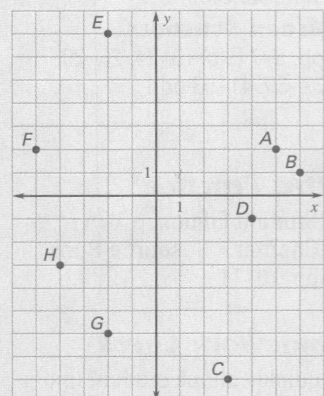

46–53.

23. Each point has a *y*-coordinate of 3. *Sample answer:* (5, 3)

24. For each point, the *x*-coordinate and *y*-coordinate are equal. *Sample answer:* (5, 5)

25. For each point, the *y*-coordinate is 2 less than half the opposite of the *x*-coordinate. *Sample answer:* (2, −3)

Link to **Careers**

LABORATORY TECHNOLOGISTS study microscopic cells, such as bacteria. The doubling period for a population of bacteria may be as short as 20 minutes.

Career Links
CLASSZONE.COM

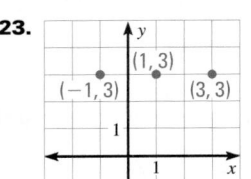 **Using Algebra** Find a pattern in the coordinates of the points. Then write the coordinates of another point in the pattern.

23.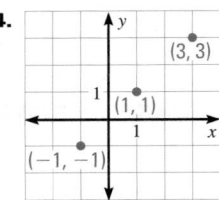

24.

25.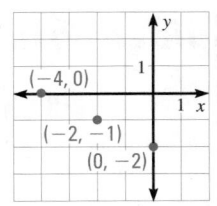

Making Predictions In Exercises 26–30, use the staircase pattern from Example 4 shown below.

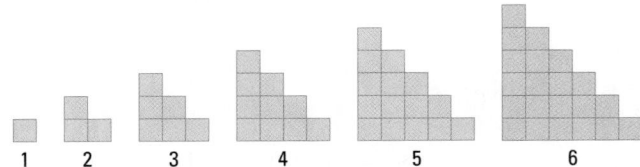

26. Find the distance around each figure. Organize your results in a table like the one shown.

Figure	1	2	3	4	5	6
Distance	4	8	?	?	?	?

        12 16 20 24

27. Use your table to describe a pattern in the distances. Each distance after the first is 4 more than the one before it.

28. Write a variable expression for the distance around the *n*th figure. 4*n*

29. Predict the distance around the tenth figure. 40

30. **You be the Judge** Will a figure in this pattern have a distance of 60? If so, which one? yes; the fifteenth figure

31. **Science Connection** Diagrams for four molecules are shown. Draw a diagram for the next two molecules in the pattern. See margin.

```
   F              F   F            F   F   F          F   F   F   F
   |              |   |            |   |   |          |   |   |   |
F—C—F        F—C—C—F       F—C—C—C—F      F—C—C—C—C—F
   |              |   |            |   |   |          |   |   |   |
   F              F   F            F   F   F          F   F   F   F
```

32. **Bacteria Growth** You are studying bacteria in biology class. The table shows the number of bacteria after *n* doubling periods. Predict the number of bacteria after 8 doubling periods. 768 billion bacteria

n (doubling periods)	0	1	2	3	4	5
Billions of bacteria	3	6	12	24	48	96

33. Braille System The Braille alphabet uses raised dots that can be read by touch. Describe a pattern that links the first ten letters and the next ten letters. Complete the missing letters. **See margin.**

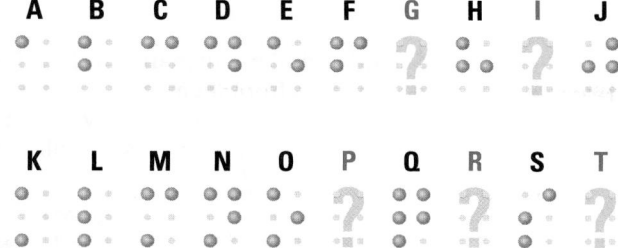

A   B   C   D   E   F   G   H   I   J

K   L   M   N   O   P   Q   R   S   T

Visualizing Patterns The first three objects in a pattern are shown. How many blocks are in the next object?

34. 16 blocks

35. 28 blocks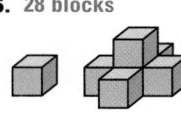

36. Multiple Choice What is the next number you expect? **C**

55, 110, 165, 220, . . .

Ⓐ 255   Ⓑ 265   Ⓒ 275   Ⓓ 440

37. Multiple Choice What is the next number you expect? **H**

1, 5, 13, 25, . . .

Ⓔ 33   Ⓕ 38   Ⓖ 41   Ⓗ 169

Problem Solving Draw a diagram to solve. *(Skills Review, p. 653)*

38. Robert, Susan, and Todd are standing in a line. How many ways can they be arranged? **6 ways**

39. A table is two feet longer than it is wide. What are five possible areas for the table? *Sample answer:* 3 square feet, 8 square feet, 15 square feet, 24 square feet, 35 square feet

Adding Decimals Find the sum. *(Skills Review, p. 655)*

40. $9.3 + 0.2$ **9.5**   **41.** $2.4 + 8.9$ **11.3**   **42.** $10.5 + 5.5$ **16**

43. $0.71 + 0.33$ **1.04**   **44.** $5.64 + 12.75$ **18.39**   **45.** $34.08 + 11.16$ **45.24**

Plotting Points Plot the points in a coordinate plane. *(Skills Review, p. 664)* **46–53. See margin.**

46. $A(5, 2)$ **47.** $B(6, 1)$ **48.** $C(3, -8)$ **49.** $D(4, -1)$

50. $E(-2, 7)$ **51.** $F(-5, 2)$ **52.** $G(-2, -6)$ **53.** $H(-4, -3)$

1.1   *Finding and Describing Patterns*   **7**

Mini-Quiz
Sketch the next figure you expect in the pattern.

1.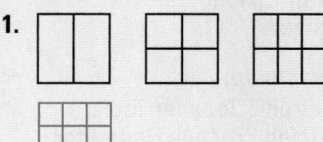

2.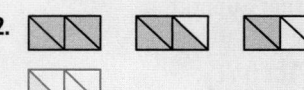

Describe a pattern in the numbers. Write the next number you expect in the pattern.

3. 100, 93, 86, 79, ... **Each number is 7 less than the previous number; 72.**

4. –43, –35, –27, –19, ... **Each number is 8 greater than the previous number; –11.**

5. Alesha's savings account balance was $55 at the end of March, $63 at the end of April, $71 at the end of May, and $79 at the end of June. Predict the amount Alesha will have in her account at the end of July. **$87**

Pacing
Suggested Number of Days

Basic: 2 days
Average: 2 days
Advanced: 2 days
Block Schedule: 1 block

Teaching Resources

📄 ***Blacklines***
(See page 1B.)

✋ ***Transparencies***
• Warm-Up with Quiz
• Answers

🖥 ***Technology***
• Electronic Teacher Tools
• Test and Practice Generator
• Online Lesson Planner
• Internet Support

Geo-Activity

Goal Use a table to make a conjecture about the number of handshakes for a group of people.

Key Discovery As the number of people increases by 1, the number of handshakes increases in a predictable pattern.

 **Teach**

―**Content and Teaching Strategies**
For background information on geometric concepts and teaching strategies related to this lesson, see pages 1E and 1F in this Teacher's Edition.

 **Inductive Reasoning**

Goal
Use inductive reasoning to make conjectures.

Key Words
• conjecture
• inductive reasoning
• counterexample

Scientists and mathematicians look for patterns and try to draw conclusions from them. A **conjecture** is an unproven statement that is based on a pattern or observation. Looking for patterns and making conjectures is part of a process called **inductive reasoning** .

Geo-Activity ◆ Making a Conjecture

Work in a group of five. You will count how many ways various numbers of people can shake hands.

2 people can shake hands 1 way.

3 people can shake hands 3 ways.

❶ Count the number of ways that 4 people can shake hands. **6 ways**

❷ Count the number of ways that 5 people can shake hands. **10 ways**

❸ Organize your results in a table like the one below.

Student Help

STUDY TIP
Copy this table in your notebook and complete it. Do not write in your textbook.⋯⋯⋯⋯⋯▶

People	2	3	4	5
Handshakes	1	3	?	?
			6	10

❹ Look for a pattern in the table. Write a conjecture about the number of ways that 6 people can shake hands.

The pattern is: Start with 1 and add 2, then 3, then 4, and so on; 15.

Much of the reasoning in geometry consists of three stages.

❶ *Look for a Pattern* Look at several examples. Use diagrams and tables to help discover a pattern.

❷ *Make a Conjecture* Use the examples to make a general conjecture. Modify it, if necessary.

❸ *Verify the Conjecture* Use logical reasoning to verify that the conjecture is true in all cases. (You will do this in later chapters.)

8 **Chapter 1** *Basics of Geometry*

SCIENTISTS use inductive reasoning as part of the scientific method. They make observations, look for patterns, and develop conjectures (*hypotheses*) that can be tested.

EXAMPLE 1 Make a Conjecture

Complete the conjecture.

Conjecture: The sum of any two odd numbers is __?__ .

Solution

Begin by writing several examples.

$1 + 1 = 2$	$5 + 1 = 6$	$3 + 7 = 10$
$3 + 13 = 16$	$21 + 9 = 30$	$101 + 235 = 336$

Each sum is even. You can make the following conjecture.

ANSWER ▶ The sum of any two odd numbers is *even*.

EXAMPLE 2 Make a Conjecture

Complete the conjecture.

Conjecture: The sum of the first n odd positive integers is __?__ .

Solution

List some examples and look for a pattern.

1	$1 + 3$	$1 + 3 + 5$	$1 + 3 + 5 + 7$

$1 = 1^2$	$4 = 2^2$	$9 = 3^2$	$16 = 4^2$

ANSWER ▶ The sum of the first n odd positive integers is n^2.

Checkpoint ✓ Make a Conjecture

Complete the conjecture based on the pattern in the examples.

1. *Conjecture:* The product of any two odd numbers is __?__ . **odd**

   EXAMPLES

$1 \times 1 = 1$	$3 \times 5 = 15$	$3 \times 11 = 33$
$7 \times 9 = 63$	$11 \times 11 = 121$	$1 \times 15 = 15$

2. *Conjecture:* The product of the numbers $(n - 1)$ and $(n + 1)$ is __?__ .
   $n^2 - 1$

   EXAMPLES

$1 \cdot 3 = 2^2 - 1$	$3 \cdot 5 = 4^2 - 1$	$5 \cdot 7 = 6^2 - 1$
$7 \cdot 9 = 8^2 - 1$	$9 \cdot 11 = 10^2 - 1$	$11 \cdot 13 = 12^2 - 1$

1.2 Inductive Reasoning **9**

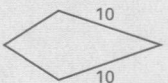 
Counterexamples Just because something is true for several examples does not *prove* that it is true in general. To prove a conjecture is true, you need to prove it is true in all cases.

A conjecture is considered false if it is *not always* true. To prove a conjecture is false, you need to find only one *counterexample*. A **counterexample** is an example that shows a conjecture is false.

EXAMPLE 3 Find a Counterexample

Show the conjecture is false by finding a counterexample.

Conjecture: The sum of two numbers is always greater than the larger of the two numbers.

Solution

Here is a counterexample. Let the two numbers be 0 and 3.

The sum is $0 + 3 = 3$, but 3 is not greater than 3.

ANSWER ▶ The conjecture is false.

EXAMPLE 4 Find a Counterexample

Show the conjecture is false by finding a counterexample.

Conjecture: All shapes with four sides the same length are squares.

Solution

Here are some counterexamples.

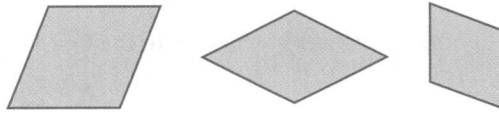

These shapes have four sides the same length, but they are not squares.

ANSWER ▶ The conjecture is false.

Checkpoint ✓ Find a Counterexample

Show the conjecture is false by finding a counterexample.

3. *Sample answer:* $7 \cdot 4 = 28$, which is an even number, but 7 is not even. The conjecture is false.

3. If the product of two numbers is even, the numbers must be even.

4. If a shape has two sides the same length, it must be a rectangle.
   See margin.

1.2 Exercises

Guided Practice

Vocabulary Check

1. Explain what a *conjecture* is. an unproven statement that is based on a pattern or an observation

2. How can you prove that a conjecture is false?
   by finding just one counterexample

Skill Check

Complete the conjecture with *odd* or *even*.

3. **Conjecture:** The difference of any two odd numbers is __?__. even

4. **Conjecture:** The sum of an odd number and an even number is __?__. odd

Show the conjecture is false by finding a counterexample.

5. *Sample answer:* 6 is divisible by 2, but it is not divisible by 4.

5. Any number divisible by 2 is divisible by 4.

6. *Sample answer:* 5 − 0 = 5 and 5 is not less than 5.

6. The difference of two numbers is less than the greater number.

7. A circle can always be drawn around a four-sided shape so that it touches all four corners of the shape. See margin.

Practice and Applications

Extra Practice

See p. 675.

8. **Rectangular Numbers** The dot patterns form rectangles with a length that is one more than the width. Draw the next two figures to find the next two "rectangular" numbers. Check drawings; 30; 42.

2   6   12   20

9. **Triangular Numbers** The dot patterns form triangles. Draw the next two figures to find the next two "triangular" numbers.
Check drawings; 15; 21.

1   3   6   10

Homework Help

Example 1: Exs. 8–16
Example 2: Exs. 8–16
Example 3: Exs. 17–19
Example 4: Exs. 17–19

Technology Use a calculator to explore the pattern. Write a conjecture based on what you observe. See margin for conjectures.

10. $101 \times 25 = $ __?__
    $101 \times 34 = $ __?__
    $101 \times 49 = $ __?__
    2525; 3434; 4949

11. $11 \times 11 = $ __?__
    $111 \times 111 = $ __?__
    $1111 \times 1111 = $ __?__
    121; 12,321; 1,234,321

12. $3 \times 4 = $ __?__
    $33 \times 34 = $ __?__
    $333 \times 334 = $ __?__
    12; 1122; 111,222

1.2 *Inductive Reasoning*   **11**

③ Apply

Assignment Guide

BASIC
Day 1: pp. 11–13 Exs. 8, 9, 13–15, 23, 24–40 even
Day 2: pp. 11–13 Exs. 17–20, 22, 25–41 odd

AVERAGE
Day 1: pp. 11–13 Exs. 8–10, 12–15, 23, 24–40 even
Day 2: pp. 11–13 Exs. 17–20, 22, 25–41 odd

ADVANCED
Day 1: pp. 11–13 Exs. 8–16, 23–41 odd
Day 2: pp. 11–13 Exs. 17–20, 21*, 22–40 even;
EC: classzone.com

BLOCK SCHEDULE
pp. 11–13 Exs. 8–15, 17–20, 22–25, 26–40 even

Extra Practice

• Student Edition, p. 675
• Chapter 1 Resource Book, pp. 19–20

Homework Check

To quickly check student understanding of key concepts, go over the following exercises:

Basic: 8, 13, 14, 17, 18
Average: 9, 10, 14, 17, 18
Advanced: 9, 12, 15, 18, 19

Teaching Tip

Exercises 10–12 are more difficult than the examples in the lesson. For Exercises 11 and 12, suggest that students represent the number of digits in the factors by the variable n. Students should try to include n in their conjectures. Consider having students continue the pattern of factors in each exercise and predict the products before using a calculator to verify the results.

7, 10–12. See Additional Answers beginning on page AA1.

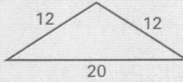
Student Help
CLASSZONE.COM

HOMEWORK HELP
Extra help with problem solving in Exs. 13–14 is at classzone.com

Making Conjectures Complete the conjecture based on the pattern you observe.

13. Conjecture: The product of an odd number and an even number is ___?___. **even**

$3 \cdot 6 = 18$	$22 \cdot 13 = 286$	$43 \cdot 102 = 4386$
$5 \cdot 12 = 60$	$-5 \cdot 2 = -10$	$254 \cdot 63 = 16{,}002$
$14 \cdot 9 = 126$	$-11 \cdot (-4) = 44$	

14. Conjecture: The sum of three consecutive integers is always three times ___?___. **the middle integer**

$3 + 4 + 5 = 3 \cdot 4$	$6 + 7 + 8 = 3 \cdot 7$	$9 + 10 + 11 = 3 \cdot 10$
$4 + 5 + 6 = 3 \cdot 5$	$7 + 8 + 9 = 3 \cdot 8$	$10 + 11 + 12 = 3 \cdot 11$
$5 + 6 + 7 = 3 \cdot 6$	$8 + 9 + 10 = 3 \cdot 9$	$11 + 12 + 13 = 3 \cdot 12$

15. Counting Diagonals In the shapes below, the *diagonals* are shown in blue. Write a conjecture about the number of diagonals of the next two shapes. *Conjecture:* The next two shapes will have 14 diagonals and 20 diagonals.

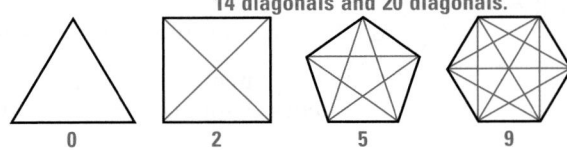
0   2   5   9

Link to
Science

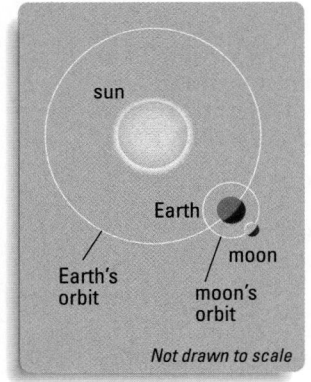

sun

Earth

moon

Earth's orbit

moon's orbit

Not drawn to scale

FULL MOONS happen when Earth is between the moon and the sun.

Application Links
CLASSZONE.COM

18. *Sample answer:* A rectangle that is 6 ft long and 4 ft wide has a perimeter of 20 ft and an area of 24 ft², but a rectangle that is 7 ft long and 3 ft wide has a perimeter of 20 ft and an area of 21 ft².

16. Moon Phases A full moon occurs when the moon is on the opposite side of Earth from the sun. During a full moon, the moon appears as a complete circle. Suppose that one year, full moons occur on these dates:

| July 21 Thursday | August 19 Friday | September 18 Sunday | October 17 Monday | November 16 Wednesday | December 15 Thursday |

Determine how many days are between these full moons and predict when the next full moon occurs. See margin for table. The next full moon should be about 30 days after December 15, or on January 14th.

Error Analysis Show the conjecture is false by finding a counterexample.

17. Conjecture: If the product of two numbers is positive, then the two numbers must both be positive. *Sample answer:* $(-4)(-5) = 20$

18. Conjecture: All rectangles with a perimeter of 20 feet have the same area. *Note:* Perimeter = 2(length + width).

19. Conjecture: If two sides of a triangle are the same length, then the third side must be shorter than either of those sides. See margin.

20. *Conjecture:* The "7" key should display the letters P, Q, and R; no, the "7" key displays P, Q, R, and S, or P, R, and S.

20. Telephone Keypad Write a conjecture about the letters you expect on the next telephone key. Look at a telephone to see whether your conjecture is correct.

21. $5x + 10 = 5(x + 2)$; since 5 is a factor of $5(x + 2)$, the sum of five consecutive integers is always divisible by 5.

21. Challenge Prove the conjecture below by writing a variable statement and using algebra.

Conjecture: The sum of five consecutive integers is always divisible by five.

$$x + (x + 1) + (x + 2) + (x + 3) + (x + 4) = \underline{\quad ? \quad}$$

Standardized Test Practice

22. Multiple Choice Which of the following is a counterexample of the conjecture below? **B**

Conjecture: The product of two positive numbers is always greater than either number.

Ⓐ 2, 2   Ⓑ $\frac{1}{2}$, 2   Ⓒ 3, 10   Ⓓ 2, −1

23. Multiple Choice You fold a large piece of paper in half four times, then unfold it. If you cut along the fold lines, how many identical rectangles will you make? **H**

Ⓕ 4   Ⓖ 8   Ⓗ 16   Ⓙ 32

Mixed Review

Describing Patterns Sketch the next figure you expect in the pattern. *(Lesson 1.1)* 24, 25. See margin.

24.

25.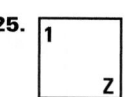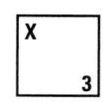

Algebra Skills

Using Integers Evaluate. *(Skills Review, p. 663)*

26. $8 + (-3)$ 5   **27.** $-2 + 9$ 7   **28.** $9 - (-1)$ 10   **29.** $-7 - 3$ −10

30. $3(-5)$ −15   **31.** $(-2)(-7)$ 14   **32.** $20 \div (-5)$ −4   **33.** $(-8) \div (-2)$ 4

Evaluating Expressions Evaluate the expression when $x = 3$. *(Skills Review, p. 670)*

34. $x + 7$ 10   **35.** $5 - x$ 2   **36.** $x - 9$ −6   **37.** $2x + 5$ 11

38. $x^2 + 6$ 15   **39.** $x^2 - 4x$ −3   **40.** $3x^2$ 27   **41.** $2x^3$ 54

1.2 *Inductive Reasoning* **13**

Assessment Resources

The Mini-Quiz below is also available on blackline (*Chapter 1 Resource Book*, p. 26) and on transparency. For more assessment resources, see:
• Chapter 1 Resource Book
• Standardized Test Practice
• Test and Practice Generator

Mini-Quiz

Complete the conjecture.

1. The sum of any three odd numbers is _?_. **odd**

2. The difference between an integer n and its opposite is _?_. **2n**

Show the conjecture is false by finding a counterexample.

3. *Conjecture:* If a positive fraction is multiplied by a positive integer, then the product is greater than the fraction.

Sample answer: $\frac{2}{3} \times 1 = \frac{2}{3}$ and $\frac{2}{3}$ is not greater than $\frac{2}{3}$.

4. *Conjecture:* If all the sides of a figure are the same length, then the figure must be a square.

Sample answer:

This triangle has three sides that are the same length and it is not a square.

24.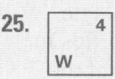

25.

	4
W	

Plan

1.3 Points, Lines, and Planes

Pacing

Suggested Number of Days

Basic: 2 days
Average: 2 days
Advanced: 2 days
Block Schedule: 1 block

Teaching Resources

 **Blacklines**

(See page 1B.)

Transparencies
• Warm-Up with Quiz
• Answers

Technology
• Electronic Teacher Tools
• Test and Practice Generator
• Online Lesson Planner
• Internet Support

Teach

Content and Teaching Strategies

For background information on geometric concepts and teaching strategies related to this lesson, see pages 1E and 1F in this Teacher's Edition.

Tips for New Teachers

Looking at diagrams and their labels can be confusing to students. On this page students need to recognize that capital letters are used to identify points, and that a line may be named by any two points on the line or by using a single lower-case letter. See the Tips for New Teachers on pp. 1–2 of the *Chapter 1 Resource Book* for additional notes about Lesson 1.3.

Goal

Use postulates and undefined terms.

Key Words

• undefined term
• point, line, plane
• postulate
• collinear, coplanar
• segment
• ray
• endpoint

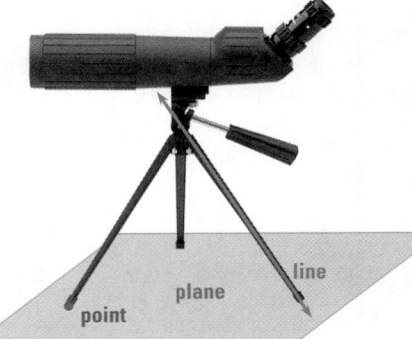

The legs of the tripod touch the table at three *points*. The legs suggest *lines*, and the table surface suggests a *plane*.

Geometry depends on a common understanding of terms such as *point*, *line*, and *plane*. Because these terms cannot be mathematically defined using other known words, they are called **undefined terms**.

A **point** has no dimension. It is represented by a small dot.

point *A* • *A*

A **line** has one dimension. It extends without end in two directions. It is represented by a line with two arrowheads.

line ℓ or $\overleftrightarrow{BC}$

A **plane** has two dimensions. It is represented by a shape that looks like a floor or wall. You have to imagine that it extends without end.

plane M or plane DEF

You need two points to describe a line, and you need three points to describe a plane, because the geometry in this book follows the two *postulates* given below. **Postulates** are statements that are accepted without further justification.

POSTULATES 1 and 2

Postulate 1 Two Points Determine a Line

Words Through any two points there is exactly one line.

Symbols Line *n* passes through points *P* and *Q*.

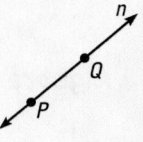

Postulate 2 Three Points Determine a Plane

Words Through any three points not on a line there is exactly one plane.

Symbols Plane *T* passes through points *A*, *B*, and *C*.

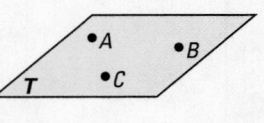

EXAMPLE 1 Name Points, Lines, and Planes

Use the diagram at the right.

  a. Name three points.

  b. Name two lines.

  c. Name two planes.

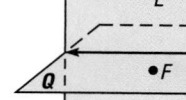

Solution

  a. *D*, *E*, and *F* are points.

  b. Line *m* and line *p*

  c. *Q* and *R* are planes.

Collinear points are points that lie on the same line.

Coplanar points are points that lie on the same plane.

Coplanar lines are lines that lie on the same plane.

Visualize It!

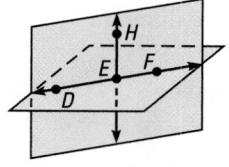

In Example 2 the points *D*, *E*, *F*, and *H* are also coplanar. The plane containing them is shown in green above.

EXAMPLE 2 Name Collinear and Coplanar Points

Use the diagram at the right.

  a. Name three points that are collinear.

  b. Name four points that are coplanar.

  c. Name three points that are not collinear.

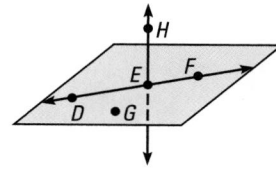

Solution

  a. Points *D*, *E*, and *F* lie on the same line. So, they are collinear.

  b. Points *D*, *E*, *F*, and *G* lie on the same plane, so they are coplanar.

  c. Points *H*, *E*, and *G* do not lie on the same line. There are many other correct answers.

Checkpoint ✓ Name Points, Lines, and Planes

Use the diagram at the right.

  1. Name two lines. *Sample answer:* line *m* and line *n*

  2. Name two planes. plane *S* and plane *T*

  3. Name three points that are collinear.
     C, *D*, and *E*

  4. Name three points that are not collinear.
     Sample answer: B, C, and *D*

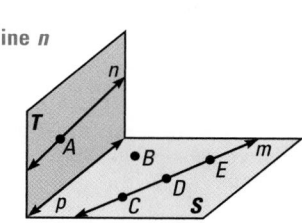

  5. Name four points that are coplanar.
     B, C, D, and *E*

  6. Name two lines that are coplanar. *p* and *n*, or *p* and *m*

Extra Example 1

Use the diagram below.

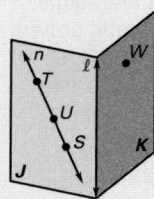

  a. Name four points. *U, T, S,* and *W* are points.

  b. Name two lines. line *ℓ* and line *n*

  c. Name two planes. *J* and *K* are planes.

Extra Example 2

Use the diagram shown in Extra Example 1 above.

  a. Name two points that are collinear. Points *S, T,* and *U* are collinear. Correct answers include *S* and *T, T* and *U,* and *S* and *U.*

  b. Name three points that are coplanar. Points *U, S,* and *T* are coplanar.

  c. Name three points that are not collinear. Points *S, T,* and *W* are not collinear. Other correct answers are *T, U, W* and *U, S, W.*

Visualize It!

Students can create their own sketches to help them remember the terms *collinear* and *coplanar*.

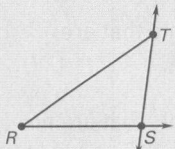
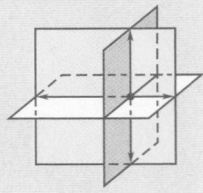
16

The line $\overleftrightarrow{AB}$ passes through *A* and *B*.

The **segment** $\overline{AB}$ consists of the **endpoints** *A* and *B*, and all points on $\overleftrightarrow{AB}$ that are between *A* and *B*.

The **ray** $\overrightarrow{AB}$ consists of the endpoint *A* and all points on $\overleftrightarrow{AB}$ that lie on the same side of *A* as *B*.

SUMMARY	LINES, SEGMENTS, and RAYS	
Word	**Symbol**	**Diagram**
line	$\overleftrightarrow{AB}$ or $\overleftrightarrow{BA}$	•A •B
segment	$\overline{AB}$ or $\overline{BA}$	endpoint •A — endpoint •B
ray	$\overrightarrow{AB}$	endpoint •A •B →
	$\overrightarrow{BA}$	← •A •B endpoint

Note that $\overleftrightarrow{AB}$ is the same as $\overleftrightarrow{BA}$. Also, $\overline{AB}$ is the same as $\overline{BA}$. However, $\overrightarrow{AB}$ is not the same as $\overrightarrow{BA}$. The two rays have different endpoints and extend in different directions.

EXAMPLE 3 Draw Lines, Segments, and Rays

Draw three noncollinear points, *J*, *K*, and *L*. Then draw $\overleftrightarrow{JK}$, $\overline{KL}$, and $\overrightarrow{LJ}$.

Solution

❶ Draw *J*, *K*, and *L*. ❷ Draw $\overleftrightarrow{JK}$. ❸ Draw $\overline{KL}$. ❹ Draw $\overrightarrow{LJ}$.

Checkpoint ✓ **Draw Lines, Segments, and Rays**

7. Draw four points as shown. **See margin.**

8. Draw the lines $\overleftrightarrow{AB}$ and $\overleftrightarrow{AC}$. Are the lines the same? Explain.

9. Draw the line segments $\overline{AC}$ and $\overline{BD}$. Are the segments the same? Explain.

10. Draw the rays $\overrightarrow{CA}$ and $\overrightarrow{CB}$. Are the rays the same? Explain.

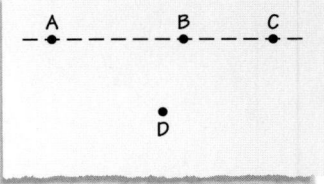

Exercises

Guided Practice

Vocabulary Check

1. Write in words how you would say each of these symbols aloud: $\overleftrightarrow{PQ}$, $\overline{PQ}$, $\overrightarrow{PQ}$, and $\overrightarrow{QP}$. line *PQ*; segment *PQ*; ray *PQ*; ray *QP*

2. Explain the difference between $\overrightarrow{PQ}$ and $\overrightarrow{QP}$. $\overrightarrow{PQ}$ has endpoint *P* and extends toward *Q*, but $\overrightarrow{QP}$ has endpoint *Q* and extends toward *P*.

Skill Check

Decide whether the statement is *true* or *false*.

3. Points *A*, *B*, and *C* are collinear. false

4. Points *A*, *B*, and *C* are coplanar. true

5. Points *B*, *C*, and *D* are coplanar. true

6. Point *C* lies on $\overleftrightarrow{AB}$. false

7. $\overleftrightarrow{AB}$ lies on plane *ABC*. true

8. $\overleftrightarrow{DE}$ lies on plane *ABC*. false

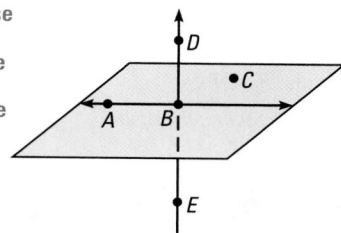

Sketch a line that contains point *S* between points *R* and *T*. Decide whether the statement is *true* or *false*.

9. $\overleftrightarrow{RS}$ is the same as $\overleftrightarrow{RT}$. true

10. $\overrightarrow{ST}$ is the same as $\overrightarrow{TS}$. false

11. $\overrightarrow{ST}$ is the same as $\overrightarrow{RT}$. false

12. $\overrightarrow{RS}$ is the same as $\overrightarrow{RT}$. true

13. $\overline{RS}$ is the same as $\overline{ST}$. false

14. $\overline{ST}$ is the same as $\overline{TS}$. true

Practice and Applications

Extra Practice

See p. 675.

Naming Points, Lines, and Planes Use the diagram at the right.

15. Name four points. any four of: *A*, *B*, *C*, *D*, *E*

16. Name two lines. $\overleftrightarrow{BC}$ and $\overleftrightarrow{DE}$

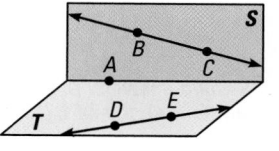

17. Name the plane that contains *A*, *B*, and *C*.
plane *S* or plane *ABC*

18. Name the plane that contains *A*, *D*, and *E*.
plane *T* or plane *ADE*

Evaluating Statements Decide whether the statement is *true* or *false*.

19. *A* lies on line ℓ.
false

20. *A*, *B*, and *C* are collinear.
false

21. *B* lies on line ℓ.
true

22. *A*, *B*, and *C* are coplanar.
true

23. *C* lies on line *m*.
true

24. *D*, *E*, and *B* are collinear.
true

25. *D* lies on line *m*.
false

26. *D*, *E*, and *B* are coplanar.
true

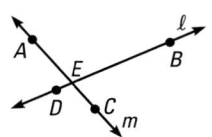

Homework Help

Example 1: Exs. 15–18
Example 2: Exs. 19–62
Example 3: Exs. 63–65

Assignment Guide

BASIC
Day 1: pp. 17–20 Exs. 15–18, 19–33 odd, 36, 38–41, 50–55, 71–81 odd
Day 2: pp. 17–20 Exs. 20–34 even, 35, 42–49, 58–60, 63, 70, Quiz 1

AVERAGE
Day 1: pp. 17–20 Exs. 15–26, 32–48 even, 54, 55, 70–82 even
Day 2: pp. 17–20 Exs. 31–51 odd, 56, 58–61, 63, 64, 66, 68, Quiz 1

ADVANCED
Day 1: pp. 17–20 Exs. 15–26, 31–51 odd, 55, 57, 70–82 even
Day 2: pp. 17–20 Exs. 34–48 even, 58–69, Quiz 1;
EC: TE p. 1D*, classzone.com

BLOCK SCHEDULE
pp. 17–20 Exs. 15–26, 28–52 even, 54–61, 63–66, 68, 70–82 even, Quiz 1

Extra Practice
• Student Edition, p. 675
• Chapter 1 Resource Book, pp. 29–30

Homework Check

To quickly check student understanding of key concepts, go over the following exercises:

Basic: 16, 29, 40, 48, 63
Average: 17, 32, 42, 51, 64
Advanced: 18, 34, 44, 55, 65

✗ Common Error

In Exercise 5, watch for students who think the points are not coplanar because the plane containing them is not drawn. Remind them that three points determine a plane and so the points are coplanar even though the plane is not shown in the figure. Similarly, in Exercise 60 on page 19, point out that point *J* and any other pair of points can determine a plane containing *J*.

Naming Collinear Points Name a point that is collinear with the given points.

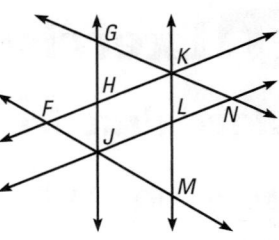

27. *F* and *H*   K   **28.** *G* and *K*   N

29. *K* and *L*   M   **30.** *M* and *J*   F

31. *J* and *N*   L   **32.** *K* and *H*   F

33. *H* and *G*   J   **34.** *J* and *F*   M

Naming Noncollinear Points Name three points that are not collinear.

35. *Sample answers: N, P, and R; N, Q, and R; P, Q, and R*

36. *T and two points on $\overleftrightarrow{RV}$; for example, R, S, T*

37. *Sample answers: W, A, and Z; W, X, and Y*

35.      **36.**   **37.**

Naming Coplanar Points Name a point that is coplanar with the given points.

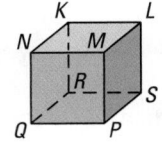

38. *A*, *B*, and *C*   D   **39.** *D*, *C*, and *F*   G

40. *G*, *A*, and *D*   H   **41.** *E*, *F*, and *G*   H

42. *A*, *B*, and *H*   E   **43.** *B*, *C*, and *F*   E

44. *A*, *B*, and *F*   G   **45.** *B*, *C*, and *G*   H

Naming Noncoplanar Points Name all the points that are *not* coplanar with the given points.

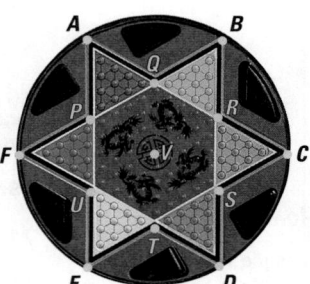

46. *N*, *K*, and *L*   **47.** *S*, *P*, and *M*
  P, Q, R, and *S*   *K, N, Q,* and *R*
48. *P*, *Q*, and *N*   **49.** *R*, *S*, and *L*
  K, L, S, and *R*   *P, Q, N,* and *M*
50. *P*, *Q*, and *R*   **51.** *R*, *K*, and *N*
  K, L, M, and *N*   *L, M, P,* and *S*
52. *P*, *S*, and *K*   **53.** *Q*, *K*, and *L*
  L, M, Q, and *R*   *M, N, R,* and *S*

Game Board In Exercises 54–57, use the game board.

54. *Sample answers: A, P, U, and E; A, Q, R, and C*

55. *Sample answers: A, Q, and B; A, B, and C*

56. *Sample answers: $\overline{RB}$, $\overline{RC}$, $\overline{RQ}$, $\overline{RA}$, $\overline{RS}$, $\overline{RD}$*

54. Name four collinear points.

55. Name three points that are not collinear.

56. Name four segments that contain point *R*.

57. $\overleftrightarrow{AD}$ divides the board in half. $\overleftrightarrow{QT}$ also divides the board in half. Name the other lines that divide the board in half. $\overleftrightarrow{BE}$, $\overleftrightarrow{CF}$, $\overleftrightarrow{RU}$, and $\overleftrightarrow{SP}$

You be the Judge In Exercises 58–62, use the diagram of the indoor tennis court.

58. Name two points that are collinear with *P*. *K* and *N*

59. Name three points that are coplanar with *P*.

60. Name two planes that contain *J*.

61. Name two planes that do not contain *J*.

62. Are the points *K* and *N* coplanar with points *J* and *Q*? Explain.

59. either three of the points *J, K, N,* and *Q,* or three of the points *K, L, M,* and *N*

60. *Sample answers:* plane *JKL* and plane *JKM*

61. *Sample answers:* plane *LKN* and plane *QNM*

62. Yes; all four points lie on the plane that contains the back wall.

Visualize It! Sketch the lines, segments, and rays. If you have geometry software, try creating your sketch using it. 63–65. See margin.

63. Draw four points *J, K, L,* and *M,* no three of which are collinear. Sketch $\overleftrightarrow{JK}$, $\overline{KL}$, $\overleftrightarrow{LM}$, and $\overrightarrow{MJ}$.

64. Draw two points, *A* and *B*. Sketch $\overrightarrow{AB}$. Add a point *C* on the ray so *B* is between *A* and *C*.

65. Draw three noncollinear points *F, G,* and *H.* Sketch $\overline{FG}$ and add a point *J* on $\overline{FG}$. Then sketch $\overrightarrow{JH}$.

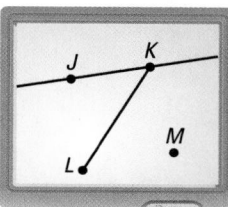

63–65. Sample answers are given.

63.
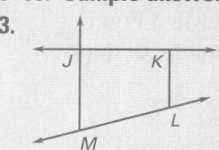

64.

65.
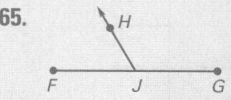

Three-Wheeled Car In Exercises 66–69, refer to the photograph of the three-wheeled car.

66. A four-wheeled car is driving slowly over uneven ground. Is it possible that only three wheels will be touching the ground at a given time? yes

67. Is it possible to draw four points that do not lie on a plane? yes

68. A three-wheeled car is driving slowly over uneven ground. Is it possible that only two wheels will be touching the ground at a given time? no

69. Is it possible to draw three points that do not lie on a plane? no

Assessment Resources

The Mini-Quiz below is also available on blackline (*Chapter 1 Resource Book*, p. 37) and on transparency. For more assessment resources, see:
- Chapter 1 Resource Book
- Standardized Test Practice
- Test and Practice Generator

Mini-Quiz

Use the diagram below.

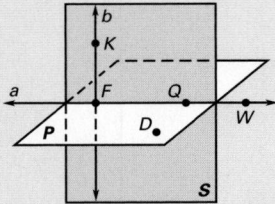

1. Name the points on plane *S*. *F, K, and Q*
2. Name two lines. *a and b*
3. Name the plane that contains point *D*. *plane P or plane DFQ*
4. Name three collinear points. *F, Q, and W*
5. Decide whether the following statement is *true* or *false*. Points *K, F,* and *D* are coplanar. *true*

1.

2.

4. *Sample answer:*
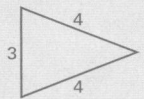

7. *Sample answer:*
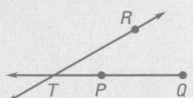

8. *Sample answer:*
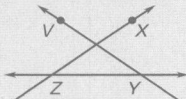

---

70. Multiple Choice Which of the statements is *false*? **D**

Ⓐ *F, G,* and *H* are collinear.

Ⓑ *C, D, K,* and *L* are coplanar.

Ⓒ *L* lies on $\overleftrightarrow{AB}$.

Ⓓ $\overrightarrow{DE}$ contains $\overline{CE}$.

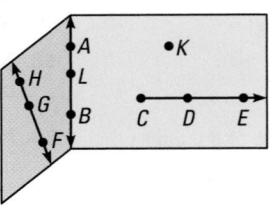

Mixed Review

Describing Number Patterns Predict the next number. *(Lesson 1.1)*

71. 6, 17, 28, 39, . . .   50

72. 9, 4, −1, −6, . . .   −11

73. 4, 20, 100, 500, . . .   2500

74. 0, 5, 15, 30, 50, . . .   75

Algebra Skills

Fractions Write the fraction as a decimal. For repeating decimals, round to the nearest hundredth. *(Skills Review, p. 657)*

75. $\frac{1}{2}$ 0.5

76. $\frac{3}{4}$ 0.75

77. $\frac{3}{5}$ 0.6

78. $\frac{4}{10}$ 0.4

79. $\frac{2}{3}$ $0.\overline{6}$; 0.67

80. $\frac{4}{3}$ $1.\overline{3}$; 1.33

81. $\frac{7}{9}$ $0.\overline{7}$; 0.78

82. $\frac{11}{2}$ 5.5

Quiz 1

1, 2. See margin.

Sketch the next figure you expect in the pattern. *(Lesson 1.1)*

1.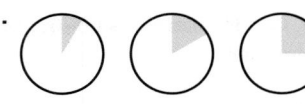

2.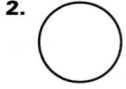

Find a counterexample to prove that the statement is false. *(Lesson 1.2)*

3. *Sample answer:* 30 is divisible by 10, but not by 20.

4. See margin.

5. *Sample answer:* 2 + 0 = 2, and 2 is not greater than 2.

6. If you fold the square along a line segment that joins two opposite corners and cut along the fold, you will create two triangles, not two rectangles.

3. If a number is divisible by 10, then it is divisible by 20.

4. Two sides of a triangle can never have the same length.

5. The sum of two numbers is always greater than either number.

6. If you fold a square piece of paper in half, then unfold it and cut along the fold, you will always create two rectangles of the same size.

Sketch the figure. *(Lesson 1.3)* 7, 8. See margin.

7. Draw three noncollinear points *P, Q,* and *R*. Sketch $\overrightarrow{QP}$. Add a point *T* on the ray so that *P* is between *Q* and *T*. Then sketch $\overleftrightarrow{RT}$.

8. Draw four points, *V, X, Y,* and *Z*, no three of which are collinear. Sketch $\overleftrightarrow{VY}$, $\overleftrightarrow{XZ}$, and $\overleftrightarrow{YZ}$.

Activity 1.4 Exploring Intersections

Question

When planes meet, what points do they have in common?

Materials
- three index cards
- scissors

Explore

1 Label two index cards as shown. Cut each card along the dotted line.

2 Slide two cards together. Answer Exercises 1–4.

3 Label and cut a third card as shown. Cut another slot in plane M and slide the cards together. Answer Exercises 5–7.

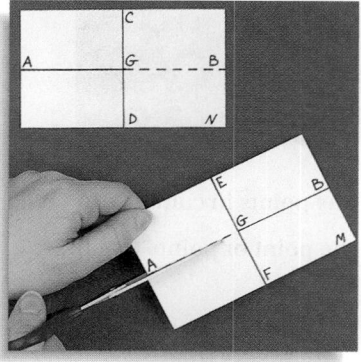

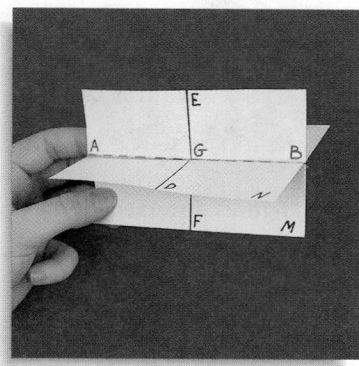

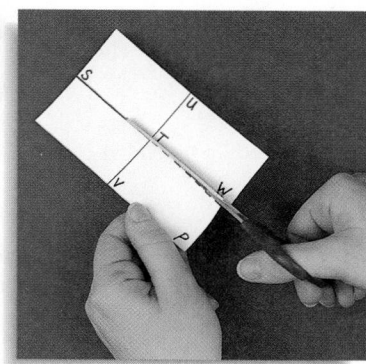

Think About It

Student Help

LOOK BACK
The index cards stand for planes, but you must imagine the planes extend without end. See p. 14.

4. yes; Two intersecting lines determine a plane.

7. The planes do not appear to meet.

1. Where do $\overleftrightarrow{AB}$ and $\overleftrightarrow{CD}$ meet? Where do $\overleftrightarrow{AB}$ and $\overleftrightarrow{EF}$ meet? *point G; point G*

2. When the cards are together, where do $\overleftrightarrow{CD}$ and $\overleftrightarrow{EF}$ meet? *point G*

3. Where do planes M and N meet? *$\overleftrightarrow{AB}$*

4. Are $\overleftrightarrow{CD}$ and $\overleftrightarrow{EF}$ coplanar? Explain.

5. Where do $\overleftrightarrow{EF}$ and $\overleftrightarrow{UV}$ meet? *point T*

6. Where do planes M and P meet? *$\overleftrightarrow{SW}$*

7. Do planes N and P meet?

Goal
Students model planes with index cards to explore how they intersect.

Materials
See the margin of the student page.

LINK TO LESSON
Refer students to this activity while discussing Example 2 on page 23. Help students recognize that the diagram in Example 2 is a two-dimensional representation of the intersecting index cards.

2 Managing the Activity

Tips for Success
Point out that the dotted lines students cut along are in different places on the two cards in Step 1.

On plane M, if students draw $\overleftrightarrow{AB}$ a little above the centerline of the card, there will be more room when sliding plane P onto plane M in Step 3.

3 Closing the Activity

KEY DISCOVERY
When two planes intersect, they intersect in a line.

Activity Assessment
Use Exercises 3 and 6 to assess student understanding.

Pacing
Suggested Number of Days

Basic: 2 days
Average: 2 days
Advanced: 2 days
Block Schedule: 1 block

Teaching Resources

☰ **Blacklines**
(See page 1B.)

📑 **Transparencies**
• Warm-Up with Quiz
• Answers

⊞ **Technology**
• Electronic Teacher Tools
• Test and Practice Generator
• Online Lesson Planner
• Internet Support

② Teach

Content and Teaching Strategies
For background information on geometric concepts and teaching strategies related to this lesson, see pages 1E and 1F in this Teacher's Edition.

Tips for New Teachers
Have students note that the postulates are numbered and that this is done just for easy reference in the textbook. Students should know the meaning of each postulate and be able to recite or write it. Giving a reason by indicating "Postulate 3" is not appropriate. See the Tips for New Teachers on pp. 1–2 of the *Chapter 1 Resource Book* for additional notes about Lesson 1.4.

1.4 Sketching Intersections

Goal
Sketch simple figures and their intersections.

Key Words
• intersect
• intersection

The photograph shows the intersection of two streets in Seattle. The intersection is the part where the two streets cross each other.

In geometry, figures **intersect** if they have any points in common.

The **intersection** of two or more figures is the point or points that the figures have in common.

> ◗ **Student Help**
>
> **READING TIP**
> The letter *P* names the point of intersection even though no dot is drawn there. ·······

POSTULATES 3 and 4

Postulate 3 Intersection of Two Lines

Words If two lines intersect, then their intersection is a point.

Symbols Lines *s* and *t* intersect at point *P*.

Postulate 4 Intersection of Two Planes

Words If two planes intersect, then their intersection is a line.

Symbols Planes *M* and *N* intersect at line *d*.

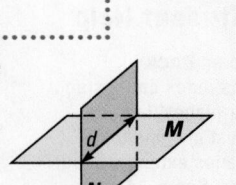

Postulates 3 and 4 are written as *if-then statements*. You will learn more about if-then statements in Lesson 2.5.

Visualize It!

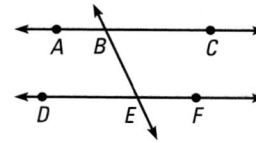

Two lines can intersect even if a diagram does not show where they intersect. The lines above intersect at a point that is not shown.

EXAMPLE 1 Name Intersections of Lines

Use the diagram at the right.

a. Name the intersection of $\overleftrightarrow{AC}$ and $\overleftrightarrow{BE}$.

b. Name the intersection of $\overleftrightarrow{BE}$ and $\overleftrightarrow{DF}$.

c. Name the intersection of $\overleftrightarrow{AC}$ and $\overleftrightarrow{DF}$.

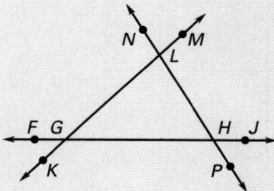

Solution

a. $\overleftrightarrow{AC}$ and $\overleftrightarrow{BE}$ intersect at point B.

b. $\overleftrightarrow{BE}$ and $\overleftrightarrow{DF}$ intersect at point E.

c. $\overleftrightarrow{AC}$ and $\overleftrightarrow{DF}$ do not appear to intersect.

EXAMPLE 2 Name Intersections of Planes

Use the diagram at the right.

a. Name the intersection of planes S and R.

b. Name the intersection of planes R and T.

c. Name the intersection of planes T and S.

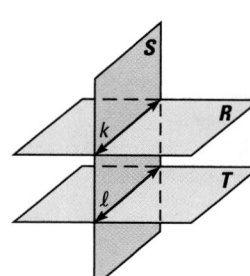

Solution

a. Planes S and R intersect at line k.

b. Planes R and T do not appear to intersect.

c. Planes T and S intersect at line ℓ.

Checkpoint ✓ Name Intersections of Lines and Planes

Use the diagram at the right.

1. Name the intersection of $\overleftrightarrow{PS}$ and $\overleftrightarrow{QR}$. point R

2. Name the intersection of $\overleftrightarrow{TV}$ and $\overleftrightarrow{QU}$.
point U

3. Name the intersection of $\overleftrightarrow{PS}$ and $\overleftrightarrow{UV}$.
They do not appear to intersect.

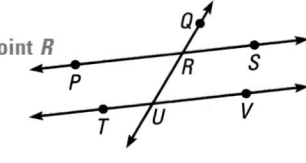

Use the diagram at the right.

4. Name the intersection of planes X and Y. line p

5. Name the intersection of planes Y and Z.
line q

6. They do not appear to intersect. **6.** Name the intersection of planes Z and X.

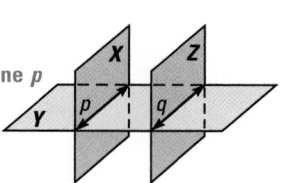

Extra Example 1

Use the diagram below.

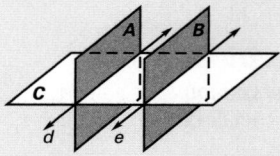

a. Name the intersection of $\overleftrightarrow{FJ}$ and $\overleftrightarrow{KM}$. point G

b. Name the intersection of $\overleftrightarrow{NP}$ and $\overleftrightarrow{KM}$. point L

c. Name the intersection of $\overleftrightarrow{NP}$ and $\overleftrightarrow{FJ}$. point H

Extra Example 2

Use the diagram below.

a. Name the intersection of planes B and C. line e

b. Name the intersection of planes A and B. The planes do not appear to intersect.

c. Name the intersection of planes A and C. line d

1.4 *Sketching Intersections* **23**

Extra Example 3

What is shown by this sketch?

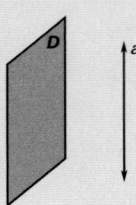

a line that does not intersect the plane

Extra Example 4

Sketch three planes that intersect in the same line.

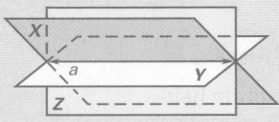

✓ Concept Check

Define *intersection*. **the point or points that two or more figures have in common**

🐢 Daily Puzzler

How many distinct planes can intersect at a single line *m*? **an infinite number**

7–9. Sample answers are given.

7.
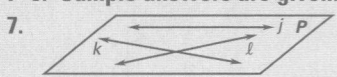

8.

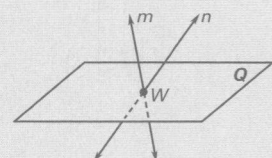

9.

Visualize It!

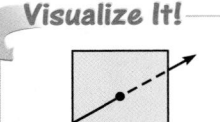

Use dashed lines to show where a line is hidden by a plane.

When you sketch lines and planes that intersect, you can use dashed lines, colored pencils, and shading to create a three-dimensional look.

EXAMPLE 3 Sketch Intersections of Lines and Planes

Sketch a plane. Then sketch each of the following.

 a. a line that is in the plane

 b. a line that does not intersect the plane

 c. a line that intersects the plane at a point

Solution

a. Draw the plane. Then draw a line in the plane.

b. Draw the plane. Then draw a line above or below the plane.

c. Draw the plane. Then draw a line that intersects the plane.

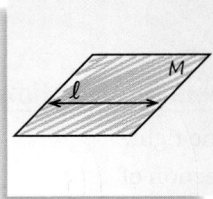

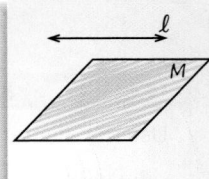

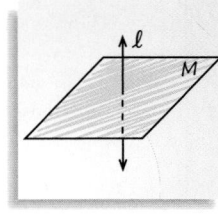

EXAMPLE 4 Sketch Intersections of Planes

Sketch two planes that intersect in a line.

Solution

❶ **Draw** one plane as if you are looking straight at it. Shade the plane.

❷ **Draw** a second plane that is horizontal. Shade this plane a different color.

❸ **Draw** the line of intersection. Use dashed lines to show where one plane is hidden.

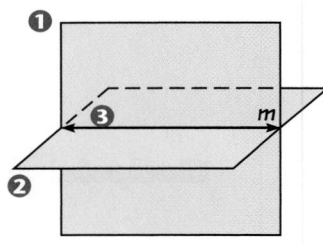

Checkpoint ✓ Sketch Intersecting Lines and Planes

 7. Sketch three lines that lie in a plane. 7–9. See margin.

 8. Sketch two lines that intersect a plane at the same point.

 9. Sketch two planes that do not intersect.

1.4 Exercises

Guided Practice

Vocabulary Check

1. Describe what an *intersection* is.
An intersection of two or more figures is the point or points that the figures have in common.

Skill Check

Decide whether the statement is *true* or *false*.

2. $\overleftrightarrow{BD}$ and $\overleftrightarrow{DC}$ intersect at point D.   true

3. $\overleftrightarrow{AB}$ and $\overleftrightarrow{BD}$ intersect at point A.   false

4. $\overleftrightarrow{BD}$ intersects plane M at point B.   false

5. $\overleftrightarrow{AB}$ and $\overleftrightarrow{DC}$ do not appear to intersect.   true

6. $\overleftrightarrow{BD}$ is the intersection of planes M and N.   false

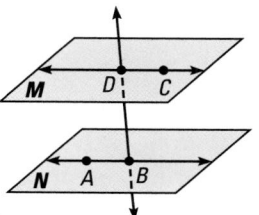

Practice and Applications

Extra Practice

See p. 676.

Everyday Intersections **What kind of geometric intersection does the photograph suggest?**

7. intersecting lines

8. the intersection of a line and a plane

9. intersecting planes

11. They do not appear to intersect.

Homework Help

Example 1: Exs. 10–15, 22–24
Example 2: Exs. 16–21, 25–27
Example 3: Exs. 28–32
Example 4: Exs. 28–32

Naming Intersections of Lines **In Exercises 10–15, use the diagram at the right.**

10. Name the intersection of $\overleftrightarrow{PQ}$ and $\overleftrightarrow{TS}$.   point R

11. Name the intersection of $\overleftrightarrow{QS}$ and $\overleftrightarrow{PT}$.

12. Name the intersection of $\overleftrightarrow{SQ}$ and $\overleftrightarrow{TR}$.   point S

13. Name the intersection of $\overleftrightarrow{RS}$ and $\overleftrightarrow{PT}$.   point T

14. Name the intersection of $\overleftrightarrow{RP}$ and $\overleftrightarrow{PT}$.   point P

15. Name the intersection of $\overleftrightarrow{RS}$ and $\overleftrightarrow{ST}$. $\overleftrightarrow{RS}$, $\overleftrightarrow{ST}$, or $\overleftrightarrow{RT}$

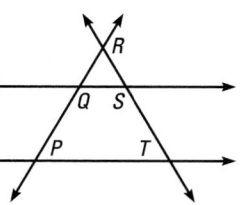

Assignment Guide

BASIC
Day 1: pp. 25–27 Exs. 7–12, 16–26 even, 35, 36, 38–40
Day 2: pp. 25–27 Exs. 28–31, 41–52

AVERAGE
Day 1: pp. 25–27 Exs. 7–19, 23–27 odd, 36–40
Day 2: pp. 25–27 Exs. 28–32, 34, 41–53 odd

ADVANCED
Day 1: pp. 25–27 Exs. 7–15 odd, 19–21, 23–26, 35–40
Day 2: pp. 25–27 Exs. 28–32, 33*, 34, 42–54 even;
EC: classzone.com

BLOCK SCHEDULE
pp. 25–27 Exs. 7–19, 23–27 odd, 28–32, 34, 36–40, 41–53 odd

Extra Practice
• Student Edition, p. 676
• Chapter 1 Resource Book, pp. 38–39

Homework Check
To quickly check student understanding of key concepts, go over the following exercises:

Basic: 10, 18, 22, 26, 28
Average: 12, 18, 23, 26, 30
Advanced: 15, 19, 21, 26, 32

26

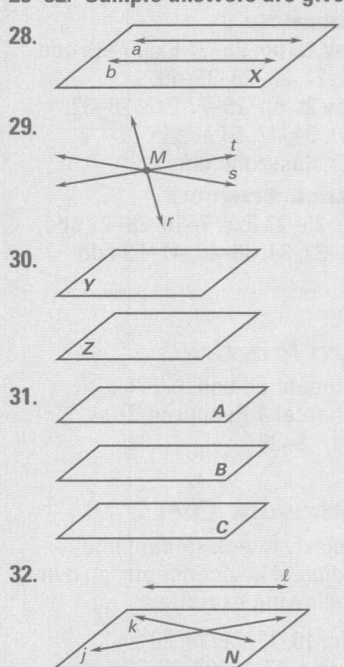

✗ Common Error
Some students may have difficulty visualizing the planes in Exercises 16–21. Help them make and label a model of the planes using index cards following the procedure discussed in Activity 1.4 on page 21.

Teaching Tip
For Exercises 28–32, point out that there may be more than one way to sketch each figure.

28–32. Sample answers are given.

28.

29.

30.

31.

32.

Naming Intersections of Planes Name the intersection of the given planes, or write *no intersection*.

16. *P* and *Q*
no intersection

17. *Q* and *R* line *ℓ*

18. *P* and *R* line *j*

19. *P* and *S* line *k*

20. *Q* and *S* line *m*

21. *R* and *S*
no intersection

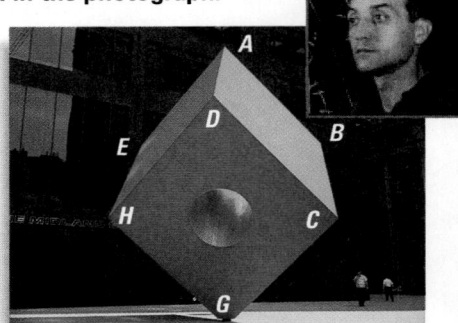

Completing Sentences Fill in the blank with the appropriate response based on the points labeled in the photograph.

22. point *B*

23. point *A*

24. point *H*

22. $\overleftrightarrow{AB}$ and $\overleftrightarrow{BC}$ intersect at __?__ .

23. $\overleftrightarrow{AD}$ and $\overleftrightarrow{AE}$ intersect at __?__ .

24. $\overleftrightarrow{HG}$ and $\overleftrightarrow{DH}$ intersect at __?__ .

25. Plane *ABC* and plane *DCG* intersect at __?__ . $\overleftrightarrow{DC}$

26. Plane *GHD* and plane *DHE* intersect at __?__ . $\overleftrightarrow{DH}$

27. Plane *EAD* and plane *BCD* intersect at __?__ . $\overleftrightarrow{AD}$

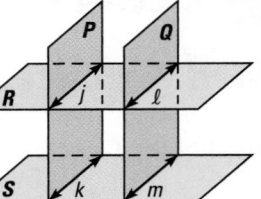

Red Cube, 1968, red painted steel, by sculptor Isamu Noguchi (above), 1904–1988. It is located on Broadway in New York City.

Visualize It! Sketch the figure described. 28–32. See margin.

28. Two lines that lie in a plane but do not intersect.

29. Three lines that intersect in a point.

30. Two planes that do not intersect.

31. Three planes that do not intersect.

32. Two lines that intersect and another line that does not intersect either one.

33. Challenge In the diagrams below, every line intersects all the other lines, but only two lines pass through each intersection point. Check sketches. Each time a line is added to a figure with *n* lines, *n* points of intersection are added.

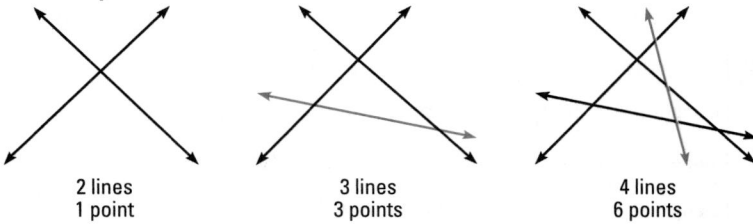

2 lines
1 point

3 lines
3 points

4 lines
6 points

Draw 5 lines that intersect in this way. Is there a pattern to the number of intersection points in the diagrams?

34. Multi-Step Problem In *perspective drawing,* lines that do not intersect in real life are represented by lines that intersect on the horizon line at a point called a *vanishing point.*

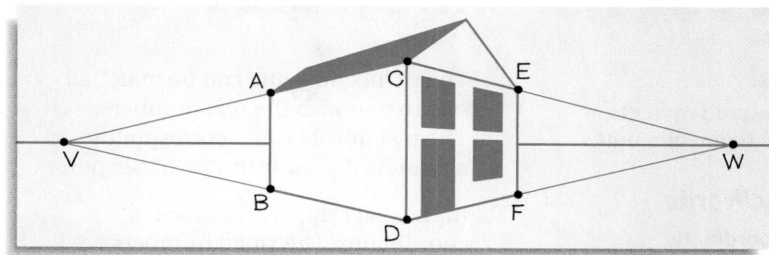

a. Name two lines that intersect at vanishing point *V.* Name two lines that intersect at vanishing point *W.* $\overleftrightarrow{AC}$, $\overleftrightarrow{BD}$; $\overleftrightarrow{CE}$, $\overleftrightarrow{DF}$

b. Trace the diagram. Draw $\overleftrightarrow{EV}$ and $\overleftrightarrow{AW}$. Label their intersection as point *G.* **b–d. See margin.**

c. Draw $\overleftrightarrow{FV}$ and $\overleftrightarrow{BW}$. Label their intersection as point *H.*

d. Use dashed lines to draw the hidden edges of the house: $\overline{GH}$, $\overline{AG}$, $\overline{BH}$, $\overline{EG}$, and $\overline{FH}$.

Mixed Review

⟨xy⟩ **Using Algebra** **Find a pattern in the coordinates of the points. Then write the coordinates of another point in the pattern.** *(Lesson 1.1)*

35. Each point has a *y*-coordinate of −3. *Sample answer:* (4, −3)

36. For each point, the *y*-coordinate is 1 more than twice the *x*-coordinate. *Sample answer:* (2, 5)

37. For each point, the *y*-coordinate is twice the opposite of the *x*-coordinate. *Sample answer:* (2, −4)

35.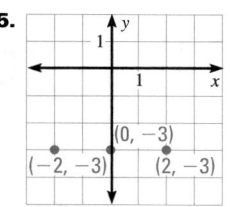

36.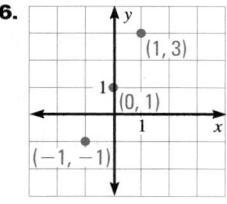

37.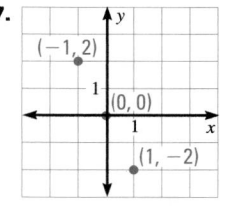

39. *Sample answers:* B, A, and C; B, A, and D; B, C, and D

40. Yes; the three points A, B, D, determine a plane, and point C is on $\overrightarrow{AD}$ so it is in the same plane.

Naming Points **Use the diagram shown.** *(Lesson 1.3)*

38. Name three collinear points. *A, C,* and *D*

39. Name three noncollinear points.

40. Are all the points coplanar? Explain.

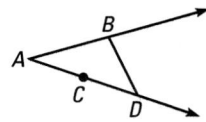

Algebra Skills

Subtracting Decimals **Find the difference.** *(Skills Review, p. 655)*

41. 13.8 − 2.4 11.4   **42.** 10.6 − 4.4 6.2   **43.** 7.5 − 3.8 3.7

44. 9.68 − 5.22 4.46   **45.** 24.72 − 16.15 8.57   **46.** 5 − 1.29 3.71

Finding Absolute Values **Evaluate.** *(Skills Review, p. 662)*

47. $|-7|$ 7   **48.** $|-3|$ 3   **49.** $|0|$ 0   **50.** $|4|$ 4

51. $|5-3|$ 2   **52.** $|12-7|$ 5   **53.** $|3-5|$ 2   **54.** $|7-12|$ 5

1.4 Sketching Intersections **27**

⟨4⟩ **Assess**

Assessment Resources
The Mini-Quiz below is also available on blackline (*Chapter 1 Resource Book,* p. 45) and on transparency. For more assessment resources, see:
- Chapter 1 Resource Book
- Standardized Test Practice
- Test and Practice Generator

Mini-Quiz
Use the diagram below.

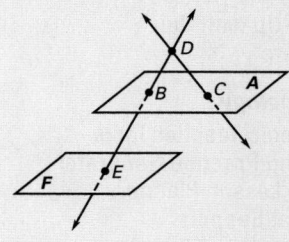

1. Name the intersection of $\overleftrightarrow{BE}$ and $\overleftrightarrow{CD}$. **point D**

2. Name the intersection of $\overleftrightarrow{BE}$ and plane *A.* **point B**

3. Name the intersection of $\overleftrightarrow{DB}$ and plane *F.* **point E**

4. Name the intersection of planes *A* and *F.* **The planes do not appear to intersect.**

5. Sketch two lines that intersect a plane at the same point. *Sample answer:*

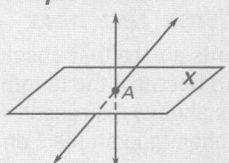

34b–d.

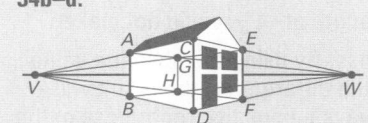

Suggested Number of Days

Basic: 2 days
Average: 2 days
Advanced: 2 days
Block Schedule: 1 block

Teaching Resources

📄 *Blacklines*
(See page 1B.)

📊 *Transparencies*
• Warm-Up with Quiz
• Answers

🖥 *Technology*
• Electronic Teacher Tools
• Test and Practice Generator
• Online Lesson Planner
• Internet Support

 **Teach**

Content and Teaching Strategies
For background information on geometric concepts and teaching strategies related to this lesson, see pages 1E and 1F in this Teacher's Edition.

Tips for New Teachers
Make students aware that absolute value is used because the length of a segment is always expressed as a positive number. In Example 1 on this page, the length of $\overline{BC}$ could have been shown as $\frac{7}{8} - 2\frac{1}{4}$, but a length of $-1\frac{3}{8}$ would not make sense. By using the absolute value of the differences, the result will always be positive regardless of the order in which the coordinates are subtracted. See the Tips for New Teachers on pp. 1–2 of the *Chapter 1 Resource Book* for additional notes about Lesson 1.5.

1.5 Segments and Their Measures

Goal
Measure segments.
Add segment lengths.

Key Words
• coordinate
• distance
• length
• between
• congruent segments

The points on a line can be matched one to one with the real numbers. The real number that corresponds to a point is the **coordinate** of the point.

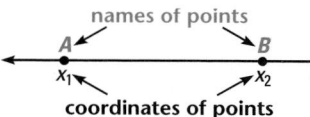

In the diagram, x_1 and x_2 are coordinates. The small numbers are *subscripts*. The coordinates are read as "x sub 1" and "x sub 2."

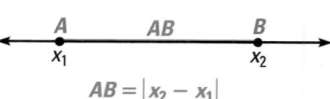

$$AB = |x_2 - x_1|$$

The **distance** between points A and B is written as AB. It is the absolute value of the difference of the coordinates of A and B. AB is also called the **length** of $\overline{AB}$.

EXAMPLE 1 Find the Distance Between Two Points

Measure the total length of the shark's tooth to the nearest $\frac{1}{8}$ inch. Then measure the length of the exposed part.

Solution

Use a ruler to measure in inches.

Student Help
SKILLS REVIEW
For help with absolute value, see p. 662.

❶ Align the zero mark of the ruler with one end of the shark's tooth.

❷ Find the length of the shark's tooth, AC.
$$AC = |2\tfrac{1}{4} - 0| = 2\tfrac{1}{4}$$

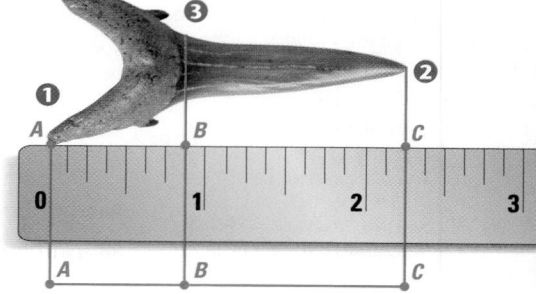

❸ Find the length of the exposed part, BC.
$$BC = |2\tfrac{1}{4} - \tfrac{7}{8}| = 1\tfrac{3}{8}$$

ANSWER ▶ The length of the shark's tooth is $2\frac{1}{4}$ inches. The length of the exposed part is $1\frac{3}{8}$ inches.

Checkpoint ✓ *Find the Distance Between Two Points*

Measure the length of the segment to the nearest $\frac{1}{8}$ inch.

1. A ●————————● B $2\frac{1}{8}$ in. **2.** C ●————————————● D $1\frac{1}{2}$ in.

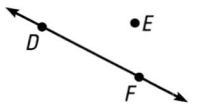

Betweenness refers to collinear points only. Point *E* is *not* between points *D* and *F*, because the points are not collinear.

When three points lie on a line, one of them is **between** the other two. In the diagram of Postulate 5 below, *B* is between *A* and *C*.

POSTULATE 5

Segment Addition Postulate

Words and Symbols

If *B* is between *A* and *C*, then *AC* = *AB* + *BC*.

If *AC* = *AB* + *BC*, then *B* is between *A* and *C*.

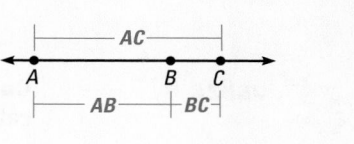

EXAMPLE 2 Find a Distance by Adding

Use the map to find the distance from Athens to Albany.

Solution

Because the three cities lie on a line, you can use the Segment Addition Postulate.

AM = 80 miles	Use map.
MB = 90 miles	Use map.
AB = *AM* + *MB*	Segment Addition Postulate
= 80 + 90	Substitute.
= 170	Add.

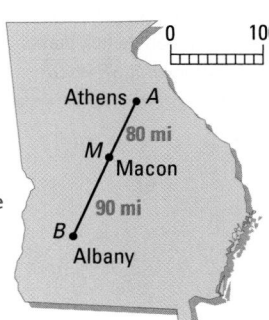

ANSWER ▶ The distance from Athens to Albany is 170 miles.

Note that in Example 2, if the cities were not collinear, you would not be able to use the Segment Addition Postulate.

 **Using Algebra**

EXAMPLE 3 Find a Distance by Subtracting

Use the diagram to find *EF*.

Solution

DF = *DE* + *EF*	Segment Addition Postulate
16 = 10 + *EF*	Substitute 16 for *DF* and 10 for *DE*.
16 − 10 = 10 + *EF* − 10	Subtract 10 from each side.
6 = *EF*	Simplify.

1.5 Segments and Their Measures **29**

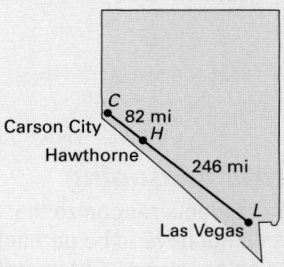
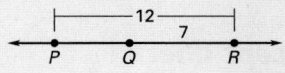

30

Extra Example 4

Are the segments shown in the coordinate plane congruent?

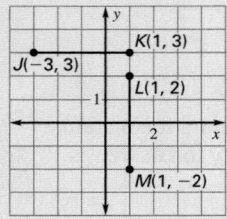

$\overline{JK}$ and $\overline{LM}$ have the same length. So, $\overline{JK} \cong \overline{LM}$.

Geometric Reasoning

Make sure students recognize that segments do not have to be parallel in order to be congruent. Congruent segments can have different orientations, overlap, intersect, and even be in different planes. Congruence refers only to length; it does not refer to orientation.

✓ Concept Check

What are *congruent segments*?
Segments that have the same length.

🦉 Daily Puzzler

Points *A, B, C,* and *D* are collinear, in that order. Find *BC* if $AC = 2x + 4$, $BC = x$, $BD = 3x + 1$, and $AD = 17$.
3

5.
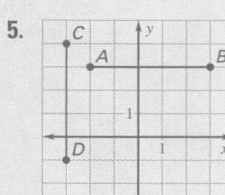

6.
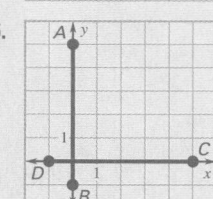

Find the length.

3. Find *AC*. 20

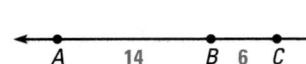

4. Find *ST*. 8

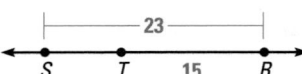

Visualize It!

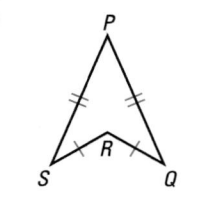

The single tick marks show that $\overline{SR} \cong \overline{QR}$. The double tick marks show that $\overline{SP} \cong \overline{QP}$.

Congruence In geometry, segments that have the same length are called **congruent segments** .

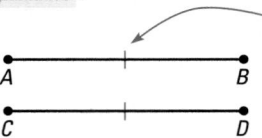

Use short tick marks to indicate congruent segments.

WORDS

The length *AB* is equal to the length *CD*.

Segment $\overline{AB}$ is congruent to segment $\overline{CD}$.

SYMBOLS

$AB = CD$

$\overline{AB} \cong \overline{CD}$

The symbol for indicating congruence is ≅.

EXAMPLE 4 Decide Whether Segments are Congruent

Are the segments shown in the coordinate plane congruent?

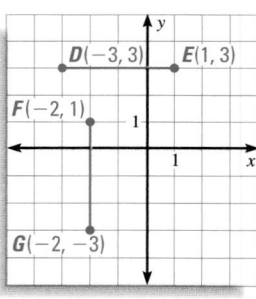

Solution

For a horizontal segment, subtract the *x*-coordinates.

$$DE = |1 - (-3)| = |4| = 4$$

For a vertical segment, subtract the *y*-coordinates.

$$FG = |-3 - 1| = |-4| = 4$$

ANSWER $\overline{DE}$ and $\overline{FG}$ have the same length. So, $\overline{DE} \cong \overline{FG}$.

Checkpoint ✓ **Decide Whether Segments are Congruent**

Plot the points in a coordinate plane. Then decide whether $\overline{AB}$ and $\overline{CD}$ are congruent.

5. $A(-2, 3)$, $B(3, 3)$, $C(-3, 4)$, $D(-3, -1)$ yes; See margin.

6. $A(0, 5)$, $B(0, -1)$, $C(5, 0)$, $D(-1, 0)$ yes; See margin.

1.5 Exercises

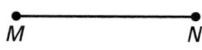
Guided Practice

Vocabulary Check

1. Is the distance between M and N the same as the length of $\overline{MN}$? **yes**

Skill Check

2. Measure the length of the line segment in Exercise 1 to the nearest millimeter. **28 mm**

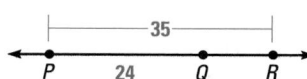

Find the length.

3. Find DF. **13**

4. Find QR. **11**

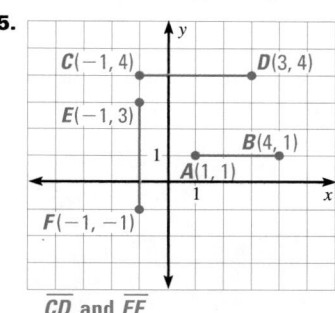
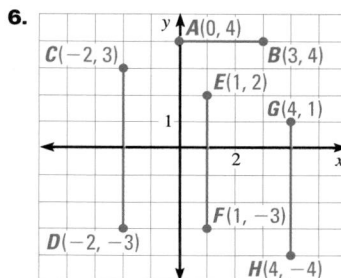

Determine which segments in the coordinate plane are congruent.

5.

$\overline{CD}$ and $\overline{EF}$

6.

$\overline{EF}$ and $\overline{GH}$

Practice and Applications

Extra Practice
See p. 676.

13. Check sketches;
$DE + EF = DF.$

14. Check sketches;
$GH + HJ = GJ.$

15. Check sketches;
$NM + MP = NP.$

16. Check sketches;
$QR + RS = QS.$

Measurement Measure the length of the segment to the nearest millimeter.

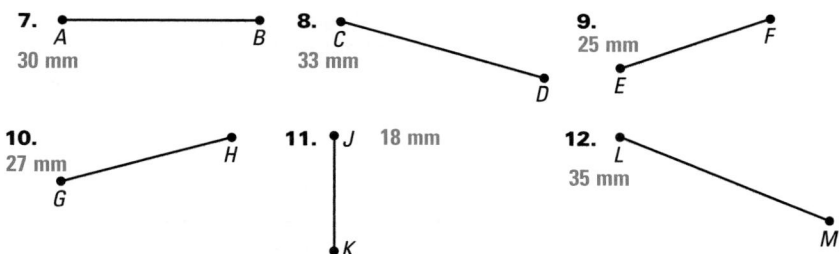

7. 30 mm

8. 33 mm

9. 25 mm

10. 27 mm

11. 18 mm

12. 35 mm

Homework Help

Example 1: Exs. 7–12
Example 2: Exs. 13–23
Example 3: Exs. 24–26
Example 4: Exs. 27–30

Visualize It! Draw a sketch of the three collinear points. Then write the Segment Addition Postulate for the points. 13–16. See margin.

13. E is between D and F.

14. H is between G and J.

15. M is between N and P.

16. R is between Q and S.

1.5 *Segments and Their Measures* **31**

Assignment Guide

BASIC
Day 1: pp. 31–33 Exs. 7–15 odd, 18–24 even, 35, 39–45 odd
Day 2: SRH p. 664 Exs. 13–20; pp. 31–33 Exs. 23–31 odd, 32–44 even

AVERAGE
Day 1: pp. 31–33 Exs. 10–16, 17–23 odd, 35–45 odd
Day 2: pp. 31–33 Exs. 25–27, 30–44 even

ADVANCED
Day 1: pp. 31–33 Exs. 10–14, 19–23, 31, 37–45 odd
Day 2: pp. 31–33 Exs. 24–32 even, 33*, 34–36, 38–44 even; EC: classzone.com

BLOCK SCHEDULE
pp. 31–33 Exs. 10–16, 17–23 odd, 25–27, 30–34 even, 35–45

Extra Practice
• Student Edition, p. 676
• Chapter 1 Resource Book, pp. 47–48

Homework Check
To quickly check student understanding of key concepts, go over the following exercises:

Basic: 7, 15, 20, 25, 27
Average: 11, 19, 21, 25, 27
Advanced: 12, 20, 21, 24, 30

Teaching Tip

Exercises 17–21 will show whether students fully understand the concept of betweenness. Any student having difficulty with these exercises should refer to the Visualize It! note on page 29.

29.

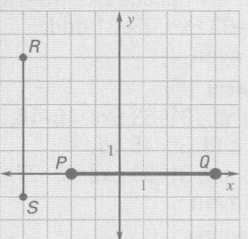

30.

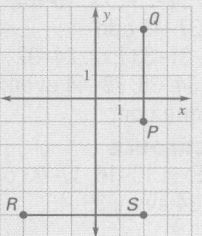

38. *Sample answer:*

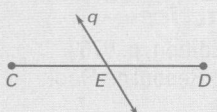

39. *Sample answer:*

You be the Judge Use the diagram to determine whether the statement is *true* or *false*.

17. *B* is between *A* and *C*. true

18. *E* is between *C* and *F*. false

19. *E* is between *D* and *H*. true

20. *D* is between *A* and *H*. true

21. *C* is between *B* and *E*. false

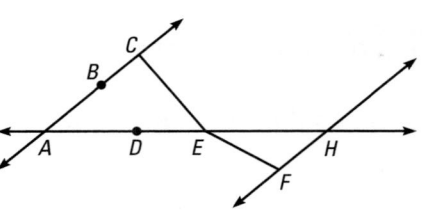

Student Help

CLASSZONE.COM

HOMEWORK HELP
Extra help with problem solving in Exs. 22–25 is at classzone.com

Segment Addition Postulate Find the length.

22. Find *PR*. 16

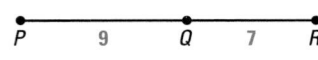

23. Find *SU*. 21

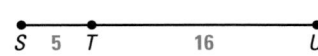

24. Find *MN*. 10

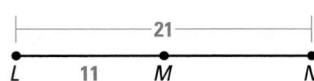

25. Find *JK*. 6

26. One fish is about $9\frac{1}{8}$ in. and the other is about $7\frac{1}{2}$ in. The difference of the lengths is $1\frac{5}{8}$ in. The total length of both fish is $16\frac{5}{8}$ in.

26. Fishing Use the photograph to determine the difference in length between the two fish and the total length of both fish.

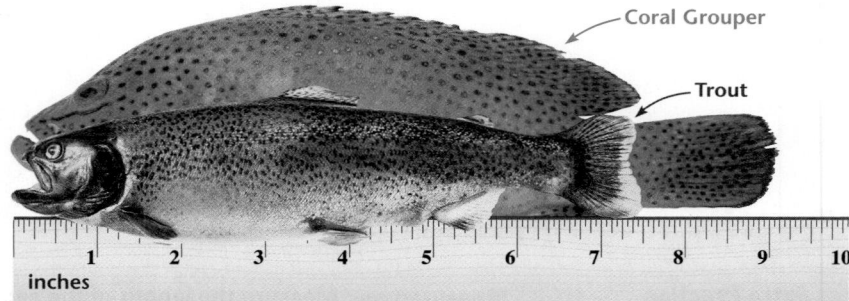

Coral Grouper

Trout

Student Help

SKILLS REVIEW
For help plotting points in a coordinate plane, see p. 664.

Congruent Segments Determine which segments in the coordinate plane are congruent.

27.

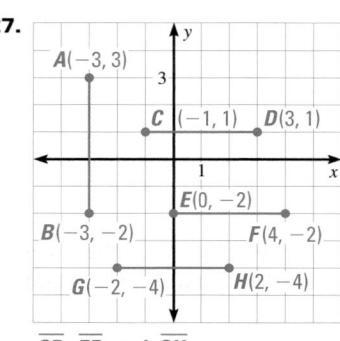

$\overline{CD}$, $\overline{EF}$, and $\overline{GH}$

28.

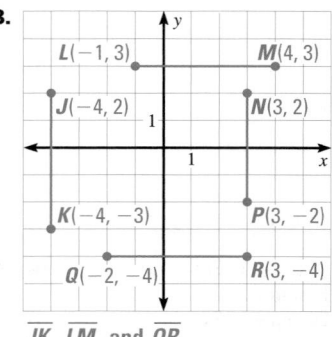

$\overline{JK}$, $\overline{LM}$, and $\overline{QR}$

32 Chapter 1 *Basics of Geometry*

Technology Plot the points in a coordinate plane. Decide whether $\overline{PQ}$ and $\overline{RS}$ are congruent. You may use geometry software.

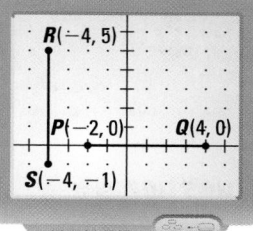

29. $P(-2, 0)$
$Q(4, 0)$
$R(-4, 5)$
$S(-4, -1)$
yes; See margin.

30. $P(2, -1)$
$Q(2, 3)$
$R(-3, -5)$
$S(2, -5)$
no; See margin.

 **Using Algebra** Write a variable expression for the length.

31. Write an expression for AC.
$8x - 1$

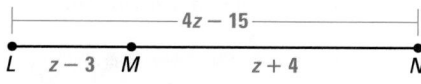
$A \quad x + 2 \quad B \qquad 7x - 3 \qquad C$

32. Write an expression for QR.
$5y + 20$
$13y + 25$

$P \quad 8y + 5 \quad Q \qquad\qquad R$

33. Challenge Find the value of z. 8

$4z - 15$
$L \quad z - 3 \quad M \qquad z + 4 \qquad N$

Standardized Test Practice

34. Multiple Choice Find the distance between the points $V(3, 5)$ and $W(3, -7)$. **D**

(A) 2   **(B)** 3   **(C)** 6   **(D)** 12

35. Multiple Choice Find the distance between the points $P(4, -1)$ and $Q(-2, -1)$. **J**

(F) -6 **(G)** -2 **(H)** 2   **(J)** 6

Mixed Review

Error Analysis Show the conjecture is false by finding a counterexample. *(Lesson 1.2)*

36. All multiples of 3 are odd. *Sample answer:* 6 is an even multiple of 3.

37. The perimeter of a rectangle can never be 17 inches.

37. Sample answer: A rectangle with a length of 5 inches and a width of 3.5 inches has a perimeter of 17 inches.

Sketching Figures Sketch the figure described. *(Lessons 1.3, 1.4)*

38. Draw two points C and D. Sketch $\overline{CD}$. Draw a line q that intersects $\overline{CD}$ at a point E. See margin.

39. Draw two lines ℓ and m that do not intersect, and a line n that intersects both lines ℓ and m. See margin.

Algebra Skills

Evaluating Expressions Evaluate the expression. *(Skills Review, p. 670)*

40. $3 + 6 \cdot 2$ 15
41. $18 \div 3 + 4 \cdot 5$ 26
42. $7 + 11 \cdot 3 - 8$ 32

43. $14 \div 7 \cdot 4$ 8
44. $(8 - 5 + 6) \div 3$ 3
45. $2(7 - 5) + 10$ 14

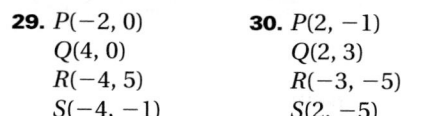 **Assess**

Assessment Resources

The Mini-Quiz below is also available on blackline (*Chapter 1 Resource Book*, p. 55) and on transparency. For more assessment resources, see:
• Chapter 1 Resource Book
• Standardized Test Practice
• Test and Practice Generator

Mini-Quiz

Measure the length of the segment to the nearest millimeter.

1.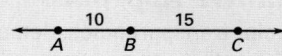
$P \qquad\qquad Q$
10 mm

2. $T \qquad\qquad U$
22 mm

Find the length.

3. Find AC. 25

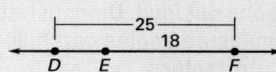
$\qquad 10 \qquad\qquad 15$
$A \qquad B \qquad\qquad C$

4. Find DE. 7

$\qquad\qquad 25 \qquad\qquad$
$\qquad\qquad 18$
$D \quad E \qquad\qquad F$

5. Are the segments shown in the coordinate plane congruent?

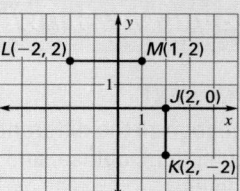

No, $\overline{JK}$ and $\overline{LM}$ are not congruent.

Planning the Activity

Goal

Students use the corners of a sheet of paper to explore how acute, right, and obtuse angles differ.

Materials

See the margin of the student page.

LINK TO LESSON

When discussing the summary at the bottom of page 36, have students use their paper models to show or form examples of each angle classification. Also, in Example 3 on page 37, students will be classifying angles by their measure.

Managing the Activity

Alternative Approach

Using color transparencies, demonstrate the activity on an overhead projector while students read the instructions out loud. Demonstrate how to measure angles carefully using a protractor.

Closing the Activity

KEY DISCOVERY

Students recognize the differences among acute, right, obtuse, and straight angles.

Activity Assessment

Use Exercises 2–4 to assess student understanding.

Activity 1.6 Kinds of Angles

Question

How do you classify angles based on their angle measures?

Materials
- paper
- scissors
- protractor

Explore

1 Tear a large corner off two pieces of paper. These are called *right angles*.

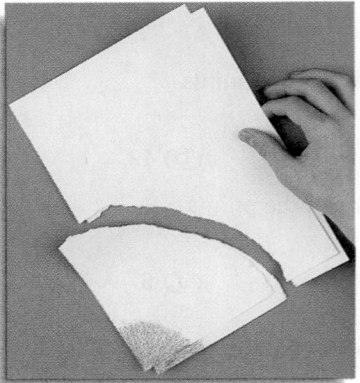

2 Cut each right angle into two smaller angles. Angles smaller than a right angle are called *acute angles*.

3 Arrange three angles to form an angle larger than a right angle. Angles larger than a right angle are called *obtuse angles*.

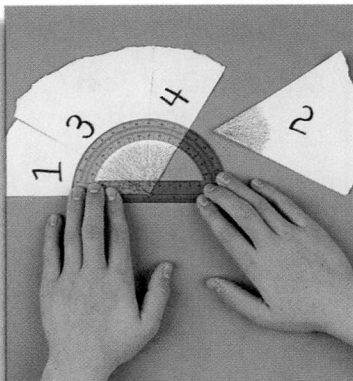

Think About It 1–4. Answers will vary.

Student Help

VOCABULARY TIP
The word *acute* means "sharp" and the word *obtuse* means "blunt." These meanings may help you remember the geometric terms.

1. Stack your angles on top of each other to compare their sizes. Arrange the angles from smallest to largest.

2. Choose two angles and place them next to each other to form a larger angle. Use a protractor to estimate the measure of the angle they form. Is the angle *acute*, *right*, or *obtuse*?

3. Choose three angles and place them next to each other to form a larger angle. Use a protractor to estimate the measure of the angle they form. Is the angle *acute*, *right*, or *obtuse*?

4. Arrange the four angles to form a *straight angle*, an angle with a measure of 180°.

1.6 Angles and Their Measures

Goal

Measure and classify angles. Add angle measures.

Key Words

- angle
- sides and vertex of an angle
- measure of an angle
- degree
- congruent angles
- acute, right, obtuse, and straight angle

An **angle** consists of two rays that have the same endpoint.

The rays are the **sides** of the angle.

The endpoint is the **vertex** of the angle.

In the photograph at the right, the sides of $\angle BAC$ are $\vec{AB}$ and $\vec{AC}$. The vertex of $\angle BAC$ is point A.

You can also write $\angle BAC$ as $\angle CAB$. Notice that the middle letter in the name of the angle is always the vertex of the angle.

You can write simply $\angle A$ if there are no other angles that have this vertex.

Mountaineers approaching Gasherbrum II, Karakorum Range, Himalayas.

EXAMPLE 1 **Name Angles**

Name the angles in the figure.

Solution

There are three different angles.

$\angle PQS$ or $\angle SQP$

$\angle SQR$ or $\angle RQS$

$\angle PQR$ or $\angle RQP$

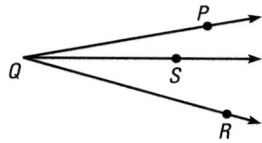

You should not name any of these angles as $\angle Q$, because all three angles have Q as their vertex. The name $\angle Q$ would not distinguish one angle from the others.

Checkpoint **Name Angles**

Name the angles in the figure.

1. $\angle S$, $\angle RST$, or $\angle TSR$

2. $\angle HMK$ or $\angle KMH$; $\angle NMK$ or $\angle KMN$; $\angle HMN$ or $\angle NMH$

3. $\angle GJC$ or $\angle CJG$; $\angle CJF$ or $\angle FJC$; $\angle GJF$ or $\angle FJG$

1.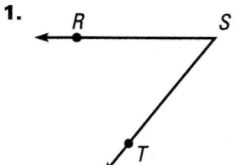

2.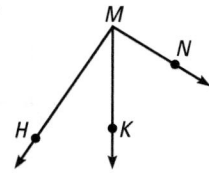

3.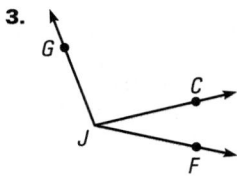

1.6 *Angles and Their Measures* **35**

---

① Plan

Pacing
Suggested Number of Days

Basic: 2 days
Average: 2 days
Advanced: 2 days
Block Schedule: 1 block

Teaching Resources

Blacklines
(See page 1B.)

Transparencies
- Warm-Up with Quiz
- Answers

Technology
- Electronic Teacher Tools
- Test and Practice Generator
- Online Lesson Planner
- Internet Support

② Teach

Content and Teaching Strategies
For background information on geometric concepts and teaching strategies related to this lesson, see pages 1E and 1F in this Teacher's Edition.

Tips for New Teachers
Stress that students say "the measure of angle *A* is equal to the measure of angle *B*" when angle measure is considered. Caution students about writing the angle symbol carefully. Some teachers have students write the angle symbol with an arc through it so it will not be confused with an inequality symbol. See the Tips for New Teachers on pp. 1–2 of the *Chapter 1 Resource Book* for additional notes about Lesson 1.6.

Extra Example 1
See next page.

36

Extra Example 1

Name the angles in the figure.

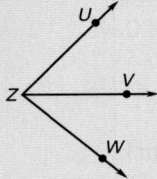

∠UZV or ∠VZU, ∠VZW or ∠WZV,
∠UZW or ∠WZU

Extra Example 2

Use a protractor to approximate the measure of ∠DEF. **110°**

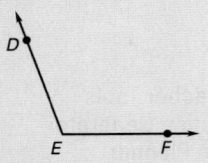

Geometric Reasoning

As you discuss congruent angles, make sure students understand that congruence refers only to the angle measure. Two angles do not need to be oriented in the same direction in order to be congruent. Also point out that the sides of an angle extend forever, so congruence of angles does not have anything to do with how "long" the sides of the angles appear on paper.

Visualize It!

Be sure students draw the right angle mark as a square, not an arc.

The **measure** of an angle is written in units called **degrees** (°). The measure of ∠A is denoted by m∠A.

EXAMPLE 2 Measure Angles

Use a protractor to approximate the measure of ∠BAC.

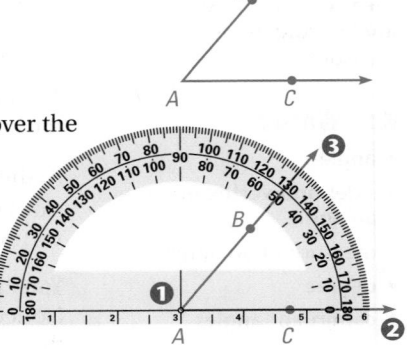

Solution

❶ Place the center of the protractor over the vertex point A.

❷ Align the protractor with one side of the angle.

❸ The second side of the angle crosses the protractor at the 50° mark. So, m∠BAC = 50°.

Two angles are **congruent angles** if they have the same measure. In the diagram below, the two angles have the same measure, so ∠DEF is congruent to ∠PQR. You can write ∠DEF ≅ ∠PQR.

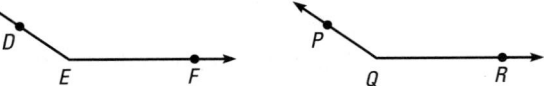

Angles are classified as **acute**, **right**, **obtuse**, or **straight**, according to their measures. Angles have measures greater than 0° and less than or equal to 180°.

Student Help

VISUAL STRATEGY
Add these words to the vocabulary pages in your notebook, as shown on p. 2.

Visualize It!

90°

A small corner mark in an angle means that the angle is a right angle.

SUMMARY   CLASSIFYING ANGLES BY THEIR MEASURES

Acute angle

A

Measure is between 0° and 90°.

Obtuse angle

A

Measure is between 90° and 180°.

Right angle

A

Measure is 90°.

Straight angle

A

Measure is 180°.

EXAMPLE 3 Classify Angles

Classify each angle.

a. $m\angle A = 130°$ **b.** $m\angle B = 90°$ **c.** $m\angle C = 45°$

Solution

a. $\angle A$ is *obtuse* because its measure is greater than 90°.

b. $\angle B$ is *right* because its measure is 90°.

c. $\angle C$ is *acute* because its measure is less than 90°.

Visualize It!

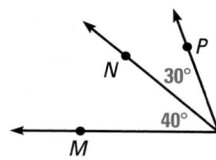

A point is in the *interior* of an angle if it is between points that lie on each side of the angle.

POSTULATE 6

Angle Addition Postulate

Words If P is in the interior of $\angle RST$, then the measure of $\angle RST$ is the sum of the measures of $\angle RSP$ and $\angle PST$.

Symbols If P is in the interior of $\angle RST$, then $m\angle RSP + m\angle PST = m\angle RST$.

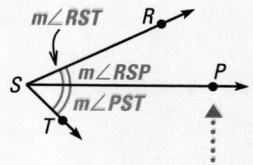

EXAMPLE 4 Add Angle Measures

Find the measure of $\angle PTM$.

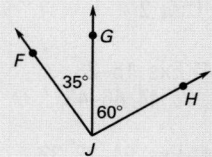

Solution

$m\angle PTM = m\angle PTN + m\angle NTM$ Angle Addition Postulate

$\qquad\quad = 30° + 40°$ Substitute 30° for $m\angle PTN$ and 40° for $m\angle NTM$.

$\qquad\quad = 70°$ Add angle measures.

ANSWER ▶ The measure of $\angle PTM$ is 70°.

Checkpoint ✓ **Add and Subtract Angle Measures**

Find the measure of $\angle ABC$.

4. 80°

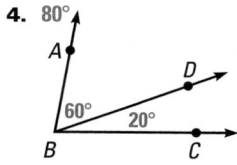

5. 130°

6. 75°

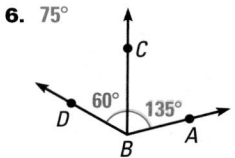

Extra Example 3

Classify each angle.

a. $m\angle G = 180°$ Angle G is straight because its measure is 180°.

b. $m\angle H = 25°$ Angle H is acute because its measure is less than 90°.

c. $m\angle J = 100°$ Angle J is obtuse because its measure is between 90° and 180°.

Extra Example 4

Find the measure of $\angle FJH$. 95°

✔ Concept Check

How do you use the measure of an angle to classify it? You use the measure to determine if the angle is acute (between 0° and 90°), right (90°), straight (180°), or obtuse (greater than 90° and less than 180°).

🧩 Daily Puzzler

$\angle JKL$ is a straight angle and $m\angle JKM = 115°$. Point H is in the interior of $\angle JKM$ and $m\angle HKL = 130°$. What is $m\angle HKM$? 65°

Assignment Guide

BASIC
Day 1: pp. 38–41 Exs. 15–23, 34, 37, 40, 41, 46–50, 53–59 odd
Day 2: pp. 38–41 Exs. 24–31, 54–60 even, Quiz 2

AVERAGE
Day 1: pp. 38–41 Exs. 15–23, 34–38 even, 40–42, 46–50, 53–59 odd
Day 2: pp. 38–41 Exs. 24–32, 44, 45, 54–60 even, Quiz 2

ADVANCED
Day 1: pp. 38–41 Exs. 15–23 odd, 35–39 odd, 42–44, 46–48, 51, 52–60 even
Day 2: pp. 38–41 Exs. 24–29, 32, 33, 45, Quiz 2; EC: TE p. 1D*, classzone.com

BLOCK SCHEDULE
pp. 38–41 Exs. 15–32, 34–38 even, 40–42, 45–50, 53–60, Quiz 2

Extra Practice
• Student Edition, p. 676
• Chapter 1 Resource Book, pp. 59–60

Homework Check

To quickly check student understanding of key concepts, go over the following exercises:

Basic: 15, 18, 21, 26, 28
Average: 16, 22, 25, 28, 30
Advanced: 19, 23, 26, 29, 32

✗ Common Error
In Exercises 18–20, watch for students who forget that the letter at the vertex must be the middle letter in the angle name.

Guided Practice

Vocabulary Check Match the angle with its classification.

A. acute **B.** obtuse **C.** right **D.** straight

1.
C
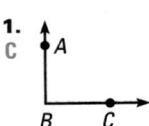

2.
D

3.
B
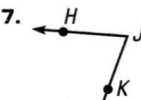

4.
A
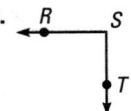

Skill Check Name the vertex and the sides of the angle. Then estimate the measure of the angle.

5. *E*; $\overrightarrow{ED}$, $\overrightarrow{EF}$; about 35°
6. *M*; $\overrightarrow{ML}$, $\overrightarrow{MN}$; about 120°
7. *J*; $\overrightarrow{JH}$, $\overrightarrow{JK}$; about 75°
8. *S*; $\overrightarrow{SR}$, $\overrightarrow{ST}$; about 90°

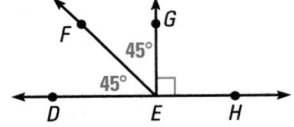

Classify the angle as *acute, right, obtuse,* or *straight*.

9. $m\angle A = 180°$ **10.** $m\angle B = 34°$ **11.** $m\angle C = 100°$ **12.** $m\angle D = 9°$
straight   acute   obtuse   acute

Use the diagram at the right to answer the questions. Explain your answers.

13. yes; $m\angle DEF = m\angle FEG$
14. yes; $m\angle DEG = m\angle HEG$

13. Is $\angle DEF \cong \angle FEG$?

14. Is $\angle DEG \cong \angle HEG$?

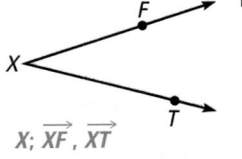

Practice and Applications

Extra Practice
See p. 676.

18. any 2 of $\angle A$, $\angle EAU$, $\angle UAE$
19. any 2 of $\angle C$, $\angle BCD$, $\angle DCB$
20. any 2 of $\angle T$, $\angle PTS$, $\angle STP$

Naming Parts of an Angle Name the vertex and the sides of the angle.

15.
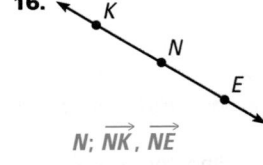
X; $\overrightarrow{XF}$, $\overrightarrow{XT}$

16.
N; $\overrightarrow{NK}$, $\overrightarrow{NE}$

17.
Q; $\overrightarrow{QR}$, $\overrightarrow{QS}$

Homework Help
Example 1: Exs. 15–20
Example 2: Exs. 21–23
Example 3: Exs. 24–26, 30–33
Example 4: Exs. 27–29

Naming Angles Write two names for the angle. See margin.

18.
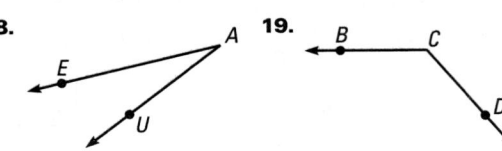

19.

20.
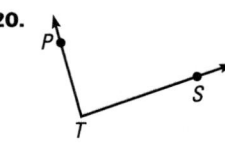

Link to Careers

SURVEYOR Surveyors use a tool called a theodolite, which can measure angles to the nearest 1/3600 of a degree.

Career Links
CLASSZONE.COM

Measuring Angles Copy the angle and use a protractor to measure it to the nearest degree. Extend the sides of the angle if necessary.

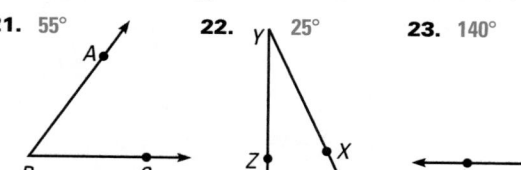

21. 55°   **22.** 25°   **23.** 140°

Classifying Angles State whether the angle appears to be *acute, right, obtuse,* or *straight.* Then estimate its measure.

24.
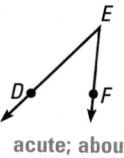
acute; about 40°

25.
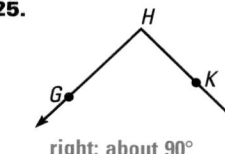
right; about 90°

26.
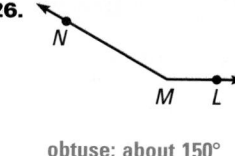
obtuse; about 150°

Angle Addition Postulate Find the measure of the angle.

27. $m\angle ABC =$ __?__ 105°   **28.** $m\angle DEF =$ __?__ 180°   **29.** $m\angle PQR =$ __?__ 140°

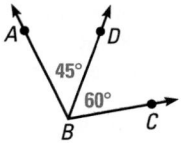
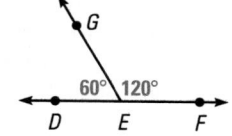
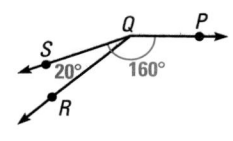

EXAMPLE   **Angles on the Coordinate Plane**

Plot the points $A(3, 0)$, $B(0, 0)$, $C(-2, 2)$ and sketch $\angle ABC$. Classify the angle.

Solution

Plot the points. Use a protractor to estimate the angle measure.

This angle has a measure of 135°.

So, $\angle ABC$ is obtuse.

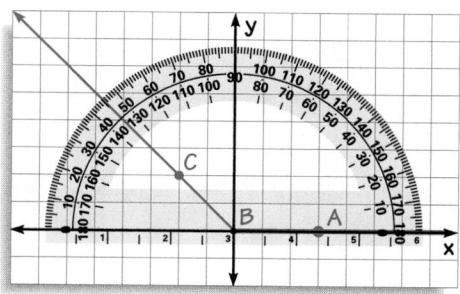

Visualize It! In Exercises 30–33, use the example above as a model. Plot the points and sketch $\angle ABC$. Classify the angle.

30. $A(3, 0)$, $B(0, 0)$, $C(0, 3)$ **31.** $A(3, 0)$, $B(0, 0)$, $C(4, -4)$

32. $A(-3, 0)$, $B(0, 0)$, $C(2, -2)$ **33.** $A(0, 4)$, $B(0, 0)$, $C(2, 2)$

30. right angle; See margin.

31. acute angle; See margin.

32. obtuse angle; See margin.

33. acute angle; See margin.

1.6 *Angles and Their Measures*   **39**

X **Common Error**

In Exercise 29, watch for students who think $m\angle PQR = 160°$. Remind them that the arc symbol implies that it is the larger angle, $\angle SQP$, that corresponds to the given angle measure.

30.
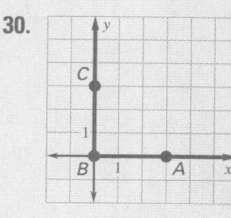

31.
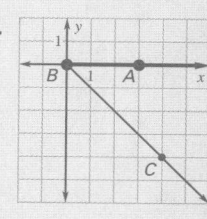

32.
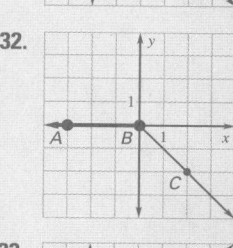

33.
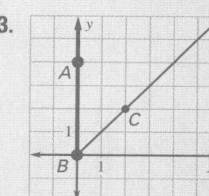

39

Link to
Geography

LONGITUDE The 0° line of longitude runs through a telescope at the Royal Observatory in Greenwich, England.

Application Links
CLASSZONE.COM

Geography For each city, estimate the measure of ∠*BOA*, where *B* is on the 0° longitude line, *O* is the North Pole, and *A* is the city.

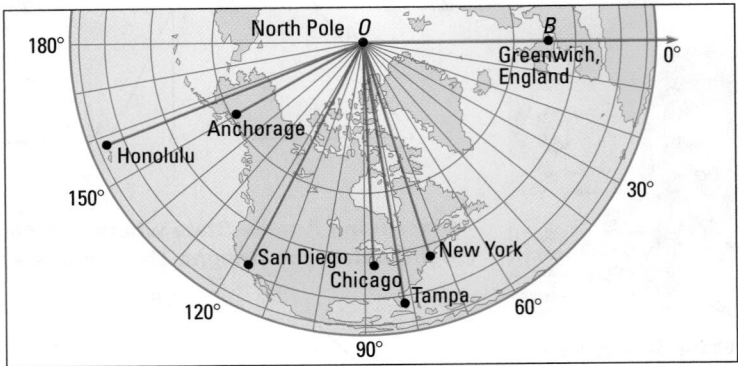

34. New York, NY 74°   **35.** Tampa, FL 82°   **36.** Chicago, IL 88°

37. San Diego, CA 117°   **38.** Honolulu, HI 158°   **39.** Anchorage, AK 150°

Airport Runways In Exercises 40–44, use the diagram of Ronald Reagan Washington National Airport and the information about runway numbering on the page facing page 1.

An airport runway number is its *bearing* (the angle measured clockwise from due north) divided by 10. Because a full circle contains 360°, runway numbers range from 1 to 36.

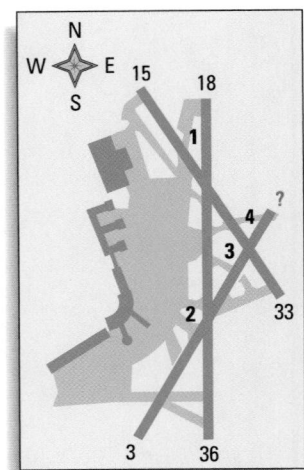

Student Help

LOOK BACK
For an example of runway numbers, see the page facing p. 1.

40. Find *m*∠1. 30°   **41.** Find *m*∠2. 150°

42. Find *m*∠3. 120°   **43.** Find *m*∠4. 60°

44. What is the number of the unlabeled runway? 21

Standardized Test Practice

45. Multi-Step Problem Fold a piece of paper in half three times and label it as shown.

 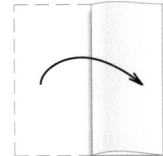 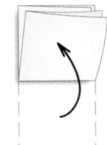  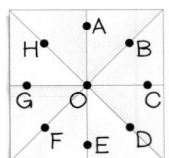

a. ∠AOB, ∠BOC, ∠COD, ∠DOE, ∠EOF, ∠FOG, ∠GOH, ∠HOA

b. ∠AOC, ∠BOD, ∠COE, ∠DOF, ∠EOG, ∠FOH, ∠GOA, ∠HOB

c. ∠AOD, ∠BOE, ∠COF, ∠DOG, ∠EOH, ∠FOA, ∠GOB, ∠HOC

a. Name eight congruent acute angles.

b. Name eight right angles.

c. Name eight congruent obtuse angles.

Mixed Review

46. one of the following:
$\overrightarrow{PM}$, $\overrightarrow{NM}$, $\overrightarrow{QM}$, $\overrightarrow{MN}$, $\overrightarrow{MQ}$

49. Check sketches;
$XY + YZ = XZ$.

50. Check sketches;
$PQ + QR = PR$.

Algebra Skills

51. Check sketches;
$AB + BC = AC$.

52. Check sketches;
$JK + KL = JL$.

Naming Rays Name the ray described. *(Lesson 1.3)*

46. Name a ray that contains *M*.
$\overrightarrow{NM}$ or $\overrightarrow{NQ}$

47. Name a ray that has *N* as an endpoint.

48. Name two rays that intersect at *P*.
$\overrightarrow{PM}$ and $\overrightarrow{PQ}$

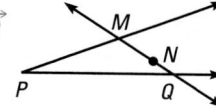

Betweenness Draw a sketch of the three collinear points. Then write the Segment Addition Postulate for the points. *(Lesson 1.5)*

49. *Y* is between *X* and *Z*.

50. *Q* is between *P* and *R*.

51. *B* is between *A* and *C*.

52. *K* is between *J* and *L*.

One-Step Equations Solve the equation. *(Skills Review p. 672)*

53. $x + 3 = 15$ 12

54. $y + 4 = 2$ −2

55. $z − 7 = 9$ 16

56. $w − 5 = −2$ 3

57. $2p = 24$ 12

58. $−9 = 3q$ −3

59. $5r = 125$ 25

60. $−12 = 6s$ −2

Quiz 2

1. *Sample answers:* $\overleftrightarrow{AB}$ and $\overleftrightarrow{EF}$, $\overleftrightarrow{AB}$ and $\overleftrightarrow{DE}$, $\overleftrightarrow{BC}$ and $\overleftrightarrow{DE}$, $\overleftrightarrow{BC}$ and $\overleftrightarrow{EF}$

2. *two of the following:* $\overleftrightarrow{AB}$, $\overleftrightarrow{BC}$, $\overleftrightarrow{BE}$

3. *two of the following:* $\overleftrightarrow{BE}$, $\overleftrightarrow{DE}$, $\overleftrightarrow{EF}$

In Exercises 1–3, use the photo shown at the right. *(Lesson 1.4)*

1. Name two lines that do not appear to intersect.

2. Name two lines that intersect at *B*.

3. Name two lines that intersect at *E*.

Judi Brown competing at the Summer Olympics.

Plot the points in a coordinate plane. Decide whether $\overline{AB}$ and $\overline{CD}$ are congruent. *(Lesson 1.5)*

4. $A(0, 0)$, $B(5, 0)$, $C(2, 4)$, $D(2, −1)$
   yes; See margin.

5. $A(−3, 2)$, $B(3, 2)$, $C(6, 0)$, $D(−6, 0)$
   no; See margin.

Classify the angle as *acute, right, obtuse,* or *straight*. *(Lesson 1.6)*

6. $m\angle Z = 90°$ right

7. $m\angle Y = 126°$ obtuse

8. $m\angle X = 180°$ straight

9. $m\angle W = 35°$ acute

10. $m\angle V = 5°$ acute

11. $m\angle U = 45°$ acute

Use the Angle Addition Postulate to find the measure of the angle. *(Lesson 1.6)*

12. Find $m\angle JKL$. 80°

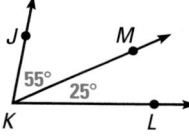

13. Find $m\angle WXY$. 75°

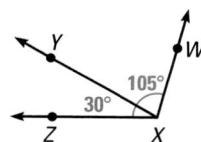

14. Find $m\angle VUW$. 60°

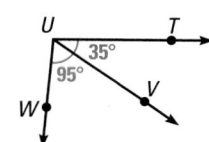

④ Assess

Assessment Resources
The Mini-Quiz below is also available on blackline (*Chapter 2 Resource Book*, p. 9) and on transparency. For more assessment resources, see:
• Chapter 1 Resource Book
• Standardized Test Practice
• Test and Practice Generator

Mini-Quiz
Use the figure below.

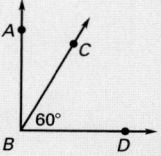

1. Use a protractor to approximate the measure of $\angle ABC$. 30°

2. Use your answer to Exercise 1 and the Angle Addition Postulate to find the measure of $\angle ABD$. 90°

3. Classify each of the three angles shown in the figure as *acute, right, obtuse,* or *straight*.
   $\angle ABC$ is acute, $\angle CBD$ is acute, and $\angle ABD$ is right.

4.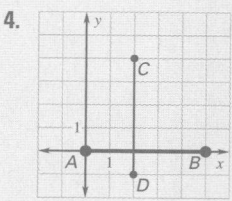

5.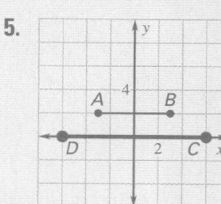

Additional Resources

The following resources are available to help review the materials in this chapter.

📖 **Chapter 1 Resource Book**
- Chapter Review Games and Activities, p. 65
- Cumulative Review, Ch. 1

Chapter  Summary and Review

VOCABULARY

- **conjecture,** *p. 8*
- **inductive reasoning,** *p. 8*
- **counterexample,** *p. 10*
- **undefined term,** *p. 14*
- **point,** *p. 14*
- **line,** *p. 14*
- **plane,** *p. 14*
- **postulate,** *p. 14*
- **collinear points,** *p. 15*
- **coplanar points,** *p. 15*
- **coplanar lines,** *p. 15*

- **segment,** *p. 16*
- **endpoint,** *p. 16*
- **ray,** *p. 16*
- **intersect,** *p. 22*
- **intersection,** *p. 22*
- **coordinate,** *p. 28*
- **distance,** *p. 28*
- **length,** *p. 28*
- **between,** *p. 29*
- **congruent segments,** *p. 30*

- **angle,** *p. 35*
- **sides of an angle,** *p. 35*
- **vertex of an angle,** *p. 35*
- **measure of an angle,** *p. 36*
- **degrees,** *p. 36*
- **congruent angles,** *p. 36*
- **acute angle,** *p. 36*
- **right angle,** *p. 36*
- **obtuse angle,** *p. 36*
- **straight angle,** *p. 36*

VOCABULARY REVIEW

Fill in the blank.

1. __?__ are lines that lie in the same plane.   coplanar lines

2. A(n) __?__ consists of two __?__ that have the same endpoint.   angle; rays

3. A(n) __?__ is an unproven statement that is based on a pattern or observation.   conjecture

4. The endpoint of the rays that form an angle is called its __?__.   vertex

5. Two or more figures __?__ if they have points in common.   intersect

6. An angle that has a measure between 0° and 90° is called a(n) __?__ angle.   acute

7. Points on the same line are __?__ points.   collinear

8. An angle that has a measure of 180° is called a(n) __?__ angle.   straight

9. Two angles that have the same measure are called __?__ angles.   congruent

10. A(n) __?__ is an example that shows a conjecture is false.   counterexample

11. An angle that has a measure between 90° and 180° is called a(n) __?__ angle.   obtuse

12. Two segments that have the same length are called __?__ segments.   congruent

1.1 FINDING AND DESCRIBING PATTERNS

Examples on pp. 3–4

> **EXAMPLE** **Describe a pattern in the numbers −7, 0, 7, 14, ...**
> **Write the next two numbers you expect in the pattern.**
>
> Each number is 7 more than the previous number.
> The next two numbers are 21 and 28.

Describe a pattern in the numbers. Write the next two numbers you expect in the pattern. 13–15. See margin.

13. 5, 14, 23, 32, . . .   **14.** 6, 18, 54, 162, . . .   **15.** 100, 90, 80, 70, . . .

Sketch the next figure you expect in the pattern. 16, 17. See margin.

16.

17.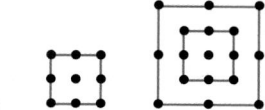

1.2 INDUCTIVE REASONING

Examples on pp. 8–10

> **EXAMPLE** **Complete the conjecture.**
>
> ***Conjecture:*** The sum of the first n even positive integers is __?__ .
>
> Write several examples and look for a pattern.
>
> | | Example | Sum | Pattern |
> |---|---|---|---|
> | First even positive integer | 2 | 2 | **1** · 2 |
> | First **2** even positive integers | 2 + 4 | 6 | **2** · 3 |
> | First **3** even positive integers | 2 + 4 + 6 | 12 | **3** · 4 |
> | First **4** even positive integers | 2 + 4 + 6 + 8 | 20 | **4** · 5 |
>
> ***Conjecture:*** The sum of the first n even positive integers is $n \cdot (n + 1)$.

18. Complete the conjecture based on the pattern you observe.

   Conjecture: The product of four consecutive numbers, plus 1, is
   always a __?__ number. squared

   $4 \cdot 5 \cdot 6 \cdot 7 + 1 = 29 \cdot 29$ $7 \cdot 8 \cdot 9 \cdot 10 + 1 = 71 \cdot 71$

   $5 \cdot 6 \cdot 7 \cdot 8 + 1 = 41 \cdot 41$ $6 \cdot 7 \cdot 8 \cdot 9 + 1 = 55 \cdot 55$

19. Show the conjecture is false by finding a counterexample.

   Conjecture: The cube of a number is always greater than
   the number. *Sample answer:* $(-2)^3 = -8$ and $-8 < -2$

---

13. Each number is 9 more than the previous number; 41; 50.

14. Each number is 3 times the previous number; 486; 1458.

15. Each number is 10 less than the previous number; 60; 50.

16.

17.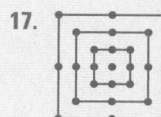

23. *Sample answer:*

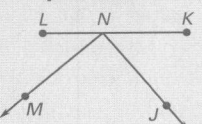

28. *Sample answer:*

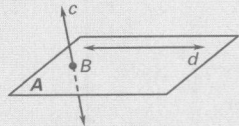

29. *Sample answer:*

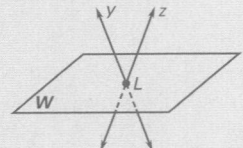

1.3   POINTS, LINES, AND PLANES

*Examples on
pp. 14–16*

> **EXAMPLES**   Decide whether the statement is *true* or *false*.
>
> **a.** C, E, and D are collinear.   **true**
>
> **b.** A, B, C, and E are coplanar.   **true**
>
> **c.** $\overline{AB}$ is a line.   **false**
>
> **d.** $\overrightarrow{EC}$ and $\overrightarrow{ED}$ are rays.   **true**

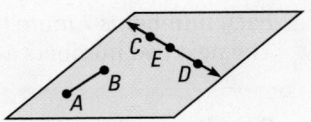

**In Exercises 20–22, use the diagram at the right
to decide whether the statement is *true* or *false*.**

20. Point A lies on line m. true

21. Point E lies on line m. false

22. Points B, C, and D are collinear. false

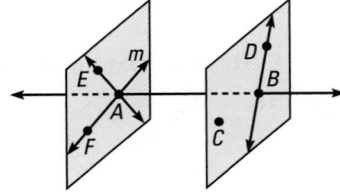

23. Draw four points J, K, L, and M, no three of which are collinear.
Sketch $\overline{LK}$ and add a point N on that line segment. Then sketch
$\overrightarrow{NJ}$ and $\overrightarrow{NM}$. **See margin.**

1.4   SKETCHING INTERSECTIONS

*Examples on
pp. 22–24*

> **EXAMPLES**   Use the diagram at the right.
>
> **a.** Name the intersection of lines $\overleftrightarrow{AT}$ and $\overleftrightarrow{BW}$.
>
>   Lines $\overleftrightarrow{AT}$ and $\overleftrightarrow{BW}$ intersect at point R.
>
> **b.** Name the intersection of planes D and E.
>
>   Planes D and E intersect at $\overleftrightarrow{BW}$.

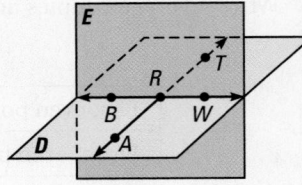

In Exercises 24–27, use the figure to fill in the blank.

24. $\overleftrightarrow{TU}$ and $\overleftrightarrow{VU}$ intersect at __?__. point U

25. Plane PQR and plane UVS intersect at __?__. $\overleftrightarrow{QS}$

26. Plane RSV and plane QUV intersect at __?__. $\overleftrightarrow{VS}$

27. Plane QSV and plane TUV intersect at __?__. $\overleftrightarrow{UV}$

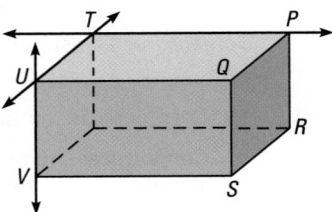

Visualize It! **Sketch the figure described.** 28, 29. See margin.

28. Two lines that are not coplanar and do not intersect

29. A plane and two lines that intersect the plane at one point

1.5   SEGMENTS AND THEIR MEASURES

Examples on pp. 28–30

EXAMPLE   Use the diagram to find *AC*.

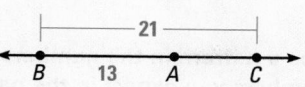

$BC = BA + AC$ Segment Addition Postulate

$21 = 13 + AC$ Substitute 21 for *BC* and 13 for *BA*.

$8 = AC$ Subtract 13 from each side.

Find the length.

30. Find *AC*.   47

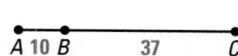

31. Find *RS*.   26

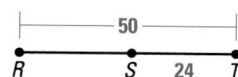

Plot the points. Decide whether $\overline{PQ}$ and $\overline{QR}$ are congruent.

32. $P(2, 3)$, $Q(2, -5)$, $R(9, -5)$
no; See margin.

33. $P(-5, 4)$, $Q(1, 4)$, $R(1, -2)$
yes; See margin.

1.6   ANGLES AND THEIR MEASURES

Examples on pp. 35–37

EXAMPLE   Find *m∠ACB*. Classify the angle.

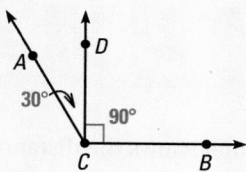

$m\angle ACB = m\angle ACD + m\angle DCB$

$= 30° + 90°$

$= 120°$

The measure of ∠ACB is 120°, so the angle is obtuse.

Copy the angle, extend its sides, and use a protractor to measure it to the nearest degree. Then classify it as *acute, right, obtuse,* or *straight*.

34.

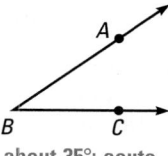

about 35°; acute

35.

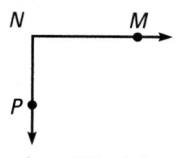

about 90°; right

36.

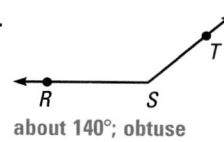

about 140°; obtuse

Find the measure of the angle.

37. *m∠DEF* = __?__   105°

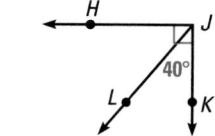

38. *m∠HJL* = __?__   50°

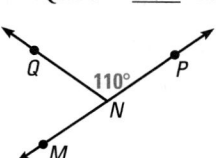

39. *m∠QNM* = __?__   70°

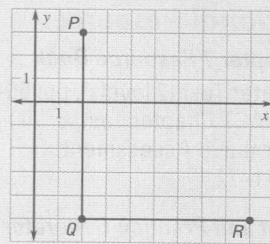

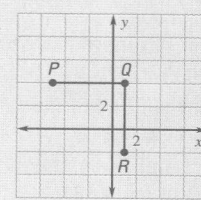

📄 **Chapter 1 Resource Book**
- Chapter Test (2 levels), pp. 66–69
- SAT/ACT Chapter Test, p. 70
- Alternative Assessment, pp. 71–72

🖥 **Test and Practice Generator**

1. Each number is 7 more than the previous number; 31; 38.

2. Each number is 17 more than the previous number; 22; 39.

3. Begin with 1 and add 4, then 8, then 12, and so on; 41; 61.

4. Each number is 0.6 more than the previous number; 5.2; 5.8.

6. See Additional Answers beginning on page AA1.

14.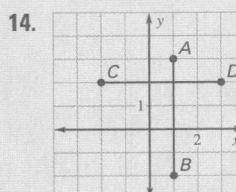

15.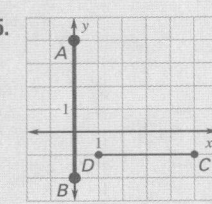

20. *Sample answers:*
∠FBE and ∠DBC are acute, ∠FBC and ∠ABD are obtuse, ∠ABC is a straight angle, and ∠FBD is a right angle.

Chapter 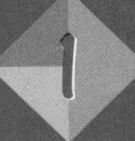 Chapter Test

Describe a pattern in the numbers. Write the next two numbers you expect in the pattern.

1–4. See margin.

1. 10, 17, 24, . . .   2. −29, −12, 5, . . .

3. 1, 5, 13, 25, . . .   4. 2.8, 3.4, 4, 4.6, . . .

5. Complete the conjecture based on the pattern you observe.

   Conjecture: The product of any two even numbers is __?__. an even number

   $2 \cdot 4 = 8$ $-4 \cdot 10 = -40$
   $4 \cdot 6 = 24$ $12 \cdot 26 = 312$
   $6 \cdot 8 = 48$ $-104 \cdot (-88) = 9152$

The first five figures in a pattern are shown. Each square in the grid is 1 unit × 1 unit.

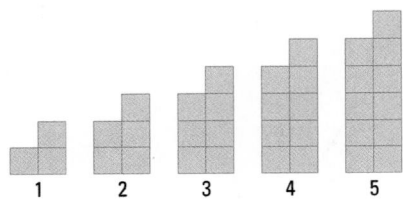

6. Make a table that shows the distance around each figure at each stage. See margin.

7. Describe the pattern of the distances and use it to predict the distance around the figure at stage 10. The distance around each figure is 2 units more than the previous distance; 26 units.

In Exercises 8–11, use the diagram below.

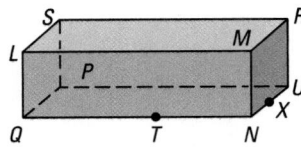

8. Name three collinear points. *Q, T, and N; or N, X, and U*

9. Name four noncoplanar points. *Sample answer: S, L, M, and N*

10. Name the intersection of $\overleftrightarrow{QT}$ and $\overleftrightarrow{UN}$. point N

11. Name the intersection of plane *LMN* and plane *QLS*. $\overleftrightarrow{QL}$

In Exercises 12 and 13, find the length.

12. Find *PR*. 17

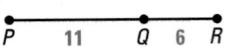

13. Find *AB*. 8

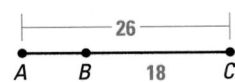

Plot the points in a coordinate plane. Then decide whether $\overline{AB}$ and $\overline{CD}$ are congruent.

14. $A(1, 3)$, $B(1, -2)$, $C(-2, 2)$, $D(3, 2)$ yes; See margin.

15. $A(0, 4)$, $B(0, -2)$, $C(5, -1)$, $D(1, -1)$ no; See margin.

In Exercises 16–20, use the diagram below.

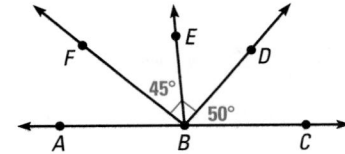

16. Name two congruent angles. ∠FBE and ∠EBD

17. Find $m\angle ABF$. 40°

18. Find $m\angle CBF$. 140°

19. Find $m\angle EBA$. 85°

20. Name an acute angle, an obtuse angle, a straight angle, and a right angle. See margin.

21. In the figure of the gymnast below, $m\angle ABD = 25°$ and $m\angle DBC = 35°$. Find the measures of ∠ABC and ∠CBE. $m\angle ABC = 60°$ $m\angle CBE = 120°$

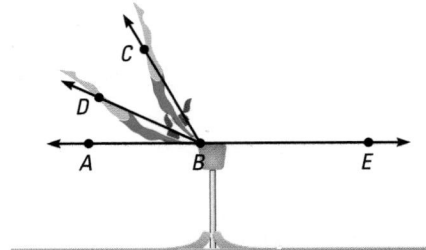

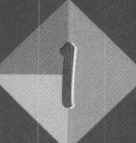

Additional Resources

📙 *Chapter 1 Resource Book*
• SAT/ACT Chapter Test, p. 70

5b. *Sample answers:*
  ∠*JBD* and ∠*EDF* are acute,
  ∠*ABJ* and ∠*GDF* are obtuse,
  ∠*ABD* is a straight angle, and
  ∠*ABC* is a right angle.

 **Test Tip**

Most standardized tests are based on concepts and skills taught in school. The best way to prepare is to keep up with your regular studies.

1. What is the next number you expect in the sequence? **C**

 4488; 44,088; 440,088; 4,400,088

  (**A**) 400,008 (**B**) 40,000,088

  (**C**) 44,000,088 (**D**) 440,000,088

2. Which statement is a counterexample to the conjecture that the square of any integer is greater than the integer? **H**

  (**F**) 4^2 is greater than 4.

  (**G**) $(-3)^2$ is greater than 3.

  (**H**) 0^2 is not greater than 0.

  (**J**) 200^2 is not greater than 200.

3. Which of the following statements is *false*? **C**

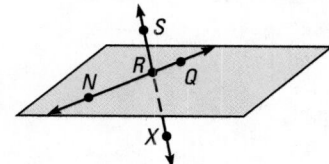

  (**A**) *R* is between *N* and *Q*.

  (**B**) *N*, *R*, and *Q* are coplanar.

  (**C**) *X*, *N*, and *R* are collinear.

  (**D**) *S*, *R*, and *X* are collinear.

4. Which of the points lies on $\overrightarrow{SR}$? **H**

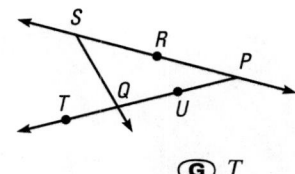

  (**F**) *Q* (**G**) *T*

  (**H**) *P* (**J**) *U*

5. Multi-Step Problem Use the figure below.

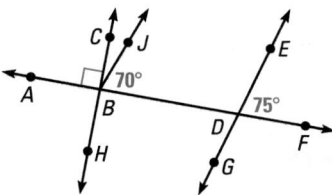

  a. Name four collinear points. *A, B, D,* and *F*

  b. Name the following types of angles: acute, obtuse, straight, and right. **See margin.**

  c. Find *m*∠*ABH*, *m*∠*GDF*, and *m*∠*ABJ*.
   m∠*ABH* = 90°; *m*∠*GDF* = 105°; *m*∠*ABJ* = 110°

6. Which of the line segments in the coordinate plane are congruent? **D**

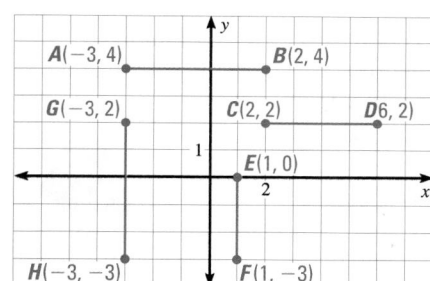

  (**A**) $\overline{AB}$ and $\overline{CD}$ (**B**) $\overline{EF}$ and $\overline{AB}$

  (**C**) $\overline{CD}$ and $\overline{GH}$ (**D**) $\overline{AB}$ and $\overline{GH}$

7. In the diagram, what is the value of *PQ*? **G**

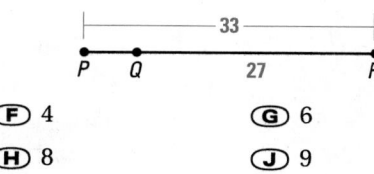

  (**F**) 4 (**G**) 6

  (**H**) 8 (**J**) 9

BraiN GaMes

Geometry Scavenger Hunt

Object of the Game To be the first team to complete the scavenger hunt.

How to Play With members of your team, search your classroom for items that fit the given description. Keep a list of objects you find. The team that finds the most items in a given time wins.

Another Way to Play The team that finds the most items that fit each category wins a point for that category. The team with the most points wins.

Or, create your own list of items to find and swap it with the list created by another team.

SCAVENGER HUNT

1. a plane
2. two planes that intersect
3. a right angle
4. a 3 inch segment
5. a 45° angle
6. a 9 inch segment
7. a pair of congruent angles that are not right angles
8. an obtuse angle
9. two planes that do not intersect
10. a pair of adjacent angles

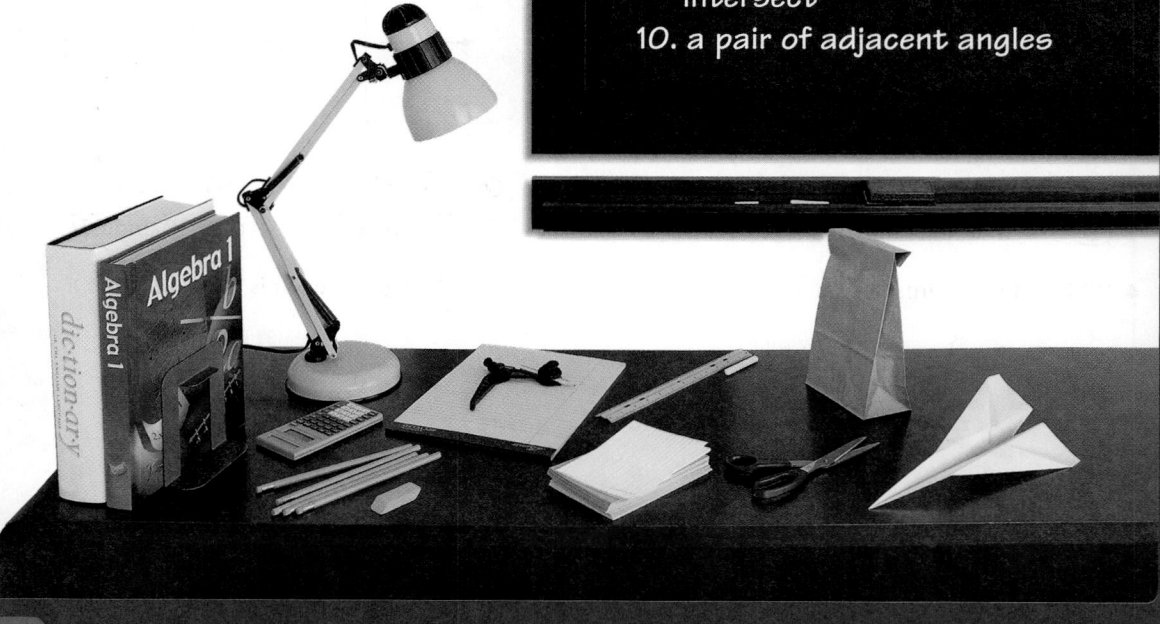

To evaluate an expression involving variables, substitute a value for each variable, and then simplify using the order of operations.

EXAMPLE 1 Evaluate Expressions

Evaluate $4x + 9y$ when $x = 3$ and $y = -5$.

Solution

$4x + 9y$	Write original expression.
$4(3) + 9(-5)$	Substitute 3 for x and -5 for y.
$12 + (-45)$	Multiply.
-33	Add.

Student Help

SKILLS REVIEW
For help using the order of operations, see p. 670.

Try These

Evaluate the expression when $x = 2$, $y = 7$, and $z = -3$.

1. $3x + 4y$ 34   **2.** $2x + 2y$ 18   **3.** $6x + y$ 19   **4.** $5x + 7y$ 59

5. $14x - 4y$ 0   **6.** $y - 2x$ 3   **7.** $5x - 3y$ -11 **8.** $11y - 4x$ 69

9. $3y + 2z$ 15   **10.** $x - 4z$ 14   **11.** $x + y + z$ 6   **12.** $-2x + 5z$ -19

EXAMPLE 2 Solve Two-Step Equations

Solve the equation $2x - 5 = 7$.

Solution

$2x - 5 = 7$	Write original equation.
$2x - 5 + 5 = 7 + 5$	Add 5 to each side.
$2x = 12$	Simplify.
$\dfrac{2x}{2} = \dfrac{12}{2}$	Divide each side by 2.
$x = 6$	Simplify.

Student Help

STUDY TIP
To check that the solution is correct, substitute 6 for x in the original equation and evaluate.
$2(x) - 5 = 7$
$2(6) - 5 = 7$
$12 - 5 = 7$ ✓

Try These

Solve the equation.

13. $3x - 2 = 13$ 5   **14.** $\dfrac{n}{8} + 2 = 4$ 16   **15.** $4 + 5w = 24$ 4   **16.** $6p - 1 = 17$ 3

17. $10 + \dfrac{y}{3} = 7$ -9 **18.** $4z + 11 = 15$ 1   **19.** $9 = -2t - 5$ -7 **20.** $1 = \dfrac{b}{4} - 6$ 28

Algebra Review **49**

Extra Example 1
Evaluate $5b - 3a$ when $a = -4$ and $b = -2$. 2

Checkpoint Exercise 1
Evaluate $-4w + 11z$ when $w = 6$ and $z = 3$. 9

Extra Example 2
Solve the equation $3c - 4 = 11$. $c = 5$

Checkpoint Exercise 2
Solve the equation $9 + \dfrac{d}{2} = -1$. $d = -20$

Pacing and Assignment Guide

REGULAR SCHEDULE

Lesson	Les. Day	Basic	Average	Advanced
2.1	Day 1	pp. 56–59 Exs. 11–24, 46–51, 53–63 odd	pp. 56–59 Exs. 11–19 odd, 20–23, 25, 42, 46–57	pp. 56–59 Exs. 13–17, 20–23, 25, 42, 46–62 even
	Day 2	SRH p. 672 Exs. 1–19 odd; pp. 56–59, Exs. 26–32, 38, 44, 45, 52–62 even	pp. 56–59 Exs. 26–40, 44, 45, 58–63	pp. 56–59 Exs. 24–34 even, 36–41, 43*, 44, 45; EC: classzone.com
2.2	Day 1	EP p. 676 Exs. 29–31; pp. 64–66 Exs. 8–22, 33, 42–47	pp. 64–66 Exs. 9–23 odd, 33–47 odd	pp. 64–66 Exs. 9–23 odd, 32*, 33–47 odd
	Day 2	SRH p. 673 Exs. 4–9, 19–21; pp. 64–66 Exs. 24–26, 28–30, 34–41	pp. 64–66 Exs. 24–31, 34–46 even	pp. 64–66 Exs. 24–31, 34–46 even; EC: classzone.com
2.3	Day 1	pp. 70–73 Exs. 8–14, 29–32, 44–59	pp. 70–73 Exs. 8–14, 29–32, 44, 46–59	pp. 70–73 Exs. 8–14 even, 29–32, 45–59
	Day 2	SRH p. 673 Exs. 13–17 odd; pp. 70–73 Exs. 15–27 odd, 28, 33, 40–42, Quiz 1	pp. 70–73 Exs. 16–28 even, 34–42 even, 45, Quiz 1	pp. 70–73 Exs. 16–28 even, 33–41 odd, 43*, Quiz 1; EC: classzone.com
2.4	Day 1	pp. 78–81 Exs. 9–19, 24–27, 60–70 even	pp. 78–81 Exs. 9–19, 23–27, 58, 64–71	pp. 78–81 Exs. 9–19 odd, 23–27, 58–71
	Day 2	pp. 78–81 Exs. 20–22, 28–35, 51–53, 59, 65–71 odd	pp. 78–81 Exs. 20–22, 28–40, 51–54, 59–63	pp. 78–81 Exs. 20–22, 28–40 even, 41–49 odd, 54–56, 57*; EC: classzone.com
2.5	Day 1	pp. 85–87 Exs. 7–12, 16, 20, 24–42 even	pp. 85–87 Exs. 7–12, 20, 21, 24–42 even	pp. 85–87 Exs. 8–12 even, 20, 21, 24–34
	Day 2	pp. 85–87 Exs. 13–19 odd, 23–43 odd	pp. 85–87 Exs. 13–19, 22, 23–43 odd	pp. 85–87 Exs. 14–18 even, 22, 23, 35–43; EC: TE p. 50D*, classzone.com
2.6	Day 1	pp. 91–94 Exs. 10–18, 28, 30–33, 36–43	pp. 91–94 Exs. 10–18, 28, 30–43	pp. 91–94 Exs. 13–18, 28, 30–43
	Day 2	pp. 91–94 Exs. 19–21, 29, 34, 35, Quiz 2	pp. 91–94 Exs. 19–23, 25, 26, 29, Quiz 2	pp. 91–94 Exs. 19–26, 27*, 29, Quiz 2; EC: classzone.com
Review	Day 1	pp. 95–97 Exs. 1–26	pp. 95–97 Exs. 1–26	pp. 95–97 Exs. 1–26
Assess	Day 1	Chapter 2 Test	Chapter 2 Test	Chapter 2 Test

YEARLY PACING   Chapter 2 Total – **14 days**   Chapters 1–2 Total – **28 days**   Remaining – **132 days**

*Challenge Exercises   EP = Extra Practice   SRH = Skills Review Handbook   EC = Extra Challenge

BLOCK SCHEDULE

Day 1	Day 2	Day 3	Day 4	Day 5	Day 6	Day 7
2.1 pp. 56–59 Exs. 11–19 odd, 20–40, 44–63	**2.2** pp. 64–66 Exs. 9–23 odd, 24–31, 33–47	**2.3** pp. 70–73 Exs. 8–14, 16–28 even, 29–32, 34–42 even, 44–59, Quiz 1	**2.4** pp. 78–81 Exs. 9–40, 51–54, 58–71	**2.5** pp. 85–87 Exs. 7–43	**2.6** pp. 91–94 Exs. 10–23, 25, 26, 28–43, Quiz 2	**Review** pp. 95–97 Exs. 1–26 **Assess** Chapter 2 Test

YEARLY PACING   Chapter 2 Total – **7 days**   Chapters 1–2 Total – **14 days**   Remaining – **66 days**

Support Materials

CHAPTER RESOURCE BOOK

CHAPTER SUPPORT

Tips for New Teachers	p. 1	Strategies for Reading Mathematics	p. 5
Parent Guide for Student Success	p. 3		

LESSON SUPPORT

	2.1	2.2	2.3	2.4	2.5	2.6
Lesson Plans (regular and block)	p. 7	p. 16	p. 26	p. 37	p. 45	p. 53
Warm-Up Exercises and Daily Quiz	p. 9	p. 18	p. 28	p. 39	p. 47	p. 55
Technology Activities & Keystrokes		p. 19	p. 29			p. 56
Practice (2 levels)	p. 10	p. 20	p. 30	p. 40	p. 48	p. 59
Reteaching with Practice	p. 12	p. 22	p. 32	p. 42	p. 50	p. 61
Quick Catch-Up for Absent Students	p. 14	p. 24	p. 34	p. 44	p. 52	p. 63
Learning Activities	p. 15					
Real-Life Applications		p. 25	p. 35			

REVIEW AND ASSESSMENT

Quizzes	pp. 36, 64	Alternative Assessment with Math Journal	p. 72
Brain Games Support	p. 65	Project with Rubric	p. 74
Chapter Review Games and Activities	p. 66	Cumulative Review	p. 76
Chapter Test (2 levels)	p. 67	Resource Book Answers	AN1
SAT/ACT Chapter Test	p. 71		

TRANSPARENCIES

	2.1	2.2	2.3	2.4	2.5	2.6
Warm-Up Exercises and Daily Quiz	p. 8	p. 9	p. 10	p. 11	p. 12	p. 13
Visualize It Transparencies		✓	✓	✓		
Answer Transparencies	✓	✓	✓	✓	✓	✓

TECHNOLOGY

- Time-Saving Test and Practice Generator
- Electronic Teacher Tools
- Geometry in Motion: Video
- Classzone.com
- Online Lesson Planner

ADDITIONAL RESOURCES

- Worked-Out Solution Key
- Resources in Spanish
- Practice Workbook with Examples

Providing Universal Access

Strategies for Strategic Learners

USE SCAFFOLDING

Geometry, Concepts and Skills already uses scaffolding to help students build the understanding they need for complex problems. Notice that many of the word problems are broken down into discrete steps. Example 3 on page 90 shows students the steps of a proof and the Checkpoint exercise following this example gives students clues to the justification for each step of a proof. Throughout the course, when students are having trouble with a particular kind of problem or proof, simplify what they are asked to do. You might first ask them to tell you aloud what they are being asked to prove, what information they already know, and what they can prove with the information they are given. Often, simply stating out loud what they know helps students clarify their thinking.

For example, in Checkpoint Exercise 5 on page 90, you could ask students questions such as the following:

In words, what are you trying to show?

Can you point out to me on the figure what you are trying to show?

What is a midpoint?

(While pointing to $\overline{AB}$) *What does this symbol mean?*

(While pointing to AB in $AB = 2 \cdot AM$) *What does this symbol mean?*

At this point, you could have students explain in their own words the five properties in the box at the top of the page. Then ask them how they could use these properties in Checkpoint Exercise 5.

PROVIDE SUFFICIENT PRACTICE

Some students need just a couple of examples in order to understand and remember a mathematical concept or procedure. Most students need more practice. Practice should be varied, so that students are assigned exercises that range from easy to difficult and so students must apply what they have learned to a variety of settings. Throughout this book, opportunities for additional practice are indicated.

CONCEPT RETENTION When students who have had a hard time with a concept finally seem to understand it, do not assume it will be retained. Come back to that concept repeatedly during the school year. Each lesson of this book includes an exercise section called "Mixed Review." Assigning these problems to students can help them retain concepts taught in earlier lessons and allow teachers to check to see what students remember and what they have forgotten. In addition, students may only know what skill to use in solving a problem when that skill is the subject of the lesson being taught. For example, students may have no difficulty applying the Pythagorean Theorem when studying the lesson where it is first taught. In later chapters, when presented with triangle problems in Mixed Review sections, students may forget that the Pythagorean Theorem can be used and may try to solve the problem using skills they have more recently acquired, which often may not be helpful.

Strategies for English Learners

ASSESS FREQUENTLY

High school is a time of change for many students. Check any available test results for your English learners. For those students new to your school this year or entering mid-year, review any standardized testing in mathematics and language arts and look for a reading fluency test. For those students who attended your school last year, look for an end-of-year math test as well. Students whose reading scores are significantly below grade level will need regular assistance in vocabulary development and in understanding the structure of the English language. Application problems will present a particular challenge. Look for support for these students from a language arts teacher, study hall or tutorial assistance, siblings, or relatives. Help English learners develop a study group of classroom peers with whom they can discuss assignments and ask questions. English learners may benefit from reading aloud to themselves quietly. Stress that they may need to read a problem over to themselves several times before they understand it. You might wish to consider rearranging students in the room so that students who wish to read quietly aloud can be separated from students who are easily distracted.

PRESENT A COURSE OVERVIEW

This course is designed so that students learn foundation skills before more complex ones. The instructional materials present mathematics in a straightforward way, and extraneous material that might distract or confuse English learners has been eliminated. Pointing out the specific, regular features of the text and giving English learners a brief overview of the course can help them organize their work in mathematics and anticipate and prepare for assignments and tests. Walk through the Table of Contents with students and point out the following regular features in each chapter that can help English learners. The Preview section of each chapter's Study Guide page includes key words for the chapter and the page numbers on which the words can be found. Each lesson begins with a list of lesson goals and key words for that lesson. Each lesson also contains student help tips in the margin, some of which pertain directly to vocabulary development.

Students may wish to prepare for each chapter or lesson by writing the list of key words and either translating them into their primary language or by making a sketch to illustrate each word's meaning. This activity could be done by English learners as a group.

OTHER FEATURES You might also point out the color coding of regular features, the appendices, any available ancillaries, and the help available on the classzone.com web site. Most search engines on the Internet have a bilingual translator readily available, and students should be encouraged to use bilingual dictionaries.

Strategies for Advanced Learners

ASSESS FREQUENTLY

This would be a good point at which to assess which students complete the review activities quickly, complete their homework with few errors, and seem ready to move at a faster pace. Advanced students may be able to move very quickly through the first four chapters. You may want to give them the Chapter Tests at the end of Chapters 2, 3, and 4. This will help you and the student determine which concepts and skills have already been mastered. If students can demonstrate mastery, substitute the challenge problems in the book or the Teacher's Edition for homework problems that are too easy. Also consider having advanced learners work on a small research project of their own on a geometry subject that interests them. Books such as *The Joy of Mathematics* by Theoni Pappas that contain dozens of intriguing math facts may spark student interest in learning more about a subject, such as fractals, topology, dynamic symmetry, or the mathematics of crystal shapes.

Alternatively, students may wish to explore some of the cross-curricular links in chapters and learn more about subjects that incorporate the mathematics in the chapter. For example, as students work through Chapter 2, they could explore questions such as the following.

(p. 55) How do you describe and plot points in 3-dimensional space? How would you adapt the Midpoint Formula to find the midpoint of a line in 3-dimensional space?

(p. 58) Find a web site that gives the latitude and longitude of any city. Think of two cities you are familiar with and estimate what location would be their midpoint. Then calculate the exact midpoint and use an atlas to check your estimate.

(p. 61) Could the wing of a hang glider be any triangular shape? Are the exact dimensions important? Do differently shaped gliders perform better under different weather conditions?

(p. 65) What is optical computing? How does a laser work?

(p. 71) Why did Mr. Calatrava use the specific angle measures he did? Do his other designs incorporate triangles?

(p. 75) How is cutting with long-bladed scissors different from cutting with short-bladed scissors? If you took the scissors shown on page 75 and doubled the length of the blades, how would that affect their cutting ability? Have you seen scissors with very short or very long blades? What are they used for?

As advanced students complete challenge problems, other students may show an interest in the problems as well. Encourage all students who show an interest to try these challenge problems. If you provide extra credit, provide it for all students in the class who complete any of the challenge problems.

Challenge Problem for use with Lesson 2.5:

Suppose 10 people are in line to buy tickets at the circus. A clown comes by and says, "When I count 3, I want everybody in line to turn around and shake hands with the person you're facing." The first person in line says, "I'll do it if the second person does." The second person says, "I'll do it if the third person does." This pattern continues to the ninth person, who says, "I'll do it." At the count of 3, how many handshakes will take place? (*1, between the ninth and tenth persons in line*)

 Content and Teaching Strategies

In Lesson 2.1, the concept of a segment bisector is introduced, leading to the definition of a midpoint of a segment, and then to the Midpoint Formula for coordinate geometry. In Lesson 2.2, the idea of bisection is applied to angles. In the third lesson, the main concept is that of classifying pairs of angles by the sum of their measures. If the sum is 90°, the angles are called *complementary*, whereas if the sum is 180°, they are *supplementary*. Lesson 2.4 introduces the topic of vertical angles, pairs of angles whose sides are respectively collinear. The linear pair relationship, a special case of supplementary angles, is also introduced. The discussion of deductive reasoning and the laws of logic in Lesson 2.5 lays the foundation for proofs, which begin in Chapter 3. Lesson 2.6 concludes the chapter with a presentation of the basic properties of equality and congruence.

Lesson 2.1

While discussing segment bisectors, the following question is often asked.

> *If one endpoint of one segment coincides with the midpoint of another, does the first segment bisect the second?*

The answer to this question is yes. Point out that to be a segment bisector, a segment, ray, line, or plane must contain the midpoint of the segment, as is the case in this situation. Another frequent question is whether one segment can bisect a second segment without the second segment bisecting the first. To check students' understanding of this possibility, you might ask students to draw a sketch of such a situation.

QUICK TIP
Since this is the first lesson to use coordinate geometry, you may want to spend some time reviewing the basics of plotting points in a coordinate plane before presenting the Midpoint Formula. Emphasize that in the Midpoint Formula each coordinate of the midpoint can be thought of as the *average* of the corresponding coordinates of the two endpoints of the segment. This will help students remember that the pairs of corresponding coordinates of the endpoints are *added*, and not subtracted in the Midpoint Formula. In both the Distance Formula and the formula for finding slope, coordinates are subtracted and confusion about which operation to use can occur.

Lesson 2.2

Similar to segment bisectors, an angle bisector divides an angle into two smaller angles of equal measure. This leads to the general conclusion that if the measure of one of the two angles formed by an angle bisector is x, then the measure of the larger bisected angle is $2x$. You may want to lead your students to this conclusion by asking them to use the results of Examples 1 and 2 to write one general statement.

If one of the angles formed by the bisector of a larger angle has measure x, then
1) the other angle formed by the bisector also has measure x, and
2) the larger (bisected) angle has measure $2x$.

To conclude the discussion of angle bisectors, you might ask your students to complete a table like the one below by entering the word *acute*, *right*, *obtuse*, or *straight* in each line.

If an angle is …	then the two angles formed by the bisector of that angle are both …
acute,	
right,	
obtuse,	
straight,	

After completing the table, point out that the measure of an angle formed by a bisector cannot be greater than 90°.

Lesson 2.3

An important point to make in this lesson is that the designation of angles as complementary or supplementary is based only on angle measure, and not on the location of the angles. Students often find it easiest to recognize *complementary angles* when a diagram shows a right angle with a ray in its interior. Stress that this is not the only situation in which complementary angles can occur. Use the following diagram in which $\angle 1$ and $\angle 2$ are complementary to show that even nonadjacent angles in a single figure can be complementary. In the figure, points A, B, and C are collinear.

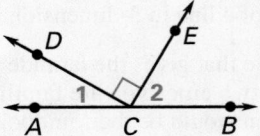

The following figure can be used to show that nonadjacent angles can be supplementary. In the figure, points A, B, and C are collinear.

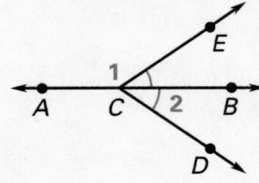

Lesson 2.4

Unlike complementary and supplementary angles, vertical angles *are* related by their position in a diagram. It is important to be able to know when angles are vertical angles, and a sure-fire test is that they must be formed by two lines, as Example 1 illustrates. Later in the course, one of the most common mistakes made by geometry students is trying to apply the Vertical Angles Theorem in cases where the two angles are not actually vertical.

QUICK TIP
Point out that a pair of vertical angles forms an "X" in a figure.

Lesson 2.5

Use Checkpoint Exercise 2 on page 83 to illustrate that many statements in mathematics which do not seem to be if-then statements actually are conditional statements. The statement of the Vertical Angles Theorem in Lesson 2.4 is another example of this fact. Ask your students to write that theorem in if-then form. (*If two angles are vertical angles, then they are congruent.*)

LOGIC It is important to emphasize that an if-then statement is a "one-way" statement. That is, if the hypothesis is true, then we can draw the stated conclusion. However, even if we know that the conclusion is true, we cannot infer that the hypothesis is also true. For example, consider the following statement.

If I am pedaling my bicycle, then the back wheel is turning.

If we know the hypothesis *I am pedaling my bicycle* is true, then the conclusion *the back wheel is turning* is true. However, just knowing that the conclusion *the back wheel is turning* is true does not mean the hypothesis is true, since the rider could be coasting on the bike. This point is also illustrated by Argument 1 of Example 4, which is not a valid argument.

Lesson 2.6

The properties of equality should be familiar to students, but the properties of congruence will be new to them, since they have probably not encountered the concept of congruence in previous math courses.

QUICK TIP
To provide a quick review of the Addition, Subtraction, Multiplication, and Division Properties of Equality, remind students that if you "do the same thing to both sides" of an equation, then an equivalent equation will result.

The property that causes the most difficulty here is the Substitution Property. Students sometimes want to use it when they are not dealing with an equation. For example, suppose $\angle 1$ and $\angle 2$ form a linear pair, and it is known that $m\angle 2 = m\angle 3$. Some students will use the Substitution Property to conclude that $\angle 1$ and $\angle 3$ form a linear pair. This is the sort of incorrect reasoning, usually in a more subtle form, that students may try to apply in solving problems later on. Emphasize that in order to apply the Substitution Property, students must substitute an algebraic expression or numerical value for another algebraic expression *in an equation*. This is one reason why a distinction must be made between a segment (which is a geometric object) and the length of the segment (which is a positive numerical value), and also between an angle and its measure.

The Transitive Property of Congruence is also sometimes misapplied by students. If two sides of a triangle are congruent to two sides of another triangle, for example, the Transitive Property cannot be used to conclude that the third sides of the two triangles are congruent. Two isosceles triangles, one with sides 6, 6, and 5 units, and the other with sides 6, 6, and 2 units, can be used to show that such a use of the property is incorrect.

Chapter Overview

Chapter Goals

Chapter 2 introduces students to special angles and their properties. These facts begin setting the foundation for students' introduction to proofs in Chapter 5. Students learn about if-then statements and the basics of logical thinking to prepare them for their study of proofs. Students will:

- Analyze segment bisectors and angle bisectors.
- Identify complementary angles, supplementary angles, vertical angles, and linear pairs.
- Use the properties of equality and congruence to justify mathematical statements.

Application Note

Baseball is based on an English game called rounders, though many people believe it was invented by Abner Doubleday, an American. The game of baseball provides an interesting array of related angle measures and segment lengths. For example, the portion of a baseball field called the *infield*, a square area with a base at each corner, contains four 90° angles. Additionally, the baseball term *fair territory* refers to the area in the interior of the 90° angle formed at home plate by the first and third baselines.

More information about strike zones is provided on the Internet at classzone.com

Application Links
CLASSZONE.COM

How is a strike zone determined?

In baseball, the *strike zone* is the area above home plate where a ball is considered a strike. The location of the strike zone is based on each player's height and batting stance.

The midpoint between the top of the batter's shoulders and the top of the batter's belt determines the top of the strike zone.

63
T
45

24

0

Learn More About It

You will learn more about strike zones in Exercises 36 and 37 on p. 58.

Who uses Segments and Angles?

KITE DESIGNER
Kite designers use geometric principles in designing and making kites. A kite's struts often bisect the angles they support. (p. 62)

ERGONOMIST
Ergonomists design offices, furniture, and equipment to improve the safety and comfort of workers. For example, drafting tables are angled so that people using them can work without injuring their backs. (p. 80)

How will you use these ideas?

- Determine the strike zone for different batting stances. (p. 58)
- Describe angle relationships in a paper airplane pattern. (p. 65)
- Predict the angle of reflection of a laser. (p. 65)
- Learn about angle relationships found in the Alamillo Bridge in Spain. (p. 71)
- Interpret the meaning of advertising slogans. (p. 86)

51

Hands-On Activities

Activities (more than 20 minutes)
2.2 Folding Angle Bisectors, p. 60
2.4 Angles and Intersecting Lines, p. 74

Geo-Activities (less than 20 minutes)
2.1 Folding a Segment Bisector, p. 53

Projects

A project covering Chapters 1–2 appears on pages 102–103 of the Student Edition. An additional project for Chapter 2 is available in the *Chapter 2 Resource Book*, pp. 74–75.

Technology

- Electronic Teacher Tools
- Test and Practice Generator
- Online Lesson Planner

Video

- **Geometry in Motion**
  There is an animation supporting Lesson 2.4.

Internet Support
CLASSZONE.COM

- Application and Career Links
  65, 71, 80, 86
- Student Help
  57, 63, 72, 79, 84, 90

PREVIEW — **What's the chapter about?**

• Analyzing segment bisectors and angle bisectors
• Identifying complementary angles, supplementary angles, vertical angles, and linear pairs
• Using properties of equality and congruence

Key Words

• midpoint, *p. 53*
• segment bisector, *p. 53*
• bisect, *p. 53*
• angle bisector, *p. 61*
• complementary angles, *p. 67*
• supplementary angles, *p. 67*

• adjacent angles, *p. 68*
• theorem, *p. 69*
• vertical angles, *p. 75*
• linear pair, *p. 75*
• if-then statement, *p. 82*
• deductive reasoning, *p. 83*

PREPARE — **Chapter Readiness Quiz**

Take this quick quiz. If you are unsure of an answer, look at the reference pages for help.

Vocabulary Check *(refer to p. 36)*

1. Suppose $m\angle ABC = 100°$. What type of angle is $\angle ABC$? **C**

Ⓐ acute   Ⓑ right   Ⓒ obtuse   Ⓓ straight

Skill Check *(refer to pp. 37, 672)*

2. In the diagram, $m\angle PQR = 165°$ and $m\angle PQS = 22°$. What is $m\angle SQR$? **G**

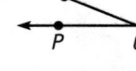

Ⓕ 121°   Ⓖ 143°   Ⓗ 153°   Ⓙ 158°

3. Which of the following is a solution of the equation $4x = -20$? **C**

Ⓐ −80   Ⓑ −16   Ⓒ −5   Ⓓ 5

VISUAL STRATEGY — **Picturing Theorems**

Visualize It! ➡

In this chapter, you will learn the first of many *theorems*.

To help you visualize a theorem, draw an example that uses specific measures.

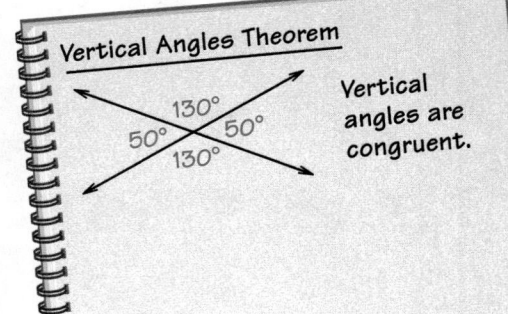

Vertical Angles Theorem

Vertical angles are congruent.

2.1 Segment Bisectors

Goal

Bisect a segment. Find the coordinates of the midpoint of a segment.

Key Words

- midpoint
- segment bisector
- bisect

Geo-Activity   Folding a Segment Bisector

1 On a piece of paper, draw $\overline{AB}$.

2 Fold the paper so that point B is on top of point A.

3 Label the point where the fold intersects $\overline{AB}$ as point M.

4 Use a ruler to measure the lengths of $\overline{AM}$, $\overline{MB}$, and $\overline{AB}$. What do you notice?

$$AM = MB = \tfrac{1}{2}AB$$

In the Geo-Activity, M is called the *midpoint* of $\overline{AB}$. The **midpoint** of a segment is the point on the segment that divides it into two congruent segments.

A **segment bisector** is a segment, ray, line, or plane that intersects a segment at its midpoint. To **bisect** a segment means to divide the segment into two congruent segments.

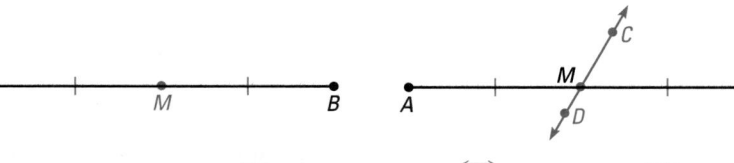

M is the midpoint of $\overline{AB}$. $\overleftrightarrow{CD}$ is a bisector of $\overline{AB}$.

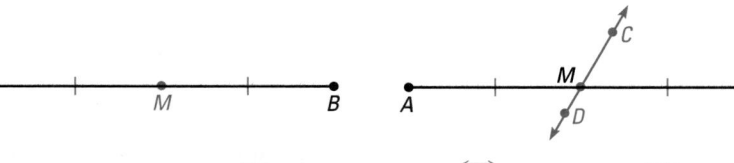

Student Help

VOCABULARY TIP
Bi- means "two," and *-sect* means "to cut." So, *bisect* means "to cut in two."

Pacing

Suggested Number of Days

Basic: 2 days
Average: 2 days
Advanced: 2 days
Block Schedule: 1 block

Teaching Resources

Blacklines
(See page 50B.)

Transparencies
- Warm-Up with Quiz
- Answers

Technology
- Electronic Teacher Tools
- Test and Practice Generator
- Online Lesson Planner
- Internet Support

Geo-Activity

Goal Use paper folding and a ruler to explore segment bisectors.

Key Discovery The midpoint of a segment divides the segment into two congruent segments.

② Teach

Content and Teaching Strategies

For background information on geometric concepts and teaching strategies related to this lesson, see pages 50E and 50F in this Teacher's Edition.

Tips for New Teachers

Students should understand that the *x*-coordinate of the midpoint is the average of the *x*-coordinates of the endpoints and likewise for the *y*-coordinate of the midpoint. See the Tips for New Teachers on pp. 1–2 of the *Chapter 2 Resource Book* for additional notes about Lesson 2.1.

Extra Example 1

K is the midpoint of $\overline{FG}$. Find *FK* and *KG*. **FK = KG = 18**

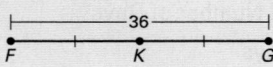

Extra Example 2

M is the midpoint of $\overline{JL}$. Find *ML* and *JL*. **ML = 7, JL = 14**

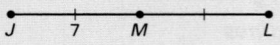

Extra Example 3

Line *s* is a segment bisector of $\overline{FH}$. Find the value of *x*. **4**

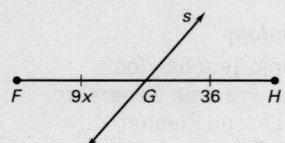

Student Help

STUDY TIP
The midpoint of a segment divides the segment in half.

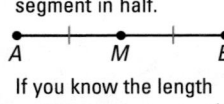

If you know the length of $\overline{AB}$, multiply *AB* by $\frac{1}{2}$ to find *AM* and *MB*.

EXAMPLE 1 Find Segment Lengths

M is the midpoint of $\overline{AB}$. Find *AM* and *MB*.

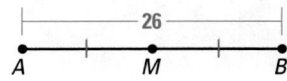

Solution

M is the midpoint of $\overline{AB}$, so *AM* and *MB* are each half the length of $\overline{AB}$.

$$AM = MB = \frac{1}{2} \cdot AB = \frac{1}{2} \cdot 26 = 13$$

ANSWER▶ *AM* = 13 and *MB* = 13.

EXAMPLE 2 Find Segment Lengths

P is the midpoint of $\overline{RS}$. Find *PS* and *RS*.

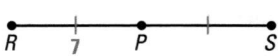

Solution

P is the midpoint of $\overline{RS}$, so *PS* = *RP*. Therefore, *PS* = 7.
You know that *RS* is twice the length of $\overline{RP}$.

$$RS = 2 \cdot RP = 2 \cdot 7 = 14$$

ANSWER▶ *PS* = 7 and *RS* = 14.

Checkpoint✓ Find Segment Lengths

1. Find *DE* and *EF*. **DE = 9; EF = 9**

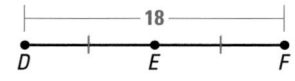

2. Find *NP* and *MP*. **NP = 11; MP = 22**

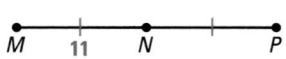

Using Algebra

EXAMPLE 3 Use Algebra with Segment Lengths

Line *ℓ* is a segment bisector of $\overline{AB}$. Find the value of *x*.

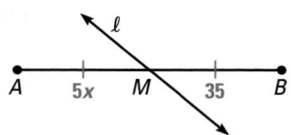

Solution

AM = *MB*	Line *ℓ* bisects $\overline{AB}$ at point *M*.
5*x* = 35	Substitute 5*x* for *AM* and 35 for *MB*.
$\frac{5x}{5} = \frac{35}{5}$	Divide each side by 5.
x = 7	Simplify.

CHECK ✓ Check your solution by substituting 7 for *x*.

$$5x = 5(7) = 35$$

Midpoints If you know the coordinates of the endpoints of a line segment in a coordinate plane, you can find the coordinates of the midpoint of the segment using the Midpoint Formula.

Student Help

READING TIP
The numbers 1 and 2 in x_1 and y_2 are called *subscripts*. You read x_1 as "x sub 1" and y_2 as "y sub 2." ·····················▶

THE MIDPOINT FORMULA

Words The coordinates of the midpoint of a segment are the averages of the x-coordinates and the y-coordinates of the endpoints.

Symbols The midpoint of the segment joining $A(x_1, y_1)$ and $B(x_2, y_2)$

is $M\left(\dfrac{x_1 + x_2}{2}, \dfrac{y_1 + y_2}{2}\right)$.

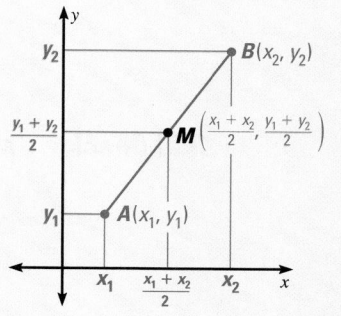

EXAMPLE 4 Use the Midpoint Formula

Find the coordinates of the midpoint of $\overline{AB}$.

a. $A(1, 2)$, $B(7, 4)$ **b.** $A(-2, 3)$, $B(5, -1)$

Solution

First make a sketch. Then use the Midpoint Formula.

Student Help

SKILLS REVIEW
For help plotting points in a coordinate plane, see p. 664.

a.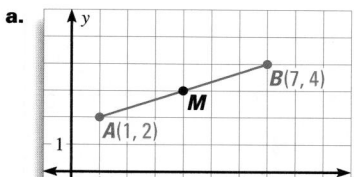

Let $(x_1, y_1) = (1, 2)$ and $(x_2, y_2) = (7, 4)$.

$$M = \left(\frac{x_1 + x_2}{2}, \frac{y_1 + y_2}{2}\right)$$

$$= \left(\frac{1 + 7}{2}, \frac{2 + 4}{2}\right)$$

$$= (4, 3)$$

b.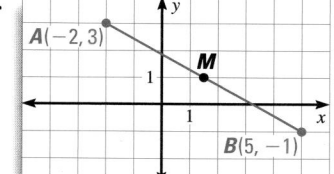

Let $(x_1, y_1) = (-2, 3)$ and $(x_2, y_2) = (5, -1)$.

$$M = \left(\frac{x_1 + x_2}{2}, \frac{y_1 + y_2}{2}\right)$$

$$= \left(\frac{-2 + 5}{2}, \frac{3 + (-1)}{2}\right)$$

$$= \left(\frac{3}{2}, 1\right)$$

Checkpoint ✓ Use the Midpoint Formula

Sketch $\overline{PQ}$. Then find the coordinates of its midpoint. $\left(-\frac{5}{2}, \frac{3}{2}\right)$

3. $P(2, 5)$, $Q(4, 3)$ $(3, 4)$ **4.** $P(0, -2)$, $Q(4, 0)$ $(2, -1)$ **5.** $P(-1, 2)$, $Q(-4, 1)$

2.1 *Segment Bisectors*   **55**

Multiple Representations
By plotting a segment and its midpoint on a coordinate plane, students can see not only that a midpoint divides a segment into two parts of equal length, but also that the coordinates of the midpoint are the means of the x-coordinates and y-coordinates of the endpoints.

Extra Example 4
Find the coordinates of the midpoint of $\overline{GH}$.

a. $G(4, 0)$, $H(-3, -1)$ $M\left(\frac{1}{2}, -\frac{1}{2}\right)$

b. $G(-2, 3)$, $H(4, -8)$ $M\left(1, -\frac{5}{2}\right)$

✔ **Concept Check**
You are given the coordinates of the endpoints of a segment. How do you find the coordinates of the midpoint of the segment? **Find the average of the x-coordinates and the average of the y-coordinates of the endpoints. These averages are the x- and y-coordinates, respectively, of the midpoint.**

🐢 **Daily Puzzler**
The midpoint of a vertical segment $\overline{RS}$ is $(-2, 5)$. Find a and b if the point R is $(-2, a)$, point S is $(-2, b)$, and $RS = 6$. **$a = 2$, $b = 8$ or $a = 8$, $b = 2$**

Assignment Guide

BASIC
Day 1: pp. 56–59 Exs. 11–24, 46–51, 53–63 odd
Day 2: SRH p. 672 Exs. 1–19 odd; pp. 56–59 Exs. 26–32, 38, 44, 45, 52–62 even

AVERAGE
Day 1: pp. 56–59 Exs. 11–19 odd, 20–23, 25, 42, 46–57
Day 2: pp. 56–59 Exs. 26–40, 44, 45, 58–63

ADVANCED
Day 1: pp. 56–59 Exs. 13–17, 20–23, 25, 42, 46–62 even
Day 2: pp. 56–59 Exs. 24–34 even, 36–41, 43*, 44, 45; EC: classzone.com

BLOCK SCHEDULE
pp. 56–59 Exs. 11–19 odd, 20–40, 44–63

Extra Practice
• Student Edition, p. 677
• Chapter 2 Resource Book, pp. 10–11

Homework Check
To quickly check student understanding of key concepts, go over the following exercises:

Basic: 18, 20, 26, 28, 31
Average: 17, 23, 29, 33, 40
Advanced: 17, 23, 28, 32, 41

Geometric Reasoning
In Exercises 11–14, a thorough discussion of why each point *M* is or is not a midpoint should help alleviate any misconceptions about midpoints.

15. *Sample answer:*

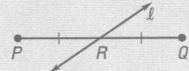

2.1 Exercises

Guided Practice

Vocabulary Check

1. In the diagram shown at the right, name the *midpoint* and a *segment bisector* of $\overline{AB}$. **C; line ℓ**

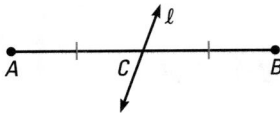

Skill Check

M is the midpoint of the segment. Find the segment lengths.

2. Find *RM* and *MS*. **3; 3**

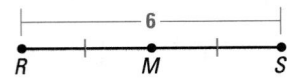

3. Find *FM* and *MG*. **25; 25**

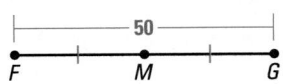

4. Find *MQ* and *PQ*. **4; 8**

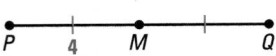

5. Find *YM* and *YZ*. **20; 40**

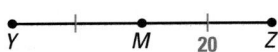

M is the midpoint of $\overline{JK}$. Find the value of the variable.

6.
47 J   x − 14   M   33   K

7.
3 J   16r   M   48   K

Find the coordinates of the midpoint of $\overline{PR}$.

8.
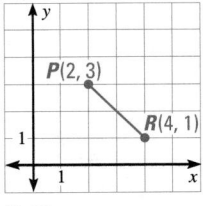
(3, 2)

9.
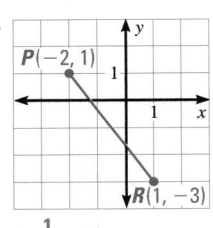
$\left(-\frac{1}{2}, -1\right)$

10.
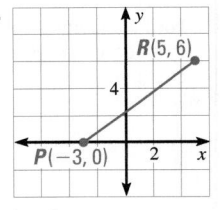
(1, 3)

Practice and Applications

Extra Practice
See p. 677.

11. No; *M* does not divide $\overline{AB}$ into two congruent segments.

12. Yes; *M* divides $\overline{AB}$ into two congruent segments.

13. No; *M* does not lie on $\overline{AB}$.

14. No; there is insufficient information to show that *M* divides $\overline{AB}$ into two congruent segments.

Recognizing Midpoints In Exercises 11–14, determine whether *M* is the midpoint of $\overline{AB}$. Explain your reasoning.

11.
A   B   M

12.
A   M   B

13.
A   M   B

14.
A   M   B

15. **Visualize It!** Sketch a line segment, $\overline{PQ}$, that is bisected by line ℓ at point *R*. **See margin.**

56 Chapter 2 *Segments and Angles*

Finding Segment Lengths *M* is the midpoint of the segment. Find the segment lengths.

16. Find *KM* and *ML*. 19; 19

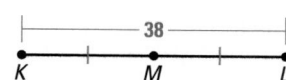

17. Find *DM* and *ME*. 41; 41

18. Find *YM* and *MZ*. 8.5; 8.5

19. Find *AM* and *MB*. 1.35; 1.35

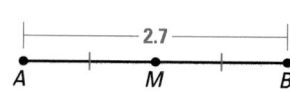

Finding Segment Lengths Line *ℓ* bisects the segment. Find the segment lengths.

20. Find *CB* and *AB*. 36; 72

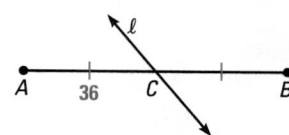

21. Find *MP* and *MN*. 15; 30

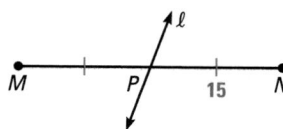

22. Find *FE* and *DE*. 29.5; 59

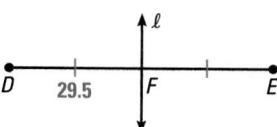

23. Find *UT* and *ST*. 3.6; 7.2

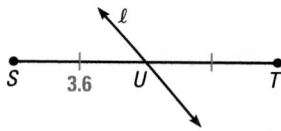

Biking The Minuteman Bikeway is a 10.5 mile bike path that runs from Arlington to Bedford, Massachusetts.

24. Caitlin and Laurie begin at opposite ends of the Minuteman Bikeway and meet at the halfway point on the path. How far does each rider bike? 5.25 miles

25. Caitlin starts on the path 4.3 miles from the Arlington end. Laurie starts on the path 3 miles from the Bedford end. How far will each rider bike before reaching the halfway point on the path?
Caitlin will bike 0.95 mi and Laurie will bike 2.25 mi.

Using Algebra Find the value of the variable.

26.

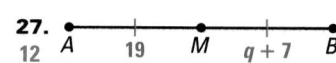

A 12 6p M 72 B

27. A 12 19 M q + 7 B

28. A 18 r − 3 M 15 B

29. A −1 4 M 2s + 6 B

Midpoint Formula Find the coordinates of the midpoint of $\overline{PQ}$.

30. $P(0, 0)$, $Q(4, 6)$ (2, 3) **31.** $P(3, 8)$, $Q(7, 6)$ (5, 7) **32.** $P(-5, 6)$, $Q(9, 7)$ $(2, \frac{13}{2})$

33. $P(-12, 0)$, $Q(6, 1)$
$(-3, \frac{1}{2})$

34. $P(-4, 4)$, $Q(4, 0)$
(0, 2)

35. $P(3, 2)$, $Q(-7, -4)$
$(-2, -1)$

2.1 *Segment Bisectors* 57

Teaching Tip

In Exercises 38–41, make sure students understand that they must calculate the coordinates as shown in the Example on this page. Stress that students should not attempt to determine the answers by simply studying the map.

Student Help

LOOK BACK
For more about baseball, see p. 50.

Strike Zone In Exercises 36 and 37, use the information below.

In baseball, the *strike zone* is the region a baseball needs to pass through for the umpire to declare it a strike if the batter does not swing. The top of the strike zone is a horizontal plane passing through the midpoint of the top of the batter's shoulders and the top of the uniform pants when the player is in a batting stance.

▶ Source: Major League Baseball

36. Find the coordinate of *T*. 51   **37.** Find the coordinate of *T*. 54

Student Help

VOCABULARY TIP
Lines of *latitude* run parallel to the Equator. Lines of *longitude* run north-south.

EXAMPLE   **Latitude-Longitude Coordinates**

Find the coordinates of the place halfway between San Francisco (37.8°N, 122.4°W) and Los Angeles (34.1°N, 118.2°W).

Solution

$$M = \left(\frac{x_1 + x_2}{2}, \frac{y_1 + y_2}{2}\right)$$

$$= \left(\frac{37.8° + 34.1°}{2}, \frac{122.4° + 118.2°}{2}\right)$$

$$= (35.95°N, 120.3°W)$$

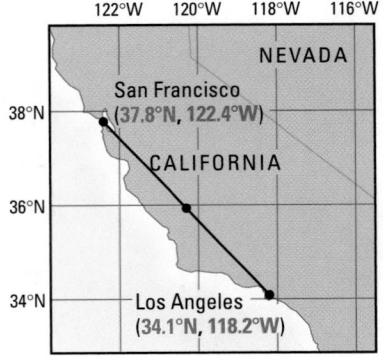

ANSWER ▶ The place halfway between San Francisco and Los Angeles has coordinates (35.95°N, 120.3°W).

Latitude-Longitude Coordinates **Find the coordinates of the place halfway between the two cities in California.**

38. (37.5°N, 121.05°W)

39. (35.75°N, 118.3°W)

40. (37.6°N, 122.25°W)

41. (36.1°N, 121°W)

38. Fresno: (36.7°N, 119.8°W)
    Napa: (38.3°N, 122.3°W)

39. Bishop: (37.4°N, 118.4°W)
    Los Angeles: (34.1°N, 118.2°W)

40. San Francisco: (37.8°N, 122.4°W)
    Palo Alto: (37.4°N, 122.1°W)

41. Santa Barbara: (34.4°N, 119.7°W)
    Oakland: (37.8°N, 122.3°W)

42. Using Midpoints In the diagram below, B is the midpoint of $\overline{AC}$, $AB = 9$, and $AD = 25$. Find CD. **7**

43. Challenge The midpoint of $\overline{AB}$ is $M(7, 5)$. The coordinates of point A are $(4, 1)$. Find the coordinates of point B. Explain. **(10, 9); explanations may vary.**

Standardized Test Practice

44. Multiple Choice T is the midpoint of $\overline{QR}$. What is the value of x? **B**

Ⓐ 17   Ⓑ 22

Ⓒ 29.5   Ⓓ 88

45. Multiple Choice What is the midpoint of the segment joining $(2, 7)$ and $(-6, 2)$? **F**

Ⓕ $\left(-2, \dfrac{9}{2}\right)$ Ⓖ $(-4, 9)$ Ⓗ $(-2, 4)$ Ⓙ $\left(\dfrac{9}{2}, -2\right)$

Mixed Review

Evaluating Statements Use the diagram at the right to determine whether the statement is *true* or *false*. *(Lessons 1.3, 1.5)*

46. Point A lies on line m. **true**

47. Point E lies on line ℓ. **true**

48. Points B, E, and C are collinear. **false**

49. Lines ℓ and m intersect at point E. **true**

50. Point E is between points B and C. **false**

51. Point F is between points A and B. **true**

Classifying Angles Name the vertex and sides of the angle. Then state whether it appears to be *acute*, *right*, *obtuse*, or *straight*. *(Lesson 1.6)*

52.
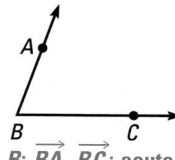
B; $\overrightarrow{BA}$, $\overrightarrow{BC}$; acute

53.
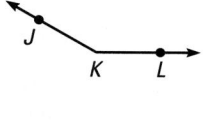
K; $\overrightarrow{KJ}$, $\overrightarrow{KL}$; obtuse

54.
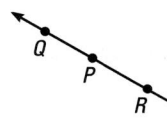
P; $\overrightarrow{PQ}$, $\overrightarrow{PR}$; straight

Algebra Skills

Evaluating Expressions Evaluate the expression. *(Skills Review, p. 670)*

55. $2 \cdot 15 + 40$ **70**

56. $120 - 35 \cdot 3$ **15**

57. $\dfrac{1}{2} \cdot 50 + 145$ **170**

58. $\dfrac{5}{4} \cdot 16 - 20$ **0**

59. $6 + 3 \cdot 5 - 2$ **19**

60. $11 \cdot 4 + 7 - 20$ **31**

61. $12 \cdot 2 - 3 \cdot 4$ **12**

62. $5 - 10 \cdot 6 + 1$ **−54**

63. $2 - (3 + 4) \cdot 5$ **−33**

2.1 *Segment Bisectors* **59**

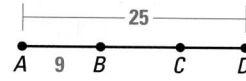

Assessment Resources

The Mini-Quiz below is also available on blackline (*Chapter 2 Resource Book*, p. 18) and on transparency. For more assessment resources, see:

• Chapter 2 Resource Book
• Standardized Test Practice
• Test and Practice Generator

Mini-Quiz

1. Line ℓ bisects $\overline{FH}$. Find GH and FH. **$GH = 27.5$, $FH = 55$**

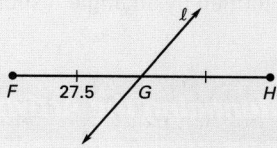

2. M is the midpoint of $\overline{AB}$. Find the value of x. **5**

3. Find the coordinates of the midpoint of the segment whose endpoints are $P(-2, -3)$ and $Q(2, 7)$. **$M(0, 2)$**

4. Julia is hiking a trail that is 3.4 miles long. She plans to stop for lunch when she reaches the halfway point on the trail. How far will she hike before stopping to eat her lunch? **1.7 mi**

Goal
Students use paper folding and a protractor to explore angle bisectors.

Materials
See the margin of the student page.

LINK TO LESSON
Relate the fold line in the activity to the definition of angle bisector presented at the beginning of Lesson 2.2. Also, in Example 1 on page 61 students find measures of the angles formed by an angle bisector.

Tips for Success
In Step 2, students must be sure the two rays are exactly aligned, one on top of the other. Be sure to have students use paper that is thin enough for them to see the rays through it when doing the folding.

KEY DISCOVERY
Any angle can be divided into two smaller congruent angles.

Activity Assessment
Use Exercises 3 and 4 to assess student understanding.

Activity ◆2.2◆ Folding Angle Bisectors

Question

How can you bisect an angle?

Materials
- protractor
- straightedge

Explore

❶ On a piece of paper, use a straightedge to draw an acute angle. Label the angle ∠ABC.

❷ Fold the paper so $\overrightarrow{BC}$ is on top of $\overrightarrow{BA}$.

❸ Draw a point D on the fold inside ∠ABC. Then use a protractor to measure ∠ABD, ∠DBC, and ∠ABC.

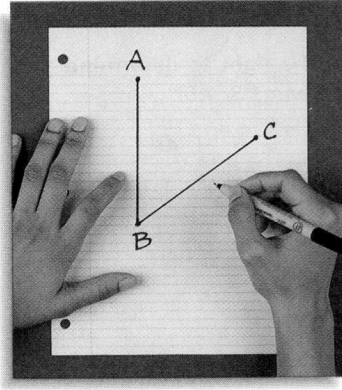  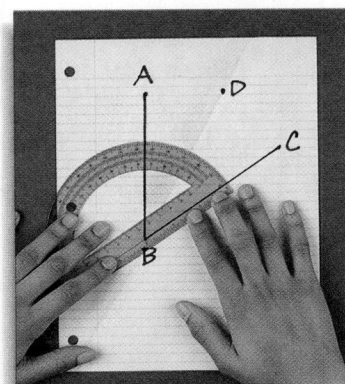

1. The measures of ∠ABD and ∠CBD are equal. The measure of ∠ABC is twice the measure of each of the smaller angles.

Student Help

LOOK BACK
For help using a protractor, see p. 36.

4. Yes; if you fold one side of the angle on top of the other side of the angle, the straight angle is divided into two congruent right angles.

Think About It

1. What do you notice about the angles you measured in Step 3?

2. Repeat Steps 1 through 3 with an obtuse angle. Compare your results with the results from Exercise 1. **The results are the same.**

3. Copy and complete:

   a. $m\angle ABD = \underline{\ ?\ } \cdot m\angle ABC$ $\frac{1}{2}$ **b.** $m\angle DBC = \underline{\ ?\ } \cdot m\angle ABC$ $\frac{1}{2}$

   c. $m\angle ABC = \underline{\ ?\ } \cdot m\angle ABD$ 2   **d.** $m\angle ABC = \underline{\ ?\ } \cdot m\angle DBC$ 2

4. **Extension** Is it possible to fold congruent angles from a straight angle? Explain your reasoning.

2.2 Angle Bisectors

Goal
Bisect an angle.

Key Words
• angle bisector

An **angle bisector** is a ray that divides an angle into two angles that are congruent. In the photograph of the hang glider, $\overrightarrow{BD}$ bisects ∠ABC because it divides the angle into two congruent angles.

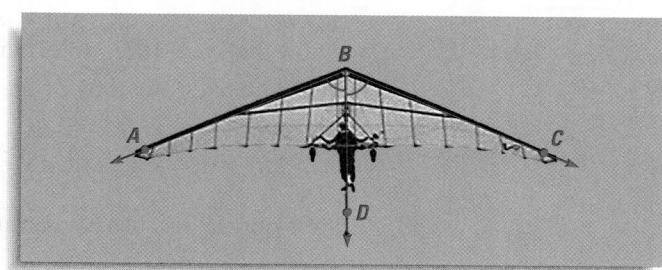

∠ABD ≅ ∠DBC

If $\overrightarrow{BD}$ bisects ∠ABC, then the measures of ∠ABD and ∠DBC are *half* the measure of ∠ABC. Also, the measure of ∠ABC is *twice* the measure of ∠ABD or ∠DBC.

Visualize It!

In this book, an arc that crosses two or more angles identifies the measure of the entire angle it crosses.

EXAMPLE 1 Find Angle Measures

$\overrightarrow{BD}$ bisects ∠ABC, and m∠ABC = 110°.
Find m∠ABD and m∠DBC.

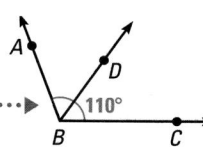

Solution

$m∠ABD = \frac{1}{2}(m∠ABC)$ $\overrightarrow{BD}$ bisects ∠ABC.

$= \frac{1}{2}(110°)$ Substitute 110° for m∠ABC.

$= 55°$ Simplify.

∠ABD and ∠DBC are congruent, so m∠DBC = m∠ABD.

ANSWER ▸ So, m∠ABD = 55°, and m∠DBC = 55°.

Checkpoint ✓ Find Angle Measures

$\overrightarrow{HK}$ bisects ∠GHJ. Find m∠GHK and m∠KHJ.

1. 26°; 26°
2. 45°; 45°
3. 80.5°; 80.5°

1.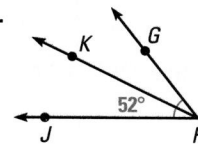

2. (figure)

3.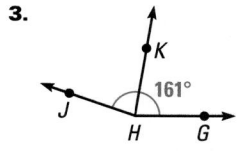

---

① Plan

Pacing
Suggested Number of Days

Basic: 2 days
Average: 2 days
Advanced: 2 days
Block Schedule: 1 block

Teaching Resources

📄 *Blacklines*
(See page 50B.)

🗲 *Transparencies*
• Warm-Up with Quiz
• Answers

🖳 *Technology*
• Electronic Teacher Tools
• Test and Practice Generator
• Online Lesson Planner
• Internet Support

② Teach

Content and Teaching Strategies
For background information on geometric concepts and teaching strategies related to this lesson, see pages 50E and 50F in this Teacher's Edition.

Tips for New Teachers
In Example 4 on page 63, point out that solving an equation for *x* is not the same as finding the measures of the angles. Checking the value of the variable by determining each angle measure is always important. See the Tips for New Teachers on pp. 1–2 of the *Chapter 2 Resource Book* for additional notes about Lesson 2.2.

Extra Example 1
See next page.

Extra Example 1

$\overrightarrow{NQ}$ bisects $\angle MNP$, and $m\angle MNP = 74°$. Find $m\angle MNQ$ and $m\angle PNQ$.

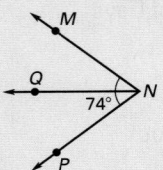

$m\angle MNQ = 37°$, $m\angle PNQ = 37°$

Extra Example 2

$\overrightarrow{AB}$ bisects $\angle FAH$, and $m\angle BAF = 38°$.

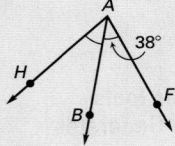

a. Find $m\angle HAB$ and $m\angle FAH$.
$m\angle HAB = 38°$, $m\angle FAH = 76°$

b. Determine whether $\angle FAH$ is *acute, right, obtuse,* or *straight.* Explain. $\angle FAH$ is acute because its measure is less than 90°.

Visualize It!

Make sure students are clear on the distinction between red arcs (used to identify congruent angles) and blue arcs (used to show the measure of an angle formed by two or more smaller angles) in this textbook. Also point out that the red arcs are drawn so they do not align with each other.

Visualize It!

In this book, matching red arcs identify congruent angles in diagrams.

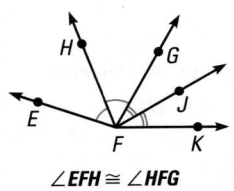

$\angle EFH \cong \angle HFG$
$\angle GFJ \cong \angle JFK$

Link to Careers

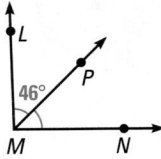

KITE DESIGNERS use geometric principles in designing and making kites. A kite's struts often bisect the angles they support.

EXAMPLE 2 Find Angle Measures and Classify an Angle

$\overrightarrow{MP}$ bisects $\angle LMN$, and $m\angle LMP = 46°$.

a. Find $m\angle PMN$ and $m\angle LMN$.

b. Determine whether $\angle LMN$ is *acute, right, obtuse,* or *straight.* Explain.

Solution

a. $\overrightarrow{MP}$ bisects $\angle LMN$, so $m\angle LMP = m\angle PMN$.
You know that $m\angle LMP = 46°$. Therefore, $m\angle PMN = 46°$.

The measure of $\angle LMN$ is twice the measure of $\angle LMP$.

$$m\angle LMN = 2(m\angle LMP) = 2(46°) = 92°$$

So, $m\angle PMN = 46°$, and $m\angle LMN = 92°$.

b. $\angle LMN$ is obtuse because its measure is between 90° and 180°.

Checkpoint ✓ Find Angle Measures and Classify an Angle

$\overrightarrow{QS}$ bisects $\angle PQR$. Find $m\angle SQP$ and $m\angle PQR$. Then determine whether $\angle PQR$ is *acute, right, obtuse,* or *straight.*

4.

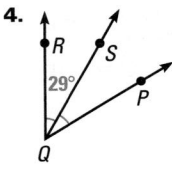

29°; 58°; acute

5.

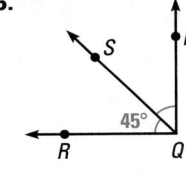

45°; 90°; right

6.

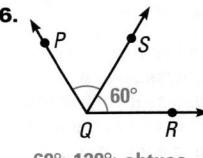

60°; 120°; obtuse

EXAMPLE 3 Use Angle Bisectors

In the kite, $\angle DAB$ is bisected by $\overrightarrow{AC}$, and $\angle BCD$ is bisected by $\overrightarrow{CA}$. Find $m\angle DAB$ and $m\angle BCD$.

Solution

$m\angle DAB = 2(m\angle BAC)$	$\overrightarrow{AC}$ bisects $\angle DAB$.
$= 2(45°)$	Substitute 45° for $m\angle BAC$.
$= 90°$	Simplify.
$m\angle BCD = 2(m\angle ACB)$	$\overrightarrow{CA}$ bisects $\angle BCD$.
$= 2(27°)$	Substitute 27° for $m\angle ACB$.
$= 54°$	Simplify.

ANSWER ▶ The measure of $\angle DAB$ is 90°, and the measure of $\angle BCD$ is 54°.

Checkpoint ✓ Use Angle Bisectors

7. $\overrightarrow{KM}$ bisects $\angle JKL$.
Find $m\angle JKM$ and $m\angle MKL$.
48°; 48°

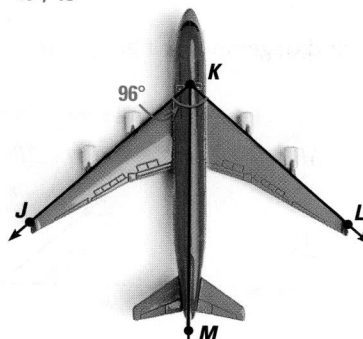

96°

8. $\overrightarrow{UV}$ bisects $\angle WUT$.
Find $m\angle WUV$ and $m\angle WUT$.
60°; 120°

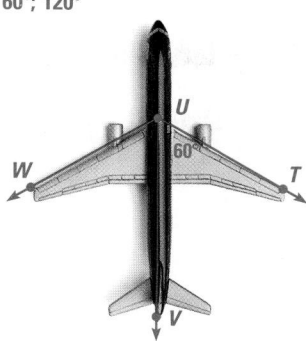

60°

Using Algebra

Student Help
CLASSZONE.COM

MORE EXAMPLES
More examples at
classzone.com

EXAMPLE 4 · Use Algebra with Angle Measures

$\overrightarrow{RQ}$ bisects $\angle PRS$. Find the value of x.

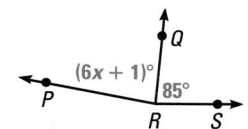

$(6x + 1)°$
85°

Solution

$m\angle PRQ = m\angle QRS$	$\overrightarrow{RQ}$ bisects $\angle PRS$.
$(6x + 1)° = 85°$	Substitute given measures.
$6x + 1 - 1 = 85 - 1$	Subtract 1 from each side.
$6x = 84$	Simplify.
$\dfrac{6x}{6} = \dfrac{84}{6}$	Divide each side by 6.
$x = 14$	Simplify.

CHECK ✓ You can check your answer by substituting 14 for x.

$$m\angle PRQ = (6x + 1)° = (6 \cdot 14 + 1)° = (84 + 1)° = 85°$$

Checkpoint ✓ Use Algebra with Angle Measures

$\overrightarrow{BD}$ bisects $\angle ABC$. Find the value of x.

9.
43

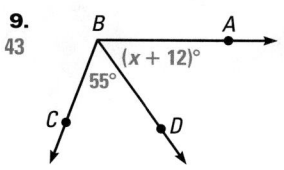

$(x + 12)°$
55°

10.
3

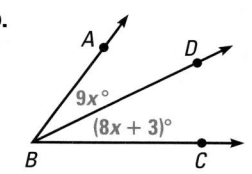

$9x°$
$(8x + 3)°$

2.2 Angle Bisectors **63**

Extra Example 3
In the award plaque, $\angle NKL$ is
bisected by $\overrightarrow{KM}$ and $\angle NML$ is
bisected by $\overrightarrow{MK}$. Find $m\angle NKL$ and
$m\angle NML$.

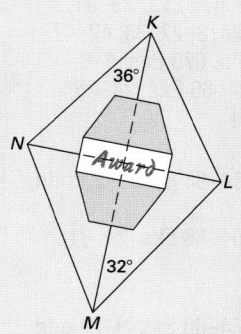

36°
Award
32°

$m\angle NKL = 72°$, $m\angle NML = 64°$

Extra Example 4
$\overrightarrow{GJ}$ bisects $\angle FGH$. Find the value
of x. 8

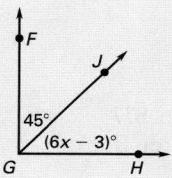

45°
$(6x - 3)°$

✓ Concept Check
A 138° angle is bisected. Give the
measures of the angles formed and
determine whether they are *acute*,
right, *obtuse*, or *straight*. The two
angles formed both measure 69°,
which means they are acute angles.

🏆 Daily Puzzler
$\overrightarrow{ZC}$ bisects $\angle AZB$, $\overrightarrow{ZD}$ bisects
$\angle AZC$, $\overrightarrow{ZE}$ bisects $\angle AZD$, and so
on. If the pattern continues and
$m\angle AZB = 128°$, find $m\angle AZH$. 2°

63

Assignment Guide

BASIC
Day 1: EP p. 676 Exs. 29–31; pp. 64–66 Exs. 8–22, 33, 42–47
Day 2: SRH p. 673 Exs. 4–9, 19–21; pp. 64–66 Exs. 24–26, 28–30, 34–41

AVERAGE
Day 1: pp. 64–66 Exs. 9–23 odd, 33–47 odd
Day 2: pp. 64–66 Exs. 24–31, 34–46 even

ADVANCED
Day 1: pp. 64–66 Exs. 9–23 odd, 32*, 33–47 odd
Day 2: pp. 64–66 Exs. 24–31, 34–46 even; EC: classzone.com

BLOCK SCHEDULE
pp. 64–66 Exs. 9–23 odd, 24–31, 33–47

Extra Practice
• Student Edition, p. 677
• Chapter 2 Resource Book, pp. 20–21

Homework Check
To quickly check student understanding of key concepts, go over the following exercises:

Basic: 10, 19, 24, 25, 28
Average: 11, 21, 24, 25, 29
Advanced: 13, 19, 24, 26, 30

✗ Common Error
In Exercises 8–13, watch for students who assign the given angle measure to one of the smaller angles rather than to ∠*PQR*. Remind students that the blue arcs each signify that the given measure applies to ∠*PQR*. You may wish to contrast these diagrams with the ones in Exercises 17–22 to further emphasize this point.

2.2 Exercises

Guided Practice

Vocabulary Check

1. What kind of geometric figure is an *angle bisector*? a ray

Skill Check

$\overrightarrow{BD}$ bisects ∠*ABC*. **Find the angle measure.**

2. Find m∠*ABD*. 20°

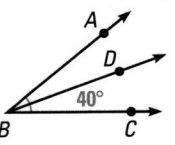

3. Find m∠*DBC*. 41°

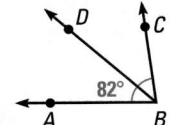

4. Find m∠*DBC*. 84°

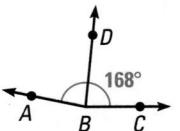

5. Find m∠*ABC*. 60°

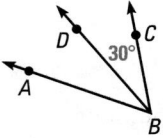

6. Find m∠*CBA*. 100°

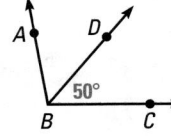

7. Find m∠*ABC*. 78°

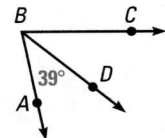

Practice and Applications

Extra Practice
See p. 677.

Finding Angle Measures $\overrightarrow{QS}$ bisects ∠*PQR*. **Find** m∠*PQS* **and** m∠*SQR*.

8.

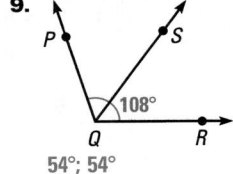

25°; 25°

9.

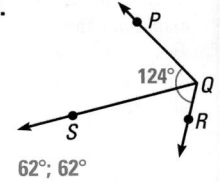

54°; 54°

10.

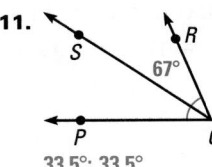

62°; 62°

11.

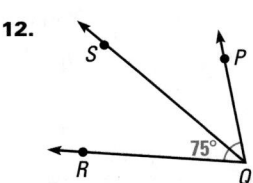

33.5°; 33.5°

12.

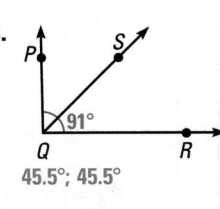

37.5°; 37.5°

13.

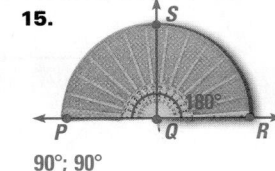

45.5°; 45.5°

Fans $\overrightarrow{QS}$ bisects ∠*PQR*. **Find** m∠*PQS* **and** m∠*SQR*.

14.

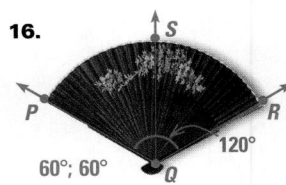

43.5°; 43.5°

15.

90°; 90°

16.

60°; 60°

Homework Help
Example 1: Exs. 8–16
Example 2: Exs. 17–22
Example 3: Exs. 24–27
Example 4: Exs. 28–30

Finding Angle Measures $\overrightarrow{BD}$ bisects ∠ABC. Find m∠ABD and m∠ABC. Then determine whether ∠ABC is *acute, right, obtuse,* or *straight.*

17.

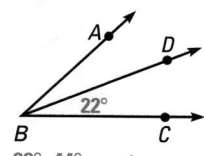

22°; 44°; acute

18.

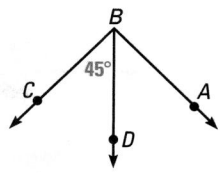

45°; 90°; right

19.

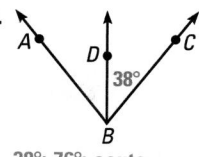

38°; 76°; acute

20.

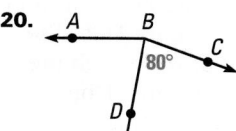

80°; 160°; obtuse

21.

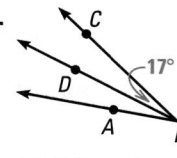

17°; 34°; acute

22.

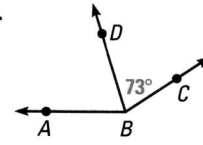

73°; 146°; obtuse

23. Paper Airplanes The diagram at the right represents an unfolded piece of paper used to make a paper airplane like the one shown below. Using the diagram, name the angles that are bisected by $\overrightarrow{AK}$.
∠JRL, ∠HAM, ∠FAN, ∠DAP

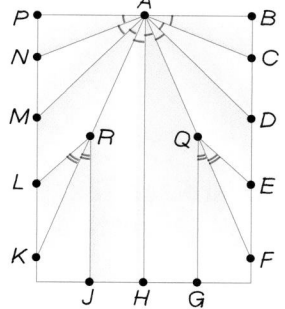

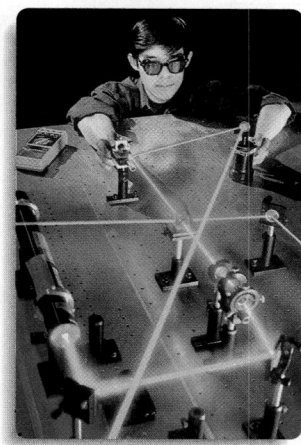
Lasers In Exercises 24–27, use the diagram below. When light is reflected by a smooth surface, the angle of incidence is equal to the angle of reflection.

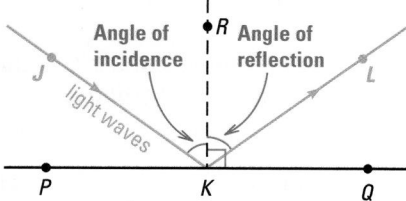

24. Name an angle bisector in the diagram. $\overrightarrow{KR}$

25. If the angle of reflection is 67°, what is the angle of incidence? 67°

26. If m∠JKL = 109°, what is the angle of reflection? 54.5°

27. **You be the Judge** Can you determine whether ∠JKP is congruent to ∠LKQ in the diagram above? Explain your reasoning.
Yes; *sample answer:* m∠PKR = m∠QKR = 90° and m∠JKR = m∠LKR, so m∠PKR − m∠JKR = m∠QKR − m∠LKR. Therefore, m∠JKP = m∠LKQ.

2.2 *Angle Bisectors* **65**

Visualize It!

Exercise 23 may be difficult for some students with the multiple sets of angles at vertex *A* and the less-than-obvious angle at vertex *R*. Suggest to students that they trace the diagram and use colored pencils to help visualize the angles that are bisected by $\overrightarrow{AK}$.

Assessment Resources

The Mini-Quiz below is also available on blackline (*Chapter 2 Resource Book*, p. 28) and on transparency. For more assessment resources, see:
- Chapter 2 Resource Book
- Standardized Test Practice
- Test and Practice Generator

Mini-Quiz

1. $\overrightarrow{BD}$ bisects $\angle ABC$. Find $m\angle ABD$ and $m\angle DBC$.

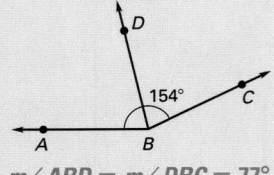

$m\angle ABD = m\angle DBC = 77°$

2. $\overrightarrow{HL}$ bisects $\angle GHJ$. Find $m\angle GHL$ and $m\angle GHJ$. Then determine whether $\angle GHJ$ is *acute, right, obtuse,* or *straight.*

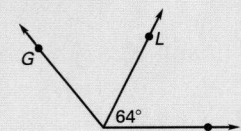

$m\angle GHL = 64°$, $m\angle GHJ = 128°$; obtuse

$\overrightarrow{QS}$ bisects $\angle PQR$. Find the value of the variable.

3.

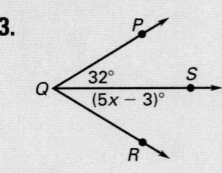

$x = 7$

4.

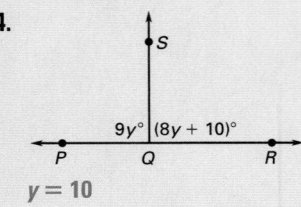

$y = 10$

 Using Algebra $\overrightarrow{KM}$ bisects $\angle JKL$. Find the value of the variable.

28.
83
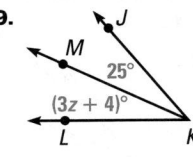

29.
7
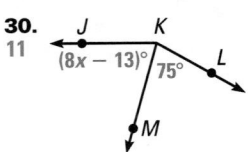

30.
11
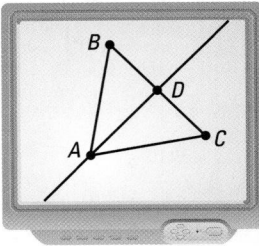

31. **Technology** Use geometry software to draw $\triangle ABC$. Construct the angle bisector of $\angle BAC$. Then find the midpoint of $\overline{BC}$. Drag any of the points. Does the angle bisector *always* pass through the midpoint of the opposite side? Does it *ever* pass through the midpoint? no; yes

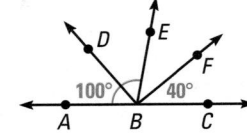

32. **Challenge** $\overrightarrow{BD}$ bisects $\angle ABE$ and $\overrightarrow{BF}$ bisects $\angle EBC$. Use the diagram shown to find $m\angle DBF$. 90°

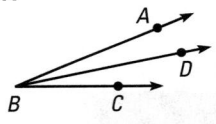

33. **Multiple Choice** In the diagram below, $\overrightarrow{BD}$ bisects $\angle ABC$ and $m\angle ABC = 23°$. What is $m\angle ABD$? A

Ⓐ $11.5°$ Ⓑ $12.5°$

Ⓒ $23°$ Ⓓ $46°$

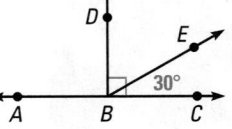

34. Each number is 8 more than the previous number; 33.

35. Each number is 5 more than the previous number; 7.

36. Numbers after the first are formed by inserting a 1 after the decimal point of the previous number; 5.11116.

37. Each number is half the previous number; 3.75.

Describing Number Patterns Describe a pattern in the numbers. Write the next number you expect in the pattern. *(Lesson 1.1)*

34. $1, 9, 17, 25, \ldots$ 35. $-13, -8, -3, 2, \ldots$

36. $5.6, 5.16, 5.116, 5.1116, \ldots$ 37. $60, 30, 15, 7.5, \ldots$

Classifying Angles Use the diagram below to classify the angle as *acute, right, obtuse,* or *straight.* *(Lesson 1.6)*

38. $\angle EBC$ acute   39. $\angle ABE$ obtuse

40. $\angle DBC$ right   41. $\angle ABC$ straight

Solving Equations Solve the equation. *(Skills Review, p. 673)*

42. $2x - 15 = 9$ 12   43. $3a + 12 = 48$ 12   44. $10 - 3y = 52$ -14

45. $5m - 11 = -46$ -7 46. $-2z + 4 = 8$ -2 47. $3 = -n + 23$ 20

2.3 Complementary and Supplementary Angles

Goal
Find measures of complementary and supplementary angles.

Key Words
• complementary angles
• supplementary angles
• adjacent angles
• theorem

Two angles are **complementary angles** if the sum of their measures is 90°. Each angle is the **complement** of the other.

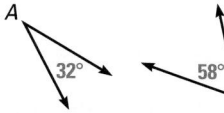

∠A and ∠B are complementary angles.
m∠A + m∠B = 32° + 58° = 90°

Two angles are **supplementary angles** if the sum of their measures is 180°. Each angle is the **supplement** of the other.

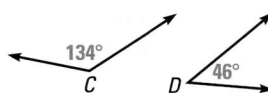

∠C and ∠D are supplementary angles.
m∠C + m∠D = 134° + 46° = 180°

Visualize It!

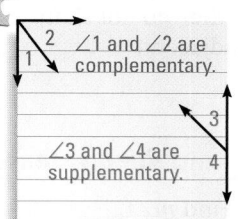

∠1 and ∠2 are complementary.

∠3 and ∠4 are supplementary.

Complementary angles make up the **C**orner of a piece of paper. Supplementary angles make up the **S**ide of a piece of paper.

EXAMPLE 1 Identify Complements and Supplements

Determine whether the angles are *complementary, supplementary,* or *neither*.

a.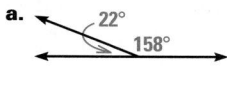

b.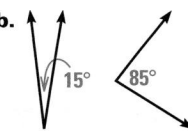

c.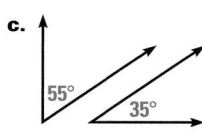

Solution

a. Because 22° + 158° = 180°, the angles are supplementary.

b. Because 15° + 85° = 100°, the angles are neither complementary nor supplementary.

c. Because 55° + 35° = 90°, the angles are complementary.

Checkpoint ✓ Identify Complements and Supplements

Determine whether the angles are *complementary, supplementary,* or *neither.*

1.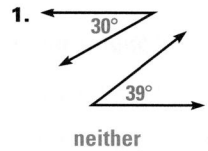
neither

2.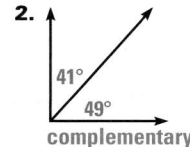
complementary

3.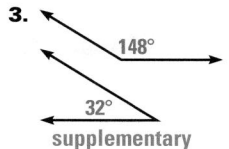
supplementary

---

① Plan

Pacing
Suggested Number of Days

Basic: 2 days
Average: 2 days
Advanced: 2 days
Block Schedule: 1 block

Teaching Resources

▤ *Blacklines*
(See page 50B.)

✎ *Transparencies*
• Warm-Up with Quiz
• Answers

▣ *Technology*
• Electronic Teacher Tools
• Test and Practice Generator
• Online Lesson Planner
• Internet Support

② Teach

Content and Teaching Strategies
For background information on geometric concepts and teaching strategies related to this lesson, see pages 50E and 50F in this Teacher's Edition.

Tips for New Teachers
Help students identify angles by using visual examples with a variety of orientations. Students with limited English proficiency can recognize the visual example and associate the description of the angle with it. If students have trouble looking at angles printed on paper, have them turn their book and view the angles from other perspectives. See the Tips for New Teachers on pp. 1–2 of the *Chapter 2 Resource Book* for additional notes about Lesson 2.3.

Extra Example 1
See next page.

Extra Example 1

State whether the angles are *complementary*, *supplementary*, or *neither*.

a.

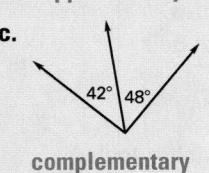

neither

b.

143° 37°

supplementary

c.

42° 48°

complementary

Extra Example 2

Tell whether the numbered angles are *adjacent* or *nonadjacent*.

a.

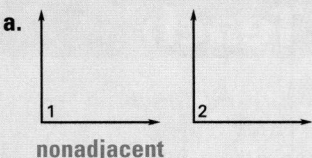

1   2

nonadjacent

b.

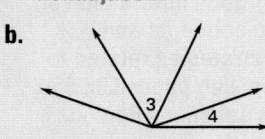

3   4

nonadjacent

c.

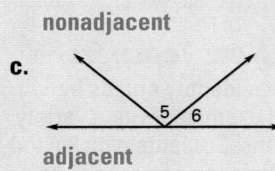

5 6

adjacent

Extra Example 3

a. ∠*F* is a complement of ∠*G*, and *m*∠*F* = 23°. Find *m*∠*G*. **67°**

b. ∠*S* is a supplement of ∠*T*, and *m*∠*S* = 95°. Find *m*∠*T*. **85°**

Two angles are **adjacent angles** if they share a common vertex and side, but have no common interior points.

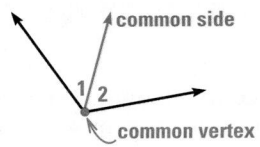

common side

∠1 and ∠2 are adjacent angles.

common vertex

EXAMPLE 2 Identify Adjacent Angles

Tell whether the numbered angles are *adjacent* or *nonadjacent*.

a.

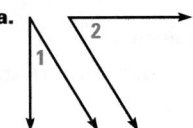

2

1

b.

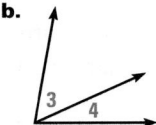

3   4

c.

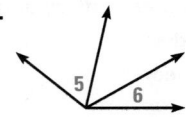

5   6

Solution

a. Because the angles do not share a common vertex or side, ∠1 and ∠2 are nonadjacent.

b. Because the angles share a common vertex and side, and they do not have any common interior points, ∠3 and ∠4 are adjacent.

c. Although ∠5 and ∠6 share a common vertex, they do not share a common side. Therefore, ∠5 and ∠6 are nonadjacent.

EXAMPLE 3 Measures of Complements and Supplements

a. ∠*A* is a complement of ∠*C*, and *m*∠*A* = 47°. Find *m*∠*C*.

b. ∠*P* is a supplement of ∠*R*, and *m*∠*R* = 36°. Find *m*∠*P*.

Solution

a. ∠*A* and ∠*C* are complements, so their sum is 90°.

$$m\angle A + m\angle C = 90°$$
$$47° + m\angle C = 90°$$
$$47° + m\angle C - 47° = 90° - 47°$$
$$m\angle C = 43°$$

b. ∠*P* and ∠*R* are supplements, so their sum is 180°.

$$m\angle P + m\angle R = 180°$$
$$m\angle P + 36° = 180°$$
$$m\angle P + 36° - 36° = 180° - 36°$$
$$m\angle P = 144°$$

Checkpoint ✓ **Measures of Complements and Supplements**

4. ∠*B* is a complement of ∠*D*, and *m*∠*D* = 79°. Find *m*∠*B*. **11°**

5. ∠*G* is a supplement of ∠*H*, and *m*∠*G* = 115°. Find *m*∠*H*. **65°**

A **theorem** is a true statement that follows from other true statements. The two theorems that follow are about complementary and supplementary angles.

Student Help

VISUAL STRATEGY
Draw examples of these theorems with specific measures, as shown on p. 52.

THEOREMS 2.1 and 2.2

2.1 Congruent Complements Theorem

Words If two angles are complementary to the same angle, then they are congruent.

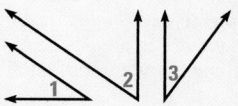

Symbols If $m\angle 1 + m\angle 2 = 90°$ and $m\angle 2 + m\angle 3 = 90°$, then $\angle 1 \cong \angle 3$.

2.2 Congruent Supplements Theorem

Words If two angles are supplementary to the same angle, then they are congruent.

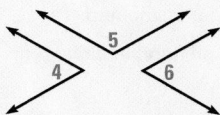

Symbols If $m\angle 4 + m\angle 5 = 180°$ and $m\angle 5 + m\angle 6 = 180°$, then $\angle 4 \cong \angle 6$.

You can use theorems in your reasoning about geometry, as shown in Example 4.

EXAMPLE 4 Use a Theorem

$\angle 7$ and $\angle 8$ are supplementary, and $\angle 8$ and $\angle 9$ are supplementary. Name a pair of congruent angles. Explain your reasoning.

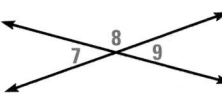

Solution

$\angle 7$ and $\angle 9$ are both supplementary to $\angle 8$. So, by the Congruent Supplements Theorem, $\angle 7 \cong \angle 9$.

Checkpoint ✓ Use a Theorem

6. In the diagram, $m\angle 10 + m\angle 11 = 90°$, and $m\angle 11 + m\angle 12 = 90°$.

Name a pair of congruent angles. Explain your reasoning.

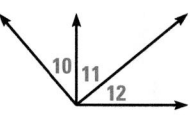

∠10 ≅ ∠12; ∠10 and ∠12 are both complementary to ∠11, so ∠10 ≅ ∠12 by the Congruent Complements Theorem.

2.3 *Complementary and Supplementary Angles*   69

Study Skills

Vocabulary Make sure students know the everyday meaning of the word *adjacent* (close to; near), and emphasize that the mathematical meaning of *adjacent angles* is more precise.

Extra Example 4

∠1 and ∠3 are complementary, and ∠3 and ∠5 are complementary. Name a pair of congruent angles. Explain your reasoning.

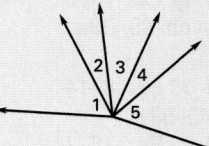

∠1 and ∠5 are both complementary to ∠3. So, by the Congruent Complements Theorem, ∠1 ≅ ∠5.

✓ Concept Check

Can a pair of angles be both complementary and supplementary? Why or why not? The sum of the measures of a pair of angles cannot be both 90° and 180°, so the pair can be either complementary or supplementary, but not both.

🧩 Daily Puzzler

Suppose ∠1 is the angle formed by the hands of a clock when it is exactly 1 o'clock. Between 12 noon and 3 o'clock, how many times will the hands of the clock form an angle complementary to ∠1? 5 times

Assignment Guide

BASIC
Day 1: pp. 70–73 Exs. 8–14, 29–32, 44–59
Day 2: SRH p. 673 Exs. 13–17 odd; pp. 70–73 Exs. 15–27 odd, 28, 33, 40–42, Quiz 1

AVERAGE
Day 1: pp. 70–73 Exs. 8–14, 29–32, 44, 46–59
Day 2: pp. 70–73 Exs. 16–28 even, 34–42 even, 45, Quiz 1

ADVANCED
Day 1: pp. 70–73 Exs. 8–14 even, 29–32, 45–59
Day 2: pp. 70–73 Exs. 16–28 even, 33–41 odd, 43*, Quiz 1; EC: classzone.com

BLOCK SCHEDULE
pp. 70–73 Exs. 8–14, 16–28 even, 29–32, 34–42 even, 44–59, Quiz 1

Extra Practice

• Student Edition, p. 677
• Chapter 2 Resource Book, pp. 30–31

Homework Check

To quickly check student understanding of key concepts, go over the following exercises:

Basic: 9, 10, 17, 23, 40
Average: 9, 10, 18, 26, 40
Advanced: 10, 14, 18, 26, 41

2.3 Exercises

Guided Practice

Vocabulary Check

1. The sum of the measures of two complementary angles is 90°. The sum of the measures of two supplementary angles is 180°.

Skill Check

3. complementary; adjacent

4. neither; nonadjacent

5. supplementary; nonadjacent

1. Explain the difference between *complementary angles* and *supplementary angles*.

2. Complete the statement: Two angles are __?__ if they share a common vertex and a common side, but have no common interior points. **adjacent**

In Exercises 3–5, determine whether the angles are *complementary*, *supplementary*, or *neither*. Also tell whether the angles are *adjacent* or *nonadjacent*.

3. (75°, 15°)   4. (90°, 110°)   5. (30°, 150°)

6. ∠A is a complement of ∠B, and m∠A = 10°. Find m∠B. **80°**

7. ∠C is a supplement of ∠D, and m∠D = 109°. Find m∠C. **71°**

Practice and Applications

Extra Practice
See p. 677.

Identifying Angles Determine whether the angles are *complementary*, *supplementary*, or *neither*. Also tell whether the angles are *adjacent* or *nonadjacent*.

8. (58° 31°)   9. (78° 102°)   10. (67° 33°)
neither; adjacent   supplementary; adjacent   neither; nonadjacent

Identifying Angles Determine whether the two angles shown on the clock faces are *complementary*, *supplementary*, or *neither*.

11. supplementary
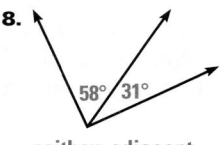

12. neither
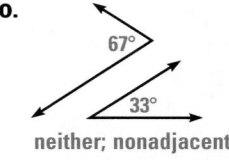

13. complementary

14. supplementary

Homework Help
Example 1: Exs. 8–14, 30–32
Example 2: Exs. 8–10
Example 3: Exs. 15–28 33, 34
Example 4: Exs. 38–42

Finding Complements Find the measure of a complement of the angle given.

15. 49°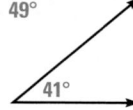
41°

16. 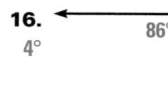 86°
4°

17. 66°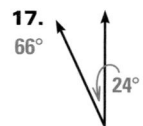
24°

18. ∠K is a complement of ∠L, and m∠K = 74°. Find m∠L. 16°

19. ∠P is a complement of ∠Q, and m∠P = 9°. Find m∠Q. 81°

Finding Supplements Find the measure of a supplement of the angle given.

20. 125° 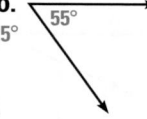 55°

21. 166°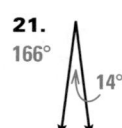
14°

22. 20°  160°

23. ∠A is a supplement of ∠B, and m∠A = 96°. Find m∠B. 84°

24. ∠P is a supplement of ∠Q, and m∠P = 7°. Find m∠Q. 173°

Finding Complements and Supplements Find the measures of a complement and a supplement of the angle.

25. m∠A = 39° 51°; 141°   **26.** m∠B = 89° 1°; 91°   **27.** m∠C = 54° 36°; 126°

28. Bridges The Alamillo Bridge in Seville, Spain, was designed by Santiago Calatrava. In the bridge, m∠1 = 58°, and m∠2 = 24°. Find the measures of the supplements of both ∠1 and ∠2. 122°; 156°

Naming Angles In the diagram, ∠QPR is a right angle.

29. Name a straight angle. ∠QPT

30. Name two congruent supplementary angles.
∠QPR, ∠RPT

31. Name two supplementary angles that are not congruent. ∠QPS, ∠SPT

32. Name two complementary angles. ∠RPS, ∠SPT

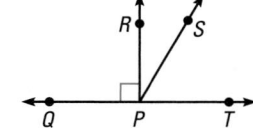

Beach Chairs Adjustable beach chairs form angles that are supplementary. Find the value of *x*.

33.
64

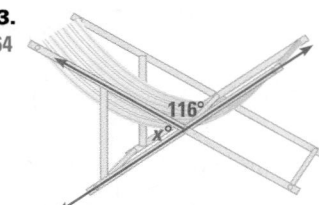

116°
x°

34.
40

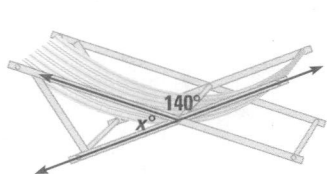

140°
x°

Student Help
CLASSZONE.COM

HOMEWORK HELP
Extra help with problem
solving in Exs. 35–37 is
at classzone.com

Using Algebra ∠*ABD* and ∠*DBC* are complementary angles. Find the value of the variable.

35.
6

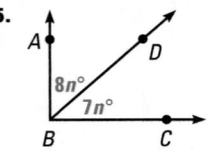

A
D
8*n*°
7*n*°
B
C

36.
5

A   B
5*x*°
13*x*°
D
C

37.
16

(3*k* + 10)° C
D
2*k*°
A   B

38. Complementary Angles ∠*ABD* and ∠*DBE* are complements, and ∠*CBE* and ∠*DBE* are complements. Can you show that ∠*ABD* ≅ ∠*CBE*? Explain.

Yes; ∠*ABD* and ∠*CBE* are both complementary to ∠*DBE*, so ∠*ABD* ≅ ∠*CBE* by the Congruent Complements Theorem.

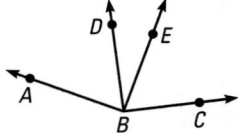

D   E
A
B   C

39. Technology Use geometry software to draw two intersecting lines. Measure three of the four angles formed. Drag the points and observe the angle measures. What theorem does this illustrate?

Congruent Supplements Theorem

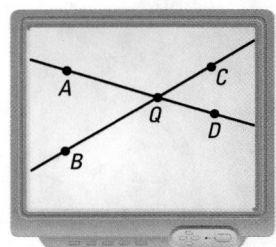

A   C
Q
D
B

Complements and Supplements Find the angle measure described.

40. ∠1 and ∠2 are both supplementary to ∠3, and *m*∠1 = 43°. Find the measure of ∠2.   43°

41. ∠4 and ∠6 are both complementary to ∠5, and *m*∠5 = 85°. Find the measure of ∠4.   5°

42. ∠*P* is supplementary to ∠*Q*, ∠*R* is supplementary to ∠*P*, and *m*∠*Q* = 60°. Find the measure of ∠*R*.   60°

43. Challenge ∠*C* and ∠*D* are supplementary angles. The measure of ∠*D* is eight times the measure of ∠*C*. Find *m*∠*C* and *m*∠*D*.
m∠*C* = 20°; *m*∠*D* = 160°

44. Multiple Choice What is the measure of a complement of a 27° angle? **B**

    (A) 53°   (B) 63°   (C) 117°   (D) 163°

45. Multiple Choice $\angle 1$ and $\angle 2$ are supplementary. Suppose that $m\angle 1 = 60°$ and $m\angle 2 = (2x + 20)°$. What is the value of x? **H**

    (F) 5   (G) 10   (H) 50   (J) 100

Mixed Review

Segment Addition Postulate **Find the length.** *(Lesson 1.5)*

46. Find *FH*. **12.7**   **47.** Find *KL*. **12**

   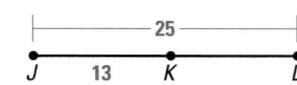

Midpoint Formula **Find the coordinates of the midpoint of $\overline{AB}$.** *(Lesson 2.1)*

48. $A(0, 0)$, $B(8, 2)$ **(4, 1)**   **49.** $A(-6, 0)$, $B(2, 4)$ **50.** $A(4, 1)$, $B(10, 3)$ **(7, 2)**
                                  (−2, 2)

51. $A(-2, 5)$, $B(-2, 7)$ **52.** $A(3, -8)$, $B(-1, 0)$ **53.** $A(-5, -9)$, $B(11, 5)$
    (−2, 6)   **(1, −4)**   **(3, −2)**

Algebra Skills

Evaluating Decimals **Evaluate.** *(Skills Review, p. 655)*

54. $2.58 + 8.04$ **10.62**   **55.** $5.17 - 1.96$ **3.21**   **56.** 1.4×3.1 **4.34**

57. 0.61×0.38 **0.2318**   **58.** $11.2 \div 1.4$ **8**   **59.** $2 \times 5.4 \times 3.9$ **42.12**

Quiz 1

1. In the diagram, *K* is the midpoint of $\overline{JL}$. Find *KL* and *JL*. *(Lesson 2.1)* **17; 34**

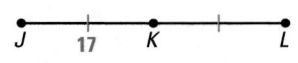

Find the coordinates of the midpoint of $\overline{AB}$. *(Lesson 2.1)*

2. $A(1, 3)$, $B(7, -1)$ **3.** $A(-4, -2)$, $B(6, 4)$ **4.** $A(-5, 3)$, $B(3, -3)$
   (4, 1)   **(1, 1)**   **(−1, 0)**

In Exercises 5–7, $\overrightarrow{KM}$ bisects $\angle JKL$. Find the angle measure. *(Lesson 2.2)*

5. Find $m\angle JKM$. **41°**   **6.** Find $m\angle JKL$. **22°**   **7.** Find $m\angle JKL$. **116°**

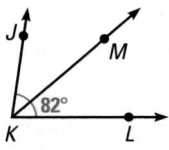      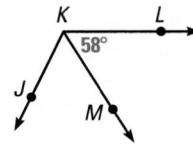

8. $\angle F$ is a supplement of $\angle G$, and $m\angle F = 101°$. Find $m\angle G$. *(Lesson 2.3)* **79°**

9. The measure of $\angle D$ is 83°. Find the measure of a complement and a supplement of $\angle D$. *(Lesson 2.3)* **7°; 97°**

2.3 *Complementary and Supplementary Angles*   **73**

Assessment Resources

The Mini-Quiz below is also available on blackline (*Chapter 2 Resource Book*, p. 39) and on transparency. For more assessment resources, see:

- Chapter 2 Resource Book
- Standardized Test Practice
- Test and Practice Generator

Mini-Quiz

Determine whether the angles are *complementary, supplementary,* or *neither.* Also tell whether the angles are *adjacent* or *nonadjacent.*

1.

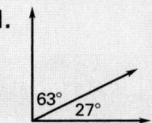

complementary; adjacent

2.

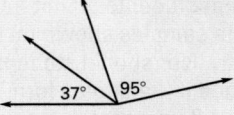

neither; nonadjacent

Find the measures of a complement and a supplement of the angle.

3. $m\angle R = 27°$ **63°; 153°**

4. $m\angle T = 11°$ **79°; 169°**

5. $\angle ABD$ and $\angle DBC$ are complementary angles. Find the value of *x*. **7**

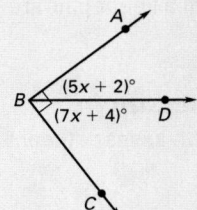

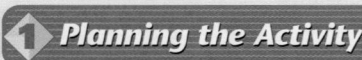

Goal
Students use a protractor to explore the measures of angles formed by two intersecting lines.

Materials
See the margin of the student page.

LINK TO LESSON
In Examples 1, 2, and 3 on pages 75 and 76, students identify pairs of angles they explored in this activity. In Example 4 on page 77, students use their discoveries in this activity to find angle measures.

Managing the Activity

Tips for Success
Point out that students do not have to model the samples shown on this page exactly. Nor should students be concerned if the angles formed by their lines do not match those of other students around them.

Closing the Activity

KEY DISCOVERY
Vertical angles are congruent and angles that form a linear pair are supplementary.

Activity Assessment
Use Exercise 3 to assess student understanding.

3. The sum of the measures of any two adjacent angles is 180°. The measures of any two nonadjacent angles are equal.

4. *Sample answer:* Three pairs of congruent angles are formed.

Activity 2.4 Angles and Intersecting Lines

Question

What is the relationship between the angles formed by two intersecting lines?

Materials
- straightedge
- protractor

Explore

❶ On a piece of paper, draw line ℓ using a straightedge. Label two points *A* and *B* on the line.

❷ Draw line *m* so that it intersects line ℓ. Label the point of intersection *E*. Label two points *C* and *D* on line *m* as shown below.

❸ Use a protractor to measure the four angles formed by the intersecting lines. Record the angle measures.

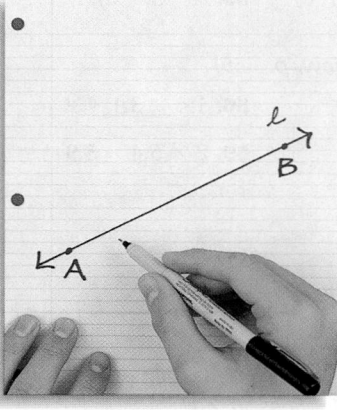 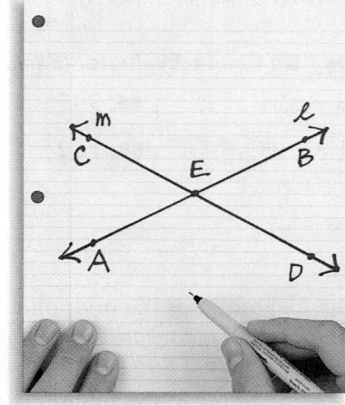 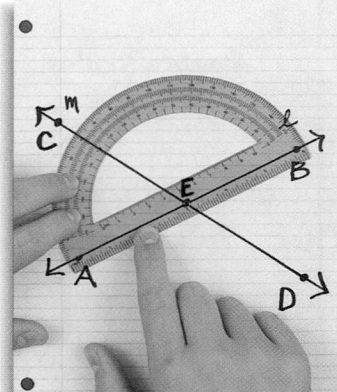

Think About It 3, 4. See margin

Student Help

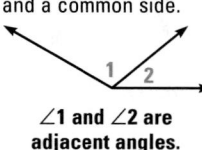
1. What do you notice about the *nonadjacent angles* you measured in Step 3? They are congruent.

2. Find the sum of the measures of any two *adjacent angles* in Step 3. What do you notice? The sum is 180°.

3. Repeat Steps 1–3 using two different lines. What do you notice about the measures of adjacent angles and nonadjacent angles?

4. **Extension** Draw a third line, *n*, that goes through point *E*. Use a protractor to measure the six angles formed by the intersecting lines. Record your results. What do you notice about the angle measures?

2.4 Vertical Angles

Goal
Find the measures of angles formed by intersecting lines.

Key Words
- vertical angles
- linear pair

Two angles are **vertical angles** if they are not adjacent and their sides are formed by two intersecting lines. The scissors show two sets of vertical angles.

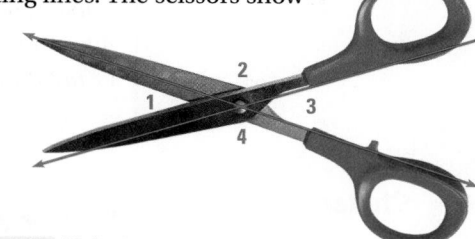

∠1 and ∠3 are vertical angles.

∠2 and ∠4 are vertical angles.

Two adjacent angles are a **linear pair** if their noncommon sides are on the same line.

∠5 and ∠6 are a linear pair.

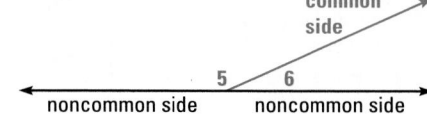

Visualize It!

You can use colored pencils to help you see pairs of vertical angles.

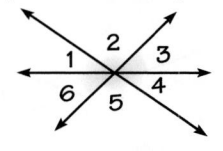

Vertical angles
∠1 and ∠4
∠2 and ∠5
∠3 and ∠6

EXAMPLE 1 Identify Vertical Angles and Linear Pairs

Determine whether the labeled angles are *vertical angles*, a *linear pair*, or *neither*.

a. 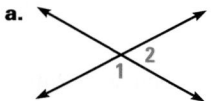 b. 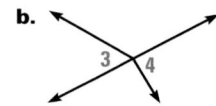 c.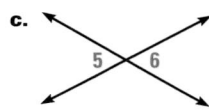

Solution

a. ∠1 and ∠2 are a linear pair because they are adjacent and their noncommon sides are on the same line.

b. ∠3 and ∠4 are neither vertical angles nor a linear pair.

c. ∠5 and ∠6 are vertical angles because they are not adjacent and their sides are formed by two intersecting lines.

POSTULATE 7

Linear Pair Postulate

Words If two angles form a linear pair, then they are supplementary.

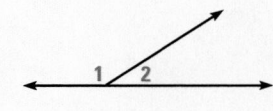

Symbols m∠1 + m∠2 = 180°

2.4 *Vertical Angles* **75**

① Plan

Pacing
Suggested Number of Days

Basic: 2 days
Average: 2 days
Advanced: 2 days
Block Schedule: 1 block

Teaching Resources

📄 *Blacklines*
(See page 50B.)

📑 *Transparencies*
• Warm-Up with Quiz
• Answers

💻 *Technology*
• Electronic Teacher Tools
• Test and Practice Generator
• Online Lesson Planner
• Internet Support

📼 *Video*
• Geometry in Motion

② Teach

Content and Teaching Strategies
For background information on geometric concepts and teaching strategies related to this lesson, see pages 50E and 50F in this Teacher's Edition.

Tips for New Teachers
Students may think that vertical angles are adjacent when viewing a diagram. It may appear to them that a pair of opposite rays of the vertical angles form a common side because the rays form a line. Be sure they understand the difference between adjacent and vertical angles. See the Tips for New Teachers on pp. 1–2 of the *Chapter 2 Resource Book* for additional notes about Lesson 2.4.

Extra Example 1
See next page.

Extra Example 1

Determine whether the labeled angles are *vertical angles*, a *linear pair*, or *neither*.

a.

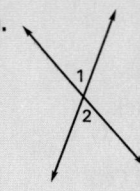

vertical angles

b.

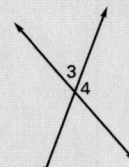

linear pair

c.

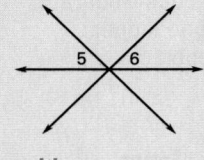

neither

Extra Example 2

Find the measure of ∠*CDF*. 73°

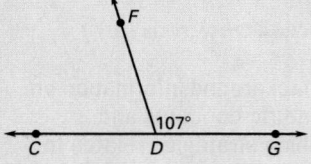

Extra Example 3

Find the measure of ∠*HJK*. 115°

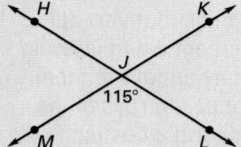

EXAMPLE **2** **Use the Linear Pair Postulate**

Find the measure of ∠*RSU*.

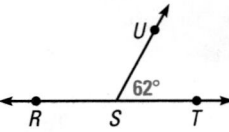

Solution

∠*RSU* and ∠*UST* are a linear pair. By the Linear Pair Postulate, they are supplementary. To find *m*∠*RSU*, subtract *m*∠*UST* from 180°.

$$m\angle RSU = 180° - m\angle UST = 180° - 62° = 118°$$

Student Help

> **VISUAL STRATEGY**
> Draw an example of this theorem with specific measures, as shown on p. 52.

THEOREM 2.3

Vertical Angles Theorem

Words Vertical angles are congruent.

Symbols ∠1 ≅ ∠3 and ∠2 ≅ ∠4.

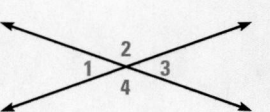

The following steps show why the Vertical Angles Theorem is true.

❶ ∠1 and ∠2 are a linear pair, so ∠1 and ∠2 are supplementary.

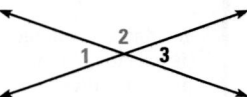

❷ ∠2 and ∠3 are a linear pair, so ∠2 and ∠3 are supplementary.

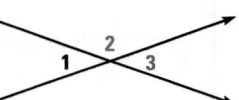

Student Help

> **LOOK BACK**
> To review the Congruent Supplements Theorem, see p. 69.

❸ ∠1 and ∠3 are supplementary to the same angle, so ∠1 is congruent to ∠3 by the Congruent Supplements Theorem.

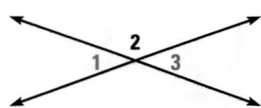

EXAMPLE **3** **Use the Vertical Angles Theorem**

Find the measure of ∠*CED*.

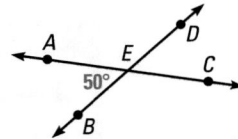

Solution

∠*AEB* and ∠*CED* are vertical angles. By the Vertical Angles Theorem, ∠*CED* ≅ ∠*AEB*, so *m*∠*CED* = *m*∠*AEB* = 50°.

STUDY TIP
When you know the measure of one vertical angle, an easy starting point is to fill in the measure of the other. ·············▶

EXAMPLE **4** **Find Angle Measures**

Find $m\angle 1$, $m\angle 2$, and $m\angle 3$.

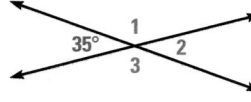

Solution

▶ $m\angle 2 = 35°$ Vertical Angles Theorem

$m\angle 1 = 180° - 35° = 145°$ Linear Pair Postulate

$m\angle 3 = m\angle 1 = 145°$ Vertical Angles Theorem

Checkpoint ✓ **Find Angle Measures**

Find $m\angle 1$, $m\angle 2$, and $m\angle 3$.

1. $m\angle 1 = 152°$; $m\angle 2 = 28°$; $m\angle 3 = 152°$

2. $m\angle 1 = 56°$; $m\angle 2 = 124°$; $m\angle 3 = 56°$

3. $m\angle 1 = 113°$; $m\angle 2 = 67°$; $m\angle 3 = 113°$

1.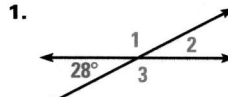

2.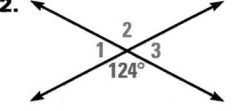

3.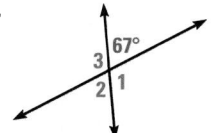

Using Algebra $_{xy}$

EXAMPLE **5** **Use Algebra with Vertical Angles**

Find the value of y.

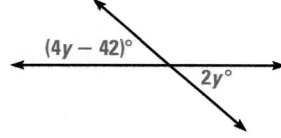

Solution

Because the two expressions are measures of vertical angles, you can write the following equation.

$(4y - 42)° = 2y°$ Vertical Angles Theorem

$4y - 42 - 4y = 2y - 4y$ Subtract $4y$ from each side.

$-42 = -2y$ Simplify.

$\dfrac{-42}{-2} = \dfrac{-2y}{-2}$ Divide each side by -2.

$21 = y$ Simplify.

Checkpoint ✓ **Use Algebra with Angle Measures**

Find the value of the variable.

4. 43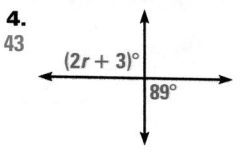

5. 16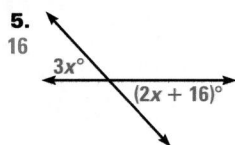

6. 5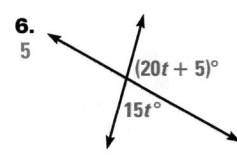

2.4 *Vertical Angles*   **77**

Extra Example 4

Find $m\angle 1$, $m\angle 2$, and $m\angle 3$.

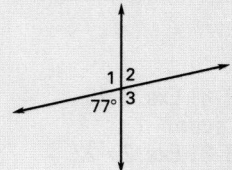

$m\angle 1 = 103°$, $m\angle 2 = 77°$, $m\angle 3 = 103°$

Extra Example 5

Find the value of x. 9

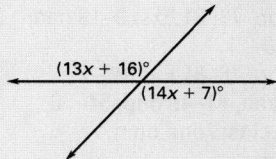

✓ Concept Check

What are some of the differences between vertical angles and linear pairs? *Sample answer:* Vertical angles have no common side, while linear pairs share a common side. Linear pairs are always supplementary, but vertical angles are only sometimes supplementary. Vertical angles are always congruent, while linear pairs may or may not be congruent.

🧩 Daily Puzzler

When two lines intersect, the measure of one of the angles they form is 20° less than three times the measure of one of the other angles formed. What are the measures of all four angles formed by the two lines? 50°, 50°, 130°, 130°

Assignment Guide

BASIC
Day 1: pp. 78–81 Exs. 9–19, 24–27, 60–70 even
Day 2: pp. 78–81 Exs. 20–22, 28–35, 51–53, 59, 65–71 odd

AVERAGE
Day 1: pp. 78–81 Exs. 9–19, 23–27, 58, 64–71
Day 2: pp. 78–81 Exs. 20–22, 28–40, 51–54, 59–63

ADVANCED
Day 1: pp. 78–81 Exs. 9–19 odd, 23–27, 58–71
Day 2: pp. 78–81 Exs. 20–22, 28–40 even, 41–49 odd, 54–56, 57*; EC: classzone.com

BLOCK SCHEDULE
pp. 78–81 Exs. 9–40, 51–54, 58–71

Extra Practice

• Student Edition, p. 678
• Chapter 2 Resource Book, pp. 40–41

Homework Check

To quickly check student understanding of key concepts, go over the following exercises:

Basic: 11, 15, 20, 30, 52
Average: 14, 19, 21, 34, 54
Advanced: 13, 19, 22, 36, 56

✗ Common Error

In Exercises 9–14, watch for students who confuse vertical angles with linear pairs. Point out that the non-shared sides of the angles in a linear pair form a line.

Guided Practice

Vocabulary Check   **Complete the statement.**

1. Two adjacent angles whose noncommon sides are on the same line are called __?__. a linear pair

2. Two angles are called __?__ if they are not adjacent and their sides are formed by two intersecting lines. vertical angles

Skill Check   **Find the measure of the numbered angle.**

3.
161°
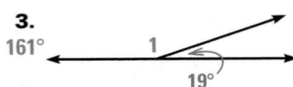
19°

4.
19°

19°

Find m∠1, m∠2, and m∠3.

5.
72° 1
3 2
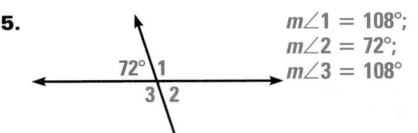

6. m∠1 = 108°;
m∠2 = 72°;
m∠3 = 108°

m∠1 = 90°;
m∠2 = 90°;
m∠3 = 90°
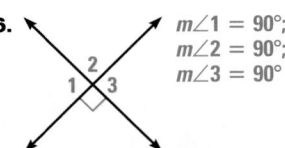

Find the value of x.

7. 4
40° (12x − 8)°
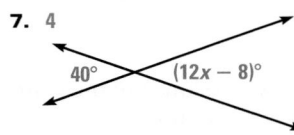

8. 10
6x° 12x°
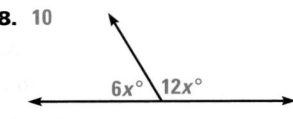

Practice and Applications

Extra Practice
See p. 678.

Vertical Angles and Linear Pairs Determine whether the angles are *vertical angles*, a *linear pair*, or *neither*.

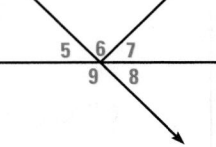

9. ∠5 and ∠6 neither
10. ∠5 and ∠9 a linear pair
11. ∠5 and ∠8 vertical angles
12. ∠6 and ∠9 neither
13. ∠8 and ∠9 a linear pair
14. ∠5 and ∠7 neither

Homework Help
Example 1: Exs. 9–14
Example 2: Exs. 15–19
Example 3: Exs. 20–22
Example 4: Exs. 28–37
Example 5: Exs. 51–56

Using the Linear Pair Postulate Find the measure of ∠1.

15.
143°
1 37°
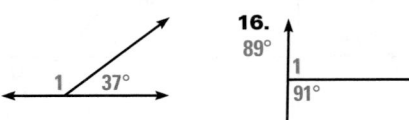

16.
89°
1
91°

17.
44°
1
136°
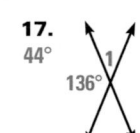

Linear Pairs Find the measure of the angle described.

18. ∠1 and ∠2 are a linear pair, and $m\angle 1 = 51°$. Find $m\angle 2$. 129°

19. ∠3 and ∠4 are a linear pair, and $m\angle 4 = 124°$. Find $m\angle 3$. 56°

Using the Vertical Angles Theorem Find the measure of ∠1.

20.
63°
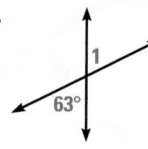
63°

21.
160°   160°
1

22.
76°
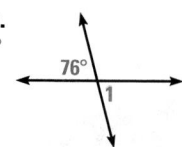
76°
1

Evaluating Statements Use the figure below to decide whether the statement is *true* or *false*.

23. If $m\angle 1 = 40°$, then $m\angle 2 = 140°$. false

24. If $m\angle 4 = 130°$, then $m\angle 2 = 50°$. true

25. ∠1 and ∠4 are a linear pair. true

26. $m\angle 1 + m\angle 4 = m\angle 3 + m\angle 2$ true

27. ∠1 and ∠4 are vertical angles. false

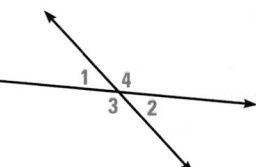

Finding Angle Measures Find $m\angle 1$, $m\angle 2$, and $m\angle 3$. 28–33. See margin.

28.
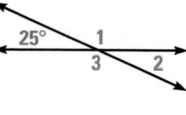
25° 1
3   2

29.
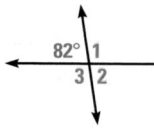
82° 1
3 2

30.
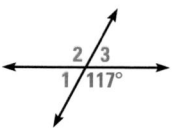
2 3
1 117°

31.
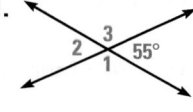
3
2   55°
1

32.
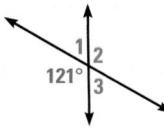
1 2
121° 3

33.
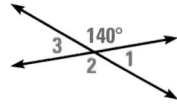
3   140°
2   1

28. $m\angle 1 = 155°$; $m\angle 2 = 25°$;
    $m\angle 3 = 155°$

29. $m\angle 1 = 98°$; $m\angle 2 = 82°$;
    $m\angle 3 = 98°$

30. $m\angle 1 = 63°$; $m\angle 2 = 117°$;
    $m\angle 3 = 63°$

31. $m\angle 1 = 125°$; $m\angle 2 = 55°$;
    $m\angle 3 = 125°$

32. $m\angle 1 = 59°$; $m\angle 2 = 121°$;
    $m\angle 3 = 59°$

33. $m\angle 1 = 40°$; $m\angle 2 = 140°$;
    $m\angle 3 = 40°$

34. $m\angle 1 = 118.1°$; $m\angle 2 = 61.9°$;
    $m\angle 3 = 118.1°$

35. $m\angle 1 = 90°$; $m\angle 2 = 90°$;
    $m\angle 3 = 90°$

36. $m\angle 1 = 53.1°$; $m\angle 2 = 126.9°$;
    $m\angle 3 = 53.1°$

Flags Each flag shown contains vertical angles. Find $m\angle 1$, $m\angle 2$, and $m\angle 3$. 34–36. See margin.

34.
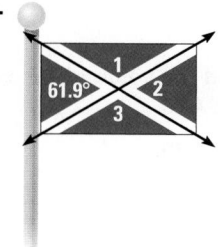
61.9° 1
      2
      3

Scotland

35.
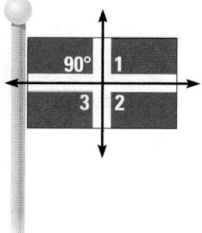
90° 1
3   2

Dominican Republic

36.

1   2
      3
126.9°

Jamaica

Teaching Tip
Exercises 41–50 involve a diagram showing many angles. Encourage students to trace out the angle given in an exercise by saying the angle name aloud while tracing along its rays with a finger.

Link to Careers

ERGONOMISTS study work conditions to improve the safety, efficiency, and comfort of workers. Drafting tables are angled so people can work at them without injuring their backs.

Career Links
CLASSZONE.COM

37. Drafting Table The legs of the drafting table form vertical angles. Find the measures of $\angle 1$, $\angle 2$, and $\angle 3$. $m\angle 1 = 95°$; $m\angle 2 = 85°$; $m\angle 3 = 95°$

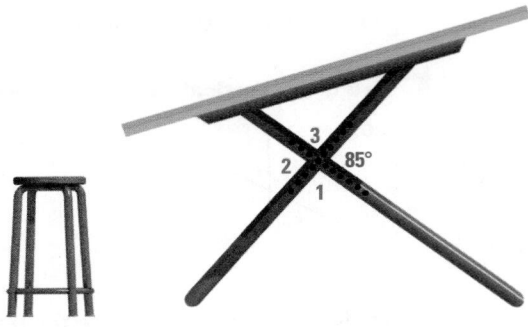

Finding Angle Measures Find $m\angle 1$, $m\angle 2$, $m\angle 3$, and $m\angle 4$. 38–40. See margin.

38.

39.

40.

Vertical Angles Use the diagram to complete the statement.

41. $\angle BGC \cong$ __?__ $\angle FGE$ **42.** $\angle AGB \cong$ __?__ $\angle DGE$

43. $\angle AGC \cong$ __?__ $\angle DGF$ **44.** $\angle CGE \cong$ __?__ $\angle BGF$

45. $m\angle AGF =$ __?__° 22 **46.** $m\angle DGE =$ __?__° 50

47. $m\angle CGE =$ __?__° 72 **48.** $m\angle BGC =$ __?__° 108

49. $m\angle DGF =$ __?__° 158 **50.** $m\angle AGD =$ __?__° 180

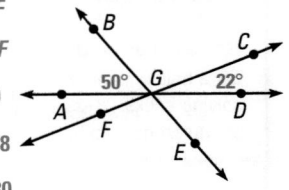

38. $m\angle 1 = 55°$; $m\angle 2 = 65°$;
    $m\angle 3 = 60°$; $m\angle 4 = 65°$

39. $m\angle 1 = 30°$; $m\angle 2 = 44°$;
    $m\angle 3 = 106°$; $m\angle 4 = 30°$

40. $m\angle 1 = 52°$; $m\angle 2 = 90°$;
    $m\angle 3 = 38°$; $m\angle 4 = 52°$

Using Algebra Find the value of the variable.

51. 79

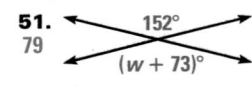

52. 20

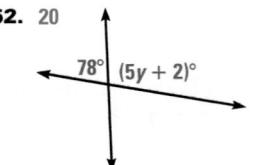

53. 58

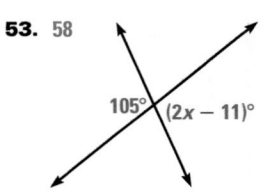

Using Algebra Find the value of the variable. Then use substitution to find $m\angle ABC$.

54. 5; 40°

55. 23; 157°

56. 4; 48°

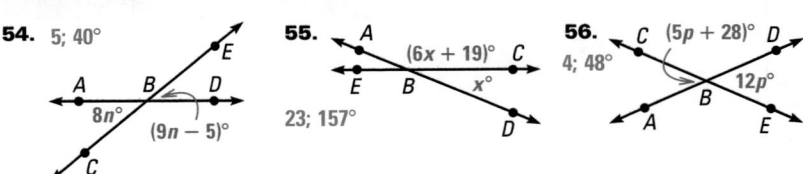

57. Challenge Find the values of x and y in the diagram below.
$x = 8; y = 7$

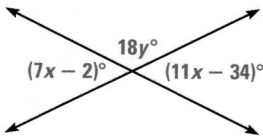

$18y°$
$(7x - 2)°$ $(11x - 34)°$

58. Visualize It! Sketch and label four angles so that $\angle 1$ and $\angle 2$ are acute vertical angles, $\angle 3$ is a right angle adjacent to $\angle 2$, and $\angle 1$ and $\angle 4$ form a linear pair. **See margin.**

Standardized Test Practice

59. Multi-Step Problem Use the diagram below.

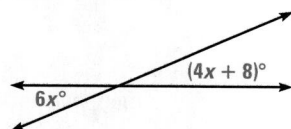

$(4x + 8)°$
$6x°$

a. Use the Vertical Angles Theorem to write an equation. $6x = 4x + 8$

b. Solve your equation to find the value of x. 4

c. Find the measures of the acute angles formed by the lines. $24°$

d. Find the measures of the obtuse angles formed by the lines. $156°$

Mixed Review

Describing Number Patterns Describe a pattern in the numbers. Write the next number you expect in the pattern. *(Lesson 1.1)*

60–63. See margin.

60. Each number is 7 more than the previous number; 32.
60. 4, 11, 18, 25, . . .
61. 3, 15, 75, 375, . . .

61. Each number is 5 times the previous number; 1875.
62. 32, 16, 8, 4, . . .
63. 404, 414, 424, 434, . . .

62. Each number is half the previous number; 2.

63. Each number is 10 more than the previous number; 444.

Congruent Segments Determine which segments in the coordinate plane are congruent. *(Lesson 1.5)*

64.
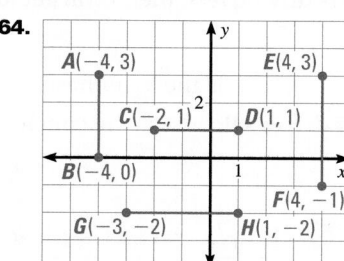
$A(-4, 3)$ $E(4, 3)$
$C(-2, 1)$ $D(1, 1)$
$B(-4, 0)$
$F(4, -1)$
$G(-3, -2)$ $H(1, -2)$

$\overline{AB} \cong \overline{CD}; \overline{EF} \cong \overline{GH}$

65.
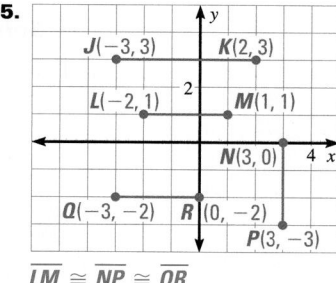
$J(-3, 3)$ $K(2, 3)$
$L(-2, 1)$ $M(1, 1)$
$N(3, 0)$
$Q(-3, -2)$ $R(0, -2)$
$P(3, -3)$

$\overline{LM} \cong \overline{NP} \cong \overline{QR}$

Algebra Skills

Simplifying Expressions Simplify the expression. *(Skills Review, p. 671)*

66. $-16x + 9x$ $\ -7x$
67. $7 + a - 2a$ $\ 7 - a$
68. $-8z^2 + 5z^2 - 4$ $\ -3z^2 - 4$

69. $6b^2 + 6b - b^2$ $\ 5b^2 + 6b$
70. $-4(t - 3) - 4t$ $\ -8t + 12$
71. $3w^2 - 1 - w^2 + 5$ $\ 2w^2 + 4$

2.4 *Vertical Angles* **81**

4 Assess

Assessment Resources
The Mini-Quiz below is also available on blackline (*Chapter 2 Resource Book*, p. 47) and on transparency. For more assessment resources, see:
• Chapter 2 Resource Book
• Standardized Test Practice
• Test and Practice Generator

Mini-Quiz
Use the diagram.

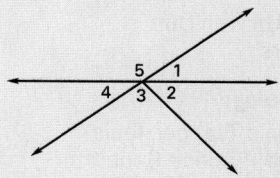

5 1
4 3 2

1. Name a linear pair. $\angle 1$ and $\angle 5$ or $\angle 4$ and $\angle 5$

2. Name a pair of vertical angles. $\angle 1$ and $\angle 4$

3. If $m\angle 1 = 35°$ and $m\angle 2 = 45°$, find $m\angle 3$, $m\angle 4$, and $m\angle 5$.
$m\angle 3 = 100°$, $m\angle 4 = 35°$, $m\angle 5 = 145°$

Use the diagram to find the value of each variable.

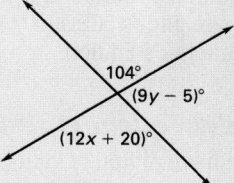

$104°$
$(9y - 5)°$
$(12x + 20)°$

4. x 7
5. y 9

58. *Sample answer:*

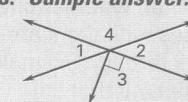

4
1 2
3

81

Plan

Pacing
Suggested Number of Days

Basic: 2 days
Average: 2 days
Advanced: 2 days
Block Schedule: 1 block

Teaching Resources

 **Blacklines**
(See page 50B.)

Transparencies
• Warm-Up with Quiz
• Answers

Technology
• Electronic Teacher Tools
• Test and Practice Generator
• Online Lesson Planner
• Internet Support

Teach

Content and Teaching Strategies
For background information on geometric concepts and teaching strategies related to this lesson, see pages 50E and 50F in this Teacher's Edition.

Tips for New Teachers
Students often have difficulty identifying hypotheses and conclusions when they begin thinking about the content of an if-then statement. Stress that identifying parts of if-then statements does not involve making a judgment on whether one agrees or disagrees with the statement. As you work through Example 2 on this page, point out that students might disagree with the statement but stress that this is irrelevant to rewriting the statement in if-then form. See the Tips for New Teachers on pp. 1–2 of the *Chapter 2 Resource Book* for additional notes about Lesson 2.5.

 **2.5 If-Then Statements and Deductive Reasoning**

Goal
Use if-then statements. Apply laws of logic.

Key Words
• if-then statement
• hypothesis
• conclusion
• deductive reasoning

An **if-then statement** has two parts. The "if" part contains the **hypothesis** . The "then" part contains the **conclusion** .

Student Help

> **VOCABULARY TIP**
> If-then statements are also called *conditional statements.*

The following sentence is an example of an if-then statement:

hypothesis

If the team wins the semi-final,
then the team will play in the championship.

conclusion

EXAMPLE 1 Identify the Hypothesis and Conclusion

Identify the hypothesis and the conclusion of the if-then statement.
    If I pass the driving test, then I will get my driver's license.

Solution

I pass the driving test is the hypothesis.
I will get my driver's license is the conclusion.

Student Help

> **STUDY TIP**
> Example 2, part (b), shows that the conclusion of a statement is not always the last part. ┈┈┈┈▶

EXAMPLE 2 Write If-Then Statements

Rewrite the statement as an if-then statement.
    a. Every game on my computer is fun to play.
    b. I will buy the CD if it costs less than $15.

Solution

    a. If a game is on my computer, then it is fun to play.
    b. If the CD costs less than $15, then I will buy it.

Deductive reasoning uses facts, definitions, accepted properties, and the laws of logic to make a logical argument. This form of reasoning differs from *inductive reasoning*, in which previous examples and patterns are used to form a conjecture.

LAWS OF LOGIC

Law of Detachment

If the hypothesis of a true if-then statement is true, then the conclusion is also true.

Law of Syllogism

If statement p, then statement q.
If statement q, then statement r. ⟩→ If these statements are true,

If statement p, then statement r. ⟵ then this statement is true.

EXAMPLE 3 Use the Law of Detachment

What can you conclude from the following true statements?

   If you wash the cotton T-shirt in hot water, then it will shrink.

   You wash the cotton T-shirt in hot water.

Solution

The hypothesis (you wash the cotton T-shirt in hot water) of a true if-then statement is true. By the Law of Detachment, the conclusion must be true.

ANSWER ▶ You can conclude that the cotton T-shirt shrinks.

Checkpoint ✓ *If-Then Statements and the Law of Detachment*

Rewrite the statement as an if-then statement. Then underline the hypothesis and circle the conclusion.

1. If <u>a teacher works at East High School</u>, then (he or she has taught for at least 5 years).

1. All teachers at East High School have taught for at least 5 years.

2. If <u>the measure of an angle is 170°</u>, then (the angle is obtuse).

2. An angle is obtuse if its measure is 170°.

What can you conclude from the given true statements?

3. If x has a value of 7, then $2x - 3$ has a value of 11. The value of x is 7. $2x - 3$ has a value of 11.

4. If you study at least two hours for the test, then you will pass the test. You study three hours for the test. You will pass the test.

Extra Example 1

Identify the hypothesis and the conclusion of the if-then statement.

If I sleep through my alarm, then I will miss the school bus.
I sleep through my alarm is the hypothesis. *I will miss the school bus* is the conclusion.

Extra Example 2

Rewrite the statement as an if-then statement.
a. Every duck on the pond is hungry. If a duck is on the pond, then it is hungry.
b. I will go running if it does not snow. If it does not snow, then I will go running.

Geometric Reasoning

Show students that any number of related if-then statements can be chained together by repeatedly applying the Law of Syllogism.

Extra Example 3

What can you conclude from the following true statements?

If you are late for your interview, you will probably not be offered the job.

You are late for your interview. You can conclude that you will probably not be offered the job.

EXAMPLE 4 Use the Law of Detachment

Which argument is correct? Explain your reasoning.

Argument 1: If two angles are vertical angles, then they are congruent. ∠1 and ∠2 are congruent. So, ∠1 and ∠2 are vertical angles.

Argument 2: If two angles are vertical angles, then they are congruent. ∠1 and ∠2 are vertical angles. So, ∠1 and ∠2 are congruent.

Solution

Argument 2 is correct. The hypothesis (two angles are vertical angles) is true, which implies that the conclusion (they are congruent) is true.

You can use the following counterexample to show that Argument 1 is false. In the diagram at the right, ∠1 ≅ ∠2, but they are not vertical angles.

Checkpoint ✓ Use the Law of Detachment

5. Which argument is correct? Explain your reasoning.

Argument 1: If two angles form a linear pair, then they are supplementary. ∠3 and ∠4 form a linear pair. So, ∠3 and ∠4 are supplementary.

Argument 2: If two angles form a linear pair, then they are supplementary. ∠3 and ∠4 are supplementary. So, ∠3 and ∠4 form a linear pair.

EXAMPLE 5 Use the Law of Syllogism

Write the statement that follows from the pair of true statements.

If the daily high temperature is 32°F or less, then the water in the pipe is frozen.

If the water in the pipe is frozen, then the pipe will break.

Solution

Use the Law of Syllogism.

If the daily high temperature is 32°F or less, then the pipe will break.

Checkpoint ✓ Use the Law of Syllogism

6. Write the statement that follows from the pair of true statements.

If the ball is thrown at the window, it will hit the window.

If the ball hits the window, then the window will break.

2.5 Exercises

Guided Practice

Vocabulary Check

Complete the statement.

1. The "if" part of an if-then statement contains the __?__ . hypothesis

2. The "then" part of an if-then statement contains the __?__ . conclusion

Skill Check

3. If two angles are adjacent angles, then they share a common side.

4. If you miss the bus, then you will be late to school.

5. The midpoint of $\overline{AB}$ is at (2, 0).

6. If the perimeter of a square is 20 feet, then the area of the square is 25 square feet.

In Exercises 3 and 4, rewrite the statement as an if-then statement.

3. Adjacent angles share a common side.

4. You will be late to school if you miss the bus.

5. What can you conclude from the following true statements?

If the endpoints of a segment have coordinates (−1, −2) and (5, 2), then the midpoint of the segment is at (2, 0).

The endpoints of $\overline{AB}$ are A(−1, −2) and B(5, 2).

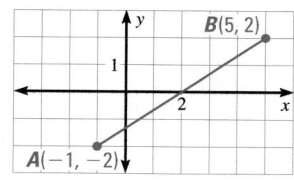

6. Use the Law of Syllogism to write the if-then statement that follows from the pair of true statements.

If the perimeter of a square is 20 feet, then the length of a side of the square is 5 feet.

If the length of a side of a square is 5 feet, then the area is 25 square feet.

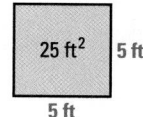

25 ft² — 5 ft — 5 ft

Practice and Applications

Extra Practice

See p. 678.

7. hypothesis: the car is running; conclusion: the key is in the ignition.

8. hypothesis: the measure of an angle is 60°; conclusion: the angle is acute.

Parts of an If-Then Statement Identify the hypothesis and the conclusion of the if-then statement.

7. If the car is running, then the key is in the ignition.

8. If the measure of an angle is 60°, then the angle is acute.

Writing If-Then Statements Rewrite the statement as an if-then statement. Then underline the hypothesis and circle the conclusion.
9–12. See margin.

9. A number divisible by 6 is also divisible by 3 and 2.

10. Eagles are fish-eating birds.

11. A shape has four sides if it is a square.

12. Two angles are supplementary if they form a linear pair.

---

3 Apply

Assignment Guide

BASIC
Day 1: pp. 85–87 Exs. 7–12, 16, 20, 24–42 even
Day 2: pp. 85–87 Exs. 13–19 odd, 23–43 odd

AVERAGE
Day 1: pp. 85–87 Exs. 7–12, 20, 21, 24–42 even
Day 2: pp. 85–87 Exs. 13–19, 22, 23–43 odd

ADVANCED
Day 1: pp. 85–87 Exs. 8–12 even, 20, 21, 24–34
Day 2: pp. 85–87 Exs. 14–18 even, 22, 23, 35–43;
EC: TE p. 50D*, classzone.com

BLOCK SCHEDULE
pp. 85–87 Exs. 7–43

Extra Practice
• Student Edition, p. 678
• Chapter 2 Resource Book, pp. 48–49

Homework Check
To quickly check student understanding of key concepts, go over the following exercises:

Basic: 7, 11, 15, 16, 17
Average: 8, 12, 14, 16, 17
Advanced: 8, 12, 14, 16, 18

X Common Error
In Exercises 11 and 12, students may think the first part of the sentence is always the hypothesis. Point out a simple example such as *We will stay home if it is raining* to illustrate that sometimes the hypothesis is written after the conclusion.

9. If a number is divisible by 6, then it is also divisible by 3 and 2.

10. If a bird is an eagle, then it eats fish.

11, 12. See Additional Answers beginning on page AA1.

Example 1: Exs. 7–12,
20, 21, 24–27
Example 2: Exs. 9–12,
20, 21, 24–27
Example 3: Exs. 13–15
Example 4: Ex. 16
Example 5: Exs. 17–19

16. Argument 1 is correct. In Argument 1, the hypothesis is true, which implies that the conclusion is true. The following counterexample shows that Argument 2 is false: The measure of ∠1 could be 60° and the measure of ∠2 could be 30°.

17. If the sun is shining, then we will have a picnic.

18. If the stereo is on, then the neighbors will complain.

19. If Chris goes to the movies, then Gabriela will go to the movies.

20. If <u>you want a great selection of used cars</u>, then you should come and see Bargain Bob's Used Cars.

Link to
Careers

ADVERTISING
COPYWRITERS work with graphic designers to develop the text for advertisements.

Career Links
CLASSZONE.COM

Using the Law of Detachment In Exercises 13–15, what can you conclude from the given true statements?

13. If two planes intersect, then their intersection is a line. Two planes are intersecting. **The intersection of the planes is a line.**

14. If x has a value of 4, then $2x$ has a value of 8. The value of x is 4.
    The value of $2x$ is 8.

15. If Central High School wins the championship, then the school will celebrate. Central High School wins the championship.
    The school will celebrate.

16. **You be the Judge** Which argument is correct? Explain your reasoning.

   Argument 1: If two angles measure 40° and 50°, then the angles are complementary. The measure of ∠1 is 40° and the measure of ∠2 is 50°. So, ∠1 and ∠2 are complementary.

   Argument 2: If two angles measure 40° and 50°, then the angles are complementary. ∠1 and ∠2 are complementary. So, $m\angle 1 = 40°$ and $m\angle 2 = 50°$.

Using the Law of Syllogism Write the if-then statement that follows from the pair of true statements.

17. If the sun is shining, then it is a beautiful day.
    If it is a beautiful day, then we will have a picnic.

18. If the stereo is on, then the volume is loud.
    If the volume is loud, then the neighbors will complain.

19. If Todd goes to the movies, then Gabriela will go to the movies.
    If Chris goes to the movies, then Todd will go to the movies.

Advertising Use the following advertising slogan: "Want a great selection of used cars? Come and see Bargain Bob's Used Cars!"

20. Rewrite the slogan as an if-then statement. Then underline the hypothesis and circle the conclusion. **See margin.**

21. Find an advertising slogan that can be written as an if-then statement. Then repeat Exercise 20 using the slogan. **Answers will vary.**

THE KOLOB ARCH is
the world's widest natural
arch. The arch, located in
Springdale, Utah, spans
310 feet.

24. If <u>you tell the truth</u>, then (you
don't have to remember anything.)

25. If <u>you want to do things</u>, then
(you first have to expect things of
yourself.)

Logical Reasoning Select the word or phrase that makes the concluding statement true.

22. The Oak Terrace apartment building does not allow dogs. Serena lives at Oak Terrace. So, Serena (*must, may, must not*) keep a dog.
    must not

23. The Kolob Arch is the world's widest natural arch. The world's widest natural arch is in Zion National Park. So, the Kolob Arch (*is, may be, is not*) in Zion National Park.   is

Quotes of Wisdom Rewrite the statement as an if-then statement. Then underline the hypothesis and circle the conclusion.   24–27. See margin.

24.
> If you tell the truth, you don't have to remember anything.
> — Mark Twain

25.
> You have to expect things of yourself before you can do them.
> — Michael Jordan

26.
> If one is lucky, a solitary fantasy can totally transform one million realities.
> — Maya Angelou

27.
> Whoever is happy will make others happy too.
> — Anne Frank

Standardized Test Practice

28. **Multiple Choice** What is the *conclusion* of the following if-then statement?   B

    If the storm passes, then our plane will take off.

    Ⓐ The storm passes.   Ⓑ Our plane will take off.

    Ⓒ Then our plane will take off.   Ⓓ None of these

Mixed Review

Classifying Angles Classify the angle as *acute, right, obtuse*, or *straight*. (Lesson 1.6)

29. $m\angle A = 20°$ acute   30. $m\angle B = 154°$ obtuse   31. $m\angle C = 90°$ right

32. $m\angle D = 7°$ acute   33. $m\angle E = 180°$ straight   34. $m\angle F = 89°$ acute

Midpoint Formula Find the coordinates of the midpoint of $\overline{AB}$. (Lesson 2.1)

35. $A(0, 0)$, $B(-2, 6)$ 36. $A(2, 3)$, $B(8, 5)$ (5, 4)   37. $A(0, -8)$, $B(6, 6)$
    $(-1, 3)$ $(3, -1)$

Algebra Skills

Solving Equations Solve the equation. (Skills Review, p. 673)

38. $4y - 2 = 4$ $\frac{3}{2}$ 39. $6 = -2t + 1$ $-\frac{5}{2}$ 40. $4x = -5 + x$ $-\frac{5}{3}$

41. $w + 4 = 3w$ 2   42. $12x + 33 = 9x$ -11 43. $14b - 17 = -3b$ 1

2.5   *If-Then Statements and Deductive Reasoning*   **87**

Mini-Quiz

1. Identify the hypothesis and the conclusion of the if-then statement.
   If the temperature begins falling, the rain will change to snow.
   hypothesis: the temperature begins falling; conclusion: the rain will change to snow

2. Rewrite the statement as an if-then statement.
   The measure of a right angle is 90°. If an angle is a right angle, then its measure is 90°.

3. What can you conclude from the given true statements.
   If two lines are parallel, then they never meet. Two lines are parallel. The two lines never meet.

4. Write the if-then statement that follows from the pair of true statements.
   If the rain becomes sleet, the roads will ice over. If the roads ice over, then school will be delayed. If the rain becomes sleet, then school will be delayed.

26. If <u>one is lucky</u>, then (a solitary) (fantasy can totally transform) (one million realities.)

27. If <u>a person is happy</u>, then (he or she will make others) (happy too.)

Pacing
Suggested Number of Days

Basic: 2 days
Average: 2 days
Advanced: 2 days
Block Schedule: 1 block

Teaching Resources

Blacklines
(See page 50B.)

Transparencies
• Warm-Up with Quiz
• Answers

Technology
• Electronic Teacher Tools
• Test and Practice Generator
• Online Lesson Planner
• Internet Support

② Teach

Content and Teaching Strategies
For background information on geometric concepts and teaching strategies related to this lesson, see pages 50E and 50F in this Teacher's Edition.

Tips for New Teachers
To refresh students' memory of the algebraic properties of equality and to help students with limited English proficiency, have students give several numeric examples of each property. This can be done prior to Example 3 on page 90. See the Tips for New Teachers on pp. 1–2 of the *Chapter 2 Resource Book* for additional notes about Lesson 2.6.

2.6 Properties of Equality and Congruence

Goal
Use properties of equality and congruence.

Key Words
• Reflexive Property
• Symmetric Property
• Transitive Property

Reflexive Property

Jean is the same height as Jean.

Symmetric Property

If Jean is the same height as Pedro, **then** Pedro is the same height as Jean.

Transitive Property

If Jean is the same height as Pedro **and** Pedro is the same height as Chris, **then** Jean is the same height as Chris.

The photos above illustrate the *Reflexive*, *Symmetric*, and *Transitive Properties* of Equality. You can use these properties in geometry with statements about equality and congruence.

Student Help

LOOK BACK
To review the difference between equality and congruence, see p. 30.

PROPERTIES OF EQUALITY AND CONGRUENCE

Reflexive Property

Equality $AB = AB$
$m\angle A = m\angle A$

Congruence $\overline{AB} \cong \overline{AB}$
$\angle A \cong \angle A$

Symmetric Property

Equality
If $AB = CD$, then $CD = AB$.
If $m\angle A = m\angle B$, then $m\angle B = m\angle A$.

Congruence
If $\overline{AB} \cong \overline{CD}$, then $\overline{CD} \cong \overline{AB}$.
If $\angle A \cong \angle B$, then $\angle B \cong \angle A$.

Transitive Property

Equality
If $AB = CD$ and $CD = EF$, then $AB = EF$.
If $m\angle A = m\angle B$ and $m\angle B = m\angle C$, then $m\angle A = m\angle C$.

Congruence
If $\overline{AB} \cong \overline{CD}$ and $\overline{CD} \cong \overline{EF}$, then $\overline{AB} \cong \overline{EF}$.
If $\angle A \cong \angle B$ and $\angle B \cong \angle C$, then $\angle A \cong \angle C$.

EXAMPLE 1 Name Properties of Equality and Congruence

Name the property that the statement illustrates.

a. If $\overline{GH} \cong \overline{JK}$, then $\overline{JK} \cong \overline{GH}$.

b. $DE = DE$

c. If $\angle P \cong \angle Q$ and $\angle Q \cong \angle R$, then $\angle P \cong \angle R$.

Solution

a. Symmetric Property of Congruence

b. Reflexive Property of Equality

c. Transitive Property of Congruence

Checkpoint ✓ *Name Properties of Equality and Congruence*

Name the property that the statement illustrates.

1. If $DF = FG$ and $FG = GH$, then $DF = GH$. **Transitive Property of Equality**

2. $\angle P \cong \angle P$ **Reflexive Property of Congruence**

3. If $m\angle S = m\angle T$, then $m\angle T = m\angle S$. **Symmetric Property of Equality**

Logical Reasoning In geometry, you are often asked to explain why statements are true. Reasons can include definitions, theorems, postulates, or properties.

EXAMPLE 2 Use Properties of Equality

In the diagram, N is the midpoint of $\overline{MP}$, and P is the midpoint of $\overline{NQ}$. Show that $MN = PQ$.

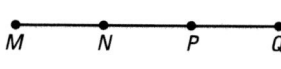

Solution

$MN = NP$	Definition of midpoint
$NP = PQ$	Definition of midpoint
$MN = PQ$	Transitive Property of Equality

Checkpoint ✓ *Use Properties of Equality and Congruence*

4. $\angle 1$ and $\angle 2$ are vertical angles, and $\angle 2 \cong \angle 3$. Show that $\angle 1 \cong \angle 3$.
Vertical Angles; Transitive

$\angle 1 \cong \angle 2$	__?__ Theorem
$\angle 2 \cong \angle 3$	Given
$\angle 1 \cong \angle 3$	__?__ Property of Congruence

2.6 *Properties of Equality and Congruence* **89**

Extra Example 1

Name the property that the statement illustrates.

a. If $\overline{ST} \cong \overline{UV}$ and $\overline{UV} \cong \overline{YZ}$, then $\overline{ST} \cong \overline{YZ}$. **Transitive Property of Congruence**

b. If $FG = HJ$, then $HJ = FG$. **Symmetric Property of Equality**

c. $\angle C \cong \angle C$ **Reflexive Property of Congruence**

Extra Example 2

In the diagram, $\overline{AB} \cong \overline{BC}$ and $\overline{BC} \cong \overline{CD}$. Show that $AB = CD$.

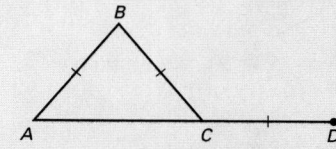

$AB = BC$ (Definition of congruent segments)
$BC = CD$ (Definition of congruent segments)
$AB = CD$ (Transitive Property of Equality)

Study Skills

Note-Taking Encourage students to copy the table of Properties of Equality and Congruence and illustrate each property with a sketch.

✗ *Common Error*

Watch for students who confuse congruence with equality. Review the table on page 88 and help them understand when to use the term *equality* and when to use the term *congruence*.

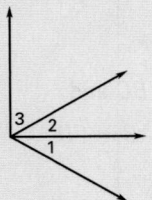
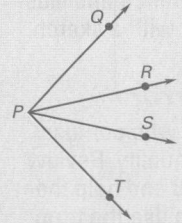
PROPERTIES OF EQUALITY

Addition Property	Example
Adding the same number to each side of an equation produces an equivalent equation.	$x - 3 = 7$ $x - 3 + 3 = 7 + 3$
Subtraction Property	**Example**
Subtracting the same number from each side of an equation produces an equivalent equation.	$y + 5 = 11$ $y + 5 - 5 = 11 - 5$
Multiplication Property	**Example**
Multiplying each side of an equation by the same nonzero number produces an equivalent equation.	$\frac{1}{4}z = 6$ $\frac{1}{4}z \cdot 4 = 6 \cdot 4$
Division Property	**Example**
Dividing each side of an equation by the same nonzero number produces an equivalent equation.	$8x = 16$ $\frac{8x}{8} = \frac{16}{8}$
Substitution Property	**Example**
Substituting a number for a variable in an equation produces an equivalent equation.	$x = 7$ $2x + 4 = 2(7) + 4$

EXAMPLE 3 Justify the Congruent Supplements Theorem

∠1 and ∠2 are both supplementary to ∠3. Show that ∠1 ≅ ∠2.

 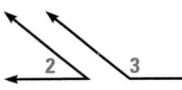

Solution

$m\angle 1 + m\angle 3 = 180°$	Definition of supplementary angles
$m\angle 2 + m\angle 3 = 180°$	Definition of supplementary angles
$m\angle 1 + m\angle 3 = m\angle 2 + m\angle 3$	Substitution Property of Equality
$m\angle 1 = m\angle 2$	Subtraction Property of Equality
∠1 ≅ ∠2	Definition of congruent angles

Checkpoint ✓ Use Properties of Equality and Congruence

5. In the diagram, M is the midpoint of $\overline{AB}$. Show that $AB = 2 \cdot AM$.

midpoint; Segment Addition; Substitution

$MB = AM$	Definition of __?__
$AB = AM + MB$	__?__ Postulate
$AB = AM + AM$	__?__ Property of Equality
$AB = 2 \cdot AM$	Distributive property

Guided Practice

Vocabulary Check | Match the statement with the property it illustrates.

1. $m\angle DEF = m\angle DEF$ **B**

2. If $\overline{PQ} \cong \overline{ST}$, then $\overline{ST} \cong \overline{PQ}$. **E**

3. $\overline{XY} \cong \overline{XY}$ **D**

4. If $\angle J \cong \angle K$ and $\angle K \cong \angle L$, then $\angle J \cong \angle L$. **F**

5. If $PQ = QR$ and $QR = RS$, then $PQ = RS$. **C**

6. If $m\angle X = m\angle Y$, then $m\angle Y = m\angle X$. **A**

A. Symmetric Property of Equality

B. Reflexive Property of Equality

C. Transitive Property of Equality

D. Reflexive Property of Congruence

E. Symmetric Property of Congruence

F. Transitive Property of Congruence

Skill Check | Name the property that the statement illustrates.

7. $\angle ABC \cong \angle ABC$ Reflexive Property of Congruence

8. If $m\angle B = m\angle D$ and $m\angle D = m\angle F$, then $m\angle B = m\angle F$.
Transitive Property of Equality

9. If $\overline{GH} \cong \overline{JK}$, then $\overline{JK} \cong \overline{GH}$.
Symmetric Property of Congruence

Practice and Applications

Extra Practice
See p. 678

Completing Statements Use the property to complete the statement.

10. *Reflexive Property of Equality:* $JK = \underline{\ ?\ }$ JK

11. *Symmetric Property of Equality:* If $m\angle P = m\angle Q$, then $\underline{\ ?\ } = \underline{\ ?\ }$.
$m\angle Q = m\angle P$

12. *Transitive Property of Equality:* If $AB = BC$ and $BC = CD$, then $\underline{\ ?\ } = \underline{\ ?\ }$. AB = CD

13. *Reflexive Property of Congruence:* $\underline{\ ?\ } \cong \angle GHJ$ $\angle GHJ$

14. *Symmetric Property of Congruence:* If $\underline{\ ?\ } \cong \underline{\ ?\ }$, then $\angle XYZ \cong \angle ABC$. $\angle ABC \cong \angle XYZ$

15. *Transitive Property of Congruence:* If $\overline{GH} \cong \overline{IJ}$ and $\underline{\ ?\ } \cong \underline{\ ?\ }$, then $\overline{GH} \cong \overline{PQ}$. $\overline{IJ} \cong \overline{PQ}$

Naming Properties Name the property that the statement illustrates.

16. If $AB = CD$, then $AB + EF = CD + EF$. Addition Property of Equality

17. If $m\angle C = 90°$, then $2(m\angle C) + 15° = 2(90°) + 15°$.
Substitution Property of Equality

18. If $XY = YZ$, then $3 \cdot XY = 3 \cdot YZ$. Multiplication Property of Equality

Homework Help

Example 1: Exs. 10–18
Example 2: Exs. 19–24
Example 3: Exs. 19–24

Assignment Guide

BASIC
Day 1: pp. 91–94 Exs. 10–18, 28, 30–33, 36–43
Day 2: pp. 91–94 Exs. 19–21, 29, 34, 35, Quiz 2

AVERAGE
Day 1: pp. 91–94 Exs. 10–18, 28, 30–43
Day 2: pp. 91–94 Exs. 19–23, 25, 26, 29, Quiz 2

ADVANCED
Day 1: pp. 91–94 Exs. 13–18, 28, 30–43
Day 2: pp. 91–94 Exs. 19–26, 27*, 29, Quiz 2;
EC: classzone.com

BLOCK SCHEDULE
pp. 91–94 Exs. 10–23, 25, 26, 28–43, Quiz 2

Extra Practice
• Student Edition, p. 678
• Chapter 2 Resource Book, pp. 59–60

Homework Check
To quickly check student understanding of key concepts, go over the following exercises:

Basic: 11, 12, 13, 17, 21
Average: 11, 12, 13, 17, 22
Advanced: 14, 15, 16, 18, 22

Geometric Reasoning

In Exercises 19–23, students begin to create simple geometric proofs although the formal introduction to proof is reserved until Chapter 5. While reviewing these exercises, you may wish to point out how each exercise begins with some given information and how each step follows logically from the ones before it until you reach your desired conclusion.

21. $PQ = RS$ (Given);
$PQ + QR = RS + QR$ (Addition Property of Equality);
$PQ + QR = PR$ (Segment Addition Postulate);
$RS + QR = QS$ (Segment Addition Postulate);
$PR = QS$ (Substitution Property of Equality)

---

19. Using Properties In the diagram, $m\angle 1 + m\angle 2 = 132°$, and $m\angle 2 = 105°$. Complete the argument to show that $m\angle 1 = 27°$.

Substitution; Subtraction

$m\angle 1 + m\angle 2 = 132°$	Given
$m\angle 2 = 105°$	Given
$m\angle 1 + 105° = 132°$	_?_ Property of Equality
$m\angle 1 = 27°$	_?_ Property of Equality

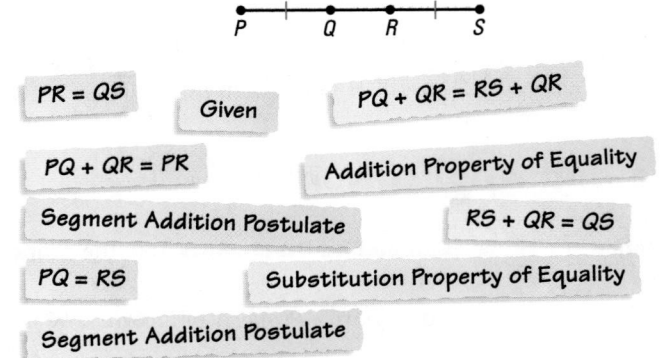

20. Using Properties of Congruence In the diagram, $\overline{AB} \cong \overline{FG}$, and $\overleftrightarrow{BF}$ bisects $\overline{AC}$ and $\overline{DG}$. Complete the argument to show that $\overline{BC} \cong \overline{DF}$. bisector; Given; bisector; Transitive

$\overline{BC} \cong \overline{AB}$	Definition of _?_
$\overline{AB} \cong \overline{FG}$	_?_
$\overline{FG} \cong \overline{DF}$	Definition of _?_
$\overline{BC} \cong \overline{DF}$	_?_ Property of Congruence

21. Unscramble the Steps In the diagram below, $PQ = RS$. Copy the diagram and arrange the statements and reasons in order to make a logical argument to show that $PR = QS$. See margin.

$PR = QS$ Given $PQ + QR = RS + QR$

$PQ + QR = PR$ Addition Property of Equality

Segment Addition Postulate $RS + QR = QS$

$PQ = RS$ Substitution Property of Equality

Segment Addition Postulate

22. Angle Addition Postulate; Angle Addition Postulate; Addition Property of Equality; Substitution Property of Equality

22. Using Properties of Equality In the diagram at the right, $m\angle WPY = m\angle XPZ$. Complete the argument to show that $m\angle WPX = m\angle YPZ$.

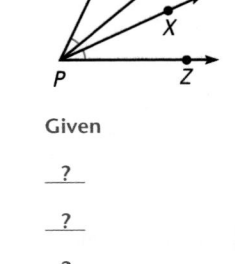

$m\angle WPY = m\angle XPZ$	Given
$m\angle WPX = m\angle WPY + m\angle YPX$	_?_
$m\angle YPZ = m\angle YPX + m\angle XPZ$	_?_
$m\angle WPY + m\angle YPX = m\angle YPX + m\angle XPZ$	_?_
$m\angle WPX = m\angle YPZ$	_?_

23. Congruent Complements Theorem Show that the Congruent Complements Theorem is true. Use Example 3 on page 90 as a model. Provide a reason for each step. *See margin.*

In the diagram, $\angle 1$ is complementary to $\angle 2$, and $\angle 3$ is complementary to $\angle 2$. Show that $\angle 1 \cong \angle 3$.

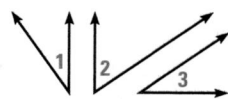

24. Error Analysis In the diagram, $\overline{SR} \cong \overline{CB}$ and $\overline{AC} \cong \overline{QR}$. Explain what is wrong with the student's argument.

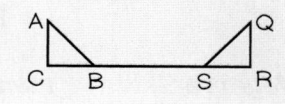

Because $\overline{SR} \cong \overline{CB}$ and $\overline{AC} \cong \overline{QR}$, then $\overline{CB} \cong \overline{AC}$ by the Transitive Property of Congruence.

Using Algebra Find the value of the variable using the given information. Provide a reason for each step. *25, 26. See margin.*

25. $AB = BC$, $BC = CD$

26. $QR = RS$, $ST = RS$

27. Challenge Fold two corners of a piece of paper so their edges match as shown at the right.

What do you notice about the angle formed by the fold lines?

Show that the angle measure is always the same. Provide a reason for each step.

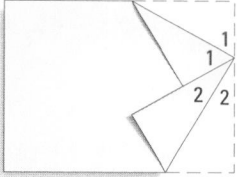

Standardized Test Practice

28. Multiple Choice Which statement illustrates the Symmetric Property of Congruence? **C**

(A) If $\overline{AD} \cong \overline{BC}$, then $\overline{DA} \cong \overline{CB}$.

(B) If $\overline{WX} \cong \overline{XY}$ and $\overline{XY} \cong \overline{YZ}$, then $\overline{WX} \cong \overline{YZ}$.

(C) If $\overline{AB} \cong \overline{GH}$, then $\overline{GH} \cong \overline{AB}$.

(D) $\overline{AB} \cong \overline{BA}$

29. Multiple Choice In the figure below, $\overline{QT} \cong \overline{TS}$ and $\overline{RS} \cong \overline{TS}$. What is the value of x? **F**

(F) 4   **(G)** 12

(H) 16   **(J)** 32

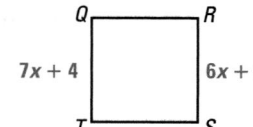

Student Help

LOOK BACK
To review the Congruent Complements Theorem, see p. 69.

24. Using the Transitive Property of Congruence, you cannot conclude anything from the given statements. If you were given $\overline{CB} \cong \overline{SR}$ and $\overline{SR} \cong \overline{AC}$, then the conclusion would be correct.

27. Because the edge of the piece of paper forms a straight angle, $m\angle 1 + m\angle 1 + m\angle 2 + m\angle 2 = 180°$. This can be simplified to $2(m\angle 1) + 2(m\angle 2) = 180°$. Factoring using the Distributive property: $2(m\angle 1 + m\angle 2) = 180°$. By the Division Property of Equality, $m\angle 1 + m\angle 2 = 90°$. Therefore, the angle is a right angle.

---

23. $m\angle 1 + m\angle 2 = 90°$ (Definition of complementary angles); $m\angle 3 + m\angle 2 = 90°$ (Definition of complementary angles); $m\angle 1 + m\angle 2 = m\angle 3 + m\angle 2$ (Substitution Property of Equality); $m\angle 1 = m\angle 3$ (Subtraction Property of Equality); $\angle 1 \cong \angle 3$ (Definition of congruent angles)

25. $AB = BC$ (Given); $BC = CD$ (Given); $AB = CD$ (Transitive Property of Equality); $AB = 3t + 1$ (Given); $CD = 7$ (Given); $3t + 1 = 7$ (Substitution Property of Equality); $3t = 6$ (Subtraction Property of Equality); $t = 2$ (Division Property of Equality)

26. $QR = RS$ (Given); $ST = RS$ (Given); $RS = ST$ (Symmetric Property of Equality); $QR = ST$ (Transitive Property of Equality); $QR = 23$ (Given); $ST = 5n - 2$ (Given); $23 = 5n - 2$ (Substitution Property of Equality); $25 = 5n$ (Addition Property of Equality); $5 = n$ (Division Property of Equality)

2.6 *Properties of Equality and Congruence*   93

The diagram for problem 29 shows a square $QRST$ with vertices Q (top left), R (top right), S (bottom right), T (bottom left). The left side is labeled $7x + 4$ and the right side is labeled $6x + 8$.

Assessment Resources

The Mini-Quiz below is also available on blackline (*Chapter 3 Resource Book*, p. 9) and on transparency. For more assessment resources, see:
• Chapter 2 Resource Book
• Standardized Test Practice
• Test and Practice Generator

Mini-Quiz

Use the property to complete the statement.

1. Transitive Property of Congruence: If $\angle F \cong \angle H$ and $\angle H \cong \angle N$, then $\underline{\ ?\ } \cong \underline{\ ?\ }$. $\angle F$, $\angle N$

2. Symmetric Property of Equality: If $JL = MN$, then $\underline{\ ?\ } \cong \underline{\ ?\ }$. MN, JL

3. Reflexive Property of Congruence: $\angle H \cong \underline{\ ?\ }$ $\angle H$

Name the property that the statement illustrates.

4. If $m\angle 3 = m\angle 4$, then $m\angle 3 - m\angle 5 = m\angle 4 - m\angle 5$.
   Subtraction Property of Equality

5. If $CD = HK$, then $\dfrac{CD}{5} = \dfrac{HK}{5}$.
   Division Property of Equality

34. *Sample answer:*

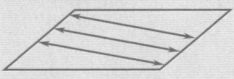

35. *Sample answer:*

36–43. **Check graphs.**

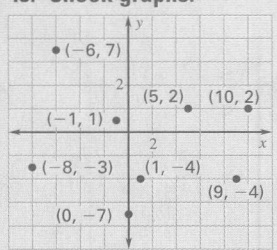

Mixed Review

Naming Collinear Points Use the diagram to name a point that is collinear with the given points. *(Lesson 1.3)*

30. G and E *C*   31. F and B *D*

32. A and D *E*   33. B and D *F*

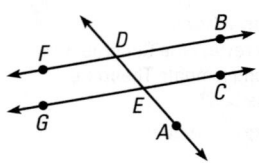

Sketching Intersections Sketch the figure described. *(Lesson 1.4)*
34, 35. See margin.

34. Three lines that do not intersect but lie in the same plane.

35. Two lines that intersect at one point, and another line that intersects both of those lines at different points.

Algebra Skills

Plotting Points Plot the point in a coordinate plane.
(Skills Review, p. 664)
36–43. See margin.

36. $(5, 2)$ 37. $(0, -7)$ 38. $(1, -4)$ 39. $(-8, -3)$

40. $(-6, 7)$ 41. $(10, 2)$ 42. $(-1, 1)$ 43. $(9, -4)$

Quiz 2

1. $m\angle 1 = 126°$; $m\angle 2 = 54°$; $m\angle 3 = 126°$

2. $m\angle 4 = 40°$; $m\angle 5 = 140°$; $m\angle 6 = 40°$

3. $m\angle 7 = 49°$; $m\angle 8 = 90°$; $m\angle 9 = 49°$; $m\angle 10 = 41°$

4. If a figure is a square, then it has four sides.

5. If $x = 5$, then the value of x^2 is 25.

6. If we charter a boat, then we will be gone all day.

7. bisector; bisector; Transitive

Find the measures of the numbered angles. *(Lesson 2.4)*

1.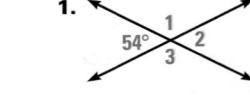

2.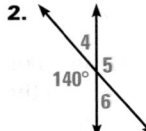

3.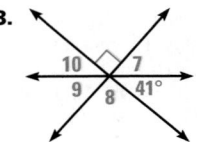

In Exercises 4 and 5, rewrite the statement as an if-then statement. *(Lesson 2.5)*

4. A square is a four-sided figure.   5. The value of x^2 is 25 if $x = 5$.

6. Use the Law of Syllogism to write the statement that follows from the pair of true statements. *(Lesson 2.5)*

   If we charter a boat, then we will go deep sea fishing.

   If we go deep sea fishing, then we will be gone all day.

7. In the diagram, $\overrightarrow{KM}$ bisects $\angle JKN$, and $\overrightarrow{KN}$ bisects $\angle MKL$. Complete the argument to show that $m\angle JKM = m\angle NKL$. *(Lesson 2.6)*

   $m\angle JKM = m\angle MKN$ Definition of $\underline{\ ?\ }$

   $m\angle MKN = m\angle NKL$ Definition of $\underline{\ ?\ }$

   $m\angle JKM = m\angle NKL$ $\underline{\ ?\ }$ Property of Equality

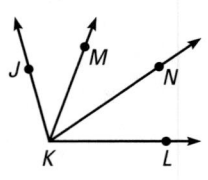

VOCABULARY

- **midpoint,** *p. 53*
- **segment bisector,** *p. 53*
- **bisect,** *p. 53*
- **angle bisector,** *p. 61*
- **complementary angles,** *p. 67*
- **complement of an angle,** *p. 67*
- **supplementary angles,** *p. 67*
- **supplement of an angle,** *p. 67*
- **adjacent angles,** *p. 68*
- **theorem,** *p. 69*
- **vertical angles,** *p. 75*
- **linear pair,** *p. 75*
- **if-then statement,** *p. 82*
- **hypothesis,** *p. 82*
- **conclusion,** *p. 82*
- **deductive reasoning,** *p. 83*

VOCABULARY REVIEW

Fill in the blank.

1. Two angles are ⎯?⎯ if the sum of their measures is 180°. supplementary

2. A(n) ⎯?⎯ is a ray that divides an angle into two congruent angles. angle bisector

3. A(n) ⎯?⎯ is a segment, ray, line, or plane that intersects a segment at its midpoint. segment bisector

4. ⎯?⎯ angles are nonadjacent angles whose sides are formed by two intersecting lines. Vertical

5. Two angles are ⎯?⎯ if they have a common vertex and side, but no common interior points. adjacent

6. Two angles are ⎯?⎯ if the sum of their measures is 90°. complementary

---

2.1 SEGMENT BISECTORS

Examples on pp. 53–55

> **EXAMPLE** *M* **is the midpoint of** $\overline{AB}$**. Find** *MB* **and** *AB*.
>
>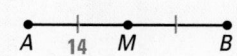
>
> *M* is the midpoint of $\overline{AB}$, so $AM = MB$. Therefore, $MB = AM = 14$.
>
> The length of $\overline{AB}$ is twice the length of $\overline{AM}$.
>
> $AB = 2 \cdot AM = 2 \cdot 14 = 28$
>
> **ANSWER** ▶ $MB = 14$ and $AB = 28$.

7. *R* is the midpoint of $\overline{PQ}$. Find *PR* and *PQ*.
5.5; 11

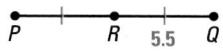

Find the coordinates of the midpoint of $\overline{FG}$.

8. $F(4, 1)$, $G(10, 9)$ $(7, 5)$

9. $F(4, -4)$, $G(0, 7)$ $\left(2, \frac{3}{2}\right)$

10. $F(-3, 1)$, $G(-6, -5)$ $\left(-\frac{9}{2}, -2\right)$

Chapter Summary and Review **95**

15. $m\angle 1 = 135°$; $m\angle 2 = 45°$; $m\angle 3 = 135°$
16. $m\angle 1 = 64°$; $m\angle 2 = 116°$; $m\angle 3 = 64°$
17. $m\angle 1 = 99°$; $m\angle 2 = 81°$; $m\angle 3 = 99°$
18. $m\angle 1 = 32°$; $m\angle 2 = 148°$; $m\angle 3 = 32°$

2.2 ANGLE BISECTORS

Examples on pp. 61–63

EXAMPLE In the diagram, $\overrightarrow{XY}$ bisects $\angle WXZ$. Find $m\angle WXY$ and $m\angle YXZ$.

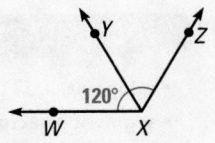

$$m\angle WXY = \frac{1}{2}(m\angle WXZ) = \frac{1}{2}(120°) = 60°$$

You know that $\angle WXY \cong \angle YXZ$, so $m\angle YXZ = 60°$ also.

$\overrightarrow{BD}$ bisects $\angle ABC$. Find the angle measure.

11. Find $m\angle ABC$. 50°   **12.** Find $m\angle ABD$. 65°   **13.** Find $m\angle DBC$. 45°

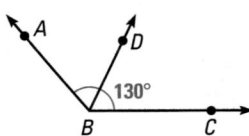   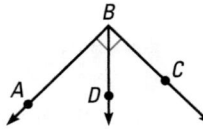   

2.3 COMPLEMENTARY AND SUPPLEMENTARY ANGLES

Examples on pp. 67–69

EXAMPLE Find the measure of a complement and a supplement of a 12° angle.

Measure of complement $= 90° - 12° = 78°$

Measure of supplement $= 180° - 12° = 168°$

14. Find the measure of a complement and a supplement of a 79° angle. 11°; 101°

2.4 VERTICAL ANGLES

Examples on pp. 75–77

EXAMPLE Find $m\angle 1$, $m\angle 2$, and $m\angle 3$.

$m\angle 2 = 100°$

$m\angle 1 = 180° - m\angle 2 = 180° - 100° = 80°$

$m\angle 3 = m\angle 1 = 80°$

Find $m\angle 1$, $m\angle 2$, and $m\angle 3$. 15–18. See margin.

15.    **16.**    **17.** 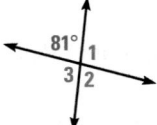   **18.**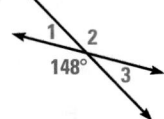

2.5 IF-THEN STATEMENTS AND DEDUCTIVE REASONING

Examples on pp. 82–84

EXAMPLE Rewrite the statement as an if-then statement. Then identify the hypothesis and the conclusion.

Our team won every game we played on our home field.

hypothesis

If our team played a game on our home field, **then** we won the game.

conclusion

In Exercises 19 and 20, write the statement as an if-then statement. Then underline the hypothesis and circle the conclusion. 19–21. See margin.

19. Every computer in the store is on sale.

20. I'll go to the mall if you come with me.

21. Use the Law of Syllogism to write the if-then statement that follows from the pair of true statements.

If Mike goes to the concert, then Jon will go to the concert.

If Jon goes to the concert, then Jeannine will go to the concert.

2.6 PROPERTIES OF EQUALITY AND CONGRUENCE

Examples on pp. 88–90

EXAMPLES Name the property that the statement illustrates.

Statement	Property
a. $\angle T \cong \angle T$	**a.** Reflexive Property of Congruence
b. If $AC = XY$ and $XY = TU$, then $AC = TU$.	**b.** Transitive Property of Equality
c. If $m\angle C = m\angle F$, then $m\angle F = m\angle C$.	**c.** Symmetric Property of Equality

Match the statement with the property it illustrates.

22. If $m\angle 2 = m\angle 3$, then $m\angle 3 = m\angle 2$. **E**   **A.** Addition Property of Equality

23. If $m\angle S = 45°$, then $m\angle S + 35° = 80°$. **A**   **B.** Transitive Property of Equality

24. If $AE = EG$ and $EG = JK$, then $AE = JK$. **B**   **C.** Multiplication Property of Equality

25. If $m\angle K = 9°$, then $3(m\angle K) = 27°$. **C**   **D.** Substitution Property of Equality

26. If $AB = 12$, then $2 \cdot AB + 3 = 2(12) + 3$. **D**   **E.** Symmetric Property of Equality

Chapter Summary and Review **97**

19. If <u>a computer is in the store</u>, then the computer is on sale.

20. If <u>you come with me</u>, then I'll go to the mall.

21. If Mike goes to the concert, then Jeannine will go to the concert.

12. If an angle has measure 60°, then the angle is an acute angle.

13. If it snows tomorrow, then Meg will be happy.

14. Reflexive Property of Equality

15. Symmetric Property of Congruence

16. Transitive Property of Equality

17. If I am in Boston, then I am in New England.

18. Angle Addition Postulate; Substitution Property of Equality

---

1. R is the midpoint of $\overline{PQ}$. Find RQ and PQ.

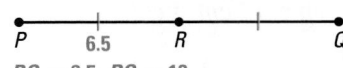

$RQ = 6.5; PQ = 13$

2. Line ℓ bisects $\overline{FG}$. Find the value of r. 15

In Exercises 3 and 4, find the coordinates of the midpoint of $\overline{AB}$.

3. $A(0, 9)$, $B(4, -3)$ (2, 3)

4. $A(-3, 5)$, $B(-7, -2)$ $\left(-5, \frac{3}{2}\right)$

5. $\overrightarrow{BD}$ bisects $\angle ABC$. Find $m\angle ABD$ and $m\angle DBC$.

70.5°; 70.5°

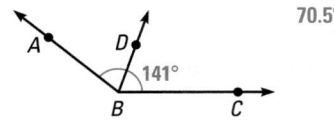

6. $\overrightarrow{KM}$ bisects $\angle JKL$. Find the value of x. 8

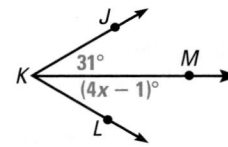

7. $\angle D$ is a supplement of $\angle E$, and $m\angle D = 29°$. Find $m\angle E$. 151°

8. $\angle F$ is a complement of $\angle G$, and $m\angle G = 76°$. Find $m\angle F$. 14°

9. $\angle 4$ and $\angle 5$ are a linear pair, and $m\angle 4 = 61°$. Find $m\angle 5$. 119°

10. Find $m\angle 1$, $m\angle 2$, and $m\angle 3$.
$m\angle 1 = 23°$; $m\angle 2 = 157°$; $m\angle 3 = 23°$

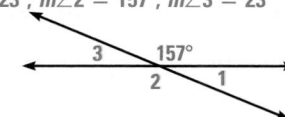

---

11. Use the diagram to find the angle measures.

a. $m\angle AHB$ 59°

b. $m\angle FHE$ 59°

c. $m\angle AHF$ 121°

d. $m\angle CHD$ 45°

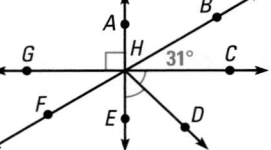

Write the statement as an if-then statement.

12. A 60° angle is an acute angle. 12, 13. See margin.

13. Meg will be happy if it snows tomorrow.

In Exercises 14–16, name the property that the statement illustrates.

14. $m\angle 2 = m\angle 2$ 14–18. See margin.

15. If $\angle 5 \cong \angle 6$, then $\angle 6 \cong \angle 5$.

16. If $AB = CD$ and $CD = FG$, then $AB = FG$.

17. Write the if-then statement that follows from the pair of true statements.

If I am in Boston, then I am in Massachusetts.

If I am in Massachusetts, then I am in New England.

18. In the diagram below, $m\angle RPQ = m\angle RPS$. Complete the argument to show that $m\angle SPQ = 2(m\angle RPQ)$.

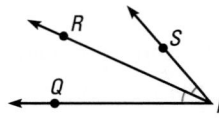

$m\angle RPQ = m\angle RPS$	Given
$m\angle SPQ = m\angle RPQ + m\angle RPS$	?
$m\angle SPQ = m\angle RPQ + m\angle RPQ$	?
$m\angle SPQ = 2(m\angle RPQ)$	Distributive property

Additional Resources

Chapter 2 Resource Book
• SAT/ACT Chapter Test, p. 71

Test Tip If you are unsure of an answer, try to eliminate some of the choices so you can make an educated guess.

8. a. *Sample answer:* ∠*CHD* and ∠*DHE*; ∠*CHB* and ∠*BHA*

   b. *Sample answer:* ∠*AHD* and ∠*DHE*; ∠*CHD* and ∠*DHG*

   c. *Sample answer:* ∠*AHD* and ∠*DHE*; ∠*CHE* and ∠*EHG*

   d. *Sample answer:* ∠*BHA* and ∠*EHF*; ∠*CHE* and ∠*AHG*

   e. 40; 90; 140

1. S is the midpoint of $\overline{RT}$, and $RS = 23$. What is RT? **D**

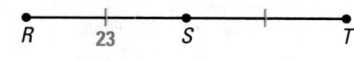

    Ⓐ 10.5   Ⓑ 11.5

    Ⓒ 23   Ⓓ 46

2. What are the coordinates of the midpoint of the segment joining $(4, -5)$ and $(-2, -1)$? **H**

    Ⓕ $(2, -6)$ Ⓖ $(1, -2)$

    Ⓗ $(1, -3)$ Ⓙ $(-8, 5)$

3. $\overrightarrow{SU}$ bisects $\angle RST$. What is the value of x? **B**

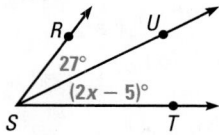

    Ⓐ 11   Ⓑ 16

    Ⓒ 49   Ⓓ 54

4. If $\angle 5$ and $\angle 6$ are both supplementary to $\angle 7$, and $m\angle 5 = 17°$, what is $m\angle 6$? **F**

    Ⓕ 17°   Ⓖ 73°

    Ⓗ 163°   Ⓙ 180°

5. Use the diagram to determine which of the following statements is *false*. **C**

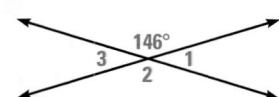

    Ⓐ $m\angle 1 = 34°$ Ⓑ $m\angle 2 = 146°$

    Ⓒ $\angle 1 \cong \angle 2$ Ⓓ $\angle 1 \cong \angle 3$

6. What is the *conclusion* of the following statement? **J**

If today is Saturday, then tomorrow is Sunday.

    Ⓕ Today is Saturday.

    Ⓖ Today is Sunday.

    Ⓗ Tomorrow is Saturday.

    Ⓙ Tomorrow is Sunday.

7. Which property is illustrated by the following statement? **B**

If $\angle 1 \cong \angle 2$, then $\angle 2 \cong \angle 1$.

    Ⓐ Symmetric Property of Equality

    Ⓑ Symmetric Property of Congruence

    Ⓒ Transitive Property of Equality

    Ⓓ Transitive Property of Congruence

8. Multi-Step Problem Use the diagram below. See margin.

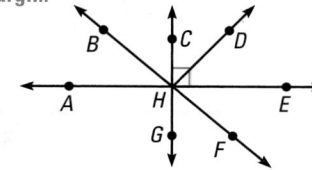

  a. Name two pairs of complementary angles.

  b. Name two pairs of supplementary angles.

  c. Name two sets of linear pairs.

  d. Name two pairs of vertical angles.

  e. Complete the statements.

    If $m\angle BHC = 50°$, then $m\angle AHB = \underline{\ ?\ }°$.

    If $m\angle CHE = 90°$, then $m\angle AHG = \underline{\ ?\ }°$.

    If $m\angle BHE = 140°$, then $m\angle AHF = \underline{\ ?\ }°$.

Teaching Tip
The Brain Games activity provides a motivating way to review selected content in the chapter. For a more comprehensive review, see the Chapter Summary and Review on pp. 95–97.

Alex: Guinea Pig named Prince

Carole: Fish named Buddy

Mark: Dog named Merlin

Sean: Rabbit named Fang

Tamara: Cat named Stripe

BraiN GaMes

Logic Puzzle

Object of the Game To use deductive reasoning to correctly determine the type and name of each student's pet.

How to Play Keep track of the clues by recording the pieces of information in a grid like the one shown. Put an X in the boxes that you can eliminate, and put a check in the boxes that you are sure of. Clue 1 of the puzzle is done for you.

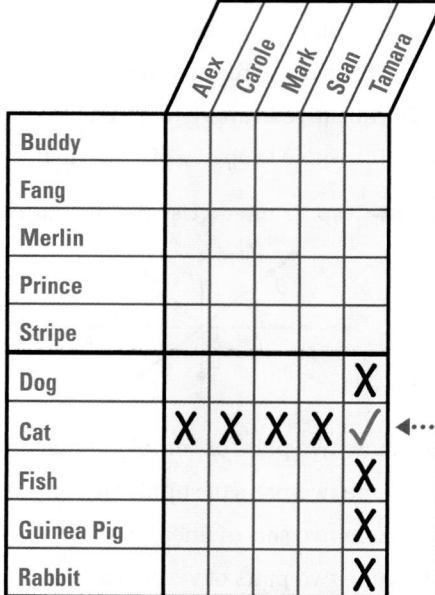

	Alex	Carole	Mark	Sean	Tamara
Buddy					
Fang					
Merlin					
Prince					
Stripe					
Dog					X
Cat	X	X	X	X	✓
Fish					X
Guinea Pig					X
Rabbit					X

Tamara has a cat, so no one else has a cat, and Tamara does not have any other pets.

Logic Puzzle

There are five students: Alex, Carole, Mark, Sean, and Tamara.

Each student has one pet: a dog, a cat, a fish, a guinea pig, or a rabbit.

The pets' names are Buddy, Fang, Merlin, Prince, and Stripe.

CLUES

1. Tamara has a cat.

2. The cat's name is Stripe.

3. Carole's pet is named Buddy.

4. Sean does not have a dog.

5. The guinea pig is not owned by Carole, Mark, or Sean.

6. Neither Mark nor Sean has a fish.

7. If Alex has a guinea pig, then its name is Prince.

8. Fang is not the name of the dog.

Like terms are terms in an expression that have the same variable raised to the same power. Constant terms, such as 2 and -5, are also considered like terms.

EXAMPLE 1 **Combine Like Terms**

Simplify the expression $6x - 2y + 10x + 7y$.

Solution

$$6x - 2y + 10x + 7y = 6x + 10x - 2y + 7y \qquad \text{Group like terms.}$$
$$= (6 + 10)x + (-2 + 7)y \qquad \text{Distributive prop.}$$
$$= 16x + 5y \qquad \text{Simplify within parentheses.}$$

Try These

Simplify the expression.

1. $7p + 15p$ $22p$
2. $r - 3r$ $-2r$
3. $3q + 6q - 2q$ $7q$
4. $2t + 4t + 7$ $6t + 7$
5. $8x + 3 - 5x$ $3x + 3$
6. $-9c - 3b + 15c$ $-3b + 6c$
7. $11y + 16z + 24z + 8y$ $19y + 40z$
8. $18f - 13g - 14f - g$ $4f - 14g$

EXAMPLE 2 **Solve Equations With Variables on Both Sides**

Solve the equation $4x - 7 = 3x + 1$.

Solution

$$4x - 7 = 3x + 1 \qquad \text{Write original equation.}$$
$$4x - 3x - 7 = 3x - 3x + 1 \qquad \text{Subtract } 3x \text{ from each side.}$$
$$x - 7 = 1 \qquad \text{Simplify.}$$
$$x - 7 + 7 = 1 + 7 \qquad \text{Add 7 to each side.}$$
$$x = 8 \qquad \text{Simplify.}$$

Student Help

STUDY TIP
Remember to check your solution by substituting the value of the variable into the original equation.

Try These

Solve the equation.

9. $4w = 3w + 10$ 10
10. $6y = 11 - 5y$ 1
11. $10p - 12 = 8p$ 6
12. $q + 14 = -6q$ -2
13. $2t + 5 = t - 9$ -14
14. $5x - 1 = 3x + 7$ 4
15. $6a - 13 = 4a + 9$ 11
16. $3r - 2 = 7r + 18$ -5
17. $12d - 7d = 2d + 6$ 2

Algebra Review **101**

Extra Example 1
Simplify the expression
$9a - 6b - 9a + 14b$. $8b$

Checkpoint Exercise 1
Simplify the expression
$10f + 7k - 11f - k$. $-f + 6k$

Extra Example 2
Solve the equation $2p + 7 = 6p - 5$.
$p = 3$

Checkpoint Exercise 2
Solve the equation
$3k + 20 = k + 46$. $k = 13$

Mathematical Goals

- Sketch rectangles, segments, and prisms.
- Represent three-dimensional objects on paper.
- Sketch points, lines, planes, and their intersections.

Managing the Project

Tips for Success

While completing Exercise 3, have small groups of students each draw the three textbooks or boxes at the same time so each student is drawing from a different angle. Students can compare their drawings to see how the difference in point of view affects how the lines of perspective are drawn.

To help students present their results, consider having photocopies of some photographs showing perspective available for use so students do not have to look for them.

Guiding Students' Work

In Step 1, point out that the horizon line does not need to be the exact same distance from the rectangle shown in the photograph. However, stress that if the horizon line is too close to the rectangle, then the perspective will be more difficult to visualize.

Chapters  1-2 Project

Drawing in Perspective

Objective

Draw objects in one-point perspective.

The photograph at the right shows one-point perspective. Lines appear to meet at a *vanishing point* at eye level, along a *horizon line*.

Materials

- paper
- pencil
- straightedge
- magazines

Drawing

❶ Draw a rectangle. Draw a horizon line first. Then pick a vanishing point.

❷ Connect corners of the rectangle to the vanishing point.

❸ Draw segments to complete a box. Erase the guide lines you used.

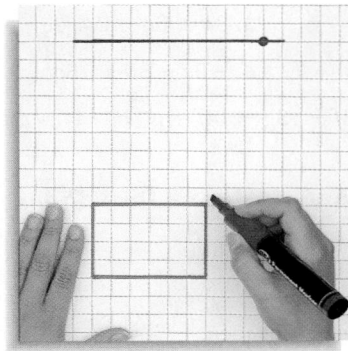 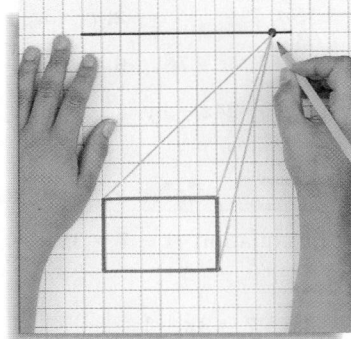 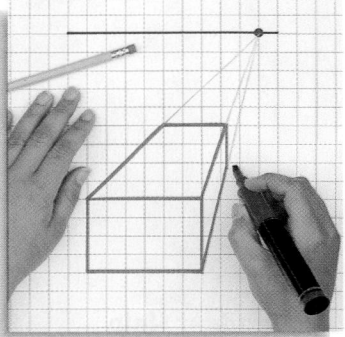

Investigation

1. Use the photograph at the top of this page. What happens to the trees as they approach the vanishing point? Do objects appear smaller or larger as they get farther away?

2. Draw three different boxes of various sizes, using the same vanishing point.

3. Place three textbooks or cereal boxes in a row on a table. Move your chair away from the table and to the left. Draw the objects. Include the vanishing point on your drawing.

Present Your Results

Make a folder of photographs and drawings that show perspective.

▶ Find photographs in newspapers or magazines that show perspective.

▶ Draw horizon lines and vanishing points on the photographs. Also draw some lines that meet at the vanishing point.

▶ Include the drawings you made in the Investigation.

▶ Write a few sentences about how the photographs and drawings demonstrate perspective.

Extension

Draw a landscape scene or street scene of your choice in one-point perspective. For example, find a straight street with buildings or a view of walkways in a park.

Project   103

Concluding the Project
Consider posting the most interesting photographs, landscapes, and street scenes on a bulletin board.

Grading the Project
A well-written project will have the following characteristics:
- Students' sketches clearly show perspective, with the horizon line and vanishing point identified.
- Students' descriptions of their photographs and drawings showing perspective demonstrate that they understand the material.

Pacing and Assignment Guide

REGULAR SCHEDULE

Lesson	Les. Day	Basic	Average	Advanced
3.1	Day 1	pp. 110–113 Exs. 9–14, 25, 26, 30–33, 42, 51–62	pp. 110–113 Exs. 9–14, 25–29, 42–50	pp. 110–113 Exs. 9–14, 30–33, 39–41*, 45–53 odd
	Day 2	EP p. 676 Exs. 22, 23; pp. 110–113 Exs. 15–19, 21–24, 36, 43–50	pp. 110–113 Exs. 15–24, 30–37, 51–62	pp. 110–113 Exs. 17–20, 23–29 odd, 34–38, 42–44, 56–62 even; EC: classzone.com
3.2	Day 1	EP p. 676 Exs. 32–34; pp. 117–120 Exs. 8–14, 26, 34–38, 44–49	pp. 117–120 Exs. 8–15, 26, 31–43	pp. 117–120 Exs. 8–16 even, 26, 27, 31–35, 37–49 odd
	Day 2	EP p. 677 Exs. 13–15; pp. 117–120 Exs. 17–23, 39–43, Quiz 1	pp. 117–120 Exs. 17–25, 44–49, Quiz 1	pp. 117–120 Exs. 20–25, 28–30, Quiz 1; EC: TE p. 104D*, classzone.com
3.3	Day 1	pp. 123–125 Exs. 9–11, 16–21, 34–48 even	pp. 123–125 Exs. 9–11, 16–25, 34–42 even	pp. 123–125 Exs. 16–28, 34–39
	Day 2	pp. 123–125 Exs. 12–15, 29–32, 35–49 odd	pp. 123–125 Exs. 12–15, 29–32, 35–41 odd, 43–49	pp. 123–125 Exs. 29–32, 33*, 40–49; EC: classzone.com
3.4	Day 1	pp. 126–127 Exs. 6–8	pp. 126–127 Exs. 6–8	pp. 126–127 Exs. 6–8
	Day 2	pp. 132–135 Exs. 16–24, 43–56	pp. 132–135 Exs. 16–26, 43–56	pp. 132–135 Exs. 16–24 even, 25–28, 43–56
	Day 3	pp. 132–135 Exs. 14, 15, 29–34, 41, 42, Quiz 2	pp. 132–135 Exs. 29–37, 39–42, Quiz 2	pp. 132–135 Exs. 29–31, 35–42, Quiz 2; EC: TE p. 104D*, classzone.com
3.5	Day 1	EP p. 678 Exs. 21–25; pp. 140–142 Exs. 6–11, 16, 19, 35–47 odd	pp. 140–142 Exs. 6–11, 16, 23, 35, 40–47	pp. 140–142 Exs. 8–11, 16, 23, 36–47
	Day 2	pp. 140–142 Exs. 12–15, 17, 25–29 odd, 36–46 even	pp. 140–142 Exs. 12–15, 17–21, 24–30, 33, 36–39	pp. 140–142 Exs. 12–15, 17–22, 27–33, 34*, 35; EC: classzone.com
3.6	Day 1	EP p. 678 Exs. 26–31; pp. 147–149 Exs. 4–9, 22–24, 29–36	pp. 147–149 Exs. 4–12, 22–24, 37–42	pp. 147–149 Exs. 4–12 even, 22–24, 27–36
	Day 2	pp. 147–149 Exs. 10–16, 19–21, 27, 28, 37–42	pp. 147–149 Exs. 13–21, 25, 27–36	pp. 147–149 Exs. 13–21, 25, 26*, 37–42; EC: classzone.com
	Day 3	pp. 150–151 Ex. 9	pp. 150–151 Ex. 9	pp. 150–151 Ex. 9
3.7	Day 1	pp. 155–159 Exs. 9–21, 49–59	pp. 155–159 Exs. 9–21, 49–59	pp. 155–159 Exs. 9–21, 46, 47, 49–59
	Day 2	SRH p. 664 Exs. 21–28; pp. 155–159 Exs. 22–27, 32–44 even, Quiz 3	pp. 155–159 Exs. 22–24, 26–44 even, 45, Quiz 3	pp. 155–159 Exs. 22, 23–43 odd, 48*, Quiz 3; EC: classzone.com
Review	Day 1	pp. 160–163 Exs. 1–32	pp. 160–163 Exs. 1–32	pp. 160–163 Exs. 1–32
Assess	Day 1	**Chapter 3 Test**	**Chapter 3 Test**	**Chapter 3 Test**

YEARLY PACING Chapter 3 Total – **18 days** Chapters 1–3 Total – **46 days** Remaining – **114 days**

*Challenge Exercises   EP = Extra Practice   SRH = Skills Review Handbook   EC = Extra Challenge

BLOCK SCHEDULE

Day 1	Day 2	Day 3	Day 4	Day 5	Day 6	Day 7	Day 8	Day 9
3.1 pp. 110–113 Exs. 9–37, 42–62	**3.2** pp. 117–120 Exs. 8–15, 17–26, 31–49, Quiz 1	**3.3** pp. 123–125 Exs. 9–25, 29–32, 34–49	**3.4** pp. 126–127 Exs. 6–8; pp. 132–135 Exs. 16–26, 43–56	**3.4 cont.** pp. 132–135 Exs. 29–37, 39–42, Quiz 2 **3.5** pp. 140–142 Exs. 6–11, 16, 23, 35, 40–47	**3.5 cont.** pp. 140–142 Exs. 12–15, 17–21, 24–30, 33, 36–39 **3.6** pp. 147–149 Exs. 4–12, 22–24, 37–42	**3.6 cont.** pp. 147–149 Exs. 13–21, 25, 27–36; pp. 150–151 Ex. 9	**3.7** pp. 155–159 Exs. 9–24, 26–44 even, 45, 49–59, Quiz 3	**Review** pp. 160–163 **Assess** Chapter 3 Test

YEARLY PACING Chapter 3 Total – **9 days** Chapters 1–3 Total – **23 days** Remaining – **57 days**

Support Materials

Providing Universal Access

Strategies for Strategic Learners

SEPARATE CONFUSING ELEMENTS

Some students grasp a concept with one or two examples, but most students need many more. If students are confused about the different properties of lines that they are learning in this chapter, select one really good problem and spend some extra time on it. Encourage students to ask questions, and as they are solving a problem, circulate around the room to see their work. Have students work on only one kind of problem before going on to the next type. Do not introduce confusing concepts on the same day—in fact, space them over several days in order to ensure that students understand one concept before you go on to the next. As part of a general weekly or monthly review, include an appropriate problem or two to remind students of concepts they found to be particularly difficult.

PROVIDE CONCRETE EXAMPLES

Look for opportunities in everyday conversation and interactions to reinforce mathematical concepts. Properties of lines, for example, lend themselves beautifully to discussions about design or construction. All students have seen parallel lines, perpendicular lines, and angles between lines in everyday life even if they didn't know the names for these concepts. Teachers can build on this understanding by discussing furniture design, interior decorating, construction, and other topics in which students have an interest. Extending the discussion of mathematics into everyday life can have the effect of extending the learning time, as students cannot help but be reminded of what they learned in mathematics class about lines and angles every time they design or build something.

RETEACH USING A DIFFERENT METHOD

Many students may find the vocabulary introduced in Lesson 3.3 particularly confusing. After you have introduced students to the four special pairs of angles, consider reviewing the material in a different way. Using space in the classroom, a hallway, the gym, or outside, construct two lines and a transversal on the ground, such as is shown in the figure at the top of page 121. You can use tape, yardsticks, rope, yarn, or even sticks to model the lines. You are going to direct pairs of students to stand in locations to represent different special pairs of angles, so the lines will need to be far enough apart so that two students can easily stand within the figure.

Now choose students and have them model pairs of angles. One way is to choose pairs of students and have them choose for themselves where to stand to correctly model the angle pair assigned to them. Another way is to place one student at a location and then direct another student to stand in an appropriate location to form a particular angle pair with the first student. For example, you could place a student at any angle and direct a second student to stand in a location to model a pair of corresponding angles. Two more students could then be directed to form a second pair of corresponding angles.

Also consider checking students' understanding by asking them to form impossible angles; for example, you could place the first student in the location of an exterior angle and then challenge a second student to form a pair of same-side interior angles with the first.

Try to repeat the activity enough times so that all students have had an opportunity to model an angle.

Strategies for English Learners

DISSECT WORD PROBLEMS

The last sentence of a word problem usually contains a verb that issues a command. It orders the reader to do something. Almost always it is followed by an additional description of what the person is to do, as in "Find the measure of $\angle A$," with *find* being the command, and *the measure of $\angle A$* identifying what is to be found. These command verbs are signals to the student that explain what the student is to do in the word problem.

COMMAND VERBS The following command verbs were compiled from sample standardized test problems, word problems found in mathematics textbooks, and state standards:

> apply, approximate, assume, calculate, choose, circle, classify, compute, convert, create, decide, demonstrate, describe, design, determine, draw, establish, estimate, examine, explain, extend, extrapolate, find, fit, identify, interpolate, interpret, investigate, make, mark, model, order, organize, perform, predict, prove, read, recognize, represent, round, select, show, sketch, solve, transform, translate, use, verify, visualize

Ask students to group the words into categories, such as words that tell you to pick from a list, words that tell you to calculate something, and so on. Allow students to make up their own categories. It is important here to let students express and discuss what they think these terms mean. Many of the words would fit into more than one category depending on the context. For example, the term *draw* can be used in the following ways:

> Draw a picture.
> Draw a conclusion.

These are very different requests, and students need to know both meanings. Through discussion, students can expose the different meanings of words and expand their concepts of what words mean. Watch for the tendency of students to confuse words that sound or look similar, for example *make* and *mark*. Encourage students to write new words in their journal,

accompanied by a picture or an explanation in their primary language. Use a dictionary and thesaurus to provide additional meanings of words, as well as synonyms and antonyms.

Suggest that students read through each chapter's review as the class begins a new chapter. Students should not concern themselves with whether or not they can solve the problems, but simply read the direction lines to make sure they understand what each direction line is asking them to do. For example, in the Chapter Summary and Review section covering Lesson 3.7 on page 163, students are asked to describe a translation using coordinate notation and to decide whether one figure is a translation of another. Students will not yet know what a translation is, but they should know the meanings of *describe* and *decide* and should not confuse the two words.

Strategies for Advanced Learners

VOCABULARY DEVELOPMENT

Most students learn vocabulary through reading. Teachers can help students build their mathematics vocabulary and understanding through the assignment of outside, independent reading. Books for young adults that discuss mathematical concepts are readily available and help students understand how mathematics is important in their daily lives. With the help of the school, district, or local library media teacher, teachers can find books that reinforce mathematical concepts, from the most simple to the most complex. Books by David Macaulay, like *The Way Things Work*, discuss mathematical and scientific terms and concepts in the context of history, architecture, and science. In the book *City*, for example, the building of an imaginary Roman city provides illustrations and explanations of area: "the rectangular area ... was seven hundred and twenty yards long by six hundred and twenty yards wide"; proportion: "no privately owned building ... could be higher than twice the width of the street on which it stood"; right angles: "the surveyors marked off the roads using an instrument called a groma to make sure all roads intersected at right angles"; perpendicular and parallel: "When weighted strings hanging from each end of the cross hung parallel to the center pole the groma was known to be perpendicular to the ground." Assigning reading that students do outside the mathematics instructional time not only builds understanding of mathematics concepts and vocabulary, but also helps students increase their reading ability. While the David Macaulay books are challenging in terms of the reading level, this strategy can be used with all students by selecting material at an appropriate reading level for each student.

Challenge Problem for use with Lesson 3.2:
In the diagram, $\overrightarrow{XB}$ bisects $\angle EXH$ and $\overleftrightarrow{AB} \perp \overleftrightarrow{CD}$. $\overleftrightarrow{EF}$ and $\overleftrightarrow{GH}$ intersect $\overleftrightarrow{AB}$ and $\overleftrightarrow{CD}$ at point X. Explain how you know that $\overrightarrow{XD}$ bisects $\angle HXF$. ($m\angle 1 = m\angle 2$, since $\overrightarrow{XB}$ bisects $\angle EXH$. Since $\overleftrightarrow{AB} \perp \overleftrightarrow{CD}$, $\angle CXB$ and $\angle BXD$ are right angles. So $\angle 1$ and $\angle 5$ are complementary and $\angle 2$ and $\angle 3$ are complementary. Then $m\angle 5 = m\angle 3$ because they are complements of equal angles. But $\angle 5$ and $\angle 4$ are vertical angles, so $m\angle 4 = m\angle 5$. By substitution, $m\angle 3 = m\angle 4$, so $\overrightarrow{XD}$ bisects $\angle HXF$.)

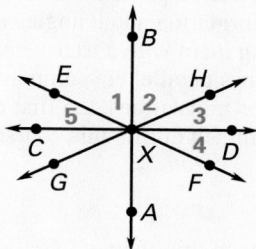

Challenge Problem for use with Lesson 3.4:
In the diagram, $j \parallel k$ and $\ell \parallel m$. Explain how you know that $\angle 1 \cong \angle 3$. (Since $\angle 1$ and $\angle 2$ are same-side interior angles formed by parallel lines, they are supplementary. The same is true for $\angle 2$ and $\angle 3$. Since $\angle 1$ and $\angle 3$ are supplements of the same angle, they are congruent.)

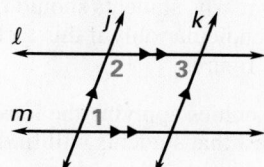

This chapter explores two of the most important relationships between pairs of lines, namely parallelism and perpendicularity. The first lesson introduces these concepts as well as the concept of skew lines (in space). Lesson 3.2 focuses on perpendicular lines and the relationship between perpendicularity and right angles. Some ramifications of both of these concepts for adjacent angles are also given. In Lesson 3.3, the angles formed by two lines and a transversal are classified according to their positions. This lesson lays the groundwork for the study of the angles formed when parallel lines are cut by a transversal, which begins in Lesson 3.4. Lesson 3.4 discusses the conclusions that can be drawn when a transversal cuts two lines that are known to be parallel. Lesson 3.5 then explores the converse question: What information about angles is needed in order to conclude that the lines forming them with a transversal are parallel? The methods for proving two lines parallel are summarized in Lesson 3.6. Finally, Lesson 3.7 introduces translations, the first of several types of transformations that will be examined during this course.

Lesson 3.1

This lesson introduces the two main ideas of this chapter, parallel lines and perpendicular lines. It may be most efficient to concentrate first on the two-dimensional case, because in a plane, any two lines that do not intersect are parallel. One point to make here is that students should not infer from a diagram that two lines are parallel simply because they appear to be. In this book, small triangles on pairs of lines indicate that the lines are parallel. There are only two means for identifying that a pair of lines are parallel; either this fact is stated directly or these triangles appear on the lines in a given figure. Similarly, students should be made aware that two lines in a figure are perpendicular only if this fact is stated or if a right angle symbol appears in the figure.

VISUALIZATION Students may have difficulties applying the idea of parallelism to oblique lines in a figure. Suggest that students will find it helpful to rotate their book so the lines are horizontal. Also, three-dimensional visualization of lines or planes in space is often difficult for students. The concept of skew lines can be particularly challenging, because without a good visualization of depth, students may have difficulty distinguishing this concept from parallelism. An effective way for students to distinguish between skew lines and parallel lines is to ask whether there is a single plane that contains both lines.

QUICK TIP
One way to help students grasp the concept of skew lines (and the other relationships introduced in this lesson) concretely is to use the walls, floor, and ceiling of the classroom to illustrate the relationship. For example, skew lines can be modeled by the line of intersection of one wall and the ceiling and the line of intersection of an adjacent wall and the floor. In addition, parallel lines can be illustrated by the lines of intersection of a single wall with its two adjacent walls. Models of perpendicular lines, a line perpendicular to a plane, and parallel planes can also be found in the classroom.

Lesson 3.2

Most students associate the concepts of perpendicular lines and right angles. Theorems 3.1 and 3.2 formally state this association. In connection with Theorem 3.2, you should emphasize the fact that if one of four angles formed by two intersecting lines is a right angle, then the other three angles are also right angles. This is stated in the Student Help note given on page 114.

MAKING CONNECTIONS Theorems 3.3 and 3.4 are easily confused. You should note that the first of these involves two right angles (the two congruent angles mentioned in the statement), while the second involves only one. Theorem 3.4 can be regarded as a special case of the Angle Addition Postulate, the case in which the given angle is a right angle.

Lesson 3.3

Emphasize that the terms defined in this lesson are based on the *relative positions* of the two angles in a diagram showing a pair of lines cut (intersected) by a transversal. It should be stressed that the lines need *not* be parallel for the angle relationships to apply. Students should be led to recognize that the use of the words *interior* and *exterior* in the terms refers to their location in relation to the pair of lines cut by the transversal.

Also make the point that in order for any of the angle relationships to hold, the angles must be formed by a pair of lines and the same transversal. In the following diagram, ∠1 and ∠2 are *not* corresponding angles because they are not formed by a pair of lines and *one* transversal. For the same reason, ∠3 and ∠4 are not alternate exterior angles.

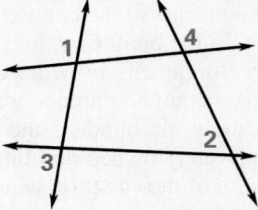

Lesson 3.4

The Corresponding Angles Postulate gives a necessary condition for two corresponding angles to be congruent. That is, pairs of corresponding angles are congruent only when the two lines cut by a transversal are parallel. The same is true of the Alternate Interior Angles Theorem and the Alternate Exterior Angles Theorem. The Same-Side Interior Angles Theorem states the necessary condition for two such angles to be supplementary.

The Visualize It notes given in the lesson are particularly helpful for visual learners. References to finding the "F" or the "Z" relationship between the angles throughout the lesson and also the remainder of the chapter is beneficial for students.

DIFFERENTIATING THEOREMS Note that the Same-Side Interior Angles Theorem is the only one whose conclusion does not involve *congruent* angles. It is easy for students to forget this fact, and to set the measures of two same-side interior angles formed by a pair of parallel lines and a transversal as equal when trying to solve an algebraic problem.

The theorems in this lesson can be useful in situations where two or more angle measures are given, either as numbers or as algebraic expressions. They provide a theoretical basis on which to set up equations, as illustrated in Examples 4 and 5 on page 131.

Lesson 3.5

The *converse* of an if-then statement is defined here and, as students will probably notice, the statements of the postulate and three theorems in this lesson are the converses of the postulate and theorems presented in Lesson 3.4. Emphasize that although all the converses are true here, it is not logically necessary that the converse of any statement be true just because the original statement is true.

READING DIAGRAMS As in previous lessons, it is important for students to be able to identify pairs of angles in a diagram that are corresponding angles, alternate interior angles, alternate exterior angles, and so on. Aside from being able to recognize the particular relationship unique to each type of angle pair, students need to be sure that the related angles are formed by only one pair of parallel lines and a single transversal.

At the conclusion of your discussion of the lesson, ask students to create two lists, one titled "If two parallel lines are cut by a transversal, what do I know?" and the other titled "What must I be able to show in order to justify that two lines cut by a transversal are parallel?"

Lesson 3.6

Geometric constructions using straightedge and compass are introduced in this lesson. As a practical matter, you might emphasize that at some point in human history, such constructions had to be relied upon to draw the right angles necessary for such pursuits as architecture, surveying, and farming. When discussing Example 1, you might point out that the validity of the construction depends on Theorem 3.12 which is stated later in this lesson.

Generally, the theorems presented in this lesson extend and relate the concepts of parallelism and perpendicularity. Example 3 illustrates how the theorems can be used in solving problems.

Lesson 3.7

In this lesson, students begin their investigation of transformations, beginning with translations. Translations can be visualized as making a copy of a figure and then sliding that copy to s new position. Translations in a coordinate plane are also discussed, with coordinate notation introduced to describe the translations. Algebraically, every translation can be described by a relation of the form

$$(x, y) \rightarrow (x + a, y + b).$$

Stress that the values of a and b can be determined easily if the coordinates of both a point and its image under the given translation are known. We say that the coordinate plane is "shifted horizontally" by a and "shifted vertically" by b. Point out that either or both of the values of a and b may be negative or zero.

Chapter Overview

Chapter Goals

Students will study the angles formed when two lines are cut by a third line, called the transversal. Throughout the chapter, they will be introduced to several theorems and postulates related to parallel and perpendicular lines. Students will:

- Identify relationships between the angles formed by two lines and a transversal, especially when the two lines are parallel.
- Use properties of parallel and perpendicular lines to find angle measures.
- Identify the angle relationships necessary for showing that two lines are parallel.
- Explore translations.

Application Note

If a sailboat's destination is upwind of its position, it must sail a series of zigzag courses that are 45° to the direction of the wind. The courses alternate between left of the direction of the wind and right of the direction of the wind. This sailing maneuver is called tacking. When a boat is ready to change its direction of tack, it can be turned into the direction of the wind, which will cause its sails to flap and the boat to lose speed. The boat's inertia then carries it into the opposite tack. If the boat's inertia is not great enough to complete the turn, the boat will stop with its sail pointing ineffectively into the wind.

More information about paths of sailboats is provided on the Internet at classzone.com

Application Links
CLASSZONE.COM

Chapter 3 Parallel and Perpendicular Lines

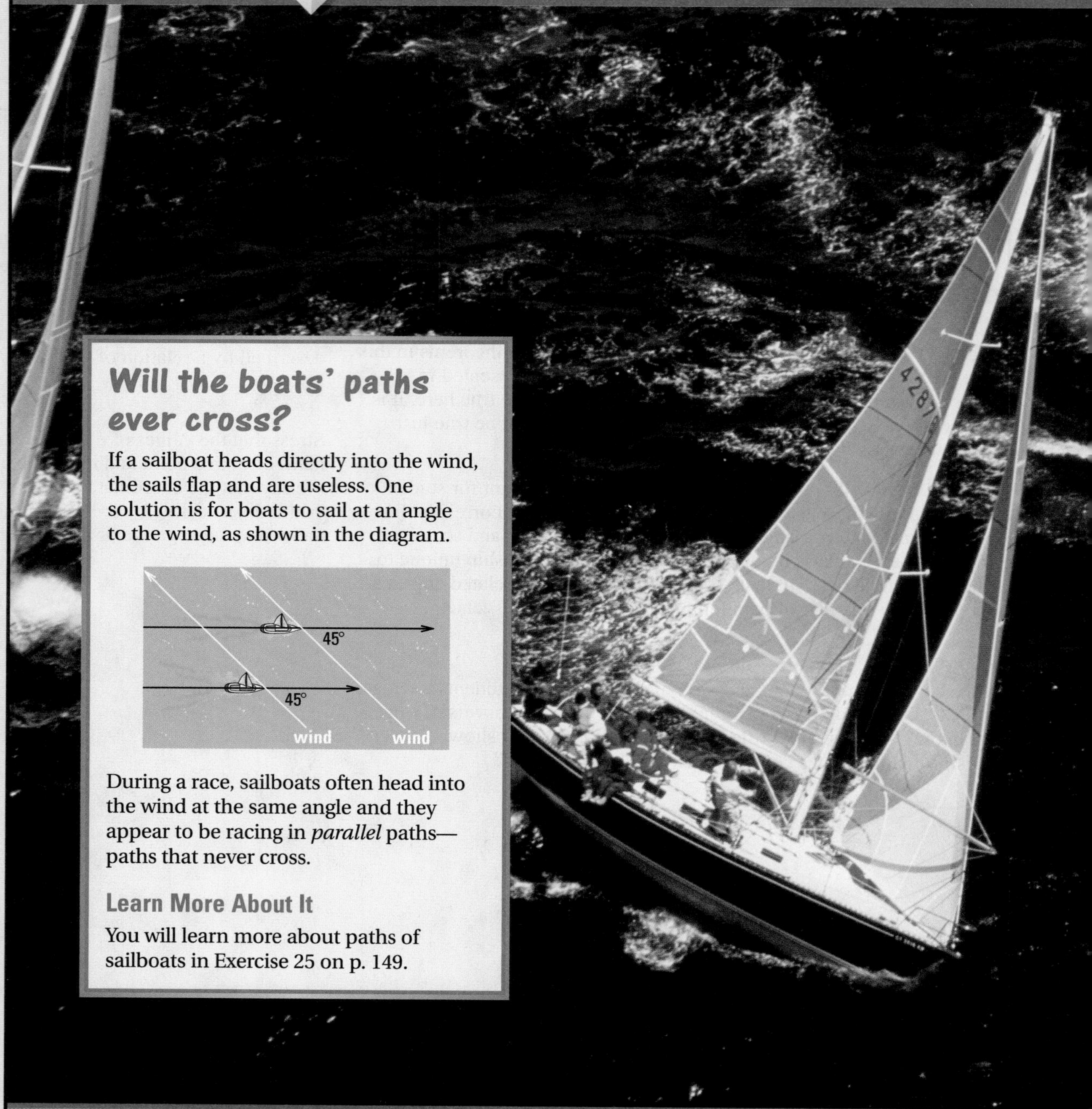

Will the boats' paths ever cross?

If a sailboat heads directly into the wind, the sails flap and are useless. One solution is for boats to sail at an angle to the wind, as shown in the diagram.

During a race, sailboats often head into the wind at the same angle and they appear to be racing in *parallel* paths—paths that never cross.

Learn More About It

You will learn more about paths of sailboats in Exercise 25 on p. 149.

Who uses Parallel and Perpendicular Lines?

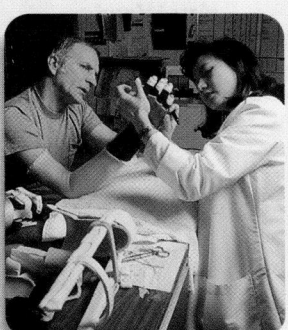

PHYSICAL THERAPIST
Physical therapists use angles to measure the range of motion of elbows, knees, and other joints. These measurements help the therapist evaluate people's injuries. (p. 134)

BICYCLE DESIGNER
The angles in a bike frame influence the position of the rider. Georgena Terry builds bicycles whose designs feature a frame geometry more suitable for women. (p. 124)

How will you use these ideas?

- Understand how escalators work. (p. 112)
- Learn how a compass and map are used in orienteering. (p. 119)
- See how surfing and kite flying are combined in the sport of kiteboarding. (p. 141)
- Understand how the strings of a guitar are related to the frets. (p. 148)

Projects

A project covering Chapters 3–4 appears on pages 228–229 of the Student Edition. An additional project for Chapter 3 is available in the *Chapter 3 Resource Book*, pp. 88–89.

Technology

- Electronic Teacher Tools
- Test and Practice Generator
- Online Lesson Planner

Video

- **Geometry in Motion**
  There is an animation supporting Lessons 3.3, 3.4, and 3.5.

Internet Support
CLASSZONE.COM

- Application and Career Links 133, 134
- Student Help 118, 133, 137, 148, 153

Diagnostic Tools

The **Chapter Readiness Quiz** can help you diagnose whether students have the following skills needed in Chapter 3:
- Identify vertical angles.
- Calculate the measures of a linear pair of angles.
- Solve multi-step equations.

Reteaching Material

This resource is available for students who need additional help with the skills on the Chapter Readiness Quiz:

📖 **Chapter 3 Resource Book**
- Reteaching with Practice (Lessons 3.1–3.7)

Additional Resources

The following resources are provided to help you prepare for the upcoming chapter and customize review materials:

📖 **Chapter 3 Resource Book**
- *Tips for New Teachers*, pp. 1–2
- *Parent Guide*, pp. 3–4
- *Lesson Plans*, pp. 7, 15, 25, 33, 42, 56, 67
- *Lesson Plans for Block Scheduling*, pp. 8, 16, 26, 34, 43, 57, 68

🖥 **Technology**
- Electronic Teaching Tools
- Online Lesson Planner
- Test and Practice Generator

> **Visualize It!**
>
> Additional suggestions for helping students visualize geometry are on pp. 107, 108, 111, 121, 128, 129, 131, 132, 140, and 148.

PREVIEW   What's the chapter about?

- Identifying relationships between lines
- Using properties of **parallel** and **perpendicular lines**
- Showing that two lines are parallel

Key Words

- **parallel lines,** *p. 108*
- **perpendicular lines,** *p. 108*
- **skew lines,** *p. 108*
- **parallel planes,** *p. 109*
- **transversal,** *p. 121*

- **corresponding angles,** *p. 121*
- **alternate interior angles,** *p. 121*
- **alternate exterior angles,** *p. 121*
- **same-side interior angles,** *p. 121*
- **converse,** *p. 136*

PREPARE   Chapter Readiness Quiz

Take this quick quiz. If you are unsure of an answer, look at the reference pages for help.

Vocabulary Check *(refer to p. 75)*

1. Name a pair of vertical angles. **B**

    (A) ∠1 and ∠4   **(B)** ∠2 and ∠4

    (C) ∠2 and ∠3   **(D)** ∠3 and ∠4

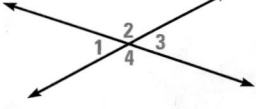

Skill Check *(refer to pp. 75, 673)*

2. The measure of ∠7 is 68°, and ∠7 and ∠8 form a linear pair. What is the measure of ∠8? **H**

   (F) 22°   **(G)** 68°   **(H)** 112°   **(J)** 80°

3. Which of the following is a solution of the equation $5x + 9 = 6x - 11$? **D**

   (A) $\frac{11}{20}$ **(B)** 2   **(C)** 11   **(D)** 20

VISUAL STRATEGY   Reading and Drawing Diagrams

Visualize It! ➡

Use dashed lines to show parts of a figure that are hidden behind other parts. The dashed line shows that line *n* is not in plane *S*.

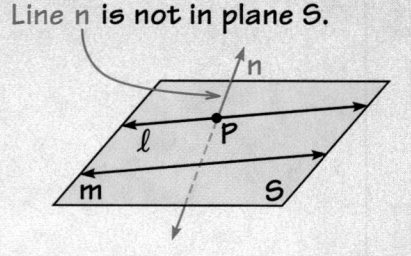

Line *n* **is not in plane S.**

Lines in Space

Question

How are lines related in space?

Materials
- pencil
- straightedge
- lined paper

Explore

❶ Use a straightedge to draw two identical rectangles.

❷ Connect the corresponding corners of the rectangles.

❸ Erase parts of "hidden" lines to form dashed lines.

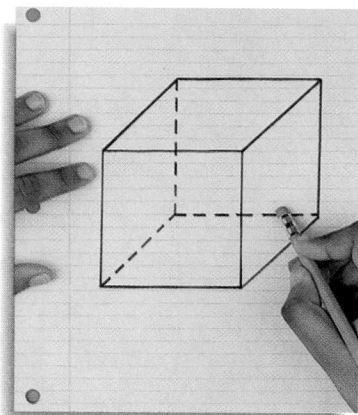

Visualize It!

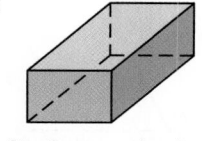

Shading your sketch will help make it look three-dimensional.

Think About It

Using your sketch from the steps above, label the corners as shown. Then extend $\overleftrightarrow{AD}$ and $\overleftrightarrow{CG}$.

1. Will $\overleftrightarrow{AD}$ and $\overleftrightarrow{CG}$ ever intersect in space? Lines that intersect on the page do not necessarily intersect in space. no

2. Determine whether the following pairs of lines will intersect in space.

a. $\overleftrightarrow{BC}$ and $\overleftrightarrow{AE}$ no **b.** $\overleftrightarrow{GH}$ and $\overleftrightarrow{DH}$ yes

c. $\overleftrightarrow{CD}$ and $\overleftrightarrow{DH}$ yes **d.** $\overleftrightarrow{AB}$ and $\overleftrightarrow{EH}$ no

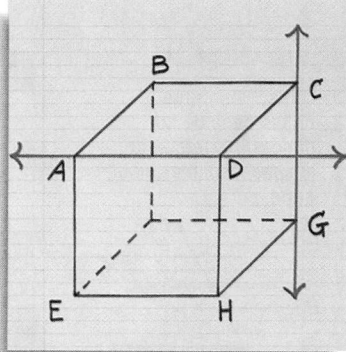

Activity **107**

---

❶ Planning the Activity

Goal
Students draw a representation of a three-dimensional figure to explore lines in space.

Materials
See the margin of the student page.

LINK TO LESSON
In Example 3 on page 109, students explore line relationships in three-dimensional space.

❷ Managing the Activity

Tips for Success
If possible, find an unmarked cardboard box or a wooden block and label the corners to match the diagram provided for Exercises 1 and 2. Use this labeled object to help students recognize that the diagram they are drawing is a representation of a three-dimensional object, not just an abstract design.

❸ Closing the Activity

KEY DISCOVERY
There are lines in space that do not intersect.

Activity Assessment
Use Exercise 2 to assess student understanding.

Suggested Number of Days

Basic: 2 days
Average: 2 days
Advanced: 2 days
Block Schedule: 1 block

Teaching Resources

📄 **Blacklines**
(See page 104B.)

📑 **Transparencies**
• Warm-Up with Quiz
• Answers

💻 **Technology**
• Electronic Teacher Tools
• Test and Practice Generator
• Online Lesson Planner
• Internet Support

② Teach

Content and Teaching Strategies
For background information on geometric concepts and teaching strategies related to this lesson, see pages 104E and 104F in this Teacher's Edition.

Tips for New Teachers

Identifying spatial relationships can be difficult for students. To help students with Checkpoint Exercises 4–6 on page 109, have an empty box and colored markers available for demonstration. Get students to suggest and agree which locations on the box should be marked as you proceed through the exercises. See the Tips for New Teachers on pp. 1–2 of the *Chapter 3 Resource Book* for additional notes about Lesson 3.1.

Goal

Identify relationships between lines.

Key Words

• parallel lines
• perpendicular lines
• skew lines
• parallel planes
• line perpendicular to a plane

Two lines are **parallel lines** if they lie in the same plane and do not intersect. On the building, lines r and s are parallel lines. You can write this as $r \parallel s$. Triangles (▸) are used to indicate that the lines are parallel.

Two lines are **perpendicular lines** if they intersect to form a right angle. Lines s and t are perpendicular lines. You can write this as $s \perp t$.

EXAMPLE 1 Identify Parallel and Perpendicular Lines

Determine whether the lines are *parallel, perpendicular,* or *neither.*

  a. n and m **b.** p and q **c.** n and p

Solution

  a. Lines n and m are parallel.

  b. Lines p and q are neither parallel nor perpendicular.

  c. Lines n and p are perpendicular.

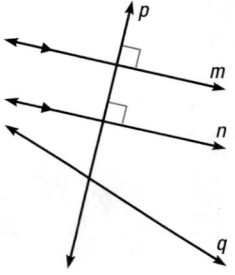

Two lines are **skew lines** if they do not lie in the same plane. Skew lines never intersect. In the diagram, lines c and b are skew lines.

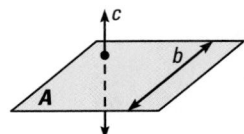

Visualize It!

Lines that intersect on the page do not necessarily intersect in space, such as lines f and g.

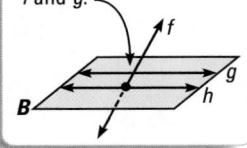

EXAMPLE 2 Identify Skew Lines

Determine whether the lines are skew.

  a. f and h

  b. f and g

Solution

  a. Lines f and h are not skew lines because they intersect.

  b. Lines f and g are skew lines.

Checkpoint ✓ *Identify Relationships Between Lines*

Use the diagram.

1. Name a pair of parallel lines. *x* and *y*

2. Name a pair of perpendicular lines. *x* and *z*

3. Name a pair of skew lines. *y* and *z*

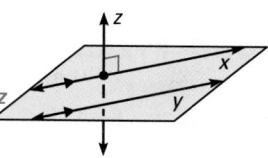

Two planes are **parallel planes** if they do not intersect.
A **line perpendicular to a plane** is a line that intersects a plane in a point and that is perpendicular to every line in the plane that intersects it.

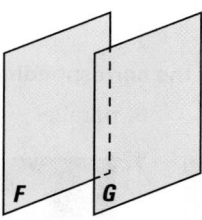

Plane *F* is parallel to plane *G*.

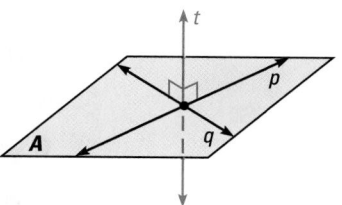

Line *t* is perpendicular to plane *A*.

EXAMPLE 3 **Identify Relationships in Space**

a. Name a plane that appears parallel to plane *B*.

b. Name a line that appears perpendicular to plane *B*.

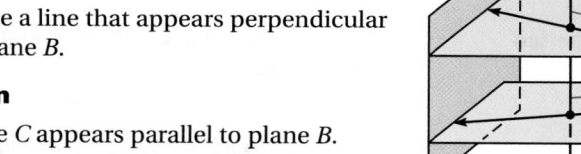

Solution

a. Plane *C* appears parallel to plane *B*.

b. Line *ℓ* appears perpendicular to plane *B*.

Checkpoint ✓ *Identify Relationships in Space*

Think of each segment in the diagram as part of a line.

4. Name a line that is skew to $\overleftrightarrow{VW}$. $\overleftrightarrow{QU}$, $\overleftrightarrow{TX}$, $\overleftrightarrow{RQ}$, or $\overleftrightarrow{ST}$

5. Name a plane that appears parallel to plane *VWX*. plane *RST*

6. Name a line that is perpendicular to plane *VWX*. $\overleftrightarrow{RV}$, $\overleftrightarrow{QU}$, $\overleftrightarrow{SW}$, or $\overleftrightarrow{TX}$

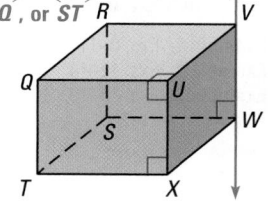

Extra Example 1

Determine whether the lines are *parallel, perpendicular,* or *neither*.

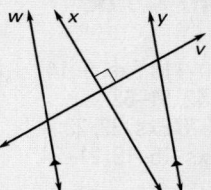

a. *v* and *x* perpendicular

b. *v* and *y* neither

c. *w* and *y* parallel

Extra Example 2

Determine whether the lines are skew.

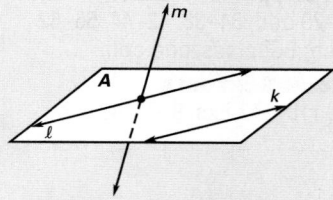

a. *k* and *m* They are skew because they are not parallel and do not intersect.

b. *ℓ* and *m* They are not skew because they intersect.

Extra Example 3

a. Name a plane that appears parallel to plane *G*. plane *F*

b. Name a line that is perpendicular to plane *F*. line *ℓ*

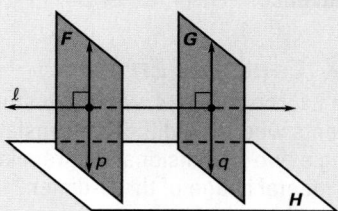

✓ **Concept Check**

What are parallel lines, perpendicular lines, and skew lines? Parallel lines lie in the same plane and do not intersect, perpendicular lines intersect to form a right angle, and skew lines do not lie in the same plane and do not intersect.

3.1 Exercises

Guided Practice

Assignment Guide

BASIC
Day 1: pp. 110–113 Exs. 9–14, 25, 26, 30–33, 42, 51–62
Day 2: EP p. 676 Exs. 22, 23; pp. 110–113 Exs. 15–19, 21–24, 36, 43–50

AVERAGE
Day 1: pp. 110–113 Exs. 9–14, 25–29, 42–50
Day 2: pp. 110–113 Exs. 15–24, 30–37, 51–62

ADVANCED
Day 1: pp. 110–113 Exs. 9–14, 30–33, 39–41*, 45–53 odd
Day 2: pp. 110–113 Exs. 17–20, 23–29 odd, 34–38, 42–44, 56–62 even; EC: classzone.com

BLOCK SCHEDULE
pp. 110–113 Exs. 9–37, 42–62

Extra Practice
• Student Edition, p. 679
• Chapter 3 Resource Book, pp. 10–11

Homework Check

To quickly check student understanding of key concepts, go over the following exercises:

Basic: 10, 13, 17, 19, 22
Average: 9, 12, 15, 23, 24
Advanced: 11, 14, 19, 24, 34

✗ Common Error

In Exercises 12–14, watch for students who have difficulty translating a two-dimensional diagram into a mental image of three-dimensional space. They may, for example, think that lines *j* and *k* in Exercise 14 are perpendicular rather than skew.

Vocabulary Check

1. How are *skew lines* and *parallel lines* alike? How are they different?

Skill Check

1. *Sample answer:* Skew lines and parallel lines are alike in that they do not intersect. They are different in that parallel lines are coplanar and skew lines are not coplanar.

Fill in the blank with ‖ or ⊥ to make the statement true.

2. Line *k* __?__ line *m*. ⊥

3. Line *m* __?__ line *ℓ*. ‖

4. Line *ℓ* __?__ line *j*. ⊥

5. Line *k* __?__ line *j*. ‖

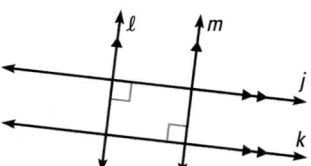

Match the photo with the corresponding description of the chopsticks.

A. skew   **B.** parallel   **C.** intersecting

6.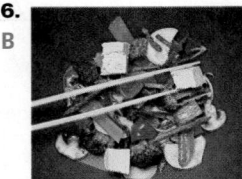
B

7.
C

8.
A

Practice and Applications

Extra Practice
See p. 679.

Line Relationships Determine whether the lines are *parallel*, *perpendicular*, or *neither*.

9. *a* and *c* neither
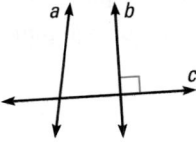

10. *q* and *s* perpendicular
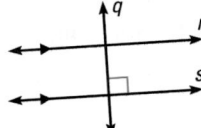

11. *y* and *z* parallel
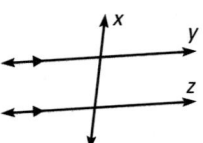

Skew Lines Determine whether the lines are skew. Explain.

12. *u* and *w*

13. *m* and *n*
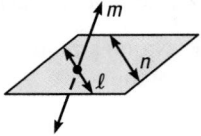

14. *j* and *k*
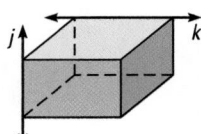

Homework Help
Example 1: Exs. 9–11
Example 2: Exs. 12–14
Example 3: Exs. 15–19, 21–24

No; *u* and *w* intersect.

Yes; *m* and *n* are not parallel and do not intersect.

Yes; *j* and *k* are not parallel and do not intersect.

Identifying Relationships In Exercises 15–19, think of each segment in the diagram as part of a line. Fill in the blank with *parallel, perpendicular,* or *skew*.

15. $\overleftrightarrow{DE}$, $\overleftrightarrow{AB}$, and $\overleftrightarrow{GC}$ appear to be __?__. parallel

16. $\overleftrightarrow{DE}$ and $\overleftrightarrow{BE}$ are __?__. perpendicular

17. $\overleftrightarrow{BE}$ and $\overleftrightarrow{GC}$ are __?__. skew

18. $\overleftrightarrow{BE}$ is __?__ to plane *DEF*. perpendicular

19. Plane *GAD* and plane *CBE* appear to be __?__. parallel

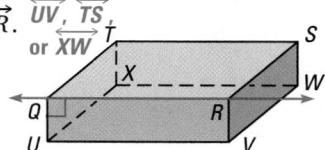

20. **Tightrope Walking** Philippe Petit sometimes uses a long pole to help him balance on the tightrope. Are the rope and the pole at the left *intersecting, perpendicular, parallel,* or *skew*? skew

Relationships in Space Think of each segment in the diagram below as part of a line. There may be more than one correct answer.

21. Name a line that appears parallel to $\overleftrightarrow{QR}$. $\overleftrightarrow{UV}$, $\overleftrightarrow{TS}$, or $\overleftrightarrow{XW}$

22. Name a line perpendicular to $\overleftrightarrow{QR}$. $\overleftrightarrow{QU}$

23. Name a line skew to $\overleftrightarrow{QR}$. $\overleftrightarrow{SW}$, $\overleftrightarrow{TX}$, $\overleftrightarrow{VW}$, or $\overleftrightarrow{UX}$

24. Name a plane that appears parallel to plane *QRS*. plane *UVW*

Visualize It! Sketch a figure that fits the description. 25–29. See margin.

25. Three lines that are parallel

26. A line that is perpendicular to two parallel lines

27. Two planes that intersect

28. A line that is perpendicular to two parallel planes

29. A line that is perpendicular to two skew lines (*Hint:* Start by sketching a figure like the one above in Exercises 21–24.)

Furniture Design In Exercises 30–33, use the photo of the chair designed by Mario Botta shown at the right.

30. Name two pairs of parallel lines.
Sample answer: h and n, k and ℓ

31. What kind of lines are *h* and *m*? parallel

32. Name two lines that are skew.
Sample answer: k and j

33. How many lines shown on the chair are perpendicular to *j*? three

Teaching Tip
After Exercises 30–33, you may wish to have students sketch their own chair design. Then have them add lines like those in the figure to their sketch and note pairs of lines that are parallel, perpendicular, and skew.

25–29. Sample answers are given.

25.

26.
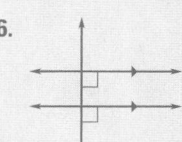

27.
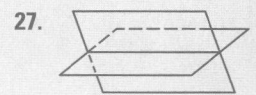

28.
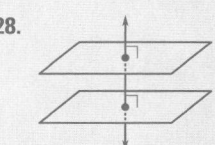

29.
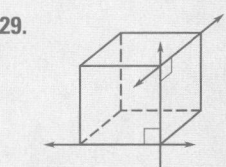

36–38. Sample answers are given.

36.

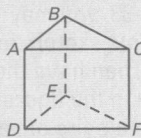

$\overline{AD}$ and $\overline{CF}$ appear parallel.

37.

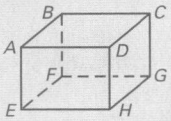

$\overline{AE}$ and $\overline{DH}$ appear parallel.

38.

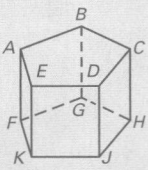

$\overline{AF}$ and $\overline{CH}$ appear parallel.

Escalators **In Exercises 34 and 35, use the following information.**

When a step on an up-escalator reaches the top, it flips over and goes back down to the bottom. On each step, let plane A be the plane you stand on.

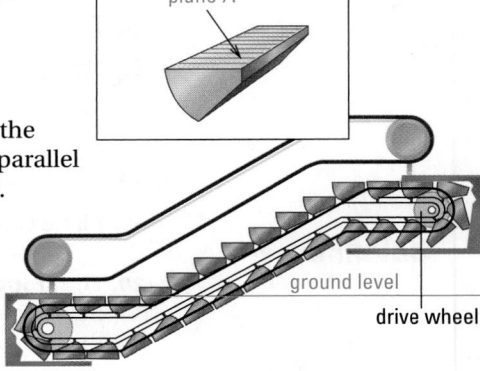

plane A

ground level

drive wheel

34. As each step moves around the escalator, is plane A always parallel to the ground level? Explain.

34. No; the step remains parallel to the ground level while it moves from the ground level to the top, but when it reaches the top, it begins to rotate and is no longer parallel to the ground level.

35. When a person is standing on plane A, is it parallel to ground level? Explain.

35. Yes; plane A remains parallel to the ground level for the entire time that the step is moving from ground level to the top.

EXAMPLE   **Sketch a Prism**

A *prism* is a three-dimensional figure with two identical faces, called *bases*, that lie in parallel planes as shown below.

a. Sketch a prism with bases that have six sides. Label the prism.

b. Name two edges that appear parallel.

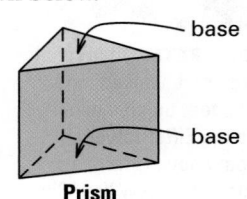

base

base

Prism

Solution

a. To draw the prism, follow these steps.

❶ *Sketch* two identical bases in parallel planes. In this case, the bases have six sides.

❷ *Connect* the bases and make hidden edges dashed. Label the prism.

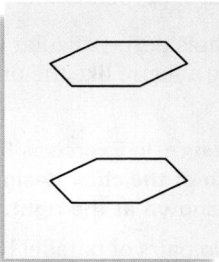

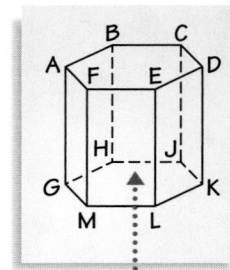

Student Help

VISUAL STRATEGY
Use dashed lines to show parts of a figure that are hidden behind other parts.

b. Edges $\overline{AG}$ and $\overline{DK}$ appear parallel.

Prisms **Sketch a prism with bases that have the given number of sides. Label the prism. Name two edges that appear parallel.**
36–38. See margin.

36. Three sides   **37.** Four sides   **38.** Five sides

Challenge Fill in the blank with *always, sometimes,* or *never.*

39. Two skew lines are __?__ parallel. never

40. Two perpendicular lines __?__ intersect. always

41. Two skew lines are __?__ coplanar. never

42. Multiple Choice Two lines are __?__ lines if they do not lie in the same plane and they do not intersect. D

Ⓐ perpendicular   Ⓑ parallel

Ⓒ coplanar   Ⓓ skew

43. Multiple Choice Use the diagram below to determine which of the following statements is *false.* Think of each segment in the diagram as part of a line. H

Ⓕ $\overleftrightarrow{QP}$ and $\overleftrightarrow{MP}$ are not parallel.

Ⓖ $\overleftrightarrow{MP}$ and $\overleftrightarrow{NR}$ are skew.

Ⓗ $\overleftrightarrow{JM}$ and $\overleftrightarrow{KS}$ are perpendicular.

Ⓙ Plane *KJM* and plane *QPM* are not parallel.

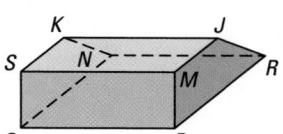

Mixed Review

If-Then Statements Identify the hypothesis and the conclusion of the if-then statement. *(Lesson 2.5)*

44. If the band plays, then each member gets $50.

45. If $m\angle 5 = 120°$, then $\angle 5$ is obtuse.

46. If there is a sale, then the store will be crowded.

47. If we can get tickets, then we'll go to the movies.

44. hypothesis: the band plays;
conclusion: each member
gets $50

45. hypothesis: $m\angle 5 = 120°$;
conclusion: $\angle 5$ is obtuse

46. hypothesis: there is a sale;
conclusion: the store will be
crowded

47. hypothesis: we can get
tickets;
conclusion: we'll go to the
movies

Properties of Congruence Use the property to complete the statement. *(Lesson 2.6)*

48. *Reflexive Property of Congruence:* __?__ ≅ $\angle XYZ$ $\angle XYZ$

49. *Symmetric Property of Congruence:* If $\angle 1 \cong \angle 2$, then __?__ ≅ __?__ .
                                          $\angle 2 \cong \angle 1$

50. *Transitive Property of Congruence:* If $\overline{AB} \cong \overline{EF}$ and $\overline{EF} \cong \overline{ST}$, then __?__ ≅ __?__ . $\overline{AB} \cong \overline{ST}$

Algebra Skills

Reciprocals Find the reciprocal. *(Skills Review, p. 656)*

51. 26 $\frac{1}{26}$ **52.** -7 $-\frac{1}{7}$ **53.** 10 $\frac{1}{10}$ **54.** $\frac{3}{8}$ $\frac{8}{3}$

Integers Evaluate. *(Skills Review, p. 663)*

55. $18 + (-3)$ 15   **56.** $-4 \div 2$ -2 **57.** $17 + (-6)$ 11   **58.** $16 - (-5)$ 21

59. $-5 + 31$ 26   **60.** $24 - 28$ -4 **61.** $(-8)(-10)$ 80   **62.** $-25 - 19$ -44

3.1 *Relationships Between Lines* **113**

Assessment Resources
The Mini-Quiz below is also available on blackline (*Chapter 3 Resource Book*, p. 17) and on transparency. For more assessment resources, see:
• Chapter 3 Resource Book
• Standardized Test Practice
• Test and Practice Generator

Mini-Quiz
Think of each segment in the diagram as part of a line. Fill in the blank with *parallel, perpendicular,* or *skew.*

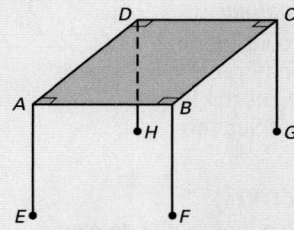

1. $\overleftrightarrow{AD}$ and $\overleftrightarrow{BC}$ are __?__ . parallel
2. $\overleftrightarrow{BC}$ and $\overleftrightarrow{CD}$ are __?__ .
    perpendicular
3. $\overleftrightarrow{BC}$ and $\overleftrightarrow{AE}$ are __?__ . skew
4. Name a line that appears to be parallel to $\overleftrightarrow{AB}$. $\overleftrightarrow{CD}$
5. Does plane *ABC* appear to be perpendicular to $\overleftrightarrow{AE}$? yes

Pacing

Suggested Number of Days

Basic: 2 days
Average: 2 days
Advanced: 2 days
Block Schedule: 1 block

Teaching Resources

 **Blacklines**
(See page 104B.)

Transparencies
• Warm-Up with Quiz
• Answers

Technology
• Electronic Teacher Tools
• Test and Practice Generator
• Online Lesson Planner
• Internet Support

Geo-Activity

Goal Use paper folding to explore angles formed by perpendicular lines.

Key Discovery Two perpendicular lines form four right angles where they intersect.

Teach

Content and Teaching Strategies
For background information on geometric concepts and teaching strategies related to this lesson, see pages 104E and 104F in this Teacher's Edition.

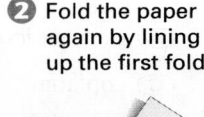 **Theorems About Perpendicular Lines**

Goal

Use theorems about perpendicular lines.

Key Words

• complementary angles
  p. 67
• perpendicular lines
  p. 108

Geo-Activity ▸ **Intersecting Lines**

 ❶ Fold a piece of paper to form a line.

 ❷ Fold the paper again by lining up the first fold.

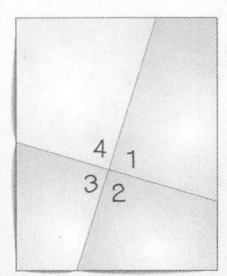 ❸ Unfold the paper. Label the angles as shown.

❹ Do the fold lines appear perpendicular? Use a protractor to measure each angle. How many right angles are formed?
Yes; all four angles are right angles.

❺ Write one or more conjectures about perpendicular lines.

Sample answer: If two lines are perpendicular, then the four angles formed are right angles.

The Geo-Activity above suggests the following theorems about perpendicular lines.

Student Help

STUDY TIP

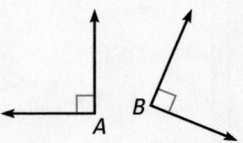

Theorem 3.2 tells you that if one right angle is marked on a pair of intersecting lines, then the other three angles are also right angles.

THEOREMS 3.1 and 3.2

Theorem 3.1

Words All right angles are congruent.

Symbols If $m\angle A = 90°$ and $m\angle B = 90°$, then $\angle A \cong \angle B$.

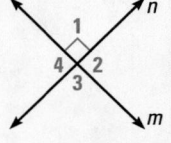

Theorem 3.2

Words If two lines are perpendicular, then they intersect to form four right angles.

Symbols If $n \perp m$, then $m\angle 1 = 90°$, $m\angle 2 = 90°$, $m\angle 3 = 90°$, and $m\angle 4 = 90°$.

EXAMPLE 1 Perpendicular Lines and Reasoning

In the diagram, $r \perp s$ and $r \perp t$. Determine whether enough information is given to conclude that the statement is true. Explain your reasoning.

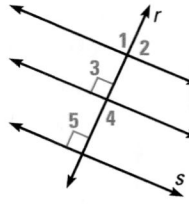

a. $\angle 3 \cong \angle 5$

b. $\angle 4 \cong \angle 5$

c. $\angle 2 \cong \angle 3$

Solution

a. Yes, enough information is given. Both angles are right angles. By Theorem 3.1, they are congruent.

b. Yes, enough information is given. Lines r and t are perpendicular. So, by Theorem 3.2, $\angle 4$ is a right angle. By Theorem 3.1, all right angles are congruent.

c. Not enough information is given to conclude that $\angle 2 \cong \angle 3$.

Checkpoint ✓ *Perpendicular Lines and Reasoning*

1. Yes; $\angle 6$ and $\angle 10$ are right angles and all right angles are congruent.

2. Yes; since $g \perp e$, $\angle 7$ is a right angle; $\angle 10$ is also a right angle; all right angles are congruent.

3. no

In the diagram, $g \perp e$ and $g \perp f$. Determine whether enough information is given to conclude that the statement is true. Explain.

1. $\angle 6 \cong \angle 10$	**2.** $\angle 7 \cong \angle 10$
3. $\angle 6 \cong \angle 8$	**4.** $\angle 7 \cong \angle 11$
5. $\angle 7 \cong \angle 9$	**6.** $\angle 6 \cong \angle 11$

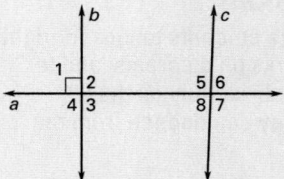

Student Help

LOOK BACK
Theorems 3.3 and 3.4 refer to *adjacent angles*. For the definition of adjacent angles, see p. 68.

4. Yes; since $g \perp e$, $\angle 7$ is a right angle; since $g \perp f$, $\angle 11$ is a right angle; all right angles are congruent.

5. no

6. Yes; $\angle 6$ is a right angle; since $g \perp f$, $\angle 11$ is a right angle; all right angles are congruent.

THEOREMS 3.3 and 3.4

Theorem 3.3

Words If two lines intersect to form adjacent congruent angles, then the lines are perpendicular.

Symbols If $\angle 1 \cong \angle 2$, then $\overleftrightarrow{AC} \perp \overleftrightarrow{BD}$.

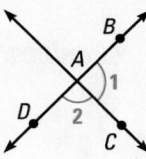

Theorem 3.4

Words If two sides of adjacent acute angles are perpendicular, then the angles are complementary.

Symbols If $\overrightarrow{EF} \perp \overrightarrow{EH}$, then $m\angle 3 + m\angle 4 = 90°$.

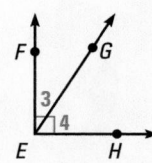

Tips for New Teachers

Help students learn to recognize parallel lines, perpendicular lines, and right angles by encouraging students to point out these features in objects they see in the classroom. Have students identify pairs of congruent angles. This activity can be applied and extended throughout this lesson and the following lessons. See the Tips for New Teachers on pp. 1–2 of the *Chapter 3 Resource Book* for additional notes about Lesson 3.2.

Visualize It!

Clarify that students do not need to mark all four right angles when two lines intersect. Marking only one right angle is sufficient.

Extra Example 1

In the diagram, $a \perp b$. Decide whether enough information is given to conclude that the statement is true. Explain your reasoning.

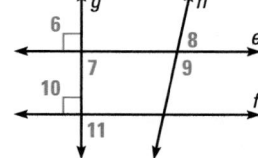

a. $\angle 1 \cong \angle 2$ Yes, enough information is given. Lines a and b are perpendicular. By Theorem 3.2, $\angle 1$ and $\angle 2$ are right angles. By Theorem 3.1, they are congruent.

b. $\angle 2 \cong \angle 6$ Not enough information is given to conclude that $\angle 2 \cong \angle 6$.

c. $\angle 7 \cong \angle 8$ Not enough information is given to conclude that $\angle 7 \cong \angle 8$.

Extra Example 2

In the window below, are $\angle FHG$ and $\angle GHD$ right angles? Explain.

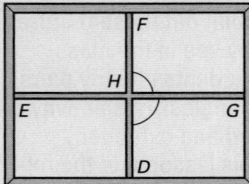

By Theorem 3.3, if two lines intersect to form adjacent congruent angles, then the lines are perpendicular. So $\overleftrightarrow{FD} \perp \overleftrightarrow{EG}$. By Theorem 3.2, perpendicular lines form four right angles. So $\angle FHG$ and $\angle GHD$ are right angles.

Extra Example 3

In the diagram, $\overleftrightarrow{HJ} \perp \overleftrightarrow{HL}$ and $m\angle LHK = 42°$. Find the value of z. **6**

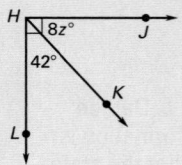

Study Skills

Encourage students to look for right angle marks on diagrams, and to immediately identify all the 90° angles they can deduce from the marks.

✓ Concept Check

What do you know about two lines intersecting to form adjacent congruent angles? **The lines are perpendicular.**

🌀 Daily Puzzler

Fold a sheet of paper in half. Fold it in half again, making this second fold perpendicular to the first. Fold the sheet in half twice more, each time perpendicular to the previous fold. How many perpendicular intersections of fold lines will you see when you unfold the paper? **9**

Link to Aviation

HELICOPTERS Main rotors of a helicopter may have two to eight blades. The blades create the helicopter's lift power.

7. $\overleftrightarrow{EH}$ and $\overleftrightarrow{FG}$ intersect to form adjacent congruent angles, so the lines are perpendicular. Perpendicular lines intersect to form 4 right angles, so $m\angle EFG = 90°$. $5x = 90$; $x = 18$.

8. $\angle BAC$ and $\angle CAD$ are adjacent acute angles and $\overrightarrow{AB} \perp \overrightarrow{AD}$, so $\angle BAC$ and $\angle CAD$ are complementary. $36° + 9y° = 90°$; $9y = 54$; $y = 6$.

9. $\angle JKM$ and $\angle MKL$ are adjacent acute angles and $\overrightarrow{KJ} \perp \overrightarrow{KL}$, so $\angle JKM$ and $\angle MKL$ are complementary. $z° + z° = 90°$; $2z = 90$; $z = 45$.

EXAMPLE 2 — Use Theorems About Perpendicular Lines

In the helicopter at the right, are $\angle AXB$ and $\angle CXB$ right angles? Explain.

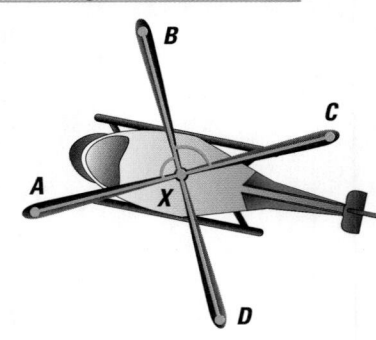

Solution

If two lines intersect to form adjacent congruent angles, as $\overline{AC}$ and $\overline{BD}$ do, then the lines are perpendicular (Theorem 3.3). So, $\overline{AC} \perp \overline{BD}$.

Because $\overline{AC}$ and $\overline{BD}$ are perpendicular, they form four right angles (Theorem 3.2). So, $\angle AXB$ and $\angle CXB$ are right angles.

EXAMPLE 3 — Use Algebra with Perpendicular Lines

In the diagram at the right, $\overrightarrow{EF} \perp \overrightarrow{EH}$ and $m\angle GEH = 30°$. Find the value of y.

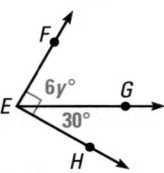

Solution

$\angle FEG$ and $\angle GEH$ are adjacent acute angles and $\overrightarrow{EF} \perp \overrightarrow{EH}$. So, $\angle FEG$ and $\angle GEH$ are complementary (Theorem 3.4).

$6y° + 30° = 90°$ $m\angle FEG + m\angle GEH = 90°$

$6y = 60$ Subtract 30 from each side.

$y = 10$ Divide each side by 6.

ANSWER ▶ The value of y is 10.

Checkpoint ✓ — Use Algebra with Perpendicular Lines

Find the value of the variable. Explain your reasoning. See margin.

7. $\angle EFG \cong \angle HFG$ **8.** $\overrightarrow{AB} \perp \overrightarrow{AD}$ **9.** $\overrightarrow{KJ} \perp \overrightarrow{KL}$, $\angle JKM \cong \angle MKL$

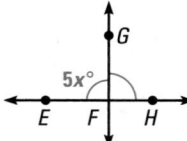   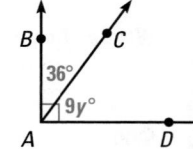   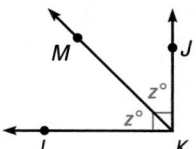

Guided Practice

Vocabulary Check

1. Complete the statement: If two lines intersect to form adjacent congruent angles, then the lines are __?__. perpendicular

Skill Check

2. If two lines are perpendicular, they intersect to form four right angles.

3. If two lines intersect to form adjacent congruent angles, then the lines are perpendicular.

4. If two sides of adjacent acute angles are perpendicular, the angles are complementary.

Write the theorem that justifies the statement about the diagram.

2. ∠5 and ∠6 are right angles.

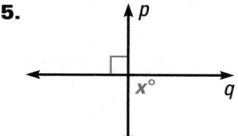

3. $j \perp k$

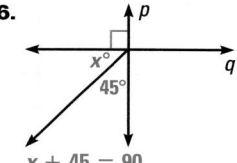

4. $m\angle 9 + m\angle 10 = 90°$

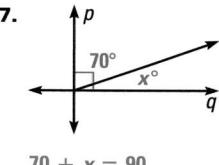

In Exercises 5–7, $p \perp q$. Write an equation to find the value of x. (Do not solve the equation.)

5.

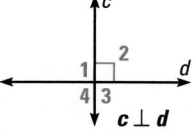

$x = 90$

6.

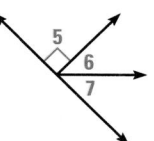

$x + 45 = 90$

7.

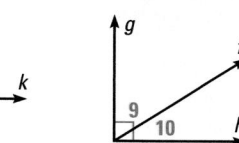

$70 + x = 90$

Practice and Applications

Extra Practice

See p. 679.

8. Yes; all right angles are congruent.

9. no

10. Yes; perpendicular lines intersect to form 4 right angles; all right angles are congruent.

11. Yes; perpendicular lines intersect to form 4 right angles; all right angles are congruent.

Homework Help

Example 1: Exs. 8–11
Example 2: Exs. 12–14
Example 3: Exs. 17–22

Perpendicular Lines and Reasoning Determine whether enough information is given to conclude that the statement is true. Explain.

8. ∠2 ≅ ∠5

9. ∠6 ≅ ∠7

10. ∠1 ≅ ∠3

11. ∠1 ≅ ∠5

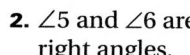

$c \perp d$

Logical Reasoning What can you conclude about ∠1 and ∠2 using the given information?

12. $\overrightarrow{BA} \perp \overleftrightarrow{BC}$

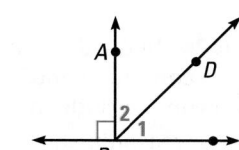

Sample answer: ∠1 and ∠2 are complementary.

13. $n \perp m$

Sample answers: ∠1 and ∠2 are right angles; ∠1 ≅ ∠2.

14. $h \perp k$

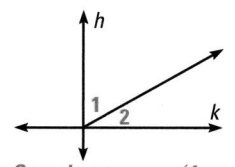

Sample answer: ∠1 and ∠2 are complementary.

3.2 *Theorems About Perpendicular Lines* **117**

X Common Error

In Exercise 26, watch for students who think the four angles must be right angles because they look like right angles in the diagram. Help them to understand that an angle measuring 89° or 91° will also look like 90° so they must use their geometry skills, not an estimate, to find angle measures.

Error Analysis Students were asked to set up an equation to find the value of *x*, given that *v* ⊥ *w*. Describe and correct any errors.

15.

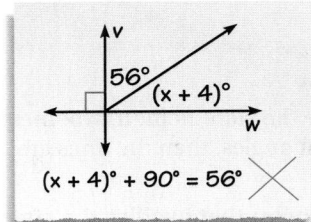

$(x + 4)° + 90° = 56°$ ✗

The equation should be $(x + 4)° + 56° = 90°$.

16.

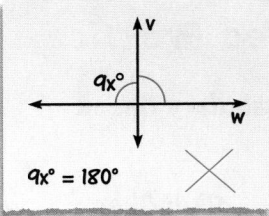

$9x° = 180°$ ✗

The equation should be $9x° = 90°$.

Student Help
CLASSZONE.COM

HOMEWORK HELP
Extra help with problem solving in Exs. 17–22 is at classzone.com

Using Algebra Find the value of *x*, given that *s* ⊥ *t*.

17. 90

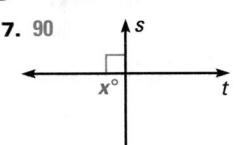

18. 30

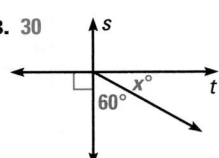

19. 35

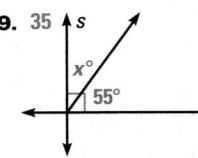

20. 35

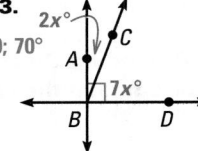

21. 15

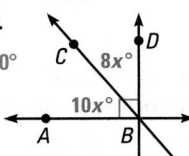

22. 6

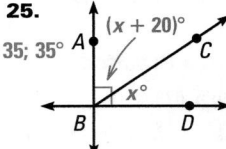

Angle Measures Find the value of *x*, given that $\overleftrightarrow{AB} \perp \overleftrightarrow{BD}$. Then use the value of *x* to find *m∠CBD*.

23.
10; 70°

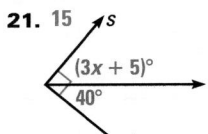

24.
5; 40°

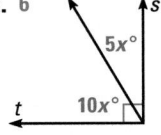

25.
35; 35°

26. No. *Sample answer:* If one of the angles is a right angle, then the crosspieces are perpendicular, so all four angles will be right angles.

27. No. Since ∠1 and ∠3 are congruent and complementary, each has measure 45°. There is no information given about the measure of ∠2, so you cannot conclude that $\overrightarrow{BA} \perp \overrightarrow{BC}$.

26. Window Repair You are fixing a window frame. You fit two strips of wood together to make the crosspieces. For the glass panes to fit, each angle formed by the crosspieces must be a right angle. Do you need to measure all four angles to be sure they are all right angles? Explain.

27. You be the Judge In the diagram shown, ∠1 and ∠3 are congruent and complementary. Can you conclude that $\overrightarrow{BA} \perp \overrightarrow{BC}$? Explain your reasoning.

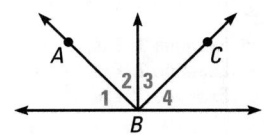

Link to
Sports

Sports In orienteering, a compass and a map are used to navigate through a wilderness area. Suppose you are in an orienteering event and you are traveling at 40° east of magnetic north, as shown below.

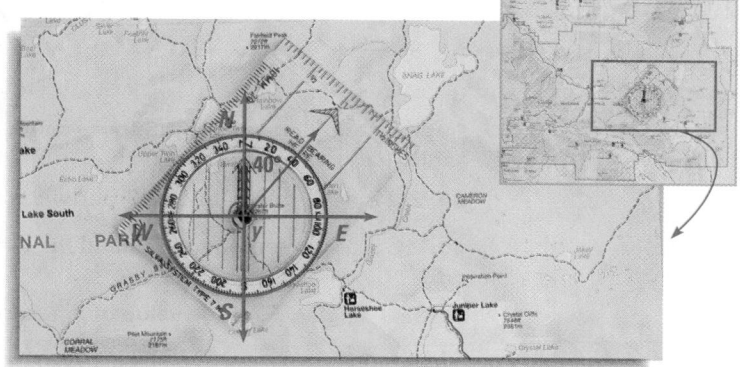

28. What can you conclude, given that $\angle NYW$ and $\angle SYW$ are congruent? Explain. $\overleftrightarrow{WE} \perp \overleftrightarrow{NS}$; if two lines intersect to form adjacent congruent angles, then the lines are perpendicular.

29. How many degrees do you need to turn to travel due east? 50°

30. How many degrees do you need to turn to travel due south from the position shown on the compass? 140°

Origami Origami is the Japanese art of folding pieces of paper into objects. The folds on the paper shown below are the basis for many objects. On the paper, $\overline{BF} \perp \overline{HD}$.

31. Yes; $\angle DJE$ and $\angle EJF$ are adjacent acute angles and $\overline{BF} \perp \overline{HD}$, so $\angle DJE$ and $\angle EJF$ are complementary.

32. Each has measure 45°.

33. No; there is no information given about $\angle AJG$ or the lines that form $\angle AJG$.

31. Are $\angle DJE$ and $\angle EJF$ complementary? Explain your reasoning.

32. If $m\angle BJC = m\angle CJD$, what are their measures?

33. Is there enough information to conclude that $\angle AJG$ is a right angle? Explain your reasoning.

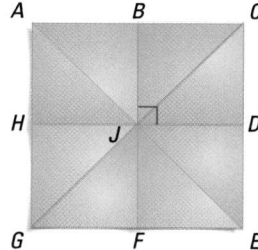

Standardized Test Practice

Multiple Choice In Exercises 34 and 35, use the diagram below.

34. Which of the following is true if $g \perp h$? B

Ⓐ $m\angle 1 + m\angle 2 > 180°$

Ⓑ $m\angle 1 + m\angle 2 < 180°$

Ⓒ $m\angle 1 + m\angle 2 = 180°$

Ⓓ None of these

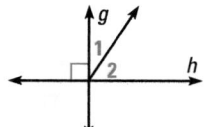

35. If $g \perp h$ and $m\angle 1 = 40°$, what is $m\angle 2$? G

Ⓕ 40°   Ⓖ 50°   Ⓗ 60°   Ⓙ 140°

Mini-Quiz
Determine whether enough information is given to conclude that the statement is true. Explain.

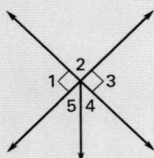

1. $\angle 1 \cong \angle 3$ Yes, all right angles are congruent.
2. $\angle 4 \cong \angle 5$ No, not enough information is given.
3. Find the value of x given that $s \perp t$. 8

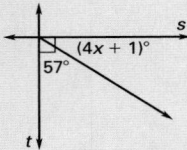

4. Angela constructs this bean pole for her beans to climb. The horizontal pieces must be at right angles to the vertical pole. Must Angela measure all four angles at each connection point to be sure they are all right angles? Explain.

No; if one angle is a right angle, then the horizontal piece and the pole are perpendicular. If they are perpendicular, then all four angles are right angles. (Theorem 3.2)

120

Mixed Review

Classifying Angles State whether the angle appears to be *acute, right, obtuse,* or *straight*. Then estimate its measure. (*Lesson 1.6*)

36.

37.

38.

obtuse; about 140°   acute; about 80°   right; about 90°

Finding Complements and Supplements Find the measure of the angle. (*Lesson 2.3*)

39. $\angle A$ is a complement of $\angle B$, and $m\angle A = 37°$. Find $m\angle B$. 53°

40. $\angle C$ is a supplement of $\angle D$, and $m\angle C = 56°$. Find $m\angle D$. 124°

Vertical Angles Find the value of x. (*Lesson 2.4*)

41. 15

80°
$(4x + 20)°$

42. 20

150° $(9x - 30)°$

43.
3

$(10x + 6)°$ $12x°$

Algebra Skills

Decimals Evaluate. (*Skills Review, p. 655*)

44. $13.6 + 9.8$ 23.4
45. $14 - 2.21$ 11.79
46. 7.4×5.9 43.66
47. $79.2 \div 9$ 8.8
48. $100 - 4.5 - 26.1$ 69.4
49. 30×11.1 333

Quiz 1

Think of each segment on the shopping bag as part of a line. There may be more than one correct answer. (*Lesson 3.1*)

1. Name two lines perpendicular to $\overleftrightarrow{FG}$. *Sample answer:* $\overleftrightarrow{BF}$, $\overleftrightarrow{CG}$
2. Name a line skew to $\overleftrightarrow{BF}$.
   Sample answer: $\overleftrightarrow{AD}$
3. Name a line that appears parallel to $\overleftrightarrow{AD}$. *Sample answer:* $\overleftrightarrow{BC}$
4. Name a line perpendicular to plane *HGC*. *Sample answer:* $\overleftrightarrow{BC}$

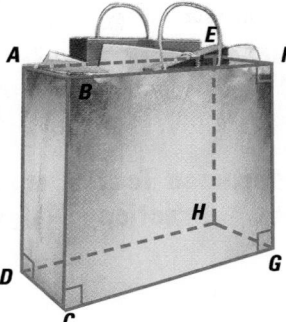

Find the value of the variable, given that $p \perp q$. (*Lesson 3.2*)

5. 67

$z°$ 23°
q

6. 11

57°
$3x°$
q

7. 34

$(3y - 12)°$
q

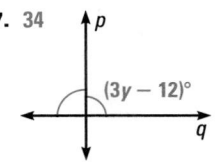

3.3 Angles Formed by Transversals

Goal

Identify angles formed by transversals.

Key Words

- transversal
- corresponding angles
- alternate interior angles
- alternate exterior angles
- same-side interior angles

A **transversal** is a line that intersects two or more coplanar lines at different points. For instance, in the diagram below, the blue line t is a transversal.

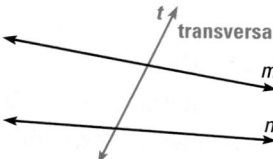

The angles formed by two lines and a transversal have special names.

Two angles are **corresponding angles** if they occupy corresponding positions.

The following pairs of angles are corresponding angles:

$\angle 1$ and $\angle 5$ $\angle 2$ and $\angle 6$

$\angle 3$ and $\angle 7$ $\angle 4$ and $\angle 8$

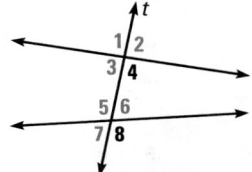

Two angles are **alternate interior angles** if they lie between the two lines on the opposite sides of the transversal.

The following pairs of angles are alternate interior angles:

$\angle 3$ and $\angle 6$

$\angle 4$ and $\angle 5$

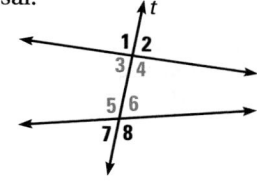

Two angles are **alternate exterior angles** if they lie outside the two lines on the opposite sides of the transversal.

The following pairs of angles are alternate exterior angles:

$\angle 1$ and $\angle 8$

$\angle 2$ and $\angle 7$

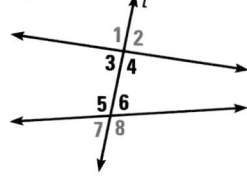

Two angles are **same-side interior angles** if they lie between the two lines on the same side of the transversal.

The following pairs of angles are same-side interior angles:

$\angle 3$ and $\angle 5$

$\angle 4$ and $\angle 6$

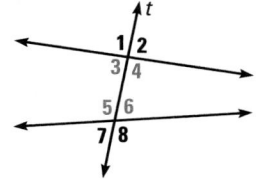

Visualize It!

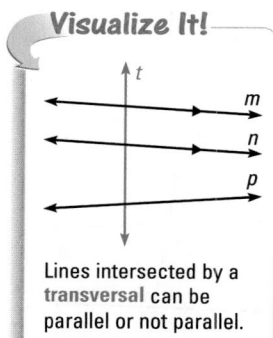

Lines intersected by a transversal can be parallel or not parallel.

3.3 *Angles Formed by Transversals* **121**

---

① Plan

Pacing

Suggested Number of Days

Basic: 2 days
Average: 2 days
Advanced: 2 days
Block Schedule: 1 block

Teaching Resources

📄 **Blacklines**
(See page 104B.)

🗂 **Transparencies**
- Warm-Up with Quiz
- Answers

💻 **Technology**
- Electronic Teacher Tools
- Test and Practice Generator
- Online Lesson Planner
- Internet Support

📼 **Video**
- **Geometry in Motion**

② Teach

Content and Teaching Strategies

For background information on geometric concepts and teaching strategies related to this lesson, see pages 104E and 104F in this Teacher's Edition.

Tips for New Teachers

Some students may be confused by all the numbers (1 through 8) shown in the textbook diagrams on this page. As an alternative to these diagrams, draw several simple, different diagrams. In each diagram show only one pair of angles (similar to the diagrams in Example 1 on page 122) and have the students identify each pair. See the Tips for New Teachers on pp. 1–2 of the *Chapter 3 Resource Book* for additional notes about Lesson 3.3.

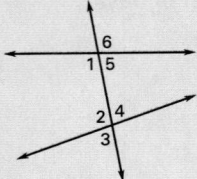
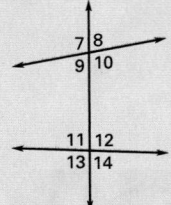

EXAMPLE 1 Describe Angles Formed by Transversals

Describe the relationship between the angles.

a. ∠1 and ∠2   b. ∠3 and ∠4   c. ∠5 and ∠6

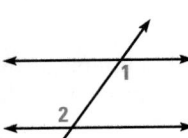   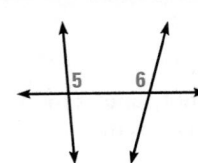

Solution

a. alternate interior angles

b. alternate exterior angles

c. same-side interior angles

Link to
Auto Racing

RACE CAR DESIGN
To maximize the speed of a
race car, the angles of the
front and rear wings can be
adjusted.

EXAMPLE 2 Identify Angles Formed by Transversals

List all pairs of angles that fit the description.

a. corresponding

b. alternate exterior

c. alternate interior

d. same-side interior

Solution

a. corresponding:
   ∠1 and ∠5   ∠2 and ∠6
   ∠3 and ∠7   ∠4 and ∠8

b. alternate exterior:
   ∠1 and ∠8   ∠3 and ∠6

c. alternate interior:
   ∠2 and ∠7   ∠4 and ∠5

d. same-side interior:
   ∠2 and ∠5   ∠4 and ∠7

Top view of car

Checkpoint ✓ Describe Angles Formed by Transversals

Describe the relationship between the angles.

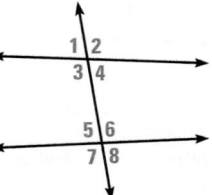

1. ∠2 and ∠7
   alternate exterior angles

2. ∠3 and ∠5
   same-side interior angles

3. ∠1 and ∠5
   corresponding angles

4. ∠4 and ∠5
   alternate interior angles

5. ∠4 and ∠8
   corresponding angles

6. ∠4 and ∠6
   same-side interior angles

Guided Practice

Vocabulary Check

1. Sketch two lines and a *transversal*. Shade a pair of *alternate interior angles*. **See margin.**

Skill Check

Match the diagram with the description of the angles.

A. alternate interior   B. corresponding   C. same-side interior

2.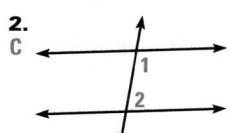
C

3.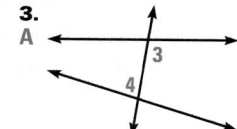
A

4.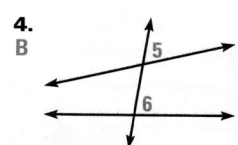
B

In Exercises 5–8, use the diagram shown below. The transversal is shown in blue.

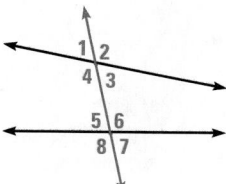

5. Name a pair of corresponding angles.
one of: ∠1 and ∠5, ∠2 and ∠6, ∠3 and ∠7, ∠4 and ∠8
6. Name a pair of alternate interior angles.
∠4 and ∠6, or ∠3 and ∠5
7. Name a pair of alternate exterior angles.
∠1 and ∠7, or ∠2 and ∠8
8. Name a pair of same-side interior angles.
∠3 and ∠6, or ∠4 and ∠5

Practice and Applications

Extra Practice
See p. 679.

9. same-side interior angles
10. alternate interior angles
11. corresponding angles

Describing Angles **Describe the relationship between ∠1 and ∠2. The transversal is shown in blue.**

9.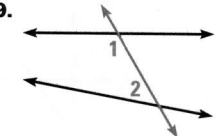

10.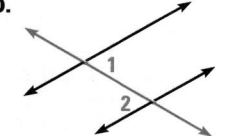

11.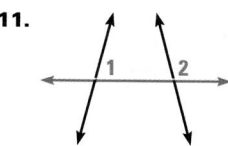

Identifying Angles **Use the diagram shown. There is more than one correct answer.**

12. Name a pair of corresponding angles.
Sample answer: ∠1 and ∠4
13. Name a pair of alternate interior angles.
Sample answer: ∠2 and ∠3
14. Name a pair of alternate exterior angles.
Sample answer: ∠1 and ∠9
15. Name a pair of same-side interior angles.
Sample answer: ∠2 and ∠4

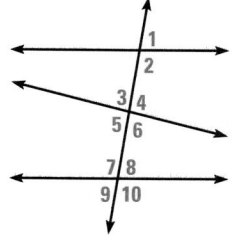

Homework Help

Example 1: Exs. 9–11, 16–28
Example 2: Exs. 12–15, 29–32

3.3 *Angles Formed by Transversals* **123**

③ Apply

Assignment Guide

BASIC
Day 1: pp. 123–125 Exs. 9–11, 16–21, 34–48 even
Day 2: pp. 123–125 Exs. 12–15, 29–32, 35–49 odd

AVERAGE
Day 1: pp. 123–125 Exs. 9–11, 16–25, 34–42 even
Day 2: pp. 123–125 Exs. 12–15, 29–32, 35–41 odd, 43–49

ADVANCED
Day 1: pp. 123–125 Exs. 16–28, 34–39
Day 2: pp. 123–125 Exs. 29–32, 33*, 40–49; EC: classzone.com

BLOCK SCHEDULE
pp. 123–125 Exs. 9–25, 29–32, 34–49

Extra Practice
• Student Edition, p. 679
• Chapter 3 Resource Book, pp. 28–29

Homework Check
To quickly check student understanding of key concepts, go over the following exercises:

Basic: 9, 11, 13, 14, 31
Average: 12, 13, 17, 24, 29
Advanced: 16, 21, 23, 26, 30

1. *Sample answer:*

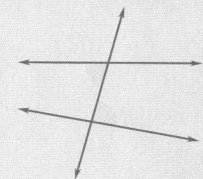

Link to
Careers

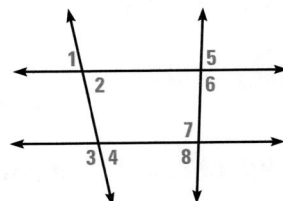

BICYCLE DESIGNER
Georgena Terry influenced the cycling world by building bicycles whose designs contain a frame geometry more suitable for women.

Describing Angles In Exercises 16–21, complete the statement using *corresponding, alternate interior, alternate exterior,* or *same-side interior.*

16. ∠6 and ∠7 are __?__ angles.
 alternate interior
17. ∠1 and ∠6 are __?__ angles.
 alternate exterior
18. ∠2 and ∠4 are __?__ angles.
 corresponding
19. ∠2 and ∠6 are __?__ angles.
 corresponding
20. ∠4 and ∠8 are __?__ angles.
 same-side interior
21. ∠4 and ∠7 are __?__ angles.
 alternate interior

22. **Bicycles** In the bicycle shown, what is the relationship between ∠7 and ∠8? same-side interior angles

Classifying Angles Use the diagram below to describe the relationship between the pair of angles.

23. ∠BCA and ∠DGJ
 alternate exterior angles
24. ∠DGJ and ∠FDE
 corresponding angles
25. ∠FDE and ∠KHL
 alternate exterior angles
26. ∠DGJ and ∠GJH
 alternate interior angles
27. ∠CGH and ∠GJH
 corresponding angles
28. ∠BCA and ∠CGH
 corresponding angles

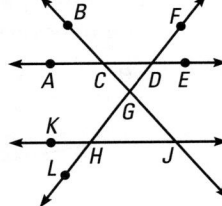

Easels Use the diagram of the easel at the right. An easel is used to display or support an artist's canvas.

29. Name two pairs of alternate exterior angles.
 ∠5 and ∠12, ∠7 and ∠10
30. Name two pairs of same-side interior angles.
 ∠6 and ∠9, ∠8 and ∠11
31. Name two pairs of alternate interior angles.
 ∠6 and ∠11, ∠8 and ∠9
32. Name three pairs of corresponding angles.
 three of: ∠5 and ∠9, ∠6 and ∠10, ∠7 and ∠11, ∠8 and ∠12

33. Challenge Sketch two lines and a transversal. Label angles 1, 2, 3, and 4 so that ∠1 and ∠2 are corresponding angles, ∠3 and ∠4 are corresponding angles, and ∠1 and ∠4 are same-side interior angles. **See margin.**

Standardized Test Practice

34. Multiple Choice In the diagram below, ∠1 and ∠3 are ____?____ angles. **B**

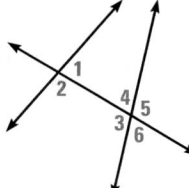

  Ⓐ corresponding

  Ⓑ alternate interior

  Ⓒ alternate exterior

  Ⓓ same-side interior

35. Multiple Choice In the diagram above, a pair of corresponding angles are ____?____. **J**

  Ⓕ ∠1 and ∠6   Ⓖ ∠3 and ∠5

  Ⓗ ∠1 and ∠2   Ⓙ None of these

Mixed Review

Line Relationships Think of each segment in the diagram as part of a line. Fill in the blank. There may be more than one correct answer. *(Lesson 3.1)*

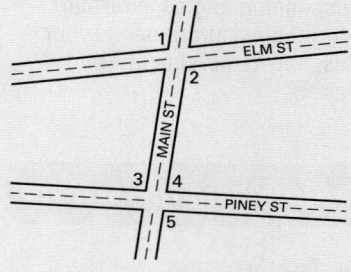

36. $\overleftrightarrow{AD}$ and ____?____ are perpendicular. $\overleftrightarrow{AB}$ or $\overleftrightarrow{AE}$

37. $\overleftrightarrow{DC}$ and ____?____ appear to be parallel. *37.* $\overleftrightarrow{AB}$, $\overleftrightarrow{GH}$, or $\overleftrightarrow{EF}$

38. $\overleftrightarrow{EH}$ and ____?____ are skew. $\overleftrightarrow{AB}$, $\overleftrightarrow{CD}$, $\overleftrightarrow{BF}$, or $\overleftrightarrow{CG}$

39. Plane *EFG* and plane ____?____ appear to be parallel. *ABC*

Perpendicular Lines Find the value of *x*, given that $\ell \perp k$. *(Lesson 3.2)*

40. 36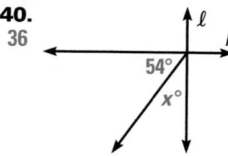

41. 24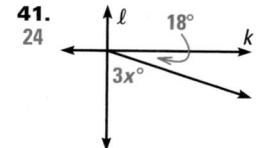

42. 18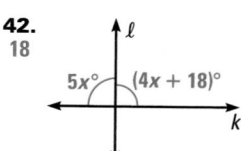

Algebra Skills

Ratios Simplify the ratio. *(Skills Review, p. 660)*

43. $\dfrac{32 \text{ ft}}{80 \text{ ft}}$ $\dfrac{2}{5}$ **44.** $\dfrac{6 \text{ yd}}{24 \text{ ft}}$ $\dfrac{3}{4}$ **45.** $\dfrac{7 \text{ ft}}{84 \text{ in.}}$ 1   **46.** $\dfrac{10 \text{ mi}}{800 \text{ ft}}$ 66

Evaluating Expressions Evaluate the expression when *y* = 4. *(Skills Review, p. 670)*

47. $y(y + 9)$ 52   **48.** $y^2 - 2y$ 8   **49.** $(y - 1)(y + 1)$ 15

④ Assess

Assessment Resources
The Mini-Quiz below is also available on blackline (*Chapter 3 Resource Book*, p. 35) and on transparency. For more assessment resources, see:
• Chapter 3 Resource Book
• Standardized Test Practice
• Test and Practice Generator

Mini-Quiz
Use the diagram below.

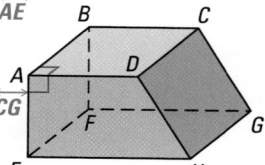

1. Which street represents a transversal? **Main Street**
2. Name a pair of corresponding angles. **∠2 and ∠5; ∠1 and ∠3**
3. Name a pair of alternate interior angles. **∠2 and ∠3**
4. Name a pair of alternate exterior angles. **∠1 and ∠5**
5. Name a pair of same-side interior angles. **∠2 and ∠4**

33. **Sample answer:**

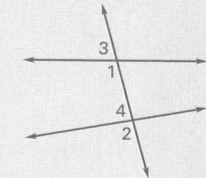

Students explore the measures of angles formed when two parallel lines are cut by a transversal.

Materials
See the margin of the student page.

LINK TO LESSON
In Examples 1–4 on pages 128–131, students apply what they discover in the activity to find measures of corresponding angles, alternate interior angles, alternate exterior angles, and same-side interior angles.

Managing the Activity

Tips for Success
Stress that at the beginning of the activity the parallel lines must be carefully and precisely drawn. If the lines are not exactly parallel, students are unlikely to make the correct discovery about the related angle pairs.

Activity 3.4 Parallel Lines and Angles

Question

What are the relationships among the angles formed when two parallel lines are cut by a transversal?

Materials
- lined paper
- straightedge
- protractor

Explore

❶ On a piece of lined paper, trace over two of the lines with a pen or pencil. Label one of the parallel lines j and label the other k.

❷ Draw a transversal intersecting the parallel lines. Label the transversal t. Label the angles as shown.

❸ Use a protractor to measure all eight angles formed by the lines. Record the angle measures on a piece of paper.

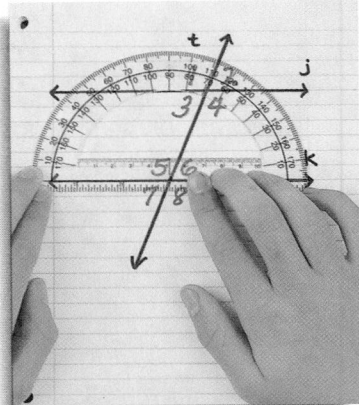

Student Help

LOOK BACK
To review definitions of corresponding, alternate interior, alternate exterior, and same-side interior angles, see p. 121.

Think About It

1. Name a pair of corresponding angles. What are their measures?
   Sample answer: ∠1 and ∠5; measures will vary, but will be equal.
2. Name a pair of alternate interior angles. What are their measures?
   ∠3 and ∠6, or ∠4 and ∠5; measures will vary, but will be equal.
3. Name a pair of alternate exterior angles. What are their measures?
   ∠1 and ∠8, or ∠2 and ∠7; measures will vary, but will be equal.
4. Name a pair of same-side interior angles. What are their measures?
   ∠3 and ∠5, or ∠4 and ∠6; measures will vary, but will have the sum 180°.

Explore

④ Draw two parallel lines *j* and *k*, and a transversal *s*. Label the angles as shown.

⑤ Use a protractor to measure all eight angles. Record the angle measures on a piece of paper.

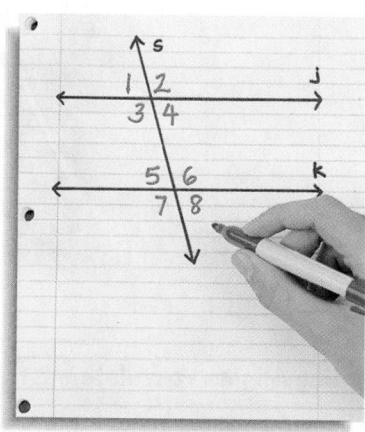

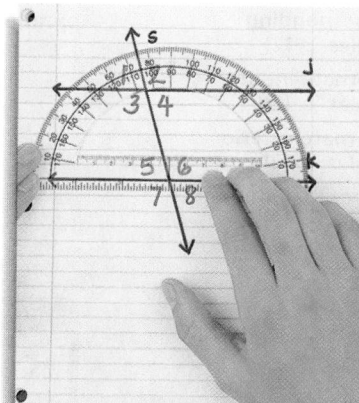

KEY DISCOVERY

If two parallel lines are cut by a transversal, then pairs of corresponding angles are congruent, pairs of alternate interior angles are congruent, pairs of alternate exterior angles are congruent, and pairs of same-side interior angles are supplementary.

Activity Assessment

Use Exercises 5–8 to assess student understanding.

5. Tables will vary. *Conjecture:* When two parallel lines are cut by a transversal, the pairs of alternate interior angles have equal measures.

6. Tables will vary. *Conjecture:* When two parallel lines are cut by a transversal, the pairs of alternate exterior angles have equal measures.

7. Tables will vary. *Conjecture:* When two parallel lines are cut by a transversal, the pairs of same-side interior angles have measures that add up to 180°.

8. Yes. When two parallel lines are cut by a transversal, the corresponding angles have equal measures, so $m\angle 2 = 90° = m\angle 6$. Therefore, $\angle 6$ is a right angle. Since lines *s* and *k* intersect to form a right angle, $s \perp k$.

⑥ Make a table like the one shown. In the angle column, list pairs of corresponding angles. Then use your list of angle measures to complete the table.

⑦ Make a conjecture about the measures of pairs of corresponding angles when two parallel lines are cut by a transversal.

Step 7. *Conjecture:* **When two parallel lines are cut by a transversal, the pairs of corresponding angles have equal measures.**

	Angle	Measure
Pair 1	$\angle 1$	?
	$\angle 5$	?
Pair 2	$\angle 2$	?
	$\angle 6$	?

Think About It

5. Repeat Steps 6 and 7 above for pairs of alternate interior angles.

6. Repeat Steps 6 and 7 above for pairs of alternate exterior angles.

7. Repeat Steps 6 and 7 above for pairs of same-side interior angles.

8. In the photo for Step 4 above, suppose $\angle 2$ is a right angle. Is line *s* perpendicular to line *k*? Explain your reasoning.

Activity **127**

Pacing
Suggested Number of Days

Basic: 3 days
Average: 3 days
Advanced: 3 days
Block Schedule: 1 block
0.5 block with 3.5

Teaching Resources

📃 **Blacklines**
(See page 104B.)

🔧 **Transparencies**
• Warm-Up with Quiz
• Answers

🖥 **Technology**
• Electronic Teacher Tools
• Test and Practice Generator
• Online Lesson Planner
• Internet Support

📼 **Video**
• **Geometry in Motion**

② Teach

Content and Teaching Strategies
For background information on geometric concepts and teaching strategies related to this lesson, see pages 104E and 104F in this Teacher's Edition.

Tips for New Teachers

Be sure students are aware of the difference between the definitions in Lesson 3.3 and the theorems in Lesson 3.4 about the angles formed by parallel lines cut by a transversal. One approach might be to explain that the definitions indicate the location of the angles and that the theorems indicate information about their measures. See the Tips for New Teachers on pp. 1–2 of the *Chapter 3 Resource Book* for additional notes about Lesson 3.4.

3.4 Parallel Lines and Transversals

Goal
Find the congruent angles formed when a transversal cuts parallel lines.

Key Words
• transversal p. 121
• corresponding angles p. 121
• alternate interior angles p. 121
• alternate exterior angles p. 121
• same-side interior angles p. 121

In the photograph of the tennis court, the angle the sideline makes with the service line is the same as the angle it makes with the base line.

This photograph illustrates a postulate about angles and parallel lines.

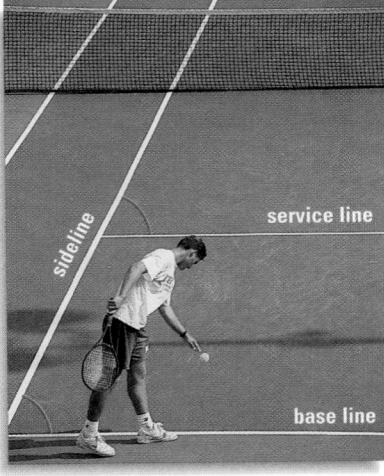
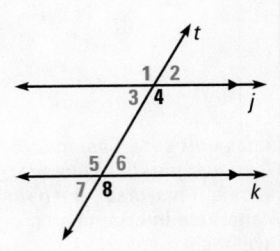

POSTULATE 8

Corresponding Angles Postulate

Words If two parallel lines are cut by a transversal, then corresponding angles are congruent.

Symbols If $j \parallel k$, then the following are true:

$$\angle 1 \cong \angle 5 \qquad \angle 2 \cong \angle 6$$
$$\angle 3 \cong \angle 7 \qquad \angle 4 \cong \angle 8$$

Visualize It!

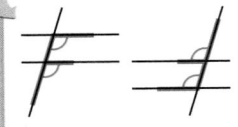

Look for angles in an F shape to help you find corresponding angles.

EXAMPLE ① Find Measures of Corresponding Angles

Find the measure of the numbered angle.

a.
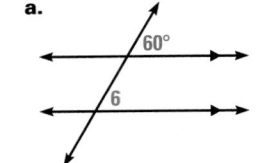

b.
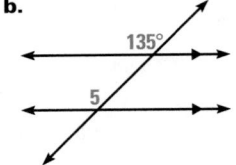

c.
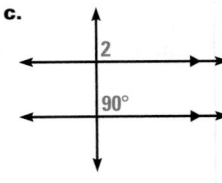

Solution

  a. $m\angle 6 = 60°$ **b.** $m\angle 5 = 135°$ **c.** $m\angle 2 = 90°$

Checkpoint ✓ *Find Measures of Corresponding Angles*

Find the measure of the numbered angle.

1.
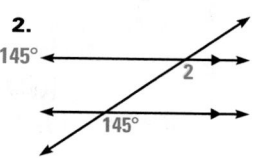
120°
120°
1

2.
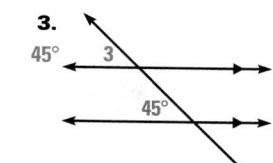
145°
2
145°

3.
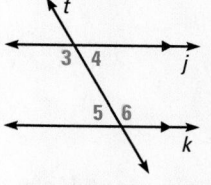
45° 3
45°

THEOREM 3.5

Alternate Interior Angles Theorem

Words If two parallel lines are cut by a transversal, then alternate interior angles are congruent.

Symbols If $j \parallel k$, then the following are true:

$\angle 3 \cong \angle 6$

$\angle 4 \cong \angle 5$

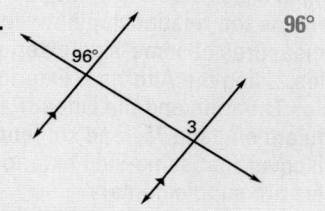

Visualize It!

Look for angles inside a Z or N shape to find alternate interior angles.

EXAMPLE 2 **Find Measures of Alternate Interior Angles**

Find the measure of $\angle PQR$.

a.
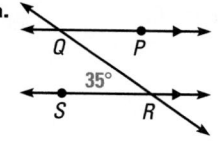
Q P
35°
S R

b.
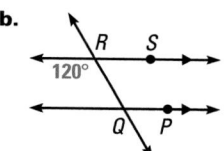
R S
120°
Q P

c.
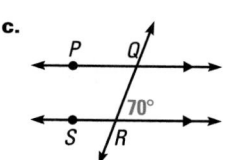
P Q
70°
S R

Solution

a. $m\angle PQR = 35°$ **b.** $m\angle PQR = 120°$ **c.** $m\angle PQR = 70°$

Checkpoint ✓ *Find Measures of Alternate Interior Angles*

Find the measure of the numbered angle.

4.
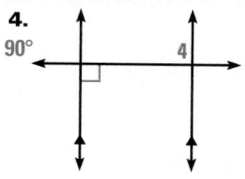
90°
4

5.
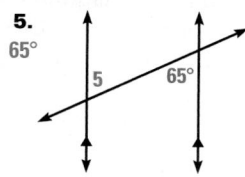
65°
5 65°

6.
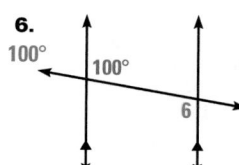
100°
100°
6

3.4 Parallel Lines and Transversals **129**

Extra Example 1

Find the measure of the numbered angle.

a.
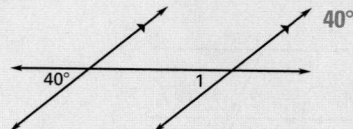
40°
40° 1

b.
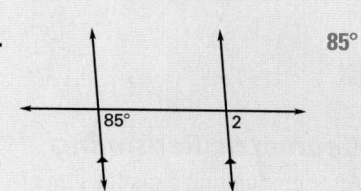
85°
85° 2

c.
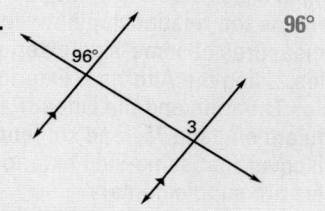
96°
96°
3

Extra Example 2

Find the measure of $\angle ABC$.

a.
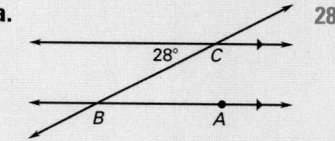
28°
28° C
B A

b.
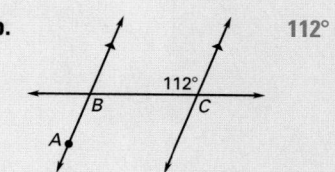
112°
112°
B C
A

c.
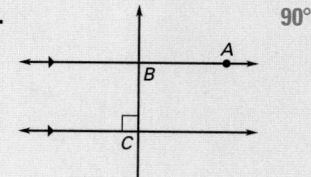
90°
A
B
C

Visualize It!

Remind students that unless lines are marked as being parallel, students cannot assume they are parallel, even though they appear to be.

129

Find the measures of ∠1 and ∠2.

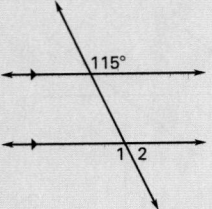

$m\angle 1 = 115°$, $m\angle 2 = 65°$

Geometric Reasoning

After completing a discussion of Example 3, ask students if they can determine the relationship between the measures of same-side exterior angles. Using the Alternate Exterior Angles Theorem and the Linear Pair Postulate on page 75, lead students to discover that same-side exterior angles are supplementary.

THEOREM 3.6

Alternate Exterior Angles Theorem

Words If two parallel lines are cut by a transversal, then alternate exterior angles are congruent.

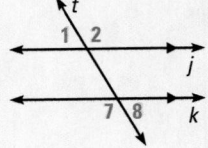

Symbols If $j \parallel k$, then the following are true:

$$\angle 1 \cong \angle 8$$
$$\angle 2 \cong \angle 7$$

EXAMPLE 3 Find Measures of Alternate Exterior Angles

Find the measures of ∠1 and ∠2.

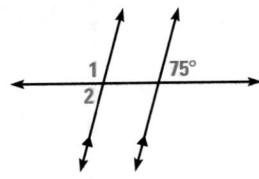

Student Help

LOOK BACK
To review linear pairs, see p. 75.

Solution

The measure of ∠2 is 75° because alternate exterior angles are congruent. The measure of ∠2 can be used to find the measure of ∠1.

$m\angle 1 + m\angle 2 = 180°$	Linear Pair Postulate
$m\angle 1 + 75° = 180°$	Substitute 75° for $m\angle 2$.
$m\angle 1 + 75° - 75° = 180° - 75°$	Subtract 75° from each side.
$m\angle 1 = 105°$	Simplify.

Checkpoint ✓ **Use Angle Relationships**

Find the measure of the numbered angle.

7.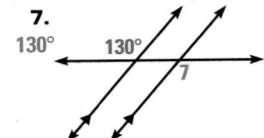

8.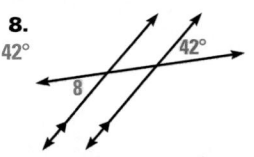

9.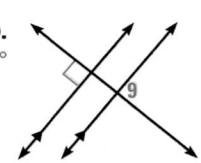

10. congruent by the Alternate Exterior Angles Theorem

11. Not congruent; the angles are a linear pair.

12. Not congruent; the angles are a linear pair.

13. congruent by the Alternate Exterior Angles Theorem

14. congruent by the Corresponding Angles Postulate

15. Not congruent; there is no special relationship between these angles.

Use the diagram below. Tell whether the angles are *congruent* or *not congruent*. Explain.

10. ∠1 and ∠8   **11.** ∠3 and ∠4

12. ∠4 and ∠2   **13.** ∠2 and ∠7

14. ∠3 and ∠7   **15.** ∠3 and ∠8

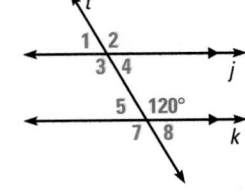

Look for angles inside a C shape to find same-side interior angles.

THEOREM 3.7

Same-Side Interior Angles Theorem

Words If two parallel lines are cut by a transversal, then same-side interior angles are supplementary.

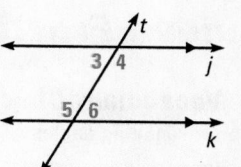

Symbols If $j \parallel k$, then the following are true:

$$m\angle 3 + m\angle 5 = 180°$$
$$m\angle 4 + m\angle 6 = 180°$$

EXAMPLE 4 **Find Measures of Same-Side Interior Angles**

Find the measure of the numbered angle.

a.
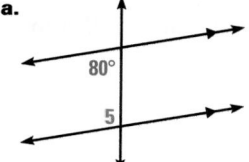

b.
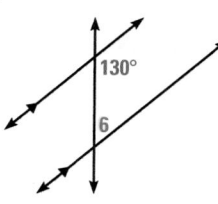

Solution

a. $m\angle 5 + 80° = 180°$
   $m\angle 5 = 100°$

b. $m\angle 6 + 130° = 180°$
   $m\angle 6 = 50°$

Using Algebra xy

EXAMPLE 5 **Use Algebra with Angle Relationships**

Find the value of x.

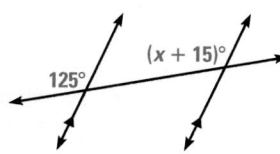

Solution

$(x + 15)° = 125°$ Corresponding Angles Postulate

$x = 110$ Subtract 15 from each side.

Checkpoint ✓ **Use Algebra with Angle Relationships**

Find the value of x.

16.
85
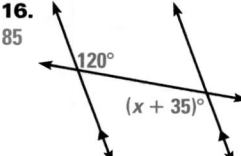

17.
104

18.
40
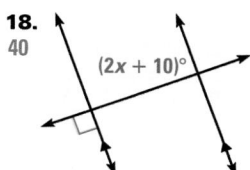

3.4 *Parallel Lines and Transversals*   131

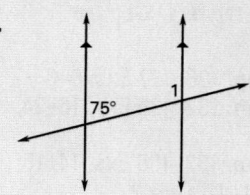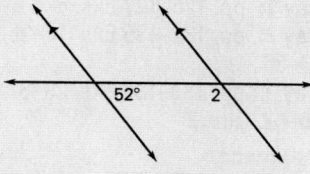
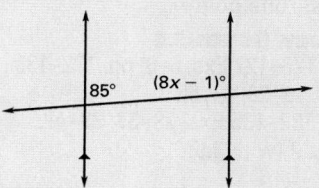
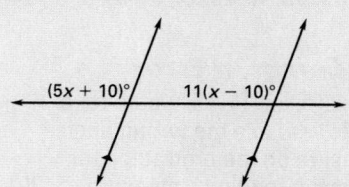

132

Assignment Guide

BASIC
Day 1: pp. 126–127 Exs. 6–8
Day 2: pp. 132–135 Exs. 16–24, 43–56
Day 3: pp. 132–135 Exs. 14, 15, 29–34, 41, 42, Quiz 2

AVERAGE
Day 1: pp. 126–127 Exs. 6–8
Day 2: pp. 132–135 Exs. 16–26, 43–56
Day 3: pp. 132–135 Exs. 29–37, 39–42, Quiz 2

ADVANCED
Day 1: pp. 126–127 Exs. 6–8
Day 2: pp. 132–135 Exs. 16–24 even, 25–28, 43–56
Day 3: pp. 132–135 Exs. 29–31, 35–42, Quiz 2; EC: TE p. 104D*, classzone.com

BLOCK SCHEDULE
pp. 126–127 Exs. 6–8; pp. 132–135 Exs. 16–26, 43–56
pp. 132–135 Exs. 29–37, 39–42, Quiz 2 (with 3.5)

Extra Practice
• Student Edition, p. 679
• Chapter 3 Resource Book, pp. 36–37

Homework Check
To quickly check student understanding of key concepts, go over the following exercises:

Basic: 17, 20, 22, 30, 33
Average: 17, 21, 23, 29, 32
Advanced: 18, 20, 24, 31, 36

✗ Common Error
In Exercises 14–15, watch for students who use the wrong angle measure on the protractor, for example, reading a measure of 120° instead of 60°. Remind students to check each angle measure by asking themselves whether the angle is acute or obtuse and thus whether its measure should be less than or greater than 90°.

3.4 Exercises

Guided Practice

Vocabulary Check
1. corresponding angles
2. alternate interior angles
3. none of these
4. corresponding angles
5. alternate exterior angles
6. corresponding angles
7. same-side interior angles
8. none of these

Tell whether the angles are corresponding angles, alternate interior angles, alternate exterior angles, same-side interior angles, or none of these.

1. $\angle 1$ and $\angle 5$ 2. $\angle 5$ and $\angle 4$
3. $\angle 2$ and $\angle 8$ 4. $\angle 6$ and $\angle 2$
5. $\angle 3$ and $\angle 6$ 6. $\angle 7$ and $\angle 3$
7. $\angle 4$ and $\angle 7$ 8. $\angle 8$ and $\angle 3$

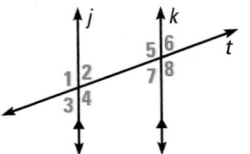

Skill Check
9. Alternate Exterior Angles Theorem
10. Alternate Interior Angles Theorem
11. Same-Side Interior Angles Theorem
12. Corresponding Angles Postulate

What postulate or theorem justifies the statement?

9. $\angle 10 \cong \angle 15$
10. $\angle 12 \cong \angle 13$
11. $m\angle 11 + m\angle 13 = 180°$
12. $\angle 9 \cong \angle 13$

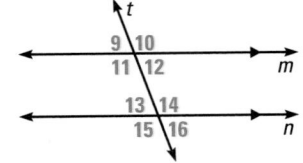

13. **Logical Reasoning** Two parallel lines are cut by a transversal so that one of the angles formed is a right angle. What can you say about the measures of all the other angles? Explain. **See margin.**

Practice and Applications

Extra Practice
See p. 679.

Homework Help
Example 1: Exs. 16–18
Example 2: Exs. 19–21, 25
Example 3: Exs. 22–24
Example 4: Exs. 29–31
Example 5: Exs. 32–37

Visualize It! Draw two parallel lines. Use a protractor to draw a transversal so that one of the angles has the given measure. Measure all the angles and write the angle measures on your drawing.
14, 15. See margin.

14.

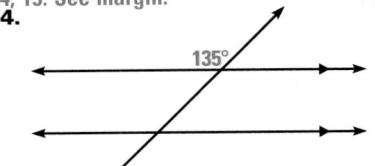

15.

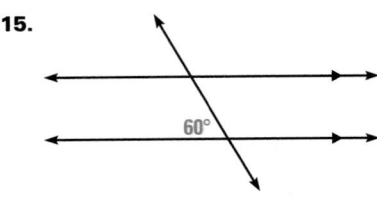

Corresponding Angles Find the measure of the numbered angle.

16.   17.   18.

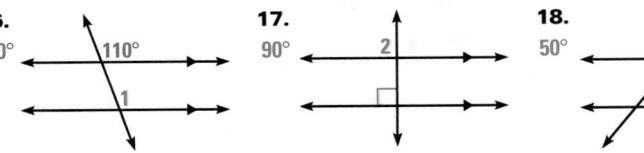

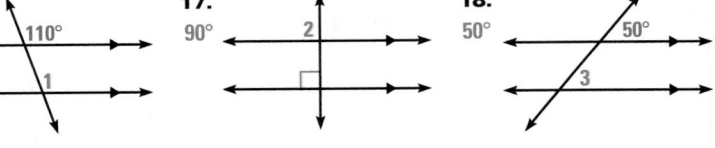

Alternate Interior Angles Find the measure of the numbered angle.

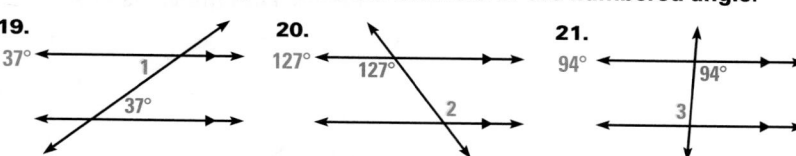

19. 37° ∠1 37°

20. 127° 127° ∠2

21. 94° 94° ∠3

Alternate Exterior Angles Find the measure of ∠*ABC*.

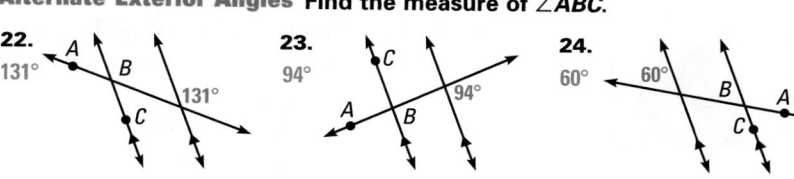

22. 131° A B C 131°

23. 94° C A B 94°

24. 60° 60° B A C

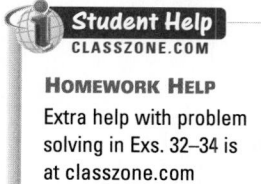
25. Rainbows When sunlight enters a drop of rain, different colors leave the drop at different angles. For red light, *m*∠2 = 42°. What is *m*∠1? Explain.

m∠1 = 42° by the Alternate Interior Angles Theorem

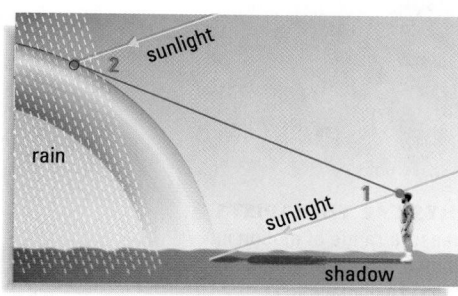

sunlight

2

rain

sunlight

1

shadow

Logical Reasoning Find *m*∠1 and *m*∠2. Explain your reasoning.

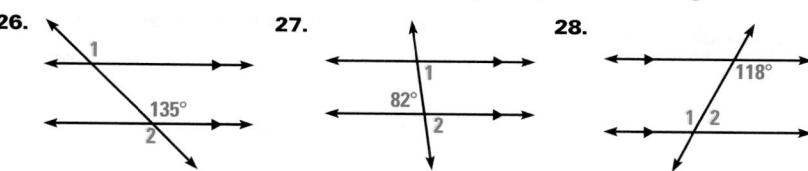

26. 1 135° 2

27. 1 82° 2

28. 118° 1 2

26. *m*∠1 = 135° by the Corresponding Angles Postulate; *m*∠2 = 135° by the Vertical Angles Theorem.

27. *m*∠1 = 82° by the Alternate Interior Angles Theorem; *m*∠2 = 82° by the Vertical Angles Theorem.

28. *m*∠1 = 118° by the Alternate Interior Angles Theorem; *m*∠2 = 62° by the Same-Side Interior Angles Theorem.

Same-Side Interior Angles Find the measure of the numbered angle.

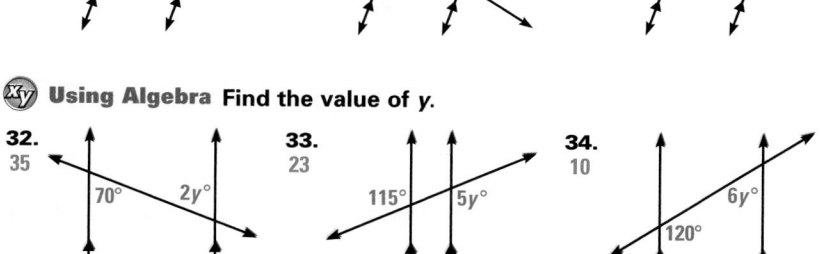

29. 131° 49° 1

30. 103° 77° 2

31. 76° 3 104°

Using Algebra Find the value of *y*.

32. 35 70° 2*y*°

33. 23 115° 5*y*°

34. 10 6*y*° 120°

Teaching Tip

If possible, invite a physics teacher to class to explain how the formation of rainbows is related to geometry.

13. The measure of all the other angles is 90°. *Sample explanation:* The two lines that form the 90° angle are perpendicular, so the other three angles formed by the two lines are also 90° angles. Since each of the other four angles is a corresponding angle to one of the right angles, the other four angles are right angles by the Corresponding Angles Postulate.

14.

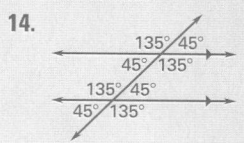

135° 45°
45° 135°
135° 45°
45° 135°

15.

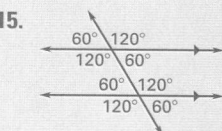

60° 120°
120° 60°
60° 120°
120° 60°

39.

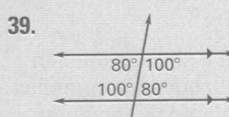

40.

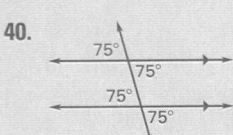

38. ∠*ABP* ≅ ∠*DEB* and
∠*PBC* ≅ ∠*BEF* by the
Corresponding Angles
Postulate. By the Angle
Addition Postulate and the
Addition Property of Equality,
∠*ABC* ≅ ∠*DEF*.

Link to
Careers

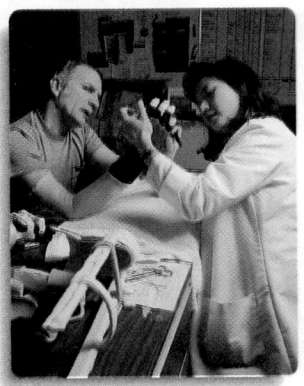

PHYSICAL THERAPISTS
measure range of motion in
elbows, knees, and other
joints. They use these
measurements as they help
people recover from injuries.

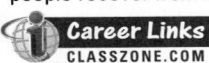
Career Links
CLASSZONE.COM

---

xy **Using Algebra** Find the value of *x*.

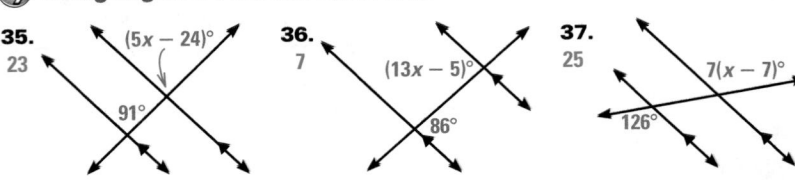

35. 23 $(5x - 24)°$ $91°$

36. 7 $(13x - 5)°$ $86°$

37. 25 $7(x - 7)°$ $126°$

38. **Physical Therapy** Sports physicians and physical therapists use a
tool called a *goniometer* to measure range of motion.

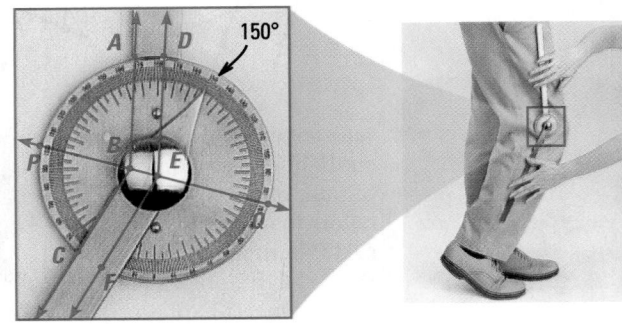

In the diagram, $\overrightarrow{BA} \parallel \overrightarrow{ED}$ and $\overrightarrow{BC} \parallel \overrightarrow{EF}$. Use the blue transversal
to explain why ∠*ABC* ≅ ∠*DEF*. See margin.

Error Analysis **A student has written some angle measures**
incorrectly. Copy the diagram and correct the errors. 39, 40. See margin.

39.

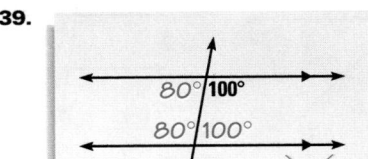

40.

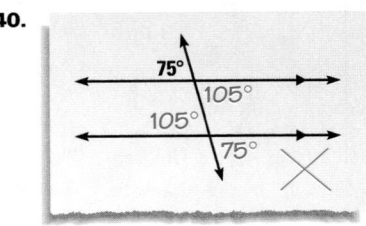

Standardized Test
Practice

41. **Multiple Choice** Which statement is *false*? C

Ⓐ $m∠2 + m∠5 = 180°$

Ⓑ $m∠5 + m∠6 = 180°$

Ⓒ $m∠6 + m∠7 = 180°$

Ⓓ $m∠3 + m∠8 = 180°$

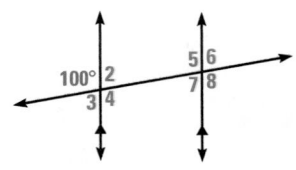

42. **Multiple Choice** Which statement about the diagram above
is *true*? J

Ⓕ ∠2 ≅ ∠4   Ⓖ ∠5 ≅ ∠7   Ⓗ ∠3 ≅ ∠8   Ⓙ ∠6 ≅ ∠3

Mixed Review

Identifying Line Relationships Fill in the blank with *parallel,* *perpendicular,* or *skew.* (*Lesson 3.1*)

43. Line j and line k are __?__. parallel

44. Line j and line m are __?__. perpendicular

45. Line k and line m are __?__. skew

46. Line m appears __?__ to plane B. perpendicular

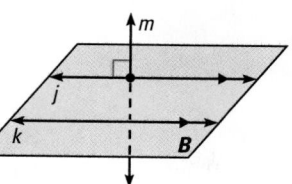

Studying Angles List all pairs of angles that fit the description. (*Lesson 3.3*)

47. ∠1 and ∠5, ∠2 and ∠6, ∠3 and ∠7, ∠4 and ∠8

48. ∠2 and ∠7, ∠4 and ∠5

49. ∠1 and ∠8, ∠3 and ∠6

50. ∠2 and ∠5, ∠4 and ∠7

47. corresponding

48. alternate interior

49. alternate exterior

50. same-side interior

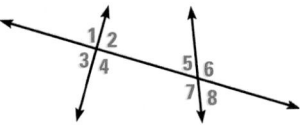

Algebra Skills

Solving Equations Solve the equation. (*Skills Review, p. 673*)

51. $3y - 4 = 20$ 8

52. $4 - 6p = 2p - 3$ $\frac{7}{8}$

53. $75 + 7x = 2x$ -15

54. $14r + 81 = -r$ -5.4

55. $12s - 5 = 7s$ 1

56. $5(z + 3) = 12$ $-\frac{3}{5}$

Quiz 2

Use the diagram to describe the relationship between the pair of angles. (*Lesson 3.3*)

1. alternate exterior angles
2. same-side interior angles
3. corresponding angles
4. alternate exterior angles
5. alternate interior angles
6. alternate interior angles

1. ∠1 and ∠8

2. ∠4 and ∠6

3. ∠6 and ∠2

4. ∠2 and ∠7

5. ∠4 and ∠5

6. ∠3 and ∠6

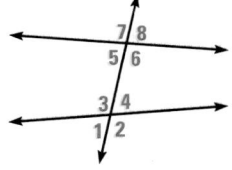

Find the measures of ∠1 and ∠2. (*Lesson 3.4*)

7.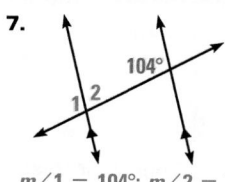

$m\angle 1 = 104°; m\angle 2 = 76°$

8.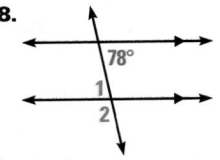

$m\angle 1 = 78°; m\angle 2 = 102°$

9.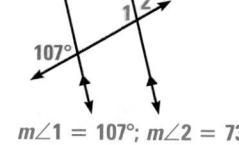

$m\angle 1 = 107°; m\angle 2 = 73°$

Find the value of x. (*Lesson 3.4*)

10. 21
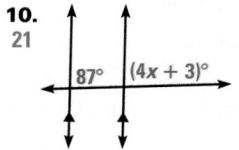

11. 26
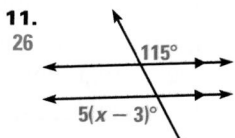

12. 11
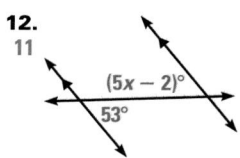

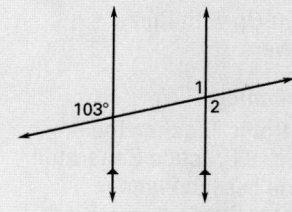

Assessment Resources

The Mini-Quiz below is also available on blackline (*Chapter 3 Resource Book*, p. 44) and on transparency. For more assessment resources, see:

- Chapter 3 Resource Book
- Standardized Test Practice
- Test and Practice Generator

Mini-Quiz

1. Find $m\angle 1$ and $m\angle 2$. Explain your reasoning.

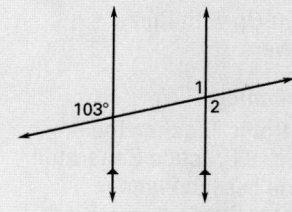

$m\angle 1 = 103°$ (corresponding angles), $m\angle 2 = 103°$ (alternate exterior angles)

2. Find $m\angle 3$ and $m\angle 4$. Explain your reasoning.

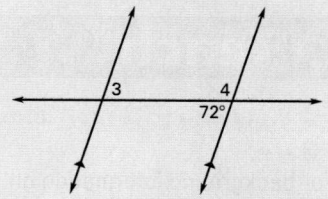

$m\angle 3 = 72°$ (alternate interior angles), $m\angle 4 = 108°$ (same-side interior angles)

3. Derek Andrews painted his family's name on a board and nailed the board to the gate in the fence at the front of his residence. The uprights (pickets) of the gate are parallel to each other. Find the value of x. 80

Pacing

Suggested Number of Days

Basic: 2 days
Average: 2 days
Advanced: 2 days
Block Schedule: 0.5 block with 3.4
0.5 block with 3.6

Teaching Resources

Blacklines
(See page 104B.)

Transparencies
• Warm-Up with Quiz
• Answers

Technology
• Electronic Teacher Tools
• Test and Practice Generator
• Online Lesson Planner
• Internet Support

Video
• Geometry in Motion

Teach

Content and Teaching Strategies

For background information on geometric concepts and teaching strategies related to this lesson, see pages 104E and 104F in this Teacher's Edition.

Tips for New Teachers

When diagrams are used in geometry, students want to draw conclusions based on the way they seem to appear. In Example 3 on page 138, the lines in each diagram appear to be parallel. Stress that only upon examination of the given information or angle measures can students accurately determine whether the lines are parallel. See the Tips for New Teachers on pp. 1–2 of the *Chapter 3 Resource Book* for additional notes about Lesson 3.5.

136

Goal

Show that two lines are parallel.

Key Words

• converse
• hypothesis p. 82
• conclusion p. 82

1. If two angles are congruent, then the two angles have the same measure; true

2. If $m\angle 3 + m\angle 4 = 90°$, then $\angle 3$ and $\angle 4$ are complementary; true

3. If $\angle 1 \cong \angle 2$, then $\angle 1$ and $\angle 2$ are right angles; false

---

Suppose two lines are cut by a transversal and a pair of corresponding angles are congruent. Are the two lines parallel? This question asks whether the *converse* of the Corresponding Angles Postulate is true.

The **converse** of an if-then statement is the statement formed by switching the hypothesis and the conclusion. Here is an example.

Statement:
If you live in Sacramento, then you live in California.

Converse:
If you live in California, then you live in Sacramento.

The converse of a true statement may or may not be true. As shown on the map at the right, if you live in California, you don't necessarily live in Sacramento; you could live in Fresno or San Diego.

EXAMPLE 1 Write the Converse of an If-Then Statement

Statement: If two segments are congruent, then the two segments have the same length.

  a. Write the converse of the true statement above.

  b. Determine whether the converse is true.

Solution

  a. *Converse:* If two segments have the same length, then the two segments are congruent.

  b. The converse is a true statement.

Checkpoint ✓ Write the Converse of an If-Then Statement

Write the converse of the true statement. Then determine whether the converse is true. 1–3. See margin.

  1. If two angles have the same measure, then the two angles are congruent.

  2. If $\angle 3$ and $\angle 4$ are complementary, then $m\angle 3 + m\angle 4 = 90°$.

  3. If $\angle 1$ and $\angle 2$ are right angles, then $\angle 1 \cong \angle 2$.

Converse of a Postulate Postulate 9 below is the converse of Postulate 8, the Corresponding Angles Postulate, which you learned in Lesson 3.4.

POSTULATE 9

Corresponding Angles Converse

Words If two lines are cut by a transversal so that corresponding angles are congruent, then the lines are parallel.

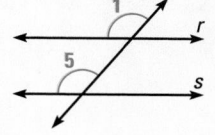

Symbols If ∠1 ≅ ∠5, then r ∥ s.

Student Help

CLASSZONE.COM

MORE EXAMPLES
More examples at classzone.com

4. Yes. Two angles are corresponding and congruent. By the Corresponding Angles Converse, the lines are parallel.

5. No. There is no information given about the angles formed where $\overleftrightarrow{SY}$ intersects $\overleftrightarrow{XZ}$.

6. Yes. *Sample answer:* Since $\overleftrightarrow{RT} \perp \overleftrightarrow{SY}$, all four angles with vertex S are right angles. Corresponding angles are both right angles, and all right angles are congruent. By the Corresponding Angles Converse, the lines are parallel.

EXAMPLE 2 Apply Corresponding Angles Converse

Is enough information given to conclude that $\overleftrightarrow{BD} \parallel \overleftrightarrow{EG}$? Explain.

a. 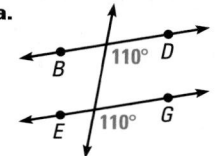 **b.** 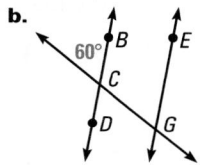 **c.**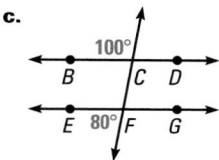

Solution

a. Yes. The two marked angles are corresponding and congruent. There is enough information to use the Corresponding Angles Converse to conclude that $\overleftrightarrow{BD} \parallel \overleftrightarrow{EG}$.

b. No. You are not given any information about the angles formed where $\overleftrightarrow{EG}$ intersects $\overleftrightarrow{CG}$.

c. Yes. You can conclude that m∠EFC = 100°. So, there is enough information to use the Corresponding Angles Converse to conclude that $\overleftrightarrow{BD} \parallel \overleftrightarrow{EG}$.

Checkpoint ✓ **Apply Corresponding Angles Converse**

Is enough information given to conclude that $\overleftrightarrow{RT} \parallel \overleftrightarrow{XZ}$? Explain.
4–6. See margin.

4. 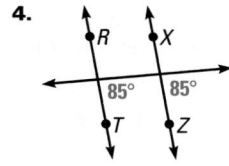 **5.** 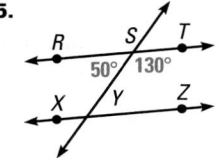 **6.**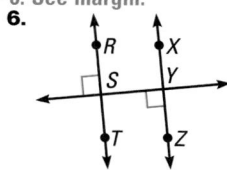

Extra Example 1

Statement: If $\overleftrightarrow{AB}$ divides $\overline{CD}$ into two congruent segments, then $\overleftrightarrow{AB}$ is a bisector of $\overline{CD}$.

a. Write the converse of the true statement above. *Converse:* If $\overleftrightarrow{AB}$ is a bisector of $\overline{CD}$, then $\overleftrightarrow{AB}$ divides $\overline{CD}$ into two congruent segments.

b. Decide whether the converse is true. The converse is a true statement.

Extra Example 2

Is enough information given to conclude that $\overleftrightarrow{JK} \parallel \overleftrightarrow{NP}$? Explain.

a.

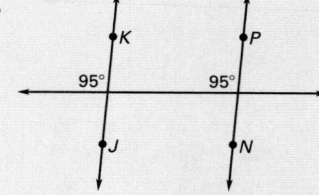

Yes. The two marked angles are corresponding and congruent. There is enough information to use the Corresponding Angles Converse to conclude that $\overleftrightarrow{JK} \parallel \overleftrightarrow{NP}$.

b.

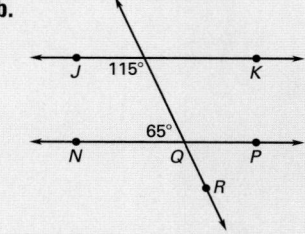

Yes. You can conclude that m∠NQR = 115°. So, there is enough information to use the Corresponding Angles Converse to conclude that $\overleftrightarrow{JK} \parallel \overleftrightarrow{NP}$.

c.

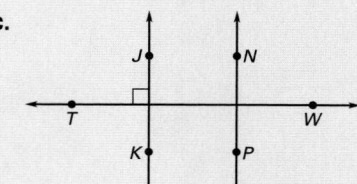

No. You are not given any information about the angles formed where $\overleftrightarrow{TW}$ intersects $\overleftrightarrow{NP}$.

137

138

Extra Example 3

Does the diagram give enough information to conclude that $m \parallel n$?

a.

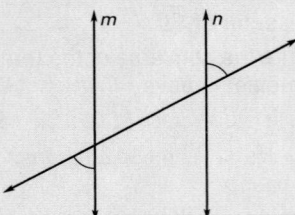

Yes. The angle congruence marks on the diagram allow you to conclude that $m \parallel n$ by the Alternate Exterior Angles Converse.

b.

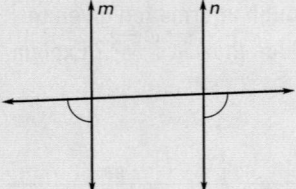

No. Not enough information is given to conclude that $m \parallel n$.

Student Help

LOOK BACK
To review the Alternate Interior Angles Theorem and the Alternate Exterior Angles Theorem, see pp. 129–130.

THEOREMS 3.8 and 3.9

3.8 Alternate Interior Angles Converse

Words If two lines are cut by a transversal so that alternate interior angles are congruent, then the lines are parallel.

Symbols If $\angle 4 \cong \angle 5$, then $r \parallel s$.

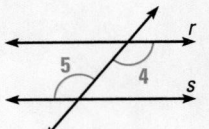

3.9 Alternate Exterior Angles Converse

Words If two lines are cut by a transversal so that alternate exterior angles are congruent, then the lines are parallel.

Symbols If $\angle 1 \cong \angle 8$, then $r \parallel s$.

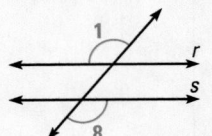

EXAMPLE 3 Identify Parallel Lines

Does the diagram give enough information to conclude that $m \parallel n$?

a.

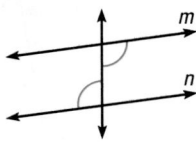

b.

Solution

a. Yes. The angle congruence marks on the diagram allow you to conclude that $m \parallel n$ by the Alternate Interior Angles Converse.

b. No. Not enough information is given to conclude that $m \parallel n$.

Checkpoint ✓ Identify Parallel Lines

7. Does the diagram give enough information to conclude that $c \parallel d$? Explain.
Yes, by the Alternate Exterior Angles Converse

THEOREM 3.10

Same-Side Interior Angles Converse

Words If two lines are cut by a transversal so that same-side interior angles are supplementary, then the lines are parallel.

Symbols If $m\angle 3 + m\angle 5 = 180°$, then $r \parallel s$.

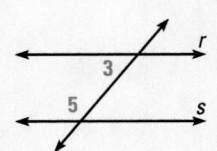

 **Using Algebra**

EXAMPLE 4 **Use Same-Side Interior Angles Converse**

Find the value of x so that $j \parallel k$.

Solution

Lines j and k are parallel if the marked angles are supplementary.

$5x° + 115° = 180°$	Supplementary angles
$5x = 65$	Subtract 115 from each side.
$x = 13$	Divide each side by 5.

ANSWER ▶ So, if $x = 13$, then $j \parallel k$.

U.S. Navy Blue Angels

Checkpoint ✓ **Use Same-Side Interior Angles Converse**

Find the value of x so that $v \parallel w$.

8.
55
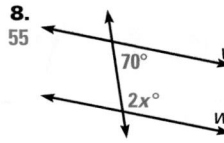

9.
30
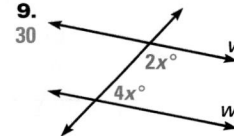

10.
68
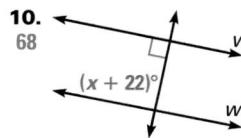

Extra Example 4
Find the value of x so that $v \parallel w$. 20

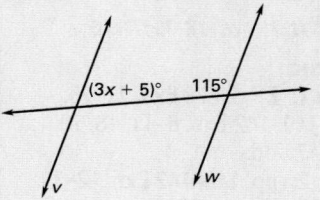

✓ Concept Check
How can you use corresponding angles to prove two lines parallel?
If a pair of corresponding angles are congruent, then the two lines are parallel.

🅐 Daily Puzzler
If $x = 95$, which lines are parallel?

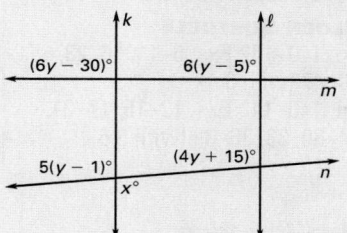

$k \parallel \ell$

3.5 Exercises

Guided Practice

Vocabulary Check

1. Describe how to form the *converse* of an if-then statement.
Switch the hypothesis and the conclusion.

2. Give an example of an if-then statement and its *converse*.
Sample answer: If a number is even, then it is divisible by 2.
Converse: If a number is divisible by 2, then the number is even.

Skill Check **Match the theorem or postulate used to explain why $p \parallel q$ with the diagram.**

3. Same-Side Interior Angles Converse   C

4. Alternate Interior Angles Converse   A

5. Corresponding Angles Converse   B

A.
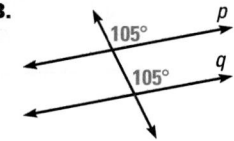

B.
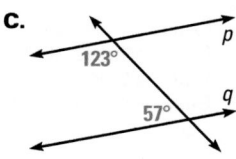

C.

③ Apply

Assignment Guide

BASIC
Day 1: EP p. 678 Exs. 21–25;
pp. 140–142 Exs. 6–11, 16, 19,
35–47 odd
Day 2: pp. 140–142 Exs. 12–15,
17, 25–29 odd, 36–46 even

AVERAGE
Day 1: pp. 140–142 Exs. 6–11,
16, 23, 35, 40–47
Day 2: pp. 140–142 Exs. 12–15,
17–21, 24–30, 33, 36–39

ADVANCED
Day 1: pp. 140–142 Exs. 8–11,
16, 23, 36–47
Day 2: pp. 140–142 Exs. 12–15,
17–22, 27–33, 34*, 35;
EC: classzone.com

BLOCK SCHEDULE
pp. 140–142 Exs. 6–11, 16, 23, 35,
40–47 (with 3.4)
pp. 140–142 Exs. 12–15, 17–21,
24–30, 33, 36–39 (with 3.6)

Extra Practice
• Student Edition, p. 680
• Chapter 3 Resource Book,
pp. 49–50

Homework Check
To quickly check student understanding of key concepts, go over
the following exercises:

Basic: 7, 10, 16, 25, 29
Average: 7, 10, 16, 18, 26
Advanced: 9, 11, 17, 21, 29

✗ Common Error
In Exercises 16–17, watch for students who cannot visualize the
solution and so do not know how to
begin. Suggest that they begin by
drawing two parallel lines, draw
any transversal, and number the
eight angles formed. Then have
them adjust the angle numbers so
they match the special angle pair
discussed in the exercise. Finally,
have students adjust the transversal until the angle measurement
condition is met.

140

Practice and Applications

Extra Practice
See p. 680.

Writing Converses In Exercises 6–9, write the converse of the true statement. Then determine whether the converse is true.

6. If two lines never intersect in a plane, then the lines are parallel.
   If two lines are parallel, then the two lines never intersect in a plane; true.
7. If $\angle 1$ and $\angle 2$ are supplementary, then $m\angle 1 + m\angle 2 = 180°$.
   If $m\angle 1 + m\angle 2 = 180°$, then $\angle 1$ and $\angle 2$ are supplementary; true.
8. If $\angle A$ measures 38°, then $\angle A$ is acute.
   If $\angle A$ is acute, then $\angle A$ measures 38°; false.
9. If $\angle B$ measures 123°, then $\angle B$ is obtuse.
   If $\angle B$ is obtuse, then $\angle B$ measures 123°; false.

10. Yes, by the Corresponding Angles Converse

11. Yes, by the Corresponding Angles Converse

12. No; there is no information given about the angles formed by the transversal and line n.

13. Yes, by the Alternate Interior Angles Converse

14. Yes, by the Alternate Exterior Angles Converse

15. No; the angles have no special relationship.

Identifying Parallel Lines Determine whether enough information is given to conclude that $m \parallel n$. Explain.

10. 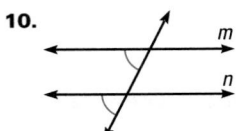   **11.** 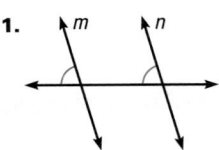   **12.**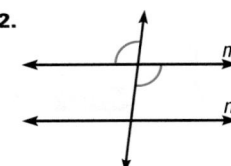

13. 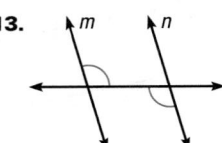   **14.** 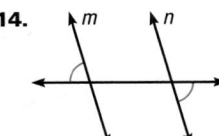   **15.**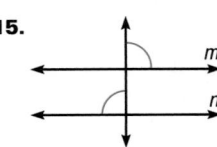

18. Yes; *Sample explanation:* $m\angle ABE = 123°$, so $\overleftrightarrow{AC} \parallel \overleftrightarrow{DF}$ by the Corresponding Angles Converse.

19. Yes; *Sample explanation:* $m\angle ABE = 143°$ by the Vertical Angles Theorem; $143° + 37° = 180°$, so $\overleftrightarrow{AC} \parallel \overleftrightarrow{DF}$ by the Same-Side Interior Angles Converse.

21. Yes; *Sample explanation:* $m\angle CBE = 115°$ by the Vertical Angles Theorem; $115° + 65° = 180°$, so $\overleftrightarrow{AC} \parallel \overleftrightarrow{DF}$ by the Same-Side Interior Angles Converse.

Visualize It! Sketch the described figure. 16, 17. See margin.

16. Draw two parallel lines and a transversal with $\angle 3$ and $\angle 4$ as congruent corresponding angles so that $m\angle 3 = m\angle 4 = 60°$.

17. Draw two parallel lines and a transversal with $\angle 5$ and $\angle 6$ as same-side interior angles so that $m\angle 5 = 45°$.

Logical Reasoning Are $\overleftrightarrow{AC}$ and $\overleftrightarrow{DF}$ parallel? Explain.

18. 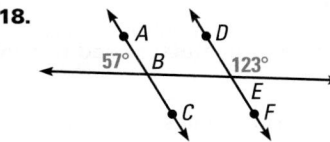   **19.**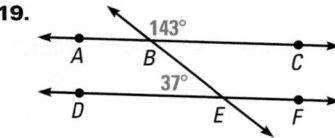

Homework Help
Example 1: Exs. 6–9
Example 2: Exs. 10–15
Example 3: Exs. 10–15
Example 4: Exs. 24–29

20. 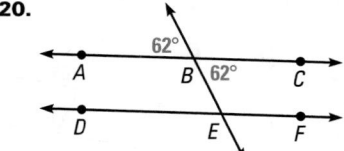   **21.**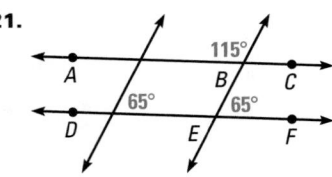

No; there is not enough information.

22. 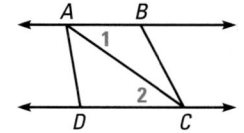 **You be the Judge** In the diagram, ∠1 ≅ ∠2. One of your classmates states that if ∠1 ≅ ∠2, then $\overleftrightarrow{AB} \parallel \overleftrightarrow{DC}$. Is your classmate right? Explain your reasoning.
Yes, by the Alternate Interior Angles Converse

23. **Kiteboarding** The lines on the photo of the kiteboarder below
20 show the angles formed between the control bar and the kite lines. Find the value of x so that n ∥ m.

🆇 **Using Algebra** Find the value of x so that m ∥ n.

24.
35
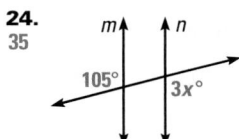

25.
21
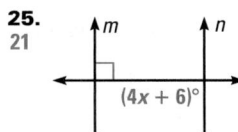

26.
59
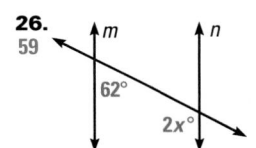

27.
11
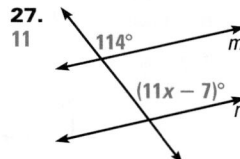

28.
24
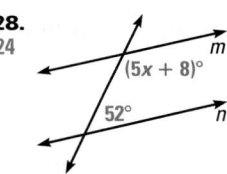

29.
10
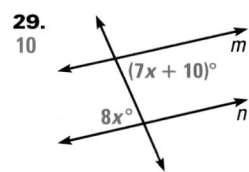

30. **Technology** Use geometry software to draw two parallel segments and a transversal. Choose a pair of alternate interior angles and construct their angle bisectors. Make a conjecture about the bisectors. Describe how to show that your conjecture is true.
See margin.

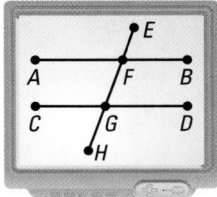

Logical Reasoning Name two parallel lines. Explain your reasoning.

31. $\overleftrightarrow{HJ} \parallel \overleftrightarrow{AB}$ by the Alternate Exterior Angles Converse.

32. $\overleftrightarrow{BC} \parallel \overleftrightarrow{DF}$ by the Same-Side Interior Angles Converse.

31.
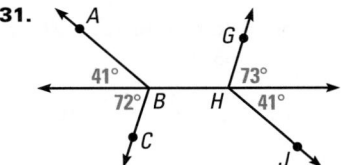

32.
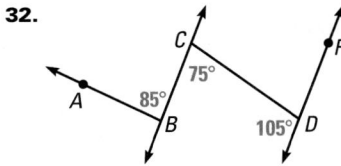

3.5 *Showing Lines are Parallel* 141

Link to
Sports

KITEBOARDING is a water sport, similar to windsurfing, in which a kite is used to help pick up speed across the water.

---

16. *Sample answer:*
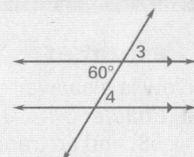

17. *Sample answer:*
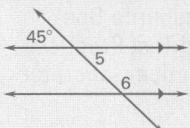

30. *Sample conjecture:* If two parallel lines are cut by a transversal, then the bisectors of the alternate interior angles are parallel.

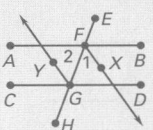

By the Alternate Interior Angles Theorem, ∠BFG ≅ ∠FGC. m∠BFG = m∠FGC and $\frac{1}{2}m\angle BFG = \frac{1}{2}m\angle FGC$ by the Multiplication Property of Equality. Since $\overrightarrow{FX}$ bisects ∠BFG, m∠1 = m∠BFG. Since $\overrightarrow{GY}$ bisects ∠FGC, m∠2 = m∠FGC. By the Substitution Property of Equality, m∠1 = m∠2. Thus ∠1 ≅ ∠2 and so $\overrightarrow{FX} \parallel \overrightarrow{GY}$ by the Alternate Interior Angles Converse.

141

Assessment Resources

The Mini-Quiz below is also available on blackline (*Chapter 3 Resource Book*, p. 58) and on transparency. For more assessment resources, see:
- Chapter 3 Resource Book
- Standardized Test Practice
- Test and Practice Generator

Mini-Quiz

Determine whether enough information is given to conclude that $m \parallel n$. Explain.

1.

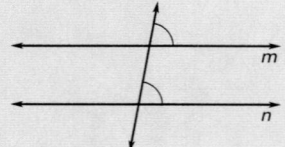

Yes. By the Corresponding Angles Converse, $m \parallel n$.

2.

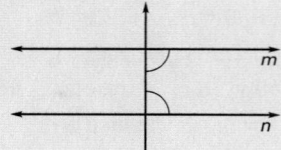

No. Not enough information is given.

Are $\overleftrightarrow{AC}$ and $\overleftrightarrow{DF}$ parallel? Explain.

3.

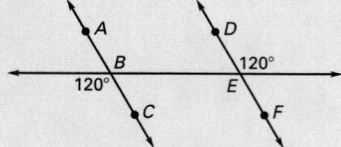

Yes. By the Alternate Exterior Angles Converse, $\overleftrightarrow{AC} \parallel \overleftrightarrow{DF}$.

4.

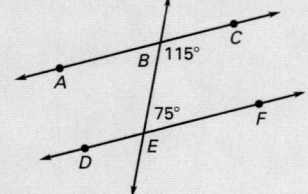

No. The same-side interior angles have measures whose sum is 190°, so the angles are not supplementary and therefore the lines are not parallel.

142

---

33. Building Stairs One way to build stairs is to attach triangular blocks to an angled support, as shown at the right. The sides of the angled support are parallel. If the support makes a 32° angle with the floor, what must $m\angle 1$ be so the top of the step will be parallel to the floor? **32°**

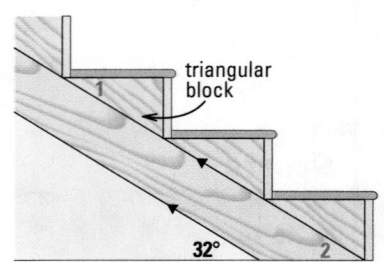

34. Challenge Using the diagram shown, find the measures of $\angle 1$ and $\angle 4$ so that $f \parallel g$.
$m\angle 1 = 40°$; $m\angle 4 = 40°$

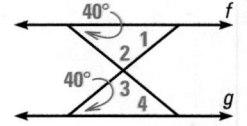

Standardized Test Practice

Multiple Choice In Exercises 35 and 36, use the diagram below.

35. If $\angle 1 \cong \angle 2$, which statement is true? **B**

(A) $r \parallel s$ (B) $q \parallel r$

(C) $s \parallel t$ (D) None of these

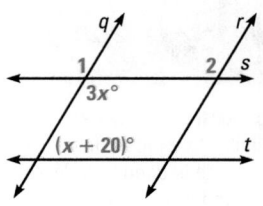

36. Find the value of x so that $s \parallel t$. **G**

(F) 10   (G) 40

(H) 50   (J) 60

Mixed Review

Perpendicular Lines What can you conclude about $\angle 1$ and $\angle 2$ using the given information? *(Lesson 3.2)*

37. $m\angle 1 = m\angle 2 = 90°$

38. $\angle 1$ and $\angle 2$ are complementary.

39. $\angle 1$ and $\angle 2$ are right angles.

37. $p \perp q$ **38.** $j \perp k$ **39.** $\angle 1 \cong \angle 2$

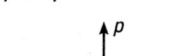   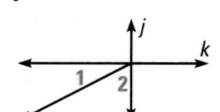   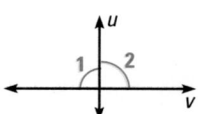

Angle Measures In the diagram below, $m \parallel n$. *(Lessons 3.3, 3.4)*

40. What kind of angles are $\angle 1$ and $\angle 5$?
alternate exterior angles

41. What kind of angles are $\angle 1$ and $\angle 2$?
corresponding angles

42. What is the measure of $\angle 4$? **72°**

43. What is the measure of $\angle 2$? **108°**

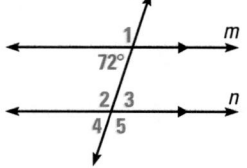

Algebra Skills

Evaluating Fractions Add or subtract. Write the answer in simplest form. *(Skills Review, p. 658)*

44. $\frac{3}{8} + \frac{7}{8}$ $\frac{5}{4}$, or $1\frac{1}{4}$ **45.** $\frac{9}{11} - \frac{2}{11}$ $\frac{7}{11}$ **46.** $\frac{7}{12} + \frac{5}{12}$ 1   **47.** $\frac{8}{9} - \frac{2}{9}$ $\frac{2}{3}$

Using Perpendicular and Parallel Lines

Goal
Construct parallel and perpendicular lines. Use properties of parallel and perpendicular lines.

Key Words
• construction

A **construction** is a geometric drawing that uses a limited set of tools, usually a compass and a straightedge (a ruler without marks).

compass

straightedge

Geo-Activity > Constructing a Perpendicular to a Line

Use the following steps to construct a perpendicular to a line in two different cases:

	Line perpendicular to a line through a point *not* on the line.	Line perpendicular to a line through a point *on* the line.
❶ Place the compass at point *P* and draw an arc that intersects line *ℓ* twice. Label the intersections *A* and *B*.	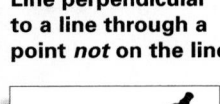	
❷ Open your compass wider. Draw an arc with center *A*. Using the same radius, draw an arc with center *B*. Label the intersection of the arcs *Q*.	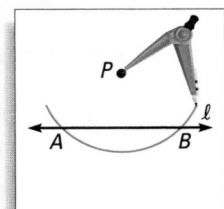	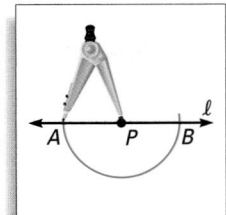
❸ Use a straightedge to draw $\overleftrightarrow{PQ}$. $\overleftrightarrow{PQ} \perp \ell$.	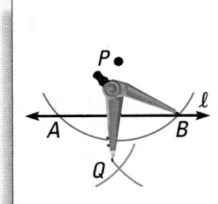	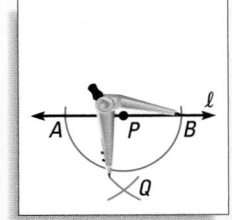

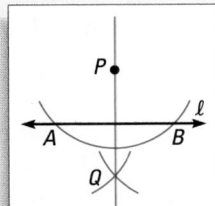
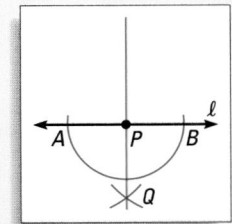

◆ Plan

Pacing
Suggested Number of Days

Basic: 3 days
Average: 3 days
Advanced: 3 days
Block Schedule: 0.5 block with 3.5 1 block

Teaching Resources

📄 **Blacklines**
(See page 104B.)

 **Transparencies**
• Warm-Up with Quiz
• Answers

🖥 **Technology**
• Electronic Teacher Tools
• Test and Practice Generator
• Online Lesson Planner
• Internet Support

Geo-Activity

Goal Use a compass and straightedge to construct a perpendicular to a line.

Key Discovery A line can be constructed perpendicular to a given line both through a point on the line and through a point not on the line.

◆ Teach

Content and Teaching Strategies
For background information on geometric concepts and teaching strategies related to this lesson, see pages 104E and 104F in this Teacher's Edition.

144

Ask students to find examples of multiple pairs of lines within the classroom that appear to be parallel or perpendicular. List some ideas on the board. Then ask students to explain how they would go about proving their conjectures. See the Tips for New Teachers on pp. 1–2 of the *Chapter 3 Resource Book* for additional notes about Lesson 3.6.

Geometric Reasoning

In Step 1 of the Geo-Activity on page 143, point out that points *A* and *B* must be the same distance away from *P* because they are both points on a circle with center *P*. Students may need a moment to clarify in their minds that all points on a circle are equidistant from its center.

Extra Example 1

What is the first step in constructing a line perpendicular to line *ℓ* that passes through a point *R* on line *ℓ*. **Place the compass at point *R* and draw an arc that intersects line *ℓ* twice. Label the intersections *A* and *B*.**

1.

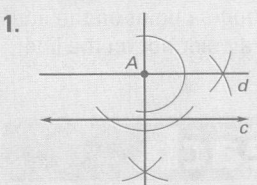

Construct a line that passes through point *P* and is parallel to line *ℓ*.

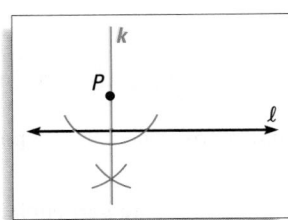

Solution

❶ **Construct** a line perpendicular to *ℓ* through *P* using the construction on the previous page. Label the line *k*.

❷ **Construct** a line perpendicular to *k* through *P* using the construction on the previous page. Label the line *j*. Line *j* is parallel to line *ℓ*.

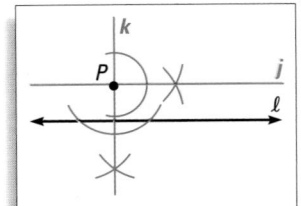

Checkpoint ✓ **Construct Parallel Lines**

1. Draw a line *c* and a point *A* not on the line. Construct a line *d* that passes through point *A* and is parallel to line *c*. See margin.

POSTULATES 10 and 11

Postulate 10 Parallel Postulate

Words If there is a line and a point not on the line, then there is exactly one line through the point parallel to the given line.

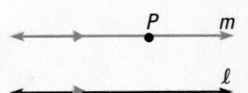

Symbols If *P* is not on *ℓ*, then there exists one line *m* through *P* such that *m* ∥ *ℓ*.

Postulate 11 Perpendicular Postulate

Words If there is a line and a point not on the line, then there is exactly one line through the point perpendicular to the given line.

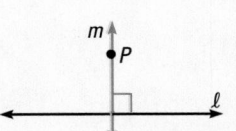

Symbols If *P* is not on *ℓ*, then there exists one line *m* through *P* such that *m* ⊥ *ℓ*.

THEOREMS 3.11 and 3.12

Theorem 3.11

Words If two lines are parallel to the same line, then they are parallel to each other.

Symbols If $q \parallel r$ and $r \parallel s$, then $q \parallel s$.

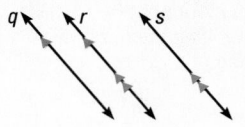

Theorem 3.12

Words In a plane, if two lines are perpendicular to the same line, then they are parallel to each other.

Symbols If $m \perp p$ and $n \perp p$, then $m \parallel n$.

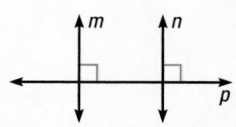

EXAMPLE 2 — Use Properties of Parallel Lines

Ladders were used to move from level to level of cliff dwellings, as shown at right. Each rung on the ladder is parallel to the rung immediately below it. Explain why $\ell \parallel p$.

Solution

You are given that $\ell \parallel m$ and $m \parallel n$. By Theorem 3.11, $\ell \parallel n$. Since $\ell \parallel n$ and $n \parallel p$, it follows that $\ell \parallel p$.

EXAMPLE 3 — Use Properties of Parallel Lines

Find the value of x that makes $\overleftrightarrow{AB} \parallel \overleftrightarrow{CD}$.

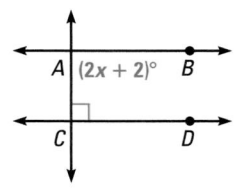

Solution

By Theorem 3.12, $\overleftrightarrow{AB}$ and $\overleftrightarrow{CD}$ will be parallel if $\overleftrightarrow{AB}$ and $\overleftrightarrow{CD}$ are both perpendicular to $\overleftrightarrow{AC}$. For this to be true $\angle BAC$ must measure 90°.

$(2x + 2)° = 90°$ $m\angle BAC$ must be 90°.

$2x = 88$ Subtract 2 from each side.

$x = 44$ Divide each side by 2.

ANSWER ▸ If $x = 44$, then $\overleftrightarrow{AB} \parallel \overleftrightarrow{CD}$.

3.6 *Using Perpendicular and Parallel Lines*   `145`

Extra Example 2

In the bookshelves shown below, each shelf is parallel to the shelf immediately above it and the top shelf is parallel to the ceiling. Explain why line c is parallel to the ceiling.

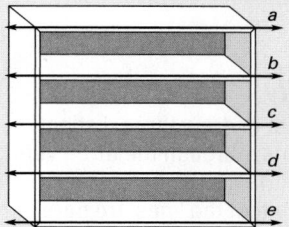

You are given that $c \parallel b$, $b \parallel a$, and a is parallel to the ceiling. By Theorem 3.11, $a \parallel c$ and therefore line c is parallel to the ceiling.

Extra Example 3

Find the value of x that makes $\overleftrightarrow{FG} \parallel \overleftrightarrow{HJ}$. 8

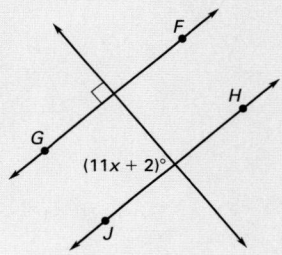

146

Concept Check

How could you use perpendicular lines to show that two lines are parallel? **By Theorem 3.12, if it can be shown that two lines in a plane are both perpendicular to a third line in the plane, then the two lines are parallel.**

Daily Puzzler

If you want to draw three distinct parallel lines through the three vertices of a triangle, how can you guarantee that the lines are parallel? *Sample answer:* **Draw them all perpendicular to one of the sides.**

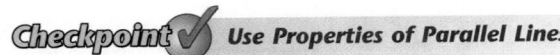 **Checkpoint** ✓ *Use Properties of Parallel Lines*

2. $a \parallel b$ and $b \parallel c$, so $a \parallel c$ (if two lines are parallel to the same line, then they are parallel to each other).

2. Use the information in the diagram to explain why $a \parallel c$.

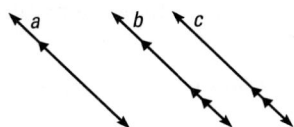

3. Find a value of x so that $d \parallel e$. **16**

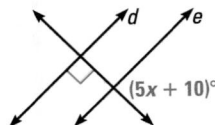

$(5x + 10)°$

You have now studied six ways to show that two lines are parallel.

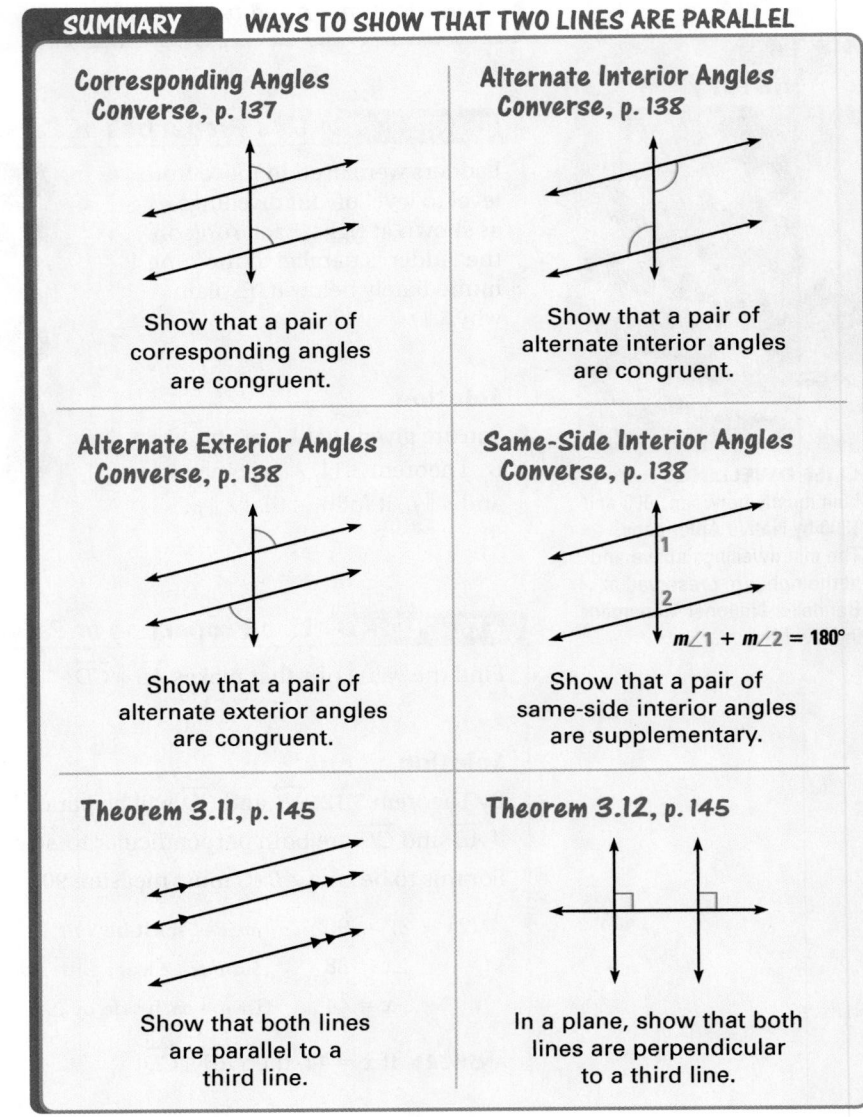

SUMMARY **WAYS TO SHOW THAT TWO LINES ARE PARALLEL**

Corresponding Angles Converse, p. 137

Show that a pair of corresponding angles are congruent.

Alternate Interior Angles Converse, p. 138

Show that a pair of alternate interior angles are congruent.

Alternate Exterior Angles Converse, p. 138

Show that a pair of alternate exterior angles are congruent.

Same-Side Interior Angles Converse, p. 138

1
2

$m\angle 1 + m\angle 2 = 180°$

Show that a pair of same-side interior angles are supplementary.

Theorem 3.11, p. 145

Show that both lines are parallel to a third line.

Theorem 3.12, p. 145

In a plane, show that both lines are perpendicular to a third line.

3.6 Exercises

Guided Practice

Vocabulary Check

1. What are the two basic tools used for a *construction*?
compass and straightedge

Skill Check

2. If two lines are parallel to the same line, then they are parallel to each other.

3. In a plane, if two lines are perpendicular to the same line, then they are parallel to each other.

Using the given information, state the theorem that you can use to conclude that *r* ∥ *s*.

2. *r* ∥ *t*, *t* ∥ *s*

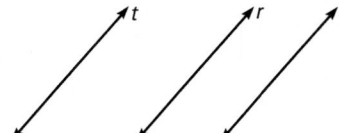

3. *r* ⊥ *t*, *t* ⊥ *s*

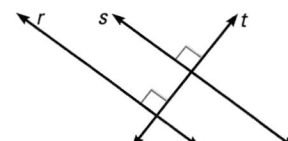

Practice and Applications

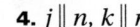

Extra Practice
See p. 680.

4. If two lines are parallel to the same line, then they are parallel to each other.

5. In a plane, if two lines are perpendicular to the same line, then they are parallel to each other.

Logical Reasoning Using the given information, state the postulate or theorem that allows you to conclude that *j* ∥ *k*.

4. *j* ∥ *n*, *k* ∥ *n*

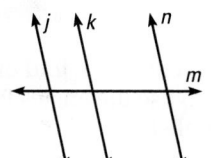

5. *j* ⊥ *n*, *k* ⊥ *n*

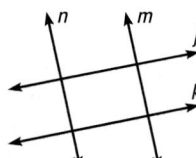

6. ∠1 ≅ ∠2

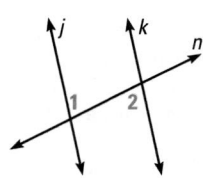

Alternate Interior Angles Converse

Showing Lines are Parallel Explain how you would show that *c* ∥ *d*. State any theorems or postulates that you would use.

7.

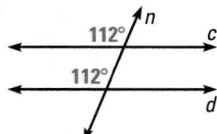

Corresponding Angles Converse

8.

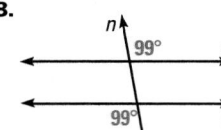

Alternate Exterior Angles Converse

9.

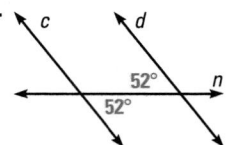

Alternate Interior Angles Converse

Homework Help

Example 1: Exs. 22–24
Example 2: Exs. 4–12
Example 3: Exs. 19–21

10.

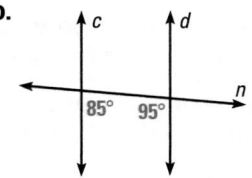

85° + 95° = 180°, so *c* ∥ *d* by the Same-Side Interior Angles Converse.

11.

In a plane, if two lines are perpendicular to the same line, then they are parallel to each other.

12.

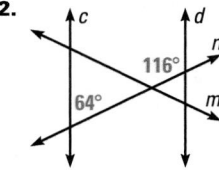

116° + 64° = 180°, so *c* ∥ *d* by the Same-Side Interior Angles Converse.

3.6 Using Perpendicular and Parallel Lines **147**

③ Apply

Assignment Guide

BASIC
Day 1: EP p. 678 Exs. 26–31; pp. 147–149 Exs. 4–9, 22–24, 29–36
Day 2: pp. 147–149 Exs. 10–16, 19–21, 27, 28, 37–42
Day 3: pp. 150–151 Ex. 9

AVERAGE
Day 1: pp. 147–149 Exs. 4–12, 22–24, 37–42
Day 2: pp. 147–149 Exs. 13–21, 25, 27–36
Day 3: pp. 150–151 Ex. 9

ADVANCED
Day 1: pp. 147–149 Exs. 4–12 even, 22–24, 27–36
Day 2: pp. 147–149 Exs. 13–21, 25, 26*, 37–42;
EC: classzone.com
Day 3: pp. 150–151 Ex. 9

BLOCK SCHEDULE
pp. 147–149 Exs. 4–12, 22–24, 37–42 (with 3.5)
pp. 147–149 Exs. 13–21, 25, 27–36; pp. 150–151 Ex. 9

Extra Practice

• Student Edition, p. 680
• Chapter 3 Resource Book, pp. 62–63

Homework Check

To quickly check student understanding of key concepts, go over the following exercises:

Basic: 5, 8, 13, 19, 23
Average: 5, 10, 14, 20, 23
Advanced: 12, 14, 16, 21, 24

✗ Common Error

In Exercises 7–12, watch for students who either omit steps or provide an inadequate explanation. Have them imagine they are explaining each answer to somebody who does not understand geometry and who needs every small step completely explained.

13. $p \parallel q$ by the Corresponding Angles Converse; $q \parallel r$ by the Same-Side Interior Angles Converse. Then, since $p \parallel q$ and $q \parallel r$, $p \parallel r$ (if two lines are parallel to the same line, then they are parallel to each other).

14. $h \parallel j$ by the Corresponding Angles Converse.

15. a and b are each perpendicular to d, so $a \parallel b$ (in a plane, if two lines are perpendicular to the same line, then they are parallel to each other); c and d are each perpendicular to a, so $c \parallel d$ (in a plane, if two lines are perpendicular to the same line, then they are parallel to each other).

16. No pairs of lines must be parallel.

17. The 8th fret is parallel to the 9th fret, and the 9th fret is parallel to the 10th fret. Therefore, the 8th fret is parallel to the 10th fret (if two lines are parallel to the same line, then they are parallel to each other).

22–24. Sample answers are given.

22.
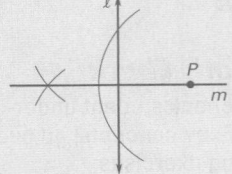

23.
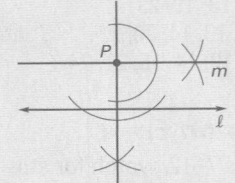

24.

26, 33–36. See Additional Answers beginning on page AA1.

Student Help
CLASSZONE.COM

HOMEWORK HELP
Extra help with problem solving in Exs. 13–16 is at classzone.com

Link to
Music

GUITARISTS press their strings against frets to play specific notes. The frets are positioned to make it easy to play scales. The frets are parallel so that the spacing between the frets is the same for all six strings.

Naming Parallel Lines In Exercises 13–16, determine which lines, if any, must be parallel. Explain your reasoning. 13–16. See margin.

13.
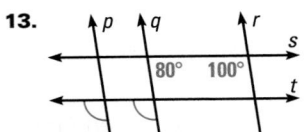

14.
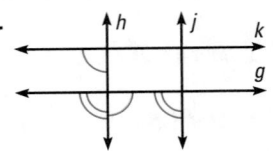

15.
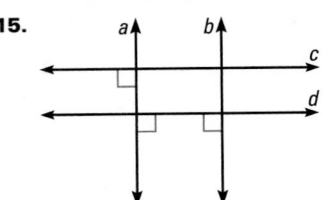

16.
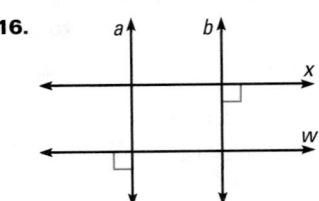

17. **Guitars** In the photo of the guitar at the right, each fret is parallel to the fret beside it. Explain why the 8th fret is parallel to the 10th fret. See margin.

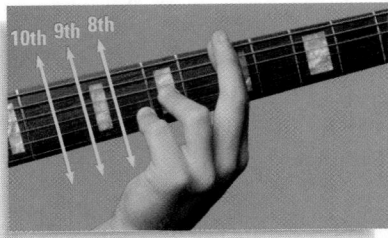

18. **Visualize It!** Make a diagonal fold on a piece of lined notebook paper. Explain how to use the angles formed to show that the lines on the paper are parallel. *Sample answer:* Show that two corresponding angles are congruent and then use the Corresponding Angles Converse.

Using Algebra Find the value of x so that $g \parallel h$.

19. 11
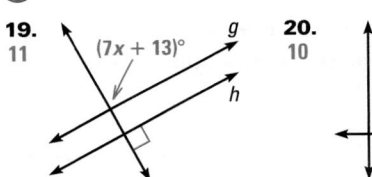
$(7x + 13)°$

20. 10
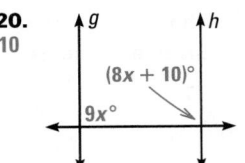
$(8x + 10)°$
$9x°$

21. 6
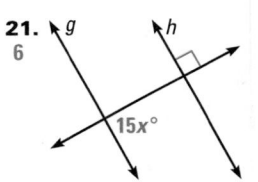
$15x°$

Constructions In Exercises 22–24, use a compass and a straightedge to construct the lines. 22–24. See margin.

22. Draw a horizontal line ℓ and choose a point P on line ℓ. Construct a line m perpendicular to line ℓ through point P.

23. Draw a vertical line ℓ and choose a point P to the right of line ℓ. Construct a line m perpendicular to line ℓ through point P.

24. Draw a horizontal line ℓ and choose a point P above line ℓ. Construct a line m parallel to line ℓ through point P.

LOOK BACK
For an example of boats sailing at an angle to the wind, see p. 104.

25. Sailing If the wind is constant, will the boats' paths ever cross? Explain.

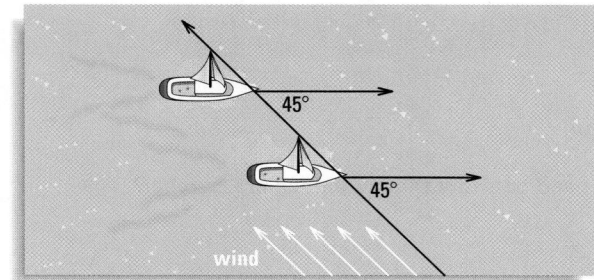

No; if the wind is constant, then the boats' paths will be parallel by the Corresponding Angles Converse. They will never cross.

26. Challenge Theorem 3.12 applies only to lines in a plane. Draw a diagram of a three-dimensional example of two lines that are perpendicular to the same line but are not parallel to each other. *See margin.*

Standardized Test Practice

27. Multiple Choice Find the value of x so that $m \parallel n$. **A**

(A) 20   (B) 25

(C) 40   (D) 90

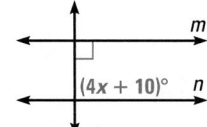

28. a. Corresponding Angles Converse

b. Same-Side Interior Angles Converse

c. 90°; $\overline{AB} \parallel \overline{EF}$ (if two lines are parallel to the same line, then they are parallel to each other); $\angle 1$ and $\angle AEF$ are same-side interior angles, so $m\angle 1 = 180° - m\angle AEF = 180° - 90° = 90°$.

28. Multi-Step Problem Use the information given in the diagram at the right.

a. Explain why $\overline{AB} \parallel \overline{CD}$.

b. Explain why $\overline{CD} \parallel \overline{EF}$.

c. What is $m\angle 1$? How do you know?

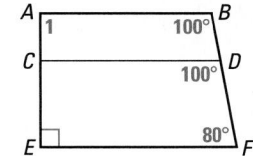

Mixed Review

Points, Lines and Planes Decide whether the statement is *true* or *false*. *(Lesson 1.3)*

29. N lies on $\overleftrightarrow{MK}$. true

30. J, K, and M are collinear. false

31. K lies in plane JML. true

32. J lies on $\overleftrightarrow{KL}$. false

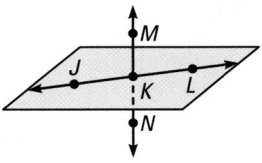

Plotting Points Plot the point in a coordinate plane. *(Skills Review, p. 664)* 33–36. See margin.

33. $A(2, 3)$ **34.** $B(-1, 6)$ **35.** $C(-4, 7)$ **36.** $D(-2, -5)$

Algebra Skills

Expressions Evaluate the expression. *(Skills Review, p. 670)*

37. $-5 \cdot 6 - 10 \div 5$ −32   **38.** $-8 + 33 - 14$ 11   **39.** $24 \div (9 + 3)$ 2

40. $4(8 - 3)^2 - 12$ 88   **41.** $48 - 3^2 \cdot 5 - 6^2$ −33   **42.** $[(1 - 8)^2 + 7] \div 8$ 7

3.6 *Using Perpendicular and Parallel Lines*   **149**

Assessment Resources

The Mini-Quiz below is also available on blackline (*Chapter 3 Resource Book*, p. 69) and on transparency. For more assessment resources, see:
• Chapter 3 Resource Book
• Standardized Test Practice
• Test and Practice Generator

Mini-Quiz

Using the given information, state the postulate or theorem that allows you to conclude that $j \parallel k$.

1. $j \parallel \ell$, $k \parallel \ell$ Theorem 3.11

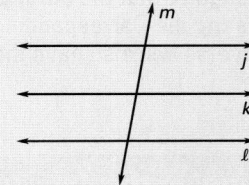

2. $j \perp p$, $k \perp p$ Theorem 3.12

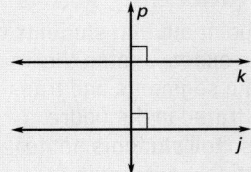

3. Draw a vertical line ℓ and choose a point P to the left of line ℓ. Construct a line m perpendicular to line ℓ through point P.

Sample answer:

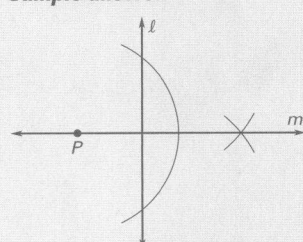

4. Find a value of x so that $g \parallel h$. **7**

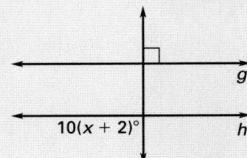

Planning the Activity

Goal
Students draw parallel lines and explore how slope can be used to prove lines parallel.

Materials
See the Technology Activity Keystrokes Masters in the *Chapter 3 Resource Book.*

LINK TO LESSON
In the Summary on page 146, students reviewed that two lines are parallel if a pair of corresponding angles are congruent. Lead students to recognize that the parallel lines illustrating the Corresponding Angles Converse will also have the same slope.

Managing the Activity

Tips for Success
In Step 1, point out that students do not need to exactly match the angle between the segments and transversal illustrated in the figure shown. Also tell students not to draw a vertical segment in Step 4 since the slope of a vertical line is undefined. An undefined slope will create a problem for students in Exercise 7.

Technology Activity **3.6** # Parallel Lines and Slope

Question

How is slope used to show that two lines are parallel?

Explore

1 Draw and label two segments and a transversal. Label the points of intersection.

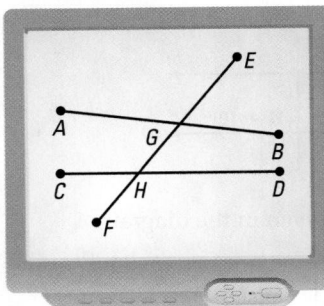

2 Measure a pair of corresponding angles.

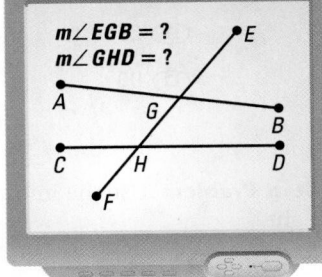

3 Drag point *B* until the two angles measured in Step 2 are congruent.

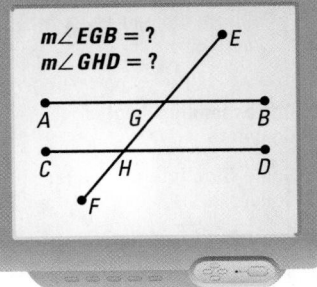

Student Help

SKILLS REVIEW
To review the slope of a line, see p. 665

Think About It

1. Are $\overline{AB}$ and $\overline{CD}$ in Step 3 parallel? What theorem does this illustrate? **Yes; Corresponding Angles Converse**

In algebra, you learned that the *slope* of a non-vertical line is the ratio of the vertical change (the rise) over the horizontal change (the run). The slope of a line can be positive or negative.

2. Measure the slopes of $\overline{AB}$ and $\overline{CD}$ in Step 3. What do you notice about the slopes? **Slopes may vary, but will be equal.**

3. Drag point *B* to a different position. Drag point *D* so that the slopes of $\overline{AB}$ and $\overline{CD}$ are equal. What are the measures of the pair of corresponding angles? **The measures of corresponding angles will vary, but will be equal.**

4. Make a conjecture about the slopes of parallel lines.
   The slopes of non-vertical parallel lines are equal.

Explore

④ Draw a non-horizontal segment $\overline{AB}$. Construct and label two points, C and D, on $\overline{AB}$.

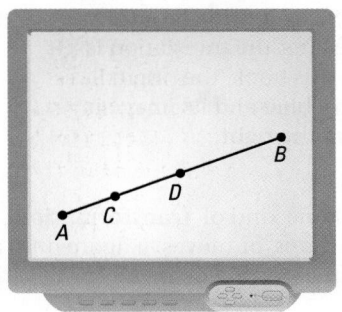

⑤ Construct two lines perpendicular to $\overline{AB}$ through points C and D.

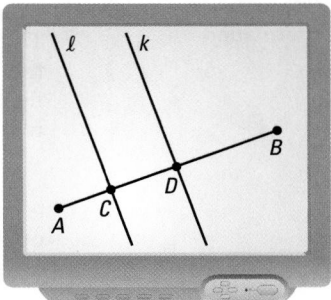

Think About It

5. What theorem allows you to conclude that the lines constructed in Step 5 are parallel? *In a plane, if two lines are perpendicular to the same line, then they are parallel to each other.*

6. Measure the slopes of the lines constructed in Step 5. Explain how to use the slopes to verify that the lines are parallel. *If the slopes are equal, then the lines must be parallel.*

7. Measure the slope of $\overline{AB}$. Multiply the slope of $\overline{AB}$ by the slope of one of the other lines. What is the result? *−1*

8. Drag point B. What happens to the calculation made in Exercise 7 as the slopes of the lines change? *The slopes change, but the product of the two slopes is still −1.*

9. Extension Construct and label point E on $\overline{AB}$. Construct line m parallel to line k through point E. What theorem allows you to conclude that lines ℓ and m are parallel? Compare the slopes of the lines to verify that they are parallel. *$\ell \parallel k$ from Exercise 6, and $k \parallel m$ by construction. Therefore, $\ell \parallel m$. (If two lines are parallel to the same line, they are parallel to each other.) The slopes of lines ℓ and m are equal.*

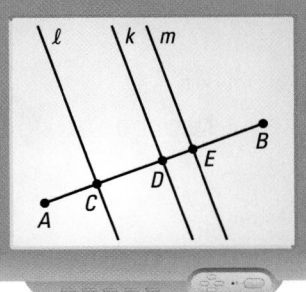

Technology Activity **151**

KEY DISCOVERY
Parallel lines have the same slope, and the product of the slopes of a pair of perpendicular lines is −1.

Activity Assessment
Use Exercises 4 and 8 to assess student understanding.

Pacing
Suggested Number of Days

Basic: 2 days
Average: 2 days
Advanced: 2 days
Block Schedule: 1 block

Teaching Resources
📄 **Blacklines**
(See page 104B.)

📑 **Transparencies**
• Warm-Up with Quiz
• Answers

🖥 **Technology**
• Electronic Teacher Tools
• Test and Practice Generator
• Online Lesson Planner
• Internet Support

Teach

Content and Teaching Strategies
For background information on geometric concepts and teaching strategies related to this lesson, see pages 104E and 104F in this Teacher's Edition.

Tips for New Teachers
The coordinate notation for a translation might be confusing to students when the values of *a* or *b* are negative. Students should think of the positive value of *a* as indicating a movement horizontally to the right and a negative value as indicating a movement horizontally to the left. Similarly, the movement associated with *b* is vertical, going up for a positive value and going down for a negative value. See the Tips for New Teachers on pp. 1–2 of the *Chapter 3 Resource Book* for additional notes about Lesson 3.7.

3.7 Translations

Goal
Identify and use translations.

Key Words
• translation
• image
• transformation

In 1996, New York City's Empire Theater was slid 170 feet up 42nd Street to a new location.

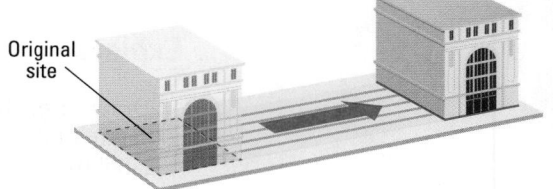

Original site

A slide is also called a **translation**. The new figure after the translation is the **image**. In this book, the original figure is given in blue and its image in red, as shown at the right.

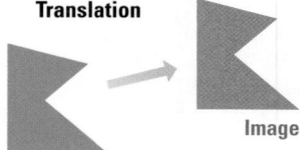

Translation

Image

A translation is one kind of **transformation**. A transformation is an operation that *maps*, or moves, a figure onto an image. You will study other transformations in Lessons 5.7, 7.6, and 11.8.

Student Help

VOCABULARY TIP
Use the following relationship to help you remember that a translation is a slide:

translation
slide

EXAMPLE 1 Compare a Figure and Its Image

Decide whether the red figure is a translation of the blue figure.

a.

b.

c.

Solution
a. Yes, this is a translation.
b. No, this is *not* a translation. The image is a mirror image of the original figure.
c. No, this is *not* a translation. The original figure is rotated.

Checkpoint ✓ Compare a Figure and Its Image

Decide whether the red figure is a translation of the blue figure.

1. no ЯR
2. yes RR
3. no ЯR

Student Help

Labeling Translations When labeling points on the image, write the prime symbol (′) next to the letter used in the original figure, as shown at the right.

In a translation, segments connecting points in the original figure to their corresponding points in the image are congruent and parallel. For example, $\overline{AA'}$ and $\overline{BB'}$ at the right are congruent and parallel.

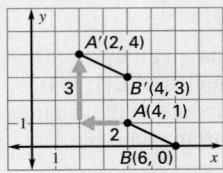

EXAMPLE 2 *Describe Translations*

Describe the translation of the segment.

Solution

Point *P* is moved 4 units to the right and 2 units down to get to point *P'*. So, every point on $\overline{PQ}$ moves 4 units to the right and 2 units down.

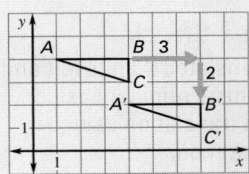

Translations in a coordinate plane can also be described using the following coordinate notation:

$$(x, y) \rightarrow (x + a, y + b)$$

Each point shifts *a* units horizontally (right or left) and *b* units vertically (up or down). When moving right or up, *add* the number of units. When moving left or down, *subtract* the number of units. Here are some examples:

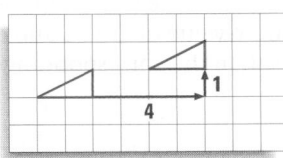

$(x, y) \rightarrow (x + 4, y + 1)$

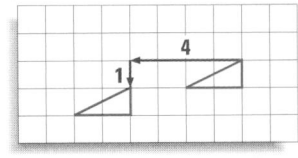

$(x, y) \rightarrow (x - 4, y - 1)$

EXAMPLE 3 *Use Coordinate Notation*

Describe the translation using coordinate notation.

Solution

Each point is moved 3 units to the left and 4 units up.

ANSWER ▶ The translation can be described using the notation $(x, y) \rightarrow (x - 3, y + 4)$.

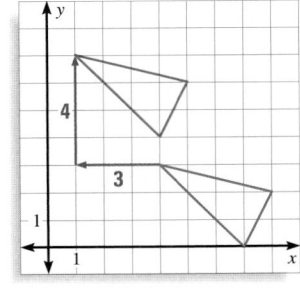

3.7 Translations **153**

Extra Example 1

How do you decide if one figure is a translation of another? A translated figure has been moved but has not changed in size, shape, or orientation.

Extra Example 2

Describe the translation of the segment.

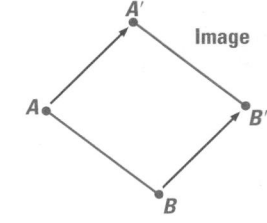

Every point on $\overline{AB}$ moves 2 units to the left and 3 units up.

Extra Example 3

Describe the translation using coordinate notation.

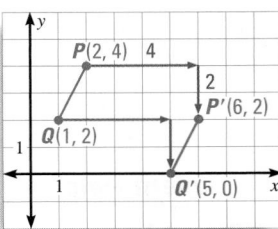

$(x, y) \rightarrow (x + 3, y - 2)$

Multiple Representations

Translations can be described in words by saying how far to the right (or left) and how for up (or down) a figure is moved. This idea is formalized using the coordinate notation $(x, y) \rightarrow (x + a, y + b)$, where a indicates the horizontal movement and b indicates the vertical movement.

Extra Example 4

Draw the triangle given by points $A(1, -1)$, $B(4, -2)$, and $C(1, -4)$. Then draw the image of the triangle after the translation given by $(x, y) \rightarrow (x - 4, y + 2)$.

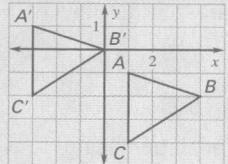

✓ Concept Check

A figure is translated 4 units left and 3 units up from its original position. Describe this translation using coordinate notation.
$(x, y) \rightarrow (x - 4, y + 3)$

😊 Daily Puzzler

A triangle has all three vertices on either the x-axis or the y-axis. After a translation, how many vertices at most can be on an axis? 2

6.
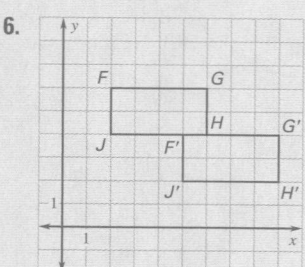

7.
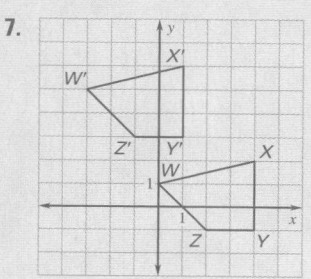

154

Describe the translation using words and coordinate notation.

4. Each point is moved 3 units to the left and 4 units down; $(x, y) \rightarrow (x - 3, y - 4)$.

5. Each point is moved 5 units to the left and 2 units up; $(x, y) \rightarrow (x - 5, y + 2)$.

4.
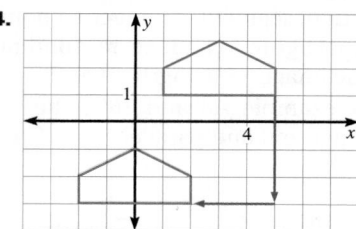

5.
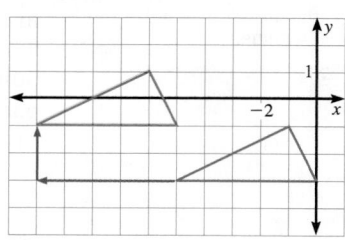

EXAMPLE 4 Draw Translated Figures

Draw the triangle given by points $A(-2, 5)$, $B(0, 7)$, and $C(3, 7)$. Then draw the image of the triangle after the translation given by $(x, y) \rightarrow (x + 2, y - 3)$.

Student Help

> **READING TIP**
> In this book, shapes are named by listing in order the labels at their corners. For example, the blue triangle in Example 4 is named $\triangle ABC$.

Solution

First, sketch $\triangle ABC$ as shown. To find points A', B', and C', start at points A, B, and C, and slide each point 2 units to the right and 3 units down.

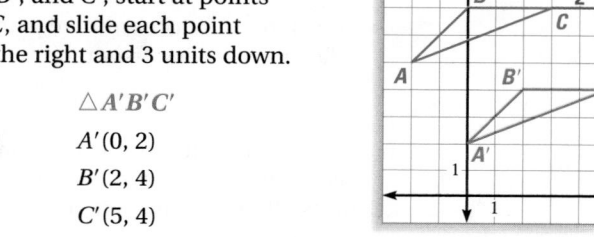

$\triangle ABC$	$\triangle A'B'C'$
$A(-2, 5)$	$A'(0, 2)$
$B(0, 7)$	$B'(2, 4)$
$C(3, 7)$	$C'(5, 4)$

Notice that each x-value of $\triangle A'B'C'$ is 2 units more than the corresponding x-value of $\triangle ABC$ and each y-value of $\triangle A'B'C'$ is 3 units less than the corresponding y-value of $\triangle ABC$.

Checkpoint ✓ Draw Translated Figures

Draw the image of the figure after the given translation. 6, 7. See margin.

6. $(x, y) \rightarrow (x + 3, y - 2)$

7. $(x, y) \rightarrow (x - 3, y + 4)$

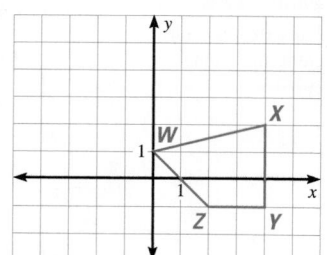

Guided Practice

Vocabulary Check

1. What is a *translation*? a slide

2. Complete the statement: A translation shows a blue triangle and a red triangle. The blue triangle is the original figure and the red triangle is the __?__. image

Skill Check

Window Frames Decide whether "opening the window" is a translation of the moving part.

3. Double hung yes

4. Casement no

5. Sliding yes

Decide whether the statement is *true* or *false*. Explain.

6. The red figure is a translation of the blue figure. true

7. To move from △*ABC* to △*A'B'C'*, shift 3 units to the right and 2 units up.
   false; shift 3 units to the left and 2 units down

8. The translation from △*ABC* to △*A'B'C'* is given by $(x, y) \rightarrow (x - 3, y - 2)$. true

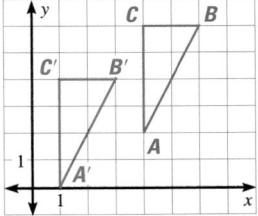

Practice and Applications

Extra Practice

See p. 680.

Compare a Figure and Its Image Decide whether the red figure is a translation of the blue figure.

9.
no

10.
yes

11.
no

12.
yes

13.
no

14.
no

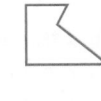

Homework Help

Example 1: Exs. 9–14
Example 2: Exs. 15–21
Example 3: Exs. 22, 23
Example 4: Exs. 38–41

③ Apply

Assignment Guide

BASIC
Day 1: pp. 155–159 Exs. 9–21, 49–59
Day 2: SRH p. 664 Exs. 21–28; pp. 155–159 Exs. 22–27, 32–44 even, Quiz 3

ADVANCED
Day 1: pp. 155–159 Exs. 9–21, 49–59
Day 2: pp. 155–159 Exs. 22–24, 26–44 even, 45, Quiz 3

ADVANCED
Day 1: pp. 155–159 Exs. 9–21, 46, 47, 49–59
Day 2: pp. 155–159 Exs. 22, 23–43 odd, 48*, Quiz 3; EC: classzone.com

BLOCK SCHEDULE
pp. 155–159 Exs. 9–24, 26–44 even, 45, 49–59, Quiz 3

Extra Practice

• Student Edition, p. 680
• Chapter 3 Resource Book, pp. 72–73

Homework Check

To quickly check student understanding of key concepts, go over the following exercises:

Basic: 10, 13, 15, 22, 38
Average: 10, 13, 17, 22, 40
Advanced: 10, 13, 20, 23, 41

Matching Translations Match the description of the translation with its diagram.

15. 4 units right and 3 units up C

16. 6 units right and 2 units down A

17. 7 units left and 1 unit up B

18. 5 units right and 2 units down D

A.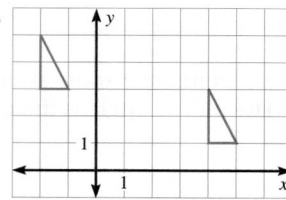

B.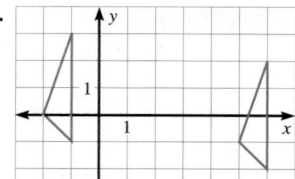

C.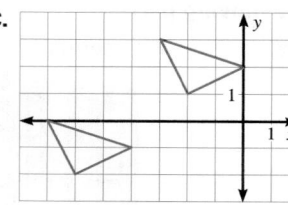

D.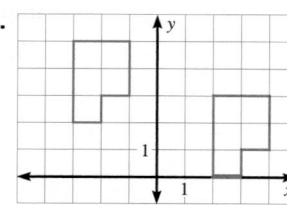

Describing Translations Describe the translation using words.

19. Each point is moved 2 units to the right and 3 units up.

20. Each point is moved 1 unit to the right and 4 units down.

21. Each point is moved 2 units to the left and 3 units up.

19.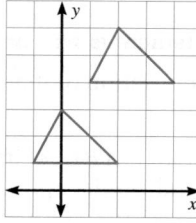

20.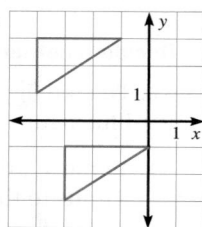

21.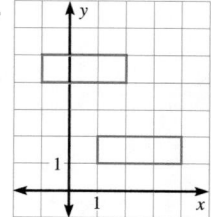

Coordinate Notation Describe the translation using coordinate notation.

22. $(x, y) \rightarrow (x + 2, y - 4)$

23. $(x, y) \rightarrow (x - 4, y + 1)$

22.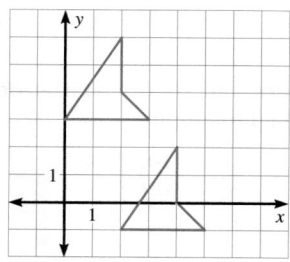

23.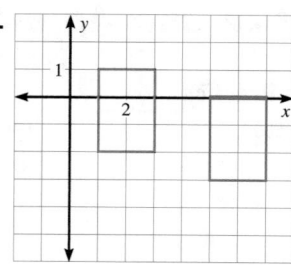

A Point and Its Image Find the image of the point using the translation $(x, y) \rightarrow (x + 4, y - 3)$.

24. $(2, 5)$ $(6, 2)$

25. $(-3, 7)$ $(1, 4)$

26. $(-1, -4)$ $(3, -7)$

27. $(4, -6)$ $(8, -9)$

28. $(0, 0)$ $(4, -3)$

29. $(-4, 3)$ $(0, 0)$

30. $(3, -4)$ $(7, -7)$

31. $(-1, -1)$ $(3, -4)$

Finding an Image Find the coordinates of P', Q', R', and S' using the given translation.

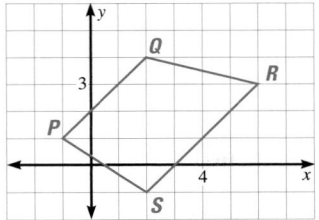

32. $(x, y) \rightarrow (x + 1, y - 4)$
   $P'(0, -3),\ Q'(3, 0),\ R'(7, -1),\ S'(3, -5)$
33. $(x, y) \rightarrow (x - 3, y + 2)$
   $P'(-4, 3),\ Q'(-1, 6),\ R'(3, 5),\ S'(-1, 1)$
34. $(x, y) \rightarrow (x + 5, y - 5)$
   $P'(4, -4),\ Q'(7, -1),\ R'(11, -2),\ S'(7, -6)$
35. $(x, y) \rightarrow (x, y - 3)$
   $P'(-1, -2),\ Q'(2, 1),\ R'(6, 0),\ S'(2, -4)$

Chess In chess, six different kinds of pieces are moved according to individual rules. The board below shows some moves for the Knight (the piece shaped like a horse).

36. Describe the translation used by the White Knight to capture the Black Pawn.
   Move 1 unit to the left and 2 units up.
37. Assume that the White Knight has taken the place of the Black Pawn. Describe the translation used by the Black Knight to move to capture the White Knight at its new location.
   Move 2 units to the right and 1 unit down.

Drawing Translated Figures Draw the image of the figure after the given translation. 38–41. See margin.

38. $(x, y) \rightarrow (x + 2, y + 1)$

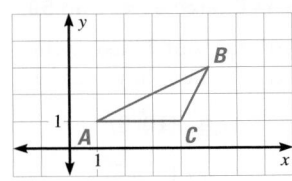

39. $(x, y) \rightarrow (x + 4, y - 5)$

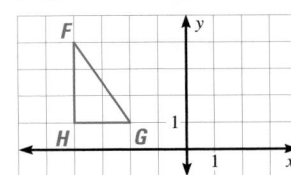

40. $(x, y) \rightarrow (x - 5, y + 3)$

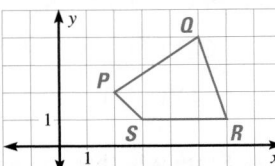

41. $(x, y) \rightarrow (x - 3, y + 8)$

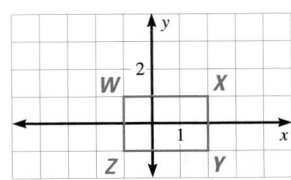

Use Points on an Image A point on an image and the translation are given. Find the corresponding point on the original figure.

42. Point on image: $(0, 3)$; translation: $(x, y) \rightarrow (x - 3, y + 2)$ $(3, 1)$

43. Point on image: $(-2, 4)$; translation: $(x, y) \rightarrow (x + 5, y - 1)$ $(-7, 5)$

44. Point on image: $(6, -1)$; translation: $(x, y) \rightarrow (x + 3, y + 7)$ $(3, -8)$

3.7 *Translations* 157

38.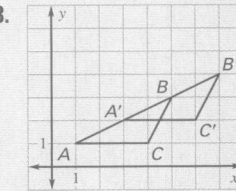

39.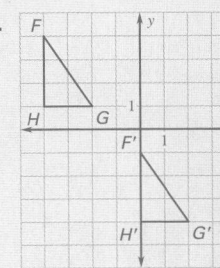

40.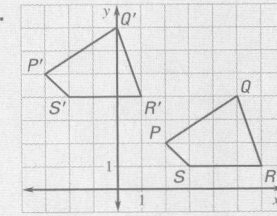

41.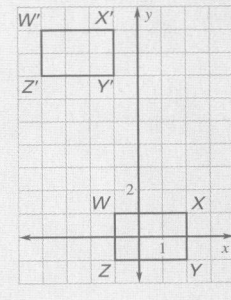

45. 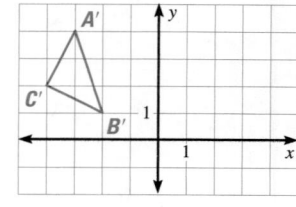 **You be the Judge** The figure on the grid shown at the right is the image after the translation $(x, y) \rightarrow (x - 6, y + 4)$. One of your classmates tells you that C on the original figure is $(2, -2)$. Do you agree? Explain your reasoning.

Yes; the image of $(2, -2)$ after the translation $(x, y) \rightarrow (x - 6, y + 4)$ is $(-4, 2)$. This is the point that is labeled C'.

Technology In Exercises 46 and 47, use geometry software to complete the steps below.

❶ Draw a triangle and translate it.

❷ Construct $\overline{JJ'}$ and $\overline{KK'}$.

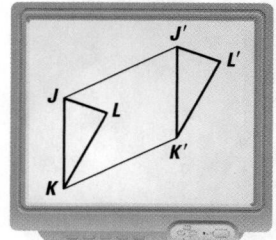

46. If two lines have the same slope, then they are parallel. Measure the slopes of $\overline{JJ'}$ and $\overline{KK'}$. Are $\overline{JJ'}$ and $\overline{KK'}$ parallel? yes

47. What should $m\angle KJJ' + m\angle K'KJ$ be? Measure the angles and check your answer. 180°

48. Challenge Point C is located at $(1, 3)$. The translation that shifts C to C' is given by $(x, y) \rightarrow (x + 5, y - 4)$. The translation that shifts C' to C'' is given by $(x, y) \rightarrow (x - 1, y + 8)$. Give the coordinate notation that describes the translation directly from C to C''. (*Hint*: Start by plotting C, C', and C''.)
$(x, y) \rightarrow (x + 4, y + 4)$

Standardized Test Practice

Multiple Choice In Exercises 49 and 50, use the diagram below.

49. Find the coordinates of T' using the translation $(x, y) \rightarrow (x - 5, y + 2)$. **A**

Ⓐ $(3, 7)$ Ⓑ $(10, 0)$

Ⓒ $(3, 5)$ Ⓓ $(-5, 7)$

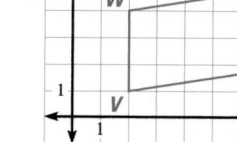

50. Find the coordinates of W' using the translation $(x, y) \rightarrow (x + 3, y - 3)$. **F**

Ⓕ $(5, 1)$ Ⓖ $(-1, 7)$

Ⓗ $(5, 7)$ Ⓙ $(-1, 1)$

Mixed Review

Classifying Angles State whether the angle appears to be *acute*, *right*, *obtuse*, or *straight*. Then estimate its measure. *(Lesson 1.6)*

51.
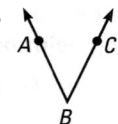
acute; about 50°

52.
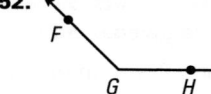
obtuse; about 135°

53.
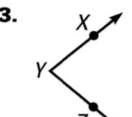
acute; about 80°

Algebra Skills

Problem Solving Use problem solving strategies to answer the question. (*Skills Review, p. 653*)

54. Your telephone company charges $.15 per minute for all long distance calls. This month you paid $12.60 for long distance calls. How many minutes did you spend on long distance calls? **84 minutes**

55. You just bought a CD single that has four tracks. In how many different orders can the songs be played? **24 orders**

Ordering Numbers Write the numbers in order from least to greatest. (*Skills Review, p. 662*)

56. −0.4, 0.5, 0, 1.0, −0.1, 0.9
  −0.4, −0.1, 0, 0.5, 0.9, 1.0

57. 3.4, −1.2, 0.7, −1.5, 0, 1.1, −4
  −4, −1.5, −1.2, 0, 0.7, 1.1, 3.4

58. 6.7, 7.6, −0.77, 6.6, −0.7, −6.7
  −6.7, −0.77, −0.7, 6.6, 6.7, 7.6

59. −6.12, 6.3, −6.8, −6.1, 6, 6.09
  −6.8, −6.12, −6.1, 6, 6.09, 6.3

Quiz 3

Determine whether enough information is given to conclude that $m \parallel n$. Explain. (*Lesson 3.5*)

1. yes, by the Alternate Exterior Angles Converse

2. No; There is no information about line m.

3. yes, by the Alternate Interior Angles Converse

4. In a plane, if two lines are perpendicular to the same line, then they are parallel to each other.

5. If two lines are parallel to the same line, then they are parallel to each other.

6. Use the Corresponding Angles Converse or Theorem 3.12 (in a plane, if two lines are perpendicular to the same line, then they are parallel to each other).

1.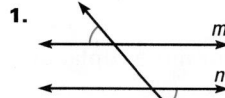

2.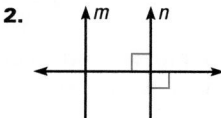

3.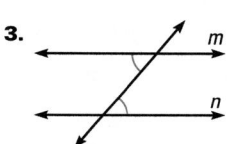

In Exercises 4–6, explain how you would show that $p \parallel q$. State any theorems or postulates that you would use. (*Lesson 3.6*)

4.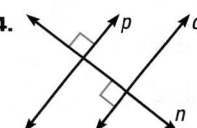

5.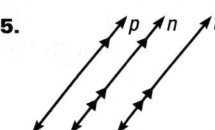

6.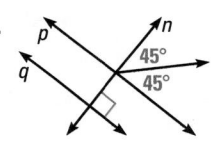

7. Draw a vertical line ℓ and construct a line m perpendicular to it through a point P to the left of line ℓ. (*Lesson 3.6*) **See margin.**

In Exercises 8 and 9, describe the translation of the figure using coordinate notation. (*Lesson 3.7*)

8.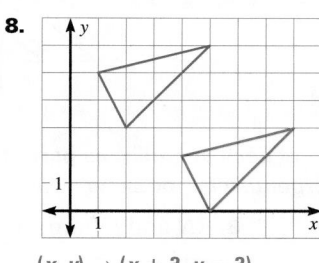

$(x, y) \rightarrow (x + 3, y - 3)$

9.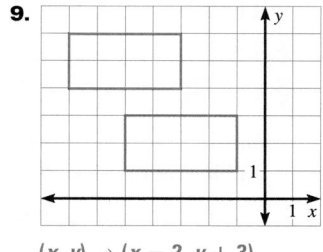

$(x, y) \rightarrow (x - 2, y + 3)$

3.7 Translations   159

Mini-Quiz

Describe the translation that maps the solid figure onto the dashed image using words and coordinate notation.

1.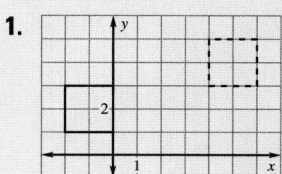

6 units right and 2 units up;
$(x, y) \rightarrow (x + 6, y + 2)$

2.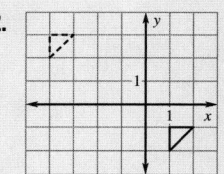

5 units left and 4 units up;
$(x, y) \rightarrow (x - 5, y + 4)$

Find the image of the point using the translation $(x, y) \rightarrow (x - 3, y + 2)$.

3. (3, 1) **(0, 3)**

4. (0, −2) **(−3, 0)**

7.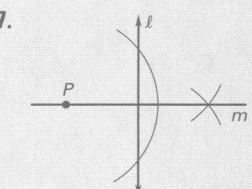

Additional Resources

The following resources are available to help review the materials in this chapter.

📖 **Chapter 3 Resource Book**
• Chapter Review Games and Activities, p. 80
• Cumulative Review, Chs. 1–3

Chapter  3 Summary and Review

VOCABULARY

- **parallel lines,** *p. 108*
- **perpendicular lines,** *p. 108*
- **skew lines,** *p. 108*
- **parallel planes,** *p. 109*
- **line perpendicular to a plane,** *p. 109*

- **transversal,** *p. 121*
- **corresponding angles,** *p. 121*
- **alternate interior angles,** *p. 121*
- **alternate exterior angles,** *p. 121*
- **same-side interior angles,** *p. 121*

- **converse,** *p. 136*
- **construction,** *p. 143*
- **translation,** *p. 152*
- **image,** *p. 152*
- **transformation,** *p. 152*

VOCABULARY REVIEW

Fill in the blank.

1. Two lines are __?__ if they lie in the same plane and do not intersect. **parallel**

2. A(n) __?__ is a line that intersects two or more coplanar lines at different points. **transversal**

3. Two lines are __?__ if they intersect to form a right angle. **perpendicular**

4. Two angles are __?__ if they lie outside two lines on opposite sides of a transversal. **alternate exterior angles**

5. A(n) __?__ is a geometric drawing that uses a limited set of tools, usually a compass and a straightedge. **construction**

6. Two planes that do not intersect are called __?__. **parallel**

3.1 RELATIONSHIPS BETWEEN LINES

Examples on pp. 108–109

EXAMPLES **Name of a pair of parallel lines, perpendicular lines, and skew lines.**

Lines *r* and *s* are parallel.

Lines *q* and *r* are perpendicular.

Lines *q* and *s* are skew.

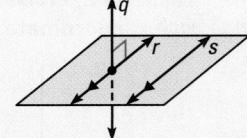

In the diagram at the right, think of each segment as part of a line. Fill in the blank with *parallel, perpendicular,* or *skew*.

7. $\overleftrightarrow{FJ}$ and $\overleftrightarrow{GH}$ appear to be __?__. **parallel**

8. $\overleftrightarrow{KN}$ and $\overleftrightarrow{JN}$ are __?__. **perpendicular**

9. $\overleftrightarrow{FK}$ and $\overleftrightarrow{HJ}$ are __?__. **skew**

10. $\overleftrightarrow{JN}$ and $\overleftrightarrow{MN}$ are __?__. **perpendicular**

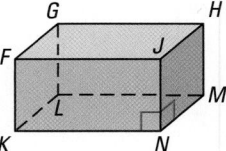

3.2 THEOREMS ABOUT PERPENDICULAR LINES

Examples on pp. 114–116

EXAMPLES In the diagram, *a* ⊥ *b* and *a* ⊥ *d*. Give the theorem that supports the statement.

a. ∠3 ≅ ∠5
All right angles are congruent. So, ∠3 ≅ ∠5. (Theorem 3.1)

b. *m*∠4 = 90°
If two lines are perpendicular, they intersect to form four right angles. So, *m*∠4 = 90°. (Theorem 3.2)

c. ∠1 and ∠2 are complementary.
If two sides of adjacent acute angles are perpendicular, then the angles are complementary. So, ∠1 and ∠2 are complementary. (Theorem 3.4)

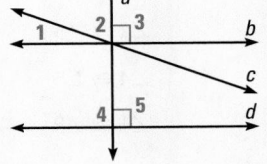

Determine whether enough information is given to conclude that the statement is true. Explain. 11–14. See margin for explanations.

11. *m*∠11 = 90° Yes

12. *m*∠9 + *m*∠10 = 90° Yes

13. *h* ⊥ *j* Yes

14. ∠9 ≅ ∠10 No

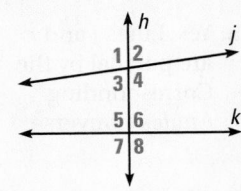

f ⊥ *g*

3.3 ANGLES FORMED BY TRANSVERSALS

Examples on pp. 121–122

EXAMPLE Name a pair of corresponding angles, alternate interior angles, alternate exterior angles, and same-side interior angles.

∠1 and ∠5 are corresponding angles.

∠3 and ∠6 are alternate interior angles.

∠1 and ∠8 are alternate exterior angles.

∠4 and ∠6 are same-side interior angles.

Complete the statement using *corresponding, alternate interior, alternate exterior*, or *same-side interior*.

15. ∠8 and ∠12 are __?__ angles. corresponding

16. ∠9 and ∠14 are __?__ angles. alternate exterior

17. ∠10 and ∠12 are __?__ angles. same-side interior

18. ∠11 and ∠12 are __?__ angles. alternate interior

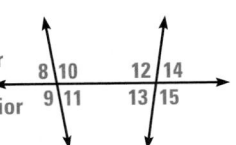

Margin:
11. *m*∠11 = 90° by the Vertical Angles Theorem
12. If two sides of adjacent acute angles are perpendicular, then the angles are complementary.
13. If two lines intersect to form adjacent congruent angles, then the lines are perpendicular.
14. The diagram shows ∠9 and ∠10 are complementary, but the actual angle measures are not given.

3.4   PARALLEL LINES AND TRANSVERSALS

Examples on pp. 128–131

EXAMPLE   In the diagram, $b \parallel c$. Name three pairs of congruent angles and one pair of supplementary angles.

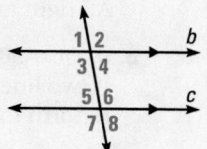

$\angle 1 \cong \angle 5$ Corresponding Angles Postulate

$\angle 4 \cong \angle 5$ Alternate Interior Angles Theorem

$\angle 1 \cong \angle 8$ Alternate Exterior Angles Theorem

$m\angle 3 + m\angle 5 = 180°$ Same-Side Interior Angles Theorem

Find the measure of the numbered angle.

19. 99°
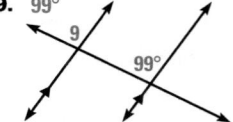

20. 63°

21. 71°
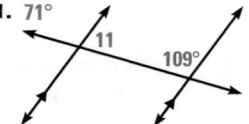

3.5   SHOWING LINES ARE PARALLEL

Examples on pp. 136–139

EXAMPLES   Is enough information given to conclude that $j \parallel k$? Explain.

a.

b.
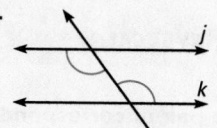

c.
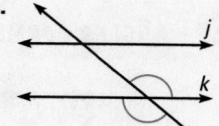

a. Yes. Lines j and k are parallel by the Corresponding Angles Converse.

b. Yes. Lines j and k are parallel by the Alternate Interior Angles Converse.

c. No. Not enough information is given.

Find the value of x so that $r \parallel s$.

22.
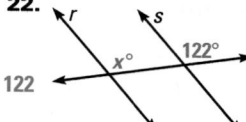

23.
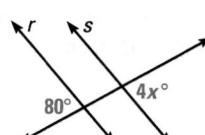

24.
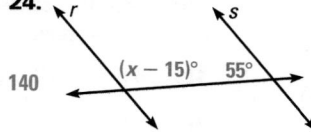

3.6 USING PERPENDICULAR AND PARALLEL LINES

Examples on pp. 143–146

> **EXAMPLE** In the diagram, $j \perp t$, $m \perp t$, and $m \parallel n$. Explain why $j \parallel n$.
>
> Because j and m lie in the same plane and are both perpendicular to t, $j \parallel m$. (Theorem 3.12)
>
> Because $j \parallel m$ and $m \parallel n$, $j \parallel n$. (Theorem 3.11)

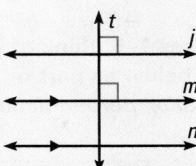

Using the given information, state the postulate or theorem that allows you to conclude that $p \parallel q$. 25–27. See margin.

25. $p \parallel r$, $q \parallel r$

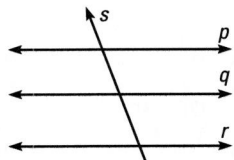

26. $p \perp s$, $q \perp s$

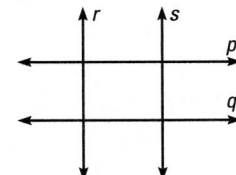

27. $\angle 1 \cong \angle 2$

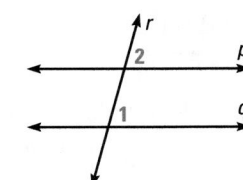

25. If two lines are parallel to the same line, then they are parallel to each other.

26. In a plane, if two lines are perpendicular to the same line, then they are parallel to each other.

27. Corresponding Angles Converse

3.7 TRANSLATIONS

Examples on pp. 152–154

> **EXAMPLE** Describe the translation using coordinate notation.
>
> Each point is moved 4 units to the left and 2 units down. So, the translation can be described using the notation $(x, y) \to (x - 4, y - 2)$.

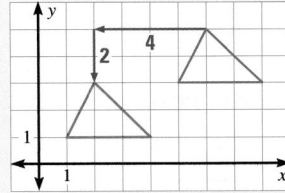

Decide whether the red figure is a translation of the blue figure.

28. no

29. yes

30. no

In Exercises 31 and 32, use the translation at the right. The original figure is blue and the image is red.

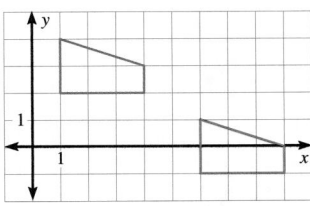

31. Describe the translation using words.
Each point is moved 5 units to the right and 3 units down.

32. Describe the translation using coordinate notation.
$(x, y) \to (x + 5, y - 3)$

Chapter Summary and Review **163**

Additional Resources

📖 **Chapter 3 Resource Book**
- Chapter Test (2 levels), pp. 81–84
- SAT/ACT Chapter Test, p. 85
- Alternative Assessment, pp. 86–87

💻 **Test and Practice Generator**

6. $\angle FGJ$ and $\angle JGH$ are complementary (or $m\angle FGJ + m\angle JGH = 90°$).

14. $m\angle 1 = 108°$ by the Corresponding Angles Postulate. $m\angle 2 = 72°$ by the Same-Side Interior Angles Theorem. $m\angle 3 = 72°$ by the Linear Pair Postulate or by the Alternate Interior Angles Theorem.

17. Yes; in a plane, if two lines are perpendicular to the same line, then they are parallel to each other.

18. Each point is moved 6 units to the right and 2 units up; $(x, y) \rightarrow (x + 6, y + 2)$.

In Exercises 1–5, think of each segment in the diagram below as part of a line. Fill in the blank with *parallel*, *perpendicular*, or *skew*.

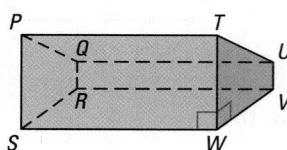

1. $\overleftrightarrow{PT}$ and $\overleftrightarrow{UV}$ are ___?___. skew

2. $\overleftrightarrow{TW}$ and $\overleftrightarrow{WV}$ are ___?___. perpendicular

3. $\overleftrightarrow{PT}$ and $\overleftrightarrow{SW}$ appear ___?___. parallel

4. Plane PQR and plane TUV appear ___?___. parallel

5. $\overleftrightarrow{TW}$ is ___?___ to plane SWV. perpendicular

6. What can you conclude about $\angle FGJ$ and $\angle JGH$, given that $\overrightarrow{FG} \perp \overleftrightarrow{GH}$? See margin.

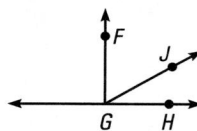

7. Find the value of x given that $c \perp d$. 38

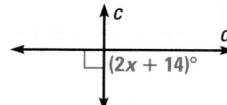

In Exercises 8–13, identify the relationship between the angles in the diagram below.

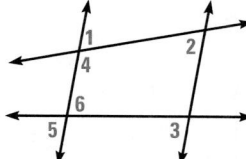

8. $\angle 1$ and $\angle 2$ alternate interior

9. $\angle 1$ and $\angle 6$ corresponding

10. $\angle 2$ and $\angle 3$ corresponding

11. $\angle 1$ and $\angle 5$ alternate exterior

12. $\angle 4$ and $\angle 2$ same-side interior

13. $\angle 5$ and $\angle 3$ corresponding

---

14. Find $m\angle 1$, $m\angle 2$, and $m\angle 3$. Explain your reasoning. See margin.

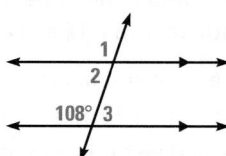

In Exercises 15 and 16, find the value of y so that $m \parallel n$.

15. 22

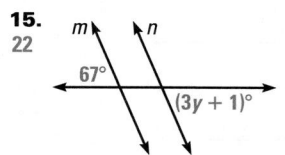

16. 32

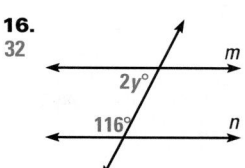

17. A carpenter wants to cut two boards to fit snugly together. The carpenter's squares are aligned along $\overline{EF}$, as shown. Are $\overline{AB}$ and $\overline{CD}$ parallel? State the theorem that justifies your answer. See margin.

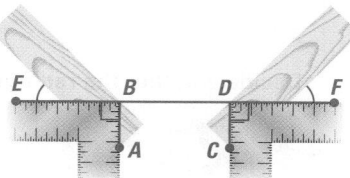

18. Describe the translation using words and coordinate notation. See margin.

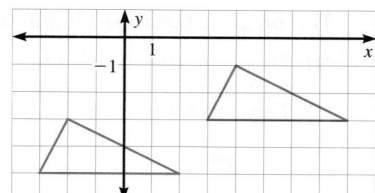

Additional Resources

📖 *Chapter 3 Resource Book*
• SAT/ACT Chapter Test, p. 85

Test Tip

Do not worry about how much time you have left or how others are doing. Concentrate on your own work.

1. Two lines are __?__ lines if they lie in the same plane and do not intersect. **B**

Ⓐ intersecting   Ⓑ parallel

Ⓒ perpendicular   Ⓓ skew

2. If two lines form adjacent congruent angles, then the lines are __?__. **G**

Ⓕ parallel   Ⓖ perpendicular

Ⓗ skew   Ⓙ collinear

3. Find the value of x, given that $p \perp q$. **A**

Ⓐ 16

Ⓑ 50

Ⓒ 90

Ⓓ 106

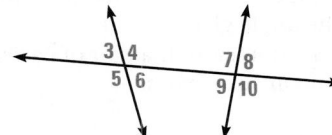

4. Which angles are alternate interior angles? **G**

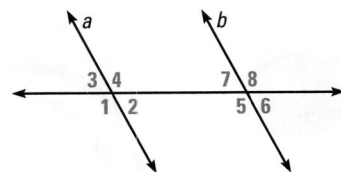

Ⓕ $\angle 3$ and $\angle 9$ Ⓖ $\angle 6$ and $\angle 7$

Ⓗ $\angle 5$ and $\angle 8$ Ⓙ All of these

5. If $a \parallel b$, which angles are congruent? **A**

Ⓐ $\angle 3$ and $\angle 6$ Ⓑ $\angle 3$ and $\angle 8$

Ⓒ $\angle 2$ and $\angle 5$ Ⓓ None of these

6. Find the value of y so that $m \parallel n$. **F**

Ⓕ 35

Ⓖ 55

Ⓗ 110

Ⓙ 180

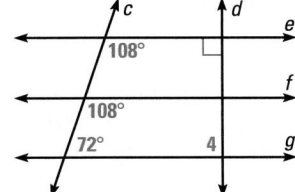

7. Multi-Step Problem Use the diagram below. See margin.

a. Explain why $e \parallel f$.

b. Explain why $f \parallel g$.

c. Explain why $e \parallel g$.

d. What is $m\angle 4$? Explain.

8. Choose the coordinate notation that describes the translation from $\triangle ABC$ to $\triangle A'B'C'$. **D**

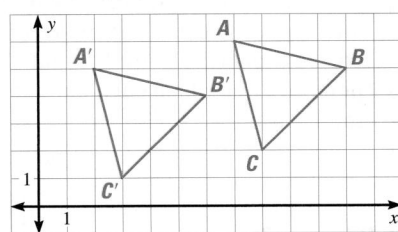

Ⓐ $(x, y) \rightarrow (x + 5, y + 1)$

Ⓑ $(x, y) \rightarrow (x + 5, y - 1)$

Ⓒ $(x, y) \rightarrow (x - 5, y + 1)$

Ⓓ $(x, y) \rightarrow (x - 5, y - 1)$

7. a. $e \parallel f$ by the Corresponding Angles Converse.
 b. $108° + 72° = 180°$, so $f \parallel g$ by the Same-Side Interior Angles Converse.
 c. $e \parallel f$ and $f \parallel g$. Therefore, $e \parallel g$: If two lines are parallel to the same line, then they are parallel to each other.
 d. $90°$; *Sample answer:* Since $d \perp e$, all four angles formed by d and e are right angles; since $e \parallel g$, $m\angle 4 = 90°$ by the Corresponding Angles Postulate.

Teaching Tip
The Brain Games activity provides a motivating way to review selected content in the chapter. For a more comprehensive review, see the Chapter Summary and Review on pp. 160–163.

BraiN GaMes

What's the Angle?

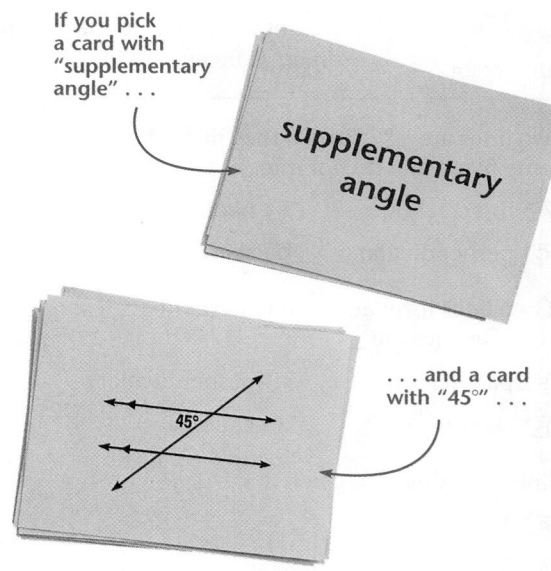

If you pick a card with "supplementary angle" . . .

supplementary angle

. . . and a card with "45°" . . .

45°

. . . your answer will be 135°.

Materials
• 6 description cards
• 26 diagram cards

Object of the Game
To correctly give the measures of angles.

Set Up
Shuffle each stack of cards. Place the description cards face down in one pile and the diagram cards face down in another pile.

How to Play

Step 1 ▶ Each player takes a turn selecting one card from each pile.

Step 2 ▶ The player calculates the measure of the angle given on the description card. If a pair of cards has no answer, say so.

Step 3 ▶ The other players determine if the answer is correct. If the answer is correct, record a point for that person. Then place each card at the bottom of its pile.

Step 4 ▶ After a set amount of time, complete the round so every player gets an equal number of turns. The player with the most points wins.

Another Way to Play
Each player has one minute to calculate as many angle measures as possible. The other players check each answer at the end of the minute. After each player has a turn, the player with the most correct answers wins.

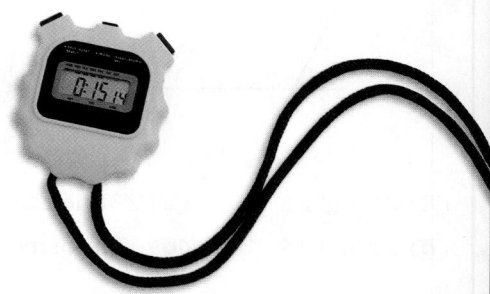

Chapter  3 Algebra Review

You must reverse the inequality sign when you multiply or divide each side of an inequality by a negative number.

Extra Example 1
Solve the inequality.
a. $3y + 5 > 20$ $y > 5$
b. $42 - 6p \leq p - 14$ $p \geq 8$

Checkpoint Exercise 1
Solve the inequality.
a. $6a + 2 < 7a$ $a > 2$
b. $22 - b \leq b$ $b \geq 11$

Extra Example 2
Evaluate the expression.
a. 5^2 25
b. $(-10)^2$ 100
c. $(\sqrt{19})^2$ 19
d. $\sqrt{49}$ 7

Checkpoint Exercise 2
a. $(-1)^2$ 1
b. 9^2 81
c. $(\sqrt{33})^2$ 33
d. $\sqrt{25}$ 5

EXAMPLE 1   Solve Inequalities

Solve the inequality.

a. $6x + 11 > 29$
b. $7 - 4y \leq y + 12$

Solution

a. $6x + 11 > 29$
$\quad\quad 6x > 18$
$\quad\quad\quad x > 3$ Divide by 6.

b. $7 - 4y \leq y + 12$
$\quad 7 - 5y \leq 12$
$\quad\quad -5y \leq 5$
$\quad\quad\quad y \geq -1$ Divide by -5.

Student Help

STUDY TIP
Both sides of the inequality have been divided by a negative number, -5, so change "≤" to "≥." ·················

Try These

Solve the inequality.

1. $x - 3 > 12$ $x > 15$ **2.** $8q + 1 < 25$ $q < 3$ **3.** $-3z + 8 \geq 20$ $z \leq -4$

4. $16 - 9c \leq -2$ $c \geq 2$ **5.** $10 - 2p \leq -4p + 4$ $p \leq -3$ **6.** $5k - 6 > 3k + 16$ $k > 11$

If $b^2 = a$, then b is a *square root* of a. Every positive number has one positive and one negative square root. The radical symbol $\sqrt{}$ indicates the positive square root.

EXAMPLE 2   Evaluate Squares and Square Roots

Evaluate the expression.

a. 7^2 **b.** $(-8)^2$ **c.** $(\sqrt{3})^2$ **d.** $\sqrt{36}$

Solution

a. $7^2 = 7 \cdot 7 = 49$
b. $(-8)^2 = (-8)(-8) = 64$
c. $(\sqrt{3})^2 = 3$
d. $\sqrt{36} = \sqrt{6^2} = 6$

Try These

Evaluate the expression.

7. 4^2 16   **8.** 14^2 196   **9.** $(-3)^2$ 9   **10.** $(-11)^2$ 121

11. $(\sqrt{5})^2$ 5   **12.** $(\sqrt{10})^2$ 10   **13.** $\sqrt{81}$ 9   **14.** $\sqrt{400}$ 20

Algebra Review   167

Additional Resources
A Cumulative Test covering
Chapters 1–3 is available in the
Chapter 3 Resource Book, pp. 92–95.

10.

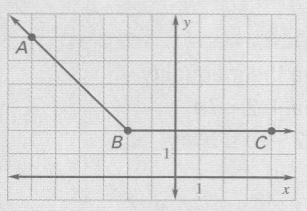

obtuse

11.

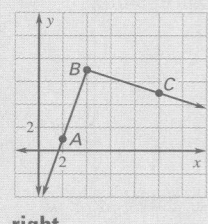

right

1. Describe a pattern in the numbers 10, 12, 15, 19, 24, ... Then
write the next number you expect in the pattern. (Lesson 1.1) You begin with 10, and add 2,
then 3, then 4, and so on; 30.

2. Show that the conjecture is false by finding a counterexample.

Conjecture: The square of a number is always greater than the
number. (Lesson 1.2) *Sample answer:* $0^2 = 0$ (or any number between 0 and 1)

In the diagram, $\overleftrightarrow{AB}$, $\overleftrightarrow{AC}$, and $\overleftrightarrow{BC}$ lie in plane *M*. (Lessons 1.3, 1.4)

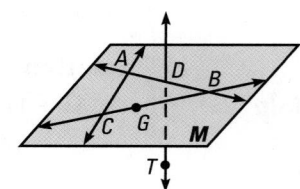

3. Name a point that is collinear with points *A* and *D*. B

4. Name a point that is not coplanar with *A*, *B*, and *C*. T

5. Name the intersection of $\overleftrightarrow{AD}$ and $\overleftrightarrow{CG}$. B

6. Name two lines that do not appear to intersect.
$\overleftrightarrow{AC}$ and $\overleftrightarrow{DT}$, or $\overleftrightarrow{BC}$ and $\overleftrightarrow{DT}$

Find the length. (Lesson 1.5)

7. Find *DE*. 16

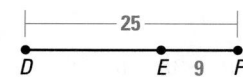

8. Find *MN*. 11

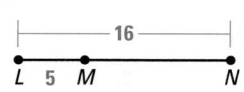

9. Find *PQ*. 7

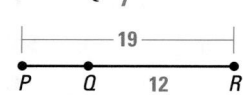

In Exercises 10 and 11, plot the points in a coordinate plane and
sketch ∠*ABC*. Classify the angle as *acute, right,* or *obtuse.* (Lesson 1.6) 10, 11. See margin.

10. $A(-6, 6)$, $B(-2, 2)$, $C(4, 2)$

11. $A(2, 1)$, $B(4, 7)$, $C(10, 5)$

12. Line ℓ bisects $\overline{AB}$. Find *AC* and *AB*. (Lesson 2.1)
$AC = 29$; $AB = 58$

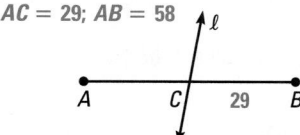

13. $\overrightarrow{KM}$ bisects ∠*JKL*. Find the value of *x*. (Lesson 2.2)
17

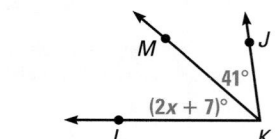

Find the value of the variable. (Lessons 2.3, 2.4)

14.
18

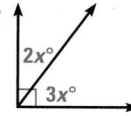

15.
55

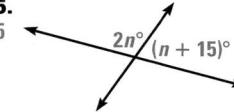

16.
10

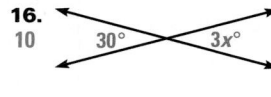

17. Rewrite the statement as an if-then statement. Then underline the
hypothesis and circle the conclusion. (Lesson 2.5)

Statement: Intersecting lines are coplanar lines.

If <u>two lines intersect</u>, then (the lines are coplanar.)

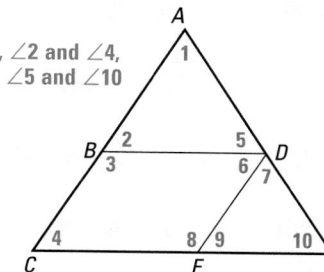
Name the property that the statement illustrates. (Lesson 2.6)

18. If $\overline{XY} \cong \overline{YZ}$, then $\overline{YZ} \cong \overline{XY}$. Symmetric Property of Congruence

19. If $m\angle P = m\angle Q$ and $m\angle Q = m\angle R$, then $m\angle P = m\angle R$. Transitive Property of Equality

20. If $AB = CD$, then $AB - 2 = CD - 2$. Subtraction Property of Equality

Sketch a figure that fits the description. (Lesson 3.1) 21, 22. See margin.

21. Line k is perpendicular to line j and line h is parallel to line k.

22. Line w lies in plane D and line v is skew to line w.

In Exercises 23–26, use the diagram at the right. (Lessons 3.3–3.5)

23. Name four pairs of corresponding angles. $\angle 1$ and $\angle 7$, $\angle 4$ and $\angle 9$, $\angle 2$ and $\angle 4$, $\angle 5$ and $\angle 10$

24. If $\overleftrightarrow{AC} \parallel \overleftrightarrow{DE}$ and $m\angle 2 = 55°$, find the measure of $\angle 6$. 55°

25. If $\overleftrightarrow{BD} \parallel \overleftrightarrow{CF}$ and $m\angle 2 = 55°$, find the measure of $\angle 4$. 55°

26. If $m\angle 3 + m\angle 6 = 180°$, which lines are parallel? Explain.
$\overleftrightarrow{AC} \parallel \overleftrightarrow{DE}$ by the Same-Side Interior Angles Converse

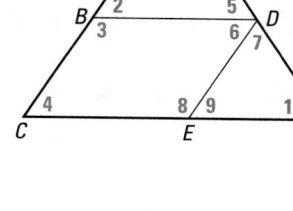

Is enough information given to conclude that $m \parallel n$? Explain.
(Lesson 3.5)

27.

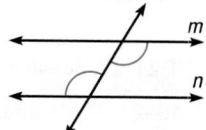

Yes, by the Alternate Interior Angles Converse

28.

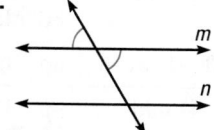

No; there is no information about line n.

29.

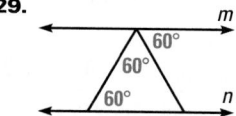

Sample answer: Yes, by the Same-Side Interior Angles Converse

Construction In Exercises 30–32, use the diagram at the right. The diagram shows two posts that support a raised deck. The posts have two parallel braces, as shown. (Lessons 3.4, 3.6)

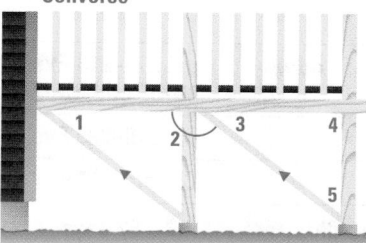

30. If $m\angle 1 = 35°$, find $m\angle 2$. 145°

31. If $m\angle 3 = 40°$, what other angle has a measure of 40°? $\angle 1$

32. Given that each post is perpendicular to the deck, explain how to show that the posts are parallel to each other.
In a plane, if two lines are perpendicular to the same line, then they are parallel to each other.

33. A segment has endpoints $A(-2, -3)$ and $B(2, 0)$. Graph $\overline{AB}$ and its image after the translation $(x, y) \rightarrow (x - 4, y + 1)$. (Lesson 3.7) See margin.

Cumulative Practice **169**

REGULAR SCHEDULE

Lesson	Les. Day	Basic	Average	Advanced
4.1	Day 1	pp. 176–178 Exs. 11–21 odd, 24–28 even, 56–58, 59–69 odd	pp. 176–178 Exs. 11–29 odd, 51–57, 60–72 even	pp. 176–178 Exs. 14–19, 23–29 odd, 48–66 even
	Day 2	pp. 176–178 Exs. 30–36, 45–49, 54, 55, 60–70 even	pp. 176–178 Exs. 30–39, 42–44, 48–50, 59–71 odd	pp. 176–178 Exs. 30–41, 43–45, 55–71 odd; EC: TE p. 170D*, classzone.com
4.2	Day 1	SRH p. 672 Exs. 2–10 even; pp. 182–184 Exs. 6–11, 15, 16, 18, 25–35 odd	pp. 182–184 Exs. 6–11, 15, 16, 18, 20, 23–35 odd	pp. 182–184 Exs. 9–11, 15–18, 20, 21, 23–25, 27–35 odd
	Day 2	SRH p. 670 Exs. 10–15; pp. 182–184 Exs. 12–14, 19, 26–34 even, Quiz 1	pp. 182–184 Exs. 12–14, 19, 22, 26–34 even, Quiz 1	pp. 182–184 Exs. 12–14, 19, 22, 26–34 even, Quiz 1; EC: TE p. 170D*, classzone.com
4.3	Day 1	SRH p. 672 Exs. 12–20 even; pp. 188–190 Exs. 7–15, 17–19, 33–37	pp. 188–190 Exs. 7–15, 17–19, 27–37	pp. 188–190 Exs. 8–10, 13–19, 27–37
	Day 2	SRH p. 673 Exs. 19–24; pp. 188–190 Exs. 20–25, 27, 28, 38–44	pp. 188–190 Exs. 20–25, 38–44	pp. 188–190 Exs. 20–25, 26*, 38–44; EC: classzone.com
4.4	Day 1	SRH p. 669 Exs. 1–4, 17–20; pp. 195–198 Exs. 8–22, 39–51 odd	pp. 195–198 Exs. 8–26, 39–55 odd	pp. 195–198 Exs. 8–22 even, 23–26, 35–37*, 39–51 odd
	Day 2	pp. 195–198 Exs. 27–32, 38–54 even, Quiz 2	pp. 195–198 Exs. 27–34, 38–54 even, Quiz 2	pp. 195–198 Exs. 27–34, 38–54 even, Quiz 2; EC: classzone.com
4.5	Day 1	pp. 203–205 Exs. 9–17, 43–53	pp. 203–205 Exs. 9–17, 37–39, 43–53 odd	pp. 203–205 Exs. 9–17, 24, 37–39, 42–45, 47–53 odd
	Day 2	pp. 203–205 Exs. 18–23, 25–36	pp. 203–205 Exs. 18–23, 25–36, 42–52 even	pp. 203–205 Exs. 18–23, 25–36, 40–41*; EC: classzone.com
4.6	Day 1	pp. 210–211 Exs. 9, 10, 18, 22–40 even	pp. 210–211 Exs. 9, 10, 18, 24–29, 30–40 even	pp. 210–211 Exs. 9, 10, 18, 24–29, 30–40 even
	Day 2	pp. 210–211 Exs. 11–16, 23–41 odd	pp. 210–211 Exs. 11–17, 19–23, 31–41 odd	pp. 210–211 Exs. 11–17, 19–23, 31–41 odd; EC: TE p. 170D*, classzone.com
4.7	Day 1	SRH p. 662 Exs. 9–12; pp. 215–218 Exs. 12–23, 26–31, 37, 38	pp. 215–218 Exs. 12–31, 37, 38, 45–48	pp. 215–218 Exs. 12–30 even, 37, 38, 45–48, 49–57 odd
	Day 2	pp. 215–218 Exs. 32–36, 44, 45–57 odd, Quiz 3	pp. 215–218 Exs. 32–36, 39, 40, 44, 49–57 odd, Quiz 3	pp. 215–218 Exs. 25, 32–36, 39–44, Quiz 3; EC: TE p. 170D*, classzone.com
Review	Day 1	pp. 219–223 Exs. 1–62	pp. 219–223 Exs. 1–62	pp. 219–223 Exs. 1–62
Assess	Day 1	Chapter 4 Test	Chapter 4 Test	Chapter 4 Test

YEARLY PACING   Chapter 4 Total – **16 days**   Chapters 1–4 Total – **62 days**   Remaining – **98 days**

*Challenge Exercises   EP = Extra Practice   SRH = Skills Review Handbook   EC = Extra Challenge

BLOCK SCHEDULE

Day 1	Day 2	Day 3	Day 4	Day 5	Day 6	Day 7	Day 8
4.1 pp. 176–178 Exs. 11–29 odd, 30–39, 42–44, 51–57, 59–72	4.2 pp. 182–184 Exs. 6–16, 18–20, 22, 23, 25–35, Quiz 1	4.3 pp. 188–190 Exs. 7–15, 17–25, 27–44	4.4 pp. 195–198 Exs. 8–34, 38–55, Quiz 2	4.5 pp. 203–205 Exs. 9–23, 25–39, 42–53	4.6 pp. 210–211 Exs. 9–41	4.7 pp. 215–218 Exs. 12–40, 44–49, 51–57 odd, Quiz 3	Review pp. 219–223 Exs. 1–62 Assess Chapter 4 Test

YEARLY PACING   Chapter 4 Total – **8 days**   Chapters 1–4 Total – **31 days**   Remaining – **49 days**

Support Materials

CHAPTER RESOURCE BOOK

CHAPTER SUPPORT

Tips for New Teachers	p. 1	Strategies for Reading Mathematics	p. 5
Parent Guide for Student Success	p. 3		

LESSON SUPPORT

	4.1	4.2	4.3	4.4	4.5	4.6	4.7
Lesson Plans (regular and block)	p. 7	p. 15	p. 26	p. 35	p. 46	p. 56	p. 64
Warm-Up Exercises and Daily Quiz	p. 9	p. 17	p. 28	p. 37	p. 48	p. 58	p. 66
Technology Activities & Keystrokes		p. 18	p. 29		p. 49		p. 67
Practice (2 levels)	p. 10	p. 19	p. 30	p. 38	p. 51	p. 59	p. 69
Reteaching with Practice	p. 12	p. 21	p. 32	p. 40	p. 53	p. 61	p. 71
Quick Catch-Up for Absent Students	p. 14	p. 23	p. 34	p. 42	p. 55	p. 63	p. 73
Learning Activities				p. 43			
Real-Life Applications		p. 24		p. 44			

REVIEW AND ASSESSMENT

Quizzes	pp. 25, 45, 74	Alternative Assessment with Math Journal	p. 81
Chapter Review Games and Activities	p. 75	Project with Rubric	p. 85
Chapter Test (2 levels)	p. 76	Cumulative Review	p. 87
SAT/ACT Chapter Test	p. 80	Resource Book Answers	AN1

TRANSPARENCIES

	4.1	4.2	4.3	4.4	4.5	4.6	4.7
Warm-Up Exercises and Daily Quiz	p. 21	p. 22	p. 23	p. 24	p. 25	p. 26	p. 27
Visualize It Transparencies	✓		✓				✓
Answer Transparencies	✓	✓	✓	✓	✓	✓	✓

TECHNOLOGY

- Time-Saving Test and Practice Generator
- Electronic Teacher Tools
- Geometry in Motion: Video
- Classzone.com
- Online Lesson Planner

ADDITIONAL RESOURCES

- Worked-Out Solution Key
- Resources in Spanish
- Practice Workbook with Examples

Providing Universal Access

Strategies for Strategic Learners

USE A VISUAL MODEL

Students learn about triangle relationships in this chapter. They begin to identify various kinds of triangles and to use the properties of triangles to solve problems involving an unknown angle measure. Help students become familiar with triangles and the new terminology by having them draw a triangle using a ruler or straightedge. Instruct them to label the three vertices A, B, and C. Ask them to identify the three sides $\overline{AB}$, $\overline{BC}$, and $\overline{AC}$. Have students measure the lengths of the sides with a ruler and find the measures of the angles using a protractor. You can use this activity to compare a variety of triangles. Since each student will draw his or her own triangle, the triangles will be of different sizes and shapes.

You might also consider asking that individual students draw a specific type of triangle, from isosceles or acute to right scalene or obtuse isosceles.

THINKING OUT LOUD

Sometimes mathematical procedures become so automatic to adults that we do not even realize all the steps we have taken. For example, in the following problem from page 196, there are several steps that some students might need to learn explicitly. Teachers can talk through their thinking while solving the problem. After students become familiar with their teacher's thinking, they can express their own thinking out loud. This gives the teacher an excellent opportunity to see where students are confused.

Problem: Find the unknown side length. Tell whether the side lengths form a Pythagorean triple.

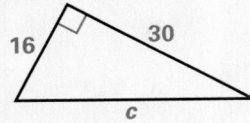

Solution: (The teacher should recite the following out loud to the class.) "Think to yourself. How do I find the length c? What do I already know? I know I have a triangle. It has a right angle, so it must be a right triangle. I know the lengths of the other two sides. What is special about right triangles? The Pythagorean Theorem applies to right triangles. What does the Pythagorean Theorem say? It says that $(\text{hypotenuse})^2 = (\text{leg})^2 + (\text{leg})^2$."

"Which side is the hypotenuse? I remember that the hypotenuse is always opposite the right angle so the side with length c is the hypotenuse. This is also the side for which I am trying to find the length. So I substitute the other two side lengths into the equation:"

$$(\text{hypotenuse})^2 = (\text{leg})^2 + (\text{leg})^2$$
$$= (16)^2 + (30)^2$$
$$= 256 + 900$$
$$= 1156$$

"To find the length of the hypotenuse, I must find the square root of each side. I use my calculator to find the square root of 1156, which is 34."

"Before I go on, I should check whether my answer is reasonable. The two legs have lengths of 16 and 30 and the hypotenuse must be longer than either of the legs. 34 is a little more than 30, so the answer seems reasonable."

"I now have the length of the hypotenuse. I go back to read the question again to be sure I have completed the problem. I see I have not. I still have to answer whether the sides form a Pythagorean triple. I remember that a Pythagorean triple is three positive integers that satisfy the Pythagorean Theorem. 16, 30, and 34 form a right triangle and satisfy the theorem, so they must be a Pythagorean triple. So the answer is yes."

Strategies for English Learners

VOCABULARY DEVELOPMENT

In math class, you might review with students the concept of *opposite*, which is a key mathematical concept and a good way for English learners to learn and remember vocabulary. Start with some simple opposites such as *up* and *down*, *high* and *low*, *tall* and *short*, and *in* and *out*, to make sure students understand the concept. Then review the following opposites, which are common and critical to the understanding of mathematics.

Give students one word from each pair and ask them to supply the opposite.

odd : even	*plus : minus*
add : subtract	*positive : negative*
multiply : divide	*true : false*
sum : difference	

MORE OPPOSITES Additional opposites that are discussed in this chapter include the pairs *equality : inequality* and *interior : exterior*.

KEY WORDS When key words are introduced in each chapter, consider whether each one has an opposite, and if so, teach it at the same time. Have students keep their lists of opposites in their notebook and encourage students to draw a picture or write an explanation in their primary language to aid in their recall of the meanings.

Strategies for Advanced Learners

DIFFERENTIATE INSTRUCTION IN TERMS OF DEPTH

The Pythagorean Theorem is a key concept in mathematics. Many geometric proofs utilize this theorem. Students should understand at this point that proofs underlie everything taught in mathematics and that a proof is just a formal display of mathematical reasoning. Some students know enough mathematics at this point to prove the Pythagorean Theorem and to understand proofs. They may be interested in looking at proofs, studying the illustrations that accompany them, and reading about the mathematicians who developed the proofs.

Students may be intrigued by the fact that there are numerous ways of showing the Pythagorean Theorem is true. This is a great opportunity to inform students that as they learn to write their own proofs often more than one method is possible. Challenge students to find a proof of the Pythagorean Theorem that both interests them and whose logic they can understand. If students are interested in history, draw the trapezoid shown below on the chalkboard and tell them it is part of a proof developed by President James Garfield, the 20th President of the United States, in 1876. Invite them to find out more about the proof.

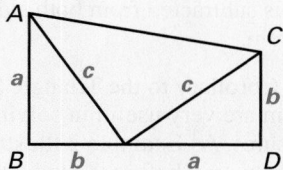

Since there are more than 200 different proofs of the Pythagorean Theorem, you might consider assigning a class project where students select a different proof of the theorem to describe and display in the classroom. As an alternative, some students may prefer to research one of the mathematicians who has a connection to this theorem, such as Euclid, Leonardo da Vinci, or Pythagoras.

For students who are more interested in experimenting on their own than researching proofs, refer them to Activity 4.4 on page 191. Ask students the following questions. Is there something special about the fact that squares are drawn in Step 2? What would happen if you drew equilateral triangles instead? Or regular hexagons? Why are squares used? Students should conclude that the square is used in the activity because it is the easiest shape for which to find the area. But the activity will work with equilateral triangles or regular hexagons, though calculating the areas of the shapes is more time-consuming.

Challenge Problem for use with Lesson 4.1:

Sketch an example of each of the following types of triangles or state that no triangle of the given type exists.

a. equiangular scalene triangle
b. acute equilateral triangle
c. obtuse right triangle

Solutions:
a. does not exist
b.

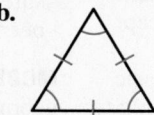

c. does not exist

Challenge Problem for use with Lesson 4.2:

In the diagram below, $m\angle A = 58°$, $m\angle ACD = 64°$, and $m\angle B = 36°$. Find $m\angle 2$. (22°)

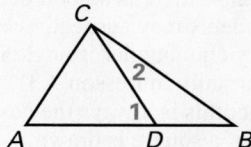

Challenge Problem for use with Lesson 4.6:

A triangle in a coordinate plane has one median whose endpoints are at (2, 5) and (5, 11). If one vertex of the triangle is at (2, 5), what are the coordinates of the centroid of the triangle? (*The coordinates of the centroid are (4, 9).*)

Challenge Problem for use with Lesson 4.7:

Name the longest segment in the diagram below. (The diagram is not drawn to scale.) Explain your reasoning. ($\overline{SR}$; *The longest segment must be the longest side of one of the two triangles. The longest side in a triangle is opposite the largest angle. In $\triangle PQS$, $\overline{SQ}$ is the longest side, but $\overline{SQ}$ is the shortest side in $\triangle SQR$. Therefore, the longest side in the diagram must be the longest side of $\triangle SQR$, which is $\overline{SR}$.*)

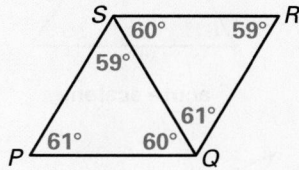

This chapter focuses on triangles and their properties. In the first lesson, students are introduced to the classification of triangles by the lengths of their sides and by the measures of their angles. Lesson 4.2 presents the Triangle Sum Theorem, as well as two implications of this fundamental and important theorem. Lesson 4.3 begins the study of special triangles, investigating isosceles triangles and equilateral triangles. This study continues with right triangles in Lesson 4.4, where the Pythagorean Theorem is introduced. The converse of this theorem is presented in Lesson 4.5, where along with two related inequalities it is used to classify triangles by relating the squares of the lengths of the sides of a triangle. The concept of the median of a triangle is introduced in Lesson 4.6, as well as the related notion of the centroid of a triangle. Lesson 4.7 presents the basic inequalities involving the sides and angles of a general triangle, leading to the statement of the Triangle Inequality Theorem.

Lesson 4.1

CLASSIFYING TRIANGLES The classification of triangles by sides and angles can be confusing for students because there is a good deal of overlap in the categories. For example, as students may suspect, every equilateral triangle is also equiangular and every equiangular triangle is also equilateral. (These facts will be stated formally in Lesson 4.3). Also, a right triangle may be isosceles, and in fact this is always the case for the two triangles formed when the diagonal of a square is drawn. Point out that any of the triangles in the Classification of Triangles by Angles summary box may be isosceles, and that an equiangular triangle is always isosceles. Stress that all but the equiangular triangle may be scalene.

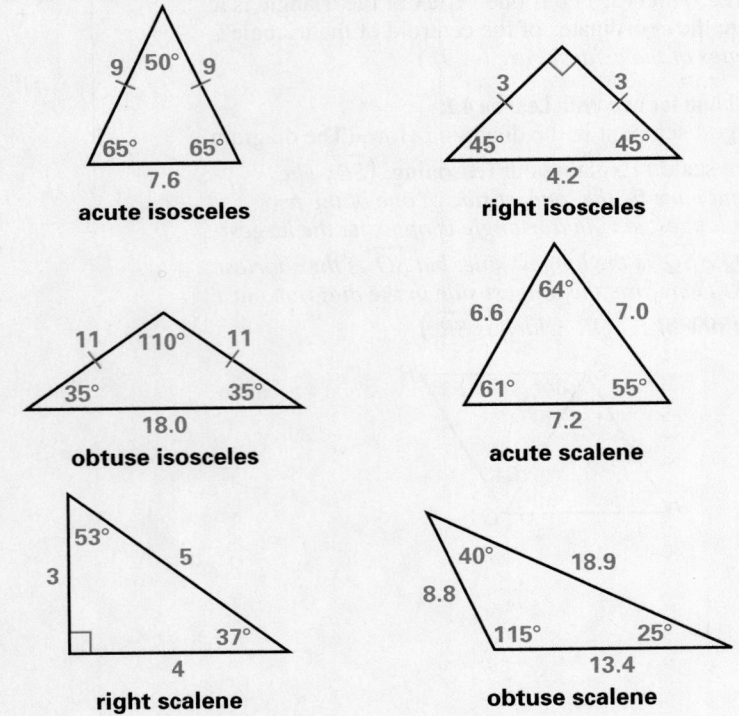

acute isosceles

right isosceles

obtuse isosceles

acute scalene

right scalene

obtuse scalene

Lesson 4.2

When first discussing exterior angles, students may become confused by the fact that two different exterior angles can be drawn at each vertex of a triangle. Point out the pairs of exterior angles at each vertex of the first triangle at the top of page 181 and have students note that the angle pairs are vertical angles. Remind them that vertical angles have the same measure, leading to the fact that either exterior angle at vertex *C* could have been shown in the figure given with the Exterior Angle Theorem.

MODELING This lesson contains one of the most important theorems in geometry, the Triangle Sum Theorem, which states that the sum of the angle measures of any triangle is 180°. You can verify the truth of this theorem by cutting a triangle out of paper or cardboard (or asking students to do this) and tearing off the three corners, which can always be arranged, in pie fashion, to form a straight angle. The theorem is also remarkable in that it states that all triangles in a plane have the same angle sum.

You can demonstrate the truth of the Exterior Angle Theorem by noting that the sum of the measures of the three angles of the triangle is 180°, as is the sum of the measures of an exterior angle and its adjacent interior angle. If these two sums are set equal to one another and the measure of the adjacent interior angle is subtracted from both sides, the result is the symbolic form of the theorem.

USING ALGEBRA Both the Corollary to the Triangle Sum Theorem and the Exterior Angle Theorem are very useful in solving algebraic problems involving angle measures. As Example 2 illustrates, the corollary must be translated into an equation before it can be useful in solving numerical problems. Note that the corollary always enables you to find the measures of all the angles of a right triangle, given the measure of one of the acute angles.

Lesson 4.3

VISUALIZING DIAGRAMS As with parallel lines, students sometimes have difficulty seeing the relationships in an isosceles triangle if the triangle is oriented on the page so that its base is not horizontal. As before, a useful hint is to rotate the book or paper on which the diagram appears until the base becomes horizontal. Another suggestion for problem-solving that involves angle measures is to label as many angles of the triangle with their measures as possible. For example, if the measure of one base angle of an isosceles triangle is given, students should immediately label the other base angle with the same measure. They can then find the measure of the third angle of the triangle using the Triangle Sum Theorem. In dealing with the lengths of sides of a triangle, students should first look to see whether the two congruent sides of an isosceles triangle are labeled with algebraic expressions. If they are, an equation can immediately be written by setting the two expressions equal to each other.

Note that sometimes you can find the angle measures of a triangle just by knowing what classification it falls in. For example, ask students to find the measures of the three angles of an isosceles right triangle.

Lesson 4.4

> **QUICK TIP**
> Use a sketch of an equilateral triangle to show students that the Pythagorean Theorem does not work for non-right triangles.

HISTORICAL CONNECTIONS The Pythagorean Theorem is one of the oldest theorems in mathematics, and arguably the most important. The Egyptians seem to have known many special cases of the theorem, as is revealed in a document known as the Rhind Papyrus. The general theorem was credited to Pythagoras (who lived in the 6th century B.C.) by Euclid (around 300 B.C.) who gave a proof of the theorem. Since then, mathematicians have found many other proofs, one of which was discovered by U.S. President James Garfield.

COMMON ERRORS Two sources of difficulty for students arise in connection with the Pythagorean Theorem. The first is that students try to apply it to non-right triangles. You should emphasize that the theorem does not work unless the triangle is known to be a right triangle. The other (and more common) error that students often make is to write down the equation incorrectly, placing the square of the length of a leg, rather than the hypotenuse, alone on one side of the equation. Therefore, it is important to urge your students to always identify which side of a given right triangle is the hypotenuse before applying the Pythagorean Theorem. Stress that there is really only one reliable way to do this, simply identify the right angle and look for the opposite side.

The Distance Formula could be considered a corollary of the Pythagorean Theorem. Students may be uncomfortable with the use of subscripts in this formula, but you should emphasize that these are merely identifiers that remind the user which pair of coordinates is associated with the first of the two points and which pair is associated with the second. Stress that it does not matter which point is thought of as the "first" and which is thought of as the "second," but it will be helpful later if students develop a consistency when subtracting.

Lesson 4.5

MODELING The converse of the Pythagorean Theorem, and its extension, stated in the box headed "Classifying Triangles" can be viewed as the means for classifying triangles by their angles—without actually finding any angle measures. You can illustrate this classification scheme by using a compass. Start with the arms of the compass at right angles to each other, and measure the distance between the free ends. Then enlarge the angle at the hinge and measure the distance again. Of course the distance between the free ends will be greater than before. Start again with the compass modeling a right angle and repeat the process but decrease the size of the angle this time. The resulting distance between the free ends will be less than it was when measured in the right-angle position.

PYTHAGOREAN TRIPLES In practical terms, when trying to identify a given triangle as a right triangle from the lengths of its sides, it may help students to memorize a few of the more common "Pythagorean triples." These are triples of whole numbers that can be the lengths of the three sides of a right triangle. Four of the more common Pythagorean triples are:

$$3, 4, 5 \qquad 5, 12, 13 \qquad 8, 15, 17 \qquad 7, 24, 25$$

If each of the numbers in a Pythagorean triple is multiplied by a whole number, the three numbers that result are also a Pythagorean triple. The triple 6, 8, 10 is an example.

Lesson 4.6

MEDIANS/CENTROID Medians are one kind of special line that can be drawn for a given triangle, and as the text indicates, they locate the physical *center of mass* of the triangle. Students may wonder, however, whether a median bisects the angle at the vertex from which it is drawn. With the exception of the median drawn to the base of an isosceles triangle, the answer to this question is no. You can demonstrate this by drawing an isosceles triangle whose base angles measure greater than 75° and drawing one of the medians to the legs (*not* the base). It should be clear that this median does not bisect the angle whose vertex is one of its endpoints.

When discussing Theorem 4.9, students may become confused about which portion of the median has length $\frac{2}{3}$ the length of the entire median, and which has length $\frac{1}{3}$ the length of the median. Emphasize that the longer portion is always the one having a vertex of the triangle as one of its endpoints. Note also that there is no general relationship between the length of one median of a triangle and the length of another, although students may guess that there are some relationships in special triangles. For example, in an isosceles triangle, the medians drawn to the congruent sides are congruent.

Lesson 4.7

ORDER The inequalities stated in Theorems 4.10 and 4.11, which are converses of each other, enable us to conclude the relative sizes of the sides of a triangle if we know the relative sizes of the angles, and vice versa. Note that the theorems give no numerical relationship between the size of the angle and the size of the opposite side (this can be done, but only by means of trigonometry). Therefore, given the length of one side of a triangle, we cannot find the measure of the opposite angle, nor can the length of a side be found given the measure of the angle opposite it.

LOGICAL REASONING The Triangle Inequality is an important theorem in any mathematical system in which the idea of "distance" is to be defined, because it is always a property that the pairwise distances between three points must satisfy. It also shows that we cannot choose just any three positive numbers and expect to form a triangle with those numbers as the lengths of the sides. In fact, given three positive numbers, we need to test the Triangle Inequality three different ways before we can be sure that a triangle can be formed from sides of those lengths. This procedure is illustrated in part c of Example 3.

Chapter Overview

Chapter Goals

In this chapter, students will learn to recognize a variety of triangles, and to compute angle measures and side lengths in them. Several fundamental theorems related to triangles will be explored. Students will:

- Classify triangles according to angle measures and side lengths.
- Use the Triangle Sum Theorem, the Base Angles Theorem, the Pythagorean Theorem, and the Triangle Inequality Theorem.
- Discover the relationship between the medians of a triangle and its centroid.

Application Note

Mathematics is often used to indirectly measure lengths that are otherwise too difficult to measure. The skywalk in the opening photo connects the Petronas Twin Towers in Kuala Lumpur, Malaysia. The skywalk is located at levels 41 and 42, and the structural support points for the skywalk are located at level 29. The supports form two congruent triangles with the skywalk. There are spherical bearings at the support points so that the skywalk does not twist, and the skywalk structure is designed so that it will not collapse if it loses its arch support. At 451.9 meters, the Petronas Twin Towers is 8.9 meters taller than the Sears Tower in Chicago.

More information about the Petronas Towers is provided on the Internet at classzone.com

Application Links
CLASSZONE.COM

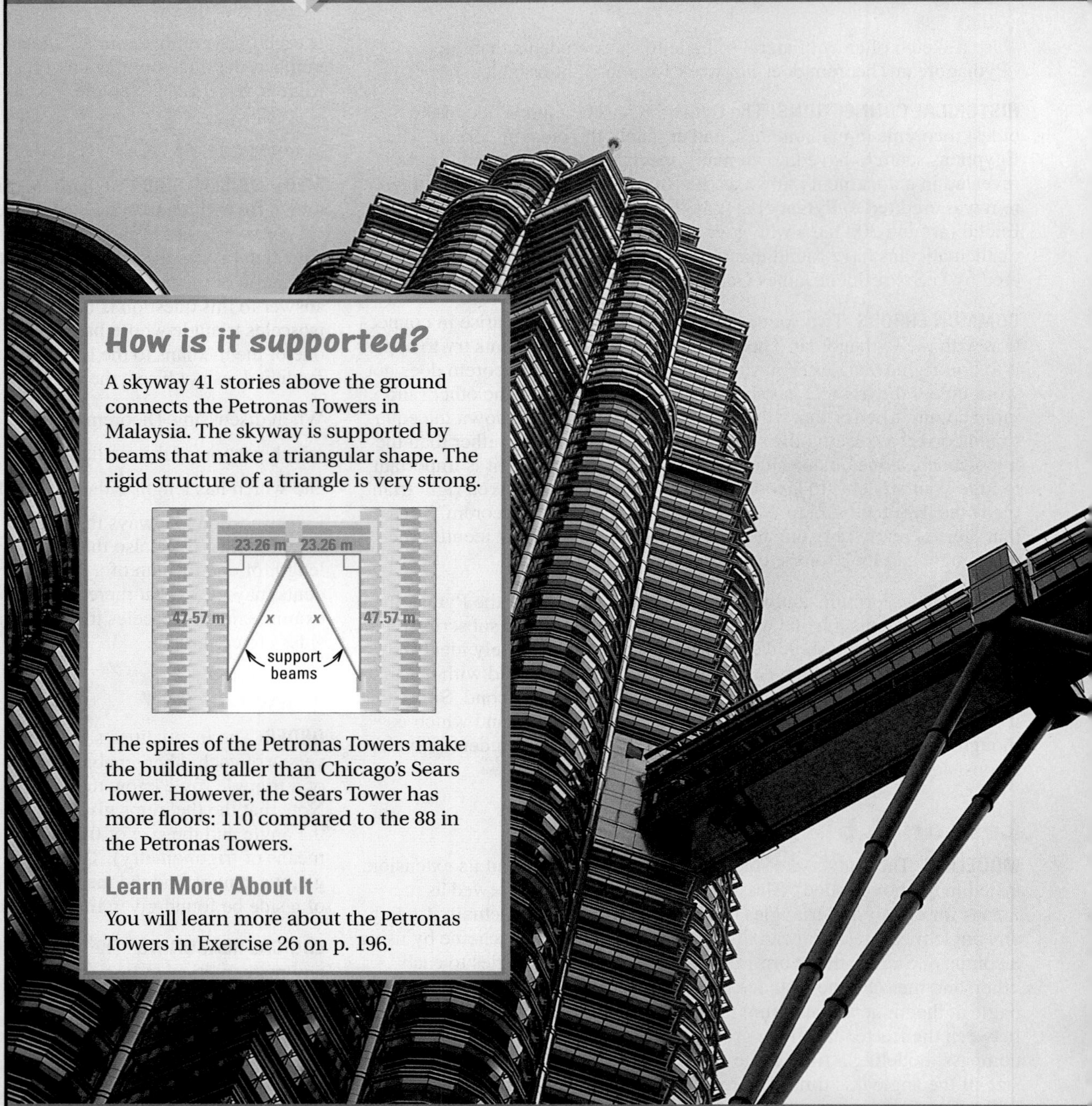

How is it supported?

A skyway 41 stories above the ground connects the Petronas Towers in Malaysia. The skyway is supported by beams that make a triangular shape. The rigid structure of a triangle is very strong.

```
           23.26 m   23.26 m

   47.57 m      x   x      47.57 m

              support
               beams
```

The spires of the Petronas Towers make the building taller than Chicago's Sears Tower. However, the Sears Tower has more floors: 110 compared to the 88 in the Petronas Towers.

Learn More About It

You will learn more about the Petronas Towers in Exercise 26 on p. 196.

Who uses Triangle Relationships?

WATER RESOURCE MANAGER
Water resource managers gather information like rainfall data and water usage to study the effects of water on the environment. They use triangular structures to minimize erosion. (p. 183)

ROCK CLIMBER
The climber is using a method of rock climbing called *top roping*. When the red and blue ropes shown are the same length, the angles they form at the top of the rock have the same measure. (p. 189)

How will you use these ideas?

* Learn more about basketball plays. (p. 177)
* Understand how triangular structures are used to prevent erosion. (p. 183)
* See how rock climbers use a safety rope. (p. 189)
* Analyze tile patterns. (p. 190)
* Investigate a baseball's path during a double play. (p. 205)

Hands-On Activities

Activities (more than 20 minutes)
4.4 Areas and Right Triangles, p. 191
4.5 Side Lengths of Triangles, p. 199
4.6 Intersecting Medians, p. 206

Geo-Activities (less than 20 minutes)
4.3 Properties of Isosceles Triangles, p. 185

Projects

A project covering Chapters 3–4 appears on pages 228–229 of the Student Edition. An additional project for Chapter 4 is available in the *Chapter 4 Resource Book*, pp. 83–84.

Technology

* Electronic Teacher Tools
* Test and Practice Generator
* Online Lesson Planner

Video

* **Geometry in Motion**
  There is an animation supporting Lessons 4.2, 4.3, 4.6, and 4.7.

CLASSZONE.COM

* Application and Career Links
  183, 189, 196
* Student Help
  174, 180, 189, 197, 202, 208, 211, 213

The **Chapter Readiness Quiz** can help you diagnose whether students have the following skills needed in Chapter 4:

• Understand angle bisectors.
• Use the Distance Formula.
• Find the midpoint of a segment.

Reteaching Material

This resource is available for students who need additional help with the skills on the Chapter Readiness Quiz:

📖 **Chapter 4 Resource Book**
• Reteaching with Practice (Lessons 4.1–4.7)

Additional Resources

The following resources are provided to help you prepare for the upcoming chapter and customize review materials:

📖 **Chapter 4 Resource Book**
• *Tips for New Teachers*, pp. 1–2
• *Parent Guide*, pp. 3–4
• *Lesson Plans*, pp. 7, 15, 26, 35, 46, 56, 64
• *Lesson Plans for Block Scheduling*, pp. 8, 16, 27, 36, 47, 57, 65

🖥 **Technology**
• Electronic Teaching Tools
• Online Lesson Planner
• Test and Practice Generator

Visualize It!

Additional suggestions for helping students visualize geometry are on pp. 178, 181, 186, 196, and 217.

PREVIEW **What's the chapter about?**

• Classifying triangles and finding their angle measures
• Using the Distance Formula, the Pythagorean Theorem, and its converse
• Showing relationships between a triangle's sides and angles

Key Words

• equilateral, isosceles, scalene triangles, *p. 173*
• equiangular, acute, right, obtuse triangles, *p. 174*
• interior, exterior angles, *p. 181*
• legs of an isosceles triangle, *p. 185*

• base angles of an isosceles triangle, *p. 185*
• hypotenuse, *p. 192*
• Pythagorean Theorem, *p. 192*
• Distance Formula, *p. 194*
• median of a triangle, *p. 207*
• centroid, *p. 208*

PREPARE **Chapter Readiness Quiz**

Take this quick quiz. If you are unsure of an answer, look at the reference pages for help.

Vocabulary Check *(refer to p. 61)*

1. In the figure shown, $\overrightarrow{BD}$ is the angle bisector of $\angle ABC$. What is the value of x?

(A) 10   (B) 15   (C) 20   (D) 30

Skill Check *(refer to pp. 30, 55)*

2. What is the distance between $P(2, 3)$ and $Q(7, 3)$? **H**

(F) 3   (G) 4   (H) 5   (J) 7

3. What is the midpoint of a segment with endpoints $A(0, 2)$ and $B(6, 4)$? **B**

(A) (3, 2)   (B) (3, 3)   (C) (4, 2)   (D) (0, 3)

VISUAL STRATEGY **Drawing Triangles**

Visualize It! ➡

When you sketch a triangle, try to make the angles roughly the correct size.

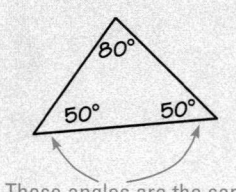

These angles are the same in an isosceles triangle.

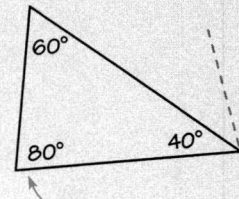

This 80° angle is twice as big as the 40° angle.

 **4.1** # Classifying Triangles

Goal

Classify triangles by their sides and by their angles.

Key Words

- equilateral, isosceles, scalene triangles
- equiangular, acute, right, obtuse triangles
- vertex

A **triangle** is a figure formed by three segments joining three noncollinear points. A triangle can be classified by its sides and by its angles.

3 noncollinear points

3 segments

SUMMARY | **CLASSIFICATION OF TRIANGLES BY SIDES**

Equilateral Triangle	Isosceles Triangle	Scalene Triangle
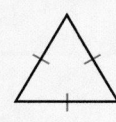	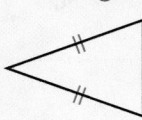	
3 congruent sides	At least 2 congruent sides	No congruent sides

---

EXAMPLE **1** **Classify Triangles by Sides**

Classify the triangle by its sides.

a. 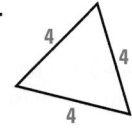 b. 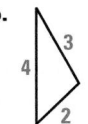 c.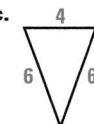

Student Help

VOCABULARY TIP
Equi- means "equal," and *-lateral* means "side." So, *equilateral* means equal sides.

Solution

a. Because this triangle has 3 congruent sides, it is equilateral.

b. Because this triangle has no congruent sides, it is scalene.

c. Because this triangle has 2 congruent sides, it is isosceles.

Checkpoint **Classify Triangles by Sides**

Classify the triangle by its sides.

1. 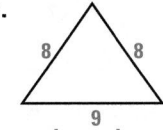   **2.** 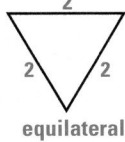   **3.**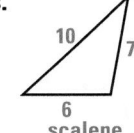
isosceles   equilateral   scalene

4.1 *Classifying Triangles* **173**

---

①Plan

Pacing
Suggested Number of Days

Basic: 2 days
Average: 2 days
Advanced: 2 days
Block Schedule: 1 block

Teaching Resources

📄 **Blacklines**
(See page 170B.)

🖨 **Transparencies**
- Warm-Up with Quiz
- Answers

💻 **Technology**
- Electronic Teacher Tools
- Test and Practice Generator
- Online Lesson Planner
- Internet Support

②Teach

Content and Teaching Strategies
For background information on geometric concepts and teaching strategies related to this lesson, see pages 170E and 170F in this Teacher's Edition.

Tips for New Teachers
Classifying triangles is usually fairly simple for students. However, students are used to thinking of equilateral and isosceles triangles as separate and distinct triangles. The idea that an equilateral triangle is isosceles may be new to them. Students need to understand the meaning of "at least 2 congruent sides" for isosceles triangles. See the Tips for New Teachers on pp. 1–2 of the *Chapter 4 Resource Book* for additional notes about Lesson 4.1.

Extra Example 1
See next page.

173

Extra Example 1

Classify the triangle by its sides.

a.

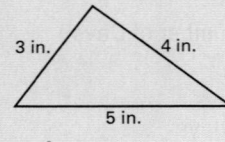

3 in. 4 in.
5 in.

scalene

b.

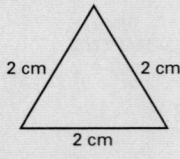

2 cm 2 cm
2 cm

equilateral

c.

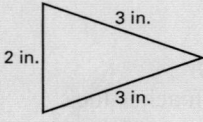

3 in.
2 in.
3 in.

isosceles

Extra Example 2

Classify the triangle by its angles and by its sides.

a.

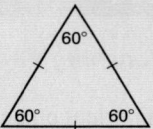

60°
60° 60°

equiangular, equilateral

b.

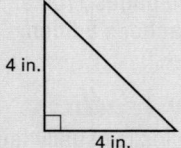

4 in.
4 in.

right, isosceles

c.

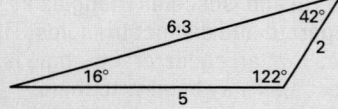

6.3 42°
  2
16° 122°
  5

obtuse, scalene

Visualize It!

Have students make a list of all the new triangle types they have encountered in this lesson. Urge students to use a table with the column headings *Triangle Type, Definition,* and *Sketch.*

174

| SUMMARY | CLASSIFICATION OF TRIANGLES BY ANGLES |

Equiangular Triangle

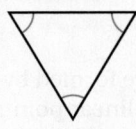

3 congruent angles

Acute Triangle

3 acute angles

Right Triangle

1 right angle

Obtuse Triangle

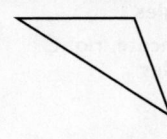

1 obtuse angle

Student Help

CLASSZONE.COM

MORE EXAMPLES
More examples at classzone.com

EXAMPLE 2 Classify Triangles by Angles and Sides

Classify the triangle by its angles and by its sides.

a.

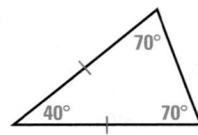

70°
40° 70°

b.

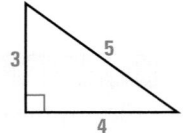

3   5
   4

c.

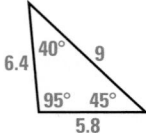

40° 9
6.4
95° 45°
  5.8

Solution

a. Because this triangle has 3 angles with measures less than 90° and 2 congruent sides, it is an *acute isosceles triangle.*

b. Because this triangle has a right angle and no congruent sides, it is a *right scalene triangle.*

c. Because this triangle has one angle greater than 90° and no congruent sides, it is an *obtuse scalene triangle.*

Checkpoint ✓ **Classify Triangles by Angles and Sides**

Classify the triangle by its angles and by its sides.

4.

72°
   36°
72°

acute isosceles triangle

5.

6.2 60° 7.6
70° 50°
   7

acute scalene triangle

6.

30° 30°
   120°

obtuse isosceles triangle

Student Help

VOCABULARY TIP
The plural of *vertex* is *vertices.*

A **vertex** of a triangle is a point that joins two sides of the triangle. The side across from an angle is the *opposite* side.

Point *B* is a vertex.

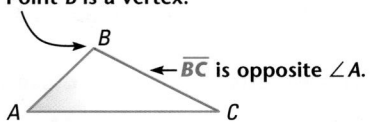

$\overline{BC}$ is opposite $\angle A$.

EXAMPLE 3 Identify the Parts of a Triangle

Name the side that is opposite the angle.

a. $\angle A$ b. $\angle B$ c. $\angle C$

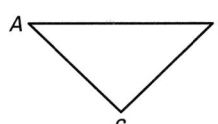

Solution

a. $\overline{BC}$ is the side that is opposite $\angle A$.

b. $\overline{AC}$ is the side that is opposite $\angle B$.

c. $\overline{AB}$ is the side that is opposite $\angle C$.

4.1 Exercises

Guided Practice

Vocabulary Check

1. What is the difference between an *obtuse triangle* and an *acute triangle*? An obtuse triangle has an obtuse angle, and an acute triangle has three acute angles.

In Exercises 2–4, use the diagram.

2. Name the side *opposite* $\angle P$. $\overline{QR}$

3. Name the side *opposite* $\angle Q$. $\overline{PR}$

4. Classify the triangle by its sides. isosceles

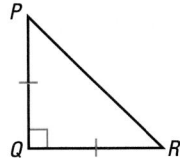

Skill Check

Classify the triangle by its sides.

5.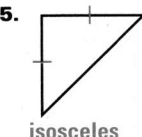
isosceles

6.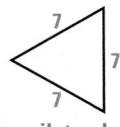
equilateral

7.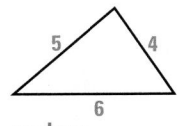
scalene

Classify the triangle by its angles.

8.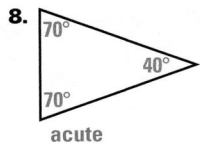
acute

9.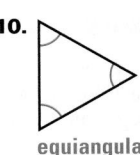
right

10.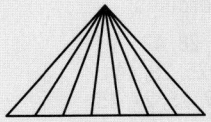
equiangular

4.1 *Classifying Triangles* 175

Multiple Representations

Remind students that an equilateral triangle can be indicated by showing identical tick marks on all three sides or by labeling all three sides with the same numerical length.

Extra Example 3

Name the side that is opposite the angle.

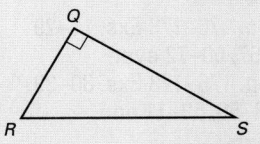

a. $\angle R$ $\overline{QS}$ is opposite $\angle R$.
b. $\angle S$ $\overline{QR}$ is opposite $\angle S$.
c. $\angle Q$ $\overline{RS}$ is opposite $\angle Q$.

✓ Concept Check

How are triangles classified by their sides? by their angles? Triangles can be equilateral (3 congruent sides), isosceles (at least 2 congruent sides), or scalene (no congruent sides). Triangles can be equiangular (3 congruent angles), acute (3 acute angles), right (1 right angle), or obtuse (1 obtuse angle).

🧩 Daily Puzzler

How many triangles are in the figure below? 28

175

3 Apply

Practice and Applications

Assignment Guide

BASIC
Day 1: pp. 176–178 Exs. 11–21 odd, 24–28 even, 56–58, 59–69 odd
Day 2: pp. 176–178 Exs. 30–36, 45–49, 54, 55, 60–70 even

AVERAGE
Day 1: pp. 176–178 Exs. 11–29 odd, 51–57, 60–72 even
Day 2: pp. 176–178 Exs. 30–39, 42–44, 48–50, 59–71 odd

ADVANCED
Day 1: pp. 176–178 Exs. 14–19, 23–29 odd, 48–66 even
Day 2: pp. 176–178 Exs. 30–41, 43–45, 55–71 odd;
EC: TE p. 170D*, classzone.com

BLOCK SCHEDULE
pp. 176–178 Exs. 11–29 odd, 30–39, 42–44, 51–57, 59–72

Extra Practice
• Student Edition, p. 681
• Chapter 4 Resource Book, pp. 10–11

Homework Check

To quickly check student understanding of key concepts, go over the following exercises:

Basic: 14, 17, 24, 28, 42
Average: 13, 17, 25, 29, 44
Advanced: 14, 17, 25, 27, 45

✕ Common Error

In Exercise 27, watch for students who do not recognize a right triangle when it is not oriented with one of the legs horizontal.

Extra Practice
See p. 681.

Classifying Triangles Classify the triangle by its sides.

11.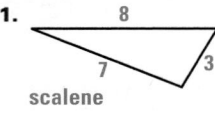
scalene

12.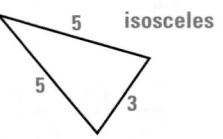
isosceles

13.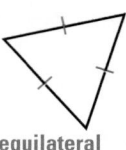
equilateral

14. 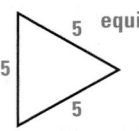 equilateral

15. 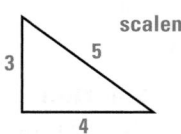 scalene

16. 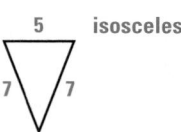 isosceles

Classifying Triangles Classify the triangle by its angles.

17.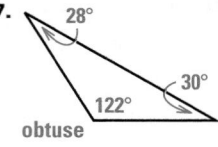
obtuse

18. 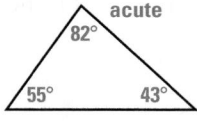 acute

19.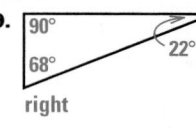
right

20. 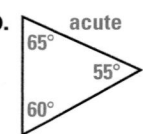 acute

21.
right

22. 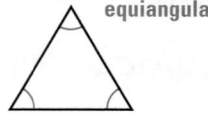 equiangular

23. An acute triangle has three acute angles, so the triangle is not an acute triangle. An obtuse triangle has one obtuse angle and two acute angles.

23. Error Analysis A student claims that the triangle is both obtuse and acute because it has an obtuse angle and an acute angle. What is wrong with his reasoning?

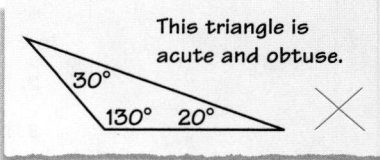

Classifying Triangles Classify the triangle by its angles and by its sides.

24.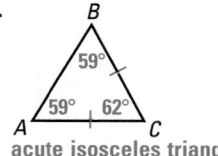
acute isosceles triangle

25. 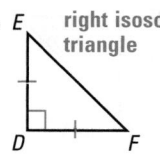 right isosceles triangle

26.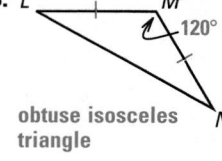
obtuse isosceles triangle
acute scalene triangle

Homework Help
Example 1: Exs. 11–16
Example 2: Exs. 17–41
Example 3: Exs. 42–47

27.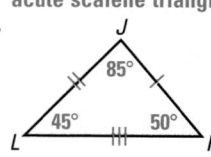
right scalene triangle

28. T 42° 42° U, 96°, V
obtuse isosceles triangle

29. J 85°, 45° 50°, L K

---

176 **Chapter 4** *Triangle Relationships*

Student Help

VISUAL STRATEGY
In Exs. 30–36 draw a sketch with side lengths or angle measures that are roughly correct, as shown on p. 172.

EXAMPLE Classify Triangles

Classify the triangle described.

a. Side lengths: 6, 8, 9

b. Angle measures: 50°, 60°, 70°

Solution

You may want to sketch the triangle.

a.

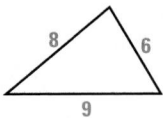

Because the triangle has three sides with different lengths, the triangle is scalene.

b.

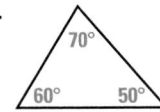

Because the triangle has three angles with measures less than 90°, the triangle is acute.

Matching Triangles In Exercises 30–36, use the example above to match the triangle description with the most specific name.

30. Side lengths: 2 cm, 3 cm, 4 cm B

31. Side lengths: 3 cm, 2 cm, 3 cm E

32. Side lengths: 4 cm, 4 cm, 4 cm A

33. Angle measures: 60°, 60°, 60° D

34. Angle measures: 30°, 60°, 90° G

35. Angle measures: 20°, 145°, 15° C

36. Angle measures: 50°, 55°, 75° F

A. Equilateral

B. Scalene

C. Obtuse

D. Equiangular

E. Isosceles

F. Acute

G. Right

Link to
Sports

BASKETBALL The triangle offense is used by many professional teams. Players are usually spaced 15 feet to 18 feet apart from each other. This provides many options for passing so a player can make a basket.

Basketball The diagram shows the position and spacing of five basketball players running the "triangle offense."

37. What type of triangle is formed by players A, B, and C? acute

38. What type of triangle is formed by players C, D, and E? right

39. What type of triangle is formed by players B, D, and E? acute

40. Which three players appear to form an obtuse triangle?
Sample answer: A, D, and E

41. Which three players appear to form a scalene triangle?
Sample answer: B, D, and E

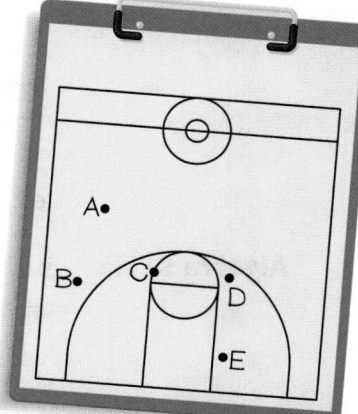

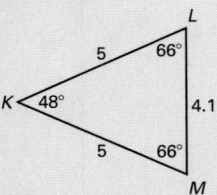
Assessment Resources

The Mini-Quiz below is also available on blackline (*Chapter 4 Resource Book*, p. 17) and on transparency. For more assessment resources, see:
• Chapter 4 Resource Book
• Standardized Test Practice
• Test and Practice Generator

Mini-Quiz

Classify the triangle by its angles and by its sides.

1.

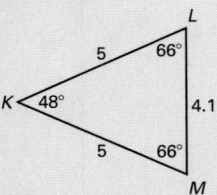

acute, isosceles

2.

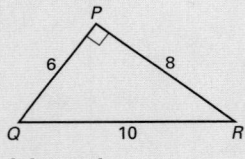

right, scalene

3. Identify which side is opposite each angle.

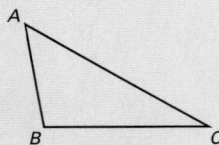

$\overline{BC}$ is opposite $\angle A$; $\overline{AB}$ is opposite $\angle C$; $\overline{AC}$ opposite $\angle B$.

4. Brittani built a garden bed frame by nailing together three equal lengths of wood to form a triangle. What type of triangle did she create? **equilateral or equiangular**

48–53. See Additional Answers beginning on page AA1.

178

Visualize It!

To help you determine the side opposite a vertex, you can draw an arrow from the vertex.

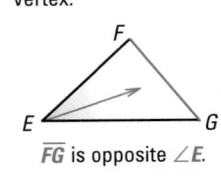

$\overline{FG}$ is opposite $\angle E$.

42. $\overline{AB}$ is opposite $\angle C$; $\overline{BC}$ is opposite $\angle A$; $\overline{AC}$ is opposite $\angle B$.

43. $\overline{DE}$ is opposite $\angle F$; $\overline{EF}$ is opposite $\angle D$; $\overline{DF}$ is opposite $\angle E$.

44. $\overline{GH}$ is opposite $\angle J$; $\overline{HJ}$ is opposite $\angle G$; $\overline{GJ}$ is opposite $\angle H$.

45. $\overline{KL}$ is opposite $\angle M$; $\overline{LM}$ is opposite $\angle K$; $\overline{KM}$ is opposite $\angle L$.

46. $\overline{NP}$ is opposite $\angle Q$; $\overline{PQ}$ is opposite $\angle N$; $\overline{NQ}$ is opposite $\angle P$.

Mixed Review

47. $\overline{RS}$ is opposite $\angle T$; $\overline{ST}$ is opposite $\angle R$; $\overline{RT}$ is opposite $\angle S$.

Identifying Parts of Triangles Identify which side is opposite each angle. **42–47. See margin.**

42. 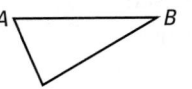 **43.** 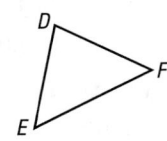 **44.**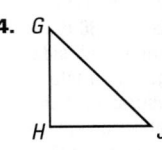

45. 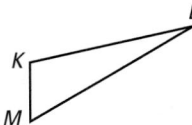 **46.** 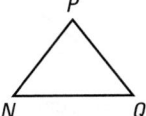 **47.**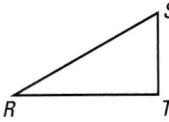

Visualize It! Draw an example of the triangle. **48–53. See margin.**

48. obtuse scalene   **49.** right isosceles   **50.** acute scalene

51. right scalene   **52.** acute isosceles   **53.** obtuse isosceles

54. Multiple Choice Which of the following terms can be used to describe a triangle with angle measures of 17°, 17°, and 146°? **C**
    (A) acute   **(B)** right   **(C)** obtuse   **(D)** equiangular

55. Multiple Choice What side is opposite $\angle C$? **F**
    (F) $\overline{AB}$ **(G)** $\overline{BC}$
    (H) $\overline{AC}$ **(J)** both $\overline{BC}$ and $\overline{AC}$

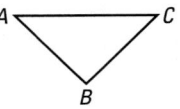

Complements and Supplements Find the value of each variable. (*Lesson 2.3*)

56. 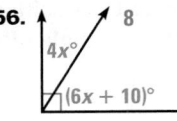   **57.** 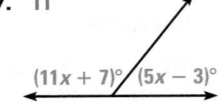   **58.**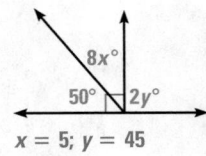

                                                                 $x = 5$; $y = 45$

Translations Find the image of the point using the translation $(x, y) \rightarrow (x - 2, y + 4)$. (*Lesson 3.7*)

59. (2, 5) (0, 9)   **60.** (1, −3) (−1, 1)   **61.** (−1, 2) (−3, 6)   **62.** (0, −5) (−2, −1)

63. (−4, −2) (−6, 2)   **64.** (0, 0) (−2, 4)   **65.** (−6, 4) (−8, 8)   **66.** (−3, −1) (−5, 3)

Algebra Skills

Solving Equations Solve the equation. (*Skills Review, p. 673*)

67. $5x - 15 = 180$ 39   **68.** $x + 2x + 36 = 180$ 48

69. $3x + 5x + 20 = 180$ 20   **70.** $-3x - (x + 8) = 180$ −47

71. $2(x - 1) - 3x + 7 = 180$ −175   **72.** $4(3x - 1) - 9x + 10 = 180$ 58

4.2 Angle Measures of Triangles

Goal
Find angle measures in triangles.

Key Words
- corollary
- interior angles
- exterior angles

The diagram below shows that when you tear off the corners of any triangle, you can place the angles together to form a straight angle.

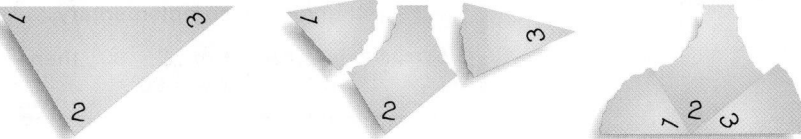

THEOREM 4.1

Triangle Sum Theorem

Words The sum of the measures of the angles of a triangle is 180°.

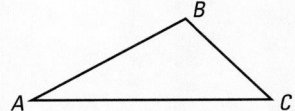

Symbols In $\triangle ABC$, $m\angle A + m\angle B + m\angle C = 180°$.

Student Help

READING TIP
Triangles are named by their vertices. $\triangle ABC$ is read "triangle *ABC*."

EXAMPLE 1 Find an Angle Measure

Given $m\angle A = 43°$ and $m\angle B = 85°$, find $m\angle C$.

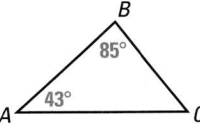

Solution

$m\angle A + m\angle B + m\angle C = 180°$	Triangle Sum Theorem
$43° + 85° + m\angle C = 180°$	Substitute 43° for $m\angle A$ and 85° for $m\angle B$.
$128° + m\angle C = 180°$	Simplify.
$128° + m\angle C - 128° = 180° - 128°$	Subtract 128° from each side.
$m\angle C = 52°$	Simplify.

ANSWER ▶ $\angle C$ has a measure of 52°.

CHECK ✓ Check your solution by substituting 52° for $m\angle C$.
$43° + 85° + 52° = 180°$

---

① Plan

Pacing
Suggested Number of Days

Basic: 2 days
Average: 2 days
Advanced: 2 days
Block Schedule: 1 block

Teaching Resources

📄 **Blacklines**
(See page 170B.)

🖐 **Transparencies**
- Warm-Up with Quiz
- Answers

🖥 **Technology**
- Electronic Teacher Tools
- Test and Practice Generator
- Online Lesson Planner
- Internet Support

📼 **Video**
- Geometry in Motion

② Teach

Content and Teaching Strategies
For background information on geometric concepts and teaching strategies related to this lesson, see pages 170E and 170F in this Teacher's Edition.

Tips for New Teachers
Some students have trouble recognizing an exterior angle. This is especially true when a triangle is drawn with a base that is not horizontal. Draw several diagrams of triangles with different orientations. Have students identify and label the exterior angles. See the Tips for New Teachers on pp. 1–2 of the *Chapter 4 Resource Book* for additional notes about Lesson 4.2.

Extra Example 1
See next page.

Extra Example 1

Given $m\angle A = 25°$ and $m\angle B = 95°$, find $m\angle C$. 60°

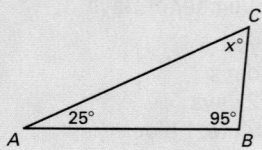

Extra Example 2

$\triangle ABC$ and $\triangle CAD$ are right triangles. Suppose $m\angle ADC = 50°$.

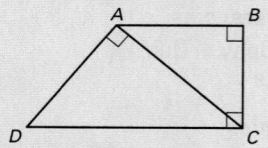

a. Find $m\angle ACD$. 40°

b. Find $m\angle BCA$. 50°

A **corollary** to a theorem is a statement that can be proved easily using the theorem. The corollary below follows from the Triangle Sum Theorem.

Student Help

LOOK BACK
For the definition of complementary angles, see p. 67.

COROLLARY

Corollary to the Triangle Sum Theorem

Words The acute angles of a right triangle are complementary.

Symbols In $\triangle ABC$, if $m\angle C = 90°$, then $m\angle A + m\angle B = 90°$.

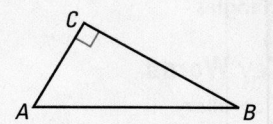

EXAMPLE 2 Find Angle Measures

$\triangle ABC$ and $\triangle ABD$ are right triangles. Suppose $m\angle ABD = 35°$.

a. Find $m\angle DAB$.

b. Find $m\angle BCD$.

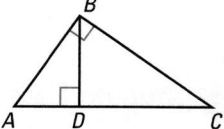

Student Help

CLASSZONE.COM

MORE EXAMPLES
More examples at classzone.com

Solution

a.

$m\angle DAB + m\angle ABD = 90°$	Corollary to the Triangle Sum Theorem
$m\angle DAB + 35° = 90°$	Substitute 35° for $m\angle ABD$.
$m\angle DAB + 35° - 35° = 90° - 35°$	Subtract 35° from each side.
$m\angle DAB = 55°$	Simplify.

b.

$m\angle DAB + m\angle BCD = 90°$	Corollary to the Triangle Sum Theorem
$55° + m\angle BCD = 90°$	Substitute 55° for $m\angle DAB$.
$m\angle BCD = 35°$	Subtract 55° from each side.

Checkpoint ✓ Find an Angle Measure

1. Find $m\angle A$. 65°

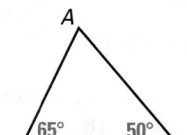

2. Find $m\angle B$. 75°

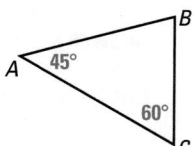

3. Find $m\angle C$. 50°

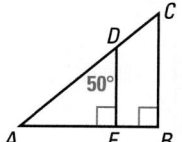

When the sides of a triangle are extended, other angles are formed. The three original angles are the **interior angles**.

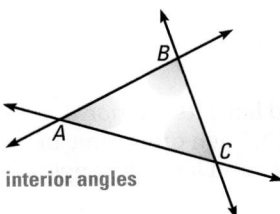

interior angles

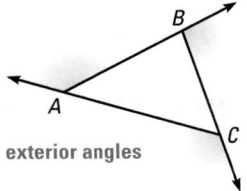

exterior angles

The angles that are adjacent to the interior angles are the **exterior angles**. It is common to show only *one* exterior angle at each vertex.

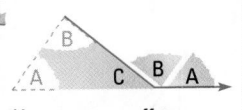
THEOREM 4.2

Exterior Angle Theorem

Words The measure of an exterior angle of a triangle is equal to the sum of the measures of the two nonadjacent interior angles.

Symbols $m\angle 1 = m\angle A + m\angle B$

EXAMPLE 3 Find an Angle Measure

Given $m\angle A = 58°$ and $m\angle C = 72°$, find $m\angle 1$.

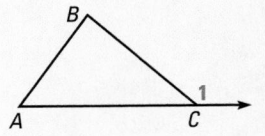

Solution

$m\angle 1 = m\angle A + m\angle C$	Exterior Angle Theorem
$= 58° + 72°$	Substitute 58° for $m\angle A$ and 72° for $m\angle C$.
$= 130°$	Simplify.

ANSWER ▶ $\angle 1$ has a measure of 130°.

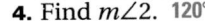

Checkpoint ✓ Find an Angle Measure

4. Find $m\angle 2$. 120°

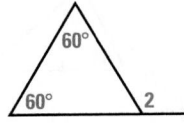

5. Find $m\angle 3$. 155°

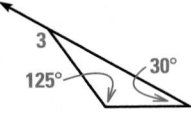

6. Find $m\angle 4$. 113°

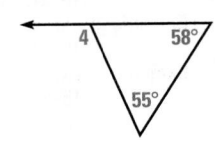

4.2 *Angle Measures of Triangles* **181**

Geometric Reasoning
Point out that an exterior angle forms a linear pair with the interior angle of the triangle that shares its vertex. Make sure students do not mistakenly think that any angle on the outside of a triangle is an exterior angle.

Extra Example 3
Given $m\angle D = 45°$ and $m\angle DFG = 100°$, find $m\angle E$. 55°

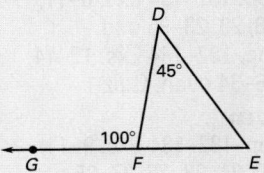

✓ Concept Check
How are the measures of interior angles and exterior angles of a triangle related? The measure of an exterior angle of a triangle is equal to the sum of the measures of the two nonadjacent interior angles. Also, the sum of the measures of an exterior angle and its adjacent interior angle is 180°.

🐢 Daily Puzzler
The ratio of the measures of the angles in a triangle is 3:2:1. Find the angle measures. 90°, 60°, 30°

Guided Practice

1. *Sample answer:*

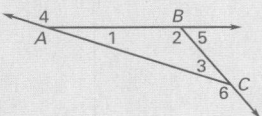

Vocabulary Check

1. Copy △*ABC* and label its *interior* angles 1, 2, and 3. Then draw three of its *exterior* angles and label the angles 4, 5, and 6. **See margin.**

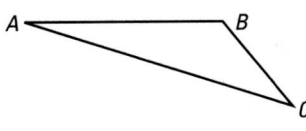

Skill Check

2. Use the diagram to determine which equation can be used to find *m∠DEF.* **A**

A. $55° + x° = 110°$ **B.** $55° + 110° = x°$

C. $55° - x° = 110°$ **D.** $55° - 110° = x°$

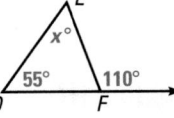

Find the value of x.

3. 61

4. 42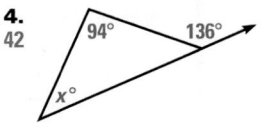

5. 35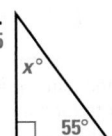

Practice and Applications

Extra Practice
See p. 681.

Finding Angle Measures Find the measure of ∠1.

6. 71°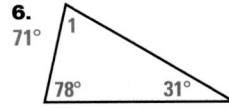

7. 110°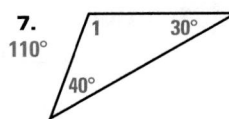

8. 52°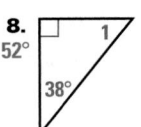

9. 60°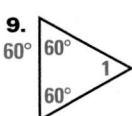

10. 37°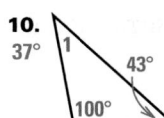

11. 45°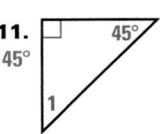

Exterior Angles Find the measure of ∠2.

12. 148°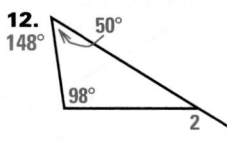

13. 139°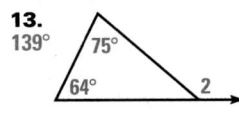

14. 43°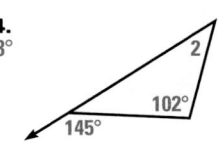

Water Resources In Exercises 15–17, use the diagram.
A structure built with rocks is used to redirect the flow of water in
a stream and increase the rate of the water's flow. Its shape is a
right triangle.

15. Identify the side opposite ∠MNL. $\overline{ML}$

16. If the measure of the upstream
angle is 37°, what is the measure of
the downstream angle? 53°

17. It is generally recommended that the
upstream angle should be between
30° and 45°. Give a range of angle
measures for the downstream angle.
45° to 60°

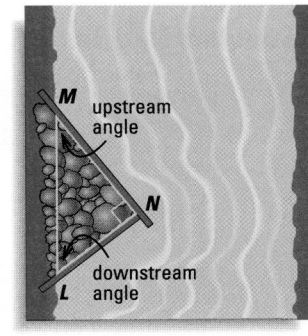

upstream
angle

downstream
angle

Using Algebra Find the value of each variable.

18.
33

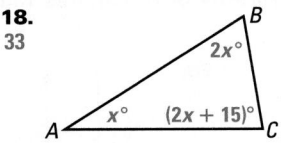

19.
20

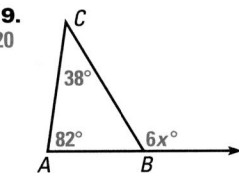

20.

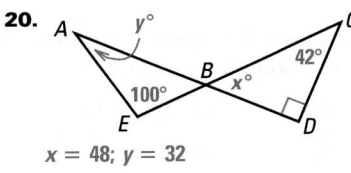

$x = 48; y = 32$

21.

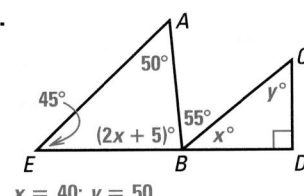

$x = 40; y = 50$

22. For any position of point C,
$m\angle PBC = m\angle BAC + m\angle BCA$.
This illustrates the Exterior
Angle Theorem.

22. Technology Use geometry software to
complete the steps below.

❶ Draw A, B, C and $\triangle ABC$.

❷ Draw $\overleftrightarrow{AB}$ and a point P on it as shown.

❸ Find $m\angle PBC$.

❹ Find $m\angle BAC + m\angle BCA$.

❺ Move point C.

What do you notice? What theorem does this demonstrate?

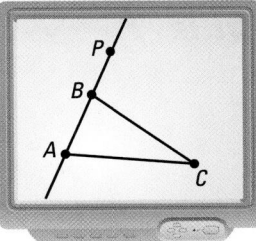

Student Help

VISUAL STRATEGY
In Ex. 23, draw a
sketch with angle
measures that are
roughly correct, as
shown on p. 172.

23. Angle Measures in a Triangle The measure of one interior angle
of a triangle is 26°. The other interior angles are congruent. Find
their measures. 77°

4 Assess

Assessment Resources

The Mini-Quiz below is also available on blackline (*Chapter 4 Resource Book*, p. 28) and on transparency. For more assessment resources, see:
• Chapter 4 Resource Book
• Standardized Test Practice
• Test and Practice Generator

Mini-Quiz

Find the measure of ∠1.

1.

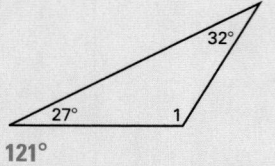

2.

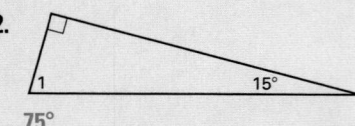

3.

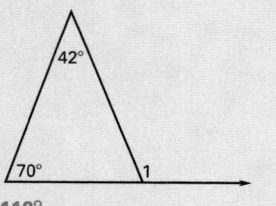

4. Find the value of the variable. **9**

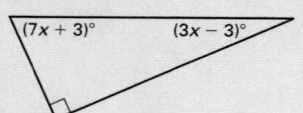

5. Raymundo is designing a logo that is formed by using two congruent right triangles as shown. The smaller acute angle of each triangle has a measure of 27°. Find the measure of each larger acute angle. **63°**

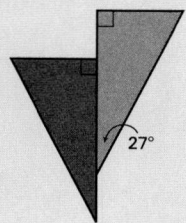

---

24. Using Algebra In △PQR, the measure of ∠P is 36°. The measure of ∠Q is five times the measure of ∠R. Find m∠Q and m∠R. *m∠Q = 120°; m∠R = 24°*

Standardized Test Practice

25. Multiple Choice Find the value of x. **C**

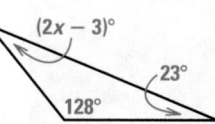

(A) 8   (B) 13

(C) 16   (D) 29

26. Multiple Choice Suppose a triangle has interior angle measures of 50°, 60°, and 70°. Which of the following is *not* an exterior angle measure? **F**

(F) 100°   (G) 110°   (H) 120°   (J) 130°

Mixed Review

Showing Lines are Parallel Explain how you would show that *m* ∥ *n*. State any theorems or postulates that you would use. *(Lesson 3.5)*

27.

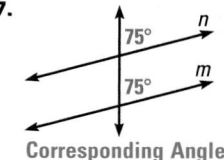

Corresponding Angles Converse

28.

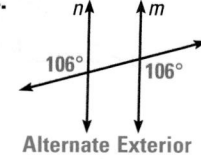

Alternate Exterior Angles Converse

29.

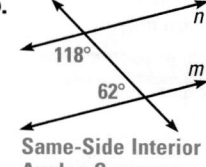

Same-Side Interior Angles Converse

Algebra Skills

Comparing Numbers Compare the two numbers. Write the answer using <, >, or =. *(Skills Review, p. 662)*

30. 1015 and 1051
1015 < 1051

31. 3.5 and 3.06
3.5 > 3.06

32. 8.09 and 8.1
8.09 < 8.1

33. 1.75 and 1.57
1.75 > 1.57

34. 0 and 0.5
0 < 0.5

35. 2.055 and 2.1
2.055 < 2.1

Quiz 1

Classify the triangle by its angles and by its sides. *(Lesson 4.1)*

1.

obtuse isosceles triangle

2.

acute scalene triangle

3.

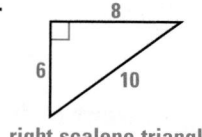

right scalene triangle

Find the measure of ∠1. *(Lesson 4.2)*

4.

5.

6.

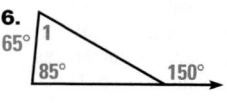

4.3 Isosceles and Equilateral Triangles

Goal
Use properties of isosceles and equilateral triangles.

Key Words
• legs of an isosceles triangle
• base of an isosceles triangle
• base angles

Geo-Activity   **Properties of Isosceles Triangles**

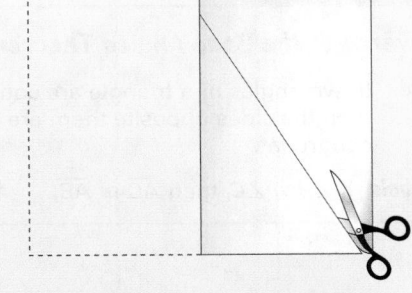

❶ Fold a sheet of paper in half. Use a straightedge to draw a line from the fold to the bottom edge. Cut along the line to form an isosceles triangle.

❷ Unfold and label the angles as shown. Use a protractor to measure ∠H and ∠K. What do you notice?
∠H and ∠K are congruent.

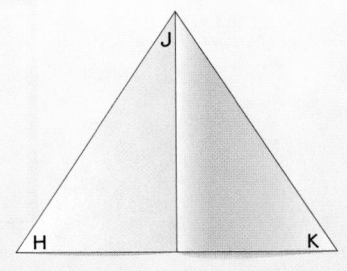

❸ Repeat Steps 1 and 2 for different isosceles triangles. What can you say about ∠H and ∠K in the different triangles?
For each isosceles triangle, ∠H and ∠K are congruent.

Student Help

VOCABULARY TIP
Isos- means "equal," and *-sceles* means "leg." So, *isosceles* means equal legs.

The Geo-Activity shows that two angles of an isosceles triangle are always congruent. These angles are opposite the congruent sides.

The congruent sides of an isosceles triangle are called **legs** .

The other side is called the **base** .

The two angles at the base of the triangle are called the **base angles** .

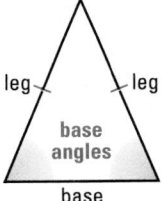

Isosceles Triangle

THEOREM 4.3

Base Angles Theorem

Words If two sides of a triangle are congruent, then the angles opposite them are congruent.

Symbols If $\overline{AB} \cong \overline{AC}$, then $\angle C \cong \angle B$.

4.3 *Isosceles and Equilateral Triangles*   **185**

① Plan

Pacing
Suggested Number of Days

Basic: 2 days
Average: 2 days
Advanced: 2 days
Block Schedule: 1 block

Teaching Resources

📄 **Blacklines**
(See page 170B.)

🔖 **Transparencies**
• Warm-Up with Quiz
• Answers

🖥 **Technology**
• Electronic Teacher Tools
• Test and Practice Generator
• Online Lesson Planner
• Internet Support

📼 **Video**
• Geometry in Motion

Geo-Activity

Goal Use paper folding to explore the measures of the base angles of an isosceles triangle.

Key Discovery The base angles of an isosceles triangle are congruent.

② Teach

Content and Teaching Strategies
For background information on geometric concepts and teaching strategies related to this lesson, see pages 170E and 170F in this Teacher's Edition.

Tips for New Teachers

Some students may need help with the vocabulary in this chapter. Ask them questions such as whether an equilateral triangle is always equiangular and whether an isosceles triangle is always acute, always obtuse, or neither. See the Tips for New Teachers on pp. 1–2 of the *Chapter 4 Resource Book* for additional notes about Lesson 4.3.

Extra Example 1

Find the measure of ∠F. **65°**

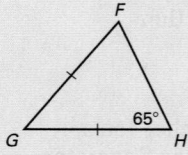

Extra Example 2

Find the value of *x*. **7**

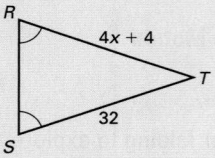

> ## Visualize It!
>
> Students who made a list of triangle types in Lesson 4.1 may wish to add some of the new facts about isosceles and equilateral triangles they learn in this lesson to their notes.

EXAMPLE **1** **Use the Base Angles Theorem**

Find the measure of ∠L.

Solution

Angle *L* is a base angle of an isosceles triangle. From the Base Angles Theorem, ∠L and ∠N have the same measure.

ANSWER ▶ The measure of ∠L is 52°.

Rock and Roll Hall of Fame, Cleveland, Ohio

THEOREM 4.4

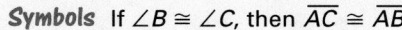

Converse of the Base Angles Theorem

Words If two angles of a triangle are congruent, then the sides opposite them are congruent.

Symbols If ∠B ≅ ∠C, then $\overline{AC} \cong \overline{AB}$.

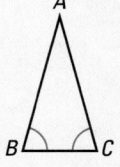

> ## Visualize It!
>
> Base angles don't have to be on the bottom of an isosceles triangle.

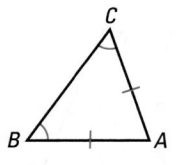

EXAMPLE **2** **Use the Converse of the Base Angles Theorem**

Find the value of *x*.

Solution

By the Converse of the Base Angles Theorem, the legs have the same length.

$DE = DF$	Converse of the Base Angles Theorem
$x + 3 = 12$	Substitute $x + 3$ for *DE* and 12 for *DF*.
$x = 9$	Subtract 3 from each side.

ANSWER ▶ The value of *x* is 9.

Checkpoint ✓ **Use Isosceles Triangle Theorems**

Find the value of y.

1. 50
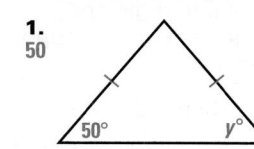

2. 9
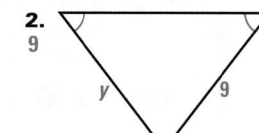

3. 12
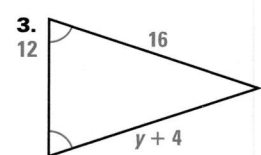

Student Help

LOOK BACK
For the definition of equilateral triangle, see p. 173.

THEOREMS 4.5 and 4.6

4.5 Equilateral Theorem

Words If a triangle is equilateral, then it is equiangular.

Symbols If $\overline{AB} \cong \overline{AC} \cong \overline{BC}$, then $\angle A \cong \angle B \cong \angle C$.

4.6 Equiangular Theorem

Words If a triangle is equiangular, then it is equilateral.

Symbols If $\angle B \cong \angle C \cong \angle A$, then $\overline{AB} \cong \overline{AC} \cong \overline{BC}$.

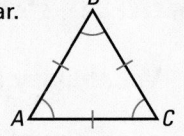

Constructing an Equilateral Triangle You can construct an equilateral triangle using a straightedge and compass.

❶ Draw $\overline{AB}$. Draw an arc with center A that passes through B.

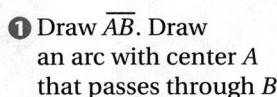

❷ Draw an arc with center B that passes through A.

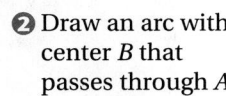

❸ The intersection of the arcs is point C. $\triangle ABC$ is equilateral.

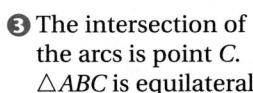

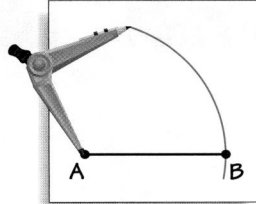
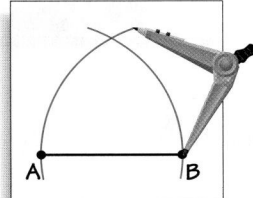
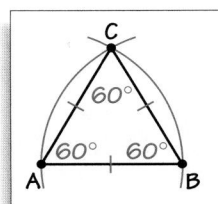

By the Triangle Sum Theorem, the measures of the three congruent angles in an equilateral triangle must add up to 180°. So, each angle in an equilateral triangle measures 60°.

Using Algebra

EXAMPLE 3 Find the Side Length of an Equiangular Triangle

Find the length of each side of the equiangular triangle.

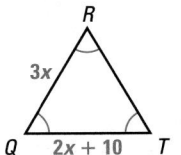

Solution

The angle marks show that $\triangle QRT$ is equiangular. So, $\triangle QRT$ is also equilateral.

$3x = 2x + 10$ — Sides of an equilateral $\triangle$ are congruent.

$x = 10$ — Subtract 2x from each side.

$3(10) = 30$ — Substitute 10 for x.

ANSWER ▶ Each side of $\triangle QRT$ is 30.

Extra Example 3
Find the length of each side of the equiangular triangle. 27

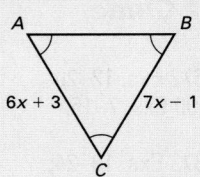

✔ **Concept Check**
If the sides of a triangle are congruent, what can be said about the angles of the triangle? **The angles opposite those sides are congruent.**

🐢 **Daily Puzzler**
Each angle of an equilateral triangle is bisected by a ray that also bisects the side opposite the angle. Six small triangles are formed in the interior of the triangle. How are these triangles related to each other? **They are congruent right triangles whose acute angles measure 30° and 60°.**

4.3 *Isosceles and Equilateral Triangles* **187**

Assignment Guide

BASIC
Day 1: SRH p. 672 Exs. 12–20 even; pp. 188–190 Exs. 7–15, 17–19, 33–37
Day 2: SRH p. 673 Exs. 19–24; pp. 188–190 Exs. 20–25, 27, 28, 38–44

AVERAGE
Day 1: pp. 188–190 Exs. 7–15, 17–19, 27–37
Day 2: pp. 188–190 Exs. 20–25, 38–44

ADVANCED
Day 1: pp. 188–190 Exs. 8–10, 13–19, 27–37
Day 2: pp. 188–190 Exs. 20–25, 26*, 38–44; EC: classzone.com

BLOCK SCHEDULE
pp. 188–190 Exs. 7–15, 17–25, 27–44

Extra Practice
• Student Edition, p. 681
• Chapter 4 Resource Book, pp. 30–31

Homework Check

To quickly check student understanding of key concepts, go over the following exercises:

Basic: 8, 10, 17, 20, 21
Average: 10, 14, 18, 20, 22
Advanced: 13, 15, 19, 22, 25

4.3 Exercises

Guided Practice

Vocabulary Check

1. What is the difference between *equilateral* and *equiangular*?
Equilateral means sides of equal length; *equiangular* means angles of equal measure.

Skill Check

Tell which sides and angles of the triangle are congruent.

2. All three sides are congruent; all three angles are congruent.

2.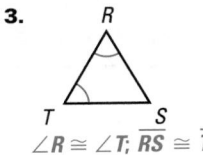

3.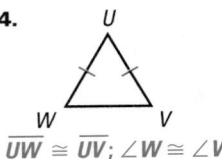
$\angle R \cong \angle T$; $\overline{RS} \cong \overline{TS}$

4.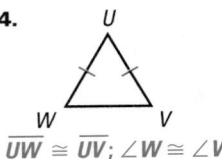
$\overline{UW} \cong \overline{UV}$; $\angle W \cong \angle V$

Find the value of *x*. Tell what theorem(s) you used.

5. **6.**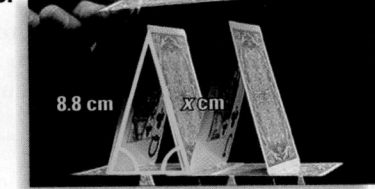

50; Base Angles Theorem   8.8; Converse of the Base Angles Theorem

Practice and Applications

Extra Practice
See p. 681.

9. 45; Corollary to the Triangle Sum Theorem and the Base Angles Theorem

Finding Measures Find the value of *x*. Tell what theorem(s) you used.

7.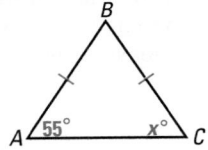
55; Base Angles Theorem

8.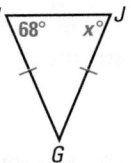
68; Base Angles Theorem

9.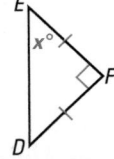
See margin.

⬣ Using Algebra Find the value of *x*.

10.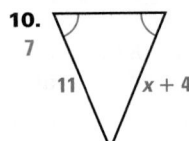
7, 11, $x + 4$

11.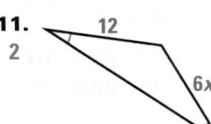
2, 12, $6x$

12.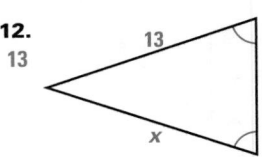
13, 13, x

Homework Help
Example 1: Exs. 7–9, 14, 15, 17–19, 27, 28
Example 2: Exs. 10–13
Example 3: Exs. 20–25

13.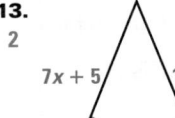
2, $7x + 5$, 19

14.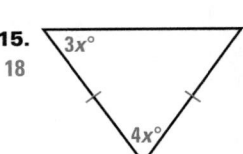
9, $(5x + 7)°$, 52°

15.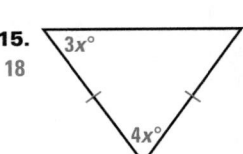
18, $3x°$, $4x°$

16. 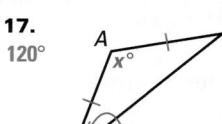 **You be the Judge** Someone in your class tells you that all equilateral triangles are isosceles triangles. Do you agree? Use theorems or definitions to support your answer.

Yes; an isosceles triangle has at least two congruent sides; since an equilateral triangle has three congruent sides, it is also isosceles.

Student Help
CLASSZONE.COM
HOMEWORK HELP
Extra help with problem solving in Exs. 17–19 is at classzone.com

Using Algebra Find the measure of ∠*A*.

17. 120°
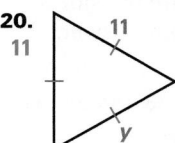

18. 80°
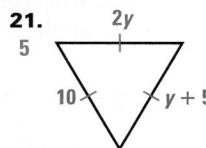

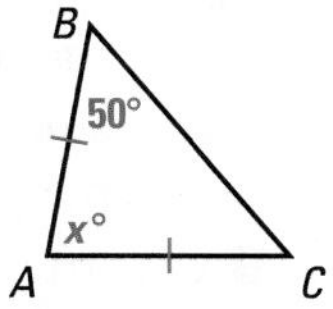

19. 90°
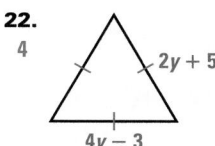

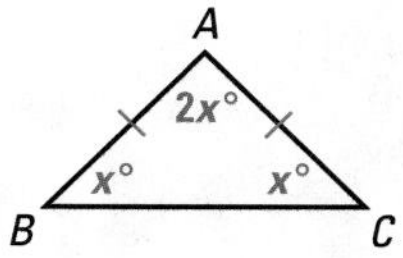

Using Algebra Find the value of *y*.

20. 11, 11
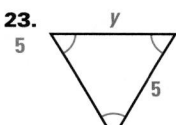

21.
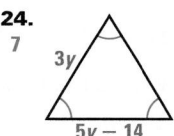

22.
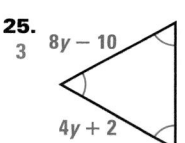

23. 5
24. 7
25. 3

26. Challenge In the diagram at the right, △*XYZ* is equilateral and the following pairs of segments are parallel: $\overline{XY}$ and $\overline{LK}$; $\overline{ZY}$ and $\overline{LJ}$; $\overline{XZ}$ and $\overline{JK}$. Describe a plan for showing that △*JKL* must be equilateral. *See margin.*

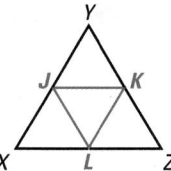

Link to
Sports

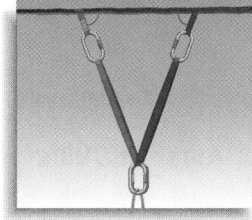

ROCK CLIMBING The climber is using a method of rock climbing called *top roping.* If the climber slips, the anchors catch the fall.

Application Links
CLASSZONE.COM

Rock Climbing In one type of rock climbing, climbers tie themselves to a rope that is supported by anchors. The diagram shows a red and a blue anchor in a horizontal slit in a rock face.

27. If the red anchor is longer than the blue anchor, are the base angles congruent?
no

28. If a climber adjusts the anchors so they are the same length, do you think that the base angles will be congruent? Why or why not?
yes; by the Base Angles Theorem

✗ Common Error
In Exercises 7–15 and 17–25, watch for students who cannot solve the problems because they overlook the symbols for congruent angles or congruent sides. Draw students' attention to the symbols and urge them to look carefully at all diagrams.

26. *Sample answer:* Since $\overline{JK} \parallel \overline{XZ}$, m∠*YJK* = 60° = m∠*X* and m∠*YKJ* = 60° = m∠*Z* by the Corresponding Angles Postulate. Since $\overline{KL} \parallel \overline{XY}$, m∠*ZLK* = 60° = m∠*X* and m∠*ZKL* = 60° = m∠*Y* by the Corresponding Angles Postulate. Since $\overline{LJ} \parallel \overline{ZY}$, m∠*XLJ* = 60° = m∠*Z* and m∠*XJL* = 60° = m∠*Y* by the Corresponding Angles Postulate. m∠*XJL* + m∠*LJK* + m∠*KJY* = 180°; 60° + m∠*LJK* + 60° = 180°; m∠*LJK* = 60°. In a similar way it can be shown that m∠*JKL* = 60° and m∠*KLJ* = 60°. △*JKL* is equiangular, so by the Equiangular Theorem △*JKL* is also equilateral.

Assessment Resources

The Mini-Quiz below is also available on blackline (*Chapter 4 Resource Book*, p. 37) and on transparency. For more assessment resources, see:
• Chapter 4 Resource Book
• Standardized Test Practice
• Test and Practice Generator

Mini-Quiz

Find the value of *x*. Tell what theorem(s) you used.

1.
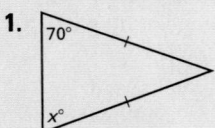

70 (Base Angles Theorem)

2.
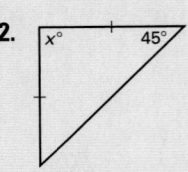

90 (Base Angles Theorem, Triangle Sum Theorem)

Find the value of *x*.

3.
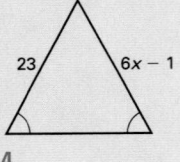

4

4.
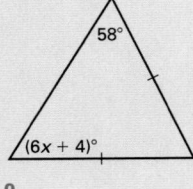

9

29.
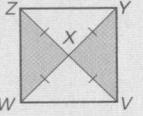

Tiles In Exercises 29–31, use the diagram at the left. In the diagram, $\overline{VX} \cong \overline{WX} \cong \overline{YX} \cong \overline{ZX}$.

29. Copy the diagram. Use what you know about side lengths to mark your diagram. **See margin.**

30. Explain why $\angle XWV \cong \angle XVW$.
$\overline{WX} \cong \overline{VX}$, so $\triangle XWV$ is isosceles. By the Base Angles Theorem, $\angle XWV \cong \angle XVW$.

31. Name four isosceles triangles.
$\triangle WXZ$, $\triangle VXW$, $\triangle YXV$, $\triangle YXZ$

32. Technology Use geometry software to complete the steps.

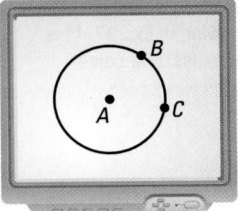

❶ Construct circle *A*.

❷ Draw points *B* and *C* on the circle.

❸ Connect the points to form $\triangle ABC$.

Is $\triangle ABC$ isosceles? Measure the sides of the triangle to check your answer.

32. Yes; $\triangle ABC$ is isosceles. (Note that two sides of the triangle are radii of the circle, and all radii of a circle have the same length.)

Standardized Test Practice

Multiple Choice In Exercises 33 and 34, use the diagram below.

33. What is the measure of $\angle EFD$? **A**

(A) 55°   (B) 65°

(C) 125°   (D) 180°

34. What is the measure of $\angle DEF$? **G**

(F) 50°   (G) 70°

(H) 125°   (J) 180°

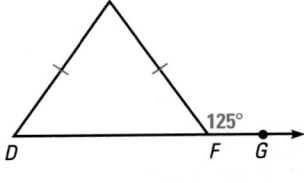

Mixed Review

Angle Bisectors $\overrightarrow{BE}$ is the angle bisector. Find $m\angle DBC$ and $m\angle ABC$. *(Lesson 2.2)*

35.

$m\angle DBC = 42°$;
$m\angle ABC = 84°$

36.

$m\angle DBC = 28°$;
$m\angle ABC = 28°$

37.

$m\angle DBC = 150°$;
$m\angle ABC = 75°$

Vertical Angles Find the value of the variable. *(Lesson 2.4)*

38.
35

$(x + 20)°$, $55°$

39.
50    $(x - 8)°$

$42°$

40.
40    $81°$

$(2x + 1)°$

Algebra Skills

Evaluating Square Roots Evaluate. *(Skills Review, p. 668)*

41. $\sqrt{49}$ 7   **42.** $\sqrt{121}$ 11   **43.** $\sqrt{1}$ 1   **44.** $\sqrt{400}$ 20

Activity 4.4 Areas and Right Triangles

Question

What is the relationship among the lengths of the sides of a right triangle?

Materials

- graph paper
- straightedge

Explore

1 On a piece of graph paper, draw a right triangle with legs that are three units each.

2 Draw a square from each side of the triangle as shown below.

3 Find the area of each square by counting each grid. Count 2 triangles as one whole square.

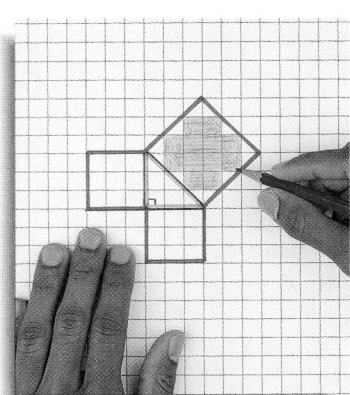

Think About It

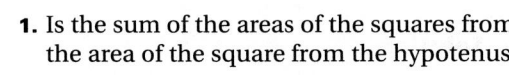

1. Is the sum of the areas of the squares from the two legs equal to the area of the square from the hypotenuse? **yes**

2. Draw another right triangle with legs that are 4 units each. Repeat Steps 1–3 and Exercise 1. Do you get the same result? **Yes, the sum of the areas from the two legs is equal to the area of the square from the hypotenuse.**

3. Extension Look at the right triangle shown at the left with legs of lengths 1 and 4. Show that the area of the square from the hypotenuse is 17. **See margin.**

4. Extension Make a conjecture about the sum of the areas of the squares from two legs and the area of the square from the hypotenuse. **When squares are drawn on the legs and the hypotenuse of a right triangle, the sum of the areas of the squares from the two legs is equal to the area of the square from the hypotenuse.**

Activity **191**

1 Planning the Activity

Goal
Students use graph paper to discover the Pythagorean Theorem by finding the areas of squares.

Materials
See the margin of the student page.

LINK TO LESSON
In Example 1 on page 192, students apply the Pythagorean Theorem to find the length of a hypotenuse of a right triangle.

2 Managing the Activity

Alternative Approach
If your classroom or a nearby hallway has a floor with square tiles, use the grid created by the flooring to demonstrate this activity for the class while students read the directions aloud. Use masking tape to show the sides of the triangle and squares on the floor.

3 Closing the Activity

KEY DISCOVERY
Students discover the Pythagorean Theorem.

Activity Assessment
Use Exercise 4 to assess student understanding.

3. See Additional Answers beginning on page AA1.

Pacing
Suggested Number of Days

Basic: 2 days
Average: 2 days
Advanced: 2 days
Block Schedule: 1 block

Teaching Resources

Blacklines
(See page 170B.)

Transparencies
• Warm-Up with Quiz
• Answers

Technology
• Electronic Teacher Tools
• Test and Practice Generator
• Online Lesson Planner
• Internet Support

Teach

Content and Teaching Strategies
For background information on geometric concepts and teaching strategies related to this lesson, see pages 170E and 170F in this Teacher's Edition.

Tips for New Teachers
Students are usually familiar with the Pythagorean Theorem by the time they are studying geometry. Some may not be aware of why the variables *a*, *b*, and *c* are typically used in the statement of the theorem. Draw a right triangle like the one shown in Theorem 4.7, label the vertices with *A*, *B*, and *C* to correspond to the labels of the sides, and tell students that this is a common way to label the vertices of a right triangle. See the Tips for New Teachers on pp. 1–2 of the *Chapter 4 Resource Book* for additional notes about Lesson 4.4.

 **4.4 The Pythagorean Theorem and the Distance Formula**

Goal
Use the Pythagorean Theorem and the Distance Formula.

Key Words
• leg
• hypotenuse
• Pythagorean Theorem
• Distance Formula

The photo shows part of twin skyscrapers in Malaysia that are connected by a skywalk. The skywalk is supported by a set of beams.

In a right triangle, the sides that form the right angle are called **legs**. The side opposite the right angle is called the **hypotenuse**.

In Exercise 26 you will find the length of the support beam using the Pythagorean Theorem.

THEOREM 4.7

The Pythagorean Theorem

Words In a right triangle, the square of the length of the hypotenuse is equal to the sum of the squares of the lengths of the legs.

$$(\text{hypotenuse})^2 = (\text{leg})^2 + (\text{leg})^2$$

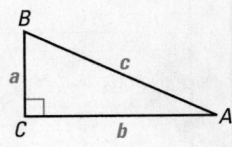

Symbols If $m\angle C = 90°$, then $c^2 = a^2 + b^2$.

EXAMPLE 1 Find the Length of the Hypotenuse

Find the length of the hypotenuse.

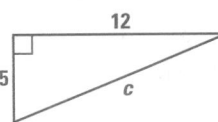

Solution

$(\text{hypotenuse})^2 = (\text{leg})^2 + (\text{leg})^2$	Pythagorean Theorem
$c^2 = 5^2 + 12^2$	Substitute.
$c^2 = 25 + 144$	Multiply.
$c^2 = 169$	Add.
$\sqrt{c^2} = \sqrt{169}$	Find the positive square root.
$c = 13$	Solve for *c*.

ANSWER ▶ The length of the hypotenuse is 13.

Student Help

SKILLS REVIEW
To review square roots, see p. 669.

EXAMPLE 2 Find the Length of a Leg

Find the unknown side length.

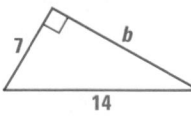

Solution

$(\text{hypotenuse})^2 = (\text{leg})^2 + (\text{leg})^2$	Pythagorean Theorem
$14^2 = 7^2 + b^2$	Substitute.
$196 = 49 + b^2$	Multiply.
$196 - 49 = 49 + b^2 - 49$	Subtract 49 from each side.
$147 = b^2$	Simplify.
$\sqrt{147} = \sqrt{b^2}$	Find the positive square root.
$12.1 \approx b$	Approximate with a calculator.

ANSWER ▶ The side length is about 12.1.

Checkpoint ✓ *Find the Lengths of the Hypotenuse and Legs*

Find the unknown side length.

1.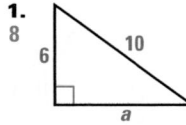

2.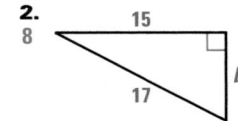

3.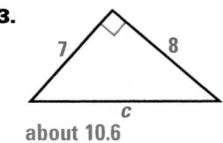
about 10.6

EXAMPLE 3 Find the Length of a Segment

Find the distance between the points
$A(1, 2)$ and $B(4, 6)$.

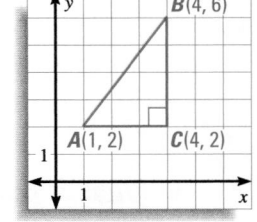

Solution

Draw a right triangle with hypotenuse $\overline{AB}$.
$BC = 6 - 2 = 4$ and $CA = 4 - 1 = 3$. Use the
Pythagorean Theorem.

$(\text{hypotenuse})^2 = (\text{leg})^2 + (\text{leg})^2$	
$(AB)^2 = 3^2 + 4^2$	Substitute.
$(AB)^2 = 9 + 16$	Multiply.
$(AB)^2 = 25$	Add.
$\sqrt{(AB)^2} = \sqrt{25}$	Find the positive square root.
$AB = 5$	Simplify.

Student Help

LOOK BACK
To review finding
distances on a
coordinate plane,
see p. 30.

Multiple Representations

As you discuss the Pythagorean Theorem, take time to show students how the diagram relates to the words and symbols. While reciting the theorem, pause to have students identify the sides in the diagram that correspond to the terms *hypotenuse* and *legs*.

Extra Example 1

Find the length of the hypotenuse.

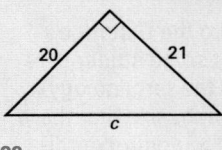

29

Extra Example 2

Find the unknown side length. **45**

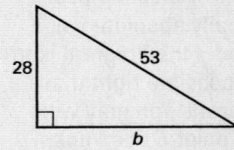

Extra Example 3

Find the distance between the points $J(-1, 7)$ and $K(5, -1)$. **10**

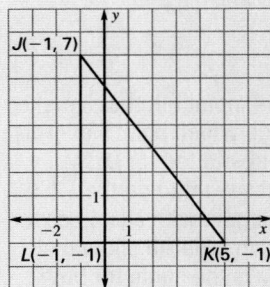

4.4 *The Pythagorean Theorem and the Distance Formula* 193

193

Distance Formula Using the steps shown in Example 3, the Pythagorean Theorem can be used to develop the **Distance Formula**, which gives the distance between two points in a coordinate plane.

THE DISTANCE FORMULA

If $A(x_1, y_1)$ and $B(x_2, y_2)$ are points in a coordinate plane, then the distance between A and B is

$$AB = \sqrt{(x_2 - x_1)^2 + (y_2 - y_1)^2}.$$

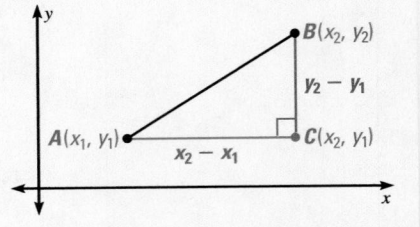

EXAMPLE 4 Use the Distance Formula

Find the distance between $D(1, 2)$ and $E(3, -2)$.

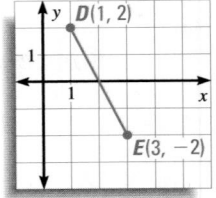

Solution

Begin by plotting the points in a coordinate plane.

$x_1 = 1$, $y_1 = 2$, $x_2 = 3$, and $y_2 = -2$.

Student Help

STUDY TIP
$\sqrt{4 + 16} \neq \sqrt{4} + \sqrt{16}$. The square root of a sum does *not* equal the sum of the square roots. You must add 4 and 16 *before* taking the square root. ·······▶

$$DE = \sqrt{(x_2 - x_1)^2 + (y_2 - y_1)^2} \qquad \text{The Distance Formula}$$

$$= \sqrt{(3 - 1)^2 + (-2 - 2)^2} \qquad \text{Substitute.}$$

$$= \sqrt{2^2 + (-4)^2} \qquad \text{Simplify.}$$

$$= \sqrt{4 + 16} \qquad \text{Multiply.}$$

$$= \sqrt{20} \qquad \text{Add.}$$

$$\approx 4.5 \qquad \text{Approximate with a calculator.}$$

ANSWER ▶ The distance between D and E is about 4.5 units.

Checkpoint ✓ Use the Distance Formula

Find the distance between the points.

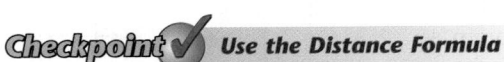

4.
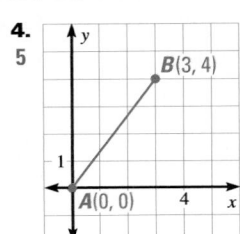

5. about 6.3
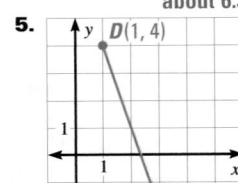

6. about 5.1
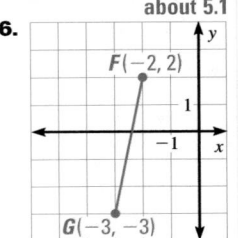

4.4 Exercises

Guided Practice

Vocabulary Check

1. Sketch a right triangle and label its vertices. Then use your triangle to state the Pythagorean Theorem. See margin.

Skill Check

Find the unknown side length. Round your answer to the nearest tenth, if necessary.

2.
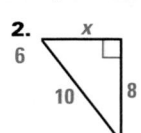

3.
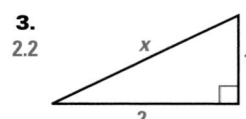

4.
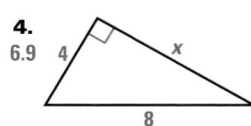

Find the distance between the points. Round your answer to the nearest tenth, if necessary.

5.
5.8
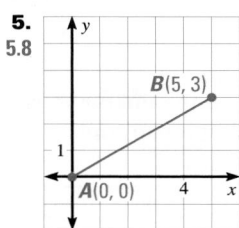

6.
5.4
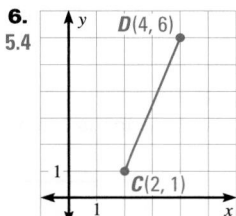

7.
6.3
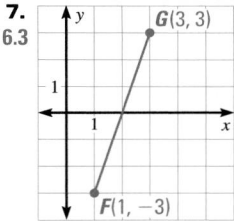

Practice and Applications

Extra Practice
See p. 681.

Finding a Hypotenuse Find the length of the hypotenuse.

8.
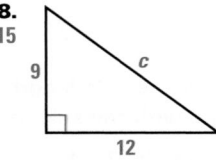

9.
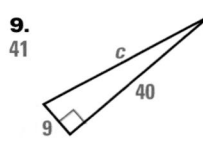

10.
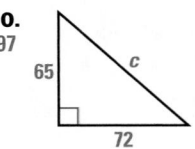

Homework Help

Example 1: Exs. 8–13, 26
Example 2: Exs. 14–22
Example 3: Exs. 27–34
Example 4: Exs. 27–34

11.
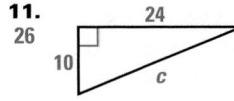

12.
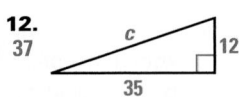

13.
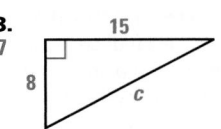

Finding a Leg Find the unknown side length.

14.

15.
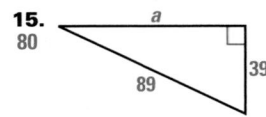

16.
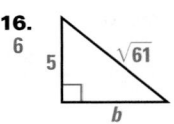

4.4   *The Pythagorean Theorem and the Distance Formula*   **195**

Assignment Guide

BASIC
Day 1: SRH p. 669 Exs. 1–4, 17–20; pp. 195–198 Exs. 8–22, 39–51 odd
Day 2: pp. 195–198 Exs. 27–32, 38–54 even, Quiz 2

AVERAGE
Day 1: pp. 195–198 Exs. 8–26, 39–55 odd
Day 2: pp. 195–198 Exs. 27–34, 38–54 even, Quiz 2

ADVANCED
Day 1: pp. 195–198 Exs. 8–22 even, 23–26, 35–37*, 39–51 odd
Day 2: pp. 195–198 Exs. 27–34, 38–54 even, Quiz 2; EC: classzone.com

BLOCK SCHEDULE
pp. 195–198 Exs. 8–34, 38–55, Quiz 2

Extra Practice
• Student Edition, p. 681
• Chapter 4 Resource Book, pp. 38–39

Homework Check
To quickly check student understanding of key concepts, go over the following exercises:

Basic: 9, 14, 18, 28, 30
Average: 10, 15, 19, 28, 31
Advanced: 12, 16, 22, 27, 32

1. *Sample answer:*

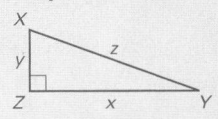

If $m\angle Z = 90°$, then $x^2 + y^2 = z^2$.

23.

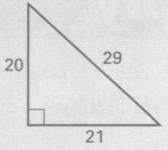

24.

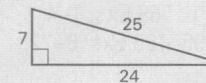

A *Pythagorean triple* is a set of three positive integers a, b, and c that satisfy the equation $c^2 = a^2 + b^2$. For example, the integers 3, 4, and 5 form a Pythagorean triple because $5^2 = 3^2 + 4^2$.

Find the length of the hypotenuse of the right triangle. Tell whether the side lengths form a Pythagorean triple.

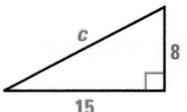

Solution

$$(\text{hypotenuse})^2 = (\text{leg})^2 + (\text{leg})^2 \qquad \text{Pythagorean Theorem}$$
$$c^2 = 8^2 + 15^2 \qquad \text{Substitute 8 and 15 for the legs.}$$
$$c^2 = 64 + 225 \qquad \text{Multiply.}$$
$$c^2 = 289 \qquad \text{Add.}$$
$$c = 17 \qquad \text{Find the positive square root.}$$

ANSWER ▶ Because the side lengths 8, 15, and 17 are integers, they form a Pythagorean triple.

Pythagorean Triples **Find the unknown side length. Tell whether the side lengths form a Pythagorean triple.**

17.

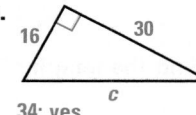

4; yes

18.

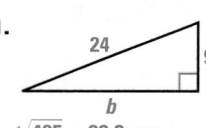

$\sqrt{52} \approx 7.2$; no

19.

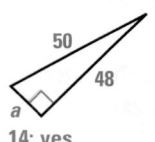

$\sqrt{170} \approx 13.0$; no

20.

16 30
c
34; yes

21.

24 9
b
$\sqrt{495} \approx 22.2$; no

22.

50 48
a
14; yes

Visualize It! **Tell whether the side lengths form a Pythagorean triple. If so, draw a right triangle with the side lengths.** See margin for sketches.

23. 21, 29, 20
yes

24. 25, 7, 24
yes

25. 5, 12, 14
no

26. Support Beam The skyscrapers shown on page 170 are connected by a skywalk with support beams. Use the diagram to find the approximate length of each support beam. about 52.95 m

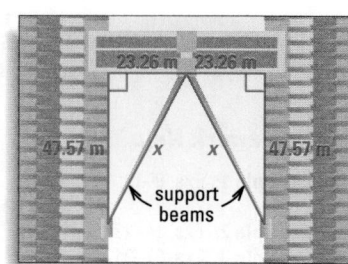

23.26 m   23.26 m

47.57 m   x   x   47.57 m

support beams

Distance Formula Find the distance between the points. Round your answer to the nearest tenth.

27.
5
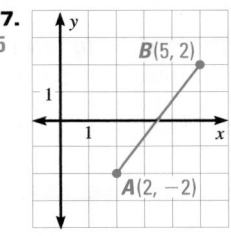

28.
8.5
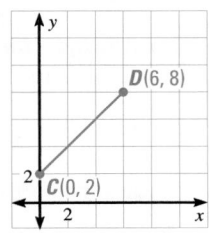

29.
6.1
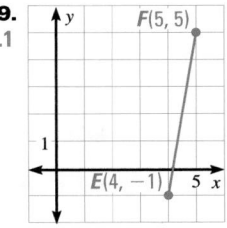

Student Help
CLASSZONE.COM

HOMEWORK HELP
Extra help with problem solving in Exs. 30–32 is at classzone.com

Congruence Graph *P, Q,* and *R.* Then use the Distance Formula to decide whether $\overline{PQ} \cong \overline{QR}$. 30–32. Check graphs.

30. $P(4, -4)$ yes
$\quad Q(1, -6)$
$\quad R(-1, -3)$

31. $P(-1, -6)$ yes
$\quad Q(-8, 5)$
$\quad R(3, -2)$

32. $P(5, 1)$ no
$\quad Q(-5, -7)$
$\quad R(-3, 6)$

Sum of Distances In Exercises 33 and 34, use the map below.
Sidewalks around the edge of a campus quadrangle connect the buildings. Students sometimes take shortcuts by walking across the grass along the pathways shown. The coordinate system shown is measured in yards.

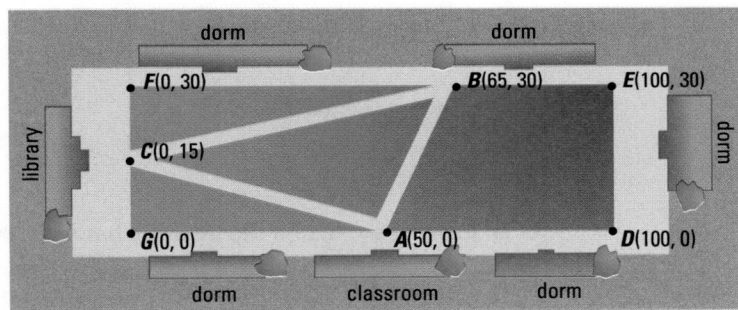

33. Find the distances from *A* to *B,* from *B* to *C,* and from *C* to *A* if you have to walk around the quadrangle along the sidewalks.
A to *B*: 115 yd; *B* to *C*: 80 yd; *C* to *A*: 65 yd

34. Find the distances from *A* to *B,* from *B* to *C,* and from *C* to *A* if you are able to walk across the grass along the pathways.
A to *B*: about 33.5 yd; *B* to *C*: about 66.7 yd; *C* to *A*: about 52.2 yd

Challenge Find the value of *x.* Use a calculator, and round your answer to the nearest tenth.

35.
13.9
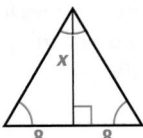

36.
7.2
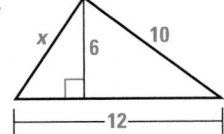

37.
7.6
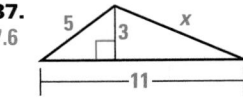

4.4 The Pythagorean Theorem and the Distance Formula 197

 Assess

Assessment Resources

The Mini-Quiz below is also available on blackline (*Chapter 4 Resource Book*, p. 48) and on transparency. For more assessment resources, see:
- Chapter 4 Resource Book
- Standardized Test Practice
- Test and Practice Generator

Mini-Quiz

Find the unknown side length. Tell whether the side lengths form a Pythagorean triple.

1.

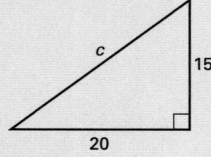

25; yes

2.

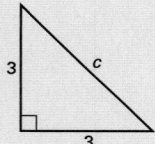

about 4.2; no

3. Find the distance between the points. Round your answer to the nearest tenth. **4.5**

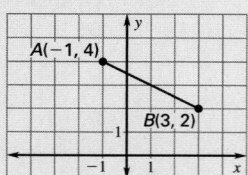

4. The diagram shows the design for a house roof. Each side of the roof is 22 feet long as shown. What is the approximate width of the house? **about 31.1 ft**

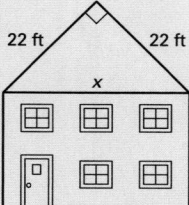

Standardized Test Practice

38. Multiple Choice What is the distance from (3, 5) to (−1, −4)? **D**

    Ⓐ $\sqrt{5}$ Ⓑ $\sqrt{17}$ Ⓒ $2\sqrt{13}$ Ⓓ $\sqrt{97}$

39. Multiple Choice Which of the following is the length of the hypotenuse of a right triangle with legs of lengths 33 and 56? **F**

    Ⓕ 65   Ⓖ 72.9   Ⓗ 85.8   Ⓙ 89

Mixed Review

Finding Absolute Values Evaluate. *(Skills Review, p. 662)*

40. $|-7|$ **7**   **41.** $|1.05|$ **1.05**   **42.** $|0|$ **0**   **43.** $|-0.02|$
                                                                          0.02

Solving Inequalities Solve the inequality. *(Algebra Review, p. 167)*

44. $x + 5 < 8$ **45.** $10 + x \geq 12$ **46.** $4x \geq 28$ **47.** $6x + 11 \leq 11$
    $x < 3$ $x \geq 2$ $x \geq 7$ $x \leq 0$

Algebra Skills

Fractions and Decimals Write the decimal as a fraction in simplest form. *(Skills Review, p. 657)*

48. 0.4 $\frac{2}{5}$ **49.** 0.08 $\frac{2}{25}$ **50.** 0.54 $\frac{27}{50}$ **51.** 0.12 $\frac{3}{25}$

52. 0.250 **53.** 0.173 **54.** $0.\overline{3}$ **55.** $0.\overline{1}$
   $\frac{1}{4}$ $\frac{173}{1000}$ $\frac{1}{3}$ $\frac{1}{9}$

Quiz 2

Find the value of x. *(Lesson 4.3)*

1.   13, $3x - 5$, 6

2.   55°, $(2x + 1)°$, 27

3. $x + 5$, $4x - 16$, 7

In Exercises 4–6, find the distance between the points. *(Lesson 4.4)*

4. $A(-2, 3)$, $B(3, 0)$ $\sqrt{34} \approx 5.8$

5. $A(-3, -1)$, $B(1, -2)$ $\sqrt{17} \approx 4.1$

6. $B(1, 2)$, $A(-1, -1)$ $\sqrt{13} \approx 3.6$

7. A device used to measure windspeed is attached to the top of a pole. Support wires are attached to the pole 5 feet above the ground. Each support wire is 6 feet long. About how far from the base of the pole is each wire attached to the ground? *(Lesson 4.4)* $\sqrt{11} \approx 3.3$ ft

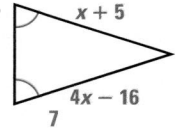

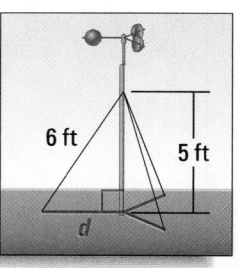

4.5 Side Lengths of Triangles

Question

What is the relationship between the side lengths of a triangle and its angle measures?

Explore

1 Draw A, B, C, and $\triangle ABC$.

2 Measure AB, BC, and CA. Calculate $(AB)^2$ and $(BC)^2 + (CA)^2$.

3 Measure $\angle ACB$.

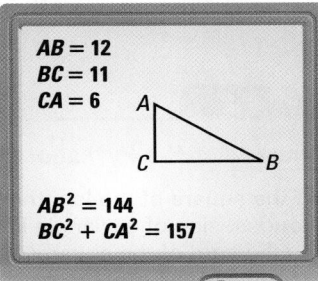

$AB = 12$
$BC = 11$
$CA = 6$

$AB^2 = 144$
$BC^2 + CA^2 = 157$

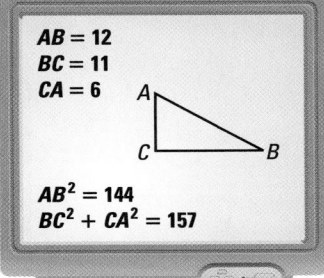

$AB = 12$
$BC = 11$
$CA = 6$

$AB^2 = 144$
$BC^2 + CA^2 = 157$

 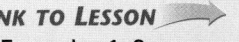
2 Managing the Activity

Tips for Success
Point out that the triangle in Step 1 does not need to be a right triangle. Encourage students to draw a scalene triangle to make their conclusions more obvious.

Think About It

Copy the table below and record the values as you work Exercises 1–3.

	AB	BC	CA	$(AB)^2$	$(BC)^2 + (CA)^2$	$m\angle ACB$
Triangle 1	?	?	?	?	?	?
Triangle 2	?	?	?	?	?	?
Triangle 3	?	?	?	?	?	?

1. Drag point A or B until $(AB)^2 = (BC)^2 + (CA)^2$. What do you notice about $m\angle ACB$? $m\angle ACB = 90°$

2. Drag point A or B until $(AB)^2 > (BC)^2 + (CA)^2$. What do you notice about $m\angle ACB$? $m\angle ACB > 90°$

3. Drag point A or B until $(AB)^2 < (BC)^2 + (CA)^2$. What do you notice about $m\angle ACB$? $m\angle ACB < 90°$

4. Can you create an acute triangle in which $(AB)^2 > (BC)^2 + (CA)^2$? no

3 Closing the Activity

KEY DISCOVERY
Triangles can be classified as acute, obtuse, or right by comparing the square of the side length of the longest side to the sum of the squares of the other two side lengths.

Activity Assessment
Use Exercises 1–3 to assess student understanding.

Pacing
Suggested Number of Days

Basic: 2 days
Average: 2 days
Advanced: 2 days
Block Schedule: 1 block

Teaching Resources

📄 **Blacklines**
(See page 170B.)

🗲 **Transparencies**
• Warm-Up with Quiz
• Answers

🖳 **Technology**
• Electronic Teacher Tools
• Test and Practice Generator
• Online Lesson Planner
• Internet Support

② Teach

---Content and Teaching---
Strategies
For background information on geometric concepts and teaching strategies related to this lesson, see pages 170E and 170F in this Teacher's Edition.

Tips for New Teachers
Stress that students should always identify the longest side of a given triangle first before applying the triangle inequalities given at the top of page 201 or even when using the Pythagorean Theorem. See the Tips for New Teachers on pp. 1–2 of the *Chapter 4 Resource Book* for additional notes about Lesson 4.5.

4.5 The Converse of the Pythagorean Theorem

Goal
Use the Converse of Pythagorean Theorem. Use side lengths to classify triangles.

Key Words
• converse p. 136

A gardener can use the Converse of the Pythagorean Theorem to make sure that the corners of a garden bed form right angles.

In the photograph, a triangle with side lengths 3 feet, 4 feet, and 5 feet ensures that the angle at one corner is a right angle.

THEOREM 4.8

The Converse of the Pythagorean Theorem

Words If the square of the length of the longest side of a triangle is equal to the sum of the squares of the lengths of the other two sides, then the triangle is a right triangle.

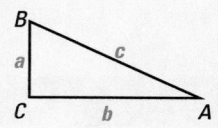

Symbols If $c^2 = a^2 + b^2$, then $\triangle ABC$ is a right triangle.

Student Help

LOOK BACK
For the definition of converse, see p. 136.
For the definition of converse, see p. 136.

EXAMPLE ① Verify a Right Triangle

Is $\triangle ABC$ a right triangle?

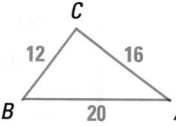

Solution

Let c represent the length of the longest side of the triangle. Check to see whether the side lengths satisfy the equation $c^2 = a^2 + b^2$.

$$c^2 \stackrel{?}{=} a^2 + b^2 \qquad \text{Compare } c^2 \text{ with } a^2 + b^2.$$
$$20^2 \stackrel{?}{=} 12^2 + 16^2 \qquad \text{Substitute 20 for } c, \text{ 12 for } a, \text{ and 16 for } b.$$
$$400 \stackrel{?}{=} 144 + 256 \qquad \text{Multiply.}$$
$$400 = 400 \qquad \text{Simplify.}$$

ANSWER ▶ It is true that $c^2 = a^2 + b^2$. So, $\triangle ABC$ is a right triangle.

Classifying Triangles You can determine whether a triangle is acute, right, or obtuse by its side lengths.

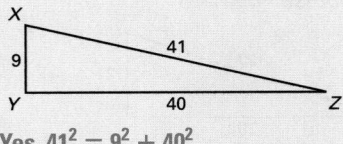
CLASSIFYING TRIANGLES

In $\triangle ABC$ with longest side c:

If $c^2 < a^2 + b^2$,   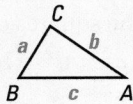   then $\triangle ABC$ is *acute*.

If $c^2 = a^2 + b^2$,   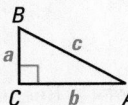   then $\triangle ABC$ is *right*.

If $c^2 > a^2 + b^2$,   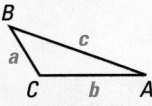   then $\triangle ABC$ is *obtuse*.

Student Help

STUDY TIP
This is the Converse of the Pythagorean Theorem.

Extra Example 2
Show that the triangle is an acute triangle.

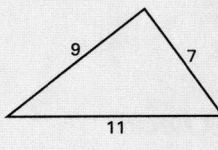

$11^2 < 9^2 + 7^2$

Extra Example 3
Show that the triangle is an obtuse triangle.

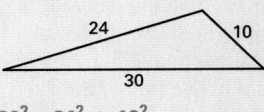

$30^2 > 24^2 + 10^2$

EXAMPLE 2 Acute Triangles

Show that the triangle is an acute triangle.

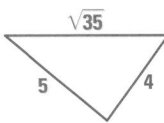

Student Help

STUDY TIP
$\sqrt{35} \approx 5.9$, so use $\sqrt{35}$ as the value of c, the longest side length of the triangle.

Solution
Compare the side lengths.

$c^2 \stackrel{?}{=} a^2 + b^2$ Compare c^2 with $a^2 + b^2$.

$(\sqrt{35})^2 \stackrel{?}{=} 4^2 + 5^2$ Substitute $\sqrt{35}$ for c, 4 for a, and 5 for b.

$35 \stackrel{?}{=} 16 + 25$ Multiply.

$35 < 41$ Simplify.

ANSWER ▶ Because $c^2 < a^2 + b^2$, the triangle is acute.

EXAMPLE 3 Obtuse Triangles

Show that the triangle is an obtuse triangle.

Solution
Compare the side lengths.

$c^2 \stackrel{?}{=} a^2 + b^2$ Compare c^2 with $a^2 + b^2$.

$(15)^2 \stackrel{?}{=} 8^2 + 12^2$ Substitute 15 for c, 8 for a, and 12 for b.

$225 \stackrel{?}{=} 64 + 144$ Multiply.

$225 > 208$ Simplify.

ANSWER ▶ Because $c^2 > a^2 + b^2$, the triangle is obtuse.

4.5 The Converse of the Pythagorean Theorem   **201**

Extra Example 4

Classify the triangle as *acute*, *right*, or *obtuse*. right

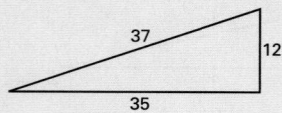

37 12
35

Extra Example 5

Classify the triangle with the given side lengths as *acute*, *right*, or *obtuse*.

a. 9, 12, 13 acute
b. 16, 30, 36 obtuse

✓ **Concept Check**

How can you determine whether a triangle is acute, right, or obtuse from its side lengths? **If the square of the length of the longest side of a triangle is greater than the sum of the squares of the lengths of the other two sides, then the triangle is an obtuse triangle. If the square of the length of the longest side of a triangle is equal to the sum of the squares of the lengths of the other two sides, then the triangle is a right triangle. If the square of the length of the longest side of a triangle is less than the sum of the squares of the lengths of the other two sides, then the triangle is an acute triangle.**

🧩 **Daily Puzzler**

Find the value(s) of *k* so that *x*, 2*x*, and *kx* are the side lengths of a right triangle. $\sqrt{3}$ or $\sqrt{5}$

Student Help
CLASSZONE.COM

MORE EXAMPLES
More examples at classzone.com

EXAMPLE 4 Classify Triangles

Classify the triangle as *acute*, *right*, or *obtuse*.

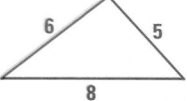

6 5
8

Solution

Compare the square of the length of the longest side with the sum of the squares of the lengths of the two shorter sides.

$$c^2 \overset{?}{=} a^2 + b^2 \qquad \text{Compare } c^2 \text{ with } a^2 + b^2.$$

$$8^2 \overset{?}{=} 5^2 + 6^2 \qquad \text{Substitute 8 for } c, \text{ 5 for } a, \text{ and 6 for } b.$$

$$64 \overset{?}{=} 25 + 36 \qquad \text{Multiply.}$$

$$64 > 61 \qquad \text{Simplify.}$$

ANSWER ▶ Because $c^2 > a^2 + b^2$, the triangle is obtuse.

EXAMPLE 5 Classify Triangles

Classify the triangle with the given side lengths as *acute*, *right*, or *obtuse*.

a. 4, 6, 7   **b.** 12, 35, 37

Solution

a. $c^2 \overset{?}{=} a^2 + b^2$ **b.** $c^2 \overset{?}{=} a^2 + b^2$

$7^2 \overset{?}{=} 4^2 + 6^2$ $37^2 \overset{?}{=} 12^2 + 35^2$

$49 \overset{?}{=} 16 + 36$ $1369 \overset{?}{=} 144 + 1225$

$49 < 52$ $1369 = 1369$

The triangle is acute.   The triangle is right.

Checkpoint ✓ Classify Triangles

Classify the triangle as *acute*, *right*, or *obtuse*. **Explain.**

1.

5 6
2

obtuse; $6^2 > 2^2 + 5^2$

2.

8 17
15

right; $17^2 = 8^2 + 15^2$

3.

7
7
7

acute; $7^2 < 7^2 + 7^2$

Use the side lengths to classify the triangle as *acute*, *right*, or *obtuse*.

4. 7, 24, 24
acute

5. 7, 24, 25
right

6. 7, 24, 26
obtuse

4.5 Exercises

❸ Apply

Guided Practice

Vocabulary Check

1. Write the Converse of the Pythagorean Theorem in your own words.

Skill Check

Determine whether the triangle is *acute, right*, or *obtuse*.

1. *Sample answer:* If you square the lengths of the two shortest sides of a triangle and add the results, and the sum is equal to the square of the length of the longest side of the triangle, then the triangle is a right triangle.

2.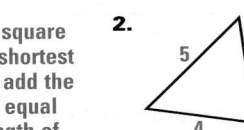
acute

3.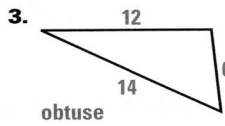
obtuse

4.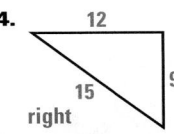
right

Match the side lengths of a triangle with the best description.

5. 2, 10, 11 **C**   **A.** right

6. 8, 5, 7 **B**   **B.** acute

7. 5, 5, 5 **D**   **C.** obtuse

8. 6, 8, 10 **A**   **D.** equiangular

Practice and Applications

Extra Practice

See p. 682.

Verifying Right Triangles Show that the triangle is a right triangle.

9.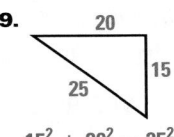
$15^2 + 20^2 = 25^2$

10.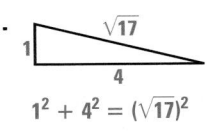
$10^2 + 24^2 = 26^2$

11.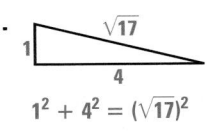
$1^2 + 4^2 = (\sqrt{17})^2$

Verifying Acute Triangles Show that the triangle is an acute triangle.

12.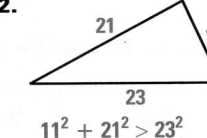
$11^2 + 21^2 > 23^2$

13.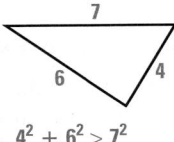
$4^2 + 6^2 > 7^2$

14.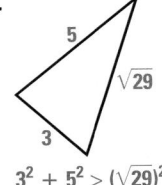
$3^2 + 5^2 > (\sqrt{29})^2$

Homework Help

Example 1: Exs. 9–11, 24
Example 2: Exs. 12–14
Example 3: Exs. 15–17
Example 4: Exs. 18–23, 37–38
Example 5: Exs. 25–36

Verifying Obtuse Triangles Show that the triangle is an obtuse triangle.

15.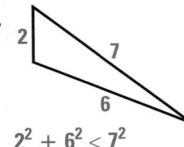
$2^2 + 6^2 < 7^2$

16.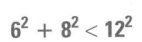
$6^2 + 8^2 < 12^2$

17.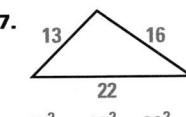
$13^2 + 16^2 < 22^2$

Assignment Guide

BASIC
Day 1: pp. 203–205 Exs. 9–17, 43–53
Day 2: pp. 203–205 Exs. 18–23, 25–36

AVERAGE
Day 1: pp. 203–205 Exs. 9–17, 37–39, 43–53 odd
Day 2: pp. 203–205 Exs. 18–23, 25–36, 42–52 even

ADVANCED
Day 1: pp. 203–205 Exs. 9–17, 24, 37–39, 42–45, 47–53 odd
Day 2: pp. 203–205 Exs. 18–23, 25–36, 40–41*;
EC: classzone.com

BLOCK SCHEDULE
pp. 203–205 Exs. 9–23, 25–39, 42–53

Extra Practice

• Student Edition, p. 682
• Chapter 4 Resource Book, pp. 51–52

Homework Check

To quickly check student understanding of key concepts, go over the following exercises:

Basic: 9, 12, 15, 18, 28
Average: 10, 13, 16, 19, 29
Advanced: 11, 14, 17, 21, 35

4.5 The Converse of the Pythagorean Theorem

Classifying Triangles Classify the triangle as *acute, right,* or *obtuse.*

18.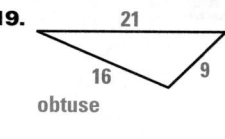
acute

19.
obtuse

20.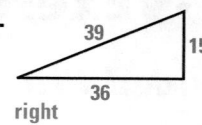
right

21.
obtuse

22.
right

23.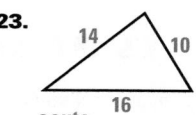
acute

Link to
History

EARLY MATHEMATICS
This photograph shows part of a Babylonian clay tablet made around 350 B.C. The tablet contains a table of numbers.

24. Early Mathematics The Babylonian tablet shown at the left contains several sets of triangle side lengths, suggesting that the Babylonians may have been aware of the relationships among the side lengths of right triangles. The side lengths in the table below show several sets of numbers from the tablet. Use a calculator to verify that each set of side lengths satisfies the Pythagorean Theorem.

$120^2 + 119^2 = 169^2$, $4800^2 + 4601^2 = 6649^2$, and $(13,500)^2 + (12,709)^2 = (18,541)^2$

a	b	c
120	119	169
4,800	4,601	6,649
13,500	12,709	18,541

Classifying Triangles Classify the triangle with the given side lengths as *acute, right,* or *obtuse.*

25. 20, 99, 101 right

26. 21, 28, 35 right

27. 26, 10, 17 obtuse

28. 7, 10, 11 acute

29. 4, $\sqrt{67}$, 9 acute

30. $\sqrt{13}$, 6, 7 right

31. 468, 595, 757 right

32. 10, 11, 14 acute

33. 4, 5, 5 acute

34. 17, 144, 145 right

35. 10, 49, 50 acute

36. $\sqrt{5}$, 5, 5.5 obtuse

Air Travel In Exercises 37 and 38, use the map below.

37. No; $714^2 < 599^2 + 403^2$, so the triangle is not a right triangle.

38. No; From Ex. 37, the triangle is not a right triangle, so the angle at Cincinnati is not a right angle. It follows that Tallahassee is not directly south of Cincinnati.

37. Use the distances given on the map to tell whether the triangle formed by the three cities is a right triangle.

38. Cincinnati is directly west of Washington, D.C. Is Tallahassee directly south of Cincinnati? Explain your answer.

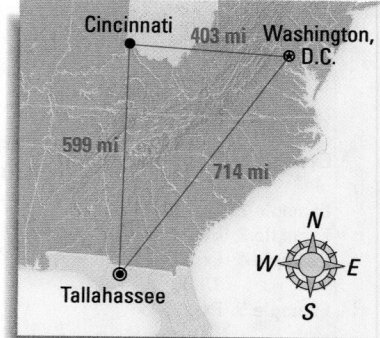

39. ⚖ *You be the Judge* A classmate tells you if you find three side lengths that form a right triangle and double each of them, the sides will form an obtuse triangle. Is your classmate correct? Explain.
See margin.

Challenge Graph points *P*, *Q*, and *R*. Connect the points to form △*PQR*. Decide whether △*PQR* is *acute*, *right*, or *obtuse*.

40. *P*(−3, 4), *Q*(5, 0), *R*(−6, −2) right

41. *P*(−1, 2), *Q*(4, 1), *R*(0, −1) acute

Standardized Test Practice

42. Multi-Step Problem A double play occurs in baseball when two outs are made on a single play. In the diagram shown, the ball is hit to the player at point *A*. A double play is made when the player at point *A* throws the ball to the player at point *B* who in turn throws it to the player at point *C*.

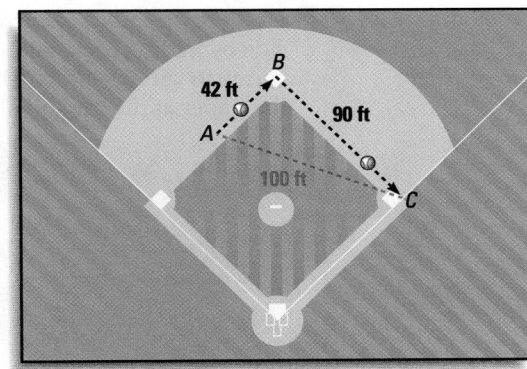

a. Use the diagram to determine what kind of triangle is formed by points *A*, *B*, and *C*. obtuse

b. What kind of triangle is formed by points *A*, *B*, and *C* if the distance between points *A* and *C* is 99 feet? acute

c. Critical Thinking Find values for *AB* and *AC* that would make △*ABC* in the diagram a right triangle if *BC* = 90 feet.
Sample answer: *AB* = 42 feet and *AC* ≈ 99.3 feet

Mixed Review

Finding Measures Find the value of *x*. *(Lesson 4.3)*

43.
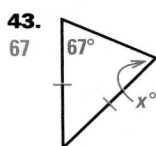

44.
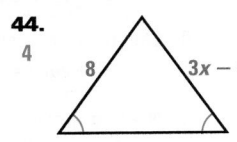

45.
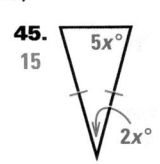

Algebra Skills

Multiplying Fractions Multiply. Write the answer as a fraction or a mixed number in simplest form. *(Skills Review, p. 659)*

46. $\frac{1}{2} \times \frac{4}{5}$ $\frac{2}{5}$

47. $\frac{3}{8} \times \frac{3}{4}$ $\frac{9}{32}$

48. $\frac{3}{11} \times \frac{11}{12}$ $\frac{1}{4}$

49. $\frac{3}{5} \times \frac{5}{9}$ $\frac{1}{3}$

50. $\frac{3}{4} \times 6$ $4\frac{1}{2}$

51. $8 \times 1\frac{3}{4}$ 14

52. $1\frac{1}{3} \times \frac{4}{9}$ $\frac{16}{27}$

53. $5\frac{1}{4} \times \frac{2}{3}$ $3\frac{1}{2}$

4.5 *The Converse of the Pythagorean Theorem* **205**

---

39. No; the doubled side lengths will also form a right triangle. *Sample answer:* Let *a*, *b*, and *c* be the lengths of the sides of a right triangle with $a^2 + b^2 = c^2$. Then $(2a)^2 + (2b)^2 = 4a^2 + 4b^2 = 4(a^2 + b^2) = 4c^2 = (2c)^2$. Since $(2a)^2 + (2b)^2 = (2c)^2$, 2*a*, 2*b*, and 2*c* are also the side lengths of a right triangle.

Mini-Quiz
Classify the triangle as *right*, *acute*, or *obtuse*.

1.
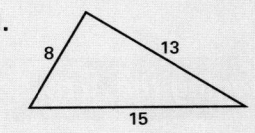
acute

2.
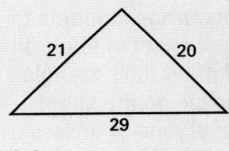
right

3.
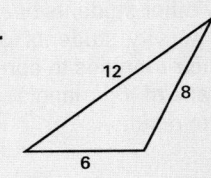
obtuse

4. Madeline makes this picture frame with a diagonal cross piece attached to the back for support. Can you tell from the dimensions whether or not the corners of the frame are right angles? Explain.

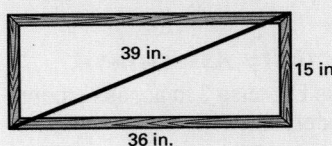

Yes, the frame corners are right angles because $36^2 + 15^2 = 39^2$.

Planning the Activity

Goal

Students use paper folding to explore the concurrence of medians of a triangle.

Materials

See the margin of the student page.

LINK TO LESSON

In Example 2 on page 208, students apply what they discover in this activity to find measures in a triangle.

2 Managing the Activity

Tips for Success

In Step 1, encourage students to use the largest sheet of paper they can find and draw their triangle as large as possible on the sheet. Encourage everyone to draw a triangle that is shaped differently from those drawn by other students near them. After the activity, students can compare their triangles to confirm that the shape of the triangle did not affect the results.

3 Closing the Activity

KEY DISCOVERY

The medians of a triangle intersect at a point that is two thirds of the distance from each vertex to the midpoint of the opposite side.

Activity Assessment

Use Exercise 2 to assess student understanding.

Activity 4.6 Intersecting Medians

Question

What is the relationship between segments formed by the intersection of the medians of a triangle?

Materials
- straightedge
- scissors
- ruler

Explore

1 On a piece of paper, draw a triangle and cut it out.

2 Find the midpoint of each side by folding each side, vertex to vertex, and pinching the paper at the middle.

3 Draw a segment from each midpoint to the vertex of the opposite angle. These segments are called *medians*.

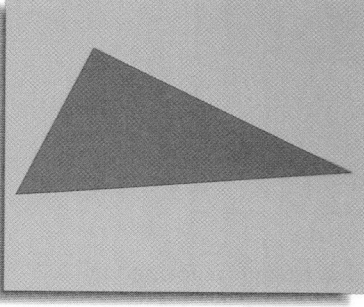  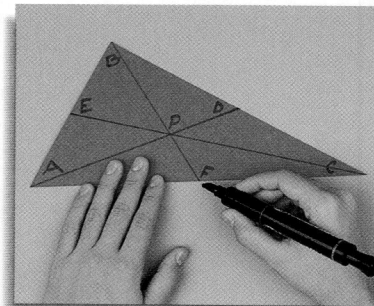

1. Answers will vary, but in each column the second and third entries should be approximately equal.

2. Yes; the distance from *P* to a vertex is equal to two thirds the distance from that vertex to the midpoint of the opposite side.

 Student Help

SKILLS REVIEW
For help with multiplying fractions, see p. 659.

3. The results are the same for any triangle.

Think About It

1. Label the triangle as shown above. Copy and complete the table.

Length of median	$AD = ?$	$BF = ?$	$CE = ?$
Length of segment from *P* to vertex	$AP = ?$	$BP = ?$	$CP = ?$
Median length multiplied by $\frac{2}{3}$	$\frac{2}{3}AD = ?$	$\frac{2}{3}BF = ?$	$\frac{2}{3}CE = ?$

2. Is there a relationship between the distance from *P* to a vertex and the distance from that vertex to the midpoint of the opposite side?

3. Extension Repeat Steps 1 through 3 with triangles of different shapes and sizes and complete a table like the one in Exercise 1. Do your results differ? Explain.

4.6 Medians of a Triangle

Goal
Identify medians in triangles.

Key Words
- median of a triangle
- centroid

A cardboard triangle will balance on the end of a pencil if the pencil is placed at a particular point on the triangle. Finding balancing points of objects is important in engineering, construction, and science.

A **median of a triangle** is a segment from a vertex to the midpoint of the opposite side.

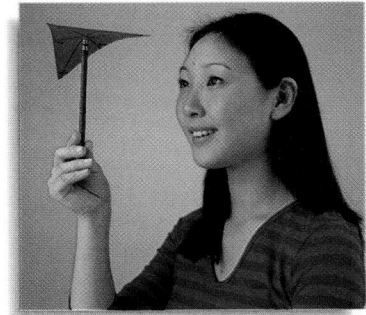

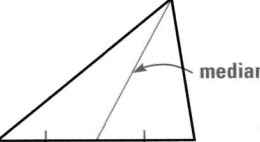 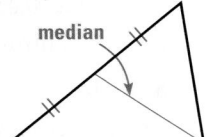 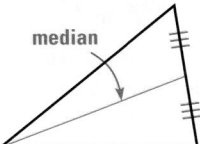

EXAMPLE ① Draw a Median

In △STR, draw a median from S to its opposite side.

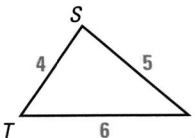

Solution

The side opposite ∠S is $\overline{TR}$.

Find the midpoint of $\overline{TR}$, and label it P. Then draw a segment from point S to point P. $\overline{SP}$ is a median of △STR.

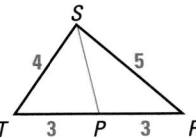

Checkpoint ✓ Draw a Median

Copy the triangle and draw a median. 1–3. See margin.

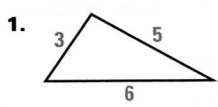 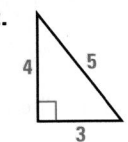 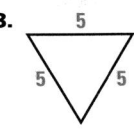

4.6 *Medians of a Triangle* **207**

① Plan

Pacing
Suggested Number of Days

Basic: 2 days
Average: 2 days
Advanced: 2 days
Block Schedule: 1 block

Teaching Resources

📄 **Blacklines**
(See page 170B.)

🖐 **Transparencies**
- Warm-Up with Quiz
- Answers

🖥 **Technology**
- Electronic Teacher Tools
- Test and Practice Generator
- Online Lesson Planner
- Internet Support

📼 **Video**
- Geometry in Motion

② Teach

Content and Teaching Strategies
For background information on geometric concepts and teaching strategies related to this lesson, see pages 170E and 170F in this Teacher's Edition.

Tips for New Teachers
Consider doing a demonstration of the centroid as the balancing point of a triangle. Cut out a few different types of triangles using stiff paper. Locate and mark the centroid. Show that the triangles will balance on the tip of a pencil. See the Tips for New Teachers on pp. 1–2 of the *Chapter 4 Resource Book* for additional notes about Lesson 4.6.

Extra Example 1
See next page.

1–3. See Additional Answers beginning on page AA1.

207

Study Skills

Vocabulary Have students look up the everyday meaning of the word *concurrent*. Relate this definition to the mathematical definitions of *concurrent* and *point of concurrence*.

Extra Example 1

In △*XYZ*, draw a median from *X* to its opposite side.

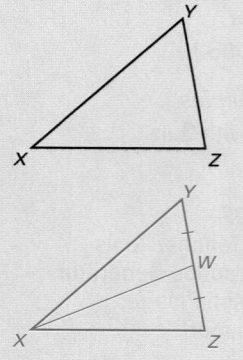

$\overline{XW}$ is a median of △*XYZ*.

Extra Example 2

Point *E* is the centroid of △*ABC* and *BF* = 24. Find *EF* and *BE*.

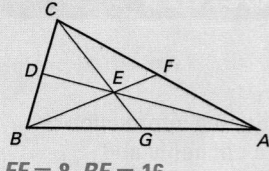

EF = 8, *BE* = 16

Extra Example 3

Point *K* is the centroid of △*FGH* and *KH* = 12. Find the length of $\overline{JH}$. 18

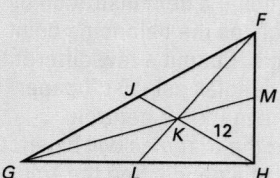

The following theorem tells you that the three medians of a triangle intersect at one point. This point is called the **centroid** of the triangle.

THEOREM 4.9

Intersection of Medians of a Triangle

Words The medians of a triangle intersect at the centroid, a point that is two thirds of the distance from each vertex to the midpoint of the opposite side.

Symbols If *P* is the centroid of △*ABC*, then

$$AP = \frac{2}{3}AD, \quad BP = \frac{2}{3}BF, \quad \text{and} \quad CP = \frac{2}{3}CE.$$

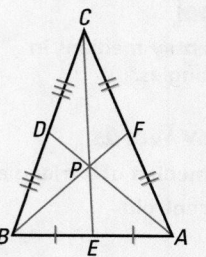

Student Help
CLASSZONE.COM

MORE EXAMPLES
More examples at classzone.com

EXAMPLE 2 Use the Centroid of a Triangle

E is the centroid of △*ABC* and *DA* = 27. Find *EA* and *DE*.

Solution

Using Theorem 4.9, you know that

$$EA = \frac{2}{3}DA = \frac{2}{3}(27) = 18.$$

Now use the Segment Addition Postulate to find *ED*.

$DA = DE + EA$	Segment Addition Postulate
$27 = DE + 18$	Substitute 27 for *DA* and 18 for *EA*.
$27 - 18 = DE + 18 - 18$	Subtract 18 from each side.
$9 = DE$	Simplify.

ANSWER ▶ $\overline{EA}$ has a length of 18 and $\overline{DE}$ has a length of 9.

EXAMPLE 3 Use the Centroid of a Triangle

P is the centroid of △*QRS* and *RP* = 10. Find the length of $\overline{RT}$.

Solution

$RP = \frac{2}{3}RT$	Use Theorem 4.9.
$10 = \frac{2}{3}RT$	Substitute 10 for *RP*.
$\frac{3}{2}(10) = \frac{3}{2}\left(\frac{2}{3}RT\right)$	Multiply each side by $\frac{3}{2}$.
$15 = RT$	Simplify.

ANSWER ▶ The median $\overline{RT}$ has a length of 15.

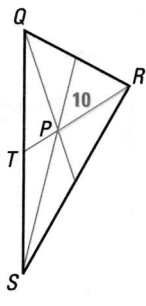

The centroid of the triangle is shown. Find the lengths.

4. Find *BE* and *ED*, given *BD* = 24.

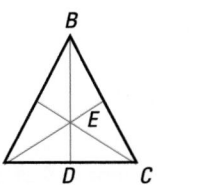

BE = 16; *ED* = 8

5. Find *JG* and *KG*, given *JK* = 4.

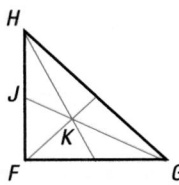

JG = 12; *KG* = 8

6. Find *PQ* and *PN*, given *QN* = 20.

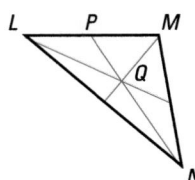

PQ = 10; *PN* = 30

4.6 Exercises

Guided Practice

Vocabulary Check

In Exercises 1 and 2, complete the statement.

1. The segment from a vertex of a triangle to the midpoint of the opposite side is a(n) __?__ . median

2. The __?__ is the point where the three medians intersect. centroid

3. Copy △*ABC*, then draw the median from point *A* to $\overline{BC}$. See margin.

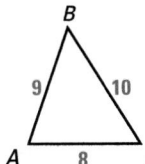

Skill Check

$\overline{BD}$ **is a median of △*ABC*. Find the length of $\overline{AD}$.**

4.

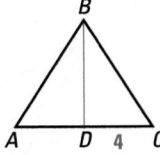

5.

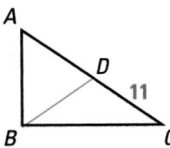

6.

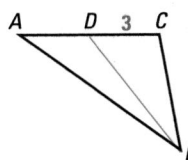

7. *T* is the centroid of △*PQR* and *PS* = 33. Find *PT* and *ST*.

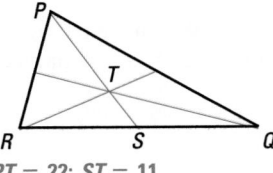

PT = 22; *ST* = 11

8. *E* is the centroid of △*ABC* and *BE* = 12. Find *BD* and *ED*.

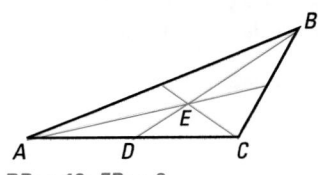

BD = 18; *ED* = 6

What is a median of a triangle and how does it relate to the centroid of the triangle? A median of a triangle is a segment from a vertex to the midpoint of the opposite side. The three medians of a triangle intersect at a point called the centroid of the triangle.

🏆 *Daily Puzzler*

Draw a triangle with two medians that are perpendicular to each other.

Sample answer:

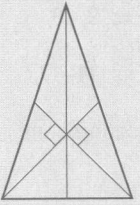

3.

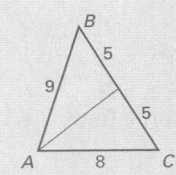

Practice and Applications

Extra Practice
See p. 682.

In Exercises 9 and 10, copy the triangle and draw the three medians of the triangle. 9, 10. See margin.

9.    **10.**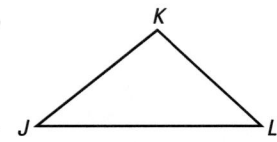

Using a Centroid *P* **is the centroid of** $\triangle LMN$. **Find** *PN* **and** *QP*.

11. $QN = 9$ **12.** $QN = 21$ **13.** $QN = 30$

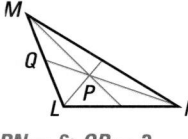   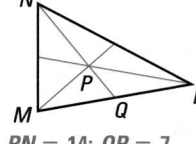   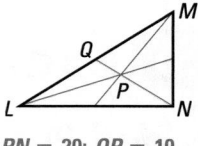

$PN = 6; QP = 3$ $PN = 14; QP = 7$ $PN = 20; QP = 10$

Using a Centroid *D* **is the centroid of** $\triangle ABC$. **Find** *CD* **and** *CE*.

14. $DE = 5$ **15.** $DE = 11$ **16.** $DE = 9$

   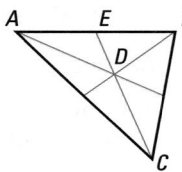   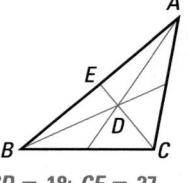

$CD = 10; CE = 15$ $CD = 22; CE = 33$ $CD = 18; CE = 27$

17. Error Analysis *D* is the centroid of $\triangle ABC$. Your friend wants to find *DE*. The median *AE* = 18. Find and correct the error. Explain your reasoning.
The equation should be $AD = \frac{2}{3}AE$, so $AD = \frac{2}{3}(18) = 12$, and $DE = 18 - 12 = 6$.

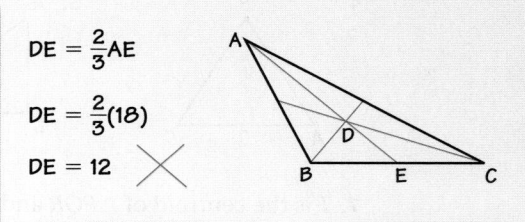

$DE = \frac{2}{3}AE$

$DE = \frac{2}{3}(18)$

$DE = 12$

Homework Help

Example 1: Exs. 9, 10, 18
Example 2: Exs. 11–13, 17
Example 3: Exs. 14–16

18. Finding a Centroid Draw a triangle and find the midpoint of each side. Then draw the three medians of your triangle. Label the centroid *P*. See margin.

Apply

Assignment Guide

BASIC
Day 1: pp. 210–211 Exs. 9, 10, 18, 22–40 even
Day 2: pp. 210–211 Exs. 11–16, 23–41 odd
AVERAGE
Day 1: pp. 210–211 Exs. 9, 10, 18, 24–29, 30–40 even
Day 2: pp. 210–211 Exs. 11–17, 19–23, 31–41 odd
ADVANCED
Day 1: pp. 210–211 Exs. 9, 10, 18, 24–29, 30–40 even
Day 2: pp. 210–211 Exs. 11–17, 19–23, 31–41 odd;
EC: TE p. 170D*, classzone.com
BLOCK SCHEDULE
pp. 210–211 Exs. 9–41

Extra Practice

• Student Edition, p. 682
• Chapter 4 Resource Book, pp. 59–60

Homework Check

To quickly check student understanding of key concepts, go over the following exercises:

Basic: 9, 11, 12, 14, 15
Average: 9, 11, 12, 14, 15
Advanced: 10, 13, 15, 16, 18

✗ Common Error

In Exercises 14–16, watch for students who feel that the centroid in these diagrams is halfway between the vertex and its opposite side. Remind students that diagrams can be misleading and that they must base their answers on calculations rather than an estimate made by just studying the diagram.

9.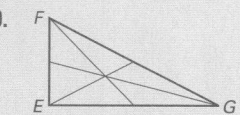

210

 **Using Algebra** Copy the graph shown.

19. Find the coordinates of Q, R, and S, the midpoints of the sides of the triangle. $Q(5, 0)$; $R(2, 2)$; $S(8, 4)$

20. Find the length of each median. $PQ = 6$; $NR = 9$; $MS = \sqrt{117} \approx 10.8$

21. Find the coordinates of the centroid. Label this point as T. $(5, 2)$

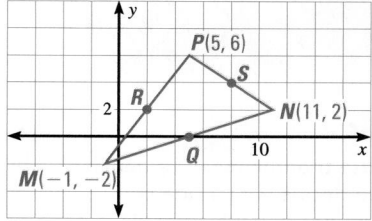

Standardized Test Practice

22. **Multiple Choice** In the figure shown, N is the centroid of $\triangle JKL$ and $KM = 36$. What is the length of $\overline{MN}$? B

Ⓐ 9   Ⓑ 12

Ⓒ 16   Ⓓ 24

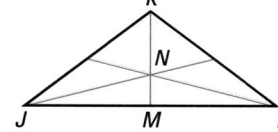

23. **Multiple Choice** In the figure shown, P is the centroid of $\triangle RST$ and $PT = 12$. What is the length of $\overline{VT}$? G

Ⓕ 8   Ⓖ 18

Ⓗ 24   Ⓙ 36

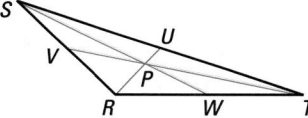

Mixed Review

Finding Angle Measures in a Triangle Find $m\angle 1$. *(Lesson 4.2)*

24.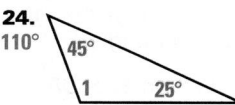

25.

26.

27.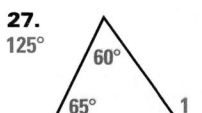

28.

29.

Algebra Skills

Writing Equivalent Fractions Write two fractions equivalent to the given fraction. *(Skills Review, p. 656)* 30–41. Sample answers are given.

30. $\frac{1}{2}$ $\frac{2}{4}$, $\frac{3}{6}$

31. $\frac{1}{5}$ $\frac{2}{10}$, $\frac{10}{50}$

32. $\frac{3}{4}$ $\frac{6}{8}$, $\frac{15}{20}$

33. $\frac{8}{14}$ $\frac{4}{7}$, $\frac{80}{140}$

34. $\frac{12}{26}$ $\frac{6}{13}$, $\frac{18}{39}$

35. $\frac{2}{20}$ $\frac{1}{10}$, $\frac{5}{50}$

36. $\frac{8}{36}$ $\frac{2}{9}$, $\frac{80}{360}$

37. $\frac{10}{45}$ $\frac{2}{9}$, $\frac{4}{18}$

38. $\frac{14}{35}$ $\frac{2}{5}$, $\frac{28}{70}$

39. $\frac{18}{24}$ $\frac{6}{8}$, $\frac{3}{4}$

40. $\frac{20}{30}$ $\frac{2}{3}$, $\frac{4}{6}$

41. $\frac{24}{33}$ $\frac{8}{11}$, $\frac{48}{66}$

4.6 *Medians of a Triangle* **211**

Mini-Quiz

1. $\overline{BD}$ is a median of $\triangle ABC$. Find the length of $\overline{AD}$. 7

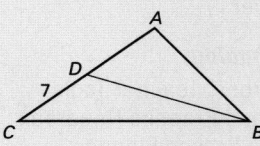

2. Point P is the centroid of $\triangle LMN$ and $QN = 45$. Find PN and QP. $PN = 30$, $QP = 15$

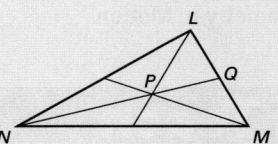

3. Point D is the centroid of $\triangle ABC$ and $DE = 14$. Find CD and CE. $CD = 28$, $CE = 42$

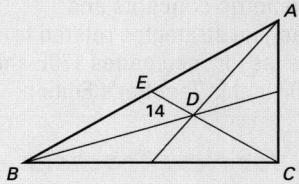

10.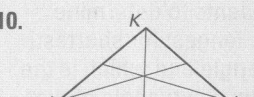

18. *Sample answer:*

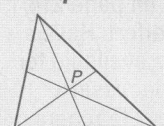

Pacing
Suggested Number of Days

Basic: 2 days
Average: 2 days
Advanced: 2 days
Block Schedule: 1 block

Teaching Resources

 **Blacklines**
(See page 170B.)

🖐 **Transparencies**
• Warm-Up with Quiz
• Answers

⊞ **Technology**
• Electronic Teacher Tools
• Test and Practice Generator
• Online Lesson Planner
• Internet Support

📼 **Video**
• **Geometry in Motion**

②Teach

Content and Teaching Strategies
For background information on geometric concepts and teaching strategies related to this lesson, see pages 170E and 170F in this Teacher's Edition.

Tips for New Teachers

With a little practice it is usually easy for students to determine which is the longest (or shortest) side of a triangle and which is the largest (or smallest) angle of that triangle. However, for problems like those in Example 3 on page 214, it takes a little more effort. Have a variety of examples available for further practice if necessary. See the Tips for New Teachers on pp. 1–2 of the *Chapter 4 Resource Book* for additional notes about Lesson 4.7.

4.7 Triangle Inequalities

Goal
Use triangle measurements to decide which side is longest and which angle is largest.

The diagrams below show a relationship between the longest and shortest sides of a triangle and the largest and smallest angles.

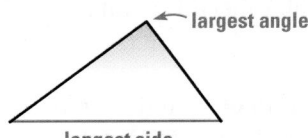
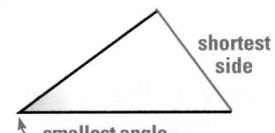

THEOREMS 4.10 and 4.11

Theorem 4.10

Words If one side of a triangle is longer than another side, then the angle opposite the longer side is larger than the angle opposite the shorter side.

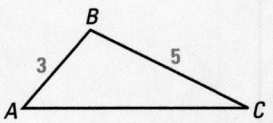

Symbols If $BC > AB$, then $m\angle A > m\angle C$.

Theorem 4.11

Words If one angle of a triangle is larger than another angle, then the side opposite the larger angle is longer than the side opposite the smaller angle.

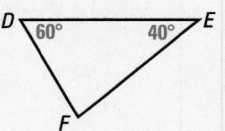

Symbols If $m\angle D > m\angle E$, then $EF > DF$.

EXAMPLE 1 Order Angle Measures

Name the angles from largest to smallest.

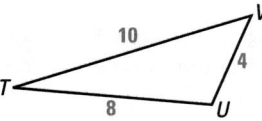

Solution

$TV > TU$, so $m\angle U > m\angle V$. Also, $TU > UV$, so $m\angle V > m\angle T$.

ANSWER ▶ The order of the angles from largest to smallest is $\angle U, \angle V, \angle T$.

Student Help
CLASSZONE.COM

MORE EXAMPLES
More examples at
classzone.com

EXAMPLE 2 Order Side Lengths

Name the sides from longest to shortest.

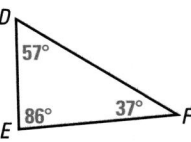

Solution

$m\angle E > m\angle D$, so $DF > FE$. Also, $m\angle D > m\angle F$, so $FE > DE$.

ANSWER ▶ The order of the sides from longest to shortest is $\overline{DF}, \overline{FE}, \overline{DE}$.

 Checkpoint ✓ *Order Angle Measures and Side Lengths*

Name the angles from largest to smallest.

1.
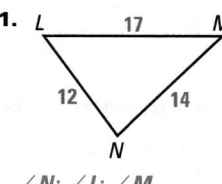
$\angle N; \angle L; \angle M$

2.
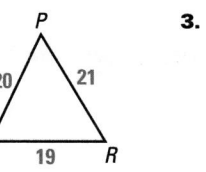
$\angle Q; \angle R; \angle P$

3.
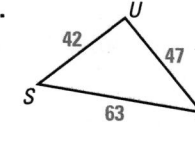
$\angle U; \angle S; \angle T$

Name the sides from longest to shortest.

4.
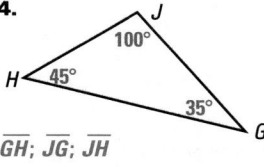
$\overline{GH}; \overline{JG}; \overline{JH}$

5.
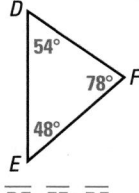
$\overline{DE}; \overline{EF}; \overline{DF}$

6.
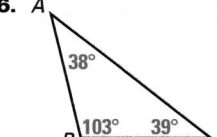
$\overline{AC}; \overline{AB}; \overline{BC}$

Segments of a Triangle Not every group of three segments can be used to form a triangle. The lengths of the segments must have the following relationship.

THEOREM 4.12

Triangle Inequality

Words The sum of the lengths of any two sides of a triangle is greater than the length of the third side.

Symbols
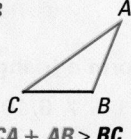

$CA + AB > \mathbf{BC}$ $AB + BC > \mathbf{CA}$ $BC + CA > \mathbf{AB}$

4.7 Triangle Inequalities **213**

Extra Example 1

Name the angles from largest to smallest. $\angle A, \angle C, \angle B$

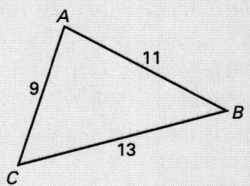

Extra Example 2

Name the sides from longest to shortest. $\overline{RT}, \overline{ST}, \overline{RS}$

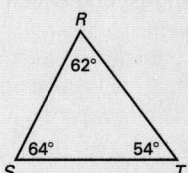

Extra Example 3

Can the side lengths form a triangle? Explain.

a. 6, 7, 15 These lengths do not form a triangle, because $6 + 7 < 15$.

b. 6, 8, 14 These lengths do not form a triangle, because $6 + 8 = 14$.

c. 6, 9, 13 These lengths do form a triangle, because $6 + 9 > 13$, $6 + 13 > 9$, and $9 + 13 > 6$.

✓ Concept Check

How can you tell from the angle measures of a triangle which side of the triangle is longest and which is shortest? The longest side is opposite the angle with the greatest measure; the shortest side is opposite the angle with the least measure.

🐢 Daily Puzzler

A triangle has sides of length x, $x + 1$, and $x + 2$, where $0 < x \le 9$. What integer value of x does NOT produce a triangle? 1

EXAMPLE 3 **Use the Triangle Inequality**

Can the side lengths form a triangle? Explain.

a. 3, 5, 9   **b.** 3, 5, 8   **c.** 3, 5, 7

Solution

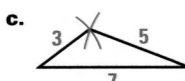

a. These lengths do not form a triangle, because $3 + 5 < 9$.

b. These lengths do not form a triangle, because $3 + 5 = 8$.

c. These lengths do form a triangle, because $3 + 5 > 7$, $3 + 7 > 5$, and $5 + 7 > 3$.

Checkpoint ✓ Use the Triangle Inequality

Can the side lengths form a triangle? Explain.

7. 5, 7, 13
No; $5 + 7 < 13$.

8. 6, 9, 12
Yes; $6 + 9 > 12$, $6 + 12 > 9$, and $9 + 12 > 6$.

9. 10, 15, 25
No; $10 + 15 = 25$.

4.7 Exercises

Guided Practice

Vocabulary Check

1. Complete the statement: The symbol ">" means ___?___ , and the symbol "<" means ___?___ . greater than; less than

Skill Check

2. Name the smallest angle of $\triangle ABC$. $\angle A$

3. Name the longest side of $\triangle ABC$. $\overline{AC}$

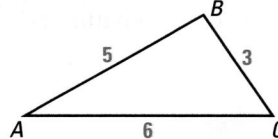

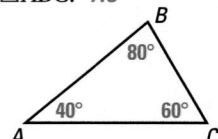

7. Yes; $6 + 10 > 15$, $6 + 15 > 10$, and $10 + 15 > 6$.

9. Yes; $7 + 8 > 13$, $7 + 13 > 8$, and $8 + 13 > 7$.

In Exercises 4 and 5, use the figure shown at the right.

4. Name the smallest and largest angles of $\triangle DEF$.
smallest, $\angle D$; largest, $\angle F$

5. Name the shortest and longest sides of $\triangle DEF$.
shortest, $\overline{EF}$; longest, $\overline{DE}$

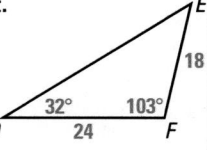

Homework Help

Example 1: Exs. 12–14,
18–24, 37, 38
Example 2: Exs. 15–17,
24–31, 37, 38
Example 3: Exs. 25,
32–36, 39–43

Can the side lengths form a triangle? Explain.

6. 1, 2, 3 No; $1 + 2 = 3$.
7. 6, 10, 15 See margin.
8. 12, 16, 30 No; $12 + 16 < 30$.

9. 7, 8, 13 See margin.
10. 4, 9, 16 No; $4 + 9 < 16$.
11. 5, 5, 10 No; $5 + 5 = 10$.

Practice and Applications

Extra Practice

See p. 682.

Comparing Angle Measures Name the smallest and largest angles of the triangle.

12.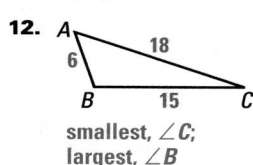
smallest, ∠C;
largest, ∠B

13.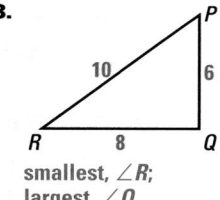
smallest, ∠R;
largest, ∠Q

14.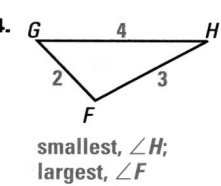
smallest, ∠H;
largest, ∠F

Comparing Side Lengths Name the shortest and longest sides of the triangle.

15.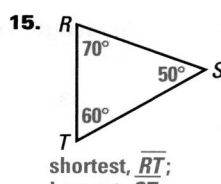
shortest, $\overline{RT}$;
longest, $\overline{ST}$

16.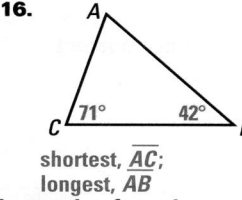
shortest, $\overline{AC}$;
longest, $\overline{AB}$

17.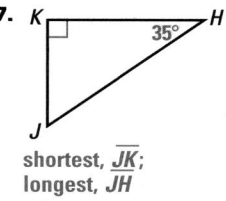
shortest, $\overline{JK}$;
longest, $\overline{JH}$

Ordering Angles Name the angles from largest to smallest.

18.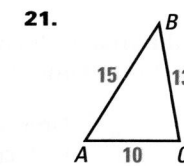
∠M; ∠K; ∠L

19.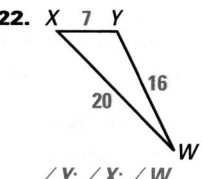
∠P; ∠Q; ∠N

20.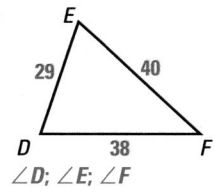
∠R; ∠S; ∠T

21.
15 13
A 10 C
∠C; ∠A; ∠B

22. X 7 Y
20 16
W
∠Y; ∠X; ∠W

23. E
29 40
D 38 F
∠D; ∠E; ∠F

24. The sides and angles cannot be labeled as shown because, for example, the longest side should be opposite the largest angle.

25. No; 3 + 5 < 9, so the side lengths do not satisfy the Triangle Inequality.

Design In Exercises 24 and 25, use the following information.
The term "kitchen triangle" refers to the imaginary triangle formed by the refrigerator, the sink, and the stove. The distances shown are measured in feet. **See margin.**

24. What is wrong with the labels on the kitchen triangle?

25. Can a kitchen triangle have the following side lengths: 9 feet, 3 feet, and 5 feet? Explain why or why not.

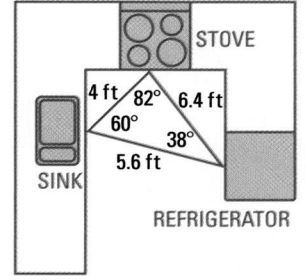

4.7 Triangle Inequalities 215

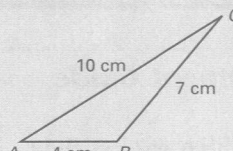
Ordering Sides Name the sides from longest to shortest.

26.
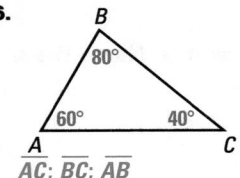
$\overline{AC}$; $\overline{BC}$; $\overline{AB}$

27.
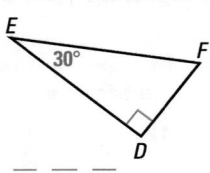
$\overline{EF}$; $\overline{DE}$; $\overline{DF}$

28.
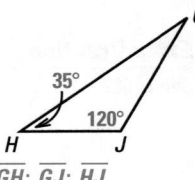
$\overline{GH}$; $\overline{GJ}$; $\overline{HJ}$

29.
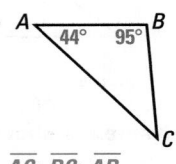
$\overline{AC}$; $\overline{BC}$; $\overline{AB}$

30.
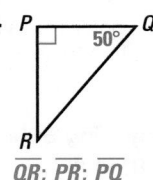
$\overline{QR}$; $\overline{PR}$; $\overline{PQ}$

31.
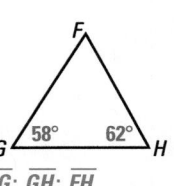
$\overline{FG}$; $\overline{GH}$; $\overline{FH}$

Error Analysis Explain why the side lengths given with the triangles are not correct.

32. $2 + 3 = 5$, so the side lengths do not satisfy the Triangle Inequality.

33. $3 + 10 < 14$, so the side lengths do not satisfy the Triangle Inequality.

32.
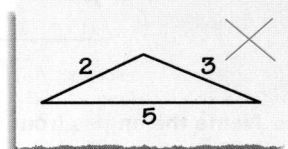

33.
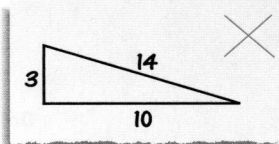

EXAMPLE *Use the Triangle Inequality*

Is it possible to draw a triangle that has side lengths of 4, 5, and 6? If so, draw the triangle.

Solution

Yes, these side lengths satisfy the Triangle Inequality: $4 + 5 > 6$, $5 + 6 > 4$, and $4 + 6 > 5$. So, it is possible to draw the triangle, as shown below.

❶ Mark $\overline{AB}$ of length 4 cm on a line. Then draw an arc of radius 5 cm with center at B.

❷ Draw an arc of radius 6 cm with center at A. Mark the intersection of the two arcs as C. $\triangle ABC$ has side lengths of 4 cm, 5 cm, and 6 cm.

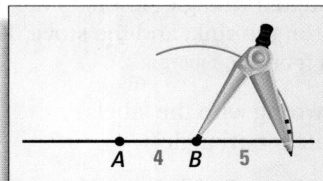

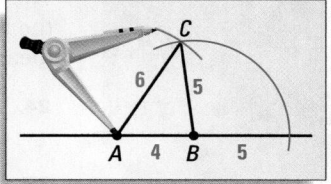

Using the Triangle Inequality Determine whether it is possible to draw a triangle with the given side lengths. If so, draw the triangle.

34. 4, 7, 10 See margin.　**35.** 10, 12, 22 no　**36.** 17, 9, 30 no

Student Help

VISUAL STRATEGY
In Exs. 37 and 38, draw a sketch with measurements that are roughly correct, as shown on p. 172.

39. The diagonal and the sidewalks along Pine St. and Union St. form a triangle. The walk along the diagonal is shorter than staying on the sidewalks by the Triangle Inequality.

Visualize It! Sketch a triangle and label it with the given angle measures and side lengths. 37, 38. See margin.

37. Angles: 59°, 46°, 75°
Sides: 13 cm, 9.7 cm, 11.5 cm

38. Angles: 135°, 15°, 30°
Sides: 7.1 cm, 2.6 cm, 5 cm

39. Taking a Shortcut Suppose you are walking south on the sidewalk of Pine Street. When you reach Pleasant Street, you cut across the empty lot to go to the corner of Oak Hill Avenue and Union Street. Explain why this route is shorter than staying on the sidewalks.

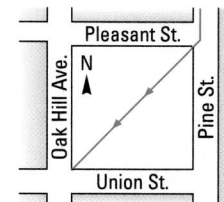

40. 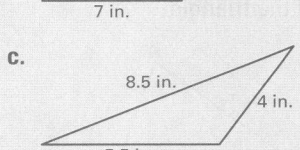 **You be the Judge** Suppose you are camping. You decide to hike 4.6 miles northwest and then turn and hike 1.8 miles east. Your friend tells you that you are about one and a half miles from camp. Is your friend right? Explain why or why not.

No; 1.5 + 1.8 < 4.6; by the Triangle Inequality, you must be more than 2.8 miles from camp.

Logical Reasoning In Exercises 41–43, use the figure shown and the given information.

By adjusting the length of the boom lines from *A* to *B*, the operator of the crane shown can raise and lower the boom. Suppose the mast $\overline{AC}$ is 50 feet long and the boom $\overline{BC}$ is 100 feet long.

41. Is the boom *raised* or *lowered* when the boom lines are shortened?
raised

42. *AB* must be less than ___?___ feet. 150

43. As the boom is raised or lowered, is ∠*ACB* ever larger than ∠*BAC*? Explain. Yes; when the boom is lowered and *AB* is greater than 100, *AB* > *BC* and so m∠*ACB* will be larger than ∠*BAC*.

Standardized Test Practice

44. Multi-Step Problem You are given an 18-inch piece of wire. You want to bend the wire to form a triangle so that the length of each side is a whole number. a–c. See margin.

a. Sketch four possible isosceles triangles and label each side length.

b. Sketch a possible acute scalene triangle.

c. Sketch a possible obtuse scalene triangle.

d. List three combinations of segment lengths with a sum of 18 that will not produce triangles. *Sample answer:* 4 in., 5 in., 9 in.; 4 in., 4 in., 10 in.; 3 in., 6 in., 9 in.

37.

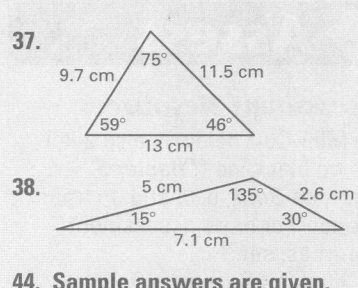

38.

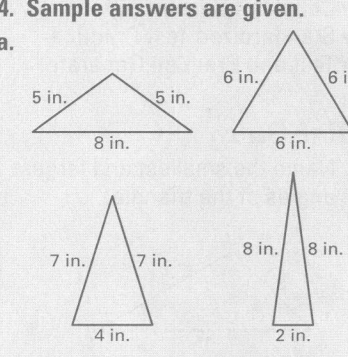

44. Sample answers are given.

a.

5 in. / 5 in. / 8 in.

6 in. / 6 in. / 6 in.

7 in. / 7 in. / 4 in.

8 in. / 8 in. / 2 in.

b.
6 in. / 5 in. / 7 in.

c.
8.5 in. / 4 in. / 5.5 in.

4.7 Triangle Inequalities **217**

Assess

Assessment Resources

The Mini-Quiz below is also available on blackline (*Chapter 5 Resource Book*, p. 9) and on transparency. For more assessment resources, see:
- Chapter 4 Resource Book
- Standardized Test Practice
- Test and Practice Generator

Mini-Quiz

1. Name the smallest and largest angles of the triangle.

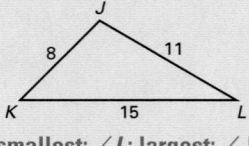

smallest: ∠L; largest: ∠J

Name the shortest and longest sides of the triangle.

2.

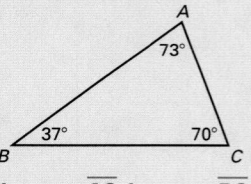

shortest: $\overline{AC}$, longest: $\overline{BC}$

3.

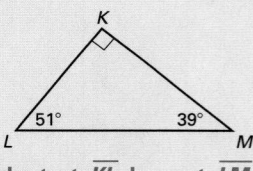

shortest: $\overline{KL}$, longest: $\overline{LM}$

4. Everett noticed that three streets in his town form a triangle. He measured each distance and made this diagram. Are his measurements correct? Explain why or why not.

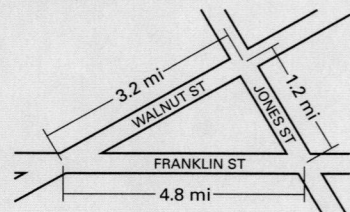

The measurements cannot be correct because 3.2 miles + 1.2 miles < 4.8 miles.

218

Mixed Review

Identifying Parts of a Triangle In Exercises 45–48, use the figure shown to complete the statement. *(Lessons 4.1, 4.3, 4.4)*

45. __?__ is the hypotenuse of △RST. $\overline{RT}$

46. In △RST, $\overline{RT}$ is the side __?__ ∠RST. **opposite**

47. The legs of △RST are __?__ and __?__. $\overline{RS}, \overline{ST}$

48. __?__ is the base of △RST. $\overline{RT}$

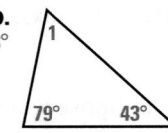

Finding Measures Find the measure of the numbered angle. *(Lesson 4.2)*

49.
58° 1
79° 43°

50.
127° 28° 2 25°
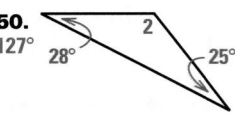

51.
34° 56° 3
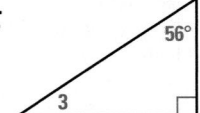

Algebra Skills

Solving Proportions Solve the proportion. *(Skills Review, p. 660)*

52. $\frac{x}{5} = \frac{6}{15}$ **2**   **53.** $\frac{18}{3} = \frac{6}{x}$ **1**   **54.** $\frac{x}{7} = \frac{6}{7}$ **6**

55. $\frac{27}{21} = \frac{9}{x}$ **7**   **56.** $\frac{5}{8} = \frac{x}{72}$ **45**   **57.** $\frac{7}{10} = \frac{49}{x}$ **70**

Quiz 3

Use the side lengths to classify the triangle as *acute, right,* or *obtuse.* *(Lesson 4.5)*

1. 6, 11, 14 **obtuse**   **2.** 15, 7, 16 **acute**   **3.** 18, 80, 82 **right**

N is the centroid of △JKL. Find KN and MN. *(Lesson 4.6)*

4. KM = 6   **5.** KM = 39   **6.** KM = 60

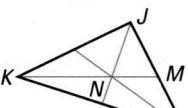
KN = 4; MN = 2

KN = 26; MN = 13

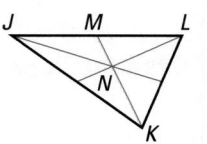
KN = 40; MN = 20

Name the sides from longest to shortest. *(Lesson 4.7)*

7.
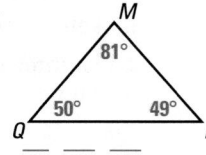
$\overline{QM}; \overline{LM}; \overline{LQ}$

8.
$\overline{PQ}; \overline{MP}; \overline{MQ}$

9.
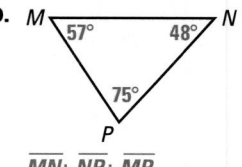
$\overline{MN}; \overline{NP}; \overline{MP}$

218 **Chapter 4** *Triangle Relationships*

Chapter 4 Summary and Review

Additional Resources

The following resources are available to help review the materials in this chapter.

📘 *Chapter 4 Resource Book*
• Chapter Review Games and Activities, p. 75
• Cumulative Review, Chs. 1–4

VOCABULARY

- **triangle**, *p. 173*
- **equilateral triangle**, *p. 173*
- **isosceles triangle**, *p. 173*
- **scalene triangle**, *p. 173*
- **equiangular triangle**, *p. 174*
- **acute triangle**, *p. 174*
- **right triangle**, *p. 174*
- **obtuse triangle**, *p. 174*

- **vertex**, *p. 175*
- **corollary**, *p. 180*
- **interior angle**, *p. 181*
- **exterior angle**, *p. 181*
- **legs of an isosceles triangle**, *p. 185*
- **base of an isosceles triangle**, *p. 185*

- **base angles of an isosceles triangle**, *p. 185*
- **legs of a right triangle**, *p. 192*
- **hypotenuse**, *p. 192*
- **Pythagorean Theorem**, *p. 192*
- **Distance Formula**, *p. 194*
- **median of a triangle**, *p. 207*
- **centroid**, *p. 208*

VOCABULARY REVIEW

Fill in the blank.

1. A(n) __?__ is a figure formed by three segments joining three noncollinear points. **triangle**

2. The side opposite the right angle is the __?__ of a right triangle. **hypotenuse**

3. A(n) __?__ to a theorem is a statement that can be proved easily using the theorem. **corollary**

4. The congruent sides of an isosceles triangle are called __?__, and the third side is called the __?__. **legs; base**

5. A point that joins two sides of a triangle is called a(n) __?__. **vertex**

6. A segment from a vertex of a triangle to the midpoint of its opposite side is called a(n) __?__. **median**

7. The point at which the medians of a triangle intersect is called the __?__ of a triangle. **centroid**

4.1 CLASSIFYING TRIANGLES

Examples on pp. 173–175

> **EXAMPLES** Classify the triangle by its angles and by its sides.
>
>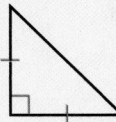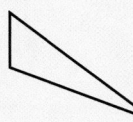
>
> **Acute isosceles** **Right isosceles** **Obtuse scalene**

Classify the triangle by its sides.

8.
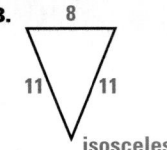
isosceles

9.
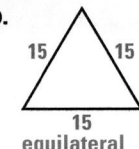
equilateral

10.
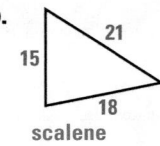
scalene

11. What kind of triangle has angle measures of 30°, 60°, and 90°? right

12. What kind of triangle has angle measures of 84°, 55°, and 41°? acute

13. What kind of triangle has side lengths of 4 feet, 8 feet, and 8 feet? isosceles

---

4.2 ANGLE MEASURES OF TRIANGLES

Examples on pp. 179–181

EXAMPLE   Given $m\angle 1 = 34°$ and $m\angle 2 = 86°$, find $m\angle 3$.

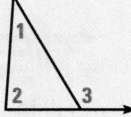

$m\angle 3 = m\angle 1 + m\angle 2$ Exterior Angle Theorem

$= 34° + 86°$ Substitute 34° for $m\angle 1$ and 86° for $m\angle 2$.

$= 120°$ Simplify.

Find $m\angle 1$.

14.
115°
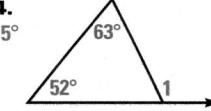
63°
52°
1

15.
142°
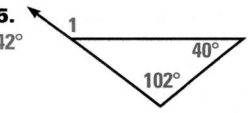
1
40°
102°

16.
137°
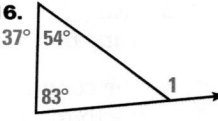
54°
83°
1

17. The measure of one interior angle of a triangle is 16°. The other interior angles are congruent. Find their measures. 82°, 82°

18. The measure of one of the interior angles of a right triangle is 31°. Find the measures of the other interior angles. 59°, 90°

4.3 ISOSCELES AND EQUILATERAL TRIANGLES

Examples on pp. 185–187

EXAMPLE   Find the value of x in the diagram.

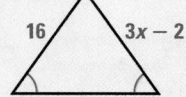
16
$3x - 2$

$3x - 2 = 16$ Converse of the Base Angles Theorem

$3x = 18$ Add 2 to each side.

$x = 6$ Divide each side by 3.

Find the value of *x*.

19.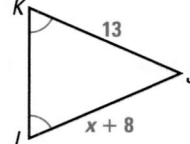
K
5
13
J
L
x + 8

20.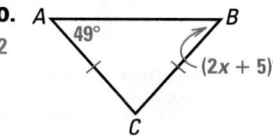
A
49°
B
22
(2*x* + 5)°
C

21.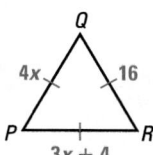
Q
4
4*x*
16
P
R
3*x* + 4

22.
K
60
x°
120°
J
L
M

23.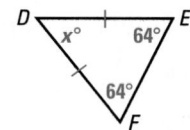
D
52
x°
64°
E
64°
F

24.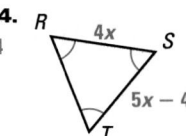
R
4
4*x*
S
5*x* − 4
T

4.4   THE PYTHAGOREAN THEOREM AND THE DISTANCE FORMULA

Examples on pp. 192–194

EXAMPLE   **Find the distance between *G*(3, 5) and *H*(7, −2).**

$$GH = \sqrt{(x_2 - x_1)^2 + (y_2 - y_1)^2}$$ Distance Formula

$$= \sqrt{(7 - 3)^2 + (-2 - 5)^2}$$ Substitute 7 for x_2, 3 for x_1, −2 for y_2, and 5 for y_1.

$$= \sqrt{4^2 + (-7)^2}$$ Simplify.

$$= \sqrt{16 + 49}$$ Multiply.

$$= \sqrt{65}$$ Add.

$$\approx 8.1$$ Approximate with a calculator.

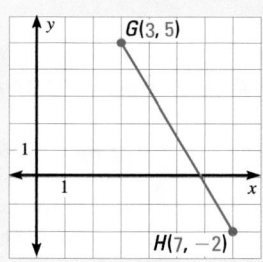

Find the unknown side length. Use a calculator to round your answer to the nearest tenth, if necessary.

25.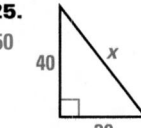
50
40
x
30

26.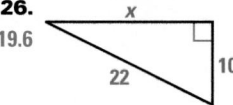
x
19.6
22
10

27.
14.8
16
6
x

Find the distance between the two points. Use a calculator to round your answer to the nearest tenth, if necessary.

28. *A*(0, 0)   5
       B(−3, 4)

29. *A*(2, 5)   9.8
       B(6, −4)

30. *A*(−8, 7)   11
       B(3, 7)

31. *A*(−4, −1)   8.1
       B(0, 6)

32. *A*(−2, −1)   7.2
       B(−6, −7)

33. *A*(8, −3)   12.2
       B(−2, 4)

34. *A*(9, 1)   13.9
       B(−3, −6)

35. *A*(5, 4)   5.4
       B(0, 6)

4.5 THE CONVERSE OF THE PYTHAGOREAN THEOREM

Examples on pp. 200–202

EXAMPLE Classify the triangle as *acute, right,* or *obtuse.*

Compare the square of the length of the longest side with the sum of the squares of the lengths of the two shorter sides.

$c^2 \stackrel{?}{=} a^2 + b^2$ Compare c^2 with $a^2 + b^2$.

$21^2 \stackrel{?}{=} 17^2 + 13^2$ Substitute 21 for c, 17 for a, and 13 for b.

$441 \stackrel{?}{=} 289 + 169$ Multiply.

$441 < 458$ Simplify.

$c^2 < a^2 + b^2$, so the triangle is acute.

Use the side lengths to classify the triangle as *acute, right,* or *obtuse.*

36. 12, 9, 15 right   **37.** 7, 11, 16 obtuse   **38.** 18, 19, 22 acute

39. 18, 42, 44 acute   **40.** 10, 3, 12 obtuse   **41.** 15, 21, 31 obtuse

4.6 MEDIANS OF A TRIANGLE

Examples on pp. 207–209

EXAMPLES Find the segment lengths.

a. D is the centroid of $\triangle ABC$ and $AE = 12$. Find AD and ED.

$AD = \frac{2}{3}AE$

$= \frac{2}{3}(12) = 8$

$AE = AD + ED$

$12 = 8 + ED$

$4 = ED$

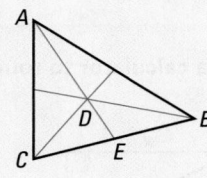

b. T is the centroid of $\triangle QRS$ and $RT = 18$. Find RU.

$RT = \frac{2}{3}RU$

$18 = \frac{2}{3}RU$

$\frac{3}{2}(18) = RU$

$27 = RU$

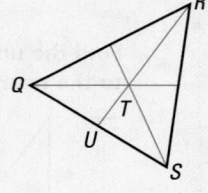

P is the centroid of $\triangle JKL$. Find KP and PM.

42. $KM = 18$

$KP = 12; PM = 6$

43. $KM = 42$

$KP = 28; PM = 14$

44. $KM = 120$

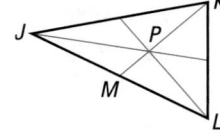

$KP = 80; PM = 40$

D is the centroid of △ABC. Find CE and DE.

45. CD = 8

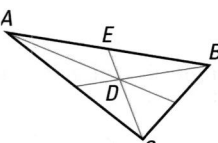

CE = 12; DE = 4

46. CD = 16

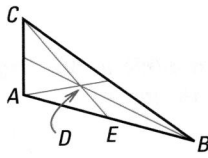

CE = 24; DE = 8

47. CD = 28

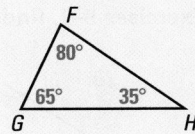

CE = 42; DE = 14

4.7 TRIANGLE INEQUALITIES

Examples on pp. 212–214

> **EXAMPLE** **Name the sides of the triangle shown from longest to shortest.**
>
> By Theorem 4.11, if one angle of a triangle is larger than another angle, then the side opposite the larger angle is longer than the side opposite the smaller angle.
>
> So, because $m\angle F > m\angle G > m\angle H$, $GH > FH > GF$. The sides from longest to shortest are $\overline{GH}$, $\overline{FH}$, and $\overline{GF}$.

Name the angles from largest to smallest.

48.

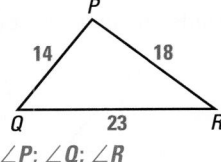

$\angle P; \angle Q; \angle R$

49.

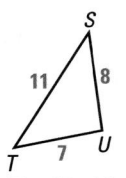

$\angle U; \angle T; \angle S$

50.

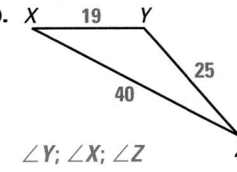

$\angle Y; \angle X; \angle Z$

Name the sides from longest to shortest.

51.

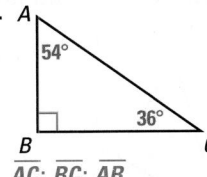

$\overline{AC}; \overline{BC}; \overline{AB}$

52.

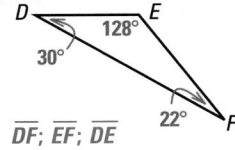

$\overline{DF}; \overline{EF}; \overline{DE}$

53.

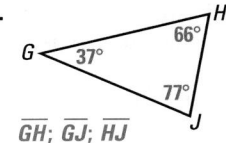

$\overline{GH}; \overline{GJ}; \overline{HJ}$

Determine whether it is possible to draw a triangle with the given side lengths. Explain your reasoning.

54. 10, 11, 20 See margin.
55. 21, 23, 25 See margin.
56. 3, 10, 15 No; 3 + 10 < 15.

57. 6, 6, 12 No; 6 + 6 = 12.
58. 13, 14, 15 See margin.
59. 2, 3, 4 See margin.

60. 4, 5, 9 No; 4 + 5 = 9.
61. 11, 11, 20
Yes; 11 + 11 > 20
and 11 + 20 > 11.
62. 14, 20, 38 No; 14 + 20 < 38.

Chapter Summary and Review **223**

Additional Resources

📖 **Chapter 4 Resource Book**
• Chapter Test (2 levels), pp. 76–79
• SAT/ACT Chapter Test, p. 80
• Alternative Assessment,
  pp. 81–82

🖥 **Test and Practice Generator**

Chapter  4 Chapter Test

Identify a triangle in the diagram that fits the given description.

1. obtuse $\triangle JLM$

2. acute $\triangle JKL$

3. equilateral $\triangle JKL$

4. isosceles $\triangle JLM$

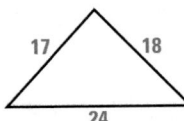

In Exercises 5–7, find the value of x.

5.
33

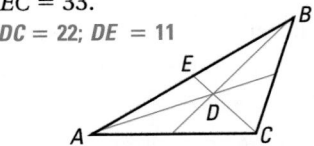

6.
11

7.
5

8. Find the length of the hypotenuse in the triangle shown. Use a calculator to round your answer to the nearest tenth. **17.2**

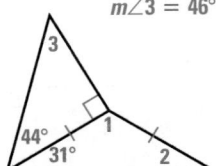

Plot the points and find the distance between them.

9. $P(0, 0)$, $Q(-6, -8)$ Check graphs; 10.

10. $P(2, 4)$, $Q(-2, 6)$ Check graphs; about 4.5.

11. Is the triangle shown below acute, right, or obtuse? **acute**

In Exercises 12 and 13, classify the triangle with the given side lengths as *acute*, *right*, or *obtuse*.

12. 6, 9, 13 obtuse   13. 12, 14, 20   obtuse

14. D is the centroid of $\triangle ABC$. Find DC and DE if $EC = 33$.
   $DC = 22$; $DE = 11$

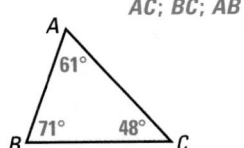

15. Name the sides from longest to shortest.
   $\overline{AC}$; $\overline{BC}$; $\overline{AB}$

16. Find the measures of all numbered angles in the figure below. $m\angle 1 = 118°$; $m\angle 2 = 31°$; $m\angle 3 = 46°$

Determine whether the side lengths can form a triangle.

17. 5, 8, 18 no   18. 20, 24, 40 yes

19. 7, 7, 14 no   20. 31, 45, 50 yes

Chapter 4 Standardized Test

Test Tip Put a mark next to questions you skip so that you will remember to come back to them later.

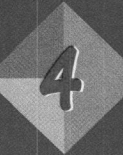

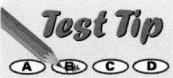

1. Classify the triangle by its angles and by its sides. D

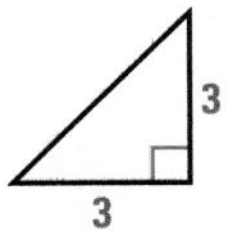

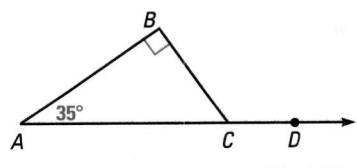

Ⓐ acute scalene

Ⓑ obtuse isosceles

Ⓒ acute equilateral

Ⓓ right isosceles

2. What is $m\angle BCD$? H

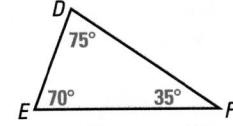

Ⓕ 35° Ⓖ 90°

Ⓗ 125° Ⓙ 180°

3. A triangle has two sides that have lengths of 14 feet and 22 feet. Which of the following lengths could *not* represent the length of the third side? A

Ⓐ 8 ft Ⓑ 12 ft

Ⓒ 14 ft Ⓓ 20 ft

4. Which of the following is the correct order of the side lengths of the triangle from longest to shortest? F

Ⓕ *EF, DF, DE*

Ⓖ *DF, EF, DE*

Ⓗ *DE, DF, EF*

Ⓙ *EF, DE, DF*

5. What is the distance between the points $J(3, 5)$ and $K(8, -2)$? C

Ⓐ $\sqrt{34}$ Ⓑ 8.5

Ⓒ $\sqrt{74}$ Ⓓ 18

6. Classify the triangle. G

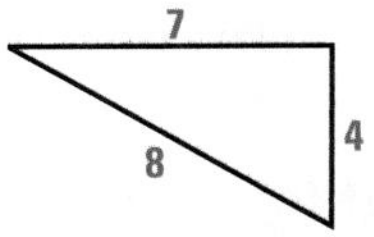

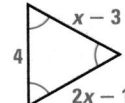

Ⓕ right

Ⓖ acute

Ⓗ equiangular

Ⓙ obtuse

7. In the triangle shown, what is the value of x? B

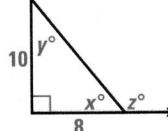

Ⓐ 4

Ⓑ 7

Ⓒ 8

Ⓓ 14

Multi-Step Problem In Exercises 8–10, use the figure shown.

8. What is the sum of x and y? 90

9. Which measure is greater, $x°$ or $y°$? $x°$

10. Which of the following is true? G

Ⓕ $z = 90$ Ⓖ $z > 90$

Ⓗ $z < 90$ Ⓙ $z = y$

Additional Resources

Chapter 4 Resource Book

- SAT/ACT Chapter Test, p. 80

BraiN GaMes

Picture it

interior angle

vertex

isosceles triangle

Materials

- 21 index cards or pieces of paper per team
- timer or watch

Object of the Game In one minute, teams guess as many vocabulary words as possible that a team member sketches.

Set Up Each team writes vocabulary words from Chapter 4 on index cards or pieces of paper. Use the vocabulary list on page 219. Place the cards in a pile with the vocabulary words facing down.

How to Play

Step 1 Each team chooses a player to sketch.

Step 2 The person who is sketching, looks at the word on the top card, without anyone else seeing it. A one minute timer is set. The sketcher draws a picture of the word on the other side of the card.

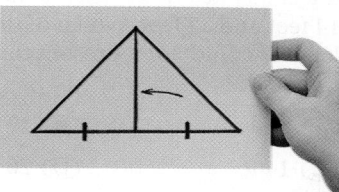

Step 3 When a team member guesses correctly, the sketcher goes to the next card.

Step 4 At the end of the minute, each card is checked to see if the word was drawn correctly. Record a point for each correct card.

Step 5 Play continues until each team member has had a turn to sketch. The team with the most points wins.

Another Way to Play Extend each turn to three minutes. Instead of just saying the vocabulary word, team members also have to write the definition on the card.

A *ratio* is a comparison of a number *a* and a nonzero number *b* using division.

EXAMPLE 1   Simplify Ratios

a. $\dfrac{8 \text{ male students}}{14 \text{ female students}}$

b. $\dfrac{16 \text{ inches}}{3 \text{ feet}}$

Solution

a. $\dfrac{8 \text{ male students}}{14 \text{ female students}} = \dfrac{8 \div 2}{14 \div 2} = \dfrac{4}{7}$

b. Use the fact that 1 foot = 12 inches to convert 3 feet to inches.

$$\dfrac{16 \text{ inches}}{3 \text{ feet}} = \dfrac{16 \text{ inches}}{3 \times 12 \text{ inches}} = \dfrac{16 \text{ inches}}{36 \text{ inches}} = \dfrac{16 \div 4}{36 \div 4} = \dfrac{4}{9}$$

Try These

Simplify the ratio.

1. $\dfrac{9 \text{ hours}}{24 \text{ hours}}$ $\dfrac{3}{8}$

2. $\dfrac{10 \text{ inches}}{2 \text{ feet}}$ $\dfrac{5}{12}$

3. $\dfrac{40 \text{ minutes}}{4 \text{ hours}}$ $\dfrac{1}{6}$

4. $\dfrac{6 \text{ pounds}}{20 \text{ ounces}}$ $\dfrac{24}{5}$

If *a* and *b* are two quantities that have different kinds of units of measure, then $\dfrac{a}{b}$ is a *rate of a per b*. A rate with 1 as its denominator is a *unit rate*. For example, \$2 per pound $\left(\dfrac{\$2}{1 \text{ lb}}\right)$ is a unit rate.

EXAMPLE 2   Find Unit Rates

Suppose you travel 336 miles using 12 gallons of gasoline. Find the unit rate in miles per gallon.

Solution

$$\text{Rate} = \dfrac{336 \text{ miles}}{12 \text{ gallons}} = \dfrac{336 \text{ miles} \div 12}{12 \text{ gallons} \div 12} = \dfrac{28 \text{ miles}}{1 \text{ gallon}} = 28 \text{ miles/gallon}$$

Try These

Find the unit rate.

5. You work 56 hours in 8 days.
   7 hours/day

6. You travel 60 miles in 2 hours.
   30 miles/hour

7. You earn \$38 in 4 hours.
   \$9.50/hour

8. You pay \$5.88 for 12 bagels.
   \$.49/bagel

Algebra Review   **227**

Mathematical Goals

- Find the centroid of a triangle and relate it to the balance point.
- Draw and identify the diagonals of a quadrilateral.
- Apply relationships between the sides and angles of a triangle.

Managing the Project

Tips for Success

Have students work in pairs to find the balance point of the triangle. One student will need to hold the hanging triangle while the partner carefully and accurately marks the lines on the triangle.

In Step 1, urge students to make the vertex holes about the same distance from each vertex.

Guiding Students' Work

In Step 3, students will need patience and a steady hand to make their triangle balance. If the triangle is not close to balancing, students should repeat Steps 1 and 2 with a new triangle as they have probably drawn their lines incorrectly.

Chapters  3-4 Project

Balancing Shapes

Objective

Find the balance points of triangles and other shapes.

Materials

- cardboard
- straightedge
- scissors
- paper punch
- string
- paper clip

How to Find a Balance Point

1 Draw a large triangle on cardboard and cut it out. Punch holes in the triangle near the vertices.

2 Tie a weight and a paper clip to the two ends of a string. Hang your triangle from the paper clip. Mark the vertical line the string makes on the triangle.

3 Repeat Step 2 with the other holes you made. Then balance the triangle on a pencil. The balance point should be close to where the lines intersect.

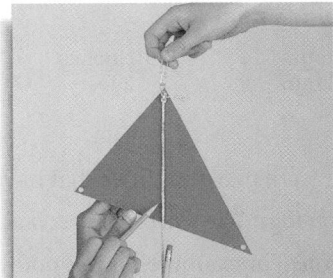

Investigation

1. Use the steps above to find the balance point of a right triangle, an equilateral triangle, and a scalene triangle. What are the lines you drew on the cardboard called? What is the balance point of a triangle called? **medians; centroid**

2. The balance point is at the intersection of the diagonals.

2. Cut out shapes like the ones below. Use the steps above to find their balance points. Then analyze the lines you drew. Make a conjecture about the balance points in relation to the diagonals.

 square
 rectangle
 parallelogram

 228   **Chapter 4** *Triangle Relationships*

Concluding the Project
Consider combining this activity with an art class and having students make colorful mobiles using colored paper or poster board shapes.

Grading the Project
A well-written project will have the following characteristics:
• The answers to Exercises 1–4 are clearly stated.
• A concise explanation of what a balance point is and how to find it is included.

3. Cut out and find the balance points of other shapes, such as those shown below. Is the balance point at the intersection of the diagonals? no

4. Cut out some shapes of your own choice. Find the balance points.

Present Your Results

Write a report about balancing shapes.

▶ Include your answers to Exercises 1–4.

▶ Explain how to find the balance points of various shapes.

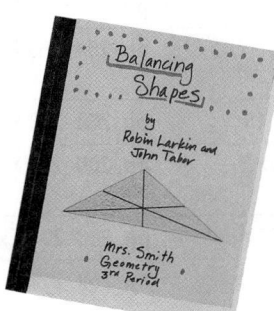

Create a mobile of the shapes you cut out.

▶ For each shape, tie a knot in a string. Thread the string through a hole at the balance point until the string stops at the knot.

▶ Hang all your shapes from one string, or hang them from different strings and tie the strings to a coat hanger.

▶ Be creative! Color or decorate the shapes.

Extension

Research the American sculptor Alexander Calder (1898–1976), creator of the first mobiles.

This mobile includes horizontal red and yellow plates that hang from their balancing points.

Project 229

229

REGULAR SCHEDULE

Lesson	Les. Day	Basic	Average	Advanced
5.1	Day 1	pp. 236–239 Exs. 14–19, 21–36	pp. 236–239 Exs. 14–36, 54–57	pp. 236–239 Exs. 14–32 even, 33–36, 58–66
	Day 2	EP pp. 679–680 Exs. 23, 26, 29, 32; pp. 236–239 Exs. 37–47, 53–63	pp. 236–239 Exs. 37–47, 51–53, 58–66	pp. 236–239 Exs. 37–57; EC: TE p. 230D*, classzone.com
5.2	Day 1	pp. 245–249 Exs. 9–20, 40, 51–58	pp. 245–249 Exs. 9–20, 32, 33, 40, 51–58	pp. 245–249 Exs. 13–20, 32, 33, 40, 51–58
	Day 2	pp. 245–249 Exs. 21–31, 41–47	pp. 245–249 Exs. 21–31, 37, 41–47	pp. 245–249 Exs. 21–31, 37, 41–50
	Day 3	pp. 245–249 Exs. 32, 34, 35, 48–50, Quiz 1	pp. 245–249 Exs. 34–36, 48–50, Quiz 1	245–249 Exs. 34–36, 38*, 39*, Quiz 1; EC: classzone.com
5.3	Day 1	pp. 254–256 Exs. 10–16, 24–30, 36–39	pp. 254–256 Exs. 10–16, 24–30, 36–39	pp. 254–256 Exs. 14–16, 27–30, 36–39
	Day 2	pp. 254–256 Exs. 17–22, 31, 46–53	pp. 254–256 Exs. 17–22, 31–33, 46–53	pp. 254–256 Exs. 14–16, 27–30, 36–39
	Day 3	pp. 254–256 Exs. 34, 35, 40–45	pp. 254–256 Exs. 34, 35, 40–45	pp. 254–256 Exs. 34, 35, 40–45; EC: TE p. 230D*, classzone.com
5.4	Day 1	pp. 260–263 Exs. 10–12, 29–31, 34–45	pp. 260–263 Exs. 10–13, 29–32, 34–45	pp. 260–263 Exs. 10–13, 29–32, 34–45
	Day 2	pp. 260–263 Exs. 15–23, 25, 32, Quiz 2	pp. 260–263 Exs. 14–26, 33, Quiz 2	pp. 260–263 Exs. 14–28, 33, Quiz 2; EC: TE p. 230D*, classzone.com
5.5	Day 1	pp. 268–271 Exs. 6–12, 14, 19, 23–35 odd	pp. 268–271 Exs. 6–14, 19, 23–35 odd	pp. 268–271 Exs. 7–14, 18, 19, 23–35 odd
	Day 2	pp. 268–271 Exs. 15–17, 20–34 even	pp. 268–271 Exs. 15–18, 20–34 even	pp. 268–271 Exs. 15–17, 20, 21*, 22–34 even; EC: classzone.com
5.6	Day 1	pp. 276–280 Exs. 7–12, 14–19, 34, 35–61 odd	pp. 276–280 Exs. 7–19, 21, 22, 34–40	pp. 276–280 Exs. 9–19, 34–36, 37–61 odd
	Day 2	pp. 276–280 Exs. 23–25, 32, 36–62 even	pp. 276–280 Exs. 20, 23, 25, 26–32, 45–61 odd	pp. 276–280 Exs. 20–33*; EC: classzone.com
5.7	Day 1	pp. 286–290 Exs. 8–17, 30, 31, 42, 44–55	pp. 286–290 Exs. 8–20, 30, 31, 42, 44–55	pp. 286–290 Exs. 11–17, 20–23, 30, 31, 42, 44–49, 53–55
	Day 2	pp. 286–290 Exs. 21–29, 33–39, 43, Quiz 3	pp. 286–290 Exs. 21–29, 32–40, 43, Quiz 3	pp. 286–290 Exs. 24–29, 32–41, 43, Quiz 3; EC: TE p. 230D*, classzone.com
Review	Day 1	pp. 291–295 Exs. 1–31	pp. 291–295 Exs. 1–31	pp. 291–295 Exs. 1–31
Assess	Day 1	Chapter 5 Test	Chapter 5 Test	Chapter 5 Test

YEARLY PACING   Chapter 5 Total – **18 days**   Chapters 1–5 Total – **80 days**   Remaining – **80 days**

BLOCK SCHEDULE

*Challenge Exercises   EP = Extra Practice   SRH = Skills Review Handbook   EC = Extra Challenge

Day 1	Day 2	Day 3	Day 4	Day 5	Day 6	Day 7	Day 8	Day 9
5.1 pp. 236–239 Exs. 14–47, 51–66	**5.2** pp. 245–249 Exs. 9–33, 37, 40–47, 51–58	**5.2 cont.** pp. 245–249 Exs. 34–36, 48–50, Quiz 1 **5.3** pp. 254–256 Exs. 10–16, 24–30, 36–39	**5.3 cont.** pp. 254–256 Exs. 17–22, 31–35, 40–53	**5.4** pp. 260–263 Exs. 10–26, 29–45, Quiz 2	**5.5** pp. 268–271 Exs. 6–20, 22–35	**5.6** pp. 276–280 Exs. 7–23, 25–32, 34–40, 45–61 odd	**5.7** pp. 286–290 Exs. 8–40, 42–55, Quiz 3	**Review** pp. 291–295 Exs. 1–31 **Assess** Chapter 5 Test

YEARLY PACING   Chapter 5 Total – **9 days**   Chapters 1–5 Total – **40 days**   Remaining – **40 days**

Support Materials

CHAPTER RESOURCE BOOK

CHAPTER SUPPORT

Tips for New Teachers	p. 1	Strategies for Reading Mathematics	p. 5
Parent Guide for Student Success	p. 3		

LESSON SUPPORT

	5.1	5.2	5.3	5.4	5.5	5.6	5.7
Lesson Plans (regular and block)	p. 7	p. 15	p. 25	p. 34	p. 45	p. 54	p. 63
Warm-Up Exercises and Daily Quiz	p. 9	p. 17	p. 27	p. 36	p. 47	p. 56	p. 65
Technology Activities & Keystrokes				p. 37		p. 57	p. 66
Practice (2 levels)	p. 10	p. 18	p. 28	p. 39	p. 48	p. 58	p. 68
Reteaching with Practice	p. 12	p. 20	p. 30	p. 41	p. 50	p. 60	p. 70
Quick Catch-Up for Absent Students	p. 14	p. 22	p. 32	p. 43	p. 52	p. 62	p. 72
Learning Activities					p. 53		
Real-Life Applications			p. 23	p. 33			

REVIEW AND ASSESSMENT

Quizzes	pp. 24, 44, 73	Alternative Assessment with Math Journal	p. 81
Brain Games Support	p. 74	Project with Rubric	p. 83
Chapter Review Games and Activities	p. 75	Cumulative Review	p. 85
Chapter Test (2 levels)	p. 76	Resource Book Answers	AN1
SAT/ACT Chapter Test	p. 80		

TRANSPARENCIES

	5.1	5.2	5.3	5.4	5.5	5.6	5.7
Warm-Up Exercises and Daily Quiz	p. 28	p. 29	p. 30	p. 31	p. 32	p. 33	p. 34
Visualize It Transparencies	✓	✓	✓	✓			✓
Answer Transparencies	✓	✓	✓	✓	✓	✓	✓

TECHNOLOGY

- Time-Saving Test and Practice Generator
- Electronic Teacher Tools
- Geometry in Motion: Video
- Classzone.com
- Online Lesson Planner

ADDITIONAL RESOURCES

- Worked-Out Solution Key
- Resources in Spanish
- Practice Workbook with Examples

Strategies for Strategic Learners

ANTICIPATE PROBLEM AREAS

Students may be confused by differences in the way figures are labeled. The letter c, for example, may be used to represent the length of a side of a triangle opposite an angle labeled C.

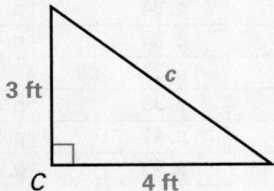

A typical problem would ask students to find the value of c.

That same side, however, might be named by its endpoints in another problem. For example, students might be asked to calculate the length of $\overline{AB}$.

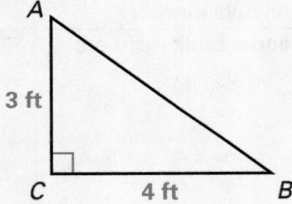

And in still another problem, algebraic expressions might be used to represent the length of the sides of the triangle.

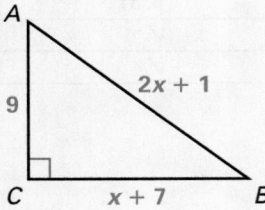

Stress that students should carefully study each figure that they encounter.

Also make sure students clearly understand the difference between a line segment and its measure. Re-emphasize that $\overline{AB}$ refers to a line segment, but AB means the length of $\overline{AB}$. Students must be encouraged to use the correct terminology when discussing segments and measures. In addition, the two symbols cannot be used the same way in writing. For example, it is perfectly correct to write a sentence such as $AB = 2 \cdot AC$, but incorrect to write $\overline{AB} = 2 \cdot \overline{AC}$. If students have difficulty understanding this, look around your classroom for two items that are related in measure, such as a window that is twice the width of a door. Announce to the class, "The width of this window is twice the width of the door." Write this statement on the chalkboard and relate it to the sentence $AB = 2 \cdot AC$.

Strategies for English Learners

VOCABULARY STUDY

The universal access suggestions for Chapter 4 suggested that students use the concept of opposites to learn and remember vocabulary words and to understand concepts in geometry such as *interior* and *exterior*. If students seem comfortable with opposites, introduce the idea of analogies. An analogy expresses a similarity between otherwise dissimilar things. Put another way, it points to the similarity of two things that are not identical. Write this simple analogy on the board and ask students to think of others, using terms used in geometry.

Perpendicular bisector is to segment as angle bisector is to   ? .

This can be written in different ways, for example,
perpendicular bisector : segment as angle bisector : angle.

Other possible examples include:

SSS is to general triangles as HL is to right triangles.
Reasons are to proofs as corresponding parts are to congruent triangles.
Included angle is to sides as included side is to angles.
Congruent angles are to arcs as congruent sides are to tick marks.

Have students make up their own analogies using geometry terms and concepts that they know.

Note that these analogies, since they are focusing on the relationships between concepts, can be translated into any language and the concepts stay the same.

Students can try analogies with shapes that are rotated or flipped, such as:

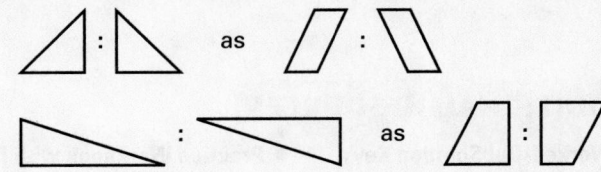

Strategies for Advanced Learners

PROVIDING WRITING OPPORTUNITIES

Have students make up their own word problems involving congruent triangles. Tell students to make the problems as interesting as possible. Advanced students can exchange and critique each other's problems. Selected problems can then be given to the rest of the class as an assignment.

DIFFERENTIATE IN TERMS OF DEPTH

Finding the areas and volumes of unusual or irregular figures often involves seeing a simpler figure, or several simpler figures, embedded in the irregular figures. To challenge advanced students' visualization skills, have them investigate the following situations.

1. Can every triangle be divided into two right triangles? Draw an example or counterexample. (*Yes. Have students try to draw one that cannot be divided this way.*)

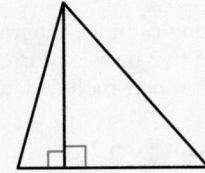

2. Can every square be divided into two isosceles triangles? If so, provide a drawing showing how to do this. (*Yes. The sides of a square are equal in length. If a line is drawn from one vertex to the opposite vertex, two of the sides of the square become two of the sides of an isosceles triangle.*)

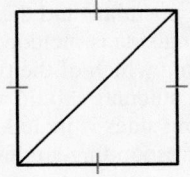

3. If three congruent circles are placed as shown at the right, will their centers always be the vertices of an equilateral triangle? (*Yes. Congruent circles have the same radius.*)

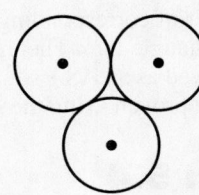

4. Can a circle always be inscribed within a square, and if so will the diameter of the circle always be equal in length to the length of a side of the square? (*Yes. Have students draw this.*)

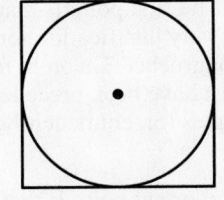

Challenge Problem for use with Lesson 5.1:
In the diagram shown, $\triangle ADB \cong \triangle EDC$, $m\angle DEC = 82°$, and $AB = 14$. State the measure of another angle and another segment in the diagram. (*$m\angle A = 82°$, $EC = 14$*)

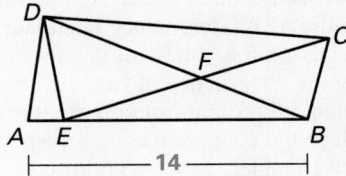

Challenge Problem for use with Lesson 5.3:
Given that $\angle D \cong \angle E$ and $\overline{DF} \cong \overline{EF}$, write a proof to show that $\triangle DFH \cong \triangle EFG$.

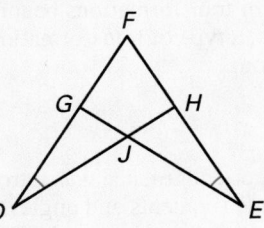

Solution:

Statements	Reasons
1. $\angle D \cong \angle E$; $\overline{DF} \cong \overline{EF}$	1. Given
2. $\angle F \cong \angle F$	2. Reflexive Property of Congruence
3. $\triangle DFH \cong \triangle EFG$	3. ASA Congruence Postulate

Challenge Problem for use with Lesson 5.4:
Given that $\overline{AC} \cong \overline{BC}$ and $\overline{CD} \perp \overline{AB}$, write a proof to show that $\triangle ACD \cong \triangle BCD$.

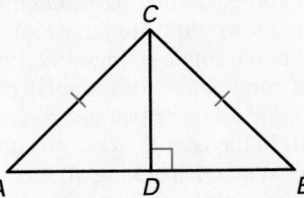

Solution:

Statements	Reasons
1. $\overline{AC} \cong \overline{BC}$; $\overline{CD} \perp \overline{AB}$	1. Given
2. $\angle ADC$ and $\angle BDC$ are right angles.	2. Perpendicular lines form right angles.
3. $\overline{CD} \cong \overline{CD}$	3. Reflexive Property of Congruence
4. $\triangle ACD \cong \triangle BCD$	4. HL Congruence Theorem

Challenge Problem for use with Lesson 5.7:
A figure in the coordinate plane is reflected in the line $y = x$. What are the new coordinates of the point with original coordinates (a, b)? (*(b, a)*)

This chapter treats the important geometric property of congruence as it relates to triangles. Lesson 5.1 introduces students to the idea of corresponding parts of two congruent triangles. Lesson 5.2 presents the first two postulates that provide a means for proving the congruence of two triangles, the SSS and SAS Congruence Postulates. The next lesson continues this theme by presenting the ASA Congruence Postulate and the AAS Congruence Theorem. In Lesson 5.4, the focus changes to right triangles and a theorem for proving the congruence of two such triangles, the HL Congruence Theorem. Lesson 5.5 presents an important use of congruent triangles, namely to deduce the congruence of corresponding sides or angles of the two congruent triangles. Lesson 5.6 introduces the Angle Bisector Theorem and the Perpendicular Bisector Theorem, both of which are used in proofs of congruent triangles. Finally, Lesson 5.7 continues the discussion of transformations begun in Lesson 3.7. The lesson focuses on reflections, a type of transformation that preserves congruence but reverses orientation.

Lesson 5.1

Although the concept of congruence was introduced in Chapter 1, there it was applied only to line segments and angles, and students may have formed the idea that congruence is just a different way of stating equality. Triangles present an example in which the idea of congruence can be more fully realized. You might start by reminding students that congruence, wherever it is encountered in geometry, carries the idea of "same size and shape." Point out that two triangles may have the same area and not be congruent.

This lesson makes the point that congruence is a powerful concept, because if we know that two triangles are congruent, we can conclude many facts about their corresponding parts. On the other hand, Example 5 shows how cumbersome it would be to have to use the definition whenever we want to prove two triangles congruent. Point out that while the use of the definition of congruence for triangles means that all three pairs of corresponding sides and all three pairs of corresponding angles must be shown to be congruent. In the next lessons, postulates and theorems are presented that allow us to avoid checking all the conditions of the definition when proving two triangles congruent.

Example 4 illustrates the usefulness of the Reflexive Property of Congruence, which many students may have regarded as so obvious they did not see a need for stating it. Emphasize that this property helps identify the corresponding sides or angles when proving the congruence of two triangles that have a common side or angle.

Lesson 5.2

This lesson presents the first of five postulates and theorems whose purpose is to simplify the task of proving two triangles congruent. Emphasize that the definition of congruent triangles contains six conditions, all of which must be verified when proving congruence from the definition. Both of the postulates in this lesson cut the number of conditions that must be shown in half.

Activity 5.2 and the Geo-Activity Copying a Triangle provide motivation for the two postulates of this lesson and it is strongly recommended that they be utilized as part of the instruction of the content in Lesson 5.2.

MAKING CONNECTIONS Emphasize that in order to use the SAS Congruence Postulate, the two corresponding angles must be *included* between the two pairs of corresponding sides. A future activity, Activity 5.4, shows why this condition is necessary, but you may want to draw diagrams like these to make this point now.

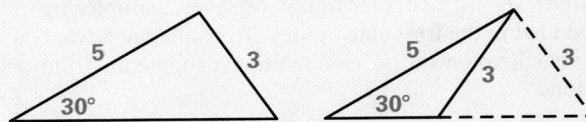

The two triangles in the diagram are clearly not congruent, even though they contain two pairs of corresponding congruent sides and a pair of congruent, but non-included, angles.

Lesson 5.3

Inform students that from a practical standpoint, the only difference between the postulate and the theorem in this lesson is that in one the corresponding side is included, while in the other it is not. Point out that when deciding which of the two to use in a specific geometric situation, the question students should ask themselves is whether each of a pair of corresponding sides is included between two angles that are congruent to a pair of corresponding angles in the other triangle.

You may want to explain why the ASA statement is a postulate while the AAS statement is a theorem. It should be clear to students, perhaps after some prompting, that in the AAS situation, we can always conclude that the third pair of corresponding angles of the two triangles are congruent using the Triangle Sum Theorem, and therefore every AAS situation can also be viewed as an ASA situation. That is, the AAS congruence statement can be proven using the ASA Congruence Postulate.

Lesson 5.4

Several points need to be made in connection with the HL Congruence Theorem. The first point is that the theorem is only useful for right triangles. Thus, any justification for the congruence of two triangles that uses the HL Congruence Theorem for the step that states the triangles are congruent must have been preceded by a step stating that both triangles are right triangles (or, equivalently, that each of the triangles contains a right angle).

Another important point is that in order to use the HL Theorem, one pair of corresponding congruent sides must be the hypotenuses of the two right triangles. Students should note that if the two pairs of sides are both legs (and the included angles are the right angles), they can use the SAS postulate to prove the triangles congruent.

Finally, note that in a proof where the HL Theorem is used as a reason for justifying congruence, there is no step stating that the two right angles are congruent, but only a step stating that two corresponding angles are in fact right angles.

Lesson 5.5

This lesson can be difficult for students, because the problems involving this material often do not mention congruent triangles in their statements (as in Example 1), yet a congruence of triangles lies at the heart of any solution.

USING DIAGRAMS You can make this hurdle less formidable for students in the following two ways.

(1) Urge them to mark all known congruent parts of the diagram with appropriate slash marks or arc symbols. This may immediately suggest a congruence of two triangles by one of the theorems or postulates of this chapter, as it does in Example 1.

(2) Ask them to list some pairs of triangles in the diagram that appear to be congruent. Symmetry will often make this task easier. Then ask them to check the given information to see if it suggests a way to prove the congruence of any of the triangle pairs they have listed. Finally, ask them to check whether the parts of the diagram that are to be proved congruent are corresponding parts of any of these triangles.

These suggestions may not do away with all students' difficulties when a diagram involves overlapping triangles. In such a situation, you should urge your students to "take the diagram apart," as is done in Examples 2 and 3. That is, they should draw a new diagram in which the overlapping triangles are separate. This often allows Method 1 above to be much more effective in suggesting a solution than it would be if applied to the original diagram.

> **QUICK TIP**
> Especially when dealing with overlapping triangles, have students begin their plans for a proof by listing all pairs of triangles that contain the two parts that are to be proven congruent. Then have students identify the pair of triangles from the list they can prove congruent.

Finally, one small detail that often confuses students at this point is the reason given for the final step in the proofs of this section: "Corresponding parts of congruent triangles are congruent." Students may ask if this is a theorem or postulate. Point out that in fact this reason is a consequence of the definition of congruent triangles.

Lesson 5.6

A common mistake students make with regard to the Angle Bisector Theorem is attempting to use the theorem to prove two segments congruent that are not perpendicular to the sides of the bisected angle. Therefore, it will be helpful to spend some time analyzing the definition of the distance from a point to a line given at the beginning of this lesson. You can drive home the point that we define this distance as the length of the *shortest* segment joining the given point with a point on the line.

You might want to point out that the converse of the Angle Bisector Theorem is also true. *If a point is equidistant from the sides of an angle, then it lies on the angle bisector.* This fact is often useful for identifying a given line through the vertex of an angle as the angle bisector. It is also true that in any triangle the bisectors of the three angles intersect in a single point. (This fact is stated for perpendicular bisectors on page 275.)

The Perpendicular Bisector Theorem usually causes fewer difficulties than the Angle Bisector Theorem. It too has a true converse, but the converse does *not* guarantee that a line containing a point equidistant from the endpoints of a segment is the perpendicular bisector of the segment, as students often assume.

Lesson 5.7

The relationship between reflection and symmetry, and especially between a line of symmetry and a line of reflection can be confusing. Emphasize that a diagram has a line of symmetry if and only if that line is a line of reflection for the two halves of the diagram that lie on either side of it. A further source of confusion is that diagrams may have more than one line of symmetry.

Chapter Goals

In this chapter, students will learn to show that two triangles are congruent. The properties of congruent triangles will then be explored. An examination of angle bisectors and perpendicular bisectors leads to a discussion of reflections and symmetry. Students will:

- Identify corresponding parts of congruent triangles.
- Show triangles are congruent using the SSS, SAS, and ASA Congruence Postulates, and the AAS and HL Congruence Theorems.
- Use angle bisectors and perpendicular bisectors to compute angle measures and segment lengths in situations involving triangles.
- Reflect figures over lines and use reflections to discover lines of symmetry in a figure.

Application Note

The game of soccer is the most popular international sport. It is played with 11 players on each team and is distinguished from other sports in that only one player, the goalkeeper, can use his or her hands. The object of the game is to propel the ball past the opponent's goalkeeper into the net. In every nation other than the United States, the sport is called *football*. The U.S. altered the name to *soccer* to distinguish this game from American football.

A soccer ball has a circumference of 27–28 inches and weighs about 14–16 ounces. A soccer field is between 100–130 yards long and between 50–100 yards wide. Normally a field is longer than it is wide. The game itself is divided into 2 halves, each of which is 45 minutes long.

More information about soccer is provided on the Internet at classzone.com

Application Links
CLASSZONE.COM

Chapter **5** Congruent Triangles

Where should a goalie stand?

A soccer goalie instinctively imagines a triangle formed by the goal posts and the ball. The best position to stand allows the goalie to reach each side of the triangle in the same amount of time.

As the ball moves and the shape of the triangle changes, the goalie's best position also changes.

Learn More About It

You will learn more about soccer in Exercises 21 and 22 on p. 278.

230

Hands-On Activities

Activities (more than 20 minutes)
5.2 Congruent Triangles, p. 240
5.4 Investigating Congruence, p. 264
5.6 Investigating Bisectors, p. 272
5.7 Investigating Reflections, p. 281

Geo-Activities (less than 20 minutes)
5.2 Copying a Triangle, p. 241
5.3 Creating Congruent Triangles, p. 250

Who uses Congruent Triangles?

FACILITIES PLANNER
Facilities planners help businesses determine the best locations for new buildings. They can help companies save money and run more efficiently. (p. 275)

TYPE DESIGNER
Type designers design fonts that appear in books, magazines, newspapers, and other materials. Erik Spiekermann has designed many fonts that are widely used today. (p. 288)

Projects

A project covering Chapters 5–6 appears on pages 352–353 of the Student Edition. An additional project for Chapter 5 is available in the *Chapter 5 Resource Book*, pp. 83–84.

⊞ Technology
• Electronic Teacher Tools
• Test and Practice Generator
• Online Lesson Planner

▭▭ Video
• **Geometry in Motion**
There is an animation supporting Lesson 5.6.

Internet Support
CLASSZONE.COM

• Application and Career Links
247, 255, 275, 277, 285, 288
• Student Help
238, 244, 247, 258, 261, 267, 270, 278

How will you use these ideas?

• See how triangles are used in sculptures. (p. 239)
• Investigate origami patterns. (p. 255)
• Identify triangles in skateboard ramps. (p. 257)
• Analyze lighting for a stage production. (p. 262)
• Look for patterns in string designs. (p. 269)
• Find out how kaleidoscope patterns are created. (p. 285)

PREVIEW   **What's the chapter about?**

• Identifying corresponding parts of congruent triangles
• Showing triangles are congruent
• Using angle bisectors and perpendicular bisectors
• Using reflections and line symmetry

Key Words

• **corresponding parts**, *p. 233*
• **congruent figures**, *p. 233*
• **proof**, *p. 243*
• **distance from a point to a line**, *p. 273*
• **equidistant**, *p. 273*
• **perpendicular bisector**, *p. 274*
• **reflection**, *p. 282*
• **line of symmetry**, *p. 284*

PREPARE   **Chapter Readiness Quiz**

Take this quick quiz. If you are unsure of an answer, look at the reference pages for help.

Vocabulary Check *(refer to p. 192)*

1. Which segment in the figure is the *hypotenuse*? D

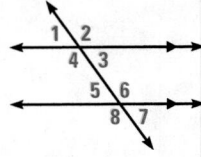

   Ⓐ $\overline{KL}$ Ⓑ $\overline{LN}$ Ⓒ $\overline{NM}$ Ⓓ $\overline{KM}$

Skill Check *(refer to pp. 61, 128)*

2. If $m\angle FGH = 100°$ and $\overrightarrow{GJ}$ bisects $\angle FGH$, what is $m\angle FGJ$?   G

   Ⓕ 25°   Ⓖ 50°   Ⓗ 90°   Ⓙ 100°

3. Use the diagram at the right to determine which of the following statements is true. D

   Ⓐ $\angle 7 \cong \angle 2$ Ⓑ $\angle 5 \cong \angle 2$

   Ⓒ $\angle 1 \cong \angle 8$ Ⓓ $\angle 3 \cong \angle 7$

VISUAL STRATEGY   **Studying Triangles**

Visualize It! ➡️   When working with overlapping triangles, you may find it helpful to draw the triangles separately.

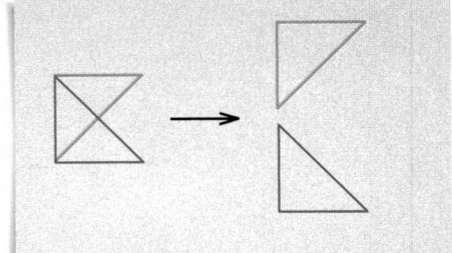

5.1 Congruence and Triangles

Goal
Identify congruent triangles and corresponding parts.

Key Words
• corresponding parts
• congruent figures

The houses below are located in San Francisco, California. The two triangles outlined on the houses have the same size and shape.

Suppose △*ABC* slides over to fit on △*DEF*. When the vertices are matched this way, ∠*A* and ∠*D* are *corresponding angles* and $\overline{AC}$ and $\overline{DF}$ are *corresponding sides*.

Corresponding angles and corresponding sides are examples of **corresponding parts** .

Figures are **congruent** if all pairs of corresponding angles are congruent and all pairs of corresponding sides are congruent.

Congruent

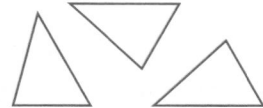

Same size and shape

Not congruent

Different sizes or shapes

EXAMPLE 1 List Corresponding Parts

Given that △*JKL* ≅ △*RST*, list all corresponding congruent parts.

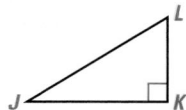 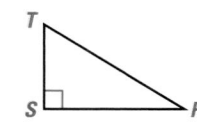

Solution
The order of the letters in the names of the triangles shows which parts correspond.

Corresponding Angles	Corresponding Sides
△*JKL* ≅ △*RST*, so ∠*J* ≅ ∠*R*.	△*JKL* ≅ △*RST*, so $\overline{JK}$ ≅ $\overline{RS}$.
△*JKL* ≅ △*RST*, so ∠*K* ≅ ∠*S*.	△*JKL* ≅ △*RST*, so $\overline{KL}$ ≅ $\overline{ST}$.
△*JKL* ≅ △*RST*, so ∠*L* ≅ ∠*T*.	△*JKL* ≅ △*RST*, so $\overline{JL}$ ≅ $\overline{RT}$.

Student Help

LOOK BACK
To review vertices of a triangle, see p. 175.

① Plan

Pacing
Suggested Number of Days

Basic: 2 days
Average: 2 days
Advanced: 2 days
Block Schedule: 1 block

Teaching Resources
📄 **Blacklines**
(See page 230B.)

👉 **Transparencies**
• Warm-Up with Quiz
• Answers

🖥 **Technology**
• Electronic Teacher Tools
• Test and Practice Generator
• Online Lesson Planner
• Internet Support

② Teach

Content and Teaching Strategies
For background information on geometric concepts and teaching strategies related to this lesson, see pages 230E and 230F in this Teacher's Edition.

Tips for New Teachers
In Example 2 on page 234, guide students to focus on the markings of the corresponding parts in the diagram. Draw additional pairs of congruent triangles for students to practice on. Draw the triangles with different placements and orientations. See the Tips for New Teachers on pp. 1–2 of the *Chapter 5 Resource Book* for additional notes about Lesson 5.1.

Extra Example 1
See next page.

Extra Example 1

Given that △PMN ≅ △EDC, list all corresponding congruent parts.

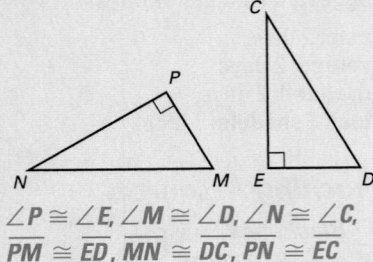

∠P ≅ ∠E, ∠M ≅ ∠D, ∠N ≅ ∠C,
PM̄ ≅ ED̄, MN̄ ≅ DC̄, PN̄ ≅ EC̄

Extra Example 2

The two triangles below are congruent.

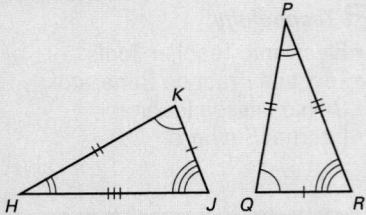

a. Identify all corresponding parts.
∠J ≅ ∠R, ∠H ≅ ∠P, ∠K ≅ ∠Q,
JK̄ ≅ RQ̄, JH̄ ≅ RP̄, HK̄ ≅ PQ̄

b. Write a congruence statement.
△HJK ≅ △PRQ

Extra Example 3

In the figure, △STU ≅ △WXY.

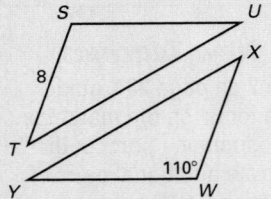

a. Find the length of WX̄. 8
b. Find m∠S. 110°

Geometric Reasoning

In this lesson, postulates and theorems that students have previously learned will be used. For instance, Example 5 utilizes the Vertical Angles Theorem (Lesson 2.4) and the Alternate Interior Angles Theorem (Lesson 3.4).

234

EXAMPLE 2 Write a Congruence Statement

The two triangles are congruent.

a. Identify all corresponding congruent parts.

b. Write a congruence statement.

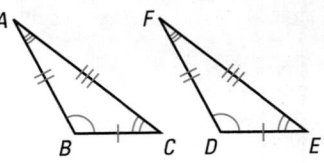

Student Help

STUDY TIP
To name the triangles in Example 2, you can start at any vertex.

△ACB ≅ △FED
△BCA ≅ △DEF
△CAB ≅ △EFD

Be sure the letters are listed in corresponding order.

Solution

a. **Corresponding Angles**

∠A ≅ ∠F
∠B ≅ ∠D
∠C ≅ ∠E

Corresponding Sides

AB̄ ≅ FD̄
BC̄ ≅ DĒ
AC̄ ≅ FĒ

b. List the letters in the triangle names so that the corresponding angles match. One possible congruence statement is
▶ △ABC ≅ △FDE.

EXAMPLE 3 Use Properties of Congruent Triangles

In the diagram, △PQR ≅ △XYZ.

a. Find the length of XZ̄.

b. Find m∠Q.

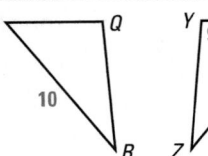 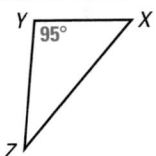

Solution

a. Because XZ̄ ≅ PR̄, you know that XZ = PR = 10.

b. Because ∠Q ≅ ∠Y, you know that m∠Q = m∠Y = 95°.

Checkpoint ✓ Name Corresponding Parts and Congruent Triangles

1. ST̄ ≅ YX̄; TŪ ≅ XZ̄;
SŪ ≅ YZ̄; ∠S ≅ ∠Y;
∠T ≅ ∠X; ∠U ≅ ∠Z

1. Given △STU ≅ △YXZ, list all corresponding congruent parts.

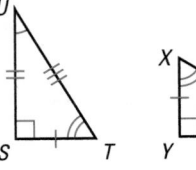 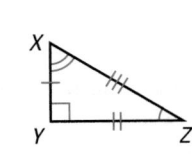

2. Given △ABC ≅ △DEF, find the length of DF̄ and m∠B. 3; 28°

3. Which congruence statement is correct? Why? B; This statement matches up the corresponding vertices in order.

A. △JKL ≅ △MNP

B. △JKL ≅ △NMP

C. △JKL ≅ △NPM

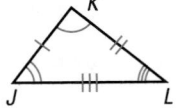 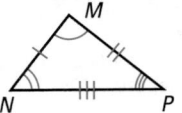

EXAMPLE 4 Determine Whether Triangles are Congruent

Use the two triangles at the right.

a. Identify all corresponding congruent parts.

b. Determine whether the triangles are congruent. If they are congruent, write a congruence statement.

Solution

a.

Corresponding Angles	Corresponding Sides
$\angle D \cong \angle G$	$\overline{DE} \cong \overline{GE}$
$\angle DEF \cong \angle GEF$	$\overline{DF} \cong \overline{GF}$
$\angle DFE \cong \angle GFE$	$\overline{EF} \cong \overline{EF}$

b. All three sets of corresponding angles are congruent and all three sets of corresponding sides are congruent, so the two triangles are congruent. A congruence statement is $\triangle DEF \cong \triangle GEF$.

EXAMPLE 5 Determine Whether Triangles are Congruent

In the figure, $\overline{HG} \parallel \overline{LK}$. Determine whether the triangles are congruent. If so, write a congruence statement.

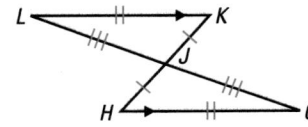

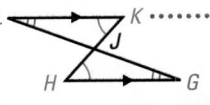
Solution

Start by labeling any information you can conclude from the figure. You can list the following angles congruent.

$\angle HJG \cong \angle KJL$	Vertical angles are congruent.
$\angle H \cong \angle K$	Alternate Interior Angles Theorem
$\angle G \cong \angle L$	Alternate Interior Angles Theorem

The congruent sides are marked on the diagram, so $\overline{HJ} \cong \overline{KJ}$, $\overline{HG} \cong \overline{KL}$, and $\overline{JG} \cong \overline{JL}$. Since all corresponding parts are congruent, $\triangle HJG \cong \triangle KJL$.

Checkpoint ✓ Determine Whether Triangles are Congruent

4. In the figure, $\overline{XY} \parallel \overline{ZW}$. Determine whether the two triangles are congruent. If they are, write a congruence statement.
yes; *Sample answer:* $\triangle XVY \cong \triangle ZVW$

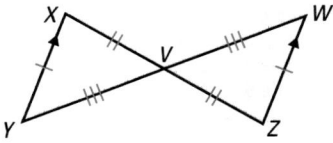

5.1 *Congruence and Triangles* **235**

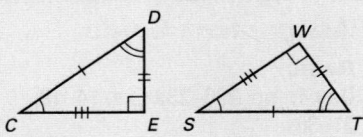
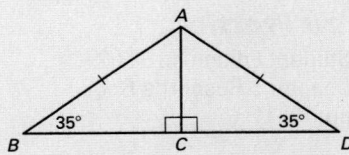

3 Apply

Assignment Guide

BASIC
Day 1: pp. 236–239 Exs. 14–19, 21–36
Day 2: EP pp. 679–680 Exs. 23, 26, 29, 32; pp. 236–239 Exs. 37–47, 53–63

AVERAGE
Day 1: pp. 236–239 Exs. 14–36, 54–57
Day 2: pp. 236–239 Exs. 37–47, 51–53, 58–66

ADVANCED
Day 1: pp. 236–239 Exs. 14–32 even, 33–36, 58–66
Day 2: pp. 236–239 Exs. 37–57; EC: TE p. 230D*, classzone.com

BLOCK SCHEDULE
pp. 236–239 Exs. 14–47, 51–66

Extra Practice

• Student Edition, p. 683
• Chapter 5 Resource Book, pp. 10–11

Homework Check

To quickly check student understanding of key concepts, go over the following exercises:

Basic: 15, 24, 33, 37, 42
Average: 21, 24, 34, 38, 43
Advanced: 16, 26, 39, 41, 45

Teaching Tip

In Exercise 20, have students put a finger under each triangle name in a pair and compare the first pair of letters, then the second pair, and then the third pair. Remind students that the congruence statement implies that corresponding pairs of letters in the triangle names also name pairs of angles that are congruent.

5.1 Exercises

Guided Practice

Vocabulary Check   Determine whether the angles or sides are *corresponding angles*, *corresponding sides*, or *neither*.

1. $\angle C$ and $\angle L$
corresponding angles

2. $\overline{AC}$ and $\overline{JK}$
neither

3. $\overline{BC}$ and $\overline{KL}$
corresponding sides

4. $\angle B$ and $\angle L$
neither

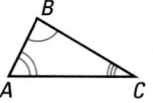 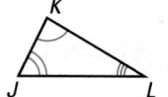

Skill Check   Given that $\triangle XYZ \cong \triangle EFD$, determine the congruent side or angle that corresponds to the side or angle.

5. $\angle Y$ $\angle F$ **6.** $\angle D$ $\angle Z$ **7.** $\overline{XZ}$ $\overline{ED}$ **8.** $\overline{FD}$ $\overline{YZ}$

In Exercises 9–12, find the missing length or angle measure, given that $\triangle LMN \cong \triangle PQR$.

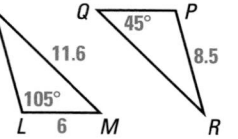

9. $m\angle P = \underline{\ ?\ }$ 105°

10. $m\angle M = \underline{\ ?\ }$ 45°

11. $QR = \underline{\ ?\ }$ 11.6

12. $LN = \underline{\ ?\ }$ 8.5

13. Determine whether the triangles are congruent. If they are, write a congruence statement.   yes; *Sample answer:* $\triangle EFG \cong \triangle KLJ$

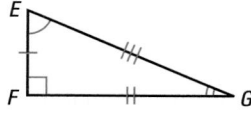 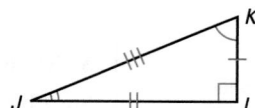

Practice and Applications

Extra Practice
See p. 683.

Corresponding Parts   Determine whether the angles or sides are *corresponding angles*, *corresponding sides*, or *neither*.

14. $\angle M$ and $\angle S$
corresponding angles

15. $\angle L$ and $\angle R$
neither

16. $\overline{MN}$ and $\overline{ST}$
neither

17. $\overline{LM}$ and $\overline{TS}$
corresponding sides

18. $\overline{RT}$ and $\overline{LN}$
corresponding sides

19. $\angle N$ and $\angle R$
corresponding angles

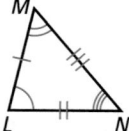 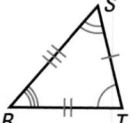

Homework Help

Example 1: Exs. 14–19, 21–24
Example 2: Exs. 21–36
Example 3: Exs. 37–40
Example 4: Exs. 42–47
Example 5: Exs. 42–47

20. Congruent Triangles   Determine which of the following is a correct congruence statement for the triangles in Exercises 14–19. There is only one correct answer.   B

A. $\triangle LNM \cong \triangle RTS$ **B.** $\triangle NML \cong \triangle RST$ **C.** $\triangle MNL \cong \triangle STR$

 Chapter 5 *Congruent Triangles*

236

Identifying Congruent Parts **In the diagram, $\triangle LMN \cong \triangle PQR$. Complete the statement with the corresponding congruent part.**

21. $\angle R \cong$ __?__ $\angle N$

22. $\overline{LN} \cong$ __?__ $\overline{PR}$

23. $\angle RPQ \cong$ __?__ $\angle NLM$

24. $\overline{QP} \cong$ __?__ $\overline{ML}$

25. $\triangle MNL \cong$ __?__ $\triangle QRP$

26. $\triangle PRQ \cong$ __?__ $\triangle LNM$

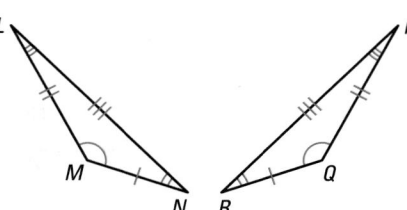

Naming Congruent Triangles **Determine whether the congruence statement correctly describes the congruent triangles shown in the necklace.**

27. $\triangle FGH \cong \triangle NML$ yes

28. $\triangle GFH \cong \triangle NLM$ no

29. $\triangle GHF \cong \triangle MLN$ yes

30. $\triangle HFG \cong \triangle LNM$ yes

31. $\triangle HGF \cong \triangle LNM$ no

32. $\triangle FHG \cong \triangle LNM$ no

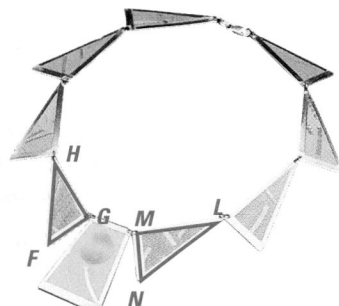

Writing Congruent Statements **In Exercises 33 and 34, the triangles are congruent. Identify all pairs of corresponding congruent parts. Then write a congruence statement.**

33. $\overline{QR} \cong \overline{TU}$; $\overline{RP} \cong \overline{US}$; $\overline{PQ} \cong \overline{ST}$; $\angle Q \cong \angle T$; $\angle R \cong \angle U$; $\angle P \cong \angle S$; *Sample answer:* $\triangle QRP \cong \triangle TUS$

34. $\overline{AB} \cong \overline{ED}$; $\overline{BC} \cong \overline{DF}$; $\overline{AC} \cong \overline{EF}$; $\angle A \cong \angle E$; $\angle B \cong \angle D$; $\angle C \cong \angle F$; *Sample answer:* $\triangle ABC \cong \triangle EDF$

33.

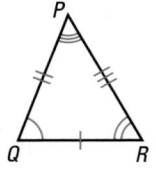

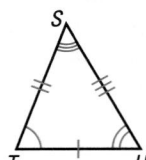

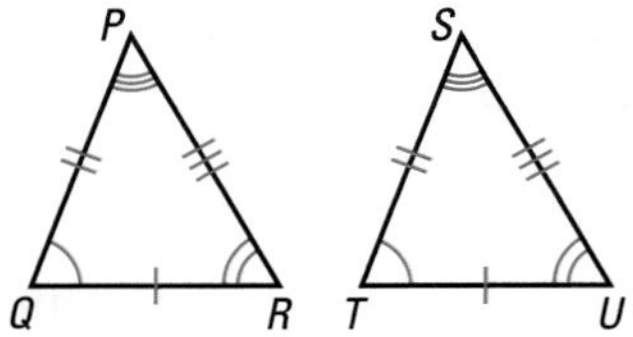

34.

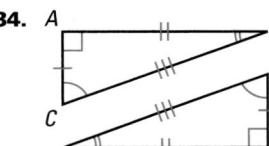

Marking Triangles **In Exercises 35 and 36, $\triangle ABC \cong \triangle DEF$. Draw the triangles on your paper. Then mark every angle and side of the triangles to show the corresponding congruent parts.**

35.

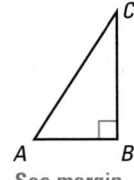

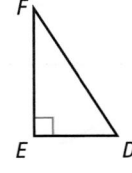

See margin.

36.

See margin.

✗ Common Error

In exercises with figures like those in Exercises 21–26, watch for students who interchange the sides with 2 and 3 tick marks or the angles with 2 and 3 arcs. Have students point to each congruency symbol and state which pair of angles or sides is indicated as being congruent.

35. ***Sample answer:***

36. ***Sample answer:***

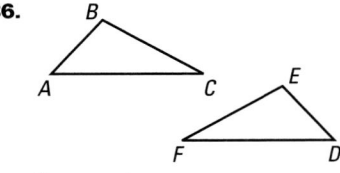

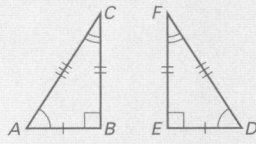

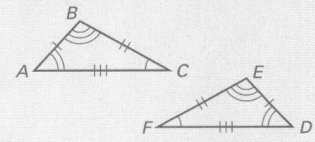

Using Congruent Triangles In Exercises 37–40, △*ABC* ≅ △*DEF*.

37. Find the length of $\overline{DF}$ and $m\angle B$. **5; 100°**

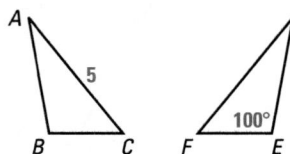

38. Find the length of $\overline{AB}$ and $m\angle F$. **14; 39°**

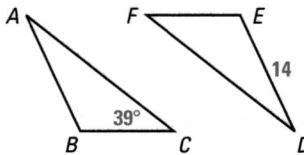

39. Find the length of $\overline{AB}$ and $m\angle F$. **6; 50°**

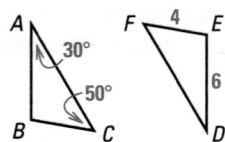

40. Find the length of $\overline{BC}$ and $m\angle D$. **5; 27°**

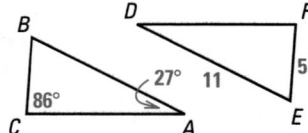

41. Writing Congruence Statements Given that △*FGH* ≅ △*TSR*, determine which congruence statement *does not* describe the triangles. **C**

   A. △*HGF* ≅ △*RST*   **B.** △*FHG* ≅ △*TRS*   **C.** △*GFH* ≅ △*SRT*

Student Help
CLASSZONE.COM

HOMEWORK HELP
Extra help with problem solving in Exs. 42–47 is at classzone.com

Congruent Triangles Determine whether the triangles are congruent. If so, write a congruence statement.

42.

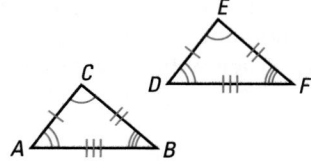

yes; *Sample answer:* △*ABC* ≅ △*DFE*

43.

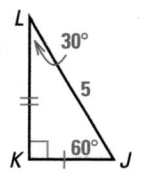

yes; *Sample answer:* △*JKL* ≅ △*PNM*

44.

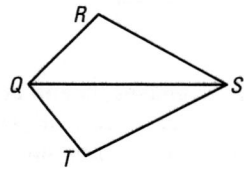

no

45.

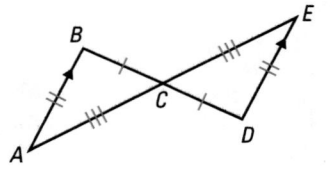

yes; *Sample answer:* △*ABC* ≅ △*EDC*

46.

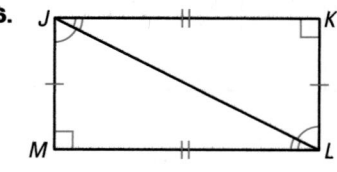

yes; *Sample answer:* △*JKL* ≅ △*LMJ*

47.

no

Link to
Art

SCULPTURE This sculpture by George Rickey, which is displayed in the Albany International Airport, uses supported triangles made from stainless steel.

48. Perpendicular lines intersect to form four right angles, and all right angles are congruent.

Standardized Test Practice

50. Yes; by the Reflexive Property of Congruence, $\overline{FJ} \cong \overline{FJ}$ so all three pairs of corresponding sides are congruent; by Exs. 48 and 49, all three pairs of corresponding angles are congruent.

Mixed Review

Algebra Skills

Sculpture The diagram below represents a view of the base of the sculpture *Four Triangles Hanging*.

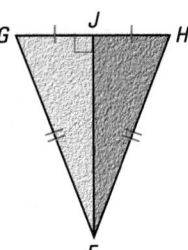

48. Explain how you know that $\angle GJF \cong \angle HJF$. **See margin.**

49. Explain how you know that $\angle JGF \cong \angle JHF$. **Base Angles Theorem**

50. Do you have enough information to show that $\triangle GFJ \cong \triangle HFJ$? Explain your reasoning. **See margin.**

Error Analysis In Exercises 51 and 52, refer to Jillian's work and the diagram below.

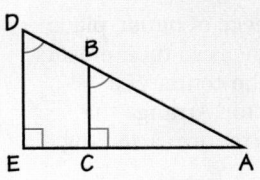

Jillian
$$\angle ABC \cong \angle ADE$$
$$\angle ACB \cong \angle AED$$
$$\angle BAC \cong \angle DAE$$

The corresponding angles are congruent so $\triangle ABC \cong \triangle ADE$.

51. How does Jillian know that $\angle BAC \cong \angle DAE$? **Reflexive Property of Congruence**

52. Is $\triangle ABC \cong \triangle ADE$? Explain your reasoning. **No; the corresponding sides are not congruent.**

53. **Multiple Choice** Which congruence statement *does not* describe the triangles at the right? **B**

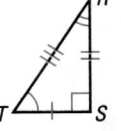

  Ⓐ $\triangle TSR \cong \triangle XYZ$ Ⓑ $\triangle RST \cong \triangle XYZ$

  Ⓒ $\triangle STR \cong \triangle YXZ$ Ⓓ $\triangle TRS \cong \triangle XZY$

Naming Properties Name the property that the statement illustrates. *(Lesson 2.6)*

54. $m\angle J = m\angle J$ **Reflexive Property of Equality**

55. $\overline{PQ} \cong \overline{QP}$ **Reflexive Property of Congruence**

56. If $\angle PQR \cong \angle FGH$, then $\angle FGH \cong \angle PQR$. **Symmetric Property of Congruence**

57. If $AB = CD$ and $CD = FG$, then $AB = FG$. **Transitive Property of Equality**

Classifying Triangles Classify the triangle with the given side lengths as *acute, right,* or *obtuse*. *(Lesson 4.5)*

58. 5, 2, 6 **obtuse**   59. 5, 9, 10 **acute**   60. 14, 5, 12 **obtuse**

Evaluating Decimals Evaluate. *(Skills Review, p. 655)*

61. $4.15 + 7.8$ **11.95**   62. $12.64 - 8.92$ **3.72**   63. $14 + 5.33$ **19.33**

64. $0.91 - 0.17$ **0.74**   65. $1.476 + 6.2 + 9.4$ **17.076**   66. $10.872 - 7.65$ **3.222**

Assessment Resources

The Mini-Quiz below is also available on blackline (*Chapter 5 Resource Book*, p. 17) and on transparency. For more assessment resources, see:
- Chapter 5 Resource Book
- Standardized Test Practice
- Test and Practice Generator

Mini-Quiz

Use the triangles below for Exercises 1–5.

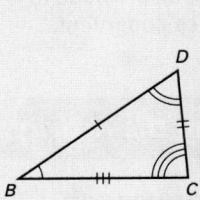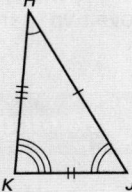

Determine whether the given angles or sides represent *corresponding angles, corresponding sides,* or *neither.*

1. $\angle B$ and $\angle H$ **corresponding angles**

2. $\overline{DB}$ and $\overline{HK}$ **neither**

Complete the statement with the corresponding congruent part.

3. $\angle J \cong \underline{\ ?\ }$ $\angle D$

4. $\overline{CB} \cong \underline{\ ?\ }$ $\overline{KH}$

5. The triangles are congruent. Identify all pairs of corresponding congruent parts. Then write a congruence statement.
$\angle B \cong \angle H, \angle D \cong \angle J, \angle C \cong \angle K,$ $\overline{BD} \cong \overline{HJ}, \overline{BC} \cong \overline{HK}, \overline{CD} \cong \overline{KJ};$ $\triangle BCD \cong \triangle HKJ$

Goal
Using pencils to model two sides of a triangle, students deduce that SAS is sufficient to prove two triangles congruent.

Materials
See the margin of the student page.

LINK TO LESSON ➤
In Example 2 on page 242, students apply what they have learned in this activity to decide whether sufficient information has been provided to prove two triangles congruent.

Alternative Approach
Using overhead markers, perform the activity on an overhead projector while students read aloud the directions. Show students how to carefully draw the sides of the triangle. Invite students to suggest ways of drawing a second triangle in Step 4.

KEY DISCOVERY
If two sides of a triangle and the angle between them are congruent to the corresponding parts of a second triangle, then the two triangles must be congruent.

Activity Assessment
Use Exercise 2 to assess student understanding.

Activity 5.2 Congruent Triangles

Question

Can you determine that two triangles are congruent without listing all of the corresponding congruent sides and angles?

Materials
- two pencils
- protractor
- straightedge

Explore

1 On a piece of paper, place two pencils so their erasers are at the center of a protractor. Arrange the pencils to form a 45° angle.

2 Mark two vertices of a triangle by pressing the pencil points to the paper. Mark the center of the protractor as the third vertex.

3 Remove the pencils and protractor and draw the sides of the triangle.

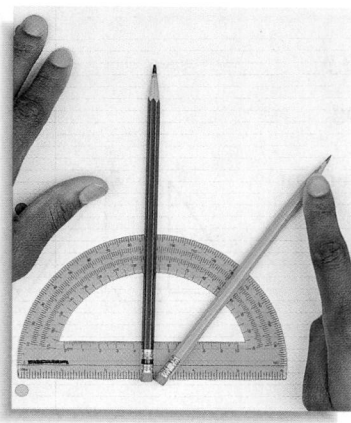
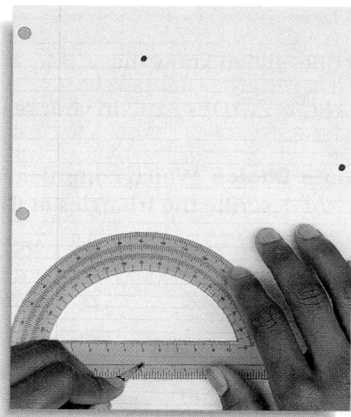
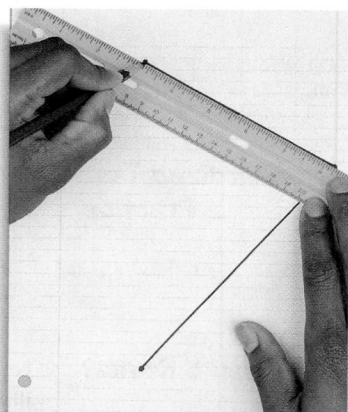

4 Repeat Steps 1–3 using the same pencils. Try to make a triangle that has a 45° angle but is *not* congruent to the one you made.

Think About It

Student Help

LOOK BACK
For help with using a protractor, see p. 36.

1. What do you notice about the triangles you made? All triangles made with the two pencils and a 45° angle appear to be congruent.
2. Based on this activity, if you know that two sides of a triangle are congruent to two sides of another triangle, what other information do you need to know to show that the triangles are congruent? You need to show that the angle formed by the two known sides of one triangle is congruent to the angle formed by the two known sides of the other triangle.

5.2 Proving Triangles are Congruent: SSS and SAS

Goal
Show triangles are congruent using SSS and SAS.

Key Words
• proof

Geo-Activity ▶ Copying a Triangle

1 Use a straightedge to draw a large triangle. Label it △ABC.

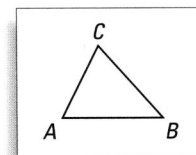

2 Open your compass to measure $\overline{AB}$ of △ABC. Use this length to construct $\overline{DE}$ so that it is congruent to $\overline{AB}$.

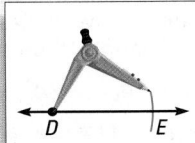

3 Open your compass to measure $\overline{AC}$. Use this length to draw an arc with the point of the compass at D.

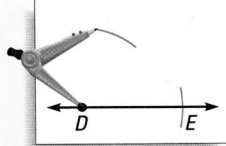

4 Open your compass to measure $\overline{BC}$. Use this length to draw an arc centered at E that intersects the arc from Step 3.

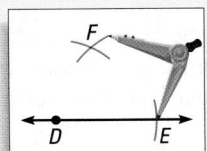

5 Label the point of intersection F. Then draw △DEF. Measure the angles of the triangles to confirm that △ABC ≅ △DEF.

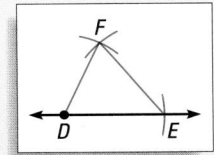

Check diagrams.

The Geo-Activity suggests the following postulate.

POSTULATE 12

Side-Side-Side Congruence Postulate (SSS)

Words If three sides of one triangle are congruent to three sides of a second triangle, then the two triangles are congruent.

Symbols If Side $\overline{MN} \cong \overline{QR}$, and
Side $\overline{NP} \cong \overline{RS}$, and
Side $\overline{PM} \cong \overline{SQ}$,
then △MNP ≅ △QRS.

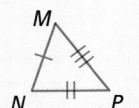
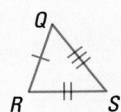

Pacing
Suggested Number of Days

Basic: 3 day
Average: 3 day
Advanced: 3 day
Block Schedule: 1 block
0.5 block with 5.3

Teaching Resources

📄 **Blacklines**
(See page 230B.)

📋 **Transparencies**
• Warm-Up with Quiz
• Answers

💻 **Technology**
• Electronic Teacher Tools
• Test and Practice Generator
• Online Lesson Planner
• Internet Support

Geo-Activity

Goal Use a compass and straight-edge to copy a triangle.

Key Discovery A triangle can be constructed with a compass and straightedge so it is congruent to a given triangle.

2 Teach

Content and Teaching Strategies
For background information on geometric concepts and teaching strategies related to this lesson, see pages 230E and 230F in this Teacher's Edition.

Tips for New Teachers

As students encounter the many ways to prove triangles congruent, they tend to focus on the abbreviations (SSS, SAS, ASA, AAS, and HL which are all in this chapter) without really understanding the relationships. Repeated practice with various diagrams can help them to recognize the needed components and to develop a higher level of accuracy. See the Tips for New Teachers on pp. 1–2 of the *Chapter 5 Resource Book* for additional notes about Lesson 5.2.

Extra Example 1

Does the diagram give enough information to show that the triangles are congruent? Explain.

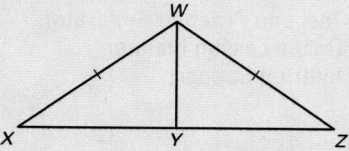

You cannot tell if $\overline{XY} \cong \overline{YZ}$ so the diagram does not give enough information to show the triangles are congruent.

Visualize It!

Students can make their own color sketches to help them visualize the included angle between two sides of a triangle.

Visualize It!

In the triangle below, ∠B is the *included angle* between sides $\overline{AB}$ and $\overline{BC}$.

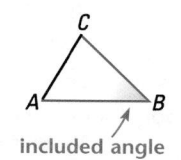

included angle

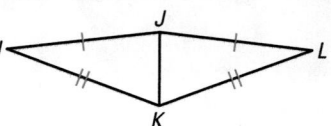

Does the diagram give enough information to show that the triangles are congruent? Explain.

Solution

From the diagram you know that $\overline{HJ} \cong \overline{LJ}$ and $\overline{HK} \cong \overline{LK}$.

By the Reflexive Property, you know that $\overline{JK} \cong \overline{JK}$.

ANSWER ▶ Yes, enough information is given. Because corresponding sides are congruent, you can use the SSS Congruence Postulate to conclude that $\triangle HJK \cong \triangle LJK$.

POSTULATE 13

Side-Angle-Side Congruence Postulate (SAS)

Words If two sides and the included angle of one triangle are congruent to two sides and the included angle of a second triangle, then the two triangles are congruent.

Symbols If Side $\overline{PQ} \cong \overline{WX}$, and

Angle ∠Q ≅ ∠X, and

Side $\overline{QR} \cong \overline{XY}$,

then $\triangle PQR \cong \triangle WXY$.

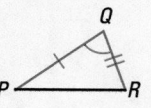 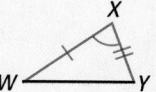

EXAMPLE 2 Use the SAS Congruence Postulate

Does the diagram give enough information to use the SAS Congruence Postulate? Explain your reasoning.

a.

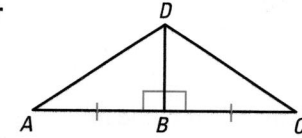

b.

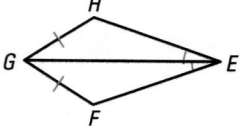

Solution

a. From the diagram, you know that $\overline{AB} \cong \overline{CB}$ and $\overline{DB} \cong \overline{DB}$.

The angle included between $\overline{AB}$ and $\overline{DB}$ is ∠ABD.

The angle included between $\overline{CB}$ and $\overline{DB}$ is ∠CBD.

Because the included angles are congruent, you can use the SAS Congruence Postulate to conclude that $\triangle ABD \cong \triangle CBD$.

b. You know that $\overline{GF} \cong \overline{GH}$ and $\overline{GE} \cong \overline{GE}$. However, the congruent angles are not included between the congruent sides, so you cannot use the SAS Congruence Postulate.

Writing Proofs A **proof** is a convincing argument that shows why a statement is true. A *two-column proof* has numbered statements and reasons that show the logical order of the argument. Each statement has a reason listed to its right.

HOW TO WRITE A PROOF

- List the given information first.

- Use information from the diagram.

- Give a reason for every statement.

- Use given information, definitions, postulates, and theorems as reasons.

- List statements in order. If a statement relies on another statement, list it later than the statement it relies on.

- End the proof with the statement you are trying to prove.

EXAMPLE 3 Write a Proof

Write a two-column proof that shows $\triangle JKL \cong \triangle NML$.

Given $\overline{JL} \cong \overline{NL}$
$\quad\quad$ L is the midpoint of $\overline{KM}$.

Prove $\triangle JKL \cong \triangle NML$

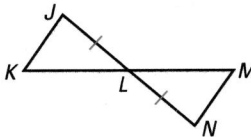

Solution

The proof can be set up in two columns. The proof begins with the given information and ends with the statement you are trying to prove.

Statements	Reasons	
S 1. $\overline{JL} \cong \overline{NL}$	1. Given	These are the given statements.
2. L is the midpoint of $\overline{KM}$.	2. Given	
A 3. $\angle JLK \cong \angle NLM$	3. Vertical Angles Theorem	This information is from the diagram.
S 4. $\overline{KL} \cong \overline{ML}$	4. Definition of midpoint	Statement 4 follows from Statement 2.
5. $\triangle JKL \cong \triangle NML$	5. SAS Congruence Postulate	Statement 5 follows from the congruences of Statements 1, 3, and 4.

Student Help

STUDY TIP
You can remind yourself about side and angle congruences by writing letters as shown.

Extra Example 2

Does the diagram give enough information to use the SAS Congruence Postulate? Explain your reasoning.

a.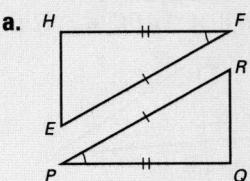

$\overline{FH} \cong \overline{PQ}$, $\overline{EF} \cong \overline{RP}$, and the included angles, $\angle F$ and $\angle P$, are congruent, so you can use the SAS Congruence Postulate.

b.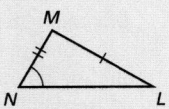

You know that $\overline{HK} \cong \overline{NM}$ and $\overline{JK} \cong \overline{LM}$. But the congruent angles, $\angle H$ and $\angle N$, are not included by the congruent sides, so you cannot use the SAS Congruence Postulate.

Extra Example 3

Given $\overline{SQ} \cong \overline{PM}$ and $\angle SQM \cong \angle PMQ$.
Prove $\triangle QSM \cong \triangle MPQ$.

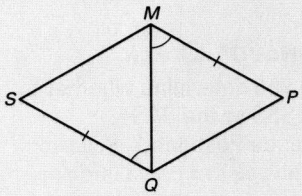

Statements (Reasons)
1. $\overline{SQ} \cong \overline{PM}$ (Given)
2. $\angle SQM \cong \angle PMQ$ (Given)
3. $\overline{QM} \cong \overline{QM}$ (Reflexive Property of Congruence)
4. $\triangle QSM \cong \triangle MPQ$ (SAS Congruence Postulate)

Extra Example 4

You are making a model of the sculpture shown in the diagram. You are told that $\overline{AC} \cong \overline{EC}$, $\overline{DC} \cong \overline{BC}$, and $\overleftrightarrow{AE} \perp \overleftrightarrow{DB}$. Write a proof to show that $\triangle ABC \cong \triangle EDC$.

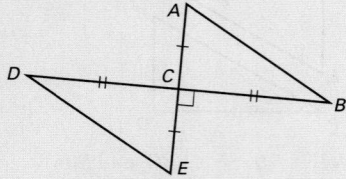

Statements (Reasons)

1. $\overline{AC} \cong \overline{EC}$ (Given)
2. $\overline{DC} \cong \overline{BC}$ (Given)
3. $\angle DCE$ and $\angle BCA$ are right angles. (If two lines are perpendicular, then they intersect to form four right angles.)
4. $\angle DCE \cong \angle BCA$ (All right angles are congruent.)
5. $\triangle ABC \cong \triangle EDC$ (SAS Congruence Postulate)

Multiple Representations

As you discuss Example 4, point out to students that it is easier to see from the diagram what information is given and how congruence can be proved than by studying the congruence statements in symbol form.

✓ Concept Check

How do you determine whether to use the SSS or the SAS Congruence Postulate? **You use SSS when you can prove three pairs of corresponding sides are congruent. You use SAS when you can prove two pairs of corresponding sides and the pair of included angles are congruent.**

🪙 Daily Puzzler

Exactly one of the following five statements about two congruent triangles is incorrect. Which is incorrect? **e**

a) $\triangle JKL \cong \triangle QRS$
b) $\angle J \cong \angle Q$
c) $\angle KLJ \cong \angle RSQ$
d) $\angle LKJ \cong \angle SRQ$
e) $\angle LJK \cong \angle SRQ$

244

EXAMPLE 4 Prove Triangles are Congruent

You are making a model of the window shown in the photo. You know that $\overline{DR} \perp \overline{AG}$ and $\overline{RA} \cong \overline{RG}$. Write a proof to show that $\triangle DRA \cong \triangle DRG$.

Solution

❶ Make a diagram and label it with the given information.

❷ Write the given information and the statement you need to prove.

Given ▷ $\overline{DR} \perp \overline{AG}$, $\overline{RA} \cong \overline{RG}$

Prove ▷ $\triangle DRA \cong \triangle DRG$

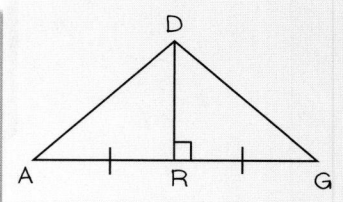

❸ Write a two-column proof. List the given statements first.

Student Help

STUDY TIP
Think about what you can say given that $\overline{DR} \perp \overline{AG}$.

Then think about what other information you can deduce from the diagram.

Statements	Reasons
1. $\overline{RA} \cong \overline{RG}$	1. Given
2. $\overline{DR} \perp \overline{AG}$	2. Given
3. $\angle DRA$ and $\angle DRG$ are right angles.	3. $\perp$ lines form right angles.
4. $\angle DRA \cong \angle DRG$	4. Right angles are congruent.
5. $\overline{DR} \cong \overline{DR}$	5. Reflexive Property of Congruence
6. $\triangle DRA \cong \triangle DRG$	6. SAS Congruence Postulate

Checkpoint ✓ Prove Triangles are Congruent

1. Fill in the missing statements and reasons.

Given ▷ $\overline{CB} \cong \overline{CE}$, $\overline{AC} \cong \overline{DC}$
Prove ▷ $\triangle BCA \cong \triangle ECD$

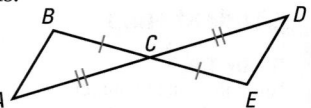

Statements	Reasons
1. $\overline{CB} \cong \overline{CE}$	1. ____?____ Given
2. ____?____ $\overline{AC} \cong \overline{DC}$	2. Given
3. $\angle BCA \cong \angle ECD$	3. ____?____ Vertical Angles Theorem
4. $\triangle BCA \cong \triangle ECD$	4. ____?____ SAS Congruence Postulate

5.2 Exercises

Guided Practice

Vocabulary Check   Use △*JKL* to name the angle included between the two sides.

1. $\overline{JK}$ and $\overline{KM}$ ∠*JKM*

2. $\overline{JK}$ and $\overline{MJ}$ ∠*J*

3. $\overline{KL}$ and $\overline{JL}$ ∠*L*

4. $\overline{KM}$ and $\overline{LM}$ ∠*KML*

5. $\overline{LK}$ and $\overline{KM}$ ∠*LKM*

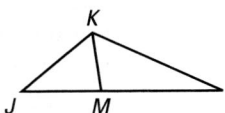

Skill Check   Decide whether enough information is given to show that the triangles are congruent. If so, tell which congruence postulate you would use.

6. △*ABC*, △*DEC*

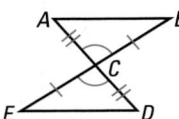

yes; SAS Congruence Postulate

7. △*FGH*, △*JKH*

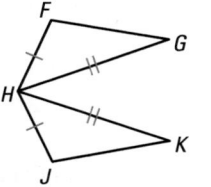

no

8. △*PQR*, △*SRQ*

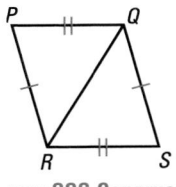

yes; SSS Congruence Postulate

Practice and Applications

Extra Practice
See p. 683.

Naming Included Angles   Use the diagram shown to name the angle included between the two sides.

9. $\overline{AB}$ and $\overline{BD}$ ∠*ABD*

10. $\overline{CD}$ and $\overline{BC}$ ∠*C*

11. $\overline{AC}$ and $\overline{CB}$ ∠*C*

12. $\overline{BA}$ and $\overline{AD}$ ∠*A*

13. $\overline{DC}$ and $\overline{BD}$ ∠*BDC*

14. $\overline{BD}$ and $\overline{BC}$ ∠*DBC*

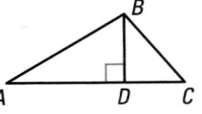

Using SSS   Decide whether enough information is given to use the SSS Congruence Postulate. Explain your reasoning.

Homework Help

Example 1: Exs. 15–17, 21–31
Example 2: Exs. 18–31
Example 3: Exs. 34–36
Example 4: Exs. 34–36

15.

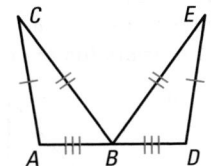

Yes; all three pairs of corresponding sides of the triangles are congruent.

16.

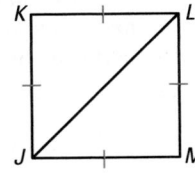

Yes; all three pairs of corresponding sides of the triangles are congruent.

17.

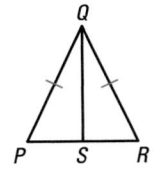

No; there is no information about $\overline{PS}$ and $\overline{RS}$.

5.2   *Proving Triangles are Congruent: SSS and SAS*   **245**

Assignment Guide

BASIC
Day 1: pp. 245–249 Exs. 9–20, 40, 51–58
Day 2: pp. 245–249 Exs. 21–31, 41–47
Day 3: pp. 245–249 Exs. 32, 34, 35, 48–50, Quiz 1

AVERAGE
Day 1: pp. 245–249 Exs. 9–20, 32, 33, 40, 51–58
Day 2: pp. 245–249 Exs. 21–31, 37, 41–47
Day 3: pp. 245–249 Exs. 34–36, 48–50, Quiz 1

ADVANCED
Day 1: pp. 245–249 Exs. 13–20, 32, 33, 40, 51–58
Day 2: pp. 245–249 Exs. 21–31, 37, 41–50
Day 3: pp. 245–249 Exs. 34–36, 38*, 39*, Quiz 1; EC: classzone.com

BLOCK SCHEDULE
pp. 245–249 Exs. 9–33, 37, 40–47, 51–58
pp. 245–249 Exs. 34–36, 48–50, Quiz 1 (with 5.3)

Extra Practice
• Student Edition, p. 683
• Chapter 5 Resource Book, pp. 18–19

Homework Check
To quickly check student understanding of key concepts, go over the following exercises:

Basic: 11, 15, 21, 22, 34
Average: 14, 22, 23, 28, 35
Advanced: 15, 25, 27, 30, 36

18. Yes; $\overline{SQ} \cong \overline{SQ}$ or the Reflexive Property of Congruence, so the triangles are congruent because two sides and the included angle of $\triangle PQS$ are congruent to two sides and the included angle of $\triangle RQS$.

19. No; the congruent angles are not included between the congruent sides.

20. Yes; $\angle JNK \cong \angle MNL$ since they are vertical angles, so the triangles are congruent by the SAS Congruence Postulate.

Using SAS Congruence Decide whether enough information is given to use the SAS Congruence Postulate. Explain your reasoning.

18.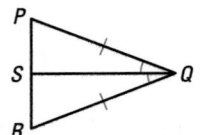

19.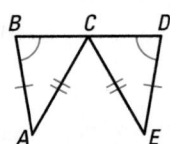

20.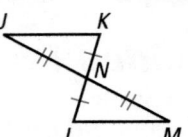

⚖️ **You be the Judge** Decide whether enough information is given to show that the triangles are congruent. If so, tell which congruence postulate you would use.

21.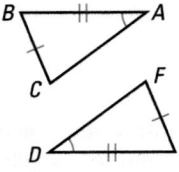

no

22.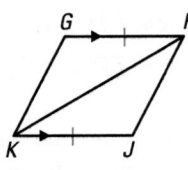

yes; SAS Congruence Postulate

23.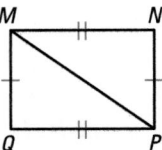

yes; SSS Congruence Postulate

24.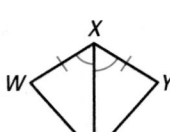

yes; SAS Congruence Postulate

25.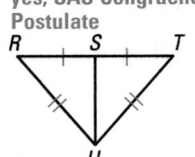

yes; SAS Congruence Postulate or SSS Congruence Postulate

26.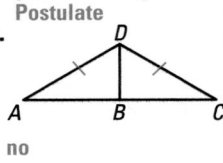

no

Textiles In Exercises 27 and 28, use the photo of the Navajo rug. In the triangles on the rug, $\overline{BC} \cong \overline{DE}$ and $\overline{AC} \cong \overline{CE}$.

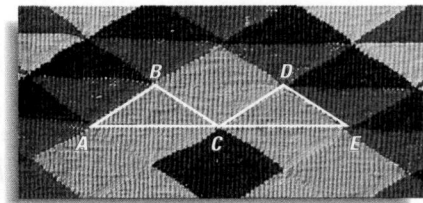

27. What additional information is needed to use the SSS Congruence Postulate to show that $\triangle ABC \cong \triangle CDE$? $\overline{AB} \cong \overline{CD}$

28. What additional information is needed to use the SAS Congruence Postulate to show that $\triangle ABC \cong \triangle CDE$? $\angle ACB \cong \angle CED$

Missing Information Determine what single piece of information you need to know in order to use either the SSS or SAS Congruence Postulate to show that the triangles are congruent.

29. $\overline{BC} \cong \overline{EF}$

30. $\overline{JK} \cong \overline{ML}$ or $\angle JLK \cong \angle MKL$

31. $\overline{PQ} \cong \overline{RQ}$ or $\angle PSQ \cong \angle RSQ$

29.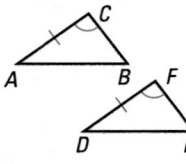

30.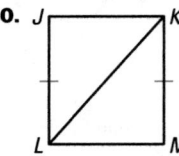

31.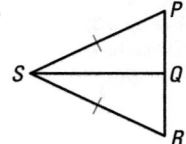

TRIANGLES To create the quilt pattern in Ex. 33, squares are cut out of two rectangular strips of fabrics sewn together.

Application Links
CLASSZONE.COM

Quilting Use the labels on the diagrams to explain how you know that the triangles are congruent. *See margin.*

32. *WXYZ* is a square.

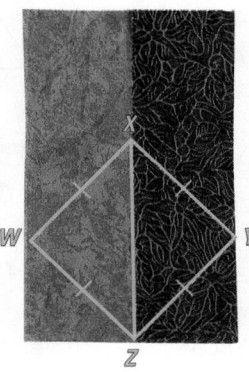

33. $\overline{AB} \parallel \overline{CD}$

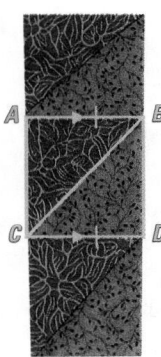

32. *Sample answer:* Since *WXYZ* is a square, all four sides of *WXYZ* are congruent. $\overline{XZ} \cong \overline{XZ}$ by the Reflexive Property, so the triangles are congruent by the SSS Congruence Postulate.

33. *Sample answer:* Since $\overline{AB} \parallel \overline{CD}$, $\angle ABC \cong \angle DCB$ by the Alternate Interior Angles Theorem. $\overline{AB} \cong \overline{DC}$ is given and $\overline{BC} \cong \overline{BC}$ by the Reflexive Property. So, the triangles are congruent by the SAS Congruence Postulate.

Student Help
CLASSZONE.COM

HOMEWORK HELP
Extra help with problem solving in Exs. 34–35 is at classzone.com

Reasoning In Exercises 34 and 35, fill in the missing statements and reasons.

34. Given ▶ $\overline{EF} \cong \overline{GH}$
         $\overline{FG} \cong \overline{HE}$
   Prove ▶ $\triangle EFG \cong \triangle GHE$

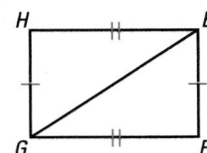

Statements	Reasons
1. $\overline{EF} \cong \overline{GH}$	**1.** Given
2. $\overline{FG} \cong \overline{HE}$	**2.** ___?___ Given
3. $\overline{GE} \cong \overline{GE}$	**3.** ___?___ Reflexive Property of Congruence
4. $\triangle EFG \cong \triangle GHE$	**4.** ___?___ SSS Congruence Postulate

35. Given ▶ $\overline{SP} \cong \overline{TP}$
         $\overline{PQ}$ bisects $\angle SPT$.
   Prove ▶ $\triangle SPQ \cong \triangle TPQ$

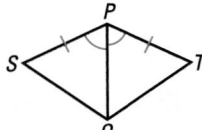

Statements	Reasons
1. $\overline{SP} \cong \overline{TP}$	**1.** Given
2. $\overline{PQ}$ bisects $\angle SPT$.	**2.** ___?___ Given
3. $\angle SPQ \cong \angle TPQ$	**3.** ___?___ Definition of angle bisector
4. ___?___ $\overline{PQ} \cong \overline{PQ}$	**4.** Reflexive Prop. of Cong.
5. $\triangle SPQ \cong \triangle TPQ$	**5.** ___?___ SAS Congruence Postulate

5.2 *Proving Triangles are Congruent: SSS and SAS*   **247**

38. *Sample answer:*
Statements (Reasons)
1. △*ABD* and △*CBD* are equilat-eral. (Given)
2. $\overline{AB} \cong \overline{BD}$ (Def. of equilateral triangle)
3. $\overline{BD} \cong \overline{BC}$ (Def. of equilateral triangle)
4. $\overline{AB} \cong \overline{BC}$ (Transitive Prop. of Congruence)
5. $\overline{AD} \cong \overline{BD}$ (Def. of equilateral triangle)
6. $\overline{BD} \cong \overline{CD}$ (Def. of equilateral triangle)
7. $\overline{AD} \cong \overline{CD}$ (Transitive Prop. of Congruence)
8. $\overline{BD} \cong \overline{BD}$ (Reflexive Prop. of Congruence)
9. △*ABD* ≅ △*CBD* (SSS Congruence Postulate)

39. *Sample answer:*
Statements (Reasons)
1. △*XYZ* is isosceles. (Given)
2. $\overline{YZ} \cong \overline{XZ}$ (Def. of isosceles triangle)
3. $\overline{ZM}$ bisects ∠*YZX*. (Given)
4. ∠*YZM* ≅ ∠*XZM* (Definition of angle bisector)
5. $\overline{ZM} \cong \overline{ZM}$ (Reflexive Prop. of Congruence)
6. △*YZM* ≅ △*XZM* (SAS Congruence Postulate)

36. Reasoning Fill in the missing statements and reasons.

Given ▸ $\overline{AC} \cong \overline{BC}$
M is the midpoint of $\overline{AB}$.

Prove ▸ △*ACM* ≅ △*BCM*

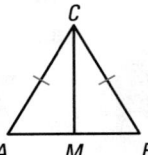

Student Help

STUDY TIP
It is helpful to label congruent sides and congruent angles in steps of a proof. ......

Statements	Reasons
S 1. $\overline{AC} \cong \overline{BC}$	1. Given
2. *M* is the midpoint of $\overline{AB}$.	2. _____?_____ Given
S 3. _____?_____ $\overline{AM} \cong \overline{BM}$	3. Definition of midpoint
S 4. $\overline{CM} \cong \overline{CM}$	4. _____?_____ Reflexive Property of Congruence
▲ 5. △*ACM* ≅ △*BCM*	5. _____?_____ SSS Congruence Postulate

37. Error Analysis Using the diagram below, Maria was asked whether it can be shown that △*ABC* ≅ △*DEF*. Explain her error.

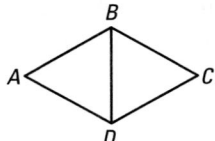

Maria
△ABC ≅ △DEF by the SAS Congruence Postulate.

The congruent angles are not included between the congruent sides.

Challenge Write a proof to show that the triangles are congruent.

38. △*ABD* and △*CBD* are equilateral. **See margin.**

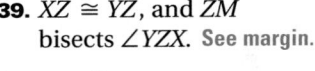

39. $\overline{XZ} \cong \overline{YZ}$, and $\overline{ZM}$ bisects ∠*YZX*. **See margin.**

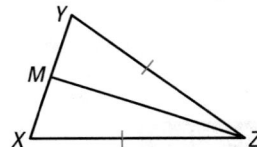

Standardized Test Practice

40. Multiple Choice In △*RST* and △*ABC*, $\overline{RS} \cong \overline{AB}$, $\overline{ST} \cong \overline{BC}$, and $\overline{TR} \cong \overline{CA}$. Which angle is congruent to ∠*T*? **C**

Ⓐ ∠*A*   Ⓑ ∠*R*   Ⓒ ∠*C*   Ⓓ ∠*B*

41. Multiple Choice In the diagram below, △*DEF* is equilateral and *G* is the midpoint of $\overline{DE}$. Which of the statements is *not* true? **G**

Ⓕ $\overline{DF} \cong \overline{EF}$ Ⓖ $\overline{DG} \cong \overline{DF}$

Ⓗ $\overline{DG} \cong \overline{EG}$ Ⓙ △*DFG* ≅ △*EFG*

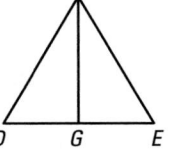

Mixed Review

Classifying Angles Use the diagram to determine whether the angles are *corresponding, alternate interior, alternate exterior,* or *same-side interior* angles. *(Lesson 3.3)*

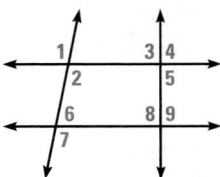

42. ∠1 and ∠5
alternate exterior angles

43. ∠2 and ∠6
same-side interior angles

44. ∠7 and ∠2
corresponding angles

45. ∠5 and ∠8
alternate interior angles

46. ∠9 and ∠4
corresponding angles

47. ∠5 and ∠9
same-side interior angles

Using the Triangle Inequality Determine whether it is possible to draw a triangle with the given side lengths. *(Lesson 4.7)*

48. 14, 8, 25 no   **49.** 20, 10, 28 yes   **50.** 16, 14, 30 no

Algebra Skills

Evaluating Square Roots Evaluate. Give the exact value if possible. If not, approximate to the nearest tenth. *(Skills Review, p. 669)*

51. $\sqrt{3}$ 1.7   **52.** $\sqrt{12}$ 3.5   **53.** $\sqrt{40}$ 6.3   **54.** $\sqrt{159}$ 12.6

55. $\sqrt{14.76}$ 3.8   **56.** $\sqrt{0.87}$ 0.9   **57.** $\sqrt{1.12}$ 1.1   **58.** $\sqrt{40.85}$ 6.4

Quiz 1

In the diagram, $\triangle ABC \cong \triangle QPR$. Complete the statement with the corresponding congruent part. *(Lesson 5.1)*

1. $\angle R \cong$ __?__ ∠C   **2.** $\overline{AB} \cong$ __?__ $\overline{QP}$

3. $\triangle BAC \cong$ __?__ $\triangle PQR$ **4.** $\triangle RPQ \cong$ __?__
$\triangle CBA$

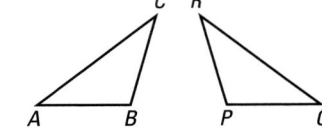

5. Write a congruence statement for the congruent triangles shown at the right. *(Lesson 5.1)*
Sample answer: $\triangle EFG \cong \triangle YXZ$

6. No; the congruent angles are not included between the congruent sides.

7. Yes; *Sample answer:* the three sides of $\triangle DGF$ are congruent to the three sides of $\triangle EGF$, so $\triangle DGF \cong \triangle EGF$ by the SSS Congruence Postulate.

8. Yes; two sides and the included angle of $\triangle JKF$ are congruent to two sides and the included angle of $\triangle NMF$, so the triangles are congruent by the SAS Congruence Postulate.

Decide whether enough information is given to show that the triangles are congruent. If so, tell which congruence postulate you would use. Explain your reasoning. *(Lesson 5.2)*

6. 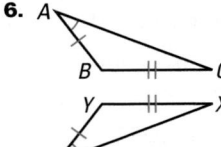   **7.** 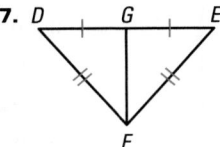   **8.**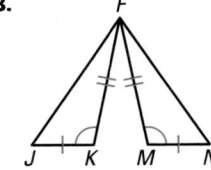

5.2 *Proving Triangles are Congruent: SSS and SAS*   **249**

④ Assess

Assessment Resources

The Mini-Quiz below is also available on blackline (*Chapter 5 Resource Book*, p. 27) and on transparency. For more assessment resources, see:
• Chapter 5 Resource Book
• Standardized Test Practice
• Test and Practice Generator

Mini-Quiz

Decide whether enough information is given to show that the triangles are congruent. If so, state the congruence postulate you would use.

1.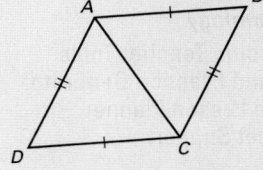

yes; SSS Congruence Postulate

2.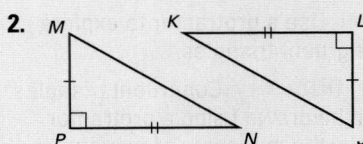

yes; SAS Congruence Postulate

Determine what information you need to know in order to use either the SSS or SAS Congruence Postulate to show that the triangles are congruent.

3.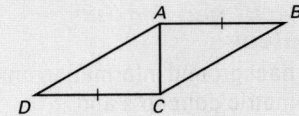

$\overline{AD} \cong \overline{CB}$ to use SSS or
$\angle ACD \cong \angle CAB$ to use SAS

4.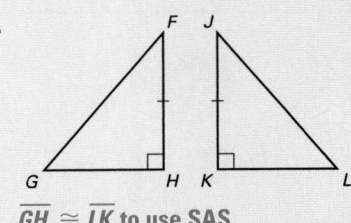

$\overline{GH} \cong \overline{LK}$ to use SAS

249

Suggested Number of Days

Basic: 3 days
Average: 3 days
Advanced: 3 days
Block Schedule: 0.5 block with 5.2
1 block

Teaching Resources

📄 *Blacklines*
(See page 230B.)

🖐 *Transparencies*
• Warm-Up with Quiz
• Answers

🖥 *Technology*
• Electronic Teacher Tools
• Test and Practice Generator
• Online Lesson Planner
• Internet Support

Geo-Activity

Goal Use a protractor to explore congruent triangles.

Key Discovery Congruent triangles can be drawn using a protractor when the measures of two angles and the length of the side between them are known.

2 Teach

Content and Teaching Strategies

For background information on geometric concepts and teaching strategies related to this lesson, see pages 230E and 230F in this Teacher's Edition.

5.3 Proving Triangles are Congruent: ASA and AAS

Goal
Show triangles are congruent using ASA and AAS.

Key Words
• vertical angles p. 75
• alternate interior angles p. 121

Geo-Activity   Creating Congruent Triangles

1 Draw a segment 3 inches long. Label the endpoints *A* and *B*.

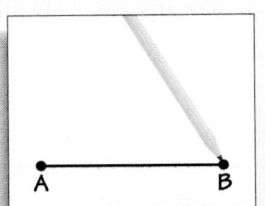

2 Draw an angle measuring 45° at point *A*.

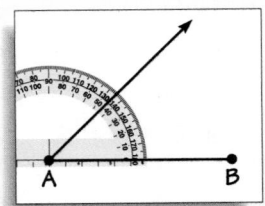

3 Draw an angle measuring 30° at point *B*.

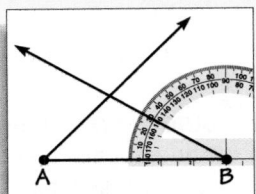

4 Label the point where the angle rays intersect as point *C*.

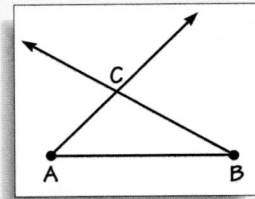

5 Compare your triangle to the triangles drawn by your classmates. Are the triangles congruent? The triangles are congruent.

The Geo-Activity above suggests the following postulate.

Visualize It!

In this triangle, $\overline{AC}$ is the *included side* between ∠*A* and ∠*C*.

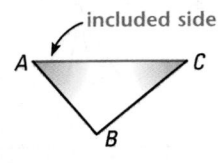

included side

POSTULATE 14

Angle-Side-Angle Congruence Postulate (ASA)

Words If two angles and the included side of one triangle are congruent to two angles and the included side of a second triangle, then the two triangles are congruent.

Symbols If Angle ∠*A* ≅ ∠*D*, and
Side $\overline{AC}$ ≅ $\overline{DF}$, and
Angle ∠*C* ≅ ∠*F*,
then △*ABC* ≅ △*DEF*.

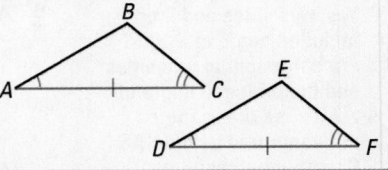

EXAMPLE 1 Determine When To Use ASA Congruence

Based on the diagram, can you use the ASA Congruence Postulate to show that the triangles are congruent? Explain your reasoning.

a.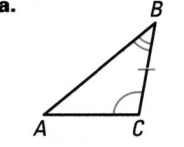

b.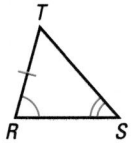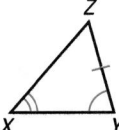

Student Help

STUDY TIP
You can use the ASA Congruence Postulate because $\overline{BC}$ and $\overline{FE}$ are *included* between the congruent angles.

Solution

a. You are given that $\angle C \cong \angle E$, $\angle B \cong \angle F$, and $\overline{BC} \cong \overline{FE}$.

▸ You can use the ASA Congruence Postulate to show that $\triangle ABC \cong \triangle DFE$.

b. You are given that $\angle R \cong \angle Y$ and $\angle S \cong \angle X$.

You know that $\overline{RT} \cong \overline{YZ}$, but these sides are not included between the congruent angles, so you cannot use the ASA Congruence Postulate.

THEOREM 5.1

Angle-Angle-Side Congruence Theorem (AAS)

Words If two angles and a non-included side of one triangle are congruent to two angles and the corresponding non-included side of a second triangle, then the two triangles are congruent.

Symbols If Angle $\angle A \cong \angle D$, and
Angle $\angle C \cong \angle F$, and
Side $\overline{BC} \cong \overline{EF}$,
then $\triangle ABC \cong \triangle DEF$.

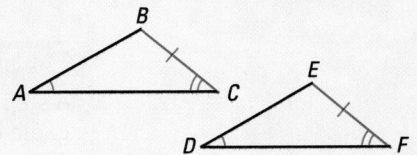

EXAMPLE 2 Determine What Information is Missing

What additional congruence is needed to show that $\triangle JKL \cong \triangle NML$ by the AAS Congruence Theorem?

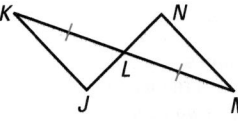

Solution

You are given $\overline{KL} \cong \overline{ML}$.

Because $\angle KLJ$ and $\angle MLN$ are vertical angles, $\angle KLJ \cong \angle MLN$.

The angles that make $\overline{KL}$ and $\overline{ML}$ the non-included sides are $\angle J$ and $\angle N$, so you need to know that $\angle J \cong \angle N$.

Tips for New Teachers

Some students may have difficulty differentiating between ASA and AAS. Stress that the order is important. Illustrate by drawing several diagrams, with different placements and orientations. Have students identify which congruence method could be used, if any. See the Tips for New Teachers on pp. 1–2 of the *Chapter 5 Resource Book* for additional notes about Lesson 5.3.

Extra Example 1

Based on the diagram, can you use the ASA Congruence Postulate to show that the triangles are congruent? Explain your reasoning.

a.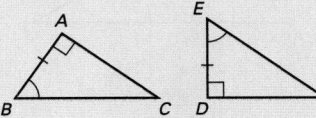

$\angle A \cong \angle D$, $\angle B \cong \angle E$, and the included sides, $\overline{AB}$ and $\overline{DE}$, are congruent so you can use the ASA Congruence Postulate to show that $\triangle ABC \cong \triangle DEF$.

b.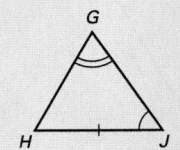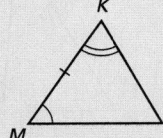

$\angle G \cong \angle K$, $\angle J \cong \angle M$, but the congruent sides, $\overline{HJ}$ and $\overline{KM}$, are not included by the congruent angles, so you cannot use the ASA Congruence Postulate.

Extra Example 2

What additional congruence is needed to show that $\triangle QPS \cong \triangle QRS$ by the AAS Congruence Theorem?

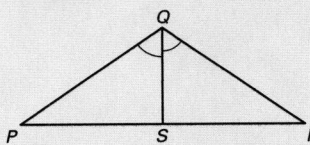

You are given $\angle PQS \cong \angle RQS$, and $\overline{QS} \cong \overline{QS}$ by the Reflexive Property of Congruence. You need to know that $\angle P \cong \angle R$ to use the AAS Congruence Theorem.

Extra Example 3

Does the diagram give enough infor-mation to show that the triangles are congruent? If so, state the postulate or theorem you would use.

a.

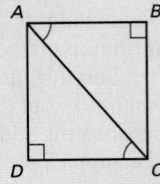

Yes; AAS Congruence Theorem

b.

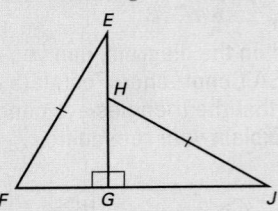

You know only that ∠*FGE* ≅ ∠*HGJ* (All right angles are ≅.) and that $\overline{FE} \cong \overline{HJ}$ (Given), so you cannot conclude that the triangles are congruent.

c.

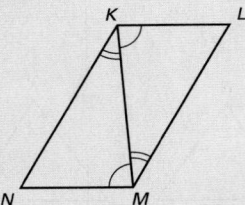

Yes; ASA Congruence Postulate

EXAMPLE 3 Decide Whether Triangles are Congruent

Does the diagram give enough information to show that the triangles are congruent? If so, state the postulate or theorem you would use.

a. 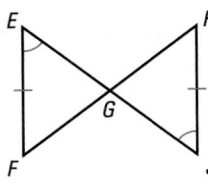   **b.** 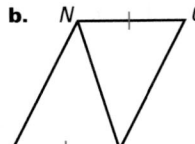   **c.**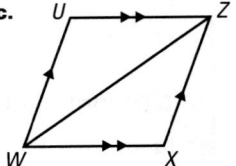

Solution

Student Help

STUDY TIP
The sides of *MNQP* look parallel but this information is not marked, so you cannot conclude that any angles are congruent. ·········▶

a. $\overline{EF} \cong \overline{JH}$ Given

  ∠*E* ≅ ∠*J*   Given

  ∠*FGE* ≅ ∠*HGJ*   Vertical Angles Theorem

Use the AAS Congruence Theorem to conclude that △*EFG* ≅ △*JHG*.

▶ **b.** Based on the diagram, you know only that $\overline{MP} \cong \overline{QN}$ and $\overline{NP} \cong \overline{NP}$. You cannot conclude that the triangles are congruent.

c. ∠*UZW* ≅ ∠*XWZ*   Alternate Interior Angles Theorem

  $\overline{WZ} \cong \overline{WZ}$ Reflexive Prop. of Congruence

  ∠*UWZ* ≅ ∠*XZW*   Alternate Interior Angles Theorem

Use the ASA Congruence Postulate to conclude that △*WUZ* ≅ △*ZXW*.

EXAMPLE 4 Prove Triangles are Congruent

A step in the Cat's Cradle string game creates the triangles shown. Prove that △*ABD* ≅ △*EBC*.

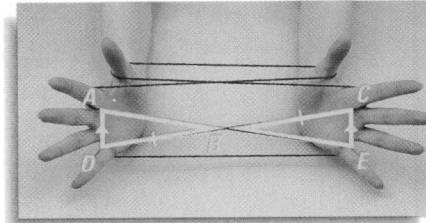

Solution

Given ▶ $\overline{BD} \cong \overline{BC}, \overline{AD} \parallel \overline{EC}$

Prove ▶ △*ABD* ≅ △*EBC*

Visualize It!

Because ∠*A* and ∠*E* are also alternate interior angles, you can show that △*ABD* ≅ △*EBC* by the AAS Congruence Theorem.

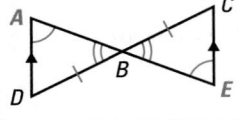

Statements	Reasons
1. $\overline{BD} \cong \overline{BC}$	**1.** Given
2. $\overline{AD} \parallel \overline{EC}$	**2.** Given
3. ∠*D* ≅ ∠*C*	**3.** Alternate Interior Angles Theorem
4. ∠*ABD* ≅ ∠*EBC*	**4.** Vertical Angles Theorem
5. △*ABD* ≅ △*EBC*	**5.** ASA Congruence Postulate

1. Complete the statement: You can use the ASA Congruence Postulate when the congruent sides are __?__ between the corresponding congruent angles. **included**

Does the diagram give enough information to show that the triangles are congruent? If so, state the postulate or theorem you would use.

2.
no
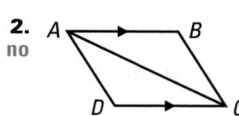

3.
no
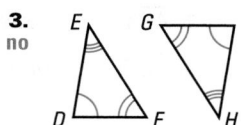

4.
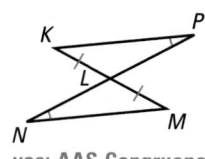
yes; AAS Congruence Theorem

5.3 Exercises

Guided Practice

Vocabulary Check

Tell whether the side is *included* or *not included* between the given angles.

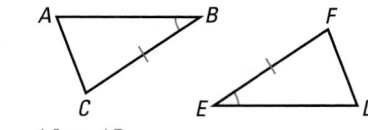

1. $\overline{FG}$ is __?__ between ∠F and ∠G. **included**

2. $\overline{GH}$ is __?__ between ∠F and ∠G. **not included**

3. $\overline{FH}$ is __?__ between ∠H and ∠G. **not included**

4. $\overline{HG}$ is __?__ between ∠H and ∠G. **included**

Skill Check

What congruence do you need to know in order to use the indicated postulate or theorem to conclude that the triangles are congruent?

5. ASA Congruence Postulate

6. AAS Congruence Theorem

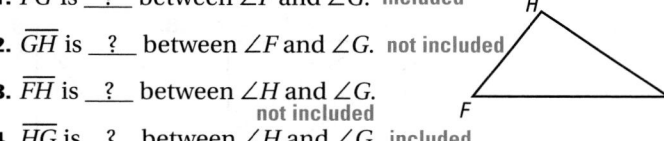

$\overline{AB} \cong \overline{DE}$

∠A ≅ ∠D

Does the diagram give enough information to show that the triangles are congruent? If so, state the postulate or theorem you would use.

7. △RST and △TQR

8. △JKL and △NML

9. △ABC and △DEF

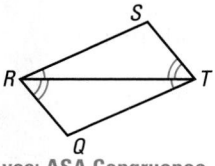
yes; ASA Congruence Postulate

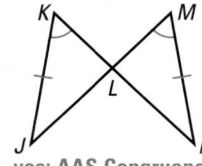
yes; AAS Congruence Theorem

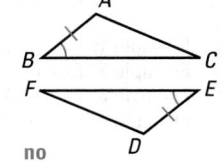
no

5.3 *Proving Triangles are Congruent: ASA and AAS* **253**

Geometric Reasoning

Before Example 4, review parallel lines and the congruent angles that are formed when parallel lines are cut by a transversal. As you begin Example 4, you may need to redraw the diagram with the parallel lines extended to help students visualize which angles are congruent.

Extra Example 4

Use the information in the diagram to prove that △FGH ≅ △GJK.

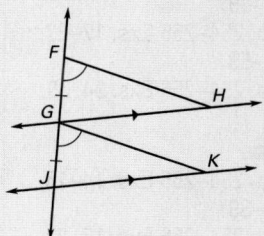

Given: $\overleftrightarrow{GH} \parallel \overleftrightarrow{JK}$, $\overline{GF} \cong \overline{JG}$, ∠GFH ≅ ∠JGK
Prove: △FGH ≅ △GJK
Statements (Reasons)
1. ∠GFH ≅ ∠JGK (Given)
2. $\overline{GF} \cong \overline{JG}$ (Given)
3. ∠FGH ≅ ∠GJK (If two parallel lines are cut by transversal, then corresponding angles are congruent.)
4. △FGH ≅ △GJK (ASA Congruence Postulate)

✔ **Concept Check**

How are the ASA Congruence Postulate and the AAS Congruence Theorem alike? How are they different? **Both postulates require that you find two pairs of congruent angles and one pair of congruent sides to prove congruence, but for the ASA Congruence Postulate the sides must be between the two angles.**

🏅 **Daily Puzzler**

The angle measures of an obtuse isosceles triangle are all whole numbers. What is the greatest possible measure of the obtuse angle? **178°**

③ Apply

Assignment Guide

BASIC
Day 1: pp. 254–256 Exs. 10–16, 24–30, 36–39
Day 2: pp. 254–256 Exs. 17–22, 31, 46–53
Day 3: pp. 254–256 Exs. 34, 35, 40–45

AVERAGE
Day 1: pp. 254–256 Exs. 10–16, 24–30, 36–39
Day 2: pp. 254–256 Exs. 17–22, 31–33, 46–53
Day 3: pp. 254–256 Exs. 34, 35, 40–45

ADVANCED
Day 1: pp. 254–256 Exs. 14–16, 27–30, 36–39
Day 2: pp. 254–256 Exs. 17–23, 31–33, 46–53
Day 3: pp. 254–256 Exs. 34, 35, 40–45; EC: TE p. 230D*, classzone.com

BLOCK SCHEDULE
pp. 254–256 Exs. 10–16, 24–30, 36–39 (with 5.2)
pp. 254–256 Exs. 17–22, 31–35, 40–53

Extra Practice

- Student Edition, p. 683
- Chapter 5 Resource Book, pp. 28–29

Homework Check

To quickly check student understanding of key concepts, go over the following exercises:

Basic: 14, 17, 24, 27, 34
Average: 15, 18, 25, 28, 34
Advanced: 16, 19, 20, 29, 35

✗ Common Error

In Exercise 20, watch for students who think the triangles can be proven congruent by SAS even though the angle is not between the two sides. Show that SSA is not proof of congruence.

254

Practice and Applications

Extra Practice
See p. 683.

14. ASA Congruence Postulate; two angles and the included side of △ABC are congruent to two angles and the included side of △DEF.

15. AAS Congruence Theorem; two angles and a non-included side of △JLM are congruent to two angles and the corresponding non-included side of △PNM.

16. ASA Congruence Postulate; two angles and the included side of △URS are congruent to two angles and the included side of △UTS.

17. Yes; SAS Congruence Postulate; two sides and the included angle of △PQR are congruent to two sides and the included angle of △TSM.

18. Yes; AAS Congruence Theorem; $\overline{BC} \cong \overline{BC}$; perpendicular lines form right angles and all right angles are congruent, so ∠BCA ≅ ∠BCD; two angles and a non-included side of △BCA are congruent to two angles and the corresponding non-included side of △BCD.

19. Yes; ASA Congruence Postulate; vertical angles are congruent, so ∠RVS ≅ ∠UVT; two angles and the included side of △RVS are congruent to two angles and the included side of △UVT.

Homework Help

Example 1: Exs. 14–22
Example 2: Exs. 24–30
Example 3: Exs. 17–22
Example 4: Exs. 34, 35

Recognizing Included Sides Tell whether the side is *included* or *not included* between the given angles.

10. $\overline{AB}$ is __?__ between ∠B and ∠BCA. not included

11. $\overline{AC}$ is __?__ between ∠BAC and ∠BCA. included

12. $\overline{AC}$ is __?__ between ∠DAC and ∠D. not included

13. $\overline{BC}$ is __?__ between ∠CAB and ∠B. not included

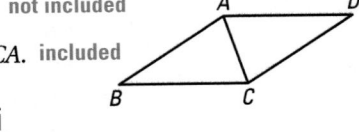

Choosing ASA or AAS Tell which postulate or theorem you would use to show that the triangles are congruent. Explain your reasoning.

14.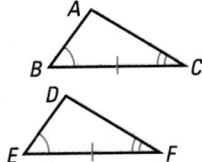

15.

16.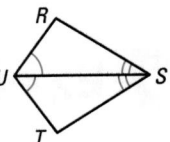

Showing Triangles are Congruent Does the diagram give enough information to show that the triangles are congruent? If so, state the postulate or theorem you would use. Explain your reasoning.

17.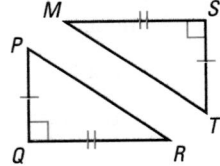

18.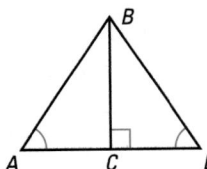

19.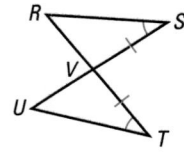

20–22. See margin.

20.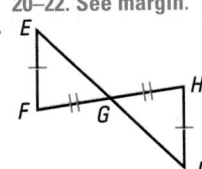

21.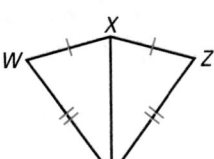

22.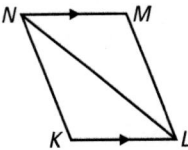

23. **Sea Plane** In the sea plane at the right, ∠L and ∠R are right angles, $\overline{KL} \cong \overline{QR}$, and ∠J ≅ ∠P.

What postulate or theorem allows you to conclude that △JKL ≅ △PQR?
AAS Congruence Theorem

Missing Information Determine what congruence is needed in order to use the indicated postulate or theorem to show that the triangles are congruent.

24. ASA Congruence Postulate $\overline{AC} \cong \overline{FD}$

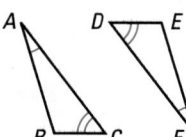

25. AAS Congruence Theorem $\angle K \cong \angle Q$

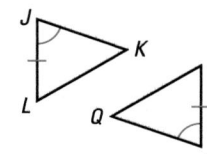

26. ASA Congruence Postulate
$\angle WZX \cong \angle YXZ$

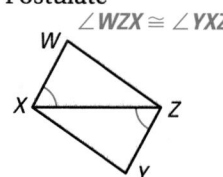

Origami The triangles below show a step in folding an origami seal. State the third congruence needed to prove that △*ABC* ≅ △*ABD* using the indicated postulate or theorem.

27. Given ▶ $\angle CBA \cong \angle DBA$, $\overline{BA} \cong \overline{BA}$
Use the AAS Congruence Theorem. $\angle C \cong \angle D$

28. Given ▶ $\angle C \cong \angle D$, $\overline{AC} \cong \overline{AD}$
Use the ASA Congruence Postulate.
$\angle CAB \cong \angle DAB$

29. Given ▶ $\angle C \cong \angle D$, $\angle CAB \cong \angle DAB$
Use the ASA Congruence Postulate. $\overline{AC} \cong \overline{AD}$

30. Given ▶ $\angle C \cong \angle D$, $\overline{AC} \cong \overline{AD}$
Use the SAS Congruence Postulate. $\overline{BC} \cong \overline{BD}$

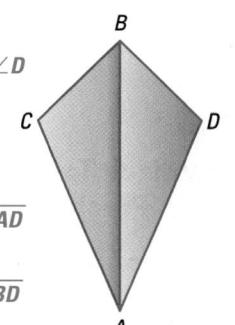

Visualize It! Use the given information to make a sketch of △*PQR* and △*STU*. Mark the triangles with the given information. 31–33. See margin.

31. $\angle Q \cong \angle T$
$\angle R \cong \angle U$
$\overline{QR} \cong \overline{TU}$

32. $\angle P \cong \angle S$
$\angle R \cong \angle U$
$\overline{PQ} \cong \overline{ST}$

33. $\angle Q \cong \angle T$
$\angle S \cong \angle P$
$\overline{QR} \cong \overline{TU}$

34. Logical Reasoning Fill in the missing statements and reasons.

Given ▶ $\overline{GF} \cong \overline{GL}$
$\overline{FH} \parallel \overline{LK}$

Prove ▶ △*FGH* ≅ △*LGK*

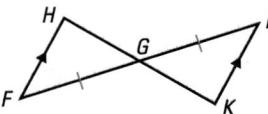

Statements	Reasons
1. $\overline{GF} \cong \overline{GL}$	**1.** Given
2. $\overline{FH} \parallel \overline{LK}$	**2.** _____?_____ Given
3. $\angle F \cong \angle L$	**3.** _____?_____ Alternate Interior Angles Theorem
4. _____?_____ $\angle HGF \cong \angle KGL$	**4.** Vertical Angles Theorem
5. △*FGH* ≅ △*LGK*	**5.** _____?_____ ASA Congruence Postulate

20. No; vertical angles are congruent, but the congruent angles are not included by the congruent sides; no postulate or theorem can be used.

21. Yes; SSS Congruence Postulate; $\overline{XY} \cong \overline{XY}$; all three sides of △*WXY* are congruent to the three sides of △*ZXY*.

22. No; one pair of alternate interior angles is congruent and one pair of sides is congruent, but this is not sufficient to show that the triangles are congruent.

31–33. Sample answers are given.

31.

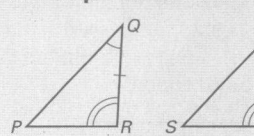

32.

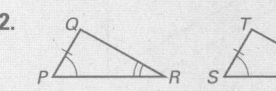

33.

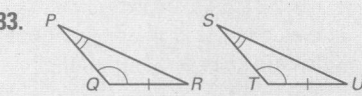

Assess

Mini-Quiz

Does the diagram give enough information to show that the triangles are congruent? If so, state the postulate or theorem you would use. Explain your reasoning.

1.

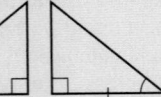

Yes; ASA Congruence Postulate; two angles and the included side of one triangle are congruent to two angles and the included side of the other triangle.

2.

Yes; AAS Cong. Thm.; two pairs of angles and a pair of non-included sides are shown congruent.

Determine what information you need in order to use the indicated postulate or theorem.

3. ASA Congruence Postulate

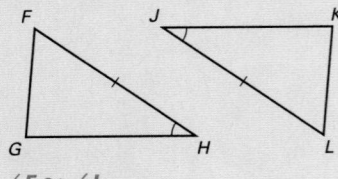

$\angle F \cong \angle L$

4. AAS Congruence Theorem

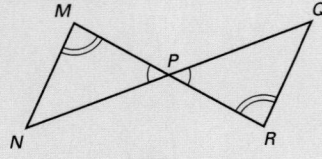

$\overline{MN} \cong \overline{RQ}$ or $\overline{NP} \cong \overline{QP}$

256

---

35. Logical Reasoning Fill in the missing statements and reasons.

Given ▶ $\overline{BC} \cong \overline{EC}$
$\overline{AB} \perp \overline{AD}$
$\overline{DE} \perp \overline{AD}$

Prove ▶ $\triangle ABC \cong \triangle DEC$

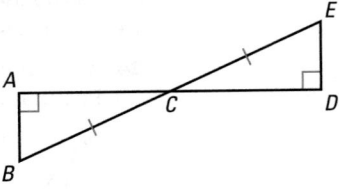

Statements	Reasons
1. ____?____ $\overline{BC} \cong \overline{EC}$	1. Given
2. $\overline{AB} \perp \overline{AD}$	2. ____?____ Given
3. $\overline{DE} \perp \overline{AD}$	3. ____?____ Given
4. $\angle A$ and $\angle D$ are right $\angle$s.	4. ____?____ Perpendicular lines form right angles.
5. ____?____ $\angle A \cong \angle D$	5. Right angles are congruent.
6. $\angle ACB \cong \angle DCE$	6. ____?____ Vertical Angles Theorem
7. $\triangle ABC \cong \triangle DEC$	7. ____?____ AAS Congruence Theorem

Standardized Test Practice

36. Multiple Choice Which theorem or postulate *cannot* be used to show that $\triangle PQR \cong \triangle SQU$? D

(A) ASA   (B) AAS
(C) SAS   (D) SSS

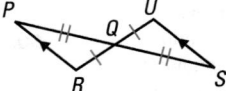

Mixed Review

Finding a Missing Length Find the unknown side length. Round your answer to the nearest tenth. *(Lesson 4.4)*

37.
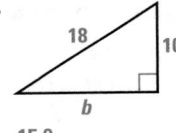
18, 10, b
15.0

38.
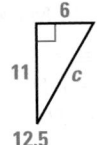
6, 11, c
12.5

39.
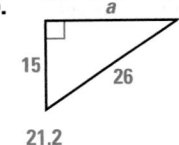
a, 15, 26
21.2

Identifying Congruent Parts In the diagram, $\triangle ABC \cong \triangle DEF$. Complete the statement. *(Lesson 5.1)*

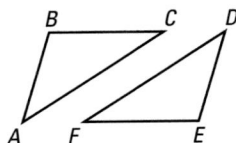

40. $\angle C \cong$ __?__ $\angle F$

41. $\overline{BA} \cong$ __?__ $\overline{ED}$

42. $\angle D \cong$ __?__ $\angle A$

43. $\overline{EF} \cong$ __?__ $\overline{BC}$

44. $\angle B \cong$ __?__ $\angle E$

45. $\overline{CB} \cong$ __?__ $\overline{FE}$

Algebra Skills

Fraction Operations Divide. Write your answer in simplest form. *(Skills Review, p. 659)*

46. $\frac{1}{2} \div 6$ $\frac{1}{12}$

47. $\frac{4}{5} \div 4$ $\frac{1}{5}$

48. $\frac{1}{6} \div \frac{2}{3}$ $\frac{1}{4}$

49. $\frac{5}{8} \div \frac{5}{16}$ 2

50. $\frac{7}{9} \div \frac{2}{7}$ $2\frac{13}{18}$

51. $\frac{3}{10} \div \frac{12}{25}$ $\frac{5}{8}$

52. $8 \div \frac{3}{4}$ $10\frac{2}{3}$

53. $\frac{4}{11} \div 12$ $\frac{1}{33}$

5.4 Hypotenuse-Leg Congruence Theorem: HL

Goal
Use the HL Congruence Theorem and summarize congruence postulates and theorems.

Key Words
• hypotenuse p. 192
• leg of a right triangle p. 192

The triangles that make up the skateboard ramp below are right triangles.

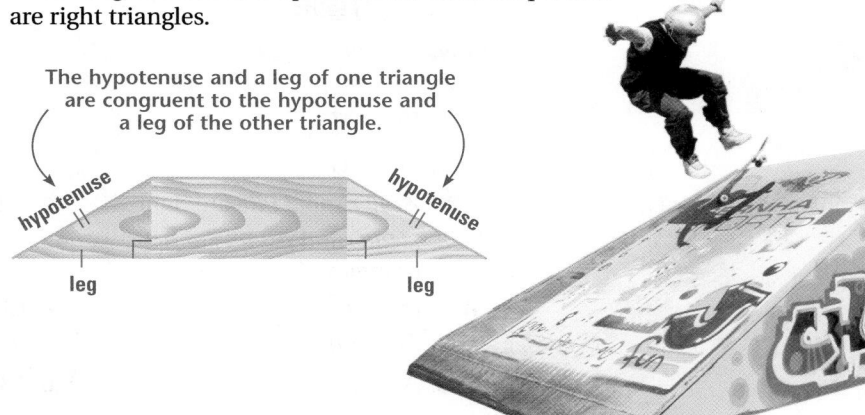

The hypotenuse and a leg of one triangle are congruent to the hypotenuse and a leg of the other triangle.

hypotenuse   hypotenuse

leg   leg

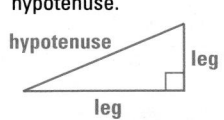

Student Help

VOCABULARY TIP
Remember that the longest side of a right triangle is called the hypotenuse.

hypotenuse
leg
leg

THEOREM 5.2

Hypotenuse-Leg Congruence Theorem (HL)

Words If the hypotenuse and a leg of a right triangle are congruent to the hypotenuse and a leg of a second right triangle, then the two triangles are congruent.

Symbols If $\triangle ABC$ and $\triangle DEF$ are right triangles, and

H $\overline{AC} \cong \overline{DF}$, and

L $\overline{BC} \cong \overline{EF}$,

then $\triangle ABC \cong \triangle DEF$.

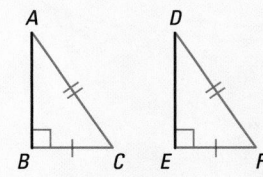

EXAMPLE 1 Determine When To Use HL

Is it possible to show that $\triangle JGH \cong \triangle HKJ$ using the HL Congruence Theorem? Explain your reasoning.

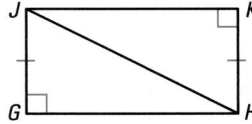

Solution

In the diagram, you are given that $\triangle JGH$ and $\triangle HKJ$ are right triangles.

By the Reflexive Property, you know $\overline{JH} \cong \overline{JH}$ (hypotenuse) and you are given that $\overline{JG} \cong \overline{HK}$ (leg). You can use the HL Congruence Theorem to show that $\triangle JGH \cong \triangle HKJ$.

5.4 Hypotenuse-Leg Congruence Theorem: HL   **257**

---

① Plan

Pacing
Suggested Number of Days

Basic: 2 days
Average: 2 days
Advanced: 2 days
Block Schedule: 1 block

Teaching Resources

Blacklines
(See page 230B.)

Transparencies
• Warm-Up with Quiz
• Answers

Technology
• Electronic Teacher Tools
• Test and Practice Generator
• Online Lesson Planner
• Internet Support

② Teach

Content and Teaching Strategies
For background information on geometric concepts and teaching strategies related to this lesson, see pages 230E and 230F in this Teacher's Edition.

Tips for New Teachers
Review the triangle congruence postulates and theorems covered in this chapter so far. Review with students why AAA and SSA are not ways to prove two triangles congruent. See the Tips for New Teachers on pp. 1–2 of the *Chapter 5 Resource Book* for additional notes about Lesson 5.4.

Extra Example 1
See next page.

Extra Example 1

Is it possible to show that △ABC ≅ △ADC using the HL Congruence Theorem? Explain your reasoning.

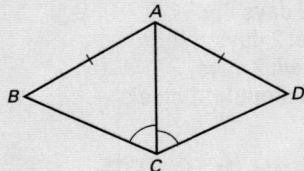

The triangles are not right triangles so they cannot be proven congruent using the HL Congruence Theorem.

Extra Example 2

Use the figure to prove that △TUV ≅ △XWV.

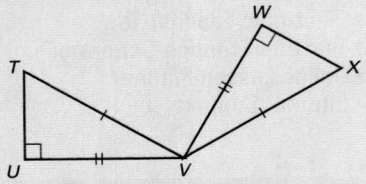

Given: $\overline{TV} \cong \overline{XV}$, $\overline{UV} \cong \overline{WV}$
Prove: △TUV ≅ △XWV
Statements (Reasons)
1. △TUV and △XWV are right triangles. (Definition of right triangle)
2. $\overline{TV} \cong \overline{XV}$ (Given)
3. $\overline{UV} \cong \overline{WV}$ (Given)
4. △TUV ≅ △XWV (HL Congruence Theorem)

Study Skills

The triangle congruence postulates and theorems have names that list the congruent parts—sides and angles—in the order you find them when looking around the triangle clockwise or counterclockwise. Point out that the names can help students recall exactly which parts of triangles they must show are congruent.

Student Help
CLASSZONE.COM

MORE EXAMPLES
More examples at classzone.com

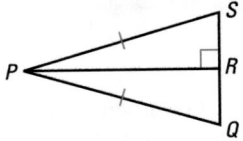

EXAMPLE 2 Use the HL Congruence Theorem

Use the diagram to prove that △PRQ ≅ △PRS.

Solution

Given ▶ $\overline{PR} \perp \overline{SQ}$
$\overline{PQ} \cong \overline{PS}$

Prove ▶ △PRQ ≅ △PRS

Statements	Reasons
1. $\overline{PR} \perp \overline{SQ}$	1. Given
2. ∠PRQ and ∠PRS are right ∡.	2. ⊥ lines form right angles.
3. △PRQ and △PRS are right triangles.	3. Definition of right triangle
H 4. $\overline{PQ} \cong \overline{PS}$	4. Given
L 5. $\overline{PR} \cong \overline{PR}$	5. Reflexive Prop. of Congruence
6. △PRQ ≅ △PRS	6. HL Congruence Theorem

SUMMARY   TRIANGLE CONGRUENCE POSTULATES AND THEOREMS

You have studied five ways to prove that △ABC ≅ △DEF.

SSS	Side	$\overline{AB} \cong \overline{DE}$	
	Side	$\overline{AC} \cong \overline{DF}$	
	Side	$\overline{BC} \cong \overline{EF}$	

SAS	Side	$\overline{AB} \cong \overline{DE}$	
	Angle	∠B ≅ ∠E	
	Side	$\overline{BC} \cong \overline{EF}$	

ASA	Angle	∠A ≅ ∠D	
	Side	$\overline{AB} \cong \overline{DE}$	
	Angle	∠B ≅ ∠E	

AAS	Angle	∠A ≅ ∠D	
	Angle	∠B ≅ ∠E	
	Side	$\overline{BC} \cong \overline{EF}$	

HL	△ABC and △DEF are right triangles.		
	Hypotenuse	$\overline{AB} \cong \overline{DE}$	
	Leg	$\overline{BC} \cong \overline{EF}$	

EXAMPLE 3 Decide Whether Triangles are Congruent

Does the diagram give enough information to show that the triangles are congruent? If so, state the postulate or theorem you would use.

a.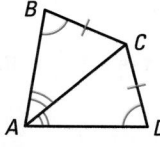

b.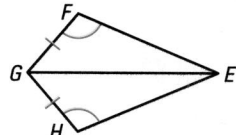

Solution

a. From the diagram, you know that $\angle BAC \cong \angle DAC$, $\angle B \cong \angle D$, and $\overline{BC} \cong \overline{DC}$. You can use the AAS Congruence Theorem to show that $\triangle BAC \cong \triangle DAC$.

b. From the diagram, you know that $\overline{FG} \cong \overline{HG}$, $\overline{EG} \cong \overline{EG}$, and $\angle EFG \cong \angle EHG$. Because the congruent angles are not included between the congruent sides, you cannot show that $\triangle FGE \cong \triangle HGE$.

Student Help

STUDY TIP
There is no SSA Congruence Theorem or Postulate, so you cannot conclude that the triangles in Example 3(b) are congruent.

EXAMPLE 4 Prove Triangles are Congruent

Use the information in the diagram to prove that $\triangle RST \cong \triangle UVW$.

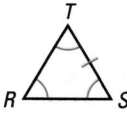 

Solution

Statements	Reasons
A 1. $\angle S \cong \angle V$	1. Given
S 2. $\overline{ST} \cong \overline{VW}$	2. Given
3. $\triangle UVW$ is equilateral.	3. Definition of equilateral triangle
4. $\angle V \cong \angle W$	4. Equilateral triangles are equiangular.
5. $\angle T \cong \angle V$	5. Given
A 6. $\angle T \cong \angle W$	6. Transitive Prop. of Congruence
7. $\triangle RST \cong \triangle UVW$	7. ASA Congruence Postulate

Checkpoint ✓ Decide Whether Triangles are Congruent

Does the diagram give enough information to show that the triangles are congruent? If so, state the postulate or theorem you would use.

1.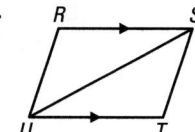

2.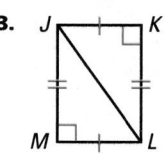

3. (diagram of triangle JKLM)

1. yes; HL Congruence Theorem

2. no

3. yes; SSS Congruence Postulate, SAS Congruence Postulate, or HL Congruence Theorem

---

Extra Example 3

Does the diagram give enough information to prove that the triangles are congruent? If so, state the postulate or theorem you would use.

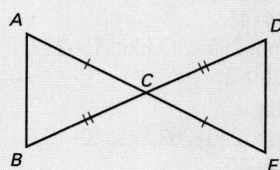

From the diagram you know that $\overline{AC} \cong \overline{EC}$ and $\overline{BC} \cong \overline{DC}$. And $\angle ACB \cong \angle ECD$ because vertical angles are congruent. You can use the SAS Congruence Postulate to prove that $\triangle ACB \cong \triangle ECD$.

Extra Example 4

Use the information in the figure to prove that $\triangle PQS \cong \triangle RQS$.

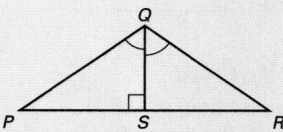

Statements (Reasons)
1. $\overline{QS} \perp \overline{PR}$ (Given)
2. $\angle PQS \cong \angle RQS$ (Given)
3. $\overline{QS} \cong \overline{QS}$ (Reflexive Property of Congruence)
4. $\angle PSQ$ and $\angle RSQ$ are right angles. (If two lines are perpendicular, then they form four right angles.)
5. $\angle PSQ \cong \angle RSQ$ (All right angles are congruent.)
6. $\triangle PQS \cong \triangle RQS$ (ASA Congruence Postulate)

✓ Concept Check

What is the HL Congruence Theorem? If the hypotenuse and leg of a right triangle are congruent to the hypotenuse and leg of a second right triangle, then the two triangles are congruent.

260

Assignment Guide

BASIC
Day 1: pp. 260–263 Exs. 10–12, 29–31, 34–45
Day 2: pp. 260–263 Exs. 15–23, 25, 32, Quiz 2

AVERAGE
Day 1: pp. 260–263 Exs. 10–13, 29–32, 34–45
Day 2: pp. 260–263 Exs. 14–26, 33, Quiz 2

ADVANCED
Day 1: pp. 260–263 Exs. 10–13, 29–32, 34–45
Day 2: pp. 260–263 Exs. 14–28, 33, Quiz 2; EC: TE p. 230D*, classzone.com

BLOCK SCHEDULE
pp. 260–263 Exs. 10–26, 29–45, Quiz 2

Extra Practice
• Student Edition, p. 683
• Chapter 5 Resource Book, pp. 39–40

Homework Check

To quickly check student understanding of key concepts, go over the following exercises:

Basic: 10, 12, 21, 29, 32
Average: 10, 18, 21, 30, 32
Advanced: 15, 17, 21, 31, 32

13. You need to know that the wires have the same length.

14. SAS Congruence Postulate; by definition of midpoint, $\overline{JK} \cong \overline{MK}$; right angles are congruent, so $\angle JKL \cong \angle MKL$; also, $\overline{KL} \cong \overline{KL}$.

5.4 Exercises

Guided Practice

Vocabulary Check

7. Yes; all three sides of △EDG are congruent to the three sides of △GFE so the triangles are congruent by the SSS Congruence Postulate.

Tell whether the segment is a *leg* or the *hypotenuse* of the right triangle.

1. $\overline{AC}$ hypotenuse
2. $\overline{BC}$ leg
3. $\overline{AB}$ leg

4. $\overline{KL}$ leg
5. $\overline{KJ}$ hypotenuse
6. $\overline{JL}$ leg

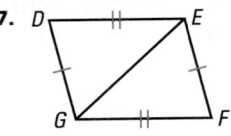
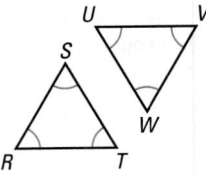

Skill Check

8. Yes; the hypotenuse and a leg of right △MNQ are congruent to the hypotenuse and a leg of right △QPM, so the triangles are congruent by the HL Congruence Theorem.

9. No; both triangles are equilateral and equiangular, but there is no information about the lengths of the sides of the triangles.

Determine whether you are given enough information to show that the triangles are congruent. Explain your answer.

7.
8.
9.

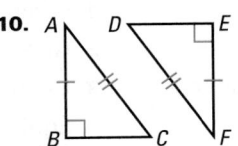
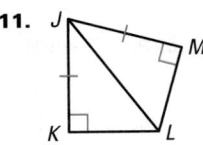
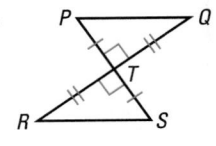

Practice and Applications

Extra Practice
See p. 683.

10. Yes; the hypotenuse and a leg of right △ABC are congruent to the hypotenuse and a leg of right △FED.

11. Yes; $\overline{JL} \cong \overline{JL}$, so the hypotenuse and a leg of right △JKL are congruent to the hypotenuse and a leg of right △JML.

12. No; there is no information given about the lengths of the hypotenuses of the right triangles.

HL Congruence Theorem Determine whether you can use the HL Congruence Theorem to show that the triangles are congruent. Explain your reasoning.

10.
11.
12.

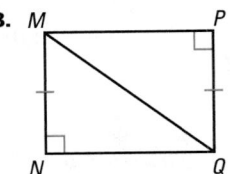

Landscaping To support a tree, you attach wires from the trunk of the tree to stakes in the ground as shown below.

13. What information do you need to know in order to use the HL Congruence Theorem to show that △JKL ≅ △MKL?
See margin.

14. Suppose K is the midpoint of $\overline{JM}$. Name a theorem or postulate you could use to show that △JKL ≅ △MKL. Explain your reasoning.
See margin.

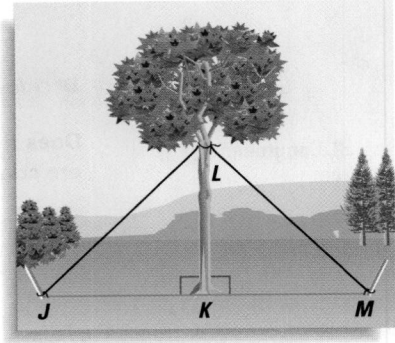

Homework Help

Example 1: Exs. 10–13, 24, 29–31
Example 2: Ex. 32
Example 3: Exs. 13–24, 29–31
Example 4: Ex. 32

15. Yes; SAS Congruence Postulate; two sides and the included angle of △ABC are congruent to two sides and the included angle of △DEF.

16. Yes; ASA Congruence Postulate; two angles and the included side of △JKL are congruent to two angles and the included side of △ZYX.

17. No; the congruent angles are not included between the congruent sides, so there is not enough information to prove the triangles congruent.

18. Yes; SSS Congruence Postulate; FH ≅ FH; all three sides of △FGH are congruent to the three sides of △FJH.

19. Yes; AAS Congruence Theorem; perpendicular lines form right angles and all right angles are congruent, so ∠UTS ≅ ∠UTV; also, UT ≅ UT; two angles and a non-included side of △STU are congruent to two angles and the corresponding non-included side of △VTU.

20. Yes; AAS Congruence Theorem; Right angles are congruent, so ∠A ≅ ∠D; also, vertical angles are congruent, so ∠ACB ≅ ∠DCE; two angles and a non-included side of △ABC are congruent to two angles and the corresponding non-included side of △DEC.

⚖ **You be the Judge** Decide whether enough information is given to show that the triangles are congruent. If so, state the theorem or postulate you would use. Explain your reasoning.

15.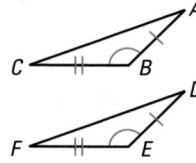

16.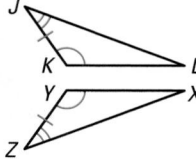

17.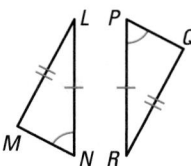

18.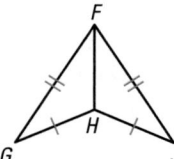

19.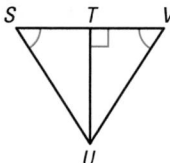

20.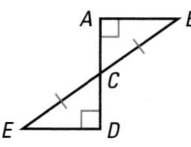

21–23. See margin.

21.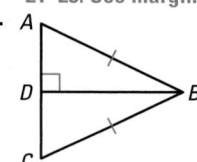

22.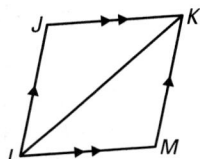

23.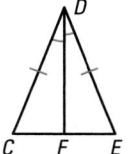

24. Logical Reasoning Three students are given the diagram shown at the right and asked which congruence postulate or theorem can be used to show that △ABC ≅ △CDA. Explain why all three answers are correct. See margin.

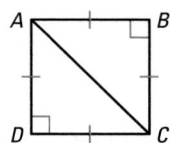

Meghan
△ABC ≅ △CDA by the SSS Congruence Postulate.

Keith
△ABC ≅ △CDA by the SAS Congruence Postulate.

Angie
△ABC ≅ △CDA by the Hypotenuse-Leg Congruence Theorem.

Visualize It! Use the given information to sketch △LMN and △STU. Mark the triangles with the given information. 25–28. See margin.

25. ∠LNM and ∠TUS are right angles. LM ≅ TS, TU ≅ LN

26. LM ⊥ MN, ST ⊥ TU, LM ≅ ST, LN ≅ SU

27. LM ⊥ MN, ST ⊥ TU, LM ≅ NM ≅ UT ≅ ST

28. ML ⊥ LN, TS ⊥ SU, LN ≅ SU, MN ≅ TU

✗ Common Error
In Exercises 15–23, watch for students who try to apply the HL Congruence Theorem to triangles that are not marked as right triangles. Remind them that a triangle must be a right triangle in order to have a hypotenuse. Stress that the hypotenuse is *not* the name of the longest side in every triangle.

21. Yes; *Sample answer:* HL Congruence Theorem; the hypotenuse and a leg (DB) of right △ABD are congruent to the hypotenuse and a leg (DB) of right △CBD.

22. Yes; ASA Congruence Postulate; when two parallel lines are cut by a transversal, alternate interior angles are congruent, so ∠JKL ≅ ∠MLK and ∠JLK ≅ ∠MKL; also LK ≅ LK; two angles and the included side of △JKL are congruent to two angles and the included side of △MLK.

23. Yes; *Sample answer:* SAS Congruence Postulate since DF ≅ DF; two sides and the included angle of △CDF are congruent to two sides and the included angle of △EDF.

24. Meghan is correct because AC ≅ AC, so all three sides of △ABC are congruent to the three sides of △CDA. Keith is correct because of the sides and angles that are marked congruent in the diagram. Angie is correct because AC ≅ AC, so the hypotenuse and a leg of right △ABC are congruent to the hypotenuse and a leg of right △CDA.

25–28. See Additional Answers beginning on page AA1.

37. Yes; AAS Congruence Theorem; two angles and a non-included side of △ABC are congruent to two angles and the corresponding non-included side of △DEF.

38. No; the congruent angles are not included between the congruent sides, so there is not enough information to prove the triangles congruent.

39. Yes; AAS Congruence Theorem; vertical angles are congruent, so ∠JLK ≅ ∠NLM; two angles and a non-included side of △JKL are congruent to two angles and the corresponding non-included side of △NML.

9. Yes; ASA Congruence Postulate or AAS Congruence Theorem; when two parallel lines are cut by a transversal, alternate interior angles are congruent, so ∠R ≅ ∠U and ∠S ≅ ∠V; to use AAS, use congruent vertical angles ∠RTS and ∠UTV along with one pair of alternate interior angles.

10. Yes; ASA Congruence Postulate or AAS Congruence Theorem; $\overline{JL} \cong \overline{JL}$; select the angles that include $\overline{JL}$ to use ASA or a pair of angles that do not include $\overline{JL}$ to use AAS.

29. ∠F ≅ ∠J, ASA Congruence Postulate; or ∠G ≅ ∠L, AAS Congruence Theorem; or $\overline{GH} \cong \overline{LK}$, SAS Congruence Postulate

30. $\overline{PQ} \cong \overline{RS}$ or $\overline{PS} \cong \overline{RQ}$, HL Congruence Theorem; or ∠PQS ≅ ∠RSQ or ∠PSQ ≅ ∠RQS, AAS Congruence Theorem

31. $\overline{WX} \cong \overline{ZX}$, ASA Congruence Postulate; or $\overline{VW} \cong \overline{YZ}$, AAS Congruence Theorem; or $\overline{VX} \cong \overline{YX}$, AAS Congruence Theorem

33. a. *Sample answer:* $\overline{BD} \cong \overline{BD}$; the hypotenuse and a leg of right △ABD are congruent to the hypotenuse and a leg of right △CBD, so △ABD ≅ △CBD by the HL Congruence Theorem.

b. Yes; *Sample answer:* It is given that $\overline{AB} \cong \overline{BC}$. By the Base Angles Theorem, ∠BAD ≅ ∠BCD. Perpendicular lines form right angles and all right angles are congruent, so ∠BDA ≅ ∠BDC. Therefore, △ABD ≅ △CBD by the AAS Congruence Theorem.

Standardized Test Practice

33. c. Yes; *Sample answer:* The hypotenuse and a leg of △ABD are congruent to the hypotenuse and a leg of △CBD, △CEG, and △FEG by the HL Congruence Theorem.

Missing Information What congruence is needed to show that the triangles are congruent? Using that congruence, tell which theorem or postulate you would use to show that the triangles are congruent.

29. 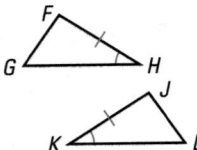 **30.** 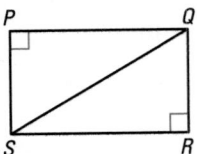 **31.**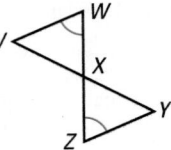

32. Logical Reasoning Fill in the missing statements and reasons.

Given ▸ $\overline{BD} \cong \overline{FD}$
 D is the midpoint of $\overline{CE}$.
 ∠BCD and ∠FED are right angles.

Prove ▸ △BCD ≅ △FED

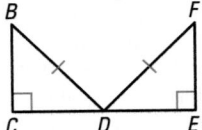

Statements	Reasons
1. $\overline{BD} \cong \overline{FD}$	1. ___?___ Given
2. ___?___ D is the midpoint of $\overline{CE}$.	2. Given
3. ___?___ $\overline{CD} \cong \overline{ED}$	3. Definition of midpoint
4. ∠BCD and ∠FED are right angles.	4. ___?___ Given
5. ___?___ are right triangles. △BCD and △FED	5. Definition of right triangle
6. △BCD ≅ △FED	6. ___?___ HL Congruence Theorem

33. Multi-Step Problem The diagram below is a plan showing the light created by two spotlights. Both spotlights are the same distance from the stage.

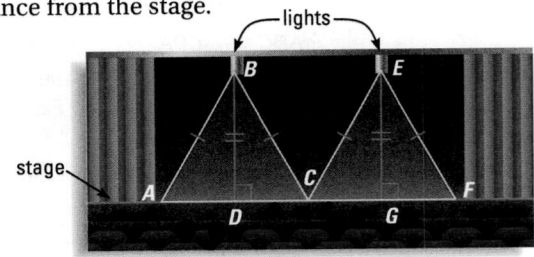

a. Show that △ABD ≅ △CBD. Tell what theorem or postulate you use and explain your reasoning.

b. Is there another way to show that △ABD ≅ △CBD? If so, tell how. Explain your reasoning.

c. Are all four right triangles in the diagram congruent? Explain your reasoning.

Mixed Review

34. m∠1 = 110° by the Alternate Interior Angles Theorem; m∠2 = 70° by the Same-Side Interior Angles Theorem (or by the Linear Pair Postulate).

35. m∠1 = 82° by the Corresponding Angles Postulate; m∠2 = 82° by the Alternate Exterior Angles Theorem (or by the Vertical Angles Theorem).

36. m∠1 = 123° by the Linear Pair Postulate; m∠2 = 57° by the Alternate Interior Angles Theorem.

Parallel Lines Find *m∠1* and *m∠2*. **Explain your reasoning.** *(Lesson 3.4)*

34.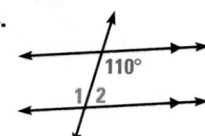

35.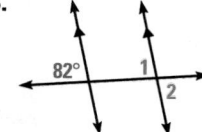

36.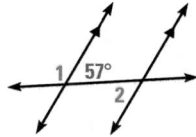

Showing Congruence Decide whether enough information is given to show that the triangles are congruent. If so, state the theorem or postulate you would use. Explain your reasoning. *(Lessons 5.2, 5.3)*

37.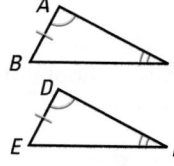

38.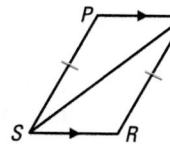

39.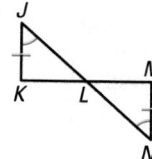

37–39. See margin.

Algebra Skills

Evaluating Expressions Evaluate. *(Skills Review, p. 670)*

40. 2 · 4 + 5 13

41. 10 − 5 · 2 0

42. 3 + 4² − 11 8

43. 7 · 2 + 6 · 3 32

44. 3 · 5 − 2 · 7 1

45. 5² − 10 · 2 5

Quiz 2

5. No; only two pairs of sides are congruent.

6. Yes; AAS Congruence Theorem; two angles and a non-included side of △JKL are congruent to two angles and the corresponding non-included side of △QRP.

7. Yes; ASA Congruence Postulate; two angles and the included side of △STU are congruent to two angles and the included side of △VUT.

8. Yes; HL Congruence Theorem; the hypotenuse and a leg of right △HEF are congruent to the hypotenuse and a leg of right △FGH.

Tell whether the theorem or postulate can be used to show that △*LMN* ≅ △*QMP*. *(Lessons 5.3, 5.4)*

1. ASA yes

2. AAS yes

3. HL no

4. SSS no

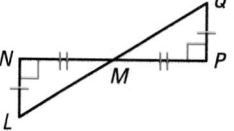

Tell whether enough information is given to show that the triangles are congruent. If so, tell which theorem or postulate you would use. Explain your reasoning. *(Lessons 5.3, 5.4)*

5.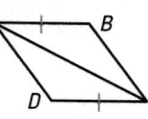

6.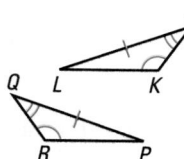

7.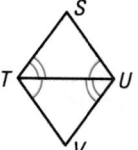

8.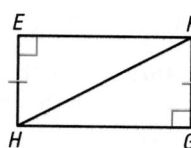

9.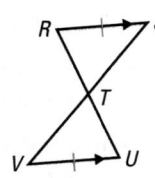

See margin.

10.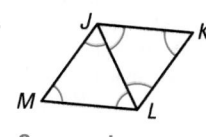

See margin.

5.4 *Hypotenuse-Leg Congruence Theorem: HL* **263**

Mini-Quiz

Determine whether you can use the HL Congruence Theorem to show that the triangles are congruent. Explain your reasoning.

1.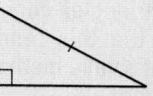

No; there is no information about the lengths of the legs of the right triangles.

2.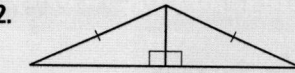

Yes; the pair of hypotenuses are shown to be congruent and a pair of legs can be shown as congruent by the Reflexive Property of Congruence, so the HL Congruence Theorem can be used.

3. Emile and his team have erected this cell phone tower and supporting guy wires. Explain how he could use the HL Congruence Theorem to show that the guy wires are attached to the ground at equal distances from the base of the tower.

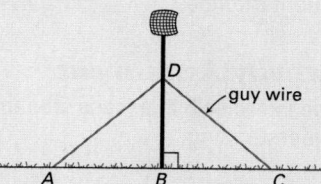

Emile could show that the guy wires have equal length and are attached to the tower at the same point. Then he would have enough information to prove △*ABD* ≅ △*CBD* by the HL Congruence Theorem. Then Emile could prove that $\overline{AB} \cong \overline{CB}$.

Planning the Activity

Goal
Students explore whether or not there is a side-side-angle congruence postulate for triangles.

Materials
See the Technology Activity Keystrokes Masters in the *Chapter 5 Resource Book*.

LINK TO LESSON

In Lesson 5.2–5.4 students learned a number of methods for proving triangles congruent. Throughout Lesson 5.5, students will use these methods to prove triangles congruent. It is important that they remember that SSA is not a valid method for proving congruence.

Managing the Activity

Tips for Success
In Step 1, instruct students not to draw $\overleftrightarrow{AB}$ close to being horizontal. The measure of ∠A should be between 30° and 60°.

Closing the Activity

KEY DISCOVERY
Two triangles can have two pairs of congruent sides and a pair of congruent angles without being congruent triangles.

Activity Assessment
Use Exercise 6 to assess student understanding.

Technology Activity

5.4 **Investigating Congruence**

Question

Is there a side-side-angle congruence postulate or theorem?

Explore

1 Draw a segment and label it $\overline{AB}$. Draw a point not on $\overleftrightarrow{AB}$ and label this point E. Construct $\overleftrightarrow{AE}$.

2 Draw a circle centered at point B that intersects $\overleftrightarrow{AE}$ in two points as shown. Label the intersection points G and H.

3 Draw $\overline{BG}$ and $\overline{BH}$. Hide the circle.

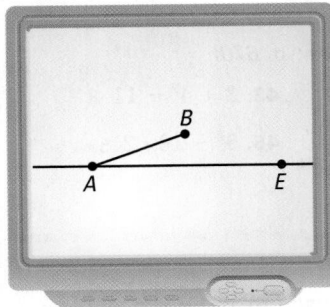

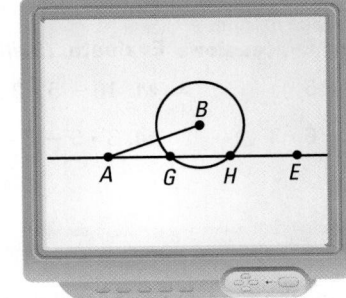

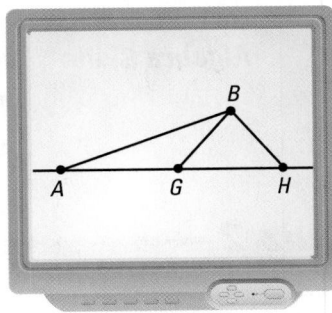

6. There is no side-side-angle congruence postulate because it is possible for two sides and a non-included angle of one triangle to be congruent to two sides and the corresponding non-included angle of another triangle without the two triangles being congruent.

Think About It

1. Find the measures of the segments and angles listed in the tables.
   Answers will vary, but *BG = BH*.

Segments

$\overline{AB}$	$\overline{AG}$	$\overline{BG}$	$\overline{BH}$	$\overline{AH}$
?	?	?	?	?

Angles

∠ABG	∠BAG	∠AGB	∠ABH	∠BAH	∠AHB
?	?	?	?	?	?

Student Help

VISUAL STRATEGY
Drawing overlapping triangles separately makes it easier to see the triangles. An example is on p. 232.

2. List the sides of △ABG that are congruent to the sides of △ABH.
   AB ≅ AB and *BG ≅ BH*
3. List the angles of △ABG that are congruent to the angles of △ABH.
   ∠*BAG ≅ ∠BAH*
4. Do you think that △ABH ≅ △ABG? Explain your reasoning.
   No; *Sample answer:* △*ABG* and △*ABH* have different shapes.
5. Sketch △ABG and △ABH separately on a piece of paper. Mark the corresponding congruences on the triangles. **Check diagrams.**

6. Explain why there is no side-side-angle congruence postulate.
   See margin.

 **Using Congruent Triangles**

Goal

Show corresponding parts of congruent triangles are congruent.

Key Words

- corresponding parts p. 233

If you know that two triangles are congruent, you can use the definition of congruent triangles from Lesson 5.1 to conclude that the corresponding parts are congruent.

If you know $\triangle ABC \cong \triangle DEF$, then you can conclude:

Corresponding Sides	Corresponding Angles
$\overline{AB} \cong \overline{DE}$	$\angle A \cong \angle D$
$\overline{BC} \cong \overline{EF}$	$\angle B \cong \angle E$
$\overline{AC} \cong \overline{DF}$	$\angle C \cong \angle F$

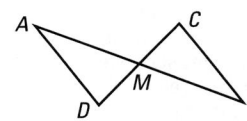

EXAMPLE **1** *Use Corresponding Parts*

In the diagram, $\overline{AB}$ and $\overline{CD}$ bisect each other at M. Prove that $\angle A \cong \angle B$.

Solution

❶ First sketch the diagram and label any congruent segments and congruent angles.

❷ Because $\angle A$ and $\angle B$ are corresponding angles in $\triangle ADM$ and $\triangle BCM$, show that $\triangle ADM \cong \triangle BCM$ to prove that $\angle A \cong \angle B$.

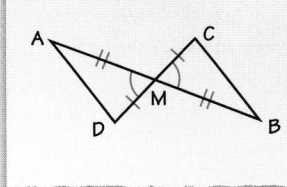

Statements	Reasons
1. $\overline{AB}$ and $\overline{CD}$ bisect each other at M.	1. Given
2. $\overline{MA} \cong \overline{MB}$	2. Definition of segment bisector
3. $\angle AMD \cong \angle BMC$	3. Vertical Angles Theorem
4. $\overline{MD} \cong \overline{MC}$	4. Definition of segment bisector
5. $\triangle ADM \cong \triangle BCM$	5. SAS Congruence Postulate
6. $\angle A \cong \angle B$	6. Corresponding parts of congruent triangles are congruent.

Student Help

LOOK BACK
For the definition of congruent figures, see p. 233.

 **Plan**

Pacing

Suggested Number of Days

Basic: 2 days
Average: 2 days
Advanced: 2 days
Block Schedule: 1 block

Teaching Resources

 Blacklines
(See page 230B.)

Transparencies
- Warm-Up with Quiz
- Answers

Technology
- Electronic Teacher Tools
- Test and Practice Generator
- Online Lesson Planner
- Internet Support

 **Teach**

Content and Teaching Strategies

For background information on geometric concepts and teaching strategies related to this lesson, see pages 230E and 230F in this Teacher's Edition.

Tips for New Teachers

Constructions and drawings provide a hands-on opportunity to explore a plan and reach valid conclusions. Overlapping figures and figures with triangles embedded in them are difficult for some students to understand. Have students practice outlining the figures in different colors or drawing them as separate figures. See the Tips for New Teachers on pp. 1–2 of the *Chapter 5 Resource Book* for additional notes about Lesson 5.5.

Extra Example 1
See next page.

265

Extra Example 1

In the diagram, $\overleftrightarrow{FK}$ bisects $\angle GHJ$ and $\angle GFJ$. Prove that $\angle G \cong \angle J$.

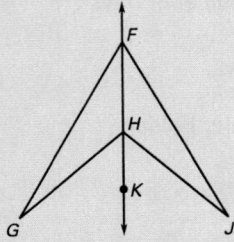

Statements (Reasons)

1. $\overleftrightarrow{FK}$ bisects $\angle GHJ$ and $\angle GFJ$. (Given)
2. $\angle GFH \cong \angle JFH$; $\angle GHK \cong \angle JHK$ (Definition of angle bisector)
3. $\angle GHF$ and $\angle GHK$ are a linear pair; $\angle JHF$ and $\angle JHK$ are a linear pair. (Definition of linear pair)
4. $\angle GHF$ and $\angle GHK$ are supplements; $\angle JHF$ and $\angle JHK$ are supplements. (Linear Pair Postulate)
5. $\angle GHF \cong \angle JHF$ (Supplements of congruent angles are congruent.)
6. $\overline{HF} \cong \overline{HF}$ (Reflexive Property of Congruence)
7. $\triangle GHF \cong \triangle JHF$ (ASA Congruence Postulate)
8. $\angle G \cong \angle J$ (Corresponding parts of congruent triangles are congruent.)

Extra Example 2

Sketch the overlapping triangles separately. Mark all congruent angles and sides. Then tell what theorem or postulate you can use to show that $\triangle ABC \cong \triangle BAD$.

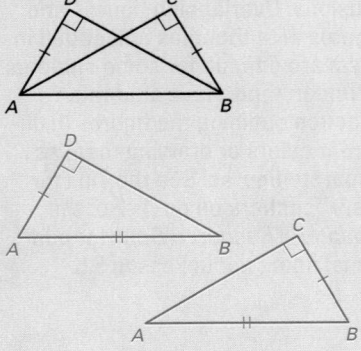

$\overline{BC} \cong \overline{AD}$ and $\overline{BA} \cong \overline{AB}$, so $\triangle ABC \cong \triangle BAD$ by the HL Congruence Theorem.

266

Diagrams often show overlapping triangles. It is usually helpful to visualize or redraw the triangles so that they do not overlap.

Original diagram

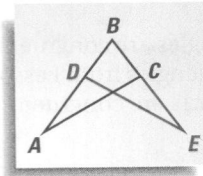

$\triangle ABC$ and $\triangle EBD$ overlap.

Redrawn diagram

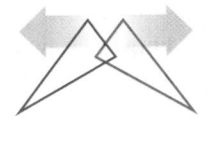

$\triangle ABC$ and $\triangle EBD$ do not overlap.

Student Help

VISUAL STRATEGY
Drawing overlapping triangles separately makes it easier to see the triangles and correctly mark congruent sides and angles, as shown on p. 232.

EXAMPLE 2 Visualize Overlapping Triangles

Sketch the overlapping triangles separately. Mark all congruent angles and sides. Then tell what theorem or postulate you can use to show $\triangle JGH \cong \triangle KHG$.

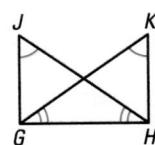

Solution

❶ Sketch the triangles separately and mark any given information. Think of $\triangle JGH$ moving to the left and $\triangle KHG$ moving to the right.

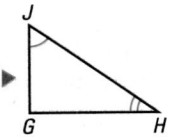 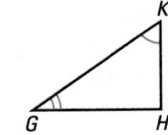

Mark $\angle GJH \cong \angle HKG$ and $\angle JHG \cong \angle KGH$.

❷ Look at the original diagram for shared sides, shared angles, or any other information you can conclude.

In the original diagram, $\overline{GH}$ and $\overline{HG}$ are the same side, so $\overline{GH} \cong \overline{HG}$.

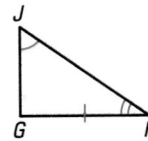 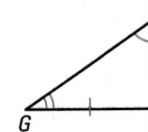

Add congruence marks to $\overline{GH}$ in each triangle.

❸ You can use the AAS Congruence Theorem to show that $\triangle JGH \cong \triangle KHG$.

Student Help
CLASSZONE.COM

MORE EXAMPLES
More examples at classzone.com

EXAMPLE **3** *Use Overlapping Triangles*

Write a proof that shows $\overline{AB} \cong \overline{DE}$.

Given ▶ $\angle ABC \cong \angle DEC$

$\overline{CB} \cong \overline{CE}$

Prove ▶ $\overline{AB} \cong \overline{DE}$

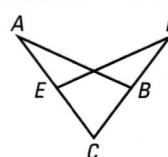

Solution

❶ Sketch the triangles separately. Then label the given information and any other information you can conclude from the diagram.

In the original diagram, $\angle C$ is the same in both triangles ($\angle BCA \cong \angle ECD$).

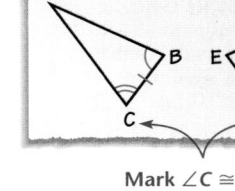

Mark $\angle C \cong \angle C$.

❷ Show $\triangle ABC \cong \triangle DEC$ to prove that $\overline{AB} \cong \overline{DE}$.

Statements	Reasons
1. $\angle ABC \cong \angle DEC$	1. Given
2. $\overline{CB} \cong \overline{CE}$	2. Given
3. $\angle C \cong \angle C$	3. Reflexive Prop. of Congruence
4. $\triangle ABC \cong \triangle DEC$	4. ASA Congruence Postulate
5. $\overline{AB} \cong \overline{DE}$	5. Corresponding parts of congruent triangles are congruent.

Checkpoint ✓ *Use Overlapping Triangles*

1. Tell which triangle congruence theorem or postulate you would use to show that $\overline{AB} \cong \overline{CD}$. **SAS**

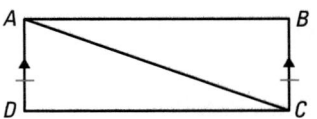

Redraw the triangles separately and label all congruences. Explain how to show that the triangles or corresponding parts are congruent.

2–3. See margin.

2. Given $\overline{KJ} \cong \overline{KL}$ and $\angle J \cong \angle L$, show $\overline{NJ} \cong \overline{ML}$.

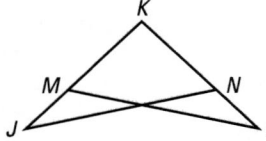

3. Given $\angle SPR \cong \angle QRP$ and $\angle Q \cong \angle S$, show $\triangle PQR \cong \triangle RSP$.

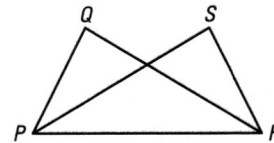

Extra Example 3

In the diagram, $\angle FGH \cong \angle JHG$ and $\angle FHG \cong \angle JGH$. Write a proof that shows $\overline{FG} \cong \overline{JH}$.

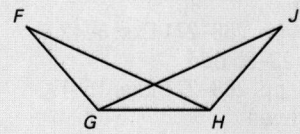

Statements (Reasons)
1. $\angle FGH \cong \angle JHG$ (Given)
2. $\overline{GH} \cong \overline{HG}$ (Reflexive Property of Congruence)
3. $\angle FHG \cong \angle JGH$ (Given)
4. $\triangle FGH \cong \triangle JHG$ (ASA Congruence Postulate)
5. $\overline{FG} \cong \overline{JH}$ (Corresponding parts of congruent triangles are congruent.)

✔ Concept Check

Why is it useful to know that two triangles are congruent? You can use the definition of congruent triangles to conclude that the corresponding parts are congruent.

🧩 Daily Puzzler

In the diagram, $\triangle PQS \cong \triangle STP$. How would you prove that $\triangle PQR \cong \triangle STR$?

Use the AAS Congruence Theorem:
$\angle PRQ \cong \angle SRT$ (Vertical angles are congruent.)
$\angle Q \cong \angle T$ (All right angles are $\cong$.)
$\overline{PQ} \cong \overline{ST}$ (Given)
$\triangle PQR \cong \triangle STR$ (AAS Congruence Theorem)

2–3. See Additional Answers beginning on page AA1.

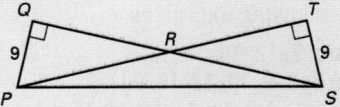

5.5 *Using Congruent Triangles* **267**

268

Guided Practice

Vocabulary Check

1. In the diagram, $\triangle ABC \cong \triangle DEF$. Why can you conclude that $\angle CBA \cong \angle FED$?
Corresponding parts of congruent triangles are congruent.

2. **Visualize It!** Tell which diagram correctly represents all the congruences in the original figure. B

Original figure   **A.**   **B.**

Skill Check

3. When two parallel lines are cut by a transversal, alternate interior angles are congruent, so $\angle VSU \cong \angle TUS$ and $\angle TSU \cong \angle VUS$; also $\overline{SU} \cong \overline{SU}$; $\triangle STU \cong \triangle UVS$ by the ASA Congruence Postulate; $\angle STU \cong \angle UVS$ since corresponding parts of congruent triangles are congruent.

4. $\angle A \cong \angle D$, $\angle ACB \cong \angle DBC$, and $\overline{BC} \cong \overline{BC}$; $\triangle ABC \cong \triangle DCB$ by the AAS Congruence Theorem; $\overline{AB} \cong \overline{DC}$ since corresponding parts of congruent triangles are congruent.

5. $\overline{LN} \cong \overline{LK}$, $\angle LNM \cong \angle LKJ$, and $\angle L \cong \angle L$; $\triangle JKL \cong \triangle MNL$ by the ASA Congruence Postulate; $\angle J \cong \angle M$ since corresponding parts of congruent triangles are congruent.

Explain how to show that the statement is true.

3. $\angle STU \cong \angle UVS$ 4. $\overline{AB} \cong \overline{DC}$ 5. $\angle J \cong \angle M$

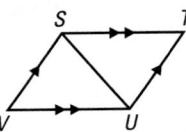   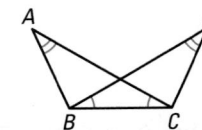   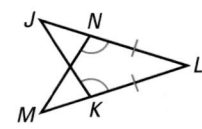

Practice and Applications

Extra Practice
See p. 684.

6. **Showing Congruence** In the diagram, $\triangle JKL \cong \triangle PQR$. Why can you conclude that $\angle K \cong \angle Q$?
Corresponding parts of congruent triangles are congruent.

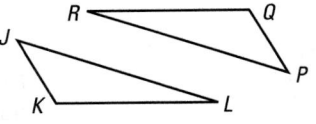

Finding Congruent Parts Tell which triangles you need to show are congruent in order to show that the statement is true.

7. $\angle A \cong \angle D$
   $\triangle ABC \cong \triangle DBC$

8. $\angle J \cong \angle N$
   $\triangle JKL \cong \triangle NML$

9. $\overline{DE} \cong \overline{BA}$
   $\triangle ABC \cong \triangle EDF$

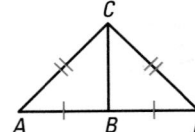   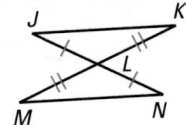   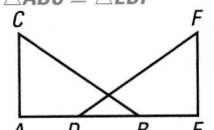

Homework Help
Example 1: Exs. 15, 16, 19
Example 2: Exs. 10–13
Example 3: Exs. 14–20

12. AAS Congruence Theorem; ∠C ≅ ∠F, ∠CBA ≅ ∠FED, and $\overline{CA} \cong \overline{FD}$.

13. ASA Congruence Postulate; ∠GJH ≅ ∠KMH, and $\overline{MH} \cong \overline{JH}$; ∠H ≅ ∠H by the Reflexive Property of Congruence.

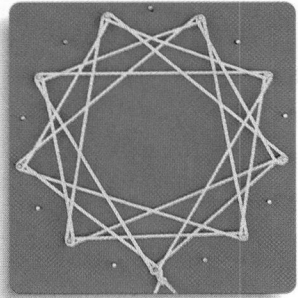

STRING DESIGNS The shape and size of a string design is determined by how many points along a circle are used to create the design.

Visualize It! Sketch the overlapping triangles separately. Mark all congruent angles and sides. Then tell what theorem or postulate you can use to show that the triangles are congruent.

10. $\overline{BC} \cong \overline{DA}$, ∠ADB ≅ ∠CBD

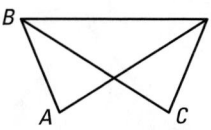

Check diagrams; SAS Congruence Postulate.

11. ∠E ≅ ∠H, $\overline{EF} \cong \overline{HJ}$

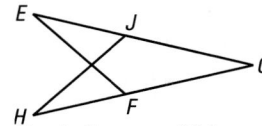

Check diagrams; AAS Congruence Theorem.

String Designs What theorem or postulate can you use to show that the triangles in the string design are congruent? Explain your reasoning. See margin.

12. △ABC ≅ △DEF

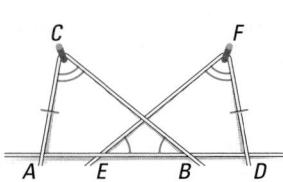

13. △GHJ ≅ △KHM

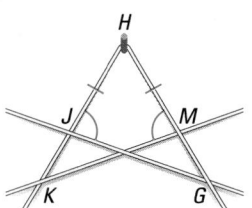

14. Logical Reasoning Fill in the missing statements and reasons.

Given ▷ $\overline{AB} \cong \overline{AE}$
∠ACB ≅ ∠ADE

Prove ▷ ∠B ≅ ∠E

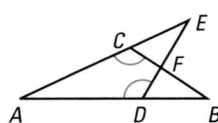

Statements	Reasons
1. $\overline{AB} \cong \overline{AE}$	**1.** _____?_____ Given
2. _____?_____ ∠ACB ≅ ∠ADE	**2.** Given
3. _____?_____ ∠A ≅ ∠A	**3.** Reflexive Prop. of Congruence
4. △ABC ≅ △AED	**4.** _____?_____ AAS Congruence Theorem
5. ∠B ≅ ∠E	**5.** _____?_____ Corresp. parts of ≅ triangles are ≅.

Finding Congruent Parts Use the information in the diagram to prove that the statement is true. 15–17. See margin.

15. ∠A ≅ ∠C

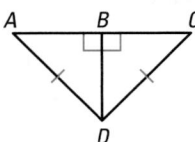

16. $\overline{JK} \cong \overline{NM}$

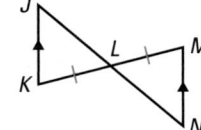

17. $\overline{QT} \cong \overline{SR}$

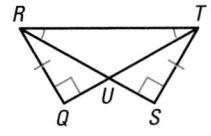

15. Statements (Reasons)
1. $\overline{AD} \cong \overline{CD}$ (Given)
2. ∠ABD and ∠CBD are right angles. (Given)
3. △ABD and △CBD are right triangles. (Def. of right triangle)
4. $\overline{BD} \cong \overline{BD}$ (Reflexive Property of Congruence)
5. △ABD ≅ △CBD (HL Congruence Theorem)
6. ∠A ≅ ∠C (Corresp. parts of ≅ triangles are ≅.)

16. Sample answer:
Statements (Reasons)
1. $\overline{KL} \cong \overline{ML}$ (Given)
2. $\overline{JK} \parallel \overline{NM}$ (Given)
3. ∠J ≅ ∠N (Alternate Interior Angles Theorem)
4. ∠K ≅ ∠M (Alternate Interior Angles Theorem)
5. △KJL ≅ △MNL (AAS Congruence Theorem)
6. $\overline{JK} \cong \overline{NM}$ (Corresp. parts of ≅ triangles are ≅.)

17. Sample answer:
Statements (Reasons)
1. $\overline{RQ} \cong \overline{TS}$ (Given)
2. ∠RTQ ≅ ∠TRS (Given)
3. ∠Q and ∠S are right angles. (Given)
4. ∠Q ≅ ∠S (Right angles are congruent.)
5. △RTQ ≅ △TRS (AAS Congruence Theorem)
6. $\overline{QT} \cong \overline{SR}$ (Corresp. parts of ≅ triangles are ≅.)

18. Argyle Patterns In the argyle pattern shown below, $\overline{UV} \cong \overline{VW} \cong \overline{XY} \cong \overline{YZ}$ and $\angle UVW \cong \angle XYZ$. Prove that $\overline{WU} \cong \overline{ZX}$. **See margin.**

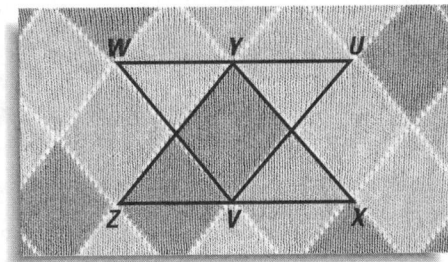

Logical Reasoning In Exercises 19 and 20, fill in the missing statements and reasons.

19. Given ▷ $\overline{BD}$ and $\overline{AE}$ bisect each other at C.
  Prove ▷ $\angle A \cong \angle E$

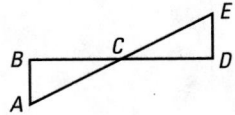

Statements	Reasons
1. $\overline{BD}$ and $\overline{AE}$ bisect each other at C.	1. _____?_____ Given
2. $\overline{BC} \cong \overline{DC}$	2. _____?_____ Def. of segment bisector
3. _____?_____ $\overline{AC} \cong \overline{EC}$	3. Def. of segment bisector
4. $\angle BCA \cong \angle DCE$	4. _____?_____ Vertical Angles Theorem
5. $\triangle ABC \cong \triangle EDC$	5. _____?_____ SAS Congruence Postulate
6. $\angle A \cong \angle E$	6. _____?_____ Corresp. parts of $\cong$ triangles are $\cong$.

20. Given ▷ $\overline{JK} \perp \overline{LK}, \overline{ML} \perp \overline{KL}$
    $\overline{JL} \cong \overline{MK}$
  Prove ▷ $\overline{JK} \cong \overline{ML}$

Statements	Reasons
1. _____?_____ $\overline{JL} \cong \overline{MK}$	1. Given
2. $\overline{JK} \perp \overline{LK}, \overline{ML} \perp \overline{KL}$	2. _____?_____ Given
3. _____?_____	3. $\perp$ lines form right angles.
4. $\triangle JKL$ and $\triangle MLK$ are right triangles.	4. _____?_____ $\perp$ lines form right $\measuredangle$; def. of right $\triangle$
5. _____?_____ $\overline{KL} \cong \overline{KL}$	5. Reflexive Prop. of Congruence
6. $\triangle JKL \cong \triangle MLK$	6. _____?_____ HL Congruence Theorem
7. _____?_____ $\overline{JK} \cong \overline{ML}$	7. _____?_____ Corresp. parts of $\cong$ triangles are $\cong$.

21. Challenge In the figure at the right, $\overline{LK} \parallel \overline{PN}$, $\angle LJK \cong \angle PMN$, and $\overline{JM} \cong \overline{ML} \cong \overline{LP}$. What theorem or postulate can be used to show that $\triangle JKL \cong \triangle MNP$? Explain.

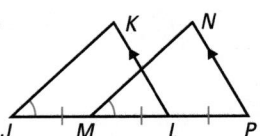

Standardized Test Practice

21. ASA Congruence Postulate; $\angle LJK \cong \angle PMN$, $\overline{JL} \cong \overline{MP}$, and $\angle JLK \cong \angle MPN$ (corresponding angles); two angles and the included side of $\triangle JKL$ are congruent to two angles and the included side of $\triangle MNP$.

22. Multiple Choice In the diagram, suppose that $\overline{AD} \cong \overline{CB}$ and $\angle BCA \cong \angle DAC$. Which triangles can you use to prove that $\angle EBA \cong \angle EDC$? **A**

ⓐ $\triangle ABC$ and $\triangle CDA$

ⓑ $\triangle ABE$ and $\triangle CDE$

ⓒ $\triangle DEB$ and $\triangle AEC$

ⓓ Not enough information

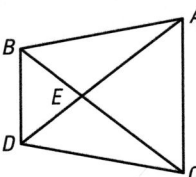

23. Multiple Choice In the diagram, suppose that $\overline{AE} \cong \overline{CE}$ and $\overline{BE} \cong \overline{DE}$. Which triangles can you use to prove that $\overline{AB} \cong \overline{CD}$? **G**

ⓕ $\triangle ABC$ and $\triangle CDA$

ⓖ $\triangle ABE$ and $\triangle CDE$

ⓗ $\triangle DEB$ and $\triangle AEC$

ⓙ Not enough information

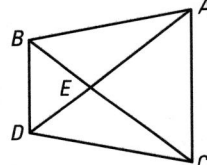

Mixed Review

Angle Bisectors In the diagram, $\overrightarrow{BD}$ bisects $\angle ABC$. Find the value of x. *(Lesson 2.2)*

24.
28
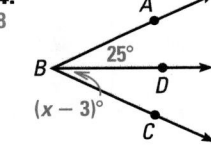

25.
18
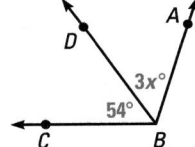

26.
6
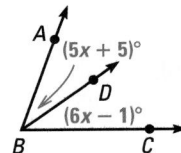

Perpendicular Lines Find the value of x, given that $p \perp q$. *(Lesson 3.2)*

27.
42
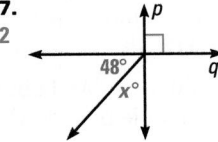

28.
13
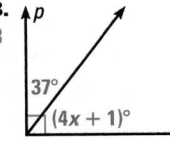

29.
11
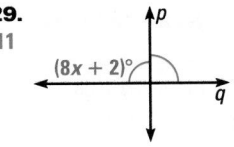

Algebra Skills

Solving Equations Solve the equation. *(Skills Review, p. 673)*

30. $x + 5 = 8$ 3

31. $7x = -63$ -9

32. $4x - 9 = 23$ 8

33. $11 + 3x = 32$ 7

34. $5x - 3x + 10 = 24$ 7

35. $x + 2x - 8 = 19$ 9

5.5 Using Congruent Triangles 271

④ Assess

Assessment Resources

The Mini-Quiz below is also available on blackline (*Chapter 5 Resource Book*, p. 56) and on transparency. For more assessment resources, see:
• Chapter 5 Resource Book
• Standardized Test Practice
• Test and Practice Generator

Mini-Quiz

1. Sketch the overlapping triangles separately. Mark all congruent angles and sides. Then tell what theorem or postulate you can use to show that the triangles are congruent.

$\overline{KL} \cong \overline{NM}, \angle KLM \cong \angle NML$

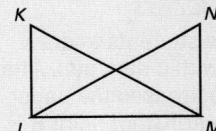

Check diagram; SAS Congruence Postulate.

2. Use the information in the diagram to show that the given statement is true.

$\overline{HJ} \cong \overline{KJ}$

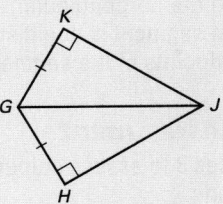

It is given that $\overline{GH} \cong \overline{GK}$, and $\overline{GJ} \cong \overline{GJ}$ by the Reflexive Property of Congruence. $\triangle GHJ$ and $\triangle GKJ$ are right triangles by definition. By the HL Congruence Theorem, $\triangle GHJ \cong \triangle GKJ$.

Therefore, $\overline{HJ} \cong \overline{KJ}$ because corresponding parts of congruent triangles are congruent.

Planning the Activity

Goal
Students use paper folding to discover the properties of a perpendicular bisector.

Materials
See the margin of the student page.

LINK TO LESSON
In Example 2 on page 274, students apply what they have learned about perpendicular bisectors to find a segment length.

Managing the Activity

Tips for Success
In Step 1, urge students to draw the segment with a dark pen. After students have creased the paper, suggest they verify that point B is directly on top of point A.

Closing the Activity

KEY DISCOVERY
Any point on the perpendicular bisector of a segment is equidistant from the endpoints of the segment.

Activity Assessment
Use Exercises 3 to assess student understanding.

Step 4. Lengths will vary, but $MA = MB$; $m\angle CMA = \angle CMB$.

Activity 5.6 Investigating Bisectors

Question

What is true about any point on the perpendicular bisector of a segment?

Materials
• ruler
• protractor

Explore

❶ On a piece of paper, draw $\overline{AB}$. Fold the paper so that point B lies directly on point A.

❷ Draw a line along the crease in the paper. Label the point where the line intersects $\overline{AB}$ as point M.

❸ Label another point on the line you made in Step 2 as point C. Draw $\overline{CA}$ and $\overline{CB}$.

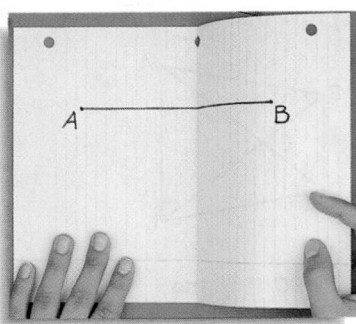
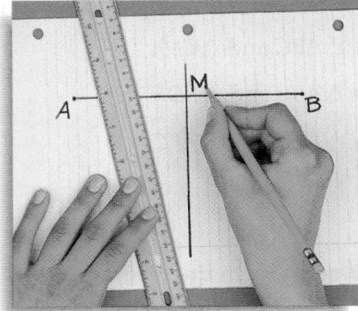
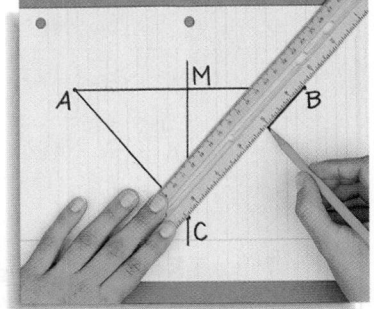

❹ Copy and complete the table at right. **See Margin.**

Segment or angle	$\overline{MA}$	$\overline{MB}$	$\angle CMA$	$\angle CMB$
Measure	?	?	?	?

Think About It

Student Help

> **LOOK BACK**
> To review segment bisectors, see p. 53.

1. Line $\overleftrightarrow{CM}$ is called the *perpendicular bisector* of $\overline{AB}$. Why do you think this describes $\overleftrightarrow{CM}$? $\overleftrightarrow{CM}$ is perpendicular to $\overline{AB}$ and $\overleftrightarrow{CM}$ bisects $\overline{AB}$.

2. Choose three other points on $\overleftrightarrow{CM}$. Label the points D, E, and F. Copy and complete the table below. What do you notice?

Point C	Point D	Point E	Point F
CA = ?	DA = ?	EA = ?	FA = ?
CB = ?	DB = ?	EB = ?	FB = ?

2. Answers will vary, but $CA = CB$, $DA = DB$, $EA = EB$, and $FA = FB$.

3. Any point on the perpendicular bisector of a segment is the same distance from the endpoints of the segment.

3. **Reasoning** What is true about any point on the perpendicular bisector of a segment?

5.6 Angle Bisectors and Perpendicular Bisectors

Goal
Use angle bisectors and perpendicular bisectors.

Key Words
- distance from a point to a line
- equidistant
- angle bisector p. 61
- perpendicular bisector

The **distance from a point to a line** is measured by the length of the perpendicular segment from the point to the line.

When a point is the same distance from one line as it is from another line, the point is **equidistant** from the two lines.

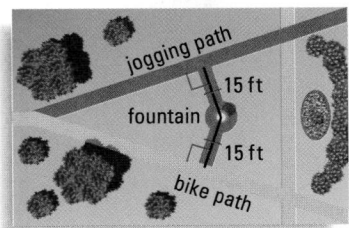

The fountain is equidistant from the jogging path and the bike path.

THEOREM 5.3

Angle Bisector Theorem

Words If a point is on the bisector of an angle, then it is equidistant from the two sides of the angle.

If 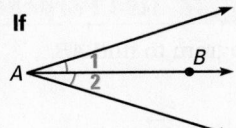 **then**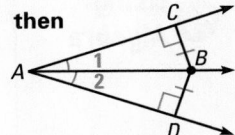

Symbols If $m\angle 1 = m\angle 2$, then $\overline{BC} \cong \overline{BD}$.

EXAMPLE 1 Use the Angle Bisector Theorem

Prove that $\triangle TWU \cong \triangle VWU$.

Given ▷ $\overrightarrow{UW}$ bisects $\angle TUV$.
$\triangle UTW$ and $\triangle UVW$ are right triangles.

Prove ▷ $\triangle TWU \cong \triangle VWU$.

Solution

Statements	Reasons
1. $\overrightarrow{UW}$ bisects $\angle TUV$.	1. Given
2. $\triangle UTW$ and $\triangle UVW$ are right triangles.	2. Given
H 3. $\overline{WU} \cong \overline{WU}$	3. Reflexive Prop. of Congruence
L 4. $\overline{WV} \cong \overline{WT}$	4. Angle Bisector Theorem
5. $\triangle TWU \cong \triangle VWU$	5. HL Congruence Theorem

---

274

Extra Example 1

In the diagram, $\overrightarrow{EG}$ bisects $\angle CEF$. Prove that $\triangle CDE \cong \triangle FDE$.

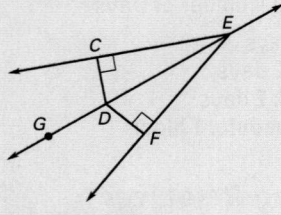

Statements (Reasons)

1. $\overrightarrow{EG}$ bisects $\angle CEF$. (Given)
2. $\angle CED \cong \angle FED$ (Definition of angle bisector)
3. $\angle DCE$ and $\angle DFE$ are right angles. (Given)
4. $\angle DCE \cong \angle DFE$ (All right angles are congruent.)
5. $\overline{DC} \cong \overline{DF}$ (Angle Bisector Theorem)
6. $\triangle CDE \cong \triangle FDE$ (AAS Congruence Theorem)

Multiple Representations

Point out that if a line is known to be a perpendicular bisector of a segment, then on a diagram you can mark a right angle where the line intersects the segment, and you can use tick marks to indicate the two congruent halves of the segment.

Extra Example 2

Use the diagram to find GH. **23**

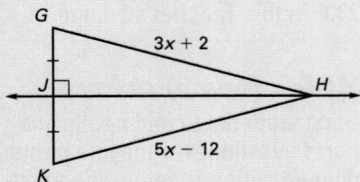

Perpendicular Bisectors A segment, ray, or line that is perpendicular to a segment at its midpoint is called a **perpendicular bisector** .

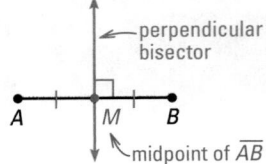

perpendicular bisector

midpoint of $\overline{AB}$

THEOREM 5.4

Perpendicular Bisector Theorem

Words If a point is on the perpendicular bisector of a segment, then it is equidistant from the endpoints of the segment.

Symbols If C is on the perpendicular bisector of $\overline{AB}$, then $\overline{CA} \cong \overline{CB}$.

If 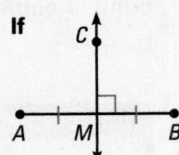   then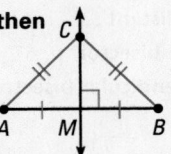

Using Algebra

EXAMPLE 2 Use Perpendicular Bisectors

Use the diagram to find AB.

Solution

In the diagram, $\overleftrightarrow{AC}$ is the perpendicular bisector of $\overline{DB}$.

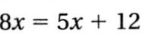

$8x = 5x + 12$ By the Perpendicular Bisector Theorem, $AB = AD$.

$3x = 12$ Subtract $5x$ from each side.

$\dfrac{3x}{3} = \dfrac{12}{3}$ Divide each side by 3.

$x = 4$ Simplify.

You are asked to find AB, not just the value of x.

ANSWER ▸ $AB = 8x = 8 \cdot 4 = 32$

Checkpoint ✓ **Use Angle Bisectors and Perpendicular Bisectors**

1. Find FH. **5**

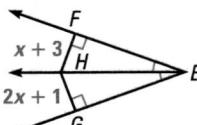

2. Find MK. **20**

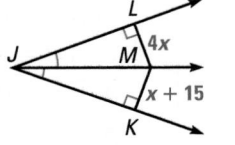

3. Find EF. **15**

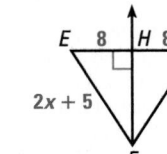

EXAMPLE 3 Use the Perpendicular Bisector Theorem

In the diagram, $\overleftrightarrow{MN}$ is the perpendicular bisector of $\overline{ST}$. Prove that $\triangle MST$ is isosceles.

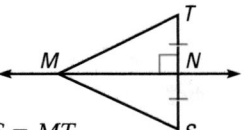

Solution

To prove that $\triangle MST$ is isosceles, show that $MS = MT$.

Statements	Reasons
1. $\overleftrightarrow{MN}$ is the $\perp$ bisector of $\overline{ST}$.	1. Given
2. $MS = MT$	2. Perpendicular Bisector Theorem
3. $\triangle MST$ is isosceles.	3. Def. of isosceles triangle

Intersecting Bisectors One consequence of the Perpendicular Bisector Theorem is that the perpendicular bisectors of a triangle intersect at a point that is equidistant from the vertices of the triangle.

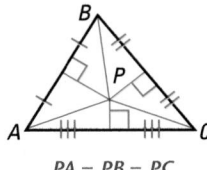

$PA = PB = PC$

Link to
Careers

FACILITIES PLANNER
By finding a location for a warehouse that is easily accessible to all its stores, a facilities planner helps a company save money and run more efficiently.

Career Links
CLASSZONE.COM

EXAMPLE 4 Use Intersecting Bisectors of a Triangle

A company plans to build a warehouse that is equidistant from each of its three stores, A, B, and C. Where should the warehouse be built?

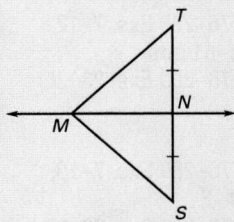

Store A
Store B
Store C

Solution

Think of the stores as the vertices of a triangle. The point where the perpendicular bisectors intersect will be equidistant from each store.

1 Trace the location of the stores on a piece of paper. Connect the points of the locations to form $\triangle ABC$.

2 Draw the perpendicular bisectors of $\overline{AB}$, $\overline{BC}$, and $\overline{CA}$. Label the intersection of the bisectors P.

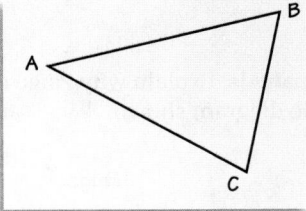

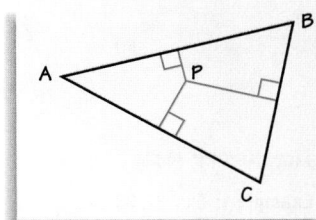

ANSWER ▶ Because P is equidistant from each vertex of $\triangle ABC$, the warehouse should be built near location P.

5.6 Angle Bisectors and Perpendicular Bisectors   **275**

Extra Example 3

Can you prove that $\triangle MST$ is an isosceles triangle if $\overleftrightarrow{MN}$ is only known to be a bisector of $\overline{ST}$?

No, if $\overleftrightarrow{MN}$ is not known to be perpendicular to $\overline{ST}$, you cannot prove that point M is equidistant from points S and T.

Extra Example 4

Where do the bisectors of the sides of an equilateral triangle intersect? Explain. Each bisector is a perpendicular bisector of a side as well as a bisector of one of the angles. The three bisectors intersect at the centroid of the triangle.

✗ Common Error

In Example 4, watch for students who confuse perpendicular bisectors with medians. Draw two large congruent obtuse triangles and label one with its perpendicular bisectors and another with its medians so students can see the difference.

✓ Concept Check

Where do the perpendicular bisectors of a triangle intersect? at a point that is equidistant from the vertices of the triangle

🧩 Daily Puzzler

Suppose you are building three new houses in a rural area. On a map, the coordinates of the three home-sites are $A(2, 5)$, $B(2, 1)$, and $C(6, 1)$. If the houses will share a well, what are the coordinates of the point where the well should be located to minimize the amount of pipe needed between the houses and the well? $(4, 3)$

275

Assignment Guide

BASIC
Day 1: pp. 276–280 Exs. 7–12, 14–19, 34, 35–61 odd
Day 2: pp. 276–280 Exs. 23–25, 32, 36–62 even

AVERAGE
Day 1: pp. 276–280 Exs. 7–19, 21, 22, 34–40
Day 2: pp. 276–280 Exs. 20, 23, 25, 26–32, 45–61 odd

ADVANCED
Day 1: pp. 276–280 Exs. 9–19, 34–36, 37–61 odd
Day 2: pp. 276–280 Exs. 20–32, 33*; EC: classzone.com

BLOCK SCHEDULE
pp. 276–280 Exs. 7–23, 25–32, 34–40, 45–61 odd

Extra Practice

• Student Edition, p. 684
• Chapter 5 Resource Book, pp. 58–59

Homework Check

To quickly check student understanding of key concepts, go over the following exercises:

Basic: 10, 14, 17, 23, 32
Average: 11, 15, 18, 23, 32
Advanced: 12, 16, 19, 24, 32

✗ Common Error

Exercise 9 highlights an error commonly made by students. Stress to all students that for PB to equal PC, both $\overline{PB}$ and $\overline{PC}$ must be perpendicular to $\overrightarrow{AP}$.

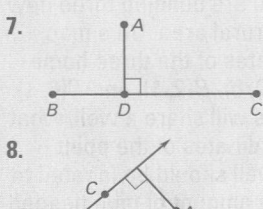

7.

8.

5.6 Exercises

Guided Practice

Vocabulary Check — **Complete the statement.**

1. If a point is on the bisector of an angle, then it is __?__ from the two sides of the angle. **equidistant**

2. If D is on the __?__ of $\overline{AB}$, then D is equidistant from A and B. **perpendicular bisector**

Skill Check — **Use the information in the diagram to find the measure.**

3. Find AD. 16

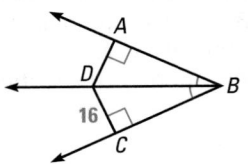

4. Find EF. 3

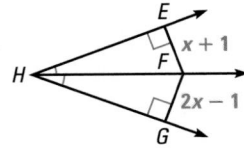

5. Find JM. 12

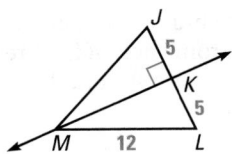

6. Find QR. 20

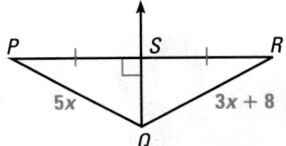

Practice and Applications

Extra Practice
See p. 684.

Visualize It! **Copy each diagram on a piece of paper. Then draw a segment that represents the distance from A to $\overrightarrow{BC}$.** 7, 8. See margin.

7.   A•

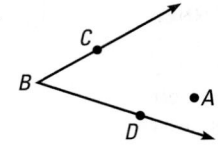

8.

9. **Error Analysis** Explain why Paige cannot make this conclusion, given the diagram shown. **Paige cannot assume that $\overline{PC} \perp \overrightarrow{AC}$.**

Homework Help

Example 1: Exs. 32, 33
Example 2: Exs. 10–12
        14–19
Example 3: Exs. 32, 33
Example 4: Exs. 23–29

> Paige
> By the Angle
> Bisector Theorem,
> $x = 7$.

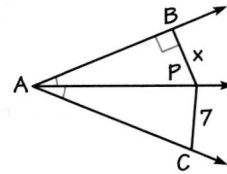

Using Algebra Find the value of *x*.

10.
10

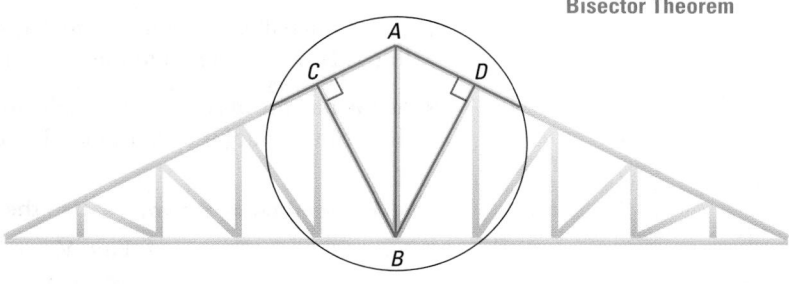

B

x + 5

D

C

15

A

11.
6

M

x   6

J   L

K

12.
2

H

3x   x + 4

E   G   F

13. Roof Trusses In the diagram of the roof truss shown below, you are given that $\overline{AB}$ bisects $\angle CAD$ and that $\angle ACB$ and $\angle ADB$ are right angles. What can you say about $\overline{BC}$ and $\overline{BD}$? Why? $\overline{BC} \cong \overline{BD}$ by the Angle Bisector Theorem

Using Bisectors Use the diagram to find the indicated measure(s).

14. Find $m\angle JKM$. 38°   **15.** Find *SV*. 18   **16.** Find *HG*. 14

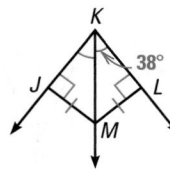

K

38°

J   L

M

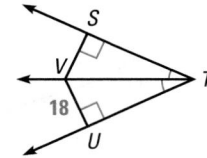

S

V

T

18

U

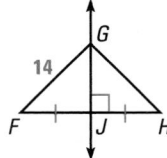

G

14

F   J   H

17. Find *LK*. 3   **18.** Find *PQ*. 5   **19.** Find *AD* and *BC*.

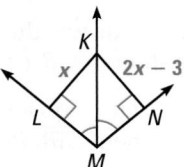

K

x   2x − 3

L   N

M

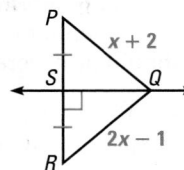

P

x + 2

S   Q

2x − 1

R

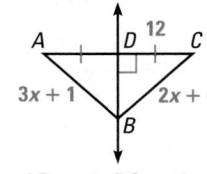

A   D   12   C

3x + 1   2x + 6

B

AD = 12; *BC* = 16

20. **Bridges** In the photo, the road is perpendicular to the support beam and $\overline{AB} \cong \overline{CB}$.

What theorem allows you to conclude that $\overline{AD} \cong \overline{CD}$? Explain. See margin.

20. Perpendicular Bisector Theorem; $\overline{AC} \perp \overline{DB}$ and $\overline{AB} \cong \overline{CB}$, so $\overline{DB}$ is the perpendicular bisector of $\overline{AC}$. Since *D* is on the perpendicular bisector of $\overline{AC}$, it is equidistant from *A* and *C*. Therefore, *AD* = *CD* and $\overline{AD} \cong \overline{CD}$.

Link to
Careers

CIVIL ENGINEERS plan and build large construction projects, such as bridges, canals, and tunnels.

Career Links
CLASSZONE.COM

5.6 Angle Bisectors and Perpendicular Bisectors   **277**

Geometric Reasoning

All students will most likely name the marked congruent segments in Exercise 25. However, some students may not name $\overline{FM} \cong \overline{HM} \cong$ $\overline{GM}$. Refer these students back to the paragraph on intersecting bisectors on page 275 and ask them to use reasoning to name any other unmarked congruent segments.

26–28.

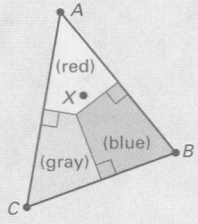

Student Help

LOOK BACK
For more about soccer, see p. 230.

Soccer One way a goalie can determine a good defensive position is to imagine a triangle formed by the goal posts and the ball.

21. When the ball is far from the goal, the goalie most likely stands on line ℓ. How is ℓ related to the goal line ($\overleftrightarrow{AC}$)?
   ℓ is the perpendicular bisector of the goal line.

22. $\overrightarrow{BG}$ should bisect $\angle ABC$ so that the goalkeeper is equidistant from the edges of the region that the goalkeeper is defending. By doing this, the goalkeeper minimizes the distance needed to reach the ball.

22. As the ball moves closer, the goalie moves from line ℓ to other places in front of the goal. How should $\overrightarrow{BG}$ relate to $\angle ABC$? Explain.

Using Perpendicular Bisectors Use the information in the diagram.

23. Find CG and AG.
   $CG = AG = 2$

24. Find VR and VQ.
   $VR = VQ = 16$

25. Name all congruent segments.

25. $\overline{FJ} \cong \overline{JG}$; $\overline{FK} \cong \overline{KH}$; $\overline{GL} \cong \overline{LH}$; $\overline{JM} \cong \overline{KM}$; $\overline{MG} \cong \overline{MH} \cong \overline{FM}$

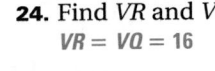
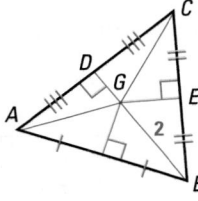
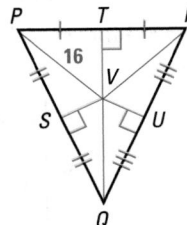
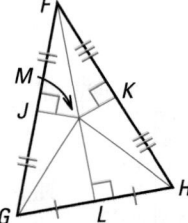

Student Help
CLASSZONE.COM

HOMEWORK HELP
Extra help with problem solving in Exs. 26–29 is at classzone.com

Analyzing a Map In Exercises 26–29, use the map shown and the following information. 26–28. See margin.
A city planner is trying to decide whether a new household at point X should be covered by fire station A, B, or C.

26. Trace the points A, B, C, and X on a piece of paper and draw the segments $\overline{AB}$, $\overline{BC}$, and $\overline{CA}$.

27. Draw the perpendicular bisectors of $\overline{AB}$, $\overline{BC}$, and $\overline{CA}$. Check that they meet at a point.

28. The perpendicular bisectors divide the town into three regions.
   Shade the region closest to fire station A red.
   Shade the region closest to fire station B blue.
   Shade the region closest to fire station C gray.

29. **Writing** In an emergency at household X, which fire station should respond? Explain your choice. The fire station at A; X is in the red region, closest to station A.

Technology In Exercises 30 and 31, use geometry drawing software to complete the steps below.

❶ Draw $\overline{AB}$. Find the midpoint of $\overline{AB}$ and label it C.

❷ Construct the perpendicular bisector of $\overline{AB}$ through C.

❸ Construct point D along the perpendicular bisector. Construct $\overline{DA}$ and $\overline{DB}$.

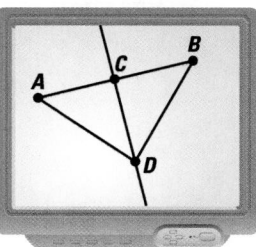

30. What is the relationship between $\overline{DA}$ and $\overline{DB}$? Measure $\overline{DA}$ and $\overline{DB}$ to verify your answer. $\overline{DA} \cong \overline{DB}$

31. Move D to another point along the perpendicular bisector. Will the relationship between $\overline{DA}$ and $\overline{DB}$ stay the same? Why?

31. Yes; by the Perpendicular Bisector Theorem, if D lies on the perpendicular bisector of $\overline{AB}$, then $\overline{DA}$ and $\overline{DB}$ will always be congruent segments.

32. Proving the Perpendicular Bisector Theorem Fill in the missing statements and reasons.

Given ▶ $\overleftrightarrow{AD}$ is the perpendicular bisector of $\overline{BC}$.

Prove ▶ $AB = AC$

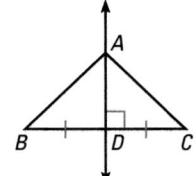

Statements	Reasons
1. $\overleftrightarrow{AD}$ is the perpendicular bisector of $\overline{BC}$.	**1.** ___?_ Given
2. $\overline{DB} \cong \overline{DC}$ $\angle ADB$ and $\angle ADC$ are right angles.	**2.** ___?___ Definition of perpendicular bisector
3. ___?___	**3.** $\perp$ lines form right angles.
4. ___?___ $\angle ADB \cong \angle ADC$	**4.** Right angles are congruent.
5. ___?___ $\overline{AD} \cong \overline{AD}$	**5.** Reflexive Prop. of Congruence
6. $\triangle ADB \cong \triangle ADC$	**6.** ___?___ SAS Congruence Postulate
7. $\overline{AB} \cong \overline{AC}$	**7.** ___?___ Corresp. parts of $\cong$ triangles are $\cong$.
8. ___?___ $AB = AC$	**8.** Def. of congruent segments

Student Help

LOOK BACK
For help with writing proofs, see p. 243.

33. Challenge Use the diagram and the information below to prove the Angle Bisector Theorem. See margin.

Given ▶ D is on the bisector of $\angle BAC$. $\overrightarrow{DB} \perp \overrightarrow{AB}, \overrightarrow{DC} \perp \overrightarrow{AC}$

Prove ▶ $\overline{DB} \cong \overline{DC}$

Hint: First prove that $\triangle ADB \cong \triangle ADC$.

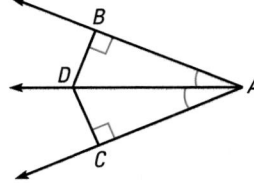

33. Statements (Reasons)
   1. D is on the bisector of $\angle BAC$. (Given)
   2. $\angle BAD \cong \angle CAD$ (Definition of angle bisector)
   3. $\overrightarrow{DB} \perp \overrightarrow{AB}, \overrightarrow{DC} \perp \overrightarrow{AC}$ (Given)
   4. $\angle ABD$ and $\angle ACD$ are right angles. (Perpendicular lines form rt. $\angle$.)
   5. $\angle ABD \cong \angle ACD$ (All right angles are congruent.)
   6. $\overline{AD} \cong \overline{AD}$ (Reflexive Property of Congruence)
   7. $\triangle ADB \cong \triangle ADC$ (AAS Congruence Theorem)
   8. $\overline{DB} \cong \overline{DC}$ (Corresp. parts of $\cong$ triangles are $\cong$.)

Assess

Assessment Resources

The Mini-Quiz below is also available on blackline (*Chapter 5 Resource Book*, p. 65) and on transparency. For more assessment resources, see:
- Chapter 5 Resource Book
- Standardized Test Practice
- Test and Practice Generator

Mini-Quiz

Find the value of *x*.

1. **3**

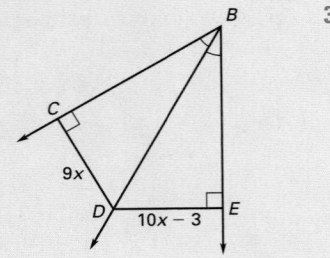

2. **7**

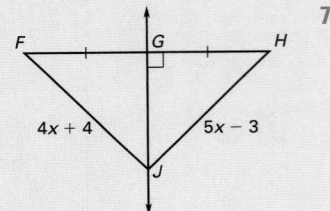

3. Find *RS*. **8**

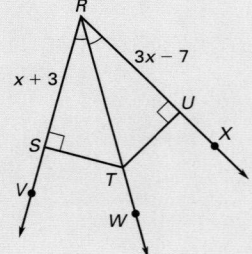

4. $\overleftrightarrow{DE}$ bisects $\angle A$ of equilateral triangle *ABC*. Explain why $\overleftrightarrow{DE}$ is a perpendicular bisector of $\overline{BC}$. **Because the triangle is equilateral, the angle bisector of $\angle A$ divides $\triangle ABC$ into two congruent right triangles. Thus $\overleftrightarrow{DE} \perp \overline{BC}$ and the point of intersection is equidistant from *B* and *C*.**

34. Multiple Choice In the figure at the right, what is *SR*? **B**

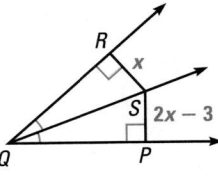

Ⓐ 2 　　　　　　Ⓑ 3

Ⓒ 4 　　　　　　Ⓓ 5

35. Multiple Choice In the figure above, what is *SP*? **G**

Ⓕ 2 　　Ⓖ 3 　　Ⓗ 4 　　Ⓙ 5

36. Multiple Choice What can you say about the figure below, in which $\overleftrightarrow{BE}$ is the perpendicular bisector of $\overline{AC}$? **D**

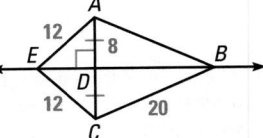

Ⓐ $AB = 20$ 　　　　Ⓑ $AC = 16$

Ⓒ $DC = 8$ 　　　　Ⓓ All of these

Mixed Review

Translations in a Coordinate Plane Find the image of the given point using the translation $(x, y) \rightarrow (x + 3, y - 6)$. *(Lesson 3.7)*

37. $(5, 1)$ $(8, -5)$　**38.** $(-2, 3)$ $(1, -3)$　**39.** $(-4, -4)$ $(-1, -10)$　**40.** $(0, -6)$ $(3, -12)$

41. $(6, 2)$ $(9, -4)$　**42.** $(2, -5)$ $(5, -11)$　**43.** $(10, 12)$ $(13, 6)$　**44.** $(-1, -1)$ $(2, -7)$

45. SAS Congruence Postulate; $DB \cong DB$ and all right angles are congruent, so two sides and the included angle of $\triangle ADB$ are congruent to two sides and the included angle of $\triangle CDB$.

46. AAS Congruence Theorem; $EF \cong EF$, so two angles and a non-included side of $\triangle HEF$ are congruent to two angles and the corresponding non-included side of $\triangle GFE$.

47. ASA Congruence Postulate; vertical angles are congruent, so two angles and the included side of $\triangle JNK$ are congruent to two angles and the included side of $\triangle LNM$.

Determining Congruent Triangles What theorem or postulate, if any, can you use to show that the triangles are congruent? Explain your reasoning. *(Lesson 5.5)*

45. 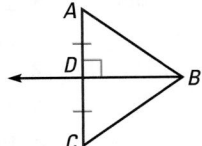　　**46.** 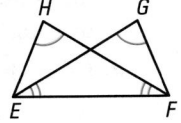　　**47.**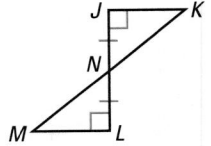

Algebra Skills

Ordering Numbers Write the numbers in order from least to greatest. *(Skills Review, p. 662)*

48. 3, −3, 0.3, −0.3, 0.6, 0
−3, −0.3, 0, 0.3, 0.6, 3

49. −0.25, 1, −0.75, 4, −1.25, 0.25
−1.25, −0.75, −0.25, 0.25, 1, 4

50. −0.4, 0.1, 0, 4.0, −0.1, −4
−4, −0.4, −0.1, 0, 0.1, 4.0

51. −3.3, 3.1, 3.8, −3.9, −3, 3.5
−3.9, −3.3, −3, 3.1, 3.5, 3.8

52. 0.55, −1, 1.1, 1, 0.5, −0.1, 0
−1, −0.1, 0, 0.5, 0.55, 1, 1.1

53. 3.2, 1, 2.1, 3.25, −2.5, 5
−2.5, 1, 2.1, 3.2, 3.25, 5

Solving Equations Solve the equation. *(Skills Review, p. 673)*

54. $4x + 3 = 11$ **2**　　**55.** $2y - 9 = -11$ **−1**　　**56.** $5d - 35 = 90$ **25**

57. $4a + 9a = 39$ **3**　　**58.** $x + 2 = 3x - 4$ **3**　　**59.** $4r - 2 = 5r + 6$ **−8**

60. $q = 2q - 9$ **9**　　**61.** $2z + 5 = 4z - 1$ **3**　　**62.** $10t + 10 = 12t$ **5**

 Chapter 5　*Congruent Triangles*

Activity 5.7 Investigating Reflections

① **Planning the Activity**

Goal
Students use paper folding to discover the properties of reflected figures.

Materials
See the margin of the student page.

LINK TO LESSON
In Example 1 on page 282, students identify reflected figures using techniques learned in this activity.

Question

What happens when a figure is reflected in a line?

Materials
- ruler
- protractor

Explore

❶ Fold a piece of paper in half. Open the paper. Draw a triangle on one side of the fold line and label the vertices X, Y, and Z.

❷ Fold the paper on the fold line and trace X, Y, and Z on the back of the paper. Open the paper, copy the points to the front, and label them X', Y', and Z'.

❸ Draw △X'Y'Z'. Then draw $\overline{XX'}$, $\overline{ZZ'}$, and $\overline{YY'}$. Label the points where $\overline{XX'}$, $\overline{ZZ'}$ and $\overline{YY'}$ intersect the fold line as A, B, and C, as shown.

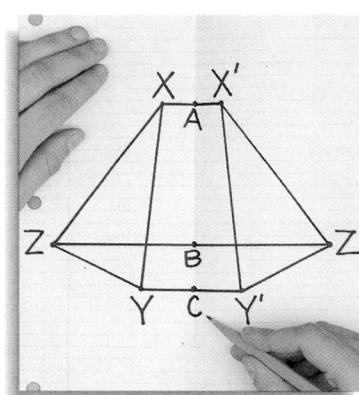

② **Managing the Activity**

Tips for Success
In Step 1, urge students to draw the triangle with a dark pen. Focus students' attention on the prime symbols. Be sure students understand the difference between the points X and X', for example.

③ **Closing the Activity**

KEY DISCOVERY
The segments connecting the corresponding points of a figure and its reflection are perpendicular to the line of reflection and are bisected by it.

Activity Assessment
Use Exercise 4 to assess student understanding.

Student Help

LOOK BACK
For help with prime notation, see p. 153.

1. Answers will vary, but XA = AX', ZB = BZ', and YC = CY'.

2. XA = AX', ZB = BZ', and YC = CY'; each pair of segments is congruent.

3. 90°; 90°; 90°; the measure of each angle is 90°.

4. The fold line is the perpendicular bisector of each segment.

Think About It

1. Copy and complete the table.

Segment	XA	AX'	ZB	BZ'	YC	CY'
Measure	?	?	?	?	?	?

2. Compare XA and AX'. Compare ZB and BZ'. Compare YC and CY'. What do you notice about these segments?

3. Copy and complete the table of angle measures. What do you notice about the angles?

Angle	∠XAB	∠ZBA	∠YCB
Measure	?	?	?

4. How does the fold line relate to $\overline{XX'}$, $\overline{YY'}$, and $\overline{ZZ'}$?

Activity **281**

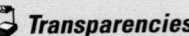

Goal
Identify and use reflections and lines of symmetry.

Key Words
• image p. 152
• reflection
• line of symmetry

A **reflection** is a transformation that creates a mirror image. The original figure is reflected in a line that is called the *line of reflection*.

PROPERTIES OF REFLECTIONS

❶ The reflected image is congruent to the original figure.

❷ The orientation of the reflected image is reversed.

❸ The line of reflection is the perpendicular bisector of the segments joining the corresponding points.

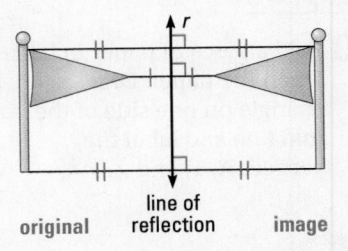

original   line of reflection   image

Visualize It!
clockwise orientation

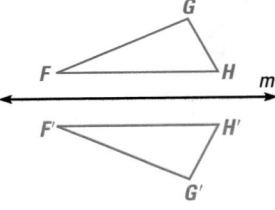

counterclockwise orientation

EXAMPLE 1 Identify Reflections

Tell whether the red triangle is the reflection of the blue triangle in line *m*.

Solution

Check to see if all three properties of a reflection are met.

❶ Is the image congruent to the original figure? Yes. ✔

❷ Is the orientation of the image reversed? Yes. ✔

　△*FGH* has a clockwise orientation.

　△*F'G'H'* has a counterclockwise orientation.

❸ Is *m* the perpendicular bisector of the segments connecting the corresponding points? Yes. ✔

To check, draw a diagram and connect the corresponding endpoints.

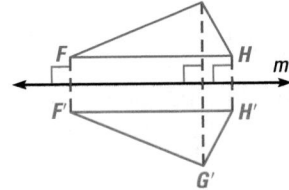

ANSWER ▶ Because all three properties are met, the red triangle is the reflection of the blue triangle in line *m*.

EXAMPLE 2 Identify Reflections

Tell whether the red triangle is the reflection of the blue triangle in line *m*.

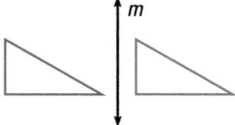

Solution

Check to see if all three properties of a reflection are met.

❶ Is the image congruent to the original figure? Yes. ✔

❷ Is the orientation of the image reversed? No.

ANSWER ▶ The red triangle is *not* a reflection of the blue triangle.

Student Help

VOCABULARY TIP
Use the following relationship to help you remember that a reflection is a flip:

reflection
flip

EXAMPLE 3 Reflections in a Coordinate Plane

a. Which segment is the reflection of $\overline{AB}$ in the *x*-axis? Which point corresponds to *A*? to *B*?

b. Which segment is the reflection of $\overline{AB}$ in the *y*-axis? Which point corresponds to *A*? to *B*?

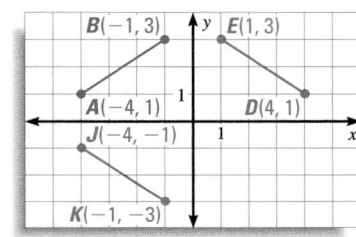

Solution

a. The *x*-axis is the perpendicular bisector of $\overline{AJ}$ and $\overline{BK}$, so the reflection of $\overline{AB}$ in the *x*-axis is $\overline{JK}$.

$A(-4, 1) \rightarrow J(-4, -1)$ *A* is reflected onto *J*.

$B(-1, 3) \rightarrow K(-1, -3)$ *B* is reflected onto *K*.

b. The *y*-axis is the perpendicular bisector of $\overline{AD}$ and $\overline{BE}$, so the reflection of $\overline{AB}$ in the *y*-axis is $\overline{DE}$.

$A(-4, 1) \rightarrow D(4, 1)$ *A* is reflected onto *D*.

$B(-1, 3) \rightarrow E(1, 3)$ *B* is reflected onto *E*.

Checkpoint ✓ Identify Reflections

Tell whether the red figure is a reflection of the blue figure. If the red figure is a reflection, name the line of reflection.

1. yes; the *x*-axis

2. no

3. yes; the *y*-axis

1.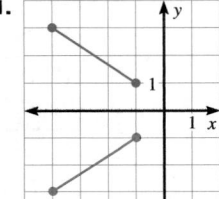

2.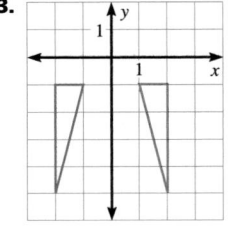

3.

Extra Example 1

Tell whether figure B is a reflection of figure A in line *m*. no

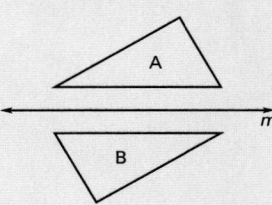

Extra Example 2

Tell whether figure B is a reflection of figure A in line *m*. yes

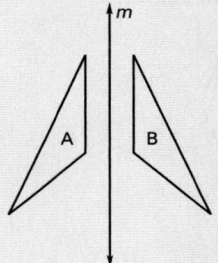

Extra Example 3

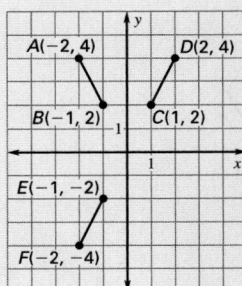

a. Which segment is the reflection of $\overline{AB}$ in the *x*-axis? Which point corresponds to *A*? to *B*?

$\overline{FE}$; *F* corresponds to *A*; *E* corresponds to *B*.

b. Which segment is a reflection of $\overline{AB}$ in the *y*-axis? Which point corresponds to *A*? to *B*?

$\overline{DC}$; *D* corresponds to *A*; *C* corresponds to *B*.

Extra Example 4

Determine the number of lines of symmetry in a rhombus. **2**

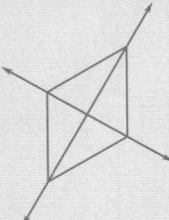

Extra Example 5

Determine the number of lines of symmetry in each figure.

a. 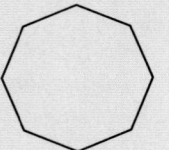   8

b. 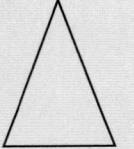   1

c. 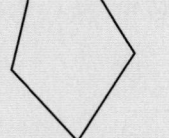   0

Symmetry In the photo, the mirror's edge creates a *line of symmetry*. A figure in the plane has a **line of symmetry** if the figure can be reflected onto itself by a reflection in the line.

A line of symmetry is a line of reflection.

Visualize It!

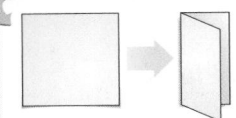

You may want to draw a shape on paper, cut it out, and then fold it to find the lines of symmetry.

EXAMPLE 4 **Determine Lines of Symmetry**

Determine the number of lines of symmetry in a square.

Solution

Think about how many different ways you can fold a square so that the edges of the figure match up perfectly.

vertical fold   **horizontal fold**   **diagonal fold**   **diagonal fold**

 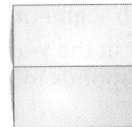  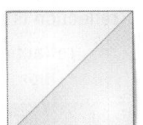

ANSWER ▶ A square has four lines of symmetry.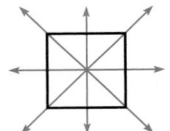

EXAMPLE 5 **Determine Lines of Symmetry**

Determine the number of lines of symmetry in each figure.

a. 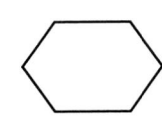   **b.**    **c.**

Solution

a. 2 lines of symmetry   **b.** no lines of symmetry   **c.** 6 lines of symmetry

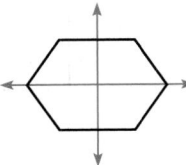   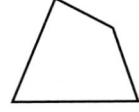   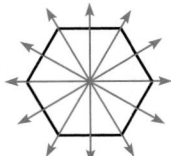

284   **Chapter 5**   *Congruent Triangles*

284

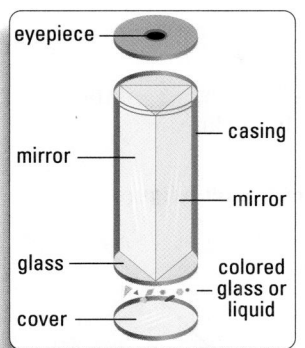

Link to
Kaleidoscopes

eyepiece

casing

mirror

mirror

glass

colored glass or liquid

cover

KALEIDOSCOPES The parts of a kaleidoscope are shown above.

Application Links
CLASSZONE.COM

EXAMPLE 6 *Use Lines of Symmetry*

Mirrors are used to create images seen through a kaleidoscope. The angle between the mirrors is $\angle A$.

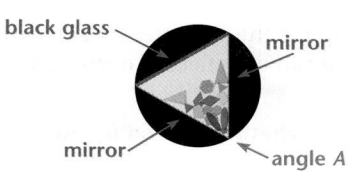

black glass — mirror

mirror

angle A

Top view

Image seen by viewer

Find the angle measure used to create the kaleidoscope design. Use the equation $m\angle A = \dfrac{180°}{n}$, where n is the number of lines of symmetry in the pattern.

a.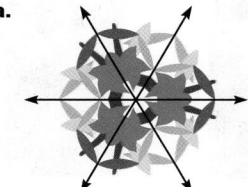

b.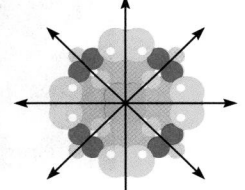

c.

Solution

a. The design has 3 lines of symmetry. So, in the formula, $n = 3$.
$$m\angle A = \frac{180°}{n} = \frac{180°}{3} = 60°$$

b. The design has 4 lines of symmetry. So, in the formula, $n = 4$.
$$m\angle A = \frac{180°}{n} = \frac{180°}{4} = 45°$$

c. The design has 6 lines of symmetry. So, in the formula, $n = 6$.
$$m\angle A = \frac{180°}{n} = \frac{180°}{6} = 30°$$

Checkpoint ✓ *Determine Lines of Symmetry*

Determine the number of lines of symmetry in the figure.

4.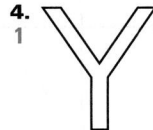
1

5.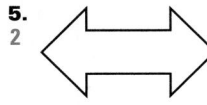
2

6.
4

5.7 Reflections and Symmetry **285**

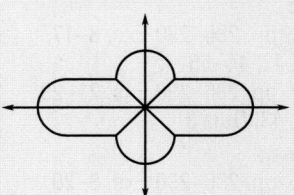

Assignment Guide

BASIC
Day 1: pp. 286–290 Exs. 8–17, 30, 31, 42, 44–55
Day 2: pp. 286–290 Exs. 21–29, 33–39, 43, Quiz 3

AVERAGE
Day 1: pp. 286–290 Exs. 8–20, 30, 31, 42, 44–55
Day 2: pp. 286–290 Exs. 21–29, 32–40, 43, Quiz 3

ADVANCED
Day 1: pp. 286–290 Exs. 11–17, 20–23, 30, 31, 42, 44–49, 53–55
Day 2: pp. 286–290 Exs. 24–29, 32–41, 43, Quiz 3;
EC: TE p. 230D*, classzone.com

BLOCK SCHEDULE
pp. 286–290 Exs. 8–40, 42–55, Quiz 3

✗ Common Error

In Exercises 11–16, watch for students who confuse the *x*- and *y*-axes. Have students hold up one hand and orient it horizontally or vertically as they say *x* or *y* to themselves out loud. Suggest that they orient their hands to model the axis being reflected over in each exercise to help them visualize the problem.

Guided Practice

Vocabulary Check

1. Complete the statement: A figure in the plane has a(n) __?__ if the figure can be reflected onto itself by a(n) __?__ in the line.
line of symmetry; reflection

Skill Check

Determine whether the red figure is a reflection of the blue figure.

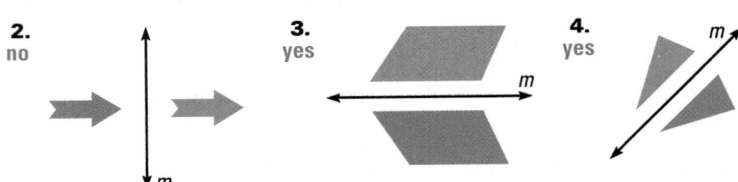

2. no   **3.** yes   **4.** yes

Flowers Determine the number of lines of symmetry in the flower.

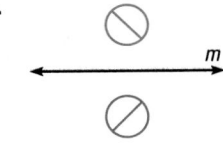

5. 3   **6.** 2   **7.** 5

Practice and Applications

Extra Practice

See p. 684.

8. Yes; all three properties of a reflection are met.

9. Yes; all three properties of a reflection are met.

10. No; the orientation is not reversed.

11. the *y*-axis

12. neither

13. the *x*-axis

Homework Help

Example 1: Exs. 8–10
Example 2: Exs. 8–10
Example 3: Exs. 11–16
Example 4: Exs. 21–29
Example 5: Exs. 21–29
Example 6: Exs. 37–39

Identifying Reflections Determine whether the figure in red is a reflection of the figure in blue. Explain why or why not.

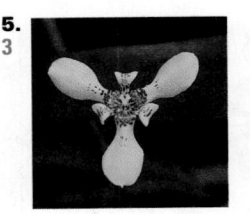

8.   **9.**   **10.**

Reflections in a Coordinate Plane Tell whether the grid shows a reflection in the *x-axis*, the *y-axis*, or *neither*.

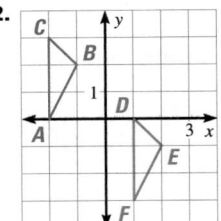

11.   **12.**   **13.**

Student Help

SKILLS REVIEW
To review coordinates,
see p. 664.

Reflections in a Coordinate Plane In Exercises 14–16, use the diagram at the right.

14. Which segment is the reflection of $\overline{AB}$ in the x-axis? Which point corresponds to A? to B?
GH; G; H

15. Which segment is the reflection of $\overline{AB}$ in the y-axis? Which point corresponds to A? to B?
CD; C; D

16. Compare the coordinates for $\overline{AB}$ with the coordinates for its reflection in the x-axis. How are the coordinates alike? How are they different?

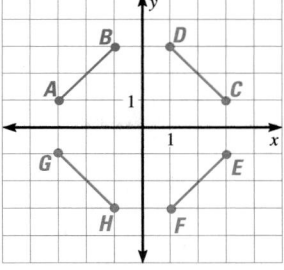

16. The coordinates are alike in that the x-coordinate of each original point is the same as the x-coordinate of the image point. The coordinates are different in that the y-coordinate of each original point is the opposite of the y-coordinate of the image point.

Visualize It! Trace the figure and draw its reflection in line *k*.
17–19. See margin.

17.

18.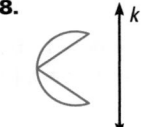

19.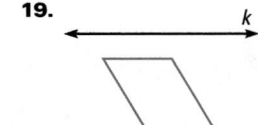

20. The triangles in sections *B* and *D*. *Sample answer:* The triangle in section *B* is a reflection of the triangle in section *A* in the horizontal fold line. The triangle in section *D* is a reflection of the triangle in section *A* in the vertical fold line.

20. Paper Folding Follow these steps.
❶ Fold a piece of paper in half, twice.
❷ Draw a triangle and cut it out.
❸ Unfold the paper and label the sections.

Which of the triangles are reflections of the triangle in section A? Explain.

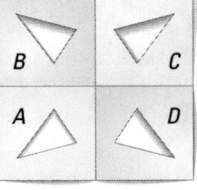

Symmetry Decide whether the line shown is a line of symmetry.

21.
no

22.
yes

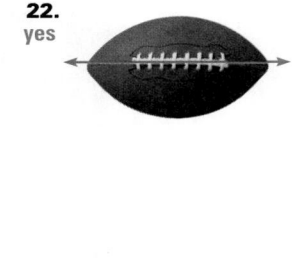

23.
yes

Lines of Symmetry Determine the number of lines of symmetry.

24.
5

25.
4

26.
2

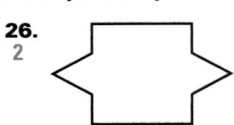

5.7 Reflections and Symmetry **287**

Geometric Reasoning

Some students may have trouble with Exercises 17–19. Refer these students to the properties of reflections box on page 282. Ask students how property 3 can be useful in doing these exercises.

17.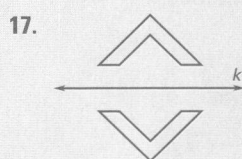

18.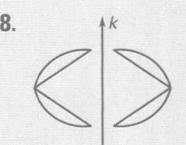

19.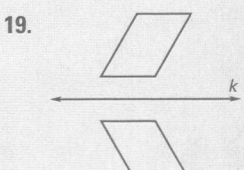

28. not all shown;

29. not all shown;

30.

33. HOOK

35. BIB

36.

WOW

You be the Judge Determine whether all lines of symmetry are shown. If not, sketch the figure and draw all the lines of symmetry.

28–29. See margin.

27.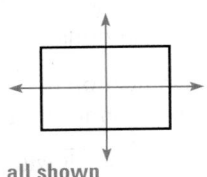

all shown

28.

29.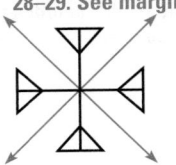

30. **Visualize It!** A piece of paper is folded in half and some cuts are made as shown. Sketch the figure that represents the piece of paper unfolded. See margin.

Type Design In Exercises 31 and 32, use the lowercase letters of the alphabet shown below.

$$a\ b\ c\ d\ e\ f\ g\ h\ i\ j\ k\ l\ m$$
$$n\ o\ p\ q\ r\ s\ t\ u\ v\ w\ x\ y\ z$$

31. Which letters are reflections of other letters? b, d, p, q

32. Draw each letter that has at least one line of symmetry and sketch its line(s) of symmetry. Which letters have one line of symmetry? Which letters have two lines of symmetry? The letters c, i, v, and w have one line of symmetry; the letters l, o, and x have two lines of symmetry.

Word Reflections Determine if the entire word has any lines of symmetry. If so, write the word and draw the line(s) of symmetry.

33. HOOK

34. NOON

35. BIB

36. WOW

Yes; see margin.   no   Yes; see margin.   Yes; see margin.

Kaleidoscope Designs Find the measure of the angle between the mirrors ($\angle A$) that produces the kaleidoscope design. Use the equation $m\angle A = \dfrac{180°}{n}$.

37. 45°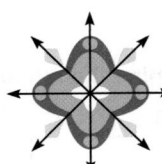

38. 90°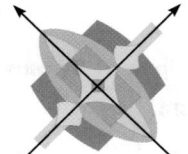

39. 60°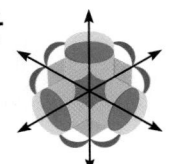

 **Using Algebra**

EXAMPLE **Show Triangles are Congruent**

Show that $\triangle ABC \cong \triangle JKL$.

Solution

Show that the corresponding sides are congruent.

For sides on a horizontal grid line, subtract the x-coordinates.

$$CA = |5 - 1| = 4$$
$$LJ = |5 - 1| = 4$$

For sides on a vertical grid line, subtract the y-coordinates.

$$BC = |4 - 2| = 2$$
$$KL = |-4 - (-2)| = |-2| = 2$$

For any other sides, use the distance formula.

$$AB = \sqrt{(5-1)^2 + (4-2)^2} = \sqrt{4^2 + 2^2} = \sqrt{20}$$

$$JK = \sqrt{(5-1)^2 + ((-4-(-2))^2} = \sqrt{4^2 + (-2)^2} = \sqrt{20}$$

By the SSS Congruence Postulate, $\triangle ABC \cong \triangle JKL$.

 **Student Help**

LOOK BACK
For help with the distance formula, see p. 194.

Showing Triangles are Congruent In Exercises 40 and 41, refer to the example above. Show that $\triangle ABC \cong \triangle DEF$.

40. $AB = 2 = DE$; $AC = 4 = DF$; $BC = \sqrt{20} = EF$; by the SSS Congruence Postulate, $\triangle ABC \cong \triangle DEF$.

41. $AB = 2 = DE$; $AC = 3 = DF$; $BC = \sqrt{13} = EF$; by the SSS Congruence Postulate, $\triangle ABC \cong \triangle DEF$.

40.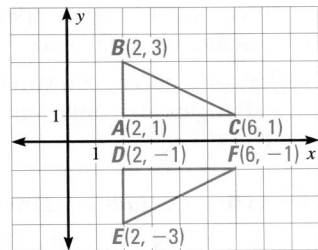

41.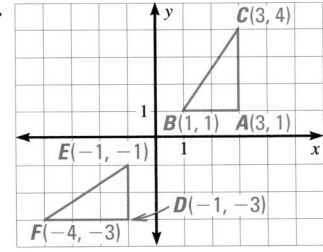

Standardized Test Practice

42. **Multiple Choice** Which triangle shows the image when $\triangle XYZ$ is reflected in the y-axis? **C**

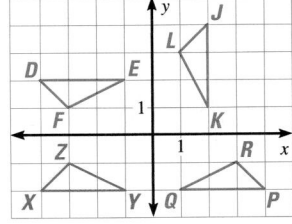

   Ⓐ $\triangle DEF$ Ⓑ $\triangle JKL$

   Ⓒ $\triangle PQR$ Ⓓ None of these

43. **Multiple Choice** How many lines of symmetry does the figure at the right have? **G**

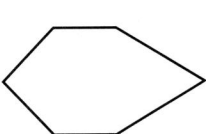

   Ⓕ 0   Ⓖ 1

   Ⓗ 2   Ⓙ 3

5.7 Reflections and Symmetry **289**

Assess

Assessment Resources
The Mini-Quiz below is also available on blackline (*Chapter 6 Resource Book*, p. 9) and on transparency. For more assessment resources, see:
• Chapter 5 Resource Book
• Standardized Test Practice
• Test and Practice Generator

Mini-Quiz
Determine whether figure B is a reflection of figure A.

1.

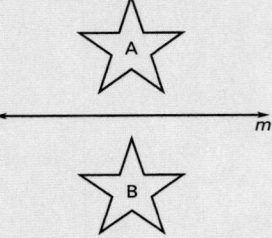

No; the orientation is not reversed.

2.

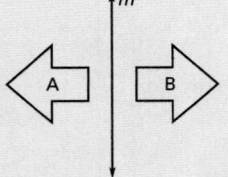

Yes; all three properties of a reflection are met.

3. Tell whether the grid shows a reflection in the *x*-axis, the *y*-axis, or *neither*. **the *y*-axis**

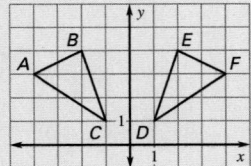

4. Find the number of lines of symmetry. **6**

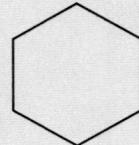

Mixed Review

Showing Lines are Parallel Find the value of *x* so that $p \parallel q$. *(Lesson 3.5)*

44.
105
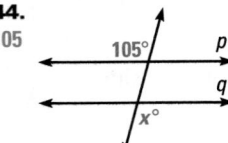

45.
72
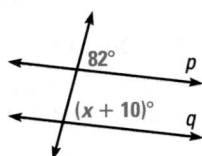

46.
31
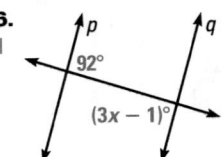

Finding Angle Measures Find the measure of $\angle 1$. *(Lesson 4.2)*

47.
67°
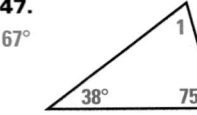

48.
99°
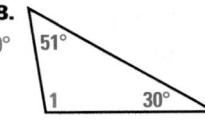

49.
46°
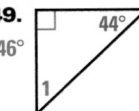

Algebra Skills

Comparing Numbers Compare the two numbers. Write the answer using >, <, or =. *(Skills Review, p. 662)*

50. 2348 and 2384   <
51. −5 and −7   >
52. 19.1 and 19.01   >

53. −11.2 and −11.238   >
54. 0.065 and 0.056   >
55. 1.011 and 1.11   <

Quiz 3

1. Sketch the overlapping triangles separately. Mark all congruent angles and sides.

Which postulate or theorem can you use to show that the triangles are congruent?
(Lesson 5.5) Check sketches; AAS Congruence Theorem.

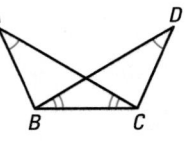

Use the diagram to find the indicated measure(s). *(Lesson 5.6)*

2. Find *DC*. **2**

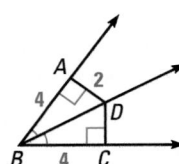

3. Find *ML* and *JK*. **ML = 9; JK = 25**

4. Find *AB*. **13**

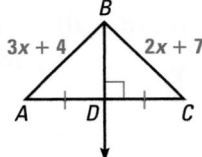

Determine the number of lines of symmetry in the figure. *(Lesson 5.7)*

5. **1**

6. **2**

7. **3**

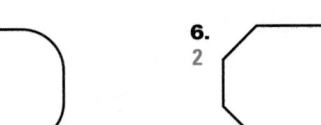
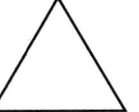

Additional Resources

The following resources are available to help review the materials in this chapter.

📖 **Chapter 5 Resource Book**
• Chapter Review Games and Activities, p. 75
• Cumulative Review, Chs. 1–5

VOCABULARY

• **corresponding parts**, *p. 233*
• **congruent figures**, *p. 233*
• **proof**, *p. 243*
• **distance from a point to a line**, *p. 273*
• **equidistant**, *p. 273*
• **perpendicular bisector**, *p. 274*
• **reflection**, *p. 282*
• **line of symmetry**, *p. 284*

VOCABULARY REVIEW

Fill in the blank.

1. When two figures are __?__, their corresponding sides and their corresponding angles are congruent. congruent

2. A(n) __?__ is a convincing argument that shows why a statement is true. proof

3. If a point is on the __?__ of a segment, then it is equidistant from the endpoints of the segment. perpendicular bisector

4. If a point is on the angle bisector of an angle, then it is __?__ from the two sides of the angle. equidistant

5. A(n) __?__ is a transformation that creates a mirror image. reflection

5.1 CONGRUENCE AND TRIANGLES

Examples on pp. 233–235

EXAMPLE In the diagram, $\triangle ABC \cong \triangle RST$. Identify all corresponding congruent parts.

Corresponding angles

$\angle A \cong \angle R$
$\angle B \cong \angle S$
$\angle C \cong \angle T$

Corresponding sides

$\overline{AB} \cong \overline{RS}$
$\overline{AC} \cong \overline{RT}$
$\overline{BC} \cong \overline{ST}$

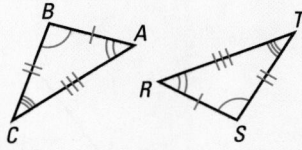

Use the triangles shown at the right to determine whether the given angles or sides represent *corresponding angles*, *corresponding sides*, or *neither*.

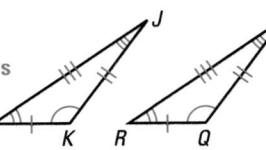

6. $\angle J$ and $\angle R$ neither
7. $\overline{KL}$ and $\overline{QR}$ corresponding sides
8. $\angle K$ and $\angle Q$ corresponding angles
9. $\overline{PQ}$ and $\overline{LJ}$ neither
10. $\overline{JK}$ and $\overline{PR}$ neither
11. $\angle R$ and $\angle L$ corresponding angles

5.2 PROVING TRIANGLES ARE CONGRUENT: SSS AND SAS

Examples on pp. 241–244

EXAMPLES Tell which congruence postulate you would use to show that the triangles are congruent.

a.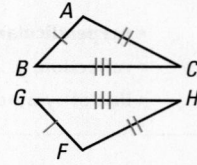

b.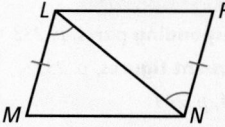

a. In the triangles shown,
S $\overline{AB} \cong \overline{FG}$,
S $\overline{BC} \cong \overline{GH}$, and
S $\overline{AC} \cong \overline{FH}$.
So, $\triangle ABC \cong \triangle FGH$ by the SSS Congruence Postulate.

b. In the triangles shown,
S $\overline{LM} \cong \overline{NP}$,
A $\angle MLN \cong \angle PNL$, and
S $\overline{LN} \cong \overline{NL}$.
So, $\triangle LMN \cong \triangle NPL$ by the SAS Congruence Postulate.

Decide whether enough information is given to show that the triangles are congruent. If so, tell which congruence postulate you would use.

12.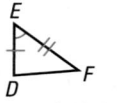

yes; SAS Congruence Postulate

13.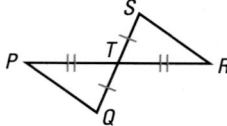

yes; SAS Congruence Postulate

14.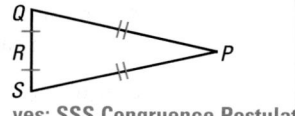

yes; SSS Congruence Postulate

5.3 PROVING TRIANGLES ARE CONGRUENT: ASA AND AAS

Examples on pp. 250–253

EXAMPLES Tell which congruence postulate or theorem you would use to show that the triangles are congruent.

a.

b.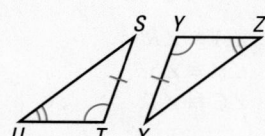

a. In the triangles shown,
A $\angle E \cong \angle K$,
S $\overline{EF} \cong \overline{KL}$, and
A $\angle F \cong \angle L$.
So, $\triangle DEF \cong \triangle JKL$ by the ASA Congruence Postulate.

b. In the triangles shown,
A $\angle U \cong \angle Z$,
A $\angle T \cong \angle Y$, and
S $\overline{ST} \cong \overline{XY}$.
So, $\triangle STU \cong \triangle XYZ$ by the AAS Congruence Theorem.

Determine what information is needed to use the indicated postulate or theorem to show that the triangles are congruent.

15. AAS Congruence Theorem

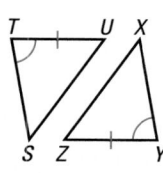

$\angle S \cong \angle X$

16. ASA Congruence Postulate

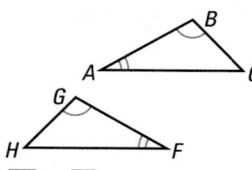

$\overline{AB} \cong \overline{FG}$

17. ASA Congruence Postulate

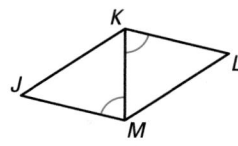

$\angle JKM \cong \angle LMK$

5.4 HYPOTENUSE-LEG CONGRUENCE THEOREM: HL

Examples on pp. 257–259

EXAMPLE Prove that $\triangle ABC \cong \triangle BFD$.

Given ▸ $\overline{CB} \perp \overline{AF}$, $\overline{DF} \perp \overline{AF}$

  B is the midpoint of $\overline{AF}$.

  $\overline{AC} \cong \overline{BD}$

Prove ▸ $\triangle ABC \cong \triangle BFD$

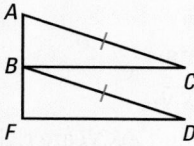

Show that the triangles are right triangles, the hypotenuses are congruent, and that corresponding legs are congruent.

Statements	Reasons
1. $\overline{CB} \perp \overline{AF}$, $\overline{DF} \perp \overline{AF}$, and B is the midpoint of $\overline{AF}$.	**1.** Given
2. $\angle CBA$ and $\angle DFB$ are right angles.	**2.** $\perp$ lines form right angles.
3. $\triangle ABC$ and $\triangle BFD$ are right triangles.	**3.** Definition of right triangle
H 4. $\overline{AC} \cong \overline{BD}$	**4.** Given
L 5. $\overline{AB} \cong \overline{BF}$	**5.** Definition of midpoint
6. $\triangle ABC \cong \triangle BFD$	**6.** HL Congruence Theorem

18. Use the information given in the diagram to fill in the missing statements and reasons to prove that $\triangle UZV \cong \triangle XYW$.

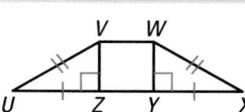

Statements	Reasons
1. $\angle UZV$ and $\angle XYW$ are right angles.	**1.** _____?_____ Given
2. $\triangle UZV$ and $\triangle XYW$ are right triangles.	**2.** _____?_____ Def. of right $\triangle$
3. _____?_____ $\overline{UV} \cong \overline{XW}$	**3.** Given
4. $\overline{UZ} \cong \overline{XY}$	**4.** _____?_____ Given
5. _____?_____ $\angle UZV \cong \angle XYW$	**5.** HL Congruence Theorem

19–21. Check sketches.

19. HL Congruence Theorem

20. AAS Congruence Theorem

21. AAS Congruence Theorem

22. Statements (Reasons)
1. ∠C and ∠D are right angles. (Given)
2. △ABC and △BAD are right triangles. (Definition of right triangle)
3. $\overline{AC} \cong \overline{BD}$ (Given)
4. $\overline{AB} \cong \overline{AB}$ (Reflexive Prop. of Congruence)
5. △ABC ≅ △BAD (HL Congruence Theorem)
6. ∠CBA ≅ ∠DAB (Corresp. parts of ≅ triangles are ≅.)

23. Statements (Reasons)
1. ∠JMN and ∠JKL are right angles. (Given)
2. ∠JMN ≅ ∠JKL (All right angles are congruent.)
3. $\overline{MN} \cong \overline{KL}$ (Given)
4. ∠J ≅ ∠J (Reflexive Prop. of Congruence)
5. △JMN ≅ △JKL (AAS Congruence Theorem)
6. $\overline{JM} \cong \overline{JK}$ (Corresp. parts of ≅ triangles are ≅.)

24. Statements (Reasons)
1. ∠PQR and ∠TSR are right angles. (Given)
2. ∠PQR ≅ ∠TSR (All right angles are congruent.)
3. $\overline{PR} \cong \overline{TR}$ (Given)
4. ∠PRQ ≅ ∠TRS (Vertical Angles Theorem)
5. △PQR ≅ △TSR (AAS Congruence Theorem)
6. $\overline{QR} \cong \overline{SR}$ (Corresp. parts of ≅ triangles are ≅.)

5.5 USING CONGRUENT TRIANGLES

Examples on pp. 265–267

EXAMPLE In the diagram, $\overline{KM} \cong \overline{KN}$ and ∠KML and ∠KNJ are right angles. Prove that $\overline{KL} \cong \overline{KJ}$.

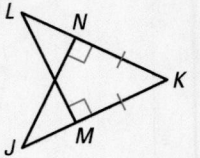

First show that △MKL ≅ △NKJ. Then use the fact that corresponding parts of congruent triangles are congruent to show that $\overline{KL} \cong \overline{KJ}$.

Sketch the triangles separately. Mark the given information and any other information you can conclude from the diagram.

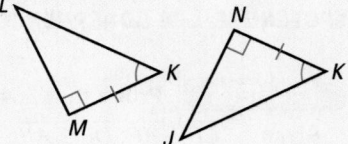

Statements	Reasons
1. $\overline{KM} \cong \overline{KN}$	1. Given
2. ∠KML and ∠KNJ are right angles.	2. Given
3. ∠KML ≅ ∠KNJ	3. Right angles are congruent.
4. ∠K ≅ ∠K	4. Reflexive Prop. of Congruence
5. △MKL ≅ △NKJ	5. ASA Congruence Postulate
6. $\overline{KL} \cong \overline{KJ}$	6. Corresponding parts of congruent triangles are congruent.

Sketch the overlapping triangles separately. Use the given information to mark all congruences. Then tell what theorem or postulate you can use to show that the triangles are congruent. 19–24. See margin.

19. $\overline{AC} \cong \overline{BD}$

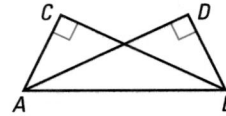

20. $\overline{MN} \cong \overline{KL}$

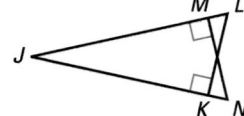

21. $\overline{PR} \cong \overline{TR}$

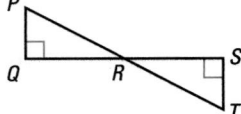

22. Use the diagram and the information given in Exercise 19 above to prove that ∠CBA ≅ ∠DAB.

23. Use the diagram and the information given in Exercise 20 above to prove that $\overline{JM} \cong \overline{JK}$.

24. Use the diagram and the information given in Exercise 21 above to prove that $\overline{QR} \cong \overline{SR}$.

5.6 ANGLE BISECTORS AND PERPENDICULAR BISECTORS

Examples on pp. 273–275

EXAMPLE Find *AB*.

By the Angle Bisector Theorem, $AB = DB$.

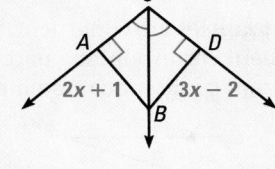

$2x + 1 = 3x - 2$ Use Angle Bisector Theorem.

$2x + 3 = 3x$ Add 2 to each side.

$3 = x$ Subtract 2x from each side.

You are asked to find *AB*, not just the value of *x*.

ANSWER ▶ $AB = 2x + 1 = 2(3) + 1 = 7$

Use the diagram to find the indicated measure.

25. Find *JM*.   3   **26.** Find *QR*.   9   **27.** Find *XY*.   7

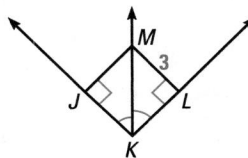 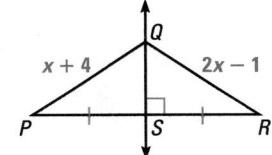 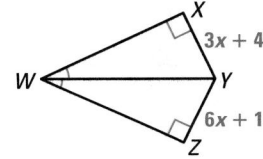

5.7 REFLECTIONS AND SYMMETRY

Examples on pp. 282–285

EXAMPLE Tell whether the red figure is a reflection of the blue figure. Then determine the number of lines of symmetry in the blue figure.

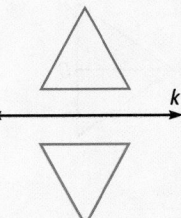

The red figure is a reflection of the blue figure in line *k*.
The blue figure has one line of symmetry.

Tell whether the red figure is a reflection of the blue figure. Then determine the number of lines of symmetry in the blue figure.

28.
yes; 1
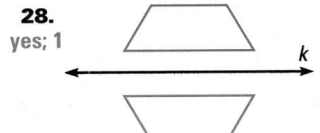

29.
no; 1
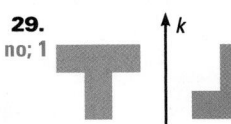

30.
yes; 1
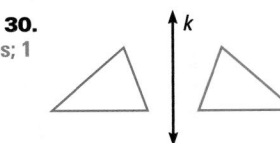

31. Draw a six-sided figure that has two lines of symmetry.   See margin.

31. *Sample answer:*

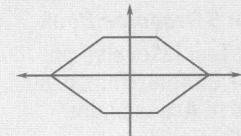

Additional Resources

📙 **Chapter 5 Resource Book**
- Chapter Test (2 levels), pp. 76–79
- SAT/ACT Chapter Test, p. 80
- Alternative Assessment, pp. 81–82

🖥 **Test and Practice Generator**

4. SSS Congruence Postulate; $\overline{BC} \cong \overline{BC}$; all three sides of △*ABC* are congruent to the three sides of △*DBC*.

5. HL Congruence Theorem; $\overline{HF} \cong \overline{HF}$; the hypotenuse and a leg of right △*HEF* are congruent to the hypotenuse and a leg of right △*HGF*.

7. Yes; the alternate interior angles are congruent and the vertical angles are congruent, so the triangles can be shown to be congruent by either the ASA Congruence Postulate or the AAS Congruence Theorem.

8. See Additional Answers beginning on page AA1.

Chapter 5 Chapter Test

1. The two triangles are congruent. List all pairs of congruent corresponding parts. Then write two different congruence statements.

$\overline{RS} \cong \overline{XY}$; $\overline{ST} \cong \overline{YZ}$; $\overline{RT} \cong \overline{XZ}$; $\angle R \cong \angle X$; $\angle S \cong \angle Y$; $\angle T \cong \angle Z$; Sample answer: △*RST* ≅ △*XYZ*

In the diagram below, △*JKL* ≅ △*STU*. Find the indicated measure.

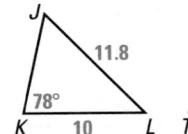 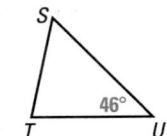

2. Find $m\angle L$. 46°   3. Find *TU*. 10

Tell which postulate or theorem you would use to show that the triangles are congruent. Explain your reasoning. 4, 5. See margin.

4.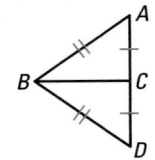

5.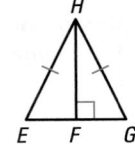

Does the diagram give enough information to show that the triangles are congruent? If so, state the postulate or theorem you would use. Explain your reasoning.

6.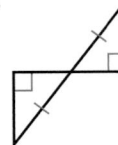

7.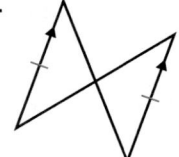

6. Yes; the vertical angles are congruent, so the triangles are congruent by the AAS Congruence Theorem.

7. See margin.

8. Given the information in the diagram, fill in the missing statements and reasons to show that $\overline{AB} \cong \overline{ED}$. **See margin.**

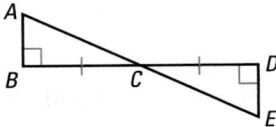

Statements	Reasons
1. $\overline{BC} \cong \overline{DC}$	1. ____?____
2. ____?____	2. Given
3. ____?____	3. Right angles are congruent.
4. $\angle ACB \cong \angle ECD$	4. ____?____
5. △*ABC* ≅ △*EDC*	5. ____?____
6. $\overline{AB} \cong \overline{ED}$	6. ____?____

Find the indicated measure.

9. Find *PR*. 8   10. Find *ST*. 5

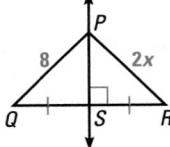 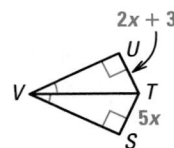

11. Tell whether the red figure is a reflection of the blue figure. If it is, tell the line of reflection. Explain your reasoning. No; the orientation of the image is not reversed.

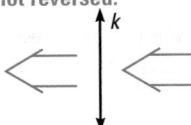

Determine the number of lines of symmetry.

12.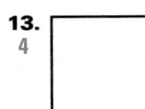
2

13.
4

Additional Resources

Chapter 5 Resource Book
• SAT/ACT Chapter Test, p. 80

Test Tip ⒶⒷⒸⒹ Draw your own sketches that show the information given in a problem. This may help you see a way to approach a problem.

1. Given that $\triangle ABC \cong \triangle FHG$, tell which of the following statements is true. **C**

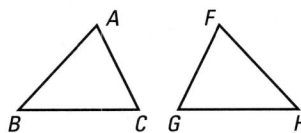

Ⓐ $\angle A \cong \angle G$ Ⓑ $\overline{AC} \cong \overline{HF}$

Ⓒ $\angle B \cong \angle H$ Ⓓ $\overline{BA} \cong \overline{FG}$

2. Based on the diagram, which theorem or postulate *cannot* be used to show that $\triangle JKL \cong \triangle NML$? **J**

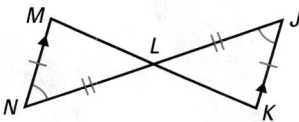

Ⓕ ASA Congruence Postulate

Ⓖ AAS Congruence Theorem

Ⓗ SAS Congruence Postulate

Ⓙ HL Congruence Theorem

3. In the diagram, $\overline{AD} \cong \overline{BC}$, $\overline{AC} \cong \overline{BD}$, and $\angle A \cong \angle B$. Decide which of the postulates or theorems listed below can be used to show that $\triangle ADC \cong \triangle BCD$. **B**

I. SAS   **II.** HL   **III.** SSS

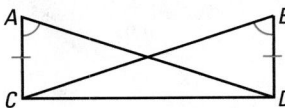

Ⓐ I only   Ⓑ I and III only

Ⓒ I, II, and III   Ⓓ None of these

4. What is the value of x? **F**

Ⓕ 3

Ⓖ 4

Ⓗ 9

Ⓙ 12

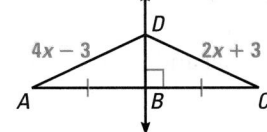

5. In the diagram above, what is AD? **C**

Ⓐ 3   Ⓑ 4   Ⓒ 9   Ⓓ 12

6. How many lines of symmetry are in the equilateral triangle below? **J**

Ⓕ 0

Ⓖ 1

Ⓗ 2

Ⓙ 3

Multi-Step Problem **In Exercises 7–9, use the diagram below.** 7, 9. See margin.

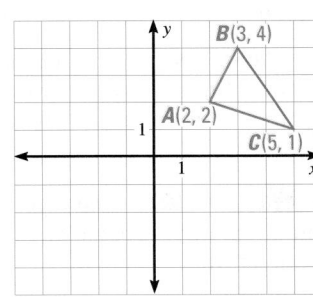

7. Sketch the reflection of the triangle in the x-axis. Label the triangle $\triangle A'B'C'$.

8. What are the coordinates of the point of the reflection that correspond to A? to B? to C?
A' is (2, −2); B' is (3, −4); C' is (5, −1).

9. How do the coordinates of $\triangle ABC$ relate to the coordinates of the reflected triangle?

7.

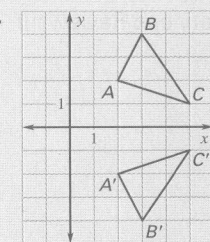

9. The x-coordinates of the points on the reflected triangles are the same as those of the corresponding points on $\triangle ABC$, while the y-coordinates are opposites.

Teaching Tip
The Brain Games activity provides a motivating way to review selected content in the chapter. For a more comprehensive review, see the Chapter Summary and Review on pp. 291–295.

BraiN ➔ GaMes

Mirror Reflections

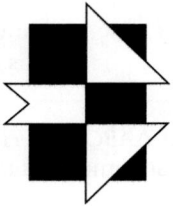

original image   target image

Materials

• mirror
• original and target images

Object of the Game To find target images made from the original image using a mirror.

How to Play There are 5 target images next to each original image. Try to create the target images.

Step 1 ▶ Place a mirror anywhere on the original image and look at the image made by the reflection. Does it match any of the target images?

Step 2 ▶ Move the mirror over the image and at different angles trying to duplicate one of the target images.

Step 3 ▶ When you create one of the target images, draw a line where you placed the mirror and label the line with the letter of the target image.

Players receive a point for each target image they find. The player with the most points after a set amount of time wins.

Another Way to Play Create your own original image and target images. Challenge a classmate to find your target images.

To find whether a point lies on a line, substitute the x- and y-values into the equation of the line and see whether the equation is true.

EXAMPLE 1 | Points on a Line

Determine whether the point lies on the line whose equation is $2x + y = 5$.

  a. (2, 1)   **b.** (−1, 3)

Solution

a. $2x + y = 5$
      $2(2) + 1 \overset{?}{=} 5$
      $4 + 1 \overset{?}{=} 5$
         $5 = 5$ ✓

b. $2x + y = 5$
      $2(-1) + 3 \overset{?}{=} 5$
      $-2 + 3 \overset{?}{=} 5$
         $1 \neq 5$

(2, 1) lies on the line.   (−1, 3) does not lie on the line.

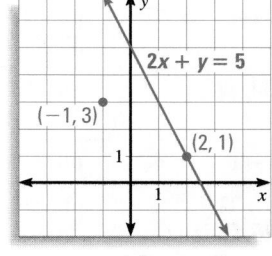

Try These

Determine whether the point lies on the line whose equation is given.

1. $3x + 2y = 13$; (3, 2) yes   **2.** $5x − 4y = 4$; (0, −1) yes

3. $y = 3x − 8$; (−4, 4) no   **4.** $−x + 6y = −2$; (4, 1) no

5. $x − 7y = 5$; (−5, 0) no   **6.** $2x − 4y = −2$; (−1, −1) no

EXAMPLE 2 | Find Slope

Find the slope of the line that passes through the points (−3, 1) and (3, −2).

Solution

Let $(x_1, y_1) = (-3, 1)$ and $(x_2, y_2) = (3, -2)$.

$$\text{slope} = \frac{y_2 - y_1}{x_2 - x_1} = \frac{-2 - 1}{3 - (-3)} = \frac{-3}{6} = -\frac{1}{2}$$

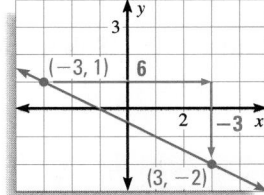

Try These

Find the slope of the line that passes through the points.

7. (3, 1) and (8, 7) $\frac{6}{5}$ **8.** (3, 4) and (3, −5) not defined   **9.** (6, −6) and (−1, 1) −1

10. (−5, 1) and (−2, 8) $\frac{7}{3}$ **11.** (−1, 4) and (4, 4) 0   **12.** (−2, −4) and (2, 6) $\frac{5}{2}$

Algebra Review   **299**

CHAPTER 6 Pacing and Assignment Guide

REGULAR SCHEDULE

Lesson	Les. Day	Basic	Average	Advanced
6.1	Day 1	pp. 306–308 Exs. 8–14, 21, 22, 30–42 even	pp. 306–308 Exs. 8–14, 21–27, 30–36 even	pp. 306–308 Exs. 8–14, 21–27, 30–42 even
	Day 2	pp. 306–308 Exs. 15–20, 29–41 odd	pp. 306–308 Exs. 15–20, 28, 29–35 odd, 37–42	pp. 306–308 Exs. 15–20, 28, 29–41 odd; EC: TE p. 300D*, classzone.com
6.2	Day 1	pp. 313–315 Exs. 14, 16–20, 22–24, 31, 32, 43, 44–54 even	pp. 313–315 Exs. 14, 16–24, 31, 32, 38–40, 43, 44–54 even	pp. 313–315 Exs. 14, 16–24, 31, 32, 38–40, 41*, 43, 44, 46
	Day 2	pp. 313–315 Exs. 13, 15, 25–30, 42, 45–55 odd	pp. 313–315 Exs. 13, 15, 25–30, 33–37, 42, 45–55 odd	pp. 313–315 Exs. 13, 15, 25–30, 33–37, 42, 47–55 odd; EC: classzone.com
6.3	Day 1	pp. 320–323 Exs. 8–13, 21, 27–40	pp. 320–323 Exs. 8–13, 20, 21, 25–38	pp. 320–323 Exs. 8–13, 20, 21, 25–34
	Day 2	EP pp. 679–680 Exs. 21, 28, 31, 34; pp. 320–323 Exs. 14–19, 25, 26, Quiz 1	pp. 320–323 Exs. 14–19, 22, 23, 39–44, Quiz 1	pp. 320–323 Exs. 14–19, 22, 23, 24*, 35–43 odd, Quiz 1; EC: classzone.com
6.4	Day 1	pp. 328–330 Exs. 7–12, 14, 15, 18, 19, 27, 28	pp. 328–330 Exs. 7–12, 14, 15, 18, 19, 27, 28	pp. 328–330 Exs. 7–12, 14, 15, 18, 19, 27–31
	Day 2	pp. 328–330 Exs. 13, 16, 17, 20–22, 29–34	pp. 328–330 Exs. 13, 16, 17, 20–22, 29–34	pp. 328–330 Exs. 13, 16, 17, 20–22, 23–26*, 32–34; EC: classzone.com
6.5	Day 1	pp. 334–336 Exs. 14–19, 35, 39–46	pp. 334–336 Exs. 14–19, 35–46	pp. 334–336 Exs. 14–19, 35–46
	Day 2	pp. 334–336 Exs. 9–13, 20–26, 34, 36–38	pp. 334–336 Exs. 9–13, 20–32, 34	pp. 334–336 Exs. 20–32, 33*, 34; EC: classzone.com
6.6	Day 1	pp. 339–341 Exs. 5–11, 20–23, 26–38 even	pp. 339–341 Exs. 5–11, 18–24, 26, 35–38	pp. 339–341 Exs. 5–11, 18–24, 25*, 26, 35–38
	Day 2	pp. 339–341 Exs. 12–17, 27–37 odd, Quiz 2	pp. 339–341 Exs. 12–17, 27–34, Quiz 2	pp. 339–341 Exs. 12–17, 27–34, Quiz 2; EC: classzone.com
Review	Day 1	pp. 342–345 Exs. 1–33	p. 342–345 Exs. 1–33	pp. 342–345 Exs. 1–33
Assess	Day 1	Chapter 6 Test	Chapter 6 Test	Chapter 6 Test

YEARLY PACING   Chapter 6 Total – **14 days**   Chapters 1–6 Total – **94 days**   Remaining – **66 days**

*Challenge Exercises   EP = Extra Practice   SRH = Skills Review Handbook   EC = Extra Challenge

BLOCK SCHEDULE

Day 1	Day 2	Day 3	Day 4	Day 5	Day 6	Day 7
6.1 pp. 306–308 Exs. 8–42	6.2 pp. 313–315 Exs. 13–40, 42–55	6.3 pp. 320–323 Exs. 8–23, 25–44, Quiz 1	6.4 pp. 328–330 Exs. 7–22, 27–34	6.5 pp. 334–336 Exs. 9–32, 34–46	6.6 pp. 339–341 Exs. 5–24, 26–38, Quiz 2	Review pp. 342–345 Exs. 1–33 Assess Chapter 6 Test

YEARLY PACING   Chapter 6 Total – **7 days**   Chapters 1–6 Total – **47 days**   Remaining – **33 days**

Support Materials

CHAPTER RESOURCE BOOK

CHAPTER SUPPORT

Tips for New Teachers	p. 1	Strategies for Reading Mathematics	p. 5
Parent Guide for Student Success	p. 3		

LESSON SUPPORT

	6.1	6.2	6.3	6.4	6.5	6.6
Lesson Plans (regular and block)	p. 7	p. 16	p. 24	p. 35	p. 46	p. 56
Warm-Up Exercises and Daily Quiz	p. 9	p. 18	p. 26	p. 37	p. 48	p. 58
Technology Activities & Keystrokes	p. 10		p. 27	p. 38	p. 49	p. 59
Practice (2 levels)	p. 11	p. 19	p. 29	p. 40	p. 51	p. 60
Reteaching with Practice	p. 13	p. 21	p. 31	p. 42	p. 53	p. 62
Quick Catch-Up for Absent Students	p. 15	p. 23	p. 33	p. 44	p. 55	p. 64
Learning Activities						
Real-Life Applications				p. 45		p. 65

REVIEW AND ASSESSMENT

Quizzes	pp. 34, 66	Project with Rubric	p. 75
Chapter Review Games and Activities	p. 67	Cumulative Review	p. 77
Chapter Test (2 levels)	p. 68	Cumulative Test for Chs. 1–6	p. 79
SAT/ACT Chapter Test	p. 72	Resource Book Answers	AN1
Alternative Assessment with Math Journal	p. 73		

TRANSPARENCIES

	6.1	6.2	6.3	6.4	6.5	6.6
Warm-Up Exercises and Daily Quiz	p. 35	p. 36	p. 37	p. 38	p. 39	p. 40
Visualize It Transparencies		✓	✓	✓	✓	
Answer Transparencies	✓	✓	✓	✓	✓	✓

TECHNOLOGY

- Time-Saving Test and Practice Generator
- Electronic Teacher Tools
- Geometry in Motion: Video
- Classzone.com
- Online Lesson Planner

ADDITIONAL RESOURCES

- Worked-Out Solution Key
- Resources in Spanish
- Practice Workbook with Examples

Providing Universal Access

Strategies for Strategic Learners

PROVIDE SCAFFOLDING

Many problems in geometry require students to perform a number of steps in a particular order. Advanced students not only remember the steps, they know when to use particular congruence theorems, they understand why they are doing what they are doing, and they know when the order of steps does or does not matter.

ASK "WHY?" Less proficient students may need extra help understanding why they take each step. When students seem overwhelmed by a multi-step problem such as a proof, have them play "The Why Game." See if they can come up with a reason for each of the steps that they take when solving a problem. Have students look at the Solution provided for Example 1 on page 332. Have them give a reason in their own words for the work done in each part of the Solution. Sample results are presented in the following table.

Solution Step	Explanation	Why?
1a. $PQRS$ is an isosceles trapezoid.	Write the given information that will be helpful in finding $m\angle R$.	The given information provides a starting point.
1b. $\angle R$ and $\angle S$ are a pair of base angles.	Write a known fact about $\angle R$.	This relates $\angle R$ to another angle whose measure is known.
1c. $m\angle R = m\angle S = 50°$	Since they are base angles of an isosceles trapezoid, they are congruent, and congruent angles have equal measures.	We want to find the measure of $\angle R$.
2a. $\overline{SR} \parallel \overline{PQ}$	Write the given information that will be helpful in finding $m\angle P$.	The given information provides a starting point.
2b. $\angle S$ and $\angle P$ are same-side interior angles.	Write a known fact about $\angle P$.	This relates $\angle P$ to $\angle S$, an angle whose measure is known.
2c. $\angle S$ and $\angle P$ are supplementary.	Because $\overline{SR} \parallel \overline{PQ}$, same-side interior angles are supplementary.	This allows us to write an equation to find the measure of $\angle P$.

(Continued in next column.)

table (continued)

Solution Step	Explanation	Why?
2d. $m\angle P = 180° - 50° = 130°$	The sum of the measures of supplementary angles is 180°.	We want to find the measure of $\angle P$.
3a. $\angle Q$ and $\angle P$ are a pair of base angles.	Write the given information that will be helpful in finding $m\angle S$.	The given information provides a starting point.
3b. $m\angle Q = m\angle P = 130°$	Since they are base angles, they are congruent.	We want to find the measure of $\angle Q$.

Geometry can be made simpler for students if they understand that they are often trying to use some given information and the theorems they have learned to prove a statement.

Strategies for Strategic, English, and Advanced Learners

VOCABULARY STUDY OF ROOTS

The analysis of Greek and Latin roots in mathematical terms and common English words is a good activity for all students because it helps them figure out the meanings of new words. Greek and Latin roots, prefixes, and suffixes occur repeatedly in mathematical and scientific terms. Students who speak another language in addition to English may find words in their other language that are derived from Greek and Latin roots. This word study can be covered in a few minutes each week during math class or in a more extensive way during a language arts class.

PREFIXES Here are several common prefixes that students will see in this geometry course:

Bi- (from the Latin for *two*)
Co-, com- (from the Latin for *with*)
Di-, dis- (from the Latin for *away from, out of, apart*)
In- (from the Latin for *in* or *not, without*)
Inter- (from the Latin for *between*)
Iso- (from the Greek for *equal*)
Multi- (from the Latin for *many*)
Per- (from the Latin for *through*, or *by means of*)
Poly- (from the Greek for *many*)
Quad- (from the Latin for *four*)
Tri- (from the Latin for *three*)

Ask students to consider how the prefixes shown on the previous page contribute to the following words.

bisect, collinear, commutative, coordinate, coplanar, diagram, diameter, dilation, distribute, inscribed, intercept, intersection, isosceles, multiple, multi-step, percent, perimeter, polynomial, polygon, polyhedra, quadrilateral, quadrant, triangle

NUMBERS Latin and Greek numbers have also influenced English mathematical words. Put these Latin and Greek words for cardinal numbers on the board and ask students to think of words that use them.

English	Cardinal Numbers in Latin	Cardinal Numbers in Greek
one	unus, una, unum	heis, miam, hen
two	duo, duae	duo
three	tres, tria	treis, tria
four	quattor	tessares, tessara
five	quinque	pente
six	sex	hex
seven	septum	hepta
eight	octo	okto
nine	novem	ennea
ten	decem	deka

Students may think of words such as *uniform* or *unicorn*, *duo*, *duet*, and *decathlon*. Many of the specific names for polygons are related to the Greek words. Spanish speakers will see that many words for numbers in Spanish are very close to the Latin number names.

Strategies for Advanced Learners

DIFFERENTIATE INSTRUCTION IN TERMS OF COMPLEXITY

Advanced learners may be interested in exploring a relatively new field of mathematics involving patterns. The concept of fractals was introduced by Polish-born mathematician Benoit Mandlebrot in the 1970s. A fractal is a geometric figure that is self-similar. That is, the appearance of any part of a fractal is similar to the whole fractal. Fractals are widely used in such fields as computer animation, geology, and biology. The following figure shows the first few stages in the formation of one fractal, the *Koch snowflake*.

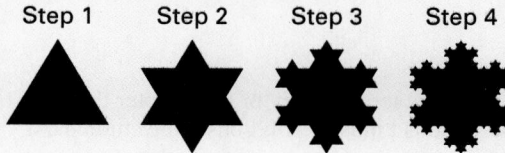

Step 1   Step 2   Step 3   Step 4

Have advanced learners investigate how the Koch snowflake is formed and how patterns are involved in, for example, the perimeter of the snowflake at each stage. Students may also be interested in studying other common fractals, such as the *Sierpinski triangle*.

Challenge Problem for use with Lesson 6.1:
Suppose a polygon with n sides has one vertex labeled V. If all the diagonals that contain vertex V are drawn, how many non-overlapping triangles are created? Give your answer in terms of n. $(n - 2)$

Challenge Problem for use with Lesson 6.3:
A parallelogram is drawn in a coordinate plane. Three of the vertices of the parallelogram are located at the points $(0, 0)$, $(a, 0)$, and (b, c), where a, b, and c are all positive numbers with $a > b$. In terms of a, b, and c, what are the coordinates of the fourth vertex of the parallelogram? (*The coordinates are $(a + b, c)$.*)

Content and Teaching Strategies

Chapter 6 carries the study of plane geometric figures beyond triangles to polygons in general, and quadrilaterals in particular. In Lesson 6.1, the most common types of polygons are discussed. The Quadrilateral Interior Angles Theorem, which is based on the Triangle Sum Theorem, is also presented. Lesson 6.2 introduces parallelograms and states several important properties related to them, while Lesson 6.3 states the converses of many of those properties. These converses detail the sufficient conditions for a quadrilateral to be a parallelogram. Lesson 6.4 discusses three special types of parallelograms, namely rhombuses, rectangles, and squares. Another type of special quadrilateral, the trapezoid, is introduced in Lesson 6.5. Finally, Lesson 6.6 synthesizes many of the facts contained in the previous lessons of the chapter, and explores ways to identify a given quadrilateral as one of the special types already studied.

Lesson 6.1

This lesson presents some of the basic terminology of the chapter that will be used to describe all the various kinds of polygons to be studied. At the start, it may be difficult for students to come to terms with the concept of a diagonal of a polygon that is not a rectangle (probably the only case with which they are familiar). Emphasize that, in general, polygons have many diagonals, and choosing any two nonconsecutive vertices uniquely determines one of them.

QUICK TIP
When asking students to draw all the diagonals of a polygon, stress that it is best to draw all possible diagonals from one vertex, then all those from a second vertex, and so on.

USING DIAGRAMS The Quadrilateral Interior Angles Theorem will perhaps be more easily remembered by students if they see that a justification for it can be offered by just drawing one diagonal of the given quadrilateral. The diagonal separates the quadrilateral into two triangles, each of which has interior angles whose measures total 180°. Students should be led to see that these angles form, either singly or in pairs, the angles of the quadrilateral. Therefore, by the Triangle Sum Theorem (of which this theorem is an extension), the sum of the interior angles of the quadrilateral is 2(180°), or 360°.

You may also want to place Theorem 6.1 into a more general context by leading students to the fact that the sum of the interior angles of an n-sided convex polygon is $(n - 2)180°$. A brief explanation of the term *convex* should be offered while making this generalization. Students can discover this fact for themselves if you ask them to draw a polygon and then draw all the diagonals *from one vertex*. By counting the number of triangles formed by the diagonals, they should be able to find the formula easily. Stress that this extension will be stated formally in Lesson 8.2.

Lesson 6.2

You might introduce this lesson by pointing out the many applications of parallelograms in the real world. One that students may be familiar with is the derailleur, or gear changer, of a multi-speed bicycle. A derailleur utilizes the shape of a parallelogram to keep the plane of the chain parallel to the plane of the rear wheel while changing gear rings. (This situation is presented in Exercise 20 of Lesson 6.3.)

COMMON ERROR Of the properties of parallelograms stated in this lesson, the most difficult for students to use properly is Theorem 6.5. Emphasize that this theorem does not say that the intersection point of the diagonals creates four congruent segments, but rather only that the two parts of each diagonal are congruent to each other.

QUICK TIP
Ask students who have difficulty understanding Theorem 6.5 to draw a large parallelogram with its diagonals shown. Have them measure the parts of the diagonals to verify the theorem.

Lesson 6.3

As stated in the Study Tip on page 316, the theorems in this lesson are the converses of the theorems of Lesson 6.2. As an activity, you might want to ask students to come up with other conditions that they believe are sufficient to show that a given quadrilateral is a parallelogram. For example, if one pair of sides are both congruent and parallel, then the given quadrilateral must be a parallelogram.

Students may again have difficulty with Theorem 6.9, the converse of Theorem 6.5 found in Lesson 6.2. Have students note that if the diagonals were drawn in the quadrilateral shown in Example 1 (which is not a parallelogram), one of the diagonals would bisect the other, but the second would not bisect the first. Remind students that in order to be certain that a quadrilateral is a parallelogram, you must know that the diagonals bisect *each other*.

QUICK TIP
To conclude the lesson, create two charts on the board, one titled "Properties of a Parallelogram" and the other "Ways to Prove a Quadrilateral is a Parallelogram." If possible, leave the charts on the board for students to view, except during assessment situations.

Lesson 6.4

Point out that the meaning of *rectangle* in a casual conversation may not agree with the mathematical definition. Often we think of a rectangle in terms of its sides, but the definition given here does not even mention side lengths. By the mathematical definition, a square is a special kind of rectangle. Point out that it is mathematically correct to say, when looking at a square, "This a rectangle." The definitions given in the text should make it clear that all the figures studied in this lesson are special types of parallelograms, but you may want to state this point explicitly for emphasis.

Some students may realize that the definitions given here contain more information than is necessary. For example, if only two consecutive sides of a parallelogram are congruent, it must be a rhombus, because of Theorem 6.2. Also, because of Theorems 6.3 and 6.4, a parallelogram with only one right angle must be a rectangle. You may want to lead students to these facts, as a way of reviewing some of the previous theorems.

COROLLARIES The corollaries given on page 326 also provide a way to review the theorems of Lesson 6.3. For example, the Rhombus Corollary can be justified by noting that if all four sides of a quadrilateral are congruent, then certainly the opposite sides are congruent in pairs, and so Theorem 6.6 applies. Similarly, the Rectangle Corollary follows from Theorem 6.7.

Finally, it may be difficult for students to keep straight the properties that apply to all parallelograms (those presented in Lesson 6.2) and those that apply only to the special types of parallelograms. If a student is in doubt, one good way to clear up their questions is to sketch and examine a parallelogram that is not one of the special types. For example, a parallelogram with one very long side and one very short side will make it clear that Theorem 6.10 does not apply to all parallelograms.

Lesson 6.5

Emphasize at the outset that trapezoids are *not* special cases of parallelograms. In fact, by the definition of a parallelogram (stated on page 310), there can be no figure that is simultaneously both a parallelogram and a trapezoid. You may also want to make sure that students understand that the term *base angles* can refer to the pair of angles with vertices at the endpoints of either base.

> **QUICK TIP**
> When drawing trapezoids throughout the discussion of the lesson, draw them in a variety of different ways. Do not always draw them with the longer base as the bottom of the figure, and avoid making them look like isosceles trapezoids unless that is the intent.

MAKING CONNECTIONS Theorems 6.12 and 6.13 (which are converses of each other) resemble the corresponding theorems about isosceles triangles (Theorems 4.3 and 4.4). Reminding students of this fact may help solidify the theorems in their minds, as well as review the previous theorems.

The fact that the length of the midsegment of a trapezoid is the average of the lengths of the bases is stated informally on page 333, but it could be regarded as a theorem.

Lesson 6.6

In order to give a schematic picture of the logic on page 337, you may want to draw a Venn diagram. A Venn diagram including all seven categories shown on page 337 would consist of one large region, representing quadrilaterals, within which there would be two separated regions, one representing parallelograms, the other representing trapezoids. The region representing trapezoids would completely contain a region labeled "Isosceles Trapezoids," while the parallelogram region would contain two overlapping regions labeled "Rhombuses" and "Rectangles." The overlap would be labeled "Squares." The diagram in Exercises 20–23 on page 340 shows a Venn diagram modeling a portion of these relationships.

This section can be used to review the many properties of various kinds of quadrilaterals that have been studied in the previous sections of this chapter. This review falls into two general categories: (1) identifying which types of quadrilaterals possess a given property, and (2) showing that a given property is sufficient to guarantee that a given quadrilateral is of one of the special types. The first of these skills is illustrated by Exercises 1–4, while the latter is shown in Examples 1–3.

KITES As a way of reviewing many of the facts of this chapter, it might be helpful to mention a type of quadrilateral that was not specifically classified in the chapter: a kite. This is a quadrilateral having two pairs of congruent consecutive sides, but not necessarily having all four sides congruent. (A kite is shown in Example 1 on page 317.) You might ask students where this quadrilateral fits into the classification scheme. You might also consider which, if any, facts that have been stated about quadrilaterals in this chapter apply to a kite. For example, the diagonals of a kite are perpendicular. You might also ask students to state conjectures about this figure that they believe are true, such as it has one pair of congruent opposite angles.

Chapter Goals

While this chapter begins with a discussion of all polygons and how to identify and classify them based on their properties, the remainder of the chapter is spent examining quadrilaterals. Students will:

- Find angle measures of quadrilaterals.
- Identify special quadrilaterals.
- Use the properties of parallelograms, rhombuses, rectangles, squares, and trapezoids to find their side lengths and angle measures.
- Investigate the midsegment of a trapezoid.

Application Note

When a photographer wants to photograph a large object from above, the necessary elevation can be achieved by using a scissors lift. The height of the lift can be quickly changed as the photographer takes pictures showing different perspectives of the object. Scissors lifts are also used when a movie scene needs to be filmed with more than one camera, and from different heights and perspectives. When the various shots are then spliced together, the result can be an enhanced sense of character movement.

More information about scissors lifts is provided on the Internet at classzone.com

Application Links
CLASSZONE.COM

Chapter 6 Quadrilaterals

How do photographers reach high places?

Scissors lifts can lift photographers, movie crews, and other people who need to reach high places.

The design of the lift ensures that as the platform is raised or lowered it is always parallel to the ground.

Learn More About It

You will learn more about scissors lifts in Exercises 34–37 on p. 314.

Who uses Quadrilaterals?

FURNITURE DESIGNER
Furniture designers use geometry and trigonometry to create designs for furniture that is structurally sound and visually interesting. (p. 329)

GEMOLOGIST
Gemologists consider the color and clarity of a gem, as well as the cut, when evaluating its value. After cutting, the facets of some gems are rectangles and trapezoids. (p. 340)

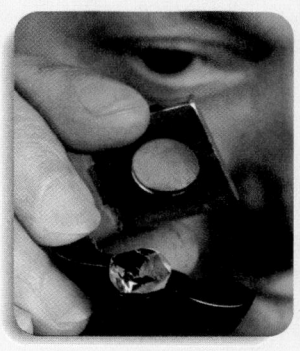

How will you use these ideas?

- Discover plant roots and stems that resemble polygons. (p. 307)
- See how photographers take overhead photos. (p. 314)
- Find out how the derailleur of a bicycle uses quadrilaterals. (p. 321)
- Predict the path of a ball on a pool table. (p. 322)
- Learn how gems are cut to enhance their sparkle. (p. 340)

301

Hands-On Activities

Activities (more than 20 minutes)
6.2 Investigating Parallelograms, p. 309
6.3 Making Parallelograms, p. 324
6.5 Midsegment of a Trapezoid, p. 331

Geo-Activities (less than 20 minutes)
6.3 Making Parallelograms, p. 316

Projects

A project covering Chapters 5–6 appears on pages 352–353 of the Student Edition. An additional project for Chapter 6 is available in the *Chapter 6 Resource Book*, pp. 75–76.

⊞ Technology

- Electronic Teacher Tools
- Test and Practice Generator
- Online Lesson Planner

▭ Video

- **Geometry in Motion**
  There is an animation supporting Lessons 6.2, 6.3, and 6.4.

Internet Support
CLASSZONE.COM

- Application and Career Links 321, 329, 340
- Student Help 307, 315, 317, 326, 336, 338

PREVIEW

What's the chapter about?

- Classifying polygons and finding angle measures of quadrilaterals
- Using properties of parallelograms, rhombuses, rectangles, squares, and trapezoids
- Identifying special quadrilaterals

Key Words

- **polygon**, *p. 303*
- **diagonal of a polygon**, *p. 303*
- **parallelogram**, *p. 310*
- **rhombus**, *p. 325*
- **rectangle**, *p. 325*
- **square**, *p. 325*
- **trapezoid**, *p. 332*
- **isosceles trapezoid**, *p. 332*
- **midsegment of a trapezoid**, *p. 333*

PREPARE

Chapter Readiness Quiz

Take this quick quiz. If you are unsure of an answer, look at the reference pages for help.

Vocabulary Check *(refer to p. 173)*

1. What type of triangle is $\triangle FGH$? **C**

   A scalene   **B** equilateral

   C isosceles   **D** None of these

Skill Check *(refer to pp. 131, 180)*

2. What is the measure of $\angle 4$? **G**

   F 45°   **G** 55°

   H 65°   **J** 125°

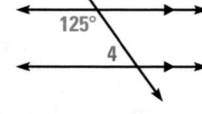

3. In $\triangle JKL$, $m\angle J = 55°$ and $m\angle K = 90°$. What is $m\angle L$? **C**

   A 15°   **B** 25°   **C** 35°   **D** 45°

VISUAL STRATEGY

Drawing Quadrilaterals

Visualize It! ➡

You can used lined paper to help you draw quadrilaterals that have at least one pair of opposite sides parallel.

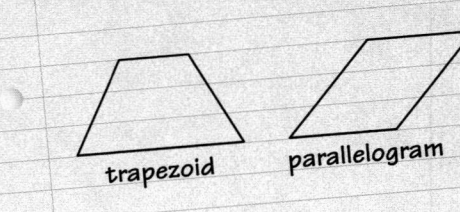

trapezoid   parallelogram

6.1 Polygons

Goal

Identify and classify polygons. Find angle measures of quadrilaterals.

Key Words

- polygon
- side of a polygon
- vertex of a polygon
- diagonal of a polygon

Each traffic sign below is an example of a *polygon*. Notice that each sign is formed with straight lines.

A **polygon** is a plane figure that is formed by three or more segments called **sides**. Each side intersects exactly two other sides at each of its endpoints. Each endpoint is a **vertex** of the polygon.

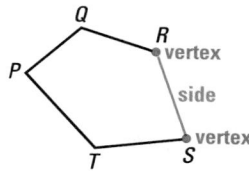

Two vertices that are the endpoints of the same side are called *consecutive* vertices. For example, in polygon *PQRST*, *R* and *S* are consecutive vertices.

A segment that joins two *nonconsecutive* vertices of a polygon is called a **diagonal**. Polygon *PQRST* has two diagonals from vertex *R*, $\overline{RP}$ and $\overline{RT}$.

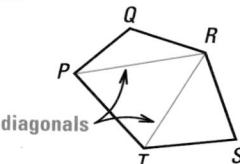

Student Help

VOCABULARY TIP
A *side* connects consecutive vertices. A *diagonal* connects nonconsecutive vertices.

EXAMPLE 1 Identify Polygons

Is the figure a polygon? Explain your reasoning.

a. 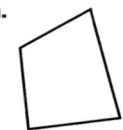 b.  c. d.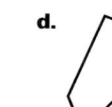

Solution

a. Yes. The figure is a polygon formed by four straight sides.

b. No. The figure is not a polygon because it has a side that is not a segment.

c. No. The figure is not a polygon because two of the sides intersect only one other side.

d. Yes. The figure is a polygon formed by six straight sides.

6.1 *Polygons* **303**

 **Plan**

Pacing
Suggested Number of Days

Basic: 2 days
Average: 2 days
Advanced: 2 days
Block Schedule: 1 block

Teaching Resources

📄 *Blacklines*
(See page 300B.)

🔧 *Transparencies*
• Warm-Up with Quiz
• Answers

💻 *Technology*
• Electronic Teacher Tools
• Test and Practice Generator
• Online Lesson Planner
• Internet Support

 **Teach**

Content and Teaching Strategies
For background information on geometric concepts and teaching strategies related to this lesson, see pages 300E and 300F in this Teacher's Edition.

Tips for New Teachers
Once students know the properties of triangles they can apply them to other figures that contain triangles, as is illustrated at the top of page 305. Students may use this approach to divide any polygon into non-overlapping triangles and determine the sum of the measures of the interior angles. See the Tips for New Teachers on pp. 1–2 of the *Chapter 6 Resource Book* for additional notes about Lesson 6.1.

Extra Example 1
See next page.

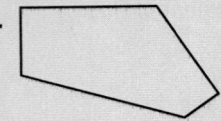

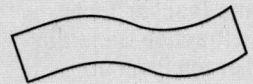
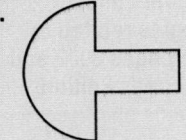

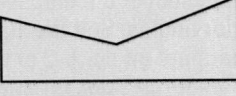
Classifying Polygons You can classify polygons by the number of sides they have. Some special types of polygons are listed below.

TYPES OF POLYGONS

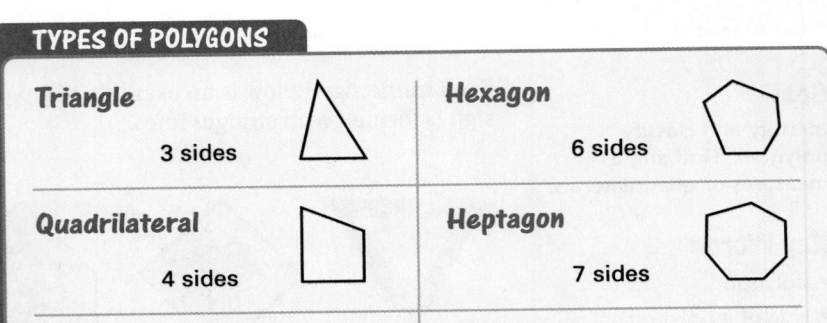

Triangle	3 sides	Hexagon	6 sides
Quadrilateral	4 sides	Heptagon	7 sides
Pentagon	5 sides	Octagon	8 sides

EXAMPLE 2 Classify Polygons

Decide whether the figure is a polygon. If so, tell what type. If not, explain why.

a.  b.  c. d.

Solution

a. The figure is a polygon with four sides, so it is a quadrilateral.

b. The figure is not a polygon because it has some sides that are not segments.

c. The figure is a polygon with five sides, so it is a pentagon.

d. The figure is not a polygon because some of the sides intersect more than two other sides.

Checkpoint ✓ **Identify and Classify Polygons**

Decide whether the figure is a polygon. If so, tell what type. If not, explain why.

1.  2.  3. 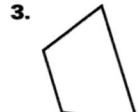 4.

yes; pentagon   yes; quadrilateral

Quadrilaterals A diagonal of a quadrilateral divides it into two triangles, each with angle measures that add up to 180°. So, the sum of the measures of the interior angles of a quadrilateral is 2 × 180°, or 360°.

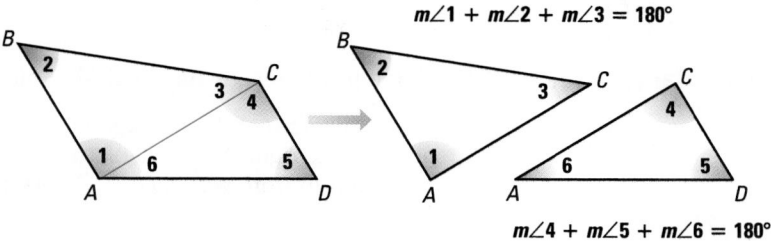

$$m\angle 1 + m\angle 2 + m\angle 3 = 180°$$

$$m\angle 4 + m\angle 5 + m\angle 6 = 180°$$

THEOREM 6.1

Quadrilateral Interior Angles Theorem

Words The sum of the measures of the interior angles of a quadrilateral is 360°.

Symbols $m\angle 1 + m\angle 2 + m\angle 3 + m\angle 4 = 360°$

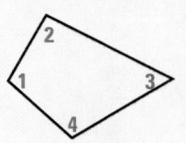

Student Help

STUDY TIP
Name a polygon by listing its vertices *consecutively* in either direction. Two names for the quadrilateral in Example 3 are *PQRS* and *RQPS*.

EXAMPLE 3 Find Angle Measures of Quadrilaterals

Find the measure of ∠S.

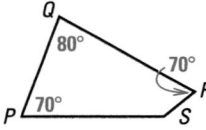

Solution

Use the fact that the sum of the measures of the interior angles of a quadrilateral is 360°.

$m\angle P + m\angle Q + m\angle R + m\angle S = 360°$	Quadrilateral Interior Angles Theorem
$70° + 80° + 70° + m\angle S = 360°$	Substitute angle measures.
$220° + m\angle S = 360°$	Simplify.
$m\angle S = 140°$	Subtract 220° from each side.

ANSWER ▶ The measure of ∠S is 140°.

Checkpoint ✓ Find Angle Measures of Quadrilaterals

Find the measure of ∠A.

5.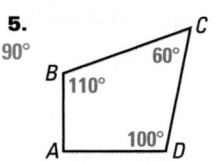

6.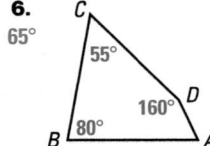

7.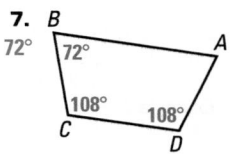

6.1 *Polygons* **305**

Geometric Reasoning
The strategy of dividing a quadrilateral into triangles and using what you know about the sum of the measures of the angles of a triangle to find the sum of the measures of the angles of the quadrilateral will be extended to polygons with more than four sides in Lesson 8.2.

Teaching Tip
Students may be skeptical that the sum of the angle measures of any quadrilateral is 360°. Have students draw several quadrilaterals of different shapes on heavy paper. Students can then cut out each shape with scissors, cut off the four corners, and arrange them with the vertices all touching to see that the sum of the four angle measures is 360°.

Extra Example 3
Find the measure of ∠V. 35°

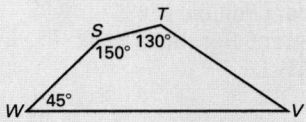

✓ Concept Check
How do you know that the figure below is a polygon?

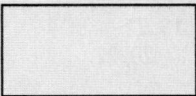

The figure is formed by four segments, each of which intersects exactly two other segments only at their endpoints.

🙂 Daily Puzzler
A polygon has at least one pair of supplementary angles and at least one pair of complementary angles. Can the polygon be a quadrilateral? If so, give an example. If not, explain why. yes; a quadrilateral with angles of 140°, 40°, 50°, and 130°

Apply

Assignment Guide

BASIC
Day 1: pp. 306–308 Exs. 8–14, 21, 22, 30–42 even
Day 2: pp. 306–308 Exs. 15–20, 29–41 odd

AVERAGE
Day 1: pp. 306–308 Exs. 8–14, 21–27, 30–36 even
Day 2: pp. 306–308 Exs. 15–20, 28, 29–35 odd, 37–42

ADVANCED
Day 1: pp. 306–308 Exs. 8–14, 21–27, 30–42 even
Day 2: pp. 306–308 Exs. 15–20, 28, 29–41 odd; EC: TE p. 300D*, classzone.com

BLOCK SCHEDULE
pp. 306–308 Exs. 8–42

Extra Practice
• Student Edition, p. 685
• Chapter 6 Resource Book, pp. 11–12

Homework Check
To quickly check student understanding of key concepts, go over the following exercises:

Basic: 8, 12, 15, 18, 21
Average: 9, 13, 16, 19, 21
Advanced: 10, 14, 17, 20, 22

Geometric Reasoning
You may wish to extend Exercise 11 by asking students what is the least number of diagonals a polygon can have (0, in a triangle) and if there is a number of diagonals that polygons cannot ever have (1).

6.1 Exercises

Guided Practice

Vocabulary Check

1. What type of polygon has 8 sides? 15 sides?   octagon; 15-gon

2. Use the diagram of the pentagon shown at the right. Name all of the *diagonals* from vertex *D*.
$\overline{DB}$ and $\overline{DA}$

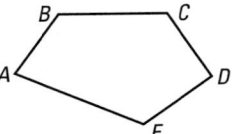

Skill Check

Is the figure a polygon? Explain your reasoning.

3.

4.
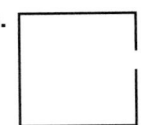

5.

Yes; the figure is a polygon formed by five straight lines.

No; one side is not a segment.

No; two of the sides intersect only one other side.

Find the measure of ∠A.

6.
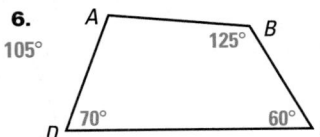

7.
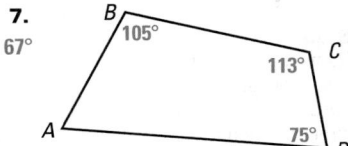

Practice and Applications

Extra Practice
See p. 685.

11. 3; in order for each side to intersect exactly two other sides at each of its endpoints, a polygon must have at least three sides; triangle.

Homework Help

Example 1: Exs. 8–10
Example 2: Exs. 8–10, 21, 24–27
Example 3: Exs. 15–20, 28

Classifying Polygons Decide whether the figure is a polygon. If so, tell what type. If not, explain why.

8.
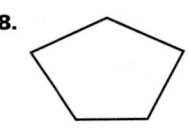

9.
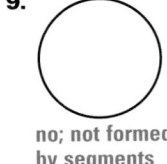

10.

yes; pentagon

no; not formed by segments

yes; 10-gon (or decagon)

11. Logical Reasoning What is the fewest number of sides a polygon can have? Explain your answer, then name the polygon.

Visualize It! Sketch the figure(s) described.   12–14. See margin.

12. Two different pentagons

13. A hexagon with three diagonals drawn from a single vertex

14. A quadrilateral with two obtuse angles

Finding Angle Measures Find the measure of ∠A.

15.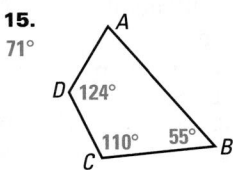
71°

16.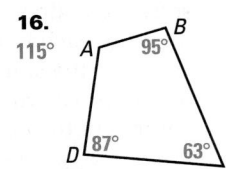
115°

17.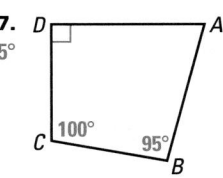
75°

Student Help
CLASSZONE.COM

HOMEWORK HELP
Extra help with problem solving in Exs. 18–20 is at classzone.com

Using Algebra Find the value of x.

18.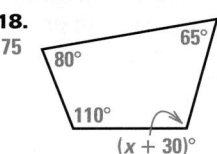
75

19.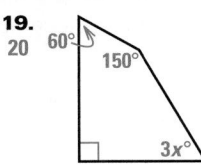
20

20.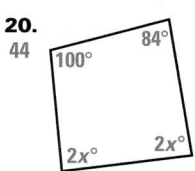
44

Parachutes Some gym classes play games using parachutes that look like the polygon below.

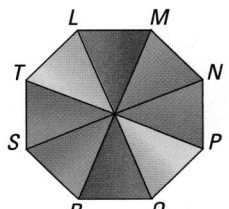

21. Tell how many sides the polygon has and what type of polygon it is. 8; octagon

22. Polygon *LMNPQRST* is one name for the polygon. State two other names using the vertices. *Sample answers: LTSRQPNM, MNPQRSTL*

23. Name all of the diagonals that have vertex *M* as an endpoint. Not all of the diagonals are shown. $\overline{MP}$, $\overline{MQ}$, $\overline{MR}$, $\overline{MS}$, $\overline{MT}$

Link to
Plants

CARAMBOLA, or star fruit, has a cross section shaped like a five-pointed star.

Plants Use the following information.
Cross sections of roots and stems often resemble polygons. Next to each cross section is the polygon it resembles. Tell how many sides each polygon has and tell what type of polygon it is.
▶ Source: *The History and Folklore of North American Wildflowers*

24. Virginia Snakeroot 6; hexagon

25. Caraway 8; octagon

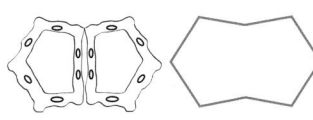

26. Fennel 5; pentagon

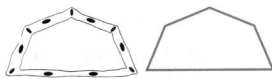

27. Poison Hemlock 17; 17-gon

✗ Common Error
Watch for students who name $\overline{ML}$ and $\overline{MN}$ as diagonals in Exercise 23. Remind these students that diagonals must pass through the interior of a polygon.

12–14. Sample answers are given.

12.

13.
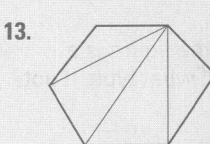

14.

Assess

Assessment Resources

The Mini-Quiz below is also available on blackline (*Chapter 6 Resource Book*, p. 18) and on transparency. For more assessment resources, see:
- Chapter 6 Resource Book
- Standardized Test Practice
- Test and Practice Generator

Mini-Quiz

Decide whether the figure is a polygon. If so, tell what type. If not, explain why.

1.

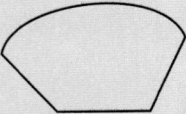

No; one side is not a segment.

2.

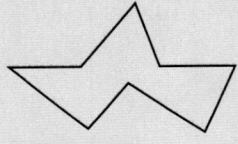

yes; octagon

Find the measure of ∠A.

3.   80°

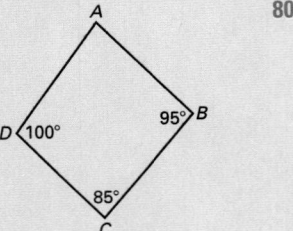

4.   108°

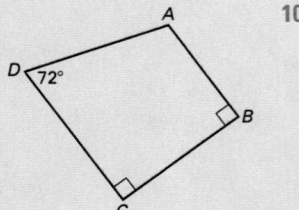

5. Find the value of x. 65

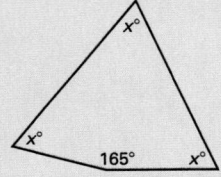

28. Technology Use geometry software to draw a quadrilateral. Measure each interior angle and calculate the sum. What happens to the sum as you drag the vertices of the quadrilateral?

The sum is 360° for any quadrilateral.

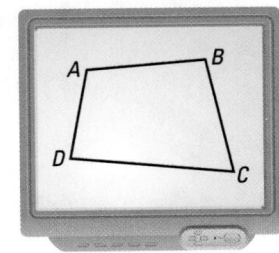

Standardized Test Practice

29. Multi-Step Problem Envelope manufacturers fold a specially-shaped piece of paper to make an envelope, as shown below.

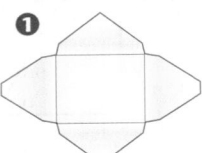 

a. How many sides are formed by the outer edges of the paper before it is folded? Name the type of polygon. 18; 18-gon

29b. Step 2: 10, 10-gon (or decagon); Step 3: 7, heptagon; Step 4: 4, quadrilateral

b. Tell how many sides are formed by the outer edges of the paper in Steps 2–4. Name the type of polygon formed after each step.

c. If the four angles of the red quadrilateral in Step 4 are congruent, then what is the measure of each angle? 90°

Mixed Review

Line Relationships Determine whether the lines are *parallel*, *perpendicular*, or *neither*. (Lesson 3.1)

30. $\overleftrightarrow{AB}$ and $\overleftrightarrow{CE}$ neither   **31.** $\overleftrightarrow{AC}$ and $\overleftrightarrow{BE}$ parallel

32. $\overleftrightarrow{AB}$ and $\overleftrightarrow{AC}$ perpendicular   **33.** $\overleftrightarrow{AC}$ and $\overleftrightarrow{CE}$ neither

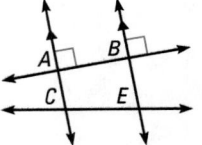

Finding Angle Measures Find the measure of the numbered angle. (Lesson 3.4)

34.   **35.**   **36.**

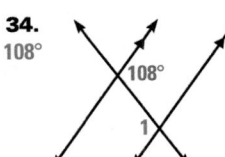   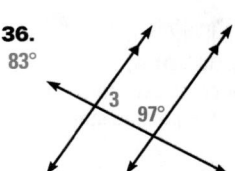

Algebra Skills

Distributive Property Use the distributive property to rewrite the expression without parentheses. (Skills Review, p. 671)

37. $4(x + 3)$ $4x + 12$ **38.** $(x - 1)6$ $6x - 6$ **39.** $-2(x - 7)$ $-2x + 14$

40. $-5(2x + 3)$ $-10x - 15$ **41.** $-3(5x - 2)$ $-15x + 6$ **42.** $(4x - 4)x$ $4x^2 - 4x$

308   **Chapter 6** *Quadrilaterals*

Activity 6.2 Investigating Parallelograms

Question

What are some of the properties of a parallelogram?

Materials

- lined paper
- tracing paper
- ruler
- protractor

Explore

1 Place a piece of tracing paper on top of a piece of lined paper. Trace two lines of the lined paper.

2 Turn the tracing paper so that the lines on the lined paper intersect the lines you drew. Trace two more lines to form a quadrilateral.

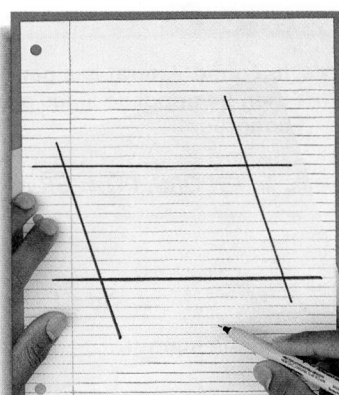

3 Use a ruler to measure the length of each side of the quadrilateral. Then use a protractor to measure each angle of the quadrilateral.

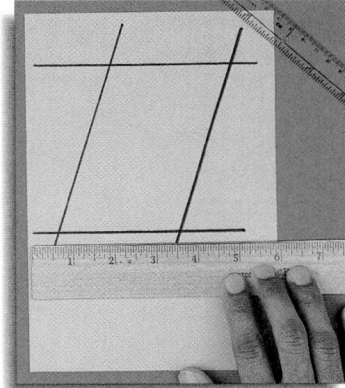

Visualize It!

In *ABCD*, $\overline{AB}$ and $\overline{DC}$ are opposite sides, and $\overline{AD}$ and $\overline{BC}$ are opposite sides.

Think About It

1. The figure you drew is called a *parallelogram*. Why do you think this type of quadrilateral is called a parallelogram? because the opposite sides are parallel

2. What do you notice about the lengths of the opposite sides of the figure you drew? The lengths of the opposite sides are equal.

3. What do you notice about the measures of the opposite angles of the figure you drew? The measures of the opposite angles are equal.

4. Based on your results from Exercises 2 and 3, complete the statements.

   a. The opposite sides of a parallelogram are __?__. congruent

   b. The opposite angles of a parallelogram are __?__. congruent

Activity 309

Pacing
Suggested Number of Days

Basic: 2 days
Average: 2 days
Advanced: 2 days
Block Schedule: 1 block

Teaching Resources

 **Blacklines**
(See page 300B.)

Transparencies
• Warm-Up with Quiz
• Answers

Technology
• Electronic Teacher Tools
• Test and Practice Generator
• Online Lesson Planner
• Internet Support

Video
• **Geometry in Motion**

Teach

Content and Teaching Strategies
For background information on geometric concepts and teaching strategies related to this lesson, see pages 300E and 300F in this Teacher's Edition.

Tips for New Teachers
Demonstrate the properties illustrated in the theorems on pages 310–312 by using a large pair of identically-labeled congruent parallelograms. Place them to show opposite sides congruent, opposite angles congruent, consecutive angles supplementary, and diagonals that bisect each other. See the Tips for New Teachers on pp. 1–2 of the *Chapter 6 Resource Book* for additional notes about Lesson 6.2.

Goal
Use properties of parallelograms.

Key Words
• parallelogram

Parallelogram lifts, like the one shown in the photograph, are used to raise heavy-duty vehicles.

A **parallelogram** is a quadrilateral with both pairs of opposite sides parallel.

The symbol □*PQRS* is read "parallelogram *PQRS*."

In □*PQRS*, $\overline{PQ} \parallel \overline{SR}$ and $\overline{QR} \parallel \overline{PS}$.

THEOREM 6.2

Words If a quadrilateral is a parallelogram, then its opposite sides are congruent.

Symbols In □*PQRS*, $\overline{PQ} \cong \overline{SR}$ and $\overline{QR} \cong \overline{PS}$.

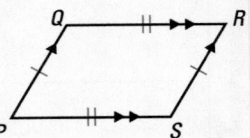

EXAMPLE 1 Find Side Lengths of Parallelograms

FGHJ is a parallelogram. Find *JH* and *FJ*.

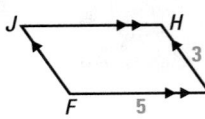

Solution

JH = *FG*	Opposite sides of a □ are congruent.
= 5	Substitute 5 for *FG*.
FJ = *GH*	Opposite sides of a □ are congruent.
= 3	Substitute 3 for *GH*.

ANSWER ▶ In □*FGHJ*, *JH* = 5 and *FJ* = 3.

Checkpoint ✓ *Find Side Lengths of Parallelograms*

1. *ABCD* is a parallelogram. Find *AB* and *AD*.
AB = 9; *AD* = 8

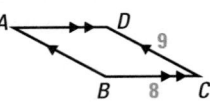

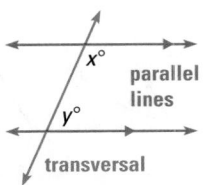

parallel lines

transversal

Consecutive angles of a parallelogram are like same-side interior angles. By Theorem 3.7, they are supplementary.

THEOREMS 6.3 and 6.4

Theorem 6.3

Words If a quadrilateral is a parallelogram, then its opposite angles are congruent.

Symbols In $\square PQRS$, $\angle P \cong \angle R$ and $\angle Q \cong \angle S$.

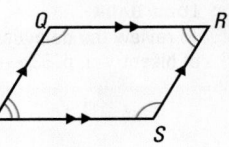

Theorem 6.4

Words If a quadrilateral is a parallelogram, then its consecutive angles are supplementary.

Symbols In $\square PQRS$, $x° + y° = 180°$.

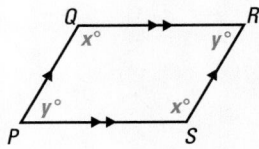

EXAMPLE 2 Find Angle Measures of Parallelograms

$PQRS$ is a parallelogram. Find the missing angle measures.

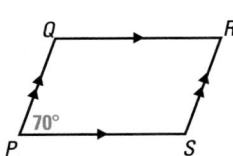

Solution

❶ By Theorem 6.3, the opposite angles of a parallelogram are congruent, so $m\angle R = m\angle P = 70°$.

❷ By Theorem 6.4, the consecutive angles of a parallelogram are supplementary.

$m\angle Q + m\angle P = 180°$ Consecutive angles of a $\square$ are supplementary.

$m\angle Q + 70° = 180°$ Substitute 70° for $m\angle P$.

$m\angle Q = 110°$ Subtract 70° from each side.

❸ By Theorem 6.3, the opposite angles of a parallelogram are congruent, so $m\angle S = m\angle Q = 110°$.

ANSWER ▶ The measure of $\angle R$ is 70°, the measure of $\angle Q$ is 110°, and the measure of $\angle S$ is 110°.

Checkpoint ✔ Find Angle Measures of Parallelograms

ABCD is a parallelogram. Find the missing angle measures.

2. $m\angle B = 120°$; $m\angle C = 60°$; $m\angle D = 120°$

3. $m\angle A = 75°$; $m\angle B = 105°$; $m\angle C = 75°$

2.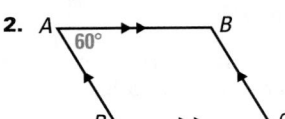

3.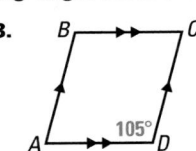

Multiple Representations

Use the diagram from Theorem 6.2 to review the symbols used on diagrams to show congruent sides and parallel sides. Make sure students recognize the difference between the symbols and relate the symbols to the terms *congruent* and *parallel*.

Extra Example 1

KLMN is a parallelogram. Find *MN* and *KN*. *MN* = 8, *KN* = 12

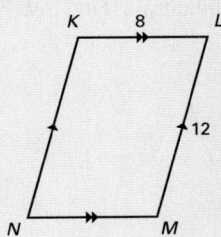

Extra Example 2

STVW is a parallelogram. Find the missing angle measures.

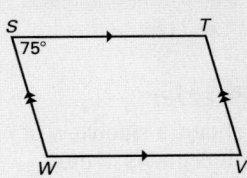

$m\angle T = 105°$; $m\angle V = 75°$; $m\angle W = 105°$

312

✗ Common Error

Watch for students who think Theorem 6.5 implies that the diagonals of a parallelogram are congruent. Explain that in a future lesson, students will learn about a special kind of parallelogram that has congruent diagonals, but stress that all parallelograms do not have this property.

Extra Example 3

JKLM is a parallelogram. Find *JN*. 8

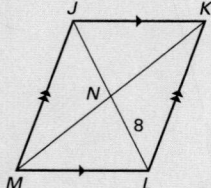

✓ Concept Check

Which parts of a parallelogram are congruent? **Both pairs of opposite sides are congruent, and both pairs of opposite angles are congruent.**

⊕ Daily Puzzler

The diagram shows a side view of a rectangular box that is in the process of being flattened. The box is 20 inches wide. When it is flattened the cardboard will be 68 inches long. What was the diagonal measure of the box before it was flattened? **52 in.**

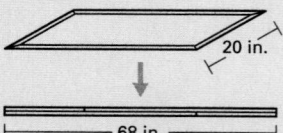

Student Help

LOOK BACK
To review the definition of bisect, see p. 53. see p. 53

THEOREM 6.5

Words If a quadrilateral is a parallelogram, then its diagonals bisect each other.

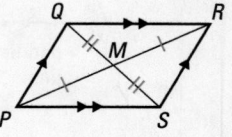

Symbols In ▱*PQRS*, $\overline{QM} \cong \overline{MS}$ and $\overline{PM} \cong \overline{MR}$.

EXAMPLE 3 Find Segment Lengths

TUVW is a parallelogram. Find *TX*.

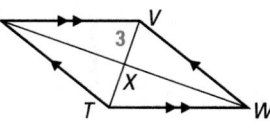

Solution

$$TX = XV \qquad \text{Diagonals of a } \square \text{ bisect each other.}$$
$$ = 3 \qquad \text{Substitute 3 for } XV.$$

SUMMARY   PROPERTIES OF PARALLELOGRAMS

Definition of parallelogram, p. 310

If a quadrilateral is a parallelogram, then both pairs of opposite sides are parallel.

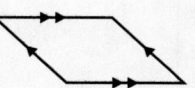

Theorem 6.2, p. 310

If a quadrilateral is a parallelogram, then its opposite sides are congruent.

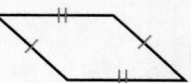

Theorem 6.3, p. 311

If a quadrilateral is a parallelogram, then its opposite angles are congruent.

Theorem 6.4, p. 311

If a quadrilateral is a parallelogram, then its consecutive angles are supplementary.

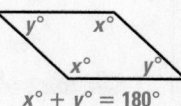

$x° + y° = 180°$

Theorem 6.5, p. 312

If a quadrilateral is a parallelogram, then its diagonals bisect each other.

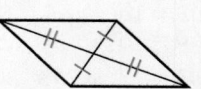

Guided Practice

Vocabulary Check

1. Complete the statement: A(n) __?__ is a quadrilateral with both pairs of opposite sides parallel.   parallelogram

Skill Check

Decide whether the figure is a parallelogram. If it is not, explain why.

2.    No; only one pair of sides is parallel.

3. 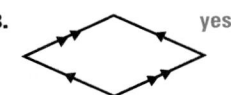   yes

Complete the statement. Give a reason for your answer.

4. $\overline{ML}$; opposite sides of a ▱ are ≅.

5. ∠KJM; opposite ∡ of a ▱ are ≅.

6. ∠LMJ; opposite ∡ of a ▱ are ≅.

7. $\overline{NL}$; diagonals of a ▱ bisect each other.

4. $\overline{JK} \cong$ __?__

5. $\angle MLK \cong$ __?__

6. $\angle JKL \cong$ __?__

7. $\overline{JN} \cong$ __?__

8. $\angle MNL \cong$ __?__   ∠KNJ; vertical ∡ are ≅.

9. $\overline{NM} \cong$ __?__ $\overline{NK}$; diagonals of a ▱ bisect each other.

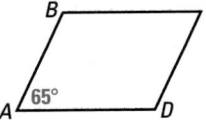

Find the measure in the parallelogram.

10. Find $m\angle C$. 65°

11. Find HK. 9

12. Find $m\angle Y$. 70°

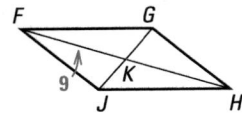

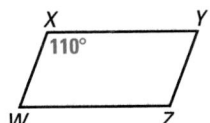

Practice and Applications

Extra Practice

See p. 685.

Congruent Segments Match the segment in ▱PQRS with a congruent one. Give a reason for your answer. 13–16. See margin.

13. $\overline{PT}$ A. $\overline{RS}$

14. $\overline{QR}$ B. $\overline{RT}$

15. $\overline{QT}$ C. $\overline{PS}$

16. $\overline{PQ}$ D. $\overline{ST}$

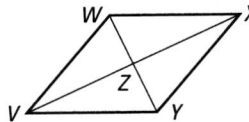

Congruent Angles Match the angle in ▱VWXY with a congruent one. Give a reason for your answer. 17–20. See margin.

17. ∠VZY   E. ∠WZX

18. ∠WVY   F. ∠VWX

19. ∠WXZ   G. ∠YVZ

20. ∠VYX   H. ∠YXW

Homework Help

Example 1: Exs. 13–16, 22–24

Example 2: Exs. 17–20, 25–27

Example 3: Exs. 13–16, 28–30

---

③ Apply

Assignment Guide

BASIC
Day 1: pp. 313–315 Exs. 14, 16–20, 22–24, 31, 32, 43, 44–54 even
Day 2: pp. 313–315 Exs. 13, 15, 25–30, 42, 45–55 odd

AVERAGE
Day 1: pp. 313–315 Exs. 14, 16–24, 31, 32, 38–40, 43, 44–54 even
Day 2: pp. 313–315 Exs. 13, 15, 25–30, 33–37, 42, 45–55 odd

ADVANCED
Day 1: pp. 313–315 Exs. 14, 16–24, 31, 32, 38–40, 41*, 43, 44, 46
Day 2: pp. 313–315 Exs. 13, 15, 25–30, 33–37, 42, 47–55 odd; EC: classzone.com

BLOCK SCHEDULE
pp. 313–315 Exs. 13–40, 42–55

Extra Practice

• Student Edition, p. 685
• Chapter 6 Resource Book, pp. 19–20

Homework Check

To quickly check student understanding of key concepts, go over the following exercises:

Basic: 13, 17, 22, 25, 28
Average: 14, 23, 26, 29, 30
Advanced: 15, 20, 27, 30, 33

13. B; diagonals of a ▱ bisect each other.
14. C; opposite sides of a ▱ are ≅.
15. D; diagonals of a ▱ bisect each other.
16. A; opposite sides of a ▱ are ≅.
17. E; vertical angles are ≅.
18. H; opposite angles of a ▱ are ≅.
19. G; if 2 ∥ lines are cut by a transversal, then alternate interior ∡ are ≅.
20. F; opposite angles of a ▱ are ≅.

313

21. **You be the Judge** *EFGH* is a parallelogram. Is $\overline{EF}$ parallel to $\overline{HG}$ or $\overline{GF}$? Explain your answer. *HG*; If you sketch parallelogram *EFGH*, you see that $\overline{EF}$ intersects $\overline{GF}$ and $\overline{EF}$ is parallel to $\overline{HG}$.

Finding Side Lengths *EFGH* is a parallelogram. Find *EF* and *FG*.

22.

EF = 8; *FG* = 7

23.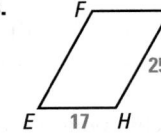

EF = 25; *FG* = 17

24.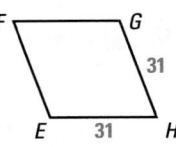

EF = 31; *FG* = 31

Finding Angle Measures *JKLM* is a parallelogram. Find the missing angle measures.

25.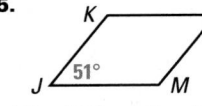

$m\angle K = 129°$; $m\angle L = 51°$; $m\angle M = 129°$

26.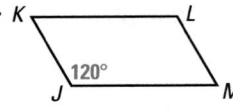

$m\angle K = 60°$; $m\angle L = 120°$; $m\angle M = 60°$

27.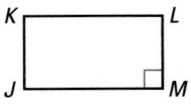

$m\angle J = 90°$; $m\angle K = 90°$; $m\angle L = 90°$

Finding Segment Lengths *ABCD* is a parallelogram. Find *DE*.

28.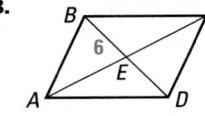

29.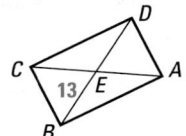

30.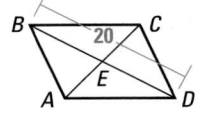

Using Algebra Find the values of *x* and *y* in the parallelogram.

31.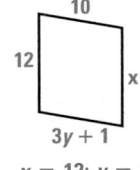

x = 12; *y* = 3

32.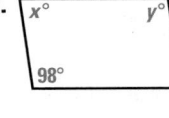

x = 82; *y* = 98

33.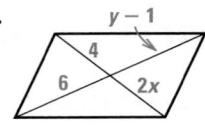

x = 2; *y* = 7

Scissors Lift Use the diagram of the scissors lift below.

34. What is $m\angle B$ when $m\angle A$ is 120°? 60°

35. Suppose you decrease $m\angle A$. What happens to $m\angle B$? $m\angle B$ increases.

36. Suppose you decrease $m\angle A$. What happens to *AD*? *AD* increases.

37. Suppose you decrease $m\angle A$. What happens to the overall height of the scissors lift? The height increases.

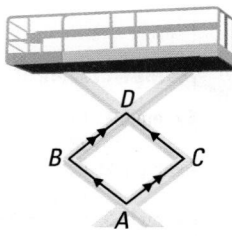

41. No; the lengths of the vertical sides of the red parallelogram are shorter than the corresponding lengths in the blue parallelogram.

Standardized Test Practice

Mixed Review

44. yes, by the Corresponding Angles Converse

45. No; same-side interior angles are not supplementary.

46. yes, by the Angle Addition Postulate and the Alternate Exterior Angles Converse

Algebra Skills

Staircases In the diagram below, the red quadrilateral and the blue quadrilateral are parallelograms.

38. Which angle in the red parallelogram is congruent to ∠1? ∠4

39. Which angles in the blue parallelogram are supplementary to ∠6? ∠5 and ∠8

40. Which postulate can be used to prove that ∠1 ≅ ∠5? Corresponding Angles Postulate

41. Challenge Is the red parallelogram congruent to the blue parallelogram? Explain your reasoning.

42. Multiple Choice Which of the following statements is *not* necessarily true about □*ABCD*? D

 Ⓐ *AE* = *CE* Ⓑ *AD* = *BC*

 Ⓒ *BE* = *DE* Ⓓ *AC* = *BD*

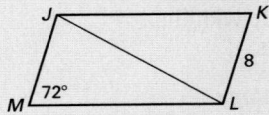

43. Multiple Choice *PQRS* is a parallelogram. What is the value of *x*? H

 Ⓕ 28 Ⓖ 34

 Ⓗ 59 Ⓙ 121

Parallel Lines Are lines *p* and *q* parallel? Explain. *(Lesson 3.5)*

44.
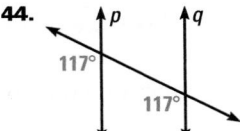

45.
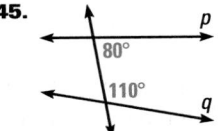

46.
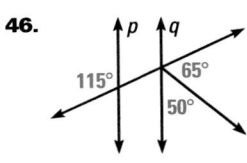

Isosceles and Equilateral Triangles Find the value of *x*. *(Lesson 4.3)*

47.
31

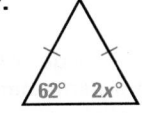

48.
7
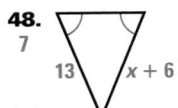

49.
5
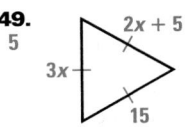

Finding Slope Find the slope of the line that passes through the points. *(Skills Review, p. 665)*

50. (1, 3) and (6, 5) $\frac{2}{5}$ **51.** (3, −8) and (7, 4) 3

52. (2, 1) and (−1, 0) $\frac{1}{3}$ **53.** (−4, 2) and (5, −1) $-\frac{1}{3}$

54. (6, −2) and (12, 14) $\frac{8}{3}$ **55.** (0, −3) and (−5, −6) $\frac{3}{5}$

6.2 Properties of Parallelograms   **315**

---

4 Assess

Assessment Resources
The Mini-Quiz below is also available on blackline (*Chapter 6 Resource Book*, p. 26) and on transparency. For more assessment resources, see:
• Chapter 6 Resource Book
• Standardized Test Practice
• Test and Practice Generator

Mini-Quiz
JKLM is a parallelogram.

1. Find *JM*. 8
2. Find the measure of ∠*K*. 72°
3. Find the measure of ∠*MJK*. 108°
4. If *ML* = 14, find *JK*. 14
5. Suppose $\overline{MK}$ is added to the diagram, with the point of intersection with $\overline{JL}$ labeled as point *Q*. What can you conclude about $\overline{JQ}$ and $\overline{LQ}$?
They are congruent.

Pacing

Suggested Number of Days

Basic: 2 days
Average: 2 days
Advanced: 2 days
Block Schedule: 1 block

Teaching Resources

📋 **Blacklines**
(See page 300B.)

✋ **Transparencies**
• Warm-Up with Quiz
• Answers

🖥 **Technology**
• Electronic Teacher Tools
• Test and Practice Generator
• Online Lesson Planner
• Internet Support

📼 **Video**
• Geometry in Motion

Geo-Activity

Goal Use straws to explore parallelograms.

Key Discovery A quadrilateral with both pairs of opposite sides congruent is a parallelogram.

②Teach

Content and Teaching Strategies
For background information on geometric concepts and teaching strategies related to this lesson, see pages 300E and 300F in this Teacher's Edition.

Tips for New Teachers

Help students recognize that the theorems in this lesson for showing a quadrilateral is a parallelogram are the converses of those given in Lesson 6.2. See the Tips for New Teachers on pp. 1–2 of the *Chapter 6 Resource Book* for additional notes about Lesson 6.3.

316

Goal

Show that a quadrilateral is a parallelogram.

Key Words

• parallelogram p. 310

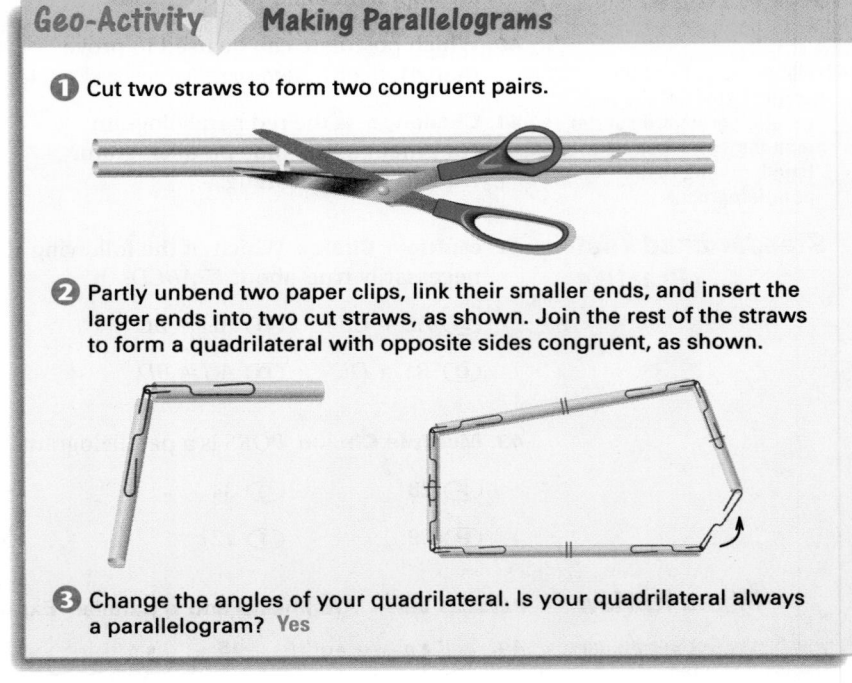

Geo-Activity ◆ **Making Parallelograms**

① Cut two straws to form two congruent pairs.

② Partly unbend two paper clips, link their smaller ends, and insert the larger ends into two cut straws, as shown. Join the rest of the straws to form a quadrilateral with opposite sides congruent, as shown.

③ Change the angles of your quadrilateral. Is your quadrilateral always a parallelogram? Yes

The Geo-Activity above describes one way to show that a quadrilateral is a parallelogram.

Student Help

STUDY TIP
The theorems in this lesson are the converses of the theorems in Lesson 6.2.

THEOREMS 6.6 and 6.7

Theorem 6.6

Words If both pairs of opposite sides of a quadrilateral are congruent, then the quadrilateral is a parallelogram.

Symbols If $\overline{PQ} \cong \overline{SR}$ and $\overline{QR} \cong \overline{PS}$, then $PQRS$ is a parallelogram.

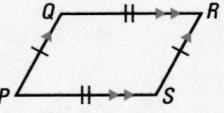

Theorem 6.7

Words If both pairs of opposite angles of a quadrilateral are congruent, then the quadrilateral is a parallelogram.

Symbols If $\angle P \cong \angle R$ and $\angle Q \cong \angle S$, then $PQRS$ is a parallelogram.

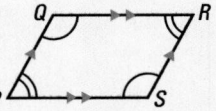

EXAMPLE 1 Use Opposite Sides

Tell whether the quadrilateral is a
parallelogram. Explain your reasoning.

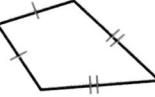

Solution

The quadrilateral is not a parallelogram. It has two pairs
of congruent sides, but opposite sides are not congruent.

EXAMPLE 2 Use Opposite Angles

Tell whether the quadrilateral is a
parallelogram. Explain your reasoning.

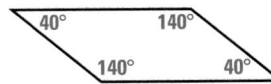

Solution

The quadrilateral is a parallelogram because both pairs of
opposite angles are congruent.

 **Use Opposite Sides and Opposite Angles**

**In Exercises 1 and 2, tell whether the quadrilateral is a parallelogram.
Explain your reasoning.**

1. Yes; both pairs of opposite
sides are congruent.

2. No; opposite angles are
not congruent.

1.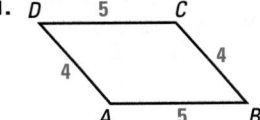

2.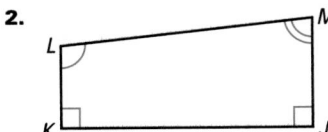

3. In quadrilateral *WXYZ*, *WX* = 15, *YZ* = 20, *XY* = 15, and *ZW* = 20.
Is *WXYZ* a parallelogram? Explain your reasoning.
No; opposite sides are not congruent.

THEOREM 6.8

Words If an angle of a quadrilateral is
supplementary to both of its
consecutive angles, then the
quadrilateral is a parallelogram.

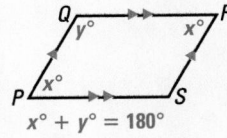

Symbols If $m\angle P + m\angle Q = 180°$ and $m\angle Q + m\angle R = 180°$,
then *PQRS* is a parallelogram.

Extra Example 1

Tell whether the quadrilateral is a
parallelogram. Explain your
reasoning.

Yes; both pairs of opposite sides are
congruent.

Extra Example 2

Tell whether the quadrilateral is a
parallelogram. Explain your
reasoning.

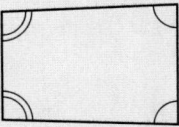

No; opposite angles are not
congruent.

318

Extra Example 3

Tell whether the quadrilateral is a parallelogram. Explain your reasoning.

a.

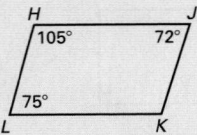

No; consecutive angles are not supplementary.

b.

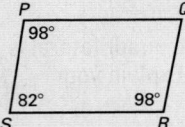

Yes; ∠S is supplementary to both of its consecutive angles.

Extra Example 4

Tell whether the quadrilateral is a parallelogram. Explain your reasoning.

a.

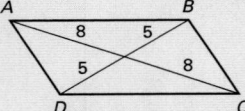

Yes; the diagonals bisect each other.

b.

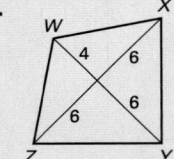

No; the diagonals do not bisect each other.

EXAMPLE 3 Use Consecutive Angles

Tell whether the quadrilateral is a parallelogram. Explain your reasoning.

a.   **b.**   **c.**

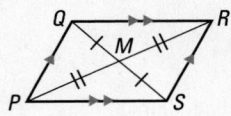

Solution

a. ∠U is supplementary to ∠T and ∠V (85° + 95° = 180°). So, by Theorem 6.8, *TUVW* is a parallelogram.

b. ∠G is supplementary to ∠F (55° + 125° = 180°), but ∠G is *not* supplementary to ∠H (55° + 120° ≠ 180°). So, *EFGH* is *not* a parallelogram.

c. ∠D is supplementary to ∠C (90° + 90° = 180°), but you are not given any information about ∠A or ∠B. Therefore, you cannot conclude that *ABCD* is a parallelogram.

THEOREM 6.9

Words If the diagonals of a quadrilateral bisect each other, then the quadrilateral is a parallelogram.

Symbols If $\overline{QM} \cong \overline{MS}$ and $\overline{PM} \cong \overline{MR}$, then *PQRS* is a parallelogram.

EXAMPLE 4 Use Diagonals

Tell whether the quadrilateral is a parallelogram. Explain your reasoning.

a.   **b.**

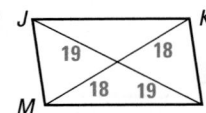
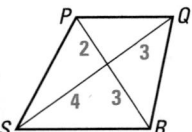

Solution

a. The diagonals of *JKLM* bisect each other. So, by Theorem 6.9, *JKLM* is a parallelogram.

b. The diagonals of *PQRS* do not bisect each other. So, *PQRS* is *not* a parallelogram.

Tell whether the quadrilateral is a parallelogram. Explain your reasoning.

4. No; opposite angles are not congruent (or consecutive angles are not supplementary).

5. Yes; one angle is supplementary to both of its consecutive angles.

6. Yes; the diagonals bisect each other.

7. No; the diagonals do not bisect each other.

4.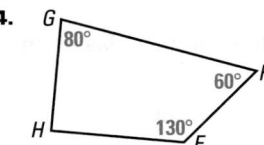

5.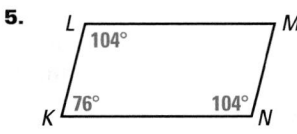

6.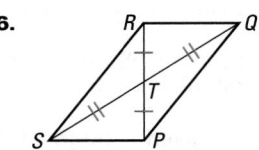

7.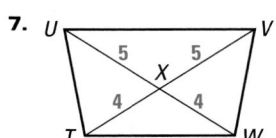

You have learned five ways to show that a quadrilateral is a parallelogram.

SUMMARY   **SHOWING A QUADRILATERAL IS A PARALLELOGRAM**

Definition of parallelogram, p. 310

Show that both pairs of opposite sides are parallel.

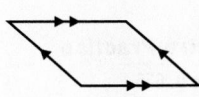

Theorem 6.6, p. 316

Show that both pairs of opposite sides are congruent.

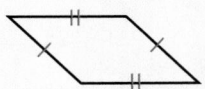

Theorem 6.7, p. 316

Show that both pairs of opposite angles are congruent.

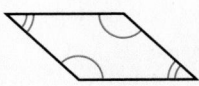

Theorem 6.8, p. 317

Show that one angle is supplementary to both of its consecutive angles.

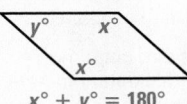

$x° + y° = 180°$

Theorem 6.9, p. 318

Show that the diagonals bisect each other.

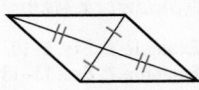

6.3 *Showing Quadrilaterals are Parallelograms*   **319**

Multiple Representations

Work slowly through the summary table. To assess students' understanding, ask them to describe how each diagram is related to its summary statement.

✓ **Concept Check**

What is the difference between a quadrilateral and a parallelogram? While a quadrilateral is any four-sided polygon, a parallelogram is a special type of quadrilateral that has both pairs of opposite sides parallel and congruent, and both pairs of opposite angles congruent.

🧩 **Daily Puzzler**

In parallelogram *ABCD*, the ratio of the measure of ∠*A* to the measure of ∠*B* is 4 : 5. What are the measures of all the angles?
$m∠A = 80°; m∠B = 100°;$
$m∠C = 80°; m∠D = 100°$

Assignment Guide

BASIC
Day 1: pp. 320–323 Exs. 8–13, 21, 27–40
Day 2: EP pp. 679–680 Exs. 21, 28, 31, 34; pp. 320–323 Exs. 14–19, 25, 26, Quiz 1

AVERAGE
Day 1: pp. 320–323 Exs. 8–13, 20, 21, 25–38
Day 2: pp. 320–323 Exs. 14–19, 22, 23, 39–44, Quiz 1

ADVANCED
Day 1: pp. 320–323 Exs. 8–13, 20, 21, 25–34
Day 2: pp. 320–323 Exs. 14–19, 22, 23, 24*, 35–43 odd, Quiz 1; EC: classzone.com

BLOCK SCHEDULE
pp. 320–323 Exs. 8–23, 25–44, Quiz 1

Extra Practice

• Student Edition, p. 685
• Chapter 6 Resource Book, pp. 29–30

Homework Check

To quickly check student understanding of key concepts, go over the following exercises:

Basic: 8, 10, 11, 14, 17
Average: 9, 12, 13, 14, 19
Advanced: 10, 13, 15, 16, 21

6.3 Exercises

Guided Practice

Vocabulary Check

1. $\overline{EF}$ is parallel to $\overline{HG}$; $\overline{EH}$ is parallel to $\overline{FG}$.

4. Every parallelogram has four sides, so every parallelogram is a quadrilateral. Some quadrilaterals have no pairs of parallel sides or one pair of parallel sides, so not every quadrilateral is a parallelogram.

In Exercises 1–3, name all the sides or angles of ▱EFGH that match the description.

1. Opposite sides are parallel.

2. Opposite angles are congruent.
   $\angle E \cong \angle G$; $\angle F \cong \angle H$

3. Consecutive angles are supplementary.
   $m\angle E + m\angle F = 180°$; $m\angle F + m\angle G = 180°$; $m\angle G + m\angle H = 180°$; $m\angle H + m\angle E = 180°$

4. Explain why every parallelogram is a quadrilateral, but not every quadrilateral is a parallelogram.

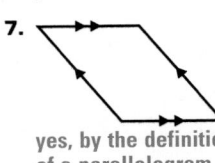

Skill Check

Decide whether you are given enough information to show that the quadrilateral is a parallelogram. Explain your reasoning.

5.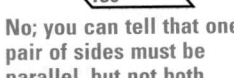

No; you can tell that one pair of sides must be parallel, but not both.

6.

No; you do not have any information about the lengths of two of the sides.

7.

yes, by the definition of a parallelogram

Practice and Applications

Extra Practice
See p. 685.

13. No; opposite angles are not congruent (or consecutive angles are not supplementary).

14. Yes; one angle is supplementary to both of its consecutive angles.

15. No; consecutive angles are not supplementary.

16. Yes; one angle is supplementary to both of its consecutive angles.

Homework Help

Example 1: Exs. 8–10
Example 2: Exs. 11–13
Example 3: Exs. 14–16
Example 4: Exs. 17–19

Using Opposite Sides Tell whether the quadrilateral is a parallelogram. Explain your reasoning.

8.

Yes; both pairs of opposite sides are congruent.

9.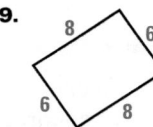

Yes; both pairs of opposite sides are congruent.

10.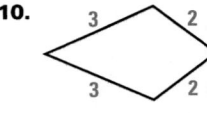

No; opposite sides are not congruent.

Using Opposite Angles Tell whether the quadrilateral is a parallelogram. Explain your reasoning.

11.

Yes; both pairs of opposite angles are congruent.

12.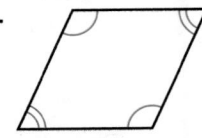

Yes; both pairs of opposite angles are congruent.

13.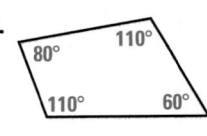

Using Consecutive Angles Tell whether the quadrilateral is a parallelogram. Explain your reasoning.

14.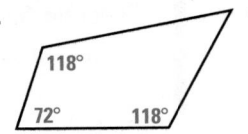

15.

16.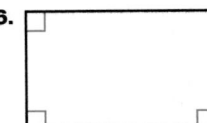

Using Diagonals Tell whether the quadrilateral is a parallelogram. Explain your reasoning.

17.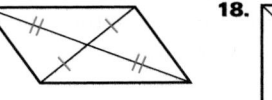

Yes; the diagonals bisect each other.

18.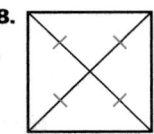

Yes; the diagonals bisect each other.

19.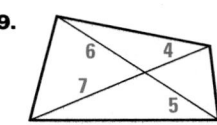

No; the diagonals do not bisect each other.

Link to
Bicycles

DERAILLEURS (named from the French word meaning "to derail") move the chain to change gears.

Application Links
CLASSZONE.COM

20. **Bicycle Gears** When you change gears on a bicycle, the derailleur moves the chain to the new gear. For the derailleur at the right, $AB = 1.8$ cm, $BC = 3.6$ cm, $CD = 1.8$ cm, and $DA = 3.6$ cm. Explain why $\overline{AB}$ and $\overline{CD}$ are always parallel when the derailleur moves. See margin.

21. **Error Analysis** What is wrong with the student's argument below? See margin.

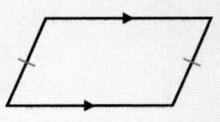

> A quadrilateral that has one pair of sides congruent and the other pair of sides parallel is always a parallelogram.

20. Both pairs of opposite sides of quadrilateral *ABCD* are congruent, so *ABCD* is a parallelogram. The opposite sides of a parallelogram are parallel, so *AB* and *CD* are always parallel.

21. The parallel sides may not be congruent so the congruent sides may not be parallel. So the quadrilateral may not be a parallelogram.

22. No; although the quadrilateral has two pairs of congruent angles, there is no information about whether the congruent angles are consecutive angles or opposite angles.

23. The diagonals of the resulting quadrilateral were drawn to bisect each other. Therefore, the resulting quadrilateral is a parallelogram.

22. **You be the Judge** Three of the interior angles of a quadrilateral have measures of 75°, 75°, and 105°. Is this enough information to conclude that the quadrilateral is a parallelogram? Explain your answer.

23. **Visualize It!** Explain why the following method of drawing a parallelogram works. State a theorem to support your answer.

❶ Use a ruler to draw a segment and its midpoint.

❷ Draw another segment so the midpoints coincide.

❸ Connect the endpoints of the segments.

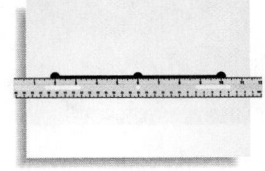
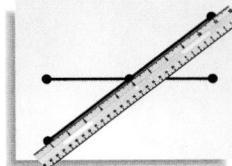
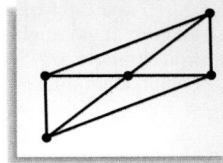

24. **Challenge** If one pair of opposite sides of a quadrilateral is both congruent and parallel, is the quadrilateral a parallelogram? Explain your reasoning. Yes; see margin.

Visualize It!
For students who answer Exercises 21 and 22 incorrectly, the best explanation may be a diagram. For both exercises an isosceles trapezoid should demonstrate the correct conclusions. For Exercise 22, make certain that it has the correct angle measures.

24. Yes; if one pair of opposite sides is congruent and parallel, the other pair of opposite sides must also be congruent and parallel.

Teaching Tip

The Example and Exercises 25–26 are coordinate geometry proofs, as were the Example and Exercises 40–41 on page 289. As before, you may wish to demonstrate some other coordinate proofs for your students.

Student Help

SKILLS REVIEW
To review the formula for finding slope, see p. 665.

25. The slope of $\overline{FG}$ = slope of $\overline{JH}$ = −3, and slope of $\overline{FJ}$ = slope of $\overline{GH}$ = 0, so the opposite sides are parallel and *FGHJ* is a parallelogram by definition.

26. The slope of $\overline{PQ} = \frac{3}{2}$ and the slope of $\overline{SR}$ = 3. Since the slopes are not equal, the sides are not parallel and *PQRS* is not a parallelogram.

EXAMPLE *Coordinate Geometry*

Use the slopes of the segments in the diagram to determine if the quadrilateral is a parallelogram.

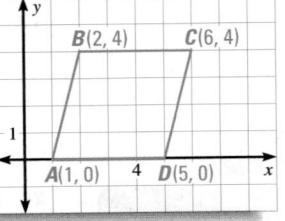
B(2, 4) C(6, 4) A(1, 0) 4 D(5, 0)

Solution

Lines and line segments are parallel if they have the same slope.

Slope of $\overline{AB}$: $\frac{4-0}{2-1} = \frac{4}{1} = 4$

Slope of $\overline{DC}$: $\frac{4-0}{6-5} = \frac{4}{1} = 4$

Slope of $\overline{BC}$: $\frac{4-4}{6-2} = \frac{0}{4} = 0$

Slope of $\overline{AD}$: $\frac{0-0}{5-1} = \frac{0}{4} = 0$

ANSWER ▶ The slopes of $\overline{AB}$ and $\overline{DC}$ are the same, so $\overline{AB} \parallel \overline{DC}$. The slopes of $\overline{BC}$ and $\overline{AD}$ are the same, so $\overline{AD} \parallel \overline{BC}$. Both pairs of opposite sides are parallel, so *ABCD* is a parallelogram.

Coordinate Geometry Use the slopes of the segments in the diagram to determine if the quadrilateral is a parallelogram. See margin.

25.
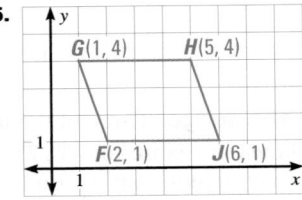
G(1, 4) H(5, 4) F(2, 1) J(6, 1)

26.
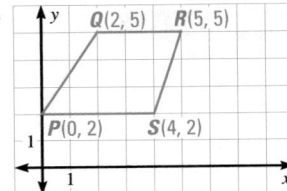
Q(2, 5) R(5, 5) P(0, 2) S(4, 2)

27a. The acute angles of a right triangle are complementary, so $m\angle AFE = 90° - 63° = 27°$.

27b. The ball bounces off each wall at the same angle at which it hit the wall, so $m\angle DFG = m\angle AFE = 27°$. $\triangle DFG$ is a right triangle, and the acute angles of a right triangle are complementary, so $m\angle FGD = 90° - m\angle DFG = 63°$.

27c. 27°; 27°

27d. $m\angle FEH = m\angle HGF = 54°$ and $m\angle EFG = m\angle GHE = 126°$; *EFGH* is a parallelogram since both pairs of opposite angles are congruent.

27. **Multi-Step Problem** Suppose you shoot a pool ball as shown below and it rolls back to where it started. The ball bounces off each wall at the same angle at which it hits the wall.

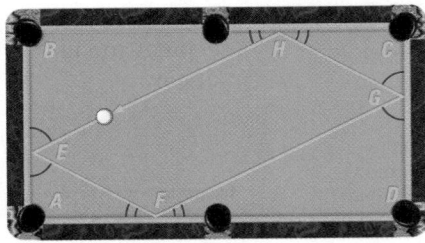

a. The ball hits the first wall at an angle of 63°. So $m\angle AEF = m\angle BEH = 63°$. Explain why $m\angle AFE = 27°$.

b. Explain why $m\angle FGD = 63°$.

c. What is $m\angle GHC$? $m\angle EHB$?

d. Find the measure of each interior angle of *EFGH*. What kind of quadrilateral is *EFGH*? How do you know?

Finding Angle Measures Find the measure of $\angle A$. *(Lesson 6.1)*

28.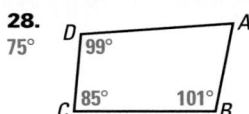
75°

29.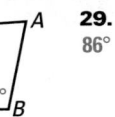
86°

30.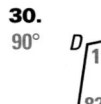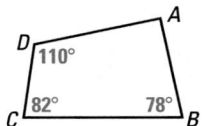

Finding Measures Find the measure in $\square$*JKLM*. *(Lesson 6.2)*

31. Find $m\angle K$. 71°

32. Find $m\angle J$. 109°

33. Find ML. 14

34. Find KL. 11

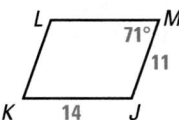

Algebra Skills

Evaluating Expressions Evaluate the expression for the given value of the variable. *(Skills Review, p. 670)*

35. $2x + 7$ when $x = 5$ 17

36. $4y - 3$ when $y = 2$ 5

37. $13 + 3m$ when $m = 3$ 22

38. $1 - b$ when $b = -10$ 11

39. $5 - 2a$ when $a = -6$ 17

40. $8c - 5$ when $c = -4$ −37

41. $12x + x^2$ when $x = -1$ −11

42. $\frac{3}{4}q^2 - 2$ when $q = 4$ 10

43. $5n^3 - 4n$ when $n = 2$ 32

44. $15p + p^2$ when $p = -3$ −36

Quiz 1

Decide whether the figure is a polygon. If so, tell what type. If not, explain why. *(Lesson 6.1)*

1.

yes; pentagon

2.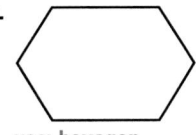

yes; hexagon

3.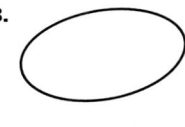

no; not formed by segments

Find the values of the variables in the parallelogram. *(Lesson 6.2)*

4.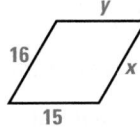

$x = 16; y = 15$

5.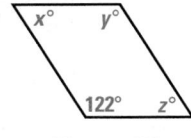

$x = 58; y = 122; z = 58$

6.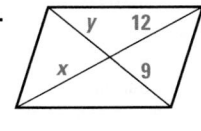

$x = 12; y = 9$

7. Yes; one angle is supplementary to both of its consecutive angles.

8. No; you can tell that one pair of sides must be parallel, but not both.

9. Yes; both pairs of opposite sides are parallel, so the quadrilateral is a parallelogram by the definition of a parallelogram.

Tell whether the quadrilateral is a parallelogram. Explain your reasoning. *(Lesson 6.3)*

7.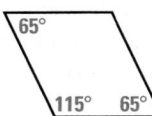

8.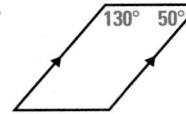

9.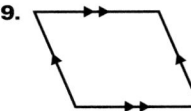

6.3 *Showing Quadrilaterals are Parallelograms* 323

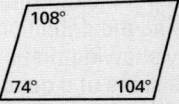 **Assess**

Assessment Resources
The Mini-Quiz below is also available on blackline (*Chapter 6 Resource Book*, p. 37) and on transparency. For more assessment resources, see:
• Chapter 6 Resource Book
• Standardized Test Practice
• Test and Practice Generator

Mini-Quiz
Tell whether the quadrilateral is a parallelogram. Explain your reasoning.

1.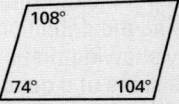

No; consecutive angles are not supplementary.

2.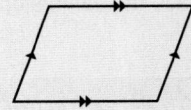

yes, by the definition of parallelogram

3.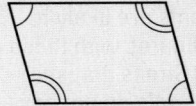

Yes; both pairs of opposite angles are congruent.

4.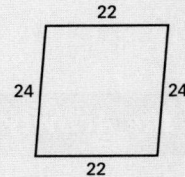

Yes; both pairs of opposite sides are congruent.

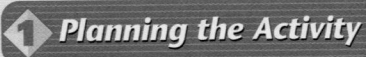
Goal

Students explore how the slopes of the opposite sides of a parallelogram are related.

Materials

See the Technology Activity Keystrokes Masters in the *Chapter 6 Resource Book*.

LINK TO LESSON ⟹

In the summary box on page 319 of Lesson 6.3, students reviewed the different ways to show that a quadrilateral is a parallelogram. This included using the definition of parallelogram by showing that both pairs of opposite sides of a quadrilateral are parallel. Help students understand that if the opposite sides of a quadrilateral are parallel, then they also have the same slope.

Managing the Activity

Tips for Success

In Step 3, students are likely to compare their figures with those of other students. Stress that students' quadrilaterals do not have to be identical. As long as both pairs of opposite angles are congruent, the specific angle measures used do not matter.

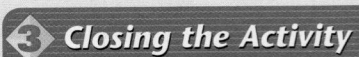
Closing the Activity

KEY DISCOVERY

Opposite sides of a parallelogram have the same slope.

Activity Assessment

Use Exercises 3 and 4 to assess student understanding.

Technology Activity ◆ **6.3** **Making Parallelograms**

Question

How can you use the angle measures in a quadrilateral to show that it is a parallelogram?

Explore

1 Draw quadrilateral *ABCD*.

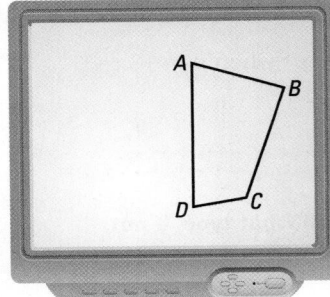

2 Measure the angles of the quadrilateral.

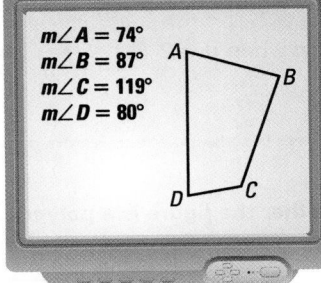

3 Drag the vertices until $m\angle A = m\angle C$ and $m\angle B = m\angle D$.

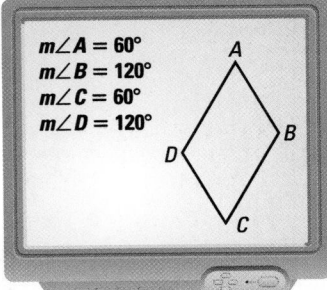

Think About It

Student Help

> **STUDY TIP**
> Recall that parallel lines and parallel segments have the same slope.

1. Find the slopes of $\overline{AB}$, $\overline{BC}$, $\overline{CD}$, and $\overline{DA}$. What do you notice about the slopes of opposite sides? *Slopes will vary; the slopes of opposite sides are equal.*
2. What do the slopes tell you about the sides of the quadrilateral? *The opposite sides of the quadrilateral are parallel.*
3. What kind of figure is quadrilateral *ABCD*? Use your results from Exercise 2 to help you. *a parallelogram by the definition of a parallelogram*
4. What theorem does this exploration illustrate? *If both pairs of opposite angles of a quadrilateral are congruent, then the quadrilateral is a parallelogram.*
5. **Extension** Draw quadrilateral *EFGH*. Draw segments $\overline{EG}$ and $\overline{FH}$. Construct a point *I* at the intersection of $\overline{EG}$ and $\overline{FH}$. Measure $\overline{EI}$, $\overline{IG}$, $\overline{FI}$, and $\overline{IH}$. Drag any of the vertices of *EFGH* so that $EI = IG$ and $FI = IH$. What do you notice? What theorem does this illustrate? *When EI = IG and FI = IH, EFGH is a parallelogram; If the diagonals of a quadrilateral bisect each other, then the quadrilateral is a parallelogram.*

 **6.4** **Rhombuses, Rectangles, and Squares**

Goal
Use properties of special types of parallelograms.

Key Words
• rhombus
• rectangle
• square

In this lesson you will study three special types of parallelograms.

A **rhombus** is a parallelogram with four congruent sides.

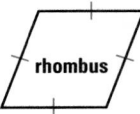

A **rectangle** is a parallelogram with four right angles.

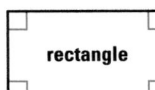

A **square** is a parallelogram with four congruent sides and four right angles.

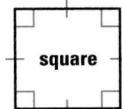

EXAMPLE **1** **Use Properties of Special Parallelograms**

In the diagram, *ABCD* is a rectangle.

a. Find *AD* and *AB*.

b. Find $m\angle A$, $m\angle B$, $m\angle C$, and $m\angle D$.

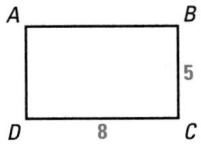

Solution

a. By definition, a rectangle is a parallelogram, so *ABCD* is a parallelogram. Because opposite sides of a parallelogram are congruent, $AD = BC = 5$ and $AB = DC = 8$.

b. By definition, a rectangle has four right angles, so $m\angle A = m\angle B = m\angle C = m\angle D = 90°$.

Checkpoint ✓ **Use Properties of Special Parallelograms**

1. In the diagram, *PQRS* is a rhombus. Find *QR*, *RS*, and *SP*. *QR* = 6, *RS* = 6, *SP* = 6

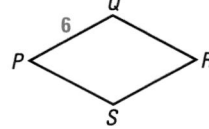

6.4 *Rhombuses, Rectangles, and Squares* **325**

Pacing
Suggested Number of Days

Basic: 2 days
Average: 2 days
Advanced: 2 days
Block Schedule: 1 block

Teaching Resources
📄 *Blacklines*
(See page 300B.)

📤 *Transparencies*
• Warm-Up with Quiz
• Answers

🖥 *Technology*
• Electronic Teacher Tools
• Test and Practice Generator
• Online Lesson Planner
• Internet Support

📼 *Video*
• Geometry in Motion

 **Teach**

─Content and Teaching─
Strategies
For background information on geometric concepts and teaching strategies related to this lesson, see pages 300E and 300F in this Teacher's Edition.

Tips for New Teachers
Consider having students make a Venn diagram illustrating the relationships of all the types of parallelograms they have learned so far. See the Tips for New Teachers on pp. 1–2 of the *Chapter 6 Resource Book* for additional notes about Lesson 6.4.

Extra Example 1
See next page.

325

Extra Example 1

In the diagram, *RSTU* is a square.

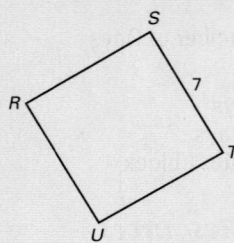

a. Find *UT*, *RU*, and *RS*.
  UT = 7, *RU* = 7, *RS* = 7

b. Find *m∠R*, *m∠S*, *m∠T*, and *m∠U*. *m∠R* = *m∠S* = *m∠T* = *m∠U* = 90°

Extra Example 2

Use the information in the diagram to name the special quadrilateral.

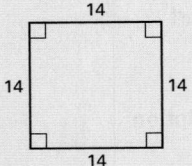

square

COROLLARIES

Rhombus Corollary

Words If a quadrilateral has four congruent sides, then it is a rhombus.

Symbols If $\overline{AB} \cong \overline{BC} \cong \overline{CD} \cong \overline{AD}$, then *ABCD* is a rhombus.

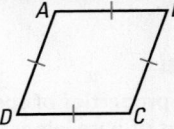

Rectangle Corollary

Words If a quadrilateral has four right angles, then it is a rectangle.

Symbols If $m\angle A = m\angle B = m\angle C = m\angle D = 90°$, then *ABCD* is a rectangle.

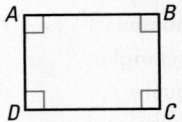

Square Corollary

Words If a quadrilateral has four congruent sides and four right angles, then it is a square.

Symbols If $\overline{AB} \cong \overline{BC} \cong \overline{CD} \cong \overline{AD}$ and $m\angle A = m\angle B = m\angle C = m\angle D = 90°$, then *ABCD* is a square.

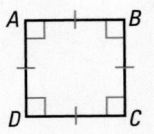

EXAMPLE 2 Identify Special Quadrilaterals

Use the information in the diagram to name the special quadrilateral.

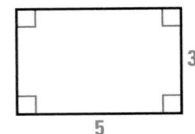

Solution

The quadrilateral has four right angles. So, by the Rectangle Corollary, the quadrilateral is a rectangle.

Because all of the sides are not the same length, you know that the quadrilateral is not a square.

Checkpoint *Identify Special Quadrilaterals*

Use the information in the diagram to name the special quadrilateral.

2. 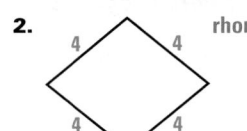 rhombus

3.  square

THEOREM 6.10

Words The diagonals of a rhombus are perpendicular.

Symbols In rhombus $ABCD$, $\overline{AC} \perp \overline{BD}$.

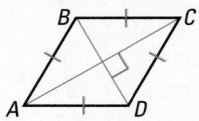

EXAMPLE 3 Use Diagonals of a Rhombus

$ABCD$ is a rhombus.
Find the value of x.

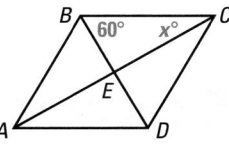

Student Help

LOOK BACK
To review the Corollary to the Triangle Sum Theorem, see p. 180.

Solution

By Theorem 6.10, the diagonals of a rhombus are perpendicular. Therefore, $\angle BEC$ is a right angle, so $\triangle BEC$ is a right triangle.

By the Corollary to the Triangle Sum Theorem, the acute angles of a right triangle are complementary. So, $x = 90 - 60 = 30$.

THEOREM 6.11

Words The diagonals of a rectangle are congruent.

Symbols In rectangle $ABCD$, $\overline{AC} \cong \overline{BD}$.

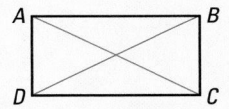

Link to
Carpentry

DOORS If a screen door is not rectangular, you can use a piece of hardware called a *turnbuckle* to shorten the longer diagonal until the door is rectangular.

EXAMPLE 4 Use Diagonals of a Rectangle

a. You nail four pieces of wood together to build a four-sided frame, as shown. What is the shape of the frame?

b. The diagonals measure 7 ft 4 in. and 7 ft 2 in. Is the frame a rectangle?

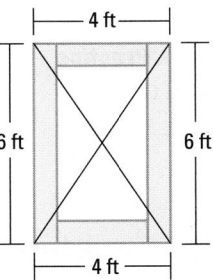

Solution

a. The frame is a parallelogram because both pairs of opposite sides are congruent.

b. The frame is not a rectangle because the diagonals are not congruent.

6.4 Rhombuses, Rectangles, and Squares **327**

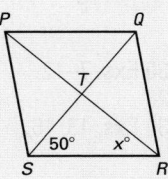
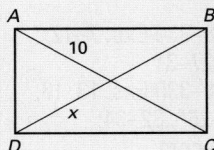

③ Apply

Assignment Guide

BASIC
Day 1: pp. 328–330 Exs. 7–12, 14, 15, 18, 19, 27, 28
Day 2: pp. 328–330 Exs. 13, 16, 17, 20–22, 29–34
AVERAGE
Day 1: pp. 328–330 Exs. 7–12, 14, 15, 18, 19, 27, 28
Day 2: pp. 328–330 Exs. 13, 16, 17, 20–22, 29–34
ADVANCED
Day 1: pp. 328–330 Exs. 7–12, 14, 15, 18, 19, 27–31
Day 2: pp. 328–330 Exs. 13, 16, 17, 20–22, 23–26*, 32–34; EC: classzone.com
BLOCK SCHEDULE
pp. 328–330 Exs. 7–22, 27–34

Extra Practice
• Student Edition, p. 686
• Chapter 6 Resource Book, pp. 40–41

Homework Check
To quickly check student understanding of key concepts, go over the following exercises:

Basic: 7, 8, 10, 13, 21
Average: 8, 9, 11, 13, 21
Advanced: 14, 17, 18, 19, 21

Checkpoint ✓ **Use Diagonals**

Find the value of x.

4. rhombus *ABCD* 90   **5.** rectangle *EFGH* 12   **6.** square *JKLM* 45

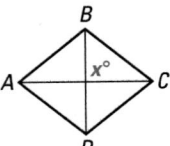 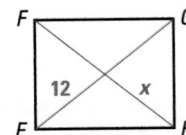 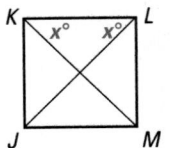

6.4 Exercises

Guided Practice

Vocabulary Check

1. What is the name for a parallelogram with four congruent sides?
rhombus

Skill Check

List all of the properties that must be true for the quadrilateral.

2. Parallelogram D   **A.** All sides are congruent.

3. Rectangle B, C, D   **B.** All angles are congruent.

4. Rhombus A, D   **C.** The diagonals are congruent.

5. Square A, B, C, D   **D.** Opposite angles are congruent.

6. *PQRS* is a rectangle. The length of $\overline{QS}$ is 12. Find *PR* and *PT*. *PR* = 12; *PT* = 6

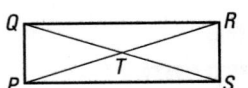

Practice and Applications

Extra Practice
See p. 686.

Using Properties **Find the measures.**

7. rhombus *ABCD*   **8.** rectangle *EFGH*   **9.** square *WXYZ*

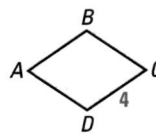 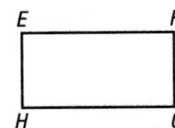 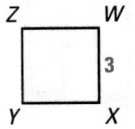

Homework Help
Example 1: Exs. 7–9
Example 2: Exs. 10–12
Example 3: Ex. 22
Example 4: Ex. 13

AB = __?__ 4 $m\angle E$ = __?__ ° 90 $m\angle W$ = __?__ ° 90

BC = __?__ 4 $m\angle F$ = __?__ ° 90   *YZ* = __?__ 3

AD = __?__ 4 $m\angle G$ = __?__ ° 90   *XY* = __?__ 3

Using Corollaries Use the information in the diagram to name the special quadrilateral.

10.
rectangle

11.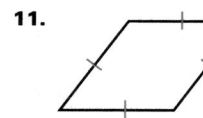
rhombus

12.
square

13. Making a Chair If you measure the diagonals of the chair frame as shown and find that they are congruent, can you conclude that the frame is rectangular? If not, what other information do you need? Explain your reasoning.
See margin.

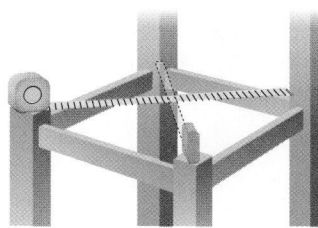

Sorting Quadrilaterals In Exercises 14–17, list each quadrilateral for which the statement is true.

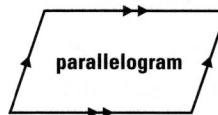
parallelogram

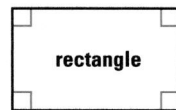
rectangle

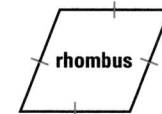
rhombus

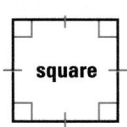
square

14. It has four right angles.
rectangle, square

15. Opposite sides are congruent.
parallelogram, rectangle, rhombus, square

16. Diagonals bisect each other.
parallelogram, rectangle, rhombus, square

17. Diagonals are perpendicular.
rhombus, square

 **Using Algebra**

EXAMPLE	Use Properties of Quadrilaterals

PQRS is a rectangle.
Find the value of *x*.

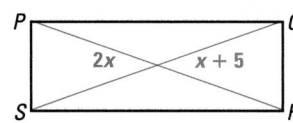

Solution

$PR = SQ$ Diagonals of a rectangle are congruent.

$2x = x + 5$ Substitute 2x for PR and x + 5 for SQ.

$x = 5$ Subtract x from each side.

Using Algebra Find the value of *x*.

18. rhombus *KLMN* 1
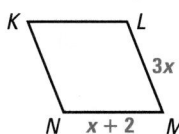

19. square *ABCD* 18
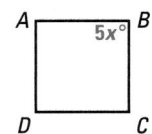

20. rectangle *EFGH* 3
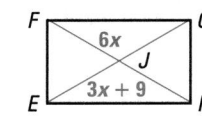

---

13. No, a quadrilateral with congruent diagonals does not have to be a rectangle; *Sample answer:* You need to know that the figure is a parallelogram, or that its diagonals bisect each other, or that all of its angles are right.

Assessment Resources

The Mini-Quiz below is also available on blackline (*Chapter 6 Resource Book*, p. 48) and on transparency. For more assessment resources, see:
• Chapter 6 Resource Book
• Standardized Test Practice
• Test and Practice Generator

Mini-Quiz

Use the information in the diagram to name the special quadrilateral.

1.

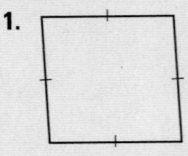

rhombus

2.

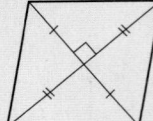

rhombus

3.

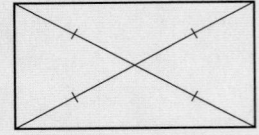

rectangle

4.

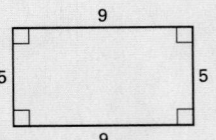

rectangle

5. Kim arranges four metersticks to make a parallelogram. Then she adjusts the metersticks so that each pair meets at a right angle. What shape has Kim formed?
square

---

21. The consecutive angles of a parallelogram are supplementary, so the two angles consecutive to ∠*J* must also be right angles. Also, opposite angles of a parallelogram are congruent, so the fourth angle must also be a right angle. By definition, a parallelogram with four right angles is a rectangle.

21. Logical Reasoning In ▱*JKLM*, ∠*J* is a right angle. Explain why ▱*JKLM* is a rectangle.

22. Using Theorems Find the value of *x* in rhombus *QRST*.   3

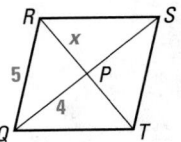

Challenge *GHJK* is a square with diagonals intersecting at *L*. Given that *GH* = 2 and *GL* = √2, complete the statement.

23. *HK* = _?_ $2\sqrt{2}$

24. m∠*KLJ* = _?_   90°

25. m∠*HJG* = _?_   45°

26. Perimeter of △*HJK* = _?_ $4 + 2\sqrt{2}$

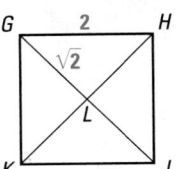

Standardized Test Practice

27. Multiple Choice In ▱*KLMN*, *KL* = *LM*. What is m∠*N*?   **D**

Ⓐ 30°   Ⓑ 45°

Ⓒ 90°   Ⓓ Cannot be determined

28. Multiple Choice In rhombus *ABCD*, *AB* = 7*x* − 3 and *CD* = 25. What is the value of *x*?   **G**

Ⓕ 3   Ⓖ 4

Ⓗ 7   Ⓙ 25

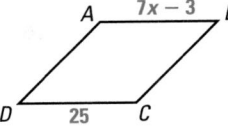

Mixed Review

Finding Angle Measures Find the measure of the numbered angle. (*Lesson 3.4*)

29.
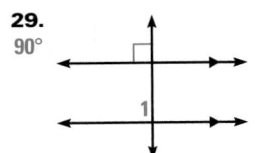

30.
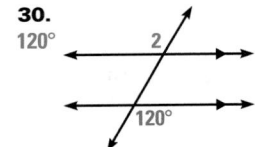

31.
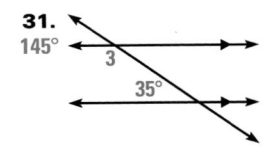

Algebra Skills

Finding Ratios Find the ratio of the length to the width for the rectangle. Write the ratio in simplest form. (*Skills Review, p. 660*)

32. $\frac{9}{5}$
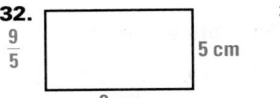

33. $\frac{2}{1}$
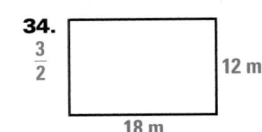

34. $\frac{3}{2}$

6.5 Midsegment of a Trapezoid

Question

What are some properties of the midsegment of a trapezoid?

Explore

1 Draw $\overleftrightarrow{AB}$. Draw a point C not on $\overleftrightarrow{AB}$ and construct a line parallel to $\overleftrightarrow{AB}$ through point C.

2 Construct a point D on the new line and draw $\overline{AD}$ and $\overline{BC}$, as shown below.

3 Construct the midpoints of $\overline{AD}$ and $\overline{BC}$. Label the points E and F. Draw $\overline{EF}$.

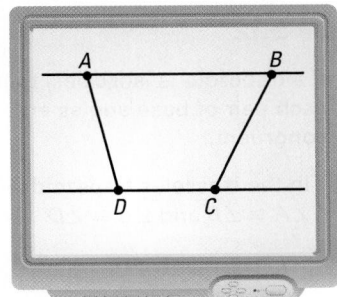

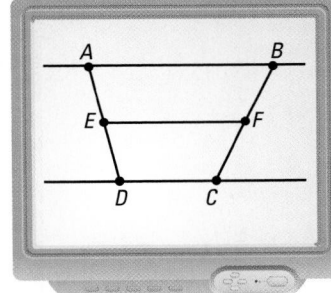

Student Help

> **VOCABULARY TIP**
> Quadrilateral *ABCD* is a *trapezoid*, a quadrilateral with exactly one pair of parallel sides. $\overline{EF}$ is called a *midsegment* of trapezoid *ABCD*.

4. The slopes may vary, but the slopes of $\overleftrightarrow{AB}$, $\overleftrightarrow{DC}$, and $\overline{EF}$ are always equal for trapezoid *ABCD*. This tells you that $\overleftrightarrow{AB}$, $\overleftrightarrow{DC}$, and $\overline{EF}$ are parallel.

Think About It

1. Measure the distances AB, DC, and EF.   Answers will vary.

2. Calculate $\frac{(AB + DC)}{2}$.   Answers will vary, but $\frac{(AB + DC)}{2}$ will be equal to EF.

3. Drag points A, B, C, and D. Do not allow $\overline{AD}$ to cross $\overline{BC}$. What do you notice about EF and $\frac{(AB + DC)}{2}$? EF and $\frac{(AB + DC)}{2}$ are always equal for trapezoid *ABCD*.

4. Measure the slopes of $\overleftrightarrow{AB}$, $\overleftrightarrow{DC}$, and $\overline{EF}$. What do you notice? What do the slopes tell you about $\overleftrightarrow{AB}$, $\overleftrightarrow{DC}$, and $\overline{EF}$?

5. **Extension** Drag the points so that $\overline{AD}$ crosses $\overline{BC}$. *ABCD* is no longer a polygon. Write an expression for EF in terms of AB and DC. $EF = \frac{|AB - DC|}{2}$

Goal
Students explore the relationship between the length of the midsegment of a trapezoid and the length of its two bases.

Materials
See the Technology Activity Keystrokes Masters in the *Chapter 6 Resource Book*.

LINK TO LESSON
In Example 2 on page 333, students apply their knowledge from this activity to find the length of the midsegment of a trapezoid given the lengths of the bases.

2 Managing the Activity

Tips for Success
In Step 1, point *C* does not need to be equidistant from points *A* and *B*.
In Step 3, point out that $\overline{EF}$ should turn out to be parallel to $\overleftrightarrow{AB}$ and $\overleftrightarrow{CD}$. If the segment is not parallel, students have found at least one of the midpoints incorrectly.

3 Closing the Activity

KEY DISCOVERY
The length of the midsegment of a trapezoid is half the sum of the lengths of the bases.

Activity Assessment
Use Exercise 3 to assess student understanding.

6.5 Trapezoids

Goal
Use properties of trapezoids.

Key Words
• trapezoid
• bases, legs, and base angles of a trapezoid
• isosceles trapezoid
• midsegment of a trapezoid

A **trapezoid** is a quadrilateral with exactly one pair of parallel sides. The parallel sides are the **bases**. The nonparallel sides are the **legs**.

A trapezoid has two pairs of **base angles**. In trapezoid $ABCD$, $\angle C$ and $\angle D$ are one pair of base angles. $\angle A$ and $\angle B$ are the other pair.

If the legs of a trapezoid are congruent, then the trapezoid is an **isosceles trapezoid**.

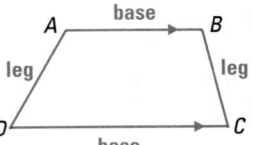

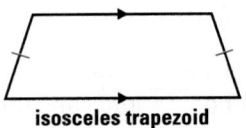

isosceles trapezoid

THEOREMS 6.12 and 6.13

Theorem 6.12

Words If a trapezoid is isosceles, then each pair of base angles are congruent.

Symbols In the isosceles trapezoid $ABCD$, $\angle A \cong \angle B$ and $\angle C \cong \angle D$.

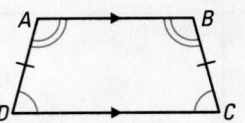

Theorem 6.13

Words If a trapezoid has a pair of congruent base angles, then it is isosceles.

Symbols In trapezoid $ABCD$, if $\angle C \cong \angle D$ then $ABCD$ is isosceles.

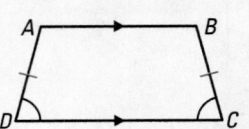

EXAMPLE 1 Find Angle Measures of Trapezoids

$PQRS$ is an isosceles trapezoid. Find the missing angle measures.

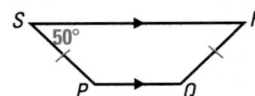

Solution

❶ $PQRS$ is an isosceles trapezoid and $\angle R$ and $\angle S$ are a pair of base angles. So, $m\angle R = m\angle S = 50°$.

❷ Because $\angle S$ and $\angle P$ are same-side interior angles formed by parallel lines, they are supplementary.
So, $m\angle P = 180° - 50° = 130°$.

❸ Because $\angle Q$ and $\angle P$ are a pair of base angles of an isosceles trapezoid, $m\angle Q = m\angle P = 130°$.

Visualize It!

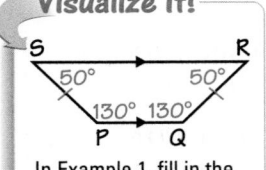

In Example 1, fill in the missing angle measures as you find them.

Checkpoint ✓ Find Angle Measures of Trapezoids

ABCD is an isosceles trapezoid. Find the missing angle measures.

1.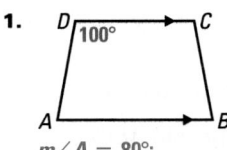

$m\angle A = 80°$;
$m\angle B = 80°$;
$m\angle C = 100°$

2.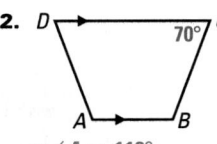

$m\angle A = 110°$;
$m\angle B = 110°$;
$m\angle D = 70°$

3.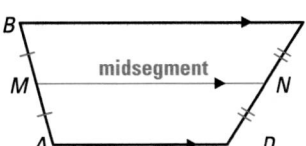

$m\angle B = 75°$;
$m\angle C = 105°$;
$m\angle D = 105°$

Student Help

VOCABULARY TIP
The midsegment of a trapezoid is sometimes called the *median* of a trapezoid.

Midsegments The **midsegment of a trapezoid** is the segment that connects the midpoints of its legs. The midsegment of a trapezoid is parallel to the bases.

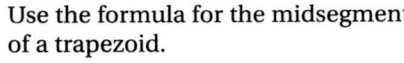

The length of the midsegment of a trapezoid is half the sum of the lengths of the bases.

$$MN = \frac{1}{2}(AD + BC)$$

EXAMPLE 2 Midsegment of a Trapezoid

Find the length of the midsegment $\overline{DG}$ of trapezoid *CEFH*.

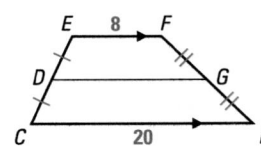

Solution

Use the formula for the midsegment of a trapezoid.

$DG = \frac{1}{2}(EF + CH)$ Formula for midsegment of a trapezoid

$ = \frac{1}{2}(8 + 20)$ Substitute 8 for *EF* and 20 for *CH*.

$ = \frac{1}{2}(28)$ Add.

$ = 14$ Multiply.

ANSWER ▶ The length of the midsegment $\overline{DG}$ is 14.

Checkpoint ✓ Midsegment of a Trapezoid

Find the length of the midsegment $\overline{MN}$ of the trapezoid.

4.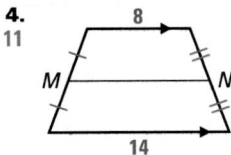

5.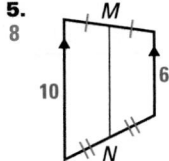

6.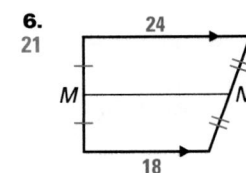

6.5 Trapezoids **333**

Extra Example 1

FGHJ is an isosceles trapezoid. Find the missing angle measures.

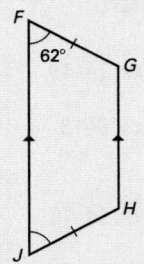

$m\angle G = 118°$; $m\angle H = 118°$;
$m\angle J = 62°$

Extra Example 2

Find the length of the midsegment $\overline{QT}$ of trapezoid *PRSU*. **10**

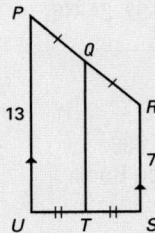

Study Skills

Note-Taking Show students some of the different shapes that a trapezoid may have, including isosceles trapezoids. Suggest that students sketch several different types of trapezoids for their notes.

✓ Concept Check

How is a trapezoid different from a parallelogram? A trapezoid has only one pair of parallel sides.

⊕ Daily Puzzler

When you join the consecutive midpoints of an isosceles trapezoid, what shape is formed? rhombus

333

Assignment Guide

BASIC
Day 1: pp. 334–336 Exs. 14–19, 35, 39–46
Day 2: pp. 334–336 Exs. 9–13, 20–26, 34, 36–38

AVERAGE
Day 1: pp. 334–336 Exs. 14–19, 35–46
Day 2: pp. 334–336 Exs. 9–13, 20–32, 34

ADVANCED
Day 1: pp. 334–336 Exs. 14–19, 35–46
Day 2: pp. 334–336 Exs. 20–32, 33*, 34; EC: classzone.com

BLOCK SCHEDULE
pp. 334–336 Exs. 9–32, 34–46

Extra Practice
• Student Edition, p. 686
• Chapter 6 Resource Book, pp. 51–52

Homework Check
To quickly check student understanding of key concepts, go over the following exercises:

Basic: 14, 17, 20, 23, 26
Average: 15, 18, 21, 24, 26
Advanced: 16, 19, 22, 25, 26

6.5 Exercises

Guided Practice

Vocabulary Check

1. Name the *bases* of trapezoid *ABCD*.
   $\overline{AB}$ and $\overline{DC}$
2. Name the *legs* of trapezoid *ABCD*.
   $\overline{AD}$ and $\overline{BC}$

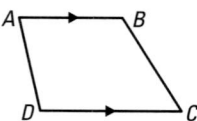

Skill Check

Decide whether the quadrilateral is a *trapezoid*, an *isosceles trapezoid*, or *neither*.

3.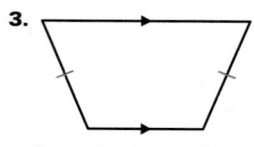
   isosceles trapezoid

4.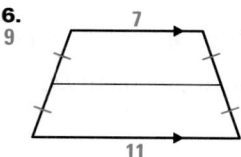
   neither

5.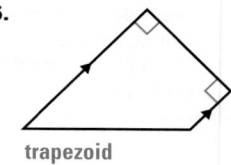
   trapezoid

Find the length of the midsegment.

6.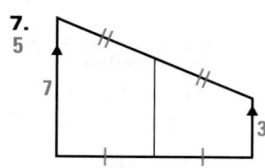
   7, 9, 11
   [midsegment]

7.
   5, 7, 3

8.
   19, 17, 15

Practice and Applications

Extra Practice
See p. 686.

Parts of a Trapezoid Match the parts of trapezoid *PQRS* with the correct description.

9. $\overline{QR}$ and $\overline{PS}$ D
10. $\angle Q$ and $\angle S$ C
11. $\angle R$ and $\angle Q$ B
12. $\overline{MN}$ E
13. $\overline{PQ}$ and $\overline{RS}$ A

A. legs
B. base angles
C. opposite angles
D. bases
E. midsegment

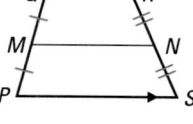

Finding Angle Measures *JKLM* is an isosceles trapezoid. Find the missing angle measures.

14.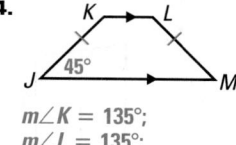
   45°
   $m\angle K = 135°$;
   $m\angle L = 135°$;
   $m\angle M = 45°$

15.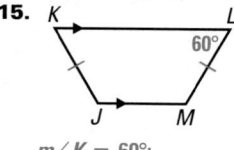
   60°
   $m\angle K = 60°$;
   $m\angle J = 120°$;
   $m\angle M = 120°$

16.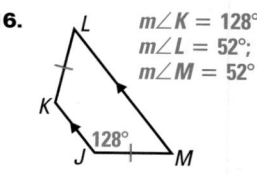
   128°
   $m\angle K = 128°$;
   $m\angle L = 52°$;
   $m\angle M = 52°$

Homework Help
Example 1: Exs. 14–19
Example 2: Exs. 20–26

Finding Angle Measure *QRST* is a trapezoid. Find the missing angle measures.

17.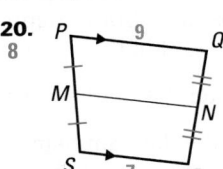

$m\angle R = 102°; m\angle T = 48°$

18.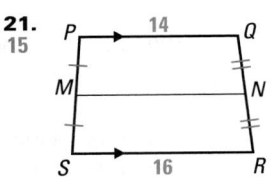

$m\angle R = 102°; m\angle S = 70°$

19.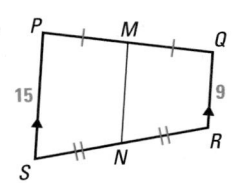

$m\angle Q = 90°; m\angle S = 30°$

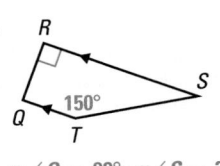

Finding Midsegments Find the length of the midsegment $\overline{MN}$ of the trapezoid.

20.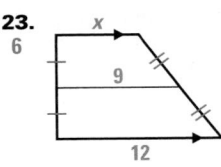

21.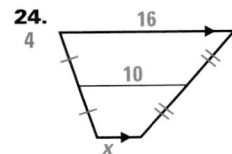

22.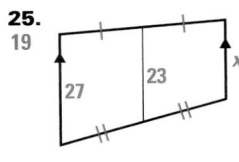

(xy) **Using Algebra** Find the value of *x*.

23.

x
6
9
12

24.

16
4
10
x

25.

19
27 23
x

26. Cake Design The top layer of the cake in the diagram at the right has a diameter of 10 inches. The bottom layer has a diameter of 22 inches. What is the diameter of the middle layer?
16 inches

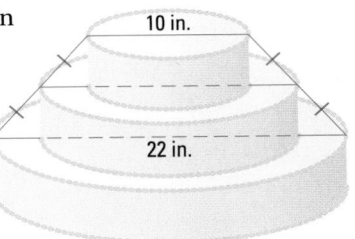

10 in.

22 in.

Coordinate Geometry The vertices of a trapezoid are *A*(2, 6), *B*(8, 6), *C*(8, 2), and *D*(4, 2).

27. Plot the vertices on a coordinate plane. Connect them to form trapezoid *ABCD*. See margin.

28. Name the bases of trapezoid *ABCD*. $\overline{AB}$ and $\overline{CD}$

29. Name the legs of trapezoid *ABCD*. $\overline{AD}$ and $\overline{BC}$

30. Find the coordinates of the midpoint of each leg. Then plot these points on the coordinate plane you drew in Exercise 27. What is the line segment that connects these two points called?
The midpoint of $\overline{AD}$ is (3, 4); the midpoint of $\overline{BC}$ is (8, 4); See diagram for Ex. 27; the midsegment.

6.5 *Trapezoids* **335**

27.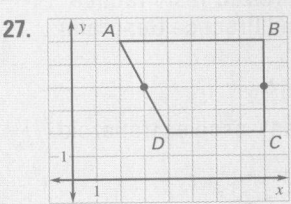

Assessment Resources

The Mini-Quiz below is also available on blackline (*Chapter 6 Resource Book*, p. 58) and on transparency. For more assessment resources, see:
- Chapter 6 Resource Book
- Standardized Test Practice
- Test and Practice Generator

Mini-Quiz

ABCD is an isosceles trapezoid.

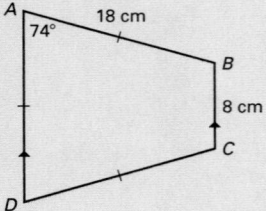

1. Find the length of $\overline{CD}$. **18 cm**
2. Find $m\angle D$. **74°**
3. Find $m\angle B$. **106°**
4. Name the legs of the trapezoid.
   $\overline{AB}$ **and** $\overline{DC}$
5. Find the length of the midsegment of the trapezoid. **13 cm**

Student Help
CLASSZONE.COM

HOMEWORK HELP
Extra help with problem solving in Exs. 31–33 is at classzone.com

32. A parallelogram; from Ex. 31, $\angle 1 \cong \angle 3$ and $\angle 2 \cong \angle 4$, so both pairs of opposite angles are congruent.

33. The opposite sides of a parallelogram are congruent, so $m + m = 9 + 5$; $2m = 9 + 5$; $m = \frac{(9+5)}{2}$; in words, the length of the midsegment is half the sum of the lengths of the bases.

Standardized Test Practice

Mixed Review

36. Yes; both pairs of opposite angles are congruent.

37. No; you can tell that one pair of sides must be parallel, but not both.

38. Yes; the diagonals bisect each other.

Algebra Skills

Visualize It! In Exercises 31–33, use the figures shown below.
The figure on the left is a trapezoid with midsegment of length m. The figure on the right is formed by cutting the trapezoid along its midsegment and rearranging the two pieces.

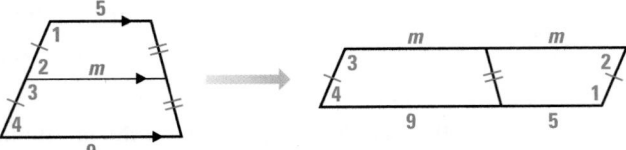

31. Which theorem or postulate from Chapter 3 can you use to show that $\angle 1 \cong \angle 3$ and $\angle 2 \cong \angle 4$ in the figure on the left?
Corresponding Angles Postulate

32. What kind of quadrilateral is on the right? Explain your answer.

33. Challenge How does the diagram help you see that the length of the midsegment is half the sum of the lengths of the bases?

34. Multiple Choice In the trapezoid at the right, what is the value of x? **C**

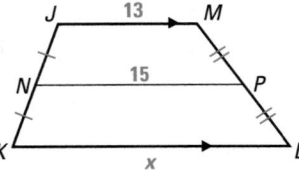

    (A) 13   (B) 15

    (C) 17   (D) 28

35. Multiple Choice Which of the following must a trapezoid have? **H**

    (F) congruent bases

    (G) diagonals that bisect each other

    (H) exactly one pair of parallel sides

    (J) a pair of congruent opposite angles

Logical Reasoning Tell whether the quadrilateral is a parallelogram. **Explain your reasoning.** *(Lesson 6.3)*

36.   **37.**   **38.**

Multiplying Multiply. Write the answer in simplest form. *(Skills Review, p. 659)*

39. $\frac{1}{2} \times 20$ **10**   **40.** $52 \times \frac{1}{4}$ **13**   **41.** $136 \times \frac{1}{8}$ **17**   **42.** $\frac{3}{4} \times 60$ **45**

43. $\frac{2}{3} \times \frac{3}{7}$ $\frac{2}{7}$ **44.** $\frac{7}{8} \times \frac{2}{14}$ $\frac{1}{8}$ **45.** $\frac{5}{6} \times \frac{1}{3}$ $\frac{5}{18}$ **46.** $\frac{4}{21} \times \frac{7}{16}$ $\frac{1}{12}$

6.6 Reasoning About Special Quadrilaterals

Goal
Identify special quadrilaterals based on limited information.

Key Words
- parallelogram p. 310
- rectangle p. 325
- rhombus p. 325
- square p. 325
- trapezoid p. 332
- isosceles trapezoid p. 332

In this chapter, you have studied six special types of quadrilaterals. The diagram below shows how these quadrilaterals are related. Each shape is a special example of the shape(s) listed above it.

Quadrilateral

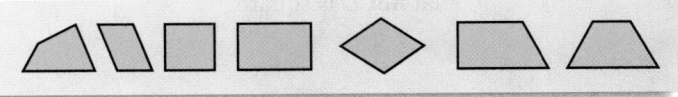

Parallelogram   **Trapezoid**

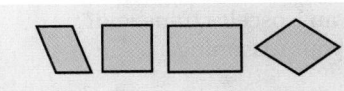 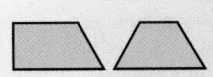

Student Help

STUDY TIP
The diagram shows that a rectangle is always a parallelogram and a quadrilateral, but it is not always a rhombus or a square.

Rectangle   **Rhombus**   **Isosceles Trapezoid**

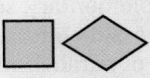 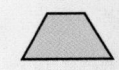 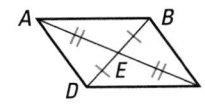

Square

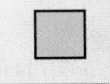

EXAMPLE 1 | Use Properties of Quadrilaterals

Determine whether the quadrilateral is a trapezoid, parallelogram, rectangle, rhombus, or square.

Solution
The diagram shows $\overline{CE} \cong \overline{EA}$ and $\overline{DE} \cong \overline{EB}$, so the diagonals of the quadrilateral bisect each other. By Theorem 6.9, you can conclude that the quadrilateral is a parallelogram.

You *cannot* conclude that $ABCD$ is a rectangle, rhombus, or square because no information about the sides or angles is given.

6.6 *Reasoning About Special Quadrilaterals* **337**

Extra Example 1

Determine whether the quadrilateral is a trapezoid, parallelogram, rectangle, rhombus, or square.

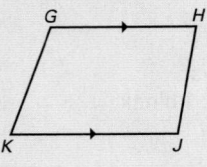

trapezoid

Extra Example 2

Are you given enough information in the diagram to conclude that *JKLM* is a square?

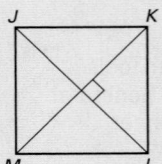

No; the diagonals are perpendicular, so the figure is a rhombus. There is no information about the lengths of the diagonals or the angle measures, so you cannot conclude that *JKLM* is a square.

Extra Example 3

Are you given enough information in the diagram to conclude that *QRST* is a rectangle?

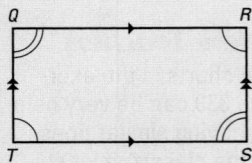

No; the figure is a parallelogram by definition, but the angle information shown is not sufficient to conclude that *QRST* is a rectangle.

✓ Concept Check

What are the six types of special quadrilaterals you have studied?
parallelogram, rectangle, rhombus, square, trapezoid, isosceles trapezoid

1. No; all four angles are right angles, so the figure is a rectangle. There is no information about the sides, so you cannot conclude that *PQRS* is a square.

2. No; both pairs of opposite angles are congruent, so the figure is a parallelogram. There is no information about the sides, so you cannot conclude that *WXYZ* is a rhombus.

3. Yes; ∠*L* and ∠*M* are supplementary, so $\overline{KL} \parallel \overline{JM}$ by the Same-Side Interior Angles Converse. Since ∠*J* and ∠*M* are not supplementary, $\overline{KJ}$ is not parallel to $\overline{LM}$. The figure has exactly one pair of parallel sides, so it is a trapezoid.

EXAMPLE 2 Identify a Rhombus

Are you given enough information in the diagram to conclude that *ABCD* is a square? Explain your reasoning.

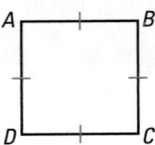

Solution

The diagram shows that all four sides are congruent. Therefore, you know that it is a rhombus. The diagram does not give any information about the angle measures, so you cannot conclude that *ABCD* is square.

EXAMPLE 3 Identify a Trapezoid

Are you given enough information in the diagram to conclude that *EFGH* is an isosceles trapezoid? Explain your reasoning.

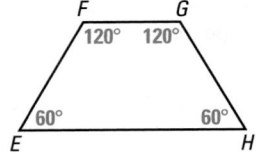

Solution

➊ *First* show that *EFGH* is a trapezoid. ∠*E* and ∠*F* are supplementary, so $\overline{FG}$ is parallel to $\overline{EH}$ by Theorem 3.10, the Same-Side Interior Angles Converse. So, *EFGH* has *at least* one pair of parallel sides.

To show that *EFGH* is a trapezoid, you must show that it has *only one* pair of parallel sides. The opposite angles of *EFGH* are not congruent, so it cannot be a parallelogram. Therefore, *EFGH* is a trapezoid.

➋ *Next* show that *EFGH* is isosceles. Because the base angles are congruent, *EFGH* is an isosceles trapezoid by Theorem 6.13.

Checkpoint ✓ Identify Quadrilaterals

Are you given enough information to conclude that the figure is the given type of special quadrilateral? Explain your reasoning. See margin.

1. A square?

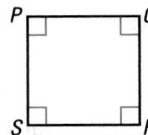

2. A rhombus?

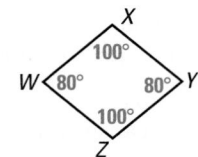

3. A trapezoid?

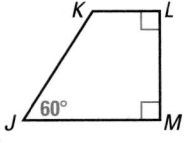

6.6 Exercises

Guided Practice

Skill Check

Copy the chart. Put a ✓ mark in the box if the shape *always* has the given property.

	Property	□	Rectangle	Rhombus	Square	Trapezoid
1.	Both pairs of opp. sides are ∥.	✓ ?	✓ ?	✓ ?	✓ ?	?
2.	Exactly 1 pair of opp. sides are ∥.	?	?	?	?	✓ ?
3.	Diagonals are perpendicular.	?	?	✓ ?	✓ ?	?
4.	Diagonals are congruent.	?	✓ ?	?	✓ ?	?

Practice and Applications

Extra Practice

See p. 686.

Properties of Quadrilaterals Copy the chart. Put a ✓ mark in the box if the shape *always* has the given property.

	Property	□	Rectangle	Rhombus	Square	Trapezoid
5.	Both pairs of opp. sides are congruent.	✓ ?	✓ ?	✓ ?	✓ ?	?
6.	Diagonals bisect each other.	✓ ?	✓ ?	✓ ?	✓ ?	?
7.	Both pairs of opp. angles are congruent.	✓ ?	✓ ?	✓ ?	✓ ?	?
8.	All sides are congruent.	?	?	✓ ?	✓ ?	?

Using Properties of Quadrilaterals Determine whether the quadrilateral is a trapezoid, parallelogram, rectangle, rhombus, or square.

9. trapezoid

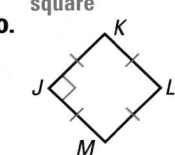

10. square

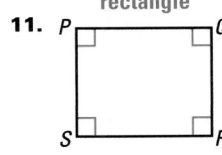

11. rectangle

6.6 *Reasoning About Special Quadrilaterals* **339**

Assignment Guide

BASIC
Day 1: pp. 339–341 Exs. 5–11, 20–23, 26–38 even
Day 2: pp. 339–341 Exs. 12–17, 27–37 odd, Quiz 2

AVERAGE
Day 1: pp. 339–341 Exs. 5–11, 18–24, 26, 35–38
Day 2: pp. 339–341 Exs. 12–17, 27–34, Quiz 2

ADVANCED
Day 1: pp. 339–341 Exs. 5–11, 18–24, 25*, 26, 35–38
Day 2: pp. 339–341 Exs. 12–17, 27–34, Quiz 2;
EC: classzone.com

BLOCK SCHEDULE
pp. 339–341 Exs. 5–24, 26–38, Quiz 2

Extra Practice
• Student Edition, p. 686
• Chapter 6 Resource Book, pp. 60–61

Homework Check
To quickly check student understanding of key concepts, go over the following exercises:

Basic: 9, 10, 12, 13, 14
Average: 9, 10, 12, 13, 14
Advanced: 13, 14, 16, 18, 19

Homework Help
Example 1: Exs. 9–11, 18, 19
Example 2: Exs. 12–17
Example 3: Exs. 12–17

340

X Common Error

In Exercises 20–23, watch for students who have difficulty interpreting the Venn diagram. Refer students to the diagram on page 337 in order to check their answers. Make sure students understand that the more general quadrilaterals are at the top of the diagram and the quadrilaterals become more specific the further down you look.

14. No; there is not enough information to determine if the figure has any parallel sides, so you cannot conclude that the figure is a trapezoid.

12. No; both pairs of opposite angles are congruent, so the figure is a parallelogram. There is no information about the sides, so you cannot conclude that the figure is a rhombus.

13. Yes; since 140° + 40° = 180°, the top and bottom sides are parallel by the Same-Side Interior Angles Converse. Since 60° + 40° ≠ 180°, the other two sides are not parallel. The figure has exactly one pair of parallel sides, so it is a trapezoid.

Link to Careers

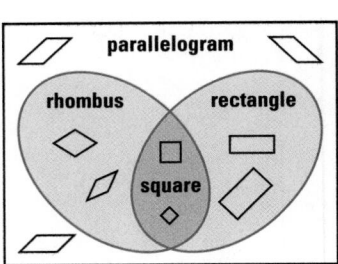

GEMOLOGISTS consider the color and clarity of a gem, as well as the cut, when evaluating its value.

Career Links
CLASSZONE.COM

15. Yes; the four angles are congruent, so the four angles are right angles by the Quadrilateral Interior Angles Theorem. A quadrilateral with four right angles is a rectangle.

16. No; there is no information about the sides or the other 3 angles, so you cannot conclude that the figure is a square.

17. Yes; both pairs of opposite sides are congruent, so the figure is a parallelogram.

Identifying Quadrilaterals Are you given enough information to conclude that the figure is the given type of special quadrilateral? Explain your reasoning. *See margin.*

12. A rhombus? **13.** A trapezoid? **14.** An isosceles trapezoid?

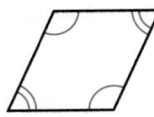 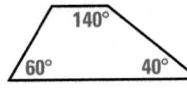 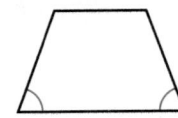

15. A rectangle? **16.** A square? **17.** A parallelogram?

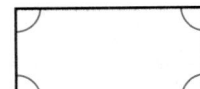  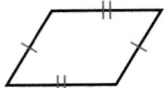

Gem Cutting Use the diagrams and the following information.
There are different ways of cutting a gem to enhance its beauty. One of the cuts used for gems is called the *step cut*. Each face of a cut gem is called a *facet*.

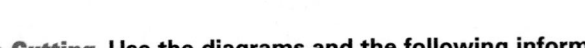

18. In *ABCD*, ∠A, ∠B, ∠C, and ∠D are all right angles. What shape is *ABCD*? **rectangle**

19. $\overline{EF}$ is parallel to $\overline{DC}$; $\overline{ED}$ and $\overline{FC}$ are congruent, but not parallel. What shape is the facet labeled *EFCD*? **isosceles trapezoid**

Using a Venn Diagram In Exercises 20–23, use the Venn diagram to decide whether the following statements are *true* or *false*.

20. All rectangles are squares. **false**

21. All squares are rectangles. **true**

22. All squares are rhombuses. **true**

23. All rhombuses are parallelograms. **true**

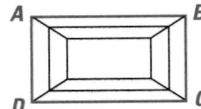

24. Technology Use geometry software to draw a triangle. Construct the midpoint of each side and connect the midpoints as shown. What type of quadrilateral is *BEFD*? Explain.
Parallelogram; $\overline{BC} \parallel \overline{DF}$ and $\overline{AB} \parallel \overline{FE}$, so the quadrilateral has two pairs of opposite sides parallel.

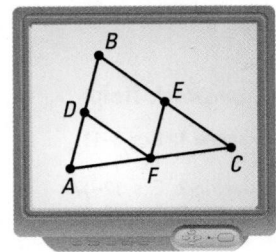

25. Challenge What type of quadrilateral is *PQRS*, with vertices *P*(2, 5), *Q*(5, 5), *R*(6, 2), and *S*(1, 2)? isosceles trapezoid

Standardized Test
Practice

26. Multiple Choice Which of the following statements is *never* true? B

    Ⓐ A rectangle is a square.

    Ⓑ A parallelogram is a trapezoid.

    Ⓒ A rhombus is a parallelogram.

    Ⓓ A parallelogram is a rectangle.

Mixed Review

Solving Proportions Solve the proportion. *(Skills Review, p. 660)*

27. $\frac{x}{3} = \frac{4}{12}$ 1   **28.** $\frac{4}{7} = \frac{x}{21}$ 12   **29.** $\frac{10}{x} = \frac{5}{8}$ 16   **30.** $\frac{3}{10} = \frac{24}{x}$ 80

31. $\frac{x}{24} = \frac{5}{12}$ 10   **32.** $\frac{3}{5} = \frac{x}{20}$ 12   **33.** $\frac{8}{x} = \frac{1}{2}$ 16   **34.** $\frac{3}{7} = \frac{21}{x}$ 49

Algebra Skills

Writing Decimals Write the fraction as a decimal. For repeating decimals, also round to the nearest hundredth for an approximation. *(Skills Review, p. 657)*

35. $\frac{1}{5}$ 0.2   **36.** $\frac{3}{8}$ 0.375   **37.** $\frac{5}{6}$ **38.** $\frac{7}{20}$ 0.35

                                                0.8333. . . ≈ 0.83

Quiz 2

Find the value of x. *(Lesson 6.4)*

1. rhombus *ABCD* 35   **2.** rectangle *FGHJ* 6   **3.** square *PQRS* 45

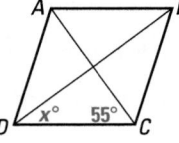   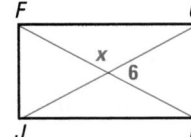   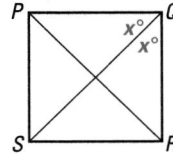

What kind of special quadrilateral is the red shape? *(Lesson 6.5)*

4.   **5.**

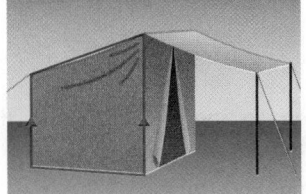   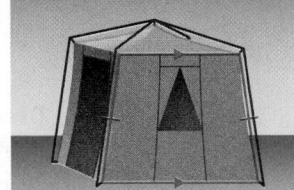

trapezoid   isosceles trapezoid

6. Which kinds of quadrilaterals can you form with four straws of the same length? You must attach the straws at their ends and cannot bend any of them. *(Lesson 6.6)* square, rectangle, rhombus, and parallelogram

---

④ Assess

Assessment Resources
The Mini-Quiz below is also available on blackline (*Chapter 7 Resource Book*, p. 9) and on transparency. For more assessment resources, see:
• Chapter 6 Resource Book
• Standardized Test Practice
• Test and Practice Generator

Mini-Quiz

1. Determine whether the quadrilateral is a trapezoid, parallelogram, rectangle, rhombus, or square.

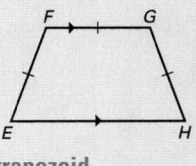

trapezoid

Are you given enough information to conclude that the figure is the given type of special quadrilateral? Explain your reasoning.

2. A parallelogram?

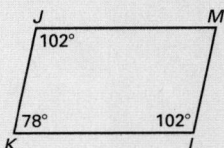

Yes; since 78° + 102° = 180°, ∠*K* is supplementary to ∠*J* and also to ∠*L*. Since one angle of the quadrilateral is supplementary to both of its consecutive angles, the quadrilateral is a parallelogram.

3. A rectangle?

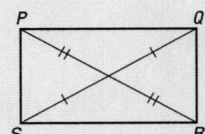

No; since the diagonals bisect each other the figure is a parallelogram, but not enough information is given to determine if ∠*P*, ∠*Q*, ∠*R*, and ∠*S* are right angles.

Additional Resources

The following resources are available to help review the materials in this chapter.

📖 **Chapter 6 Resource Book**
• Chapter Review Games and Activities, p. 67
• Cumulative Review, Chs. 1–6

Chapter  **6** **Summary and Review**

VOCABULARY

• **polygon**, *p. 303*
• **side of a polygon**, *p. 303*
• **vertex of a polygon**, *p. 303*
• **diagonal of a polygon**, *p. 303*
• **parallelogram**, *p. 310*

• **rhombus**, *p. 325*
• **rectangle**, *p. 325*
• **square**, *p. 325*
• **trapezoid**, *p. 332*
• **bases of a trapezoid**, *p. 332*

• **legs of a trapezoid**, *p. 332*
• **base angles of a trapezoid**, *p. 332*
• **isosceles trapezoid**, *p. 332*
• **midsegment of a trapezoid**, *p. 333*

VOCABULARY REVIEW

Fill in each blank.

1. A(n) __?__ is a quadrilateral with both pairs of opposite sides parallel. **parallelogram**

2. A(n) __?__ is a plane figure that is formed by three or more segments called sides. **polygon**

3. A(n) __?__ is a parallelogram with four congruent sides. **rhombus**

4. A(n) __?__ is a parallelogram with four right angles. **rectangle**

5. A parallelogram with four congruent sides and four right angles is a(n) __?__. **square**

6. A(n) __?__ is a quadrilateral with exactly one pair of sides parallel. **trapezoid**

7. Each endpoint of the sides of a polygon is called a(n) __?__. **vertex**

8. A segment that joins two nonconsecutive vertices of a polygon is called a(n) __?__. **diagonal**

9. The nonparallel sides of a trapezoid are called its __?__ and the parallel sides are called its __?__. **legs; bases**

6.1 **POLYGONS**

Examples on pp. 303–305

EXAMPLE **Decide whether the figure is a polygon. If so, tell what type. If not, explain why.**

The figure is a polygon because it has straight sides and each side intersects exactly two other sides at each of its endpoints.

The polygon has five sides, so it is a pentagon.

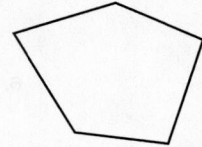

Decide whether the figure is a polygon. If so, tell what type. If not, explain why.

10.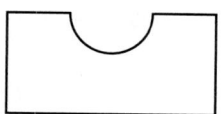

No; one of the sides is not a segment.

11.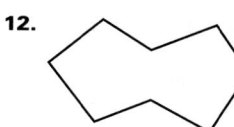

yes; hexagon

12.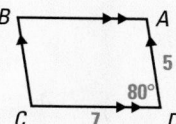

yes; octagon

Sketch the figure(s) described. 13, 14. See margin.

13. Two different heptagons

14. A quadrilateral with one right angle

6.2 PROPERTIES OF PARALLELOGRAMS

Examples on pp. 310–312

> **EXAMPLES** **ABCD is a parallelogram. Find the measure.**
>
> **a.** Find *AB* and *BC*.
>
> By Theorem 6.2, the opposite sides of a parallelogram are congruent. So, $AB = DC = 7$, and $BC = AD = 5$.
>
> **b.** Find $m\angle A$.
>
> By Theorem 6.4, the consecutive angles of a parallelogram are supplementary.
>
> $m\angle A + m\angle D = 180°$ Consec. $\angle$s of a $\square$ are supplementary.
>
> $m\angle A + 80° = 180°$ Substitute 80° for $m\angle D$.
>
> $m\angle A = 100°$ Subtract 80° from each side.

Find the measures in the parallelogram.

15. Find *BC* and *DC*.

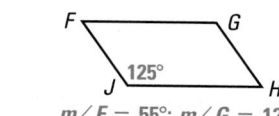

BC = 4; DC = 8

16. Find $m\angle F$ and $m\angle G$.

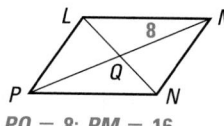

$m\angle F = 55°$; $m\angle G = 125°$

17. Find *PQ* and *PM*.

PQ = 8; PM = 16

6.3 SHOWING QUADRILATERALS ARE PARALLELOGRAMS

Examples on pp. 316–319

> **EXAMPLE** **Tell whether JKLM is a parallelogram. Explain your reasoning.**
>
>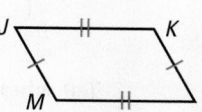
>
> By Theorem 6.6, *JKLM* is a parallelogram because both pairs of opposite sides are congruent.

13. *Sample answer:*

14. *Sample answer:*

18. By the Same-Side Interior Angles Converse, $\overline{BC} \parallel \overline{AD}$. Since .55° + 105° ≠ 180°, $\overline{AB}$ and $\overline{CD}$ are not parallel. The figure has exactly one pair of parallel sides, so it is not a parallelogram.

19. *EFGH* is a parallelogram because both pairs of opposite sides are congruent. (If both pairs of opposite sides of a quadrilateral are congruent, then the quadrilateral is a parallelogram.)

20. *JKLM* is a parallelogram because the diagonals bisect each other. (If the diagonals of a quadrilateral bisect each other, then the quadrilateral is a parallelogram.)

Tell whether the quadrilateral is a parallelogram. Explain your reasoning. 18–20. See margin.

18.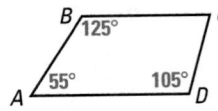

19.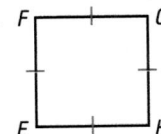

20.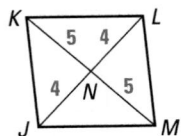

6.4 RHOMBUSES, RECTANGLES, AND SQUARES

Examples on pp. 325–327

EXAMPLES Use the information in the diagram to name the special quadrilateral.

a. *ABCD* is a quadrilateral with four congruent sides. By the Rhombus Corollary, *ABCD* is a rhombus.

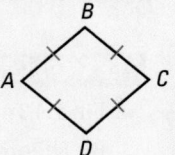

b. *JKLM* is a quadrilateral with four right angles. By the Rectangle Corollary, *JKLM* is a rectangle.

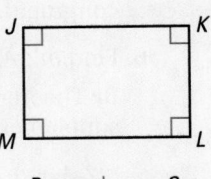

c. *PQRS* is a quadrilateral with four congruent sides. By the Rhombus Corollary, *PQRS* is a rhombus. A rhombus is a type of parallelogram, so its opposite angles are congruent.

$$m\angle Q = m\angle S = 90° \qquad m\angle R = m\angle P = 90°$$

PQRS has four congruent sides and four right angles. By definition, *PQRS* is a square.

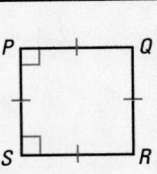

Find the values of x and y.

21. square *ABCD*

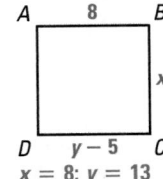

$x = 8; y = 13$

22. rectangle *FGHJ*

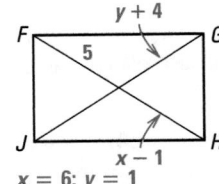

$x = 6; y = 1$

23. rhombus *KLMN*

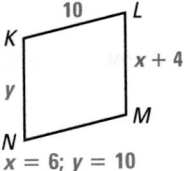

$x = 6; y = 10$

Tell whether the statement is *true* or *false*.

24. All rhombuses are squares. false

25. All rectangles are parallelograms. true

26. All rhombuses are rectangles. false

27. All squares are parallelograms. true

6.5 TRAPEZOIDS

Examples on pp. 332–333

EXAMPLES   *ABCD* is an isosceles trapezoid. Find the measure.

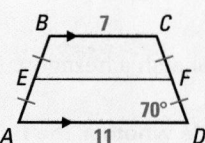

a. Find $m\angle A$.

ABCD is an isosceles trapezoid, so the base
angles are congruent. Therefore $m\angle A = m\angle D = 70°$.

b. Find *EF*.

The length of the midsegment $\overline{EF}$ is half the sum of the bases.

$$EF = \tfrac{1}{2}(BC + AD) = \tfrac{1}{2}(7 + 11) = \tfrac{1}{2}(18) = 9$$

Find the measure in the trapezoid.

28. Find $m\angle P$.   69°

29. Find $m\angle C$.   71°

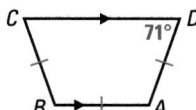

30. Find *TU*.   11

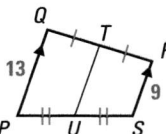

6.6 REASONING ABOUT SPECIAL QUADRILATERALS

Examples on pp. 337–338

EXAMPLE   Are you given enough information in the
diagram to conclude that *FGHJ* is a trapezoid?
Explain your reasoning.

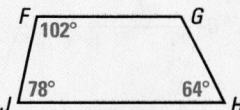

Because $m\angle F + m\angle J = 102° + 78° = 180°$, $\angle F$ and $\angle J$
are supplementary. The same-side interior angles are
supplementary. So, by Theorem 3.10, $\overline{FG} \parallel \overline{JH}$.

Because $m\angle J + m\angle H = 78° + 64° = 142°$, $\angle J$ and $\angle H$ are *not*
supplementary, so $\overline{FJ}$ is *not* parallel to $\overline{GH}$.

FGHJ has *exactly* one pair of parallel sides, so it is a trapezoid.

**Are you given enough information to conclude that the figure is the
given type of special quadrilateral? Explain your reasoning.**   31–33. See margin.

31. A rhombus?

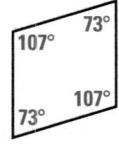

32. A trapezoid?

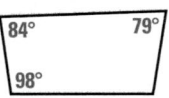

33. A rectangle?

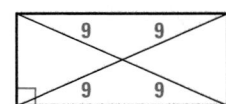

---

31. No; both pairs of opposite angles
are congruent, so the figure is a
parallelogram. There is no information about the sides, so you
cannot conclude that the figure
is a rhombus.

32. No; neither pair of consecutive
angles is supplementary, so
neither pair of opposite sides is
parallel and the figure is not a
trapezoid.

33. Yes; *Sample answer:* The diagonals bisect each other, so the figure is a parallelogram. One angle
of the figure is given as a right
angle. Consecutive angles of a
parallelogram are supplementary, so the other three angles
can be shown to be right angles.
Because the figure is a parallelogram with four right angles, it is
a rectangle by definition.

Additional Resources

📖 **Chapter 6 Resource Book**
• Chapter Test (2 levels), pp. 68–71
• SAT/ACT Chapter Test, p. 72
• Alternative Assessment, pp. 73–74

🖥 **Test and Practice Generator**

1. *Sample answer:*

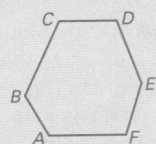

7. Yes; the figure is a parallelogram because both pairs of opposite sides are congruent.

8. No; the opposite angles are not congruent, so the figure is not a parallelogram.

9. Yes; the diagonals bisect each other.

10. Yes; one of the angles is supplementary to both of its consecutive angles.

15. square

16. trapezoid (if $\overline{BC}$ is not parallel to $\overline{AD}$)

17. Yes; because both pairs of opposite sides are parallel, the quadrilateral is a parallelogram by definition. Consecutive angles of a parallelogram are supplementary, so the other three angles can be shown to be right angles. Since the figure is a parallelogram with four right angles, it is a rectangle by definition.

Chapter  6 Chapter Test

1. Sketch a hexagon. Label its vertices. **See margin.**

Decide whether the figure is a polygon. If so, tell what type. If not, explain why.

2.

yes; heptagon

3.

No; two of the sides intersect only one other side.

4. Find $m\angle D$. **120°**

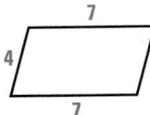

In □ABCD, find the values of x and y.

5.
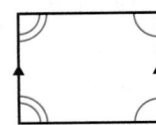
$x = 61; y = 7$

6.
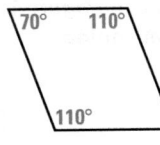
$x = 110; y = 100$

Tell whether the quadrilateral is a parallelogram. Explain your reasoning. 7–10. See margin.

7.
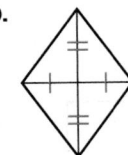

8.

9.
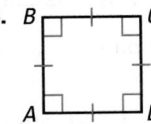

10.
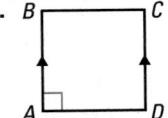

---

11. What other information do you need to conclude that the figure is a rectangle?

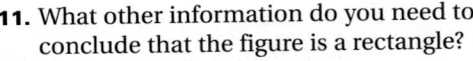

The four angles are right angles.

12. In rhombus *JKLM*, find the values of *a* and *b*.

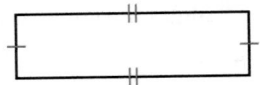

$a = 11; b = 4$

13. *PQRS* is a trapezoid. Find $m\angle S$ and $m\angle Q$.

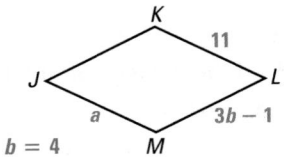

$m\angle S = 112°; m\angle Q = 58°$

14. Find the length of the midsegment of trapezoid *STUV*. **14**

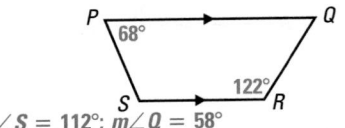

Determine whether the quadrilateral is a trapezoid, parallelogram, rectangle, rhombus, or square. 15, 16. See margin.

15.
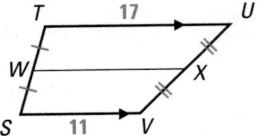

16.

17. Are you given enough information in the diagram to conclude that *EFGH* is a rectangle? Explain your reasoning. **See margin.**

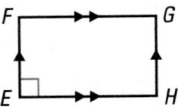

Additional Resources

📘 *Chapter 6 Resource Book*
• SAT/ACT Chapter Test, p. 72

✏️ *Test Tip* Diagrams may not be drawn to scale. Use only information that is given or marked on diagrams.

1. What type of polygon is shown below? **A**

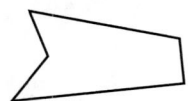

   (A) pentagon   (B) hexagon

   (C) heptagon   (D) octagon

2. Find the value of *x*. **J**

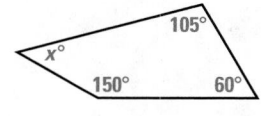

   (F) 22   (G) 40

   (H) 44   (J) 45

3. The diagonals of a parallelogram ___?___. **C**

   (A) are congruent   (B) are perpendicular

   (C) bisect each other (D) are parallel

4. *FGHJ* is a rectangle. What is *FG*? **F**

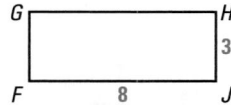

   (F) 3   (G) 5

   (H) 8   (J) 11

5. *ABCD* is a trapezoid. Which of the following statements is true? **A**

   (A) $m\angle C = 56°$

   (B) $m\angle A = 75°$

   (C) $\angle A$ and $\angle C$ are supplementary.

   (D) $\angle C$ and $\angle D$ are supplementary.

6. Which of the following terms could *not* be used to describe the figure below? **J**

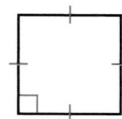

   (F) parallelogram   (G) rhombus

   (H) square   (J) trapezoid

7. *WXYZ* is an isosceles trapezoid. What are the values of *x* and *y*? **D**

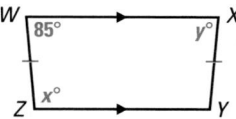

   (A) $x = 85, y = 95$ (B) $x = 100, y = 85$

   (C) $x = 105, y = 85$ (D) $x = 95, y = 85$

8. *STUV* is a rhombus. What are the values of *x* and *y*? **H**

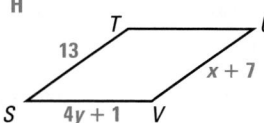

   (F) $x = 3, y = 3$ (G) $x = 3, y = 6$

   (H) $x = 6, y = 3$ (J) $x = 6, y = 4$

9. What is the length of the midsegment of trapezoid *DEFG*? **D**

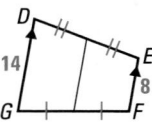

   (A) 8   (B) 9

   (C) 10   (D) 11

BraiN GaMes

Polyominoes

Materials
- graph paper
- scissors

Object of the Game To be the team to find the most number of ways to connect five squares edge to edge.

Set Up Cut squares out of graph paper.

How to Play

Step 1 Arrange 5 squares edge to edge in as many ways as possible. Do not include any congruent figures. If a figure is a rotation or a reflection of one of the other figures, it is a congruent figure.

Step 2 Draw a picture of each figure on graph paper.

Step 3 Work until your teacher tells you time is up. Each team receives a point for each figure it finds. A team loses a point if they have congruent figures or they have a figure in which the squares are not arranged edge to edge.

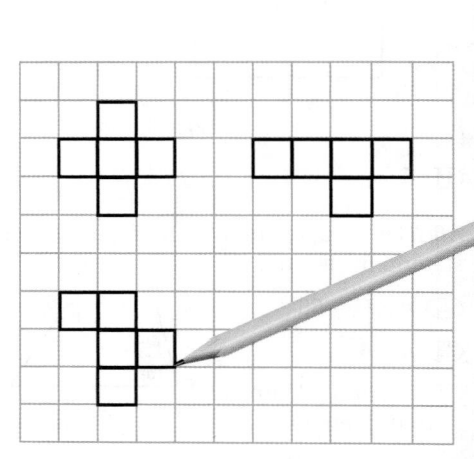

Another Way to Play After you have found all of the possible figures, cut out the drawings and arrange them so that they fit together as a large rectangle.

A *proportion* is an equation that states that two ratios are equal. A property that is helpful when solving a proportion is the *cross product property*.

$$\text{If } \frac{a}{b} = \frac{c}{d}, \text{ then } ad = bc.$$

EXAMPLE **Solve Proportions**

Solve the proportion.

a. $\dfrac{x}{18} = \dfrac{5}{6}$ **b.** $\dfrac{3}{8} = \dfrac{y+2}{32}$

Solution

a.

$\dfrac{x}{18} = \dfrac{5}{6}$	Write original proportion.
$x \cdot 6 = 18 \cdot 5$	Cross product property
$6x = 90$	Multiply.
$\dfrac{6x}{6} = \dfrac{90}{6}$	Divide each side by 6.
$x = 15$	Simplify.

b.

$\dfrac{3}{8} = \dfrac{y+2}{32}$	Write original proportion.
$3 \cdot 32 = 8(y+2)$	Cross product property
$96 = 8y + 16$	Multiply and distribute.
$80 = 8y$	Subtract 16 from each side.
$\dfrac{80}{8} = \dfrac{8y}{8}$	Divide each side by 8.
$10 = y$	Simplify.

Student Help

STUDY TIP
Check your solutions by substituting the value of the variable into the original equation.

Try These

Solve the proportion.

1. $\dfrac{3}{4} = \dfrac{t}{40}$ 30 **2.** $\dfrac{4}{9} = \dfrac{x}{63}$ 28 **3.** $\dfrac{2}{5} = \dfrac{6}{w}$ 15 **4.** $\dfrac{5}{b} = \dfrac{8}{15}$ $\dfrac{75}{8} = 9.375$

5. $\dfrac{7}{2} = \dfrac{7}{y}$ 2 **6.** $\dfrac{35}{x} = \dfrac{5}{14}$ 98 **7.** $\dfrac{d}{4} = \dfrac{11}{14}$ $\dfrac{22}{7} = 3\dfrac{1}{7}$ **8.** $\dfrac{54}{h} = \dfrac{9}{10}$ 60

9. $\dfrac{3z}{10} = \dfrac{7}{20}$ $1\dfrac{1}{6}$ **10.** $\dfrac{4}{7} = \dfrac{c+2}{21}$ 10 **11.** $\dfrac{8}{15} = \dfrac{40}{r+1}$ 74 **12.** $\dfrac{1}{6} = \dfrac{4}{v+5}$ 19

Algebra Review **349**

Extra Example

Solve the proportion.

a. $\dfrac{k}{18} = \dfrac{7}{42}$ 3

b. $\dfrac{52}{13} = \dfrac{28}{5+g}$ 2

Checkpoint Exercise

Solve the proportion.

a. $\dfrac{5r}{80} = \dfrac{22}{88}$ 4

b. $\dfrac{s+5}{12} = \dfrac{19}{38}$ 1

1. Describe the pattern in the numbers 2, 11, 20, 29, ... Then write the next two numbers you expect in the pattern. **(Lesson 1.1)** Each number is 9 more than the previous number; 38, 47.

2. Plot the points $A(-4, 0)$, $B(5, 0)$, $C(3, 6)$, and $D(3, -2)$ in a coordinate plane. Then decide whether $\overline{AB}$ and $\overline{CD}$ are congruent. **(Lesson 1.5)** no

Find the coordinates of the midpoint of $\overline{FG}$. **(Lesson 2.1)**

3. $F(-2, 5)$, $G(0, -7)$ $(-1, -1)$ **4.** $F(1, 4)$, $G(7, -2)$ (4, 1)   **5.** $F(-4, -2)$, $G(6, 4)$ (1, 1)

$\overrightarrow{BD}$ **bisects $\angle ABC$. Find the value of x. Then determine whether $\angle ABC$ is *acute, right, obtuse,* or *straight.*** **(Lesson 2.2)**

6.
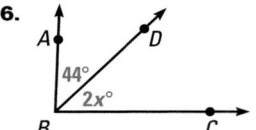
22; acute

7.
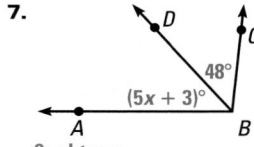
9; obtuse

8.
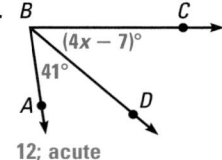
12; acute

9. $\angle P$ is a complement of $\angle Q$, and $m\angle P = 28°$. Find $m\angle Q$. **(Lesson 2.3)** 62°

10. $\angle Y$ is a supplement of $\angle Z$, and $m\angle Y = 146°$. Find $m\angle Z$. **(Lesson 2.3)** 34°

Use the diagram below to complete the statement using *corresponding, alternate interior, alternate exterior, same-side interior,* or *vertical.* **(Lessons 2.4, 3.3)**

11. $\angle 1$ and $\angle 4$ are __?__ angles. same-side interior

12. $\angle 3$ and $\angle 5$ are __?__ angles.
alternate exterior

13. $\angle 1$ and $\angle 7$ are __?__ angles. alternate interior

14. $\angle 4$ and $\angle 6$ are __?__ angles. vertical

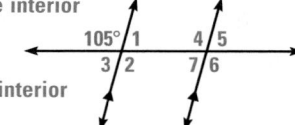

Use the diagram above to find the measure of the angle. **(Lesson 3.4)**

15. $m\angle 4 = $ __?__ ° 105   **16.** $m\angle 7 = $ __?__ ° 75   **17.** $m\angle 5 = $ __?__ ° 75   **18.** $m\angle 2 = $ __?__ ° 105

Find the value of x. **(Lessons 4.2, 4.3)**

19.
58
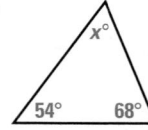

20.
3
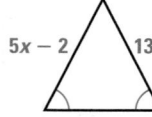

21.
48
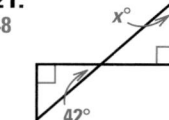

22.
132
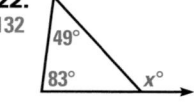

350

Find the distance between A and B. (Lesson 4.4)

23. $A(1, 4)$, $B(-3, 1)$ 5

24. $A(-2, 1)$, $B(3, 13)$ 13

25. $A(3, -4)$, $B(9, 4)$ 10

Can the side lengths form a triangle? Explain. (Lesson 4.7)

26. 6, 7, 8
yes; $6 + 7 > 8$, $7 + 8 > 6$; $8 + 6 > 7$

27. 3, 6, 12 no; $3 + 6 < 12$

28. 4, 10, 14 no; $4 + 10 = 14$

Does the diagram give enough information to show that the triangles are congruent? If so, state the theorem or postulate you would use.
(Lessons 5.2, 5.3, 5.4)

29. $\triangle ABC$ and $\triangle DEC$
no

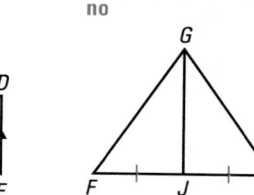

30. $\triangle FJG$ and $\triangle HJG$
no

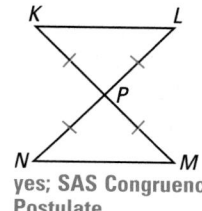

31. $\triangle KPL$ and $\triangle NPM$

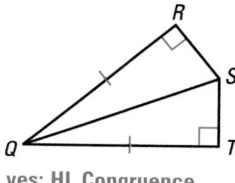

yes; SAS Congruence Postulate

32. $\triangle QRS$ and $\triangle QTS$

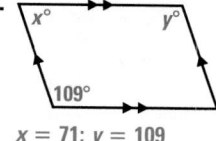

yes; HL Congruence Theorem

33. Sketch a square. How many lines of symmetry does a square have?
(Lessons 5.7, 6.4) See margin.

Find the values of x and y. (Lesson 6.2)

34.

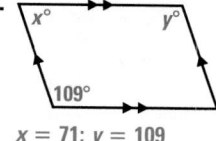

$x = 71$; $y = 109$

35. $x = 18$; $y = \frac{5}{2}$

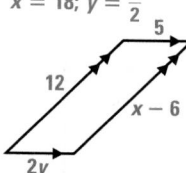

36.

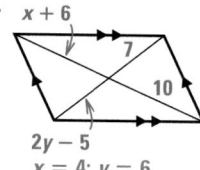

$x = 4$; $y = 6$

37. In quadrilateral $EFGH$, $m\angle E = 90°$, $m\angle F = 90°$, and $m\angle G = 67°$. What special kind of quadrilateral must $EFGH$ be? Explain your reasoning. (Lesson 6.6) A trapezoid; by the Quadrilateral Interior Angles Theorem, $m\angle H = 113°$; since $\angle E$ and $\angle F$ are supplementary, $\overline{EH} \parallel \overline{FG}$, but the other two sides are not parallel; a quadrilateral with exactly one pair of parallel sides is a trapezoid.

Billboard Supports The two ten-foot posts that support a vertical billboard form an angle of 115° with level ground, as shown.
38–40. See margin.

38. What theorem or postulate could you use to show that $\triangle ABC$ is congruent to $\triangle DEF$? (Lesson 5.3)

39. Are $\overleftrightarrow{AB}$ and $\overleftrightarrow{DE}$ parallel? Explain. (Lesson 3.6)

40. What type of quadrilateral must $ADEB$ be? Explain your reasoning. (Lesson 6.6)

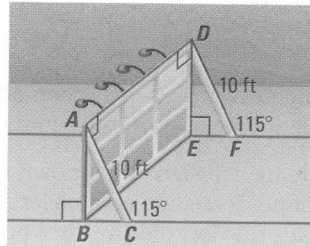

351

33. *Sample answer:*

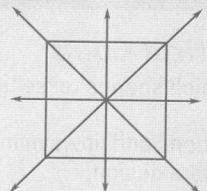

4

38. $AC = DF$, $m\angle ACB = 65° = m\angle DFE$, and $m\angle ABC = 90° = m\angle DEF$, so $\triangle ABC \cong \triangle DEF$ by the AAS Congruence Theorem.

39. Yes; $\overleftrightarrow{AB}$ and $\overleftrightarrow{DE}$ are both perpendicular to $\overleftrightarrow{BE}$ and the three lines are coplanar, so $\overleftrightarrow{AB} \parallel \overleftrightarrow{DE}$.

40. a rectangle; *Sample answer:* It is given that $\overline{AB} \perp \overline{AD}$ and that $\overline{DE} \perp \overline{AD}$. $\overline{AB}$ is perpendicular to the level ground, so $\overline{AB}$ is perpendicular to every line in the plane of the ground that intersects $\overline{AB}$; therefore, $\overline{AB} \perp \overline{BE}$, and so $\angle ABE$ is a right angle. Similarly, $\angle DEB$ is a a right angle. Each angle of quadrilateral $ADEB$ is a right angle, so $ADEB$ must be a rectangle by the Rectangle Corollary.

Mathematical Goals

- Identify which shapes tessellate a plane.
- Use reflections and line symmetry to create a design.
- Classify and identify polygons, and use the properties of squares, triangles, and hexagons.

Managing the Project

Tips for Success

Point out that the triangles shown at the top of this page are equilateral, and that the hexagons created using hexagon grid paper are special hexagons whose sides are all the same length and whose angles all have the same measure, 120°.

Guiding Students' Work

In Step 2, students should not copy the design shown in the book. Point out that they will learn more if they each make a unique design so the designs can be compared. Inform them that any design with symmetry will work, but suggest that they not make their design too complicated. Help them to see that adding color to their design in Step 3 will help make it distinctive.

Creating Tessellations

Objective

Materials

- square, triangle, and hexagon grid paper
- straightedge
- colored pencils or markers

Use regular tessellations to create geometric designs.

A tessellation, or tiling, of a plane is a collection of tiles that fill the plane with no gaps or overlaps. You can tessellate a plane with squares, triangles, or hexagons. You can use these tessellations to create other geometric designs that tessellate.

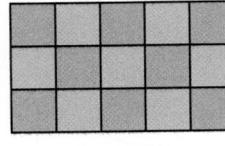

squares

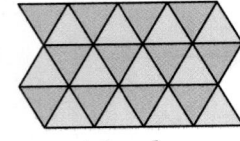

triangles

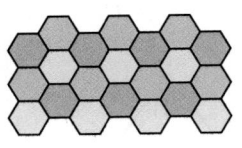

hexagons

Drawing

❶ Choose a basic shape that tessellates. Draw lines of symmetry.

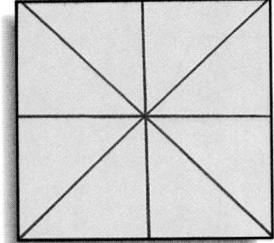

❷ Draw a design that has symmetry.

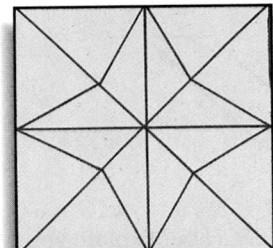

❸ Color the design.

Investigation

1. The design in Step 3 is copied four times to tessellate an area. What type of polygon is in the center? octagon

2. Create a design of your own using a square as the basic shape. Tessellate an area large enough to see the overall pattern. How did the lines of symmetry you drew help you create a design? *Sample answer:* Seeing the lines of symmetry is helpful when making a symmetric design.

Investigation

3. Create a design of your own using a triangle as the basic shape. Tessellate a large enough area to see an overall pattern. What shapes do you see before and after your design is tessellated?

Sample answer: triangles; hexagons

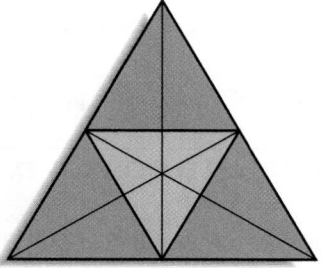

Present Your Results

Write a report to present your results.

▶ Make a book of designs that tessellate.

▶ Include the designs you copied and created in the Investigation and any additional designs you create.

▶ On one page show the steps for creating the design using one basic shape. On another page show the result when this design is tessellated.

Extension

Think of a way in which you would like to use a design that tessellates. Examples include designs in fabric, upholstery, car seat covers, bracelets, belts, wallpaper, and flooring. Then create a design and show how you would use it.

Concluding the Project

If possible, consider inviting to class an interior decorator, fabric designer, or similar person who works with fabric patterns. Invite the guest to look at the students' designs and discuss what kinds of patterns would be used for different decorating applications.

Grading the Project

A well-written project will have the following characteristics:

• A variety of designs are carefully drawn, with sufficient replication of each design to show that it does tesselate the plane.
• Detailed answers are given to the questions in the Investigations.

CHAPTER 7 Pacing and Assignment Guide

REGULAR SCHEDULE

Lesson	Les. Day	Basic	Average	Advanced
7.1	Day 1	SRH p. 661 Exs. 7–12; pp. 361–363 Exs. 13–27, 32, 33, 35, 54, 59–61	pp. 361–363 Exs. 17–37, 54, 56–58	pp. 361–363 Exs. 21–29, 32–36, 54–60 even
	Day 2	SRH p. 661 Exs. 16–21; pp. 361–363 Exs. 38–45, 50–53, 55–58, 62–67	pp. 361–363 Exs. 38–45, 47–53, 55, 59–67	pp. 361–363 Exs. 42–45, 46*, 47–51, 55–67 odd; EC: classzone.com
7.2	Day 1	SRH p. 656 Exs. 6–10; pp. 369–371 Exs. 8, 9, 11–16, 22, 23, 34–37	SRH p. 661 Exs. 22–27; pp. 369–371 Exs. 17–20, 24–27, 38–41	pp. 369–371 Exs. 8–16 even, 22, 23, 29–34
	Day 2	SRH p. 661 Exs. 22–27; pp. 369–371 Exs. 17–20, 24–27, 38–41	pp. 369–371 Exs. 17–21, 24–27, 35–41	pp. 369–371 Exs. 17–21 odd, 24–27, 28*, 35–41; EC: classzone.com
7.3	Day 1	pp. 375–378 Exs. 7–15, 39–53	pp. 375–378 Exs. 7–15, 27–29, 37, 39–49	pp. 375–378 Exs. 10–15, 27–29, 37, 39–49
	Day 2	SRH p. 661 Exs. 28–30; pp. 375–378 Exs. 16–26, 38, Quiz 1	pp. 375–378 Exs. 16–26, 30, 38, Quiz 1	pp. 375–378 Exs. 21–26, 30, 31–36*, 38, Quiz 1; EC: classzone.com
7.4	Day 1	pp. 382–385 Exs. 5–13, 33–49 odd	pp. 382–385 Exs. 5–13, 33–37, 42–49	pp. 382–385 Exs. 8–13, 33–45
	Day 2	pp. 382–385 Exs. 15–26, 32–48 even	pp. 382–385 Exs. 14–26, 29–32, 38–41	pp. 382–385 Exs. 14–20, 22–26 even, 27–32; EC: TE p. 354D*, classzone.com
7.5	Day 1	pp. 390–392 Exs. 10–19, 40–52 even	pp. 390–392 Exs. 14–19, 40, 44–52	pp. 390–392 Exs. 14–19, 30, 31*, 32*, 40–43
	Day 2	pp. 390–392 Exs. 20–29, 33–37, 41–51 odd	pp. 390–392 Exs. 20–29, 33–39, 41–43	pp. 390–392 Exs. 20–28 even, 33–39, 44–47; EC: classzone.com
7.6	Day 1	pp. 396–398 Exs. 6–9, 21–29	pp. 396–398 Exs. 6–10, 15, 21–29	pp. 396–398 Exs. 6–10, 15, 21–29
	Day 2	pp. 396–398 Exs. 11–14, 17–20, Quiz 2	pp. 396–398 Exs. 11–14, 16–20, Quiz 2	pp. 396–398 Exs. 11–14, 16–20, Quiz 2; EC: TE p. 354D*, classzone.com
Review	Day 1	pp. 400–403 Exs. 1–26	pp. 400–403 Exs. 1–26	pp. 400–403 Exs. 1–26
Assess	Day 1	Chapter 7 Test	Chapter 7 Test	Chapter 7 Test

YEARLY PACING   Chapter 7 Total – **14 days**   Chapters 1–7 Total – **108 days**   Remaining – **52 days**

*Challenge Exercises   EP = Extra Practice   SRH = Skills Review Handbook   EC = Extra Challenge

BLOCK SCHEDULE

Day 1	Day 2	Day 3	Day 4	Day 5	Day 6	Day 7
7.1 pp. 361–363 Exs. 17–45, 47–67	**7.2** pp. 369–371 Exs. 8–27, 29–41	**7.3** pp. 375–378 Exs. 7–30, 37–49, Quiz 1	**7.4** pp. 382–385 Exs. 5–26, 29–49	**7.5** pp. 390–392 Exs. 14–29, 33–52	**7.6** pp. 396–398 Exs. 6–29, Quiz 2	**Review** pp. 400–403 Exs. 1–26 **Assess** Chapter 7 Test

YEARLY PACING   Chapter 7 Total – **7 days**   Chapters 1–7 Total – **54 days**   Remaining – **26 days**

Support Materials

CHAPTER RESOURCE BOOK

CHAPTER SUPPORT

Tips for New Teachers	p. 1	Strategies for Reading Mathematics	p. 5
Parent Guide for Student Success	p. 3		

LESSON SUPPORT

	7.1	7.2	7.3	7.4	7.5	7.6
Lesson Plans (regular and block)	p. 7	p. 16	p. 28	p. 38	p. 50	p. 59
Warm-Up Exercises and Daily Quiz	p. 9	p. 18	p. 30	p. 40	p. 52	p. 61
Technology Activities & Keystrokes		p. 19		p. 41	p. 53	p. 62
Practice (2 levels)	p. 10	p. 22	p. 31	p. 45	p. 54	p. 64
Reteaching with Practice	p. 12	p. 24	p. 33	p. 47	p. 56	p. 66
Quick Catch-Up for Absent Students	p. 14	p. 26	p. 35	p. 49	p. 58	p. 68
Learning Activities			p. 36			
Real-Life Applications	p. 15	p. 27				

REVIEW AND ASSESSMENT

Quizzes	pp. 37, 69	Alternative Assessment with Math Journal	p. 77
Brain Games Support	p. 70	Project with Rubric	p. 79
Chapter Review Games and Activities	p. 71	Cumulative Review	p. 81
Chapter Test (2 levels)	p. 72	Resource Book Answers	AN1
SAT/ACT Chapter Test	p. 76		

TRANSPARENCIES

	7.1	7.2	7.3	7.4	7.5	7.6
Warm-Up Exercises and Daily Quiz	p. 41	p. 42	p. 43	p. 44	p. 45	p. 46
Visualize It Transparencies	✓	✓	✓	✓		✓
Answer Transparencies	✓	✓	✓	✓	✓	✓

TECHNOLOGY

- Time-Saving Test and Practice Generator
- Electronic Teacher Tools
- Geometry in Motion: Video
- Classzone.com
- Online Lesson Planner

ADDITIONAL RESOURCES

- Worked-Out Solution Key
- Resources in Spanish
- Practice Workbook with Examples

Strategies for Strategic Learners

ANTICIPATE PROBLEM AREAS

The concepts of similar polygons and congruent polygons may be confused by students. Especially confusing to some students is the existence of both a SSS Congruence Postulate and a SSS Similarity Theorem, as well as a SAS Congruence Postulate and a SAS Similarity Theorem. Remind students that if two figures can be shown to be identical, they are congruent.

REVISIT CONGRUENT FIGURES
Ask students to recall that if two figures are congruent, then their corresponding angles are congruent and their corresponding sides are congruent. Two congruent triangles, for example, have three pairs of congruent angles and three pairs of congruent sides. However one does not need that much information to decide that two triangles are congruent. Remind students that knowing any one of the following is enough to determine that two triangles are congruent.

Three pairs of sides are congruent.
Two pairs of corresponding angles and their included sides are congruent.
Two pairs of corresponding sides and their included angles are congruent.

Have students verify these statements by drawing pairs of triangles using just the information given in each statement.

SIMILAR FIGURES
Point out that similar figures are proportional in terms of their side lengths. This means that, in general, one shape cannot be superimposed exactly on the other figure the way congruent figures can, because similar figures are seldom the same size. Suggest that it is as if you had two congruent figures and then shrunk or enlarged one of them.

Emphasize that two figures are similar if their corresponding angles are congruent and the ratios of the lengths of their corresponding sides are equal. Just as for congruent triangles, it is not necessary to show that all three pairs of angles are congruent and that all three pairs of corresponding sides are proportional. Any one of the following is sufficient for determining that two triangles are similar.

Two angles of one triangle are congruent to two angles of the other triangle. (AA Similarity Postulate)
All three pairs of corresponding sides are proportional.
One pair of corresponding angles are congruent and the corresponding sides that form these angles have lengths that are proportional.

Again, have students demonstrate that these statements are true by drawing triangles, one bigger than the other, that satisfy the conditions.

Strategies for English Learners

BUILD ON WHAT STUDENTS KNOW

English learners who have arrived recently from other countries may have had adequate or excellent prior schooling in mathematics. They may be familiar with algorithms that are not commonly used in this country. Encourage recent immigrants to share alternate forms of describing mathematical concepts or algorithms and provide an opportunity for the whole class to understand and discuss these alternative methods. Such a discussion not only promotes respect for cultural diversity and gives the teacher a glimpse into the student's thinking, but it also provides all students with an opportunity to build a deeper understanding of mathematics.

USE A PICTURE OR MODEL

Students who are not yet fluent in English can investigate the concepts of similarity and congruence, as well as the relationship between different figures, by cutting out paper models of the figures. This is a fun activity for all students, and helps provide a concrete visualization of abstract concepts. Have students do the following.

On a sheet of 8.5-by-11 inch paper, draw a dashed line along one of the diagonals of the rectangle and cut along the diagonal to divide the sheet of paper into two right triangles. Then have students rotate and flip the shapes until the two triangles can be arranged one on top of the other. This illustrates the concept of congruence.

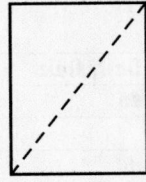

Now have the students take one of their two right triangles and draw the dashed line shown in the figure below. Have them divide the triangle into two smaller right triangles by cutting along the dashed line.

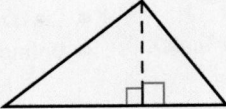

Students will then have three right triangles that are similar to each other, but not congruent. You can illustrate the concept of proportion by having students measure the lengths of the sides of the triangles and comparing the measures of the corresponding sides. Finally, have students measure the acute angles of each triangle to verify that the corresponding angles of the triangles are congruent.

Strategies for Advanced Learners

USE CROSS-CURRICULAR CONNECTIONS

Throughout this textbook there are suggestions, called "links," to connect mathematics to other subject areas. History, science, language arts, music, art, and other subjects all offer rich opportunities to combine studies in two or more disciplines. For example, links to history give the teacher an opportunity to encourage advanced students to investigate the state of the thinking in mathematics in each culture. The history and mathematics teachers could collaborate on assignments that ask students to investigate the contributions of each culture to mathematics and how mathematical ideas and practices spread through trade and military conquests.

When the mathematics teacher needs to work with a group of students to reteach a concept that has already been mastered by the advanced students, those advanced students can be working on a long-term project or investigation across subject areas. In this way, they can extend their knowledge of mathematics, use their time productively, and be ready to rejoin the class when the teacher moves on to new material. Teachers should be careful, however, not to misuse cross-curricular challenge investigations. Cross-curricular investigations should contribute to the student's understanding of mathematics and provide an opportunity for the teacher to instruct at a level characteristic of a more difficult mathematics class. Many investigations are very appropriate for homework, especially when the teacher knows that homework appropriate for the majority of the class is too easy for a particular student or group of students.

Possible cross-curricular challenge investigations include the following. The expected learning outcomes for students are described in italics within the parentheses:

1. Major mathematical discoveries of a particular time period. (*Expected outcomes: (1) Students will better understand a mathematical concept by looking at the evolution of that concept over time; (2) Students will understand that in mathematics, new discoveries are built on old ones.*) Each period of history has made significant contributions to the area of mathematics, with the exception perhaps of the Dark Ages in Europe. In history class, students might be challenged to make a list or timeline of major mathematical contributions of civilizations in whatever historical period they are studying. In language arts class they could write about one or more of these contributions, and in mathematics class they could share their findings orally with the class.

2. Ancient mathematical procedures that are useful today. (*Expected outcome: Students will understand how large artworks, like a frieze on an ancient building, were constructed from a small model or a sketch using similarity.*) Greek and Roman art and architecture offer many examples of artwork that can be investigated.

Challenge Problem for use with Lesson 7.4:

In the diagram, $\frac{EC}{EA} = \frac{ED}{EB}$. Explain how you know that $\frac{CD}{AB} = \frac{ED}{EB}$.

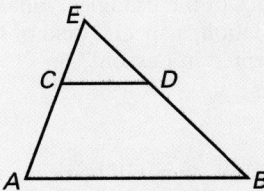

Solution: Since $\angle E \cong \angle E$ by the Reflexive Property of Congruence, $\triangle ECD \sim \triangle EAB$ by the SAS Similarity Theorem. Therefore, by the definition of similar triangles, $\frac{CD}{AB} = \frac{ED}{EB}$.

Challenge Problem for use with Lesson 7.6:

In the diagram, $\overline{WX}$ is the image of $\overline{YZ}$ under a dilation with center C. Suppose $CY = YW$, $YW = YZ$, and $WX = 20$. Find YZ. (*10*)

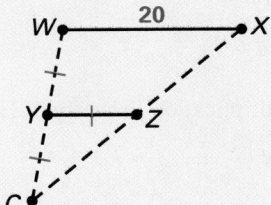

This chapter investigates properties of geometric figures that are pre-served under a change of scale. Lesson 7.1 lays the algebraic groundwork that will be needed for dealing with such scale changes, mainly the manipulation of ratios and proportions. In Lesson 7.2, the definition of similarity as it relates to polygons is given, along with some of its imme-diate implications. Lesson 7.3 narrows the discussion of similarity to triangles, with the introduction of the Angle-Angle Similarity Postulate. In Lesson 7.4, two theorems for justifying the similarity of triangles, the Side-Side-Side Similarity Theorem and the Side-Angle-Side Similarity Theorem, are stated. Lesson 7.5 extends the discussion of proportions in the definition of similarity to cover lengths of segments that are related to a pair of similar triangles but are not sides of the triangles. Finally, the relationship between similarity and a dilation, a special kind of transfor-mation that preserves similarity, is presented in Lesson 7.6.

Lesson 7.1

This lesson reviews the basic algebraic skills that will be needed for deal-ing with linear measurements in similar figures. Mastering the terminol-ogy at the outset will help students acquire these skills, so you should make sure they understand that a ratio is written as a fraction, while a proportion is an equation stating that two ratios have the same value.

> **QUICK TIP**
> The most important point to stress in this lesson is that all the rules for solving for the value of the unknown in a proportion, are applications of the properties of equality studied in Lesson 2.6. A quick review of these properties will be helpful to all students.

Point out that the Cross Product Property

$$\frac{a}{b} = \frac{c}{d} \Rightarrow ad = bc$$

is actually just an application of the Multiplication Property of Equality:

$$\frac{a}{b} = \frac{c}{d} \Rightarrow bd \cdot \frac{a}{b} = bd \cdot \frac{c}{d} \Rightarrow ad = bc.$$

Another important point to make is the one stated in the Study Tip on page 357: When simplifying any ratio of two measurements, make sure the units of the measurements in the numerator and denominator are the same. This rule can be extended to cover proportions, but here you might note that answers will be correct as long as either (1) the two numerators have the same units and the two denominators have the same units, or (2) the two measurements in one ratio have the same units, and the two measurements in the other ratio have the same units.

Finally, note that when a proportion is stated using colons or in words, it is important to write it using fractions consistently. That is, "a is to b as c is to d" must be written as $\frac{a}{b} = \frac{c}{d}$, and not, for example, as $\frac{a}{b} = \frac{d}{c}$.

Lesson 7.2

In this lesson, the formal definition of similar polygons is presented. Note that, in general, all the conditions in the definition are necessary. You can emphasize this point by asking students to name two figures whose corre-sponding sides are proportional but which are not similar (a rhombus and a square provide an example), and two figures whose corresponding angles are congruent but are not similar (for example, a rectangle and a square).

ORGANIZING PROPORTIONS The most important practical advice you can give students for dealing with proportions that arise from similar poly-gons is that the terms must be arranged correctly. One way to ensure this is to promote a scheme like the following:

$$\begin{array}{cc} \downarrow & \downarrow \\ \rightarrow \dfrac{a}{b} & = \dfrac{c}{d} \end{array} \begin{array}{l} \leftarrow \text{same polygon} \\ \leftarrow \text{same polygon} \end{array}$$

corr. corr.
sides sides

This diagram indicates that in each ratio the numerator and denominator represent corresponding sides of the two polygons, while both numerators are parts of the same polygon, as are both denominators. Stress that it does not matter which polygon contributes the two numerators and which contributes the two denominators. The arrangement of terms in the pro-portion lends itself well to the concept of scale factor, which is then just the value of either ratio.

Finally, point out that Theorem 7.1 about the perimeters of similar poly-gons is a specific instance of a more general principle: In similar poly-gons, any two corresponding linear measurements have the same ratio as two corresponding side lengths. For example, as students may conjecture, the lengths of corresponding medians of similar triangles have the same ratio as the lengths of any two corresponding sides.

Lesson 7.3

The Angle-Angle Similarity Postulate is analogous to the postulates and theorems of Chapter 5 for proving triangles congruent. The advantage of this postulate is that it allows us to deduce proportions involving the lengths of the sides of triangles from information about only two pairs of corresponding angles. But emphasize that the postulate only applies to triangles.

Several geometric situations involving the Angle-Angle Similarity Postulate crop up again and again, and you may want to make note of them. Two of the more common such situations are: (1) a triangle with a segment that joins points on two sides and that is parallel to the third side (as in Example 2), and (2) two triangles with one pair of corresponding angles that are vertical angles, and whose sides opposite these angles are parallel (as in Exercises 4, 12, and 26).

Lesson 7.4

This lesson presents two more methods for proving triangles similar, the SSS Similarity Theorem and the SAS Similarity Theorem. Emphasize that in order to use the SAS Similarity Theorem, the congruent angles must be *included* between the two pairs of corresponding sides. If they are not, similarity is not assured. A counterexample, in which the angles are not included by the proportional sides, is shown below. It should be obvious to students that the triangles are not similar.

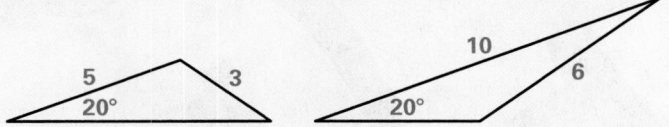

Lesson 7.5

Note that the Triangle Proportionality Theorem is an extension of the proportions that occur in the definition of similar triangles, because two of the segments $(\overline{TQ}$ and $\overline{US}$ in the diagram shown with the theorem$)$ are not sides of any triangle. Also, note that many problems can be solved with the Triangle Proportionality Theorem or by using the definition. For example, an alternative method of solving the problem posed in Example 2 is to use the following proportion, based on the definition of similar triangles.

$$\frac{y}{20} = \frac{3}{3+9}, \text{ or } \frac{y}{20} = \frac{3}{12}$$

Encourage students to take a moment to study a problem first in order to decide the method of solution that will be easiest for them before setting up their equations.

MAKING CONNECTIONS This lesson provides another opportunity to make a connection to previously-learned theorems. Note that the Midsegment Theorem can be justified as a result of the SAS Similarity Theorem, and also that it relates to the statement about the length of the midsegment of a trapezoid in Lesson 6.5 (when the triangle is thought of as a trapezoid with one base of zero length).

Lesson 7.6

The study of transformations of the plane begun in Lesson 3.7 continues here. Dilations "stretch" a figure by a scale factor k, which must be a positive number. Note that if $k < 1$, then the dilation *shrinks* the figure. Therefore, although in English the word *dilation* means an enlargement, in mathematics it can also refer to a reduction. This fact may confuse some students, who might have a tendency to assume that the larger polygon in a diagram is always the image of the smaller. Make sure they understand that the smaller polygon may be the image of the larger.

Dilations are especially relevant to the concept of similarity, because the image of a polygon under a dilation is always a polygon similar to the original polygon. This fact can be easily illustrated in a drawing that contains the original polygon, its image, and the rays drawn from the center of dilation through corresponding vertices of the two polygons. The Converse of the Triangle Proportionality Theorem then shows that corresponding sides of the original polygon and its image are parallel, and from this it is easy to see that the two polygons are similar.

A popular mechanical device, before computer graphics tools were available, was called a *pantograph*. This was a drawing tool that allowed the user to trace over a given picture or diagram and in the process produce a similar, but larger, copy of the image. Its design was based on some of the theorems of this chapter. If your school has one of these devices, you might want to borrow it to illustrate the effect of a dilation.

Chapter Overview

Chapter Goals

In this chapter, the concept of similarity will be explored. After studying ratios and proportions, students use them in connection with similar polygons. Extensive work with similar triangles is accomplished before the chapter concludes with a lesson on dilations. Students will:

- Use ratios and solve proportions, especially as they relate to similar polygons.
- Identify similar polygons, and use postulates and theorems to show that two triangles are similar.
- Identify and draw dilations of polygons.

Application Note

When a large mural is created from a drawing, it is common to overlay the drawing with a grid. The outline of the drawing in each section of the grid can be traced and then enlarged using a photocopier. The enlarged pieces can then be used to create the mural. Most copiers have settings that allow you to enlarge the drawing by a percentage. You can use repeated enlarging to enlarge by a greater percentage.

More information about scale drawings is provided on the Internet at classzone.com

Application Links
CLASSZONE.COM

How does a drawing become a mural?

Murals are often created by enlarging an original drawing. People use projection machines, grids, and other methods to make sure that all parts of a mural are in proportion to the original drawing.

original figure   enlarged figure

Learn More About It

You will learn more about enlarging drawings in Exercise 21 on p. 370.

Who uses Similarity?

MAP MAKER
Map makers analyze photographs, satellite images, and other data to create maps. Maps may be designed for travel, tourism, weather forecasting, and geological exploration. (p. 360)

ARTIST
Painters, photographers, and sculptors use proportions to enlarge and reduce the size of an original art piece. Proportions are also used to draw human figures and make perspective drawings. (p. 370)

How will you use these ideas?

- Estimate distances on a map. (p. 360)
- Compare the dimensions of television screens. (p. 371)
- Analyze a hockey pass. (p. 374)
- Calculate the height of a flag pole. (p. 377)
- See how similarity appears in fractals. (p. 391)

Projects
A project covering Chapters 7–8 appears on pages 468–469 of the Student Edition. An additional project for Chapter 7 is available in the *Chapter 7 Resource Book*, pp. 79–80.

Technology
- Electronic Teacher Tools
- Test and Practice Generator
- Online Lesson Planner

Video
- **Geometry in Motion**
  There is an animation supporting Lesson 7.5.

Internet Support
CLASSZONE.COM

- Application and Career Links
  360, 370, 377, 391
- Student Help
  363, 366, 376, 383, 387, 397

Diagnostic Tools

The **Chapter Readiness Quiz** can help you diagnose whether students have the following skills needed in Chapter 7:

• Write a congruence statement for two triangles.
• Find the length of the midsegment of a trapezoid.
• Simplify fractions.

Reteaching Material

This resource is available for students who need additional help with the skills on the Chapter Readiness Quiz:

📙 **Chapter 7 Resource Book**
• Reteaching with Practice (Lessons 7.1–7.6)

Additional Resources

The following resources are provided to help you prepare for the upcoming chapter and customize review materials:

📙 **Chapter 7 Resource Book**
• *Tips for New Teachers*, pp. 1–2
• *Parent Guide*, pp. 3–4
• *Lesson Plans*, pp. 7, 16, 28, 38, 50, 59
• *Lesson Plans for Block Scheduling*, pp. 8, 17, 29, 39, 51, 60

🖥 **Technology**
• Electronic Teaching Tools
• Online Lesson Planner
• Test and Practice Generator

Visualize It!

Additional suggestions for helping students visualize geometry are on pp. 359, 373, 391, 394, and 396.

PREVIEW   What's the chapter about?

• Using **ratios** and **proportions**
• Identifying **similar polygons** and showing that triangles are similar
• Identifying and drawing **dilations**

Key Words

• ratio of *a* to *b*, *p. 357*
• proportion, *p. 359*
• similar polygons, *p. 365*
• scale factor, *p. 366*
• midsegment of a triangle, *p. 389*
• dilation, *p. 393*

PREPARE   Chapter Readiness Quiz

Take this quick quiz. If you are unsure of an answer, look at the reference pages for help.

Vocabulary Check *(refer to pp. 234, 241)*

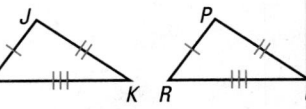

1. Which *congruence statement* is correct for the triangles at the right? **C**

(A) $\triangle LKJ \cong \triangle QRP$ (B) $\triangle JLK \cong \triangle PQR$

(C) $\triangle LJK \cong \triangle RPQ$ (D) $\triangle KLJ \cong \triangle QPR$

Skill Check *(refer to pp. 333, 656)*

2. What is the value of *x*? **G**

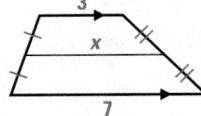

(F) 4   (G) 5

(H) 5.5   (J) 6

3. Which of the following fractions can be simplified to $\frac{2}{7}$? **B**

(A) $\frac{8}{18}$ (B) $\frac{6}{21}$ (C) $\frac{12}{28}$ (D) $\frac{18}{33}$

VISUAL STRATEGY   Separating Triangles

Visualize It! ➡   To help you see triangle relationships more clearly, sketch overlapping triangles separately and label them with their measures.

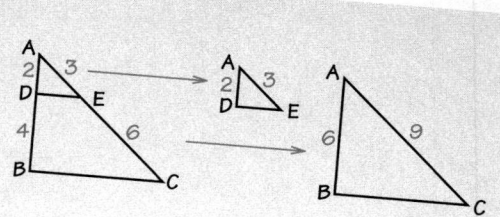

7.1 Ratio and Proportion

Goal
Use ratios and proportions.

Key Words
- ratio of *a* to *b*
- proportion
- means
- extremes

In 2000, Nomar Garciaparra of the Boston Red Sox won the American League batting title for the second straight year, with a *batting average* of .372. He was at bat 529 times and had 197 hits. A baseball player's batting average is calculated using a *ratio*.

$$\frac{\text{hits}}{\text{times at bat}} = \frac{197}{529} \approx 0.372$$

A **ratio** is a comparison of a number *a* and a nonzero number *b* using division.

The ratio of *a* to *b* can be written in three ways: as the fraction $\frac{a}{b}$ (or an equivalent decimal), as $a : b$, or as "*a* to *b*." A ratio is usually written in simplest form.

EXAMPLE 1 Simplify Ratios

Simplify the ratio.

a. 60 cm : 200 cm

b. $\dfrac{3\text{ ft}}{18\text{ in.}}$

Solution

a. 60 cm : 200 cm can be written as the fraction $\dfrac{60\text{ cm}}{200\text{ cm}}$.

$$\frac{60\text{ cm}}{200\text{ cm}} = \frac{60 \div 20}{200 \div 20} \qquad \text{Divide numerator and denominator by their greatest common factor, 20.}$$

$$= \frac{3}{10} \qquad \text{Simplify. } \tfrac{3}{10} \text{ is read as "3 to 10."}$$

b. $\dfrac{3\text{ ft}}{18\text{ in.}} = \dfrac{3 \cdot 12\text{ in.}}{18\text{ in.}} \qquad$ Substitute 12 in. for 1 ft.

$$= \frac{36\text{ in.}}{18\text{ in.}} \qquad \text{Multiply.}$$

$$= \frac{36 \div 18}{18 \div 18} \qquad \text{Divide numerator and denominator by their greatest common factor, 18.}$$

$$= \frac{2}{1} \qquad \text{Simplify. } \tfrac{2}{1} \text{ is read as "2 to 1."}$$

358

Extra Example 1

Simplify the ratio.

a. 35 m : 700 m $\frac{1}{20}$

b. $\frac{12 \text{ ft}}{5 \text{ yd}}$ $\frac{4}{5}$

Extra Example 2

In the diagram, *DE* : *EF* is 1 : 2 and *DF* = 45. Find *DE* and *EF*.

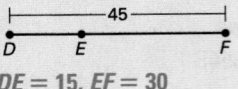

DE = 15, *EF* = 30

Extra Example 3

The perimeter of a rectangle is 70 centimeters. The ratio of the length to the width is 3 : 2. Find the length and the width of the rectangle.   **length: 21 cm; width: 14 cm**

EXAMPLE 2 *Use Ratios*

In the diagram, *AB* : *BC* is 4 : 1 and *AC* = 30. Find *AB* and *BC*.

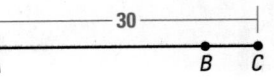

Solution

Let *x* = *BC*. Because the ratio of *AB* to *BC* is 4 to 1, you know that *AB* = 4*x*.

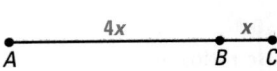

$AB + BC = AC$	Segment Addition Postulate
$4x + x = 30$	Substitute 4*x* for *AB*, *x* for *BC*, and 30 for *AC*.
$5x = 30$	Add like terms.
$x = 6$	Divide each side by 5.

To find *AB* and *BC*, substitute 6 for *x*.

$$AB = 4x = 4 \cdot 6 = 24 \qquad BC = x = 6$$

ANSWER ▶ So, *AB* = 24 and *BC* = 6.

EXAMPLE 3 *Use Ratios*

The perimeter of a rectangle is 80 feet. The ratio of the length to the width is 7 : 3. Find the length and the width of the rectangle.

Solution

The ratio of length to width is 7 to 3. You can let the length $\ell = 7x$ and the width $w = 3x$.

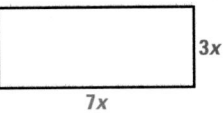

Student Help

SKILLS REVIEW
To review the formula for the perimeter of a rectangle, see p. 674.

▶ $2\ell + 2w = P$	Formula for the perimeter of a rectangle
$2(7x) + 2(3x) = 80$	Substitute 7*x* for ℓ, 3*x* for *w*, and 80 for *P*.
$14x + 6x = 80$	Multiply.
$20x = 80$	Add like terms.
$x = 4$	Divide each side by 20.

To find the length and width of the rectangle, substitute 4 for *x*.

$$\ell = 7x = 7 \cdot 4 = 28 \qquad w = 3x = 3 \cdot 4 = 12$$

ANSWER ▶ The length is 28 feet, and the width is 12 feet.

Checkpoint *Use Ratios*

1. In the diagram, *EF* : *FG* is 2 : 1 and *EG* = 24. Find *EF* and *FG*.   *EF* = 16; *FG* = 8

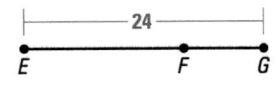

2. length, 24 ft; width, 18 ft

2. The perimeter of a rectangle is 84 feet. The ratio of the length to the width is 4 : 3. Find the length and the width of the rectangle.

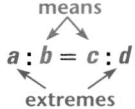

Visualize It!

means

$$a : b = c : d$$

extremes

The **means** are the inner terms, b and c, and the **extremes** are the outer terms, a and d.

Solving Proportions An equation that states that two ratios are equal is called a **proportion** .

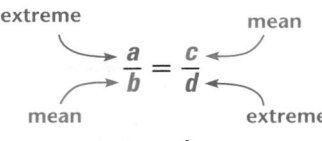

extreme   mean

$$\frac{a}{b} = \frac{c}{d}$$

mean   extreme

proportion

The numbers b and c are the **means** of the proportion. The numbers a and d are the **extremes** of the proportion. You can solve a proportion using the *cross product property*.

CROSS PRODUCT PROPERTY

Words In a proportion, the product of the extremes is equal to the product of the means.

$$ad = bc$$

$$\frac{a}{b} \diagdown \frac{c}{d}$$

Symbols If $\frac{a}{b} = \frac{c}{d}$, then $ad = bc$.

Using Algebra

EXAMPLE 4 Solve a Proportion

Solve the proportion $\frac{5}{3} = \frac{y + 2}{6}$.

Solution

$\frac{5}{3} = \frac{y + 2}{6}$	Write original proportion.
$5 \cdot 6 = 3(y + 2)$	Cross product property
$30 = 3y + 6$	Multiply and use distributive property.
$30 - 6 = 3y + 6 - 6$	Subtract 6 from each side.
$24 = 3y$	Simplify.
$\frac{24}{3} = \frac{3y}{3}$	Divide each side by 3.
$8 = y$	Simplify.

CHECK ✓ Check your solution by substituting 8 for y.

$$\frac{y + 2}{6} = \frac{8 + 2}{6} = \frac{10}{6} = \frac{5}{3}$$

Checkpoint ✓ *Solve a Proportion*

Solve the proportion.

3. $\frac{3}{x} = \frac{6}{8}$ 4

4. $\frac{5}{3} = \frac{15}{y}$ 9

5. $\frac{m + 2}{5} = \frac{14}{10}$ 5

7.1 *Ratio and Proportion*   **359**

Teaching Tip
When discussing the Cross Product Property, point out that the order of the ratios in a proportion does not matter. Demonstrate this by reversing the order of the ratios $\frac{a}{b}$ and $\frac{c}{d}$ and applying the property. Stress that the result, $cb = da$, is equivalent to the equality $ad = bc$.

Extra Example 4
Solve the proportion $\frac{2}{7} = \frac{8}{w - 2}$.
$w = 30$

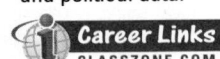
EXAMPLE **5** **Write and Solve a Proportion**

Use the map of Texas to estimate the distance between Dallas and Houston.

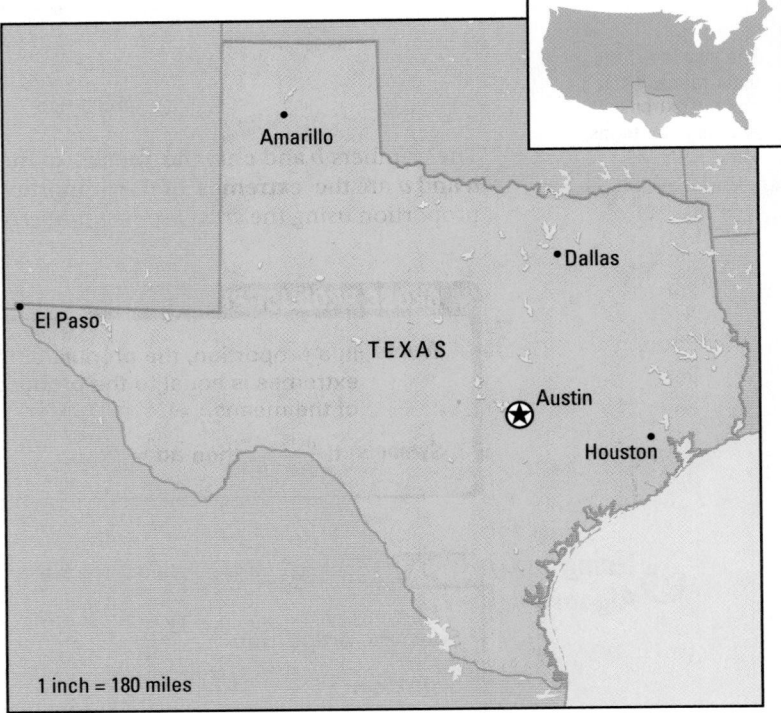

Amarillo

•Dallas

•El Paso

TEXAS

⊛ Austin

Houston

1 inch = 180 miles

Solution

From the scale on the map, you know that 1 inch represents 180 miles.

On the map, the distance between Dallas and Houston is $1\frac{1}{4}$ inches, which can be written as 1.25 inches.

Let x represent the actual distance between Dallas and Houston. You can write the following proportion.

$$\frac{1\text{ in.}}{180\text{ mi}} = \frac{1.25\text{ in.}}{x\text{ mi}}$$ Write a proportion.

$$1 \cdot x = 180 \cdot 1.25$$ Cross product property

$$x = 225$$ Simplify.

ANSWER ▶ The distance between Dallas and Houston is about 225 miles.

 **Write and Solve a Proportion**

Use the map of Texas and a ruler to estimate the distance between the two cities.

6. about 360 miles

7. about 146 miles

6. El Paso and Amarillo   **7.** Houston and Austin

Guided Practice

Vocabulary Check

1. Give an example of a *ratio*. *Sample answer:* The teacher-pupil ratio at the elementary school is 1 to 19.

2. Identify the *means* and the *extremes* of the proportion $\frac{a}{b} = \frac{3}{4}$. The means are *b* and 3; the extremes are *a* and 4.

Skill Check

In Exercises 3–6, simplify the ratio.

3. $\frac{12 \text{ m}}{10 \text{ m}}$ $\frac{6}{5}$

4. $\frac{21 \text{ pencils}}{35 \text{ pencils}}$ $\frac{3}{5}$

5. $\frac{2 \text{ years}}{8 \text{ months}}$ $\frac{3}{1}$

6. $\frac{2 \text{ yd}}{16 \text{ ft}}$ $\frac{3}{8}$

7. In the diagram, $FG : GH = 4 : 3$ and $FH = 56$. Find FG and GH.
$FG = 32$; $GH = 24$

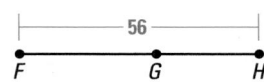

8. The perimeter of a rectangle is 27. The ratio of the length to the width is $2 : 1$. Find the length and width of the rectangle. length, 9; width 4.5

Solve the proportion.

9. $\frac{x}{2} = \frac{5}{10}$ 1

10. $\frac{3}{t} = \frac{12}{20}$ 5

11. $\frac{8}{5} = \frac{24}{y}$ 15

12. $\frac{x+1}{3} = \frac{14}{21}$ 1

Practice and Applications

Extra Practice
See p. 687.

Writing Ratios **A track team won 8 meets and lost 2. Find the ratio.**

13. wins to losses $\frac{4}{1}$

14. wins to the number of track meets $\frac{4}{5}$

15. losses to wins $\frac{1}{4}$

16. losses to the number of track meets $\frac{1}{5}$

Simplifying Ratios **Simplify the ratio.**

17. $\frac{16 \text{ lb}}{2 \text{ lb}}$ $\frac{8}{1}$

18. $\frac{14 \text{ in.}}{42 \text{ in.}}$ $\frac{1}{3}$

19. $\frac{35 \text{ mi}}{12 \text{ mi}}$ $\frac{35}{12}$

20. $\frac{6 \text{ days}}{4 \text{ weeks}}$ $\frac{3}{14}$

21. $\frac{3 \text{ ft}}{12 \text{ in.}}$ $\frac{3}{1}$

22. $\frac{6 \text{ yd}}{10 \text{ ft}}$ $\frac{9}{5}$

23. $\frac{60 \text{ cm}}{1 \text{ m}}$ $\frac{3}{5}$

24. $\frac{400 \text{ m}}{0.5 \text{ km}}$ $\frac{4}{5}$

Finding Ratios **Find the ratio of the length to the width of the rectangle. Then simplify the ratio.**

Homework Help

Example 1: Exs. 13–31
Example 2: Exs. 32–34
Example 3: Exs. 35–37
Example 4: Exs. 38–45
Example 5: Exs. 47–53

25.

16 mm

20 mm

$\frac{20 \text{ mm}}{16 \text{ mm}}, \frac{5}{4}$

26.

7 cm

7 cm

$\frac{7 \text{ cm}}{7 \text{ cm}}, \frac{1}{1}$

27.

12 in.

2 ft

$\frac{2 \text{ ft}}{12 \text{ in.}}$ or $\frac{24 \text{ in.}}{12 \text{ in.}}, \frac{2}{1}$

7.1 *Ratio and Proportion* **361**

Assignment Guide

BASIC
Day 1: SRH p. 661 Exs. 7–12; pp. 361–363 Exs. 13–27, 32, 33, 35, 54, 59–61
Day 2: SRH p. 661 Exs. 16–21; pp. 361–363 Exs. 38–45, 50–53, 55–58, 62–67

AVERAGE
Day 1: pp. 361–363 Exs. 17–37, 54, 56–58
Day 2: pp. 361–363 Exs. 38–45, 47–53, 55, 59–67

ADVANCED
Day 1: pp. 361–363 Exs. 21–29, 32–36, 54–60 even
Day 2: pp. 361–363 Exs. 42–45, 46*, 47–51, 55–67 odd; EC: classzone.com

BLOCK SCHEDULE
pp. 361–363 Exs. 17–45, 47–67

Extra Practice
• Student Edition, p. 687
• Chapter 7 Resource Book, pp. 10–11

Homework Check
To quickly check student understanding of key concepts, go over the following exercises:

Basic: 21, 32, 35, 39, 50
Average: 22, 33, 37, 42, 53
Advanced: 24, 34, 36, 45, 51

Finding Ratios Use the number line to find the ratio of the segment lengths.

28. $AB : CD$ $\frac{2}{3}$ **29.** $BD : CF$ $\frac{1}{1}$ **30.** $BF : AD$ $\frac{11}{9}$ **31.** $CF : AB$ $\frac{7}{2}$

Using Ratios Find the segment lengths.

32. In the diagram, $FG : GH$ is $1 : 3$ and $FH = 12$. Find FG and GH. FG = 3; GH = 9

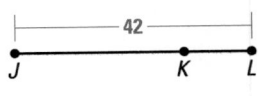

33. In the diagram, $JK : KL$ is $5 : 2$ and $JL = 42$. Find JK and KL. JK = 30; KL = 12

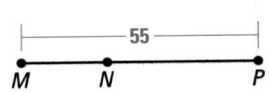

34. In the diagram, $MN : NP$ is $4 : 7$ and $MP = 55$. Find MN and NP. MN = 20; NP = 35

Using Ratios Sketch the rectangle described and label its sides using Example 3 as a model. Then find the length and the width of the rectangle. 35–37. Check sketches.

35. The perimeter of a rectangle is 110 inches. The ratio of the length to the width is $4 : 1$. length, 44 inches; width, 11 inches

36. The perimeter of a rectangle is 84 feet. The ratio of the width to the length is $2 : 5$. length, 30 feet; width, 12 feet

37. The perimeter of a rectangle is 132 meters. The ratio of the length to the width is $8 : 3$. length, 48 meters; width, 18 meters

Using Algebra In Exercises 38–45, solve the proportion.

38. $\frac{1}{2} = \frac{x}{22}$ 11   **39.** $\frac{6}{a} = \frac{2}{5}$ 15   **40.** $\frac{9}{4} = \frac{36}{z}$ 16   **41.** $\frac{b}{35} = \frac{8}{5}$ 56

42. $\frac{r+7}{60} = \frac{4}{15}$ 9   **43.** $\frac{s-1}{6} = \frac{1}{2}$ 4   **44.** $\frac{5}{3} = \frac{t+8}{18}$ 22   **45.** $\frac{12}{5x-3} = \frac{6}{11}$ 5

46. Challenge Solve the proportion $\frac{x}{4} = \frac{25}{x}$. 10, −10

Babe Ruth's Bat A sculpture of Babe Ruth's 35-inch bat is 120 feet long. Round your answers to the nearest tenth of an inch.

47. How long is the sculpture in inches? 1440 inches

48. The diameter of the sculpture near the base is 9 feet. Estimate the corresponding diameter of Babe Ruth's bat. about 2.6 inches

49. The diameter of the handle of the sculpture is 3.5 feet. Estimate the diameter of the handle of Babe Ruth's bat. about 1.0 inches

Link to
Sculpture

BABE RUTH'S BAT
A free-standing sculpture patterned after Babe Ruth's bat stands outside a sports museum in Louisville, Kentucky.

Student Help
CLASSZONE.COM

HOMEWORK HELP
Extra help with problem
solving in Exs. 50–53 is
at classzone.com

Using a Map Use the map of North Carolina and a ruler to estimate
the distance between the two cities.

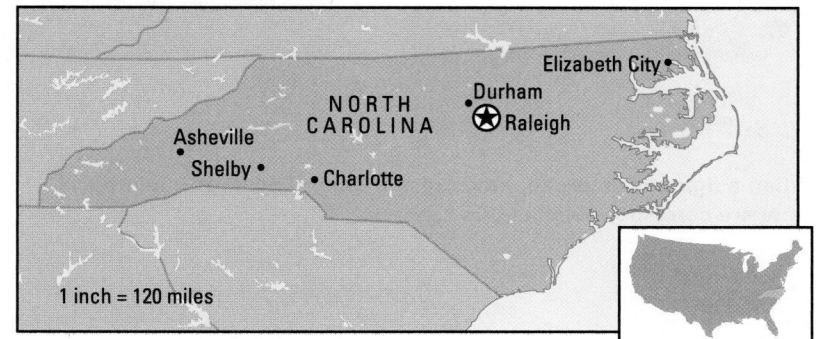

1 inch = 120 miles

50. Charlotte and Durham
  about 120 miles

51. Shelby and Elizabeth City
  about 300 miles

52. Elizabeth City and Asheville
  about 360 miles

53. Asheville and Shelby
  about 60 miles

**Standardized Test
Practice**

54. Multiple Choice The perimeter of a bedroom is 88 feet.
The ratio of the width to the length is 5 : 6. What are the
dimensions of the bedroom? B

  (A) 10 ft × 12 ft
  (B) 20 ft × 24 ft
  (C) 5 ft × 6 ft
  (D) 22 ft × 44 ft

55. Multiple Choice Solve the proportion $\frac{x+7}{6} = \frac{5}{3}$. H

  (F) 2
  (G) 2.5
  (H) 3
  (J) 10

Mixed Review

Identifying Congruent Parts In the diagram, $\triangle ABC \cong \triangle RQP$.
Complete the statement with the corresponding congruent part.
(Lesson 5.1)

56. $\angle Q \cong$ __?__ $\angle B$

57. $\angle C \cong$ __?__ $\angle P$

58. $\overline{AB} \cong$ __?__ $\overline{RQ}$

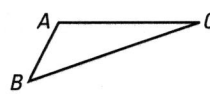
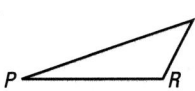

Identifying Polygons Decide whether the figure is a polygon.
If so, tell what type. If not, explain why. *(Lesson 6.1)*

59.
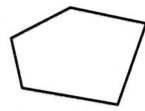
yes; pentagon

60.
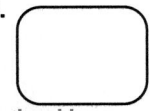
No; the sides are
not segments.

61.

yes; octagon

Algebra Skills

Decimal Operations Evaluate. *(Skills Review, p. 655)*

62. 4.22 + 1.07 5.29

63. 8.36 + 9.98 18.34

64. 7.2 − 2.4 4.8

65. 3.5 × 5.6 19.6

66. 7.35 ÷ 0.15 49

67. 15.12 ÷ 1.26 12

7.1 Ratio and Proportion 363

---

4 Assess

Assessment Resources
The Mini-Quiz below is also avail-
able on blackline (*Chapter 7
Resource Book*, p. 18) and on trans-
parency. For more assessment
resources, see:
  • Chapter 7 Resource Book
  • Standardized Test Practice
  • Test and Practice Generator

Mini-Quiz
Simplify the ratio.

1. $\frac{20\ m}{32\ m}$ $\frac{5}{8}$

2. $\frac{40\ min}{2\ h}$ $\frac{1}{3}$

3. In the diagram, $BC : CD$ is 5 : 7
and $BD = 48$. Find BC and CD.

$BC = 20$, $CD = 28$

Solve the proportion.

4. $\frac{4}{5} = \frac{x}{35}$ $x = 28$

5. $\frac{12}{7} = \frac{96}{y+2}$ $y = 54$

363

① Planning the Activity

Goal

Students explore the relationship between the measures of corresponding angles and the lengths of corresponding sides in similar figures.

Materials

See the margin of the student page.

LINK TO LESSON ⟹

In Example 1 on page 365, students show that they understand similar figures by identifying the corresponding sides and angles of two similar triangles.

② Managing the Activity

Tips for Success

Suggest that students find all the measures first before calculating any of the ratios. Point out that students should find the ratio of photo 1 to photo 2 since the activity deals with enlargement rather than reduction.

③ Closing the Activity

KEY DISCOVERY

When a figure is enlarged, the corresponding angles are congruent and the corresponding side lengths are proportional.

Activity Assessment

Use Exercises 1 and 2 to assess student understanding.

Activity 7.2 — Conjectures About Similarity

Question

When a figure is enlarged, how are corresponding angles related? How are corresponding lengths related?

Materials
- ruler
- calculator
- protractor

Explore

Photo 1 is an enlargement of Photo 2.

① Use a ruler to find the length of $\overline{AB}$ in each photo. Then use a calculator to find the ratio of AB in Photo 1 to AB in Photo 2. Round to the nearest tenth.

② Use a protractor to find $m\angle1$ in each photo. Then find the ratio of $m\angle1$ in Photo 1 to $m\angle1$ in Photo 2.

③ Continue finding measurements in the photos and record your results in a table like the one shown below.

Photo 1

Measurement	Photo 1	Photo 2	Ratio
AB	6 cm	5 cm	1.2
AF	? 4 cm	? 3.3 cm	? 1.2
CD	? 2.5 cm	? 2.1 cm	? 1.2
$m\angle1$	? 130°	? 130°	? 1
$m\angle2$	? 40°	? 40°	? 1

Photo 2

Student Help

LOOK BACK
For help measuring segments and angles, see pp. 28 and 36.

Think About It

1. Make a conjecture about the relationship between corresponding lengths when a figure is enlarged.
   When a figure is enlarged, pairs of corresponding lengths have the same ratio.
2. Make a conjecture about the relationship between corresponding angles when a figure is enlarged.
   When a figure is enlarged, corresponding angles have the same measure.
3. Suppose an angle in Photo 2 has a measure of 35°. What is the measure of the corresponding angle in Photo 1? 35°
4. **Extension** Suppose a segment in Photo 1 is 5 centimeters long. What is the measure of the corresponding segment in Photo 2? about 4.2 cm

7.2 Similar Polygons

① **Plan**

Pacing
Suggested Number of Days

Basic: 2 days
Average: 2 days
Advanced: 2 days
Block Schedule: 1 block

Teaching Resources

📄 **Blacklines**
(See page 354B.)

✒ **Transparencies**
• Warm-Up with Quiz
• Answers

▦ **Technology**
• Electronic Teacher Tools
• Test and Practice Generator
• Online Lesson Planner
• Internet Support

Goal
Identify similar polygons.

Key Words
• similar polygons
• scale factor

In geometry, two figures that have the same shape are called *similar*.

Two polygons are **similar polygons** if corresponding angles are congruent and corresponding side lengths are proportional.

In the diagram, *ABCD* is similar to *EFGH*. The symbol ~ indicates similarity. So, you can write *ABCD* ~ *EFGH*. When you refer to similar polygons, list their corresponding vertices in the same order.

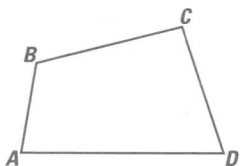
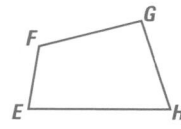

ABCD ~ EFGH

Corresponding Angles	Ratios of Corresponding Sides
$\angle A \cong \angle E$	$\dfrac{EF}{AB} = \dfrac{FG}{BC} = \dfrac{GH}{CD} = \dfrac{HE}{DA}$
$\angle B \cong \angle F$	
$\angle C \cong \angle G$	
$\angle D \cong \angle H$	

Student Help

VOCABULARY TIP
$\triangle PRQ \sim \triangle STU$ is called a *similarity statement*.

EXAMPLE ① Use Similarity Statements

▶ $\triangle PRQ \sim \triangle STU$.

  a. List all pairs of congruent angles.

  b. Write the ratios of the corresponding sides in a statement of proportionality.

  c. Check that the ratios of corresponding sides are equal.

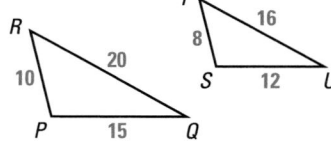

Solution

  a. $\angle P \cong \angle S$, $\angle R \cong \angle T$, and $\angle Q \cong \angle U$.

  b. $\dfrac{ST}{PR} = \dfrac{TU}{RQ} = \dfrac{US}{QP}$

  c. $\dfrac{ST}{PR} = \dfrac{8}{10} = \dfrac{4}{5}$, $\dfrac{TU}{RQ} = \dfrac{16}{20} = \dfrac{4}{5}$, and $\dfrac{US}{QP} = \dfrac{12}{15} = \dfrac{4}{5}$.

  The ratios of corresponding sides are all equal to $\dfrac{4}{5}$.

② **Teach**

Content and Teaching Strategies
For background information on geometric concepts and teaching strategies related to this lesson, see pages 354E and 354F in this Teacher's Edition.

Tips for New Teachers
When discussing Theorem 7.1, a question may arise concerning the ratio of areas of similar figures. To explore that idea, use a number of squares of different sizes. List the ratios of corresponding side lengths, the ratio of the perimeters, and the ratio of the areas to show that the ratio of the areas is the square of the ratio of corresponding side lengths. See the Tips for New Teachers on pp. 1–2 of the *Chapter 7 Resource Book* for additional notes about Lesson 7.2.

Extra Example 1
See next page.

7.2 *Similar Polygons*    365

366

Multiple Representations

While discussing the similar quadrilaterals shown above Example 1 on page 365, point out that the congruency statements describe the pairs of congruent angles in the two figures. Then discuss each of the ratios and have students point to the sides in the quadrilaterals as you read aloud the ratios.

Extra Example 1

$JKLM \sim NPQR$.

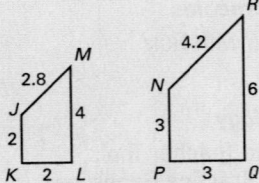

a. List all pairs of congruent angles. $\angle J \cong \angle N$, $\angle K \cong \angle P$, $\angle L \cong \angle Q$, and $\angle M \cong \angle R$

b. Write the ratios of the corresponding sides in a statement of proportionality.
$$\frac{JK}{NP} = \frac{KL}{PQ} = \frac{LM}{QR} = \frac{MJ}{RN}$$

c. Check that the ratios of the corresponding sides are equal.
$\frac{JK}{NP} = \frac{2}{3}$, $\frac{KL}{PQ} = \frac{2}{3}$, $\frac{LM}{QR} = \frac{2}{3}$, and and $\frac{MJ}{RN} = \frac{2}{3}$. The ratios of corresponding sides are all equal to $\frac{2}{3}$.

Geometric Reasoning

Some students may quickly deduce that if the scale factor is less than 1, the second figure is smaller. Inform students that reductions, enlargements, and how they relate to the value of scale factors are covered in depth in Lesson 7.6.

Scale Factor If two polygons are similar, then the ratio of the lengths of two corresponding sides is called the **scale factor**.

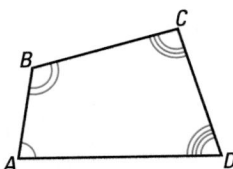 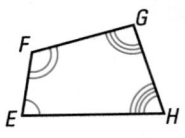

$$\text{scale factor of } EFGH \text{ to } ABCD = \frac{EF}{AB} = \frac{FG}{BC} = \frac{GH}{CD} = \frac{HE}{DA}$$

Student Help
CLASSZONE.COM
MORE EXAMPLES
More examples at classzone.com

EXAMPLE 2 Determine Whether Polygons are Similar

Determine whether the triangles are similar. If they are similar, write a similarity statement and find the scale factor of Figure B to Figure A.

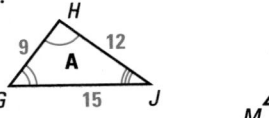 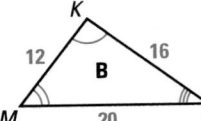

Solution

❶ Check whether the corresponding angles are congruent.

From the diagram, you can see that $\angle G \cong \angle M$, $\angle H \cong \angle K$, and $\angle J \cong \angle L$. Therefore, the corresponding angles are congruent.

❷ Check whether the corresponding side lengths are proportional.

$$\frac{MK}{GH} = \frac{12}{9} = \frac{12 \div 3}{9 \div 3} = \frac{4}{3}$$

$$\frac{KL}{HJ} = \frac{16}{12} = \frac{16 \div 4}{12 \div 4} = \frac{4}{3}$$

All three ratios are equal, so the corresponding side lengths are proportional.

$$\frac{LM}{JG} = \frac{20}{15} = \frac{20 \div 5}{15 \div 5} = \frac{4}{3}$$

ANSWER ▶ By definition, the triangles are similar. $\triangle GHJ \sim \triangle MKL$.

The scale factor of Figure B to Figure A is $\frac{4}{3}$.

Checkpoint ✓ **Determine Whether Polygons are Similar**

Determine whether the polygons are similar. If they are similar, write a similarity statement and find the scale factor of Figure B to Figure A.

1. yes; $\triangle XYZ \sim \triangle DEF$; $\frac{3}{2}$

2. no $\left(\frac{9}{6} \neq \frac{12}{10}\right)$

1.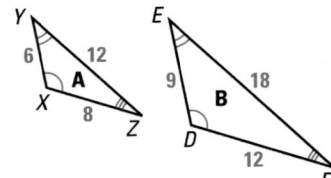

2.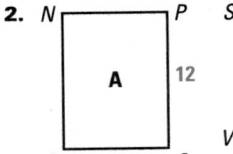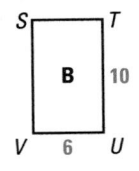

EXAMPLE 3 Use Similar Polygons

$\triangle RST \sim \triangle GHJ$.
Find the value of x.

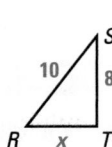
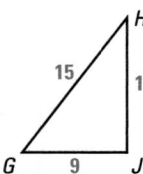

Solution

Because the triangles are similar, the corresponding side lengths are proportional. To find the value of x, you can use the following proportion.

$\dfrac{GH}{RS} = \dfrac{JG}{TR}$ Write proportion.

$\dfrac{15}{10} = \dfrac{9}{x}$ Substitute 15 for GH, 10 for RS, 9 for JG, and x for TR.

$15 \cdot x = 10 \cdot 9$ Cross product property

$15x = 90$ Multiply.

$\dfrac{15x}{15} = \dfrac{90}{15}$ Divide each side by 15.

$x = 6$ Simplify.

Student Help

STUDY TIP
You can find x using other proportions that include TR. Another proportion you can use is $\dfrac{ST}{HJ} = \dfrac{TR}{JG}$.

EXAMPLE 4 Perimeters of Similar Polygons

The outlines of a pool and the patio around the pool are similar rectangles.

a. Find the ratio of the length of the patio to the length of the pool.

b. Find the ratio of the perimeter of the patio to the perimeter of the pool.

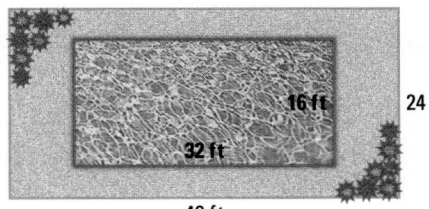

Solution

a. The ratio of the length of the patio to the length of the pool is

$$\frac{\text{length of patio}}{\text{length of pool}} = \frac{48 \text{ feet}}{32 \text{ feet}} = \frac{48 \div 16}{32 \div 16} = \frac{3}{2}.$$

b. The perimeter of the patio is $2(24) + 2(48) = 144$ feet.

The perimeter of the pool is $2(16) + 2(32) = 96$ feet.

The ratio of the perimeter of the patio to the perimeter of the pool is

$$\frac{\text{perimeter of patio}}{\text{perimeter of pool}} = \frac{144 \text{ feet}}{96 \text{ feet}} = \frac{144 \div 48}{96 \div 48} = \frac{3}{2}.$$

Extra Example 2

Determine whether the triangles are similar. If they are similar, write a similarity statement and find the scale factor of Figure B to Figure A.

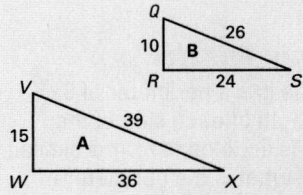

By definition, the triangles are similar: $\triangle QRS \sim \triangle VWX$. The scale factor of Figure B to Figure A is $\dfrac{2}{3}$.

Extra Example 3

$ABCD \sim KLMN$. Find the value of x.

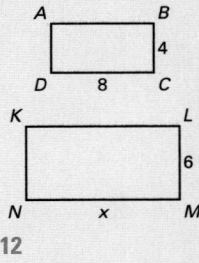

12

Extra Example 4

A triangular work of art and the frame around it are similar triangles.

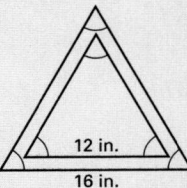

a. Find the ratio of the length of one side of the art to the length of one side of the frame. $\dfrac{3}{4}$

b. Find the ratio of the perimeter of the art to the perimeter of the frame. $\dfrac{3}{4}$

Concept Check

What are similar polygons?
Two polygons are similar if corresponding angles are congruent and corresponding side lengths are proportional.

Daily Puzzler

A triangle has a perimeter of $6x$. If the length of each side of the triangle is doubled to form a similar triangle, what is the perimeter of the larger triangle? **12x**

In Example 4 on the previous page, notice that the ratio of the perimeters of the similar figures is equal to the ratio of the side lengths. This observation is generalized in the following theorem.

THEOREM 7.1

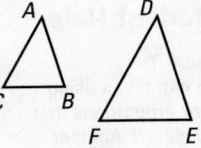

Perimeters of Similar Polygons

Words If two polygons are similar, then the ratio of their perimeters is equal to the ratio of their corresponding side lengths.

Symbols If $\triangle ABC \sim \triangle DEF$, then $\dfrac{DE + EF + FD}{AB + BC + CA} = \dfrac{DE}{AB} = \dfrac{EF}{BC} = \dfrac{FD}{CA}$.

 **Use Similar Polygons**

In the diagram, $\triangle PQR \sim \triangle STU$.

3. Find the value of x. **18**

4. Find the ratio of the perimeter of $\triangle STU$ to the perimeter of $\triangle PQR$. $\frac{1}{2}$

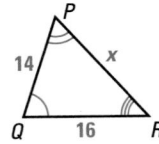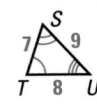

7.2 Exercises

Guided Practice

Vocabulary Check

1. If two triangles are *congruent*, must they be *similar*? Explain.

2. If two triangles are *similar*, must they be *congruent*? Explain.
   No; two similar triangles are congruent only if their scale factor is $\frac{1}{1}$.

Skill Check

1. Yes; congruent triangles have congruent corresponding angles and corresponding side lengths have the ratio $\frac{1}{1}$.

4. *Sample answer:*
$\dfrac{LM}{AB} = \dfrac{MN}{BC} = \dfrac{NL}{CA}$

In Exercises 3–6, $\triangle ABC \sim \triangle LMN$.

3. List all pairs of congruent angles.
   $\angle A \cong \angle L; \angle B \cong \angle M; \angle C \cong \angle N$

4. Write the ratios of the corresponding sides in a statement of proportionality.

5. Find the scale factor of $\triangle LMN$ to $\triangle ABC$. $\frac{3}{4}$

6. Find the value of x. **36**

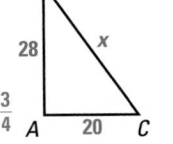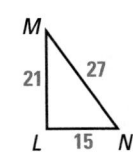

7. Are the two rectangles shown at the right similar? Explain your reasoning.
   no; $\frac{7}{15} \neq \frac{4}{10}$

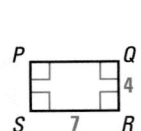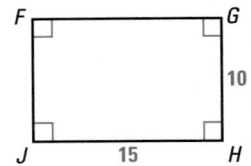

Practice and Applications

Extra Practice

See p. 687.

8. Sample answer:

$\angle P \cong \angle D; \angle Q \cong \angle E;$
$\angle R \cong \angle F; \dfrac{PQ}{DE} = \dfrac{QR}{EF} = \dfrac{RP}{FD}$

9. Sample answer:

$\angle A \cong \angle Q; \angle B \cong \angle R;$
$\angle C \cong \angle S; \angle D \cong \angle T;$
$\angle E \cong \angle U;$
$\dfrac{AB}{QR} = \dfrac{BC}{RS} = \dfrac{CD}{ST} = \dfrac{DE}{TU} = \dfrac{EA}{UQ}$

10. $\angle F \cong \angle J; \angle G \cong \angle K;$
$\angle H \cong \angle L;$
$\dfrac{FG}{JK} = \dfrac{GH}{KL} = \dfrac{FH}{JL}$

16. yes; $\triangle RST \sim \triangle YZX; \dfrac{2}{3}$

Using Similarity Statements List all pairs of congruent angles. Then write the ratios of the corresponding sides in a statement of proportionality.

8. $\triangle PQR \sim \triangle DEF$

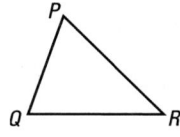

9. $ABCDE \sim QRSTU$

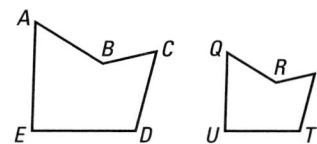

10. Error Analysis $\triangle FGH \sim \triangle JKL$. A student was asked to list all pairs of congruent angles and write the ratios of the corresponding sides in a statement of proportionality. Copy the diagram and correct the student's errors.

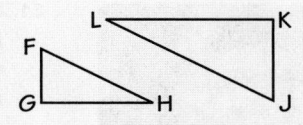

$\angle F \cong \angle J, \angle G \cong \angle L, \angle H \cong \angle K$

$\dfrac{FG}{JK} = \dfrac{KL}{GH} = \dfrac{FH}{JL}$ ✕

Determining Similarity Determine whether the polygons are similar. If they are similar, write a similarity statement and find the scale factor of Figure B to Figure A.

11.

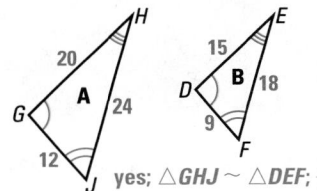

yes; $\triangle GHJ \sim \triangle DEF; \dfrac{3}{4}$

12.

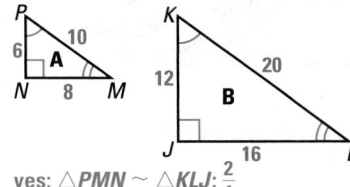

yes; $\triangle PMN \sim \triangle KLJ; \dfrac{2}{1}$

13.

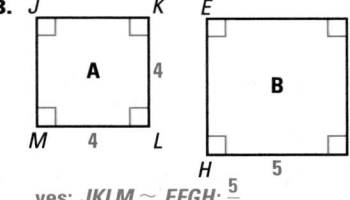

yes; $JKLM \sim EFGH; \dfrac{5}{4}$

14.

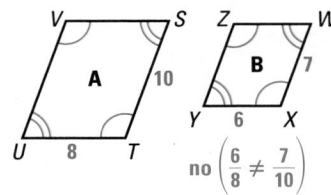

no $\left(\dfrac{6}{8} \neq \dfrac{7}{10} \right)$

15.

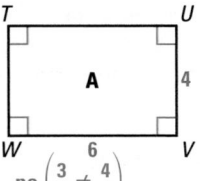

no $\left(\dfrac{3}{4} \neq \dfrac{4}{6} \right)$

16.

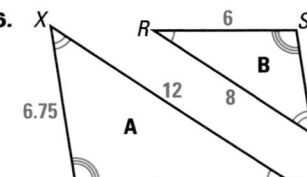

Homework Help

Example 1: Exs. 8–10, 22, 23
Example 2: Exs. 11–16, 29
Example 3: Exs. 17–21, 24, 25
Example 4: Exs. 26, 27

❸ Apply

Assignment Guide

BASIC
Day 1: SRH p. 656 Exs. 6–10; pp. 369–371 Exs. 8, 9, 11–16, 22, 23, 34–37
Day 2: SRH p. 661 Exs. 22–27; pp. 369–371 Exs. 17–20, 24–27, 38–41

AVERAGE
Day 1: pp. 369–371 Exs. 8–16, 22, 23, 29–34
Day 2: pp. 369–371 Exs. 17–21, 24–27, 35–41

ADVANCED
Day 1: pp. 369–371 Exs. 8–16 even, 22, 23, 29–34
Day 2: pp. 369–371 Exs. 17–21 odd, 24–27, 28*, 35–41; EC: classzone.com

BLOCK SCHEDULE
pp. 369–371 Exs. 8–27, 29–41

Extra Practice

• Student Edition, p. 687
• Chapter 7 Resource Book, pp. 22–23

Homework Check

To quickly check student understanding of key concepts, go over the following exercises:

Basic: 8, 12, 17, 22, 26
Average: 9, 14, 19, 23, 26
Advanced: 10, 16, 21, 23, 27

Using Similar Polygons The two polygons are similar. Find the values of *x* and *y*.

17.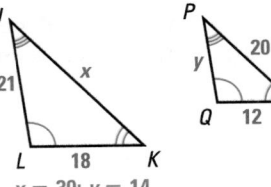
x = 30; y = 14

18.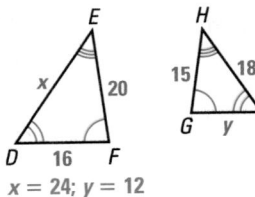
x = 24; y = 12

19.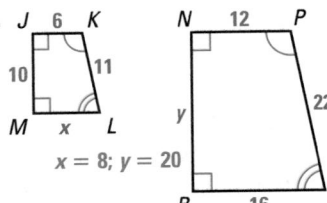
x = 8; y = 20

20.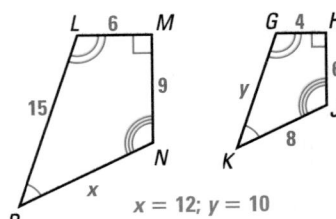
x = 12; y = 10

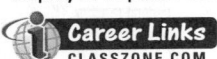
21. Mural Alejandro Romero created the mural, *Chicago Federation of Labor*, by enlarging the 56 in. by 21 in. sketch shown below. Romero used a scale factor of about 3.5. What are the dimensions of the mural in inches? In feet?   196 in. by 73.5 in.; 16 ft 4 in. by 6 ft 1½ in.

Using Similar Polygons In Exercises 22–26, use the diagram below, where *FGHJ* ~ *KLMN*.

22. Find m∠N. 90°

23. Find m∠F. 80°

24. Find the value of *x*. 6

25. Find the value of *y*. 15

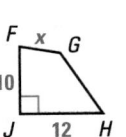 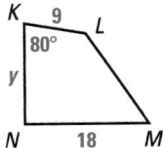

26. Find the ratio of the perimeter of *KLMN* to the perimeter of *FGHJ*. $\frac{3}{2}$

27. Perimeters of Similar Polygons △*QRS* is similar to △*XYZ*. The ratio of one side of △*XYZ* to the corresponding side of △*QRS* is 6 : 7. What is the ratio of the perimeter of △*XYZ* to the perimeter of △*QRS*? 6 : 7

28. Challenge △*JKL* is similar to △*STU*. The ratio of *ST* to *JK* is 5 to 2. The perimeter of △*STU* is 35 feet. Find the perimeter of △*JKL*. 14 feet

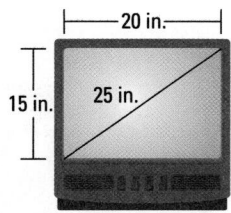

Link to

Technology

← 20 in. →

15 in.   25 in.

TELEVISION SCREENS
have sizes that are based on
the length of their diagonal.
The *aspect ratio* of a 25-inch
standard definition television
screen is 20 in. : 15 in., or 4 : 3.

29. No; the ratio $\frac{\text{length}}{\text{length}} = \frac{4}{16} = \frac{1}{4}$
and the ratio $\frac{\text{width}}{\text{width}} = \frac{3}{9} = \frac{1}{3}$;
$\frac{1}{4} \neq \frac{1}{3}$.

---

29. Television Screens The *aspect ratio* of a television screen is the
length-to-width ratio of the screen. A standard definition television has
an aspect ratio of 4 : 3. A high definition projection television has an
aspect ratio of 16 : 9. Are the television screens similar rectangles?
See margin.

EXAMPLE Logical Reasoning

Are two isosceles trapezoids *always, sometimes,* or *never* similar?

Solution

It is possible to sketch two
similar isosceles trapezoids.

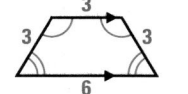 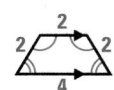

It is also possible to sketch
two isosceles trapezoids that
are *not* similar.

 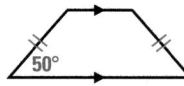

ANSWER ▶ Therefore, two isosceles trapezoids are sometimes similar.

Logical Reasoning Are the polygons *always, sometimes,* or
never similar?

30. Two isosceles triangles sometimes **31.** Two rhombuses sometimes

32. Two equilateral triangles always **33.** A right and an isosceles triangle
                                                        sometimes

Standardized Test Practice

34. Multiple Choice $\triangle DEF \sim \triangle MNP$. Which statement may be *false*? B

A $\angle E \cong \angle N$ **B** $\angle P \cong \angle D$

C $\frac{MN}{DE} = \frac{NP}{EF}$ **D** $\frac{MN}{DE} = \frac{PM}{FD}$

Mixed Review

Showing Triangles are Congruent Does the diagram give enough
information to show that the triangles are congruent? If so, state
the postulate or theorem you would use. *(Lessons 5.3, 5.4)*

35.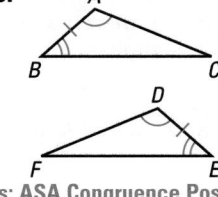

36.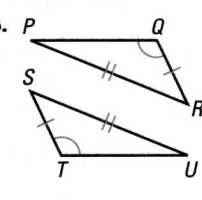
no

37.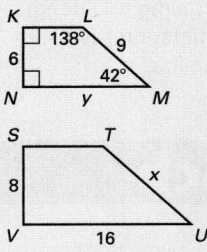
yes; HL Congruence 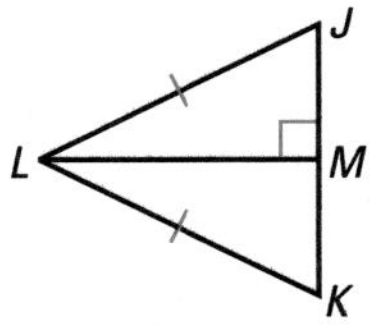
Theorem or AAS
Congruence Theorem

yes; ASA Congruence Postulate

Algebra Skills

Writing Equivalent Fractions Write two equivalent fractions.
(Skills Review, p. 656) 38–41. Sample answers are given.

38. $\frac{1}{3}$ $\frac{2}{6}$, $\frac{3}{9}$ **39.** $\frac{2}{5}$ $\frac{4}{10}$, $\frac{20}{50}$ **40.** $\frac{4}{7}$ $\frac{8}{14}$, $\frac{20}{35}$ **41.** $\frac{9}{4}$ $\frac{18}{8}$, $\frac{54}{24}$

7.2 Similar Polygons 371

---

4 Assess

Assessment Resources

The Mini-Quiz below is also avail-
able on blackline (*Chapter 7
Resource Book*, p. 30) and on trans-
parency. For more assessment
resources, see:
• Chapter 7 Resource Book
• Standardized Test Practice
• Test and Practice Generator

Mini-Quiz

1. Determine whether the triangles
   are similar. If they are similar,
   write a similarity statement and
   find the scale factor of Figure B
   to Figure A.

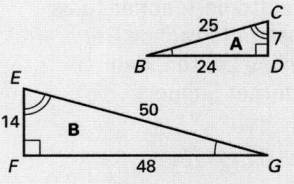

$\triangle BCD \sim \triangle GEF$. The scale factor
of Figure B to Figure A is 2.

In the diagram, *KLMN ~ STUV*.

2. Find $m\angle T$. 138°
3. Find $m\angle V$. 90°
4. Find the value of *x*. 12
5. Find the value of *y*. 12

371

1 Plan

Pacing
Suggested Number of Days

Basic: 2 days
Average: 2 days
Advanced: 2 days
Block Schedule: 1 block

Teaching Resources

📄 **Blacklines**
(See page 354B.)

✍ **Transparencies**
• Warm-Up with Quiz
• Answers

💻 **Technology**
• Electronic Teacher Tools
• Test and Practice Generator
• Online Lesson Planner
• Internet Support

Geo-Activity

Goal Use a protractor and ruler to draw similar triangles using just the measures of two corresponding angles.

Key Discovery Having two congruent angles is sufficient for two triangles to be similar.

2 Teach
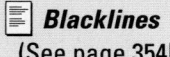

─ **Content and Teaching Strategies**
For background information on geometric concepts and teaching strategies related to this lesson, see pages 354E and 354F in this Teacher's Edition.

7.3 Showing Triangles are Similar: AA

Goal
Show that two triangles are similar using the AA Similarity Postulate.

Key Words
• similar polygons p. 365

Step 5. Yes; the corresponding angles are congruent and the corresponding side lengths are proportional.

Geo-Activity ▸ **Angles and Similar Triangles**

❶ Use a protractor to draw a triangle that has a 40° angle and a 60° angle. Label the triangle △ABC.

❷ Use a protractor to draw a larger triangle that has a 40° and a 60° angle. Label this triangle △DEF.

❸ Use a protractor to measure the third angle of each triangle. It should measure 80°. Does it?
   yes

❹ Use a ruler to measure the lengths of the sides of both triangles. Record your results.
   Answers will vary.

❺ Are the triangles similar? Explain your reasoning.

POSTULATE 15

Angle-Angle Similarity Postulate (AA)

Words If two angles of one triangle are congruent to two angles of another triangle, then the two triangles are similar.

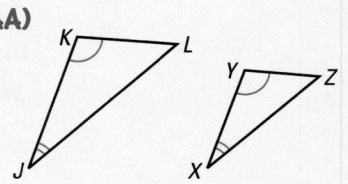

Symbols If ∠K ≅ ∠Y and ∠J ≅ ∠X, then △JKL ~ △XYZ.

This postulate allows you to say that two triangles are similar if you know that two pairs of angles are congruent. In other words, you don't need to compare all of the side lengths and angle measures to show that two triangles are similar.

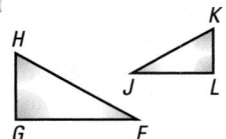

Use colored pencils to show congruent angles. This will help you write similarity statements.

EXAMPLE 1 Use the AA Similarity Postulate

Determine whether the triangles are similar. If they are similar, write a similarity statement. Explain your reasoning.

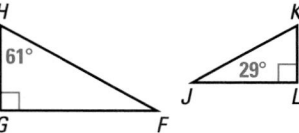

Solution

If two pairs of angles are congruent, then the triangles are similar.

❶ $\angle G \cong \angle L$ because they are both marked as right angles.

❷ Find $m\angle F$ to determine whether $\angle F$ is congruent to $\angle J$.

$m\angle F + 90° + 61° = 180°$	Triangle Sum Theorem
$m\angle F + 151° = 180°$	Add.
$m\angle F = 29°$	Subtract 151° from each side.

Both $\angle F$ and $\angle J$ measure 29°, so $\angle F \cong \angle J$.

ANSWER ▶ By the AA Similarity Postulate, $\triangle FGH \sim \triangle JLK$.

EXAMPLE 2 Use the AA Similarity Postulate

Are you given enough information to show that $\triangle RST$ is similar to $\triangle RUV$? Explain your reasoning.

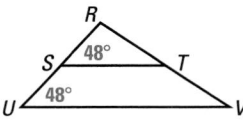

Student Help

VISUAL STRATEGY
Redraw overlapping triangles as two separate triangles, as shown on p. 356.

Solution

Redraw the diagram as two triangles: $\triangle RUV$ and $\triangle RST$.

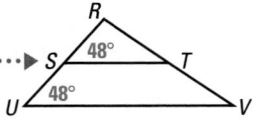 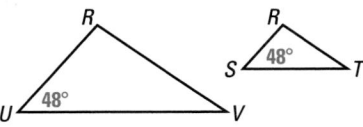

From the diagram, you know that both $\angle RST$ and $\angle RUV$ measure 48°, so $\angle RST \cong \angle RUV$. Also, $\angle R \cong \angle R$ by the Reflexive Property of Congruence. By the AA Similarity Postulate, $\triangle RST \sim \triangle RUV$.

Checkpoint ✓ Use the AA Similarity Postulate

Determine whether the triangles are similar. If they are similar, write a similarity statement.

1.
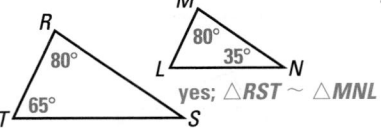
yes; $\triangle RST \sim \triangle MNL$

2.
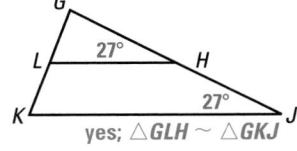
yes; $\triangle GLH \sim \triangle GKJ$

7.3 *Showing Triangles are Similar: AA* 373

Tips for New Teachers
A discussion of the Geo-Activity on page 372 leads nicely into Postulate 15. Lead students to the conclusion that they have created similar triangles by knowing only that two angles of one triangle are congruent to two angles of another triangle. See the Tips for New Teachers on pp. 1–2 of the *Chapter 7 Resource Book* for additional notes about Lesson 7.3.

Extra Example 1
Determine whether the triangles are similar. If they are similar, write a similarity statement. Explain your reasoning.

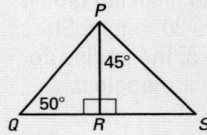

$m\angle S = 45°$ and $m\angle QPR = 40°$ by the Triangle Sum Theorem. The triangles do not have two congruent angles and therefore are not similar.

Extra Example 2
Are you given enough information to show that $\triangle UVY$ is similar to $\triangle UWX$? Explain your reasoning.

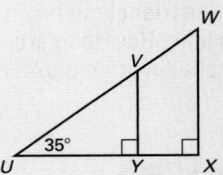

From the diagram, you know that $\angle UYV \cong \angle UXW$ because they are both marked as right angles. Also, $\angle U \cong \angle U$ by the Reflexive Property of Congruence. By the AA Similarity Postulate, $\triangle UVY \sim \triangle UWX$.

Visualize It!
Students may wish to trace similar figures using colored pencils for corresponding segments to help them identify the correct pairs of corresponding parts when writing similarity statements.

374

Geometric Reasoning

Point out that two similar triangles have corresponding congruent angles even if one triangle is much larger than the other. Some students may occasionally think that the angles in a very large figure must have greater angle measures than those of the similar smaller figure. On the classroom floor, use masking tape to model a 30° angle. Ask students to imagine that the other two ends of the tape are the vertices of a triangle. Now extend the masking tape as far as possible and ask students if the measure of the 30° angle changed when the sides of the angle were made longer. Point out that no matter how long each piece of masking tape is, they still meet at a 30° angle. Stress that the measure of the angles do not change when a triangle is enlarged.

Extra Example 3

At camp, Celeste bends a thin stick 2.5 inches long to form an isosceles triangle with a base of 1 inch. She shines her flashlight at the triangle and notices that it forms a shadow on the side of her tent. The shadow is similar to the triangle formed by the stick. She estimates the base of the shadow of the triangle to be about 6 inches long. How long are the legs of the shadow triangle?
about 4.5 in.

✓ Concept Check

Explain how to use the AA Similarity Postulate. If two angles of one triangle can be shown congruent to two angles of another triangle, then the two triangles are similar.

🌑 Daily Puzzler

△ABC ~ △DEF and each side of △DEF is four times the length of its corresponding side of △ABC. What is the ratio of the perimeter of △ABC to the perimeter of △DEF?
1 : 4

 **Using Algebra**

Student Help

> **STUDY TIP**
> In problems like Example 3, you must show that the triangles are similar before you can write and solve the proportion. ·········

EXAMPLE 3 **Use Similar Triangles**

A hockey player passes the puck to a teammate by bouncing the puck off the wall of the rink, as shown below. According to the laws of physics, the angles that the path of the puck makes with the wall are congruent. How far from the wall will the teammate pick up the pass?

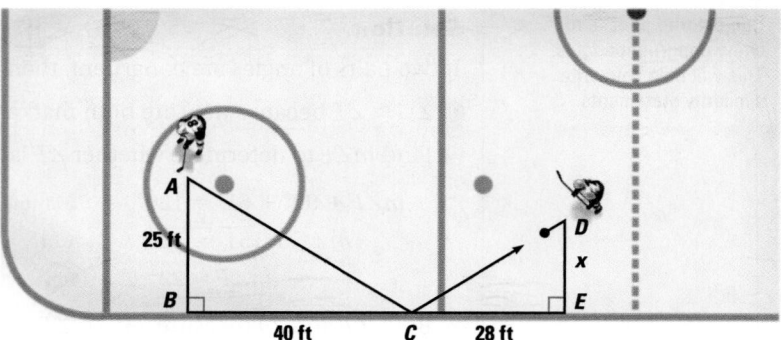

Solution

From the diagram, you know that $\angle B \cong \angle E$. From the laws of physics given in the problem, $\angle ACB \cong \angle DCE$. Therefore, $\triangle ABC \sim \triangle DEC$ by the AA Similarity Postulate.

$\dfrac{DE}{AB} = \dfrac{EC}{BC}$	Write a proportion.
$\dfrac{x}{25} = \dfrac{28}{40}$	Substitute x for DE, 25 for AB, 28 for EC, and 40 for BC.
$x \cdot 40 = 25 \cdot 28$	Cross product property
$40x = 700$	Multiply.
$\dfrac{40x}{40} = \dfrac{700}{40}$	Divide each side by 40.
$x = 17.5$	Simplify.

ANSWER ▶ The teammate will pick up the pass 17.5 feet from the wall.

Checkpoint ✓ **Use Similar Triangles**

Write a similarity statement for the triangles. Then find the value of the variable.

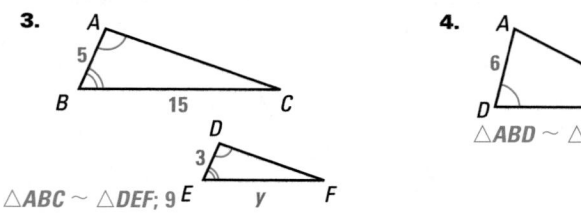

3. △ABC ~ △DEF; 9

4. △ABD ~ △EBC; 3

Guided Practice

Vocabulary Check

1. Complete the statement: If two angles of one triangle are congruent to two angles of another triangle, then __?__.

the triangles are similar

Skill Check

Determine whether the triangles are similar. If they are similar, write a similarity statement. Explain your reasoning.

3. Yes; ∠D ≅ ∠G and ∠F ≅ ∠H, so △DEF ~ △GJH by the AA Similarity Postulate.

4. Yes; ∠L ≅ ∠P and ∠LMK ≅ ∠PMN, so △LMK ~ △PMN by the AA Similarity Postulate.

5. Yes; ∠Q ≅ ∠Q and ∠QRT ≅ ∠QSU, so △QRT ~ △QSU by the AA Similarity Postulate.

2.
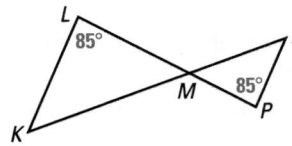
No; only one pair of angles is congruent.

3.
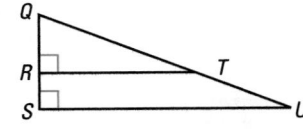

4.
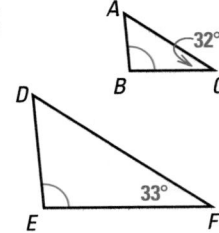

5.
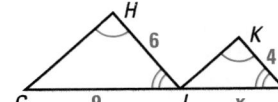

6. Write a similarity statement for the triangles. Then find the value of *x*.
△HJG ~ △KLJ; 6

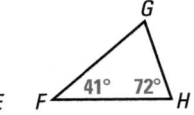

Practice and Applications

Extra Practice

See p. 687.

11. yes; △XYZ ~ △GFH

Homework Help

Example 1: Exs. 7–12, 31
Example 2: Exs. 13–15, 37
Example 3: Exs. 16–19, 21–26, 30–36

Using the AA Similarity Postulate **Determine whether the triangles are similar. If they are similar, write a similarity statement.**

7.
no

8.
no

9.
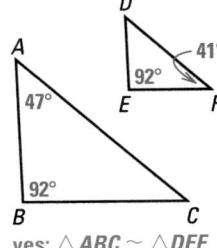
yes; △ABC ~ △DEF

10.
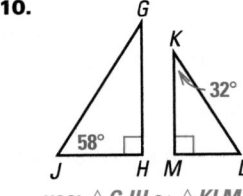
yes; △GJH ~ △KLM

11.
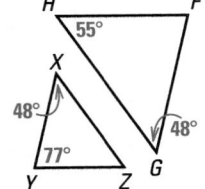

12.
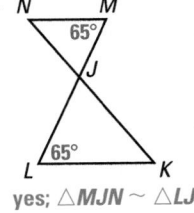
yes; △MJN ~ △LJK

7.3 *Showing Triangles are Similar: AA* **375**

③ Apply

Assignment Guide

BASIC
Day 1: pp. 375–378 Exs. 7–15, 39–53
Day 2: SRH p. 661 Exs. 28–30; pp. 375–378 Exs. 16–26, 38, Quiz 1

AVERAGE
Day 1: pp. 375–378 Exs. 7–15, 27–29, 37, 39–49
Day 2: pp. 375–378 Exs. 16–26, 30, 38, Quiz 1

ADVANCED
Day 1: pp. 375–378 Exs. 10–15, 27–29, 37, 39–49
Day 2: pp. 375–378 Exs. 21–26, 30, 31–36*, 38, Quiz 1; EC: classzone.com

BLOCK SCHEDULE
pp. 375–378 Exs. 7–30, 37–49, Quiz 1

Extra Practice
• Student Edition, p. 687
• Chapter 7 Resource Book, pp. 31–32

Homework Check
To quickly check student understanding of key concepts, go over the following exercises:

Basic: 7, 10, 13, 16, 21
Average: 8, 10, 15, 17, 23
Advanced: 11, 14, 24, 26, 37

✗ Common Error
In Exercises 11 and 12, watch for students who have difficulty identifying corresponding sides and angles when one figure of a pair is rotated. Have students trace one of the two figures, rotate the tracing, and then mark the corresponding sides and angles on the tracing. Students should then use the other original triangle to determine if similarity exists.

X *Common Error*

To avoid making errors in setting up proportions in situations like Exercise 25, students may want to redraw overlapping triangles separately, labeling each one individually. The four terms of the proportion will then be easier to read from the diagrams. In Exercise 25, a common error is to write the proportion $\frac{10}{14} = \frac{15}{x}$, when the correct proportion is $\frac{10}{14} = \frac{25}{x}$.

Teaching Tip

In Exercises 27–29, encourage students to make sketches of the triangles before deciding whether each statement is true or false.

13. Yes; $\angle U \cong \angle U$ and $\angle UVW \cong \angle S$, so $\triangle UVW \sim \triangle UST$ by the AA Similarity Postulate.

15. Yes; *Sample answer:* $\overline{DE} \parallel \overline{BA}$, so $\angle CED \cong \angle A$ and $\angle CDE \cong \angle B$; $\triangle CDE \sim \triangle CBA$ by the AA Similarity Postulate.

Student Help
CLASSZONE.COM

HOMEWORK HELP
Extra help with problem solving in Exs. 21–26 is at classzone.com

Using the AA Similarity Postulate **Determine whether you can show that the triangles are similar. If they are similar, write a similarity statement. Explain your reasoning.**

13.

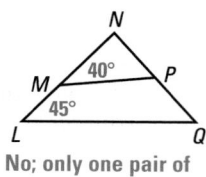

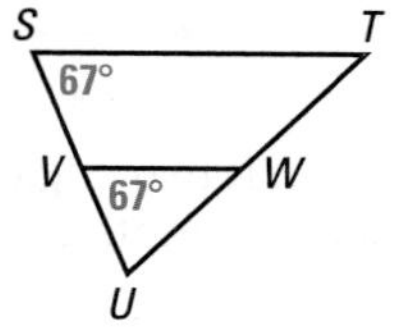

14.

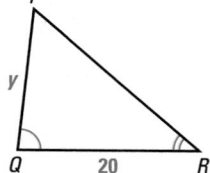

No; only one pair of angles is congruent.

15.

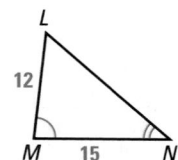

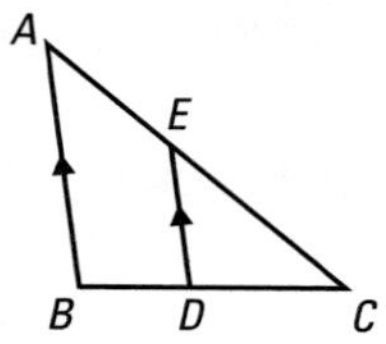

Similar Triangles **Use the diagram to complete the statement.**

16. $\triangle PQR \sim$ __?__ $\triangle LMN$

17. $\frac{LM}{PQ} = \frac{?}{QR}$ MN

18. $\frac{12}{y} = \frac{15}{?}$ 20

19. $y =$ __?__ 16

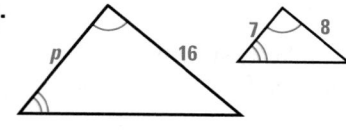

20. The scale factor of $\triangle LMN$ to $\triangle PQR$ is __?__. $\frac{3}{4}$

Using Similar Triangles **Find the value of the variable.**

21. 14

22. 36

23. 24

24. 24

25. 35

26. 20

Logical Reasoning **Decide whether the statement is *true* or *false*.**

27. If an acute angle of a right triangle is congruent to an acute angle of another right triangle, then the triangles are similar. true

28. Some equilateral triangles are not similar. false

29. All isosceles triangles with a 40° vertex angle are similar. true

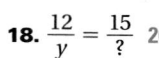

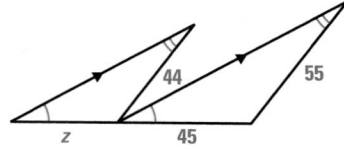

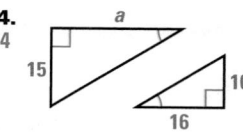

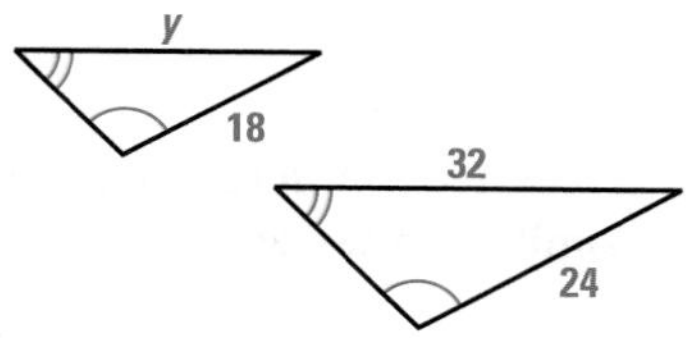

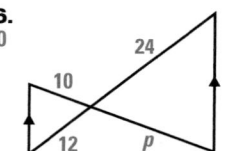

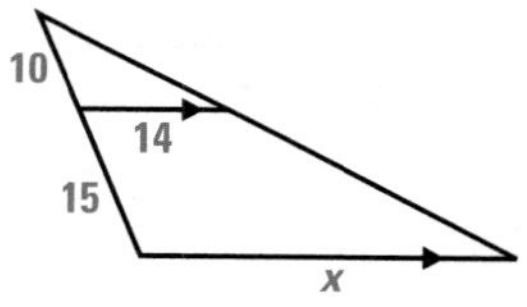

30. Unisphere To estimate the height of the Unisphere, you place a mirror on the ground and stand where you can see the top of the model in the mirror, as shown in the diagram. Write and solve a proportion to estimate the height of the Unisphere.   140 ft

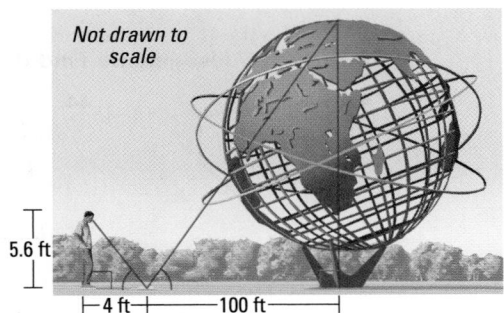

Not drawn to scale

5.6 ft

├─4 ft─┤   ─100 ft─┤

Challenge *ABCD* is a trapezoid, *AB* = 8, *AE* = 6, *EC* = 15, and *DE* = 10. Complete the statement.

31. $\triangle ABE \sim \underline{\quad ? \quad}$ $\triangle CDE$

32. $\dfrac{AB}{? \ CD} = \dfrac{AE}{? \ CE} = \dfrac{BE}{? \ DE}$

33. $\dfrac{6}{? \ 15} = \dfrac{8}{? \ x}$

34. $\dfrac{15}{? \ 6} = \dfrac{10}{? \ y}$

35. $x = \underline{\quad ? \quad}$ 20

36. $y = \underline{\quad ? \quad}$ 4

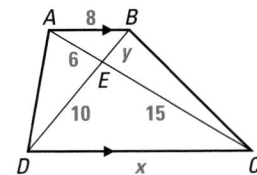

37. Meredith is right; $m\angle ACB = 55°$, so $\angle ACB \cong \angle ADE$; $\angle A \cong \angle A$, so $\triangle ABC \sim \triangle AED$ by the AA Similarity Postulate.

37. **You be the Judge** Meredith claims that the triangles shown at the right are similar. Brian thinks that they are not similar. Who is right? Explain your reasoning.

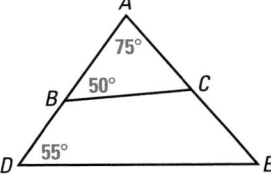

38. a. Julia and the flagpole are both perpendicular to the ground, and the overlapping triangles share a common angle. Therefore, the overlapping triangles are similar by the AA Similarity Postulate.

b. $\dfrac{5}{7} = \dfrac{x}{35}$

c. **25 feet**

38. Multi-Step Problem Julia uses the shadow of a flagpole to estimate its height. She stands so that the tip of her shadow coincides with the tip of the flagpole's shadow as shown.

a. Explain why the two overlapping triangles in the diagram are similar.

b. Using the similar triangles, write a proportion that models the situation.

c. Solve the proportion to calculate the height of the flagpole.

5 ft

7 ft   28 ft

Assessment Resources

The Mini-Quiz below is also available on blackline (*Chapter 7 Resource Book*, p. 40) and on transparency. For more assessment resources, see:
- Chapter 7 Resource Book
- Standardized Test Practice
- Test and Practice Generator

Mini-Quiz

Determine whether the triangles are similar. If they are similar, write a similarity statement.

1.

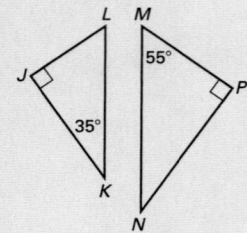

$\triangle JKL \sim \triangle PNM$

2.

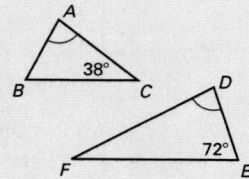

$\triangle ABC$ is not similar to $\triangle DEF$.

3. Find the value of *x*. 15

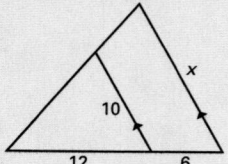

46–53. *Sample answer:*

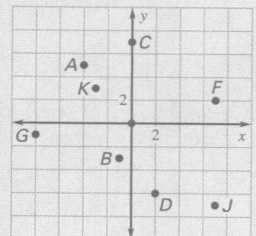

Mixed Review

Congruent Triangles In the diagram, $\triangle FGH \cong \triangle RST$. Complete the statement. *(Lesson 5.1)*

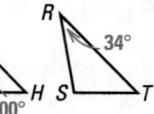

39. $m\angle F = $ __?__ ° 34
40. $m\angle T = $ __?__ ° 46
41. $\overline{GH} \cong$ __?__ $\overline{ST}$
42. $\triangle TSR \cong$ __?__ $\triangle HGF$

Trapezoid Midsegments Find the value of *x*. *(Lesson 6.5)*

43. 17

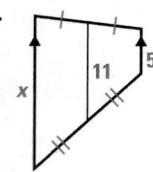

44. 10

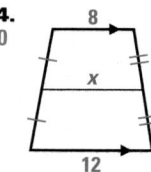

45. 15

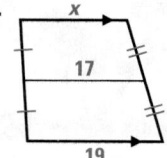

Algebra Skills

Plotting Points Plot the points in a coordinate plane. *(Skills Review, p. 664)* 46–53. See margin.

46. $A(-4, 5)$
47. $B(-1, -3)$
48. $C(0, 7)$
49. $D(2, -6)$
50. $F(7, 2)$
51. $G(-8, -1)$
52. $J(7, -7)$
53. $K(-3, 3)$

Quiz 1

Solve the proportion. *(Lesson 7.1)*

1. $\dfrac{x}{16} = \dfrac{3}{4}$ 12

2. $\dfrac{5}{8} = \dfrac{25}{y}$ 40

3. $\dfrac{11}{2} = \dfrac{z+3}{6}$ 30

The two polygons are similar. Find the value of the variable. *(Lesson 7.2)*

4. 10

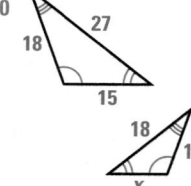

5. 5

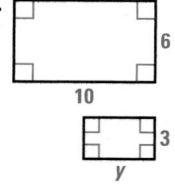

6. 20

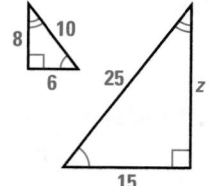

Determine whether the triangles are similar. If they are similar, write a similarity statement. Explain your reasoning. *(Lesson 7.3)*

7. Yes; $\angle A \cong \angle F$ and $\angle B \cong \angle G$, so $\triangle ABC \sim \triangle FGH$ by the AA Similarity Postulate.

8. Yes; $\angle J \cong \angle R$ and $\angle K \cong \angle P$, so $\triangle JKL \sim \triangle RPQ$ by the AA Similarity Postulate.

9. Yes; vertical angles are congruent, so $\angle SUT \cong \angle WUV$; also $\angle T \cong \angle V$, so $\triangle STU \sim \triangle WVU$ by the AA Similarity Postulate.

7.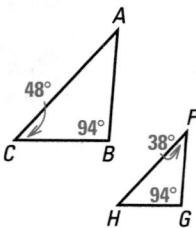

8.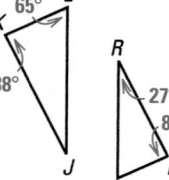

9.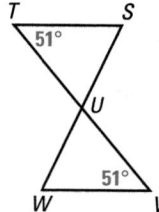

 **7.4** # Showing Triangles are Similar: SSS and SAS

Goal
Show that two triangles are similar using the SSS and SAS Similarity Theorems.

Key Words
• similar polygons p. 365

The triangles in the Navajo rug look similar. To show that they are similar, you can use the definition of similar polygons or the AA Similarity Postulate.

In this lesson, you will learn two new methods to show that two triangles are similar.

THEOREM 7.2

Side-Side-Side Similarity Theorem (SSS)

Words If the corresponding sides of two triangles are proportional, then the triangles are similar.

Symbols If $\dfrac{FG}{AB} = \dfrac{GH}{BC} = \dfrac{HF}{CA}$, then $\triangle ABC \sim \triangle FGH$.

EXAMPLE 1 Use the SSS Similarity Theorem

Determine whether the triangles are similar. If they are similar, write a similarity statement and find the scale factor of Triangle B to Triangle A.

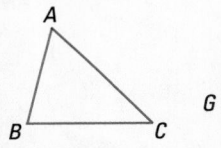 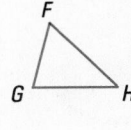

Solution

Find the ratios of the corresponding sides.

$\dfrac{SU}{PR} = \dfrac{6}{12} = \dfrac{6 \div 6}{12 \div 6} = \dfrac{1}{2}$

$\dfrac{UT}{RQ} = \dfrac{5}{10} = \dfrac{5 \div 5}{10 \div 5} = \dfrac{1}{2}$

$\dfrac{TS}{QP} = \dfrac{4}{8} = \dfrac{4 \div 4}{8 \div 4} = \dfrac{1}{2}$

All three ratios are equal. So, the corresponding sides of the triangles are proportional.

ANSWER ▸ By the SSS Similarity Theorem, $\triangle PQR \sim \triangle STU$. The scale factor of Triangle B to Triangle A is $\dfrac{1}{2}$.

7.4 *Showing Triangles are Similar: SSS and SAS* **379**

① Plan

Pacing
Suggested Number of Days

Basic: 2 days
Average: 2 days
Advanced: 2 days
Block Schedule: 1 block

Teaching Resources

📄 *Blacklines*
(See page 354B.)

🪧 *Transparencies*
• Warm-Up with Quiz
• Answers

🖥 *Technology*
• Electronic Teacher Tools
• Test and Practice Generator
• Online Lesson Planner
• Internet Support

② Teach

Content and Teaching Strategies
For background information on geometric concepts and teaching strategies related to this lesson, see pages 354E and 354F in this Teacher's Edition.

Tips for New Teachers
Stress the importance of writing the proportions first and then substituting measurements for the lengths of segments as shown in Examples 1 and 2 on pages 379–380. Students should always check to make sure they have identified the correct pair of corresponding parts in their proportions or ratios. This practice should be continued throughout the rest of the chapter. See the Tips for New Teachers on pp. 1–2 of the *Chapter 7 Resource Book* for additional notes about Lesson 7.4.

Extra Example 1
See next page.

Extra Example 1

Determine whether the triangles are similar. If they are similar, write a similarity statement and find the scale factor of Triangle B to Triangle A.

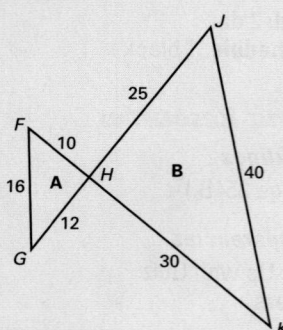

By the SSS Similarity Theorem, $\triangle FGH \sim \triangle JKH$. The scale factor of Triangle B to Triangle A is $\frac{5}{2}$.

Extra Example 2

All three figures below are rectangles. Is either *GHJK* or *LMNP* similar to *CDEF*?

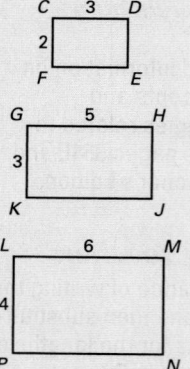

Because the ratios of all corresponding sides are not equal, *CDEF* and *GHJK* are not similar. Because the ratios of all corresponding sides are equal, *CDEF* and *LMNP* are similar.

Student Help

STUDY TIP
When using the SSS Similarity Theorem, compare the shortest sides, the longest sides, and then the remaining sides.

EXAMPLE 2 Use the SSS Similarity Theorem

Is either $\triangle DEF$ or $\triangle GHJ$ similar to $\triangle ABC$?

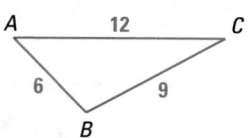 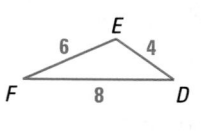 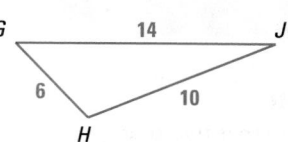

Solution

❶ Look at the ratios of corresponding sides in $\triangle ABC$ and $\triangle DEF$.

Shortest sides	Longest sides	Remaining sides
$\dfrac{DE}{AB} = \dfrac{4}{6} = \dfrac{2}{3}$	$\dfrac{FD}{CA} = \dfrac{8}{12} = \dfrac{2}{3}$	$\dfrac{EF}{BC} = \dfrac{6}{9} = \dfrac{2}{3}$

ANSWER ▶ Because all of the ratios are equal, $\triangle ABC \sim \triangle DEF$.

❷ Look at the ratios of corresponding sides in $\triangle ABC$ and $\triangle GHJ$.

Shortest sides	Longest sides	Remaining sides
$\dfrac{GH}{AB} = \dfrac{6}{6} = \dfrac{1}{1}$	$\dfrac{JG}{CA} = \dfrac{14}{12} = \dfrac{7}{6}$	$\dfrac{HJ}{BC} = \dfrac{10}{9}$

ANSWER ▶ Because the ratios are not equal, $\triangle ABC$ and $\triangle GHJ$ are not similar.

Checkpoint ✔ **Use the SSS Similarity Theorem**

Determine whether the triangles are similar. If they are similar, write a similarity statement.

1.

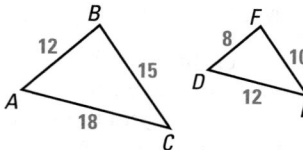

yes; $\triangle ABC \sim \triangle DFE$

2.

no

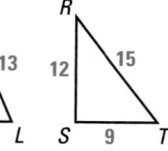

Student Help

LOOK BACK
To review included angles, see p. 242.

THEOREM 7.3

Side-Angle-Side Similarity Theorem (SAS)

Words If an angle of one triangle is congruent to an angle of a second triangle and the lengths of the sides that include these angles are proportional, then the triangles are similar.

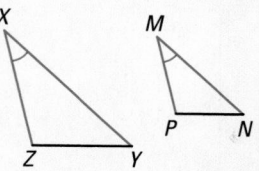

Symbols If $\angle X \cong \angle M$ and $\dfrac{PM}{ZX} = \dfrac{MN}{XY}$, then $\triangle XYZ \sim \triangle MNP$.

EXAMPLE **3** **Use the SAS Similarity Theorem**

Determine whether the triangles are
similar. If they are similar, write a
similarity statement.

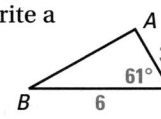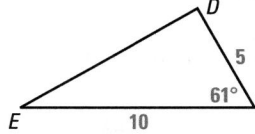

Solution

∠C and ∠F both measure 61°, so ∠C ≅ ∠F.

Compare the ratios of the side lengths that include ∠C and ∠F.

Shorter sides $\frac{DF}{AC} = \frac{5}{3}$ **Longer sides** $\frac{FE}{CB} = \frac{10}{6} = \frac{5}{3}$

The lengths of the sides that include ∠C and ∠F are proportional.

ANSWER ▶ By the SAS Similarity Theorem, △ABC ~ △DEF.

EXAMPLE **4** *Similarity in Overlapping Triangles*

Show that △VYZ ~ △VWX.

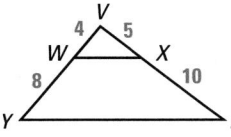

Solution

Separate the triangles, △VYZ and △VWX, and label the side lengths.

Student Help

VISUAL STRATEGY
Redraw overlapping
triangles as two
separate triangles,
as shown on p. 356. ┈┈┈┈┈▶

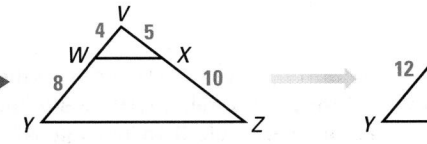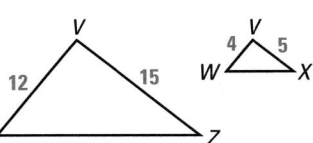

∠V ≅ ∠V by the Reflexive Property of Congruence.

Shorter sides

$\frac{VW}{VY} = \frac{4}{4+8} = \frac{4}{12} = \frac{1}{3}$

Longer sides

$\frac{XV}{ZV} = \frac{5}{5+10} = \frac{5}{15} = \frac{1}{3}$

The lengths of the sides that include ∠V are proportional.

ANSWER ▶ By the SAS Similarity Theorem, △VYZ ~ △VWX.

Checkpoint ✓ *Use the SAS Similarity Theorem*

3. No; ∠H ≅ ∠M but $\frac{8}{6} \neq \frac{12}{8}$.

4. Yes; ∠P ≅ ∠P, $\frac{PQ}{PS} = \frac{3}{6} = \frac{1}{2}$,
and $\frac{PR}{PT} = \frac{5}{10} = \frac{1}{2}$; the lengths of
the sides that include ∠P are
proportional, so △PQR ~ △PST
by the SAS Similarity Theorem.

Determine whether the triangles are similar. If they are similar,
write a similarity statement. Explain your reasoning.

3.

4.

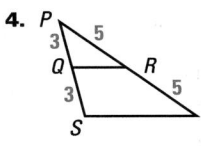

7.4 *Showing Triangles are Similar: SSS and SAS* **381**

Extra Example 3

Determine whether the triangles
are similar. If they are similar, write
a similarity statement.

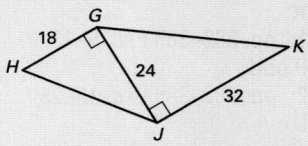

By the SAS Similarity Theorem,
△GHJ ~ △JGK.

Extra Example 4

Show that △LMQ ~ △LNP.

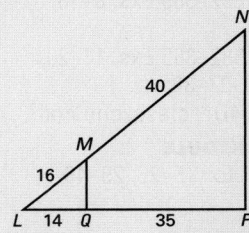

The lengths of the sides that include
∠L are proportional: $\frac{LM}{LN} = \frac{16}{56} = \frac{2}{7}$
and $\frac{LQ}{LP} = \frac{14}{49} = \frac{2}{7}$. So, by the SAS
Similarity Theorem, △LMQ ~ △LNP.

✓ Concept Check

Explain two ways to use corresponding sides to prove two triangles similar. If the ratios of the
lengths of all three pairs of corresponding sides can be shown to be
equal, then the SSS Similarity
Theorem can be used to justify that
the triangles are similar. If the ratios
of the lengths of two pairs of corresponding sides can be shown to be
equal and the angles between those
pairs of sides can be shown to be
congruent, then the SAS Similarity
Theorem can be used to justify that
the triangles are similar.

Assignment Guide

BASIC
Day 1: pp. 382–385 Exs. 5–13, 33–49 odd
Day 2: pp. 382–385 Exs. 15–26, 32–48 even

AVERAGE
Day 1: pp. 382–385 Exs. 5–13, 33–37, 42–49
Day 2: pp. 382–385 Exs. 14–26, 29–32, 38–41

ADVANCED
Day 1: pp. 382–385 Exs. 8–13, 33–45
Day 2: pp. 382–385 Exs. 14–20, 22–26 even, 27–32;
EC: TE p. 354D*, classzone.com

BLOCK SCHEDULE
pp. 382–385 Exs. 5–26, 29–49

Extra Practice

• Student Edition, p. 688
• Chapter 7 Resource Book, pp. 45–46

Homework Check

To quickly check student understanding of key concepts, go over the following exercises:

Basic: 5, 7, 11, 15, 19
Average: 7, 9, 12, 15, 19
Advanced: 8, 13, 17, 20, 26

✗ Common Error

In Exercises 9 and 10, watch for students who have difficulty identifying the corresponding sides due to the orientations of the figures. Suggest that students find the ratio of the lengths of the shortest sides first, then the ratio of the lengths of the longest sides, and finally the lengths of the remaining sides. Students will find that this suggestion is useful for Exercises 13, 17, and 18 as well.

Guided Practice

Vocabulary Check

1. If two sides of a triangle are proportional to two sides of another triangle, can you conclude that the triangles are similar?

Skill Check

In Exercises 2 and 3, determine whether the triangles are similar. If they are similar, write a similarity statement.

1. No; to do so you must also know that the included angles are congruent.

2. Yes, $\triangle ABC \sim \triangle FHG$ by the SSS Similarity Theorem.

4. $\triangle LMN$ is not similar to $\triangle ABC$ because the ratios of corresponding sides are not equal. $\triangle ABC \sim \triangle XYZ$ by the SSS Similarity Theorem because the ratios of corresponding sides are all equal to $\frac{3}{4}$.

2.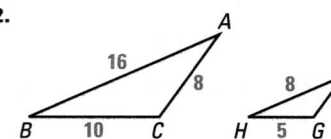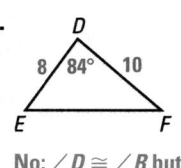

3.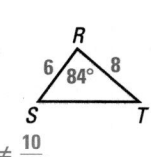

No; $\angle D \cong \angle R$ but $\frac{8}{6} \neq \frac{10}{8}$.

4. Is either $\triangle LMN$ or $\triangle XYZ$ similar to $\triangle ABC$? Explain.

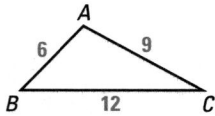 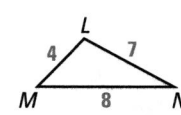 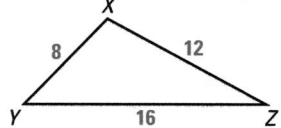

Practice and Applications

Extra Practice

See p. 688.

SSS Similarity Theorem Determine whether the two triangles are similar. If they are similar, write a similarity statement and find the scale factor of Triangle B to Triangle A.

5.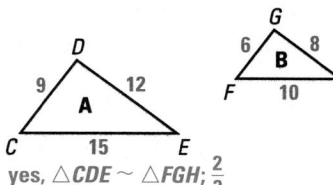
yes, $\triangle CDE \sim \triangle FGH$; $\frac{2}{3}$

6.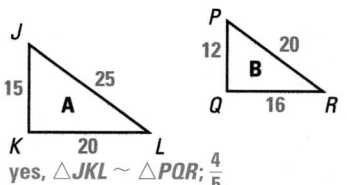
yes, $\triangle JKL \sim \triangle PQR$; $\frac{4}{5}$

7. no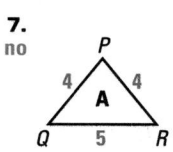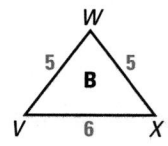

8.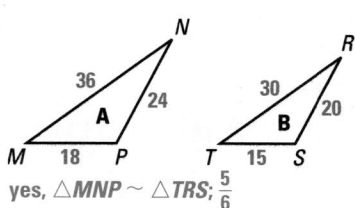
yes, $\triangle MNP \sim \triangle TRS$; $\frac{5}{6}$

Homework Help

Example 1: Exs. 5–10, 21–26
Example 2: Exs. 11–13
Example 3: Exs. 14–18, 21–26
Example 4: Exs. 19, 20, 26–29

9.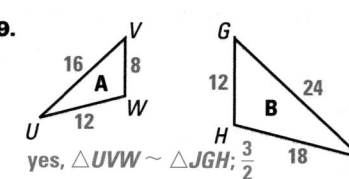
yes, $\triangle UVW \sim \triangle JGH$; $\frac{3}{2}$

10. no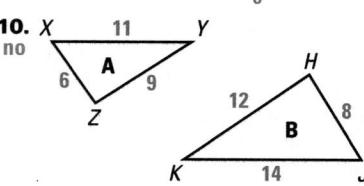

11. △RST is not similar to △ABC because the ratios of corresponding sides are not equal. △ABC ~ △XYZ by the SSS Similarity Theorem because the ratios of corresponding sides are all equal to $\frac{3}{2}$.

12. △RST is not similar to △ABC because the ratios of corresponding sides are not equal. △ABC ~ △XYZ by the SSS Similarity Theorem because the ratios of corresponding sides are all equal to $\frac{2}{5}$.

13. △RST is not similar to △ABC because the ratios of corresponding sides are not equal. △ABC ~ △XYZ by the SSS Similarity Theorem because the ratios of corresponding sides are all equal to $\frac{3}{4}$.

SSS Similarity Theorem Is either △RST or △XYZ similar to △ABC? Explain your reasoning.

11.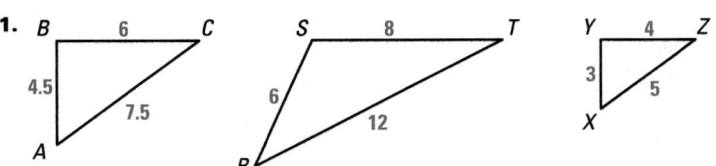

12.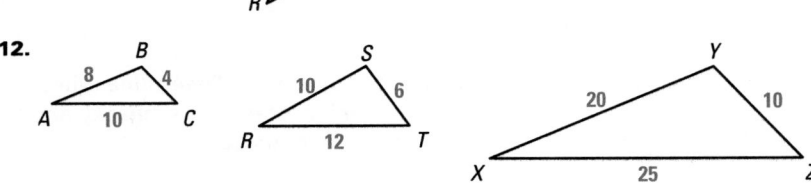

13.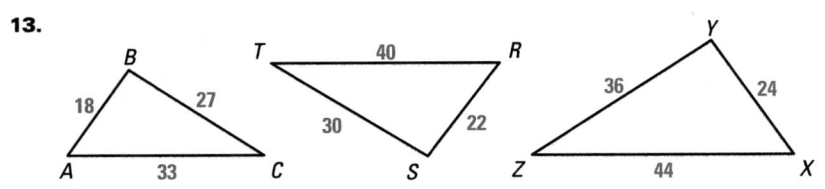

14. **A-Frame Building** Suppose you are constructing an A-frame home that is modeled after a ski lodge. The ski lodge and home are shown below. Are the triangles similar? Explain your reasoning.
The triangles are similar by the SAS Similarity Theorem.

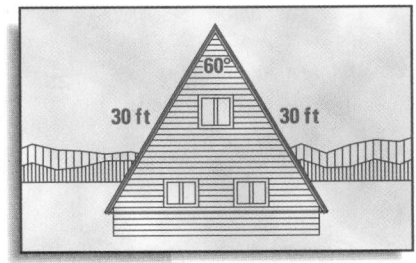

Student Help
CLASSZONE.COM

HOMEWORK HELP
Extra help with problem solving in Exs. 15–18 is at classzone.com

SAS Similarity Theorem Determine whether the two triangles are similar. If they are similar, write a similarity statement.

15.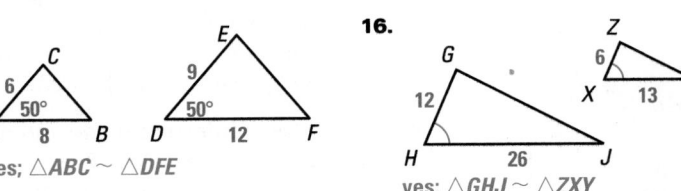

yes; △ABC ~ △DFE

16.

yes; △GHJ ~ △ZXY

17. no

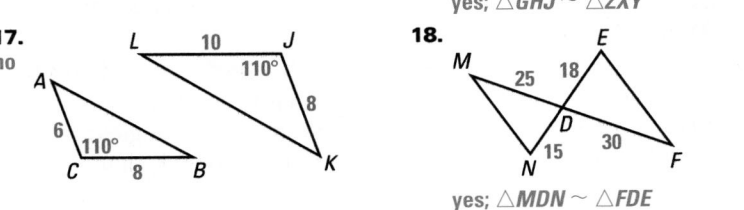

18.

yes; △MDN ~ △FDE

7.4 *Showing Triangles are Similar: SSS and SAS* **383**

Teaching Tip

For Exercises 21–26, suggest that students look again at the given information and decide which of AA, SSS, or SAS they could use to determine if the triangles are similar. For example, in Exercises 21 and 24 students should use SSS since all three side lengths are given for both triangles.

Student Help

VISUAL STRATEGY
Redraw overlapping triangles as two separate triangles, as shown on p. 356.

19. $\angle A \cong \angle A$, $\dfrac{AD}{AB} = \dfrac{9}{30} = \dfrac{3}{10}$, and $\dfrac{AE}{AC} = \dfrac{6}{20} = \dfrac{3}{10}$; the lengths of the sides that include $\angle A$ are proportional, so $\triangle ADE \sim \triangle ABC$ by the SAS Similarity Theorem.

20. $\angle J \cong \angle J$, $\dfrac{JM}{JK} = \dfrac{10}{14} = \dfrac{5}{7}$, and $\dfrac{JN}{JL} = \dfrac{20}{28} = \dfrac{5}{7}$; the lengths of the sides that include $\angle J$ are proportional, so $\triangle JMN \sim \triangle JKL$ by the SAS Similarity Theorem.

21. Yes; $\triangle JKL \sim \triangle XYZ$ (or $\triangle YXZ$) by the SSS Similarity Theorem.

22. Yes; $\triangle GHJ \sim \triangle LKM$ by the AA Similarity Postulate.

27. $\dfrac{AE}{AC} = \dfrac{AD}{AB}$ or $\dfrac{AE}{AC} = \dfrac{DE}{BC}$

29. *Sample answer:* Jon is correct. $\dfrac{6}{27} = \dfrac{4}{18}$, so the triangles are similar when $x = 6$.

Link to
Sports

SHUFFLEBOARD is played on a long flat court. Players earn points by using sticks called *cues* to push circular disks onto a scoring area at the opposite end of the court.

Overlapping Triangles Show that the overlapping triangles are similar. Then write a similarity statement.

19.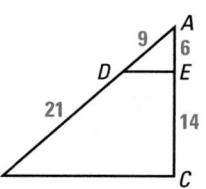

20.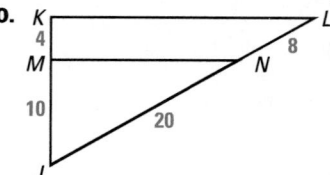

Determining Similarity Determine whether the triangles are similar. If they are similar, state the similarity and the postulate or theorem that justifies your answer.

21.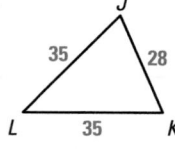

22.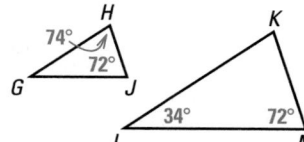

23. no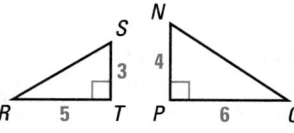

24. no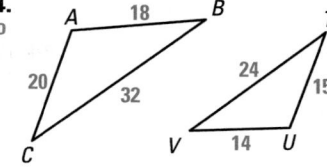

25.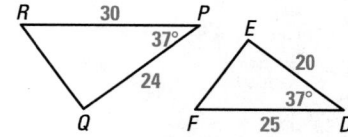

Yes; $\triangle PQR \sim \triangle DEF$ by the SAS Similarity Theorem.

26.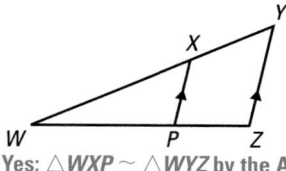

Yes; $\triangle WXP \sim \triangle WYZ$ by the AA Similarity Postulate.

Shuffleboard In the portion of a shuffleboard court shown, $\dfrac{AD}{AB} = \dfrac{DE}{BC}$.

27. What piece of information do you need in order to show that $\triangle ADE \sim \triangle ABC$ using the SSS Similarity Theorem? See margin.

28. What piece of information do you need in order to show that $\triangle ADE \sim \triangle ABC$ using the SAS Similarity Theorem? $\angle ADE \cong \angle ABC$

29. **You be the Judge** Jon claims that $\triangle SUV$ is similar to $\triangle SRT$ when $x = 6$. Dave believes that the triangles are similar when $x = 5$. Who is right? Explain your reasoning. See margin.

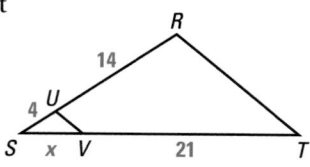

Technology In Exercises 30 and 31, use geometry software to complete the steps below.

❶ Draw △ABC.

❷ Construct a line perpendicular to $\overline{AB}$ through C. Label the intersection D.

❸ Measure $\overline{CA}$, $\overline{CD}$, $\overline{CB}$, and $\overline{BD}$.

❹ Calculate the ratios $\dfrac{CA}{CD}$ and $\dfrac{CB}{BD}$.

❺ Drag point C until $\dfrac{CA}{CD} = \dfrac{CB}{BD}$.

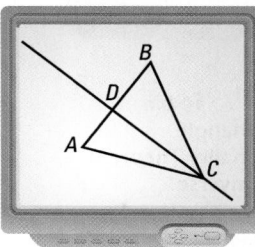

30. For what measure of ∠ACB are △ABC and △CBD similar? 90°

31. What theorem supports your answer to Exercise 30?
SAS Similarity Theorem

Standardized Test Practice

32. Multiple Choice Which method can be used to show that the two triangles at the right are similar? C

(A) AA   (B) SSS

(C) SAS   (D) Cannot be shown

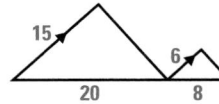

33. Multiple Choice In the diagram, △MNP ~ △RST. Find the value of x. G

(F) 20   (G) 24

(H) 30   (J) 32

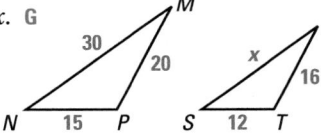

Mixed Review

37. Yes; *Sample answer:* $\overrightarrow{TV}$ bisects ∠STU, so ∠STV ≅ ∠UTV; ∠TSV ≅ ∠TUV and $\overline{TV} \cong \overline{TV}$, so △STV ≅ △UTV by the AAS Congruence Theorem; corresponding parts of congruent triangles are congruent, so ∠TVS ≅ ∠TVU.

Using Bisectors In the diagram below, $\overrightarrow{TV}$ bisects ∠STU. *(Lesson 5.6)*

34. $\overline{ST} \cong \underline{\ ?\ }$ $\overline{UT}$

35. ∠VTU ≅ $\underline{\ ?\ }$ ∠VTS

36. m∠STU = $\underline{\ ?\ }$ 54°

37. Is ∠TVS congruent to ∠TVU? Explain your reasoning.

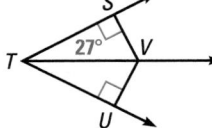

Solving Proportions Solve the proportion. *(Lesson 7.1)*

38. $\dfrac{b}{12} = \dfrac{5}{6}$ 10   **39.** $\dfrac{24}{y} = \dfrac{4}{9}$ 54   **40.** $\dfrac{5}{8} = \dfrac{c}{56}$ 35   **41.** $\dfrac{5}{2} = \dfrac{60}{a}$ 24

Algebra Skills

Writing Decimals as Fractions Write the decimal as a fraction in simplest form. *(Skills Review, p. 657)*

42. 0.4 $\dfrac{2}{5}$ **43.** 0.25 $\dfrac{1}{4}$ **44.** 0.64 $\dfrac{16}{25}$ **45.** 0.88 $\dfrac{22}{25}$

46. 0.26 $\dfrac{13}{50}$ **47.** 0.55 $\dfrac{11}{20}$ **48.** 0.7 $\dfrac{7}{10}$ **49.** 0.34 $\dfrac{17}{50}$

---

④ Assess

Assessment Resources

The Mini-Quiz below is also available on blackline (*Chapter 7 Resource Book*, p. 52) and on transparency. For more assessment resources, see:
• Chapter 7 Resource Book
• Standardized Test Practice
• Test and Practice Generator

Mini-Quiz

Determine whether the triangles are similar. If they are similar, state the similarity and the postulate or theorem that justifies your answer.

1.

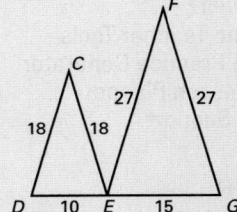

By the SSS Similarity Theorem, △CDE ~ △FEG.

2.

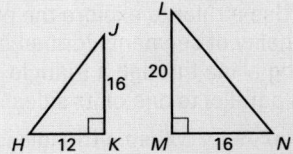

△HJK is not similar to △NLM.

3.

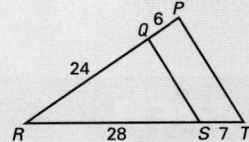

By the SAS Similarity Theorem, △RQS ~ △RPT.

Pacing

Suggested Number of Days

Basic: 2 days
Average: 2 days
Advanced: 2 days
Block Schedule: 1 block

Teaching Resources

📄 **Blacklines**
(See page 354B.)

📃 **Transparencies**
• Warm-Up with Quiz
• Answers

🖥 **Technology**
• Electronic Teacher Tools
• Test and Practice Generator
• Online Lesson Planner
• Internet Support

📼 **Video**
• Geometry in Motion

Geo-Activity

Goal Use a ruler to explore the pro-portionality of segments formed by drawing a line through a triangle that is parallel to one of its sides.

Key Discovery When a triangle is cut by a line parallel to a side of the triangle, it divides the other two sides proportionally.

 **Teach**

-**Content and Teaching Strategies**
For background information on geometric concepts and teaching strategies related to this lesson, see pages 354E and 354F in this Teacher's Edition.

 **7.5** # Proportions and Similar Triangles

Goal
Use the Triangle Proportionality Theorem and its converse.

Key Words
• midsegment of a triangle

4. When a line parallel to one side of a triangle cuts the other two sides, then the ratios of the segment lengths of the divided sides are equal.

Geo-Activity ▶ **Investigating Proportional Segments**

❶ Draw a triangle. Label its vertices *A*, *B*, and *C*. Make sure that each side is at least 4 cm. Draw a point on $\overline{AB}$. Label the point *D*.

❷ Draw a line through *D* parallel to $\overline{AC}$. Label the intersection of the line and $\overline{BC}$ as point *E*.

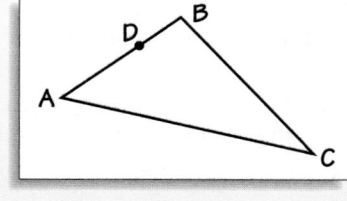

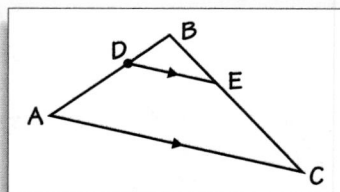

❸ Measure $\overline{BD}$, $\overline{DA}$, $\overline{BE}$, and $\overline{EC}$ in centimeters. Then calculate the ratios $\frac{BD}{DA}$ and $\frac{BE}{EC}$. Measures will vary, but $\frac{BD}{DA} = \frac{BE}{EC}$.

❹ Make a conjecture about the ratios of segment lengths of a triangle's sides when the triangle is cut by a line parallel to the triangle's third side.

Proportionality Suppose that a point *P* lies on $\overline{GH}$ and a point *Q* lies on $\overline{JK}$. If $\frac{GP}{PH} = \frac{JQ}{QK}$, then we say that $\overline{GH}$ and $\overline{JK}$ are *divided proportionally*.

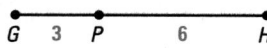

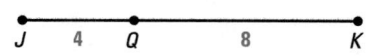

In the Geo-Activity above, $\overline{DE}$ divides $\overline{AB}$ and $\overline{CB}$ proportionally.

THEOREM 7.4

Triangle Proportionality Theorem

Words If a line parallel to one side of a triangle intersects the other two sides, then it divides the two sides proportionally.

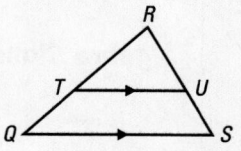

Symbols In △*QRS*, if $\overline{TU} \parallel \overline{QS}$, then $\frac{RT}{TQ} = \frac{RU}{US}$.

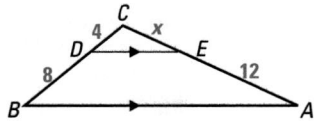
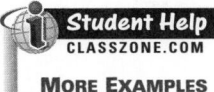
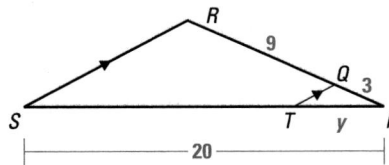
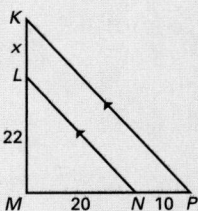
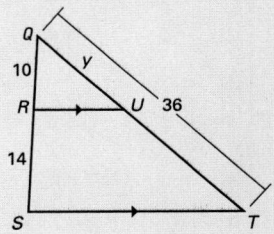

Extra Example 3

Given the diagram, determine whether $\overline{CD}$ is parallel to $\overline{EF}$.

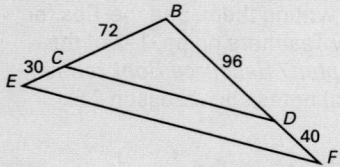

$\dfrac{BC}{CE} = \dfrac{72}{30} = \dfrac{12}{5}$ and $\dfrac{BD}{DF} = \dfrac{96}{40} = \dfrac{12}{5}$;

because $\dfrac{12}{5} = \dfrac{12}{5}$, $\overline{CD}$ is parallel to $\overline{EF}$.

Converse of the Triangle Proportionality Theorem

Words If a line divides two sides of a triangle proportionally, then it is parallel to the third side.

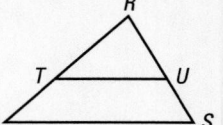

Symbols In $\triangle QRS$, if $\dfrac{RT}{TQ} = \dfrac{RU}{US}$, then $\overline{TU} \parallel \overline{QS}$.

EXAMPLE 3 Determine Parallels

Given the diagram, determine whether $\overline{MN}$ is parallel to $\overline{GH}$.

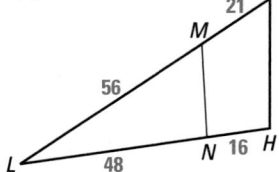

Solution

Find and simplify the ratios of the two sides divided by $\overline{MN}$.

$$\frac{LM}{MG} = \frac{56}{21} = \frac{8}{3} \qquad\qquad \frac{LN}{NH} = \frac{48}{16} = \frac{3}{1}$$

ANSWER ▶ Because $\dfrac{8}{3} \ne \dfrac{3}{1}$, $\overline{MN}$ is not parallel to $\overline{GH}$.

Checkpoint ✓ **Find Segment Lengths and Determine Parallels**

Find the value of the variable.

1.
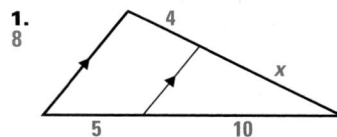

2.
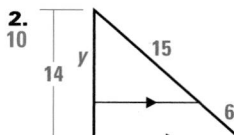

Given the diagram, determine whether $\overline{QR}$ is parallel to $\overline{ST}$. Explain.

3.
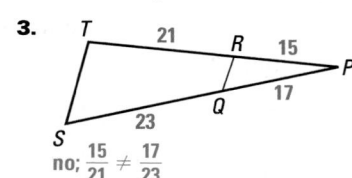

no; $\dfrac{15}{21} \ne \dfrac{17}{23}$

4.
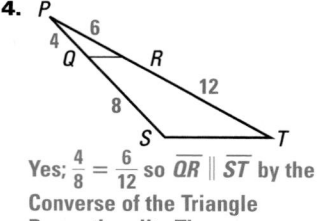

Yes; $\dfrac{4}{8} = \dfrac{6}{12}$ so $\overline{QR} \parallel \overline{ST}$ by the Converse of the Triangle Proportionality Theorem.

A **midsegment of a triangle** is a segment that connects the midpoints of two sides of a triangle. The following theorem about midsegments is a special case of the Triangle Proportionality Theorem.

THEOREM 7.6

The Midsegment Theorem

Words The segment connecting the midpoints of two sides of a triangle is parallel to the third side and is half as long.

Symbols In $\triangle ABC$, if $CD = DA$ and $CE = EB$, then $\overline{DE} \parallel \overline{AB}$ and $DE = \frac{1}{2}AB$.

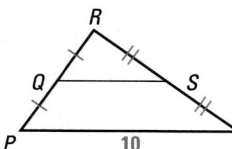

EXAMPLE 4 Use the Midsegment Theorem

Find the length of $\overline{QS}$.

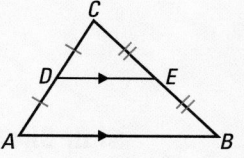

Solution

From the marks on the diagram, you know S is the midpoint of $\overline{RT}$, and Q is the midpoint of $\overline{RP}$. Therefore, $\overline{QS}$ is a midsegment of $\triangle PRT$. Use the Midsegment Theorem to write the following equation.

$$QS = \frac{1}{2}PT = \frac{1}{2}(10) = 5$$

ANSWER ▶ The length of $\overline{QS}$ is 5.

Checkpoint ✓ *Use the Midsegment Theorem*

Find the value of the variable.

5. 8
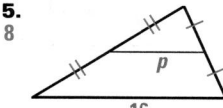

6. 28
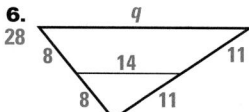

7. Use the Midsegment Theorem to find the perimeter of $\triangle ABC$. 24

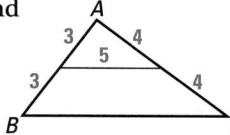

7.5 *Proportions and Similar Triangles* 389

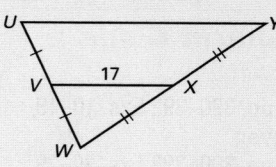
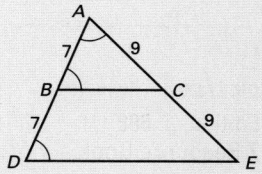

Assignment Guide

BASIC
Day 1: pp. 390–392 Exs. 10–19, 40–52 even
Day 2: pp. 390–392 Exs. 20–29, 33–37, 41–51 odd

AVERAGE
Day 1: pp. 390–392 Exs. 14–19, 40, 44–52
Day 2: pp. 390–392 Exs. 20–29, 33–39, 41–43

ADVANCED
Day 1: pp. 390–392 Exs. 14–19, 30, 31*, 32*, 40–43
Day 2: pp. 390–392 Exs. 20–28 even, 33–39, 44–47;
EC: classzone.com

BLOCK SCHEDULE
pp. 390–392 Exs. 14–29, 33–52

Extra Practice

• Student Edition, p. 688
• Chapter 7 Resource Book, pp. 54–55

Homework Check

To quickly check student understanding of key concepts, go over the following exercises:

Basic: 14, 20, 22, 24, 35
Average: 16, 21, 22, 25, 36
Advanced: 19, 20, 22, 26, 37

✗ Common Error

In Exercises 18 and 19, watch for students who have difficulty setting up the proportion. Suggest that students review how the proportion was set up in Example 2 on page 387.

Guided Practice

Vocabulary Check

Complete the statement.

1. The __?__ Theorem states that if a line divides two sides of a triangle proportionally, then it is __?__ to the third side.
   Converse of the Triangle Proportionality; parallel

2. A __?__ of a triangle is a segment that connects the midpoints of two sides of a triangle. midsegment

Skill Check

Copy and complete the proportion using the diagram below.

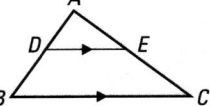

3. $\dfrac{AD}{DB} = \dfrac{?}{EC}$ AE

4. $\dfrac{?}{DA} = \dfrac{CE}{EA}$ BD

5. $\dfrac{AD}{?} = \dfrac{AE}{AC}$ AB

6. $\dfrac{BD}{BA} = \dfrac{CE}{?}$ CA

Find the value of the variable.

7.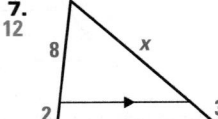
12

8.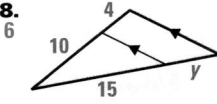
6

9.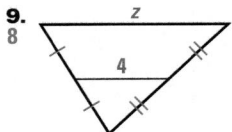
8

Practice and Applications

Extra Practice

See p. 688.

ⓧⓨ Using Algebra Solve the proportion.

10. $\dfrac{2}{3} = \dfrac{m}{36}$ 24

11. $\dfrac{t}{2} = \dfrac{5}{12}$ $\dfrac{5}{6}$

12. $\dfrac{21}{y} = \dfrac{7}{18}$ 54

13. $\dfrac{27}{r} = \dfrac{3}{4}$ 36

Finding Segment Lengths Find the value of the variable.

14.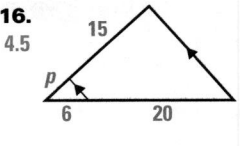
14

15.
18

16.
4.5

Homework Help

Example 1: Exs. 14–19
Example 2: Exs. 14–19
Example 3: Exs. 20–23
Example 4: Exs. 24–29, 33–37

17.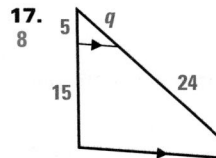
8

18.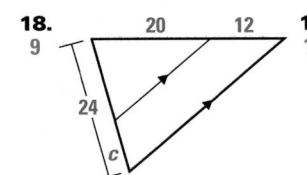
9

19.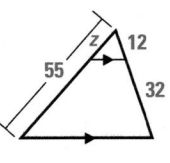
15

Determining Parallels Given the diagram, determine whether $\overline{QS}$ is parallel to $\overline{PT}$.

20.
yes
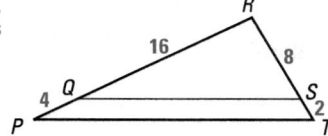

21.
yes
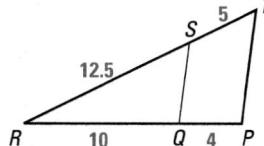

22.
no
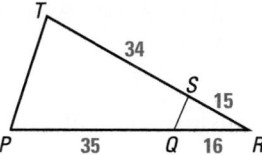

23.
yes
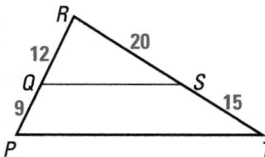

Using the Midsegment Theorem Find the value of the variable.

24.
2
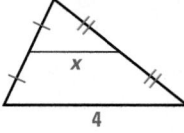

25.
22
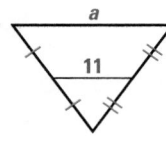

26.
9
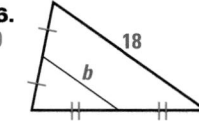

27.
10
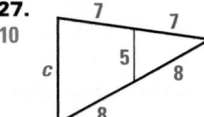

28.
$\frac{1}{2}$
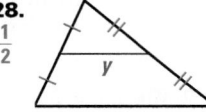

29.
13.5
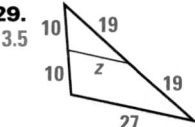

Visualize It! The design below approximates a fractal. Begin with an equilateral triangle. Shade the triangle formed by the three midsegments. Continue the process for each unshaded triangle.

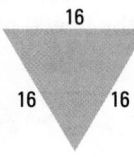
Stage 0

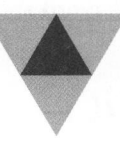
Stage 1

Stage 2

Stage 3

30. Find the perimeter of the dark blue triangle in Stage 1. 24

31. Challenge Find the total perimeter of all the dark blue triangles in Stage 2. 60

32. Challenge Find the total perimeter of all the dark blue triangles in Stage 3. 114

7.5 Proportions and Similar Triangles   391

Mini-Quiz
Solve the proportion.

1. $\frac{4}{9} = \frac{x}{45}$ 20

2. $\frac{49}{n} = \frac{7}{12}$ 84

Find the value of the variable.

3.   32
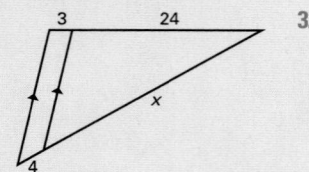

4.   11
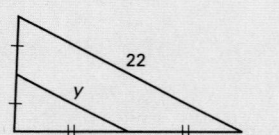

5. Given the diagram, determine whether $\overline{QS}$ is parallel to $\overline{PT}$.

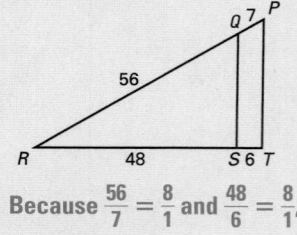

Because $\frac{56}{7} = \frac{8}{1}$ and $\frac{48}{6} = \frac{8}{1}$, $\overline{QS} \parallel \overline{PT}$.

41. ~~OHIO~~

43. ~~BOOK~~

Midsegment Theorem Use the diagram below to complete the statement.

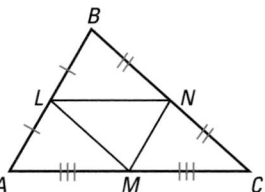

33. $\overline{LM} \parallel$ __?__ $\overline{BC}$

34. $\overline{AB} \parallel$ __?__ $\overline{MN}$

35. If $AC = 15$, then $LN =$ __?__ .   7.5

36. If $MN = 7.4$, then $AB =$ __?__ .   14.8

37. If $NC = 9.5$, then $LM =$ __?__ .   9.5

Technology In Exercises 38 and 39, use geometry software to complete the steps below.

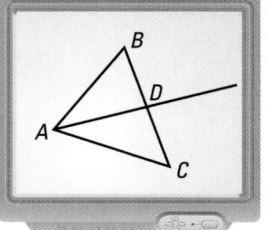

❶ Draw $\triangle ABC$.

❷ Construct the angle bisector of $\angle A$.

❸ Construct the intersection of the angle bisector and $\overline{BC}$. Label it D.

❹ Measure $\overline{DC}$, $\overline{AC}$, $\overline{DB}$, and $\overline{AB}$. Then calculate the ratios $\frac{DC}{AC}$ and $\frac{DB}{AB}$.

38. Drag one or more of the triangle's vertices. What do you notice about the ratios as the shape changes?   The ratios remain equal.

39. Complete the conjecture: If a ray bisects an angle of a triangle, then it divides the opposite side into segments whose lengths are __?__ to the lengths of the other two sides.   proportional

Standardized Test Practice

40. Multiple Choice What is the value of x?   A

(A) 21   (B) 24

(C) 32   (D) 42

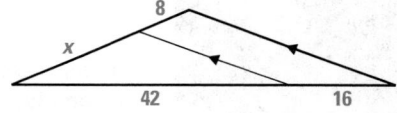

Mixed Review

Reflections Determine if the entire word has any lines of symmetry. If so, write the word and draw the line(s) of symmetry. (*Lesson 5.7*)

41–43. See margin for drawings.

41. yes

42. no
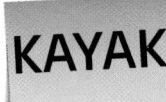

43. yes

Algebra Skills

Finding Slope Find the slope of the line that passes through the points. (*Skills Review, p. 665*)

44. (0, 2) and (4, 8) $\frac{3}{2}$

45. (1, 2) and (3, 4)   1

46. (5, 2) and (5, 3)   undefined

47. (−5, 6) and (1, 2) $-\frac{2}{3}$

48. (−4, 4) and (2, 0) $-\frac{2}{3}$

49. (3, 7) and (−1, −3) $\frac{5}{2}$

50. (5, 3) and (−1, 1) $\frac{1}{3}$

51. (0, −4) and (3, 5)   3

52. (−3, 2) and (6, −5) $-\frac{7}{9}$

7.6 Dilations

Goal
Identify and draw dilations.

Key Words
- dilation
- reduction
- enlargement

Student Help

LOOK BACK
To review other types of transformations, see pp. 152 and 282.

Geo-Activity — **Drawing a Dilation**

❶ Draw a triangle. Label it △PQR. Choose a point C outside the triangle.

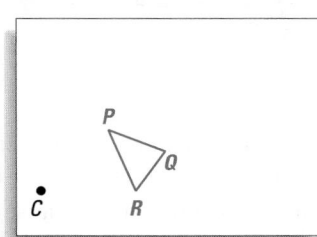

❷ Use a straightedge to draw lines from C through P, Q, and R.

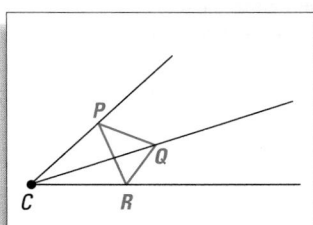

❸ Measure $\overrightarrow{CP}$ and locate a point P' on $\overrightarrow{CP}$ such that $CP' = 2 \cdot CP$. Locate Q' and R' the same way.

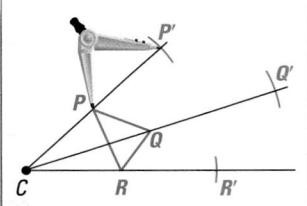

❹ Connect the points P', Q', and R' to form △P'Q'R'.

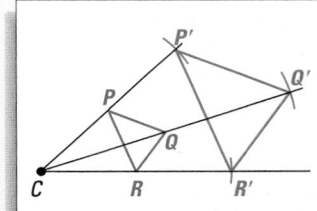

Step 5. $\frac{P'R'}{PR} = \frac{2}{1}$, $\frac{R'Q'}{RQ} = \frac{2}{1}$, and $\frac{Q'P'}{QP} = \frac{2}{1}$; by the SSS Similarity Theorem, $\triangle PQR \sim \triangle P'Q'R'$.

❺ Calculate the ratios $\frac{P'R'}{PR}$, $\frac{R'Q'}{RQ}$, and $\frac{Q'P'}{QP}$. Then show that △PQR is similar to △P'Q'R'.

❻ What is the scale factor of △P'Q'R' to △PQR? How does it compare to the ratio $\frac{CP'}{CP}$? $\frac{2}{1}$; the ratios are equal.

A **dilation** is a transformation with center C and scale factor k that maps each point P to an image point P' so that P' lies on $\overrightarrow{CP}$ and $CP' = k \cdot CP$.

As you saw in the Geo-Activity above, a dilation maps a figure onto a similar figure, called the image.

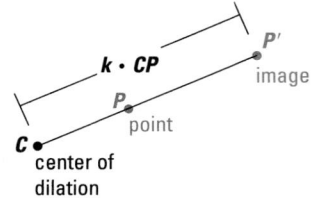

7.6 Dilations 393

---

❶ Plan

Pacing
Suggested Number of Days

Basic: 2 days
Average: 2 days
Advanced: 2 days
Block Schedule: 1 block

Teaching Resources

📄 **Blacklines**
(See page 354B.)

📃 **Transparencies**
- Warm-Up with Quiz
- Answers

🖥 **Technology**
- Electronic Teacher Tools
- Test and Practice Generator
- Online Lesson Planner
- Internet Support

Geo-Activity

Goal Use a compass and straightedge to create a dilation and explore the measures of the two figures.

Key Discovery Two similar triangles, one of which is a dilation of the other, have corresponding side lengths whose ratio is the scale factor of the dilation.

❷ Teach

Content and Teaching Strategies
For background information on geometric concepts and teaching strategies related to this lesson, see pages 354E and 354F in this Teacher's Edition.

Extra Example 1

Tell whether the dilation is a *reduction* or an *enlargement*.

a.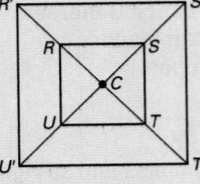

enlargement

b.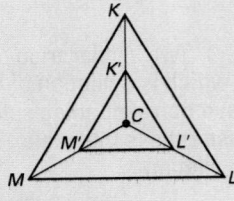

reduction

Extra Example 2

Find the scale factor of the dilation.

a.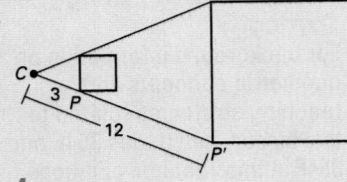

4

b.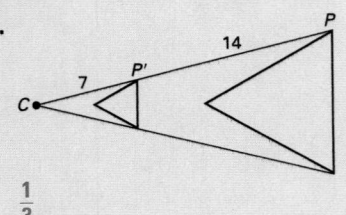

$\frac{1}{3}$

394

Visualize It!

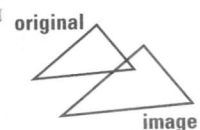

original

image

For transformations in this book, the original figure is blue and the image is red.

Student Help

STUDY TIP

If you know the scale factor k of a dilation, you can tell the type of dilation.

If $k < 1$, then the dilation is a reduction.

If $k > 1$, then the dilation is an enlargement.

Types of Dilations If the image is smaller than the original figure, then the dilation is a **reduction**. If the image is larger than the original figure, then the dilation is an **enlargement**.

EXAMPLE 1 Identify Dilations

Tell whether the dilation is a *reduction* or an *enlargement*.

a.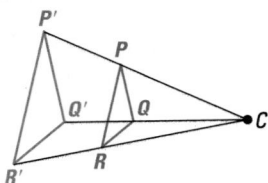

b.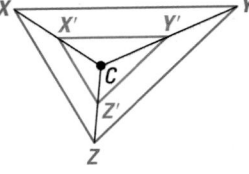

Solution

a. The dilation is an enlargement because the image ($\triangle P'Q'R'$) is larger than the original figure ($\triangle PQR$).

b. The dilation is a reduction because the image ($\triangle X'Y'Z'$) is smaller than the original figure ($\triangle XYZ$).

Scale Factor The *scale factor* of a dilation is the ratio of CP' to CP. This is also equal to the scale factor of $\triangle P'Q'R'$ to $\triangle PQR$.

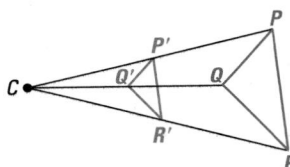

$$\text{scale factor} = k = \frac{CP'}{CP} = \frac{P'Q'}{PQ}$$

EXAMPLE 2 Find Scale Factors

Find the scale factor of the dilation.

a.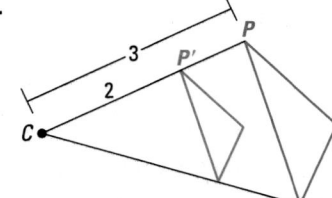

b.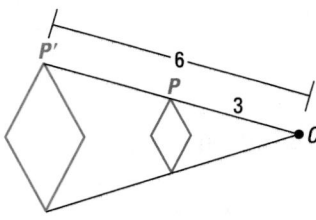

Solution

Find the ratio of CP' to CP.

a. scale factor $= \frac{CP'}{CP} = \frac{2}{3}$

b. scale factor $= \frac{CP'}{CP} = \frac{6}{3} = 2$

Checkpoint ✓ *Identify Dilations and Find Scale Factors*

Tell whether the dilation is a *reduction* or an *enlargement*. Then find the scale factor of the dilation.

1.

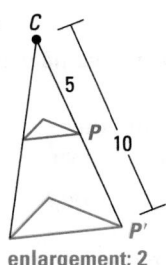

enlargement; 2

2.

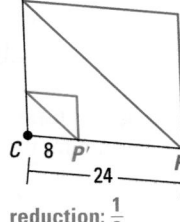

reduction; $\frac{1}{3}$

3.

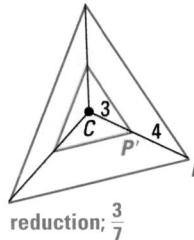

reduction; $\frac{3}{7}$

EXAMPLE 3 **Dilations and Similar Figures**

$\triangle P'Q'R'$ is the image of $\triangle PQR$ after a reduction. Find the value of x.

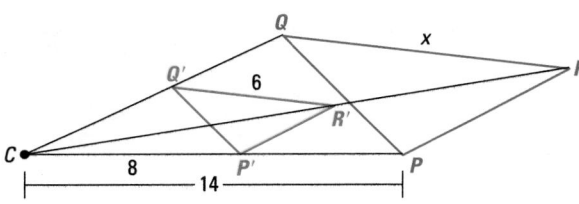

Student Help

STUDY TIP
Recall that the scale factor of the dilation equals the ratio of the corresponding sides of the figures. ......▶

Solution

$$\frac{CP'}{CP} = \frac{Q'R'}{QR} \qquad \text{Write a proportion.}$$

$$\frac{8}{14} = \frac{6}{x} \qquad \text{Substitute 8 for } CP', \text{ 14 for } CP, \text{ 6 for } Q'R', \text{ and } x \text{ for } QR.$$

$$8 \cdot x = 14 \cdot 6 \qquad \text{Cross product property}$$

$$8x = 84 \qquad \text{Multiply.}$$

$$x = 10.5 \qquad \text{Divide each side by 8.}$$

Checkpoint ✓ *Dilations and Similar Figures*

The red figure is the image of the blue figure after a dilation. Find the value of the variable.

4.

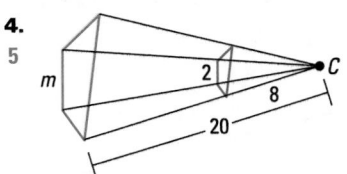

5.

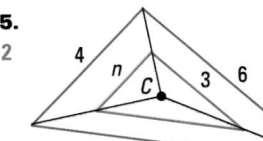

7.6 Dilations **395**

Geometric Reasoning

If you have access to a photocopier, draw a rectangle exactly 2 inches by 4 inches on a sheet of paper. Photocopy the sheet on different enlargement and reduction settings, noting which settings you used. Convert the settings to scale factors and list the scale factors on the board or overhead. Show students the original rectangle and have them match the photocopies to the scale factors. Students can also measure the photocopied figures to check how accurately the photocopier enlarges or reduces.

Extra Example 3

$\triangle S'T'U'$ is the image of $\triangle STU$ after a reduction. Find the value of x. **21**

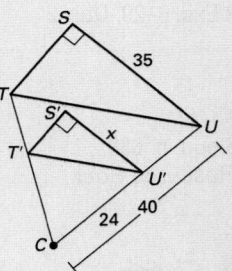

✓ **Concept Check**

How do you tell from the scale factor whether a dilation is a reduction or an enlargement? **If the scale factor is less than 1, the dilation is a reduction. If the scale factor is greater than 1, the dilation is an enlargement.**

🔒 **Daily Puzzler**

Suppose you have a photocopier that does enlargements where the image is 50% larger than the original. Explain how you can use repeated copying to make an enlargement where the image is 125% larger than the original.
Sample answer: Copy the image one time to get a scale factor of 1.5. Then copy the copy again to get a scale factor of 1.5 · 1.5 = 2.25. This image is 125% larger than the original.

395

7.6 Exercises

Assignment Guide

BASIC
Day 1: pp. 396–398 Exs. 6–9, 21–29
Day 2: pp. 396–398 Exs. 11–14, 17–20, Quiz 2
AVERAGE
Day 1: pp. 396–398 Exs. 6–10, 15, 21–29
Day 2: pp. 396–398 Exs. 11–14, 16–20, Quiz 2
ADVANCED
Day 1: pp. 396–398 Exs. 6–10, 15, 21–29
Day 2: pp. 396–398 Exs. 11–14, 16–20, Quiz 2; EC: TE p. 354D*, classzone.com
BLOCK SCHEDULE
pp. 396–398 Exs. 6–29, Quiz 2

Extra Practice

• Student Edition, p. 688
• Chapter 7 Resource Book, pp. 64–65

Homework Check

To quickly check student understanding of key concepts, go over the following exercises:

Basic: 6, 7, 8, 11, 13
Average: 7, 8, 10, 11, 13
Advanced: 8, 9, 10, 12, 16

✗ Common Error

In Exercise 9, watch for students who forget that $CP' = 8 + 28$. Have students review Checkpoint Exercise 3 on page 395 or work the exercise in class.

Guided Practice

Vocabulary Check

1. Complete: In a *dilation*, every image is __?__ to the original figure.
similar

Skill Check

2. She found $\frac{CP}{CP'}$ instead of $\frac{CP'}{CP}$.

2. Error Analysis Katie found the scale factor of the dilation shown to be $\frac{1}{2}$. What did Katie do wrong?

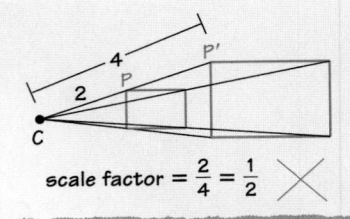

scale factor $= \frac{2}{4} = \frac{1}{2}$ ✗

3. Is the dilation shown a reduction or an enlargement? How do you know?
Enlargement; the red image is larger than the blue original figure.

Tell whether the dilation is a *reduction* or an *enlargement*.

4.

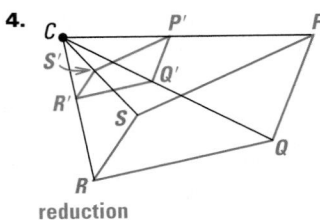

reduction

5.

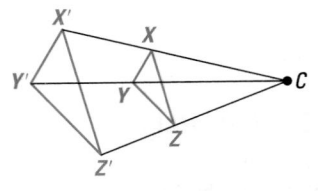

enlargement

Practice and Applications

Extra Practice

See p. 688.

Identifying Dilations Tell whether the dilation is a *reduction* or an *enlargement*. Then find its scale factor.

6.

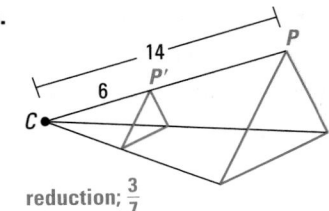

reduction; $\frac{3}{7}$

7.

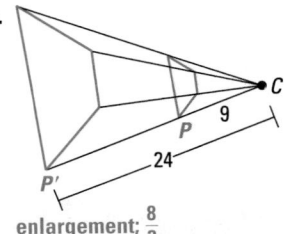

enlargement; $\frac{8}{3}$

8.

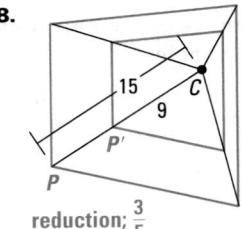

reduction; $\frac{3}{5}$

9.

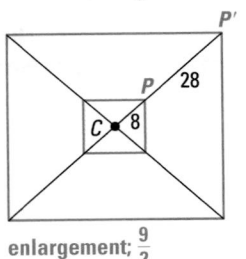

enlargement; $\frac{9}{2}$

Homework Help

Example 1: Exs. 6–9, 11–14
Example 2: Exs. 6–9, 15
Example 3: Exs. 11–14, 16

10. Visualize It! Sketch quadrilateral *EFGH*. Then sketch a dilation of *EFGH* with center *C* and scale factor 3. See margin.

Dilations and Similar Figures The red figure is the image of the blue figure after a dilation. Tell whether the dilation is a *reduction* or an *enlargement*. Then find the value of the variable.

11.
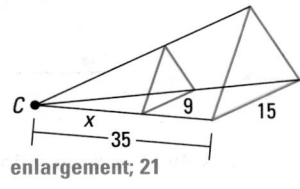
enlargement; 21

12.
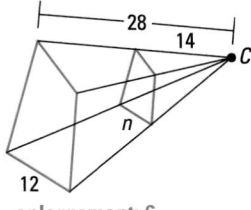
enlargement; 6

13.
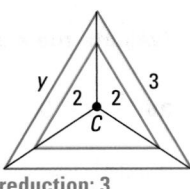
reduction; 3

14.
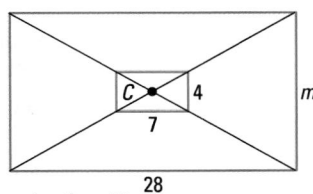
reduction; 16

10. *Sample answer:*
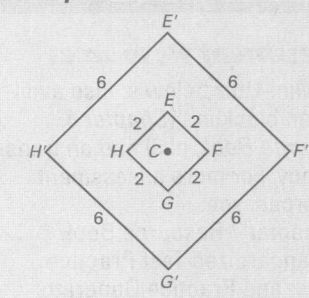

Student Help
CLASSZONE.COM

HOMEWORK HELP
Extra help with problem solving in Exs. 15–16 is at classzone.com

Flashlight Image In Exercises 15 and 16, use the following information and the diagram below.

The flashlight in the photograph below has a removable clear cap with a picture of a bug on it. When the flashlight is on, the bug is projected onto a wall. This situation models a dilation.

15. What is the scale factor of the dilation? 4

16. Write and solve a proportion to find the height of the bug projected onto the wall. 8 cm

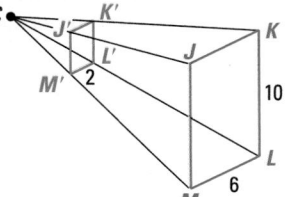

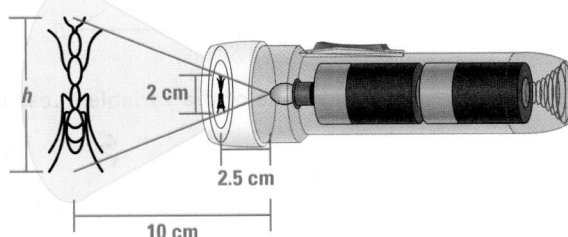

Standardized Test Practice

17. Multiple Choice The length of a side of $J'K'L'M'$ is what percent of the length of the corresponding side of $JKLM$? C

(A) 3%

(B) 12%

(C) $33\frac{1}{3}$%

(D) 300%

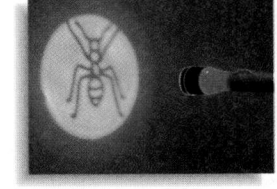

7.6 Dilations **397**

Assessment Resources

The Mini-Quiz below is also available on blackline (*Chapter 8 Resource Book*, p. 9) and on transparency. For more assessment resources, see:
- Chapter 7 Resource Book
- Standardized Test Practice
- Test and Practice Generator

Mini-Quiz

1. Tell whether the dilation is a *reduction* or *enlargement*. Then find its scale factor.

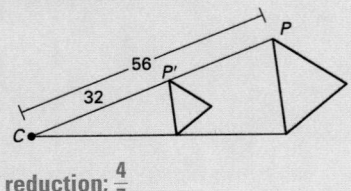

reduction; $\frac{4}{7}$

The figure with vertex P' is the image of the figure with vertex P after a dilation. Tell whether the dilation is a *reduction* or an *enlargement*. Then find the value of the variable.

2.

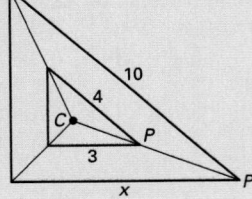

enlargement; $x = 7.5$

3.

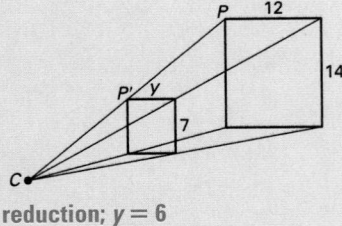

reduction; $y = 6$

Mixed Review

Finding Angle Measures Find the measure of $\angle 1$. *(Lesson 4.2)*

18.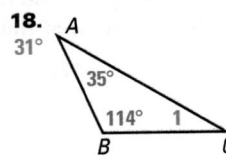

19.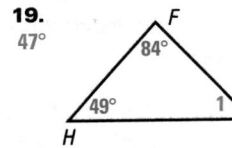

20.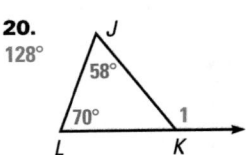

Solving Proportions Solve the proportion. *(Lesson 7.1)*

21. $\frac{d}{7} = \frac{12}{28}$ 3

22. $\frac{35}{10} = \frac{7}{x+1}$ 1

23. $\frac{9}{2} = \frac{t+3}{12}$ 51

Algebra Skills

Evaluating Expressions Evaluate the expression when $x = -3$. *(Skills Review, p. 670)*

24. $x^2 + 8$ 17

25. $-4x - 9$ 3

26. $(x + 2)(x - 5)$ 8

27. $11 - \frac{15}{x}$ 16

28. $2x^3$ -54

29. $4x^2 + 3x - 1$ 26

Quiz 2

Determine whether the triangles are similar. If they are similar, write a similarity statement. *(Lesson 7.4)*

1.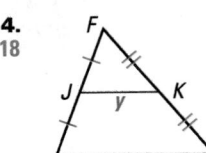

yes (by the SAS Similarity Theorem); $\triangle ABC \sim \triangle FGH$

2.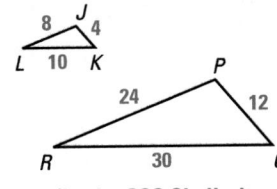

yes (by the SSS Similarity Theorem); $\triangle JKL \sim \triangle PQR$

Find the value of the variable. *(Lesson 7.5)*

3. 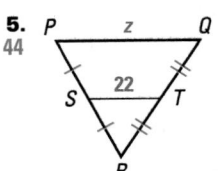 8

4. 18

5. 44

Tell whether the dilation is a *reduction* or an *enlargement*. Then find its scale factor. *(Lesson 7.6)*

6.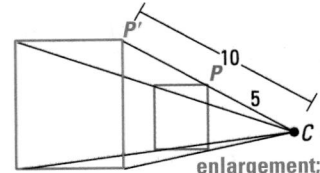

enlargement; 2

7.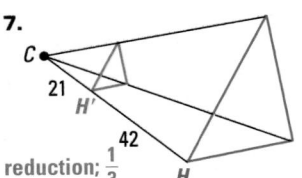

reduction; $\frac{1}{3}$

Question

In a dilation in a coordinate plane, how are the coordinates of the original figure and its image related?

Explore

1 Create a set of axes. Draw a triangle, △*CDE*.

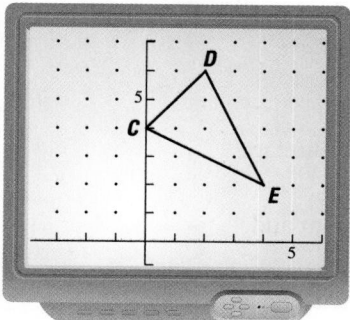

2 Dilate △*CDE* with center (0, 0) using a scale factor of 0.5. This creates △*C'D'E'*.

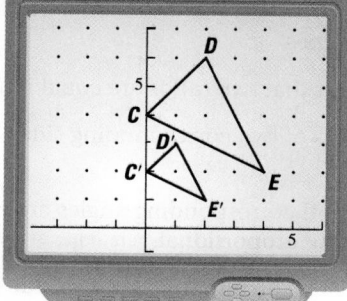

3 Find the coordinates of points *C*, *C'*, *D*, *D'*, *E*, and *E'*.

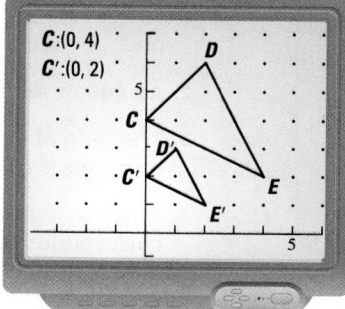

Think About It

Student Help

KEYSTROKE HELP
If your software has a Snap to Grid feature, use it to plot points at integer coordinates.

1. The image of each point (*x*, *y*) is the point (0.5*x*, 0.5*y*).

2. For a dilation with center at (0, 0) and scale factor *k*, the image of point *P*(*x*, *y*) is the point *P'*(*kx*, *ky*).

3. Predictions may vary; the coordinates of the vertices of the image are *X'*(−3, 9), *Y'*(6, 6), and *Z'*(0, −3).

1. Drag any of the vertices. Compare the *x*- and *y*-coordinates of each vertex of △*CDE* with the *x*- and *y*-coordinates of the corresponding vertex of △*C'D'E'*.

2. Repeat Steps 1–3 and Exercise 1 several times using different scale factors. Make a conjecture about the relationship between the scale factor of the dilation and the coordinates of the vertices of the original figure and the image.

3. △*XYZ* is shown at the right. Suppose you dilate △*XYZ* using the origin as the center of dilation and a scale factor of 3.

Use your conjecture from Exercise 2 to predict the coordinates of the vertices of the image. Check your results using the geometry software.

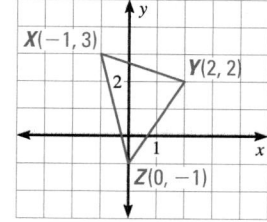

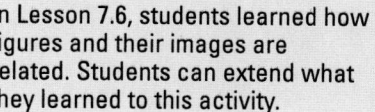

Goal
Students explore the relationship between the coordinates of a figure and its dilated image.

Materials
See the Technology Activity Keystrokes Masters in the *Chapter 7 Resource Book*.

LINK TO LESSON
In Lesson 7.6, students learned how figures and their images are related. Students can extend what they learned to this activity.

2 Managing the Activity

Tips for Success
In Step 1, students' triangles do not have to match the one pictured in the text. However, the activity will work best if their triangles are 3 or 4 units away from the origin so there is room for the dilated image between the origin and the original figure. If students have difficulty making a conjecture in Exercise 2, have them restart the activity using a triangle located entirely in the first quadrant with vertices whose coordinates are all *even* numbers. This will insure that the coordinates of the vertices of the image will be integers.

3 Closing the Activity

KEY DISCOVERY
When you dilate a figure with center (0, 0), the coordinates of the vertices of the image can be found by multiplying the coordinates of the vertices of the original figure by the scale factor.

Activity Assessment
Use Exercises 2 and 3 to assess student understanding.

Chapter  7 Summary and Review

VOCABULARY

• **ratio of *a* to *b*,** *p. 357*
• **proportion,** *p. 359*
• **means of a proportion,** *p. 359*
• **extremes of a proportion,** *p. 359*

• **similar polygons,** *p. 365*
• **scale factor,** *p. 366*
• **midsegment of a triangle,** *p. 389*

• **dilation,** *p. 393*
• **reduction,** *p. 394*
• **enlargement,** *p. 394*

VOCABULARY REVIEW

Fill in each blank.

1. In the proportion $\frac{f}{g} = \frac{h}{j}$, __?__ and __?__ are the means, and __?__ and __?__ are the extremes. *g; h; f; j*

2. An equation that states that two ratios are equal is a(n) __?__. proportion

3. The ratio of the lengths of two corresponding sides of two similar polygons is called the __?__. scale factor

4. Two polygons are __?__ if corresponding angles are congruent and corresponding sides are proportional. similar

5. A segment that connects the midpoints of two sides of a triangle is called the __?__. midsegment

6. A(n) __?__ is a transformation that maps a figure onto a similar figure. dilation

7. If the image of a dilation is larger than the original figure, then the dilation is a(n) __?__. enlargement

---

7.1 RATIO AND PROPORTION

Examples on pp. 357–360

EXAMPLE Solve the proportion $\frac{2}{5} = \frac{4}{x + 6}$.

$\frac{2}{5} = \frac{4}{x + 6}$ Write original proportion.

$2(x + 6) = 5 \cdot 4$ Cross product property

$2x + 12 = 20$ Use distributive property and multiply.

$2x = 8$ Subtract 12 from each side.

$x = 4$ Divide each side by 2.

Solve the proportion.

8. $\frac{y}{18} = \frac{1}{3}$ 6

9. $\frac{4}{7} = \frac{24}{s}$ 42

10. $\frac{2}{3} = \frac{z-6}{15}$ 16

11. $\frac{x+4}{40} = \frac{3}{5}$ 20

7.2 SIMILAR POLYGONS

Examples on pp. 365–368

> **EXAMPLE** **ABCD ~ EFGH. Find the value of x.**
>
> $\frac{EF}{AB} = \frac{EH}{AD}$ Write proportion.
>
> $\frac{x}{20} = \frac{20}{25}$ Substitute.
>
> $25x = 400$ Cross product property
>
> $x = 16$ Divide each side by 25.

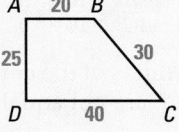 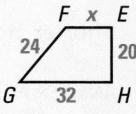

The polygons are similar. Find the value of the variable.

12.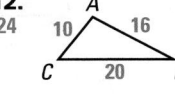

13.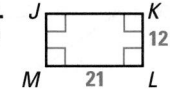

14.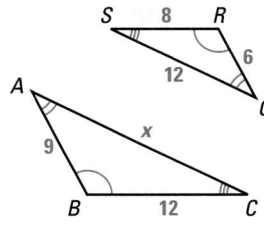

7.3 SHOWING TRIANGLES ARE SIMILAR: AA

Examples on pp. 372–374

> **EXAMPLE** **Determine whether the triangles below are similar. If they are similar, write a similarity statement.**
>
> $\angle A \cong \angle D$ because they both measure 55°, and $\angle B \cong \angle E$ because they are both right angles.
>
> Therefore, by the AA Similarity Postulate, the triangles are similar. $\triangle ABC \sim \triangle DEF$.

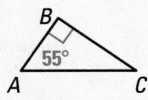 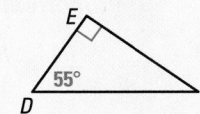

Determine whether the triangles are similar. If they are similar, write a similarity statement. 15, 17. See margin.

15.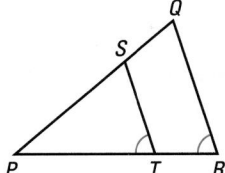

16. no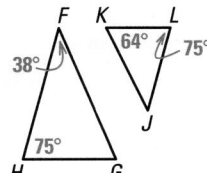

17.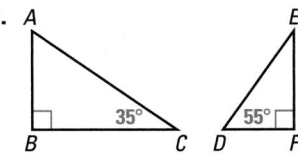

15. $\angle P \cong \angle P$ and $\angle STP \cong \angle QRP$, so $\triangle PST \sim \triangle PQR$ by the AA Similarity Postulate.

17. $\angle B \cong \angle F$ and $\angle A \cong \angle D$, so $\triangle ABC \sim \triangle DFE$ by the AA Similarity Postulate.

18. $\angle U \cong \angle X$, $\frac{SU}{VX} = \frac{24}{18} = \frac{4}{3}$, and $\frac{UT}{XW} = \frac{16}{12} = \frac{4}{3}$; the lengths of the sides that include the congruent angles are proportional, so $\triangle SUT \sim \triangle VXW$ by the SAS Similarity Theorem.

19. $\angle QSR \cong \angle UST$, $\frac{QS}{US} = \frac{35}{28} = \frac{5}{4}$, and $\frac{RS}{TS} = \frac{30}{24} = \frac{5}{4}$; the lengths of the sides that include the congruent angles are proportional, so $\triangle QRS \sim \triangle UTS$ by the SAS Similarity Theorem.

7.4 SHOWING TRIANGLES ARE SIMILAR: SSS AND SAS

Examples on pp. 379–381

EXAMPLE Determine whether the triangles at the right are similar. If they are similar, write a similarity statement.

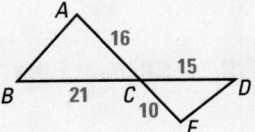

$\angle ACB \cong \angle ECD$, because vertical angles are congruent.

To apply the SAS Similarity Theorem, compare the ratios of the corresponding sides that include $\angle ACB$ and $\angle ECD$.

Shorter sides $\frac{CE}{AC} = \frac{10}{16} = \frac{5}{8}$ **Longer sides** $\frac{CD}{BC} = \frac{15}{21} = \frac{5}{7}$

ANSWER ▸ The ratios of the corresponding sides that include $\angle ACB$ and $\angle DCE$ are not equal. Therefore, the triangles are not similar.

Determine whether the triangles are similar. If they are similar, write a similarity statement. 18, 19. See margin.

18.
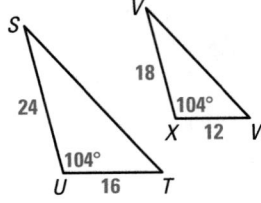

19.
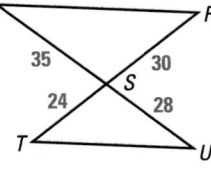

20.
no
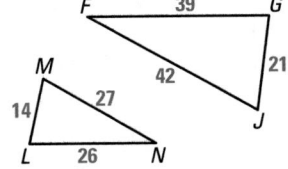

7.5 PROPORTIONS AND SIMILAR TRIANGLES

Examples on pp. 386–389

EXAMPLE Find the value of *x*.

Because $\overline{NM}$ is parallel to $\overline{KL}$, you can use the Triangle Proportionality Theorem to find the value of *x*.

$\frac{JN}{NK} = \frac{JM}{ML}$ Triangle Proportionality Theorem

$\frac{12}{20} = \frac{x}{25}$ Substitute.

$12 \cdot 25 = 20 \cdot x$ Cross product property

$300 = 20x$ Multiply.

$15 = x$ Divide each side by 20.

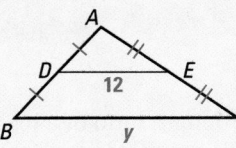

EXAMPLE   **Find the value of *y*.**

$DE = \frac{1}{2}BC$ Midsegment Theorem

$12 = \frac{1}{2}y$ Substitute.

$24 = y$ Multiply each side by 2.

Find the value of the variable.

21.

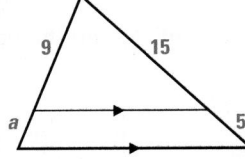

22.

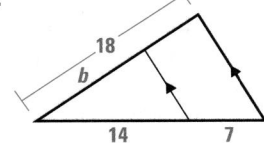

23.

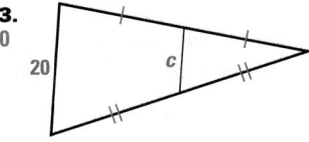

7.6   DILATIONS

Examples on
pp. 393–395

EXAMPLES   **Tell whether the dilation is a *reduction* or an *enlargement*. Then find the scale factor of the dilation.**

a.

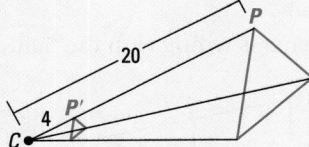

b.

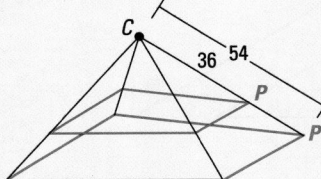

The image is smaller than the original figure, so the dilation is a reduction.

The image is larger than the original figure, so the dilation is an enlargement.

Scale factor $= \dfrac{CP'}{CP} = \dfrac{4}{20} = \dfrac{1}{5}$

Scale factor $= \dfrac{CP'}{CP} = \dfrac{54}{36} = \dfrac{3}{2}$

Tell whether the dilation is a *reduction* or an *enlargement*. Then find the scale factor of the dilation.

24.

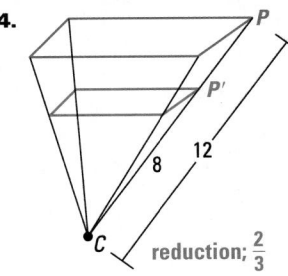

reduction; $\frac{2}{3}$

25.

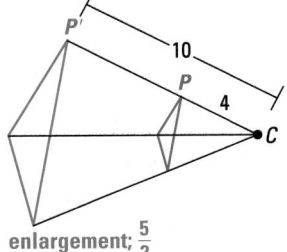

enlargement; $\frac{5}{2}$

26.

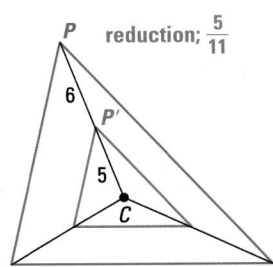

reduction; $\frac{5}{11}$

Additional Resources

Chapter 7 Resource Book
- Chapter Test (2 levels), pp. 72–75
- SAT/ACT Chapter Test, p. 76
- Alternative Assessment, pp. 77–78

Test and Practice Generator

Chapter  7 Chapter Test

1. The height-to-length ratio of a brick wall is 3 : 20. The wall is 40 feet long. How high is it? **6 feet**

Solve the proportion.

2. $\frac{2}{9} = \frac{x}{27}$ **6**

3. $\frac{x-3}{35} = \frac{4}{7}$ **23**

Determine whether the polygons are similar. If they are similar, write a similarity statement.

4. **no**
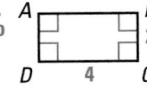

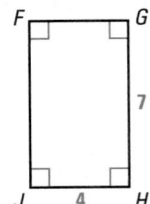

5.
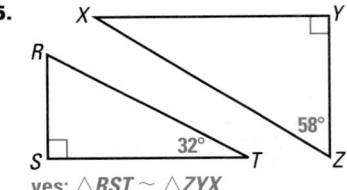
yes; △RST ~ △ZYX

6.
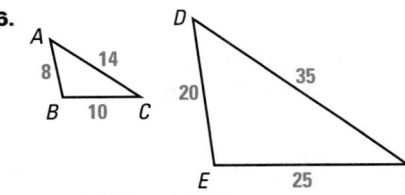
yes; △ABC ~ △DEF

7. **no**
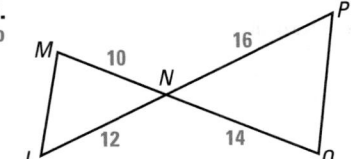

8. *QRST* is similar to *WXYZ*. The ratio of *QR* to *WX* is 3 : 7. What is the ratio of the perimeter of *QRST* to the perimeter of *WXYZ*? **3 : 7**

9. Find *TQ* in the figure below. **18**
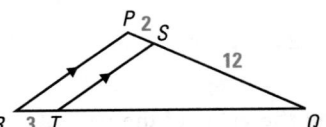

In Exercises 10 and 11, find the value of the variable.

10.
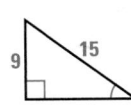 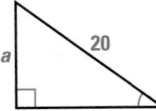

11.
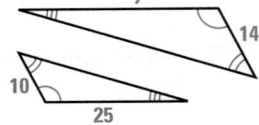

12. What is wrong with the figure below?
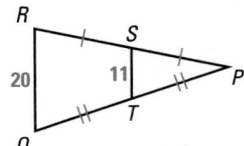
The midsegment of a triangle should be half as long as the side to which it is parallel.

Determine whether the dilation is a *reduction* or an *enlargement*. Then find its scale factor.

13.
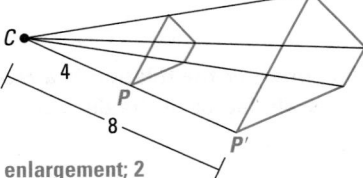
enlargement; 2

14.
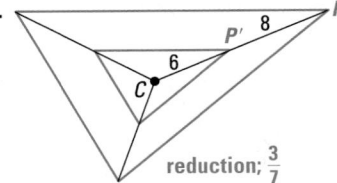
reduction; $\frac{3}{7}$

Test Tip Scan all the questions in the section you are working on. This can help you see easy questions and plan how to spend your time.

1. Simplify the ratio $\frac{16 \text{ ft}}{4 \text{ yd}}$. **B**

   (A) $\frac{1}{4}$ (B) $\frac{4}{3}$ (C) $\frac{4}{1}$ (D) $\frac{3}{4}$

2. $\triangle ABC$ is similar to $\triangle DEF$. What are the values of x and y? **G**

   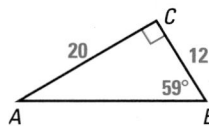 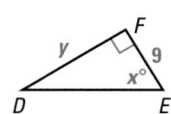

   (F) $x = 31, y = 15$ (G) $x = 59, y = 15$
   (H) $x = 59, y = 16$ (J) $x = 31, y = 12$

3. Which of the following statements is *false*? **D**

   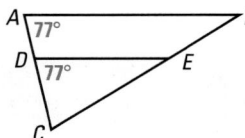

   (A) $\frac{CD}{CA} = \frac{CE}{CB}$ (B) $\angle DEC \cong \angle ABC$

   (C) $\triangle ACB \sim \triangle DCE$ (D) $\triangle BAC \sim \triangle DEC$

4. What is the value of x in the diagram? **H**

   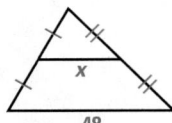

   (F) 8 (G) 12 (H) 24 (J) 96

5. Solve the proportion $\frac{x + 4}{18} = \frac{11}{6}$. **B**

   (A) 24 (B) 29 (C) 33 (D) 37

6. You want to enlarge a drawing of a car that is 4 inches long. Your enlargement will be 12 inches long. What is the scale factor of the enlargement to the drawing? **F**

   (F) 3 to 1 (G) 4 to 1
   (H) 1 to 3 (J) 1 to 4

7. Determine whether the dilation is a reduction or enlargement. Then find its scale factor. **B**

   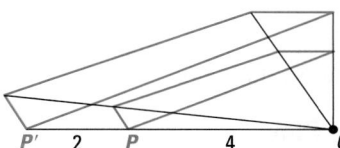

   (A) enlargement; $\frac{2}{3}$ (B) enlargement; $\frac{3}{2}$

   (C) reduction; $\frac{2}{3}$ (D) reduction; $\frac{3}{2}$

8. **Multi-Step Problem** The Greek mathematician Thales calculated the height of the Great Pyramid in Egypt by placing a rod at the tip of the pyramid's shadow and using similar triangles as shown. **See margin.**

   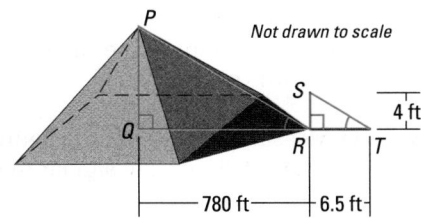

   Not drawn to scale

   a. Explain why $\triangle PQR$ is similar to $\triangle SRT$.
   b. Find the scale factor of $\triangle PQR$ to $\triangle SRT$.
   c. Find the height of the Great Pyramid.

8. a. $\angle Q \cong \angle SRT$ and $\angle PRQ \cong \angle STR$, so $\triangle PQR \sim \triangle SRT$ by the AA Similarity Postulate.

   b. $\frac{120}{1}$

   c. 480 feet

Teaching Tip
The Brain Games activity provides a motivating way to review selected content in the chapter. For a more comprehensive review, see the Chapter Summary and Review on pp. 400–403.

BraiN GaMes

Tangrams

Materials
• 7 tangram pieces
• scissors
• ruler

Object of the Game To find as many ways as possible to use some or all of the tangram pieces to make shapes that are similar to individual pieces.

Set Up If your tangram pieces are printed on paper, carefully cut them out.

How to Play

Step 1 ▶ Look at the smallest triangle. Find tangram pieces that are similar to the triangle.

Step 2 ▶ Use more than one piece to make triangles that are similar to the original triangle. Trace each figure to record your answers.

Step 3 ▶ Look at the parallelogram. Use more than one piece to make parallelograms that are similar to the original parallelogram. Trace each figure to record your answers.

Step 4 ▶ Measure the sides of the figures to check if they are similar to the original figure.

Another Way to Play Use more than one piece to create a figure then try to create figures similar to it.

When simplifying radical expressions, you can use the *Product Property of Radicals*, which states that $\sqrt{ab} = \sqrt{a} \cdot \sqrt{b}$, where $a \geq 0$ and $b \geq 0$.

EXAMPLE 1   Simplify Radical Expressions

Simplify the expression $\sqrt{20}$.

Solution

Look for perfect square factors to remove from the radicand.

$$\sqrt{20} = \sqrt{4 \cdot 5} \qquad \text{Factor using perfect square factor.}$$
$$= \sqrt{4} \cdot \sqrt{5} \qquad \text{Product Property of Radicals}$$
$$= 2\sqrt{5} \qquad \text{Simplify: } \sqrt{4} = 2.$$

Try These

Simplify the expression.

1. $\sqrt{12}$ $2\sqrt{3}$ **2.** $\sqrt{63}$ $3\sqrt{7}$ **3.** $\sqrt{44}$ $2\sqrt{11}$ **4.** $\sqrt{125}$ $5\sqrt{5}$

5. $\sqrt{150}$ $5\sqrt{6}$ **6.** $\sqrt{80}$ $4\sqrt{5}$ **7.** $\sqrt{48}$ $4\sqrt{3}$ **8.** $\sqrt{72}$ $6\sqrt{2}$

EXAMPLE 2   Use Formulas

Solve for ℓ in the formula for the perimeter of a rectangle, $P = 2\ell + 2w$.

Solution

$$P = 2\ell + 2w \qquad \text{Write the formula.}$$
$$P - 2w = 2\ell \qquad \text{Subtract 2w from each side.}$$
$$\frac{P - 2w}{2} = \ell \qquad \text{Divide each side by 2.}$$

ANSWER ▶ The formula, when solved for ℓ, is $\ell = \dfrac{P - 2w}{2}$.

Try These

Solve for the indicated variable in the formula.

9. Solve for h: $A = bh$ $h = \dfrac{A}{b}$ **10.** Solve for B: $V = \frac{1}{3}Bh$ $B = \dfrac{3V}{h}$

11. Solve for P: $S = 2B + Ph$ **12.** Solve for b_2: $A = \frac{1}{2}(b_1 + b_2)h$
$$P = \frac{S - 2B}{h} \qquad\qquad\qquad b_2 = \frac{2A}{h} - b_1$$

Algebra Review   **407**

Extra Example 1
Simplify the expression $\sqrt{18}$. $3\sqrt{2}$

Checkpoint Exercise 1
Simplify the expression $\sqrt{28}$. $2\sqrt{7}$

Extra Example 2
Solve for a in the Pythagorean Theorem, $a^2 + b^2 = c^2$.
$a = \sqrt{c^2 - b^2}$

Checkpoint Exercise 2
Solve for ℓ in the formula
$S = \pi r^2 + \pi r\ell$. $\ell = \dfrac{S - \pi r^2}{\pi r}$

Pacing and Assignment Guide

REGULAR SCHEDULE

Lesson	Les. Day	Basic	Average	Advanced
8.1	Day 1	EP p. 685 Exs. 1–3; pp. 413–415 Exs. 10–21, 33–40, 41–51 odd	pp. 413–415 Exs. 10–27, 31–40, 41–51 odd	pp. 413–415 Exs. 10–12, 13–25 odd, 27–40, 42–50 even; EC: TE p. 408D*, classzone.com
8.2	Day 1	pp. 421–423 Exs. 8–20, 26–28, 39–51 odd	pp. 421–423 Exs. 8–20, 26–28, 39–51 odd	pp. 421–423 Exs. 14–20, 26–28, 39–51 odd
	Day 2	EP p. 685 Exs. 4–6; pp. 421–423 Exs. 21–25, 31–33, 40–50 even, Quiz 1	pp. 421–423 Exs. 21–25, 29–34, 40–50 even, Quiz 1	pp. 421–423 Exs. 21–25, 29–34, 35–38*, 40, Quiz 1; EC: classzone.com
8.3	Day 1	pp. 427–429 Exs. 8–16, 20–22, 24–28, 34–46	pp. 427–429 Exs. 8–30, 34–46	pp. 427–429 Exs. 8–16 even, 17–19, 22–26, 30–34, 35–45 odd; EC: TE p. 408D*, classzone.com
8.4	Day 1	EP p. 681 Exs. 16–18; pp. 434–437 Exs. 5–13, 16–20, 26, 27, 32–35, 36–44 even, Quiz 2	pp. 434–437 Exs. 5–19 odd, 21–27, 32–35, 37–43 odd, Quiz 2	pp. 434–437 Exs. 6–20 even, 21–35, 37–43 odd, Quiz 2; EC: TE p. 408D*, classzone.com
8.5	Day 1	pp. 442–445 Exs. 8–13, 18–22, 32, 35, 40–52 even	pp. 442–445 Exs. 8–22, 40–56 even	pp. 442–445 Exs. 8–22, 38*, 40–56 even
	Day 2	pp. 442–445 Exs. 23–28, 34, 37, 41–55 odd	pp. 442–445 Exs. 23–37, 41–55 odd	pp. 442–445 Exs. 23–37, 39*, 41–55 odd; EC: classzone.com
8.6	Day 1	SRH p. 671 Exs. 1–4; pp. 448–450 Exs. 8–13, 15, 18–22, 26–28, 32–42	pp. 448–450 Exs. 8–25, 29–40	pp. 448–450 Exs. 11–17, 20–38; EC: TE p. 408D*, classzone.com
8.7	Day 1	SRH p. 669 Exs. 5–8; pp. 456–459 Exs. 10–20, 47–55	pp. 456–459 Exs. 10–23, 41–46, 47–55 odd	pp. 456–459 Exs. 10–20 even, 21–23, 41–55 odd
	Day 2	pp. 456–459 Exs. 24–27, 31–35, 40–46, Quiz 3	pp. 456–459 Exs. 24–29, 31–40, Quiz 3	pp. 456–459 Exs. 26–29, 30*, 32–40, Quiz 3; EC: classzone.com
Review	Day 1	pp. 460–463 Exs. 1–40	pp. 460–463 Exs. 1–40	pp. 460–463 Exs. 1–40
Assess	Day 1	Chapter 8 Test	Chapter 8 Test	Chapter 8 Test

YEARLY PACING   Chapter 8 Total – **12 days**   Chapters 1–8 Total – **120 days**   Remaining – **40 days**

*Challenge Exercises   EP = Extra Practice   SRH = Skills Review Handbook   EC = Extra Challenge

BLOCK SCHEDULE

Day 1	Day 2	Day 3	Day 4	Day 5	Day 6
8.1 pp. 413–415 Exs. 10–27, 31–40, 41–51 odd **8.2** pp. 421–423 Exs. 8–20, 26–28, 39–51 odd	**8.2 cont.** pp. 421–423 Exs. 21–25, 29–34, 40–50 even, Quiz 1 **8.3** pp. 427–429 Exs. 8–30, 34–46	**8.4** pp. 434–437 Exs. 5–19 odd, 21–27, 32–35, 37–43 odd, Quiz 2 **8.5** pp. 442–445 Exs. 8–22, 40–56 even	**8.5 cont.** pp. 442–445 Exs. 23–37, 41–55 odd **8.6** pp. 448–450 Exs. 8–25, 29–40	**8.7** pp. 456–459 Exs. 10–29, 31–46, 47–55 odd, Quiz 3	**Review** pp. 460–463 Exs. 1–40 **Assess** Chapter 8 Test

YEARLY PACING   Chapter 8 Total – **6 days**   Chapters 1–8 Total – **60 days**   Remaining – **20 days**

Support Materials

CHAPTER RESOURCE BOOK

CHAPTER SUPPORT

Tips for New Teachers	p. 1	Strategies for Reading Mathematics	p. 5
Parent Guide for Student Success	p. 3		

LESSON SUPPORT

	8.1	8.2	8.3	8.4	8.5	8.6	8.7
Lesson Plans (regular and block)	p. 7	p. 15	p. 25	p. 33	p. 44	p. 55	p. 66
Warm-Up Exercises and Daily Quiz	p. 9	p. 17	p. 27	p. 35	p. 46	p. 57	p. 68
Technology Activities & Keystrokes				p. 36	p. 47	p. 58	
Practice (2 levels)	p. 10	p.18	p. 28	p. 38	p. 50	p. 60	p. 69
Reteaching with Practice	p. 12	p. 20	p. 30	p. 40	p. 52	p. 62	p. 71
Quick Catch-Up for Absent Students	p. 14	p. 22	p. 32	p. 42	p. 54	p. 64	p. 73
Learning Activities		p. 23					
Real-Life Applications						p. 65	p. 74

REVIEW AND ASSESSMENT

Quizzes	pp. 24, 43, 75	Alternative Assessment with Math Journal	p. 84
Brain Games Support	p. 76	Project with Rubric	p. 86
Chapter Review Games and Activities	p. 78	Cumulative Review	p. 88
Chapter Test (2 levels)	p. 79	Resource Book Answers	AN1
SAT/ACT Chapter Test	p. 83		

TRANSPARENCIES

	8.1	8.2	8.3	8.4	8.5	8.6	8.7
Warm-Up Exercises and Daily Quiz	p. 47	p. 48	p. 49	p. 50	p. 51	p. 52	p. 53
Visualize It Transparencies	✓	✓	✓	✓			✓
Answer Transparencies	✓	✓	✓	✓	✓	✓	✓

TECHNOLOGY

- Time-Saving Test and Practice Generator
- Electronic Teacher Tools
- Geometry in Motion: Video
- Classzone.com
- Online Lesson Planner

ADDITIONAL RESOURCES

- Worked-Out Solution Key
- Resources in Spanish
- Practice Workbook with Examples

Strategies for Strategic Learners

PROVIDE MODELS AND ILLUSTRATIONS

This textbook is well organized to address the needs of pupils who need additional scaffolding. Notice that students (1) are asked to draw figures, (2) recognize geometric figures in common objects, and (3) are provided with examples to which they can return. In addition to what is already provided in the book for Lessons 8.3–8.7, teachers can use graph paper to help students understand the idea of measuring area in square units. Providing hands-on cardboard models of various plane figures and asking students to find their areas will help many learners conceptualize the content of these lessons. Consider asking students to bring in other models they can use to estimate or calculate area.

PROVIDE SUFFICIENT PRACTICE

Many of the topics covered in this chapter can be related to daily situations that students encounter. For example, some students will have heard discussions about the square footage of a house or apartment. Carpet is sold in square feet and housing costs depend on the size of the home. Students may think of other examples, such as insect foggers that are designed for specific square footage areas.

Before discussing Lesson 8.3, ask students to draw a sketch representing the floor plan of their home, or even an imaginary home. Students' sketches should look something like the one below, with each room labeled with its length and width.

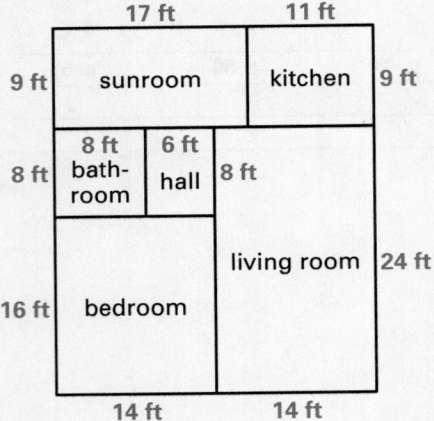

Once students have worked Examples 4 and 5 in Lesson 8.3, they can modify their sketches to create additional examples by removing some of the measures, as shown in the figure at the top of the next column.

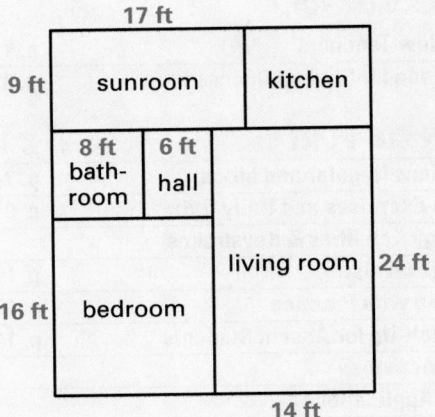

Students can then exchange sketches and try to find the area of the partner's home.

Strategies for English Learners

VOCABULARY STUDY OF ROOTS

Many mathematical terms share a Greek or Latin root. Recognizing the root can help students understand the meaning of new words. Have students look at the following table of Greek and Latin roots.

Greek or Latin Root	Mathematical Terms
congruere (Latin, *to agree*)	*congru*ent, *congru*ence
polys (Greek, *many*)	*poly*gon, *poly*hedron
hedra (Greek, *surface*)	poly*hedr*on, octa*hedr*al
circum (Latin, *around*)	*circum*ference
circulus (Latin, *circle* or *ring*)	*circ*le, *circul*ar
ferre (Latin, *bear, carry*)	circum*fere*nce

Encourage students to look for the similarities in words, and to watch for prefixes, suffixes, or roots that may give them clues to a word's meaning.

Strategies for Advanced Learners

RECORD UNANSWERED QUESTIONS

One characteristic of experts in any field is that they identify the unanswered questions in their field. Mathematicians are no exception, as there are questions which have intrigued them for centuries. Encourage advanced students to write down and

keep a record of their unanswered questions. These questions can help guide them as they continue to study mathematics. Some of their questions will be explained in future math courses.

Students may be intrigued to learn that there are substantial cash prizes offered for solutions to some famous unsolved mathematical problems. For example, in 2000, the Clay Mathematics Institute announced million-dollar awards for solutions to each of seven classic problems, including a geometry puzzle called *The Hodge Conjecture*.

DIFFERENTIATE INSTRUCTION IN TERMS OF DEPTH

By now students are familiar with the concept of the arithmetic mean of two numbers, which they know simply as the mean of the numbers. But students may not be familiar with a *geometric mean*. Advanced students may be able to see the parallel between the following problems.

1. Find the length x of a side of a square that has the same perimeter as a rectangle with sides of lengths 2 inches and 8 inches. Then write a general formula for the length x of a side of a square that has the same perimeter as a rectangle with sides of length a and b.

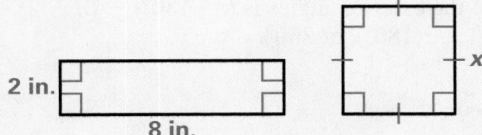

The perimeter of the rectangle is $2(2) + 2(8) = 4 + 16$, or 20 inches. So the side length of a square is $\frac{20}{4} = 5$ inches.

If the lengths of the sides of the rectangle are a and b, the perimeter is $2a + 2b$, so the length of a side of the square is $\frac{2a + 2b}{4} = \frac{a + b}{2}$. This is the formula for the **arithmetic mean** of two numbers a and b.

2. Find the length x of a side of a square that has the same area as a rectangle with sides of lengths 2 inches and 8 inches.

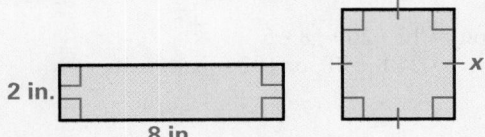

In order to solve this problem, students will have to note that if the area of a square is x^2, the length of a side of the square is the positive number x. The area of the rectangle is

$2(8) = 16$ square inches. Since $4^2 = 16$, the length of a side of the square is 4 inches.

In order to write a general formula for the length x of a side of a square that has the same area as a rectangle with sides of lengths a and b, students need to be familiar with square roots and the square root symbol. The formula is $x = \sqrt{ab}$. Note that this is also the formula for the **geometric mean** of two positive numbers a and b.

As students progress in mathematics, they will learn more about arithmetic and geometric means. The purpose of the Challenge problems at this point is to pique their interest and give them a basic understanding of more complicated relationships that will be studied in depth at a later time.

Challenge Problem for use with Lesson 8.1:

A certain kind of regular polygon can be divided into non-overlapping equilateral triangles, all of which share a common vertex. How many sides does the polygon have? *(6)*

Challenge Problem for use with Lesson 8.3:

A rectangular swimming pool that is 20 feet wide is to have a brick walk of a uniform width of 4 feet built on two sides, as shown. What must be the length x of the pool if the area of the walk is 224 square feet? *(32 ft)*

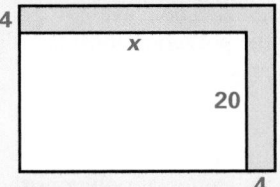

Challenge Problem for use with Lesson 8.4:

Find the area, to the nearest tenth, of an equilateral triangle whose sides are each 6 units long. (*Hint:* Use the fact that the height of an equilateral triangle bisects the base of the triangle.) *(15.6 square units)*

Challenge Problem for use with Lesson 8.6:

The legs of trapezoid *PQRS* are extended so that they meet at point *T*. Suppose the measurements are as shown. What is the area of trapezoid *PQRS*? *(256 square units)*

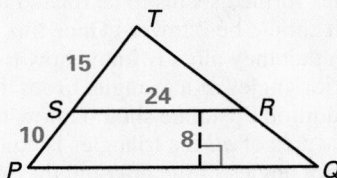

Content and Teaching Strategies

Finding the areas of polygons and circles is the focus of this chapter. Lesson 8.1 introduces the regular polygon and explores its characteristics. In Lesson 8.2, facts about the exterior and interior angles of a polygon are presented, generalizing some of the facts previously stated with regard to triangles and quadrilaterals. Lesson 8.3 begins the study of finding areas of the most common special kinds of polygons, focusing on squares and rectangles. Lesson 8.4 carries this study on to the important special case of triangles, while the area of parallelograms is taken up in Lesson 8.5. The last special kind of polygon whose area formula is presented in this chapter is the trapezoid, the topic of Lesson 8.6. Finally, the area and circumference formulas for a circle, as well as a method for finding the area of a sector of a circle, are given in Lesson 8.7.

Lesson 8.1

This lesson introduces the concept of a regular polygon. Make sure students understand that a regular polygon must have all its interior angles congruent, as well as all its sides congruent. To stress this point, you might ask students to think of examples in which all the sides of a polygon are congruent but not all the angles, and vice versa. Rhombuses are an example for which the sides but not necessarily the angles are congruent, while rectangles are an example of a polygon whose angles are congruent, but not necessarily its sides. Due to these facts, neither of these types of quadrilaterals can be called regular.

The distinction between concave and convex polygons is important, and it is helpful for students to be able to recognize a concave polygon when they see one. After this lesson, most of the polygons considered in this book will be convex.

Lesson 8.2

In this lesson, some of the theorems about angles of polygons that were introduced in the special cases of triangles and quadrilaterals back in Chapters 4 and 6 are generalized to include polygons with an arbitrary number of sides. Before starting this study, students should be able to identify which are the interior and exterior angles of a polygon in a given diagram. Remind them that an exterior angle is formed by extending *only one* side of the polygon.

It may help students remember the Polygon Interior Angles Theorem if you have students do the following activity. Have students draw an *n*-sided polygon (it's best if students draw a variety of polygons) and then draw all the diagonals from one vertex of that polygon. Once this is done, ask them how the number of triangles formed seems to be related to the number of sides of their polygon. (It should be 2 fewer.) Once this relationship is established, remind them that they already know how to find the sum of the measures of the interior angles of a triangle. From there, several applications of the Angle Addition Postulate should allow them to see that the sum of all the angle measures of *all* the triangles is equal to the sum of the measures of the interior angles of the polygon they drew. Note that in the case of a regular polygon, we can find the measure of *one* interior angle of the polygon, but since the angles of a non-regular polygon do not have equal measures, the formula does not allow us to do so for non-regular polygons.

> **QUICK TIP**
> You might want to lead students to acknowledge that the expression $\frac{(n-2) \cdot 180°}{n}$ can be used to find the measure of each interior angle of a regular convex polygon with *n* sides.

The Polygon Exterior Angles Theorem can be made more meaningful to students if you ask them to imagine that they are walking in a clockwise direction along the perimeter of the polygon in the diagram that accompanies the theorem. Each time they come to a vertex, they will need to turn to their right through a certain angle (an exterior angle at that vertex) to walk the next side of the polygon. If they continue doing this all the way back to their starting point, they will naturally have turned through a total of 360°. It follows that the sum of the exterior angles must be 360°.

USING ALGEBRA An alternative way to justify the Polygon Exterior Angles Theorem is to point out that at each vertex the sum of the interior angle measure and the measure of an exterior angle at that vertex is 180°. Therefore, the sum of the measures of the interior and exterior angles of an *n*-sided polygon—one of each at each vertex—is $n \cdot 180°$. Since the sum of the measures of the interior angles is $(n-2) \cdot 180°$, then the sum of the measures of the exterior angles is $(n \cdot 180°) - (n-2) \cdot 180° = (180n)° - (180n)° + 2(180°)$, or 360°.

Lesson 8.3

VISUALIZING COMPLEX POLYGONS The idea of dividing a complex polygon into rectangles in order to find its area (illustrated in Example 4) is an important one for practical applications of the formulas in this lesson. Mention that in some situations it may be easier to use subtraction than addition in order to find an area, as shown in the following example.

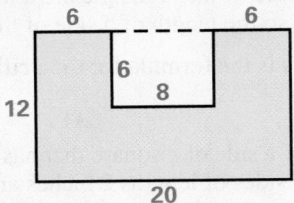

$$\text{Area of polygon} = 12 \cdot 20 - 8 \cdot 6$$
$$= 240 - 48, \text{ or } 192 \text{ square units}$$

Lesson 8.4

Even if you decide not to carry out Activity 8.4 that precedes this lesson, it is a good idea to go over the results. Particularly, have students notice that the height of the rectangle that encloses the triangle is the same as the height of the triangle, and the length of the rectangle is also the base of the triangle.

When discussing the area formula for a triangle, be sure to note that for a right triangle the area can always be found easily if the lengths of the legs are known. One of these lengths can be the base of the triangle while the other can be the height.

Emphasize that the paragraph and figures at the bottom of page 431 make the point that triangles can have the same area without being congruent.

Theorem 8.3 about the areas of similar polygons may be difficult for students to grasp because they often assume that the ratio of areas of similar polygons is equal to the scale factor, rather than to the square of the scale factor. As a memory tool, suggest that students make a connection between the *square* units of an area and the *square* of the scale factor for similar polygons.

Lesson 8.5

COMMON ERRORS Two mistakes often occur in connection with the formula for the area of a parallelogram. The first of these is that students often inadvertently include a factor of $\frac{1}{2}$ in the formula, confusing this formula with the formula for finding the area of a triangle. You can try to prevent this mistake by pointing out that a parallelogram is divided by one of its diagonals into two congruent triangles, each having the same base and height as the parallelogram. So the total area is $\frac{1}{2}bh + \frac{1}{2}bh = bh$.

The second mistake involves the misconception that the area of a parallelogram is given by the product of the lengths of two consecutive sides. One way to convince students that this thinking is incorrect is to use a model of a parallelogram with hinged corners that can be collapsed, making its height arbitrarily small. Students should recognize that as the height of the parallelogram changes, its area changes but the lengths of its sides do not. Stress that this means it is not possible to express the area of the parallelogram by a formula that involves only the lengths of the sides.

Lesson 8.6

Many students, even those who are strong in mathematics, have a tendency to avoid the trapezoid formula when one leg of the trapezoid is perpendicular to the bases. In this case the trapezoid can be subdivided by a vertical segment into a triangle and a rectangle. Students then use the area formulas for a triangle and a rectangle to find the area of the trapezoid. While the method is mathematically sound, there are two drawbacks: (1) students try to use the method when the trapezoid does not have a leg perpendicular to the bases, and (2) they miscalculate the base of the triangle, which is found by subtracting the length of one base from the length of the other.

The algebra required by a problem such as the one shown in Example 2 can be daunting for some students. Emphasize that there is usually more than one correct approach once the equation is set up. For example, since the area formula involves the fraction $\frac{1}{2}$, some students may want to multiply both sides by 2 in order to clear both sides of fractions as soon as possible, as is done in the example. Others may feel more comfortable finding the product of $\frac{1}{2}$ and 8 initially, to produce 4 (of course this works best when the height is an even integer). Remind students that doing the algebra step by step, being sure to always do the same thing to both sides, with the goal of isolating the unknown quantity on one side of the equation, is the approach to take.

Lesson 8.7

Many students find the symbol π somewhat mysterious. The simple fact is that this number is *defined* by the circumference formula (given in the box on page 452). The activity preceding this lesson suggests a way to show that this definition leads to the area formula on page 453. Remind students that π is an irrational number and that the value shown on a calculator when the π key is pressed is only an approximation.

> **QUICK TIP**
> Watch for students who are confused about which formula to use when finding the circumference or area of a circle. Point out that the area of a figure is given in *square* units and that the area formula for circles involves the *square* of the radius, r.

The method of finding the area of a sector shown in Example 4 should remind students of the proportions they studied in connection with similar triangles.

Chapter Goals

In this chapter, students learn to classify polygons as convex, concave, and regular. They build on this knowledge to explore angle measures in polygons. The remainder of the chapter is spent deriving the area formulas for several special polygons and using the formulas to find the areas of these polygons. Students will:

- Find measures of interior and exterior angles of polygons.
- Find the area of squares, rectangles, triangles, parallelograms, and trapezoids.
- Find the circumference and area of circles.

Application Note

Students may be fascinated by these hexagonal basaltic columns. Ask them to think of other examples of naturally occurring polygons. A common example is the hexagonal-shaped honeycomb. Also, flower blossoms often contain pentagon and hexagon designs. Spider webs are another common example. The crystals of many common substances, such as salt, exhibit polygonal shapes.

More information about rock formations is provided on the Internet at classzone.com

Application Links
CLASSZONE.COM

Chapter 8 Polygons and Area

Where do hexagons appear in nature?

Basaltic columns are geological formations that result from rapidly cooling lava. Most basaltic columns are hexagonal (six-sided) shapes.

The Giant's Causeway on the Irish coast features hexagonal columns that reach a height of 82 feet.

7.8 in.

9 in.

Learn More About It

You will learn more about rock formations in Exercise 31 on p. 436.

Who uses Polygons and Area?

FLOORING INSTALLER
Flooring installers are hired by flooring contractors to install, repair, or replace floor coverings, such as carpet and hardwood. (p. 414)

LANDSCAPE ARCHITECT
Landscape architects use engineering knowledge and creativity to design outdoor environments, such as parks, gardens, and recreational areas. (p. 457)

How will you use these ideas?

- Measure the angles of home plate on a baseball field. (p. 422)
- Find the area of a judo mat. (p. 428)
- Estimate the cost of planting a cornfield maze. (p. 429)
- See how cranberries are harvested. (p. 456)
- Determine the area covered by a lawn sprinkler. (p. 457)

409

Hands-On Activities

Activities (more than 20 minutes)
8.2 Angle Sum of Polygons, p. 416
8.4 Finding Area of Triangles, p. 430
8.4 Area of Similar Triangles, p. 438
8.7 Finding Area of Circles, p. 451

Geo-Activities (less than 20 minutes)
8.5 Exploring the Area of a Parallelogram, p. 439

Projects

A project covering Chapters 7–8 appears on pages 468–469 of the Student Edition. An additional project for Chapter 8 is available in the *Chapter 8 Resource Book*, pp. 86–87.

⊞ Technology

- Electronic Teacher Tools
- Test and Practice Generator
- Online Lesson Planner

📼 Video

- **Geometry in Motion**
  There is an animation supporting Lesson 8.2.

Internet Support
CLASSZONE.COM

- Application and Career Links
  436, 457
- Student Help
  412, 421, 425, 441, 449, 457

Diagnostic Tools

The **Chapter Readiness Quiz** can help you diagnose whether students have the following skills needed in Chapter 8:
- Recognize characteristics of polygons.
- Use angle measures of polygons.

Reteaching Material

This resource is available for students who need additional help with the skills on the Chapter Readiness Quiz:

📘 *Chapter 8 Resource Book*
- Reteaching with Practice (Lessons 8.1–8.7)

Additional Resources

The following resources are provided to help you prepare for the upcoming chapter and customize review materials:

📘 *Chapter 8 Resource Book*
- *Tips for New Teachers*, pp. 1–2
- *Parent Guide*, pp. 3–4
- *Lesson Plans*, pp. 7, 15, 25, 33, 44, 55, 66
- *Lesson Plans for Block Scheduling*, pp. 8, 16, 26, 34, 45, 56, 67

🖥 *Technology*
- Electronic Teaching Tools
- Online Lesson Planner
- Test and Practice Generator

Visualize It!

Additional suggestions for helping students visualize geometry are on pp. 414, 419, 422, 426, 427, 428, 432, 435, 449, and 454.

Chapter  8 Study Guide

PREVIEW — What's the chapter about?

- Finding measures of interior and exterior angles of polygons
- Finding the area of squares, rectangles, triangles, parallelograms, and trapezoids
- Finding the circumference and area of circles

Key Words

- convex, *p. 411*
- concave, *p. 411*
- equilateral, *p. 412*
- equiangular, *p. 412*
- regular, *p. 412*

- area, *p. 424*
- radius, *p. 452*
- diameter, *p. 452*
- circumference, *p. 452*
- sector, *p. 454*

PREPARE — Chapter Readiness Quiz

Take this quick quiz. If you are unsure of an answer, look at the reference pages for help.

Vocabulary Check *(refer to p. 325)*

1. Which term does *not* describe the polygon below? **A**

    (A) rhombus   **(B)** quadrilateral

    (C) rectangle   **(D)** parallelogram

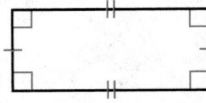

Skill Check *(refer to p. 305)*

2. What is the sum of the measures of the interior angles of a quadrilateral? **H**

    (F) 90°   **(G)** 180°   **(H)** 360°   **(J)** 540°

VISUAL STRATEGY — Breaking Polygons into Parts

Visualize It! ➡ You can break a complex polygon into simpler parts whose areas you know.

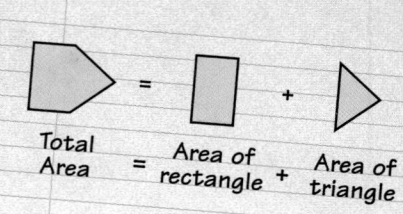

8.1 Classifying Polygons

Goal
Describe polygons.

Key Words
- convex
- concave
- equilateral
- equiangular
- regular

The Iranian tile pattern at the right shows several polygons. These polygons can be classified as *convex* or *concave*.

A polygon is **convex** if no line that contains a side of the polygon passes through the interior of the polygon. A polygon that is not convex is called **concave** .

Convex polygon

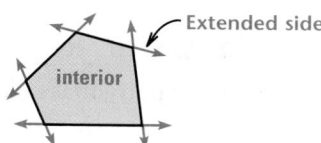
Extended side
interior

None of the extended sides pass through the interior.

Concave polygon

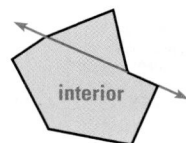
interior

At least one extended side passes through the interior.

EXAMPLE 1   Identify Convex and Concave Polygons

Decide whether the polygon is *convex* or *concave*.

a.

b.

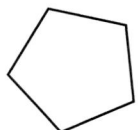

Solution

a. At least one extended side passes through the interior. So, the polygon is concave.

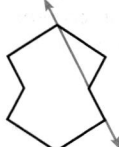

b. None of the extended sides pass through the interior. So, the polygon is convex.

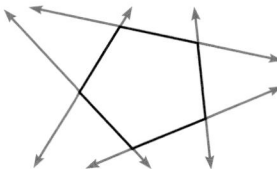

---

① Plan

Pacing
Suggested Number of Days
Basic: 1 day
Average: 1 day
Advanced: 1 day
Block Schedule: 0.5 block with 8.2

Teaching Resources

 Blacklines
(See page 408B.)

 **Transparencies**
• Warm-Up with Quiz
• Answers

Technology
• Electronic Teacher Tools
• Test and Practice Generator
• Online Lesson Planner
• Internet Support

② Teach

Content and Teaching Strategies
For background information on geometric concepts and teaching strategies related to this lesson, see pages 408E and 408F in this Teacher's Edition.

Tips for New Teachers
Students have a tendency to confuse the terms *convex* and *concave* for polygons. Provide several more polygonal shapes for students to copy. Have them extend the sides of each polygon as shown in Example 1 on this page. See the Tips for New Teachers on pp. 1–2 of the *Chapter 8 Resource Book* for additional notes about Lesson 8.1.

Extra Example 1
See next page.

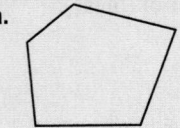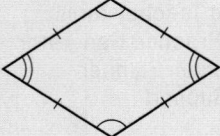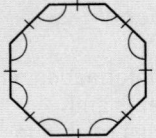
A polygon is **equilateral** if all of its sides are congruent. A polygon is **equiangular** if all of its interior angles are congruent. A polygon is **regular** if it is both equilateral and equiangular.

Equilateral

All sides are congruent.

Equiangular

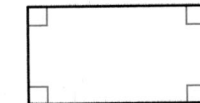

All angles are congruent.

Regular Polygons

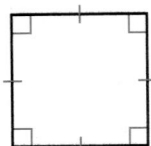 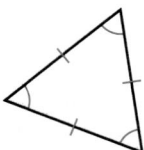 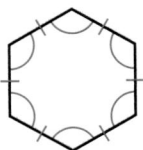

All sides are congruent and all angles are congruent.

EXAMPLE 2 Identify Regular Polygons

Decide whether the polygon is regular. Explain your answer.

a. 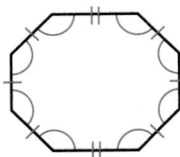 **b.** 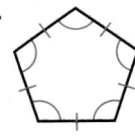 **c.**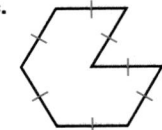

Solution

a. Although the polygon is equiangular, it is not equilateral. So, the polygon is not regular.

b. Because the polygon is both equilateral and equiangular, it is regular.

c. Although the polygon is equilateral, it is not equiangular. So, the polygon is not regular.

 **Describe Polygons**

Decide whether the polygon is *convex* or *concave*. Then decide whether the polygon is regular. Explain your answer.

1.  **2.**  **3.**

Guided Practice

Vocabulary Check

1. Sketch a *concave* polygon. See margin.

2. Describe the difference between an *equilateral* polygon and an *equiangular* polygon. All the *sides* of an equilateral polygon are congruent; all the *angles* of an equiangular polygon are congruent.

3. What is a *regular* polygon?
A regular polygon is both equilateral and equiangular.

Skill Check

Decide whether the polygon shown in black is *convex* or *concave*.

4.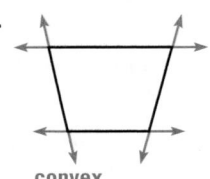
convex

5.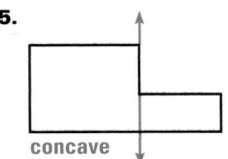
concave

6.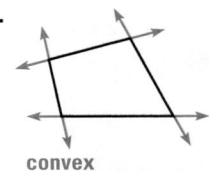
convex

Match the polygon with the description.

A. concave   **B.** equilateral   **C.** convex equiangular

7.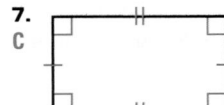
C

8.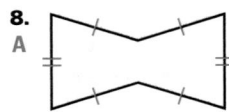
A

9.
B

Practice and Applications

Extra Practice
See p. 689.

Convex and Concave Polygons Decide whether the polygon is *convex* or *concave.*

10.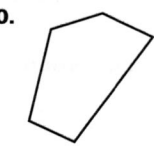
convex

11.
concave

12.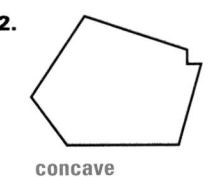
concave

Equilateral and Equiangular Polygons Decide whether the polygon is *equilateral, equiangular,* or *neither.*

Homework Help
Example 1: Exs. 10–12
Example 2: Exs. 13–21

13.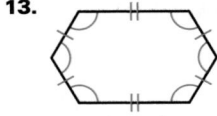
equiangular

14.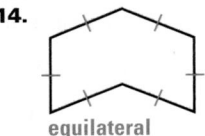
equilateral

15.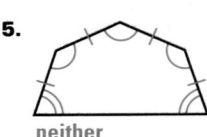
neither

8.1 *Classifying Polygons*   **413**

Assignment Guide

BASIC
Day 1: EP p. 685 Exs. 1–3;
pp. 413–415 Exs. 10–21, 33–40,
41–51 odd
AVERAGE
Day 1: pp. 413–415 Exs. 10–27,
31–40, 41–51 odd
ADVANCED
Day 1: pp. 413–415 Exs. 10–12,
13–25 odd, 27–40, 42–50 even;
EC: TE p. 408D*, classzone.com
BLOCK SCHEDULE
pp. 413–415 Exs. 10–27, 31–40,
41–51 odd (with 8.2)

Extra Practice
• Student Edition, p. 689
• Chapter 8 Resource Book,
  pp. 10–11

Homework Check
To quickly check student understanding of key concepts, go over the following exercises:

Basic: 10, 11, 16, 17, 18
Average: 10, 11, 16, 17, 21
Advanced: 11, 12, 17, 19, 21

1. *Sample answer:*

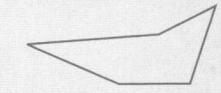

413

29–32. Sample answers are given.

29.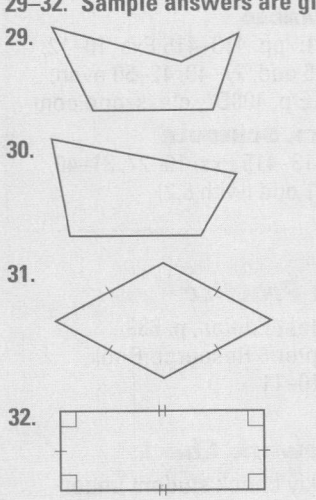

30.

31.

32.

Regular Polygons Decide whether the polygon is regular. Explain your answer.

16.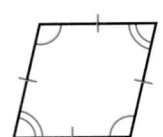

not regular (equilateral but not equiangular)

17.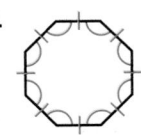

regular (both equilateral and equiangular)

18.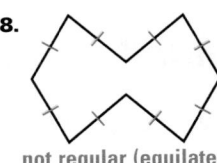

not regular (equilateral but not equiangular)

19.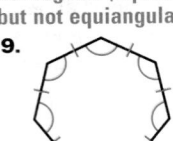

regular (both equilateral and equiangular)

20.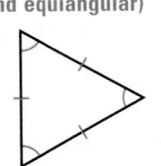

regular (both equilateral and equiangular)

21.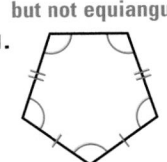

not regular (equiangular but not equilateral)

Flooring Decide whether the polygon outlined in red in the floor pattern is *convex* or *concave*.

22.

concave

23.

concave

24.

convex

Link to
Careers

FLOORING INSTALLERS are hired by flooring contractors to install, repair, or replace floor coverings such as carpet and hardwood.

Web site Icons Use the polygons outlined on the website icons shown below.

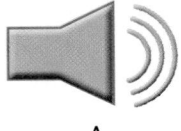
A

B

C

27. Yes. The sum of the measures of the angles of a triangle is 180°, or 3(60°). So, the missing angle measure is 60°. All three angles are congruent, so the triangle is equiangular.

28. Yes. Opposite sides of a parallelogram are congruent, so all four sides are congruent and the parallelogram is equilateral.

25. Which polygons are convex? Which polygons are concave? **C; A, B**

26. Do any of the polygons appear to be regular? Explain.
No. *Sample answer:* None of the polygons appear to be both equilateral and equiangular.

Logical Reasoning Answer the question about the polygon.

27. Is the triangle equiangular? Explain.

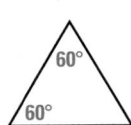
60°
60°

28. Is the parallelogram equilateral? Explain.

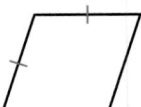

Student Help

LOOK BACK
To review the names given to polygons, see p. 304. For example, a *pentagon* has five sides.

Visualize It! Decide whether it is possible to sketch a polygon that fits the description. If so, sketch it. 29–32. See margin for sketches.

29. A concave pentagon **yes**

30. A convex quadrilateral **yes**

31. A polygon that is equilateral but not equiangular **yes**

32. A polygon that is equiangular but not equilateral **yes**

Finding Perimeters In Exercises 33–35, the polygons are equilateral. Find the perimeter of the polygon.

33.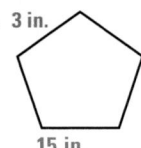
3 in.

15 in.

34.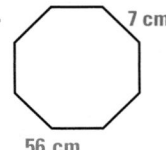
7 cm

56 cm

35.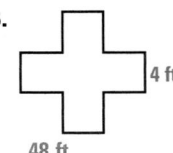
4 ft

48 ft

xy Using Algebra In Exercises 36–38, the polygons are regular. Find the value of *x*.

36.
60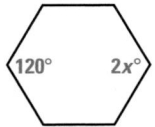
120°   2x°

37.
6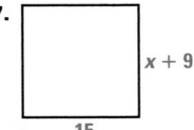
x + 9

15

38.
13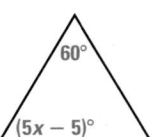
60°

(5x − 5)°

Standardized Test Practice

Multiple Choice In Exercises 39 and 40, use the terms below.

  I. Equilateral   **II.** Equiangular   **III.** Convex   **IV.** Concave

39. Which of the terms best describe the polygon below? A

  Ⓐ I and III   Ⓑ III only

  Ⓒ I and IV   Ⓓ I, II, and III

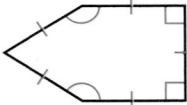

40. Which of the terms best describe the polygon below? H

  Ⓕ I and III   Ⓖ I, II, and IV

  Ⓗ I and IV   Ⓙ IV only

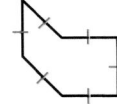

Mixed Review

Classifying Polygons Decide whether the figure is a polygon. If so, tell what type. *(Lesson 6.1)*

41.
yes; pentagon

42.
yes; quadrilateral

43.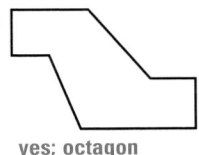
yes; octagon

Algebra Skills

Evaluating Powers Evaluate the expression. *(Skills Review, p. 668)*

44. 5^2 25   **45.** $(-4)^2$ 16   **46.** 6^3 216   **47.** 2^5 32

Evaluating Radicals Evaluate. Give the exact value if possible. If not, approximate to the nearest tenth. *(Skills Review, p. 668)*

48. $\sqrt{36}$ 6   **49.** $\sqrt{1}$ 1   **50.** $\sqrt{169}$ 13   **51.** $\sqrt{5}$ 2.2

8.1 *Classifying Polygons* **415**

Mini-Quiz
Decide whether the polygon is *convex* or *concave*.

1.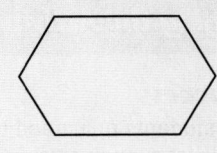
concave

2.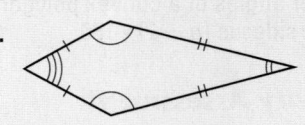
convex

Decide whether the polygon is *equilateral, equiangular,* or *neither*.

3.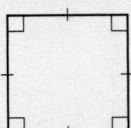
equilateral

4.
neither

5. Decide whether the polygon is regular. Explain your answer.

Because the polygon is both equilateral and equiangular, it is regular.

Goal

Students measure the interior angles of polygons to discover the relationship between the sum of the measures of the angles and the number of sides of the polygon.

Materials

See the margin of the student page.

LINK TO LESSON

In Example 1 on page 417, students find the sum of the angle measures of a heptagon using the formula discovered in this activity.

2 Managing the Activity

Tips for Success

Point out that students only need to draw diagonals from one vertex. Students may be tempted to connect all the vertices in the polygon.

3 Closing the Activity

KEY DISCOVERY

The sum of the measures of the interior angles of a convex polygon with n sides is $(n - 2) \cdot 180°$.

Activity Assessment

Use Exercise 2 to assess student understanding.

Question

What is the sum of the measures of the interior angles of a polygon with a given number of sides?

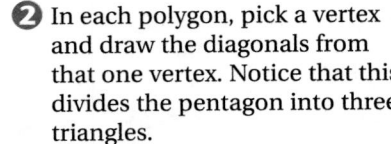

Materials

- paper
- pencil
- straightedge

Explore

1 Using your straightedge, draw convex polygons with three sides, four sides, five sides, and six sides. A pentagon has been drawn below.

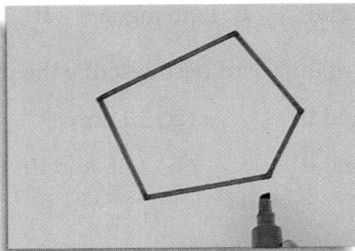

2 In each polygon, pick a vertex and draw the diagonals from that one vertex. Notice that this divides the pentagon into three triangles.

3 Copy and complete the table below. Use the fact that the sum of the measures of the interior angles of a triangle is 180°.

Polygon	Number of sides	Number of triangles	Sum of measures of interior angles
Triangle	3	1	$1 \cdot 180° = 180°$
Quadrilateral	? 4	? 2	$2 \cdot 180° \overset{?}{=} 360°$
Pentagon	? 5	? 3	$3 \cdot 180° \overset{?}{=} 540°$
Hexagon	? 6	? 4	$4 \cdot 180° \overset{?}{=} 720°$

Think About It

1. Look for a pattern in the last column of the table. What is the sum of the measures of the interior angles of a heptagon (7-sided polygon)? of an octagon (8-sided polygon)? Put the answers in your table. $5 \cdot 180° = 900°$; $6 \cdot 180° = 1080°$

2. Write an expression for the sum of the measures of the interior angles of any convex polygon with n sides. $(n - 2) \cdot 180°$

8.2 Angles in Polygons

Goal

Find the measures of interior and exterior angles of polygons.

Key Words

- interior angle p. 181
- exterior angle p. 181

The definitions for interior angles and exterior angles can be extended to include angles formed in any polygon. In the diagrams shown below, interior angles are red, and exterior angles are blue.

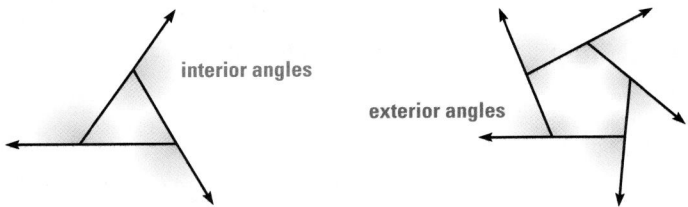

interior angles

exterior angles

Activity 8.2 suggests the *Polygon Interior Angles Theorem* shown below.

THEOREM 8.1

Polygon Interior Angles Theorem

Words The sum of the measures of the interior angles of a convex polygon with n sides is $(n - 2) \cdot 180°$.

Symbols $m\angle 1 + m\angle 2 + \cdots + m\angle n = (n - 2) \cdot 180°$

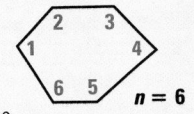

$n = 6$

EXAMPLE ① Use the Polygon Interior Angles Theorem

Find the sum of the measures of the interior angles of a convex heptagon.

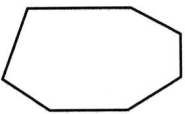

Solution

A heptagon has 7 sides. Use the Polygon Interior Angles Theorem and substitute 7 for n.

$(n - 2) \cdot 180° = (7 - 2) \cdot 180°$ Substitute 7 for n.

$= 5 \cdot 180°$ Simplify.

$= 900°$ Multiply.

ANSWER ▶ The sum of the measures of the interior angles of a convex heptagon is 900°.

Pacing
Suggested Number of Days

Basic: 2 days
Average: 2 days
Advanced: 2 days
Block Schedule: 0.5 block with 8.1
                    0.5 block with 8.3

Teaching Resources

📄 *Blacklines*
(See page 408B.)

📑 *Transparencies*
- Warm-Up with Quiz
- Answers

💻 *Technology*
- Electronic Teacher Tools
- Test and Practice Generator
- Online Lesson Planner
- Internet Support

 *Video*
- Geometry in Motion

② Teach

Content and Teaching Strategies
For background information on geometric concepts and teaching strategies related to this lesson, see pages 408E and 408F in this Teacher's Edition.

Tips for New Teachers
The Activity on page 416 gives students experience discovering patterns. Students should recognize that the number of sides and the number of interior angles are the same in a polygon. Point out that the formulas in Theorems 8.1 and 8.2 can be used to find the number of sides in a polygon as well as the angle measures. See the Tips for New Teachers on pp. 1–2 of the *Chapter 8 Resource Book* for additional notes about Lesson 8.2.

Extra Example 1
See next page.

Extra Example 1

Find the sum of the measures of the interior angles of a convex hexagon. **720°**

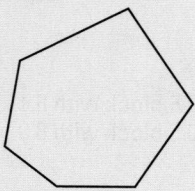

Extra Example 2

Find the measure of ∠*A* in the diagram. **90°**

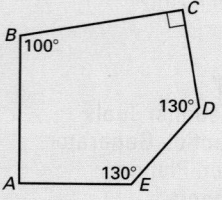

Geometric Reasoning

In Example 3, some students may initially think there is insufficient information provided to solve the problem. Be sure they understand why the problem can be solved. You might want to repeat the problem drawing a different sized regular octagon in order to show students that the size of the octagon has no effect on the interior angle measures.

Extra Example 3

Find the measure of an interior angle of a regular hexagon. **120°**

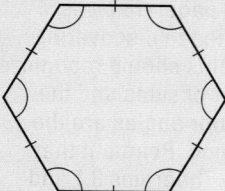

EXAMPLE **2** **Find the Measure of an Interior Angle**

Find the measure of ∠*A* in the diagram.

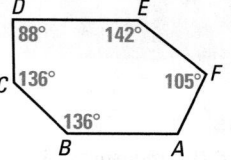

Solution

The polygon has 6 sides, so the sum of the measures of the interior angles is:

$$(n - 2) \cdot 180° = (6 - 2) \cdot 180° = 4 \cdot 180° = 720°.$$

Add the measures of the interior angles and set the sum equal to 720°.

$136° + 136° + 88° + 142° + 105° + m\angle A = 720°$	The sum is 720°.
$607° + m\angle A = 720°$	Simplify.
$m\angle A = 113°$	Subtract 607°.

ANSWER ▶ The measure of ∠*A* is 113°.

Student Help

SKILLS REVIEW
To review solving an equation, see p. 671.

EXAMPLE **3** **Interior Angles of a Regular Polygon**

Find the measure of an interior angle of a regular octagon.

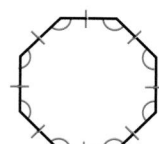

Solution

The sum of the measures of the interior angles of any octagon is:

$$(n - 2) \cdot 180° = (8 - 2) \cdot 180° = 6 \cdot 180° = 1080°.$$

Because the octagon is regular, each angle has the same measure. So, divide 1080° by 8 to find the measure of one interior angle.

$$\frac{1080°}{8} = 135°$$

ANSWER ▶ The measure of an interior angle of a regular octagon is 135°.

Checkpoint ✓ **Find Measures of Interior Angles**

In Exercises 1–3, find the measure of ∠*G*.

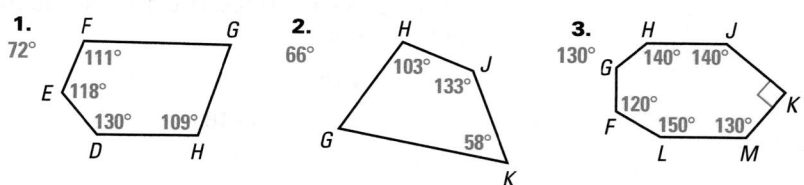

4. Find the measure of an interior angle of a regular polygon with twelve sides. **150°**

Exterior Angles The diagrams below show that the sum of the measures of the exterior angles of the convex polygon is 360°.

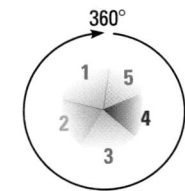

❶ Shade one exterior angle at each vertex.

❷ Cut out the exterior angles.

❸ Arrange the exterior angles to form 360°.

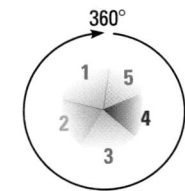

Visualize It!

A circle contains two straight angles. So, there are 180° + 180°, or 360°, in a circle.

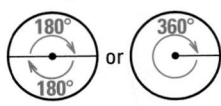

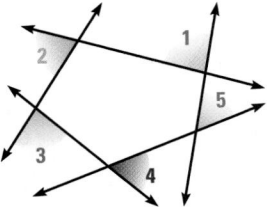

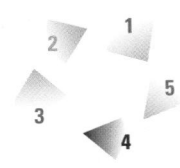

The sum of the measures of the exterior angles of a convex polygon does not depend on the number of sides that the polygon has.

THEOREM 8.2

Polygon Exterior Angles Theorem

Words The sum of the measures of the exterior angles of a convex polygon, one angle at each vertex, is 360°.

Symbols $m\angle 1 + m\angle 2 + \cdots + m\angle n = 360°$

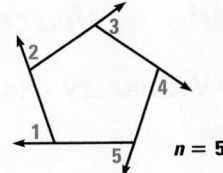

n = 5

Using Algebra

EXAMPLE 4 Find the Measure of an Exterior Angle

Find the value of *x*.

Solution

Using the Polygon Exterior Angles Theorem, set the sum of the measures of the exterior angles equal to 360°.

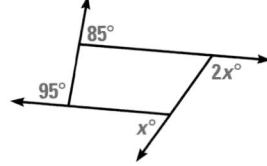

$95° + 85° + 2x° + x° = 360°$	Polygon Exterior Angles Theorem
$180 + 3x = 360$	Combine like terms.
$3x = 180$	Subtract 180 from each side.
$x = 60$	Divide each side by 3.

ANSWER ▶ The value of *x* is 60.

Common Error

In Checkpoint Exercise 5, watch for students who omit the right angle in their calculations.

✓ **Concept Check**

What do you know about the measures of the interior angles of a convex polygon? **The sum of the measures of the angles is $(n - 2) \cdot 180°$, where n is the number of sides of the polygon.**

🐢 **Daily Puzzler**

One interior angle of a convex polygon with 16 sides measures 157.5°. Is this a regular polygon? Explain. *Sample answer:* Not necessarily; in a regular polygon with 16 sides, each interior angle does measure 157.5°. However, it is possible for only one of the angles to measure 157.5° while the sum of the measures of all the angles is 2520°. Also, all of the interior angles could measure 157.5° and the polygon would still not be regular if its sides are not all the same length.

Checkpoint ✓ *Find Exterior Angle Measures*

Find the value of x.

5.
91
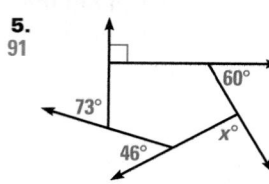

6.
21
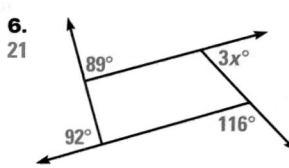

7.
125
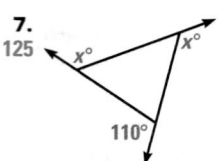

8.
30
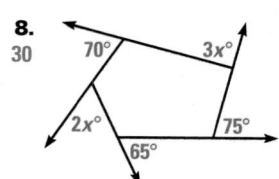

8.2 **Exercises**

Guided Practice

Vocabulary Check

1. Name an *interior angle* and an *exterior angle* of polygon *ABCDE* shown at the right.
Sample answer: ∠DCB, ∠BCG

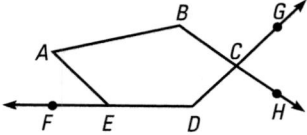

Skill Check

2. Write the formula that is used to find the sum of the measures of the interior angles of any convex polygon with *n* sides.
$(n - 2) \cdot 180°$

3. Use your answer from Exercise 2 to find the sum of the measures of the interior angles of a convex polygon with 9 sides. **1260°**

Write an equation to find the measure of ∠1. Do not solve the equation.

4.
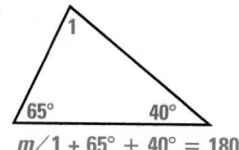
$m\angle 1 + 65° + 40° = 180°$

5.
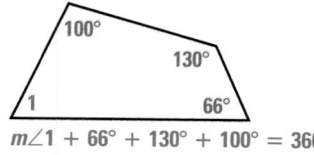
$m\angle 1 + 66° + 130° + 100° = 360°$

6.
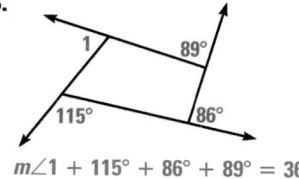
$m\angle 1 + 115° + 86° + 89° = 360°$

7.
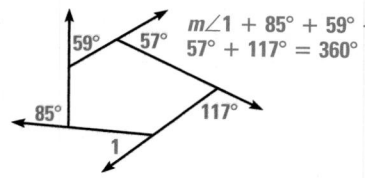
$m\angle 1 + 85° + 59° + 57° + 117° = 360°$

Practice and Applications

Extra Practice

See p. 689.

Sum of Interior Angle Measures Find the sum of the measures of the interior angles of the convex polygon.

8.
720°
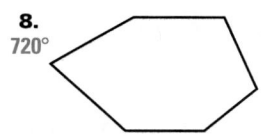

9.
1080°
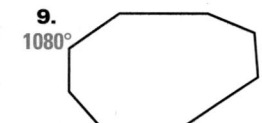

10.
360°

Polygons with *n* Sides Find the sum of the measures of the interior angles of the convex polygon with *n* sides.

11. $n = 10$ 1440°   **12.** $n = 15$ 2340°   **13.** $n = 20$ 3240°

14. $n = 30$ 5040°   **15.** $n = 52$ 9000°   **16.** $n = 100$ 17,640°

Interior Angle Measures Find the measure of $\angle A$.

17.
99°
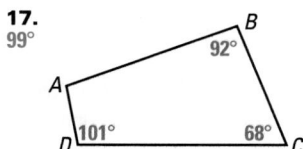

18.
135°
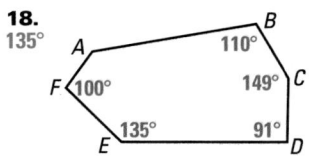

19.
127°
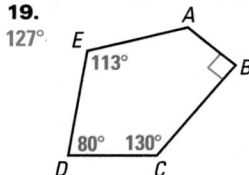

20.
70°
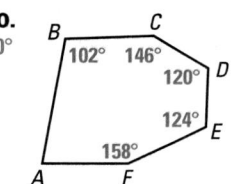

Student Help
CLASSZONE.COM

HOMEWORK HELP
Extra help with problem solving in Exs. 21–23 is at classzone.com

Interior Angles of Regular Polygons Find the measure of an interior angle of the regular polygon.

21.
120°
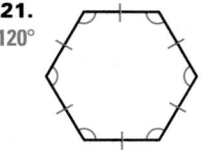

22.
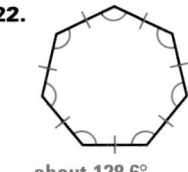
about 128.6°

23.
60°
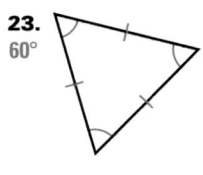

Using Algebra Find the value of *x*.

24.
46
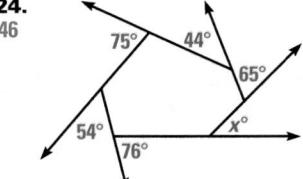

25.
32
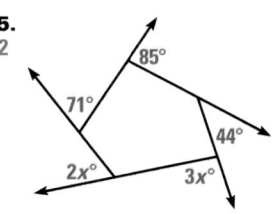

Homework Help

Example 1: Exs. 8–16
Example 2: Exs. 17–20
Example 3: Exs. 21–23
Example 4: Exs. 24, 25

③ Apply

Assignment Guide

BASIC
Day 1: pp. 421–423 Exs. 8–20, 26–28, 39–51 odd
Day 2: EP p. 685 Exs. 4–6; pp. 421–423 Exs. 21–25, 31–33, 40–50 even, Quiz 1

AVERAGE
Day 1: pp. 421–423 Exs. 8–20, 26–28, 39–51 odd
Day 2: pp. 421–423 Exs. 21–25, 29–34, 40–50 even, Quiz 1

ADVANCED
Day 1: pp. 421–423 Exs. 14–20, 26–28, 39–51 odd
Day 2: pp. 421–423 Exs. 21–25, 29–34, 35–38*, 40, Quiz 1; EC: classzone.com

BLOCK SCHEDULE
pp. 421–423 Exs. 8–20, 26–28, 39–51 odd (with 8.1)
pp. 421–423 Exs. 21–25, 29–34, 40–50 even, Quiz 1 (with 8.3)

Extra Practice
• Student Edition, p. 689
• Chapter 8 Resource Book, pp. 18–19

Homework Check

To quickly check student understanding of key concepts, go over the following exercises:

Basic: 8, 17, 21, 24, 26
Average: 11, 18, 21, 24, 26
Advanced: 15, 19, 22, 25, 26

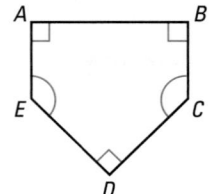

Link to
Sports

BASEBALL Home plate is used to find the placement of the foul lines on the playing field.

Baseball A home plate for a baseball field is a pentagon as shown.

26. Is the polygon regular? Explain why or why not. **No; it is not equilateral or equiangular.**

27. What is the sum of the interior angles? **540°**

28. Find the measures of ∠C and ∠E.
$m\angle C = m\angle E = 135°$

Visualize It! Cut a strip of lined paper and tie it into an overhand knot as shown. Gently flatten the knot to form a pentagon. The pentagon should be regular.

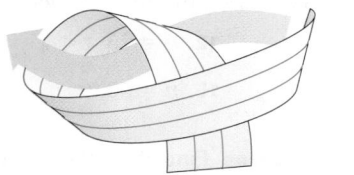

29. Find the measure of an interior angle of a regular pentagon. **108°**

30. Measure the interior angles of your knot with a protractor. Use your answer to Exercise 29 to determine whether your pentagon is regular. **Answers will vary.**

Logical Reasoning Select the word that makes the statement true.

31. The sum of the measures of the exterior angles of a convex polygon, one at each vertex, is (*always, sometimes, never*) 360°. **always**

32. The sum of the measures of the interior angles of a convex octagon is (*always, sometimes, never*) 1440°. **never**

33. The measures of the exterior angles of a convex polygon are (*always, sometimes, never*) equal. **sometimes**

34. Outdoor Furniture You are constructing a regular hexagonal wooden bench like the one shown. On the bench, ∠1 ≅ ∠2. Find the measures of ∠1 and ∠2 so that you know what angle to use to cut the pieces of wood.
$m\angle 1 = m\angle 2 = 60°$

Challenge Find the number of sides of the regular polygon with the given exterior angle measure.

35. 60° **6**   **36.** 20° **18**   **37.** 72° **5**   **38.** 10° **36**

Multiple Choice In Exercises 39 and 40, use the diagram below.

39. What is the value of *x*? **B**

(A) 18   (B) 78

(C) 117   (D) 198

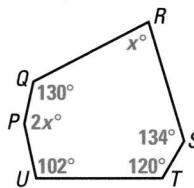

40. What is the measure of ∠*P*? **H**

(F) 39°   (G) 78°   (H) 156°   (J) 234°

Mixed Review

Using a Centroid *E* is the centroid of △*ABC*. Find *BE* and *ED*.
(Lesson 4.6)

41. *BD* = 6

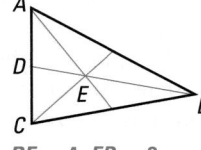

BE = 4, *ED* = 2

42. *BD* = 33

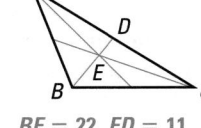

BE = 22, *ED* = 11

43. *BD* = 52

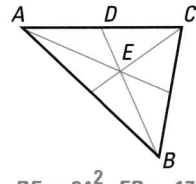

$BE = 34\frac{2}{3}$, $ED = 17\frac{1}{3}$

Algebra Skills

Absolute Value Evaluate. *(Skills Review, p. 662)*

44. $|-5|$ 5   **45.** $|11|$ 11   **46.** $|-48|$ 48   **47.** $|0|$ 0

48. $|13.2|$ 13.2   **49.** $|-2|$ 2   **50.** $|-0.001|$ 0.001   **51.** $|-1.11|$ 1.11

Quiz 1

Decide whether the polygon is regular. Explain your answer.
(Lesson 8.1)

1.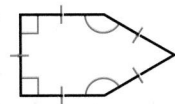

not regular (equilateral
but not equiangular)

2.

regular (both equilateral
and equiangular)

3.

not regular (not
equilateral and not
equiangular)

Find the measure of ∠1. *(Lesson 8.2)*

4.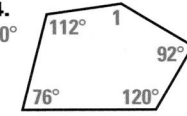

5.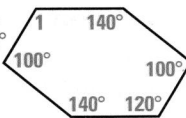

6.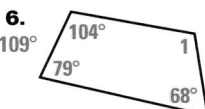

7.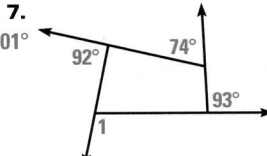

8.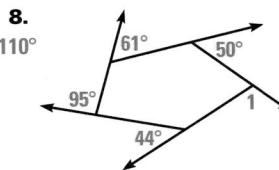

8.2 *Angles in Polygons*   423

---

4 Assess

Assessment Resources
The Mini-Quiz below is also available on blackline (*Chapter 8 Resource Book*, p. 27) and on transparency. For more assessment resources, see:
• Chapter 8 Resource Book
• Standardized Test Practice
• Test and Practice Generator

Mini-Quiz

1. Find the measure of ∠*A*. 85°

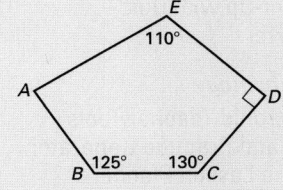

2. A regular polygon has 11 sides. Find the sum of the measures of the interior angles. 1620°

3. Find the measure of one interior angle of a regular polygon with 12 sides. 150°

4. What is the measure of each exterior angle of a regular pentagon? 72°

Goal
Find the area of squares and rectangles.

Key Words
• area
• square p. 325
• rectangle p. 325

Can you tell which of the rectangles below covers more surface?

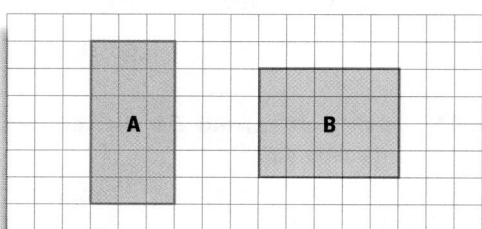

Rectangle A is made up of 18 squares while rectangle B is made up of 20 squares. So, rectangle B covers more *area*. The amount of surface covered by a figure is its **area** .

Area is measured in square units such as square inches (in.^{2}) and square meters (m^2).

AREA OF A SQUARE

Words Area = (side)^2

Symbols $A = s^2$

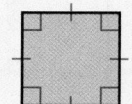

EXAMPLE **1** **Find the Area of a Square**

Find the area of the square.

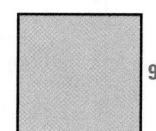

9 ft

Solution
Use the formula for the area of a square and substitute 9 for *s*.

$A = s^2$ Formula for the area of a square

$= 9^2$ Substitute 9 for *s*.

$= 81$ Simplify.

ANSWER ▶ The area of the square is 81 square feet.

AREA OF A RECTANGLE

Words Area = (base)(height)

Symbols $A = bh$

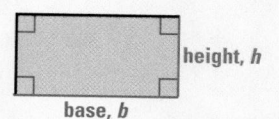

height, h

base, b

Student Help

CLASSZONE.COM

MORE EXAMPLES
More examples at classzone.com

EXAMPLE 2 Find the Area of a Rectangle

Find the area of the rectangular pool.

Solution

Use the formula for the area of a rectangle. Substitute 24 for b and 16 for h.

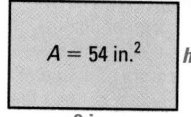

16 ft

24 ft

$A = bh$ Formula for the area of a rectangle

$= 24 \cdot 16$ Substitute 24 for b and 16 for h.

$= 384$ Multiply.

ANSWER ▶ The area of the pool is 384 square feet.

Using Algebra

EXAMPLE 3 Find the Height of a Rectangle

The rectangle has an area of 54 square inches. Find its height.

Solution

$A = 54$ in.^{2} h

9 in.

Use the formula for the area of a rectangle and substitute 54 for A and 9 for b.

$A = bh$ Formula for the area of a rectangle

$54 = 9h$ Substitute 54 for A and 9 for b.

$6 = h$ Divide each side by 9.

ANSWER ▶ The height of the rectangle is 6 inches.

Checkpoint ✓ *Area of Squares and Rectangles*

Find the area of the quadrilateral.

1.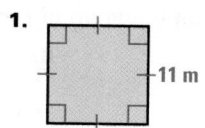
11 m

121 m^2

2.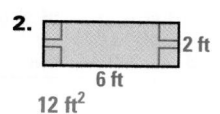
2 ft
6 ft
12 ft^2

3.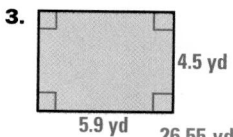
4.5 yd
5.9 yd   26.55 yd^2

4. A rectangle has an area of 52 square meters and a height of 4 meters. Find the length of its base. **13 m**

8.3 Area of Squares and Rectangles **425**

Extra Example 1

Find the area of the square. **400 in.^{2}**

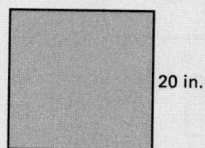

20 in.

Extra Example 2

Find the area of the rectangle.

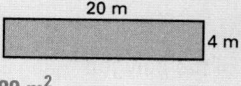
20 m
4 m

80 m^2

Extra Example 3

The rectangle has an area of 98 square centimeters. Find the length of its base. **14 cm**

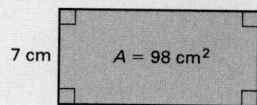
7 cm $A = 98$ cm^2

Visualize It!

The formula for the area of a square is merely the first area formula in this chapter. Suggest that students begin making a table of area formulas, complete with sketches, that they can add to during the course of the chapter.

Extra Example 4

Find the dimensions of rectangles A and B.

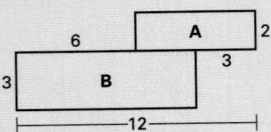

Rectangle A has height 2 units and base 6 units. Rectangle B has height 3 units and base 9 units.

Extra Example 5

Find the area of the polygon made up of rectangles. **131 in.²**

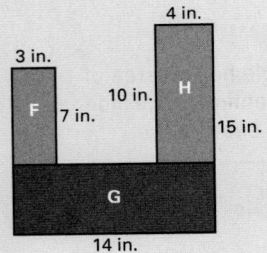

✓ Concept Check

How do you find the area of a rectangle? **Multiply the base by the height.**

To find the area of a complex polygon, divide the polygon into smaller regions whose areas you can find.

EXAMPLE **4** **Divide a Complex Polygon into Rectangles**

Find the dimensions of rectangles A and B.

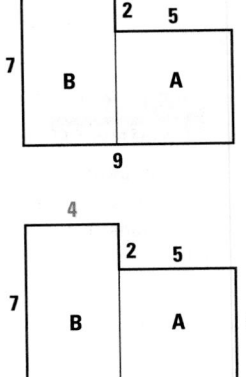

Solution

Rectangle A
The base is 5 units.
Because rectangle B is 2 units taller than rectangle A, the height of rectangle A is
7 − 2 = 5 units.

Rectangle B
The height is 7 units.
The base of rectangle B is the total of both bases minus the base of rectangle A,
or **9 − 5 = 4** units.

Visualize It!

Labels on diagrams are centered on the segment with which they correspond. In Example 5, the 9 cm label refers to a side of the polygon, not just the height of rectangle G.

EXAMPLE **5** **Find the Area of a Complex Polygon**

Find the area of the polygon made up of rectangles.

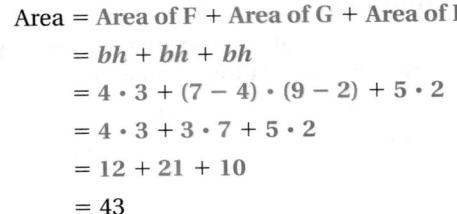

Solution

Add the areas of the rectangles.

$$\text{Area} = \text{Area of F} + \text{Area of G} + \text{Area of H}$$
$$= bh + bh + bh$$
$$= 4 \cdot 3 + (7 - 4) \cdot (9 - 2) + 5 \cdot 2$$
$$= 4 \cdot 3 + 3 \cdot 7 + 5 \cdot 2$$
$$= 12 + 21 + 10$$
$$= 43$$

ANSWER ▶ The total area of the polygon is 43 square centimeters.

Checkpoint ✓ **Polygons Made Up of Rectangles**

Find the area of the polygon made up of rectangles.

5. **30 in.²**

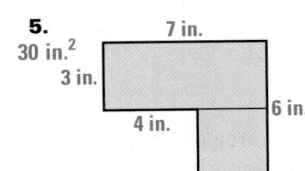

6. **54 m²**

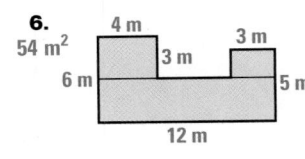

8.3 Exercises

Guided Practice

Vocabulary Check

1. What kind of quadrilateral has opposite sides parallel, opposite sides congruent, and four right angles? **rectangle**

Skill Check

Match the figure with the corresponding area equation.

A. $A = x^2$ **B.** $A = 2x^2$ **C.** $A = 4x^2$

2.
C
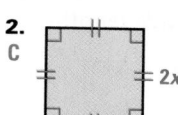
2x

3.
A
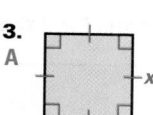
x

4.
B
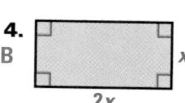
x
2x

Determine whether the statement about the diagram is _true_ or _false_. Explain your answer.

5. True; the area of the polygon is Area A + Area B + Area C.

5. To find the area of the entire polygon, add the areas of the three rectangles.

6. The height of rectangle A is 1 unit.
False; the height is $7 - 3 - 2 = 2$ units.

7. The height of rectangle C is 5 units.
True; the height is $3 + 2 = 5$ units.

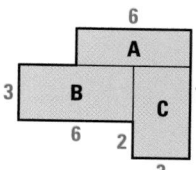

Practice and Applications

Extra Practice

See p. 689.

Area of a Square Find the area of the square.

8.
36 ft²

6 ft

9.
144 m²
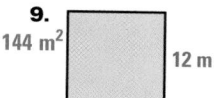
12 m

10.
400 in.²
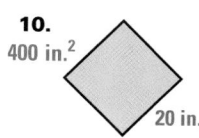
20 in.

Area of a Rectangle Find the area of the rectangle.

11.
10 cm²
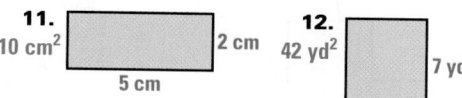
2 cm
5 cm

12.
42 yd²
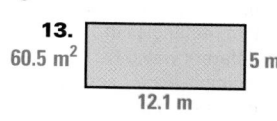
7 yd
6 yd

13.
60.5 m²
12.1 m
5 m

Homework Help

Example 1: Exs. 8–10, 14
Example 2: Exs. 11–13, 15, 16
Example 3: Exs. 20–22
Example 4: Exs. 24–26
Example 5: Exs. 27–30

Visualize It! Sketch the figure and find its area. **14–16. See margin.**

14. A square with side lengths of 2.2 centimeters

15. A rectangle with a base of 4 meters and a height of 11 meters

16. A rectangle with a base of 13 feet and a height of 8 feet

8.3 *Area of Squares and Rectangles*   **427**

Apply

Assignment Guide

BASIC
Day 1: pp. 427–429 Exs. 8–16, 20–22, 24–28, 34–46

AVERAGE
Day 1: pp. 427–429 Exs. 8–30, 34–46

ADVANCED
Day 1: pp. 427–429 Exs. 8–16 even, 17–19, 22–26, 30–34, 35–45 odd; EC: TE p. 408D*, classzone.com

BLOCK SCHEDULE
pp. 427–429 Exs. 8–30, 34–46 (with 8.2)

Extra Practice
• Student Edition, p. 689
• Chapter 8 Resource Book, pp. 28–29

Homework Check
To quickly check student understanding of key concepts, go over the following exercises:

Basic: 8, 11, 20, 24, 27
Average: 9, 12, 21, 25, 28
Advanced: 14, 16, 22, 26, 30

14.
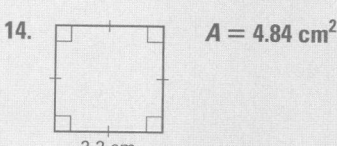
$A = 4.84$ cm²
2.2 cm

15.
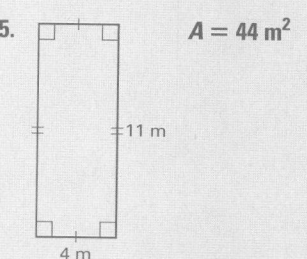
$A = 44$ m²
11 m
4 m

16.
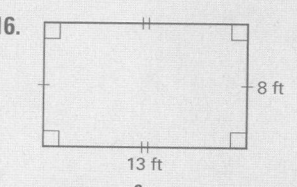
8 ft
13 ft
$A = 104$ ft²

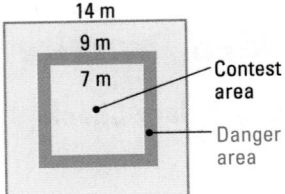

X Common Error

In Exercises 24–30, watch for students who misread the diagrams. In particular, watch for students who misread a measure as applying to only part of a side instead of correctly reading that it is the measure of the entire side. Refer those students to the Visualize It note on page 426 to help them read the diagrams correctly.

Judo The dimensions of the squares on a judo mat are given in the diagram.

17. Find the area of the entire mat. 196 m²

18. Find the area of the contest area. 49 m²

19. Find the area of the contest area including the danger area. 81 m²

Using Algebra In Exercises 20–22, *A* gives the area of the rectangle. Find the missing side length.

20.
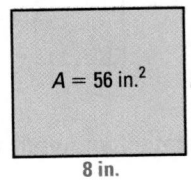
$A = 56$ in.² *h*
8 in.
h = 7 in.

21.
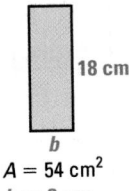
18 cm
b
$A = 54$ cm²
b = 3 cm

22.
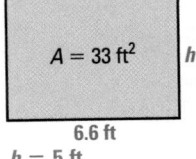
$A = 33$ ft² *h*
6.6 ft
h = 5 ft

23. *You be the Judge* The perimeter of a square is 28 feet. Can you conclude that the area of the square is 49 square feet? Explain.
Yes. Since the perimeter is 28 feet, each side measures 28 ÷ 4 = 7 feet.
Area = 7² = 49 square feet.

Dividing a Polygon Find the dimensions of the rectangle.

24. Rectangle A
b = 5 ft, *h* = 3 ft
25. Rectangle B
b = 3 ft, *h* = 6 ft
26. Rectangle C
b = 4 ft, *h* = 7 ft

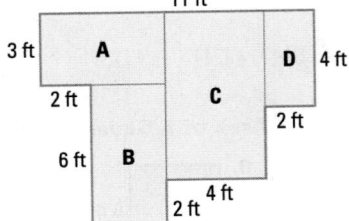

Visualize It!

In Exs. 27–30, the polygons can be divided into rectangles in different ways. For example, Ex. 27 can be divided as follows:

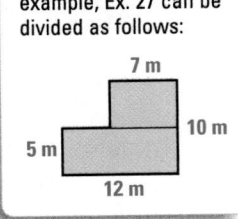

Area of Complex Polygons Find the area of the polygon made up of rectangles.

27.
95 m²
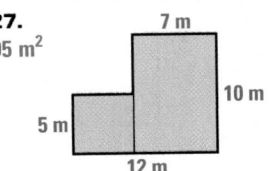

28.
47 in.²
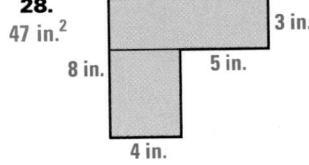

29.
116 yd²
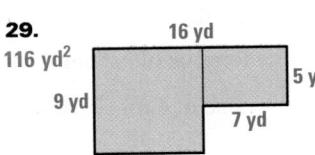

30.
152 cm²
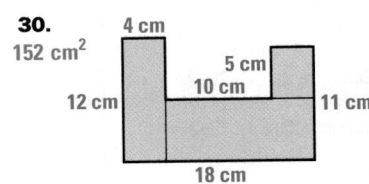

Maize Maze Brett Herbst transforms cornfields into mazes. His maze in Utah, shown at the right, is in the shape of Utah.

31. What is the area covered by the maze, which is made up of two rectangles?
207,500 ft²

32. How many acres does the maze cover? (1 acre = 43,560 square feet)
about 4.76 acres

33. Suppose corn seed costs $34 per acre and fertilizer costs $57 per acre. How much will it cost to seed and fertilize a field with the same dimensions as the maze? $433.16

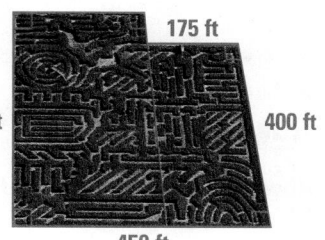

175 ft
500 ft
400 ft
450 ft

Standardized Test Practice

34. Multi-Step Problem The polygon below is made up of rectangles.

a. Write an expression for the area of the polygon. 13x

b. Suppose the area is 65 square units. Find the value of x. 5

c. Using your results from part (b), sketch the figure and label all of its dimensions. See margin.

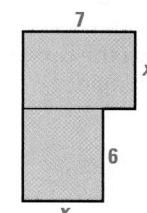

7
x
6
x

Mixed Review

Congruent Parts Use the diagram of parallelogram *ABCD*. Match the segment or angle with a congruent one. Give a reason for your answer. *(Lesson 6.2)*

35. $\overline{CE}$ **A.** $\overline{AB}$

36. $\overline{CD}$ **B.** $\angle ADC$

37. $\angle ABD$ **C.** $\overline{AE}$

38. $\angle CBA$ **D.** $\angle CDB$

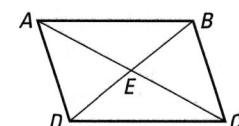

35. C; Diagonals of a parallelogram bisect each other.

36. A; Opposite sides of a parallelogram are congruent.

37. D; If two parallel lines are cut by a transversal, then alternate interior angles are congruent.

38. B; Opposite angles of a parallelogram are congruent.

Determining Similarity Determine whether the triangles are similar. If so, state the similarity and the postulate or theorem that justifies your answer. *(Lesson 7.4)*

39.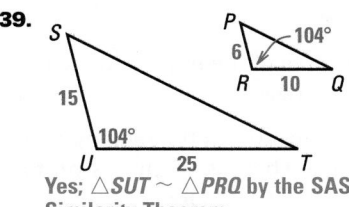

Yes; △*SUT* ~ △*PRQ* by the SAS Similarity Theorem.

40.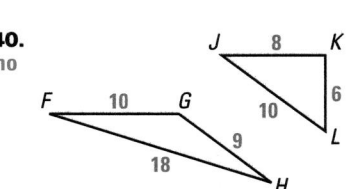
no

Algebra Skills

Comparing Numbers Compare the two numbers. Write the answer using <, >, or =. *(Skills Review, p. 662)*

41. 8 and −18
8 > −18

42. 2459 and 2495
2459 < 2495

43. −10 and 0
−10 < 0

44. −1.12 and −1.01

45. 2.44 and 2.044

46. −0.75 and −0.7

44. −1.12 < −1.01
45. 2.44 > 2.044
46. −0.75 < −0.7

---

4 Assess

Assessment Resources

The Mini-Quiz below is also available on blackline (*Chapter 8 Resource Book*, p. 35) and on transparency. For more assessment resources, see:
- Chapter 8 Resource Book
- Standardized Test Practice
- Test and Practice Generator

Mini-Quiz

Find the area of each figure.

1.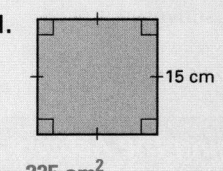
15 cm
225 cm²

2.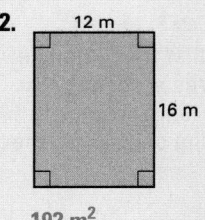
12 m
16 m
192 m²

3.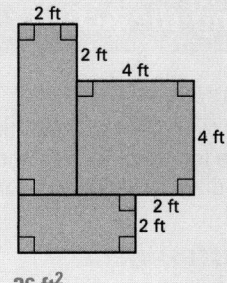
2 ft
2 ft
4 ft
4 ft
2 ft
2 ft
36 ft²

4. A rectangular mural has an area of 504 square feet and is 12 feet high. What is the length of the base of the mural? 42 ft

34c.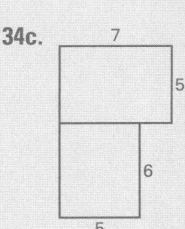
7
5
6
5

Goal
Students compare the area of a triangle to the area of a rectangle to discover the formula for the area of a triangle.

Materials
See the margin of the student page.

LINK TO LESSON
In Example 1 on page 432, students find the area of a triangle using the area formula discovered in this activity.

Managing the Activity

Tips for Success
In Step 1, stress that the rectangle must touch all three vertices of the triangle, with the longest side of the triangle forming one side of the rectangle.

Closing the Activity

KEY DISCOVERY
The area of a triangle is one half the product of the length of a base and its corresponding height.

Activity Assessment
Use Exercise 4 to assess student understanding.

Activity 8.4 Finding Area of Triangles

Question

How do you find the area of a triangle?

Materials
- grid paper
- colored pencils or markers
- scissors

Explore

❶ Draw a triangle on grid paper. Then draw a rectangle that encloses the triangle. Write down the dimensions of the rectangle.

❷ Cut out the rectangle. Then cut the triangles out of the rectangle.

❸ Try to cover the large triangle with the two smaller triangles.

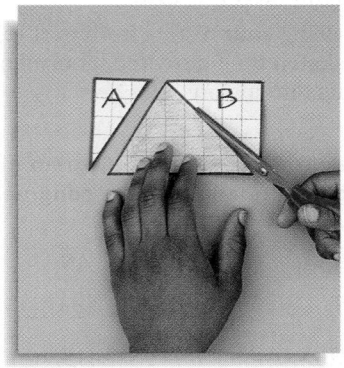

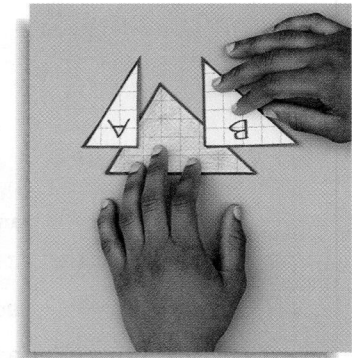

Think About It

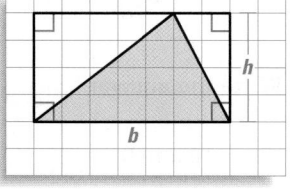

1. Do the two smaller triangles cover the same area as the large triangle? yes

2. How is the area of the large triangle related to the area of the original rectangle? The area of the triangle is half the area of the rectangle.

3. Use the dimensions of the rectangle in Step 1 to find the area of the rectangle. Then use your answer to find the area of the large triangle. Answers will vary.

4. Use the diagram at the left to write a rule for finding the area of a triangle given its base b and its height h. $A = \frac{1}{2}bh$

5. Find the area of a triangle with a base of 14 feet and a height of 6 feet. 42 ft^2

 **Area of Triangles**

Goal
Find the area of triangles.

Key Words
- height of a triangle
- base of a triangle

The amount of material needed to make the sail at the right is determined by the area of the triangular sail.

The *height* and *base* of a triangle are used to find its area.

The **height of a triangle** is the perpendicular segment from a vertex to the line containing the opposite side. The opposite side is called the **base of the triangle** . The terms *height* and *base* are also used to represent the segment lengths.

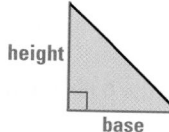

In a right triangle, a leg is a height.

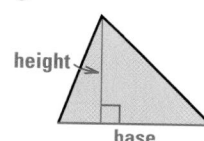

A height can be inside the triangle.

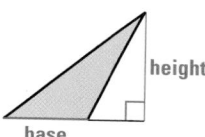

A height can be outside the triangle.

As shown in Activity 8.4, the area of a triangle is found using a base and its corresponding height.

AREA OF A TRIANGLE

Words Area = $\frac{1}{2}$ (base)(height)

Symbols $A = \frac{1}{2}bh$

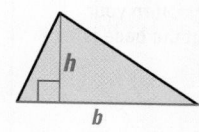

Triangles with the Same Area Triangles can have the same area without necessarily being congruent. For example, all of the triangles below have the same area but they are not congruent.

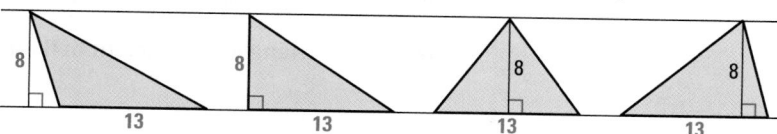

Pacing
Suggested Number of Days

Basic: 1 day
Average: 1 day
Advanced: 1 day
Block Schedule: 0.5 block with 8.5

Teaching Resources

📄 *Blacklines*
(See page 408B.)

📑 *Transparencies*
- Warm-Up with Quiz
- Answers

💻 *Technology*
- Electronic Teacher Tools
- Test and Practice Generator
- Online Lesson Planner
- Internet Support

②Teach

Content and Teaching Strategies
For background information on geometric concepts and teaching strategies related to this lesson, see pages 408E and 408F in this Teacher's Edition.

Tips for New Teachers
Some students may not be convinced that the formula for the area of a triangle applies for triangles of all shapes and sizes. Encourage students to repeat the activity on page 430 using each of the four triangles at the bottom of this page. See the Tips for New Teachers on pp. 1–2 of the *Chapter 8 Resource Book* for additional notes about Lesson 8.4.

Geometric Reasoning

For triangles where the vertical height dimension is located outside the triangle (as shown on page 431), stress that the area students calculate is that of the original triangle and not the larger triangle whose side is formed by the height.

Extra Example 1

Find the area of the right triangle.

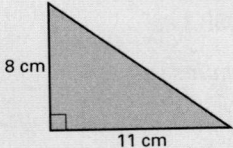

44 cm²

Extra Example 2

Find the area of the triangle.

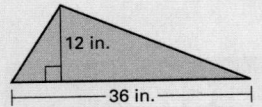

216 in.²

Extra Example 3

Find the height of the triangle, given that its area is 27 square meters. **6 m**

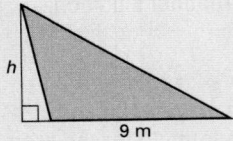

Visualize It!

To help you determine the base and height of a tilted triangle, turn your book so that the base is horizontal.

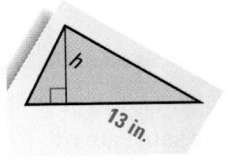

EXAMPLE **1** **Find the Area of a Right Triangle**

Find the area of the right triangle.

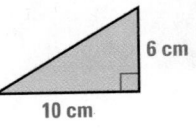

Solution

Use the formula for the area of a triangle. Substitute 10 for b and 6 for h.

$A = \frac{1}{2}bh$ Formula for the area of a triangle

$= \frac{1}{2}(10)(6)$ Substitute 10 for b and 6 for h.

$= 30$ Simplify.

ANSWER ▶ The triangle has an area of 30 square centimeters.

EXAMPLE **2** **Find the Area of a Triangle**

Find the area of the triangle.

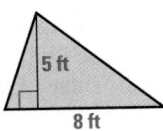

Solution

$A = \frac{1}{2}bh$ Formula for the area of a triangle

$= \frac{1}{2}(8)(5)$ Substitute 8 for b and 5 for h.

$= 20$ Simplify.

ANSWER ▶ The triangle has an area of 20 square feet.

EXAMPLE **3** **Find the Height of a Triangle**

Find the height of the triangle, given that its area is 39 square inches.

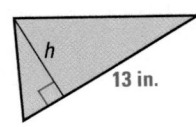

Solution

$A = \frac{1}{2}bh$ Formula for the area of a triangle

$39 = \frac{1}{2}(13)h$ Substitute 39 for A and 13 for b.

$78 = 13h$ Multiply each side by 2.

$6 = h$ Divide each side by 13.

ANSWER ▶ The triangle has a height of 6 inches.

In Exercises 1–3, find the area of the triangle.

1.

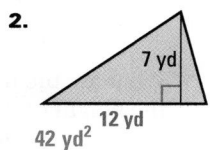

8 in. 9 in.
36 in.²

2.

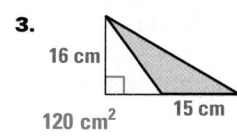

7 yd 12 yd
42 yd²

3.

16 cm
15 cm
120 cm²

4. A triangle has an area of 84 square inches and a height of 14 inches. Find the base. 12 in.

EXAMPLE **4** **Areas of Similar Triangles**

a. Find the ratio of the areas of the similar triangles.

b. Find the scale factor of △ABC to △DEF and compare it to the ratio of their areas.

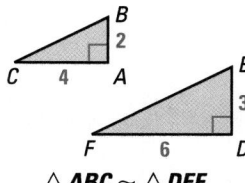

△ **ABC ~** △ **DEF**

Solution

a. Area of △ABC = $\frac{1}{2}bh = \frac{1}{2}(4)(2) = 4$ square units

   Area of △DEF = $\frac{1}{2}bh = \frac{1}{2}(6)(3) = 9$ square units

   Ratio of areas = $\frac{\text{Area of } \triangle ABC}{\text{Area of } \triangle DEF} = \frac{4}{9}$

Student Help

LOOK BACK
To review scale factor, see p. 366.

b. The scale factor of △ABC to △DEF is $\frac{2}{3}$.

   The ratio of the areas is the square of the scale factor: $\frac{2^2}{3^2} = \frac{4}{9}$.

The relationship in Example 4 is generalized for all similar polygons in the following theorem.

THEOREM 8.3

Areas of Similar Polygons

Words If two polygons are similar with a scale factor of $\frac{a}{b}$, then the ratio of their areas is $\frac{a^2}{b^2}$.

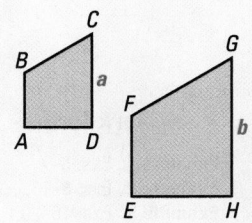

Symbols If ABCD ~ EFGH with a scale factor of $\frac{a}{b}$, then $\frac{\text{Area of } ABCD}{\text{Area of } EFGH} = \frac{a^2}{b^2}$.

Extra Example 4

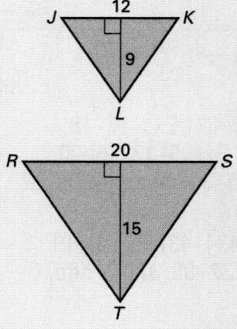

a. Find the ratio of the areas of the similar triangles. $\frac{9}{25}$

b. Find the scale factor of △JKL to △RST and compare it to the ratio of their areas. $\frac{3}{5}$; The ratio of the areas is the square of the scale factor: $\frac{3^2}{5^2} = \frac{9}{25}$.

✓ Concept Check

How do you find the area of a triangle? Multiply one half the length of the base by the height.

🧩 **Daily Puzzler**

Suppose the area of the largest triangle in the figure below is *x*. What is the total area of the shaded regions in the figure? $\frac{x}{3}$

Apply

Assignment Guide

BASIC
Day 1: EP p. 681 Exs. 16–18;
pp. 434–437 Exs. 5–13, 16–20,
26, 27, 32–35, 36–44 even, Quiz 2

AVERAGE
Day 1: pp. 434–437 Exs. 5–19
odd, 21–27, 32–35, 37–43 odd,
Quiz 2

ADVANCED
Day 1: pp. 434–437 Exs. 6–20
even, 21–35, 37–43 odd, Quiz 2;
EC: TE p. 408D*, classzone.com

BLOCK SCHEDULE
pp. 434–437 Exs. 5–19 odd,
21–27, 32–35, 37–43 odd, Quiz 2
(with 8.5)

Extra Practice

• Student Edition, p. 690
• Chapter 8 Resource Book,
  pp. 38–39

Homework Check

To quickly check student understanding of key concepts, go over
the following exercises:

Basic: 5, 8, 11, 16, 26
Average: 7, 9, 11, 17, 26
Advanced: 6, 10, 12, 20, 27

✗ Common Error

In Exercises 5–13, watch for students whose answers are exactly
half of the correct answer. This is a
common error that occurs when
students try to find the area of a triangle mentally. They first halve the
measure of the base, multiply this
measure by the height, and then
multiply the result by $\frac{1}{2}$. Urge
students to write down each step
of their calculations, rather than
trying to find the areas mentally.

8.4 Exercises

Guided Practice

Vocabulary Check

1. What are the measures of the base and
the height of the shaded triangle at the
right? **base = 9 ft, height = 5 ft**

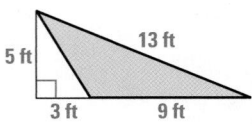

Skill Check

**The triangle has a horizontal base of 15 units and a height of 7 units.
Sketch the triangle and label its base and its height.** 2–4. See margin.

2.    **3.**    **4.**

Practice and Applications

Extra Practice
See p. 690.

Area of a Right Triangle Find the area of the right triangle.

5. 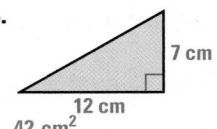   **6.** 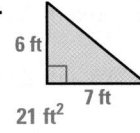   **7.**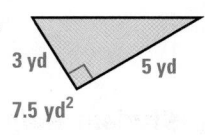

Finding Area In Exercises 8–13, find the area of the triangle.

8. 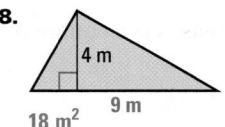   **9.**   **10.**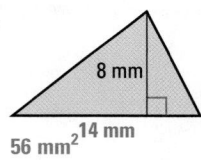

11. 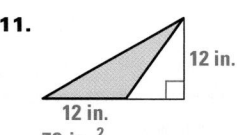   **12.** 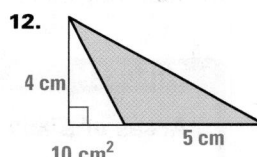   **13.**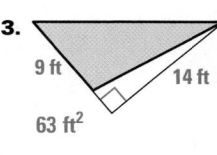

Homework Help

Example 1: Exs. 5–7
Example 2: Exs. 8–13
Example 3: Exs. 16–20
Example 4: Exs. 26, 27

14. **You be the Judge** In the triangle
at the right, Trisha says the base is 15
and the height is 4. Luis says that the
base is 5 and the height is 12. Who is
right? Explain your reasoning.
They are both right. *Sample answer:* The
sides measuring 15 and 5 can each be a
base of the triangle, and 4 and 12 would be
their corresponding heights.

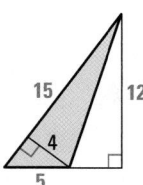

15. Visualize It! Draw three different triangles that each have an area of 24 square units. **See margin.**

Using Algebra In Exercises 16–18, *A* gives the area of the triangle. Find the missing measure.

16. $A = 22$ ft^2

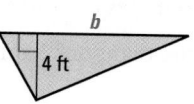

b

4 ft

b = 11 ft

17. $A = 63$ cm^2

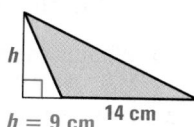

h

h = 9 cm   14 cm

18. $A = 80$ m^2

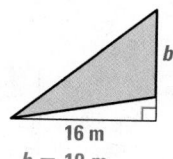

b

16 m

b = 10 m

19. Finding the Height A triangle has an area of 78 square inches and a base of 13 inches. Find the height. **12 in.**

20. Finding the Base A triangle has an area of 135 square meters and a height of 9 meters. Find the base. **30 m**

Tiles In Exercises 21 and 22, use the diagram of the tile pattern.

2 in.

4 in.

21. Find the area of one triangular tile. **4 in.^{2}**

22. The tiles are being used to make a rectangular border that is 4 inches high and 48 inches long. How many tiles are needed for the border? (*Hint:* Start by finding the area of the border.) **48 tiles**

Complex Polygons Find the area of the polygon by using the triangles and rectangles shown.

23.

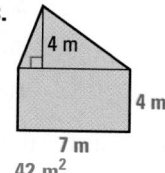

4 m

4 m

7 m

42 m^2

24.

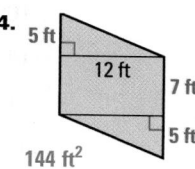

5 ft

12 ft

7 ft

5 ft

144 ft^2

25.

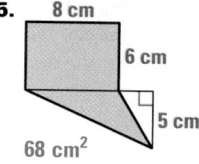

8 cm

6 cm

5 cm

68 cm^2

Areas of Similar Triangles In Exercises 26 and 27, the triangles are similar. Find the scale factor of △*PQR* to △*XYZ*. Then find the ratio of their areas.

26.

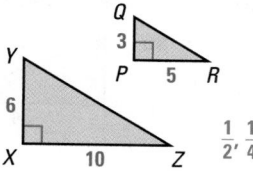

Q
3
P   5   *R*

Y
6
X   10   *Z*

$\frac{1}{2}, \frac{1}{4}$

27.

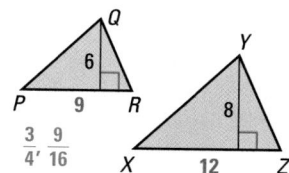

Q
6
P   9   *R*

Y
8
X   12   *Z*

$\frac{3}{4}, \frac{9}{16}$

2.

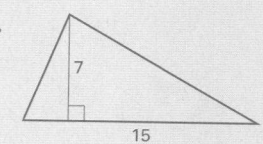

7

15

3.

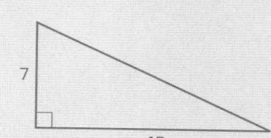

7

15

4.

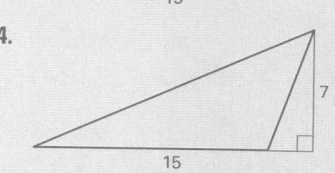

7

15

15. Sample answer:

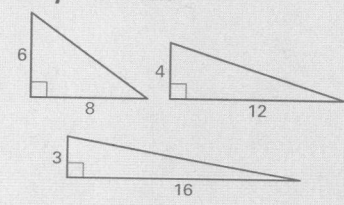

6

8

4

12

3

16

29.

Area of a Regular Octagon In Exercises 28–30, use the regular octagon at the right.

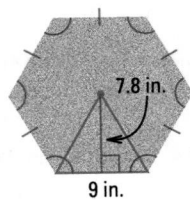

28. Find the area of △GXF in the octagon.
 9.6 square units

29. Copy the diagram. To form congruent triangles, connect the following pairs of vertices: A and E, B and F, C and G, D and H. How many triangles are formed?
 See margin; 8 triangles.

30. What is the area of the octagon? Explain.
 76.8 square units, since 8(9.6) = 76.8

31. Rock Formations Many basaltic columns are hexagonal. The top of one of these columns is a regular hexagon as shown below. Find its area. (Another photograph of basaltic columns is on page 408.) 210.6 in.²

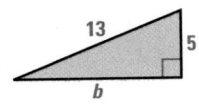

9 in.

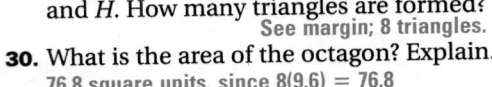
EXAMPLE *Using the Pythagorean Theorem*

Find the area of the triangle.

13   5
b

Solution

First, find the base. Use the Pythagorean Theorem to find the value of *b*.

$(\text{hypotenuse})^2 = (\text{leg})^2 + (\text{leg})^2$ Pythagorean Theorem

$(13)^2 = (5)^2 + (b)^2$ Substitute.

$169 = 25 + b^2$ Simplify.

$144 = b^2$ Subtract 25 from each side.

$12 = b$ Find the positive square root.

Use 12 as the base in the formula for the area of a triangle.

$A = \frac{1}{2}bh = \frac{1}{2}(12)(5) = 30$ square units

Student Help

LOOK BACK
To review the Pythagorean Theorem, see pp. 192 and 193.

Using the Pythagorean Theorem Find the area of the triangle.

32.

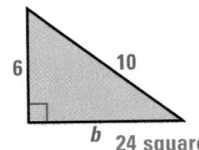

6   10
b 24 square units

33.

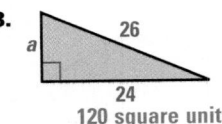

a   26
24
120 square units

34.

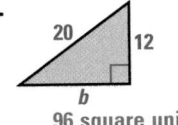

20   12
b
96 square units

35. Multiple Choice Given that the area of the triangle is 99 square meters, what is the height of the triangle? B

(A) 4.5 m (B) 9 m

(C) 11 m (D) 22 m

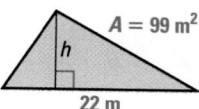

Mixed Review

Trapezoids Find the value of *x* in the trapezoid. *(Lesson 6.5)*

36.
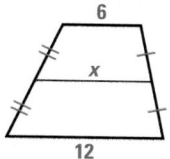

37.
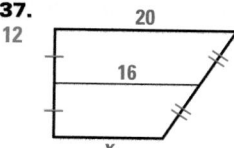

38.
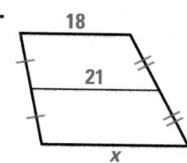

Algebra Skills

Naming Coordinates Give the coordinates of the point.
(Skills Review, p. 664)

39. A (−3, 3) **40.** B (2, 1)

41. C (6, 2) **42.** D (−2, −2)

43. E (3, −1) **44.** F (6, −2)

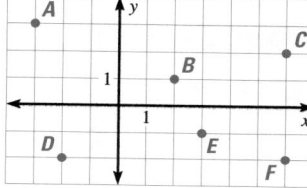

Quiz 2

Find the area of the polygon. *(Lessons 8.3, 8.4)*

1.
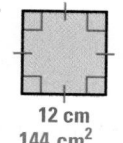
12 cm
144 cm²

2.
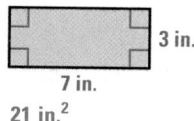
3 in.
7 in.
21 in.²

3.
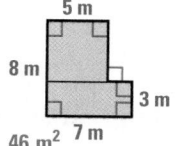
5 m
8 m
3 m
46 m² 7 m

4.
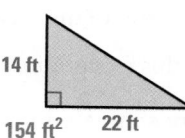
14 ft
22 ft
154 ft²

5.
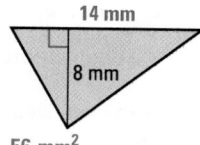
14 mm
8 mm
56 mm²

6.
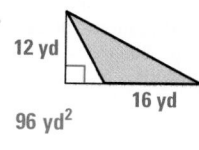
12 yd
16 yd
96 yd²

In Exercises 7–9, A gives the area of the polygon. Find the missing measure. *(Lessons 8.3, 8.4)*

7. $A = 48$ in.²
$b = 8$ in.
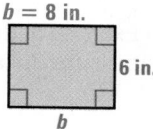
6 in.
b

8. $A = 90$ m²
$h = 12$ m
h
15 m

9. $A = 63$ cm²
$b = 14$ cm
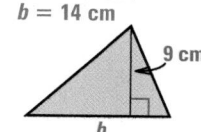
9 cm
b

8.4 *Area of Triangles*   437

Assessment Resources
The Mini-Quiz below is also available on blackline (*Chapter 8 Resource Book*, p. 46) and on transparency. For more assessment resources, see:
• Chapter 8 Resource Book
• Standardized Test Practice
• Test and Practice Generator

Mini-Quiz
Find the area of the triangle.

1.
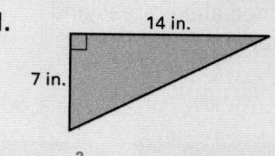
14 in.
7 in.
49 in.²

2.
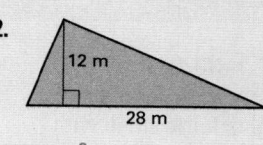
12 m
28 m
168 m²

In Exercises 3 and 4, *A* gives the area of the triangle. Find the missing measure.

3. $A = 65$ ft² 10 ft
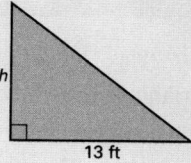
h
13 ft

4. $A = 27$ cm² 9 cm
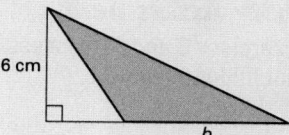
6 cm
b

5. The triangular sail of a boat has a base 5 feet long and the sail has an area of 20 square feet. What is the height of the sail?
8 ft

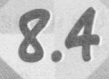
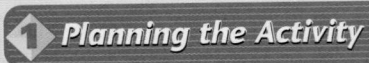

Goal
Students explore how the ratio of the areas of two similar triangles is related to the scale factor.

Materials
See the Technology Activity Keystrokes Masters in the *Chapter 8 Resource Book*.

LINK TO LESSON
This activity reinforces what students learned about areas and scale factors in Example 4 of Lesson 8.4.

Tips for Success
In Step 1, students should make their triangles fairly small with point *D* close to the triangle so that there will be room on the screen for the dilation in Step 2.

KEY DISCOVERY
If two similar triangles have a scale factor of $\frac{a}{b}$, then the ratio of the areas of the triangles is $\frac{a^2}{b^2}$.

Activity Assessment
Use Exercises 3 and 5 to assess student understanding.

Question

How are the areas of similar triangles related to the scale factor of the triangles?

Explore

1 Draw a triangle. Construct point *D* outside the triangle.

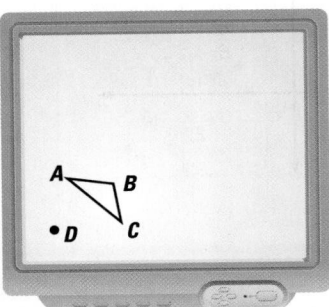

2 Dilate the triangle by a scale factor of 3 using *D* as the center.

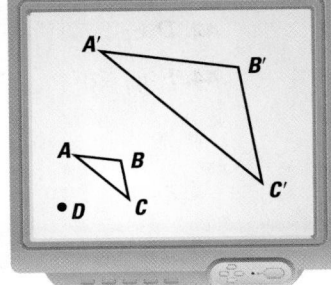

3 Construct the interior of each triangle. Measure the area of each triangle.

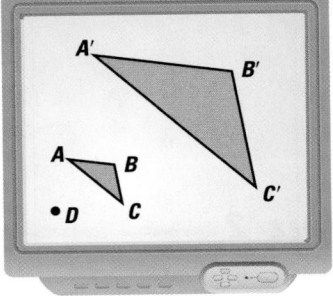

Think About It

Student Help

LOOK BACK
To review the definition of scale factor, see p. 366.

3. If two polygons are similar with a scale factor of $\frac{a}{b}$, then the ratio of their areas is $\frac{a^2}{b^2}$.

1. Divide the area of $\triangle A'B'C'$ by the area of $\triangle ABC$. How is the ratio of the areas related to the scale factor?
   The ratio of the areas is the square of the scale factor.

2. What happens to the ratio of the areas as you drag point *A*?
   It stays the same.

3. What theorem does this exploration illustrate?

Extension Use geometry software to draw a triangle. Construct the interior of the triangle.

4. What is the area of the triangle? Answers will vary.

5. The scale factor of this triangle to a larger triangle is 2.7. What is the area of the larger triangle? Answers will vary, but the area of the larger triangle will be 2.7^2, or 7.29 times the area of the original triangle.

6. To check your answer to Exercise 5, dilate the original triangle by a scale factor of 2.7 and measure the area of the larger triangle.
   Check students' work.

8.5 Area of Parallelograms

Goal
Find the area of parallelograms.

Key Words
- base of a parallelogram
- height of a parallelogram
- parallelogram p. 310
- rhombus p. 325

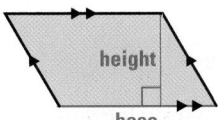 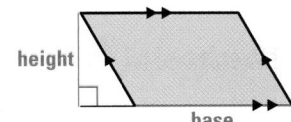

Geo-Activity ▶ **Exploring the Area of a Parallelogram**

❶ Use a straightedge to draw a line through one of the vertices of an index card.
1., 2. Check students' work.

5
3

❷ Cut out the triangle. Tape the triangle to the opposite side to form a parallelogram.

5
3

❸ How does the area of the parallelogram compare to the area of the rectangular index card? How do their bases compare? How do their heights compare? area of parallelogram = area of rectangular index card; bases are equal; heights are equal.

❹ Write a conjecture about the formula for the area of a parallelogram. *Sample answer:* The area of a parallelogram is the product of the base and the height.

Either pair of parallel sides of a parallelogram are called the **bases of a parallelogram**.

The shortest distance between the bases of a parallelogram is called the **height of a parallelogram**. The segment that represents the height is perpendicular to the bases.

height
base

A height can be inside the parallelogram.

height
base

A height can be outside the parallelogram.

As the Geo-Activity suggests, the area of a parallelogram is found using a base and its corresponding height.

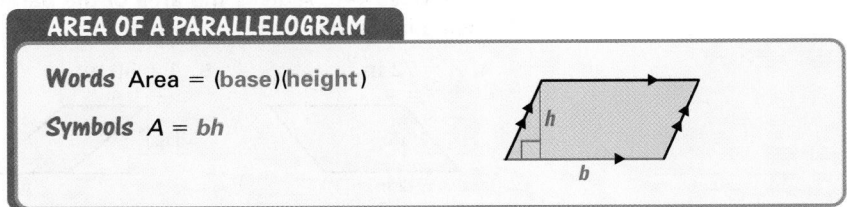

AREA OF A PARALLELOGRAM

Words Area = (base)(height)

Symbols $A = bh$

h
b

8.5 *Area of Parallelograms* **439**

❶ Plan

Pacing
Suggested Number of Days

Basic: 2 days
Average: 2 days
Advanced: 2 days
Block Schedule: 0.5 block with 8.4
                    0.5 block with 8.6

Teaching Resources
📄 **Blacklines**
(See page 408B.)

📑 **Transparencies**
- Warm-Up with Quiz
- Answers

💻 **Technology**
- Electronic Teacher Tools
- Test and Practice Generator
- Online Lesson Planner
- Internet Support

Geo-Activity

Goal Use a paper model to explore the relationship between the area formulas for parallelograms and rectangles.

Key Discovery The area of a parallelogram is the same as the area of a rectangle that has the same base and height as the parallelogram.

❷ Teach

Content and Teaching Strategies
For background information on geometric concepts and teaching strategies related to this lesson, see pages 408E and 408F in this Teacher's Edition.

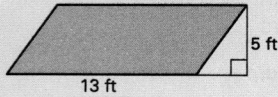
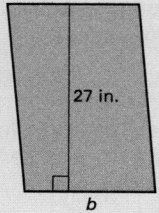
EXAMPLE 1 Find the Area of a Parallelogram

Find the area of the parallelogram.

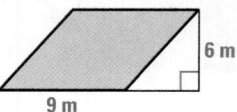

6 m

9 m

Solution

Use the formula for the area of a parallelogram. Substitute 9 for *b* and 6 for *h*.

$$A = bh \qquad \text{Formula for the area of a parallelogram}$$
$$= (9)(6) \qquad \text{Substitute 9 for } b \text{ and 6 for } h.$$
$$= 54 \qquad \text{Multiply.}$$

ANSWER ▶ The parallelogram has an area of 54 square meters.

Using Algebra *xy*

EXAMPLE 2 Find the Height of a Parallelogram

Find the height of the parallelogram given that its area is 78 square feet.

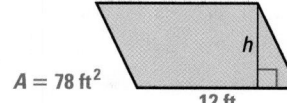

$A = 78 \text{ ft}^2$

h

12 ft

Solution

$$A = bh \qquad \text{Formula for the area of a parallelogram}$$
$$78 = 12h \qquad \text{Substitute 78 for } A \text{ and 12 for } b.$$
$$6.5 = h \qquad \text{Divide each side by 12.}$$

ANSWER ▶ The parallelogram has a height of 6.5 feet.

Checkpoint ✓ **Area of Parallelograms**

Find the area of the parallelogram.

1.

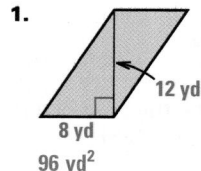

12 yd

8 yd

96 yd²

2.

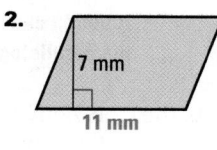

7 mm

11 mm

77 mm²

3.

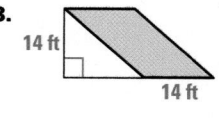

14 ft

14 ft

196 ft²

In Exercises 4–6, *A* gives the area of the parallelogram. Find the missing measure.

4. $A = 72 \text{ in.}^2$
$h = 6 \text{ in.}$

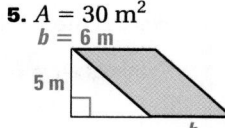

h

12 in.

5. $A = 30 \text{ m}^2$
$b = 6 \text{ m}$

5 m

b

6. $A = 28 \text{ cm}^2$
$h = 4 \text{ cm}$

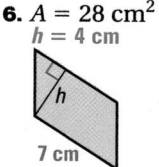

h

7 cm

AREA OF A RHOMBUS

Words Area = $\frac{1}{2}$(product of diagonals)

Symbols $A = \frac{1}{2}d_1d_2$

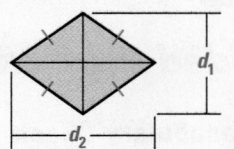

Area of a Rhombus The formula for the area of a rhombus can be justified using the area of a triangle. A specific case is given below.

The diagonals divide a rhombus into four congruent right triangles. So, the area of the rhombus is 4 times the area of one of the right triangles.

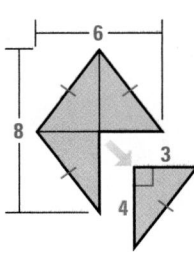

Area of 1 triangle = $\frac{1}{2}bh = \frac{1}{2}(3)(4) = 6$ square units

Area of 4 triangles = 4(6) = 24 square units

Notice that $\frac{1}{2}d_1d_2$, or $\frac{1}{2}(6)(8)$, also equals 24 square units.

Student Help
CLASSZONE.COM

MORE EXAMPLES
More examples at classzone.com

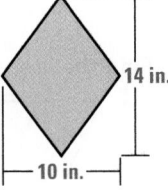

EXAMPLE 3 Find the Area of a Rhombus

Find the area of the rhombus.

a.

14 in.

10 in.

b.

6 m
9 m   9 m
6 m

Solution

a. $A = \frac{1}{2}d_1d_2$

$= \frac{1}{2}(14)(10)$

$= 70$

The area of the rhombus is 70 square inches.

Solution

b. $A = \frac{1}{2}d_1d_2$

$= \frac{1}{2}(6 + 6)(9 + 9)$

$= \frac{1}{2}(12)(18)$

$= 108$

The area of the rhombus is 108 square meters.

8.5 Area of Parallelograms   **441**

Extra Example 3
Find the area of the rhombus.

a.

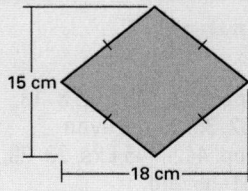

15 cm

18 cm

135 cm^2

b.

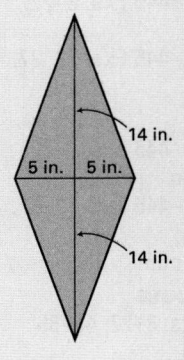

14 in.

5 in.   5 in.

14 in.

140 in.^{2}

✓ Concept Check

How is finding the area of a parallelogram different from finding the area of a rhombus? The area of a parallelogram is found by multiplying the base and the height. The area of a rhombus is given by half the product of the lengths of its diagonals.

🧩 Daily Puzzler

Eight congruent rectangles form a large rectangle whose area is 120 square units, as shown in the figure below. What is the perimeter of the large rectangle? 46 units

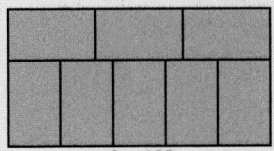

A = 120

8.5 Exercises

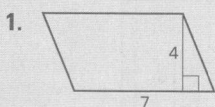
Guided Practice

Vocabulary Check

1. The parallelogram at the right has a base of 7 units and a height of 4 units. Sketch the parallelogram and label its base and height. *See margin.*

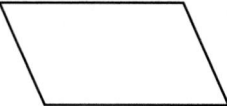

Skill Check

Match the quadrilateral with the corresponding area equation.

A. $A = \frac{1}{2}(6)(10)$ **B.** $A = (13)(6)$ **C.** $A = (10)(6)$

2. C
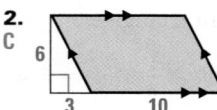

3. A
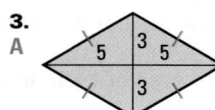

4. B
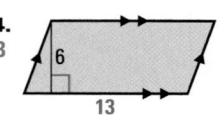

Find the measures of the diagonals of the rhombus.

5.
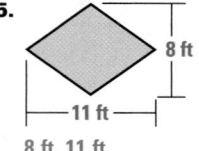
8 ft, 11 ft

6.
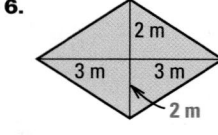
4 m, 6 m

7.
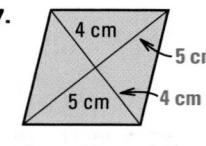
8 cm, 10 cm

Practice and Applications

Extra Practice

See p. 690.

Area of Parallelograms Find the area of the parallelogram.

8.
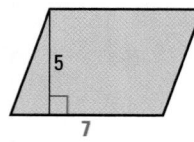
35 square units

9.
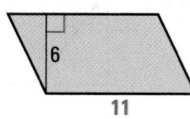
66 square units

10.
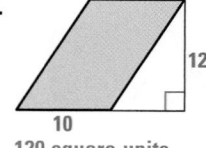
120 square units

11.
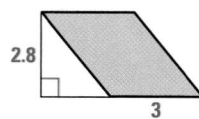
8.4 square units

12.
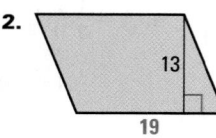
247 square units

13.
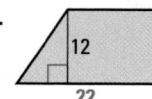
264 square units

14. **You be the Judge** One of your classmates states that it is possible to draw a parallelogram that is different from the one at the right, but has the same area. Do you agree or disagree? Explain your reasoning. *Agree. Sample answer: Other parallelograms with area 42 ft² can be drawn, for example with base 21 ft and height 2 ft.*

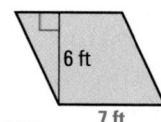

Link to
Crafts

STAINED GLASS The area of a stained-glass design can be used to estimate the amount of glass needed to complete the piece.

16. The base is 4, not 7. So, the area is (4)(5), or 20 square units.

17. The formula is $A = bh$, not $A = \frac{1}{2}bh$. So, the area is (14)(11), or 154 square units.

---

15. Stained Glass The piece of stained glass at the right is made up of eight congruent parallelograms. Each parallelogram has a base of 8 centimeters and a height of 3 centimeters. Find the area of the entire piece. **192 cm²**

Error Analysis In Exercises 16 and 17, students were asked to find the area of the parallelogram. Describe and correct any errors. See margin.

16.
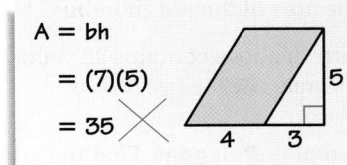

$A = bh$

$= (7)(5)$

$= 35$

17.

$A = \frac{1}{2}bh$

$= \frac{1}{2}(14)(11)$

$= 77$

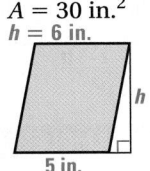

Using Algebra In Exercises 18–20, **A** gives the area of the parallelogram. Find the missing measure.

18. $A = 88 \text{ cm}^2$
$b = 11$ cm
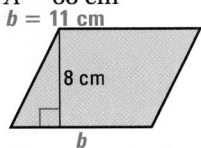

19. $A = 30 \text{ in.}^2$
$h = 6$ in.
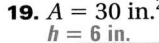

20. $A = 39 \text{ ft}^2$
$b = 6.5$ ft
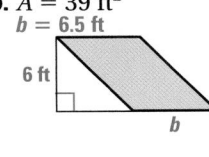

21. A parallelogram has a base of 13 meters and an area of 104 square meters. Find the height. **8 m**

22. A parallelogram has a height of 12 feet and an area of 132 square feet. Find the base. **11 ft**

Area of Rhombuses Find the area of the rhombus.

23.
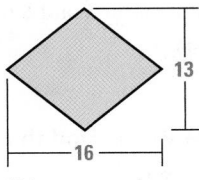
104 square units

24.
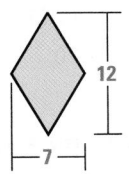
42 square units

25.
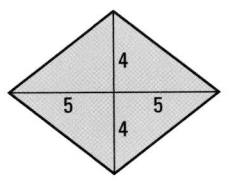
40 square units

26.
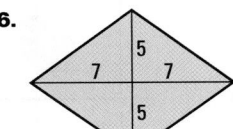
70 square units

27.
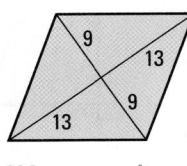
234 square units

28.
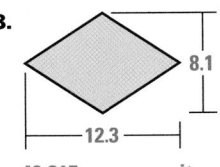
49.815 square units

8.5 *Area of Parallelograms* **443**

Geometric Reasoning

It is possible for students to answer Exercise 14 correctly, but still use incorrect reasoning to reach the correct conclusion. For example, students may fixate on the base and height and say that a rectangle with the same base of 7 feet and a height of 6 feet is a different parallelogram with the same area. Although this is correct, students may think this is the only other parallelogram with an area of 42 square feet. To convince them otherwise, draw parallelograms with bases and heights of 21 feet and 2 feet, and 14 feet and 3 feet, respectively, and ask students to calculate the areas of these figures.

✗ Common Error

In Exercises 23–28, watch for students whose answers are exactly double the correct answers. They have likely forgotten to multiply by $\frac{1}{2}$ when finding the area of each rhombus. Review the formula for the area of a rhombus given on page 441.

443

Link to

Quilts

PENROSE RHOMBUSES

Many quilts feature designs made up of rhombuses. The quilt shown above consists entirely of Penrose Rhombuses.

Penrose Rhombuses The rhombuses shown below are named after the British physicist Roger Penrose. He found that these rhombuses can be used to cover a surface without any gaps and without having to repeat the same pattern. An example is shown below.

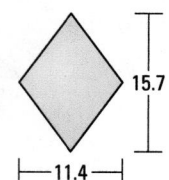 15.7

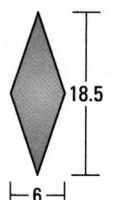 18.5

29. Find the area of the yellow rhombus. **89.49 square units**

30. Find the area of the red rhombus. **55.5 square units**

31. The pattern above contains 32 yellow and 21 red rhombuses. Find its area. **4029.18 square units**

Area of Complex Polygons Find the area of the entire polygon by adding the areas of the smaller polygons.

32.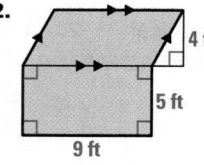

81 ft²

33.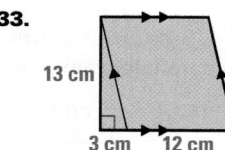

175.5 cm²

34.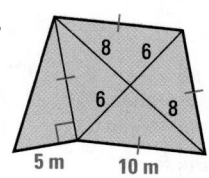

121 m²

EXAMPLE **Area on the Coordinate Plane**

Find the area of the parallelogram.

Solution

Count the squares to find the base and the height. The base is 4 units and the height is 5 units.

$A = bh$ Formula for area of a parallelogram

$= (4)(5)$ Substitute 4 for b and 5 for h.

$= 20$ Multiply.

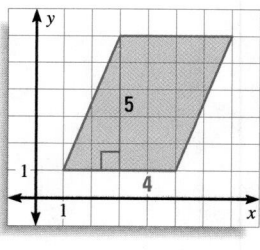

Area on the Coordinate Plane Find the area of the parallelogram.

35.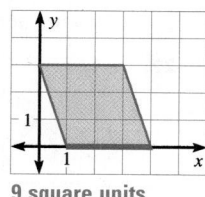

9 square units

36.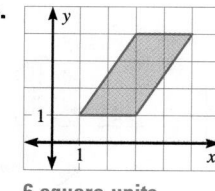

6 square units

37.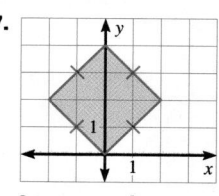

8 square units

Challenge Find the area of the given quadrilateral.

38. Parallelogram *ABCD*
216 square units

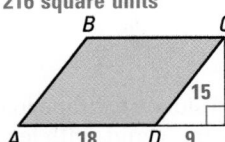

39. Rhombus *FGHJ*   480 square units

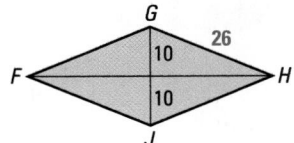

Standardized Test Practice

40. Multiple Choice What is the area of the parallelogram?   B

(A) 12 cm²   (B) 15 cm²

(C) 18 cm²   (D) 30 cm²

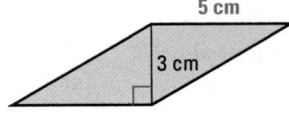

41. Multiple Choice What is the area of the rhombus?   F

(F) 42 ft²   (G) 63 ft²

(H) 72 ft²   (J) 84 ft²

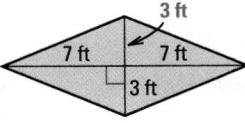

42. Multiple Choice What is the area of the shaded polygon?   C

(A) 30 in.²   (B) 40 in.²

(C) 50 in.²   (D) 60 in.²

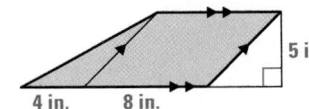

Mixed Review

Similar Triangles Find the value of *x*. *(Lesson 7.5)*

43.
1

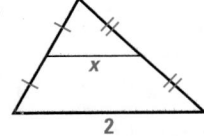

44.
32

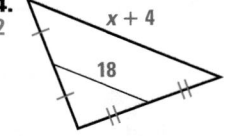

45.
8

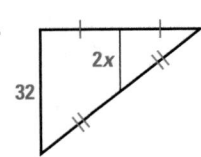

Convex and Concave Polygons Decide whether the polygon is *convex* or *concave*. *(Lesson 8.1)*

46.

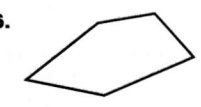

convex

47.

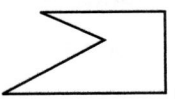

concave

48.

convex

Algebra Skills

Simplifying Radicals Simplify the expression. *(Skills Review, p. 668)*

49. $\sqrt{12}$ $2\sqrt{3}$ **50.** $\sqrt{45}$ $3\sqrt{5}$ **51.** $\sqrt{52}$ $2\sqrt{13}$ **52.** $\sqrt{80}$ $4\sqrt{5}$

53. $\sqrt{300}$ $10\sqrt{3}$ **54.** $\sqrt{125}$ $5\sqrt{5}$ **55.** $\sqrt{180}$ $6\sqrt{5}$ **56.** $\sqrt{432}$ $12\sqrt{3}$

8.5 Area of Parallelograms   **445**

Assessment Resources

The Mini-Quiz below is also available on blackline (*Chapter 8 Resource Book*, p. 57) and on transparency. For more assessment resources, see:
• Chapter 8 Resource Book
• Standardized Test Practice
• Test and Practice Generator

Mini-Quiz

1. Find the area of the parallelogram.

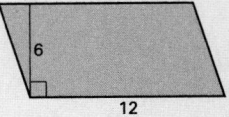

72 square units

2. Find the area of the rhombus.

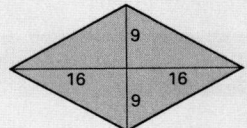

288 square units

3. The area of the parallelogram is 27 square meters. Find the height. 9 m

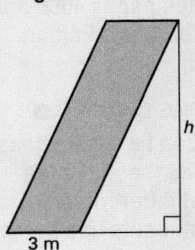

4. A parallelogram has a height of 8 feet and an area of 96 square feet. Find the base. 12 ft

Plan

Pacing
Suggested Number of Days

Basic: 1 day
Average: 1 day
Advanced: 1 day
Block Schedule: 0.5 block with 8.5

Teaching Resources

Blacklines
(See page 408B.)

Transparencies
• Warm-Up with Quiz
• Answers

Technology
• Electronic Teacher Tools
• Test and Practice Generator
• Online Lesson Planner
• Internet Support

Teach

Content and Teaching Strategies
For background information on geometric concepts and teaching strategies related to this lesson, see pages 408E and 408F in this Teacher's Edition.

Tips for New Teachers
Any trapezoid can be cut diagonally into two triangles that have the same height but different base lengths. Have students use the formula for the area of a triangle to show the total area of the trapezoid as the sum of the areas of the two triangles. Assist them in using algebra to simplify their result to derive the formula for the area of a trapezoid. See the Tips for New Teachers on pp. 1–2 of the *Chapter 8 Resource Book* for additional notes about Lesson 8.6.

8.6 Area of Trapezoids

Goal
Find the area of trapezoids.

Key Words
• trapezoid p. 332
• base of a trapezoid p. 332
• height of a trapezoid

Recall that the parallel sides of a trapezoid are called the *bases* of the trapezoid, with lengths denoted by b_1 and b_2.

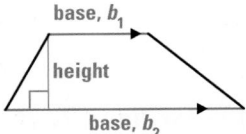

The shortest distance between the bases is the **height of the trapezoid** .

Suppose that two congruent trapezoids with bases b_1 and b_2 and height h are arranged to form a parallelogram as shown.

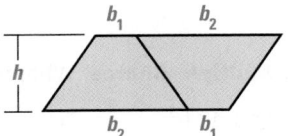

Student Help

<blockquote>
LOOK BACK
To review more about trapezoids, see p. 332.
</blockquote>

The area of the parallelogram is $h(b_1 + b_2)$. Because the two trapezoids are congruent, the area of one of the trapezoids is half the area of the parallelogram.

AREA OF A TRAPEZOID

Words Area $= \frac{1}{2}$(height)(sum of bases)

Symbols $A = \frac{1}{2}h(b_1 + b_2)$

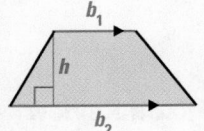

EXAMPLE **1** **Find the Area of a Trapezoid**

Find the area of the trapezoid.

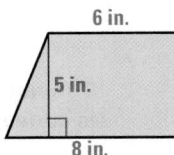

Solution

$A = \frac{1}{2}h(b_1 + b_2)$ Formula for the area of a trapezoid

$= \frac{1}{2}(5)(6 + 8)$ Substitute 5 for h, 6 for b_1, and 8 for b_2.

$= \frac{1}{2}(5)(14)$ Simplify within parentheses.

$= 35$ Simplify.

ANSWER ▶ The area of the trapezoid is 35 square inches.

 **Area of Trapezoids**

Find the area of the trapezoid.

1.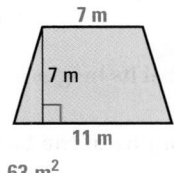
7 m

7 m

11 m

63 m²

2.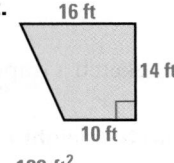
16 ft

14 ft

10 ft

182 ft²

3.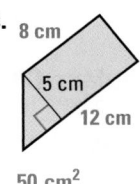
8 cm

5 cm

12 cm

50 cm²

EXAMPLE 2 **Use the Area of a Trapezoid**

Find the value of b_2 given that the area of the trapezoid is 96 square meters.

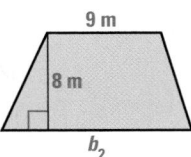
9 m

8 m

A = 96 m²

b_2

Student Help

STUDY TIP
The equation is easier to solve without the fraction. So, multiply each side by 2 before distributing.

Solution

$A = \frac{1}{2}h(b_1 + b_2)$	Formula for the area of a trapezoid
$96 = \frac{1}{2}(8)(9 + b_2)$	Substitute 96 for A, 8 for h, and 9 for b_1.
$192 = 8(9 + b_2)$	Multiply each side by 2.
$192 = 72 + 8b_2$	Use the distributive property.
$120 = 8b_2$	Subtract 72 from each side.
$15 = b_2$	Divide each side by 8.

ANSWER The value of b_2 is 15 meters.

 **Use the Area of a Trapezoid**

A gives the area of the trapezoid. Find the missing measure.

4. $A = 77$ ft² $h = 7$ ft

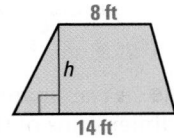
8 ft

h

14 ft

5. $A = 39$ cm² $b_1 = 5$ cm

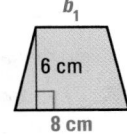
b_1

6 cm

8 cm

6. $A = 84$ in.² $b_2 = 8$ in.

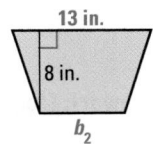
13 in.

8 in.

b_2

7. A trapezoid has an area of 294 square yards. Its height is 14 yards and the length of one base is 30 yards. Find the length of the other base. **12 yards**

Extra Example 1
Find the area of the trapezoid.

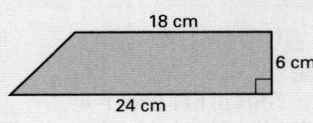
18 cm

6 cm

24 cm

126 cm²

Extra Example 2
Find the value of h given that the area of the trapezoid is 120 square feet. **12 ft**

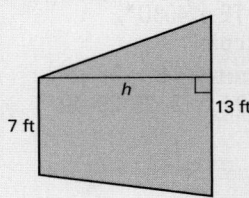
h

13 ft

7 ft

Visualize It!
Remind students to continue to add area formulas and sketches to the table they began in Lesson 8.3.

✓ Concept Check
Explain how to find the area of a trapezoid. *Sample answer:* First find the sum of the two bases of the trapezoid. Then multiply this sum by the height of the trapezoid. Finally, multiply this product by $\frac{1}{2}$.

③ Apply

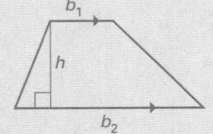
8.6 Exercises

Guided Practice

Vocabulary Check **1.** Sketch a trapezoid. Label its height h and its bases b_1 and b_2. See margin.

Skill Check **Find the height and the lengths of the bases of the trapezoid.**

2.
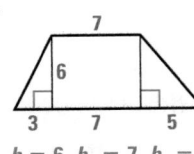
$h = 6$, $b_1 = 7$, $b_2 = 15$

3.
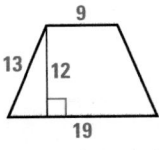
$h = 12$, $b_1 = 9$, $b_2 = 19$

4.
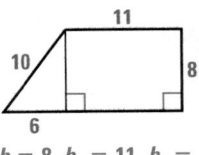
$h = 8$, $b_1 = 11$, $b_2 = 17$

Match the trapezoid with the equation used to find the height.

A. $A = \frac{1}{2}h(5 + 13)$ **B.** $A = \frac{1}{2}h(8 + 13)$ **C.** $A = \frac{1}{2}h(5 + 8)$

5.
C
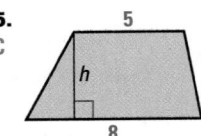

6.
A
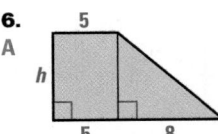

7.
B
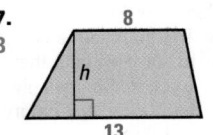

Practice and Applications

Area of a Trapezoid **Find the area of the trapezoid.**

8.
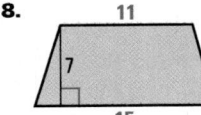
91 square units

9.
6
8
10
64 square units

10.
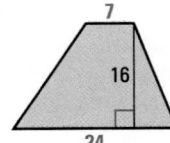
248 square units

11.
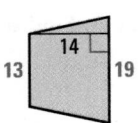
224 square units

12.
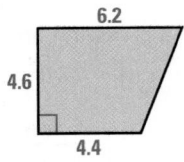
24.38 square units

13.
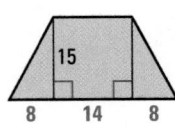
330 square units

14. 🔨 **You be the Judge** A classmate states
that if you double the dimensions of the
trapezoid at the right, then its area doubles.
Do you agree? Explain your answer.
Sample answer: No. The area is quadrupled:
$\frac{1}{2} \cdot 6(6 + 9) = 45$ and $\frac{1}{2} \cdot 12(12 + 18) = 180$; $4(45) = 180$.

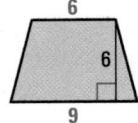

15. Visualize It! Draw three different trapezoids with a height of 5 units and bases of 3 units and 7 units. Then find the areas of the trapezoids. What do you notice? *See margin; the areas are equal.*

Student Help

LOOK BACK
To review the midsegment of a trapezoid, see p. 333.

Technology In Exercises 16 and 17, use geometry software.

① Draw a trapezoid.

② Draw the midsegment.

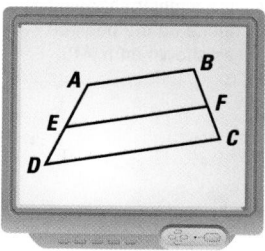

16. Find the length of the midsegment and the height of the trapezoid. Multiply the two measures. *Answers will vary.*

17. Find the area of the trapezoid. How does the area compare to your answer for Exercise 16? *Answers will vary; they are the same.*

Using Algebra In Exercises 18–20, A gives the area of the trapezoid. Find the missing measure.

18. $A = 135$ cm² $b_1 = 9$ cm

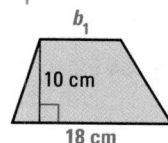

19. $A = 132$ in.² $h = 11$ in.

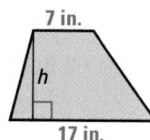

20. $A = 198$ m² $b_2 = 18$ m

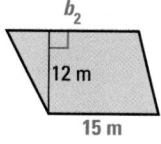

Student Help
CLASSZONE.COM

HOMEWORK HELP
Extra help with problem solving in Exs. 21–22 is at classzone.com

21. A trapezoid has an area of 50 square units. The lengths of the bases are 10 units and 15 units. Find the height. *4 units*

22. A trapezoid has an area of 24 square units. The height is 3 units and the length of one of the bases is 5 units. Find the length of the other base. *11 units*

Bridges In Exercises 23–25, use the following information.
The roof on the bridge below, consists of four sides: two congruent trapezoids and two congruent triangles.

Doe River Covered Bridge in Elizabethton, Tennessee

23. Find the combined area of the two trapezoids.
about 3897 ft²

24. Use the diagram at the right to find the combined area of the two triangles.
about 285 ft²

25. What is the area of the entire roof?
about 4182 ft²

Detail of roof

8.6 *Area of Trapezoids* **449**

Geometric Reasoning
Exercises 16 and 17 demonstrate that another way to find the area of a trapezoid is to multiply the length of the midsegment times the height. You may wish to ask students what the formula for finding the length of the midsegment is, given the lengths of the bases of a trapezoid.

15.

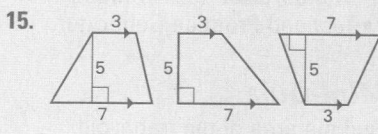

Assessment Resources

The Mini-Quiz below is also available on blackline (*Chapter 8 Resource Book*, p. 68) and on transparency. For more assessment resources, see:
- Chapter 8 Resource Book
- Standardized Test Practice
- Test and Practice Generator

Mini-Quiz

Find the area of the trapezoid.

1.

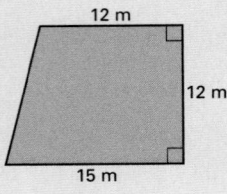

162 m²

2.

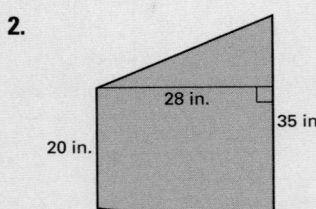

770 in.²

3. A trapezoid has an area of 125 square centimeters. The lengths of the bases are 8 centimeters and 17 centimeters. Find the height. **10 cm**

4. A trapezoid has an area of 190 square feet, a height of 20 feet, and the measure of one base is 10 feet. Find the measure of the other base. **9 ft**

Student Help

> **VISUAL STRATEGY**
> To find the area of a complex polygon, you can add the areas of the simpler shapes that make up the polygon, as shown on p. 410.

Windows Find the area of the window.

26.

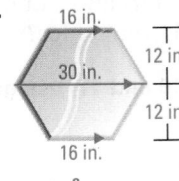

552 in.²

27.

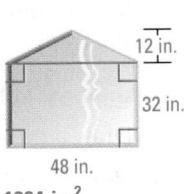

1824 in.²

28.

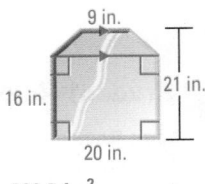

392.5 in.²

Using the Pythagorean Theorem Find the height using the Pythagorean Theorem and a calculator. Then find the area of the trapezoid.

29.

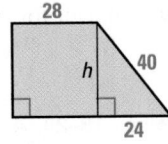

1280 square units

30.

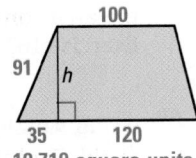

10,710 square units

31.

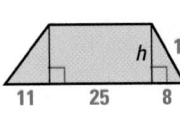

517.5 square units

Standardized Test Practice

32. Multiple Choice What is the area of the trapezoid? **B**

Ⓐ 25 in.²   Ⓑ 42 in.²

Ⓒ 68 in.²   Ⓓ 84 in.²

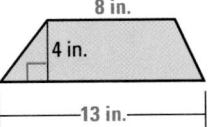

33. Multiple Choice What is the area of the trapezoid? **H**

Ⓕ 88 ft²   Ⓖ 128 ft²

Ⓗ 152 ft²   Ⓙ 176 ft²

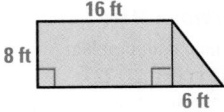

Mixed Review

Finding Area Match the region with a formula for its area. Use each formula exactly once. *(Lessons 8.3–8.6)*

34. Region 1 **A**   **A.** $A = s^2$

35. Region 2 **E**   **B.** $A = \frac{1}{2}d_1d_2$

36. Region 3 **C**   **C.** $A = \frac{1}{2}bh$

37. Region 4 **B**   **D.** $A = \frac{1}{2}h(b_1 + b_2)$

38. Region 5 **D**   **E.** $A = bh$

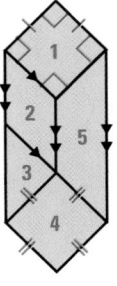

Algebra Skills

Fraction Operations Add or subtract. Write the answer as a fraction in simplest form. *(Skills Review, p. 658)*

39. $\frac{3}{8} + \frac{5}{8}$ **1**   **40.** $\frac{5}{9} - \frac{2}{9}$ **$\frac{1}{3}$**   **41.** $\frac{3}{4} + \frac{1}{12}$ **$\frac{5}{6}$**   **42.** $\frac{4}{7} - \frac{1}{5}$ **$\frac{13}{35}$**

Activity 8.7 Finding Area of Circles

Question

How do you find the area of a circle using the radius?

Explore

1 Use a compass to draw a circle on a piece of paper. Cut the circle out. Fold the circle in half, four times.

2 Cut the circle along the fold lines to divide the circle into 16 equal wedges.

3 Arrange the wedges to form a shape resembling a parallelogram. The base and height of the parallelogram are labeled.

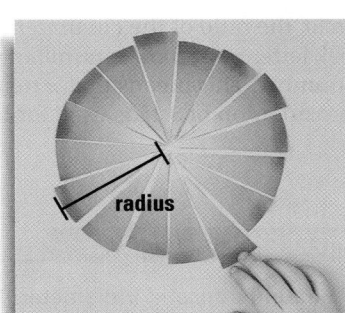

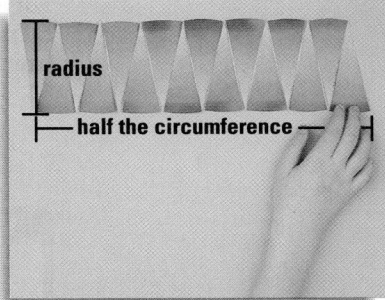

radius

radius

half the circumference

Think About It

Student Help

VOCABULARY TIP
The *circumference* is the distance around a circle.

1. The area of the parallelogram in Step 3 can be used to approximate the area of the original circle. Write a verbal expression for the area of the parallelogram. *The area of the parallelogram is the product of half the circumference and the radius.*

2. The ratio of the circumference C to the diameter d of a circle is denoted by the Greek letter π (or *pi*), which is approximately equal to 3.14. So, $C = \pi d$. Because the diameter is twice the radius r of the circle,

$$C = \pi d = \pi(2r) = 2\pi r.$$

Use the fact that $C = 2\pi r$ to write your verbal expression from Exercise 1 using variables. Simplify the expression. $A = \frac{1}{2}(2\pi r) \cdot r,$ or $A = \pi r^2$

3. Find the area of the circle at the right. Use 3.14 as an approximation for π. *about 50.24 square units*

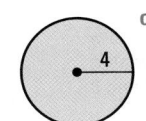

4

---

1 **Planning the Activity**

Goal
Students cut a circle into wedges in order to derive the formula for the area of a circle.

Materials
See the margin of the student page.

LINK TO LESSON
In Example 2 on page 453, students find the area of a circle using the formula $A = \pi r^2$ derived in this activity.

2 **Managing the Activity**

Alternative Approach
If you have access to a large sheet of paper and a large compass, consider demonstrating the activity for the class while volunteers read the instructions aloud and assist with Steps 2 and 3.

3 **Closing the Activity**

KEY DISCOVERY
The area of a circle is the square of the radius multiplied by the constant pi (π).

Activity Assessment
Use Exercise 2 to assess student understanding.

Pacing
Suggested Number of Days

Basic: 2 days
Average: 2 days
Advanced: 2 days
Block Schedule: 1 block

Teaching Resources

📄 **Blacklines**
(See page 408B.)

🔖 **Transparencies**
• Warm-Up with Quiz
• Answers

💻 **Technology**
• Electronic Teacher Tools
• Test and Practice Generator
• Online Lesson Planner
• Internet Support

Teach

Content and Teaching Strategies
For background information on geometric concepts and teaching strategies related to this lesson, see pages 408E and 408F in this Teacher's Edition.

Tips for New Teachers
Students have had some experience finding measures for radii, diameters, area, and circumference, but this may be the first time that they have worked with a sector of a circle. They can think of a sector of a circle as a piece of pie. See the Tips for New Teachers on pp. 1–2 of the *Chapter 8 Resource Book* for additional notes about Lesson 8.7.

8.7 Circumference and Area of Circles

Goal
Find the circumference and area of circles.

Key Words
• circle
• center
• radius
• diameter
• circumference
• central angle
• sector

A **circle** is the set of all points in a plane that are the same distance from a given point, called the **center** of the circle. A circle with center *P* is called "circle *P*," or ⊙*P*.

The distance from the center to a point on the circle is the **radius**. The plural of radius is *radii*.

The distance across the circle, through the center, is the **diameter**. The diameter *d* is twice the radius *r*. So, $d = 2r$.

The **circumference** of a circle is the distance around the circle.

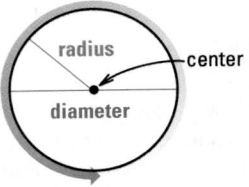

For any circle, the ratio of the circumference to its diameter is denoted by the Greek letter π, or *pi*. The number π is 3.14159 . . . , which is an irrational number. This means that π neither terminates nor repeats. So, an approximation of 3.14 is used for π.

CIRCUMFERENCE OF A CIRCLE

Words Circumference = π(diameter)
                            = 2π(radius)

Symbols $C = \pi d$ or $C = 2\pi r$

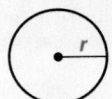

EXAMPLE 1 Find the Circumference of a Circle

Find the circumference of the circle.

Solution

$C = 2\pi r$	Formula for the circumference
$= 2\pi(4)$	Substitute 4 for *r*.
$= 8\pi$	Simplify.
$\approx 8(3.14)$	Use 3.14 as an approximation for π.
$= 25.12$	Multiply.

ANSWER ▶ The circumference is about 25 inches.

Student Help

STUDY TIP
When simplifying an expression involving π, substitute 3.14 for π. You can also use the π key on your calculator, as in Examples 2 and 3.

Checkpoint ✓ *Find the Circumference of a Circle*

Find the circumference of the circle. Round your answer to the nearest whole number.

1.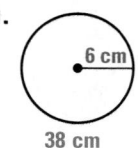
6 cm
38 cm

2.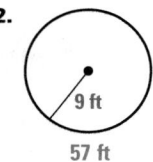
9 ft
57 ft

3.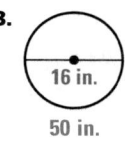
16 in.
50 in.

AREA OF A CIRCLE

Words Area $= \pi(\text{radius})^2$

Symbols $A = \pi r^2$

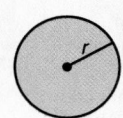
r

EXAMPLE 2 **Find the Area of a Circle**

Find the area of the circle.

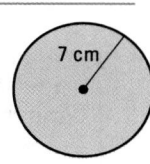
7 cm

Solution

$A = \pi r^2$ Formula for the area of a circle

$= \pi(7)^2$ Substitute 7 for r.

$= 49\pi$ Simplify.

≈ 153.94 Use a calculator.

ANSWER ▶ The area is about 154 square centimeters.

Student Help

KEYSTROKE HELP
If your calculator has π written above a key, use the following keystrokes to simplify 49π:

49 ☒ 2nd π

EXAMPLE 3 **Use the Area of a Circle**

Find the radius of a circle with an area of 380 square feet.

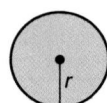
r
$A = 380 \text{ ft}^2$

Solution

$A = \pi r^2$ Formula for the area of a circle

$380 = \pi r^2$ Substitute 380 for A.

$120.96 \approx r^2$ Divide each side by π. Use a calculator.

$11 \approx r$ Take the positive square root.

ANSWER ▶ The radius is about 11 feet.

8.7 *Circumference and Area of Circles*   **453**

Extra Example 1
Find the circumference of the circle. **about 38 cm**

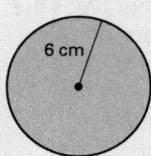
6 cm

Extra Example 2
Find the area of the circle.

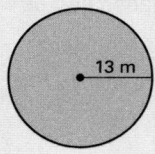
13 m

about 531 m²

Extra Example 3
Find the radius of a circle with an area of 255 square centimeters. **about 9 cm**

Multiple Representations
Point out that only one measurement, the radius, is needed to find the circumference or area of a circle. Point to the symbolic form of the area formula and ask students what each symbol represents. Make sure students are clear that π is a constant and not a measurement of the circle. Then have students look at the figures illustrating the formulas to confirm that only r is represented in them.

454

✗ Common Error

Throughout this lesson, watch for students whose answers for area problems are exactly four times the correct answer. They are calculating the area of the circles using the diameter instead of the radius. Invite students to think of a mnemonic they could use to help avoid making this error.

Extra Example 4

Find the area of the shaded sector.

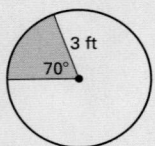

about 5.5 ft²

✓ Concept Check

How do you find the circumference and area of a circle if you know the diameter? **Find the radius, which is half the diameter. The circumference is the diameter multiplied by π. The area is π multiplied by the square of the radius.**

🐢 Daily Puzzler

To the nearest meter, what is the side length of a square that encloses the same area as a circle with radius 13 meters? **23 m**

 **Find the Area of a Circle**

Find the area of the circle. Round your answer to the nearest whole number.

4.
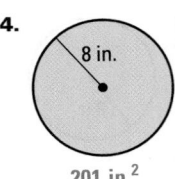
8 in.

201 in.²

5.
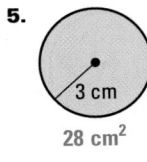
3 cm

28 cm²

6.
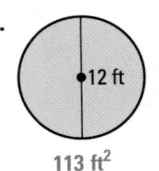
12 ft

113 ft²

Student Help

VOCABULARY TIP
The term *radius* is also used to name a segment that connects the center of a circle to a point on the circle. Two such radii are used to determine a *sector* of a circle.

Central Angles An angle whose vertex is the center of a circle is a **central angle** of the circle.

A region of a circle determined by two radii and a part of the circle is called a **sector** of the circle.

Because a sector is a portion of a circle, the following proportion can be used to find the area of a sector.

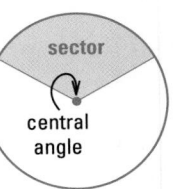
sector

central angle

$$\frac{Area\ of\ sector}{Area\ of\ entire\ circle} = \frac{Measure\ of\ central\ angle}{Measure\ of\ entire\ circle}$$

EXAMPLE 4 Find the Area of a Sector

Find the area of the blue sector.

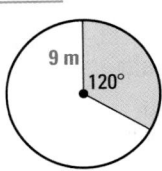
9 m
120°

Solution

❶ First find the area of the circle.

$$A = \pi r^2 = \pi(9)^2 \approx 254.47$$

The area of the circle is about 254 square meters.

❷ Then find the area of the sector. Let x equal the area of the sector.

$$\frac{Area\ of\ sector}{Area\ of\ entire\ circle} = \frac{Measure\ of\ central\ angle}{Measure\ of\ entire\ circle}$$

Visualize It!

A circle contains two straight angles. So, there are 180° + 180°, or 360°, in a circle. ⋯⋯⋯⋯

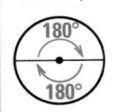
180°
180°

or

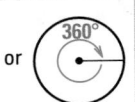
360°

$\dfrac{x}{254} = \dfrac{120°}{360°}$	Substitute.
$360x = 254 \cdot 120$	Cross product property
$360x = 30,480$	Simplify.
$\dfrac{360x}{360} = \dfrac{30,480}{360}$	Divide each side by 360.
$x \approx 84.67$	Simplify.

ANSWER ▶ The area of the sector is about 85 square meters.

In Exercises 7 and 8, *A* represents the area of the entire circle and *x* represents the area of the blue sector. Complete the proportion used to find *x*. Do not solve the proportion.

7. $A = 22 \text{ m}^2$

$$\frac{x}{?} = \frac{180°}{?}$$

$$\frac{x}{22} = \frac{180}{360}$$

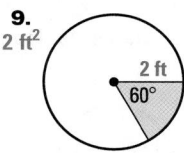

8. $A = 28 \text{ ft}^2$

$$\frac{x}{?} = \frac{?}{360°}$$

$$\frac{x}{28} = \frac{170}{360}$$

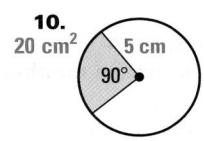

Find the area of the blue sector. Round your answer to the nearest whole number.

9.
2 ft^2

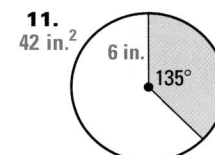

10.
20 cm^2

11.
42 in.^2

8.7 Exercises

Guided Practice

Vocabulary Check

1. Sketch a circle. Sketch and label a *radius* and a *diameter* of the circle. **See margin.**

2. Describe how to find the *circumference* of a circle given its radius.
Multiply the radius by 2. Then multiply by π using the π key on a calculator or using 3.14 as an approximation.

Skill Check Copy and complete the table below.

	Radius, *r*	Diameter, *d*	
3.	?	14 in.	7 in.
4.	11 cm	?	22 cm
5.	3.5 m	?	7 m
6.	?	1 ft	0.5 ft

Write an equation for the area *A* or the circumference *C* by filling in the missing number.

7. $C = 2\pi \boxed{?}$
8

8. $A = \boxed{?} (3)^2$
π

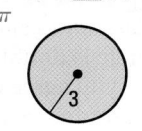

9. $A = \pi (\boxed{?})^2$
7

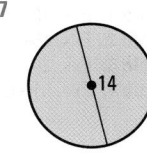

Assignment Guide

BASIC
Day 1: SRH p. 669 Exs. 5–8; pp. 456–459 Exs. 10–20, 47–55
Day 2: pp. 456–459 Exs. 24–27, 31–35, 40–46, Quiz 3

AVERAGE
Day 1: pp. 456–459 Exs. 10–23, 41–46, 47–55 odd
Day 2: pp. 456–459 Exs. 24–29, 31–40, Quiz 3

ADVANCED
Day 1: pp. 456–459 Exs. 10–20 even, 21–23, 41–55 odd
Day 2: pp. 456–459 Exs. 26–29, 30*, 32–40, Quiz 3; EC: classzone.com

BLOCK SCHEDULE
pp. 456–459 Exs. 10–29, 31–46, 47–55 odd, Quiz 3

Extra Practice
• Student Edition, p. 690
• Chapter 8 Resource Book, pp. 69–70

Homework Check
To quickly check student understanding of key concepts, go over the following exercises:

Basic: 10, 15, 19, 24, 33
Average: 11, 16, 19, 27, 34
Advanced: 12, 20, 28, 35, 39

1.
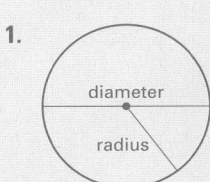

455

Practice and Applications

Extra Practice

See p. 690.

Finding Circumference Find the circumference of the circle. Round your answer to the nearest whole number.

10.
44 cm
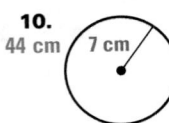
7 cm

11.
31 m
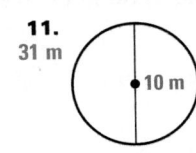
10 m

12.
13 in.
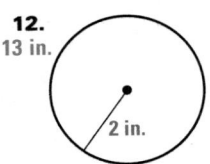
2 in.

13. A circle with a radius of 13 yards 82 yd

14. A circle with a diameter of 15 meters 47 m

Finding Area Find the area of the circle. Round your answer to the nearest whole number.

15.
50 ft²
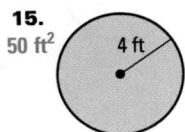
4 ft

16.
3 cm²
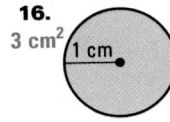
1 cm

17.
113 in.²
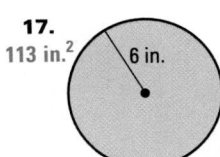
6 in.

18.
254 m²
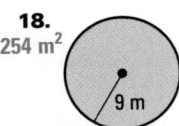
9 m

19.
201 ft²
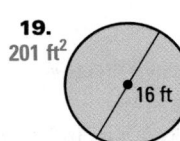
16 ft

20.
284 yd²
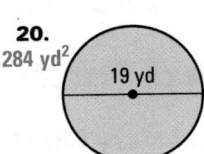
19 yd

21. Cranberries To harvest cranberries, the field is flooded so that the berries float. What area of cranberries can be gathered into a circular region with a radius of 5.5 meters? about 95 m²

22. Error Analysis A student was asked to find the area of the circle below. Describe any errors. The radius is 6, not 12.

23. *Sample answer:* Disagree. The area of the circle with radius 5 is about 79. The area of the circle with radius 10 is about 314. Since 314 ÷ 79 ≈ 4, the area is quadrupled, not doubled.

$$A = \pi r^2$$
$$\approx (3.14)(12)^2$$
$$= 452.16$$
12

23. **You be the Judge** One of your classmates states that if the radius of a circle is doubled, then its area doubles. Do you agree or disagree? Explain your reasoning using the circles at the right.

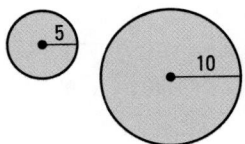
5
10

(xy) **Using Algebra** Use the area *A* of the circle to find the radius *r*.
Round your answer to the nearest whole number.

24. $A = 13$ in.² 2 in.   **25.** $A = 531$ m² 13 m   **26.** $A = 154$ ft² 7 ft

    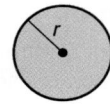   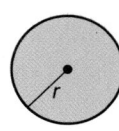

27. A circle has an area of 50 square units. What is its radius? about 4 units

28. A circle has an area of 452 square units. What is its diameter?
about 24 units

29. A circle has an area of 28 square units. What is its diameter?
about 6 units

30. Challenge The circle at the right has an
area of 78.5 square yards. What is its
circumference? about 31 yd

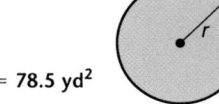

$A = 78.5$ yd²

Sectors and Proportions In Exercises 31 and 32, *A* represents the
area of the entire circle and *x* represents the area of the blue sector.
Complete the proportion used to find *x*. Do not solve the proportion.

31. $A = 12$ ft²

$$\frac{x}{?} = \frac{?}{360°}$$

$$\frac{x}{12} = \frac{60}{360}$$

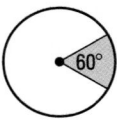

32. $A = 198$ yd²

$$\frac{x}{?} = \frac{100°}{?}$$

$$\frac{x}{198} = \frac{100}{360}$$

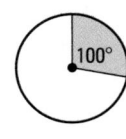

Area of Sectors Find the area of the blue sector. Start by finding the
area of the circle. Round your answer to the nearest whole number.

33.
61 m²   10 m
70°

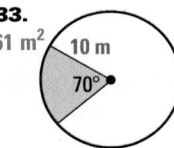

34.
35 ft²   6 ft
110°

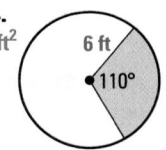

35.
25 cm²   45° 8 cm

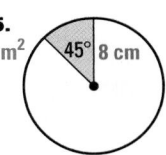

Landscaping The diagram shows the area of a lawn covered by a
water sprinkler. Round your answer to the nearest whole number.

36. What is the area of the lawn that
is covered by the sprinkler? 285 ft²

37. Suppose the water pressure is
weakened so that the radius is
12 feet. What is the area of lawn
that will be covered? 182 ft²

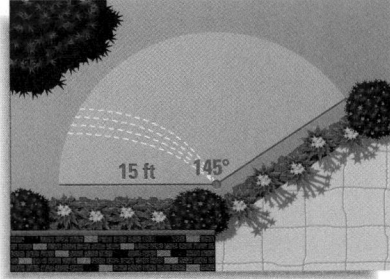

15 ft   145°

In Exercises 28 and 29, students
begin with a known area and are
asked to find the diameter. Some
students may use the formula
$A = \pi r^2$ and stop after finding the
value of *r*. Remind these students
that $d = 2r$ and that they must
multiply their value for the radius
by 2 to find the diameter.

✗ Common Error

EXAMPLE **Find Complex Areas**

Find the area of the shaded region.

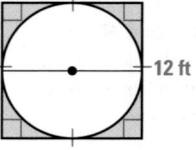
12 ft

Solution

❶ First find the area of the outer region, which is a square.

$$\text{Area of square} = s^2 = (12)^2 = 144 \text{ ft}^2$$

❷ Then find the area of the inner region, which is a circle. The diameter of the circle is 12 feet, so the radius is 6 feet.

$$\text{Area of circle} = \pi r^2 \approx \pi(6)^2 \approx 113 \text{ ft}^2$$

❸ Finally, subtract the area of the circle from the area of the square.

$$\text{Area of shaded region} = \text{Area of square} - \text{Area of circle}$$
$$\approx 144 - 113$$
$$= 31$$

ANSWER ▶ The area of the shaded region is about 31 square feet.

Finding Complex Areas Use the method in the example above to find the area of the shaded region. Round your answer to the nearest whole number.

38.
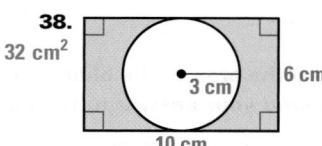
32 cm²
3 cm
6 cm
10 cm

39.
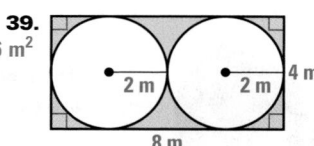
6 m²
2 m 2 m
4 m
8 m

40. Multi-Step Problem Earth has a radius of about 3960 miles at the equator. Suppose a cable is wrapped around Earth at the equator, as shown in the diagram.

a. Find the length of the cable by finding the circumference of Earth. (Assume that Earth is perfectly round. Use 3.14 for π.)

b. Suppose 10 miles is added to the cable length in part (a). Use this length as the circumference of a new circle. Find the radius of the larger circle. (Use 3.14 for π.)

c. Suppose you are standing at the equator. How far are you from the cable in part (b)?

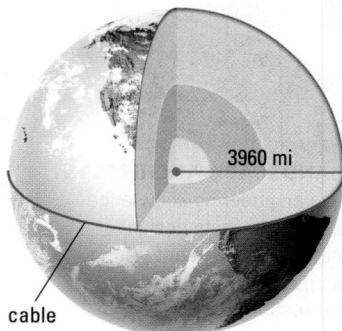
3960 mi
cable

Mixed Review

Identifying Quadrilaterals Use the information in the diagram to name the special quadrilateral. *(Lesson 6.6)*

41.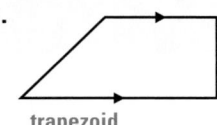
trapezoid

42.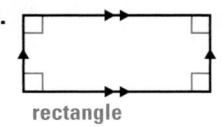
rectangle

43.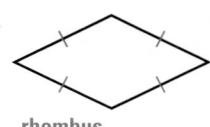
rhombus

Squares and Rectangles Find the area of the polygon. *(Lesson 8.3)*

44.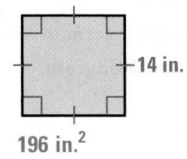
14 in.
196 in.^{2}

45.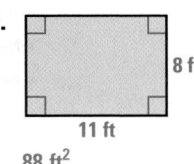
8 ft
11 ft
88 ft^2

46.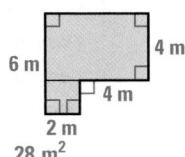
4 m
6 m
4 m
2 m
28 m^2

Algebra Skills

Integers Evaluate the expression. *(Skills Review, p. 663)*

47. $-4 + 7$ 3

48. $-5 - (-5)$ 0

49. $(-4)(-11)$ 44

50. $12 - (-12)$ 24

51. $33 \div (-3)$ -11

52. $14 + (-3) + 4$ 15

53. $-6 - 10$ -16

54. $(-5)(-13)$ 65

55. $22 - (-9)$ 31

Quiz 3

Find the area of the figure. In Exercises 5 and 6, round your answer to the nearest whole numbers. *(Lessons 8.5–8.7)*

1.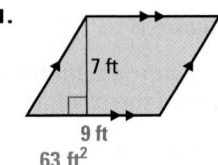
7 ft
9 ft
63 ft^2

2.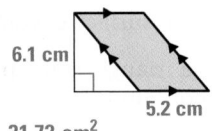
6.1 cm
5.2 cm
31.72 cm^2

3.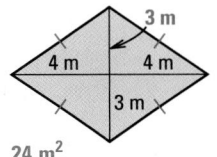
3 m
4 m   4 m
3 m
24 m^2

4.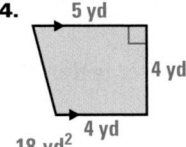
5 yd
4 yd
4 yd
18 yd^2

5.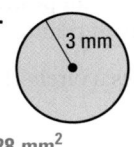
3 mm
28 mm^2

6.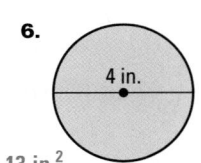
4 in.
13 in.^{2}

In Exercises 7–9, *A* gives the area of the figure. Find the missing measure. In Exercise 9, round your answer to the nearest whole number. *(Lessons 8.5–8.7)*

7. $A = 195$ cm^2
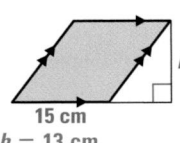
h
15 cm
h = 13 cm

8. $A = 57$ ft^2
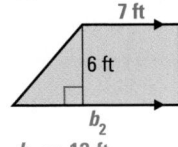
7 ft
6 ft
b_2
$b_2 = 12$ ft

9. $A = 201$ in.^{2}
r
r ≈ 8 in.

8.7   Circumference and Area of Circles   459

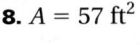

4 **Assess**

Assessment Resources
The Mini-Quiz below is also available on blackline (*Chapter 8 Resource Book*, p. 9) and on transparency. For more assessment resources, see:
- Chapter 8 Resource Book
- Standardized Test Practice
- Test and Practice Generator

Mini-Quiz
Find the circumference and area of the circle. Round your answers to the nearest whole number.

1. A circle with a radius of 3 inches   circumference: 19 in.; area: 28 in.^{2}

2. A circle with a diameter of 13 meters   circumference: 41 m; area: 133 m^2

3. Find the area of the shaded sector. Start by finding the area of the circle. Round your answer to the nearest whole number.

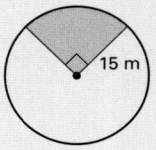
15 m

177 m^2

4. A circular swimming pool has a circumference of 63 feet. What is the radius of the pool, to the nearest foot?   10 ft

Additional Resources

The following resources are available to help review the materials in this chapter.

📖 **Chapter 8 Resource Book**
- Chapter Review Games and Activities, p. 78
- Cumulative Review, Chs. 1–8

Chapter  8 Summary and Review

VOCABULARY

- **convex,** *p. 411*
- **concave,** *p. 411*
- **equilateral,** *p. 412*
- **equiangular,** *p. 412*
- **regular,** *p. 412*
- **area,** *p. 424*
- **height of a triangle,** *p. 431*

- **base of a triangle,** *p. 431*
- **base of a parallelogram,** *p. 439*
- **height of a parallelogram,** *p. 439*
- **height of a trapezoid,** *p. 446*
- **circle,** *p. 452*

- **center of a circle,** *p. 452*
- **radius,** *p. 452*
- **diameter,** *p. 452*
- **circumference,** *p. 452*
- **central angle,** *p. 454*
- **sector,** *p. 454*

VOCABULARY REVIEW

Fill in the blank.

1. A polygon is __?__ if no line that contains a side of the polygon passes through the interior of the polygon. convex

2. A polygon is __?__ if all of its interior angles are congruent. equiangular

3. If a polygon is equilateral and equiangular, then it is __?__. regular

4. The amount of surface covered by a figure is its __?__. area

5. The __?__ of a triangle is the perpendicular segment from a vertex to the line containing the opposite side of the triangle. height

6. Either pair of parallel sides of a parallelogram are called the __?__. bases

7. A(n) __?__ is the set of all points in a plane that are the same distance from a given point. circle

8. The distance across a circle through its center is the __?__. diameter

---

8.1 CLASSIFYING POLYGONS

Examples on pp. 411–412

> **EXAMPLE** **Decide whether the polygon is regular. Explain your answer.**
>
> Although the polygon is equilateral, it is not equiangular. So, the polygon is not regular.

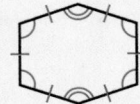

Decide whether the polygon is regular. Explain your answer. 9–12. See margin.

9.

10.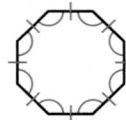

11.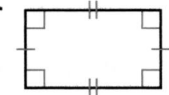

12.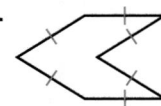

9. not regular (not equilateral and not equiangular)
10. regular (both equilateral and equiangular)
11. not regular (equiangular but not equilateral)
12. not regular (equilateral but not equiangular)

8.2 ANGLES IN POLYGONS

Examples on pp. 417–420

EXAMPLE   **Find the measure of ∠A in the diagram.**

The polygon has 5 sides, so the sum of the measures of the interior angles is:

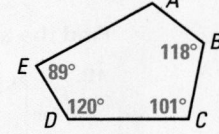

$(n - 2) \cdot 180° = (5 - 2) \cdot 180° = 3 \cdot 180° = 540°.$

Add the measures of the interior angles and set the sum equal to 540°.

$89° + 120° + 101° + 118° + m\angle A = 540°$ The sum is 540°.

$428° + m\angle A = 540°$ Simplify.

$m\angle A = 112°$ Subtract 428° from each side.

Find the measure of ∠1.

13.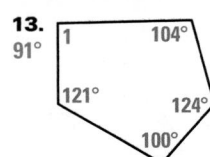

14.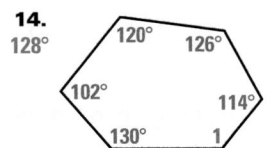

15.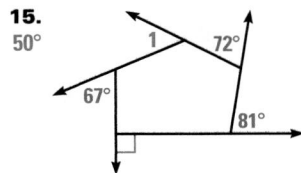

8.3 AREA OF SQUARES AND RECTANGLES

Examples on pp. 424–426

EXAMPLE   **Find the area of the rectangle.**

$A = bh$ Formula for the area of a rectangle

$= (8)(5)$ Substitute 8 for *b* and 5 for *h*.

$= 40 \text{ cm}^2$ Multiply.

Find the area of the polygon.

16.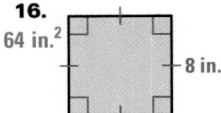
64 in.²   8 in.

17.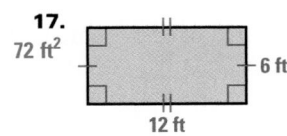
72 ft²   6 ft   12 ft

18.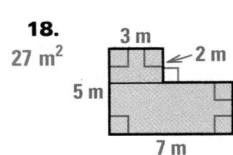
27 m²   3 m   2 m   5 m   7 m

8.4   AREA OF TRIANGLES

Examples on pp. 431–433

EXAMPLE   Find the area of the triangle.

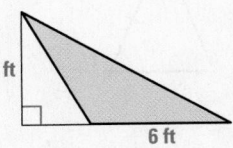

$$A = \frac{1}{2}bh$$ Formula for the area of a triangle

$$= \frac{1}{2}(6)(5)$$ Substitute 6 for b and 5 for h.

$$= 15 \text{ ft}^2$$ Multiply.

Find the area of the triangle.

19.
44 m² 8 m
11 m

20.
3 ft² 2 ft   3 ft
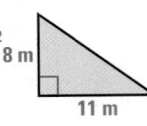

21.
58.5 cm²
9 cm
13 cm

22.
46.75 in.² 11 in.
$8\frac{1}{2}$ in.
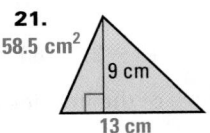

23. A triangle has an area of 117 square feet and a height of 9 feet. Find the base.   **26 ft**

24. A triangle has an area of 81 square meters and a base of 18 meters. Find the height.   **9 m**

8.5   AREA OF PARALLELOGRAMS

Examples on pp. 439–441

EXAMPLE   Find the area of the parallelogram.

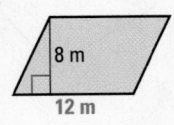

$$A = bh$$ Formula for the area of a parallelogram

$$= (12)(8)$$ Substitute 12 for b and 8 for h.

$$= 96 \text{ m}^2$$ Multiply.

Find the area of the parallelogram.

25.
56 ft²
7 ft
8 ft
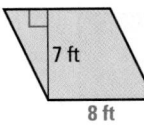

26.
60 cm²
10 cm
6 cm
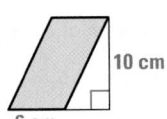

27.
22.5 in.²
5 in.
9 in.
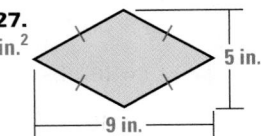

28. A parallelogram has an area of 135 square inches and a base of 15 inches. Find the height.   **9 in.**

29. A parallelogram has an area of 121 square meters and a height of 11 meters. Find the base.   **11 m**

8.6 AREA OF TRAPEZOIDS

Examples on pp. 446–447

EXAMPLE **Find the area of the trapezoid.**

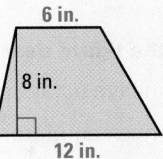
6 in.
8 in.
12 in.

$A = \frac{1}{2}h(b_1 + b_2)$ Formula for the area of a trapezoid

$= \frac{1}{2}(8)(6 + 12)$ Substitute 8 for h, 6 for b_1, and 12 for b_2.

$= \frac{1}{2}(8)(18)$ Simplify within parentheses.

$= 72$ in.^{2} Multiply.

Find the area of the trapezoid.

30. 56 mm^2
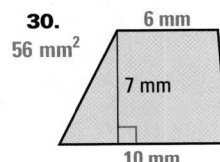
6 mm
7 mm
10 mm

31. 36 ft^2
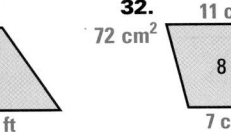
4 ft
6 ft
8 ft

32. 72 cm^2
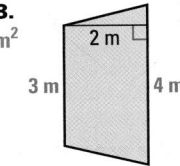
11 cm
8 cm
7 cm

33. 7 m^2
2 m
3 m
4 m

34. A trapezoid has an area of 100 square yards. The lengths of the bases are 9 yards and 11 yards. Find the height. **10 yd**

8.7 CIRCUMFERENCE AND AREA OF CIRCLES

Examples on pp. 452–455

EXAMPLE **Find the circumference and the area of the circle.**

9 m

$C = 2\pi r$

$= 2\pi(9)$

$= 18\pi$

≈ 57 m^2

$A = \pi r^2$

$= \pi(9)^2$

$= 81\pi$

≈ 254 m^2

Find the circumference and the area of the circle. Round your answer to the nearest whole number.

35. $C \approx 82$ in.; $A \approx 531$ in.^{2}
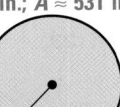
13 in.

36. $C \approx 6$ m; $A \approx 3$ m^2
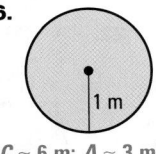
1 m

37. $C \approx 25$ ft; $A \approx 50$ ft^2
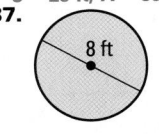
8 ft

38. $C \approx 44$ cm; $A \approx 154$ cm^2
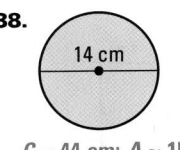
14 cm

39. A circle has a circumference of 50 square feet. Find the radius. **about 8 ft**

40. A circle has an area of 380 square centimeters. Find the radius. **about 11 cm**

Chapter Summary and Review **463**

1–4. Sample answers are given.

1.

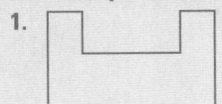

2.

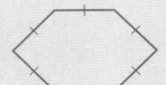

3.

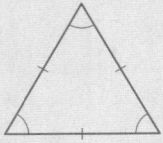

4.

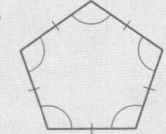

Chapter  8 · Chapter Test

Sketch the figure described. 1–4. See margin.

1. A concave octagon

2. A convex equilateral hexagon

3. A regular triangle

4. A regular pentagon

Find the measure of ∠1.

5.

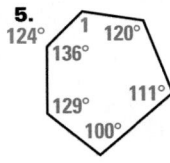

6.

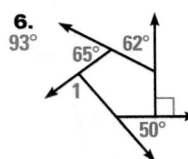

Find the area of the polygon.

7. 49 ft²

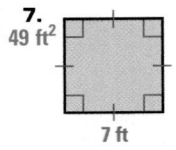

7 ft

8. 20 cm²

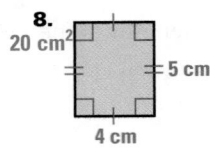

5 cm
4 cm

9. 14 m²

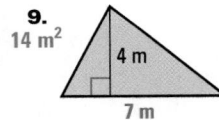

4 m
7 m

10. 104 in.²

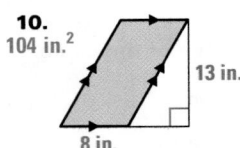

13 in.
8 in.

11. 60 mm²

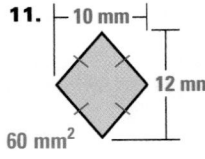

10 mm
12 mm

12. 45 yd²

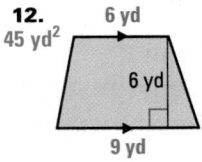

6 yd
6 yd
9 yd

13. 45 m²

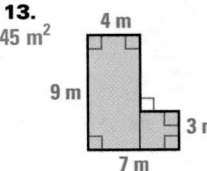

4 m
9 m
3 m
7 m

14. 29.5 ft²

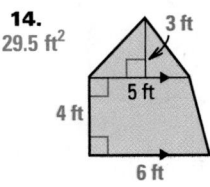

3 ft
5 ft
4 ft
6 ft

 464 Chapter 8 *Polygons and Area*

Find the circumference and the area of the circle. Round your answer to the nearest whole number.

15.

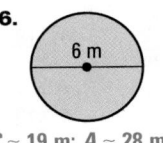

2 in.

$C \approx 13$ in.; $A \approx 13$ in.²

16.

6 m

$C \approx 19$ m; $A \approx 28$ m²

Find the area of the blue sector. Round your answer to the nearest whole number.

17. 29 ft²

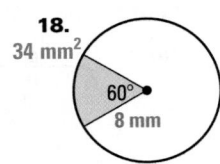

5 ft
130°

18. 34 mm²

60°
8 mm

19. Find the area of the shaded region. Round your answer to the nearest whole number. 79 cm²

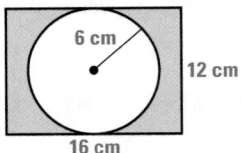

6 cm
12 cm
16 cm

In Exercises 20–23, *A* gives the area of the figure. Find the missing measure. In Exercise 23, round your answer to the nearest whole number.

20. $A = 80$ m² $b = 10$ m   **21.** $A = 104$ in.² $h = 13$ in.

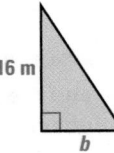

16 m
b

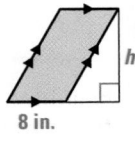

h
8 in.

22. $A = 126$ ft² $b_1 = 7$ ft   **23.** $A = 50$ cm² $r \approx 4$ cm

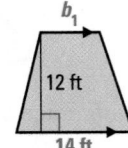

b_1
12 ft
14 ft

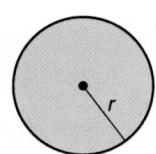

r

Test Tip   Go back and check your answers. Try to use a method other than the one you originally used, to avoid making the same mistake twice.

1. What is the measure of ∠C? D

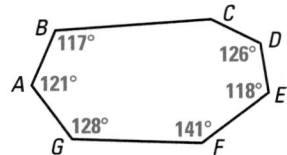

Ⓐ 125°   Ⓑ 132°

Ⓒ 144°   Ⓓ 149°

2. What is the value of *x*? F

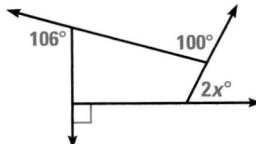

Ⓕ 32   Ⓖ 58

Ⓗ 64   Ⓙ 138

3. What is the area of the polygon made up of rectangles? B

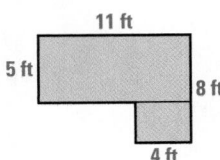

Ⓐ 28 ft²   Ⓑ 67 ft²

Ⓒ 72 ft²   Ⓓ 87 ft²

4. A triangle has an area of 54 square inches. What is the base of the triangle, given that its height is 9 inches? J

Ⓕ 6 in.   Ⓖ 8 in.

Ⓗ 10 in.   Ⓙ 12 in.

5. What is the area of the polygon? C

Ⓐ 78 cm²

Ⓑ 84 cm²

Ⓒ 120 cm²

Ⓓ 156 cm²

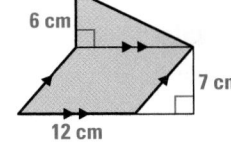

6. What is the area of the rhombus? H

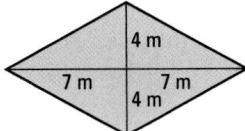

Ⓕ 22 m²   Ⓖ 28 m²

Ⓗ 56 m²   Ⓙ 112 m²

7. What is the area of the trapezoid? A

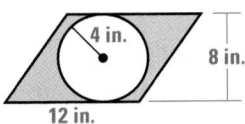

Ⓐ 100 yd²   Ⓑ 130 yd²

Ⓒ 170 yd²   Ⓓ 200 yd²

Multi-Step Problem In Exercises 8–10, use the diagram below.

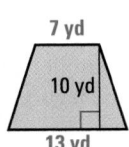

8. Find the area of the parallelogram. 96 in.²

9. Find the area of the circle. Round your answer to the nearest whole number. 50 in.²

10. Find the area of the shaded region. 46 in.²

Teaching Tip
The Brain Games activity provides a motivating way to review selected content in the chapter. For a more comprehensive review, see the Chapter Summary and Review on pp. 460–463.

BraiN GaMes

Sum of Parts

Materials
- 3 puzzles
- scissors
- pen or pencil

Object of the Game To be the first team to solve all three of the puzzles.

Set Up For each puzzle, carefully cut out the smaller pieces.

How to Play

Step 1 ▶ Use the 12 small rhombuses to completely cover the hexagon.

Step 2 ▶ Use the 8 trapezoids to completely cover the hexagon.

Step 3 ▶ Use the 6 small concave hexagons to completely cover the hexagon.

Another Way to Play Create your own puzzles by copying the hexagon and cutting it into smaller pieces. Challenge members of your team to solve your puzzle.

An expression like 7^3 is called a *power*. The *exponent* 3 represents the number of times the *base* 7 is used as a factor: $7^3 = 7 \cdot 7 \cdot 7 = 343$.

EXAMPLE 1 Evaluate Powers

Evaluate the expression.

a. 8^2 **b.** $(-5)^3$ **c.** $(-3)^1$ **d.** 2^5

Solution

a. $8^2 = 8 \cdot 8 = 64$ **b.** $(-5)^3 = (-5) \cdot (-5) \cdot (-5) = -125$

c. $(-3)^1 = -3$ **d.** $2^5 = 2 \cdot 2 \cdot 2 \cdot 2 \cdot 2 = 32$

Try These

Evaluate the expression.

1. 2^4 16 **2.** 9^3 729 **3.** $(-3)^5$ −243 **4.** $(-1)^4$ 1 **5.** 10^3 1000 **6.** 3^6 729

To solve percent problems, use the equation $a = p \cdot b$, where a is the number being compared to the base, p is the percent, and b is the base number.

EXAMPLE 2 Use Percents

a. What is 36% of 150? **b.** 24 is what percent of 80?

Solution

a. $a = p \cdot b$

$a = (36\%) \cdot 150$

$a = 0.36 \cdot 150$

$a = 54$

b. $a = p \cdot b$

$24 = p \cdot 80$

$\dfrac{24}{80} = \dfrac{80p}{80}$

$0.3 = p$

$30\% = p$

ANSWER ▶ 36% of 150 is 54. *ANSWER* ▶ 24 is 30% of 80.

Try These

7. What is 60% of 85? 51 **8.** What is 4% of 200? 8

9. 18 is what percent of 40? 45% **10.** 48 is what percent of 300? 16%

Algebra Review **467**

Mathematical Goals

- Find and use scale factors.
- Use ratios and proportions.
- Find the area of squares, rectangles, and circles.

Managing the Project

Tips for Success

Some students may feel overwhelmed by the amount of data in this project. Urge students to slow down, read carefully, and work one step at a time, rather than reading the entire project first and becoming intimidated by all the measurement data. You might wish to have students work in 4-person teams with each team member handling one of the first four steps in the Investigation, and also one of the four rows in the table given in Step 5. Students should work together to create their presentation, and also to work on the Extension.

Alternate Approach

If you have access to a large area with square floor tiles (or some other gridded surface), consider having students make their scale drawings using these tiles. Students can mark the tiles with colored yarn, string, or tape.

Designing a Park

Objective

Use similarity and area in park design.

Suppose your town has decided to plan a new park, and you are in charge of designing it. To present ideas for approval, you'll need to have a scale drawing of the park and all of its parts. The area set aside for the park is 800 feet wide and 1000 feet long.

Materials
- paper
- ruler
- poster board

Investigation

1 Make a scale drawing of the outline of the park on a poster, showing the park as a rectangle that is 8 inches wide and 10 inches long. What is the scale factor for the drawing? 100 ft : 1 in.

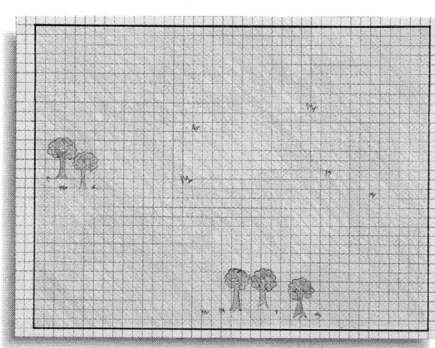

2 The town wants a playing field to be 200 feet wide and 400 feet long. What are the field's dimensions on your scale drawing? What is the actual area? the area on your scale drawing? 2 in. by 4 in.; 80,000 ft^2; 8 in.^{2}

3 You set aside a place for dogs to play. You sketch the Free Run Meadow as a 2 inch by 2 inch square on the poster. What is the Meadow area on the scale drawing? What will its actual area be? 4 in.^{2}; 40,000 ft^2

4 The Parks Department decides to have a circular pool, with radius of 50 feet. What is the approximate area covered by the pool? What is the length of the radius on your scale drawing? 7850 ft^2; 0.5 in.

Chapter 8 *Polygons and Area*

5 Use your answers in Steps 1–4 to copy and complete the table. See margin.

	Actual Dimensions	Actual Area	Dimensions on Scale Drawing	Area on Scale Drawing
Entire Park	?	?	?	?
Playing Field	?	?	?	?
Free Run Meadow	?	?	?	?
Pool	?	?	?	?

Present Your Results

Make a poster that displays your scale drawing and completed table.

▶ Include at least two other areas you want in the park, such as a community garden, a running track, or a soccer field.

▶ Expand your table to include these new areas.

Extension

Design another special park for your city, such as a playground or a picnic area. Find dimensions for swings, picnic tables, etc., and present a scale drawing of your plan.

Concluding the Project
Discuss with students ways in which a scale drawing is useful. Ask students where they have seen scale drawings before, such as the map at a mall, state park, or theme park. Consider extending the discussion by asking students if they can clarify what characteristics of scale drawings make the drawing more useful or less useful. Some students may recall reading a map that was confusing because there was no scale given, or because it was a poor representation of the actual area.

Grading the Project
A well-written project will have the following characteristics:
• The values in the table are correctly calculated.
• Students' drawings match the table and include all the required features.

Step 5.
Entire Park: 800 ft by 1000 ft; 800,000 ft^2; 8 in. by 10 in.; 80 in.^{2}
Playing Field: 200 ft by 400 ft; 80,000 ft^2; 2 in. by 4 in.; 8 in.^{2}
Free Run Meadow: 200 ft by 200 ft; 40,000 ft^2; 2 in. by 2 in.; 4 in.^{2}
Pool: radius = 50 ft; 7850 ft^2; radius = 0.5 in.; 0.785 in.^{2}

REGULAR SCHEDULE

Lesson	Les. Day	Basic	Average	Advanced
9.1	Day 1	pp. 477–480 Exs. 17–29 odd, 33–41 odd, 43–45, 52–58 even, 59–79 odd	pp. 477–480 Exs. 20–32, 33–51 odd, 52–55, 59–79 odd	pp. 477–480 Exs. 20–32, 34–50 even, 54–61, 62–80 even; EC: TE p. 470D*, classzone.com
9.2	Day 1	SRH p. 674 Exs. 1–4; pp. 487–490 Exs. 8–19, 21–23, 33–35, 48–52	pp. 487–490 Exs. 8–23, 33–35, 49–52	pp. 487–490 Exs. 12–20 even, 21–23, 33–35, 49–58
	Day 2	EP p. 690 Exs. 33–35; pp. 487–490 Exs. 24–31, 36–41, 47, 53–58	pp. 487–490 Exs. 24–32, 36–44, 47, 48, 53–58	pp. 487–490 Exs. 24–32, 36–45, 46*, 47, 48; EC: classzone.com
9.3	Day 1	pp. 495–499 Exs. 9–20, 32, 36, 47–55	pp. 495–499 Exs. 9–22, 32, 33, 36, 41, 42, 47–55	pp. 495–499 Exs. 12–22, 32, 33, 36, 41, 42, 50–55
	Day 2	EP p. 690 Exs. 22–28; pp. 495–499 Exs. 23–28, 34, 37, 46, 56–61, Quiz 1	pp. 495–499 Exs. 23–31, 34, 35, 37–40, 46, 59–61, Quiz 1	pp. 495–499 Exs. 23–31 odd, 34, 35, 37–40, 43–45*, 46, 59–61, Quiz 1; EC: classzone.com
9.4	Day 1	pp. 503–507 Exs. 8–24, 43–46, 51–57	pp. 503–507 Exs. 8–26, 44–46, 51–54	pp. 503–507 Exs. 11–21 odd, 22–26, 44–46, 51–60
	Day 2	pp. 503–507 Exs. 27–38, 42, 50, 58–63	pp. 503–507 Exs. 27–43, 50, 55–63	pp. 503–507 Exs. 27–43, 47–49*, 50; EC: classzone.com
9.5	Day 1	pp. 513–516 Exs. 9–16, 20, 22–25, 45–51	pp. 513–516 Exs. 9–16, 20, 22–25, 27, 29, 45–57 odd	pp. 513–516 Exs. 12–16, 20, 22–25, 27, 45–51
	Day 2	pp. 513–516 Exs. 17–19, 21, 26, 30–38, 52–57	pp. 513–516 Exs. 17–19, 21, 26, 28, 30–38, 46–56 even	pp. 513–516 Exs. 17–19, 28–32, 36–44, 54–57; EC: TE p. 470D*, classzone.com
9.6	Day 1	pp. 520–523 Exs. 8–10, 12–17, 23–28, 37–39, 43–45, 46–60 even, Quiz 2	pp. 520–523 Exs. 11–25, 33–41, 46, 47–61 odd, Quiz 2	pp. 520–523 Exs. 14–22, 25–36, 39–46, 57–61 odd, Quiz 2; EC: p 470D*, classzone.com
Review	Day 1	pp. 524–527 Exs. 1–29	pp. 524–527 Exs. 1–29	pp. 524–527 Exs. 1–29
Assess	Day 1	Chapter 9 Test	Chapter 9 Test	Chapter 9 Test

YEARLY PACING   Chapter 9 Total – **12 days**   Chapters 1–9 Total – **132 days**   Remaining – **28 days**

*Challenge Exercises   EP = Extra Practice   SRH = Skills Review Handbook   EC = Extra Challenge

BLOCK SCHEDULE

Day 1	Day 2	Day 3	Day 4	Day 5	Day 6
9.1 pp. 477–480 Exs. 20–32, 33–51 odd, 52–55, 59–61, 62–68 even **9.2** pp. 487–490 Exs. 8–23, 33–35, 49–52	**9.2 cont.** pp. 487–490 Exs. 24–32, 36–44, 47, 48 **9.3** pp. 495–499 Exs. 9–22, 32, 33, 36, 41, 42, 47–55	**9.3 cont.** pp. 495–499 Exs. 23–31, 34, 35, 37–40, 46, 59–61, Quiz 1 **9.4** pp. 503–507 Exs. 8–26, 44–46, 51–54	**9.4 cont.** pp. 503–507 Exs. 27–43, 50, 55–63 **9.5** pp. 513–516 Exs. 9–16, 20, 22–25, 27, 29, 45–57 odd	**9.5 cont.** pp. 513–516 Exs. 17–19, 21, 26, 28, 30–38, 46–56 even **9.6** pp. 520–523 Exs. 11–25, 33–41, 46, 47–61 odd, Quiz 2	**Review** pp. 524–527 Exs. 1–29 **Assess** Chapter 9 Test

YEARLY PACING   Chapter 9 Total – **6 days**   Chapters 1–9 Total – **66 days**   Remaining – **14 days**

Support Materials

CHAPTER RESOURCE BOOK

CHAPTER SUPPORT

Tips for New Teachers	p. 1	Strategies for Reading Mathematics	p. 5
Parent Guide for Student Success	p. 3		

LESSON SUPPORT

	9.1	9.2	9.3	9.4	9.5	9.6
Lesson Plans (regular and block)	p. 7	p. 15	p. 24	p. 36	p. 44	p. 53
Warm-Up Exercises and Daily Quiz	p. 9	p. 17	p. 26	p. 38	p. 46	p. 55
Technology Activities & Keystrokes			p. 27			p. 56
Practice (2 levels)	p.10	p. 18	p. 30	p. 39	p. 47	p. 57
Reteaching with Practice	p. 12	p. 20	p. 32	p. 41	p. 49	p. 59
Quick Catch-Up for Absent Students	p. 14	p. 22	p. 34	p. 43	p. 51	p. 61
Learning Activities						
Real-Life Applications		p. 23			p. 52	

REVIEW AND ASSESSMENT

Quizzes	pp. 35, 62	Alternative Assessment with Math Journal	p. 71
Brain Games Support	p. 63	Project with Rubric	p. 73
Chapter Review Games and Activities	p. 65	Cumulative Review	p. 75
Chapter Test (2 levels)	p. 66	Cumulative Test for Chs. 1–9	p. 77
SAT/ACT Chapter Test	p. 70	Resource Book Answers	AN1

TRANSPARENCIES

	9.1	9.2	9.3	9.4	9.5	9.6
Warm-Up Exercises and Daily Quiz	p. 54	p. 55	p. 56	p. 57	p. 58	p. 59
Visualize It Transparencies	✓	✓	✓		✓	
Answer Transparencies	✓	✓	✓	✓	✓	✓

TECHNOLOGY

- Time-Saving Test and Practice Generator
- Electronic Teacher Tools
- Geometry in Motion: Video
- Classzone.com
- Online Lesson Planner

ADDITIONAL RESOURCES

- Worked-Out Solution Key
- Resources in Spanish
- Practice Workbook with Examples

Providing Universal Access

Strategies for Strategic Learners

ANTICIPATE PROBLEM AREAS

Some students have difficulty differentiating between area and volume, especially when they are asked to find the surface area and volume of three-dimensional objects.

For some students, finding the volume of a small box by filling it with unit cubes helps them internalize the concept of volume. Other students may respond better to the same activity on a larger scale—filling a small room with cardboard boxes representing unit cubes. Such an activity is described below.

1) Prepare cardboard boxes before class; between 6 and 9 boxes should be sufficient. Ideally, the boxes should be cubes 1 yard on each side. You may be able to find a small group of students who would enjoy making these cubes using cardboard and tape.

2) Choose a small room or area in a hallway that would be suitable for this activity. The ideal space would be clear of furniture or other items and have a ceiling low enough that students could estimate its height in feet. Although your classroom is likely the most convenient location, it is also likely to be so full of furniture that it will be difficult for students to visualize the volume of the empty space contained within the room.

3) Ask students to first estimate the volume of the room. Ask students if they know what volume means and use students' explanations to help those who are unsure. Then introduce the cubes and explain that volume means how many of these cubes would fit in the room. Have students begin stacking cubes in a corner until students can estimate the height of the ceiling. Then have students lay the cubes side by side along the walls to estimate the length and width of the room. Draw a scale diagram on the board illustrating the number of cubes and lead students to recognize that it is not necessary to completely fill the room in order to determine the volume of the room.

4) Conclude the activity by asking students to explain in their own words the difference between surface area and volume. If there is time, you can have students use the measures you recorded on the board to calculate the surface area of the room they "measured" with their cubes.

The act of stacking and laying cubes in a space seems to help some students internalize that volume is a three-dimensional measure. In the process, they can be led to derive the formula for the volume of a rectangular prism.

Draw students' attention to the fact that the calculations for surface area and for volume use the same measures (namely, the length, width, and height), but that the measures are used in different ways.

Strategies for English Learners

VOCABULARY DEVELOPMENT

Certain letters are commonly (but not always) used in formulas to represent specific things. Review with your English learners common abbreviations (variables) used in geometry formulas.

area (A)	perimeter (P)
base (b)	radius (r)
base area (B)	slant height (ℓ)
circumference (C)	surface area (S)
diameter (d)	volume (V)
height or altitude (h)	width (w)
length (ℓ)	

Note that some letters are commonly used to represent more than one measure, and students will need to look at the situation in order to understand the abbreviation being used. Ask students to tell you what the variables represent in the following common formulas.

$d = 2r$ (d is diameter; r is radius.)

$A = \frac{1}{2}bh$ (A is area; b is base; h is height.)

$A = \pi r^2$ (A is area; r is radius.)

$C = \pi d$ (C is circumference; d is diameter.)

$A = \ell w$ (A is area; ℓ is length; w is width.)

You can also have students review the common abbreviations of units of measurement.

Strategies for Advanced Learners

DIFFERENTIATE INSTRUCTION IN TERMS OF COMPLEXITY

Throughout geometry, students will use properties of figures and formulas. They are already familiar with some formulas, such as the Distance Formula, the Pythagorean Theorem, and the formula for the interior angles of a polygon.

Have students skim the chapter to find and list formulas they have learned, or will learn. Have them make a chart with an example showing how to apply each formula and display this information for use by the entire class. This activity serves two purposes: (1) it helps advanced students organize their thinking, anticipate what they will learn in geometry, and communicate key mathematics concepts; and (2) it serves as a review for the entire class.

DIFFERENTIATE INSTRUCTION IN TERMS OF DEPTH

The following activity extends the material presented in this chapter to a more complex subject in mathematics. Students can work on this activity during class while you work with the rest of the class on aspects of this chapter that Advanced learners have already mastered.

1) Have students bring a tennis ball or similar sized ball to class. Provide each student with a few rubber bands. Students will also need a protractor.

2) Review the definitions of *plane* and *line*. Explain that these definitions are true only for the type of geometry you have been studying, called *Euclidean* geometry. Geometry on a sphere (called *spherical geometry*) differs from Euclidean geometry. In spherical geometry, a *plane* is a spherical surface and a *line* is a special kind of circle, called a *great circle*, on that surface.

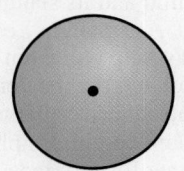 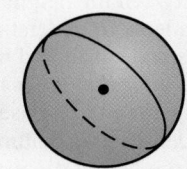

a **plane** in the geometry of a sphere   a **line** in the geometry of a sphere

3) Have students put a rubber band around the center of their tennis ball. Point out that the surface of the ball is a plane and the rubber band is a great circle, that is, a circle on the sphere whose center is the center of the sphere. Explain that circles on the sphere that do not have the same center as the sphere are not considered lines.

Have students answer the following questions using their tennis balls, rubber bands, and protractors.

- If a point lies on a line, does the point that is opposite it on the sphere also lie on the line? (*yes*)
- Is it possible to draw two lines on the sphere that do not intersect? Explain. (*No; any two great circles must intersect.*)
- Is it possible to draw two lines on the sphere that intersect to form right angles? (*yes*)
- Make a triangular region enclosed by three lines. Measure the angles of the triangle with a protractor. What do you notice about the sum of the measures of the angles of the triangle? (*The sum is greater than 180°.*)
- Make an equiangular triangular region enclosed by three lines. What are the angle measures? (*The measures will be between 60° and 180°.*)

- Is it possible to draw a 60°-60°-60° triangle on a sphere? a 90°-90°-90° triangle? a 120°-120°-120° triangle? (*no; yes; yes*)
- What is the range of values for the sum of the measures of the angles of a triangle on a sphere? (*The sum of the measures of the angles is between 180° and 540°.*)
- Summarize how *plane* and *line* are viewed differently in Euclidean geometry and spherical geometry. Include drawings with markings of lines that intersect on a sphere. Summarize your results about the angles of triangles on a sphere.

Encourage interested students to research how a spherical surface, such as the map on a globe, becomes distorted when it is represented as a flat surface. Students may also wish to investigate other non-Euclidean geometries, such as elliptic geometry and hyperbolic geometry. You may also be able to recommend to them a simple introduction to topology.

Challenge Problem for use with Lesson 9.1:
Suppose a prism and a pyramid have congruent bases. Which polyhedron has more faces? How many more? (*The prism has one more face than the pyramid.*)

Challenge Problem for use with Lesson 9.5:
A square pyramid and a cone have the same height, and a side of the base of the square pyramid is the same length as the diameter of the cone. What is the ratio of the volume of the cone to the volume of the pyramid? Leave your answer in terms of π. $\left(\dfrac{\pi}{4}\right)$

Challenge Problem for use with Lesson 9.6:
The surface area of a sphere is 9π square units. What is the volume of the sphere? Leave your answer in terms of π. $\left(\dfrac{9\pi}{2}\right)$

This chapter investigates the question of how to find the surface area and volume of important types of solids. Lesson 9.1 introduces the concept of a polyhedron, along with some essential terminology. The topic of surface area is taken up in Lesson 9.2, beginning with prisms and cylinders. In Lesson 9.3, the study of surface area is extended to pyramids and cones. Then volume is introduced in Lesson 9.4, again using prisms and cylinders as the first solids examined. Lesson 9.5 extends the study of volume to pyramids and cones. Finally, in Lesson 9.6, both the volume and surface area of spheres are explored.

Lesson 9.1

The study of solid figures is always a challenge for geometry students, even though we live in a three-dimensional world. The reasons for this difficulty appear to be twofold: (1) students have not yet had extensive experience in analyzing three-dimensional situations mathematically, and (2) problems involving solids are presented on a printed page and thus the solids are necessarily shown in a two-dimensional form that is often hard for students to decipher. Pyramids and cones can be especially difficult for students to visualize from their two-dimensional representations.

USING MODELS There are ways you can try to overcome these obstacles. One way is to use a computer drawing program that allows students to manipulate three-dimensional images. Another way is to make some cardboard models of solids (or have your students make some using nets provided to them) and to use them for class demonstrations. Also, it is important to be consistent in your terminology when discussing solids in class, and insist that your students do so as well. For example, the distinction between an *edge* and a *face* of a polyhedron is critical, and it is a good idea to avoid the word "side" when discussing polyhedra, which could refer to either feature.

It will also help students if they are able to draw their own diagrams of polyhedra, and you should take advantage of the instruction and practice that are given in this lesson. Drawing their own diagrams will certainly aid students throughout the chapter in their visualization of three-dimensional figures.

Lesson 9.2

There may be some ambiguity about the meaning of certain terms when students are dealing with the surface area of a polyhedron. The words *base* and *height*, for example, may refer to the dimensions of a face (as in the base and height of a rectangle) or they may refer to dimensions of the polyhedron itself, such as the area of the base of a prism and its height. It will be helpful to students seeing this material for the first time if you are careful to specify, when using these terms in a discussion, the sense in which they are being used. Point out to students that the formulas stated in this chapter consistently uses capital letters to denote area measurements and lowercase letters to denote linear measurements.

It may help students remember the surface area formula for a prism if you derive it in a special case, using numerical measurements. For example, for a rectangular prism whose base measures 3 feet by 5 feet and whose height is 7 feet, you could write the following.

$$\text{Surface area} = 3 \times 5 + 3 \times 5 + (3 \times 7 + 5 \times 7 + 3 \times 7 + 5 \times 7)$$
$$= 2(3 \times 5) + (3 + 5 + 3 + 5)7$$
$$= 2(\text{area of base}) + \text{perimeter} \times \text{height}$$

QUICK TIP
Make sure students understand that the formula for the surface area of a cylinder can be viewed as a special case of the formula for the surface area of a prism. The bases of a cylinder are circles whose "perimeter" is their circumference.

Lesson 9.3

The distinction between the height of a pyramid and its slant height may not be easy for students to understand at first. As stated in Lesson 9.1, computer graphics and cardboard models may help enlighten students. Although the term *lateral edge* (a segment connecting the vertex of a pyramid with one vertex of its base) is not introduced in the text, students may mistakenly identify this feature as the slant height. Emphasize that the slant height is the height of any triangle that forms a face of the pyramid. A second definition of slant height is given in the Visualize It feature on page 491. (Stress that we are only considering *regular pyramids* in this lesson, which allows slant height to be well-defined.) Stressing that the faces of a pyramid are always triangles, may help students remember the factor $\frac{1}{2}$ in the formula for the surface area of a pyramid.

USING THE PYTHAGOREAN THEOREM It should be clear to students from the examples in the lesson that the Pythagorean Theorem is an important tool for finding the linear measurements of a pyramid that are needed to calculate its surface area. A common mistake when using the Pythagorean Theorem for this purpose is the incorrect placement of the height of the pyramid as the length of the hypotenuse rather than the slant height. (This can also occur with cones.) This mistake can almost always be avoided if students are willing to make a careful sketch of the pyramid or cone, clearly showing these segments and indicating the right angle between the height of the pyramid or cone and its base.

Although the Student Help note on page 493 asks students to recall the formula for the area of a sector, it may not be clear how this formula helps in deriving the formula for the lateral area of a cone. The following discussion can be presented to illuminate this point. (The symbol θ represents the measure of the central angle of the sector of the circle.)

Lateral area of cone = area of a sector of a circle with radius ℓ

$$= \frac{\theta}{360} \cdot \pi \ell^2$$

$$= \frac{\theta}{360} \cdot 2\left(\frac{\pi \ell^2}{2}\right)$$

$$= \frac{\theta}{360} \cdot (2\pi \ell)\left(\frac{\ell}{2}\right)$$

$$= \left(\frac{\theta}{360}\right)(2\pi \ell) \cdot \left(\frac{\ell}{2}\right)$$

But $\left(\frac{\theta}{360}\right)(2\pi \ell)$ is the arc length of the sector, which equals $2\pi r$, the circumference of the base of the cone. Therefore,

Lateral area of cone $= (2\pi r) \cdot \left(\frac{\ell}{2}\right)$, or $\pi r \ell$.

Lesson 9.4

Having discussed prisms and cylinders in Lesson 9.2, the formulas in this lesson should present few difficulties for students. The most common mistake made by students is the use of a surface area formula where a volume formula should have been used, and vice versa. The best way to avoid this error may be to call attention to the fact that, in general, all area formulas involve the product of two linear measurements while all volume formulas involve the product of three linear measurements (not necessarily all distinct). This is also a good opportunity to emphasize that the units for surface area are square units and the units for volume are cubic units.

Lesson 9.5

The "volume puzzle" shown on page 510 can help students understand why there is factor of $\frac{1}{3}$ in the formula for the volume of a pyramid. You can construct a cardboard model of this puzzle, or have your students make one for themselves. A net for each of the three pieces is shown below.

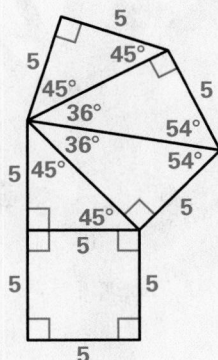

The most common mistake students make when using these formulas is to substitute $\frac{1}{2}$ for $\frac{1}{3}$, indicating their confusion of these formulas with the corresponding surface area formulas.

Lesson 9.6

The sphere formulas may be difficult for students to remember, because they are not as intuitive as some of the other surface area and volume formulas. It would be difficult to justify these formulas using the mathematical knowledge that students possess at this time, but you can offer the following intuitive picture of the *relationship* between the volume and surface area formulas, which may help them remember both.

VISUALIZING MODELS/DERIVING FORMULAS Imagine a sphere formed by a large number of cones whose vertices all lie at the center of the sphere and whose bases combine to form the surface of the sphere. As the radius of the base of these cones becomes smaller, the total volume of these cones approximates the volume of the sphere whose radius is the height of each cone. To find the combined volume of these cones, we use the formula $V = \frac{1}{3}Bh$ for each cone, making the combined volume approximately $\frac{1}{3}$(sum of the bases of the cones)(height of each cone). Since the height of each cone is approximately the radius of the sphere, and since the total area of the bases of all the cones approximates the surface area of the sphere, the volume of the sphere will be approximately $\frac{1}{3}$(surface area of the sphere)(radius of the sphere).

Volume of sphere $= \frac{1}{3}(4\pi r^2)r$, or $\frac{4}{3}\pi r^3$

If students can remember the two formulas, their only mistake may be using the surface area formula when the volume formula is required, and vice versa.

Surface Area and Volume

Chapter Overview

Chapter Goals

This chapter explores solid figures, both polyhedra (prisms and pyramids) and solids that are not polyhedra (cylinders, cones, and spheres). Students will:

- Identify and name solid figures.
- Find the surface area and volume of prisms, cylinders, pyramids, cones, and spheres.

Application Note

The first planetariums were mechanical models of the solar system. They were also called *orreries*, after Charles Boyle, the fourth Earl of Orrery in Ireland. The clockwork-driven orreries showed the planets from Mercury to Saturn revolving around the sun.

Today, planetariums use a projection system inside a dome to show the sun, the moon, the planets, and the stars. The projection system produces images of the moon and planets in correct positions and phases. Other astronomical features, such as galaxies and eclipses, can also be superimposed onto the dome.

More information about planetariums is provided on the Internet at classzone.com

Application Links
CLASSZONE.COM

What shapes are used in a planetarium?

Planetariums project images of the night sky onto the inside of a dome.

The dome of New York City's Hayden Planetarium is in the top half of the sphere visible in this photograph.

The sphere is built inside a glass cube that houses the Rose Center for Earth and Space at the American Museum of Natural History.

Learn More About It

You will learn more about the planetarium in Exercises 33–36 on p. 522.

Who uses Surface Area and Volume?

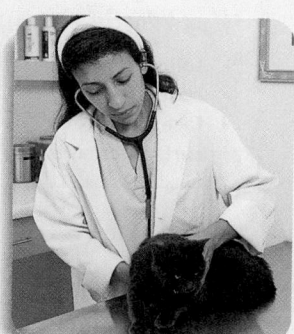

VETERINARIAN
Veterinarians provide treatment for sick and injured animals. They also offer preventive care for healthy animals. (p. 497)

AQUARIUM DIVER
Aquarium divers take care of fish and plants in an aquarium. They monitor the volume of water in a tank and insure that the tank is not overcrowded. (p. 505)

How will you use these ideas?

- Analyze the shape of the first music recordings. (p. 489)
- Determine the amount of material needed to make various objects. (p. 497)
- Compare the volumes of swimming pools. (p. 505)
- Calculate the amount of food in a pet feeder. (p. 515)
- Find the amount of glass in the cube of the Rose Center for Earth and Space. (p. 522)

Hands-On Activities

Activities (more than 20 minutes)
9.2 Investigating Surface Area, pp. 481–482
9.5 Investigating Volume, pp. 508–509

Projects

A project covering Chapters 9–10 appears on pages 584–585 of the Student Edition. An additional project for Chapter 9 is available in the *Chapter 9 Resource Book*, pp. 73–74.

Technology

- Electronic Teacher Tools
- Test and Practice Generator
- Online Lesson Planner

Internet Support
CLASSZONE.COM

- Application and Career Links 497, 521
- Student Help 486, 494, 497, 504, 506, 514, 515, 518, 522

PREVIEW

What's the chapter about?

• Identifying and naming solids
• Finding the surface area of solids
• Finding the volume of solids

Key Words

• solid, *p. 473*
• base, *p. 473*
• face, *p. 474*
• surface area, *p. 483*

• lateral area, *p. 484*
• height, slant height, *p. 491*
• volume, *p. 500*
• hemisphere, *p. 517*

PREPARE

Chapter Readiness Quiz

Take this quick quiz. If you are unsure of an answer, look at the reference pages for help.

Skill Check *(refer to pp. 192, 431, 453)*

1. What is the length of the hypotenuse in the triangle shown? **A**

   (A) 13 (B) 17
   (C) 19 (D) 169

   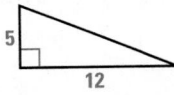

2. What is the area of the triangle shown above? **G**

   (F) 15 units^2 (G) 30 units^2 (H) 60 units^2 (J) 120 units^2

3. What is the approximate area of the circle? **C**

   (A) 18.8 in.^{2} (B) 37.7 in.^{2}
   (C) 113 in.^{2} (D) 226 in.^{2}

   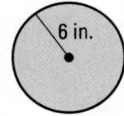

VISUAL STRATEGY

Drawing Three Dimensional Figures

Visualize It! ➡

Here is a helpful method for drawing three dimensional figures.

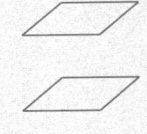

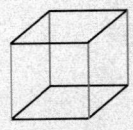
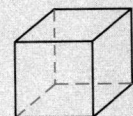

Draw the bases.

Connect the bases using vertical lines.

Erase so that hidden edges are dashed lines.

9.1 Solid Figures

Key Words
- solid
- polyhedron
- base
- face
- edge

The three-dimensional shapes on this page are examples of *solid figures*, or **solids**. When a solid is formed by polygons, it is called a **polyhedron**.

Polyhedra Prisms and pyramids are examples of polyhedra. To name a prism or pyramid, use the shape of the *base*.

Rectangular prism

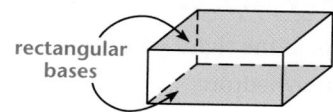

rectangular bases

Triangular pyramid

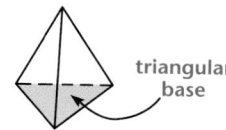

triangular base

The two **bases** of a prism are congruent polygons in parallel planes.

The **base** of a pyramid is a polygon.

Not Polyhedra Solids with curved surfaces, like the cylinder, cone, and sphere shown below, are not polyhedra.

Cylinder   **Cone**   **Sphere**

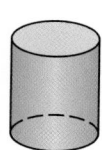 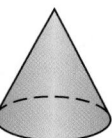 

Student Help

VOCABULARY TIP
Poly- means "many" and *-hedron* is Greek for "side" or "face." A polyhedron is a figure with many faces. The plural of polyhedron is *polyhedra*.

EXAMPLE 1 **Identify and Name Polyhedra**

Tell whether the solid is a polyhedron. If so, identify the shape of the bases. Then name the solid.

a. 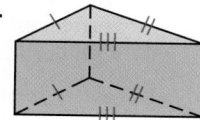   b.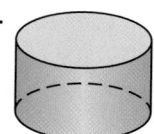

Solution

a. The solid is formed by polygons so it is a polyhedron. The bases are congruent triangles in parallel planes. This figure is a triangular prism.

b. A cylinder has a curved surface, so it is not a polyhedron.

---

① **Plan**

Pacing
Suggested Number of Days

Basic: 1 day
Average: 1 day
Advanced: 1 day
Block Schedule: 0.5 block with 9.2

Teaching Resources

📄 **Blacklines**
(See page 470B.)

 **Transparencies**
- Warm-Up with Quiz
- Answers

🖥 **Technology**
- Electronic Teacher Tools
- Test and Practice Generator
- Online Lesson Planner
- Internet Support

② **Teach**

Content and Teaching Strategies
For background information on geometric concepts and teaching strategies related to this lesson, see pages 470E and 470F in this Teacher's Edition.

Tips for New Teachers
Have models similar to those in the summary table on page 475 available for students to look at and touch. Wood or plastic models may be available in your math, art, or science departments. See the Tips for New Teachers on pp. 1–2 of the *Chapter 9 Resource Book* for additional notes about Lesson 9.1.

Extra Example 1
See next page.

Extra Example 1

Tell whether the solid is a polyhedron. If so, identify the shape of the
bases. Then name the solid.

a.

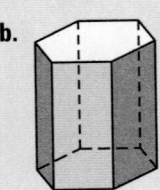

The solid is a polyhedron. The
bases are congruent rectangles
in parallel planes. The solid is a
rectangular prism.

b.

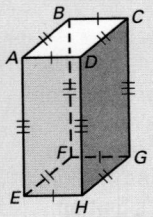

The solid is a polyhedron. The
bases are congruent hexagons
in parallel planes. The solid is a
hexagonal prism.

Extra Example 2

Use the diagram.

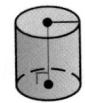

a. Name the polyhedron.
   rectangular prism

b. Count the number of faces and
   edges. 6 faces, 12 edges

c. List any congruent faces and
   congruent edges.
   congruent faces: *ABCD* ≅ *EFGH*,
   DCGH ≅ *ABFE*, *ADHE* ≅ *BCGF*;
   congruent edges: $\overline{AD}$ ≅ $\overline{BC}$ ≅
   $\overline{EH}$ ≅ $\overline{FG}$, $\overline{AB}$ ≅ $\overline{DC}$ ≅ $\overline{EF}$ ≅ $\overline{HG}$,
   $\overline{AE}$ ≅ $\overline{DH}$ ≅ $\overline{CG}$ ≅ $\overline{BF}$

Extra Example 3

Sketch a rectangular pyramid.

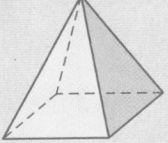

STUDY TIP
In this book, all of the
solids are *right solids*.
This means that the
segment representing
the *height* is
perpendicular to the
base(s).

Parts of a Polyhedron To avoid confusion, the word *side* is not used
when describing polyhedra. Instead, the plane surfaces are called
faces and the segments joining the vertices are called **edges** .

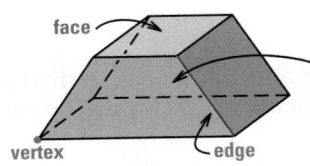

face

The trapezoidal faces of this polyhedron
are the bases.

vertex   edge

EXAMPLE 2 Find Faces and Edges

Use the diagram at the right.

a. Name the polyhedron.

b. Count the number of faces and edges.

c. List any congruent faces and
   congruent edges.

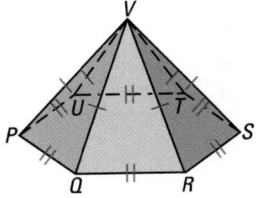

Solution

a. The polyhedron is a hexagonal pyramid.

b. The polyhedron has 7 faces and 12 edges.

c. Using the markings on the diagram, you can conclude the following:

Congruent faces

△*PQV* ≅ △*QRV* ≅ △*RSV* ≅
△*STV* ≅ △*TUV* ≅ △*UPV*

Congruent edges

$\overline{PQ}$ ≅ $\overline{QR}$ ≅ $\overline{RS}$ ≅ $\overline{ST}$ ≅ $\overline{TU}$ ≅ $\overline{UP}$
$\overline{PV}$ ≅ $\overline{QV}$ ≅ $\overline{RV}$ ≅ $\overline{SV}$ ≅ $\overline{TV}$ ≅ $\overline{UV}$

Student Help

VISUAL STRATEGY
For more help drawing
three dimensional
figures, see p. 472.

EXAMPLE 3 Sketch a Polyhedron

Sketch a triangular prism.

Solution

❶ Draw the
   triangular bases.

❷ Connect the
   corresponding
   vertices of the bases
   with vertical lines.

❸ Partially erase the
   hidden lines to create
   dashed lines. Shade
   the prism.

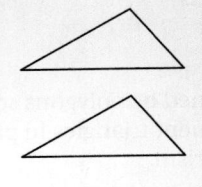 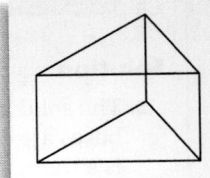 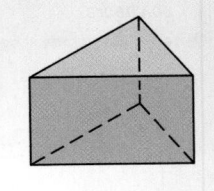

Checkpoint ✓ Identify and Sketch Polyhedra

Tell whether the solid is a polyhedron. If so, identify the shape of the base(s). Then name the solid.

1.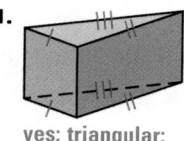
yes; triangular;
triangular prism

2.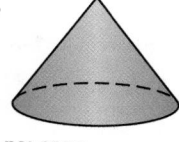
no; cone

3.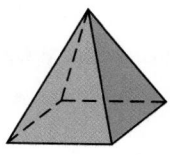
yes; rectangular;
rectangular pyramid

4. Copy the partial drawing of a triangular pyramid. Then complete the drawing of the pyramid. *See margin.*

TYPES OF SOLIDS

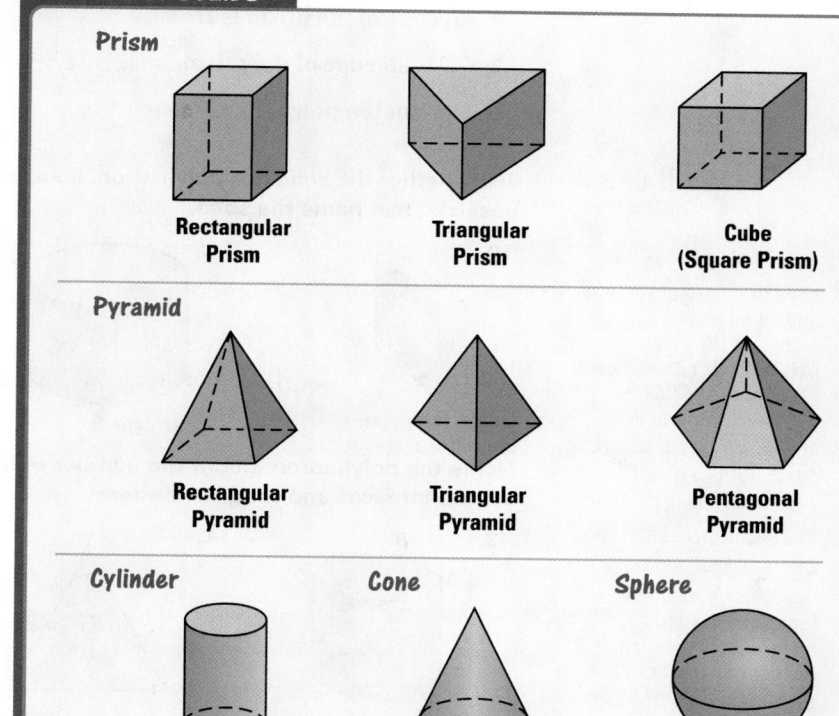

Prism

Rectangular Prism   Triangular Prism   Cube (Square Prism)

Pyramid

Rectangular Pyramid   Triangular Pyramid   Pentagonal Pyramid

Cylinder   Cone   Sphere

Student Help

STUDY TIP
In this book, the faces of a pyramid, not including the base, are congruent isosceles triangles, unless otherwise noted.

Visualize It!

Show students how to draw the circular bases of cylinders and cones as ellipses. Also show them how to connect the bases with vertical lines to complete their sketches.

Teaching Tip

Make sure students are correctly interpreting the sketches that represent three-dimensional objects. Point out that the perspective of a sketch distorts shapes. For example, a cylinder with a circular base is drawn with an oval base but should be interpreted as a circle.

Geometric Reasoning

Point out that the base of a solid may not always be the part that the shape appears to be resting on. Draw a cylinder oriented different ways to show that the cylinder still has a base even though it is laying on its side or at an angle.

Study Skills

Vocabulary Throughout this chapter, stress that students will be expected to know the names of all the solids shown on this page. Encourage students to sketch each of these solids and write notes about the properties of each solid in order to help them learn the names of the solids.

✓ Concept Check

What is the difference between a solid figure and a polyhedron?
A solid figure is a three-dimensional shape; a polyhedron is a solid figure formed by polygons.

4.

Assignment Guide

BASIC
Day 1: pp. 477–480 Exs. 17–29 odd, 33–41 odd, 43–45, 52–58 even, 59–79 odd

AVERAGE
Day 1: pp. 477–480 Exs. 20–32, 33–51 odd, 52–55, 59–79 odd

ADVANCED
Day 1: pp. 477–480 Exs. 20–32, 34–50 even, 54–61, 62–80 even; EC: TE p. 470D*, classzone.com

BLOCK SCHEDULE
pp. 477–480 Exs. 20–32, 33–51 odd, 52–55, 59–61, 62–68 even (with 9.2)

Extra Practice

• Student Edition, p. 691
• Chapter 9 Resource Book, pp. 10–11

Homework Check

To quickly check student understanding of key concepts, go over the following exercises:

Basic: 17, 18, 33, 37, 43
Average: 22, 32, 35, 37, 47
Advanced: 24, 32, 34, 38, 50

9.1 Exercises

Guided Practice

Vocabulary Check

In Exercises 1–3, match the solid with its name.

A. prism   **B.** pyramid   **C.** cylinder

1.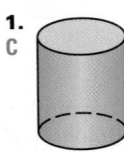
C

2.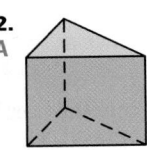
A

3.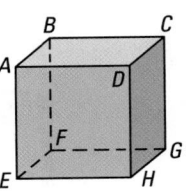
B

In Exercises 4–9, tell whether the statement is *true* or *false*. Refer to the prism below, if necessary.

4. *CDHG* is a face of the prism.   true

5. A prism has only one base.   false

6. *ABCD* and *EFGH* are possible bases of the prism.   true

7. An edge of the prism is *H*.   false

8. $\overline{GC}$ is an edge of the prism.   true

9. A prism is a polyhedron.   true

Skill Check

Tell whether the solid is a polyhedron. If so, identify the shape of the base(s). Then name the solid.

10.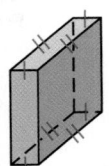
yes; rectangular; rectangular prism

11.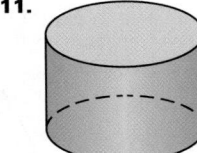
no; cylinder

12.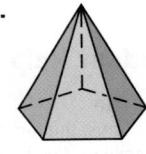
yes; pentagonal; pentagonal pyramid

Name the polyhedron. Count the number of faces and edges. List any congruent faces and congruent edges.

13. hexagonal prism; 8 faces and 18 edges; congruent faces: *ABVU* ≅ *BCWV* ≅ *CDXW* ≅ *DEYX* ≅ *EFZY* ≅ *FAUZ* and *ABCDEF* ≅ *UVWXYZ*; congruent edges: $\overline{AB} \cong \overline{BC} \cong \overline{CD} \cong \overline{DE} \cong \overline{EF} \cong \overline{FA} \cong \overline{UV} \cong \overline{VW} \cong \overline{WX} \cong \overline{XY} \cong \overline{YZ} \cong \overline{ZU}$ and $\overline{AU} \cong \overline{BV} \cong \overline{CW} \cong \overline{DX} \cong \overline{EY} \cong \overline{FZ}$

14. triangular pyramid; 4 faces and 6 edges; congruent faces: △*PQR* ≅ △*PQS*; congruent edges: $\overline{QR} \cong \overline{QS}$ and $\overline{PR} \cong \overline{PS}$

15. cube or square prism; 6 faces and 12 edges; congruent faces: *JKLM* ≅ *TUVW* ≅ *JKUT* ≅ *KLVU* ≅ *LMWV* ≅ *MJTW*

13.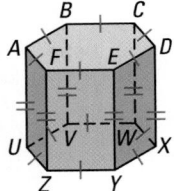

14.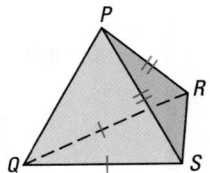

15.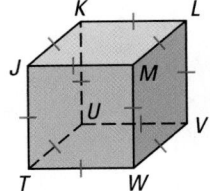

16. **Visualize It!** How many faces and edges does a box of cereal have?   6 faces and 12 edges

Practice and Applications

Extra Practice

See p. 691.

Name Bases and Solids Tell whether the solid is a polyhedron. If so, identify the shape of the base(s). Then name the solid.

17.
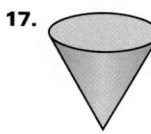
no; cone

18.
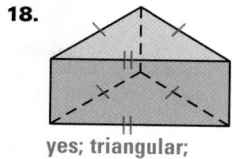
yes; triangular; triangular prism

19.
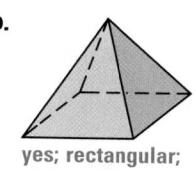
yes; rectangular; rectangular pyramid

Logical Reasoning Tell whether the statement is *true* or *false*.

20. A rectangular pyramid has two bases. false

21. A triangular prism has two bases. true

22. The bases of a prism are congruent polygons. true

23. A cone has two bases. false

24. A sphere is a polyhedron. false

Identify Solids Match the solid with its name.

A. cone   **B.** pyramid   **C.** cylinder

D. rectangular prism   **E.** cube   **F.** sphere

25.
F

26.
D

27.
A

28.
E

29.
B

30.
C

32. A pyramid with a square base has one square face and four triangular faces, while a triangular prism has two congruent triangular faces and three rectangular faces.

Homework Help

Example 1: Exs. 17–19, 25–35
Example 2: Exs. 36–38
Example 3: Exs. 43–51

Error Analysis Julie incorrectly identified the solid below as a pyramid with a square base.

31. Correctly identify the solid.
triangular prism

32. What would you say to Julie to help her tell the difference between this solid and a pyramid?

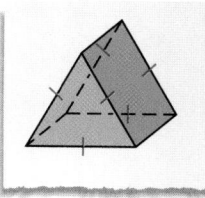

Teaching Tip

Most students should be able to complete Exercises 43–51, although the sphere in Exercise 51 may cause some difficulty since it is the only solid that has no base. For students who do have trouble with these exercises, refer them to Example 3 on page 474 and the chart on page 475. Also, the Visualize It note on page 517 discusses a useful method for sketching a sphere.

43.

44.

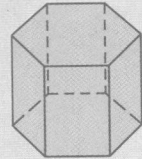

45.

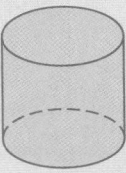

46–51. Sample answers are given.

46.

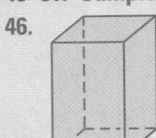

47.

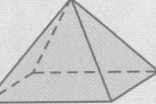

48.

49.

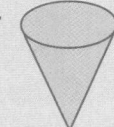

36. rectangular pyramid; 5 faces and 8 edges; congruent faces: △ABE ≅ △ACD and △ADE ≅ △ACB; congruent edges: $\overline{AB} \cong \overline{AC} \cong \overline{AD} \cong \overline{AE}$, $\overline{BC} \cong \overline{ED}$, and $\overline{BE} \cong \overline{CD}$

37. triangular prism; 5 faces and 9 edges; congruent faces: △FGH ≅ △JKL; congruent edges: $\overline{FG} \cong \overline{JK}$, $\overline{GH} \cong \overline{KL}$, $\overline{FH} \cong \overline{JL}$, $\overline{FJ} \cong \overline{GK} \cong \overline{HL}$

38. pentagonal pyramid; 6 faces and 10 edges; congruent faces: △NPQ ≅ △NQR ≅ △NRS ≅ △NST ≅ △NTP; congruent edges: $\overline{NP} \cong \overline{NQ} \cong \overline{NR} \cong \overline{NS} \cong \overline{NT}$ and $\overline{PQ} \cong \overline{QR} \cong \overline{RS} \cong \overline{ST} \cong \overline{TP}$

40. False; cylinders, cones, and spheres are not polyhedra because they have curved surfaces.

41. False; if a prism is not rectangular, then the bases are the congruent faces that are not rectangular.

Student Help

STUDY TIP
Use a pencil when drawing solids so that you can erase hidden lines easily.

Identify Polyhedra Tell whether the solid is a polyhedron. If so, identify the shape of the base(s). Then name the solid.

33.

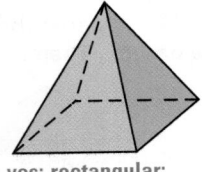

yes; rectangular; rectangular pyramid

34.

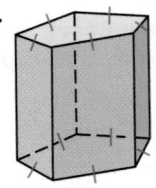

yes; pentagonal; pentagonal prism

35. no

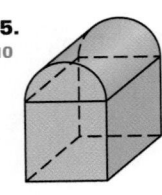

Counting Faces and Edges Name the polyhedron. Then count the number of faces and edges. List any congruent faces and congruent edges.

36.

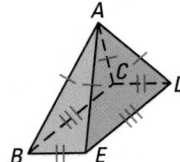

37.

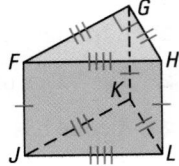

38.

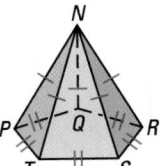

Logical Reasoning Determine whether the statement is *true* or *false*. Explain your reasoning.

39. Prisms, pyramids, cylinders, cones, and spheres are all solids.
True; they are all three-dimensional shapes.

40. Prisms, pyramids, cylinders, cones, and spheres are all polyhedra.

41. Every face of a prism is also a base of the prism.

42. Every base of a prism is also a face of the prism.
True; a prism has six faces, of which two are bases.

Visualize It! Copy the partial drawing. Then complete the drawing of the solid. 43–45. See margin.

43. square pyramid   **44.** hexagonal prism   **45.** cylinder

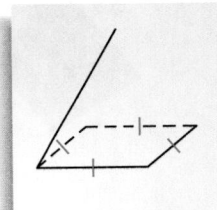
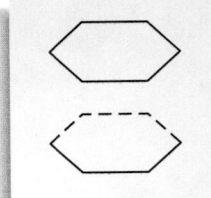

Sketching Solids Sketch the solid described. 46–51. See margin.

46. rectangular prism   **47.** rectangular pyramid

48. cube   **49.** cone

50. cylinder   **51.** sphere

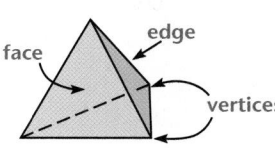
EXAMPLE   Euler's Formula

Mathematician Leonhard Euler proved that the number of faces (F), vertices (V), and edges (E) of a polyhedron are related by the formula $F + V = E + 2$.

Use Euler's Formula to find the number of vertices on the *tetrahedron* shown.

face   edge

vertices

Solution
The tetrahedron has 4 faces and 6 edges.

$F + V = E + 2$	Write Euler's Formula.
$4 + V = 6 + 2$	Substitute 4 for F and 6 for E.
$4 + V = 8$	Simplify.
$V = 8 - 4$	Subtract 4 from each side.
$V = 4$	Simplify.

ANSWER ▶ The tetrahedron has 4 vertices.

Platonic Solids A Platonic solid has faces that are congruent, regular polygons. Use the example above to find the number of vertices on the Platonic solid.

52. cube
   6 faces, 12 edges **8 vertices**

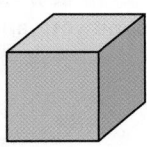

53. octahedron
   8 faces, 12 edges **6 vertices**

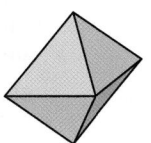

54. dodecahedron
   12 faces, 30 edges **20 vertices**

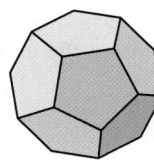

55. icosahedron
   20 faces, 30 edges **12 vertices**

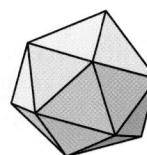

 Using Algebra Use Euler's Formula to find the number of faces, edges, or vertices. Use the example above as a model.

56. A prism has 5 faces and 6 vertices. How many edges does it have?
   9 edges

57. A pyramid has 12 edges and 7 vertices. How many faces does it have? **7 faces**

58. A prism has 8 faces and 12 vertices. How many edges does it have? **18 edges**

9.1 *Solid Figures*   479

4 Assess

Assessment Resources

The Mini-Quiz below is also available on blackline (*Chapter 9 Resource Book*, p. 17) and on transparency. For more assessment resources, see:
• Chapter 9 Resource Book
• Standardized Test Practice
• Test and Practice Generator

Mini-Quiz

Tell whether the solid is a polyhedron. If so, identify the shape of the bases. Then name the solid.

1.

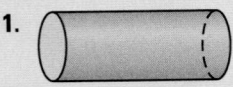

The solid is not a polyhedron.

2.

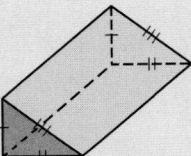

The solid is a polyhedron with congruent triangular bases in parallel planes, so it is a triangular prism.

3. Name the polyhedron. Count the number of faces and edges and then list the congruent faces and congruent edges.

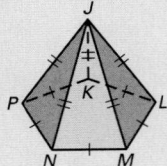

pentagonal pyramid; 6 faces, 10 edges; congruent faces: $\triangle PJN \cong \triangle NJM \cong \triangle MJL \cong \triangle LJK \cong \triangle KJP$; congruent edges: $\overline{PJ} \cong \overline{NJ} \cong \overline{MJ} \cong \overline{LJ} \cong \overline{KJ}$, $\overline{PN} \cong \overline{NM} \cong \overline{ML} \cong \overline{LK} \cong \overline{KP}$

Standardized Test Practice

59. Multiple Choice How many faces does the prism below have? **D**

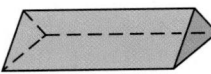

(A) 2   (B) 3

(C) 4   (D) 5

60. Multiple Choice How many edges does the pyramid at the right have? **F**

(F) 6   (G) 5

(H) 4   (J) 3

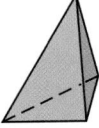

61. Multiple Choice How many vertices does the pyramid above have? **C**

(A) 6   (B) 5   (C) 4   (D) 3

Mixed Review

Finding Measures of Squares and Rectangles **Use the given information to find the missing measure.** *(Lesson 8.3)*

62. A square has a side length of 9 centimeters. Find its area. **81 cm²**

63. A rectangle has a height of 4 meters and a base length of 7 meters. Find its area. **28 m²**

64. A rectangle has an area of 60 square inches and a height of 6 inches. Find the length of its base. **10 in.**

65. A square has an area of 169 square feet. Find its side length. **13 ft**

Finding Circumference and Area of a Circle **Find the circumference and the area of the circle. Round your answers to the nearest whole number.** *(Lesson 8.7)*

66.

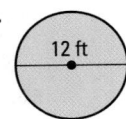

12 ft

38 ft; 113 ft²

67.

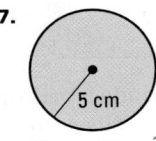

5 cm

31 cm; 79 cm²

68.

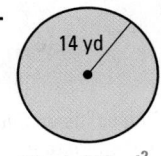

14 yd

88 yd; 616 yd²

Algebra Skills

Evaluating Expressions **Evaluate the expression.** *(Skills Review, p. 670)*

69. $92 - (12 + 39)$ **41**   **70.** $8 + 4 \cdot 3 - 5$ **15**   **71.** $(7 - 5) \cdot 14$ **28**

72. $10 - (5 - 2)^2 + 8$ **9**   **73.** $14 + 4^2 - 26$ **4**   **74.** $3(10 - 3)^2$ **147**

Substituting and Simplifying Expressions **Evaluate the expression when $\ell = 3$, $h = 5$, and $w = 2$. Write your answer in terms of π, if appropriate.** *(Skills Review, p. 674)*

75. $\ell \cdot w \cdot h$ **30**   **76.** $2\ell + 2w + 2h$ **20**   **77.** $2\pi h$ **10π**

78. $\pi w^2 h$ **20π**   **79.** $\pi \ell^2$ **9π**   **80.** $2\ell + 2w$ **10**

Activity 9.2 ▸ Investigating Surface Area

Materials
- graph paper
- scissors

Question

How can you find the surface area of a solid figure?

Explore

1 The diagram below shows a pattern, or *net*, for making a solid. Copy the net on graph paper. Draw the dashed lines as shown. Then label the sections as shown.

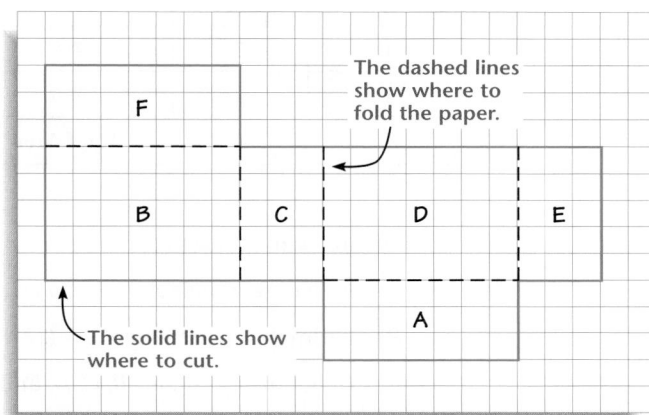

The dashed lines show where to fold the paper.

The solid lines show where to cut.

F

B   C   D   E

A

2 Cut out the net along the solid lines.

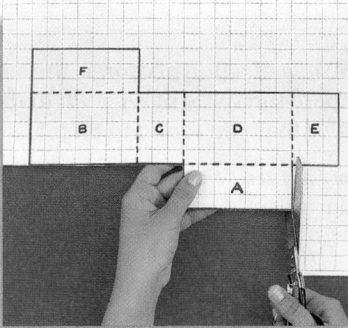

3 Fold the net along the dashed lines to form a polyhedron.

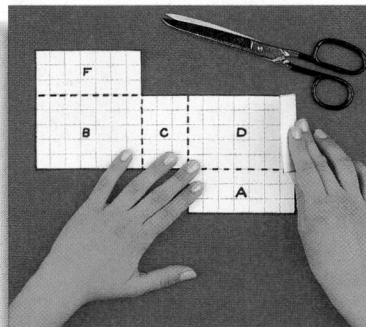

4 What is the name of the polyhedron formed in Step 3? rectangular prism

continued on next page

Activity 481

481

KEY DISCOVERY

The surface area of a polyhedron is the sum of the areas of its faces.

Activity Assessment

Use Exercises 3 and 4 to assess student understanding.

5 Use the net you made to complete the table below. Each square on the graph paper measures 1 unit by 1 unit.

Rectangle	Length	Width	Area
A	? 7	? 3	? 21
B	? 7	? 5	? 35
C	? 3	? 5	? 15
D	? 7	? 5	? 35
E	? 3	? 5	? 15
F	? 7	? 3	? 21
		Total Area	? 142

6 Find the following measures.

A = Area of rectangle $A = (\ell \times w) =$ _?_   21 square units

P = Perimeter of rectangle $A = (2\ell + 2w) =$ _?_   20 units

h = Height of rectangles B, C, D, and $E =$ _?_   5 units

Think About It

1. Use the values for A, P, and h from Step 6 above to find the value of $2A + Ph$. 142

2. The *surface area* of a polyhedron is the sum of the areas of its faces. Compare the value of $2A + Ph$ to the total of the areas of all the rectangles that you found in Step 5. They are equal.

3. **Make a Conjecture** Write a formula for finding the surface area of a rectangular prism. Surface area = $2A + Ph$, where A is the area of a base, P is the perimeter of the base, and h is the height of the prism.

4. **Extension** Draw the net of another rectangular prism with different dimensions. Cut out and fold the net to make sure that it forms a rectangular prism. Use the formula you wrote in *Think About It* Question 3 to calculate the surface area. Answers will vary.

9.2 Surface Area of Prisms and Cylinders

Goal
Find the surface areas of prisms and cylinders.

Key Words
- prism
- surface area
- lateral face
- lateral area
- cylinder

A **prism** is a polyhedron with two congruent faces that lie in parallel planes.

To visualize the surface area of a prism, imagine unfolding it so that it lies flat. The flat representation of the faces is called a *net*.

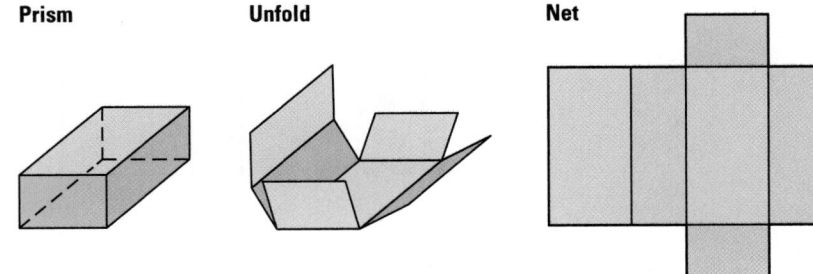

Prism   Unfold   Net

The **surface area** of a polyhedron is the sum of the areas of its faces. The surface area of a prism is equal to the area of its net.

EXAMPLE 1 Use the Net of a Prism

Find the surface area of the rectangular prism.

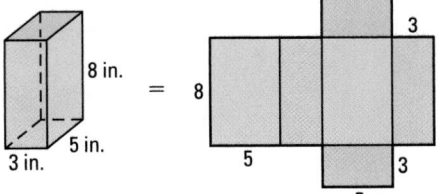

Solution
Add the areas of all the rectangles that form the faces of the prism.

Congruent Faces	Dimensions	Area of Face
Left face and right face	8 in. by 5 in.	$8 \times 5 = 40$ in.^{2}
Front face and back face	8 in. by 3 in.	$8 \times 3 = 24$ in.^{2}
Top face and bottom face	3 in. by 5 in.	$3 \times 5 = 15$ in.^{2}

Student Help

STUDY TIP
Remember that area is measured in square units, such as ft^2, m^2, and in.^{2}

Add the areas of all the faces to get the surface area.

$S = 40 + 40 + 24 + 24 + 15 + 15$ Add the area of all six faces.

$\quad = 158$ Simplify.

ANSWER ▶ The surface area of the prism is 158 square inches.

Pacing
Suggested Number of Days

Basic: 2 days
Average: 2 days
Advanced: 2 days
Block Schedule: 0.5 block with 9.1
                       0.5 block with 9.3

Teaching Resources

📄 **Blacklines**
(See page 470B.)

📠 **Transparencies**
- Warm-Up with Quiz
- Answers

💻 **Technology**
- Electronic Teacher Tools
- Test and Practice Generator
- Online Lesson Planner
- Internet Support

② Teach

Content and Teaching Strategies
For background information on geometric concepts and teaching strategies related to this lesson, see pages 470E and 470F in this Teacher's Edition.

Tips for New Teachers
Consider having nets of a few solids available to help students with limited English proficiency. Make nets like those on this page and on pages 484, 488, and 489. This should help students to make a connection between the formulas and the models for the surface areas of prisms. See the Tips for New Teachers on pp. 1–2 of the *Chapter 9 Resource Book* for additional notes about Lesson 9.2.

Extra Example 1
See next page.

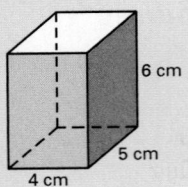
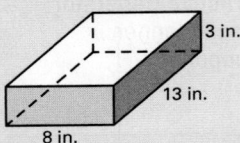
Lateral Faces and Area The **lateral faces** of a prism are the faces of the prism that are not bases. **Lateral area** is the sum of the areas of the lateral faces.

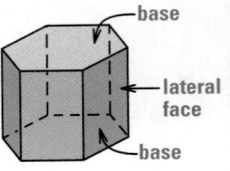

base
lateral face
base

Surface Area of a Prism One way to find the surface area of a prism is to use the method in Activity 9.2, summarized below.

Surface area	=	2(area of base)	+	area of lateral faces
	=	2B	+	Ph

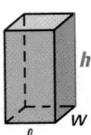

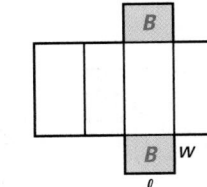

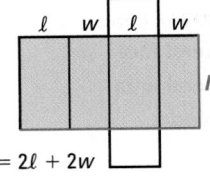

$P = 2\ell + 2w$

Student Help

STUDY TIP
The formula at the right works for any prism, regardless of the shape of its bases.

SURFACE AREA OF A PRISM

Words Surface area =
2(area of base) + (perimeter of base)(height)

Symbols $S = 2B + Ph$

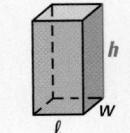

EXAMPLE 2 Find Surface Area of a Prism

Find the surface area of the prism.

5 m
3 m
2 m
4 m

Visualize It!

It may help you to visualize the bases and the height if you redraw the solid in Example 2 so that it lies on a base.

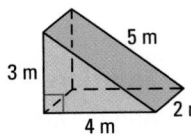

5
2
3
4

Solution

❶ Find the area of a triangular base.

$$B = \frac{1}{2} \cdot 4 \cdot 3 = 2 \cdot 3 = 6$$

❷ Find the perimeter of a base.

$$P = 3 + 4 + 5 = 12$$

❸ Find the height of the prism. In the diagram, $h = 2$.

❹ Use the formula for surface area of a prism.

$S = 2B + Ph$	Formula for the surface area of a prism
$= 2 \cdot 6 + 12 \cdot 2$	Substitute 6 for B, 12 for P, and 2 for h.
$= 12 + 24$	Multiply.
$= 36$	Add.

ANSWER ▶ The surface area of the prism is 36 square meters.

Find the surface area of the prism.

1.

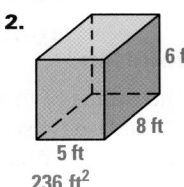

3 in.

6 in.

2 in.

72 in.²

2.

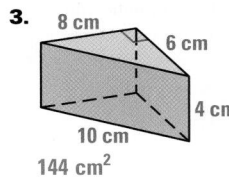

6 ft

8 ft

5 ft

236 ft²

3.

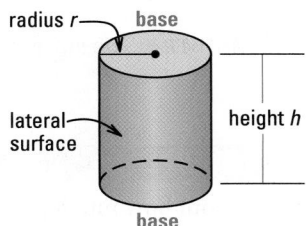

8 cm

6 cm

4 cm

10 cm

144 cm²

Surface Area of a Cylinder A **cylinder** is a solid with two congruent circular bases that lie in parallel planes. The *lateral area* of a cylinder is the area of the curved surface.

radius *r* — base

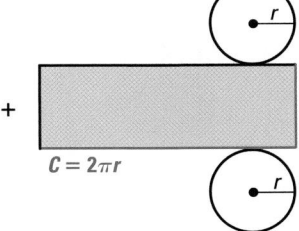

height *h*

lateral surface

base

The radius of the cylinder is the radius of a base.

The height of the cylinder is the perpendicular distance between the bases.

Visualize It!

The label below represents the lateral surface of a cylinder. When unwrapped, the label is a rectangle.

The diagram below shows how to find the surface area of a cylinder.

Surface area	=	2(area of base)	+	lateral area
	=	$2\pi r^2$	+	$(2\pi r)h$

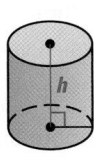

$B = \pi r^2$

$C = 2\pi r$

r

h

SURFACE AREA OF A CYLINDER

Words Surface area =
2(area of base) + (circumference of base)(height)

Symbols $S = 2B + Ch$
$= 2\pi r^2 + 2\pi rh$

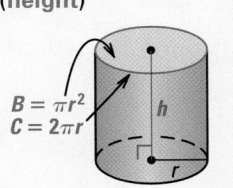

$B = \pi r^2$
$C = 2\pi r$

h

r

9.2 *Surface Area of Prisms and Cylinders* **485**

Extra Example 3

Find the surface area of the cylinder. Round your answer to the nearest whole number.

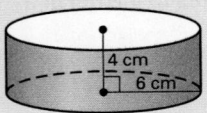

about 377 cm²

Extra Example 4

About how much wallpaper is needed to cover a cylindrical column that reaches from floor to ceiling in a room that is 9 feet high if the radius of the column is 1.5 feet? about 85 ft²

✓ Concept Check

Give the formulas for the surface areas of a prism and a cylinder. How are these formulas alike?
prism: $S = 2B + Ph$, cylinder: $S = 2\pi r^2 + 2\pi rh$; Both formulas involve adding 2 times the area of a base to the lateral area of the solid.

🐢 Daily Puzzler

If you "stretch" a prism so that its length doubles, but its base remains unchanged, does its surface area double? Explain or give an example.
No; for example, a square prism with a base edge of 5 and a height of 4 has a surface area of 130. Changing the height to 8 changes the surface area to 210, and 210 ≠ 2 · 130.

Student Help

LOOK BACK
See p. 452 to review calculations with π.

Student Help
CLASSZONE.COM

MORE EXAMPLES
More examples at classzone.com

EXAMPLE 3 Find Surface Area of a Cylinder

Find the surface area of the cylinder. Round your answer to the nearest whole number.

Solution

The radius of the base is 3 feet and the height 4 feet. Use these values in the formula for surface area of a cylinder.

$$S = 2\pi r^2 + 2\pi rh \qquad \text{Write the formula for surface area.}$$
$$= 2\pi(3^2) + 2\pi(3)(4) \qquad \text{Substitute 3 for } r \text{ and 4 for } h.$$
$$= 18\pi + 24\pi \qquad \text{Simplify.}$$
$$= 42\pi \qquad \text{Add.}$$
$$\approx 132 \qquad \text{Multiply.}$$

ANSWER ▶ The surface area is about 132 square feet.

EXAMPLE 4 Find Lateral Area

About how much plastic is used to make a straw that has a diameter of 5 millimeters and a height of 195 millimeters?

Solution

The straw is a cylinder with no bases. Use the formula for the surface area of a cylinder, but do not include the areas of the bases. The diameter is 5 millimeters. So the radius is 5 ÷ 2 = 2.5.

$$\text{Lateral area} = 2\pi rh \qquad \text{Surface area formula without bases.}$$
$$= 2\pi(2.5)(195) \qquad \text{Substitute 2.5 for } r \text{ and 195 for } h.$$
$$= 975\pi \qquad \text{Simplify.}$$
$$\approx 3063 \qquad \text{Multiply.}$$

ANSWER ▶ The straw is made with about 3063 square millimeters of plastic.

Checkpoint ✓ Find Surface Area of Cylinders

Find the area described. Round your answer to the nearest whole number.

4. surface area 151 in.²

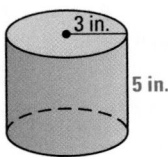

5. surface area 603 ft²

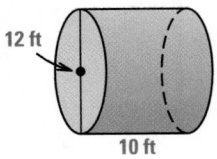

6. lateral area 13 m²

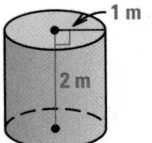

Guided Practice

Vocabulary Check

Tell whether the statement is *true* or *false*.

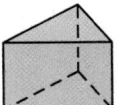

1. The solid is a triangular prism. true

2. The blue face is a lateral face of the solid. false

3. The red face is a lateral face of the solid. true

4. The blue face is a base of the solid. true

Skill Check

Find the surface area of the solid. If necessary, round your answer to the nearest whole number.

5.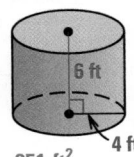
10 in.
6 in. 4 in.
248 in.²

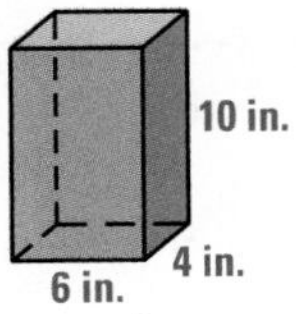

6.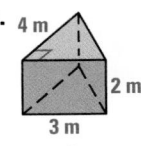
6 ft
4 ft
251 ft²

7. 4 m
2 m
3 m
36 m²

Practice and Applications

Extra Practice

See p. 691.

Identifying Parts of a Prism In Exercises 8–10, use the diagram at the right.

8. What is the height of the solid? 5 in.

9. What is the area of a base? 6 in.²

10. What is the perimeter of a base? 10 in.

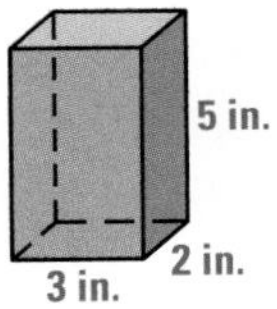

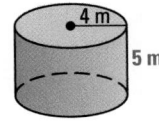
5 in.
3 in. 2 in.

Identifying Parts of a Cylinder In Exercises 11–13, use the diagram at the right.

11. What is the height of the solid? 5 m

12. What is the area of a base? about 50 m²

13. What is the circumference of a base? about 25 m

4 m
5 m

Homework Help

Example 1: Exs. 17–20, 28, 29, 31
Example 2: Exs. 17–20, 28, 29, 31
Example 3: Exs. 24–27, 30
Example 4: Exs. 39–44

Identifying Parts of a Prism In Exercises 14–16, use the diagram at the right.

14. What is the height of the solid? 15 ft

15. What is the area of a base? 24 ft²

16. What is the perimeter of a base? 24 ft

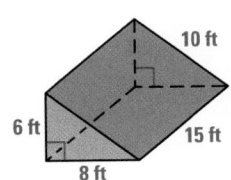
10 ft
6 ft 15 ft
8 ft

③ **Apply**

Assignment Guide

BASIC
Day 1: SRH p. 674 Exs. 1–4; pp. 487–490 Exs. 8–19, 21–23, 33–35, 48–52
Day 2: EP p. 690 Exs. 33–35; pp. 487–490 Exs. 24–31, 36–41, 47, 53–58

AVERAGE
Day 1: pp. 487–490 Exs. 8–23, 33–35, 49–52
Day 2: pp. 487–490 Exs. 24–32, 36–44, 47, 48, 53–58

ADVANCED
Day 1: pp. 487–490 Exs. 12–20 even, 21–23, 33–35, 49–58
Day 2: pp. 487–490 Exs. 24–32, 36–45, 46*, 47, 48; EC: classzone.com

BLOCK SCHEDULE
pp. 487–490 Exs. 8–23, 33–35, 49–52 (with 9.1)
pp. 487–490 Exs. 24–32, 36–44, 47, 48 (with 9.3)

Extra Practice

• Student Edition, p. 691
• Chapter 9 Resource Book, pp. 18–19

Homework Check

To quickly check student understanding of key concepts, go over the following exercises:

Basic: 17, 21, 24, 28, 39
Average: 18, 22, 25, 30, 40
Advanced: 20, 23, 26, 29, 41

Surface Area of a Prism Find the surface area of the prism.

17.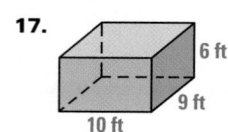
6 ft
9 ft
10 ft
408 ft²

18.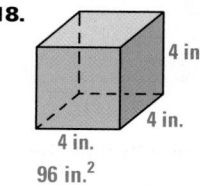
4 in.
4 in.
4 in.
96 in.²

19.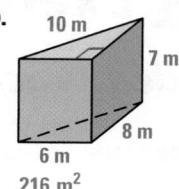
10 m
7 m
8 m
6 m
216 m²

20. **You be the Judge**

The height of Prism A is twice the height of Prism B. Is the surface area of Prism A twice the surface area of Prism B? Explain.

Prism A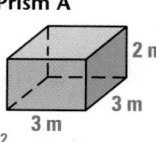
2 m
3 m
3 m

Prism B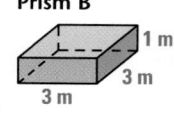
1 m
3 m
3 m

No; the surface area of Prism A is 42 m², and the surface area of Prism B is 30 m².

Analyzing Nets Name the solid that can be folded from the net.

21.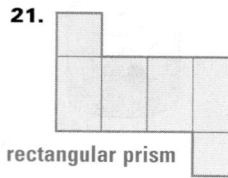

rectangular prism

22.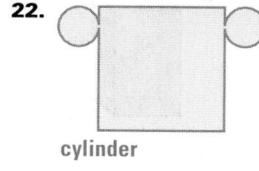

cylinder

23.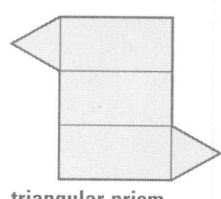

triangular prism

Finding Surface Area of a Cylinder Find the surface area of the cylinder. Round your answer to the nearest whole number.

24.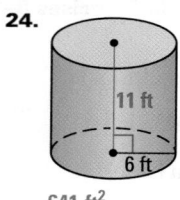
11 ft
6 ft
641 ft²

25.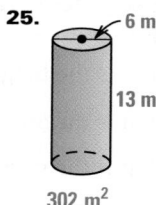
6 m
13 m
302 m²

26.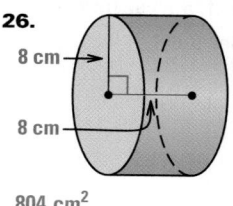
8 cm
8 cm
804 cm²

Finding Surface Area Find the surface area of the solid. Round your answer to the nearest whole number.

27. 377 in.²
5 in.
7 in.

28. Cube with 16 mm sides   1536 mm²
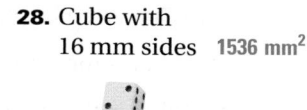

29.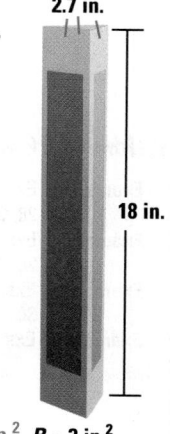
2.7 in.
18 in.

30. 24 in.²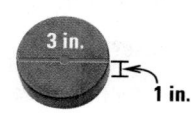
3 in.
1 in.

31.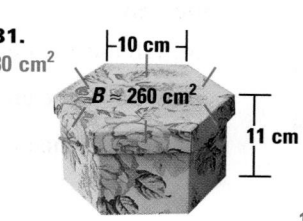
1180 cm²
10 cm
B ≈ 260 cm²
11 cm

152 in.² B ≈ 3 in.²

32. The radius of the cylinder is 6 in., not 12 in. Also, $12^2 \neq 24$. The surface area is $2\pi(6^2) + 2\pi(6)(10) = 192\pi \approx 603$ in.^{2}

32. Error Analysis Juanita is trying to find the surface area of the cylinder shown below. What did she do wrong?

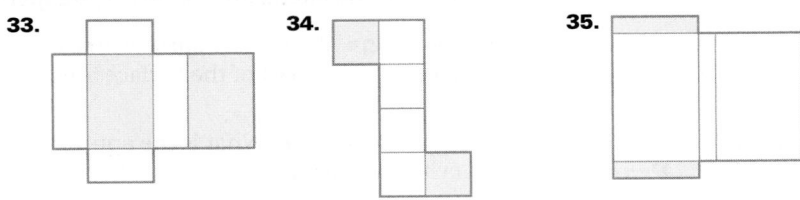

$$S = 2\pi(12^2) + 2\pi(12)(10)$$
$$= 2\pi(24) + 2\pi(120)$$
$$= 288\pi$$
$$\approx 905 \text{ in.}^2$$

Visualize It! Sketch the solid that results after the net has been folded. Use the shaded faces as bases. 33–35. See margin.

33. **34.** **35.**

Visualize It! Sketch the solid described and find its surface area.
36–38. See margin.
36. A rectangular prism with a height of 10 feet, a length of 3 feet, and a width of 6 feet.

37. A cylinder with a radius of 3 meters and a height of 7 meters.

38. A triangular prism with a height of 5 inches and a base that is a right triangle with legs of 8 inches and 6 inches.

Finding Lateral Area Find the lateral area of the solid. If necessary, round your answer to the nearest whole number.

39.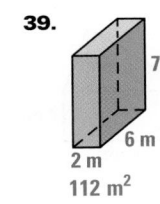
7 m
6 m
2 m
112 m^2

40.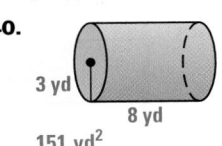
3 yd
8 yd
151 yd^2

41.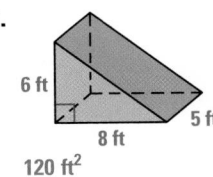
6 ft
8 ft
5 ft
120 ft^2

42. Clothes Rod Find the lateral area of a clothes rod that has a radius of 2 centimeters and a height of 90 centimeters. 1131 cm^2

43. Wax Cylinder Records In the late 1800's, a standard-sized cylinder record was about 2 inches in diameter and 4 inches long. Find the lateral area of the cylinder described. 25 in.^{2}

44. Compact Discs A standard compact disc has an outer radius of 60 millimeters and a height of 1.2 millimeters. Find the lateral area of the disc. 452 mm^2

33.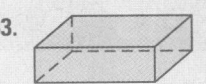

34.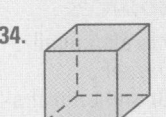

35.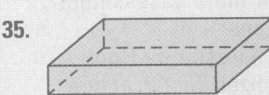

36.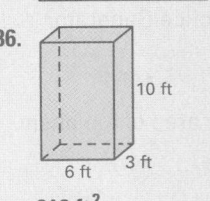
10 ft
6 ft
3 ft
216 ft^2

37.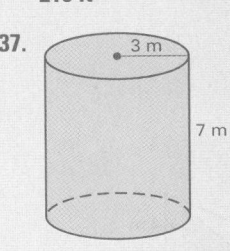
3 m
7 m
about 188 m^2

38.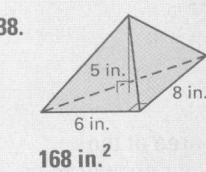
5 in.
8 in.
6 in.
168 in.^{2}

9.2 *Surface Area of Prisms and Cylinders* **489**

Mini-Quiz

Find the surface area of the prism.

1.

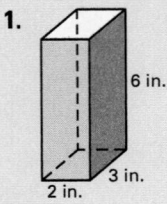

6 in.
3 in.
2 in.

72 in.²

2.

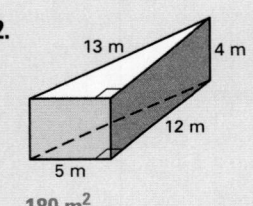

13 m
4 m
12 m
5 m

180 m²

Find the surface area of the cylinder. Round your answer to the nearest whole number.

3.

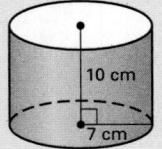

10 cm
7 cm

748 cm²

4.

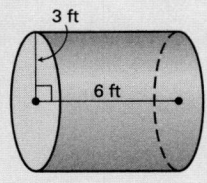

3 ft
6 ft

170 ft²

Architecture **In Exercises 45 and 46 use the following information.**
Suppose a skyscraper is a prism that is 415 meters tall and each base is a square that measures 64 meters on a side.

45. What is the lateral area of this skyscraper?   106,240 m²

46. Challenge What is the surface area of this skyscraper? (*Hint*: The ground is not part of the surface area of the skyscraper.)   110,336 m²

Standardized Test Practice

47. Multiple Choice What is the approximate surface area of the cylinder shown? B

(A) 502 cm²   (B) 628 cm²

(C) 785 cm²   (D) 1570 cm²

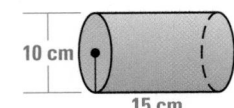

10 cm
15 cm

48. Multiple Choice What is the height of the prism shown if the surface area is 104 square feet? H

(F) 4 feet   (G) 5 feet

(H) 6 feet   (J) 7 feet

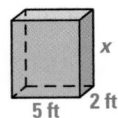

x
5 ft   2 ft

Mixed Review

Finding Area **Find the area of the polygon made up of rectangles and triangles.** (*Lessons 8.3, 8.4*)

49.

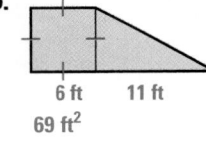

6 ft   11 ft

69 ft²

50.

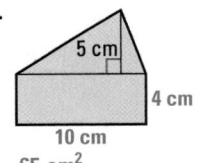

5 cm
4 cm
10 cm

65 cm²

51.

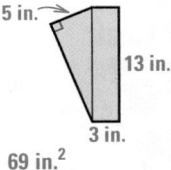

5 in.
13 in.
3 in.

69 in.²

52. A parallelogram has a height of 5 feet and an area of 70 square feet. Find the base. (*Lesson 8.5*) 14 ft

Algebra Skills

Solving Proportions **Solve the proportion.** (*Skills Review, p. 660*)

53. $\frac{6}{x} = \frac{2}{5}$ 15

54. $\frac{1}{10} = \frac{4}{x}$ 40

55. $\frac{3}{5} = \frac{x}{35}$ 21

56. $\frac{4}{7} = \frac{16}{x}$ 28

57. $\frac{33}{x} = \frac{11}{13}$ 39

58. $\frac{x}{32} = \frac{5}{8}$ 20

Surface Area of Pyramids and Cones

Goal

Find the surface areas of pyramids and cones.

Key Words

- pyramid
- height of a pyramid
- slant height of a pyramid
- cone
- height of a cone
- slant height of a cone

The Rainforest Pyramid on Galveston Island in Texas is one example of how pyramids are used in architecture.

The base of a **pyramid** is a polygon and the lateral faces are triangles with a common vertex.

This greenhouse is home to plants, butterflies, and bats.

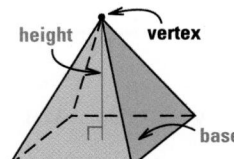

Height

The **height of a pyramid** is the perpendicular distance between the vertex and base.

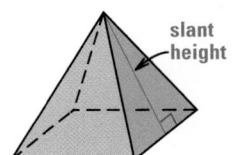

Slant height

The **slant height of a pyramid**, represented by the letter ℓ, is the height of any of its lateral faces.

Visualize It!

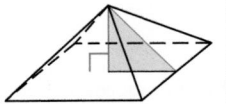

For a square pyramid, the slant height is the length of the hypotenuse of a right triangle formed by the height of the pyramid and half of the length of the base.

EXAMPLE 1 Find the Slant Height

Find the slant height of the Rainforest Pyramid. Round your answer to the nearest whole number.

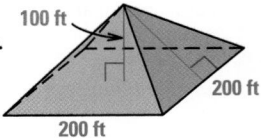

100 ft · 200 ft · 200 ft

Solution

To find the slant height, use the right triangle formed by the height and half of the base.

$$(\text{slant height})^2 = (\text{height})^2 + \left(\frac{1}{2}\,\text{side}\right)^2 \qquad \text{Use the Pythagorean Theorem.}$$

$$= 100^2 + \left(\frac{1}{2} \cdot 200\right)^2 \qquad \text{Substitute 100 for height and 200 for base side length.}$$

$$= 10{,}000 + 10{,}000 \qquad \text{Simplify.}$$

$$\text{slant height} = \sqrt{20{,}000} \qquad \text{Take the positive square root.}$$

$$\approx 141.42 \qquad \text{Use a calculator.}$$

ANSWER The slant height is about 141 feet.

---

① Plan

Pacing
Suggested Number of Days

Basic: 2 days
Average: 2 days
Advanced: 2 days
Block Schedule: 0.5 block with 9.2
0.5 block with 9.4

Teaching Resources

📄 *Blacklines*
(See page 470B.)

🔗 *Transparencies*
- Warm-Up with Quiz
- Answers

🖥 *Technology*
- Electronic Teacher Tools
- Test and Practice Generator
- Online Lesson Planner
- Internet Support

② Teach

Content and Teaching Strategies
For background information on geometric concepts and teaching strategies related to this lesson, see pages 470E and 470F in this Teacher's Edition.

Tips for New Teachers
Students may confuse slant height with height for a pyramid. Stress that the slant height is the altitude of a lateral face. Consider having them draw diagrams that include both the pyramid and its net, like those shown at the top of page 492. Drawing and labeling both diagrams should help them avoid confusion. See the Tips for New Teachers on pp. 1–2 of the *Chapter 9 Resource Book* for additional notes about Lesson 9.3.

Extra Example 1
See next page.

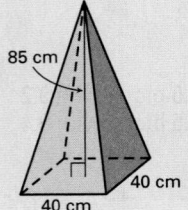

85 cm

40 cm

40 cm

about 87 cm

Extra Example 2

Find the surface area of the pyramid shown in Extra Example 1.
about 8560 cm²

Surface Area of a Pyramid The diagrams show the surface area of a pyramid with a square base.

Surface area	=	area of base	+	area of lateral faces
	=	B	+	$\frac{1}{2}P\ell$

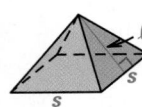

= 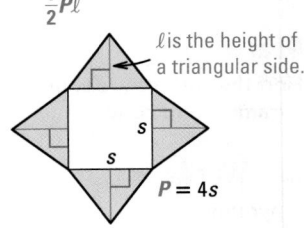 +

ℓ is the height of a triangular side.

$P = 4s$

Student Help

STUDY TIP
A *regular pyramid* has a regular polygon for a base. The slant height, ℓ, is the same on all of the lateral faces of a regular pyramid.

SURFACE AREA OF A PYRAMID

Words Surface area =
(area of base) + $\frac{1}{2}$(perimeter of base)(slant height)

Symbols $S = B + \frac{1}{2}P\ell$

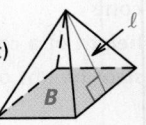

EXAMPLE **2** **Find Surface Area of a Pyramid**

Find the surface area of the pyramid.

Solution

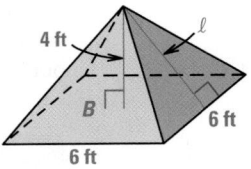

4 ft

B

6 ft

6 ft

❶ Find the area of the base.

$B = 6 \times 6 = 36$

Student Help

SKILLS REVIEW
For help with area and perimeter, see p. 674.

❷ Find the perimeter of the base.

$P = 6 + 6 + 6 + 6 = 24$

❸ Find the slant height.

(slant height)² = (height)² + ($\frac{1}{2}$side)²	Use the Pythagorean Theorem.
= 4² + 3²	Substitute. Half of 6 is 3.
= 16 + 9	Simplify powers.
= 25	Simplify.
slant height = $\sqrt{25}$ = 5	Take positive square root.

❹ Substitute values into the formula for surface area of a pyramid.

$S = B + \frac{1}{2}P\ell$	Write the formula for surface area.
= 36 + $\frac{1}{2}$(24)(5)	Substitute.
= 96	Simplify.

ANSWER ▶ The surface area of the pyramid is 96 square feet.

Find the surface area of the pyramid.

1.

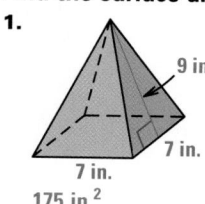

9 in.

7 in.

7 in.

175 in.²

2.

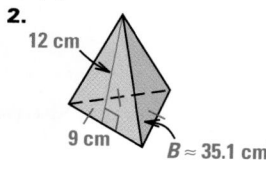

12 cm

9 cm

$B \approx 35.1 \text{ cm}^2$

about 197.1 cm²

3.

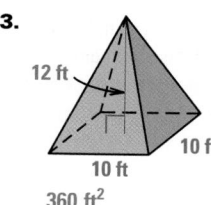

12 ft

10 ft

10 ft

360 ft²

Surface Area of a Cone A **cone** has a circular base and a vertex that is not in the same plane as the base. In a right cone, the height meets the base at its center. In this lesson, only right cones are shown.

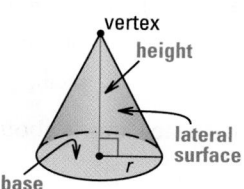

vertex

height

lateral surface

base

r

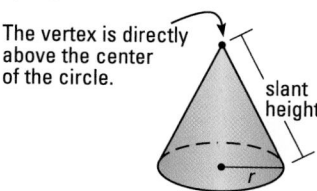

The vertex is directly above the center of the circle.

slant height

r

The **height of a cone** is the perpendicular distance between the vertex and the base.

The **slant height of a cone** is the distance between the vertex and a point on the base edge.

Student Help

LOOK BACK
To review how to find the area of a sector, see p. 454.

The diagrams show the surface area of a cone.

Surface area	=	area of base	+	area of sector
	=	B	+	$\pi r \ell$

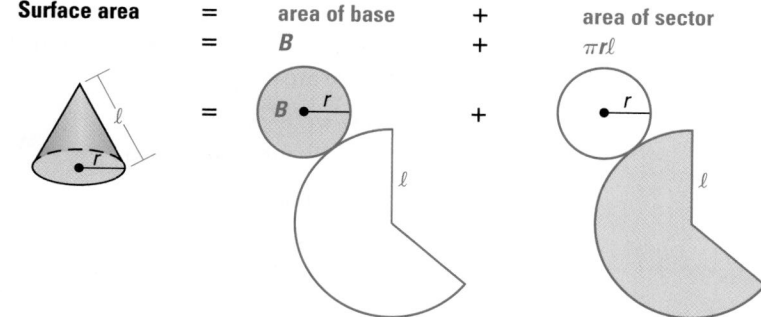

ℓ

r

=

B • r

+

r

ℓ

ℓ

SURFACE AREA OF A CONE

Words Surface area = (area of base) + (area of sector)

= (area of base) + π(radius of base)(slant height)

Symbols $S = B + \pi r \ell$

$= \pi r^2 + \pi r \ell$

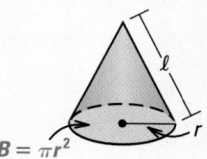

ℓ

$B = \pi r^2$

r

Teaching Tip
As you discuss the definition of a right cone, draw a cone with a height that does not meet the base at its center so students can see how it differs from the right cones shown in this lesson.

Geometric Reasoning
Show how the formula for the surface area of a right cone, $S = \pi r^2 + \pi r \ell$, can be derived from that for a square pyramid, $S = B + \frac{1}{2}P\ell$. The term πr^2 corresponds to *B* because it is the area of the circular base of the cone. The circumference of the circle is $2\pi r$, so $\frac{1}{2}P\ell$ is $\frac{1}{2}(2\pi r)\ell$, or $\pi r \ell$.

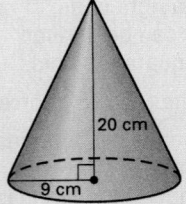
EXAMPLE 3 Find Surface Area of a Cone

Find the surface area of the cone to the nearest whole number.

a.

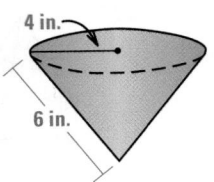

4 in.

6 in.

b.

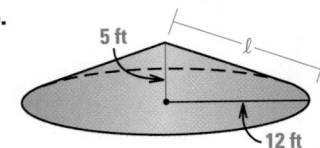

5 ft

ℓ

12 ft

Solution

a. The radius of the base is 4 inches and the slant height is 6 inches.

$S = \pi r^2 + \pi r \ell$ Write the formula for surface area of a cone.

$= \pi(4)^2 + \pi(4)(6)$ Substitute 4 for r and 6 for ℓ.

$= 40\pi$ Simplify $16\pi + 24\pi$.

≈ 126 Multiply.

ANSWER ▶ The surface area is about 126 square inches.

b. *First* find the slant height.

Using Algebra

$(\text{slant height})^2 = r^2 + h^2$ Use the Pythagorean Theorem.

$= (12)^2 + (5)^2$ Substitute 12 for r and 5 for h.

$= 169$ Simplify $144 + 25$.

$\text{slant height} = \sqrt{169}$ Find the positive square root.

$= 13$ Simplify.

Next substitute 12 for r and 13 for ℓ in the formula for surface area.

$S = \pi r^2 + \pi r \ell$ Write the formula for surface area.

$= \pi(12)^2 + \pi(12)(13)$ Substitute.

$= 300\pi$ Simplify $144\pi + 156\pi$.

≈ 942 Multiply.

ANSWER ▶ The surface area is about 942 square feet.

Checkpoint ✓ Find Surface Area of Cones

Find the surface area of the cone to the nearest whole number.

4.
691 in.²

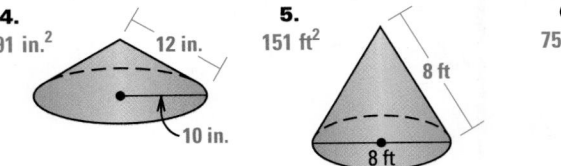

12 in.

10 in.

5.
151 ft²

8 ft

8 ft

6.
75 cm²

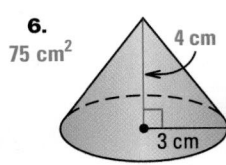

4 cm

3 cm

Guided Practice

Vocabulary Check   **Complete the statement using *height* or *slant height*.**

1. The red line segment is the ___?___ of the pyramid. height

2. The blue line segment is the ___?___ of the pyramid. slant height

3. The height of the lateral faces of the pyramid is the ___?___. slant height

4. The ___?___ of a pyramid is the perpendicular distance between the vertex and base. height

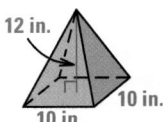

Skill Check   5. Find the slant height of the pyramid shown at the right. 13 in.

12 in.
10 in.
10 in.

Find the surface area of the solid.

6.
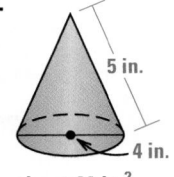
5 m
$B \approx 3.9 \text{ m}^2$
3 m
about 26 m²

7.
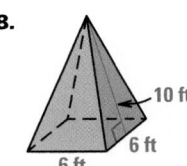
5 in.
4 in.
about 44 in.²

8.
10 ft
6 ft
6 ft
156 ft²

Practice and Applications

Extra Practice
See p. 691.

Recognizing Slant Height Tell whether the red line segment is the *height* or the *slant height*.

9.
height

10.
height

11.
slant height

Homework Help

Example 1: Exs. 9–14
Example 2: Exs. 15–20
Example 3: Exs. 23–28

Finding Slant Height Find the slant height of the solid.

12.
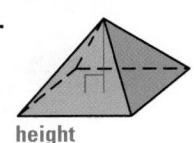
10 m   8 m
12 m
12 m

13.
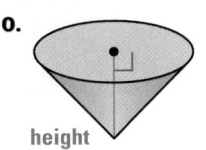
15 mm   18 mm
12 mm

14.
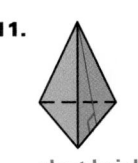
17 in.
15 in.
8 in.

Assignment Guide

BASIC
Day 1: pp. 495–499 Exs. 9–20, 32, 36, 47–55
Day 2: EP p. 690 Exs. 22–28; pp. 495–499 Exs. 23–28, 34, 37, 46, 56–61, Quiz 1

AVERAGE
Day 1: pp. 495–499 Exs. 9–22, 32, 33, 36, 41, 42, 47–55
Day 2: pp. 495–499 Exs. 23–31, 34, 35, 37–40, 46, 59–61, Quiz 1

ADVANCED
Day 1: pp. 495–499 Exs. 12–22, 32, 33, 36, 41, 42, 50–55
Day 2: pp. 495–499 Exs. 23–31 odd, 34, 35, 37–40, 43–45*, 46, 59–61, Quiz 1;
EC: classzone.com

BLOCK SCHEDULE
pp. 495–499 Exs. 9–22, 32, 33, 36, 41, 42, 47–55 (with 9.2)
pp. 495–499 Exs. 23–31, 34, 35, 37–40, 46, 59–61, Quiz 1 (with 9.4)

Extra Practice
• Student Edition, p. 691
• Chapter 9 Resource Book, pp. 30–31

Homework Check
To quickly check student understanding of key concepts, go over the following exercises:

Basic: 9, 12 ,15, 23, 26
Average: 10, 13, 16, 24, 27
Advanced: 14, 18, 20, 27, 30

Surface Area of a Pyramid In Exercises 15–20, find the surface area of the pyramid.

15.

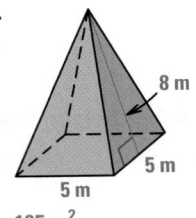

8 m

5 m

5 m

105 m²

16.

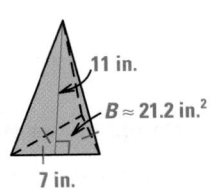

11 in.

$B \approx 21.2$ in.²

7 in.

about 136.7 in.²

17.

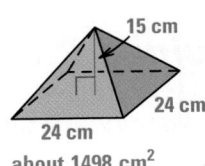

15 cm

24 cm

24 cm

about 1498 cm²

18.

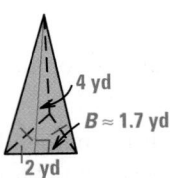

4 yd

$B \approx 1.7$ yd²

2 yd

about 13.7 yd²

19.

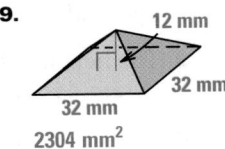

12 mm

32 mm

32 mm

2304 mm²

20.

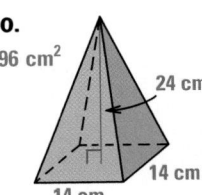

896 cm²

24 cm

14 cm

14 cm

21. The slant height is the hypotenuse of a right triangle formed by the height of the pyramid and half the length of the base. Since the hypotenuse of a right triangle is always the longest side of the triangle, the slant height is always greater than the height.

22. He used the height instead of the slant height, ℓ.
$S = 40^2 + \frac{1}{2}(160)(25) = 3600$ m².

21. Logical Reasoning Explain why the slant height of a pyramid must be greater than the height of the pyramid.

22. Error Analysis Jamie is trying to find the surface area of the pyramid below. His solution is shown. What did he do wrong in his calculations?

$S = 40^2 + \frac{1}{2}(160)(15)$

$= 1600 + 1200$

$= 2800$ m²   ✗

15 m

40 m

40 m

Surface Area of a Cone Find the surface area of the cone. Round your answer to the nearest whole number.

23.

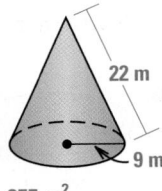

22 m

9 m

877 m²

24.

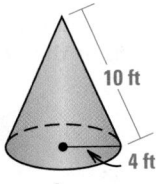

10 ft

4 ft

176 ft²

25.

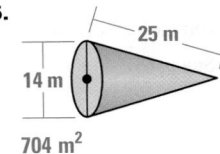

25 m

14 m

704 m²

26.

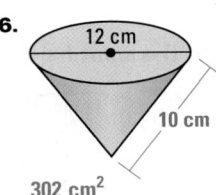

12 cm

10 cm

302 cm²

27.

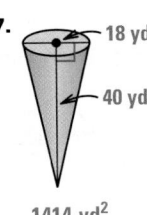

18 yd

40 yd

1414 yd²

28.

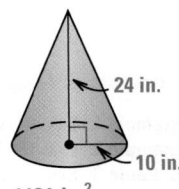

24 in.

10 in.

1131 in.²

Finding Lateral Area Find the lateral area of the object.

29.

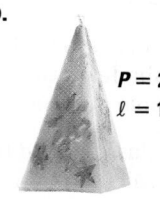

$P = 28$ cm
$\ell = 14$ cm

196 cm²

30.

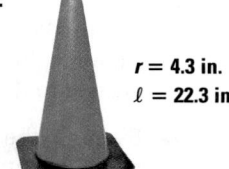

$r = 4.3$ in.
$\ell = 22.3$ in.

about 301.25 in.²

31.

GO TEAM!

$d = 8$ in.
$\ell = 14$ in.

about 176 in.²

Student Help
CLASSZONE.COM

HOMEWORK HELP
Extra help with problem solving in Exs. 32–35 is at classzone.com

Visualize It! Sketch the described solid and find its surface area. If necessary, round your answer to the nearest whole number.

32–35. See margin.

32. A pyramid has a square base with congruent edges of 12 meters and a height of 8 meters.

33. A pyramid has a triangular base with congruent edges of 8 feet and a slant height of 13 feet. The base area is 27.7 square feet.

34. A cone has a diameter of 6 yards and a slant height of 7 yards.

35. A cone has a radius of 10 inches and a height of 14 inches.

Using Nets Name the solid that can be folded from the net. Then find its surface area.

36.

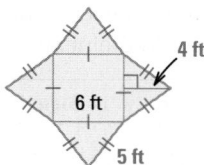

4 ft
6 ft
5 ft

square pyramid; 84 ft²

37.

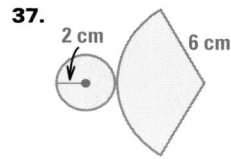

2 cm   6 cm

cone; about 50 cm²

Link to
Careers

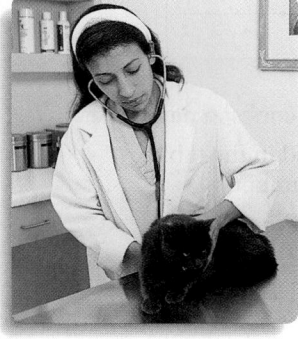

VETERINARIANS provide treatment for sick and injured animals. They also offer preventive care, such as vaccinations, for healthy animals.

Career Links
CLASSZONE.COM

Veterinary Medicine A cone-shaped collar, called an Elizabethan collar, is used to prevent pets from aggravating a healing wound.

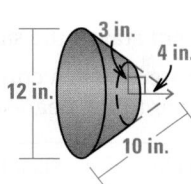

3 in.
4 in.
12 in.
10 in.

Diagram of a collar

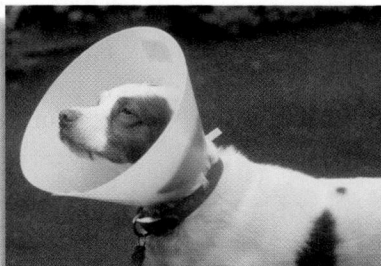

38. Find the lateral area of the entire cone shown above.   about 188 in.²

39. Find the lateral area of the small cone that has a radius of 3 inches and a height of 4 inches.   about 47 in.²

40. Use your answers to Exercises 38 and 39 to find the amount of material needed to make the Elizabethan collar shown.   about 141 in.²

9.3 *Surface Area of Pyramids and Cones*   **497**

32.

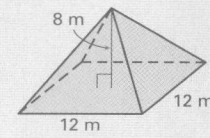

8 m
12 m
12 m

384 m²

33.

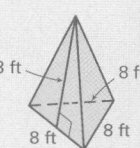

13 ft   8 ft
8 ft   8 ft

184 ft²

34.

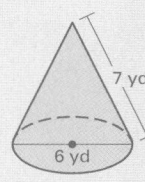

7 yd
6 yd

about 94 yd²

35.

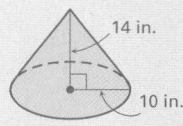

14 in.
10 in.

about 855 in.²

Link to

Lamp Design

LAMPSHADES Many
lampshades are shaped like
cones or pyramids. This lamp
was designed by architect and
designer Frank Lloyd Wright.

Lampshades In Exercises 41 and 42, refer to the lampshade with a
square base shown at the right.

41. Use the Pythagorean Theorem to find the
slant height of the lampshade. Round your
answer to the nearest whole number. **23 cm**

42. Estimate the amount of glass needed to
make the lampshade by calculating the
lateral area of the pyramid. **about 1288 cm²**

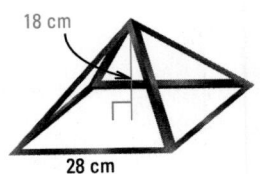

Challenge Find the surface area of the combined solids. (*Hint*: Find
the surface area of each solid and add them together. In each
calculation, remember to omit the surface where the solids connect.)

43.

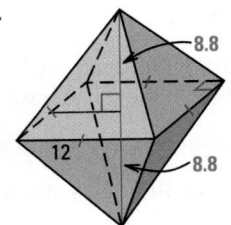

about 511 square units

44.

about 101 square units

45.

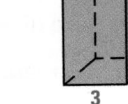

about 264 square units

**Standardized Test
Practice**

46. **Multi-Step Problem** Complete parts (a)–(e).

a. Find the surface area of each cone described in the table.
Round your answer to the nearest whole number.

	Radius	Slant Height	Surface Area
Cone A	3 ft	6 ft	? 85 ft²
Cone B	3 ft	8 ft	? 104 ft²
Cone C	3 ft	10 ft	? 123 ft²

b. What measurement stayed the same in the cones in part (a)? the radius

c. Find the surface area of each cone described in the table.
Round your answer to the nearest whole number.

	Radius	Slant Height	Surface Area
Cone D	2 ft	8 ft	? 63 ft²
Cone E	4 ft	8 ft	? 151 ft²
Cone F	6 ft	8 ft	? 264 ft²

the slant height

d. What measurement stayed the same in the cones in part (c)?

e. Compare the measurements you found in the two tables. Which
measurement has a greater influence on surface area? Why?
The radius; the radius is used in computing both the base area and the
lateral surface area, and is squared when computing the base area, while
the slant height affects only the lateral surface area.

Mixed Review

Evaluating Expressions Evaluate the expression for the given value of the variable. *(Skills Review, p. 670)*

47. $3x^2$ when $x = 6$ **108**
48. $x^2 + 6$ when $x = 4$ **22**
49. $2x^2 - 3$ when $x = 3$ **15**

50. $4x^2 - 10$ when $x = 2$ **6**
51. $x^2 + 4x$ when $x = 6$ **60**
52. $3x^2 + 5x$ when $x = 5$ **100**

Finding Areas of Sectors Find the area of the green sector given the area of the circle. Round your answer to the nearest whole number. *(Lesson 8.7)*

53. $A = 180$ m² **50 m²**
54. $A = 114$ ft² **19 ft²**
55. $A = 258$ cm² **86 cm²**

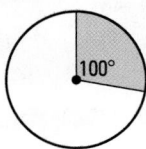 100°
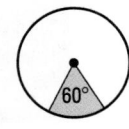 60°
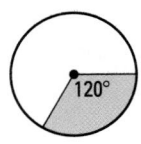 120°

Algebra Skills

Simplifying Expressions Simplify. *(Skills Review, p. 670)*

56. $-2x + 3 + 6x$ **4x + 3**
57. $4y + 5 - 4y - 1$ **4**
58. $4 - (2x - 1) + x$ **5 − x**

59. $7x - 2 - (3x + 2)$ **4x − 4**
60. $x - 9x + 6x$ **−2x**
61. $10c - (5 - 3c)$ **13c − 5**

Quiz 1

Identify the shape of the base(s) of the solid and name the solid. Then tell if the solid is a polyhedron. If so, count the number of faces of the polyhedron. *(Lesson 9.1)*

1.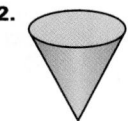
triangular; triangular prism; yes; 5 faces

2.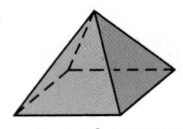
circular; cone; no

3.
rectangular; rectangular pyramid; yes; 5 faces

Find the surface area of the figure. If necessary, round your answer to the nearest whole number. *(Lessons 9.2, 9.3)*

4. 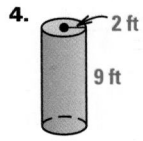 2 ft, 9 ft
138 ft²

5. 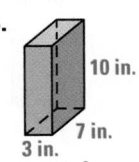 10 in., 7 in., 3 in.
242 in.²

6. 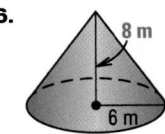 8 m, 6 m
302 m²

7. 12 in., 5 in., 8 in., 13 in.
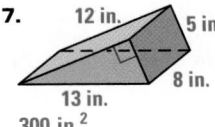
300 in.²

8. 9 m, 10 m, 10 m
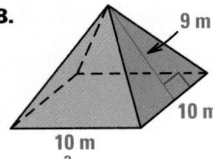
280 m²

9. 5 cm, 7 cm
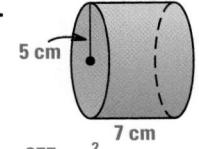
377 cm²

9.3 Surface Area of Pyramids and Cones **499**

④ Assess

Assessment Resources

The Mini-Quiz below is also available on blackline (*Chapter 9 Resource Book*, p. 38) and on transparency. For more assessment resources, see:
• Chapter 9 Resource Book
• Standardized Test Practice
• Test and Practice Generator

Mini-Quiz

Find the surface area of the solid. If necessary, round your answer to the nearest whole number.

1. 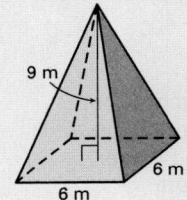 9 m, 6 m, 6 m
150 m²

2. 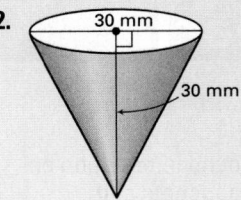 30 mm, 30 mm
2287 mm²

3. Find the lateral area of a right cone with radius 5 inches and slant height 22 inches. Round your answer to the nearest whole number. **346 in.²**

499

Pacing
Suggested Number of Days

Basic: 2 days
Average: 2 days
Advanced: 2 days
Block Schedule: 0.5 block with 9.3
0.5 block with 9.5

Teaching Resources

 **Blacklines**
(See page 470B.)

Transparencies
• Warm-Up with Quiz
• Answers

Technology
• Electronic Teacher Tools
• Test and Practice Generator
• Online Lesson Planner
• Internet Support

Teach

Content and Teaching Strategies
For background information on geometric concepts and teaching strategies related to this lesson, see pages 470E and 470F in this Teacher's Edition.

Tips for New Teachers
Consider having some type of model available for demonstration similar to the model in Example 1 on this page. You could use a set of smaller boxes and a larger one, blocks of wood and their container, or a box of sugar cubes. There are many possible models that can give students a real-world example of volume. See the Tips for New Teachers on pp. 1–2 of the *Chapter 9 Resource Book* for additional notes about Lesson 9.4.

 **9.4**

Volume of Prisms and Cylinders

Goal
Find the volumes of prisms and cylinders.

Key Words
• prism p. 483
• cylinder p. 485
• volume

The amount of water in an aquarium is an example of *volume*. The **volume** of a solid is the number of cubic units contained in its interior.

EXAMPLE 1 Find the Volume of a Rectangular Prism

Find the volume of the box by determining how many unit cubes fit in the box.

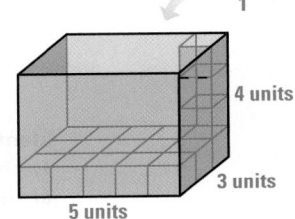

Solution
The base is 5 units by 3 units. So, 3 • 5, or 15 unit cubes are needed to cover the base layer.

There are 4 layers. Each layer has 15 cubes. So, the total number of cubes is 4 • 15, or 60.

ANSWER ▶ The volume of the box is 60 cubic units.

Student Help

READING TIP
Volume is measured in cubic units, such as ft^3, read as "cubic feet."

Volume of a Prism The process used in Example 1 can be used to determine the volume of any prism.

Volume of prism = **area of base** × **height**

$$= B \times h$$

VOLUME OF A PRISM

Words Volume = (area of base)(height)

Symbols $V = Bh$

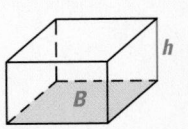

EXAMPLE 2 Find the Volume of a Prism

Find the volume of the prism.

a.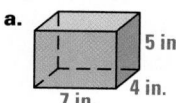
5 in.
4 in.
7 in.

b.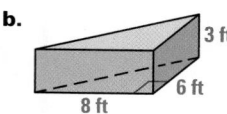
3 ft
6 ft
8 ft

Solution

a. $V = Bh$ Write the formula for volume of a prism.

$= (7 \cdot 4) \cdot 5$ Area of rectangular base $= \ell \cdot w = 7 \cdot 4$.

$= 140$ Simplify.

ANSWER ▶ The volume is 140 cubic inches.

b. $V = Bh$ Write the formula for volume of a prism.

$= \left(\dfrac{1}{2} \cdot 8 \cdot 6\right) \cdot 3$ Area of triangular base $= \dfrac{1}{2} \cdot 8 \cdot 6$.

$= 72$ Simplify.

▶ **ANSWER** ▶ The volume is 72 cubic feet.

Student Help

STUDY TIP
Because you are multiplying three units of measure when you find volume, your answer will always be in cubic units.
ft $\times$ ft $\times$ ft $=$ ft^3

Checkpoint ✔ **Find Volume of Prisms**

Find the volume of the prism.

1.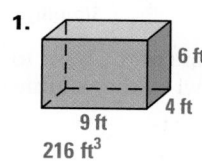
6 ft
4 ft
9 ft
216 ft^3

2.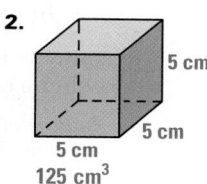
5 cm
5 cm
5 cm
125 cm^3

3.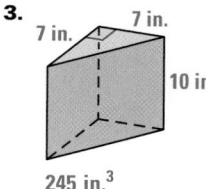
7 in.
7 in.
10 in.
245 in.^{3}

Volume of a Cylinder The method for finding the volume of a cylinder is the same for finding the volume of a prism.

Volume of cylinder = **area of base** × **height**

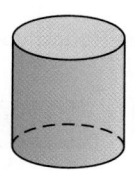
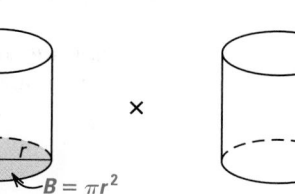

$=$ × h

$B = \pi r^2$

9.4   *Volume of Prisms and Cylinders*   **501**

Extra Example 1

Find the volume of the box by determining how many unit cubes fit in the box.

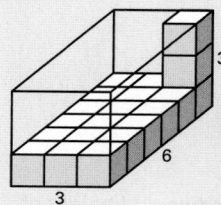
3
6
3

The volume of the box is 54 cubic units.

Extra Example 2

Find the volume of the prism.

a.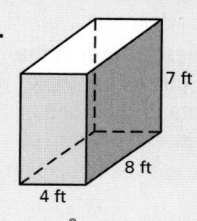
7 ft
8 ft
4 ft
224 ft^3

b.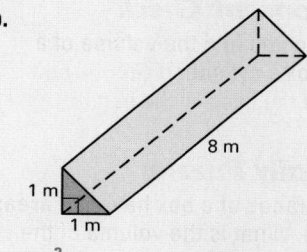
8 m
1 m
1 m
4 m^3

Extra Example 3

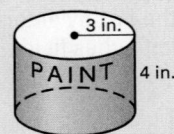

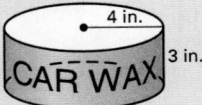

a. How do the radius and height of the paint can compare to the radius and height of the can of car wax? **The height of the paint can is the same as the radius of the can of car wax and the height of the can of car wax is the same as the radius of the paint can.**

b. How many times greater is the volume of the can of car wax than the volume of the paint can? **The volume of the can of car wax is $\frac{4}{3}$ times the volume of the paint can.**

✓ Concept Check
How do you find the volume of a prism or a cylinder? **Multiply the area of the base by the height.**

🐞 Daily Puzzler
Three faces of a box have the areas shown. What is the volume of the box? **1536 cubic units**

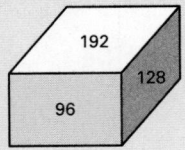

VOLUME OF A CYLINDER

Words Volume = (area of base)(height)

Symbols $V = Bh$
$= \pi r^2 h$

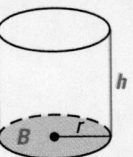

EXAMPLE **3** **Compare Volumes of Cylinders**

a. How do the radius and height of the mug compare to the radius and height of the dog bowl?

b. How many times greater is the volume of the bowl than the volume of the mug?

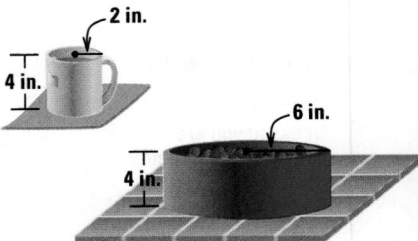

Solution

a. The radius of the mug is 2 inches and the radius of the dog bowl is 6 inches. The radius of the bowl is three times the radius of the mug. The height of the mug is the same as the height of the bowl.

b.

Volume of mug		**Volume of dog bowl**
$V = \pi r^2 h$	Write the formula for volume.	$V = \pi r^2 h$
$= \pi(2^2)(4)$	Substitute for r and for h.	$= \pi(6^2)(4)$
$= 16\pi$	Simplify.	$= 144\pi$

To compare the volume of the bowl to the volume of the mug, divide the volume of the bowl by the volume of the mug.

$$\frac{\text{Volume of bowl}}{\text{Volume of mug}} = \frac{144\pi}{16\pi} = 9$$

ANSWER ▶ The volume of the bowl is nine times the volume of the mug.

Checkpoint ✓ **Find Volume of Cylinders**

Find the volume of the cylinder. Round your answer to the nearest whole number.

4. 38 ft^3
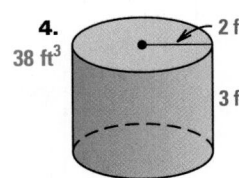

5. 16 in.^3
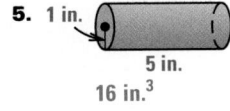

6. 126 m^3
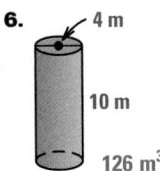

9.4 Exercises

Guided Practice

Vocabulary Check — Based upon the units, tell whether the number is a measure of *surface area* or *volume*.

1. 5 ft^3 volume

2. 7 yd^2 surface area

3. 3 m^2 surface area

4. 2 cm^3 volume

Skill Check — **Candles** Find the volume of the candle.

5.
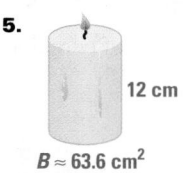
12 cm
$B \approx 63.6 \text{ cm}^2$
about 763.2 cm^3

6.
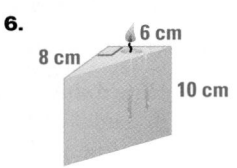
6 cm
8 cm
10 cm
240 cm^3

7.
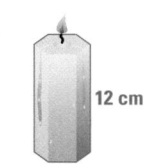
12 cm
$B \approx 23.4 \text{ cm}^2$
about 280.8 cm^3

Practice and Applications

Extra Practice
See p. 692.

Using Unit Cubes Find the number of unit cubes that will fit in the box. Explain your reasoning.

8.
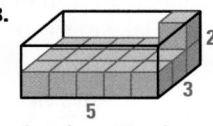
2
3
5
30 unit cubes; 15 unit cubes per layer and 2 layers in all

9.
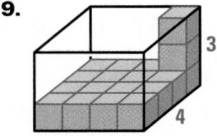
3
4
4
48 unit cubes; 16 unit cubes per layer and 3 layers in all

10.
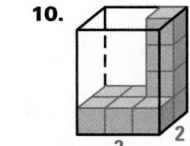
4
2
3
24 unit cubes; 6 unit cubes per layer and 4 layers in all

Volume of a Prism Find the volume of the prism.

11.
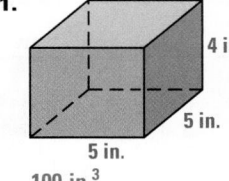
4 in.
5 in.
5 in.
100 in.^3

12.
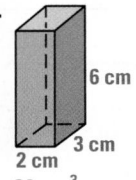
6 cm
2 cm
3 cm
36 cm^3

13.
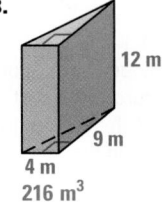
12 m
9 m
4 m
216 m^3

Volume of a Cube In Exercises 14–16, you are given the length of each side of a cube. Sketch the cube and find its volume.

14–16. Check sketches.

14. 3 meters
27 m^3

15. 7 feet
343 ft^3

16. 10 centimeters
1000 cm^3

Visualize It! In Exercises 17 and 18, make a sketch of the solid. Then find its volume. 17, 18. See margin.

17. A prism has a square base with 4 meter sides and a height of 7 meters.

18. A prism has a rectangular base that is 3 feet by 6 feet and a height of 8 feet.

Homework Help
Example 1: Exs. 8–10
Example 2: Exs. 11–18
Example 3: Exs. 27–40

9.4 Volume of Prisms and Cylinders   **503**

③ Apply

Assignment Guide

BASIC
Day 1: pp. 503–507 Exs. 8–24, 43–46, 51–57
Day 2: pp. 503–507 Exs. 27–38, 42, 50, 58–63

AVERAGE
Day 1: pp. 503–507 Exs. 8–26, 44–46, 51–54
Day 2: pp. 503–507 Exs. 27–43, 50, 55–63

ADVANCED
Day 1: pp. 503–507 Exs. 11–21 odd, 22–26, 44–46, 51–60
Day 2: pp. 503–507 Exs. 27–43, 47–49*, 50; EC: classzone.com

BLOCK SCHEDULE
pp. 503–507 Exs. 8–26, 44–46, 51–54 (with 9.3)
pp. 503–507 Exs. 27–43, 50, 55–63 (with 9.5)

Extra Practice
• Student Edition, p. 692
• Chapter 9 Resource Book, pp. 39–40

Homework Check
To quickly check student understanding of key concepts, go over the following exercises:

Basic: 8, 11, 19, 27, 30
Average: 9, 12, 17, 20, 31
Advanced: 15, 18, 21, 32, 42

✗ Common Error
In Exercise 13, watch for students who calculate volume by finding $4 \times 9 \times 12$. Remind them that the base is a triangle, so the area of the base is $\frac{1}{2} \times 4 \times 9$.

17.
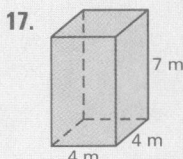
7 m
4 m
4 m
112 m^3

18.
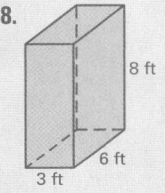
8 ft
3 ft
6 ft
144 m^3

HOMEWORK HELP
Extra help with problem
solving in Exs. 19–21 is
at classzone.com

Finding Volume Find the volume of the combined prisms.

19.
3 ft
8 ft
2 ft
2 ft 5 ft
78 ft³

20.
6 in.
2 in.
1 in.
4 in.
2 in. 1 in.
20 in.³

21.
2 m
5 m
4 m
10 m
7 m
350 m³

Shopping In Exercises 22–24, use the information about the sizes of
the cereal boxes shown below.

22. Find the volume of each box of cereal.
smaller box: 160 in.³; bigger box: 640 in.³

23. How many small boxes
of cereal do you have to
buy to equal the amount
of cereal in a large box? 4 boxes

24. Which box gives you the
most cereal for your
money? Explain.
The bigger box. Four times the volume of
the smaller box should cost $8, not $6.

$2.00
Cereal
10 in.
8 in. 2 in.

$6.00
Cereal
16 in.
10 in. 4 in.

Link to
Civil Engineering

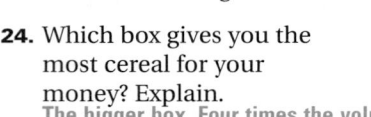

SOO LOCKS The first locks
system between Lake Superior
and Lake Huron was built
around 1797. Today, four locks
are available in the Soo Locks
system.

Soo Locks In Exercises 25 and 26, use the information below.
Lake Superior is about 22 feet higher than Lake Huron. In order for
ships to safely pass from one lake to the other, they must go through
one of the four Soo Locks.

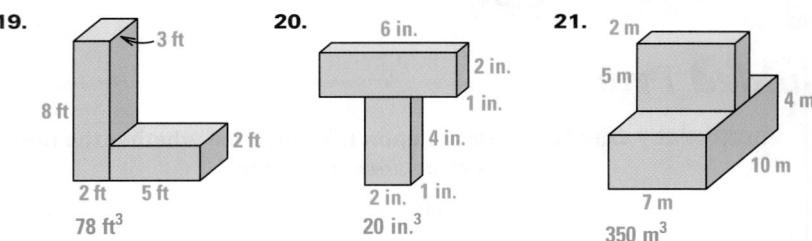

Top View
Lake
Huron
Lake
Superior
lower gates   upper gates

Side View
80 ft
22 ft
Lake
Huron
800 ft
Lake
Superior

Not drawn to scale

25. Water is added to the MacArthur Lock until the height is increased by
22 feet. To find the amount of water added to the lock, find the volume
of a rectangular prism with a length of 800 feet, a width of 80 feet, and a
height of 22 feet. 1,408,000 ft³

26. How many gallons of water are added to the MacArthur Lock
to raise the ship to the level of Lake Superior? Use the fact that
1 ft³ ≈ 7.5 gal. 10,560,000 gal

Volume of a Cylinder Find the volume of the cylinder. Round your answer to the nearest whole number.

27.
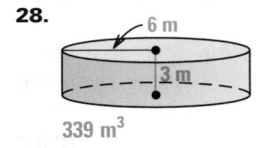
4 in.
9 in.
452 in³

28.
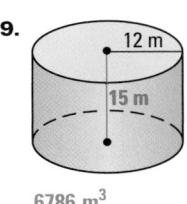
6 m
3 m
339 m³

29.
12 m
15 m
6786 m³

Swimming Pools In Exercises 30–32, find the volume of the pool. Round your answer to the nearest whole number. Then compare the volumes of the pools to answer Exercise 33.

30.
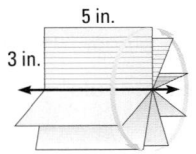
20 ft
4 ft
1257 ft³

31.
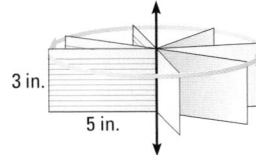
24 ft
4 ft
1810 ft³

32.
15 ft
3 ft
530 ft³

33. Which pool above requires the least amount of water to fill it?
the pool in Exercise 32

Visualize It! In Exercises 34 and 35, use the information below.
Suppose that a 3-inch by 5-inch index card is rotated around a horizontal line and a vertical line to produce two different solids.

5 in.
3 in.

3 in.
5 in.

34. Find the volume of each solid.

35. Which solid has a greater volume? Explain your reasoning.

Aquariums In Exercises 36 and 37, use the information below.
The Giant Ocean Tank at the New England Aquarium is a cylinder that is 23 feet deep and 40 feet in diameter as shown.

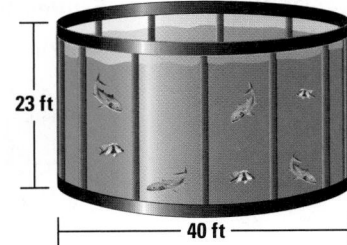
23 ft
40 ft

36. Find the volume of the tank.
about 28,903 ft³

37. How many gallons of water are needed to fill the tank?
(1 ft³ ≈ 7.5 gal)   about 216,773 gal

38. Personal Aquariums To avoid overcrowding in a personal aquarium, you should buy one fish for every gallon of water (231 in.³ ≈ 1 gal). About how many fish should be in an aquarium that is a rectangular prism measuring 20 inches wide, 10 inches long, and is filled with water to a height of 11 inches? 9 or 10 fish

34. horizontal line: about 141 in.³;
   vertical line: about 236 in.³

35. The solid on the right; the solid with the vertical line of rotation has a volume that is almost twice the volume of the solid with the horizontal line of rotation.

Link to
Careers

AQUARIUM DIVER In addition to feeding and taking care of the fish and the plants in an aquarium, divers make sure that the tank does not get too crowded.

Teaching Tip
For students who have difficulty with the visualization required in Exercises 34 and 35, point out that the 3-inch dimension of the index card models the radius in one cylinder but the height in the other. The 5-inch dimension of the card models the height and then the radius in the cylinders, respectively.

9.4 *Volume of Prisms and Cylinders*   505

⚖ You be the Judge In Exercises 39 and 40, use the cartons shown.

39. Find the volume of each carton of ice cream.
    about 785 cm³; about 6283 cm³

40. Terry assumes that because the dimensions doubled, the jumbo carton contains twice as much ice cream as the regular carton. Is Terry right? Explain your reasoning.
    No; the answer to Exercise 39 shows that the jumbo carton contains about 8 times as much ice cream as the regular carton.

EXAMPLE Find Volume

Find the volume of the passenger car of the Space Spiral at Cedar Point Amusement Park in Sandusky, Ohio.

Solution

The passenger car is a cylinder with a "hole" in it. To find the volume, subtract the volume of the hole from the volume of the larger cylinder.

$$\text{Volume of larger cylinder} = \pi r^2 h$$
$$= \pi(10^2)(14)$$
$$\approx 4398$$

$$\text{Volume of "hole"} = \pi r^2 h$$
$$= \pi(4^2)(14)$$
$$\approx 704$$

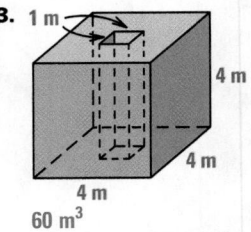

ANSWER ▶ The volume of the passenger car is about $4398 - 704 = 3694$ cubic feet.

Student Help
CLASSZONE.COM

HOMEWORK HELP
Extra help with problem solving in Exs. 41–43 is at classzone.com

Finding Volume In Exercises 41–43, find the volume of the solid.

41.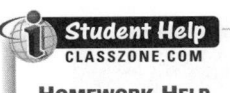
    2 in., 1 in., 8 in., 6 in., 2 in.
    92 in.³

42.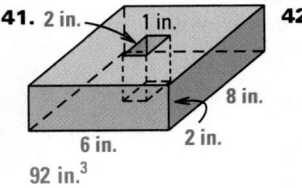
    8 ft, 3 ft, 10 ft
    about 1728 ft³

43.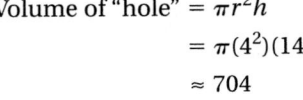
    1 m, 4 m, 4 m, 4 m
    60 m³

🅧🅨 Using Algebra Write an expression for the volume of the solid in terms of x.

44. 12x
    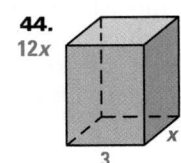
    4, x, 3

45. $7x^2$
    x, 2x, 7

46. $15x^3$
    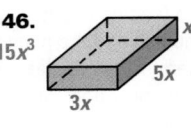
    x, 5x, 3x

Challenge In Exercises 47–49, find the missing dimension(s). If necessary, round your answer to the nearest whole number.

47. A cylinder has a volume of 100.48 cubic inches and a diameter of 4 inches. Find the height of the cylinder. 8 in.

48. A cylinder has a volume of 1538.6 cubic feet and a height of 10 feet. Find the radius of the cylinder. 7 ft

49. The length of a rectangular prism is twice its width. The height of the prism equals the width. Find the dimensions of the prism, given that the volume is 54 cubic inches. 6 in. long, 3 in. wide, 3 in. high

Standardized Test Practice

50. Multiple Choice What is the approximate volume of the cylinder shown at the right? C

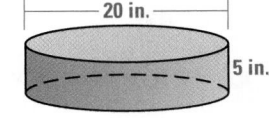

Ⓐ 100 in.³ Ⓑ 785 in.³

Ⓒ 1570 in.³ Ⓓ 6280 in.³

51. Multiple Choice The volume of the prism shown at the right is 168 cubic feet. What is the height of the prism? H

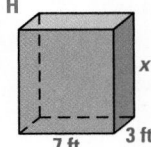

Ⓕ 6 feet Ⓖ 7 feet

Ⓗ 8 feet Ⓙ 9 feet

Mixed Review

Using the Pythagorean Theorem Find the unknown side length. Round your answer to the nearest tenth. *(Lesson 4.4)*

52. 9.7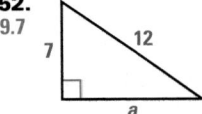

53. 4.1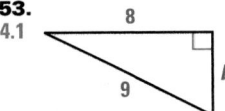

54. 11.3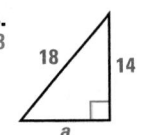

Surface Area Find the surface area of the solid. If necessary, round your answer to the nearest whole number. *(Lessons 9.2, 9.3)*

55.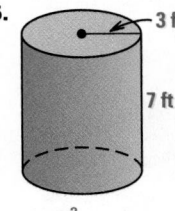

188 ft²

56.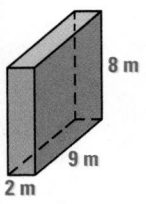

212 m²

57.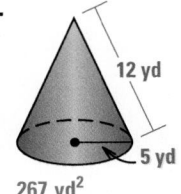

267 yd²

Algebra Skills

Solving Equations Solve the equation. *(Skills Review, p. 672)*

58. $x - 7 = 0$ 7

59. $m - 1 = -12$ −11

60. $10 + c = -3$ −13

61. $\frac{3}{4}b = 24$ 32

62. $-14d = 2$ $-\frac{1}{7}$

63. $6n = 102$ 17

Assessment Resources

The Mini-Quiz below is also available on blackline (*Chapter 9 Resource Book*, p. 46) and on transparency. For more assessment resources, see:
• Chapter 9 Resource Book
• Standardized Test Practice
• Test and Practice Generator

Mini-Quiz

Find the volume of the prism.

1.
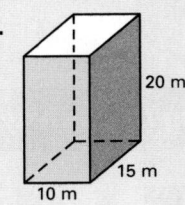

3000 m³

2. 9 ft 12 ft
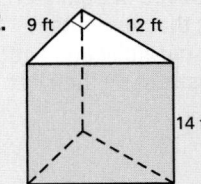
14 ft

756 ft³

3. Find the volume of the cylinder. Round your answer to the nearest whole number.

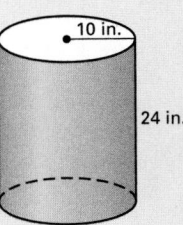

10 in.

24 in.

7540 in.³

4. Find the volume of a cube with side length 14 meters. 2744 m³

Students use nets to create solids to compare the volumes of prisms and pyramids.

Materials
See the margin of the student page.

LINK TO LESSON
In Example 1 on page 510, students find the volume of a pyramid using the volume formula derived in this activity.

2 Managing the Activity

Tips for Success
In Step 1, point out that the net on the right is composed of equilateral triangles. Stress that students should measure carefully and accurately when constructing their net.

Activity 9.5 Investigating Volume

Question

How does the volume of a pyramid relate to the volume of a prism with the same base?

Materials
- poster board
- ruler
- scissors
- tape
- unpopped popcorn, uncooked rice, or dried beans

Explore

1 Use a ruler to draw the two nets shown below on poster board. Be sure to draw the dashed lines on the net as shown.

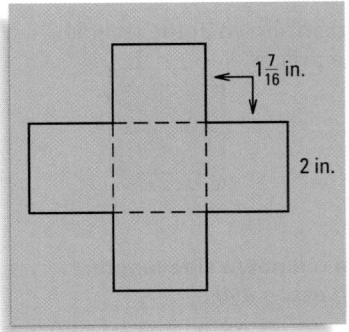

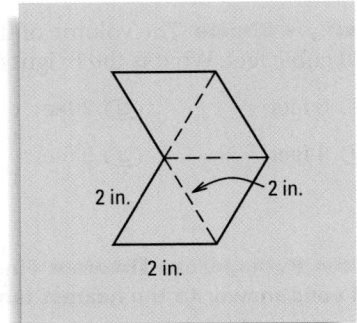

2 Cut out the nets.

3 Fold along the dashed lines to form an open prism and an open pyramid.

4 Tape each solid to hold it in place, making sure that the edges do not overlap.

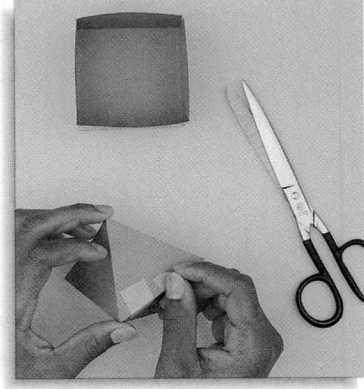

KEY DISCOVERY
The volume of a pyramid is one-third the volume of a prism with the same base and height.

Activity Assessment
Use Exercises 5 and 6 to assess student understanding.

Think About It

1. Compare the area of the base of the pyramid to the area of the base of the prism. Fitting the pyramid inside the prism will help. What do you notice? The base areas are the same.

2. Compare the heights of the solids. What do you notice?
   The heights are the same.

3. Which solid do you think has the greater volume? Why?
   The rectangular prism; the pyramid fits inside the prism with extra room around the pyramid.

Explore

5 Fill the pyramid to the top with unpopped popcorn, uncooked rice, or dried beans. Pour the contents of the pyramid into the prism.

6 Repeat Step 5 until the prism is full. Count the number of times you empty the contents of the pyramid into the prism.

Think About It

4. How many times did you empty the contents of the pyramid into the prism? What does this tell you about the volume of the pyramid and the prism? 3 times; the volume of the prism is 3 times the volume of the pyramid.

5. Complete each sentence with the appropriate number.

   Volume of the prism = __?__ × Volume of the pyramid 3

   Volume of the pyramid = __?__ × Volume of the prism $\frac{1}{3}$

6. Use your results from Exercise 5 to write a formula for the volume of a pyramid.

   Volume of a prism: $V = Bh$

   Volume of a pyramid: $V =$ __?__ Bh $\frac{1}{3}$

Activity 509

Pacing
Suggested Number of Days

Basic: 2 days
Average: 2 days
Advanced: 2 days
Block Schedule: 0.5 block with 9.4
　　　　　　　　　0.5 block with 9.6

Teaching Resources

📄 *Blacklines*
(See page 470B.)

📀 *Transparencies*
• Warm-Up with Quiz
• Answers

💻 *Technology*
• Electronic Teacher Tools
• Test and Practice Generator
• Online Lesson Planner
• Internet Support

② Teach

Content and Teaching Strategies
For background information on geometric concepts and teaching strategies related to this lesson, see pages 470E and 470F in this Teacher's Edition.

Tips for New Teachers
Students sometimes think that the volume of a pyramid or a cone is $\frac{1}{2}$ the volume of a prism or a cylinder, respectively, when the solids have the same height. It is a matter of their visual perception when they look at models or diagrams. Assure them that the relationship is $\frac{1}{3}$. This can be demonstrated by using hollow models and filling them with sand (sugar, salt, or water may work as well). Practice before you attempt the demonstration in class. See the Tips for New Teachers on pp. 1–2 of the *Chapter 9 Resource Book* for additional notes about Lesson 9.5.

510

9.5 Volume of Pyramids and Cones

Goal
Find the volumes of pyramids and cones.

Key Words
• **pyramid** p. 491
• **cone** p. 493
• **volume** p. 500

In the puzzle below, you can see that the square prism can be made using three congruent pyramids. The volume of each pyramid is one-third the volume of the prism.

Volume Puzzle

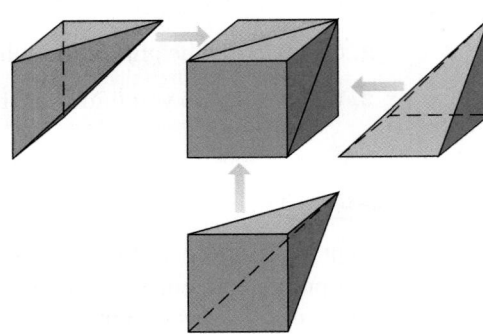

VOLUME OF A PYRAMID

Words Volume = $\frac{1}{3}$(area of base)(height)

Symbols $V = \frac{1}{3}Bh$

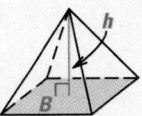

Student Help

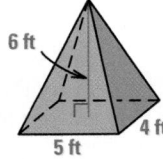

STUDY TIP
In Example 1(b), 6 m is the height of the triangular base and 8 m is the height of the pyramid.

EXAMPLE 1 Find the Volume of a Pyramid

Find the volume of the pyramid.

a.

6 ft
4 ft
5 ft

b.
8 m
6 m
7 m

Solution

a. $V = \frac{1}{3}Bh$　　Write the formula for volume.

$= \frac{1}{3}(5 \cdot 4)(6)$　　Substitute.

$= 40$　　Simplify.

ANSWER ▶ The volume is 40 cubic feet.

b. $V = \frac{1}{3}Bh$

$= \frac{1}{3}\left(\frac{1}{2} \cdot 7 \cdot 6\right)(8)$

$= 56$

ANSWER ▶ The volume is 56 cubic meters.

Find the volume of the pyramid.

1.

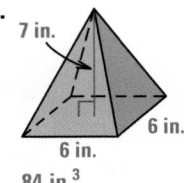
7 in.
6 in.
6 in.
84 in.^{3}

2.

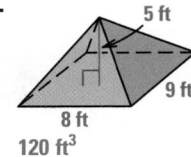
5 ft
9 ft
8 ft
120 ft^3

3.

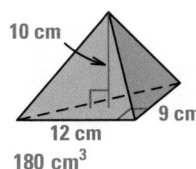
10 cm
9 cm
12 cm
180 cm^3

Student Help

LOOK BACK
For help with relating
the volume of a pyramid
to the volume of a
prism, see p. 508.

Volume of a Cone The volume of a cone is related to the volume
of a cylinder in the same way that the volume of a pyramid is related
to the volume of a prism.

$$\frac{1}{3}Bh \quad + \quad \frac{1}{3}Bh \quad + \quad \frac{1}{3}Bh \quad = \quad Bh$$

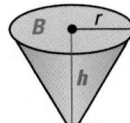 + 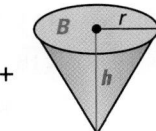 + 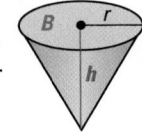 =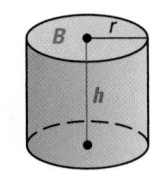

VOLUME OF A CONE

Words Volume = $\frac{1}{3}$(area of base)(height)

Symbols $V = \frac{1}{3}Bh$

$= \frac{1}{3}\pi r^2 h$

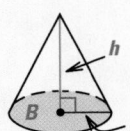

EXAMPLE **2** **Find the Volume of a Cone**

Find the volume of the cone. Round your
answer to the nearest whole number.

12 cm
8 cm

Solution
The radius of the cone is $r = 8$ cm.
The height of the cone is $h = 12$ cm.

$V = \frac{1}{3}\pi r^2 h$ Write the formula for volume of a cone.

$= \frac{1}{3}\pi(8^2)(12)$ Substitute 8 for r and 12 for h.

≈ 804 Multiply.

ANSWER The volume is about 804 cubic centimeters.

 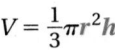
Extra Example 1
Find the volume of the pyramid.

a.

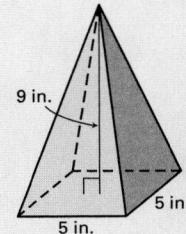
9 in.
5 in.
5 in.

75 in.^{3}

b.

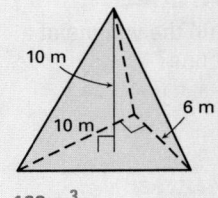
10 m
10 m
6 m

100 m^3

Extra Example 2
Find the volume of the cone. Round
your answer to the nearest whole
number. **about 236 ft^3**

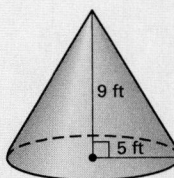
9 ft
5 ft

What is the volume of the cone?

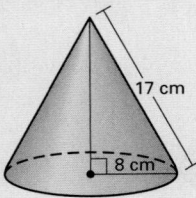

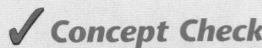

about 1005 cm³

✓ Concept Check

How do you find the volume of a pyramid or a cone? **Multiply one third of the area of the base by the height.**

🕮 Daily Puzzler

A cone with radius 3 inches has a volume of 103.7 cubic inches. Find the height of the cone to the nearest inch. **11 in.**

 **Using Algebra**

EXAMPLE 3 Find the Volume of a Cone

What is the volume of the cone shown at the right?

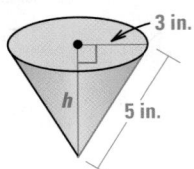

Solution

You are given the slant height of the cone. You need to find the height of the cone before you can find the volume.

❶ Find the height.

$(\text{leg})^2 + (\text{leg})^2 = (\text{hypotenuse})^2$	Use the Pythagorean Theorem.
$3^2 + h^2 = 5^2$	Substitute.
$9 + h^2 = 25$	Simplify.
$h^2 = 25 - 9$	Subtract 9 from each side.
$h^2 = 16$	Simplify.
$h = \sqrt{16}$	Take the positive square root.
$h = 4$	Simplify.

❷ Find the volume.

$V = \frac{1}{3}\pi r^2 h$	Write the formula for volume.
$= \frac{1}{3}\pi(3^2)(4)$	Substitute 3 for r and 4 for h.
≈ 38	Multiply.

ANSWER ▶ The volume is about 38 cubic inches.

Student Help

LOOK BACK
For help with the Pythagorean Theorem, see p. 192.

Checkpoint ✓ Find the Volume of a Cone

Find the volume of the cone. Round your answer to the nearest whole number.

4.

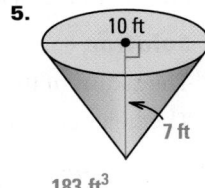

236 in.³

5.

10 ft

7 ft

183 ft³

6.

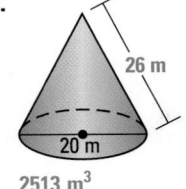

2513 m³

7. Find the volume of a cone with a height of 6 inches and a diameter of 8 inches. **101 in.³**

8. Find the volume of a cone with a slant height of 17 feet and a diameter of 16 feet. **1005 ft³**

9.5 Exercises

Guided Practice

Vocabulary Check

Match the solid with its volume formula. Use each formula once.

A. Pyramid   **B.** Cone   **C.** Prism   **D.** Cylinder

1. $V = \pi r^2 h$ D   **2.** $V = Bh$ C   **3.** $V = \frac{1}{3}\pi r^2 h$ B   **4.** $V = \frac{1}{3}Bh$ A

Skill Check

Find the volume of the solid. If necessary, round your answer to the nearest whole number.

5.
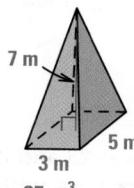
7 m
5 m
3 m
35 m³

6.
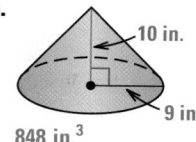
10 in.
9 in.
848 in.³

7.
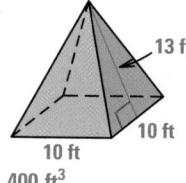
13 ft
10 ft
10 ft
400 ft³

8. Find the volume of a cone with a slant height of 15 inches and a radius of 9 inches. Leave your answer in terms of π.   324π in.³

Practice and Applications

Extra Practice

See p. 692.

Find Base Areas Find the area of the base of the solid.

9.
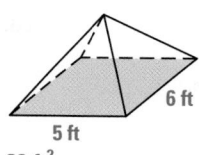
6 ft
5 ft
30 ft²

10.
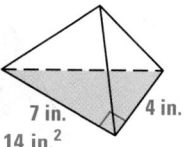
7 in.
4 in.
14 in.²

11.
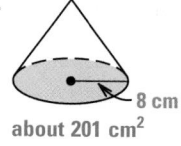
8 cm
about 201 cm²

Volume of a Pyramid Find the volume of the pyramid.

12.
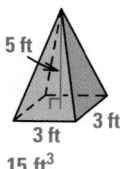
5 ft
3 ft
3 ft
15 ft³

13.
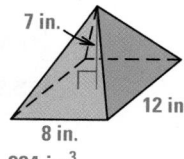
7 in.
12 in.
8 in.
224 in.³

14.
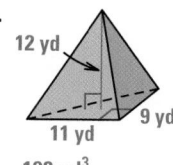
12 yd
9 yd
11 yd
198 yd³

Volume of a Pyramid Find the volume.

15. Find the volume of a pyramid with a base area of 48 square feet and a height of 5 feet.   80 ft³

16. Find the volume of a pyramid with a height of 3 inches and a square base with side lengths of 4 inches.   16 in.³

Homework Help

Example 1: Exs. 12–16, 20–25
Example 2: Exs. 17–19, 21, 26
Example 3: Exs. 30–32

③ Apply

Assignment Guide

BASIC
Day 1: pp. 513–516 Exs. 9–16, 20, 22–25, 45–51
Day 2: pp. 513–516 Exs. 17–19, 21, 26, 30–38, 52–57

AVERAGE
Day 1: pp. 513–516 Exs. 9–16, 20, 22–25, 27, 29, 45–57 odd
Day 2: pp. 513–516 Exs. 17–19, 21, 26, 28, 30–38, 46–56 even

ADVANCED
Day 1: pp. 513–516 Exs. 12–16, 20, 22–25, 27, 45–51
Day 2: pp. 513–516 Exs. 17–19, 28–32, 36–44, 54–57;
EC: TE p. 470D*, classzone.com

BLOCK SCHEDULE
pp. 513–516 Exs. 9–16, 20, 22–25, 27, 29, 45–57 odd (with 9.4)
pp. 513–516 Exs. 17–19, 21, 26, 28, 30–38, 46–56 even (with 9.6)

Extra Practice

• Student Edition, p. 692
• Chapter 9 Resource Book, pp. 47–48

Homework Check

To quickly check student understanding of key concepts, go over the following exercises:

Basic: 12, 17, 20, 30, 33
Average: 13, 18, 22, 31, 34
Advanced: 15, 19, 28, 29, 39

Volume of a Cone Find the volume of the cone. Round your answer to the nearest whole number.

17.

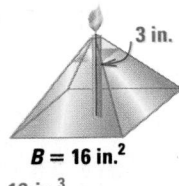

10 yd

3 yd

94 yd³

18.

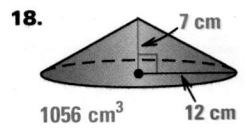

7 cm

1056 cm³   12 cm

19.

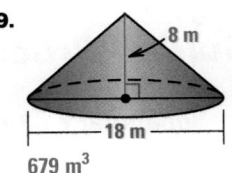

8 m

18 m

679 m³

Finding Volume Find the volume of the object. Round your answer to the nearest whole number.

20.

3 in.

B = 16 in.²

16 in.³

21.

5 cm

8 cm

209 cm³

22.

h = 144 m

B = 46,535 m²

2,233,680 m³

Logical Reasoning In Exercises 23 and 24, use a pyramid that has a height of 8 feet and a square base with a side length of 6 feet.

23. How does the volume of the pyramid change if the base stays the same and the height is doubled? The volume is doubled.

24. How does the volume of the pyramid change if the height stays the same and the side length of the base is doubled?
The volume is multiplied by 4.

Error Analysis In Exercises 25 and 26, explain the student's error and correct it.

25. The student used the slant height instead of the height of the pyramid. The volume should be $\frac{1}{3}(10 \cdot 10)(12) = 400$ in.³

26. The student used the diameter instead of the radius of the base. The volume should be 100.53 ft³.

25.

13 in.

10 in.

10 in.

$V = \frac{1}{3}(10 \cdot 10)(13)$

$= \frac{1}{3}(100)(7)$

≈ 433.33 in.³

26.

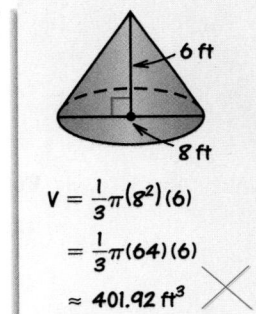

6 ft

8 ft

$V = \frac{1}{3}\pi(8^2)(6)$

$= \frac{1}{3}\pi(64)(6)$

≈ 401.92 ft³

Student Help
CLASSZONE.COM

HOMEWORK HELP
Extra help with problem solving in Exs. 27–29 is at classzone.com

Finding Dimensions Find the missing dimension.

27. A pyramid has a volume of 20 cubic inches and the area of the base is 15 square inches. What is the height of the pyramid? 4 in.

28. A cone has a volume of 8π cubic feet and a height of 6 feet. What is the radius of the base? 2 ft

29. A pyramid with a square base has a volume of 120 cubic meters and a height of 10 meters. What is a side length of the base? 6 m

Finding Volume with Slant Height Find the volume of the solid. If necessary, round your answer to the nearest whole number.

30.

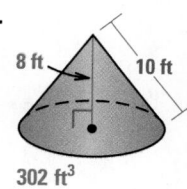

8 ft   10 ft

302 ft³

31.

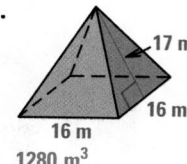

17 m

16 m

16 m

1280 m³

32. 14 in.

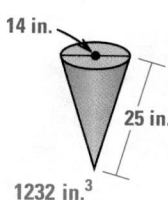

25 in.

1232 in.³

Student Help

STUDY TIP
Recall that the radius of the sector that forms a cone is the slant height of the cone. This is shown on p. 493.

Using Nets In Exercises 33–35, use the net to sketch the solid. Then find the volume of the solid. Round your answer to the nearest whole number. 33–35. See margin.

33.

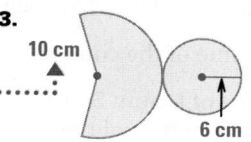

10 cm

6 cm

34.

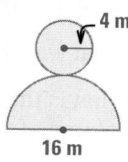

4 m

16 m

35.

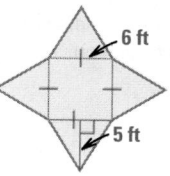

6 ft

5 ft

Popcorn A movie theater serves a small size of popcorn in a conical container and a large size of popcorn in a cylindrical container.

36. What is the volume of the small container? What is the volume of the large container?
about 57 in.³; about 170 in.³

37. How many small containers of popcorn do you have to buy to equal the amount of popcorn in a large container? 3

38. Which container gives you more popcorn for your money? Explain your reasoning.

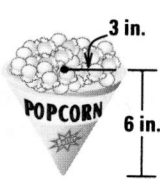

3 in.

POPCORN

6 in.

$2.00

3 in.

Butter

POPCORN

6 in.

$4.00

38. The large container; *Sample answer:* You need to buy three small containers for $6 to equal the amount of popcorn in the large container, which costs $4.

Student Help

CLASSZONE.COM

HOMEWORK HELP
Extra help with problem solving in Exs. 39–41 is at classzone.com

Pet Feeder In Exercises 39–41, use the diagram of the automatic pet feeder.

39. Calculate the amount of food that can be placed in the feeder. (*Hint:* Add the volume of the cylinder and the volume of the cone.)
about 173.4 in.³

40. If a cat eats 1 cup of food each day, how much food does the cat eat in five days? Express your answer in cubic inches. (1 cup ≈ 14.4 in.³) 72 in.³

41. Will the feeder hold enough food for the five days described in Exercise 40? Explain.
Yes; the feeder holds about 173.4 in.³ and only 72 in.³ are needed for five days.

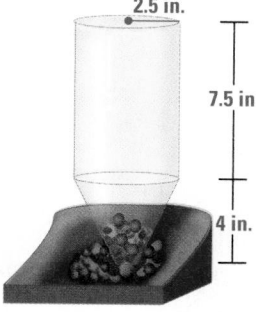

2.5 in.

7.5 in.

4 in.

33.

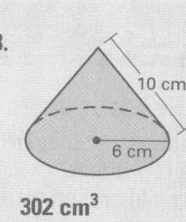

10 cm

6 cm

302 cm³

34.

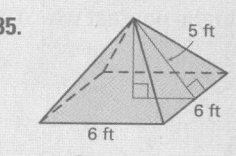

8 m

4 m

116 m³

35.

5 ft

6 ft

6 ft

48 ft³

9.5 *Volume of Pyramids and Cones* **515**

Assessment Resources

The Mini-Quiz below is also available on blackline (*Chapter 9 Resource Book*, p. 55) and on transparency. For more assessment resources, see:
- Chapter 9 Resource Book
- Standardized Test Practice
- Test and Practice Generator

Mini-Quiz

Find the volume of the solid. Round your answer to the nearest whole number.

1.

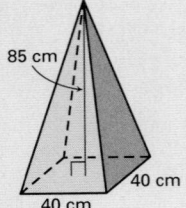

45,333 cm³

2.

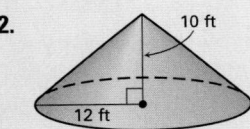

1508 ft³

3. a cone with height 12 inches and a base of radius 7 inches 616 in.³

4. A pyramid has a volume of 72 cubic feet and the area of the base is 36 square feet. What is the height of the pyramid? 6 ft

---

Link to
Career

GEOLOGISTS collect and interpret data about volcanoes to help predict when a volcano will erupt.

Standardized Test Practice

Mixed Review

Algebra Skills

Volcanoes In Exercises 42–44, use the information below.

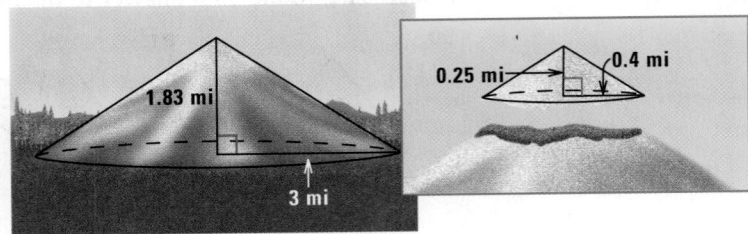

Before 1980, Mount St. Helens was a cone-shaped volcano.

In 1980, Mount St. Helens erupted, destroying the cone-shaped tip.

42. What was the volume of Mount St. Helens before 1980? about 17.25 mi³

43. What was the volume of the cone-shaped tip that was destroyed? about 0.04 mi³

44. What is the volume of Mount St. Helens today? (*Hint*: Subtract the volume of the tip from the volume before 1980.) about 17.21 mi³

45. Multiple Choice A pyramid has a height of 9 yards and a volume of 96 cubic yards. Which of these are possible dimensions for its rectangular base? **A**

(A) 4 yards by 8 yards

(B) 2 yards by 8 yards

(C) 6 yards by 7 yards

(D) 5 yards by 8 yards

Finding Circumference and Area Find the circumference and the area of the circle. Round your answer to the nearest whole number. (*Lesson 8.7*)

46.

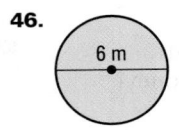

6 m

$C \approx 19$ m; $A \approx 28$ m²

47.

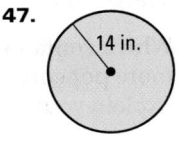

14 in.

$C \approx 88$ in.; $A \approx 616$ in.²

48.

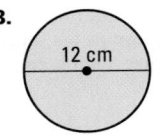

12 cm

$C \approx 38$ cm; $A \approx 113$ cm²

Surface Area Find the surface area of the figure. If necessary, round your answer to the nearest whole number. (*Lessons 9.2, 9.3*)

49.

7 in.

8 in.

5 in.

262 in.²

50.

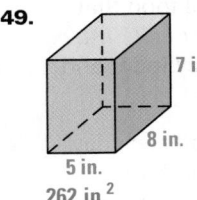

5 ft

9 ft

440 ft²

51.

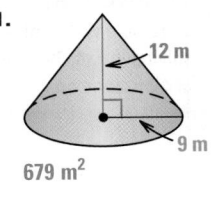

12 m

9 m

679 m²

Slope Plot the points and draw a line through them. Then tell if the slope is *positive*, *negative*, *zero*, or *undefined*. (*Skills Review, p. 665*)

52–57. Check graphs.

52. (0, 0) and (4, 3) positive

53. (5, 5) and (1, −2) positive

54. (−3, 2) and (6, 2) zero

55. (−2, −2) and (0, 4) positive

56. (−4, −1) and (−4, 7) undefined

57. (−3, 5) and (4, −1) negative

 **9.6** # Surface Area and Volume of Spheres

Goal
Find surface areas and volumes of spheres.

Key Words
• sphere
• hemisphere

A globe is an example of a *sphere*. A **sphere** is the set of all points in space that are the same distance from a point, the center of the sphere.

A geometric plane passing through the center of a sphere divides it into two **hemispheres**. The globe is divided into the Northern Hemisphere and the Southern Hemisphere.

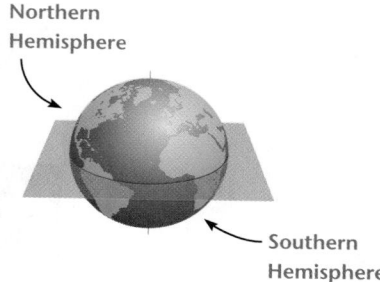

Northern Hemisphere

Southern Hemisphere

The globe is divided into two hemispheres.

SURFACE AREA OF A SPHERE

Words Surface area = $4\pi(\text{radius})^2$

Symbols $S = 4\pi r^2$

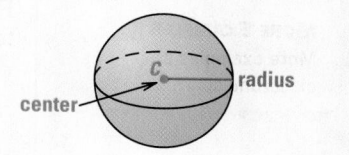

center

radius

C

Visualize It!

To sketch a sphere, draw a circle and its center. Then draw an oval to give the sphere dimension.

EXAMPLE 1 Find the Surface Area of a Sphere

Find the surface area of the sphere. Round your answer to the nearest whole number.

a.

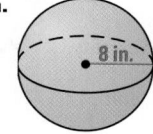

8 in.

b.

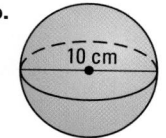

10 cm

Solution

a. The radius is 8 inches, so $r = 8$.

$$S = 4\pi r^2$$
$$= 4 \cdot \pi \cdot 8^2$$
$$\approx 804$$

The surface area is about 804 square inches.

b. The diameter is 10 cm, so the radius is $\frac{10}{2} = 5$. So, $r = 5$.

$$S = 4\pi r^2$$
$$= 4 \cdot \pi \cdot 5^2$$
$$\approx 314$$

The surface area is about 314 square centimeters.

9.6 *Surface Area and Volume of Spheres* **517**

---

① Plan

Pacing
Suggested Number of Days

Basic: 1 day
Average: 1 day
Advanced: 1 day
Block Schedule: 0.5 block with 9.5

Teaching Resources

📄 **Blacklines**
(See page 470B.)

🖨 **Transparencies**
• Warm-Up with Quiz
• Answers

💻 **Technology**
• Electronic Teacher Tools
• Test and Practice Generator
• Online Lesson Planner
• Internet Support

② Teach

Content and Teaching Strategies
For background information on geometric concepts and teaching strategies related to this lesson, see pages 470E and 470F in this Teacher's Edition.

Tips for New Teachers
Show students how to draw a sphere by first drawing a circle and then adding an ellipse. Similar to the diagrams on this page, half of the ellipse should be solid and half should be dashed to create a three-dimensional perspective. See the Tips for New Teachers on pp. 1–2 of the *Chapter 9 Resource Book* for additional notes about Lesson 9.6.

Extra Example 1
See next page.

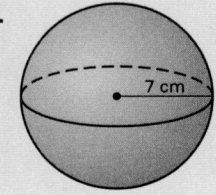
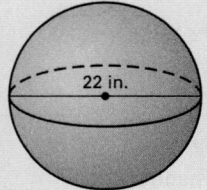
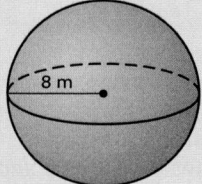
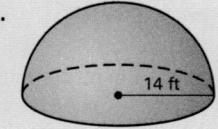
 **Checkpoint** ✓ *Find the Surface Area of a Sphere*

Find the surface area of the sphere. Round your answer to the nearest whole number.

1.

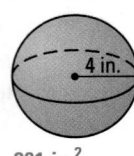

201 in.²

2.

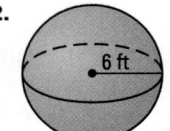

452 ft²

3.

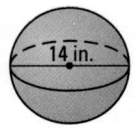

616 in.²

VOLUME OF A SPHERE

Words Volume = $\frac{4}{3}\pi(\text{radius})^3$

Symbols $V = \frac{4}{3}\pi r^3$

EXAMPLE 2 **Find the Volume of a Sphere**

Find the volume of the sphere or hemisphere. Round your answer to the nearest whole number.

a.

b.

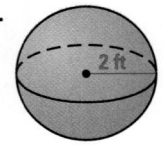

Solution

a. $V = \frac{4}{3}\pi r^3$ Write the formula for volume of a sphere.

$= \frac{4}{3} \cdot \pi \cdot 2^3$ Substitute 2 for r.

$= \frac{32}{3}\pi$ Simplify. $2^3 = 2 \cdot 2 \cdot 2 = 8$

≈ 34 Multiply.

ANSWER ▶ The volume is about 34 cubic feet.

b. A hemisphere has half the volume of a sphere.

$V = \frac{1}{2}\left(\frac{4}{3}\pi r^3\right)$ Write the formula for $\frac{1}{2}$ the volume of a sphere.

$= \frac{1}{2} \cdot \left(\frac{4}{3} \cdot \pi \cdot 5^3\right)$ Substitute 5 for r.

$= \frac{250}{3}\pi$ Simplify. $5^3 = 5 \cdot 5 \cdot 5 = 125$

≈ 262 Multiply.

ANSWER ▶ The volume is about 262 cubic inches.

EXAMPLE **3** **Find the Volume of a Sphere**

Estimate the volume of air in a beach ball that has a 12 inch diameter. Round your answer to the nearest whole number.

Solution

$$V = \frac{4}{3}\pi r^3$$ Write volume formula.

$$= \frac{4}{3} \cdot \pi \cdot 6^3$$ Substitute $\frac{12}{2} = 6$ for r.

$$= 288\pi$$ Simplify.

$$\approx 905$$ Multiply.

ANSWER ▶ The volume of air in the ball is about 905 cubic inches.

 **Find the Volume of a Sphere**

Find the volume to the nearest whole number.

4.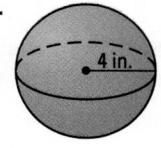

4 in.

268 in.³

5.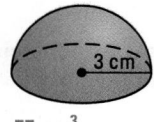

3 cm

57 cm³

6.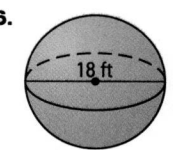

18 ft

3054 ft³

9.6 Exercises

Guided Practice

Vocabulary Check
1. Explain the difference between a *sphere* and a *hemisphere*.
   A sphere can be divided into two hemispheres, so a hemisphere is half a sphere.

Skill Check
Find the surface area to the nearest whole number.

2.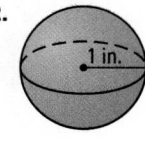

1 in.

13 in.²

3.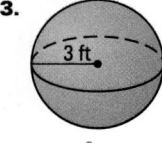

3 ft

113 ft²

4.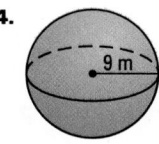

9 m

1018 m²

Find the volume to the nearest whole number.

5.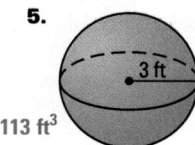

3 ft

113 ft³

6.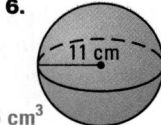

11 cm

5575 cm³

7.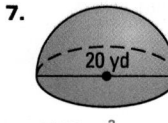

20 yd

2094 yd³

9.6 Surface Area and Volume of Spheres 519

✗ Common Error
Throughout this lesson, watch for students who confuse the surface area and volume formulas. Invite students to think of mnemonics they can use to distinguish between the two formulas.

Extra Example 3
Estimate the volume of rubber in a solid ball that has a 6-centimeter diameter. Round your answer to the nearest whole number.
about 113 cm³

✓ Concept Check
How do you find the surface area and volume of a sphere?
For surface area, compute four times the product of π and the square of the radius. For volume, compute four-thirds of the product of π and the cube of the radius of the sphere.

🐢 Daily Puzzler
Eight congruent spheres are packed into a cube with edge length x so that each sphere is touching three faces of the cube and three other spheres as shown. What is the ratio of the total volume of the spheres to the volume of the cube? $\frac{\pi}{6}$

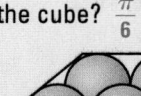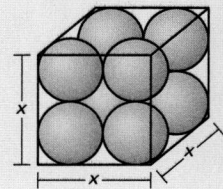

Assignment Guide

BASIC
Day 1: pp. 520–523 Exs. 8–10,
12–17, 23–28, 37–39, 43–45,
46–60 even, Quiz 2

AVERAGE
Day 1: pp. 520–523 Exs. 11–25,
33–41, 46, 47–61 odd, Quiz 2

ADVANCED
Day 1: pp. 520–523 Exs. 14–22,
25–36, 39–46, 57–61 odd, Quiz 2;
EC: p 470D*, classzone.com

BLOCK SCHEDULE
pp. 520–523 Exs. 11–25, 33–41,
46, 47–61 odd, Quiz 2 (with 9.5)

Extra Practice

- Student Edition, p. 692
- Chapter 9 Resource Book,
  pp. 57–58

Homework Check

To quickly check student understanding of key concepts, go over
the following exercises:

Basic: 8, 15, 23, 26, 37
Average: 13, 19, 24, 38, 40
Advanced: 17, 25, 34, 39, 41

Extra Practice
See p. 692.

Find Surface Area of a Sphere Find the surface area of the sphere.
Round your answer to the nearest whole number.

8.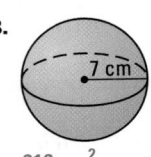
7 cm

616 cm^2

9.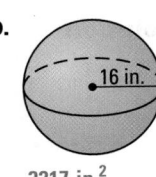
16 in.

3217 in.^{2}

10.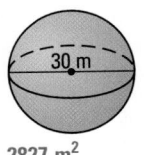
30 m

2827 m^2

11. Bob wrote V rather than S for surface area, used the wrong formula, used the diameter rather than the radius, and wrote the answer in cubic units rather than square units. $S = 4\pi r^2 = 4\pi(5^2) = 100\pi \approx 314 \text{ mm}^2$.

11. Error Analysis Bob is asked to find the surface area of a sphere with a diameter of 10 millimeters. Explain and correct his error(s).

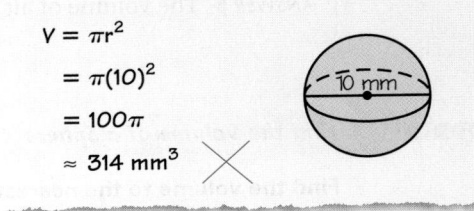

$$V = \pi r^2$$
$$= \pi(10)^2$$
$$= 100\pi$$
$$\approx 314 \text{ mm}^3 \quad \times$$

10 mm

Sports In Exercises 12–17, estimate the surface area of the ball.
Round your answer to the nearest whole number.

12. Soccer ball 232 in.^{2} **13.** Tennis ball 137 cm^2 **14.** Bowling ball 1493 cm^2

$r = 4.3$ in.

$r = 3.3$ cm

$r = 10.9$ cm

15. Golf ball 9 in.^{2} **16.** Basketball 284 in.^{2} **17.** Softball 290 cm^2

$d = 1.7$ in.

$d = 9.5$ in.

$d = 9.6$ cm

Homework Help

Example 1: Exs. 8–17
Example 2: Exs. 23–28,
37–39
Example 3: Exs. 23–28,
37–39

18. **You be the Judge** Julie thinks that if you double the radius of the sphere shown at the right, the surface area will double. Is she right? Explain your reasoning.
No; if you double the radius, the surface area will quadruple from 36π to 144π.

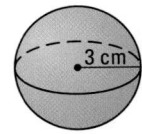
3 cm

ASTRONOMERS study the planets, stars, and solar system. Powerful telescopes are used to collect information about astronomical objects.

Application Links
CLASSZONE.COM

21. The surface area of Earth is about 13.4 times the surface area of the moon.

Astronomy In Exercises 19–22, use the information about Earth and its moon given in the photo.

19. Find the surface area of Earth.
about 197,000,000 mi^2

20. Find the surface area of Earth's moon. about 14,700,000 mi^2

21. Compare the surface areas of Earth and its moon. **See margin.**

22. About 70% of Earth's surface is water. How many square miles of water are on Earth's surface?
about 138,000,000 mi^2

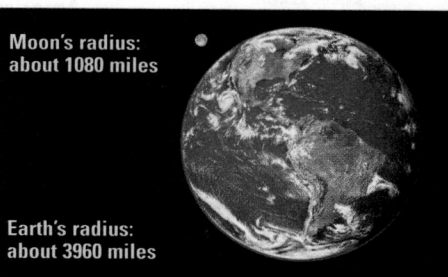
Moon's radius: about 1080 miles
Earth's radius: about 3960 miles

Finding Volume of a Sphere Find the volume of the sphere. Round your answer to the nearest whole number.

23.
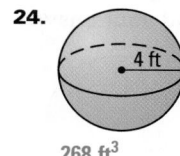
8 m
2145 m^3

24.
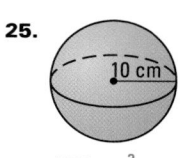
4 ft
268 ft^3

25.
10 cm
4189 cm^3

26.
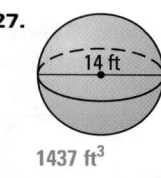
22 yd
5575 yd^3

27.
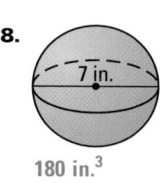
14 ft
1437 ft^3

28.
7 in.
180 in.^{3}

Technology Use formulas to create a spreadsheet like the one shown. Then answer Exercises 29–32.

	A	B	C		
	Radius, r	Surface area, $4\pi r^2$	Surface area of new sphere / Surface area of original sphere		
1					
2	3	113.1	1		
3	6	452.4	4		
4	9	?	1017.9	?	9
5	12	?	1809.6	?	16

29. How many times greater is the surface area of a sphere if the radius is doubled? tripled? quadrupled?
4 times greater; 9 times greater; 16 times greater

30. Explain why the surface area changes by a greater amount than the radius. because the radius is squared in the formula $S = 4\pi r^2$

31. How many times greater do you think the volume of a sphere will be if the radius is doubled? tripled?

31. Answers may vary. The volume is multiplied by 8 when the radius is doubled, and the volume is multiplied by 27 when the radius is tripled.

32. Create a spreadsheet for the volume of a sphere. Then answer Exercises 29 and 30 for the volume of a sphere.
8 times greater; 27 times greater; 64 times greater; because the radius is cubed in the formula $V = \frac{4}{3}\pi r^3$

9.6 *Surface Area and Volume of Spheres* **521**

X Common Error

Exercises 18 and 29–32 again point out a very common error that students make; namely that doubling or tripling dimensions doubles or triples the surface area and volume of a solid. A thorough discussion of Exercises 29–32 should help correct this common misconception.

X Common Error

In Exercises 37–39, watch for students who find the volume of the entire sphere, rather than the hemisphere. Remind them to multiply by $\frac{1}{2}$.

Student Help

LOOK BACK
See pp. 470–471 for more information about The Rose Center for Earth and Space.

Spheres in Architecture In Exercises 33–36, refer to the information below about The Rose Center for Earth and Space at New York City's American Museum of Natural History.

The sphere has a diameter of 87 feet. The glass cube surrounding the sphere is 95 feet long on each edge.

33. Find the surface area of the sphere.
about 23,779 ft²

34. Find the volume of the sphere. about 344,791 ft³

35. Find the volume of the glass cube. 857,375 ft³

36. Find the approximate amount of glass used to make the cube. (*Hint*: Do not include the ground or roof in your calculations.) 36,100 ft²

Finding Volume of a Hemisphere Find the volume of the hemisphere. Round your answer to the nearest whole number.

37.
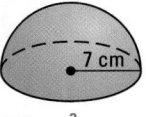
7 cm
718 cm³

38.
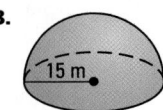
15 m
7069 m³

39.
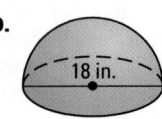
18 in.
1527 in.³

Composite Solids Find the volume of the solid. Round your answer to the nearest whole number.

40.
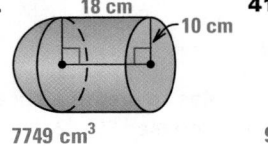
18 cm
10 cm
7749 cm³

41.
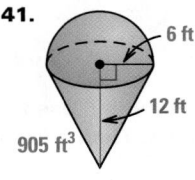
6 ft
12 ft
905 ft³

42.
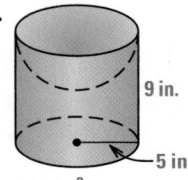
9 in.
5 in.
445 in.³

Student Help
CLASSZONE.COM

HOMEWORK HELP
Extra help with problem solving in Exs. 43–45 is at classzone.com

Architecture The entrance to the Civil Rights Institute in Birmingham, Alabama, includes a hemisphere that has a radius of 25.3 feet.

25.3 ft

43. Find the volume of the hemisphere.
about 33,917 ft³

44. Find the surface area of the hemisphere, not including its base. about 4022 ft²

45. The walls of the hemisphere are 1.3 feet thick. So, the rounded surface inside the building is a hemisphere with a radius of 24 feet. Find its surface area, not including its base. about 3619 ft²

Standardized Test Practice

46. Multiple Choice What is the approximate surface area of the sphere shown? A

A 3217 in.²　　**B** 4287 in.²

C 12,861 in.²　　**D** 17,149 in.²

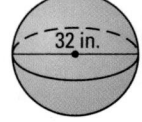
32 in.

Mixed Review

Surface Area Find the surface area of the solid. If necessary, round your answer to the nearest whole number. *(Lessons 9.2, 9.3)*

47. A cone has a height of 12 meters and a base radius of 3 meters. 145 m²

48. A pyramid has a slant height of 3 feet and a square base that measures 4 feet on a side. 40 ft²

49. A cylinder has a radius of 9 centimeters and a height of 9 centimeters. 1018 cm²

Simplifying Radicals Evaluate. Give the exact value if possible. Otherwise, approximate to the nearest tenth. *(Skills Review, p. 668)*

50. $\sqrt{6}$ 2.4   **51.** $\sqrt{18}$ 4.2   **52.** $\sqrt{77}$ 8.8   **53.** $\sqrt{400}$ 20

54. $\sqrt{256}$ 16   **55.** $\sqrt{99}$ 9.9   **56.** $\sqrt{40}$ 6.3   **57.** $\sqrt{120}$ 11.0

Algebra Skills

Using Formulas Find the missing length using the given information. *(Skills Review, p. 674)*

58. A rectangle is 6 feet wide and 11 feet long. Find the perimeter. 34 ft

59. A square has an area of 100 square inches. Find the perimeter. 40 in.

60. Find the width of a rectangle with a length of 8 meters and an area of 40 square meters. 5 m

61. The perimeter of a square is 44 yards. Find the side length. 11 yd

Quiz 2

Find the volume of the solid. If necessary, round your answer to the nearest whole number. *(Lessons 9.4, 9.5, 9.6)*

1.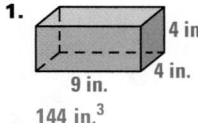
4 in.
4 in.
9 in.
144 in.³

2.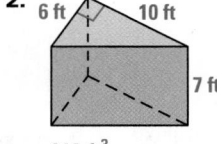
6 ft   10 ft
7 ft
210 ft³

3.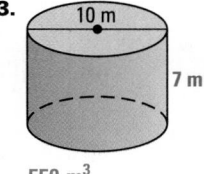
10 m
7 m
550 m³

4.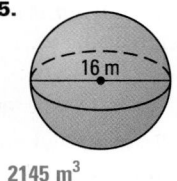
12 ft
10 ft
10 ft
400 ft³

5.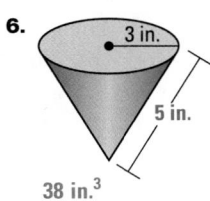
16 m
2145 m³

6.
3 in.
5 in.
38 in.³

7. Sketch a cylinder with a radius of 4 inches and a height of 4 inches. Then find its volume. *(Lesson 9.4)* Check sketch; about 201 in.³

8. Sketch a sphere with a radius of 9 centimeters. Then find its surface area. *(Lesson 9.6)* Check sketch; about 1018 cm².

9.6 Surface Area and Volume of Spheres   **523**

④ Assess

Assessment Resources

The Mini-Quiz below is also available on blackline (*Chapter 10 Resource Book*, p. 9) and on transparency. For more assessment resources, see:
- Chapter 9 Resource Book
- Standardized Test Practice
- Test and Practice Generator

Mini-Quiz

Round all answers to the nearest whole number.

1. Find the surface area and the volume of the sphere.

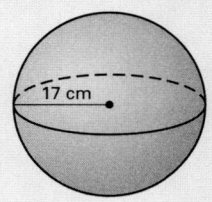
17 cm

3632 cm²; 20,580 cm³

2. Find the volume of the hemisphere.

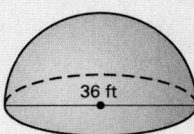
36 ft

12,215 ft³

3. Find the surface area of a large beach ball with diameter 4 feet. 50 ft²

4. Find the volume of a hemisphere with radius 6 centimeters. 452 cm³

Additional Resources

The following resources are available to help review the materials in this chapter.

📖 **Chapter 9 Resource Book**
- Chapter Review Games and Activities, p. 65
- Cumulative Review, Chs. 1–9

Chapter  Summary and Review

VOCABULARY

- **solid**, *p. 473*
- **polyhedron**, *p. 473*
- **base**, *p. 473*
- **face**, *p. 474*
- **edge**, *p. 474*
- **prism**, *p. 483*
- **surface area**, *p. 483*

- **lateral face**, *p. 484*
- **lateral area**, *p. 484*
- **cylinder**, *p. 485*
- **pyramid**, *p. 491*
- **height of a pyramid**, *p. 491*
- **slant height of a pyramid**, *p. 491*

- **cone**, *p. 493*
- **height of a cone**, *p. 493*
- **slant height of a cone**, *p. 493*
- **volume**, *p. 500*
- **sphere**, *p. 517*
- **hemisphere**, *p. 517*

VOCABULARY REVIEW

Fill in the blank.

1. Polyhedra are named by the shape of their __?__. **base(s)**

2. The __?__ of a polyhedron is the sum of the areas of its faces. **surface area**

3. A(n) __?__ is a polyhedron with two congruent faces that lie in parallel planes. **prism**

4. The __?__ of a prism are the faces of the prism that are not bases. **lateral faces**

5. The __?__ of a solid is the number of cubic units contained in its interior. **volume**

9.1 SOLID FIGURES

Examples on pp. 473–475

EXAMPLE   Tell whether the solid is a polyhedron. If so, identify the shape of the base(s) and then name the solid.

The solid is formed by polygons so it is a polyhedron. There are two congruent triangular bases. This is a triangular prism.

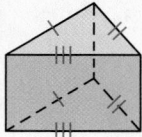

Tell whether the solid is a polyhedron. If so, identify the shape of the base(s) and then name the solid.

6. 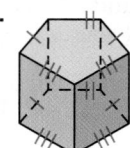   yes; pentagonal; pentagonal prism

7.   no

8. yes; hexagonal; hexagonal pyramid

9.2 SURFACE AREA OF PRISMS AND CYLINDERS

Examples on pp. 483–486

EXAMPLES Find the surface area to the nearest whole number.

a.

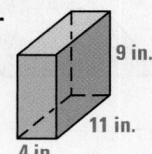

9 in.
11 in.
4 in.

$S = 2B + Ph$

$= 2(11 \cdot 4) + (2 \cdot 4 + 2 \cdot 11)(9)$

$= 2(44) + 270$

$= 358 \text{ in.}^2$

b.

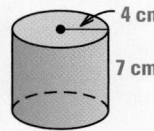

4 cm
7 cm

$S = 2\pi r^2 + 2\pi rh$

$= 2\pi(4^2) + 2\pi(4)(7)$

$= 88\pi$

$\approx 276 \text{ cm}^2$

Find the surface area to the nearest whole number.

9.

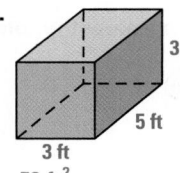

3 ft
5 ft
3 ft
78 ft²

10.

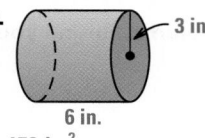

3 in.
6 in.
170 in.²

11.

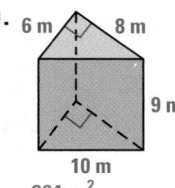

6 m   8 m
9 m
10 m
264 m²

9.3 SURFACE AREA OF PYRAMIDS AND CONES

Examples on pp. 491–494

EXAMPLES Find the surface area to the nearest whole number.

a.

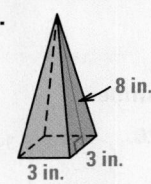

8 in.
3 in.   3 in.

$S = B + \frac{1}{2}P\ell$

$= (3 \cdot 3) + \frac{1}{2}(4 \cdot 3)(8)$

$= 9 + \frac{1}{2}(12)(8)$

$= 57 \text{ in.}^2$

b.

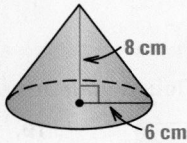

8 cm
6 cm

$\ell^2 = 6^2 + 8^2 = 100$ Find the slant height.

$\ell = \sqrt{100} = 10$

$S = \pi r^2 + \pi r\ell$ Formula for surface area.

$= \pi(6)^2 + \pi(6)(10)$

$= 96\pi$

$\approx 302 \text{ cm}^2$

Find the surface area of the solid. Round your answer to the nearest whole number.

12. 95 m²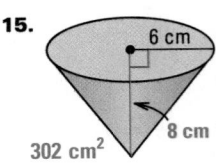

7 m
5 m
5 m

13. 478 in.²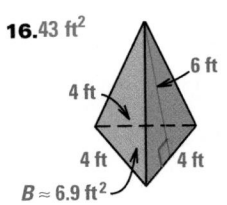

11 in.
8 in.

14. 96 ft²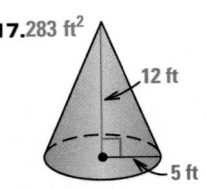

4 ft
6 ft
6 ft

15.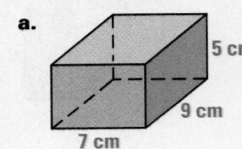

6 cm
8 cm
302 cm²

16. 43 ft²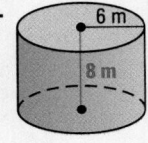

6 ft
4 ft
4 ft   4 ft
$B \approx 6.9 \text{ ft}^2$

17. 283 ft²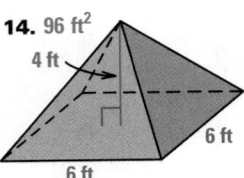

12 ft
5 ft

9.4 VOLUME OF PRISMS AND CYLINDERS

Examples on pp. 500–502

EXAMPLES Find the volume. Round your answer to the nearest whole number.

a.

5 cm
9 cm
7 cm

$V = Bh$

$= (7 \cdot 9)(5)$

$= (63)(5)$

$= 315 \text{ cm}^3$

b.

6 m
8 m

$V = \pi r^2 h$

$= \pi(6^2)(8)$

$= 288\pi$

$\approx 905 \text{ m}^3$

Find the volume. Round your answer to the nearest whole number.

18.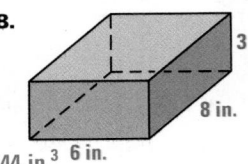

3 in.
8 in.
144 in.³ 6 in.

19.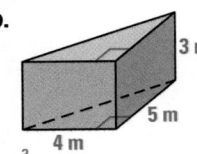

3 m
5 m
4 m
30 m³

20.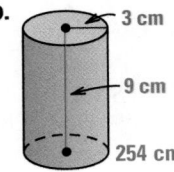

3 cm
9 cm
254 cm³

21.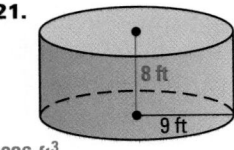

8 ft
9 ft
2036 ft³

22.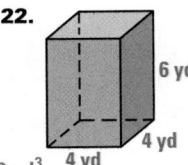

6 yd
4 yd
96 yd³ 4 yd

23.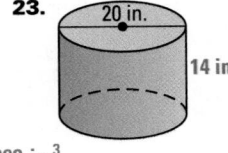

20 in.
14 in.
4398 in.³

9.5 VOLUME OF PYRAMIDS AND CONES

Examples on pp. 510–512

EXAMPLES **Find the volume. Round your answer to the nearest whole number.**

a.

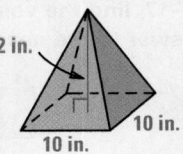

12 in.

10 in.
10 in.

$$V = \frac{1}{3}Bh$$

$$= \frac{1}{3}(10 \cdot 10)(12) = 400 \text{ in.}^3$$

b.

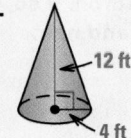

12 ft

4 ft

$$V = \frac{1}{3}\pi r^2 h$$

$$= \frac{1}{3}\pi(4^2)(12) = 64\pi \approx 201 \text{ ft}^3$$

Find the volume. Round your answer to the nearest whole number.

24. 30 m³

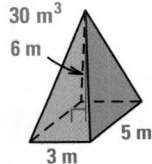

6 m

5 m
3 m

25.

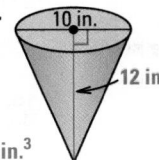

10 in.

12 in.

314 in.³

26. 1018 cm³

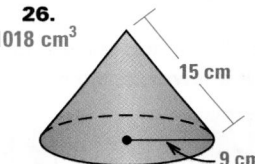

15 cm

9 cm

9.6 SURFACE AREA AND VOLUME OF SPHERES

Examples on pp. 517–519

EXAMPLE **Find the surface area and the volume of the sphere. Round your answer to the nearest whole number.**

6 in.

Surface area = $4\pi r^2$

$= 4\pi(6^2)$

$= 144\pi$

$\approx 452 \text{ in.}^2$

Volume = $\frac{4}{3}\pi r^3$

$= \frac{4}{3}\pi(6^3)$

$= 288\pi$

$\approx 905 \text{ in.}^3$

Find the surface area and the volume of the sphere. Round your answer to the nearest whole number.

27.

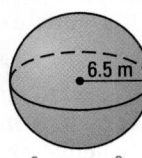

6.5 m

531 m²; 1150 m³

28.

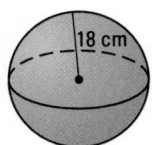

18 cm

4072 cm²; 24,429 cm³

29.

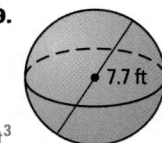

7.7 ft

186 ft²; 239 ft³

Chapter Summary and Review **527**

Tell whether the solid is a polyhedron. If so, identify the shape of the base(s) and name the solid.

1.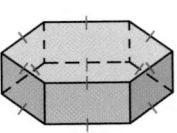

yes; hexagonal; hexagonal prism

2.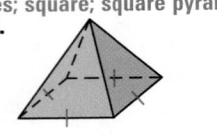

yes; square; square pyramid

3.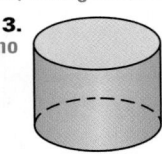

no

4.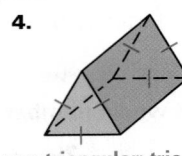

yes; triangular; triangular prism

In Exercises 5–10, find the surface area of the solid. Round your answer to the nearest whole number.

5. 202 ft² 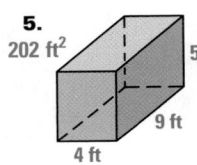 5 ft, 9 ft, 4 ft

6. 54 in.² 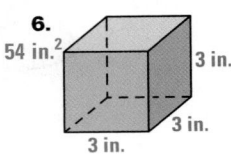 3 in., 3 in., 3 in.

7. 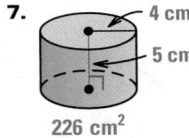 4 cm, 5 cm

226 cm²

8. 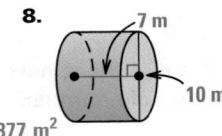 7 m, 10 m

377 m²

9. 452 in.² 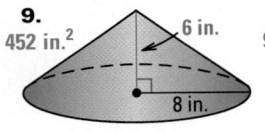 6 in., 8 in.

10. 96 yd² 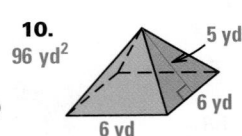 5 yd, 6 yd, 6 yd

11. Name the solid that is represented by the net below. Then find its surface area and volume. Round your answers to the nearest whole number. cylinder; 245 ft²; 283 ft³

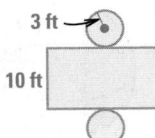 3 ft, 10 ft

In Exercises 12–17, find the volume of the solid. Round your answer to the nearest whole number.

12. 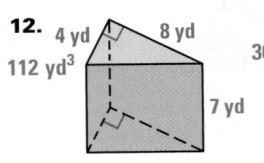 4 yd, 8 yd, 7 yd

112 yd³

13. 36 in.³ 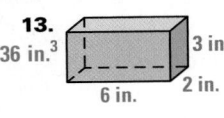 3 in., 6 in., 2 in.

14. 108 m³ 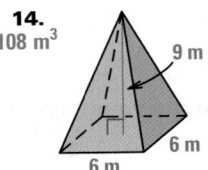 9 m, 6 m, 6 m

15. 63 ft³ 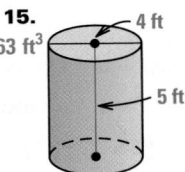 4 ft, 5 ft

16. 3054 ft³ 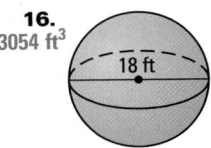 18 ft

17. 21 cm³ 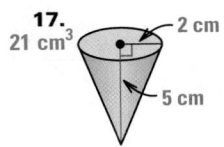 2 cm, 5 cm

18. How much does the volume of a cylinder increase if the radius doubles and the height stays the same? The volume is multiplied by 4.

19. How much does the volume of a cylinder increase if the height doubles and the radius stays the same? The volume doubles.

20. How much does the volume of a sphere increase if the radius doubles?
The volume is multiplied by 8.

In Exercises 21 and 22, use the aquarium below.

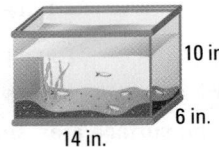 10 in., 6 in., 14 in.

21. If you fill the aquarium to a height of 8 inches, what is the volume of water in the aquarium? 672 in.³

22. How much glass is used to make the aquarium? (Do not include the top of the aquarium in your calculations.) 484 in.³

Additional Resources

📖 **Chapter 9 Resource Book**
• SAT/ACT Chapter Test, p. 70

10. The volume of the prism section is 3 times the volume of the pyramid section. *Sample answer:* The prism and the pyramid have the same base and height; the formulas for the volumes of a prism and a pyramid tell us that the volume of the prism is 3 times the volume of the pyramid.

Test Tip Read each question carefully to avoid missing preliminary steps. Do not look at the answers until you are sure you understand the question.

1. Which term correctly describes the solid shown below? **B**

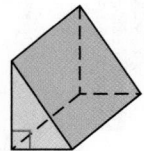

(A) square pyramid

(B) triangular prism

(C) triangular pyramid

(D) rectangular prism

2. What is the approximate surface area of the solid shown below? **G**

(F) 87.9 in.^{2}

(G) 113 in.^{2}

(H) 276 in.^{2}

(J) 352 in.^{2}

4 in.

7 in.

3. What is the approximate surface area of a sphere with a diameter of 8 centimeters? **C**

(A) 50 cm^2 (B) 100 cm^2

(C) 201 cm^2 (D) 268 cm^2

4. What is the volume of the pyramid shown? **G**

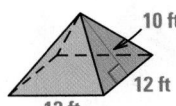

10 ft

12 ft

12 ft

(F) 120 ft^3 (G) 384 ft^3

(H) 480 ft^3 (J) 1440 ft^3

5. Suppose the length of each side of the cube is doubled. How many times larger is the surface area of the new cube? **C**

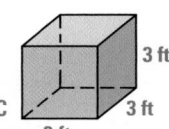

3 ft
3 ft
3 ft

(A) 2   (B) 3

(C) 4   (D) 8

6. A cone has a diameter of 12 inches and a volume of 48π cubic inches. Find the height. **J**

(F) 1 in.   (G) 1.3 in.

(H) 3 in.   (J) 4 in.

7. What is the approximate volume of the solid shown below? **B**

(A) 213 ft^3

(B) 269 ft^3

(C) 288 ft^3

(D) 307 ft^3

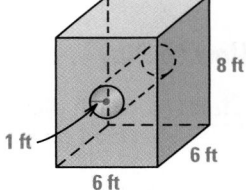

8 ft

1 ft

6 ft

6 ft

Multi-Step Problem Use the solid below.

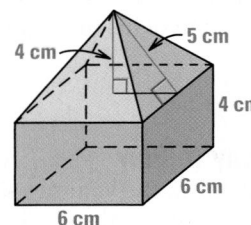

4 cm

5 cm

4 cm

6 cm

6 cm

8. Find the volume of the solid. **192 cm^3**

9. Find the surface area of the solid. **192 cm^2**

10. How does the volume of the prism section of the solid relate to the volume of the pyramid section of the solid? Explain your answer. See margin.

Teaching Tip
The Brain Games activity provides a motivating way to review selected content in the chapter. For a more comprehensive review, see the Chapter Summary and Review on pp. 524–527.

BraiN GaMes

Volume War

Materials
• 40 volume cards

Object of the Game To get more cards than your opponent.

Set Up Mix up the cards and deal them so that all players have the same number of cards. Put any extra cards aside. Each player makes a pile of their cards, face down.

How to Play

Step 1 Each player turns over the top card on his or her pile of cards.

Step 2 Compare the volume of each card. The player with the greatest volume takes all of the cards that are face up and puts them on the bottom of his or her pile of cards.

Step 3 If two or more of the cards have the same volume, the players who turned those cards over each turn a new card and the winner takes all of the face up cards.

Step 4 Continue playing until one player has all the cards or the most cards in a set amount of time.

Another Way to Play Make a set of your own cards. None of the volumes should be greater than 36 cubic units.

530

EXAMPLE 1 Use a Calculator

Use a calculator to evaluate $\frac{3}{\sqrt{2}}$. Round to the nearest tenth.

Solution

Calculator keystrokes	Display	Rounded value
3 ÷ √ 2 ENTER	2.121320344	2.1

ANSWER ▶ So, $\frac{3}{\sqrt{2}} \approx 2.1$.

Try These

Use a calculator to evaluate. Round to the nearest tenth.

1. $\frac{1}{\sqrt{5}}$ 0.4
2. $\frac{7}{\sqrt{2}}$ 4.9
3. $\frac{3}{\sqrt{10}}$ 0.9
4. $\frac{9}{\sqrt{17}}$ 2.2
5. $\frac{10}{\sqrt{7}}$ 3.8

6. $\frac{\sqrt{5}}{\sqrt{11}}$ 0.7
7. $\frac{\sqrt{19}}{\sqrt{3}}$ 2.5
8. $\frac{4\sqrt{21}}{7\sqrt{5}}$ 1.2
9. $6 - \frac{6}{\sqrt{6}}$ 3.6
10. $5 + \frac{2}{\sqrt{3}}$ 6.2

The *quotient property of radicals* states the following:

$\sqrt{\frac{a}{b}} = \frac{\sqrt{a}}{\sqrt{b}}$, where $a \geq 0$ and $b > 0$.

EXAMPLE 2 Use the Quotient Property

Simplify the expression.

a. $\sqrt{\frac{25}{4}}$

b. $\sqrt{\frac{45}{100}}$

Solution

a. $\sqrt{\frac{25}{4}} = \frac{\sqrt{25}}{\sqrt{4}} = \frac{5}{2}$

b. $\sqrt{\frac{45}{100}} = \frac{\sqrt{45}}{\sqrt{100}} = \frac{\sqrt{9 \cdot 5}}{\sqrt{100}} = \frac{\sqrt{9} \cdot \sqrt{5}}{\sqrt{100}} = \frac{3\sqrt{5}}{10}$

Try These

Simplify the expression.

11. $\sqrt{\frac{25}{49}}$ $\frac{5}{7}$
12. $\sqrt{\frac{9}{64}}$ $\frac{3}{8}$
13. $\sqrt{\frac{81}{121}}$ $\frac{9}{11}$
14. $\sqrt{\frac{1}{36}}$ $\frac{1}{6}$
15. $\sqrt{\frac{49}{16}}$ $\frac{7}{4}$

16. $\sqrt{\frac{7}{16}}$ $\frac{\sqrt{7}}{4}$
17. $\sqrt{\frac{18}{81}}$ $\frac{\sqrt{2}}{3}$
18. $\sqrt{\frac{8}{100}}$ $\frac{\sqrt{2}}{5}$
19. $\sqrt{\frac{147}{400}}$ $\frac{7\sqrt{3}}{20}$
20. $\sqrt{\frac{128}{144}}$ $\frac{2\sqrt{2}}{3}$

9. SSS Congruence Postulate; $\overline{AB} \cong \overline{DC}$ (Given); $\overline{BC} \cong \overline{BC}$ (Reflexive Property of Congruence); $\overline{AC} \cong \overline{DB}$ (Given)

10. AAS Congruence Theorem; $\angle K \cong \angle M$ (Given); $\angle JLK \cong \angle NLM$ (Vertical Angles Theorem); $\overline{JL} \cong \overline{NL}$ (Given)

11. HL Congruence Theorem; $\overline{PR} \cong \overline{ST}$ (Given); $\overline{QR} \cong \overline{RT}$ (Given)

1. In a coordinate plane, plot the points $A(4, 6)$, $B(-1, 3)$, and $C(5, -1)$ and sketch $\angle ABC$. Then classify the angle. **(Lesson 1.6)** Check sketches; acute.

2. Find $m\angle 1$, $m\angle 2$, and $m\angle 3$ in the diagram shown at the right. **(Lesson 2.4)** $m\angle 1 = 54°$, $m\angle 2 = 54°$, $m\angle 3 = 126°$

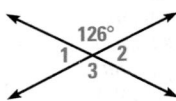

Find the measure of each numbered angle. (Lessons 3.2, 3.4)

3.

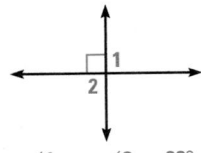

$m\angle 1 = m\angle 2 = 90°$

4.

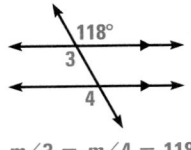

$m\angle 3 = m\angle 4 = 118°$

5.

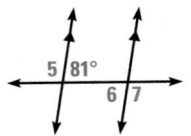

$m\angle 5 = 99°$, $m\angle 6 = 81°$, $m\angle 7 = 99°$

Use the triangle shown at the right. (Lesson 4.2, 4.7)

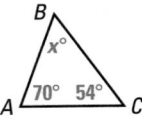

6. Find the value of x. 56

7. Classify the triangle by its angles. acute

8. Write the sides of the triangle from longest to shortest. $\overline{BC}$, $\overline{AC}$, $\overline{AB}$

Is it possible to show that the triangles are congruent? If so, state the postulate or theorem you would use. Explain your reasoning.
(Lessons 5.2–5.4) 9–11. See margin.

9.

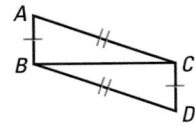

10.

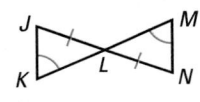

11.

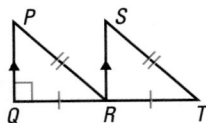

Use *always, sometimes,* or *never* to complete the statement.
(Lessons 2.3, 5.7, 6.4)

12. Two acute angles are __?__ complementary. sometimes

13. The sides of a rhombus are __?__ congruent. always

14. A rectangle __?__ has exactly one line of symmetry. never

15. Two squares are __?__ congruent. sometimes

Find the value of the variable(s). (Lessons 6.1, 6.4, 6.5)

16. *PQST* is a trapezoid. $x = 10$

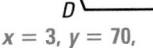

17. *ABCD* is a rhombus.

$x = 3$, $y = 70$, $z = 110$

18. *JKLM* is a rectangle.

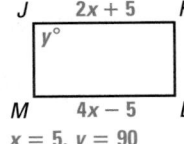

$x = 5$, $y = 90$

532

Use the diagram shown at the right. (Lessons 7.3–7.5)

19. Which theorem or postulate could you use to show that
△*ABC* ~ △*ADE*? **AA Similarity Postulate**

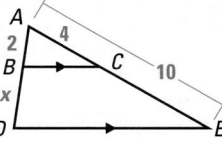

20. Complete the proportion $\frac{DB}{DA} = \frac{?}{EA}$. **EC**

21. Find the value of *x*. **3**

Use the polygon shown at the right. (Lessons 6.1, 8.1, 8.2)

22. Classify the polygon by its sides. **pentagon**

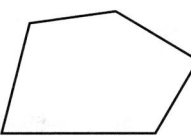

23. Is the polygon convex or concave? Explain your reasoning.
Convex; *Sample answer:* **none of the extended sides pass through the interior.**

24. What is the sum of the measures of the interior angles of the
polygon? What is the sum of its exterior angles, one at each vertex?
540°; 360°

Tiles The tile pattern shown at the right contains regular octagonal
tiles. (Lessons 6.6, 8.2)

25. What is the measure of an interior angle of one of the regular
octagonal tiles? **135°**

26. What type of shape are the yellow figures? Explain your reasoning. **See margin.**

Find the area of the polygon. (Lessons 8.3–8.6)

27.
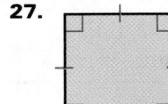
196 m² 14 m

28.
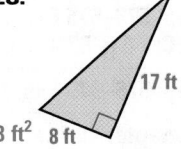
17 ft
68 ft² 8 ft

29.
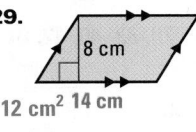
8 cm
112 cm² 14 cm

30.
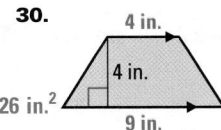
4 in.
4 in.
26 in.² 9 in.

**Find the surface area and the volume of the solid. If necessary, round
your answer to the nearest whole number.** (Lessons 9.2–9.5)

31.
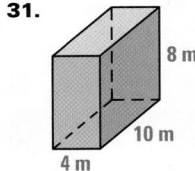
8 m
10 m
4 m
304 m², 320 m³

32.
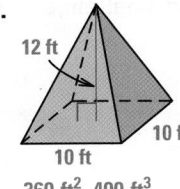
12 ft
10 ft
10 ft
360 ft², 400 ft³

33.
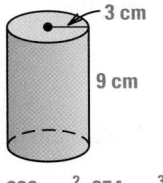
3 cm
9 cm
226 cm², 254 cm³

34.
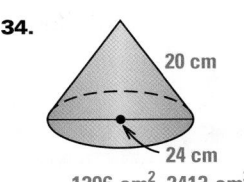
20 cm
24 cm
1206 cm², 2413 cm³

**The sphere shown at the right has a diameter of 32 inches. Round
your answer to the nearest whole number.** (Lesson 9.6)

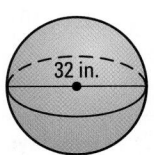
32 in.

35. Find the surface area of the sphere. **3217 in.²**

36. What is half the volume of the sphere? **8579 in.³**

26. Square; *Sample answer:* The sum
of the measures of one interior
angle in each of two octagons
and in one of the yellow figures
is 360°. So, each interior angle of
the yellow figure must measure
90° (135° + 135° + 90° = 360°).
Since the sides of the yellow
figure are congruent, the yellow
figure is a square.

REGULAR SCHEDULE

Lesson	Les. Day	Basic	Average	Advanced
10.1	Day 1	SRH p. 669 Exs. 9–16; pp. 539–541 Exs. 12–25, 54, 55, 63–68	pp. 539–541 Exs. 16–25, 54, 55, 63–68	pp. 539–541 Exs. 16–25, 54, 55, 63–68
	Day 2	pp. 539–541 Exs. 26–43, 56–62	pp. 539–541 Exs. 26–51, 56–62	pp. 539–541 Exs. 29–52, 53*, 56–62; EC: classzone.com
10.2	Day 1	pp. 545–547 Exs. 6–17, 35–45	pp. 545–547 Exs. 6–18, 38–45, 47–53 odd	pp. 545–547 Exs. 6–18, 39–53 odd
	Day 2	SRH p. 669 Exs. 37–40; pp. 545–547 Exs. 19–27 odd, 32–34, 46–53	pp. 545–547 Exs. 19–31, 34–37	pp. 545–547 Exs. 19–37; EC: TE p. 534D*, classzone.com
10.3	Day 1	pp. 552–555 Exs. 11–22, 45–56	pp. 552–555 Exs. 11–22, 37–39, 43, 45–56	pp. 552–555 Exs. 14–16, 20–22, 37–41, 43, 45–56
	Day 2	pp. 552–555 Exs. 23–28, 30–35, 43, 44, Quiz 1	pp. 552–555 Exs. 23–36, 41, 44, Quiz 1	pp. 552–555 Exs. 26–36, 42*, 44, Quiz 1; EC: classzone.com
10.4	Day 1	pp. 560–562 Exs. 10–23, 44–49	pp. 560–562 Exs. 10–23, 44–49	pp. 560–562 Exs. 10–23, 44–49
	Day 2	pp. 560–562 Exs. 24–29, 31–43	pp. 560–562 Exs. 24–43	pp. 560–562 Exs. 24–43; EC: TE p. 534D*, classzone.com
10.5	Day 1	pp. 566–568 Exs. 11–16, 39–48	pp. 566–568 Exs. 11–16, 33–35, 39–48	pp. 566–568 Exs. 11–16, 33–35, 36*, 39–48
	Day 2	pp. 566–568 Exs. 17–30, 38, 49–51	pp. 566–568 Exs. 17–32, 37, 38, 49–51	pp. 566–568 Exs. 17–32, 37, 38, 49–51; EC: classzone.com
10.6	Day 1	pp. 573–575 Exs. 12–23, 41, 43–54	pp. 573–575 Exs. 12–23, 30–32, 41, 43–54	pp. 573–575 Exs. 12–23, 30–32, 41–53 odd
	Day 2	pp. 573–575 Exs. 24–29, 33–39, 42, Quiz 2	pp. 573–575 Exs. 24–29, 33–40, 42, Quiz 2	pp. 573–575 Exs. 24–29, 33–40, 42, Quiz 2; EC: TE p. 534D*, classzone.com
Review	Day 1	pp. 576–579 Exs. 1–60	pp. 576–579 Exs. 1–60	pp. 576–579 Exs. 1–60
Assess	Day 1	Chapter 10 Test	Chapter 10 Test	Chapter 10 Test

YEARLY PACING Chapter 10 Total – **14 days**   Chapters 1–10 Total – **146 days**   Remaining – **14 days**

*Challenge Exercises   EP = Extra Practice   SRH = Skills Review Handbook   EC = Extra Challenge

BLOCK SCHEDULE

Day 1	Day 2	Day 3	Day 4	Day 5	Day 6	Day 7
10.1 pp. 539–541 Exs. 16–51, 54–68	**10.2** pp. 545–547 Exs. 6–31, 34–45, 47–53 odd	**10.3** pp. 552–555 Exs. 11–39, 41, 43–56, Quiz 1	**10.4** pp. 560–562 Exs. 10–49	**10.5** pp. 566–568 Exs. 11–35, 37–51	**10.6** pp. 573–575 Exs. 12–54, Quiz 2	Review pp. 576–579 Exs. 1–60 Assess Chapter 10 Test

YEARLY PACING Chapter 10 Total – **7 days**   Chapters 1–10 Total – **73 days**   Remaining – **7 days**

Support Materials

CHAPTER RESOURCE BOOK

CHAPTER SUPPORT

Tips for New Teachers	p. 1	Strategies for Reading Mathematics	p. 5
Parent Guide for Student Success	p. 3		

LESSON SUPPORT

	10.1	10.2	10.3	10.4	10.5	10.6
Lesson Plans (regular and block)	p. 7	p. 15	p. 24	p. 34	p. 45	p. 55
Warm-Up Exercises and Daily Quiz	p. 9	p. 17	p. 26	p. 36	p. 47	p. 57
Technology Activities & Keystrokes		p. 18		p. 37	p. 48	
Practice (2 levels)	p.10	p. 19	p. 27	p. 40	p. 49	p. 58
Reteaching with Practice	p. 12	p. 21	p. 29	p. 42	p. 51	p. 60
Quick Catch-Up for Absent Students	p. 14	p. 23	p. 31	p. 44	p. 53	p. 62
Learning Activities					p. 54	
Real-Life Applications			p. 32			p. 63

REVIEW AND ASSESSMENT

Quizzes	pp. 33, 64	Alternative Assessment with Math Journal	p. 74
Brain Games Support	p. 65	Project with Rubric	p. 76
Chapter Review Games and Activities	p. 68	Cumulative Review	p. 78
Chapter Test (2 levels)	p. 69	Resource Book Answers	AN1
SAT/ACT Chapter Test	p. 73		

TRANSPARENCIES

	10.1	10.2	10.3	10.4	10.5	10.6
Warm-Up Exercises and Daily Quiz	p. 60	p. 61	p. 62	p. 63	p. 64	p. 65
Visualize It Transparencies	✓	✓	✓	✓	✓	
Answer Transparencies	✓	✓	✓	✓	✓	✓

TECHNOLOGY

- Time-Saving Test and Practice Generator
- Electronic Teacher Tools
- Geometry in Motion: Video
- Classzone.com
- Online Lesson Planner

ADDITIONAL RESOURCES

- Worked-Out Solution Key
- Resources in Spanish
- Practice Workbook with Examples

Providing Universal Access

Strategies for Strategic Learners

REVIEW PREVIOUSLY LEARNED CONCEPTS

USING CALCULATORS Chapter 10 presumes that students have spent some time using a calculator. Review with students some of the keystrokes they will encounter in this chapter. By now, many students will have learned the importance of careful and accurate keystroking. Ask students to name some of the errors they have made with calculators. Answers may include hitting the key next to the one you wanted to press, and forgetting to hit the 2nd or ALPHA function keys before selecting a second function or a letter. Encourage students to practice with a calculator before beginning the exercises in each lesson. Students with large fingers or long fingernails may need to spend extra time practicing hitting the keys accurately.

Another source of frustration for students when using calculators occurs when they try to fix a keystroke error. Many calculators have one key that cancels just the most recent keystroke and other key that erases the entire current entry so the whole keystroke sequence must be re-entered. Encourage students to become familiar with how to fix keystroke errors using the correct cancel keys so they do not waste time repeating keystrokes unnecessarily.

Also encourage students to practice basic calculations on their calculator. Students can try the calculations shown in the Examples of a lesson using their calculator. Doing this allows them to check that their answers match those shown in the textbook.

SQUARES AND SQUARE ROOTS Finding square roots is another concept that students should have learned in earlier mathematics classes. By now students should be able to immediately recognize perfect squares. If students have not fully acquired this skill, have them list the squares of the numbers from 1 through 20 on an index card and then write the square root corresponding to each perfect square beside it. Throughout this chapter, students will be finding square roots and checking that their answers are reasonable. The check portion will be considerably easier and faster if students have memorized the perfect squares.

Strategies for English Learners

DISSECT WORD PROBLEMS

Word problems are a big challenge for English learners. For an English learner, a word problem can look something like the first two lines of Lewis Carroll's "Jabberwocky," (*'Twas brillig, and the slithy toves/Did gyre and gimble in the wabe.*) where the only words you know are *and*, *the*, *did*, and *in*.

In some cases the teacher can turn the word problem into a picture problem.

Word problem: The length of a rectangular prism is twice its width. The height of the prism equals the width. Find the dimensions of the prism given that the volume is 54 cubic inches.

Picture problem:

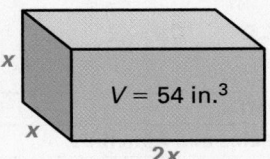

Reducing the complexity of word problems to avoid specialized vocabulary and idioms allows English learners to focus on the mathematics. The teacher may want to organize lessons for English learners to start first with problems that have few words. Several times a week, however, teachers and students should focus on word problems.

SIGNAL WORDS

Fortunately most word problems follow a standard format, and there are clues provided by "signal words," words of high frequency that signal what will follow or what came before. Most word problems also contain numbers, and usually you must do something with the numbers to solve the problems. Analyzing the units in a problem can provide a clue as to the answer that is needed. English learners can study the techniques that help good problem solvers solve problems, so that they can use these strategies too.

Examine the structure of a word problem with your students, choosing one from a previous chapter so that students will already be familiar with at least some of the vocabulary. For example, use Exercise 38 from page 505 in Chapter 9:

38. Personal Aquariums To avoid overcrowding in a personal aquarium, you should buy one fish for every gallon of water. ($231 \text{ in.}^3 \approx 1 \text{ gal}$) About how many fish should be in an aquarium that is 20 inches wide, 10 inches long, and is filled with water to a height of 11 inches?

Point out the features of word problems shown in the following chart:

Heading (optional) *Personal Aquariums*	A word or short phrase at the beginning gives you a clue about something in the problem. In this case, it gives you a context for the problem but is not needed to solve the problem. In other cases the title may give you a clue as to the type of problem (for example, Multiple Choice, see page 507) or the concept or skill you will use (for example, Logical Reasoning, see page 514).
First sentence *To avoid overcrowding in a personal aquarium, you should buy one fish for every gallon of water. (231 in.^{3} ≈ 1 gal)*	Sometimes the first sentence is a background sentence that provides only context information, but usually it provides at least one numerical fact. In this case, it gives the facts that you should have one fish for every gallon of water and that 231 in.^{3} is about 1 gal. Sometimes this first sentence will provide *all* the facts you will need to work with.
Second sentence *About how many fish should be in an aquarium that is 20 inches wide, 10 inches long, and is filled with water to a height of 11 inches?*	The last sentence or group of sentences often contains one of the following words or phrases: *who, what, when, where, why, how many, how much, which, how long,* or a command verb like *write, evaluate, find, solve.* The signal word or words (in this case, *how many*) are usually followed closely by a description of what you are supposed to provide in your answer, in this case, the appropriate number of fish for this aquarium.

Have students find other word problems in Chapter 10 that follow this basic pattern. Also look for variations, such as a word problem followed by answers in a multiple choice format.

Strategies for Advanced Learners

SUBSTITUTE CHALLENGE PROBLEMS

Grouping advanced students together to work on challenge problems can be good for the whole class. While the advanced students work together on problems that provide them with an interesting challenge, the teacher can work with a smaller group of students who need pre-teaching or re-teaching. Alternately, at times the teacher will want to provide some students with extra time to practice necessary skills, allowing him or her to work with advanced students to review what they have been working on independently or to extend their learning. Depending on the class, teachers may want to use the challenge problems with the whole class.

Challenge Problem for use with Lesson 10.2:
The outer edges of the diagram at the right form a square with side length 12. If every triangle in the diagram is a 45°-45°-90° triangle, what is the value of *x*? Write your answer in radical form. $(3\sqrt{2})$

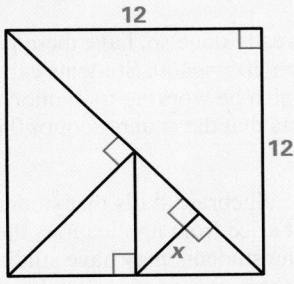

Challenge Problem for use with Lesson 10.4:
In the diagram, $\angle Q$ is a right angle, $m\angle QPS = 32°$, $m\angle RPS = 24°$, and $PQ = 14$. Find *RS* to the nearest tenth of a unit. (*12.0*)

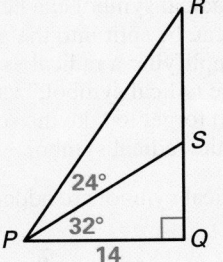

Challenge Problem for use with Lesson 10.6:
A kite flying at point *C* is held by two strings, 48 feet long and 34 feet long respectively, whose other ends are at points *A* and *C* on the ground. The string from point *A* makes an angle of 35° with the ground. What is $m\angle B$ to the nearest degree? (*54°*)

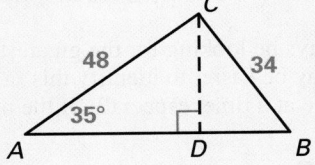

This chapter uses tools developed in previous chapters, especially the properties of similarity, to explore right triangles and their role in trigonometry. Lesson 10.1 lays the algebraic groundwork that will be needed for working with radical expressions throughout the chapter. Lessons 10.2 and 10.3 explore the two most important special types of right triangles, 45°-45°-90° triangles and 30°-60°-90° triangles. The study of trigonometry is begun in Lesson 10.4, where the tangent function is introduced. This study is continued in Lesson 10.5 with the introduction of the other two basic trigonometric functions: sine and cosine. The ideas of the previous two lessons are used in Lesson 10.6 to solve right triangles, that is, to find all the unknown side lengths and angle measures of a given right triangle.

Lesson 10.1

> **QUICK TIP**
> If students have not already done so, have them prepare a list of the perfect squares through $20^2 = 400$. Students can refer to their lists as necessary, but should also be working to memorize the values. You may want to remind students that the square root of 0 is 0 and the square root of 1 is 1.

This lesson provides the algebraic skills that students need in order to deal with quantities that arise from applications the Pythagorean Theorem. Although some students may have studied radical expressions and how to manipulate them in a previous math course, it will be helpful to review several common errors, the following two in particular.

1) If an expression under a radical symbol can be factored into two positive numbers, the radical can be split into the product of two radicals. This fact is used when simplifying a radical expression by "taking a number out from under the radical symbol," as illustrated in Example 4. A common mistake is to forget to take the square root when bringing a number out from under the radical symbol.

2) If expressions under a radical symbol are added, no simplification can be done. For example,
$$\sqrt{9 + 16} = \sqrt{25} = 5 \quad \text{and} \quad \sqrt{9} + \sqrt{16} = 3 + 4 = 7.$$
Therefore,
$$\sqrt{9 + 16} \neq \sqrt{9} + \sqrt{16}.$$

Students should always be looking for the greatest perfect-square factor of the radicand. It may be easier to identify this factor if they take perfect-square factors out one at a time, especially if the number is large.

Lesson 10.2

USING PATTERNS Students always find problems like the one illustrated in Example 1 easier to solve than those of the kind illustrated in Example 4. This is due to the fact that in Example 4 they must divide both sides of

an equation by the radical $\sqrt{2}$ rather than simply substitute, as is done in Example 1. Emphasize that Theorem 10.1 provides a pattern that can always be used to solve 45°-45°-90° triangles. A given side length is simply equated with the corresponding symbolic side length in the model triangle given with the theorem. Then algebra is used, if necessary, to solve for the value of the variable x in the model triangle. Once this value is known, the other side lengths can be easily determined.

Lesson 10.3

The 30°-60°-90° triangle is inherently more difficult for students to work with than the 45°-45°-90° studied in the previous lesson, for two related reasons:

1) There are three different symbolic quantities that represent the side lengths of the 30°-60°-90° triangle, rather than the two shown for 45°-45°-90° triangles.

2) Students must remember which quantity goes opposite which acute angle. In particular, it is often difficult for them to keep straight whether x or $x\sqrt{3}$ is the measure of the side opposite the 30° angle.

One way to deal with the second difficulty is to urge students to draw an accurate sketch of the right triangle on their papers, even if one is provided in the problem. By drawing the triangle so that it does *not* appear to be an isosceles right triangle, students are forced to make a distinction between a smaller angle (30°) and a larger one (60°), and in the process, they should recognize that the side opposite the larger angle must be longer (according to Theorem 4.11). Since $x\sqrt{3} > x$, the expression opposite the 60° angle must be $x\sqrt{3}$, and x must be the length of the side opposite the 30° angle.

IDENTIFYING A PATTERN The pattern for Theorem 10.2 can be easily derived by dividing an equilateral triangle, each of whose sides has length $2x$, into two congruent 30°-60°-90° triangles by drawing one altitude. It then can be seen that one side of the equilateral triangle is divided into two corresponding sides of the 30°-60°-90° triangles, each of length x, and it is not hard to calculate the length of the third (common) side by the Pythagorean Theorem. This derivation will not be of theoretical importance to students, but it may aid their memorization of the pattern shown in Theorem 10.2.

As before, the model triangle of the theorem should be used to set up an equation, by putting a given side length on one side and the corresponding symbolic lengths on the other, and then solving for x. Once this is done, the values of all three sides of the triangle, which are expressed in terms of x, can be calculated.

Lesson 10.4

This lesson begins the study of trigonometry, which continues throughout the remainder of the chapter. It is important to emphasize at the outset that right triangle trigonometry is based on the principle of similarity. As Activity 10.4 illustrates, any two right triangles with, say, one acute angle that measures 40° are similar. Therefore, the ratio of the length of the leg opposite the 40° angle to the length of the other leg is the same regardless of the size of the triangle.

> **QUICK TIP**
> To simplify students' introduction to trigonometry, use a 45°-45°-90° triangle to find the tangent ratio the first time. Then use a 30°-60°-90° triangle. Doing so will help to make a connection between the content of Lesson 10.4 and that of Lessons 10.2 and 10.3.

Students often have trouble performing the algebraic steps needed to solve an equation like the one that occurs in Method 1 of the Solution of Example 4,

$$\tan 35° = \frac{4}{x},$$

where the variable occurs in the denominator of the ratio. For students who have difficulty with such equations, Method 2 of the Solution shows how to avoid such an equation. On the other hand, this situation may not always be avoidable when students encounter the sine and cosine functions in the next lesson, so it is important that they can solve this equation by Method 1, as shown.

Finally, it is important to note that in this lesson and in the forthcoming lessons of this chapter we will be dealing exclusively with right triangles. Stress to students that if the methods given here are applied to non-right triangles, they will give incorrect results. This is the point of Exercise 30 on page 561, which you should assign or discuss in class.

Lesson 10.5

Students may become confused as to whether the sine or the cosine ratio should be used in the solution of a given problem. You should stress that in a typical problem, two sides and one acute angle of a right triangle are involved (at least, we can deal with only two sides and one angle at a time). Two of these will have given values and the value of the third will be unknown. In a problem that calls for sine or cosine, one of the two sides will be the hypotenuse, and the other should be characterized in terms of its relationship to the angle involved, either as the side opposite or adjacent to that angle. Once this is done, it is just a question of remembering that the sine ratio is used when dealing with the side opposite the angle whose measure is known, while the cosine ratio is used when dealing with the adjacent side.

RELATING SINE AND COSINE Note that any problem in which the sine ratio is involved can be converted to a problem in which the cosine ratio is involved, by simply replacing the given angle with its complement (the other acute angle of the right triangle). Although there is very little practical advantage to doing this in most instances, it may be well to mention the fact to your students and use the occasion to explain (as a way of helping them organize their thoughts) that the prefix "*co-*" in cosine is associated with the idea of complement.

Lesson 10.6

Point out that this lesson deals with right triangle situations where the unknown measure is not a side length, but rather the measure of one of the acute angles.

USING A CALCULATOR One of the most important practical ideas conveyed in this lesson is the use of the inverse trigonometric keys on the calculator. These are almost always accessed by means of a function key ("2nd" or "INV"), which must be pressed before the main key ("sin," "cos," or "tan"). The use of these keys can be summarized by saying that they are used to solve an equation of the form

> a trigonometric ratio of x = a number.

Examples 1 and 3 illustrate the use of these keys in the simplest situations. In dealing with more complex situations, such as in Example 4, emphasize that students should think in terms of reducing the problem to an equation of the above form first. Then they can use their calculators.

Chapter Overview

Chapter Goals

After an introduction to simplifying square roots, the chapter investigates the relationship between the side lengths of 45°-45°-90° triangles and the side lengths of 30°-60°-90° triangles. The remainder of the chapter is spent discussing the use of the three main trigonometric ratios. Students will:
- Simplify radicals.
- Find the sine, cosine, and tangent ratios of the acute angles in a right triangle.
- Solve right triangles.

Application Note

The first space shuttle launch was on April 12, 1981, and today, dozens of launches later, space shuttle flights seemingly have become almost routine. During landings, the space shuttle behaves more like a glider than an airplane because it does not use its engines like an airplane when landing. Also, unlike an airplane, the pilot of the space shuttle cannot abort a landing and "go around" for another attempt, so the calculation of the glide angle is critical. During space shuttle missions, students with Internet access can use web sites such as spaceflight.nasa.gov/realdata to locate the exact path of the space shuttle over or near their city. Using this web site 1–2 days prior to a space shuttle landing, students can choose Landing Ground Tracks to view the flight path that the space shuttle will follow when it lands back on Earth.

More information about calculating glide paths is provided on the Internet at classzone.com

Application Links
CLASSZONE.COM

Chapter **10** # Right Triangles and Trigonometry

At what angle does the space shuttle land?

The diagram shows the *glide angle* at which the space shuttle approaches Earth during landing.

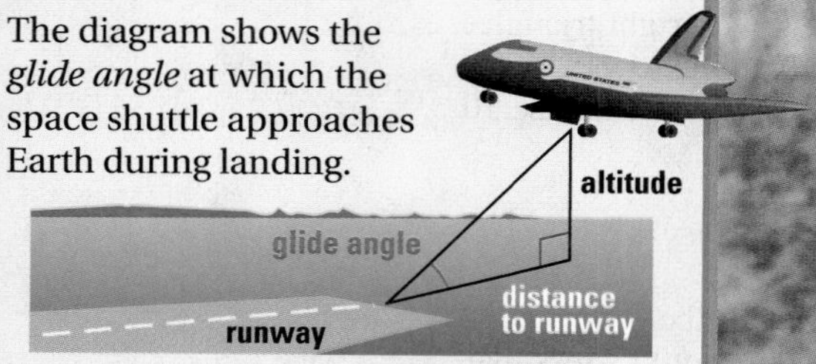

altitude

glide angle

distance to runway

runway

To calculate the glide angle, you need to know the altitude, the distance to the runway, and *trigonometric ratios*, which you will study in this chapter.

Learn More About It

You will learn more about calculating glide angles in Exercise 33 on p. 574.

Hands-On Activities

Activities (more than 20 minutes)
10.3 Special Right Triangles, p. 548
10.4 Right Triangle Ratio, p. 556

Geo-Activities (less than 20 minutes)
10.2 Exploring an Isosceles Right Triangle, p. 542

Projects

A project covering Chapters 9–10 appears on pages 584–585 of the Student Edition. An additional project for Chapter 10 is available in the *Chapter 10 Resource Book*, pp. 76–77.

 Technology

• Electronic Teacher Tools
• Test and Practice Generator
• Online Lesson Planner

Internet Support
CLASSZONE.COM

• Application and Career Links
  553, 559, 561, 574
• Student Help
  540, 543, 551, 562, 567, 570

Who uses Right Triangles and Trigonometry?

PERSONAL TRAINER
Personal trainers develop fitness programs suited to an individual's abilities and goals. They study anatomy, nutrition, and physiology. (p. 553)

FORESTER
Foresters manage and protect forests. To determine the height of a tree, a forester may use trigonometry. (p. 559)

How will you use these ideas?

• Analyze jewelry designs. (p. 546)
• Determine correct positions for certain exercises. (p. 553)
• Measure the height of a tree. (p. 559)
• Plan a skateboard ramp. (p. 567)
• Calculate the glide angle of the space shuttle. (p. 574)

Diagnostic Tools

The **Chapter Readiness Quiz** can help you diagnose whether students have the following skills needed in Chapter 10:
- Classify triangles.
- Find angle measures in triangles.
- Use the Pythagorean Theorem.

Reteaching Material

This resource is available for students who need additional help with the skills on the Chapter Readiness Quiz:

📖 **Chapter 10 Resource Book**
- Reteaching with Practice (Lessons 10.1–10.6)

Additional Resources

The following resources are provided to help you prepare for the upcoming chapter and customize review materials:

📖 **Chapter 10 Resource Book**
- *Tips for New Teachers*, pp. 1–2
- *Parent Guide*, pp. 3–4
- *Lesson Plans*, pp. 7, 15, 24, 34, 45, 55
- *Lesson Plans for Block Scheduling*, pp. 8, 16, 25, 35, 46, 56

🖥 **Technology**
- Electronic Teaching Tools
- Online Lesson Planner
- Test and Practice Generator

Visualize It!

Additional suggestions for helping students visualize geometry are on pp. 549, 554, and 567.

Chapter  10 Study Guide

PREVIEW — **What's the chapter about?**

- Simplifying square roots
- Finding the side lengths of 45°-45°-90° triangles and 30°-60°-90° triangles
- Finding the sine, cosine, and tangent of acute angles
- Solving right triangles

Key Words

- **radical,** *p. 537*
- **radicand,** *p. 537*
- **45°-45°-90° triangle,** *p. 542*
- **30°-60°-90° triangle,** *p. 549*
- **trigonometric ratio,** *p. 557*

- **leg opposite an angle,** *p. 557*
- **leg adjacent to an angle,** *p. 557*
- **tangent,** *p. 557*
- **sine,** *p. 563*
- **cosine,** *p. 563*
- **solve a right triangle,** *p. 569*

PREPARE — **Chapter Readiness Quiz**

Take this quick quiz. If you are unsure of an answer, look at the reference pages for help.

Vocabulary Check *(refer to p. 201)*

1. What type of triangle is shown at the right? **B**

(A) acute   **(B)** right   **(C)** obtuse   **(D)** isosceles

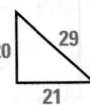

Skill Check *(refer to pp. 180, 192)*

2. What is the value of x in the triangle shown at the right? **G**

(F) 20   **(G)** 30   **(H)** 50   **(J)** 60

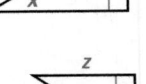

3. What is the approximate value of z in the triangle shown at the right? **B**

(A) 4.0   **(B)** 7.5   **(C)** 10.3   **(D)** 16.0

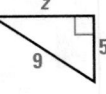

VISUAL STRATEGY — **Labeling Triangles**

Visualize It! ➡ When labeling a right triangle, it may be helpful to use colored pencils. Use one color for the triangle, a second for its angle measures, and a third for its segment lengths.

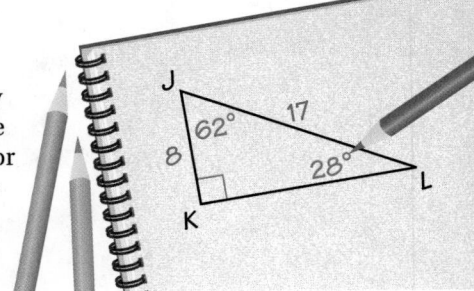

10.1 Simplifying Square Roots

Goal
Simplify square roots.

Key Words
- radical
- radicand

Square roots are written with a radical symbol $\sqrt{}$. An expression written with a radical symbol is called a *radical expression*, or **radical**. The number or expression inside the radical symbol is the **radicand**.

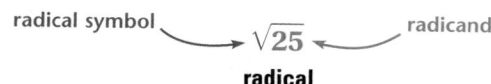

radical symbol $\longrightarrow$ $\sqrt{25}$ $\longleftarrow$ radicand

radical

The radical symbol always indicates the nonnegative square root of a number. For example, $\sqrt{25} = 5$ because $5^2 = 25$.

EXAMPLE 1 Use a Calculator to Find Square Roots

Find the square root of 52. Round your answer to the nearest tenth. Check that your answer is reasonable.

Solution

Calculator keystrokes	Display	Rounded value
52 $\sqrt{}$ or $\sqrt{}$ 52 **ENTER**	7.21110	$\sqrt{52} \approx 7.2$

This is reasonable, because 52 is between the perfect squares 49 and 64. So, $\sqrt{52}$ should be between $\sqrt{49}$ and $\sqrt{64}$, or 7 and 8. The answer 7.2 is between 7 and 8.

EXAMPLE 2 Find Side Lengths

Use the Pythagorean Theorem to find the length of the hypotenuse to the nearest tenth.

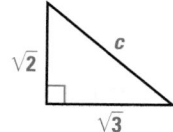

Student Help

> **STUDY TIP**
> Recall that for any number $a \geq 0$, $(\sqrt{a})^2 = a$.

Solution

$$a^2 + b^2 = c^2 \qquad \text{Write Pythagorean Theorem.}$$
$$(\sqrt{2})^2 + (\sqrt{3})^2 = c^2 \qquad \text{Substitute } \sqrt{2} \text{ for } a \text{ and } \sqrt{3} \text{ for } b.$$
$$2 + 3 = c^2 \qquad \text{Simplify.}$$
$$5 = c^2 \qquad \text{Add.}$$
$$\sqrt{5} = c \qquad \text{Take the square root of each side.}$$
$$2.2 \approx c \qquad \text{Use a calculator.}$$

10.1 Simplifying Square Roots **537**

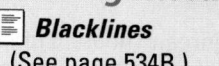 **Plan**

Pacing
Suggested Number of Days

Basic: 2 days
Average: 2 days
Advanced: 2 days
Block Schedule: 1 block

Teaching Resources

📄 *Blacklines*
(See page 534B.)

🔧 *Transparencies*
- Warm-Up with Quiz
- Answers

🖥 *Technology*
- Electronic Teacher Tools
- Test and Practice Generator
- Online Lesson Planner
- Internet Support

Teach

┌ **Content and Teaching Strategies** ┐
For background information on geometric concepts and teaching strategies related to this lesson, see pages 534E and 534F in this Teacher's Edition.

Tips for New Teachers
Students must be careful when using formulas or equations that involve squaring radicals, as in Example 2 on this page. It is easy for students to mistakenly write $(\sqrt{2})^2 = 4$ and $(\sqrt{3})^2 = 9$. Remind students that when squaring a radical all they need to do is "drop" the radical symbol. See the Tips for New Teachers on pp. 1–2 of the *Chapter 10 Resource Book* for additional notes about Lesson 10.1.

Extra Examples 1 and 2
See next page.

538

Extra Example 1

Find the square root of 74. Round your answer to the nearest tenth. Check that your answer is reasonable. $\sqrt{74} \approx 8.6$; This is reasonable because 74 is between the perfect squares 64 and 81. So $\sqrt{74}$ should be between $\sqrt{64}$ and $\sqrt{81}$, or 8 and 9. The answer 8.6 is between 8 and 9.

Extra Example 2

Use the Pythagorean Theorem to find the length of the hypotenuse to the nearest tenth. 3.3

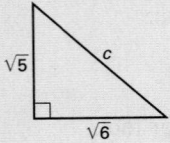

Extra Example 3

Multiply the radicals. Then simplify if possible.
a. $\sqrt{5} \cdot \sqrt{6}$ $\sqrt{30}$
b. $\sqrt{3} \cdot \sqrt{12}$ 6

Extra Example 4

Simplify the radical expression.
a. $\sqrt{28}$ $2\sqrt{7}$
b. $\sqrt{54}$ $3\sqrt{6}$

✓ Concept Check

How do you use the Product Property of Radicals to simplify radical expressions? **You simplify a radical expression by finding the perfect square factors of the radicand and then simplifying these factors.**

Student Help

SKILLS REVIEW
To review the Product Property of Radicals, see p. 669.

Multiplying Radicals You can use the *Product Property of Radicals* to multiply radical expressions.

$$\sqrt{a} \cdot \sqrt{b} = \sqrt{ab}, \text{ where } a \geq 0 \text{ and } b \geq 0.$$

EXAMPLE 3 **Multiply Radicals**

Multiply the radicals. Then simplify if possible.

a. $\sqrt{3} \cdot \sqrt{7}$ **b.** $\sqrt{2} \cdot \sqrt{8}$

Solution

a. $\sqrt{3} \cdot \sqrt{7} = \sqrt{3 \cdot 7}$ **b.** $\sqrt{2} \cdot \sqrt{8} = \sqrt{2 \cdot 8}$
$\phantom{\textbf{a.} \sqrt{3} \cdot \sqrt{7}} = \sqrt{21}$ $\phantom{\textbf{b.} \sqrt{2} \cdot \sqrt{8}} = \sqrt{16}$
$\phantom{\textbf{b.} \sqrt{2} \cdot \sqrt{8} = \sqrt{16}} = 4$

Simplifying Radicals You can also use the *Product Property of Radicals* to simplify radical expressions.

$$\sqrt{ab} = \sqrt{a} \cdot \sqrt{b}, \text{ where } a \geq 0 \text{ and } b \geq 0.$$

To factor the radicand, look for perfect square factors.

EXAMPLE 4 **Simplify Radicals**

Simplify the radical expression.

a. $\sqrt{12}$ **b.** $\sqrt{45}$

Student Help

STUDY TIP
When you factor a radicand, write the perfect square factors first.

Solution

a. $\sqrt{12} = \sqrt{4 \cdot 3}$ **b.** $\sqrt{45} = \sqrt{9 \cdot 5}$
$\phantom{\textbf{a.} \sqrt{12}} = \sqrt{4} \cdot \sqrt{3}$ $\phantom{\textbf{b.} \sqrt{45}} = \sqrt{9} \cdot \sqrt{5}$
$\phantom{\textbf{a.} \sqrt{12}} = 2\sqrt{3}$ $\phantom{\textbf{b.} \sqrt{45}} = 3\sqrt{5}$

Checkpoint ✓ **Evaluate, Multiply, and Simplify Radicals**

Find the square root. Round your answer to the nearest tenth. Check that your answer is reasonable.

1. $\sqrt{27}$ **2.** $\sqrt{46}$ **3.** $\sqrt{8}$ **4.** $\sqrt{97}$

1. 5.2; $\sqrt{25} < \sqrt{27} < \sqrt{36}$, so $5 < 5.2 < 6$.
2. 6.8; $\sqrt{36} < \sqrt{46} < \sqrt{49}$, so $6 < 6.8 < 7$.
3. 2.8; $\sqrt{4} < \sqrt{8} < \sqrt{9}$, so $2 < 2.8 < 3$.
4. 9.8; $\sqrt{81} < \sqrt{97} < \sqrt{100}$, so $9 < 9.8 < 10$.

Multiply the radicals. Then simplify if possible.

5. $\sqrt{3} \cdot \sqrt{5}$ $\sqrt{15}$ **6.** $\sqrt{11} \cdot \sqrt{6}$ $\sqrt{66}$ **7.** $\sqrt{3} \cdot \sqrt{27}$ 9   **8.** $5\sqrt{3} \cdot \sqrt{3}$ 15

Simplify the radical expression.

9. $\sqrt{20}$ $2\sqrt{5}$ **10.** $\sqrt{8}$ $2\sqrt{2}$ **11.** $\sqrt{75}$ $5\sqrt{3}$ **12.** $\sqrt{112}$ $4\sqrt{7}$

Guided Practice

Vocabulary Check

1. What is the *radicand* in the expression $\sqrt{25}$? 25

Match the radical expression with its simplified form.

2. $\sqrt{36}$ D   **A.** $\sqrt{6}$

3. $\sqrt{3} \cdot \sqrt{2}$ A   **B.** $3\sqrt{2}$

4. $\sqrt{3} \cdot \sqrt{6}$ B   **C.** $4\sqrt{2}$

5. $\sqrt{32}$ C   **D.** 6

Skill Check

Use the figure shown at the right.

6. Use the Pythagorean Theorem to find the length of the hypotenuse in radical form. $2\sqrt{5}$

7. Use a calculator to find the length of the hypotenuse to the nearest tenth. 4.5

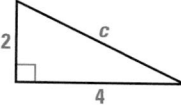

12. 3.6; $\sqrt{9} < \sqrt{13} < \sqrt{16}$, so $3 < 3.6 < 4$.

13. 2.4; $\sqrt{4} < \sqrt{6} < \sqrt{9}$, so $2 < 2.4 < 3$.

Simplify the expression.

8. $\sqrt{49}$ 7   9. $\sqrt{28}$ $2\sqrt{7}$ 10. $\sqrt{72}$ $6\sqrt{2}$ 11. $\sqrt{54}$ $3\sqrt{6}$

Practice and Applications

Extra Practice

See p. 693.

14. 9.5; $\sqrt{81} < \sqrt{91} < \sqrt{100}$, so $9 < 9.5 < 10$.

15. 5.8; $\sqrt{25} < \sqrt{34} < \sqrt{36}$, so $5 < 5.8 < 6$.

16. 10.3; $\sqrt{100} < \sqrt{106} < \sqrt{121}$, so $10 < 10.3 < 11$.

17. 12.2; $\sqrt{144} < \sqrt{148} < \sqrt{169}$, so $12 < 12.2 < 13$.

18. 7.9; $\sqrt{49} < \sqrt{62} < \sqrt{64}$, so $7 < 7.9 < 8$.

19. 13.6; $\sqrt{169} < \sqrt{186} < \sqrt{196}$, so $13 < 13.6 < 14$.

Finding Square Roots Find the square root. Round your answer to the nearest tenth. Check that your answer is reasonable.

12. $\sqrt{13}$ 13. $\sqrt{6}$ 14. $\sqrt{91}$ 15. $\sqrt{34}$

16. $\sqrt{106}$ 17. $\sqrt{148}$ 18. $\sqrt{62}$ 19. $\sqrt{186}$

Pythagorean Theorem Find the length of the hypotenuse. Write your answer in radical form.

20.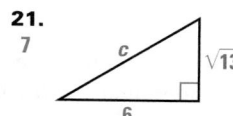

21.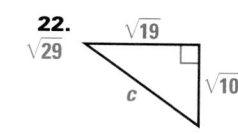

22. (figure with $\sqrt{19}$, $\sqrt{29}$, c, $\sqrt{10}$)

Homework Help

Example 1: Exs. 12–19
Example 2: Exs. 20–25
Example 3: Exs. 26–34
Example 4: Exs. 35–45

Pythagorean Theorem Find the missing side length of the right triangle. Round your answer to the nearest tenth.

23. 4.1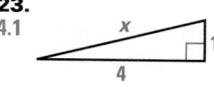

24. 5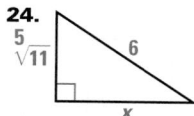

25. 2.8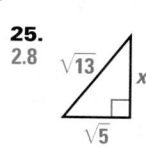

③ Apply

Assignment Guide

BASIC
Day 1: SRH p. 669 Exs. 9–16; pp. 539–541 Exs. 12–25, 54, 55, 63–68
Day 2: pp. 539–541 Exs. 26–43, 56–62

AVERAGE
Day 1: pp. 539–541 Exs. 16–25, 54, 55, 63–68
Day 2: pp. 539–541 Exs. 26–51, 56–62

ADVANCED
Day 1: pp. 539–541 Exs. 16–25, 54, 55, 63–68
Day 2: pp. 539–541 Exs. 29–52, 53*, 56–62; EC: classzone.com

BLOCK SCHEDULE
pp. 539–541 Exs. 16–51, 54–68

Extra Practice
• Student Edition, p. 693
• Chapter 10 Resource Book, pp. 10–11

Homework Check
To quickly check student understanding of key concepts, go over the following exercises:

Basic: 16, 20, 26, 32, 38
Average: 17, 24, 30, 33, 42
Advanced: 19, 25, 31, 34, 43

Study Skills
Note-Taking Suggest that students make a list of all perfect squares from 1^2 to 20^2 to help them quickly recognize and use these values in this exercise set.

540

Multiplying Radicals Multiply the radicals. Then simplify if possible.

26. $\sqrt{7} \cdot \sqrt{2}$ $\sqrt{14}$

27. $\sqrt{5} \cdot \sqrt{5}$ 5

28. $\sqrt{3} \cdot \sqrt{11}$ $\sqrt{33}$

29. $2\sqrt{5} \cdot \sqrt{7}$ $2\sqrt{35}$

30. $\sqrt{10} \cdot 4\sqrt{3}$ $4\sqrt{30}$

31. $\sqrt{11} \cdot \sqrt{22}$ $11\sqrt{2}$

EXAMPLE *Square a Radical*

Evaluate the expression.

    a. $(3\sqrt{7})^2$ **b.** $(2\sqrt{11})^2$

Solution

  a. $(3\sqrt{7})^2 = 3\sqrt{7} \cdot 3\sqrt{7}$ **b.** $(2\sqrt{11})^2 = 2\sqrt{11} \cdot 2\sqrt{11}$

           $= 3 \cdot 3 \cdot \sqrt{7} \cdot \sqrt{7}$ $= 2 \cdot 2 \cdot \sqrt{11} \cdot \sqrt{11}$

           $= 9 \cdot 7$ $= 4 \cdot 11$

           $= 63$ $= 44$

Squaring Radicals Evaluate the expression. Use the example above as a model.

32. $(6\sqrt{5})^2$ 180

33. $(5\sqrt{3})^2$ 75

34. $(7\sqrt{2})^2$ 98

Simplifying Radicals Simplify the radical expression.

35. $\sqrt{18}$ $3\sqrt{2}$

36. $\sqrt{50}$ $5\sqrt{2}$

37. $\sqrt{48}$ $4\sqrt{3}$

38. $\sqrt{60}$ $2\sqrt{15}$

39. $\sqrt{56}$ $2\sqrt{14}$

40. $\sqrt{125}$ $5\sqrt{5}$

41. $\sqrt{200}$ $10\sqrt{2}$

42. $\sqrt{162}$ $9\sqrt{2}$

43. $\sqrt{44}$ $2\sqrt{11}$

⚖ **You be the Judge** Determine whether the expression can be simplified further. If so, explain how you would do so.

44. Yes; simplify $\sqrt{20}$ as $2\sqrt{5}$, then $\sqrt{80} = 2 \cdot 2\sqrt{5} = 4\sqrt{5}$.

44. $\sqrt{80} = \sqrt{4 \cdot 20}$

        $= \sqrt{4} \cdot \sqrt{20}$

        $= 2\sqrt{20}$

45. $\sqrt{8} \cdot \sqrt{12} = \sqrt{8 \cdot 12}$

   no $= \sqrt{4 \cdot 2 \cdot 4 \cdot 3}$

               $= 4\sqrt{6}$

Area Formula Use the area formula $A = lw$ to find the area of the rectangle. Round your answer to the nearest tenth.

46.
11.8
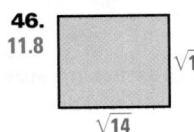

47.
62.0
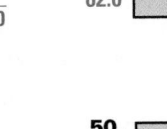

48.
124.7

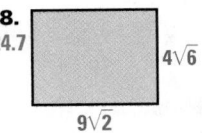

ℹ **Student Help**
CLASSZONE.COM

HOMEWORK HELP
Extra help with problem solving in Exs. 46–51 is at classzone.com

49.
48
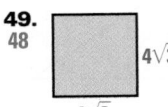

50.
135.8
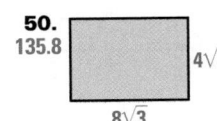

51.
15.9
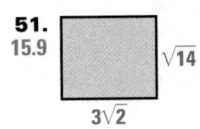

52. Area of an Equilateral Triangle The area of an equilateral triangle with side length *s* is given by the formula

$$A = \frac{1}{4}s^2\sqrt{3}.$$

The flower bed shown is an equilateral triangle with a side length of 30 feet. Find its area.
$225\sqrt{3} \approx 389.7 \text{ ft}^2$

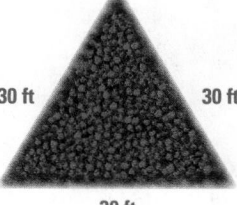
30 ft   30 ft

30 ft

53. Challenge An equilateral triangle has an area of 1 square meter. What is the length of each side? Round your answer to the nearest centimeter. **152 centimeters**

Standardized Test Practice

54. Multiple Choice Which number is a perfect square? **C**

Ⓐ 44   Ⓑ 110   Ⓒ 169   Ⓓ 500

55. Multiple Choice $\sqrt{220}$ is between which two integers? **H**

Ⓕ 12 and 13   Ⓖ 13 and 14   Ⓗ 14 and 15   Ⓙ 15 and 16

56. Multiple Choice Which of the following expressions could *not* be used to represent the length of the hypotenuse in the triangle shown at the right? **D**

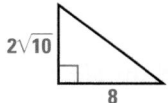
$2\sqrt{10}$

8

Ⓐ $2\sqrt{26}$ Ⓑ $\sqrt{104}$ Ⓒ about 10.2   Ⓓ $6\sqrt{10}$

Mixed Review

Finding Angle Measures Find the measure of ∠1. *(Lesson 4.2)*

57.

45°
45°   1

58.
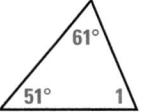
68°
61°
51°   1

59.
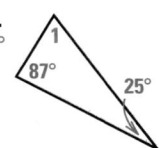
68°
1
87°
25°

Isosceles Triangles Find the value of *x*. *(Lesson 4.3)*

60.

13   9   x − 4

61.
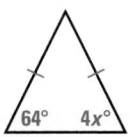
16
64°   4x°

62.
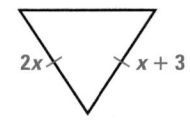
3
2x   x + 3

Algebra Skills

Distributive Property Use the distributive property to rewrite the expression without parentheses. *(Skills Review, p. 671)*

63. $x(x + 5)$ $x^2 + 5x$ **64.** $4(2x − 1)$ $8x − 4$ **65.** $x(3x + 4)$ $3x^2 + 4x$

66. $5x(x + 2)$ $5x^2 + 10x$ **67.** $−3(1 − x)$ $−3 + 3x$ **68.** $2x − (x − 6)$ $x + 6$

10.1 *Simplifying Square Roots*   541

Mini-Quiz
1. Find the square root of 110. Round your answer to the nearest tenth. Check that your answer is reasonable.
10.5; $\sqrt{100} < \sqrt{110} < \sqrt{121}$, so $10 < 10.5 < 11$.

2. Multiply $\sqrt{2} \cdot \sqrt{32}$. Then simplify if possible. **8**

Simplify the radical expression.
3. $\sqrt{45}$ $3\sqrt{5}$
4. $\sqrt{75}$ $5\sqrt{3}$

5. Find the length of the hypotenuse of a right triangular roof support whose legs measure $\sqrt{13}$ meters and 4 meters. Round your answer to the nearest tenth of a meter. **5.4 m**

Pacing

Suggested Number of Days

Basic: 2 days
Average: 2 days
Advanced: 2 days
Block Schedule: 1 block

Teaching Resources

📄 **Blacklines**
(See page 534B.)

🖐 **Transparencies**
• Warm-Up with Quiz
• Answers

🖥 **Technology**
• Electronic Teacher Tools
• Test and Practice Generator
• Online Lesson Planner
• Internet Support

Geo-Activity

Goal Use paper folding to discover the side and angle relationships of an isosceles right triangle.

Key Discovery In a right triangle with angle measures of 45°, 45°, and 90°, the lengths of the two legs are equal and the hypotenuse is about 1.4 times the length of each leg.

②Teach

Content and Teaching Strategies
For background information on geometric concepts and teaching strategies related to this lesson, see pages 534E and 534F in this Teacher's Edition.

Goal

Find the side lengths of 45°-45°-90° triangles.

Key Words

• 45°-45°-90° triangle
• isosceles triangle p. 173
• leg of a right triangle p. 192
• hypotenuse p. 192

Student Help

LOOK BACK
To review the Pythagorean Theorem, see p. 192.

Geo-Activity ▷ Exploring an Isosceles Right Triangle

❶ Fold a large piece of paper so the top lines up with one side.

❷ Measure the angles of the triangle formed.
45°, 45°, 90°

❸ Measure the legs of the triangle.
Answers may vary.

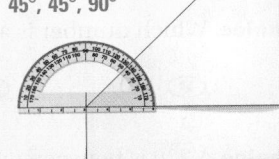

❹ Use the Pythagorean Theorem to predict the length of the hypotenuse.
Answers may vary, but will be √2 ≈ 1.4 times the answer in Step 3.

❺ Measure the hypotenuse to verify your answer in Step 4.
The answer should be the same as in Step 4.

A right triangle with angle measures of 45°, 45°, and 90° is called a **45°-45°-90° triangle**. You can use the Pythagorean Theorem to find the length of the hypotenuse of any 45°-45°-90° triangle.

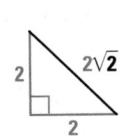

$$\sqrt{2^2 + 2^2} = \sqrt{4 + 4}$$
$$= \sqrt{4 \cdot 2}$$
$$= \sqrt{4} \cdot \sqrt{2}$$
$$= 2\sqrt{2}$$

$$\sqrt{3^2 + 3^2} = \sqrt{9 + 9}$$
$$= \sqrt{9 \cdot 2}$$
$$= \sqrt{9} \cdot \sqrt{2}$$
$$= 3\sqrt{2}$$

$$\sqrt{4^2 + 4^2} = \sqrt{16 + 16}$$
$$= \sqrt{16 \cdot 2}$$
$$= \sqrt{16} \cdot \sqrt{2}$$
$$= 4\sqrt{2}$$

THEOREM 10.1

45°-45°-90° Triangle Theorem

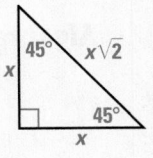

Words In a 45°-45°-90° triangle, the length of the hypotenuse is the length of a leg times √2.

Symbols hypotenuse = leg · √2

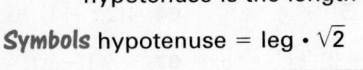

EXAMPLE 1 Find Hypotenuse Length

Find the length x of the hypotenuse in the
45°-45°-90° triangle shown at the right.

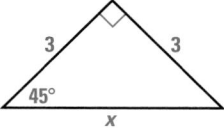

Solution

By the 45°-45°-90° Triangle Theorem, the length of the
hypotenuse is the length of a leg times $\sqrt{2}$.

hypotenuse = leg · $\sqrt{2}$ 45°-45°-90° Triangle Theorem

         = 3 · $\sqrt{2}$ Substitute.

ANSWER ▶ The length of the hypotenuse is $3\sqrt{2}$.

EXAMPLE 2 Find Leg Length

Find the length x of each leg in the
45°-45°-90° triangle shown at the right.

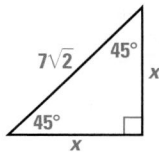

Solution

By the 45°-45°-90° Triangle Theorem, the length of the
hypotenuse is the length of a leg times $\sqrt{2}$.

hypotenuse = leg · $\sqrt{2}$ 45°-45°-90° Triangle Theorem

     $7\sqrt{2} = x\sqrt{2}$ Substitute.

     $\dfrac{7\sqrt{2}}{\sqrt{2}} = \dfrac{x\sqrt{2}}{\sqrt{2}}$ Divide each side by $\sqrt{2}$.

       $7 = x$ Simplify.

ANSWER ▶ The length of each leg is 7.

Student Help

READING TIP
The expression $x\sqrt{2}$ is
equivalent to $\sqrt{2}x$.

Checkpoint ✓ *Find Hypotenuse and Leg Lengths*

Find the value of x.

1.
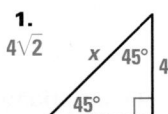

2.
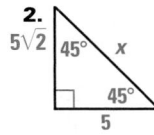

3.
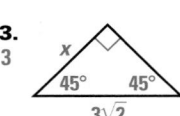

4.
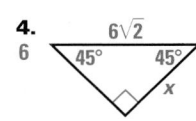

10.2 *45°-45°-90° Triangles*   **543**

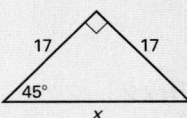
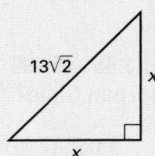

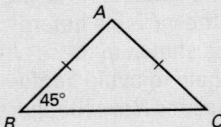
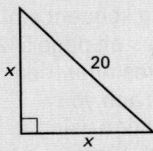
544

Using Algebra

EXAMPLE 3 Identify 45°-45°-90° Triangles

Determine whether there is enough information to conclude that the triangle is a 45°-45°-90° triangle. Explain your reasoning.

Solution

By the Triangle Sum Theorem, $x° + x° + 90° = 180°$.

So, $2x° = 90°$, and $x = 45$.

ANSWER ▶ Since the measure of each acute angle is 45°, the triangle is a 45°-45°-90° triangle.

Example 3 shows that whenever a right triangle has congruent acute angles, it is a 45°-45°-90° triangle.

Student Help

LOOK BACK
To review the Base Angles Theorem, see p. 185.

EXAMPLE 4 Find Leg Length

Show that the triangle is a 45°-45°-90° triangle. Then find the value of x.

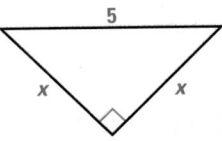

Solution

The triangle is an isosceles right triangle. By the Base Angles Theorem, its acute angles are congruent. From the result of Example 3, this triangle must be a 45°-45°-90° triangle.

You can use the 45°-45°-90° Triangle Theorem to find the value of x.

hypotenuse = leg · $\sqrt{2}$	45°-45°-90° Triangle Theorem
$5 = x\sqrt{2}$	Substitute.
$\dfrac{5}{\sqrt{2}} = \dfrac{x\sqrt{2}}{\sqrt{2}}$	Divide each side by $\sqrt{2}$.
$\dfrac{5}{\sqrt{2}} = x$	Simplify.
$3.5 \approx x$	Use a calculator to approximate.

5. The triangle is an isosceles right triangle. By the Base Angles Theorem, its acute angles are congruent. From the result of Example 3, the triangle is a 45°-45°-90° triangle. $x = \dfrac{8}{\sqrt{2}} \approx 5.7$.

Checkpoint ✓ Find Leg Lengths

6. The triangle has congruent acute angles. By Example 3, the triangle is a 45°-45°-90° triangle. $x = \dfrac{12}{\sqrt{2}} \approx 8.5$.

Show that the triangle is a 45°-45°-90° triangle. Then find the value of x. Round your answer to the nearest tenth.

5.
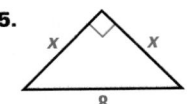

6.
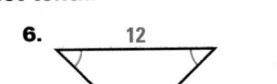

10.2 Exercises

Guided Practice

Vocabulary Check

1. How many congruent sides does an *isosceles right triangle* have? 2

2. How many congruent angles does an isosceles right triangle have? What are the measures of the three angles? 2; 45°, 45°, and 90°

Skill Check

Find the value of x in the 45°-45°-90° triangle. Write your answer in radical form.

3.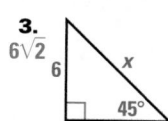
$6\sqrt{2}$, 6, x, 45°

4.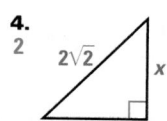
2, $2\sqrt{2}$, x, x

5.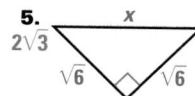
x, $2\sqrt{3}$, $\sqrt{6}$, $\sqrt{6}$

Practice and Applications

Extra Practice
See p. 693.

Finding Hypotenuse Lengths Find the length of the hypotenuse in the 45°-45°-90° triangle. Write your answer in radical form.

6.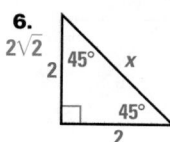
$2\sqrt{2}$, 2, 45°, x, 45°, 2

7.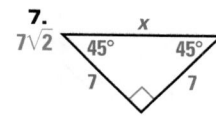
$7\sqrt{2}$, x, 45°, 45°, 7, 7

8.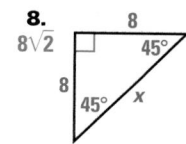
8, $8\sqrt{2}$, 45°, 8, 45°, x

9.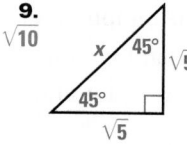
$\sqrt{10}$, x, 45°, $\sqrt{5}$, 45°, $\sqrt{5}$

10.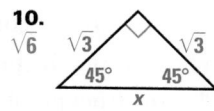
$\sqrt{6}$, $\sqrt{3}$, $\sqrt{3}$, 45°, 45°, x

11.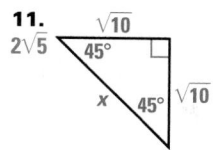
$\sqrt{10}$, $2\sqrt{5}$, 45°, x, 45°, $\sqrt{10}$

Finding Leg Lengths Find the length of a leg in the 45°-45°-90° triangle.

12.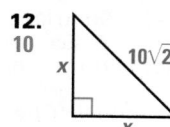
10, x, $10\sqrt{2}$, x

13.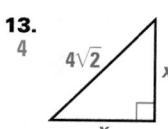
4, $4\sqrt{2}$, x, x

14.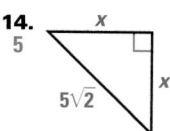
x, 5, x, $5\sqrt{2}$

15.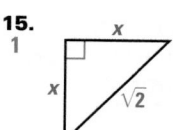
1, x, x, $\sqrt{2}$

16.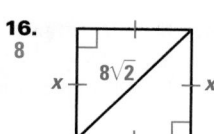
8, x, $8\sqrt{2}$, x

17.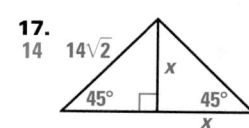
14, $14\sqrt{2}$, x, 45°, 45°, x

Homework Help

Example 1: Exs. 6–11, 18
Example 2: Exs. 12–17
Example 3: Exs. 19–27
Example 4: Exs. 22–27

③ Apply

Assignment Guide

BASIC
Day 1: pp. 545–547 Exs. 6–17, 35–45
Day 2: SRH p. 669 Exs. 37–40; pp. 545–547 Exs. 19–27 odd, 32–34, 46–53

AVERAGE
Day 1: pp. 545–547 Exs. 6–18, 38–45, 47–53 odd
Day 2: pp. 545–547 Exs. 19–31, 34–37

ADVANCED
Day 1: pp. 545–547 Exs. 6–18, 39–53 odd
Day 2: pp. 545–547 Exs. 19–37; EC: TE p. 534D*, classzone.com

BLOCK SCHEDULE
pp. 545–547 Exs. 6–31, 34–45, 47–53 odd

Extra Practice
• Student Edition, p. 693
• Chapter 10 Resource Book, pp. 19–20

Homework Check
To quickly check student understanding of key concepts, go over the following exercises:

Basic: 7, 9, 12, 19, 23
Average: 8, 11, 15, 20, 26
Advanced: 9, 13, 17, 21, 27

✗ Common Error

In Exercises 22–27, watch for students who multiply the hypotenuse by $\sqrt{2}$ instead of dividing by $\sqrt{2}$.

28. $\triangle ADB, \triangle ACD, \triangle BCD$; since D is on the perpendicular bisector of $\overline{AB}$, $\overline{AD} \cong \overline{DB}$ by the Perpendicular Bisector Theorem; $m\angle ADB = 90°$, so $\triangle ADB$ is an isosceles right triangle; by the Base Angles Theorem, $\angle A \cong \angle B$; from the result of Example 3 on p. 544, $\triangle ADB$ is a 45°-45°-90° triangle. $m\angle ACD = 90°$, so by the Triangle Sum Theorem, $m\angle ADC = 45°$ and so $\triangle ACD$ is a 45°-45°-90° triangle. $m\angle BCD = 90°$, so by the Triangle Sum Theorem, $m\angle BDC = 45°$ and so $\triangle BCD$ is a 45°-45°-90° triangle.

31. If the hypotenuse has length $5\sqrt{2}$, the legs should each have length 5. If the legs have length $\sqrt{5}$, then the hypotenuse has length $\sqrt{5} \cdot \sqrt{2}$, or $\sqrt{10}$.

19. No; you cannot determine the measures of the other two angles.

20. Yes; the triangle has congruent acute angles, so by Example 3 on p. 544, the triangle is a 45°-45°-90° triangle.

21. No; the triangle is isosceles, but there is no information about the angle measures.

22. The triangle is an isosceles right triangle. By the Base Angles Theorem, its acute angles are congruent. From the result of Example 3, the triangle is a 45°-45°-90° triangle. $x = \dfrac{4}{\sqrt{2}} \approx 2.8$.

23. The triangle has congruent acute angles. By Example 3, the triangle is a 45°-45°-90° triangle. $x = \dfrac{9}{\sqrt{2}} \approx 6.4$.

24. By the Triangle Sum Theorem, the third angle measures 90°. So the triangle is a 45°-45°-90° triangle. $x = \dfrac{32}{\sqrt{2}} \approx 22.6$.

25. By Example 3, $x = 45$ and the triangle is a 45°-45°-90° triangle. $y = \dfrac{8}{\sqrt{2}} \approx 5.7$.

26. The triangle is an isosceles right triangle. By the Base Angles Theorem, its acute angles are congruent. From the result of Example 3, the triangle is a 45°-45°-90° triangle. $x = \dfrac{20}{\sqrt{2}} \approx 14.1$.

27. By the Triangle Sum Theorem, the third angle measures 45°. So the triangle is a 45°-45°-90° triangle. $x = \dfrac{35}{\sqrt{2}} \approx 24.7$.

29. Lengths may vary, but AC, CB, and CD are all equal. Since $\triangle ACD$ and $\triangle BCD$ are 45°-45°-90° triangles, they are isosceles triangles by the Converse of the Base Angles Theorem. Therefore, $AC = CD$ and $CD = BC$.

30. Lengths may vary, but $AD = DB$ and each is equal to $\sqrt{2} \approx 1.4$ times the length in Exercise 29.

18. Jewelry Use a calculator to find the length x of the earring shown at the right. Round your answer to the nearest tenth.
2.0 centimeters

1.4 cm
1.4 cm

You be the Judge Determine whether there is enough information to conclude that the triangle is a 45°-45°-90° triangle. Explain your reasoning.

19.
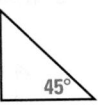
45°

20.

21.
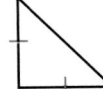

Finding Leg Lengths Show that the triangle is a 45°-45°-90° triangle. Then find the value of each variable. Round to the nearest tenth.

22.
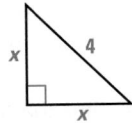
x
4
x

23.
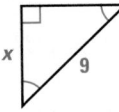
x
9

24.
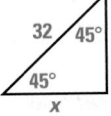
32
45°
45°
x

25.
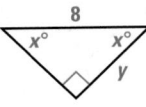
8
$x°$
$x°$
y

26.
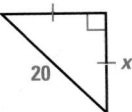
20
x

27.
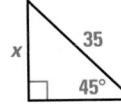
x
35
45°

Technology In Exercises 28–30, use geometry software.

❶ Draw $\overline{AB}$ and construct its midpoint, C.

❷ Construct the perpendicular bisector of $\overline{AB}$.

❸ Construct point D on the bisector and construct $\overline{AD}$ and $\overline{DB}$.

❹ Measure $\angle ADB$. Drag point D until $m\angle ADB = 90°$.

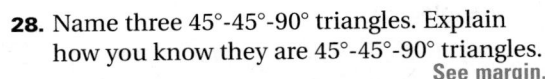
D
A
C
B

28. Name three 45°-45°-90° triangles. Explain how you know they are 45°-45°-90° triangles.
See margin.

29. Measure $\overline{AC}$, $\overline{CB}$, and $\overline{CD}$. What do you notice? Explain.

30. Predict the measures of $\overline{AD}$ and $\overline{DB}$. Check your answer by measuring the segments.

31. Error Analysis A student labels a 45°-45°-90° triangle as shown. Explain and correct the error.
See margin.

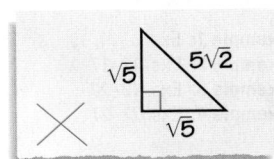
$\sqrt{5}$
$5\sqrt{2}$
$\sqrt{5}$

546 **Chapter 10** *Right Triangles and Trigonometry*

34. a. $x = 15$; $m\angle A = 45°$;
   $m\angle B = 45°$; $m\angle C = 90°$

b. $b = 12$; $c = 12\sqrt{2}$

c. Check using the
   Pythagorean Theorem:
   $12^2 + 12^2 \stackrel{?}{=} (12\sqrt{2})^2$
   $144 + 144 \stackrel{?}{=} 144 \cdot 2$
   $288 = 288$

Standardized Test Practice

Mixed Review

Algebra Skills

Quilt Design The quilt design in the photo is based on the pattern in the diagram below. Use the diagram in Exercises 32 and 33.

"Wheel of Theodorus,"
by Diana Venters

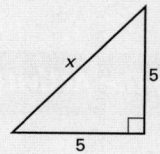

32. Working from left to right, use the Pythagorean Theorem in each right triangle to find the values of r, s, t, u, v, and w.
$r = \sqrt{2}$; $s = \sqrt{3}$; $t = 2$; $u = \sqrt{5}$; $v = \sqrt{6}$; $w = \sqrt{7}$

33. Identify any 45°-45°-90° triangles in the figure.
the right triangle with legs of length 1 and
hypotenuse of length $r = \sqrt{2}$

34. Multi-Step Problem Use the triangle shown below. **See margin.**

a. Find the value of x. Then find $m\angle A$, $m\angle B$, and $m\angle C$.

b. Find the values of b and c.

c. Use the Pythagorean Theorem or the 45°-45°-90° Triangle Theorem to justify your answers in part (b).

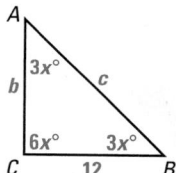

Classifying Triangles Classify the triangle as *acute*, *right*, or *obtuse*. *(Lesson 4.5)*

35.
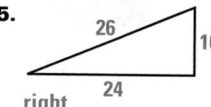
right

36.
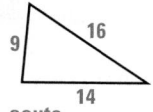
acute

37.
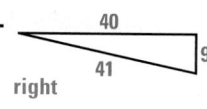
right

Simplifying Radicals Simplify the radical expression. *(Lesson 10.1)*

38. $\sqrt{24}$ $2\sqrt{6}$ **39.** $\sqrt{63}$ $3\sqrt{7}$ **40.** $\sqrt{52}$ $2\sqrt{13}$ **41.** $\sqrt{64}$ 8

42. $\sqrt{80}$ $4\sqrt{5}$ **43.** $\sqrt{196}$ 14 **44.** $\sqrt{250}$ $5\sqrt{10}$ **45.** $\sqrt{117}$ $3\sqrt{13}$

Writing Fractions as Decimals Write the fraction as a decimal. For repeating decimals, also round to the nearest hundredth. *(Skills Review, p. 657)*

46. $\frac{9}{10}$ 0.9 **47.** $\frac{3}{5}$ 0.6 **48.** $\frac{2}{3}$ $0.666... \approx 0.67$ **49.** $\frac{33}{100}$ 0.33

50. $\frac{4}{9}$ $0.444... \approx 0.44$ **51.** $\frac{3}{20}$ 0.15 **52.** $\frac{47}{50}$ 0.94 **53.** $\frac{1}{6}$ $0.1666... \approx 0.17$

10.2 *45°-45°-90° Triangles* **547**

Mini-Quiz
Find the length of the hypotenuse in the 45°-45°-90° triangle. Write your answer in radical form.

1.

$5\sqrt{2}$

2.
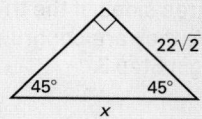
44

3. Find the length of a leg in the 45°-45°-90° triangle. Round your answer to the nearest tenth. **12.7**
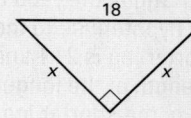

Goal

Students explore the relationship between side measures in a 30°-60°-90° triangle.

Materials

See the margin of the student page.

LINK TO LESSON

In Examples 2 and 3 on page 550, students find the lengths of the hypotenuse and longer leg of a 30°-60°-90° triangle using the ratios they discover in this activity.

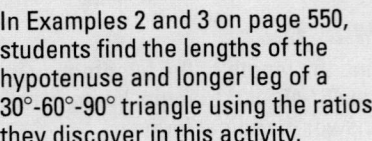

Tips for Success

After Step 2, suggest that students measure the three sides of the triangle to be sure they are congruent before they begin Step 3.

KEY DISCOVERY

In a 30°-60°-90° triangle, the ratio of the length of the hypotenuse to the length of the shorter leg is 2 : 1 and the ratio of the length of the longer leg to the length of the shorter leg is about 1.73 : 1.

Activity Assessment

Use Exercise 5 to assess student understanding.

Activity 10.3 Special Right Triangles

Question

What is special about the ratios of the side lengths in a triangle with angle measures 30°, 60°, and 90°?

Materials
- compass
- centimeter ruler
- calculator

Explore

1 Draw a segment at least 10 centimeters long. Label it $\overline{AB}$. Set your compass opening to AB. Draw arcs with center A and center B.

2 Label the intersection of the arcs C. Draw equilateral $\triangle ABC$. Use your ruler to locate the midpoint of $\overline{AB}$. Label it D.

3 Draw $\overline{CD}$. $\triangle ACD$ has angle measures of 30°, 60°, and 90°. Measure AC, AD, and CD to the nearest millimeter.

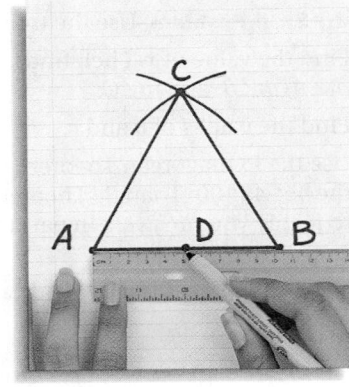
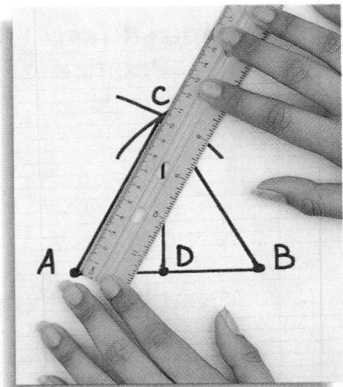

Think About It

In Exercises 1–3, AC, AD, and CD are the lengths in centimeters of the sides of triangles similar to the one you constructed. Copy and complete the table. In Exercise 4, use the values from your drawing.

Student Help

LOOK BACK
To review ratios of side lengths in similar triangles, see p. 365.

	AC	AD	CD	$\frac{AC}{AD}$	$\frac{CD}{AD}$	
1.	10	5	8.7	? 2	?	1.74
2.	20	10	17.3	? 2	?	1.73
3.	50	25	43.3	? 2	?	1.732
4.	?	?	?	?	?	

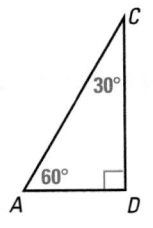

4. Lengths will vary, but
$AC = 2 \cdot AD$ and
$CD = \sqrt{3} \cdot AD \approx 1.73 \cdot AD$.

5. The ratio $\frac{AC}{AD}$ is 2 : 1 and the ratio $\frac{CD}{AD}$ is approximately 1.73 : 1.

5. What do you notice about the ratios $\frac{AC}{AD}$ and $\frac{CD}{AD}$ for $\triangle ACD$ with $m\angle A = 60°$, $m\angle C = 30°$, and $m\angle D = 90°$?

10.3 30°-60°-90° Triangles

Goal
Find the side lengths of 30°-60°-90° triangles.

Key Words
• 30°-60°-90° triangle

A right triangle with angle measures of 30°, 60°, and 90° is called a **30°-60°-90° triangle**.

Activity 10.3 shows that the ratio of the length of the hypotenuse of a 30°-60°-90° triangle to the length of the shorter leg is 2 : 1.

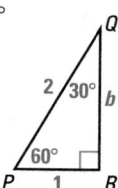

EXAMPLE 1 Find Leg Length

In the diagram above, △PQR is a 30°-60°-90° triangle with PQ = 2 and PR = 1. Find the value of b.

Solution

You can use the Pythagorean Theorem to find the value of b.

$(\text{leg})^2 + (\text{leg})^2 = (\text{hypotenuse})^2$	Write the Pythagorean Theorem.
$1^2 + b^2 = 2^2$	Substitute.
$1 + b^2 = 4$	Simplify.
$b^2 = 3$	Subtract 1 from each side.
$b = \sqrt{3}$	Take the square root of each side.

Visualize It!

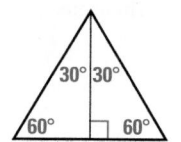

An equilateral triangle can be divided into two 30°-60°-90° triangles.

Because all 30°-60°-90° triangles are similar, the ratio of the length of the longer leg to the length of the shorter leg is always $\sqrt{3}$: 1. This result is summarized in the theorem below.

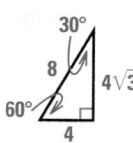 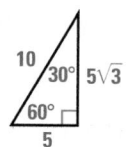 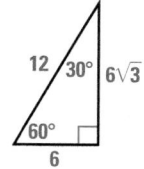

THEOREM 10.2

30°-60°-90° Triangle Theorem

Words In a 30°-60°-90° triangle, the hypotenuse is twice as long as the shorter leg, and the longer leg is the length of the shorter leg times $\sqrt{3}$.

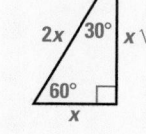

Symbols Hypotenuse = 2 · shorter leg
Longer leg = shorter leg · $\sqrt{3}$

---

① Plan

Pacing
Suggested Number of Days

Basic: 2 days
Average: 2 days
Advanced: 2 days
Block Schedule: 1 block

Teaching Resources

 **Blacklines**
(See page 534B.)

Transparencies
• Warm-Up with Quiz
• Answers

Technology
• Electronic Teacher Tools
• Test and Practice Generator
• Online Lesson Planner
• Internet Support

② Teach

Content and Teaching Strategies
For background information on geometric concepts and teaching strategies related to this lesson, see pages 534E and 534F in this Teacher's Edition.

Tips for New Teachers
Expect students to be skeptical that Theorem 10.2 applies for different sizes of 30°-60°-90° triangles. Use the Visualize It note in the margin of this page to help see that 30°-60°-90° triangles can be made from equilateral triangles of any size. Draw several different size 30°-60°-90° triangles and label the shorter leg and hypotenuse. Have half the students find the length of the longer leg using Theorem 10.2 and the other half using the Pythagorean Theorem. See the Tips for New Teachers on pp. 1–2 of the *Chapter 10 Resource Book* for additional notes about Lesson 10.3.

Extra Example 1
See next page.

550

Extra Example 1

In the diagram below, △*ABC* is a 30°-60°-90° triangle with *AB* = 20 and *AC* = 10. Find the value of *a*.

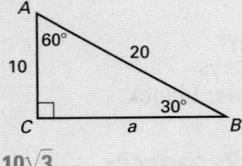

$10\sqrt{3}$

Extra Example 2

In the 30°-60°-90° triangle below, the length of the shorter leg is given. Find the length of the hypotenuse. **46**

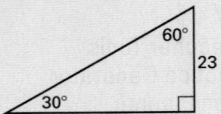

Extra Example 3

In the 30°-60°-90° triangle below, the length of the shorter leg is given. Find the length of the longer leg. $12\sqrt{3}$

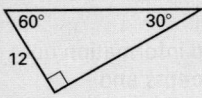

Study Skills

Note-Taking Suggest that students write an explanation of how to find the unknown length of the sides of a 30°-60°-90°, given different combinations of side lengths.

EXAMPLE 2 Find Hypotenuse Length

In the 30°-60°-90° triangle at the right, the length of the shorter leg is given. Find the length of the hypotenuse.

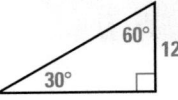

Solution

The hypotenuse of a 30°-60°-90° triangle is twice as long as the shorter leg.

hypotenuse = 2 • shorter leg	30°-60°-90° Triangle Theorem
= 2 • 12	Substitute.
= 24	Simplify.

ANSWER ▶ The length of the hypotenuse is 24.

EXAMPLE 3 Find Longer Leg Length

In the 30°-60°-90° triangle at the right, the length of the shorter leg is given. Find the length of the longer leg.

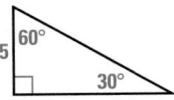

Solution

The length of the longer leg of a 30°-60°-90° triangle is the length of the shorter leg times $\sqrt{3}$.

longer leg = shorter leg • $\sqrt{3}$	30°-60°-90° Triangle Theorem
= 5 • $\sqrt{3}$	Substitute.

ANSWER ▶ The length of the longer leg is $5\sqrt{3}$.

In a 30°-60°-90° triangle, the longer leg is opposite the 60° angle, and the shorter leg is opposite the 30° angle.

Checkpoint ✓ Find Lengths in a Triangle

Find the value of x. Write your answer in radical form.

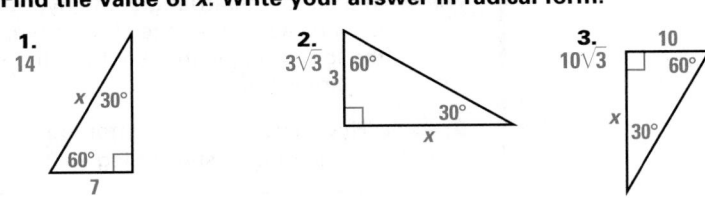

EXAMPLE 4 Find Shorter Leg Length

In the 30°-60°-90° triangle at the right, the length of the longer leg is given. Find the length x of the shorter leg. Round your answer to the nearest tenth.

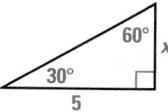

Solution

The length of the longer leg of a 30°-60°-90° triangle is the length of the shorter leg times $\sqrt{3}$.

longer leg = shorter leg $\cdot \sqrt{3}$	30°-60°-90° Triangle Theorem
$5 = x \cdot \sqrt{3}$	Substitute.
$\dfrac{5}{\sqrt{3}} = x$	Divide each side by $\sqrt{3}$.
$2.9 \approx x$	Use a calculator.

ANSWER ▶ The length of the shorter leg is about 2.9.

Student Help

CLASSZONE.COM

MORE EXAMPLES
More examples at classzone.com

EXAMPLE 5 Find Leg Lengths

In the 30°-60°-90° triangle at the right, the length of the hypotenuse is given. Find the length x of the shorter leg and the length y of the longer leg.

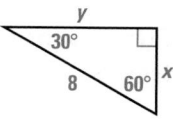

Solution

Use the 30°-60°-90° Triangle Theorem to find the length of the shorter leg. Then use that value to find the length of the longer leg.

Shorter leg	**Longer leg**
hypotenuse = 2 $\cdot$ shorter leg	longer leg = shorter leg $\cdot \sqrt{3}$
$8 = 2 \cdot x$	$y = 4 \cdot \sqrt{3}$
$4 = x$	$y = 4\sqrt{3}$

ANSWER ▶ The length of the shorter leg is 4.
The length of the longer leg is $4\sqrt{3}$.

Checkpoint ✓ **Find Leg Lengths**

Find the value of each variable. Round your answer to the nearest tenth.

4.
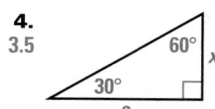

5.
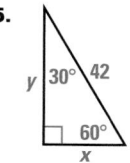

$x = 21; y = 21\sqrt{3} \approx 36.4$

10.3 *30°-60°-90° Triangles* 551

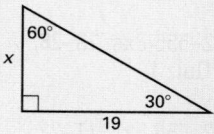
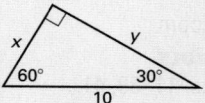
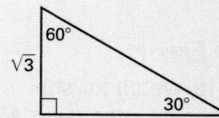

10.3 **Exercises**

Guided Practice

Vocabulary Check

1. Name two special right triangles by their angle measures.
   45°-45°-90° triangle and 30°-60°-90° triangle

Skill Check

Use the diagram to tell whether the equation is *true* or *false*.

2. $t = 7\sqrt{3}$ true   3. $t = \sqrt{3}h$ false   4. $h = 2t$ false

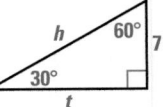

5. $h = 14$ true   6. $7 = \dfrac{h}{2}$ true   7. $7 = \dfrac{t}{\sqrt{3}}$ true

Find the value of each variable. Write your answers in radical form.

8.
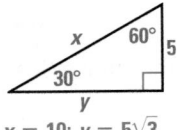
$x = 10$; $y = 5\sqrt{3}$

9.

10.
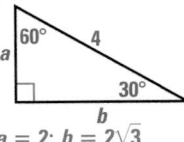
$a = 2$; $b = 2\sqrt{3}$

Practice and Applications

Extra Practice

See p. 693.

Finding Hypotenuse Lengths Find the length of the hypotenuse.

11.

12.

13.

14.

15.

16.

Finding Leg Lengths Find the length of the longer leg of the triangle. Write your answer in radical form.

17.

18.

19.

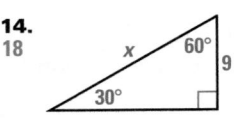

Homework Help

Example 2: Exs. 11–16
Example 3: Exs. 17–22
Example 4: Exs. 23–28
Example 5: Exs. 29–35

20.

21.

22.
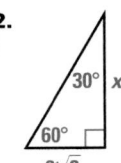

Assignment Guide

BASIC
Day 1: pp. 552–555 Exs. 11–22, 45–56
Day 2: pp. 552–555 Exs. 23–28, 30–35, 43, 44, Quiz 1

AVERAGE
Day 1: pp. 552–555 Exs. 11–22, 37–39, 43, 45–56
Day 2: pp. 552–555 Exs. 23–36, 41, 44, Quiz 1

ADVANCED
Day 1: pp. 552–555 Exs. 14–16, 20–22, 37–41, 43, 45–56
Day 2: pp. 552–555 Exs. 26–36, 42*, 44, Quiz 1;
EC: classzone.com

BLOCK SCHEDULE
pp. 552–555 Exs. 11–39, 41, 43–56, Quiz 1

Extra Practice
• Student Edition, p. 693
• Chapter 10 Resource Book, pp. 27–28

Homework Check

To quickly check student understanding of key concepts, go over the following exercises:

Basic: 14, 20, 26, 30, 33
Average: 15, 21, 27, 33, 37
Advanced: 16, 22, 28, 35, 37

✗ **Common Error**

In Exercises 11–16, watch for students who think the hypotenuse is a multiple of $\sqrt{3}$ in a 30°-60°-90° triangle. Students may mistakenly think that $\sqrt{3} > 2$ and that it therefore corresponds to the longest side, the hypotenuse. Have students use a calculator to find the value of $\sqrt{3}$ and compare it to 2.

552

Finding Leg Lengths Find the length of the shorter leg of the triangle. Round your answer to the nearest tenth.

23. 2.3
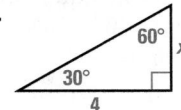

24. 4.0
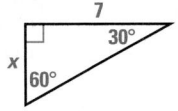

25. 10.4
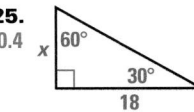

26. 5.8
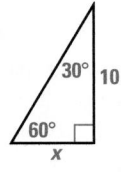

27. 14
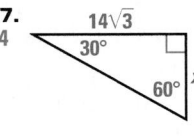

28. 6.9
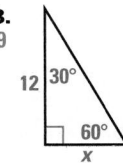

29. Tipping Platform A tipping platform is a ramp used to unload trucks as shown below. What is the height of an 60 foot ramp when it is tipped up to a 30° angle? **30 feet**

ramp

60 ft

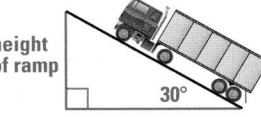

height of ramp

30°

Finding Leg Lengths Find the value of each variable. Write your answers in radical form.

30.
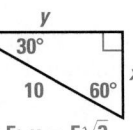
$x = 5$; $y = 5\sqrt{3}$

31.
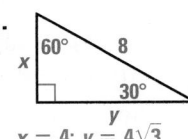
$x = 4$; $y = 4\sqrt{3}$

32.
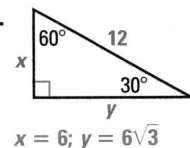
$x = 6$; $y = 6\sqrt{3}$

33.
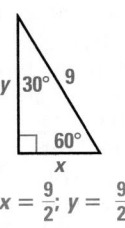
$x = \dfrac{9}{2}$; $y = \dfrac{9}{2}\sqrt{3}$

34.
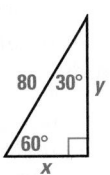
$x = 40$; $y = 40\sqrt{3}$

35.
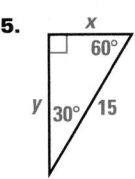
$x = \dfrac{15}{2}$; $y = \dfrac{15}{2}\sqrt{3}$

36. Fitness A "crunch" is the type of sit-up shown in the photo at the right. A personal trainer tells you that in doing a crunch, your back and shoulders should be lifted to an angle of about 30°. If your shoulder-to-waist length is 18 inches, how high should your shoulders be lifted?
9 inches

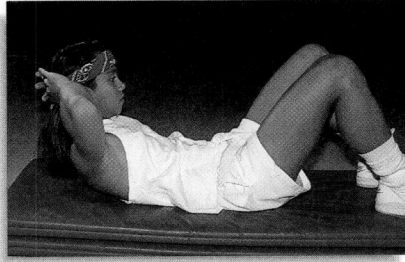

EXAMPLE Find Area Using 30°-60°-90° Triangles

The road sign is shaped like an equilateral triangle with side lengths of 36 inches. Estimate the area of the sign.

Solution

Divide the triangle into two 30°-60°-90° triangles.

The length of the shorter leg of each triangle is 18 inches. The length of the longer leg of each triangle is $18\sqrt{3}$ inches, by the 30°-60°-90° Triangle Theorem.

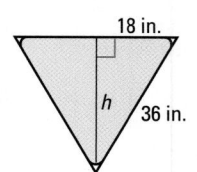

Use the formula for the area of a triangle.

Student Help

LOOK BACK
To review the area of a triangle, see p. 431. ·········

▶ Area $= \frac{1}{2}bh = \frac{1}{2} \cdot 36 \cdot 18\sqrt{3} \approx 561.18$

ANSWER ▶ The area of the sign is about 561 square inches.

Finding Area Find the area of each triangle. Use the example above as a model.

37.

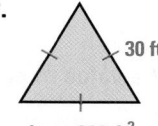

30 ft

about 390 ft²

38. 18 in.

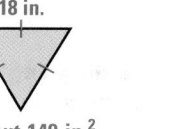

about 140 in.²

39. 7 cm

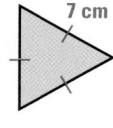

about 21 cm²

40. The area of a triangle is given by $A = \frac{1}{2}bh$. If an equilateral triangle with sides of length s is divided into two 30°-60°-90° triangles, then the length of the base is s and the height is $\frac{s}{2}\sqrt{3}$. Then $A = \frac{1}{2}bh = \frac{1}{2} \cdot s \cdot \frac{s}{2}\sqrt{3} = \frac{1}{4} \cdot s^2 \cdot \sqrt{3}$.

40. 🔑 **Using Algebra** Use the figure in the example above to explain why the area A of an equilateral triangle with side length s is given by the formula $A = \frac{1}{4} \cdot s^2 \cdot \sqrt{3}$.

41. Visualize It! A 30°-60°-90° triangle has a shorter leg length of 15 centimeters. Sketch the triangle and find the length of the hypotenuse and the length of the longer leg in radical form.
Check sketches; hypotenuse, 30 cm; longer leg, $15\sqrt{3}$ cm.

42. Challenge The side length of the hexagonal nut shown at the right is 1 centimeter. Find the value of x. (*Hint*: Use the fact that a regular hexagon can be divided into six congruent equilateral triangles.) **1.73 cm**

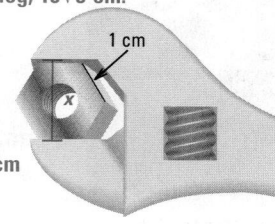

1 cm

Standardized Test Practice

43. Multiple Choice Which triangle is labeled correctly? **C**

Ⓐ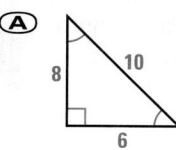

Ⓑ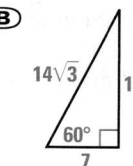

Ⓒ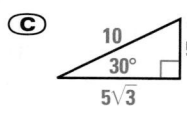

Ⓓ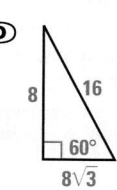

44. Multiple Choice Find the perimeter of the triangle shown below to the nearest tenth of a centimeter. **F**

    Ⓕ 28.4 cm   Ⓖ 30 cm

    Ⓗ 31.2 cm   Ⓙ 41.6 cm

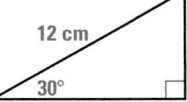

12 cm   30°

Mixed Review

Writing Ratios A football team won 10 games and lost 6 games. Find the ratio. *(Lesson 7.1)*

45. wins to losses $\frac{5}{3}$ **46.** losses to wins $\frac{3}{5}$

47. wins to the number of games $\frac{5}{8}$ **48.** losses to the number of games $\frac{3}{8}$

Similar Triangles Determine whether the triangles are similar. If they are similar, write a similarity statement. *(Lessons 7.3, 7.4)*

49.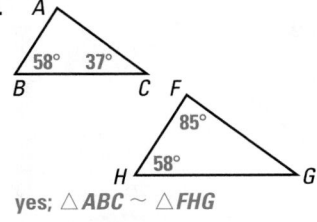

yes; $\triangle ABC \sim \triangle FHG$

50.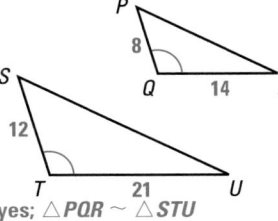

yes; $\triangle PQR \sim \triangle STU$

Algebra Skills

Evaluating Expressions Evaluate the expression when $x = -4$. *(Skills Review, p. 670)*

51. $5x + 4$ -16 **52.** $10x - 1$ -41 **53.** $x^2 - 7$ 9

54. $(x + 3)(x - 3)$ 7 **55.** $2x^2 - x + 1$ 37 **56.** $5x^2 + 2x - 3$ 69

Quiz 1

Multiply the radical expression. Then simplify if possible. *(Lesson 10.1)*

1. $\sqrt{8} \cdot \sqrt{3}$ $2\sqrt{6}$ **2.** $\sqrt{2} \cdot \sqrt{15}$ $\sqrt{30}$ **3.** $\sqrt{8} \cdot \sqrt{18}$ 12 **4.** $\sqrt{80} \cdot \sqrt{5}$ 20

Simplify the radical expression. *(Lesson 10.1)*

5. $\sqrt{27}$ $3\sqrt{3}$ **6.** $\sqrt{176}$ $4\sqrt{11}$ **7.** $\sqrt{52}$ $2\sqrt{13}$ **8.** $\sqrt{180}$ $6\sqrt{5}$

Find the value of each variable. Write your answer in radical form. *(Lessons 10.2, 10.3)*

9.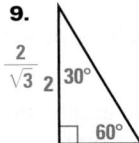

10.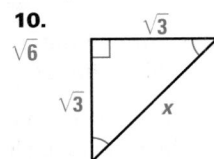

11.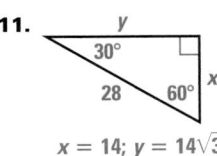

$x = 14$; $y = 14\sqrt{3}$

10.3   *30°-60°-90° Triangles*   **555**

Assessment Resources

The Mini-Quiz below is also available on blackline (*Chapter 10 Resource Book*, p. 36) and on transparency. For more assessment resources, see:

• Chapter 10 Resource Book
• Standardized Test Practice
• Test and Practice Generator

Mini-Quiz

Find the value of each variable. Round your answer to the nearest tenth, if necessary.

1.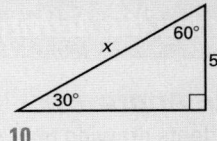

60°   x   5   30°   10

2.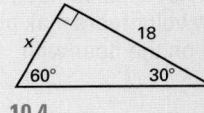

x   18   60°   30°   10.4

3.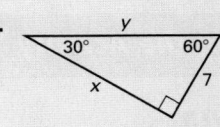

30°   y   60°   x   7

$x \approx 12.1$, $y = 14$

Goal
Students discover the tangent ratio by measuring the leg lengths of similar right triangles.

Materials
See the margin of the student page.

LINK TO LESSON

In Example 1 on page 557, students use a ratio to find the tangent of an angle.

Managing the Activity

Alternative Approach
Instead of students drawing their triangles, if you have a tiled floor, supervise a few volunteers making a large triangle on the floor with masking tape.

Closing the Activity

KEY DISCOVERY
The ratio of the leg lengths in a right triangle is the same regardless of the size of the triangle.

Activity Assessment
Use Exercise 3 to assess student understanding.

Activity 10.4 Right Triangle Ratio

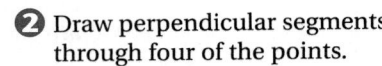
- centimeter ruler
- protractor
- calculator

Question

Does the size of similar right triangles affect the ratio of their leg lengths?

Explore

❶ Draw a 40° angle. Mark points every 5 cm along one side.

❷ Draw perpendicular segments through four of the points.

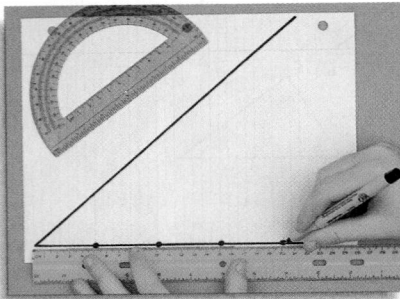

Think About It

1. There are four similar triangles in your drawing. Measure the legs and complete a table like the one below.

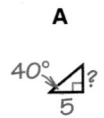 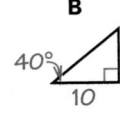 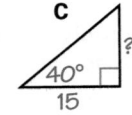 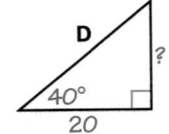

Triangle	longer leg	shorter leg	shorter leg / longer leg
A	5 cm	4.2 cm	0.84
B	10 cm	? 8.4 cm	? 0.84
C	15 cm	? 12.6 cm	? 0.84
D	20 cm	? 16.8 cm	? 0.84

2. Compare the ratios of the leg lengths in the last column of your table. What do you notice? Each ratio is 0.84.

3. Answers will vary, but the ratios in the fourth column will always be equal; the ratio of leg lengths depends on the measures of its angles, not the size of the right triangle.

3. Repeat using a different acute angle. Based on your results, does the ratio of leg lengths depend on the size of a right triangle or the measures of its angles?

10.4 Tangent Ratio

Goal
Find the tangent of an acute angle.

Key Words
- trigonometric ratio
- leg opposite an angle
- leg adjacent to an angle
- tangent

How can you find the height of the tree in the photograph at the right? It is too tall to be measured directly in any simple manner.

You can determine the height using a *trigonometric ratio*. A **trigonometric ratio** is a ratio of the lengths of two sides of a right triangle.

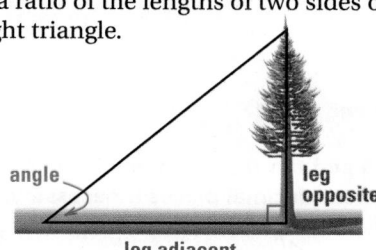

For any acute angle of a right triangle, there is a **leg opposite** the angle and a **leg adjacent** to the angle. The ratio of these legs is the **tangent** of the angle.

Student Help
READING TIP
The tangent of angle *A* is written as "tan *A*."

TANGENT RATIO

For any acute angle A of a right triangle:

$$\tan A = \frac{\text{leg opposite } \angle A}{\text{leg adjacent to } \angle A} = \frac{a}{b}$$

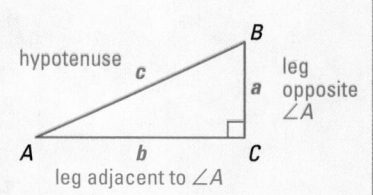

EXAMPLE 1 Find Tangent Ratio

Find tan *S* and tan *R* as fractions in simplified form and as decimals rounded to four decimal places.

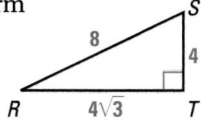

Solution

$$\tan S = \frac{\text{leg opposite } \angle S}{\text{leg adjacent to } \angle S} = \frac{4\sqrt{3}}{4} = \sqrt{3} \approx 1.7321$$

$$\tan R = \frac{\text{leg opposite } \angle R}{\text{leg adjacent to } \angle R} = \frac{4}{4\sqrt{3}} = \frac{1}{\sqrt{3}} \approx 0.5774$$

10.4 *Tangent Ratio* **557**

---

1 Plan

Pacing
Suggested Number of Days

Basic: 2 days
Average: 2 days
Advanced: 2 days
Block Schedule: 1 block

Teaching Resources

📄 **Blacklines**
(See page 534B.)

🖐 **Transparencies**
- Warm-Up with Quiz
- Answers

💻 **Technology**
- Electronic Teacher Tools
- Test and Practice Generator
- Online Lesson Planner
- Internet Support

2 Teach

Content and Teaching Strategies
For background information on geometric concepts and teaching strategies related to this lesson, see pages 534E and 534F in this Teacher's Edition.

Tips for New Teachers
Point out to students that the angle symbol is not used when writing ratios or equations involving trigonometric ratios. While looking at Example 2 on page 558, explain that the value of the trigonometric ratio is frequently expressed as a decimal approximation, rounded to the nearest ten thousandth (four decimal places). See the Tips for New Teachers on pp. 1–2 of the *Chapter 10 Resource Book* for additional notes about Lesson 10.4.

Extra Example 1
See next page.

558

Geometric Reasoning

When discussing the definition of the tangent ratio on page 557, point out to students that to find the tangent ratio for $\angle A$ (or $\angle B$) in the triangle, you do not need to know the length of the hypotenuse.

Extra Example 1

Find tan G and tan J as fractions in simplified form and as decimals. Round to four decimal places if necessary.

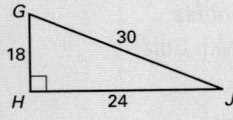

$\tan G = \frac{24}{18} \approx 1.3333$; $\tan J = \frac{18}{24} = 0.75$

Extra Example 2

Approximate tan 62° to four decimal places. 1.8807

Extra Example 3

Use a tangent ratio to find the value of x. Round your answer to the nearest tenth. 10.0

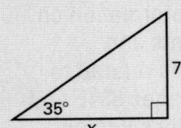

Tangent Function You can use the TAN function on a calculator to approximate the tangent of an angle. You can also use the table of trigonometric ratios on page 705.

Student Help

KEYSTROKE HELP
When calculating the tangent of an angle, be sure your calculator is in DEGREE mode.

EXAMPLE 2 Use a Calculator for Tangent

Approximate tan 74° to four decimal places.

Solution

Calculator keystrokes	Display	Rounded value
74 [TAN] *or* [TAN] 74 [ENTER]	3.487414444	3.4874

Checkpoint ✓ Find Tangent Ratio

Find tan S and tan R as fractions in simplified form and as decimals. Round to four decimal places if necessary.

1.

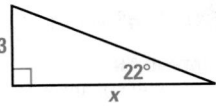

2.

1. $\tan S = \frac{3}{4} = 0.75$;

   $\tan R = \frac{4}{3} \approx 1.3333$

2. $\tan S = \frac{5}{12} \approx 0.4167$;

   $\tan R = \frac{12}{5} = 2.4$

Use a calculator to approximate the value to four decimal places.

3. tan 35° 0.7002   **4.** tan 85° 11.4301   **5.** tan 10° 0.1763

EXAMPLE 3 Find Leg Length

Use a tangent ratio to find the value of x. Round your answer to the nearest tenth.

Solution

$$\tan 22° = \frac{\text{opposite leg}}{\text{adjacent leg}}$$ Write the tangent ratio.

$$\tan 22° = \frac{3}{x}$$ Substitute.

$$x \cdot \tan 22° = 3$$ Multiply each side by x.

$$x = \frac{3}{\tan 22°}$$ Divide each side by tan 22°.

$$x \approx \frac{3}{0.4040}$$ Use a calculator or table to approximate tan 22°.

$$x \approx 7.4$$ Simplify.

In Example 3, the unknown, x, was in the *denominator* of the ratio. If you prefer to use a ratio in which the unknown is in the *numerator*, use the tangent ratio for the other acute angle in the triangle.

Student Help

STUDY TIP
Given the measure of one acute angle of a right triangle, use the Corollary to the Triangle Sum Theorem to find the measure of the other angle.········

EXAMPLE 4 Find Leg Length

Use two different tangent ratios to find the value of x to the nearest tenth.

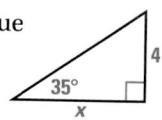

Solution

First, find the measure of the other acute angle: $90° - 35° = 55°$.

Method 1

$$\tan 35° = \frac{\text{opposite leg}}{\text{adjacent leg}}$$

$$\tan 35° = \frac{4}{x}$$

$$x \cdot \tan 35° = 4$$

$$x = \frac{4}{\tan 35°}$$

$$x \approx \frac{4}{0.7002}$$

$$x \approx 5.7$$

Method 2

$$\tan 55° = \frac{\text{opposite leg}}{\text{adjacent leg}}$$

$$\tan 55° = \frac{x}{4}$$

$$4 \tan 55° = x$$

$$4(1.4281) \approx x$$

$$x \approx 5.7$$

ANSWER The two methods yield the same answer: $x \approx 5.7$.

Link to
Careers

FORESTERS Foresters manage and protect forests. To determine the height of a tree, a forester may use trigonometry.

Career Links
CLASSZONE.COM

EXAMPLE 5 Estimate Height

You stand 45 feet from the base of a tree and look up at the top of the tree as shown in the diagram. Use a tangent ratio to estimate the height of the tree to the nearest foot.

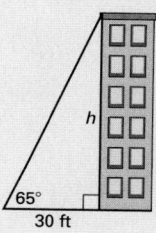

Solution

$$\tan 59° = \frac{\text{opposite leg}}{\text{adjacent leg}}$$ Write ratio.

$$\tan 59° = \frac{h}{45}$$ Substitute.

$$45 \tan 59° = h$$ Multiply each side by 45.

$$45(1.6643) \approx h$$ Use a calculator or table to approximate $\tan 59°$.

$$74.9 \approx h$$ Simplify.

ANSWER The tree is about 75 feet tall.

10.4 *Tangent Ratio* **559**

Extra Example 4
Use two different tangent ratios to find the value of x to the nearest tenth. **13.3**

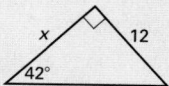

Extra Example 5
You stand 30 feet from the base of a building and look up at the top of the building as shown in the diagram. Use a tangent ratio to estimate the height of the building to the nearest foot. **64 ft**

✗ Common Error
Throughout this lesson, watch for students who confuse the hypotenuse with the adjacent side as they write ratios. Suggest that before students write their ratios, they point to and name each side of the triangle they are working with to help them use the sides correctly.

 **Checkpoint** ✓ **Find Side Length**

Write two equations you can use to find the value of *x*.

6. $\tan 44° = \frac{8}{x}$ and $\tan 46° = \frac{x}{8}$

7. $\tan 37° = \frac{4}{x}$ and $\tan 53° = \frac{x}{4}$

8. $\tan 59° = \frac{5}{x}$ and $\tan 31° = \frac{x}{5}$

6.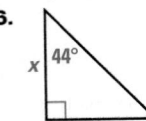

7.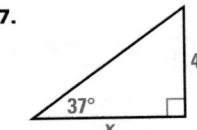

8.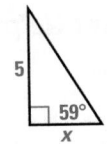

Find the value of *x*. Round your answer to the nearest tenth.

9. 10.4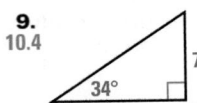

10. 12.6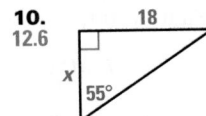

11. 34.6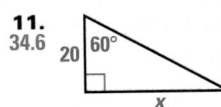

10.4 Exercises

Guided Practice

Vocabulary Check

1. Name the acute angles in △*DEF*. ∠*D* and ∠*E*

2. Identify the leg opposite ∠*D* and the leg adjacent to ∠*D*.
   $\overline{EF}$ is opposite ∠*D*, and $\overline{DF}$ is adjacent to ∠*D*.

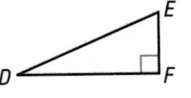

Skill Check

Find tan *A* as a fraction in simplified form and as a decimal. Round to four decimal places if necessary.

3.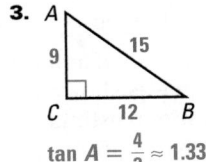

$\tan A = \frac{4}{3} \approx 1.3333$

4.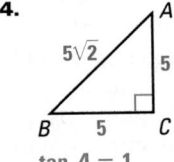

$\tan A = 1$

5.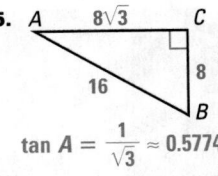

$\tan A = \frac{1}{\sqrt{3}} \approx 0.5774$

Use a calculator to approximate the value to four decimal places.

6. tan 25° 0.4663   **7.** tan 62° 1.8807   **8.** tan 80° 5.6713   **9.** tan 43° 0.9325

Practice and Applications

Extra Practice
See p. 693.

Finding Tangent Ratios **Find tan *A*. Write your answer as a fraction.**

10. $\frac{8}{15}$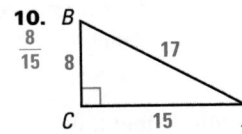

11. $\sqrt{3}$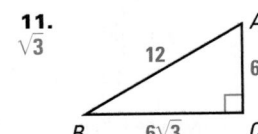

12. $\frac{2}{3}$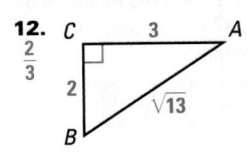

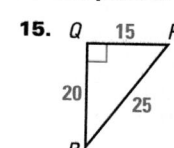
Finding Tangent Ratios Find tan *P* and tan *R* as fractions in simplified form and as decimals rounded to four decimal places. See margin.

13.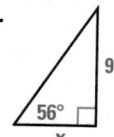

14.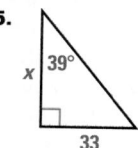

15.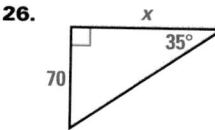

13. $\tan P = \dfrac{7}{24} \approx 0.2917$;

$\tan R = \dfrac{24}{7} \approx 3.4286$

14. $\tan P = \dfrac{12}{35} \approx 0.3429$;

$\tan R = \dfrac{35}{12} \approx 2.9167$

15. $\tan P = \dfrac{4}{3} \approx 1.3333$;

$\tan R = \dfrac{3}{4} = 0.75$

Using a Calculator Use a calculator to approximate the value to four decimal places.

16. tan 28° 0.5317 **17.** tan 54° 1.3764 **18.** tan 5° 0.0875 **19.** tan 89° 57.2900

20. tan 67° 2.3559 **21.** tan 40° 0.8391 **22.** tan 12° 0.2126 **23.** tan 83° 8.1443

Using Tangent Ratios Write two equations you can use to find the value of *x*. Then find the value of *x* to the nearest tenth. See margin.

24.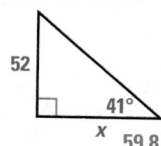

25.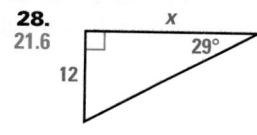

26.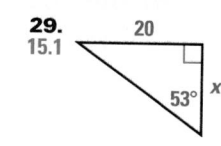

24. $\tan 56° = \dfrac{9}{x}$; $\tan 34° = \dfrac{x}{9}$; 6.1

25. $\tan 39° = \dfrac{33}{x}$; $\tan 51° = \dfrac{x}{33}$; 40.8

26. $\tan 35° = \dfrac{70}{x}$; $\tan 55° = \dfrac{x}{70}$; 100.0

Finding Leg Lengths Find the value of *x* to the nearest tenth.

27. 52, 41°, *x* 59.8

28. 21.6, 12, 29°, *x* 24.7

29. 20, 15.1, 53°, *x*

30. Error Analysis To find the length of $\overline{BC}$ in the diagram at the right, a student wrote $\tan 55° = \dfrac{18}{BC}$. Explain the student's error.

△*ABC* is not a right triangle, so you cannot use the tangent ratio.

31. Water Slide A water slide makes an angle of about 13° with the ground. The slide extends horizontally about 58.2 meters as shown below. Find the height *h* of the slide to the nearest tenth of a meter. 13.4 m

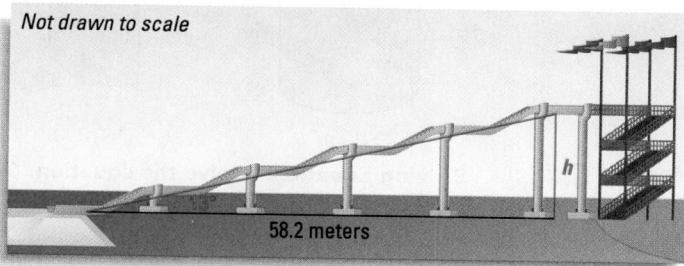

Not drawn to scale

58.2 meters

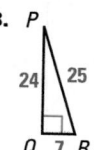

Assessment Resources

The Mini-Quiz below is also available on blackline (*Chapter 10 Resource Book*, p. 47) and on transparency. For more assessment resources, see:
• Chapter 10 Resource Book
• Standardized Test Practice
• Test and Practice Generator

Mini-Quiz

Find tan *A* and tan *B* as fractions in simplified form and as decimals. Round to four decimal places if necessary.

1.

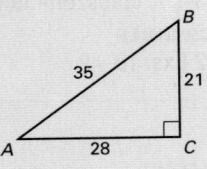

$\tan A = \dfrac{3}{4} = 0.75$,

$\tan B = \dfrac{4}{3} \approx 1.3333$

2.

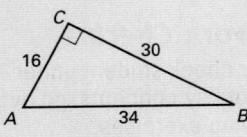

$\tan A = \dfrac{15}{8} = 1.875$,

$\tan B = \dfrac{8}{15} \approx 0.5333$

3. Find the value of *x* to the nearest tenth. 42.9

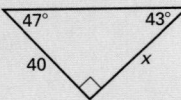

4. You stand 15 feet from the base of a water tower and look up at an angle of 78° at the top of the tower. Use a tangent ratio to estimate the height of the tower to the nearest foot. 71 ft

Finding Side Lengths Find the value of *x* to the nearest tenth.

32.

33.

34.

35.
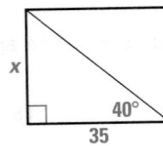

36.
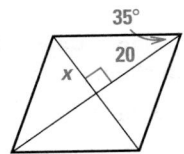

37.
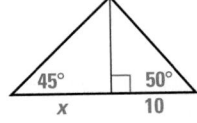

38. Surveying To find the distance *d* from a house on shore to a house on an island, a surveyor measures from the house on shore to point *B*, as shown in the diagram. An instrument called a *transit* is used to find the measure of ∠*B*. Find the distance *d* to the nearest tenth of a meter.
36.0 m

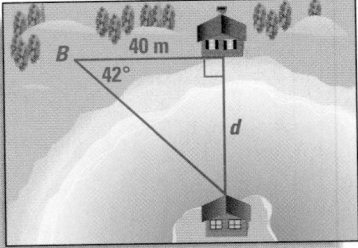

Standardized Test Practice

39. Multiple Choice Which expression can be used to find the value of *x* in the triangle shown? C

Ⓐ $x = 10 \tan 38°$ Ⓑ $x = \dfrac{\tan 38°}{10}$

Ⓒ $x = \dfrac{10}{\tan 38°}$ Ⓓ $x = \dfrac{10}{\tan 52°}$

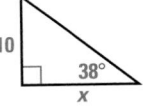

40. Multiple Choice What is the approximate value of *y* in the triangle shown? F

Ⓕ 7.2   Ⓖ 8.4

Ⓗ 9.3   Ⓙ 10.1

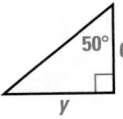

Mixed Review

Volume of Solids Find the volume of the solid. If necessary, round your answer to the nearest whole number. *(Lessons 9.4, 9.5)*

41.
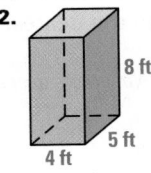
1018 m³

42.
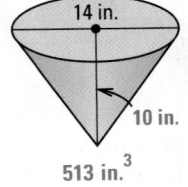
160 ft³

43.

513 in.³

Algebra Skills

Solving Equations Solve the equation. *(Skills Review, p. 673)*

44. $8x - 10 = 3x$ 2   **45.** $4(x + 3) = 32$ 5   **46.** $3x - 7 - x = 11$ 9

47. $6x + 5 = 3x - 4$ −3   **48.** $2 - x = 4x + 22$ −4   **49.** $5x - 18 = 2x + 21$ 13

10.5 Sine and Cosine Ratios

① Plan

Pacing
Suggested Number of Days

Basic: 2 days
Average: 2 days
Advanced: 2 days
Block Schedule: 1 block

Teaching Resources

📄 **Blacklines**
(See page 534B.)

✋ **Transparencies**
• Warm-Up with Quiz
• Answers

🖥 **Technology**
• Electronic Teacher Tools
• Test and Practice Generator
• Online Lesson Planner
• Internet Support

Goal
Find the sine and cosine of an acute angle.

Key Words
• sine
• cosine

Two other trigonometric ratios are **sine** and **cosine** . These are abbreviated as *sin* and *cos*. Unlike the tangent ratio, these ratios involve the hypotenuse of a right triangle.

SINE AND COSINE RATIOS

For any acute angle A of a right triangle:

$$\sin A = \frac{\text{leg opposite } \angle A}{\text{hypotenuse}} = \frac{a}{c}$$

$$\cos A = \frac{\text{leg adjacent to } \angle A}{\text{hypotenuse}} = \frac{b}{c}$$

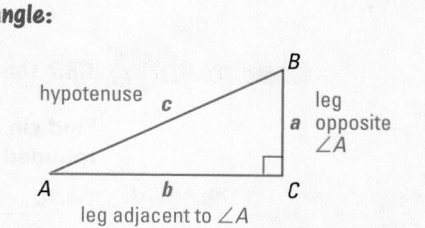

EXAMPLE ① Find Sine and Cosine Ratios

Find sin *A* and cos *A*.

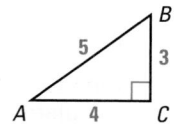

Solution

$$\sin A = \frac{\text{leg opposite } \angle A}{\text{hypotenuse}}$$ Write ratio for sine.

$$= \frac{3}{5}$$ Substitute.

$$\cos A = \frac{\text{leg adjacent to } \angle A}{\text{hypotenuse}}$$ Write ratio for cosine.

$$= \frac{4}{5}$$ Substitute.

✔ Checkpoint Find Sine and Cosine Ratios

Find sin *A* and cos *A*.

1. $\sin A = \frac{15}{17}$; $\cos A = \frac{8}{17}$

2. $\sin A = \frac{24}{25}$; $\cos A = \frac{7}{25}$

3. $\sin A = \frac{4}{5}$; $\cos A = \frac{3}{5}$

1.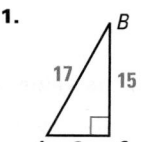

2.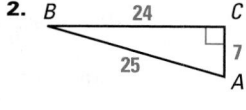

3.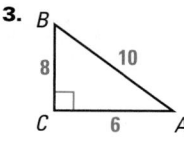

10.5 *Sine and Cosine Ratios* **563**

② Teach

Content and Teaching Strategies
For background information on geometric concepts and teaching strategies related to this lesson, see pages 534E and 534F in this Teacher's Edition.

Tips for New Teachers
The hypotenuse of a right triangle is usually the easiest side to identify in a diagram. Students tend to make mistakes when trying to identify the other two sides by confusing which is opposite and which is adjacent. Stress that students identify the hypotenuse first. Then they should focus on a particular angle in the right triangle and identify the side opposite that angle. The remaining side must be the adjacent one, if everything else was identified correctly. See the Tips for New Teachers on pp. 1–2 of the *Chapter 10 Resource Book* for additional notes about Lesson 10.5.

Extra Example 1
See next page.

Extra Example 1

Find sin E and cos E.

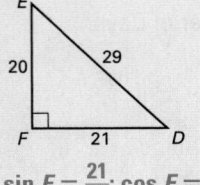

$\sin E = \dfrac{21}{29}$; $\cos E = \dfrac{20}{29}$

Extra Example 2

Find sin K and cos K. Write your answers as fractions and as decimals rounded to four decimal places.

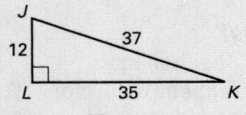

$\sin K = \dfrac{12}{37} \approx 0.3243$;

$\cos K = \dfrac{35}{37} \approx 0.9459$

Extra Example 3

Use a calculator to approximate sin 14° and cos 14°. Round your answers to four decimal places.
$\sin 14° \approx 0.2419$; $\cos 14° \approx 0.9703$

EXAMPLE 2 Find Sine and Cosine Ratios

Find sin A and cos A. Write your answers as fractions and as decimals rounded to four decimal places.

Solution

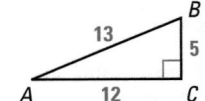

$$\sin A = \frac{\text{leg opposite } \angle A}{\text{hypotenuse}} = \frac{5}{13} \approx 0.3846$$

$$\cos A = \frac{\text{leg adjacent to } \angle A}{\text{hypotenuse}} = \frac{12}{13} \approx 0.9231$$

Checkpoint ✓ Find Sine and Cosine Ratios

Find sin A and cos A. Write your answers as fractions and as decimals rounded to four decimal places.

4. $\sin A = \dfrac{40}{41} \approx 0.9756$;

   $\cos A = \dfrac{9}{41} \approx 0.2195$

5. $\sin A = \dfrac{\sqrt{2}}{2} \approx 0.7071$;

   $\cos A = \dfrac{\sqrt{2}}{2} \approx 0.7071$

6. $\sin A = \dfrac{\sqrt{39}}{8} \approx 0.7806$;

   $\cos A = \dfrac{5}{8} = 0.625$

4.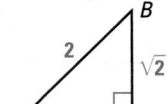
5.
6.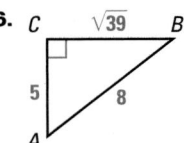

Sine and Cosine Functions You can use the SIN and COS functions on a calculator to approximate the sine and cosine of an angle. You can also use the table of trigonometric ratios on page 705.

EXAMPLE 3 Use a Calculator for Sine and Cosine

Use a calculator to approximate sin 74° and cos 74°. Round your answers to four decimal places.

Solution

Calculator keystrokes	Display	Rounded value
74 [SIN] *or* [SIN] 74 [ENTER]	0.961261696	0.9613
74 [COS] *or* [COS] 74 [ENTER]	0.275637356	0.2756

Checkpoint ✓ Use a Calculator for Sine and Cosine

Use a calculator to approximate the value to four decimal places.

7. sin 43° 0.6820   8. cos 43° 0.7314   9. sin 15° 0.2588   10. cos 15° 0.9659

11. cos 72° 0.3090   12. sin 72° 0.9511   13. cos 90° 0   14. sin 90° 1

EXAMPLE 4 Find Leg Lengths

Find the lengths of the legs of the triangle. Round your answers to the nearest tenth.

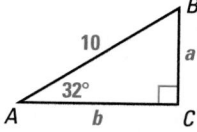

Solution

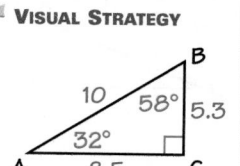

Student Help

VISUAL STRATEGY

You can label side lengths and angle measures in different colors. See p. 536. · · · · · · · · ·

$$\sin A = \frac{\text{leg opposite } \angle A}{\text{hypotenuse}} \qquad \cos A = \frac{\text{leg adjacent to } \angle A}{\text{hypotenuse}}$$

$$\sin 32° = \frac{a}{10} \qquad\qquad \cos 32° = \frac{b}{10}$$

$$10(\sin 32°) = a \qquad\qquad 10(\cos 32°) = b$$

$$10(0.5299) \approx a \qquad\qquad 10(0.8480) \approx b$$

$$5.3 \approx a \qquad\qquad\qquad 8.5 \approx b$$

▶ **ANSWER** ▶ In the triangle, BC is about 5.3 and AC is about 8.5.

Checkpoint ✔ Find Leg Lengths

Find the lengths of the legs of the triangle. Round your answers to the nearest tenth.

15.

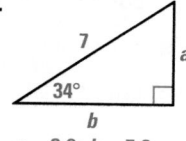

$a \approx 3.9; b \approx 5.8$

16.

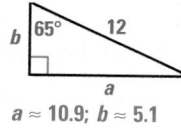

$a \approx 10.9; b \approx 5.1$

17.

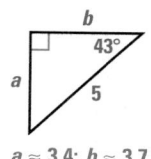

$a \approx 3.4; b \approx 3.7$

SUMMARY   TRIGONOMETRIC RATIOS

For any acute angle A of a right triangle:

Tangent of ∠A

$$\tan A = \frac{\text{leg opposite } \angle A}{\text{leg adjacent to } \angle A} = \frac{a}{b}$$

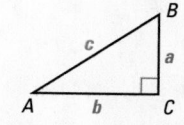

Sine of ∠A

$$\sin A = \frac{\text{leg opposite } \angle A}{\text{hypotenuse}} = \frac{a}{c}$$

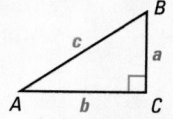

Cosine of ∠A

$$\cos A = \frac{\text{leg adjacent to } \angle A}{\text{hypotenuse}} = \frac{b}{c}$$

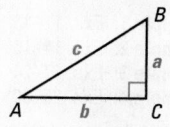

10.5 *Sine and Cosine Ratios* **565**

Extra Example 4

Find the lengths of the legs of the triangle. Round your answers to the nearest tenth.

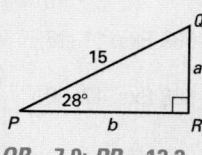

$QR \approx 7.0; PR \approx 13.2$

Study Skills

Note-Taking Suggest that students write notes summarizing the three trigonometric ratios and think of mnemonics they can use to help remember the definitions of the ratios.

✔ Concept Check

Which sides are used for the trigonometric ratios sine, cosine, and tangent?

The sine ratio is $\frac{\text{opposite side}}{\text{hypotenuse}}$,

the cosine ratio is $\frac{\text{adjacent side}}{\text{hypotenuse}}$,

and the tangent ratio is $\frac{\text{opposite side}}{\text{adjacent side}}$.

🐢 Daily Puzzler

Find $\tan A$ and the ratio $\frac{\sin A}{\cos A}$ for any right triangle. What do you notice? $\tan A = \frac{\sin A}{\cos A}$

10.5 Exercises

Guided Practice

Vocabulary Check — Use the diagram shown at the right to match the trigonometric ratios.

1. $\cos D$ B

2. $\sin D$ C

3. $\tan D$ A

A. $\dfrac{EF}{DE}$

B. $\dfrac{DE}{DF}$

C. $\dfrac{EF}{DF}$

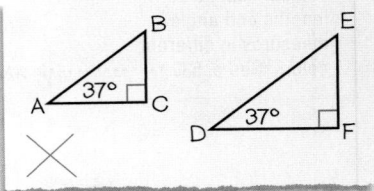

4. **Error Analysis** A student says that $\sin D > \sin A$ because the side lengths of $\triangle DEF$ are greater than the side lengths of $\triangle ABC$. Explain why the student is incorrect.

The value of sin 37° is constant, so sin D = sin A. The sine of an acute angle of a right triangle depends on the measure of the angle, not on the size of the triangle.

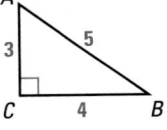

Skill Check — In Exercises 5–10, use the diagram shown below to find the trigonometric ratio.

5. $\sin A$ $\dfrac{4}{5} = 0.8$

6. $\cos A$ $\dfrac{3}{5} = 0.6$

7. $\tan A$ $\dfrac{4}{3} \approx 1.3333$

8. $\sin B$ $\dfrac{3}{5} = 0.6$

9. $\cos B$ $\dfrac{4}{5} = 0.8$

10. $\tan B$ $\dfrac{3}{4} = 0.75$

14. $\sin P = \dfrac{12}{37} \approx 0.3243$;

$\cos P = \dfrac{35}{37} \approx 0.9459$

Practice and Applications

Extra Practice
See p. 694.

15. $\sin P = \dfrac{6\sqrt{2}}{11} \approx 0.7714$;

$\cos P = \dfrac{7}{11} \approx 0.6364$

16. $\sin P = \dfrac{1}{3} \approx 0.3333$;

$\cos P = \dfrac{2\sqrt{2}}{3} \approx 0.9428$

Homework Help
Example 1: Exs. 11–13
Example 2: Exs. 14–16
Example 3: Exs. 17–24
Example 4: Exs. 25–30

Finding Sine and Cosine Ratios Find sin A and cos A. Write your answers as fractions in simplest form.

11.
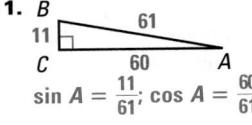
$\sin A = \dfrac{11}{61}$; $\cos A = \dfrac{60}{61}$

12.
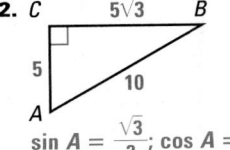
$\sin A = \dfrac{\sqrt{3}}{2}$; $\cos A = \dfrac{1}{2}$

13.
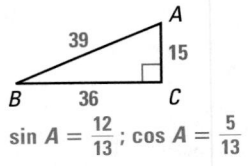
$\sin A = \dfrac{12}{13}$; $\cos A = \dfrac{5}{13}$

Finding Sine and Cosine Ratios Find sin P and cos P. Write your answers as fractions in simplest form and as decimals rounded to four decimal places. 14–16. See margin.

14.
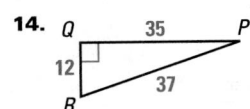

15.
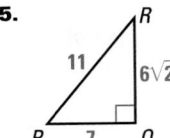

16.
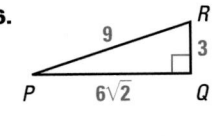

Assignment Guide

BASIC
Day 1: pp. 566–568 Exs. 11–16, 39–48
Day 2: pp. 566–568 Exs. 17–30, 38, 49–51

AVERAGE
Day 1: pp. 566–568 Exs. 11–16, 33–35, 39–48
Day 2: pp. 566–568 Exs. 17–32, 37, 38, 49–51

ADVANCED
Day 1: pp. 566–568 Exs. 11–16, 33–35, 36*, 39–48
Day 2: pp. 566–568 Exs. 17–32, 37, 38, 49–51;
EC: classzone.com

BLOCK SCHEDULE
pp. 566–568 Exs. 11–35, 37–51

Extra Practice
• Student Edition, p. 694
• Chapter 10 Resource Book, pp. 49–50

Homework Check
To quickly check student understanding of key concepts, go over the following exercises:

Basic: 11, 14, 17, 18, 25
Average: 12, 15, 19, 20, 28
Advanced: 13, 16, 23, 24, 30

Calculator Use a calculator to approximate the value to four decimal places.

17. sin 40° 0.6428 **18.** cos 23° 0.9205 **19.** sin 80° 0.9848 **20.** cos 5° 0.9962

21. sin 59° 0.8572 **22.** cos 61° 0.4848 **23.** sin 90° 1 **24.** cos 77° 0.2250

Finding Leg Lengths Find the lengths of the legs of the triangle. Round your answers to the nearest tenth.

25.
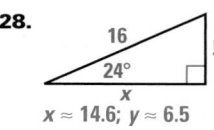
x ≈ 5.8; y ≈ 5.6

26.
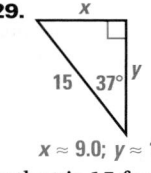
x ≈ 8.2; y ≈ 11.3

27.
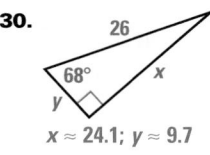
x ≈ 5.3; y ≈ 9.6

28.
16, 24°, x
x ≈ 14.6; y ≈ 6.5

29.
x, 15, 37°, y
x ≈ 9.0; y ≈ 12.0

30.
26, 68°, x, y
x ≈ 24.1; y ≈ 9.7

Student Help
CLASSZONE.COM
HOMEWORK HELP
Extra help with problem solving in Exs. 25–30 is at classzone.com

31. Visualize It! A ladder that is 15 feet long is leaning against a wall. The ladder makes an angle of 70° with the ground. Make a sketch. Then determine how high up the wall the ladder reaches. Round your answer to the nearest foot. Check sketches; 14 ft.

Link to
Skateboarding

SKATEBOARD RAMPS built with curves, called half pipes, allow freestyle skateboard riders to perform acrobatic maneuvers.

32. Skateboard Ramp You are constructing a skateboarding ramp like the one shown below. Your ramp will be 8 feet long and the ramp angle will be about 22°. Find the lengths of the legs of the triangles that support the ramp. Round your answers to the nearest inch. 36 in. and 89 in.

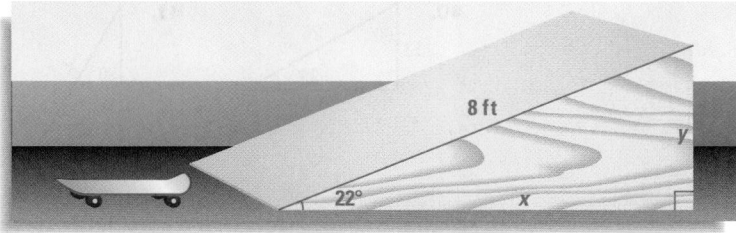

Technology In Exercises 33–35, use geometry software.

❶ Draw $\overline{AB}$.

❷ Construct a perpendicular to $\overline{AB}$ through B.

❸ Add point C on the perpendicular.

❹ Draw $\overline{AC}$.

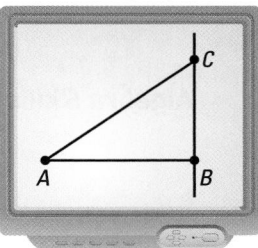

33. Find $m\angle A$, sin A, and cos A. Answers will vary.

34. Calculate $(\sin A)^2 + (\cos A)^2$. 1

35. Drag point C. What do you notice? $(\sin A)^2 + (\cos A)^2 = 1$.

X *Common Error*
In Exercises 25–30, students will be using both the sine and cosine formulas and may confuse them. Suggest that they jot down the definitions of the ratios on an index card to refer to as they work these exercises.

Mini-Quiz

1. Find sin *K* and cos *K*. Write your answers as fractions in simplest form and as decimals rounded to four decimal places.

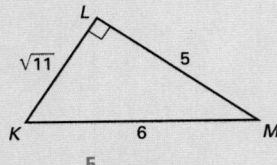

$\sin K = \frac{5}{6} \approx 0.8333;$

$\cos K = \frac{\sqrt{11}}{6} \approx 0.5528$

2. Find the lengths *x* and *y* of the legs of the triangle. Round your answers to the nearest tenth.

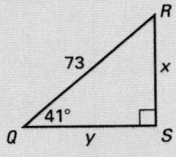

$x \approx 47.9;\ y \approx 55.1$

3. A ramp is built with a ramp angle of 25° as shown in the figure below. The end of the ramp is 15 feet above the ground. What is the length of the inclined surface of the ramp? Round your answer to the nearest tenth.

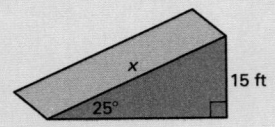

35.5 ft

36. See Additional Answers beginning on page AA1.

36. Challenge Let *A* be any acute angle of a right triangle. Use the definitions of sin *A*, cos *A*, and tan *A* to prove the following result. **See margin.**

$$\frac{\sin A}{\cos A} = \tan A$$

37. **You be the Judge** One student uses the ratio $\sin 42° = \frac{r}{34}$ to find the length of $\overline{ST}$. Another student uses the ratio $\cos 48° = \frac{r}{34}$.

Assuming the students make no errors in calculation, who will get the correct answer? **Both students will get the correct answer.**

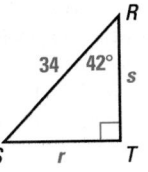

Standardized Test Practice

38. Multiple Choice Use the diagram below. Which expression could be used to find *CD*? C

(A) 8(cos 25°) (B) 8(sin 25°)

(C) $\frac{8}{\sin 25°}$ (D) $\frac{8}{\cos 25°}$

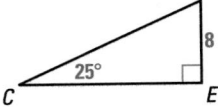

39. Multiple Choice Which statement *cannot* be true? H

(F) sin *A* = 0.55 (G) sin *A* = 0.61

(H) sin *A* = 1.2 (J) sin *A* = 0.4869

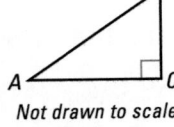

Not drawn to scale

Mixed Review

Finding Leg Lengths Find the value of each variable. Write your answer in radical form. *(Lesson 10.3)*

40.
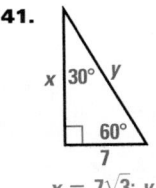

41.
$x = 7\sqrt{3};\ y = 14$

42.
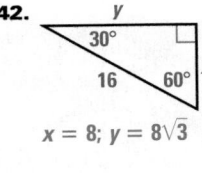

$x = 8;\ y = 8\sqrt{3}$

Using a Calculator Use a calculator to approximate the value to four decimal places. *(Lesson 10.4)*

43. tan 32° 0.6249 **44.** tan 88° 28.6363 **45.** tan 56° 1.4826

46. tan 24° 0.4452 **47.** tan 17° 0.3057 **48.** tan 49° 1.1504

Algebra Skills

Ordering Numbers Write the numbers in order from least to greatest. *(Skills Review, p. 662)*

49. −0.8, 1.8, −8, 0.08, −18, 0, −1.8 −18, −8, −1.8, −0.8, 0, 0.08, 1.8

50. 2641, 2146, 2614, 2416, 2164, 2461 2146, 2164, 2416, 2461, 2614, 2641

51. −0.56, −0.47, −0.61, −0.5, −0.6 −0.61, −0.6, −0.56, −0.5, −0.47

10.6 Solving Right Triangles

Goal
Solve a right triangle.

Key Words
- tangent p. 557
- sine p. 563
- cosine p. 563
- solve a right triangle
- inverse tangent
- inverse sine
- inverse cosine

To **solve a right triangle** means to find the measures of both acute angles and the lengths of all three sides. Suppose you know the lengths of the *legs* of a right triangle. How would you find the measures of the *angles*?

In the triangle at the right, the legs have lengths 7 and 10, so the tangent of $\angle A$ is $\frac{7}{10}$, or 0.7.

You can use the table of trigonometric ratios on page 705 to find the measure of $\angle A$. Or you can use the **inverse tangent** function ($\tan^{-1} x$) of a scientific calculator to find the angle measure.

Look for 0.7 in the tangent column.

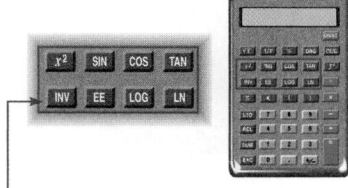

The angle with a tangent of 0.7 has a measure of about 35°.

On this calculator, you press [INV] then [TAN] to get the inverse tangent.

Student Help

READING TIP
The phrase "$\tan^{-1} z$" is read as "the inverse tangent of z." ············

INVERSE TANGENT

For any acute angle A of a right triangle:

▶ If $\tan A = z$, then $\tan^{-1} z = m\angle A$.

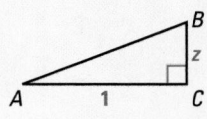

EXAMPLE 1 Use Inverse Tangent

Use a calculator to approximate the measure of $\angle A$ to the nearest tenth of a degree.

Solution

Since $\tan A = \frac{8}{10} = 0.8$, $\tan^{-1} 0.8 = m\angle A$.

Expression	Calculator keystrokes	Display
$\tan^{-1} 0.8$	0.8 [INV] [TAN] *or* 	38.65980825

ANSWER ▶ Because $\tan^{-1} 0.8 \approx 38.7°$, $m\angle A \approx 38.7°$.

10.6 *Solving Right Triangles* **569**

---

① Plan

Pacing
Suggested Number of Days

Basic: 2 days
Average: 2 days
Advanced: 2 days
Block Schedule: 1 block

Teaching Resources

📄 *Blacklines*
(See page 534B.)

✏ *Transparencies*
• Warm-Up with Quiz
• Answers

💻 *Technology*
• Electronic Teacher Tools
• Test and Practice Generator
• Online Lesson Planner
• Internet Support

② Teach

―**Content and Teaching Strategies**
For background information on geometric concepts and teaching strategies related to this lesson, see pages 534E and 534F in this Teacher's Edition.

Tips for New Teachers

Students need to practice interpreting and using the inverse trigonometric functions on their calculators. Give them decimal values and a function and ask them to find the angle measure. You should also have them practice using just the trigonometric tables on page 705. See the Tips for New Teachers on pp. 1–2 of the *Chapter 10 Resource Book* for additional notes about Lesson 10.6.

Extra Example 1
See next page.

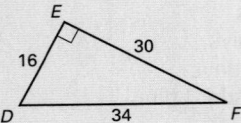
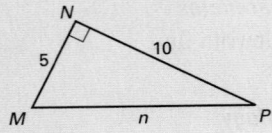

EXAMPLE 2 Solve a Right Triangle

Find each measure to the nearest tenth.

a. c **b.** $m\angle B$ **c.** $m\angle A$

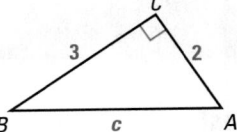

Solution

a. Use the Pythagorean Theorem to find c.

$(\text{hypotenuse})^2 = (\text{leg})^2 + (\text{leg})^2$	Pythagorean Theorem
$c^2 = 3^2 + 2^2$	Substitute.
$c^2 = 13$	Simplify.
$c = \sqrt{13}$	Find the positive square root.
$c \approx 3.6$	Use a calculator to approximate.

b. Use a calculator to find $m\angle B$.

Since $\tan B = \dfrac{2}{3} \approx 0.6667$, $m\angle B \approx \tan^{-1} 0.6667 \approx 33.7°$.

c. $\angle A$ and $\angle B$ are complementary, so $m\angle A \approx 90° - 33.7° = 56.3°$.

Checkpoint ✓ **Use Inverse Tangent**

$\angle A$ is an acute angle. Use a calculator to approximate the measure of $\angle A$ to the nearest tenth of a degree.

1. $\tan A = 3.5$ **74.1°**   **2.** $\tan A = 2$ **63.4°**   **3.** $\tan A = 0.4402$ **23.8°**

Find the measure of $\angle A$ to the nearest tenth of a degree.

4.
29.1°
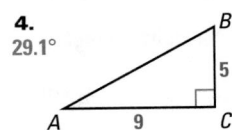

5.
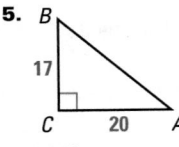
40.4°

6.
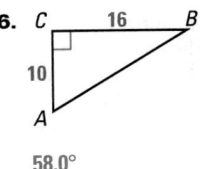
58.0°

Inverse Sine and Inverse Cosine A scientific calculator has **inverse sine** ($\sin^{-1} x$) and **inverse cosine** ($\cos^{-1} x$) functions. Use these inverse functions if you are given the lengths of one leg and the hypotenuse.

On this calculator, you press
INV **SIN** and **INV** **COS**
to get the inverse functions. ⎯⎯

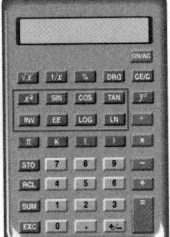

INVERSE SINE AND INVERSE COSINE

For any acute angle A of a right triangle:

If sin $A = y$, then sin$^{-1} y = m\angle A$.

If cos $A = x$, then cos$^{-1} x = m\angle A$.

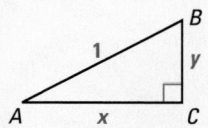

Student Help

STUDY TIP
To use the table of ratios on p. 705 to approximate sin^{-1} 0.55, find the number closest to 0.55 in the sine column, then read the angle measure at its left.

EXAMPLE 3 Find the Measures of Acute Angles

$\angle A$ is an acute angle. Use a calculator to approximate the measure of $\angle A$ to the nearest tenth of a degree.

a. sin $A = 0.55$ **b.** cos $A = 0.48$

Solution

a. Since sin $A = 0.55$, $m\angle A = \sin^{-1} 0.55$.

$\sin^{-1} 0.55 \approx 33.36701297$, so $m\angle A \approx 33.4°$.

b. Since cos $A = 0.48$, $m\angle A = \cos^{-1} 0.48$.

$\cos^{-1} 0.48 \approx 61.31459799$, so $m\angle A \approx 61.3°$.

Student Help

VISUAL STRATEGY

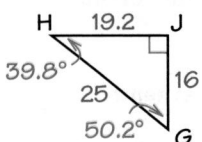

The triangle in Example 4 can be labeled in color, as suggested on p. 536.

EXAMPLE 4 Solve a Right Triangle

Solve $\triangle GHJ$ by finding each measure. Round decimals to the nearest tenth.

a. $m\angle G$ **b.** $m\angle H$ **c.** g

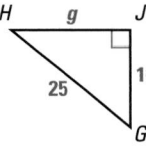

Solution

a. Since cos $G = \dfrac{16}{25} = 0.64$, $m\angle G = \cos^{-1} 0.64$.

$\cos^{-1} 0.64 \approx 50.2081805$, so $m\angle G \approx 50.2°$.

b. $\angle G$ and $\angle H$ are complementary.

$m\angle H = 90° - m\angle G \approx 90° - 50.2° = 39.8°$

c. Use the Pythagorean Theorem to find g.

$(\text{leg})^2 + (\text{leg})^2 = (\text{hypotenuse})^2$	Pythagorean Theorem
$16^2 + g^2 = 25^2$	Substitute.
$256 + g^2 = 625$	Simplify.
$g^2 = 369$	Subtract 256 from each side.
$g = \sqrt{369}$	Find the positive square root.
$g \approx 19.2$	Use a calculator to approximate.

10.6 *Solving Right Triangles*   **571**

Extra Example 3

$\angle P$ is an acute angle. Use a calculator to approximate the measure of $\angle P$ to the nearest tenth of a degree.

a. sin $P = 0.37$ 21.7°

b. cos $P = 0.28$ 73.7°

Extra Example 4

Solve $\triangle ABC$ by finding each measure. Round decimals to the nearest tenth.

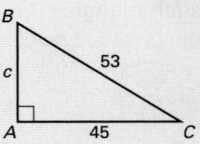

a. $m\angle B$ 58.1°

b. $m\angle C$ 31.9°

c. c 28

✗ Common Error

Throughout this lesson, watch for students who set up the equations incorrectly when they use trigonometric ratios to find side lengths.

Concept Check

When would you use the inverse tangent function? **when you know the lengths of the sides of a right triangle but not the measures of the acute angles and you want to find the measures of the angles**

Daily Puzzler

In right triangle ABC, $\sin A = \frac{3}{5}$ and $\sin B = \frac{4}{5}$. Give the possible side lengths of three such triangles. **The side lengths could be 3, 4, and 5; 6, 8, and 10; 9, 12, and 15; or any other multiples of 3, 4, and 5.**

∠A is an acute angle. Use a calculator to approximate the measure of ∠A to the nearest tenth of a degree.

7. $\sin A = 0.5$ **30°**　　**8.** $\cos A = 0.92$ **23.1°**　　**9.** $\sin A = 0.1149$ **6.6°**

10. $\cos A = 0.5$ **60°**　　**11.** $\sin A = 0.25$ **14.5°**　　**12.** $\cos A = 0.45$ **63.3°**

Solve the right triangle. Round decimals to the nearest tenth.

13.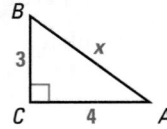

$x = 5$; $m\angle A \approx 36.9°$; $m\angle B \approx 53.1°$

14.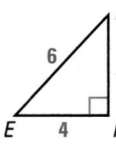

$y \approx 4.5$; $m\angle D \approx 41.8°$; $m\angle E \approx 48.2°$

15.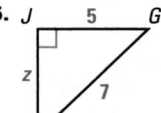

$z \approx 4.9$; $m\angle G \approx 44.4°$; $m\angle H \approx 45.6°$

10.6 Exercises

Guided Practice

Vocabulary Check

1. Explain what is meant by *solving a right triangle*. **finding the missing lengths of the sides and the missing measures of the acute angles**

Skill Check

Tell whether the statement is *true* or *false*.

2. You can solve a right triangle given only the lengths of two sides. **true**

3. You can solve a right triangle given only the measure of one acute angle. **false**

Find the value of x. Round your answer to the nearest tenth.

4.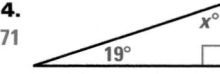

5.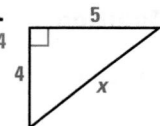

6.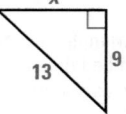

Calculator ∠A is an acute angle. Use a calculator to approximate the measure of ∠A to the nearest tenth of a degree.

7. $\tan A = 5.4472$ **79.6°**　**8.** $\sin A = 0.8988$ **64.0°**　**9.** $\cos A = 0.3846$ **67.4°**

Solve the right triangle. Round your answers to the nearest tenth.

10.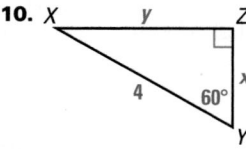

$x = 2$; $y \approx 3.5$; $m\angle X = 30°$

11.

$d \approx 7.2$; $m\angle D \approx 31.0°$; $m\angle E \approx 59.0°$

Practice and Applications

Extra Practice

See p. 694.

Calculator ∠A is an acute angle. Use a calculator to approximate the measure of ∠A to the nearest tenth of a degree.

12. $\tan A = 0.5$ 26.6°   **13.** $\tan A = 1.0$ 45°   **14.** $\tan A = 2.5$ 68.2°

15. $\tan A = 0.2311$ 13.0° **16.** $\tan A = 1.509$ 56.5° **17.** $\tan A = 4.125$ 76.4°

18. Use the Pythagorean Theorem; 73.

19. *Sample answer:* Use the inverse tangent function; 48.9°.

20. *Sample answer:* Use the fact that ∠Q and ∠S are complements; 41.1°.

Solving a Triangle Tell what method you would use to solve for the indicated measure. Then find the measure to the nearest tenth.

18. QS **19.** $m\angle Q$ **20.** $m\angle S$

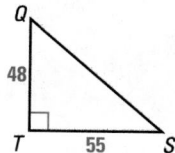

Inverse Tangent Use the Pythagorean Theorem to find the length of the hypotenuse. Then use the inverse tangent to find the measure of ∠A to the nearest tenth of a degree.

21.
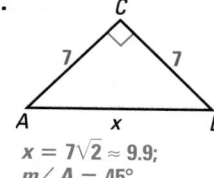
$x = 7\sqrt{2} \approx 9.9$;
$m\angle A = 45°$

22.
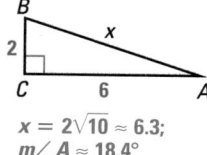
$x = 2\sqrt{10} \approx 6.3$;
$m\angle A \approx 18.4°$

23.
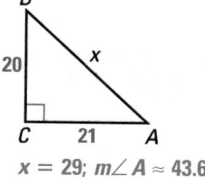
$x = 29$; $m\angle A \approx 43.6°$

Calculator ∠A is an acute angle. Use a calculator to approximate the measure of ∠A to the nearest tenth of a degree.

24. $\sin A = 0.75$ 48.6°   **25.** $\cos A = 0.1518$ 81.3° **26.** $\sin A = 0.6$ 36.9°

27. $\cos A = 0.45$ 63.3°   **28.** $\cos A = 0.1123$ 83.6° **29.** $\sin A = 0.6364$ 39.5°

Ramps In Exercises 30–32, use the information about ramps.
The Uniform Federal Accessibility Standards require that the measure of the angle used in a wheelchair ramp be less than or equal to 4.76°.

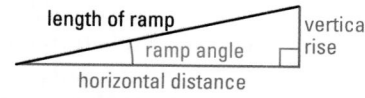

30. A ramp has a length of 20 feet and a vertical rise of 2.5 feet. Find the ramp's horizontal distance and the measure of its ramp angle. Does this ramp meet the standards? about 19.8 ft; about 7.2°; no

Homework Help

Example 1: Exs. 12–17
Example 2: Exs. 18–23
Example 3: Exs. 24–29
Example 4: Exs. 34–39

31. Suppose a ramp has a vertical rise of 4 feet. Give an example of a possible length of the ramp that meets the standards. *Sample answer:* 48 ft (Any length greater than or equal to 48.2 ft will meet the standards.)

32. Measurement Measure the horizontal distance and the vertical rise of a ramp near your home or school. Find the measure of the ramp angle. Does the ramp meet the standards? Explain. Answers will vary.

10.6 Solving Right Triangles **573**

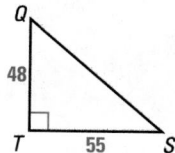

③ Apply

Assignment Guide

BASIC
Day 1: pp. 573–575 Exs. 12–23, 41, 43–54
Day 2: pp. 573–575 Exs. 24–29, 33–39, 42, Quiz 2

AVERAGE
Day 1: pp. 573–575 Exs. 12–23, 30–32, 41, 43–54
Day 2: pp. 573–575 Exs. 24–29, 33–40, 42, Quiz 2

ADVANCED
Day 1: pp. 573–575 Exs. 12–23, 30–32, 41–53 odd
Day 2: pp. 573–575 Exs. 24–29, 33–40, 42, Quiz 2;
EC: TE p. 534D*, classzone.com

BLOCK SCHEDULE
pp. 573–575 Exs. 12–54, Quiz 2

Extra Practice
• Student Edition, p. 694
• Chapter 10 Resource Book, pp. 58–59

Homework Check

To quickly check student understanding of key concepts, go over the following exercises:

Basic: 12, 21, 24, 27, 34
Average: 15, 22, 25, 26, 35
Advanced: 17, 23, 28, 29, 36

573

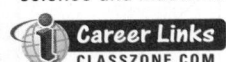
33. Space Shuttle The glide angle of a space shuttle is the angle indicated in the photo. During the shuttle's approach to Earth, the glide angle changes. When the shuttle's altitude is about 15.7 miles, its horizontal distance to the runway is about 59 miles. Find the measure of the glide angle. Round your answer to the nearest tenth. **14.9°**

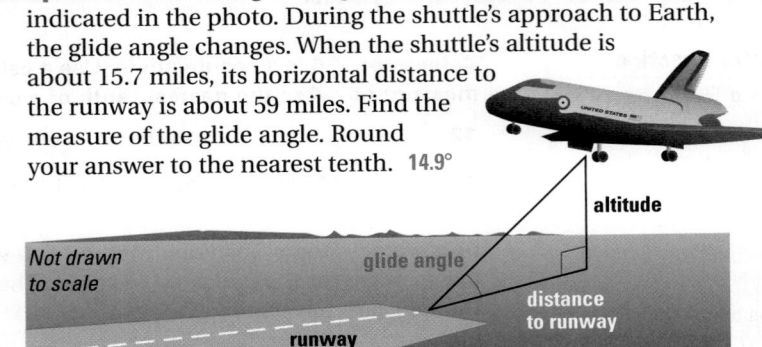

Inverse Sine and Inverse Cosine Find the measure of ∠A to the nearest tenth of a degree.

34. 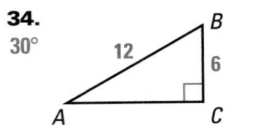 30°

35. 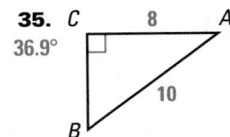 36.9°

36. 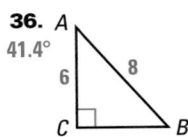 41.4°

Solving Right Triangles Solve the right triangle. Round decimals to the nearest tenth.

37.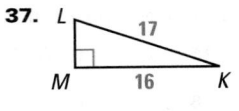

38.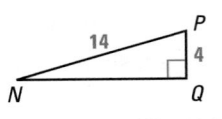

39.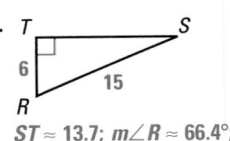

$LM \approx 5.7$; $m\angle L \approx 70.3°$;
$m\angle K \approx 19.7°$

$NQ \approx 13.4$; $m\angle N \approx 16.6°$;
$m\angle P \approx 73.4°$

$ST \approx 13.7$; $m\angle R \approx 66.4°$;
$m\angle S \approx 23.6°$

40. **You be the Judge** Each of the expressions $\sin^{-1}\frac{BC}{AC}$, $\cos^{-1}\frac{AB}{AC}$, and $\tan^{-1}\frac{BC}{AB}$ can be used to approximate $m\angle A$. Which expression would you choose? Explain your choice. $\tan^{-1}\frac{BC}{AB}$; the diagram gives the lengths AB and BC, but does not give the length AC.

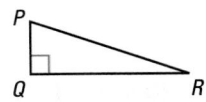

41. Multiple Choice Which additional information would *not* be enough to solve △PQR? **B**

Ⓐ $m\angle P$ and PR Ⓑ $m\angle P$ and $m\angle R$

Ⓒ PQ and PR Ⓓ $m\angle P$ and PQ

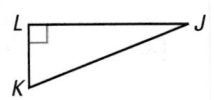

42. Multiple Choice Which expression is correct? **G**

Ⓕ $\sin^{-1}\frac{JL}{JK} = m\angle J$ Ⓖ $\tan^{-1}\frac{KL}{JL} = m\angle J$

Ⓗ $\cos^{-1}\frac{JL}{JK} = m\angle K$ Ⓙ $\sin^{-1}\frac{JL}{KL} = m\angle K$

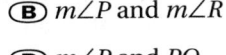

Circumference and Area of Circles Find the circumference and the area of the circle. Round your answers to the nearest whole number. *(Lesson 8.7)*

43.

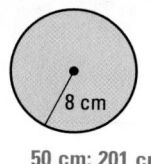

50 cm; 201 cm^2

44.

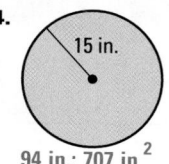

15 in.

94 in.; 707 in.^{2}

45.

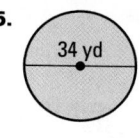

34 yd

107 yd; 908 yd^2

Volume of Solids Find the volume of the solid. Round your answers to the nearest whole number. *(Lesson 9.6)*

46.

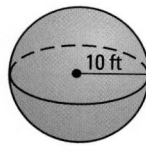

10 ft

4189 ft^3

47.

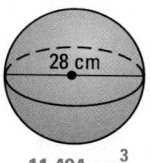

28 cm

11,494 cm^3

48.

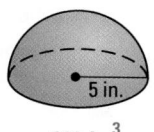

5 in.

262 in.^{3}

Algebra Skills

Decimal Operations Evaluate. *(Skills Review, p. 655)*

49. $0.36 + 0.194$ 0.554

50. $\$8.42 - \$2.95$ $5.47

51. 7×4.65 32.55

52. $55.40 \div 0.04$ 1385

53. $700 \div 0.35$ 2000

54. $\$22.50 \times 0.08$ $1.80

Quiz 2

Find the value of each variable. Round the results to the nearest tenth. *(Lessons 10.4, 10.5)*

1.

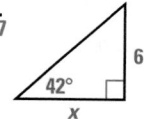

6.7, 6, 42°, x

2.

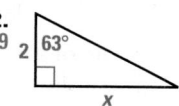

3.9, 2, 63°, x

3.

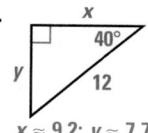

x, 40°, y, 12

$x \approx 9.2$; $y \approx 7.7$

4.

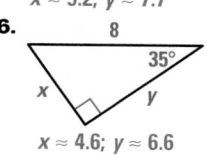

19, x, 21°, y

$x \approx 6.8$; $y \approx 17.7$

5.

14, 7.8, 29°, x

6.

8, 35°, x, y

$x \approx 4.6$; $y \approx 6.6$

Use a calculator to approximate the value to four decimal places. *(Lessons 10.4, 10.5)*

7. $\tan 72°$ 3.0777

8. $\sin 52°$ 0.7880

9. $\cos 36°$ 0.8090

Solve the right triangle. Round decimals to the nearest tenth. *(Lesson 10.6)*

10.

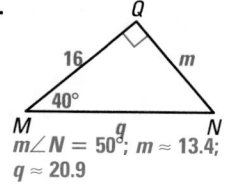

Q, 16, m, 40°, M, q, N

$m\angle N = 50°$; $m \approx 13.4$; $q \approx 20.9$

11.

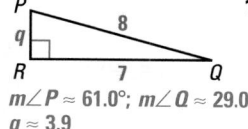

P, q, 8, R, 7, Q

$m\angle P \approx 61.0°$; $m\angle Q \approx 29.0°$; $q \approx 3.9$

12.

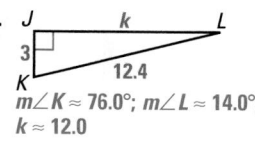

J, k, L, 3, K, 12.4

$m\angle K \approx 76.0°$; $m\angle L \approx 14.0°$; $k \approx 12.0$

10.6 Solving Right Triangles **575**

Mini-Quiz

$\angle A$ is an acute angle. Use a calculator to approximate the measure of $\angle A$ to the nearest tenth of a degree.

1. $\sin A = 0.3404$ 19.9°

2. $\cos A = 0.301$ 72.5°

3. $\tan A = 4.8288$ 78.3°

4. Solve the right triangle. Round your answers to the nearest tenth.

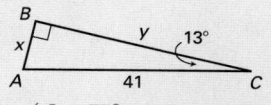

B, x, y, 13°, A, 41, C

$m\angle A = 77°$; $x \approx 9.2$; $y \approx 39.9$

Additional Resources

The following resources are available to help review the materials in this chapter.

Chapter 10 Resource Book
• Chapter Review Games and Activities, p. 68
• Cumulative Review, Chs. 1–10

VOCABULARY

• **radical**, *p. 537*
• **radicand**, *p. 537*
• **45°-45°-90° triangle**, *p. 542*
• **30°-60°-90° triangle**, *p. 549*
• **trigonometric ratio**, *p. 557*

• **leg opposite an angle**, *p. 557*
• **leg adjacent to an angle**, *p. 557*
• **tangent**, *p. 557*
• **sine**, *p. 563*
• **cosine**, *p. 563*

• **solve a right triangle**, *p. 569*
• **inverse tangent**, *p. 569*
• **inverse sine**, *p. 570*
• **inverse cosine**, *p. 570*

VOCABULARY REVIEW

Fill in the blank.

1. A(n) __?__ is an expression written with a radical symbol. radical

2. The number or expression inside the radical symbol is the __?__. radicand

3. A(n) __?__ is a ratio of the lengths of two sides of a right triangle. trigonometric ratio

4. A right triangle with side lengths 9, $9\sqrt{3}$, and 18 is a(n) __?__ triangle. 30°-60°-90° triangle

5. To __?__ a right triangle means to determine the lengths of all three sides of the triangle and the measures of both acute angles. solve

6. A right triangle with side lengths 4, 4, and $4\sqrt{2}$ is a(n) __?__ triangle. 45°-45°-90° triangle

7. If $\angle F$ is an acute angle of a right triangle, then

$$\underline{\quad?\quad} \text{ of } \angle F = \frac{\text{leg opposite } \angle F}{\text{leg adjacent to } \angle F}.$$ tangent

8. If $\angle F$ is an acute angle of a right triangle, then

$$\underline{\quad?\quad} \text{ of } \angle F = \frac{\text{leg adjacent } \angle F}{\text{hypotenuse}}.$$ cosine

9. If $\angle F$ is an acute angle of a right triangle, then

$$\underline{\quad?\quad} \text{ of } \angle F = \frac{\text{leg opposite } \angle F}{\text{hypotenuse}}.$$ sine

---

10.1 SIMPLIFYING SQUARE ROOTS

Examples on pp. 537–538

EXAMPLES Use the Product Property of Radicals to simplify the expression.

a. $\sqrt{7} \cdot \sqrt{5}$

$\sqrt{7} \cdot \sqrt{5} = \sqrt{7 \cdot 5} = \sqrt{35}$

b. $\sqrt{52}$

$\sqrt{52} = \sqrt{4 \cdot 13} = 2\sqrt{13}$

Multiply the radicals. Then simplify if possible.

10. $\sqrt{13} \cdot \sqrt{13}$ 13

11. $\sqrt{2} \cdot \sqrt{72}$ 12

12. $\sqrt{7} \cdot \sqrt{10}$ $\sqrt{70}$

13. $\left(4\sqrt{7}\right)^2$ 112

14. $\sqrt{3} \cdot \sqrt{19}$ $\sqrt{57}$

15. $\sqrt{5} \cdot \sqrt{5}$ 5

16. $\sqrt{6} \cdot \sqrt{18}$ $6\sqrt{3}$

17. $\left(3\sqrt{11}\right)^2$ 99

Simplify the radical expression.

18. $\sqrt{27}$ $3\sqrt{3}$

19. $\sqrt{72}$ $6\sqrt{2}$

20. $\sqrt{150}$ $5\sqrt{6}$

21. $\sqrt{68}$ $2\sqrt{17}$

22. $\sqrt{108}$ $6\sqrt{3}$

23. $\sqrt{80}$ $4\sqrt{5}$

24. $\sqrt{7500}$ $50\sqrt{3}$

25. $\sqrt{507}$ $13\sqrt{3}$

Use the formula $A = \ell w$ to find the area of the rectangle. Round your answer to the nearest tenth.

26. 31.6
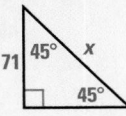
$2\sqrt{5}$
$5\sqrt{2}$

27. 50.9
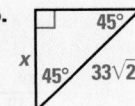
$3\sqrt{3}$
$4\sqrt{6}$

28. 8.9
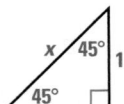
$\sqrt{10}$
$2\sqrt{2}$

10.2   45°-45°-90° TRIANGLES

Examples on pp. 542–544

EXAMPLES **Find the value of x.**

a.
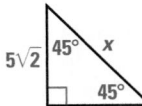

By the 45°-45°-90° Triangle Theorem, the length of the hypotenuse is the length of a leg times $\sqrt{2}$, so $x = 71\sqrt{2}$.

b.
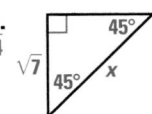

By the 45°-45°-90° Triangle Theorem, the length of the hypotenuse is the length of a leg times $\sqrt{2}$, so $33\sqrt{2} = x\sqrt{2}$, and $x = 33$.

Find the length of the hypotenuse in the 45°-45°-90° triangle. Write your answer in radical form.

29. $15\sqrt{2}$
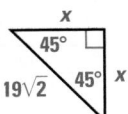
x 45° 15 45°

30. 10
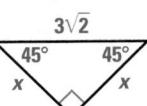
$5\sqrt{2}$ 45° x 45°

31. $\sqrt{14}$
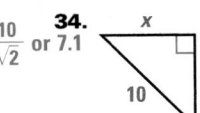
45° $\sqrt{7}$ 45° x

Find the length of each leg in each 45°-45°-90° triangle. Write your answer in radical form or as a decimal to the nearest tenth.

32. 19
x
$19\sqrt{2}$ 45° 45° x

33. 3
$3\sqrt{2}$
45° 45°
x x

34. $\dfrac{10}{\sqrt{2}}$ or 7.1
x
10 x

39. $\tan A = \dfrac{8}{15} \approx 0.5333;$

    $\tan B = \dfrac{15}{8} = 1.875$

40. $\tan A = \dfrac{1}{\sqrt{3}} \approx 0.5774;$

    $\tan B = \sqrt{3} \approx 1.7321$

41. $\tan A = \dfrac{3}{2} = 1.5;$

    $\tan B = \dfrac{2}{3} \approx 0.6667$

10.3   30°-60°-90° TRIANGLES

Examples on pp. 549–551

EXAMPLES   Find the value of each variable.

a.
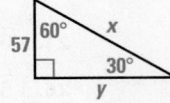

b.
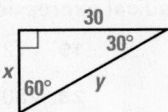

By the 30°-60°-90° Triangle Theorem, the length of the hypotenuse is twice the length of the shorter leg, so $x = 2(57) = 114$.

The length of the longer leg is the length of the shorter leg times $\sqrt{3}$, so $y = 57\sqrt{3}$.

By the 30°-60°-90° Triangle Theorem, the length of the longer leg is the length of the shorter leg times $\sqrt{3}$,

so $30 = x\sqrt{3}$ and $x = \dfrac{30}{\sqrt{3}} \approx 17.3$.

Then $y = 2x \approx 34.6$.

Find the value of each variable. Write your answers in radical form or as a decimal to the nearest tenth.

35.
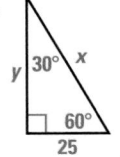

$x = 50;\ y = 25\sqrt{3} \approx 43.3$

36.
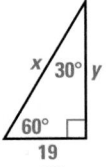

$x = 38;\ y = 19\sqrt{3} \approx 32.9$

37.
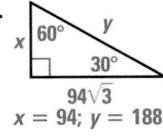

$x = 94;\ y = 188$

38.
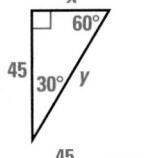

$x = \dfrac{45}{\sqrt{3}} \approx 26.0;\ y \approx 52.0$

10.4   TANGENT RATIO

Examples on pp. 557–559

EXAMPLE   Find tan *A* and tan *B* as fractions and as decimals.

$\tan A = \dfrac{\text{leg opposite to } \angle A}{\text{leg adjacent to } \angle A} = \dfrac{21}{20} = 1.05$

$\tan B = \dfrac{\text{leg opposite to } \angle B}{\text{leg adjacent to } \angle B} = \dfrac{20}{21} \approx 0.9524$

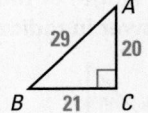

Find tan *A* and tan *B* as fractions in simplest form and as decimals. Round your answers to four decimal places if necessary. 39–41. See margin.

39.
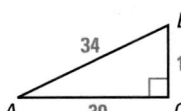

40.
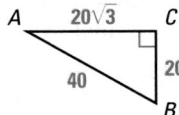

41.
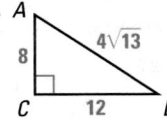

Approximate the value to four decimal places.

42. $\tan 17°$ 0.3057   43. $\tan 81°$ 6.3138   44. $\tan 36°$ 0.7265   45. $\tan 24°$ 0.4452

10.5 SINE AND COSINE RATIOS

Examples on pp. 563–565

> **EXAMPLE** Find sin A and cos A as fractions and as decimals.
>
> $\sin A = \dfrac{\text{leg opposite } \angle A}{\text{hypotenuse}} = \dfrac{20}{29} \approx 0.6897$
>
> $\cos A = \dfrac{\text{leg adjacent to } \angle A}{\text{hypotenuse}} = \dfrac{21}{29} \approx 0.7241$
>
>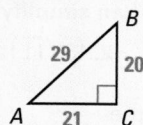

Find sin A and cos A as fractions in simplest form and as decimals. Round your answers to four decimal places if necessary. 46–48. See margin.

46.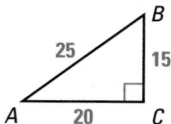
47.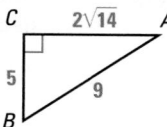
48.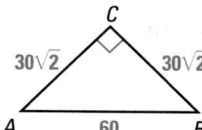

Approximate the value to four decimal places.

49. sin 57° 0.8387
50. sin 12° 0.2079
51. cos 31° 0.8572
52. cos 75° 0.2588

53. Find the lengths of the legs of the triangle. Round your answers to the nearest tenth. $x \approx 5.2$; $y \approx 8.6$

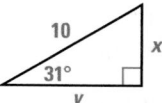

10.6 SOLVING RIGHT TRIANGLES

Examples on pp. 569–571

> **EXAMPLE** Solve $\triangle XYZ$.
>
> By the Pythagorean Theorem, $y^2 = 5^2 + 7^2 = 25 + 49 = 74$, so $y \approx 8.6$.
>
> $\tan X = \dfrac{7}{5} = 1.4$, so $m\angle X = \tan^{-1} 1.4 \approx 54.5°$.
>
> Since $\angle X$ and $\angle Z$ are complementary,
> $m\angle Z = 90° - m\angle X \approx 90° - 54.5° = 35.5°$.
>
>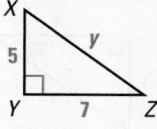

$\angle A$ is an acute angle. Use a calculator to approximate the measure of $\angle A$ to the nearest tenth of a degree.

54. tan A = 3.2145
72.7°
55. sin A = 0.0888
5.1°
56. cos A = 0.2243
77.0°
57. tan A = 1.2067
50.4°

Solve the right triangle. Round decimals to the nearest tenth. 58–60. See margin.

58.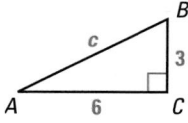
59.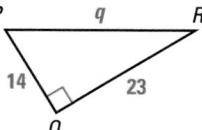
60.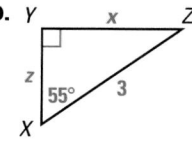

Chapter Summary and Review 579

---

46. $\sin A = \dfrac{3}{5} = 0.6$; $\cos A = \dfrac{4}{5} = 0.8$

47. $\sin A = \dfrac{5}{9} \approx 0.5556$;
$\cos A = \dfrac{2\sqrt{14}}{9} \approx 0.8315$

48. $\sin A = \dfrac{\sqrt{2}}{2} \approx 0.7071$;
$\cos A = \dfrac{\sqrt{2}}{2} \approx 0.7071$

58. $c \approx 6.7$; $m\angle A \approx 26.6°$;
$m\angle B \approx 63.4°$

59. $q \approx 26.9$; $m\angle P \approx 58.7°$;
$m\angle R \approx 31.3°$

60. $m\angle Z = 35°$; $x \approx 2.5$; $z \approx 1.7$

Additional Resources

 **Chapter 10 Resource Book**
- Chapter Test (2 levels), pp. 69–72
- SAT/ACT Chapter Test, p. 73
- Alternative Assessment, pp. 74–75

Test and Practice Generator

Chapter 10 Chapter Test

Multiply the radicals. Then simplify if possible.

1. $\sqrt{15} \cdot \sqrt{7}$ $\sqrt{105}$ **2.** $(3\sqrt{11})^2$ 99

3. Find the length of the hypotenuse of the triangle at the right. Write your answer in radical form. $4\sqrt{2}$

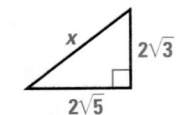

Use the diagram below to match the segment with its length.

4. $\overline{DC}$ D   **A.** 12

5. $\overline{AB}$ A   **B.** $6\sqrt{2}$

6. $\overline{BC}$ B   **C.** $6\sqrt{3}$

7. $\overline{AD}$ C   **D.** 6

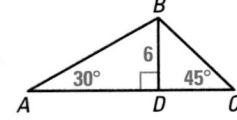

8. The hypotenuse of a 45°-45°-90° triangle has length 47. Find the length of each leg to the nearest tenth.   33.2

Find the value of each variable to the nearest tenth.

9.
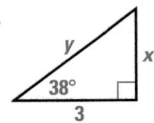
$x \approx 2.3; y \approx 3.8$

10.
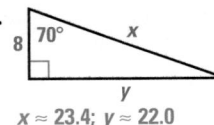
$x \approx 23.4; y \approx 22.0$

11.
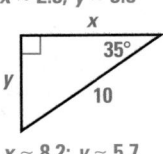
$x \approx 8.2; y \approx 5.7$

12.
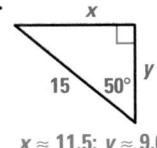
$x \approx 11.5; y \approx 9.6$

Use a calculator to approximate the value to four decimal places.

13. $\tan 70°$ 2.7475   **14.** $\cos 14°$ 0.9703

15. $\tan 31°$ 0.6009   **16.** $\sin 26°$ 0.4384

17. $\cos 30°$ 0.8660   **18.** $\tan 45°$ 1

19. $\sin 5°$ 0.0872   **20.** $\tan 10°$ 0.1763

$\angle A$ is an acute angle. Use a calculator to approximate the measure of $\angle A$ to the nearest tenth of a degree.

21. $\tan A = 5.2$ 79.1°   **22.** $\tan A = 7$ 81.9°

23. $\sin A = 0.3091$ 18.0°   **24.** $\sin A = 0.5318$ 32.1°

25. $\cos A = 0.6264$ 51.2°   **26.** $\cos A = 0.3751$ 68°

Solve the right triangle. Round decimals to the nearest tenth.

27.
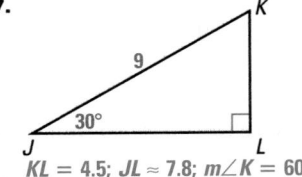
$KL = 4.5; JL \approx 7.8; m\angle K = 60°$

28.
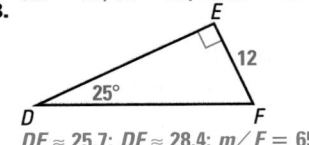
$DE \approx 25.7; DF \approx 28.4; m\angle F = 65°$

29.
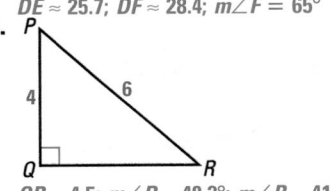
$QR \approx 4.5; m\angle P \approx 48.2°; m\angle R \approx 41.8°$

30. In the diagram of the roller coaster, $\triangle XYZ$ is a right triangle with $XZ = 320$ and $YZ = 180$. Find the measure of $\angle X$ to the nearest tenth of a degree.   29.4°

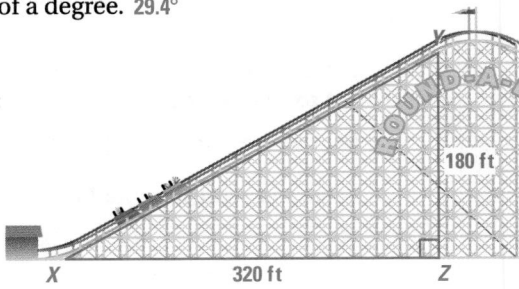

Additional Resources

📘 **Chapter 10 Resource Book**
• SAT/ACT Chapter Test, p. 73

 **Test Tip**

Memorize important ideas like the Pythagorean Theorem, the Midpoint Theorem, and formulas for area and volume.

1. Simplify the expression $\sqrt{124}$. **A**

   (A) $2\sqrt{31}$ (B) 12

   (C) $2\sqrt{62}$ (D) $4\sqrt{31}$

2. What is the value of x in the triangle shown below? **J**

   (F) 7

   (G) $\sqrt{21}$

   (H) $2\sqrt{7}$

   (J) $\sqrt{14}$

3. For $\angle J$ in the figure shown below, which statement is *true*? **B**

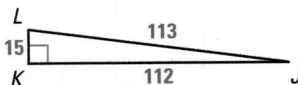

   (A) $\tan J = \dfrac{15}{113}$ (B) $\tan J = \dfrac{15}{112}$

   (C) $\tan J = \dfrac{112}{113}$ (D) $\tan J = \dfrac{112}{15}$

4. $\angle W$ is an acute angle and $\sin W = 0.8170$. Approximate the measure of $\angle W$. **J**

   (F) $0.01°$ (G) $35.2°$

   (H) $39.2°$ (J) $54.8°$

5. Which of the following statements about the figure below is *true*? **C**

   (A) $SR = 48\sqrt{2}$

   (B) $QR = 96$

   (C) $PS = 48\sqrt{3}$

   (D) $PQ = 96\sqrt{3}$

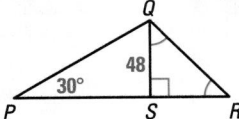

6. Find the measure of $\angle A$ to the nearest tenth of a degree. **G**

   (F) $28.6°$

   (G) $33.1°$

   (H) $37.4°$

   (J) $56.9°$

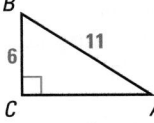

7. Which of the following statements about the triangle shown below is *true*? **B**

   (A) $m\angle P = 48°$

   (B) $PR \approx 11.3$

   (C) $RQ \approx 8.6$

   (D) None of these

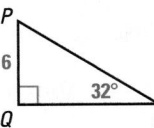

8. The distance from a point P on the ground to a point R at the base of a cliff is 30 meters. In the figure below, the measure of $\angle P$ is $72°$. Which of the following gives the approximate height h of the cliff? **J**

   (F) 60.8 m

   (G) 78.6 m

   (H) 90.4 m

   (J) 92.3 m

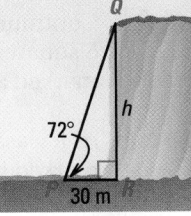

9. Which of the following gives the approximate values of x and y for the triangle shown? **D**

   (A) $x \approx 18.8, y \approx 20.1$

   (B) $x \approx 14.4, y \approx 19.2$

   (C) $x \approx 12.6, y \approx 18.5$

   (D) $x \approx 18.8, y \approx 20.5$

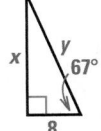

Chapter Standardized Test **581**

Teaching Tip

The Brain Games activity provides a motivating way to review selected content in the chapter. For a more comprehensive review, see the Chapter Summary and Review on pp. 576–579.

BraiN ⤳ GaMes

Right Triangle Bingo

Materials

- bingo card
- chips (optional)
- calculator

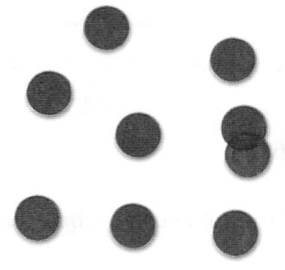

Object of the Game   To be the first to get 5 answers in a row.

Set Up   The bingo card has a five-by-five grid and 30 answers listed. Choose 24 of the answers and use them to fill in the empty squares on the card.

Step 1 ▶ Your teacher (or a student assigned to be the caller) will read a problem.

Step 2 ▶ Solve the problem.

Step 3 ▶ If the answer is on your card, mark it with a chip or use a pencil to circle the number.

Step 4 ▶ Continue playing until someone has 5 squares in a row, column, or diagonal marked and says "BINGO."

Step 5 ▶ As a class, check that the answers marked have been called.

B	I	N	G	O
6	4.70	5.03	26.15	.36
5.26	$7\sqrt{2}$	0.59	0.21	0.85
11.92	1.73	FREE	28	0.05
10.46	10	0.87	19.42	1.50
12	25	0.97	$3\sqrt{3}$	0.90

Another Way to Play   The first one to get a T, X, or Z marked on the bingo card wins. Or use 30 questions submitted by students and write the answers on the board. Create a new bingo card and fill it in with the answers on the board. Play like above.

 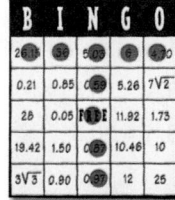 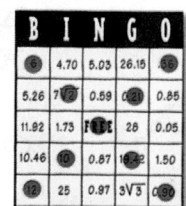  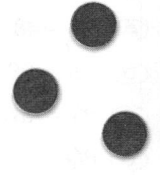

EXAMPLE **Solve Systems of Equations by Substitution**

Solve the linear system using substitution.

$-x + y = 8$	Equation 1
$2x + 3y = -1$	Equation 2

Solution

❶ Solve for y in Equation 1 because it is easy to isolate y.

$-x + y = 8$	Original Equation 1
$y = x + 8$	Revised Equation 1

❷ Substitute $x + 8$ for y in Equation 2 and solve for x.

$2x + 3y = -1$	Original Equation 2
$2x + 3(x + 8) = -1$	Substitute $x + 8$ for y in Equation 2.
$5x + 24 = -1$	Simplify and combine like terms.
$5x = -25$	Subtract 24 from both sides and simplify.
$x = -5$	Divide each side by 5 and simplify.

❸ Substitute -5 for x in the revised Equation 1 and solve for y.

$$y = x + 8 = -5 + 8 = 3$$

❹ Check that $(-5, 3)$ is a solution by substituting -5 for x and 3 for y in each of the original equations.

Equation 1: $-x + y = 8$ Equation 2: $2x + 3y = -1$

$-(-5) + 3 \stackrel{?}{=} 8$ $2(-5) + 3(3) \stackrel{?}{=} -1$

$5 + 3 = 8$ ✓ $-10 + 9 = -1$ ✓

ANSWER ▶ The solution is $(-5, 3)$.

Try These

Use the substitution method to solve the linear system.

1. $x + y = 1$
$4x + 5y = 7$ $(-2, 3)$

2. $x + 2y = 9$
$3x - y = -1$ $(1, 4)$

3. $3x + y = 3$
$7x + 2y = 1$ $(-5, 18)$

4. $x - y = -4$
$x + y = 16$ $(6, 10)$

5. $-x + y = 1$
$2x + y = 4$ $(1, 2)$

6. $6x - y = 2$
$4x + 3y = -6$ $(0, -2)$

7. $2x + 3y = 5$
$x - 4y = -3$ $(1, 1)$

8. $-3x - 2y = -5$
$-x + 3y = -9$ $(3, -2)$

9. $5x + 2y = 7$
$2x - 4y = 22$ $(3, -4)$

Mathematical Goals

- Use proportions and similar triangles to find unknown measures.
- Use the tangent ratio to find the side lengths of right triangles.

Managing the Project

Tips for Success

This project is best done by students working in pairs, with one student standing in place while their partner marks the location. Then the partners can easily measure the distances. Alternatively, one partner can use the mirror method while the other partner uses the shadow method. However, the project can be completed alone if some students prefer to work without a partner.

Guiding Students' Work

When using the mirror method, instruct students to place the mirror so that their shadow or the object's shadow does not make it hard to view the mirror. When students are using the shadow method, make sure everyone understands that the tops of the two shadows must coincide.

Indirect Measurement

Objective

Use indirect measurement to estimate the heights of objects.

Materials
- measuring tape
- mirror
- protractor

How to Estimate the Height of a Tall Object

You can use similar triangles to estimate the height of a tall object.

Mirror Method

Place a mirror on the ground between yourself and the object. Step backward until you can see the top of the object in the mirror. Use similar triangles to estimate the height y of the object.

Shadow Method

Stand so that the top of your shadow coincides with the top of the object's shadow. You can use this statement to find the height y.

$$\frac{x}{\text{Your shadow}} = \frac{y}{\text{Object's shadow}}$$

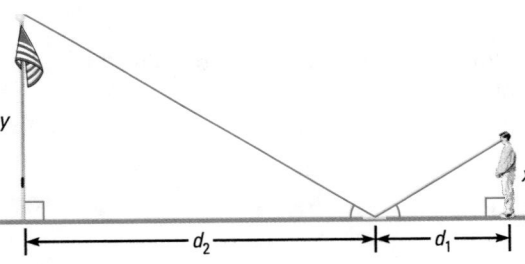

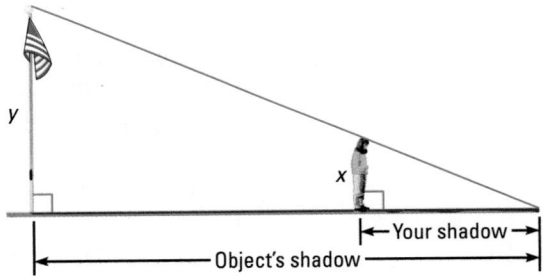

Investigation

❶ Use the mirror method to estimate the heights of three objects outside, such as a tree or a building. Record your data in a table like the one below.

	Your distance from mirror, d_1	Your height, x	Distance from object to mirror, d_2	Height of object, y
Object 1	?	?	?	?
Object 2	?	?	?	?
Object 3	?	?	?	?

2 Use the shadow method to estimate the heights of the same three objects outside. Record your data in a table like the one below.

	Your height, x	Length of your shadow	Height of object, y	Length of object's shadow
Object 1	?	?	?	?
Object 2	?	?	?	?
Object 3	?	?	?	?

3 Compare the heights you found using the two methods.

Indirect Measurement

by
Jim Russsell
and
Robert Pike

Present Your Results

Write a report about the mirror and shadow methods of indirect measurement.

▶ Include your answers to Steps 1–3 of the Investigation.

▶ Use similar triangles to explain why each method works. Provide a diagram with each explanation.

▶ Compare the two methods. Which do you prefer? Describe any advantages or disadvantages of the methods.

Extension

Try another method for estimating height. Use the diagram at the right and follow the steps below.

▶ Stand and look at the top of the object.

▶ Use a protractor to estimate the value of a.

▶ Measure your distance b from the object.

▶ Use the tangent ratio to find c:

$$\tan a = \frac{\text{opposite side}}{\text{adjacent side}} = \frac{c}{b}$$

▶ Estimate the height of the object, $c + d$.

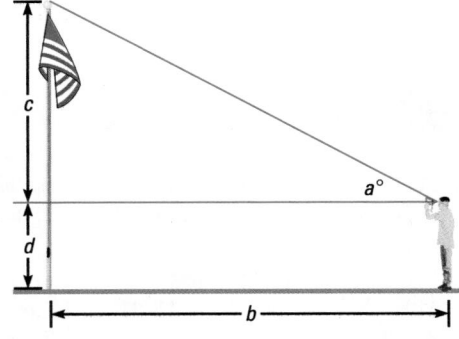

Concluding the Project
If possible, consider inviting a surveyor or similar person to class. Ask the speaker, or have students research and share with the class, information on how measuring devices such as transits, sextants, and theodolites are used. For each object whose height was measured indirectly by several students, have these students compare their results and work to arrive at a consensus about the height of the object.

Grading the Project
A well-written project will have the following characteristics:
• The measured values in the tables are reasonable and the calculated values are correct.
• Students use similar triangles to correctly explain how each method works.
• Students' written comparisons of the two methods demonstrate that they have tried both methods and thought about them critically.

Pacing and Assignment Guide

REGULAR SCHEDULE

Lesson	Les. Day	Basic	Average	Advanced
11.1	Day 1	pp. 591–593 Exs. 13–35 odd, 40–44, 48, 49–67 odd	pp. 591–593 Exs. 14–34 even, 40–44, 48–53, 56–68 even	pp. 591–593 Exs. 21–35 odd, 36–39, 43–52, 53–67 odd; EC: TE p. 586D*, classzone.com
11.2	Day 1	pp. 598–600 Exs. 7–24, 29, 30–44 even	pp. 598–600 Exs. 7–27, 29, 30, 31–45 odd	pp. 598–600 Exs. 7–12, 17–21, 25–27, 28*, 29, 30–44 even; EC: classzone.com
11.3	Day 1	pp. 604–607 Exs. 13–39 odd, 43–51 odd, 55–61 odd, Quiz 1	pp. 604–607 Exs. 14–38 even, 40–42, 44–52 even, 55–62, Quiz 1	pp. 604–607 Exs. 19–42, 43–53 odd, 54*, 55–60, Quiz 1; EC: classzone.com
11.4	Day 1	pp. 611–612 Exs. 6–8, 15, 16, 19, 24–29	pp. 611–612 Exs. 6–8, 15, 16, 19, 21, 24–29	pp. 611–612 Exs. 15, 16, 19, 21, 22, 24–29
	Day 2	pp. 611–612 Exs. 9–14, 17, 18, 20, 23, 30–32	pp. 611–612 Exs. 9–14, 17, 18, 20, 23, 30–32	pp. 611–612 Exs. 9–14, 17, 18, 20, 23, 30–32; EC: TE p. 586D*, classzone.com
11.5	Day 1	pp. 617–619 Exs. 9–20, 39, 50–55	pp. 617–619 Exs. 12–26, 39, 50–55	pp. 617–619 Exs. 12–17, 21–27, 39, 50–55
	Day 2	pp. 617–619 Exs. 28–30, 32–36, 40–49	pp. 617–619 Exs. 28–38, 40–49	pp. 617–619 Exs. 28–38, 40–49; EC: TE p. 586D*, classzone.com
11.6	Day 1	pp. 623–625 Exs. 8–17, 31–38	pp. 623–625 Exs. 8–17, 31–38	pp. 623–625 Exs. 12–18, 31–38
	Day 2	pp. 623–625 Exs. 20–25, 27–30, Quiz 2	pp. 623–625 Exs. 19–30, Quiz 2	pp. 623–625 Exs. 19–30, Quiz 2; EC: TE p. 586D*, classzone.com
11.7	Day 1	pp. 630–632 Exs. 5–27, 28–34 even, 41, 42–50 even	pp. 630–632 Exs. 9–35 odd, 36–38, 41–51	pp. 630–632 Exs. 8–34 even, 36–38, 39*, 40*, 41–46; EC: classzone.com
11.8	Day 1	pp. 636–639 Exs. 10–21, 38–45	pp. 636–639 Exs. 10–21, 31, 32, 38–45	pp. 636–639 Exs. 10–21, 31–36, 38–45
	Day 2	pp. 636–639 Exs. 22–24, 27–30, 37, Quiz 3	pp. 636–639 Exs. 22–30, 37, Quiz 3	pp. 636–639 Exs. 22–30, 37, Quiz 3; EC: TE p. 586D*, classzone.com
Review	Day 1	pp. 641–645 Exs. 1–56	pp. 641–645 Exs. 1–56	pp. 641–645 Exs. 1–56
Assess	Day 1	Chapter 11 Test	Chapter 11 Test	Chapter 11 Test

YEARLY PACING   Chapter 11 Total – **14 days**   Chapters 1–11 Total – **160 days**   Remaining – **0 days**

*Challenge Exercises   EP = Extra Practice   SRH = Skills Review Handbook   EC = Extra Challenge

BLOCK SCHEDULE

Day 1	Day 2	Day 3	Day 4	Day 5	Day 6	Day 7
11.1 pp. 591–593 Exs. 14–34 even, 40–44, 48–53, 56–68 even	**11.3** pp. 604–607 Exs. 14–38 even, 40–42, 44–52 even, 55–62, Quiz 1	**11.4 cont.** pp. 611–612 Exs. 9–14, 17, 18, 20, 23, 30–32	**11.5 cont.** pp. 617–619 Exs. 28–38, 40–49	**11.6 cont.** pp. 623–625 Exs. 19–30, Quiz 2	**11.8** pp. 636–639 Exs. 10–32, 37–45, Quiz 3	**Review** pp. 641–645 Exs. 1–56
11.2 pp. 598–600 Exs. 7–27, 29, 30, 31–45 odd	**11.4** pp. 611–612 Exs. 6–8, 15, 16, 19, 21, 24–29	**11.5** pp. 617–619 Exs. 12–26, 39, 50–55	**11.6** pp. 623–625 Exs. 8–17, 31–38	**11.7** pp. 630–632 Exs. 9–35 odd, 36–38, 41–51		**Assess** Chapter 11 Test

YEARLY PACING   Chapter 11 Total – **7 days**   Chapters 1–11 Total – **80 days**   Remaining – **0 days**

Support Materials

CHAPTER RESOURCE BOOK

CHAPTER SUPPORT

Tips for New Teachers	p. 1	Strategies for Reading Mathematics	p. 5
Parent Guide for Student Success	p. 3		

LESSON SUPPORT

	11.1	11.2	11.3	11.4	11.5	11.6	11.7	11.8
Lesson Plans (regular and block)	p. 7	p. 15	p. 24	p. 34	p. 42	p. 50	p. 62	p. 73
Warm-Up Exercises and Daily Quiz	p. 9	p. 17	p. 26	p. 36	p. 44	p. 52	p. 64	p. 75
Technology Activities & Keystrokes						p. 53	p. 65	p. 76
Practice (2 levels)	p. 10	p. 18	p. 27	p. 37	p. 45	p. 56	p. 67	p. 78
Reteaching with Practice	p. 12	p. 20	p. 29	p. 39	p. 47	p. 58	p. 69	p. 80
Quick Catch-Up for Absent Students	p. 14	p. 22	p. 31	p. 41	p. 49	p. 60	p. 71	p. 82
Learning Activities							p. 72	
Real-Life Applications		p. 23	p. 32					p. 83

REVIEW AND ASSESSMENT

Quizzes	pp. 33, 61, 84	Alternative Assessment with Math Journal	p. 92
Brain Games Support	p. 85	Project with Rubric	p. 94
Chapter Review Games and Activities	p. 86	Cumulative Review	p. 96
Chapter Test (2 levels)	p. 87	Cumulative Test for Chs. 1–11	p. 98
SAT/ACT Chapter Test	p. 91	Resource Book Answers	AN1

TRANSPARENCIES

	11.1	11.2	11.3	11.4	11.5	11.6	11.7	11.8
Warm-Up Exercises and Daily Quiz	p. 66	p. 67	p. 68	p. 69	p. 70	p. 71	p. 72	p. 73
Visualize It Transparencies	✓		✓				✓	✓
Answer Transparencies	✓	✓	✓	✓	✓	✓	✓	✓

TECHNOLOGY

- Time-Saving Test and Practice Generator
- Electronic Teacher Tools
- Geometry in Motion: Video
- Classzone.com
- Online Lesson Planner

ADDITIONAL RESOURCES

- Worked-Out Solution Key
- Resources in Spanish
- Practice Workbook with Examples

Providing Universal Access

Strategies for Strategic Learners

DIFFERENTIATE INSTRUCTION IN TERMS OF PACING

Geometry is very difficult for students who have not mastered the basic skills necessary. If students are struggling at this point, it is probably because they do not really understand the key concepts that have been covered thus far, such as angle relationships, areas, perimeters, and volumes. Nothing is to be gained by rushing these students through; they may complete the course with a passing grade and (if following the traditional sequence) may even do well in some areas of Algebra 2, but if they do not have a fundamental understanding of geometry, they will struggle with trigonometry and algebra problems that build on geometry concepts.

A better strategy is to slow the pace down for these students and analyze which key concepts are not well understood. The teacher could then use the Test and Practice Generator available for this textbook to help diagnose students' weak areas. If students did not understand square roots, they would not have understood the last chapter. If students do not understand that if two ratios are equal then their cross products are also equal, they will not be able to solve problems involving the properties of arcs discussed in Lesson 11.3.

Students who are falling behind need more mathematics instructional time. Work with the student and his or her parents and with the school principal and district office to see if one of the following schedule changes is possible so that the time spent learning mathematics is extended. Use the additional time to go back over essential concepts from previous chapters.

> **Tutorial**
> **Study hall**
> **Summer school**
> **Extend the school year into the summer**

If none of these time extensions is an option, students who will complete the course with a D or lower should be strongly encouraged to repeat the course. Repeating a course is something most adults become comfortable with. Many of us take a course over or repeat a course we took some time ago because we have forgotten the material. But for a student, course repetition is usually seen as a failure. Repeating a math course should be presented to students as a viable option for ensuring that students have the proper foundation for future study. College counselors often look favorably on a student who repeated a difficult subject and showed improvement the second time. They see this as a sign of maturity and effort. Students may find that the course is much easier the second time because they are building on those concepts and skills they did learn. This can provide a welcome relief if the student has a difficult class schedule. Repeating a course can demonstrate to

students that persistence pays off. And most of all, research has shown that self-esteem grows when students tackle a challenge and succeed.

Strategies for English Learners

VOCABULARY STUDY OF ROOTS

Many mathematical terms stem from a common root word, usually of Latin origins. Recognizing the Latin prefixes and roots can help students comprehend sophisticated mathematics vocabulary. Consider the following roots and words:

Latin Root	Mathematical Terms
aequus (even)	equal, equality, equate, equation, unequal, inequality
ex (out) and premere (to press)	express, expression
extraneous (foreign, strange)	extra, extraneous, extrapolate
numerus (number)	number, numeral, numeration, numerical, numerator
radicare (to root)	radical, radicand
ratus (to think) and ration (to reason)	ratio, rational, irrational
simplex (simple)	simplify
solvere (solve, loosen, dissolve)	solve, solution, solvable
value	value, evaluate
variare (various, diverse)	vary, variance, variable, variation

Encourage students to look at the similarity between new words and known words. Recognizing, for example, that *irrational* has the word *ratio* in it, plus the prefix *ir-*, which generally means *not*, will help students remember that an irrational number is one that cannot be expressed as the ratio of two integers.

Strategies for Advanced Learners

DIFFERENTIATE INSTRUCTION IN TERMS OF COMPLEXITY

Mathematics is an integral part of the study of other subjects, such as science, economics, and history. Teachers at this grade level may want to work together to develop interesting and more challenging assignments that combine the objectives for other courses into short projects or investigations that can be

substituted for easier assignments. These assignments can be a welcome alternative to having advanced students wait until the rest of the class catches up. The assignments should be designed so that students can work relatively independently, with a group, or with the science or history teacher. They can then be used while the mathematics teacher focuses on review or reteaching with the rest of the class.

USE CROSS-CURRICULAR CONNECTIONS

Many states have adopted standards in the core subject areas. As states implement a standards-based education, assignments that simultaneously address standards across several subject areas maximize the efficient use of instructional time. A discussion between mathematics and history teachers, for example, can result in the formulation of topics for advanced exploration that students could pursue in either mathematics class or in history class.

The links to other subjects in this textbook suggest areas for students to apply and extend their learning of mathematics. For example, on page 606, students learn about the Leaning Tower of Pisa. The teacher can suggest that a group of students research more about this tower. When was it built? Why is it leaning? How are engineers saving the tower? Why have previous attempts to save the tower failed? Suggest that students make a timeline of the tower's history and rescue attempts.

Challenge Problem for use with Lesson 11.1:
Draw a line and label a point P on the line. Draw at least four circles that are tangent to the line at point P. Describe the figure that contains the centers of all the circles you drew. (*a line perpendicular to the original line at point P*)

Challenge Problem for use with Lesson 11.4:
In the diagram below, both circles are centered at P and $\overline{AB}$ is tangent to the smaller circle at point D. Explain how you know that $AD = DB$. (*Since $\overline{AB}$ is tangent to the smaller circle, it is perpendicular to radius $\overline{PD}$. Therefore, the diameter $\overline{EC}$ of the larger circle bisects chord $\overline{AB}$. By the definition of segment bisector, $AD = DB$.*)

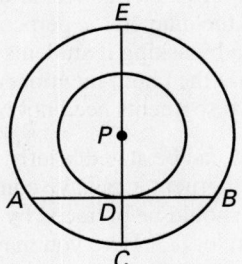

Challenge Problem for use with Lesson 11.5:
In the diagram, $CDEFGHJ$ is a regular 7-sided polygon inscribed in a circle. Explain how you know that the measures of the numbered angles are all equal. (*Since the sides of the polygon are congruent and are chords of the circle, the arcs $\overarc{CD}$, $\overarc{DE}$, and so on are all congruent. Since each numbered angle is an inscribed angle, the measure of each of these angles is half that of its intercepted arc. Since these arcs are all congruent, the numbered angles are all congruent.*)

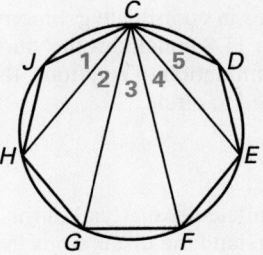

Challenge Problem for use with Lesson 11.6:
In the diagram, $m\overarc{AB} = 20°$, $m\overarc{CD} = 40°$, $CE = 6$, $XB = 6$, and $XD = 12$. Find AX. (*6*)

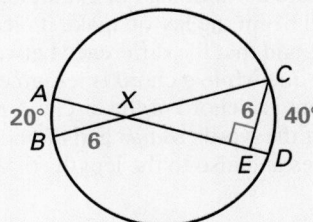

Challenge Problem for use with Lesson 11.8:
What special kind of quadrilateral has rotational symmetry but does *not* have any lines of symmetry? (*a parallelogram that is neither a rectangle nor a rhombus*)

Content and Teaching Strategies

This chapter explores the geometry of the circle and its connections to coordinate geometry, trigonometry, and transformational geometry. Lesson 11.1 introduces the important terminology that will be used throughout the chapter. Lesson 11.2 examines tangents to circles and their properties. In Lesson 11.3 the focus is on the relationship between circular arcs and angles formed by two radii, while relationships between chords and arcs are explored in Lesson 11.4. In Lesson 11.5, attention turns to the relationship between circles and inscribed angles and polygons. Lesson 11.6 looks at two chords intersecting inside a circle, in particular the angles formed by two such chords and the segments into which they divide each other. The role of circles in coordinate geometry is explored in Lesson 11.7. Finally, Lesson 11.8 concludes the study of transformations of the plane with an examination of rotations, the type of transformation that is most closely related to circles.

Lesson 11.1

It will be important for students to assimilate the material of this lesson thoroughly, so they will be able to understand the discussions in the chapter that inevitably involve the terms introduced here. Students should be encouraged to memorize these terms.

VOCABULARY Although the concepts introduced in this lesson are basic, some confusion can easily arise about the definitions. For example, you should stress that the definitions which begin the lesson make it clear that a *diameter* is a special case of a *chord*, and that the difference between a *secant* and a *chord* is that a secant is a line while a chord is a segment. Students may notice that part of a secant is a chord of the circle. A more subtle source of confusion is the use of the words *radius* and *diameter* to refer to both the segments themselves and also to the <u>lengths</u> of those segments.

Point out that two circles which are tangent to the same line at the same point (as in Example 3), are said to be tangent to each other.

Lesson 11.2

Theorem 11.1 and its converse, Theorem 11.2, are important because they provide a way of thinking about the concept of tangency in terms of the familiar concept of perpendicularity. When the tangency of a line is stated in a problem, the idea that the given line is perpendicular to the radius drawn to the point of tangency is the way this given information should almost always be interpreted.

> **QUICK TIP**
> The diagram shown in Theorem 11.3 is a useful picture for students to have in mind when a problem involves two tangents meeting at a point outside a circle. In such a situation, if you draw the line connecting the point of intersection with the center of the circle, students may be able to discover for themselves that this line bisects the angle formed by the two tangents.

Lesson 11.3

The distinction between major and minor arcs is a necessary one, because minor arcs are the only ones that have central angles directly associated with them (since angles have been defined in this book as always having measure less than or equal to 180°). Point out that the measure of a major arc is determined by the central angle associated with its related minor arc. Emphasize that major arcs are named using three points, the two endpoints and one other point on the arc. Note also that a *semicircle* is an arc of exactly 180°, and not, as students sometimes believe, just any arc whose measure is greater than 90°.

MAKING CONNECTIONS Students should have no trouble understanding the Arc Addition Postulate due to its similarity to the Segment Addition Postulate and the Angle Addition Postulate. However note that unlike the Angle Addition Postulate, the Arc Addition Postulate can be applied to two arcs whose measures have a sum greater than 180°.

An important concept for students in this lesson is the distinction between the measure of an arc and arc length. First of all, stress that the measure of an arc is given in degrees (and is therefore a measure of rotation), while arc length is given in linear units.

Point out that the fraction in the arc length formula on page 603 is just the fraction of the whole circle that the arc represents. Finally, the concept of congruent arcs requires that two arcs have the same measure in degrees and the same arc length. Stress that this makes sense in terms of the general meaning of congruent as having the "same size and same shape."

Lesson 11.4

Theorem 11.4 is an important and useful theorem for gaining information about the lengths of segments in circles from information about the angle at which they intersect.

> **QUICK TIP**
> Note that if the word "radius" is substituted for the word "diameter" everywhere in Theorem 11.4, the theorem will still be true, since any radius can be extended to produce a diameter.

USING LOGIC Students may notice that Theorem 11.5 is not quite a logical converse of Theorem 11.4. A more likely converse would read: "If a diameter bisects a chord, then the diameter is perpendicular to the chord." You might initiate a discussion by asking if students think that this is a true statement. It is not, because the chord mentioned could be another diameter, in which case the two segments need not be perpendicular.

The two parts of Theorem 11.6 can be stated briefly as "Congruent chords have congruent arcs," and "Congruent arcs have congruent chords." Both of these statements, however, should be prefaced by the words, "In the same circle or in congruent circles... ." Here you may want to remind students that in order for two arcs to be congruent, they must have the same measure and the same length; they must be in the same circle or in congruent circles.

Lesson 11.5

Theorem 11.7 may cause confusion in applications because students are not sure on which side of the equation the factor $\frac{1}{2}$ belongs. An easy way for them to answer this question for themselves is to draw a quick sketch of an inscribed angle and its associated central angle (whose measure is the same as the measure of the intercepted arc), as shown in the diagram for Theorem 11.7. It should be clear from such a diagram that the central angle is the larger of the two angles, and so $\frac{1}{2}$ belongs on the side of the equation containing the measure of the arc and not the measure of the inscribed angle.

Theorem 11.8 could be regarded as a corollary to Theorem 11.7, since it is a special case where the arc intercepted by the inscribed angle is a semicircle. Theorem 11.9 might also be regarded as a corollary of Theorem 11.7.

Lesson 11.6

Both of the theorems in this lesson can be proven by connecting two end-points, one from each of the two chords, with a segment and connecting the other two endpoints with another segment, thus forming two triangles that are similar by the AA Similarity Theorem. Theorem 11.10 is then the result of an application of Theorem 11.7 combined with the Exterior Angle Theorem, and Theorem 11.11 follows from the definition of similar triangles.

The algebra work in problems like the one shown in Example 2 may pose challenges for some students. Remind students that a step-by-step approach, in which each step is justified by one of the properties of equality, should be utilized. Urge students to write out their work carefully.

Lesson 11.7

Working with the equations of circles can present difficulties for students, especially if the center of the given circle has one or two negative coordinates. Students often make the mistake of disregarding either the negative sign of the formula or the one attached to a coordinate in such a situation. (Note that this is one of the errors presented in Exercise 38.) For this reason, it will be important to go over Example 2 carefully in class. Note also that, in the reverse situation, where students are asked to identify the center and the radius of a circle given its equation, rewriting an expression like $(x + 3)^2$ as $(x - (-3))^2$ should help avoid any errors.

At this point, equations referring to coordinate geometry objects may have lost some of their immediacy for students, and it may be a good idea to review the basic fact that a point (x, y) is on the graph of an equation if and only if its coordinates make the equation true. You can review this concept effectively by taking a particular equation such as $(x - 1)^2 + (y + 2)^2 = 25$ and asking whether various coordinate pairs lie on the graph of this circle. Some good choices might be $(5, 1)$ (*yes*), $(3, 4)$ (*no*), and $(4, -6)$ (*yes*). Such an exercise will be more meaningful when students work on Exercises 28–35, which generalize this idea.

Lesson 11.8

The final lesson of the book should not present many conceptual obstacles for students, but one of these may be the relationship between a rotation and symmetry. The connection is that if a figure looks the same after a certain rotation, then it has *rotational symmetry*, and it may very well happen that more than one rotation will map the figure onto itself. Students may wonder whether a reflection can demonstrate rotational symmetry, since this transformation seems to resemble rotation through a 180° angle. Stress that the answer to this question is no, since a reflection over the symmetry line changes the orientation of a figure, while rotation does not change orientation. This can be easily demonstrated with a few simple examples, such as a kite-shaped figure that can be mapped onto itself by a reflection but which has no rotational symmetry.

MODELING In connection with rotational symmetry, you may want to ask students to bring in objects that demonstrate such symmetry. Hubcaps are a good example; almost all have rotational symmetry. Referring to each object, you can then ask students which rotational measures demonstrate the symmetry.

Chapter Goals

Students study and use properties involving tangents, secants, and chords of a circle. After discussions about central angles, inscribed angles, and inscribed polygons, the chapter concludes with lessons on the equations of circles and rotations. Students will:

- Identify parts of a circle, such as arcs, and study their properties.
- Identify inscribed angles and intersecting chords in circles, and use the properties related to them.
- Write and graph equations of circles.
- Identify rotational symmetry and rotations in a plane.

Application Note

Fireworks displays are associated with various celebrations throughout the year. Some fireworks can be seen from a distance because they reach a height of 1000 to 1300 feet before they explode and can remain visible for several seconds after they burst. Explosives include ingredients such as potassium nitrate, charcoal, and sulfur. A fuse is also attached that determines how high the firework travels.

More information about fireworks is provided on the Internet at classzone.com

Application Links
CLASSZONE.COM

Chapter 11 Circles

How far away can fireworks be seen?

If you watch fireworks as you sail out to sea on a clear night, the fireworks will gradually disappear over the horizon. When the ship is at point E, the fireworks are no longer visible.

3960 mi

Learn More About It

You will learn more about fireworks in Exercise 28 on p. 600.

Who uses Circles?

EMT
Some emergency medical technicians train for wilderness emergencies. They use the geometry of circles to find people trapped by avalanches. (p. 612)

GRAPHIC DESIGNER
Graphic designers may create symbols to represent a company or organization. These symbols often appear on packaging, stationery, and Web sites. (p. 638)

How will you use these ideas?

- Calculate how far global positioning systems can transmit signals. (p. 599)
- Find the time difference between cities. (p. 605)
- Design a logo for the Internet. (p. 624)
- Determine how cell phone towers transmit calls. (p. 631)
- Investigate patterns of rotational symmetry in art. (p. 638)

587

Projects

A project for Chapter 11 is available in the *Chapter 11 Resource Book*, pp. 94–95.

 *Technology*

- Electronic Teacher Tools
- Test and Practice Generator
- Online Lesson Planner

Internet Support
CLASSZONE.COM

- Application and Career Links 599, 609, 612, 638
- Student Help 590, 599, 606, 614, 618, 621, 630, 635

The **Chapter Readiness Quiz** can help you diagnose whether students have the following skills needed in Chapter 11:

- Use the Converse of the Pythagorean Theorem.
- Find the coordinates of the endpoints of a line segment after a reflection.
- Solving multi-step equations.

Reteaching Material

This resource is available for students who need additional help with the skills on the Chapter Readiness Quiz:

📘 **Chapter 11 Resource Book**
- Reteaching with Practice (Lessons 11.1–11.8)

Additional Resources

The following resources are provided to help you prepare for the upcoming chapter and customize review materials:

📘 **Chapter 11 Resource Book**
- *Tips for New Teachers*, pp. 1–2
- *Parent Guide*, pp. 3–4
- *Lesson Plans*, pp. 7, 15, 24, 34, 42, 50, 62, 73
- *Lesson Plans for Block Scheduling*, pp. 8, 16, 25, 35, 43, 51, 63, 74

🖥 *Technology*
- Electronic Teaching Tools
- Online Lesson Planner
- Test and Practice Generator

Visualize It!

Additional suggestions for helping students visualize geometry are on pp. 592, 599, 602, 611, 616, 633, and 637.

Chapter  11 Study Guide

PREVIEW

What's the chapter about?

- Identifying **parts of a circle** and studying their properties
- Writing **equations** of circles
- Identifying **rotations** in a plane

Key Words

- **chord**, *p. 589*
- **secant**, *p. 589*
- **tangent**, *p. 589*
- **minor arc**, *p. 601*
- **major arc**, *p. 601*

- **arc length**, *p. 603*
- **inscribed angle**, *p. 614*
- **intercepted arc**, *p. 614*
- **rotation**, *p. 633*
- **rotational symmetry**, *p. 634*

PREPARE

Chapter Readiness Quiz

Take this quick quiz. If you are unsure of an answer, look at the reference pages for help.

Vocabulary Check *(refer to p. 200)*

1. Which of the following represents side lengths of a right triangle? **D**

   (**A**) 5, 8, 10   (**B**) 12, 15, 20   (**C**) 4, 12, 13   (**D**) 16, 30, 34

Skill Check *(refer to pp. 283, 673)*

2. A segment has endpoints $C(2, 1)$ and $D(5, 4)$. What are the endpoints of the segment after it is reflected in the y-axis? **F**

   (**F**) $C'(-2, 1)$ (**G**) $C'(2, -1)$ (**H**) $C'(-2, -1)$ (**J**) $C'(-1, 2)$
       $D'(-5, 4)$ $D'(5, -4)$ $D'(-5, -4)$ $D'(-4, 5)$

3. What is the value of x in the equation $5x - 8 = 2x + 7$? **D**

   (**A**) $\frac{1}{3}$ (**B**) 2   (**C**) 3   (**D**) 5

VISUAL STRATEGY

Marking Diagrams

Visualize It! ➤

As you solve exercises, copy diagrams and add information as you learn it.

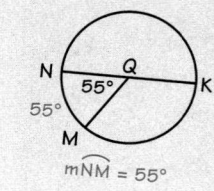

11.1 Parts of a Circle

Goal
Identify segments and lines related to circles.

Key Words
- chord
- diameter p. 452
- radius p. 452
- secant
- tangent
- point of tangency

The diagrams below show special segments and lines of a circle.

A **chord** is a segment whose endpoints are points on a circle.

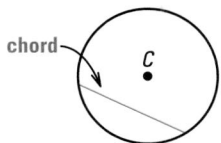
chord

A *diameter* is a chord that passes through the center of a circle.

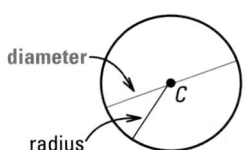
diameter
radius

A **secant** is a line that intersects a circle in two points.

secant

A **tangent** is a line in the plane of a circle that intersects the circle in exactly one point. The point is called a **point of tangency**.

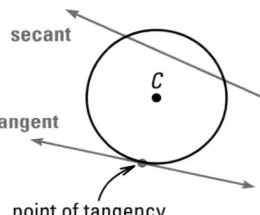
tangent
point of tangency

Student Help

STUDY TIP
To identify a circle, you can name the point that is the center of the circle. In Example 1, *C* is the center, so the circle is called ⊙*C*.

EXAMPLE 1 **Identify Special Segments and Lines**

Tell whether the line or segment is best described as a *chord*, a *secant*, a *tangent*, a *diameter*, or a *radius* of ⊙*C*.

a. $\overline{AD}$

b. $\overline{HB}$

c. $\overleftrightarrow{EG}$

d. $\overrightarrow{JK}$

Solution

a. $\overline{AD}$ is a diameter because it passes through the center *C* and its endpoints are points on the circle.

b. $\overline{HB}$ is a chord because its endpoints are on the circle.

c. $\overleftrightarrow{EG}$ is a tangent because it intersects the circle in exactly one point.

d. $\overrightarrow{JK}$ is a secant because it intersects the circle in two points.

11.1 *Parts of a Circle* **589**

Pacing
Suggested Number of Days

Basic: 1 day
Average: 1 day
Advanced: 1 day
Block Schedule: 0.5 block with 11.2

Teaching Resources

Blacklines
(See page 586B.)

Transparencies
- Warm-Up with Quiz
- Answers

Technology
- Electronic Teacher Tools
- Test and Practice Generator
- Online Lesson Planner
- Internet Support

Teach

Content and Teaching Strategies
For background information on geometric concepts and teaching strategies related to this lesson, see pages 586E and 586F in this Teacher's Edition.

Tips for New Teachers
Secants and tangents are defined as lines on this page. Students have a tendency to write them as segments when identifying them in relation to a circle. Stress that they are lines and not segments when discussing the definitions. Also emphasize the Visualize It note on page 592. See the Tips for New Teachers on pp. 1–2 of the *Chapter 11 Resource Book* for additional notes about Lesson 11.1.

Extra Example 1
See next page.

Extra Example 1

Tell whether the line or segment is best described as a *chord*, a *secant*, a *tangent*, a *diameter*, or a *radius* of ⊙A.

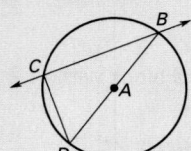

a. $\overline{AB}$ radius
b. $\overline{BD}$ diameter
c. $\overleftrightarrow{BC}$ secant
d. $\overline{CD}$ chord

Extra Example 2

Identify a chord, a secant, a diameter, a tangent, two radii, the center, and a point of tangency.

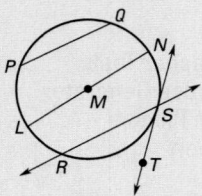

chord: $\overline{PQ}$; secant: $\overleftrightarrow{RS}$; diameter: $\overline{LN}$; tangent: $\overleftrightarrow{ST}$; radii: $\overline{LM}$, $\overline{MN}$; center: *M*; point of tangency: *S*

Extra Example 3

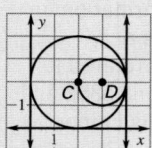

a. Name the coordinates of the center of each circle.
 C(2, 2), *D*(3, 2)

b. Name the coordinates of the intersection of the circles. (4, 2)

c. What is the line that is tangent to both circles? Name the coordinates of the point of tangency.
 the line *x* = 4; (4, 2)

d. What is the length of the diameter of ⊙*D*? What is the length of the radius of ⊙*C*? 2; 2

✓ Concept Check

Explain the difference between a chord and a secant. A chord is a segment whose endpoints are on a circle. A secant is a line that intersects a circle in two points.

590

Student Help

VOCABULARY TIP
The plural of *radius* is *radii*.

Student Help
CLASSZONE.COM

MORE EXAMPLES
More examples at classzone.com

EXAMPLE 2 Name Special Segments, Lines, and Points

Identify a chord, a secant, a tangent, a diameter, two radii, the center, and a point of tangency.

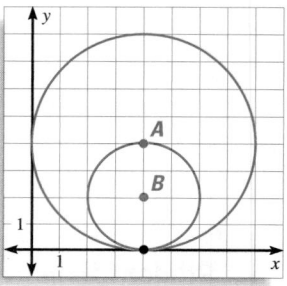

Solution

$\overline{AB}$ is a chord. $\overleftrightarrow{HJ}$ is a secant.

$\overleftrightarrow{FG}$ is a tangent. $\overline{DE}$ is a diameter.

$\overline{DC}$ is a radius. $\overline{CE}$ is a radius.

C is the center.   *K* is a point of tangency.

EXAMPLE 3 Circles in Coordinate Geometry

When a circle lies in a coordinate plane, you can use coordinates to describe particular points of the circle.

a. Name the coordinates of the center of each circle.

b. Name the coordinates of the intersection of the two circles.

c. What is the line that is tangent to both circles? Name the coordinates of the point of tangency.

d. What is the length of the diameter of ⊙*B*? What is the length of the radius of ⊙*A*?

Solution

a. The center of ⊙*A* is *A*(4, 4). The center of ⊙*B* is *B*(4, 2).

b. The intersection of the two circles is the point (4, 0).

c. The *x*-axis is tangent to both circles. The point of tangency is (4, 0).

d. The diameter of ⊙*B* is 4. The radius of ⊙*A* is 4.

Checkpoint ✓ Parts of a Circle

1. $\overline{DE}$ is a chord (as are $\overleftrightarrow{JK}$ and $\overleftrightarrow{FH}$); $\overleftrightarrow{JK}$ is a secant; $\overleftrightarrow{AC}$ is a tangent; $\overline{FH}$ is a diameter; $\overline{FG}$ is a radius (as is $\overline{GH}$); *G* is the center; *B* is the point of tangency.

1. Identify a chord, a secant, a tangent, a diameter, a radius, the center, and a point of tangency.

2. In Example 3, name the coordinates of the point of tangency of the *y*-axis to ⊙*A*. (0, 4)

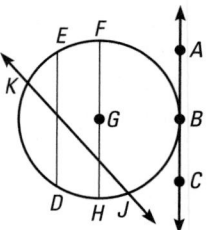

Guided Practice

Vocabulary Check

1. Sketch a circle. Then sketch and label a *radius*, a *diameter*, a *chord*, and a *tangent*. **See margin.**

Skill Check

Match the part of the circle with the term that best describes it.

2. $\overline{GH}$ B   **A.** Center

3. M E   **B.** Chord

4. $\overline{JM}$ D   **C.** Diameter

5. J A   **D.** Radius

6. $\overline{MH}$ C   **E.** Point of tangency

7. $\overleftrightarrow{GH}$ F   **F.** Secant

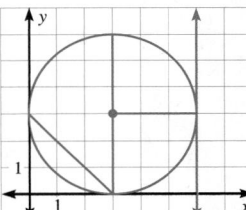

Use the circle to name the coordinates of the points.

8. center (3, 3)

9. endpoints of a diameter
   (3, 0) and (3, 6)

10. a point of tangency
    (0, 3), (3, 0), or (6, 3)

11. endpoints of a chord that
    is not a diameter
    (0, 3) and (3, 0)

12. endpoints of a radius
    (3, 3) and one of the following: (3, 0), (6, 3), (3, 6)

Practice and Applications

Extra Practice
See p. 695.

Finding Radii The diameter of a circle is given. Find the radius.

13. $d = 15$ cm   14. $d = 6.5$ in.   15. $d = 3$ ft   1.5 ft   16. $d = 8$ m   4 m
    7.5 cm   3.25 in.

Finding Diameters The radius of a circle is given. Find the diameter.

17. $r = 26$ in.   52 in.   18. $r = 62$ ft   124 ft   19. $r = 8.7$ m   20. $r = 4.4$ cm
                                              17.4 m   8.8 cm

Identifying Terms Name the term that best describes the given line, segment, or point.

21. $\overline{CD}$ chord   22. $\overleftrightarrow{FG}$ tangent

23. $\overline{EC}$ diameter   24. $\overline{AB}$ radius

25. H 26. A center of the circle
    point of tangency

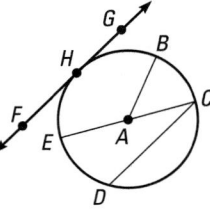

Homework Help

Example 1: Exs. 21–32
Example 2: Exs. 33–39
Example 3: Exs. 40–47

③ Apply

Assignment Guide

BASIC
Day 1: pp. 591–593 Exs. 13–35 odd, 40–44, 48, 49–67 odd

AVERAGE
Day 1: pp. 591–593 Exs. 14–34 even, 40–44, 48–53, 56–68 even

ADVANCED
Day 1: pp. 591–593 Exs. 21–35 odd, 36–39, 43–52, 53–67 odd; EC: TE p. 586D*, classzone.com

BLOCK SCHEDULE
pp. 591–593 Exs. 14–34 even, 40–44, 48–53, 56–68 even (with 11.2)

Extra Practice
• Student Edition, p. 695
• Chapter 11 Resource Book, pp. 10–11

Homework Check
To quickly check student understanding of key concepts, go over the following exercises:

Basic: 27, 29, 31, 33, 43
Average: 28, 30, 32, 34, 44
Advanced: 27, 29, 31, 35, 45

1. *Sample answer:*

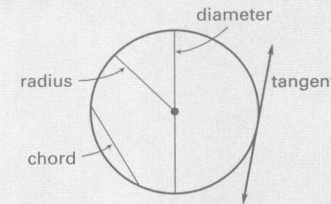

Common Error

In Exercises 36 and 37, watch for students who confuse chords and secants. Remind them that a chord is a segment whose endpoints are on a circle. A secant is a line that passes through a circle.

33. $\overline{EG}$ is a chord (as is $\overline{EF}$); $\overleftrightarrow{EG}$ is a secant; $\overline{EF}$ is a diameter; $\overline{CE}$ is a radius (as are $\overline{CF}$ and $\overline{CG}$); D is a point of tangency.

34. $\overline{MN}$ is a chord (as is $\overline{JL}$); $\overleftrightarrow{MN}$ is a secant; $\overline{JL}$ is a diameter; $\overline{KR}$ is a radius (as are $\overline{KJ}$ and $\overline{KL}$); U is a point of tangency.

35. $\overline{LM}$ is a chord (as is $\overline{PN}$); $\overleftrightarrow{LM}$ is a secant; $\overline{PN}$ is a diameter; $\overline{QR}$ is a radius (as are $\overline{QP}$ and $\overline{QN}$); K is a point of tangency.

Visualize It!

A chord is a segment.
A secant is a line.

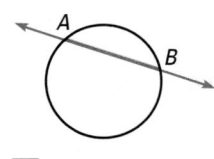

$\overline{AB}$ is a chord.
$\overleftrightarrow{AB}$ is a secant.

Identifying Terms Tell whether the line or segment is best described as a *chord*, a *secant*, a *tangent*, a *diameter*, or a *radius*.

27. $\overline{PZ}$ chord

28. $\overline{RT}$ radius

29. $\overline{ST}$ diameter

30. $\overleftrightarrow{PZ}$ secant

31. $\overleftrightarrow{VW}$ tangent

32. $\overleftrightarrow{TU}$ secant

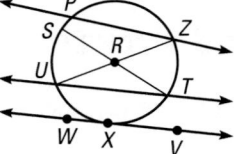

Identifying Terms Identify a chord, a secant, a diameter, a radius, and a point of tangency. 33–35. See margin.

33.

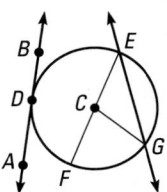

34.

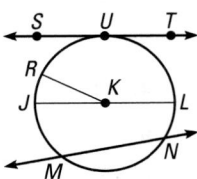

35.

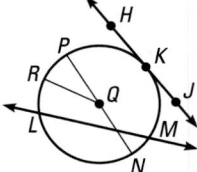

Link to Geography

Island Map The diagram shows the layout of the streets on Mexcaltitán Island.

36. Name two secants. $\overleftrightarrow{FA}$ and $\overleftrightarrow{EB}$

37. Name two chords.
   any two of $\overline{GD}$, $\overline{HC}$, $\overline{FA}$, and $\overline{EB}$

38. Is the diameter of the circle longer than $\overline{HC}$? Explain. Yes; the diameter of a circle is the longest chord of a circle.

39. Can you draw a line through three of the given points that is tangent to the circle?
   Yes; $\overleftrightarrow{JK}$ is a tangent through J, G, and K.

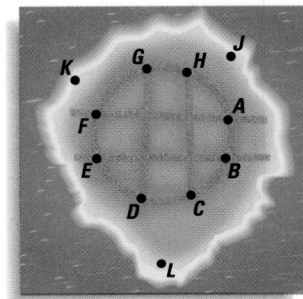

Coordinate Geometry Use the diagram below.

40. What are the coordinates of the center of $\odot A$? of $\odot B$? (2, 2); (6, 2)

41. What is the length of the radius of $\odot A$? of $\odot B$? 2; 2

42. Name the coordinates of the intersection of the two circles. (4, 2)

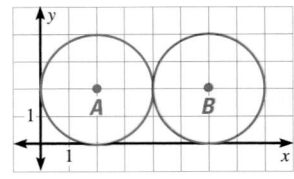

Coordinate Geometry Name the coordinates of the center of each circle, identify the point of intersection of the circles, and identify a line that is tangent to both circles.

43. $\odot A$: (3, 2); $\odot B$: (3, 3); intersection: (3, 0); tangent line: x-axis

44. $\odot A$: (3, 3); $\odot B$: (2, 3); intersection: (0, 3); tangent line: y-axis

43.

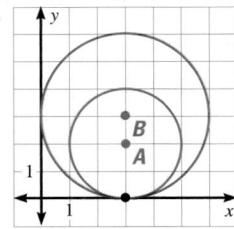

44.

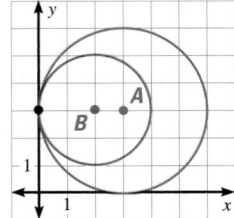

Coordinate Geometry Use the diagram below.

45. What are the lengths of the radius and the diameter of the circle? 2; 4

46. Find the length of the chord $\overline{AB}$.
$2\sqrt{2} \approx 2.8$

47. Copy the diagram and sketch a tangent that passes through *A*.
See margin.

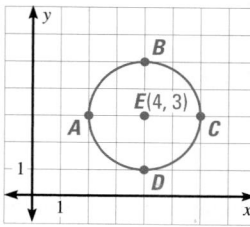

Standardized Test Practice

In Exercises 48 and 49, use the diagram below.

48. Multiple Choice Which of the following is a secant? D

   (A) $\overleftrightarrow{EF}$ (B) $\overleftrightarrow{GH}$

   (C) $\overline{AB}$ (D) $\overrightarrow{EF}$

49. Multiple Choice Which of the following is a tangent? G

   (F) $\overrightarrow{EF}$ (G) $\overleftrightarrow{GH}$

   (H) $\overline{AB}$ (J) $\overline{AC}$

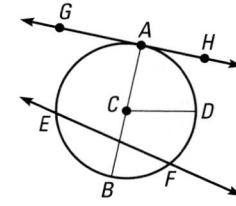

Mixed Review

50. HL Congruence Theorem; $\overline{BC} \cong \overline{BC}$ and $\overline{BA} \cong \overline{BD}$ in right triangles *BAC* and *BDC*, so $\triangle BAC \cong \triangle BDC$ by the HL Congruence Theorem.

51. SSS Congruence Postulate; three sides of $\triangle JKL$ are congruent to three sides of $\triangle PQR$, so $\triangle JKL \cong \triangle PQR$ by the SSS Congruence Postulate.

52. ASA Congruence Postulate; $\angle P \cong \angle T$, $\overline{PR} \cong \overline{TR}$, and $\angle PRQ \cong \angle TRS$, so $\triangle PQR \cong \triangle TSR$ by the ASA Congruence Postulate.

Congruent Triangles Tell which theorem or postulate you can use to show that the triangles are congruent. Explain your reasoning. *(Lessons 5.2–5.4)*

50.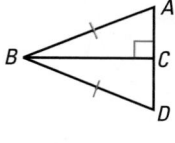

51.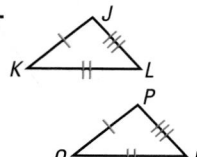

52.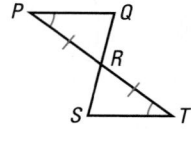

Coordinate Geometry Plot the points and draw the quadrilateral. Use the slopes of the segments to determine whether the quadrilateral is a parallelogram. *(Lesson 6.3)*

53. $A(0, 0)$, $B(1, 3)$, $C(5, 3)$, $D(4, 0)$
Check graphs; yes.

54. $P(2, 1)$, $Q(0, 5)$, $R(2, 5)$, $S(4, 1)$
Check graphs; yes.

Algebra Skills

Simplifying Radicals Find the square root. Round your answer to the nearest tenth. *(Lesson 10.1)*

55. $\sqrt{32}$ 5.7   **56.** $\sqrt{81}$ 9   **57.** $\sqrt{40}$ 6.3   **58.** $\sqrt{104}$ 10.2

59. $\sqrt{98}$ 9.9   **60.** $\sqrt{192}$ 13.9   **61.** $\sqrt{250}$ 15.8   **62.** $\sqrt{242}$ 15.6

Solving Equations Solve the equation. *(Skills Review, p. 673)*

63. $2x + 5 = 19$ 7   **64.** $7x - 7 = 14$ 3   **65.** $5x + 9 = 4$ −1

66. $3x - 10 = 20$ 10   **67.** $12 - 8x = 84$ −9   **68.** $4x + 3 = 23$ 5

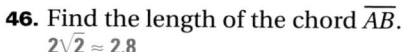 **Assess**

Assessment Resources
The Mini-Quiz below is also available on blackline (*Chapter 11 Resource Book*, p. 17) and on transparency. For more assessment resources, see:
• Chapter 11 Resource Book
• Standardized Test Practice
• Test and Practice Generator

Mini-Quiz
Name the term that best describes the given line or segment.

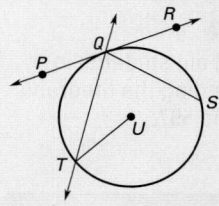

1. $\overline{TU}$ radius
2. $\overline{QS}$ chord
3. $\overleftrightarrow{QT}$ secant
4. $\overrightarrow{PR}$ tangent
5. Name the coordinates of the center of each circle, identify the point of intersection of the circles, and identify a line that is tangent to both circles.

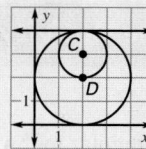

$C(2, 3)$ and $D(2, 2)$; $(2, 4)$; the line $y = 4$

47.

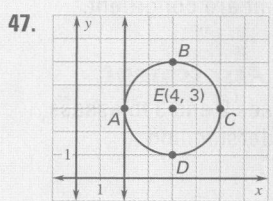

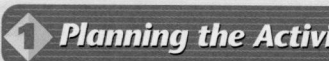

Activity 11.2 Tangents and Circles

Question

What is the relationship between a tangent and a circle?

Materials
- compass
- ruler
- protractor

Explore

1 Use your compass to draw a circle. Label the center *C*. Draw a point outside your circle. Label this point *P*.

2 Draw two lines tangent to ⊙*C* from point *P*. Label the points of tangency *M* and *N*. Draw the radii $\overline{CM}$ and $\overline{CN}$.

3 Use a ruler to measure the *tangent segments $\overline{MP}$ and $\overline{NP}$*. Use a protractor to measure ∠*CMP* and ∠*CNP*.

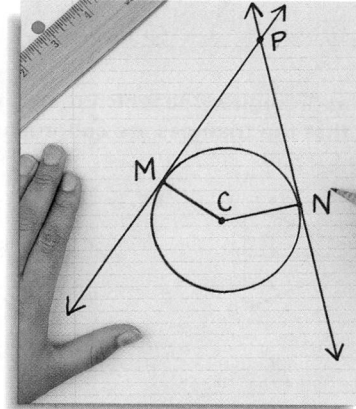

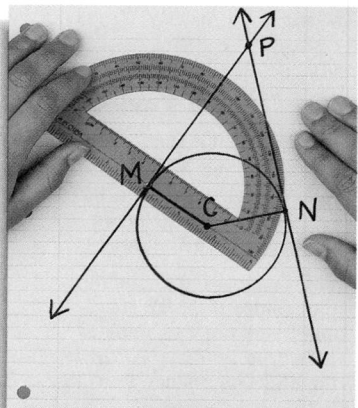

1. You get the same results for each circle: *MP* = *NP* and m∠*CMP* = m∠*CNP* = 90°.

> #### Student Help
>
> **LOOK BACK**
> To review the definition of conjecture, see p. 8.

Think About It

1. Repeat Steps 1–3 for three circles of different sizes. Compare your results. Do your results depend on the size of the circle?

2. Make a conjecture about the lengths of the two tangent segments drawn to a circle from the same exterior point.
   The lengths are equal.

3. Make a conjecture about the angle formed by a tangent and the radius drawn to the point of tangency.
   The angle is a right angle.

11.2 Properties of Tangents

Goal
Use properties of a tangent to a circle.

Key Words
- point of tangency p. 589
- perpendicular p. 108
- tangent segment

A discus thrower spins around in a circle one and a half times, then releases the discus. The discus forms a path tangent to the circle.

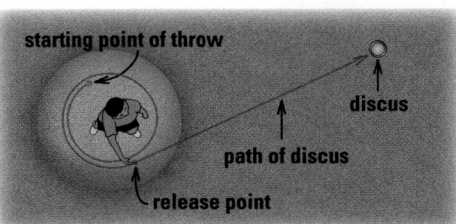

starting point of throw
discus
path of discus
release point

THEOREMS 11.1 and 11.2

Theorem 11.1

Words If a line is tangent to a circle, then it is perpendicular to the radius drawn at the point of tangency.

Symbols If ℓ is tangent to $\odot C$ at B, then $\ell \perp \overline{CB}$.

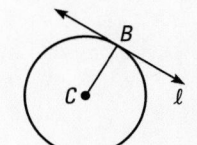

Theorem 11.2

Words In a plane, if a line is perpendicular to a radius of a circle at its endpoint on the circle, then the line is tangent to the circle.

Symbols If $\ell \perp \overline{CB}$, then ℓ is tangent to $\odot C$ at B.

Student Help

VOCABULARY TIP
Tangent is based on a Latin word meaning "to touch."

EXAMPLE 1 Use Properties of Tangents

$\overrightarrow{AC}$ is tangent to $\odot B$ at point C. Find BC.

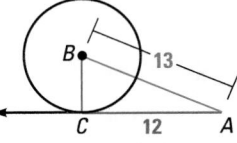

Solution
$\overline{BC}$ is a radius of $\odot B$, so you can apply Theorem 11.1 to conclude that $\overline{BC}$ and $\overrightarrow{AC}$ are perpendicular.

So, $\angle BCA$ is a right angle, and $\triangle BCA$ is a right triangle. To find BC, use the Pythagorean Theorem.

$(BA)^2 = (BC)^2 + (AC)^2$	Pythagorean Theorem
$13^2 = (BC)^2 + 12^2$	Substitute 13 for BA and 12 for AC.
$169 = (BC)^2 + 144$	Multiply.
$25 = (BC)^2$	Subtract 144 from each side.
$5 = BC$	Find the positive square root.

11.2 *Properties of Tangents* **595**

① Plan

Pacing
Suggested Number of Days

Basic: 1 day
Average: 1 day
Advanced: 1 day
Block Schedule: 0.5 block with 11.1

Teaching Resources

📋 **Blacklines**
(See page 586B.)

🖐 **Transparencies**
- Warm-Up with Quiz
- Answers

💻 **Technology**
- Electronic Teacher Tools
- Test and Practice Generator
- Online Lesson Planner
- Internet Support

② Teach

Content and Teaching Strategies
For background information on geometric concepts and teaching strategies related to this lesson, see pages 586E and 586F in this Teacher's Edition.

Tips for New Teachers
The statement of Theorem 11.3 uses the word *segments* because the theorem applies to the part of each tangent line from the inter-section of the tangents to the point of tangency. Stress the definition of *tangent segment* to help alleviate any confusion about the terminology used. See the Tips for New Teachers on pp. 1–2 of the *Chapter 11 Resource Book* for addi-tional notes about Lesson 11.2.

Extra Example 1
See next page.

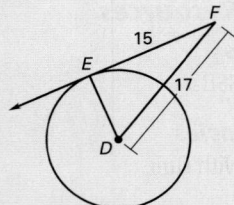
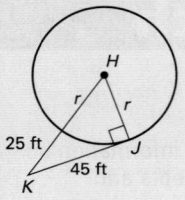
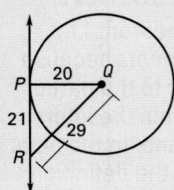

EXAMPLE 2 **Find the Radius of a Circle**

You are standing at C, 8 feet from a silo. The distance to a point of tangency is 16 feet. What is the radius of the silo?

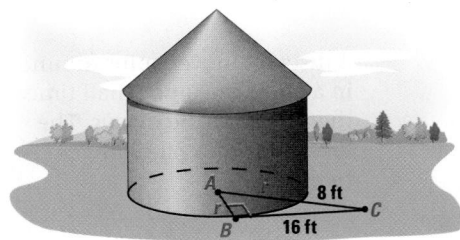

Solution

Tangent $\overleftrightarrow{BC}$ is perpendicular to radius $\overline{AB}$ at B, so $\triangle ABC$ is a right triangle. So, you can use the Pythagorean Theorem.

$(AC)^2 = (AB)^2 + (BC)^2$	Pythagorean Theorem
$(r + 8)^2 = r^2 + 16^2$	Substitute $r + 8$ for AC, r for AB, and 16 for BC.
$r^2 + 16r + 64 = r^2 + 256$	$(r + 8)(r + 8) = r^2 + 16r + 64$
$16r + 64 = 256$	Subtract r^2 from each side.
$16r = 192$	Subtract 64 from each side.
$r = 12$	Divide each side by 16.

ANSWER ▶ The radius of the silo is 12 feet.

You can use the Converse of the Pythagorean Theorem to show that a line is tangent to a circle.

EXAMPLE 3 **Verify a Tangent to a Circle**

How can you show that $\overleftrightarrow{EF}$ must be tangent to $\odot D$?

Solution

Use the Converse of the Pythagorean Theorem to determine whether $\triangle DEF$ is a right triangle.

$(DF)^2 \stackrel{?}{=} (DE)^2 + (EF)^2$	Compare $(DF)^2$ with $(DE)^2 + (EF)^2$.
$15^2 \stackrel{?}{=} 9^2 + 12^2$	Substitute 15 for DF, 9 for DE, and 12 for EF.
$225 \stackrel{?}{=} 81 + 144$	Multiply.
$225 = 225$	Simplify.

$\triangle DEF$ is a right triangle with right angle E. So, $\overline{EF}$ is perpendicular to $\overline{DE}$. By Theorem 11.2, it follows that $\overleftrightarrow{EF}$ is tangent to $\odot D$.

Tangent Segment A **tangent segment** touches a circle at one of the segment's endpoints and lies in the line that is tangent to the circle at that point.

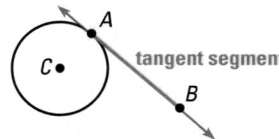

Activity 11.2, on page 594, shows that tangent segments from the same exterior point are congruent.

THEOREM 11.3

Words If two segments from the same point outside a circle are tangent to the circle, then they are congruent.

Symbols If $\overline{SR}$ and $\overline{ST}$ are tangent to $\odot P$ at points R and T, then $\overline{SR} \cong \overline{ST}$.

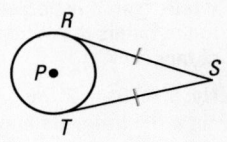

EXAMPLE 4 Use Properties of Tangents

$\overline{AB}$ is tangent to $\odot C$ at B.
$\overline{AD}$ is tangent to $\odot C$ at D.
Find the value of x.

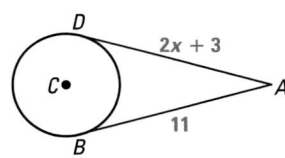

Solution

$$AD = AB$$ Two tangent segments from the same point are congruent.

$$2x + 3 = 11$$ Substitute $2x + 3$ for AD and 11 for AB.

$$2x = 8$$ Subtract 3 from each side.

$$x = 4$$ Divide each side by 2.

Checkpoint ✓ Use Properties of Tangents

$\overline{CB}$ and $\overline{CD}$ are tangent to $\odot A$. Find the value of x.

1.

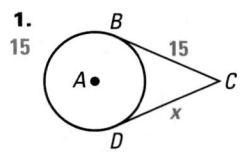

2.

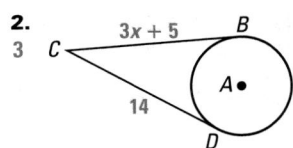

11.2 *Properties of Tangents* **597**

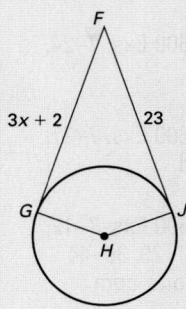

Assignment Guide

BASIC
Day 1: pp. 598–600 Exs. 7–24, 29, 30–44 even

AVERAGE
Day 1: pp. 598–600 Exs. 7–27, 29, 30, 31–45 odd

ADVANCED
Day 1: pp. 598–600 Exs. 7–12, 17–21, 25–27, 28*, 29, 30–44 even; EC: classzone.com

BLOCK SCHEDULE
pp. 598–600 Exs. 7–27, 29, 30, 31–45 odd (with 11.1)

Extra Practice

- Student Edition, p. 695
- Chapter 11 Resource Book, pp. 18–19

Homework Check

To quickly check student understanding of key concepts, go over the following exercises:

Basic: 7, 10, 17, 20, 22
Average: 8, 11, 18, 21, 23
Advanced: 9, 12, 19, 21, 25

11.2 Exercises

Guided Practice

Vocabulary Check

1. Complete the statement: In the diagram at the right, $\overleftrightarrow{AB}$ is __?__ to $\odot C$, and point B is the __?__.
    tangent; point of tangency

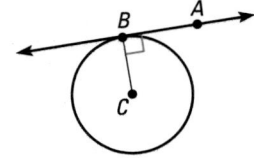

Skill Check

2. 90°; If a line is tangent to a circle, then it is perpendicular to the radius drawn at the point of tangency.

3. No; $5^2 + 5^2 \neq 7^2$, so $\triangle ABD$ is not a right triangle and $\overline{AB}$ is not perpendicular to $\overline{BD}$. Therefore, $\overline{BD}$ is not tangent to $\odot C$.

2. In the diagram below, $\overleftrightarrow{XY}$ is tangent to $\odot C$ at point P. What is $m\angle CPX$? Explain.

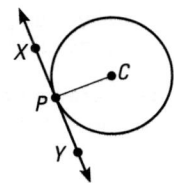

3. In the diagram below, $AB = BD = 5$ and $AD = 7$. Is $\overline{BD}$ tangent to $\odot C$? Explain.

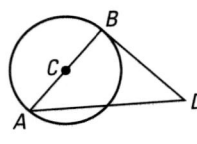

$\overline{AB}$ is tangent to $\odot C$ at A and $\overline{DB}$ is tangent to $\odot C$ at D. Find the value of x.

4.
4

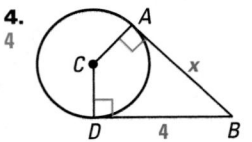

5.
2

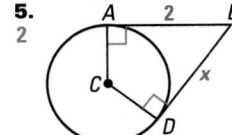

6.
5

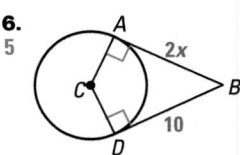

Practice and Applications

Extra Practice
See p. 695.

Finding Segment Lengths $\overleftrightarrow{AB}$ is tangent to $\odot C$. Find the value of r.

7.
3

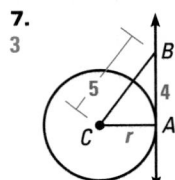

8.
8

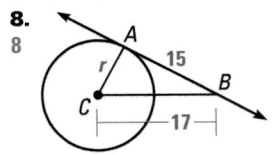

9.
15

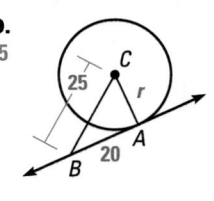

Finding Segment Lengths $\overline{AB}$ and $\overline{AD}$ are tangent to $\odot C$. Find the value of x.

Homework Help

Example 1: Exs. 7–9, 27
Example 2: Exs. 13–19
Example 3: Exs. 20–21
Example 4: Exs. 10–12, 22–26

10.
7

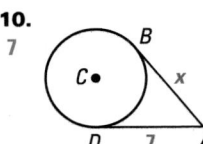

11.
15

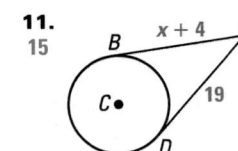

12.
5

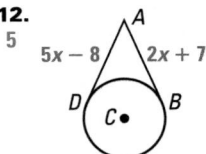

20. No; $5^2 + 15^2 \neq 17^2$, so $\triangle ABC$ is not a right triangle and $\overline{AB}$ is not perpendicular to $\overline{AC}$. Therefore, $\overline{AB}$ is not tangent to $\odot C$.

21. No; $5^2 + 14^2 \neq 15^2$, so $\triangle ABC$ is not a right triangle and $\overline{AB}$ is not perpendicular to $\overline{AC}$. Therefore, $\overline{AB}$ is not tangent to $\odot C$.

22. $\overline{AB} \cong \overline{AD}$, $\overline{BC} \cong \overline{DC}$

23. $\angle ABC \cong \angle ADC$; $\angle DAC \cong \angle BAC$; $\angle BCA \cong \angle DCA$.

24. $\triangle ABC \cong \triangle ADC$

Using Algebra Square the binomial.

13. $(x + 2)^2$
$x^2 + 4x + 4$

14. $(x + 4)^2$
$x^2 + 8x + 16$

15. $(x + 7)^2$
$x^2 + 14x + 49$

16. $(x + 12)^2$
$x^2 + 24x + 144$

Finding the Radius of a Circle $\overline{AB}$ is tangent to $\odot C$. Find the value of r.

17.
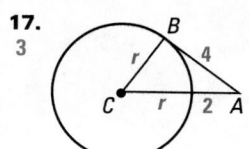

18.
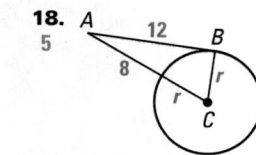

19.
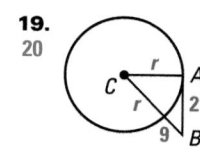

Verifying Tangents Tell whether $\overline{AB}$ is tangent to $\odot C$. Explain your reasoning.

20.
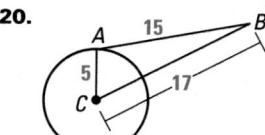

21.
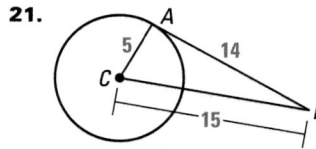

Finding Congruent Parts In Exercises 22–24, $\overline{AB}$ and $\overline{AD}$ are tangent to $\odot C$.

22. Name all congruent segments.

23. Name all congruent angles.

24. Name two congruent triangles.

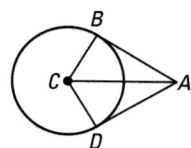

Visualize It! In Exercises 25 and 26, $\odot L$ has radii $\overline{LJ}$ and $\overline{LK}$ that are perpendicular. $\overline{KM}$ and $\overline{JM}$ are tangent to $\odot L$. **25, 26. See margin.**

25. Sketch $\odot L$, $\overline{LJ}$, $\overline{LK}$, $\overline{KM}$, and $\overline{JM}$.

26. Is $\triangle JLM$ congruent to $\triangle KLM$? Explain your reasoning.

27. Global Positioning System GPS satellites orbit 12,500 miles above Earth. Because GPS signals can't travel through Earth, a satellite can transmit signals only as far as points A and C from point B. Find BA and BC to the nearest mile. $BA = BC \approx 15{,}977$ miles

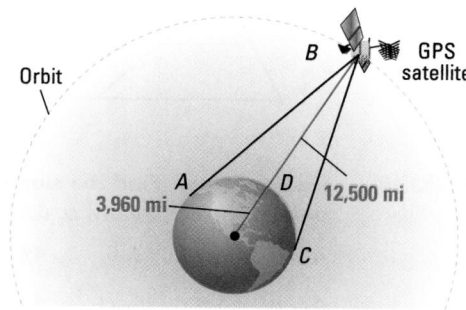

25.
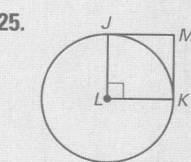

26. Yes. Explanations may vary.

Sample answer: $\overline{JM} \cong \overline{KM}$ since tangent segments from the same point are congruent. $\overline{LJ} \cong \overline{LK}$ since all the radii of a circle are congruent. $\overline{LM} \cong \overline{LM}$ by the Reflexive Property of Congruence. Then $\triangle JLM \cong \triangle KLM$ by the SSS Congruence Postulate.

Assessment Resources
The Mini-Quiz below is also available on blackline (*Chapter 11 Resource Book*, p. 26) and on transparency. For more assessment resources, see:
• Chapter 11 Resource Book
• Standardized Test Practice
• Test and Practice Generator

Mini-Quiz
$\vec{CA}$ is tangent to $\odot F$ at B. $\vec{CE}$ is tangent to $\odot F$ at D.

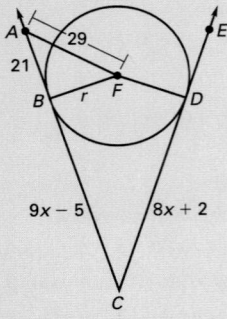

1. Find the value of r. **20**
2. Find the value of x. **7**
3. You are standing 6 feet from the edge of a large circular fountain. The distance to a point of tangency is 12 feet. What is the radius of the fountain? **9 ft**

Student Help

LOOK BACK
For more about fireworks, see p. 586.

28. Challenge You are cruising away from a fireworks show over a bay at point A. The highest point of the fireworks is point F. When your ship reaches point D, you can no longer see the fireworks over the horizon. You are standing at point E. $\overline{FE}$ is tangent to Earth at B. Find FE. Round your answer to the nearest mile. **49 miles**

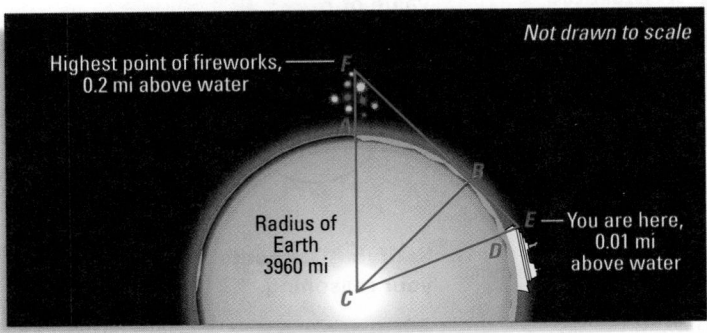

Not drawn to scale
Highest point of fireworks, — F
0.2 mi above water
B
Radius of Earth 3960 mi
E — You are here, 0.01 mi above water
D
C

Standardized Test Practice

29. Multiple Choice In the diagram below, $\overline{EF}$ and $\overline{EG}$ are tangent to $\odot C$. What is the value of x? **D**
Ⓐ -4 Ⓑ -1
Ⓒ 1 Ⓓ 4

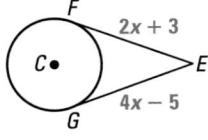

F
$2x + 3$
C • E
$4x - 5$
G

30. Multiple Choice In the diagram below, $\overline{SR}$ is tangent to $\odot P$. Find the radius of $\odot P$. **G**
Ⓕ 18 Ⓖ 27
Ⓗ 36 Ⓙ 45

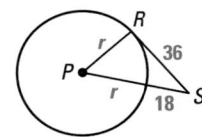

R
r 36
P •
r 18 S

Mixed Review

Using the Triangle Inequality Can the side lengths form a triangle? Explain. *(Lesson 4.7)*

31. yes; $5 + 11 > 14$, $5 + 14 > 11$, and $11 + 14 > 5$.
32. no; $8 + 14 < 23$.
33. yes; $15 + 3 > 13$, $15 + 13 > 3$, and $3 + 13 > 15$.
34. yes; $18 + 25 > 9$, $18 + 9 > 25$, and $9 + 25 > 18$.
35. no; $3 + 7 = 10$.
36. no; $22 + 6 < 29$.

31. 5, 11, 14   **32.** 8, 14, 23   **33.** 15, 3, 13
34. 18, 25, 9   **35.** 10, 3, 7   **36.** 22, 6, 29

Using the Midsegment Theorem Find the value of x. *(Lesson 7.5)*

37. 10

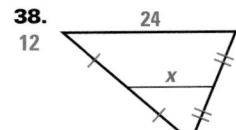

5
x

38. 12

24
x

39. 19

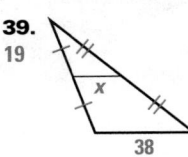

x
38

Algebra Skills

Finding Slope Find the slope of the line that passes through the points. *(Skills Review, p. 665)*

40. (0, 0) and $(-3, 6)$ -2 **41.** (2, 4) and (8, 0) $-\dfrac{2}{3}$ **42.** (1, 5) and $(-2, 1)$ $\dfrac{4}{3}$

43. (0, -3) and (4, 7) $\dfrac{5}{2}$ **44.** (-1, 6) and (4, -5) $-\dfrac{11}{5}$ **45.** (-7, 2) and (-1, 4) $\dfrac{1}{3}$

 **Arcs and Central Angles**

Goal
Use properties of arcs of circles.

Key Words
- minor arc
- major arc
- semicircle
- congruent circles
- congruent arcs
- arc length

Any two points A and B on a circle C determine a *minor arc* and a *major arc* (unless the points lie on a diameter).

If the measure of $\angle ACB$ is less than 180°, then A, B, and all the points on $\odot C$ that lie in the interior of $\angle ACB$ form a **minor arc**.

Points A, B, and all the points on $\odot C$ that do not lie on $\widehat{AB}$ form a **major arc**.

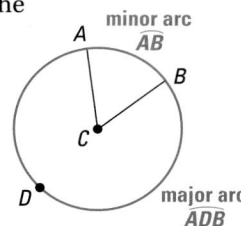

You name an arc by its endpoints. Use one other point on a major arc as part of its name to distinguish it from the minor arc.

The *measures* of a minor arc and a major arc depend on the central angle of the minor arc.

Student Help

LOOK BACK
For the definition of a central angle, see p. 454.

The **measure of a minor arc** is the measure of its central angle.

The **measure of a major arc** is the difference of 360° and the measure of the related minor arc.

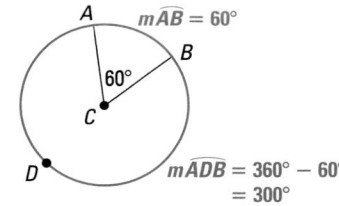

A **semicircle** is an arc whose central angle measures 180°. A semicircle is named by three points. Its measure is 180°.

EXAMPLE 1 Name and Find Measures of Arcs

Name the red arc and identify the type of arc. Then find its measure.

a.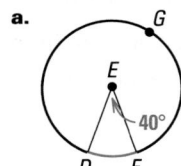

b.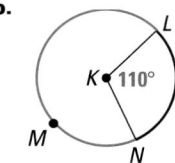

Solution
- a. $\widehat{DF}$ is a minor arc. Its measure is 40°.
- b. $\widehat{LMN}$ is a major arc. Its measure is $360° - 110° = 250°$.

---

① Plan

Pacing
Suggested Number of Days

Basic: 1 day
Average: 1 day
Advanced: 1 day
Block Schedule: 0.5 block with 11.4

Teaching Resources

📄 **Blacklines**
(See page 586B.)

📘 **Transparencies**
- Warm-Up with Quiz
- Answers

💻 **Technology**
- Electronic Teacher Tools
- Test and Practice Generator
- Online Lesson Planner
- Internet Support

② Teach

Content and Teaching Strategies
For background information on geometric concepts and teaching strategies related to this lesson, see pages 586E and 586F in this Teacher's Edition.

Tips for New Teachers
Let students know that three letters are used only when naming major arcs and semicircles. They should be aware that the order of the letters also indicates the direction of movement around the circle for all arcs. See the Tips for New Teachers on pp. 1–2 of the *Chapter 11 Resource Book* for additional notes about Lesson 11.3.

Extra Example 1
See next page.

Extra Example 1

Identify each arc as a major or minor arc. Then find its measure.

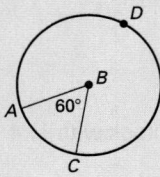

$\overarc{AC}$ is a minor arc. Its measure is 60°. $\overarc{ADC}$ is a major arc. Its measure is 360° − 60° = 300°.

Extra Example 2

Find the measure of $\overarc{ACD}$. 195°

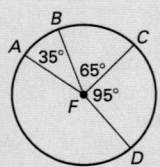

Extra Example 3

Find the measures of $\overarc{AB}$ and $\overarc{CD}$. Are the arcs congruent?

a.

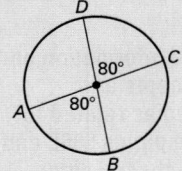

$\overarc{AB}$ and $\overarc{CD}$ are in the same circle. Because $m\overarc{AB} = m\overarc{CD} = 80°$, $\overarc{AB} \cong \overarc{CD}$.

b.

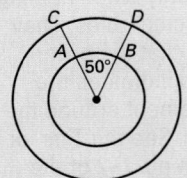

$\overarc{AB}$ and $\overarc{CD}$ are not in the same circle or in congruent circles. Therefore, although $m\overarc{AB} = m\overarc{CD} = 50°$, the arcs are not congruent.

Visualize It!

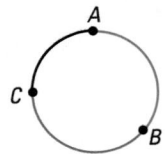

Arcs of a circle are *adjacent* if they intersect only at their endpoints.

$\overarc{AB}$ and $\overarc{BC}$ are adjacent.

POSTULATE 16

Arc Addition Postulate

Words The measure of an arc formed by two adjacent arcs is the sum of the measures of the two arcs.

Symbols $m\overarc{ACB} = m\overarc{AC} + m\overarc{CB}$

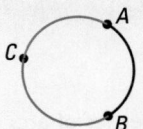

EXAMPLE 2 Find Measures of Arcs

Find the measure of $\overarc{GEF}$.

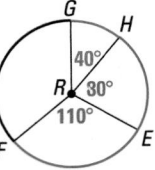

Solution

$$m\overarc{GEF} = m\overarc{GH} + m\overarc{HE} + m\overarc{EF}$$
$$= 40° + 80° + 110°$$
$$= 230°$$

Two circles are **congruent circles** if they have the same radius. Two arcs of the same circle or of congruent circles are **congruent arcs** if they have the same measure.

EXAMPLE 3 Identify Congruent Arcs

Find the measures of the blue arcs. Are the arcs congruent?

a.

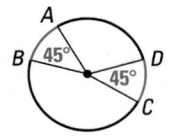

b.

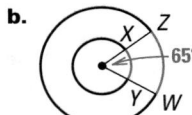

Solution

a. Notice that $\overarc{AB}$ and $\overarc{DC}$ are in the same circle. Because $m\overarc{AB} = m\overarc{DC} = 45°$, $\overarc{AB} \cong \overarc{DC}$.

b. Notice that $\overarc{XY}$ and $\overarc{ZW}$ are not in the same circle or in congruent circles. Therefore, although $m\overarc{XY} = m\overarc{ZW} = 65°$, $\overarc{XY} \not\cong \overarc{ZW}$.

---

1. $m\overarc{BC} = 58°$; $m\overarc{EF} = 58°$; yes

2. $m\overarc{BC} = 58°$; $m\overarc{CD} = 72°$; no

3. $m\overarc{CD} = 72°$; $m\overarc{DE} = 72°$; yes

4. $m\overarc{BFE} = 158°$; $m\overarc{CBF} = 158°$; yes

Checkpoint ✓ **Identify Congruent Arcs**

Find the measures of the arcs. Are the arcs congruent?

1. $\overarc{BC}$ and $\overarc{EF}$

2. $\overarc{BC}$ and $\overarc{CD}$ 1–4. See margin.

3. $\overarc{CD}$ and $\overarc{DE}$

4. $\overarc{BFE}$ and $\overarc{CBF}$

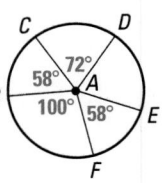

Student Help

SKILLS REVIEW
To review finding circumference of a circle, see p. 674.

Arc Length An **arc length** is a portion of the circumference of a circle. You can write a proportion to find *arc length*.

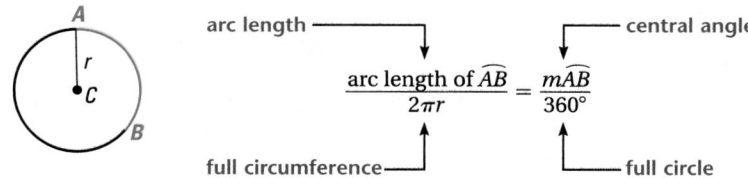

arc length ————→
————— central angle

$$\frac{\text{arc length of } \widehat{AB}}{2\pi r} = \frac{m\widehat{AB}}{360°}$$

full circumference ————
———— full circle

ARC LENGTH

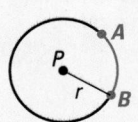

Words In a circle, the ratio of the length of a given arc to the circumference is equal to the ratio of the measure of the arc to 360°.

Symbols Arc length of $\widehat{AB} = \dfrac{m\widehat{AB}}{360°} \cdot 2\pi r$

EXAMPLE 4 Find Arc Lengths

Find the length of the red arc.

a.
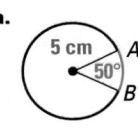

b.
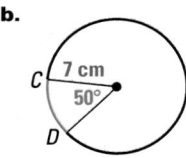

c.
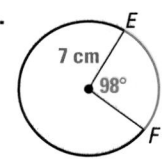

Student Help

STUDY TIP
You can substitute 3.14 as an approximation of π or use a calculator.

Solution

a. Arc length of $\widehat{AB} = \dfrac{50°}{360°} \cdot 2\pi(5) \approx 4.36$ centimeters

b. Arc length of $\widehat{CD} = \dfrac{50°}{360°} \cdot 2\pi(7) \approx 6.11$ centimeters

c. Arc length of $\widehat{EF} = \dfrac{98°}{360°} \cdot 2\pi(7) \approx 11.97$ centimeters

Checkpoint ✓ **Find Arc Lengths**

Find the length of the red arc. Round your answer to the nearest hundredth.

5.
4.19 in.
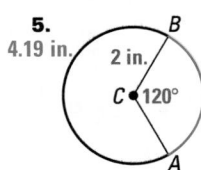

6.
12.57 ft
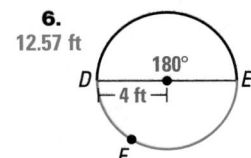

7.
9.42 cm
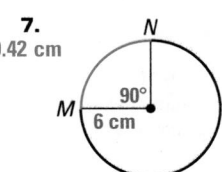

11.3 *Arcs and Central Angles*   **603**

Multiple Representations

As you introduce arc length, draw a large circle with radius *r* on the board. Ask students to give the length of the circumference. ($2\pi r$) Erase half the circle and draw a diameter to complete the half-circle. Ask students to give the arc length of half the circumference. (πr) Erase half again and draw a radius to make a quarter-circle. Again, ask students the arc length of one quarter of the circumference. $\left(\dfrac{1}{2}\pi r\right)$ Lead students from these expressions to the formula for arc length.

Extra Example 4

Find the length of the minor arc.

a.
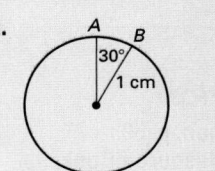
about 0.52 cm

b.
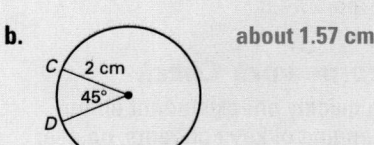
about 1.57 cm

c.
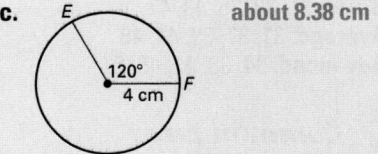
about 8.38 cm

✓ Concept Check

How do you find the length of an arc of a circle? Divide the measure of the arc by 360° and multiply the result by the circumference of the circle.

🧩 Daily Puzzler

A clock face has a diameter of 12 inches. The tip of the minute hand traces the circumference of the clock. How many inches does the tip of the minute hand travel each hour? about 37.7 in.

Assignment Guide

BASIC
Day 1: pp. 604–607 Exs. 13–39 odd, 43–51 odd, 55–61 odd, Quiz 1

AVERAGE
Day 1: pp. 604–607 Exs. 14–38 even, 40–42, 44–52 even, 55–62, Quiz 1

ADVANCED
Day 1: pp. 604–607 Exs. 19–42, 43–53 odd, 54*, 55–60, Quiz 1; EC: classzone.com

BLOCK SCHEDULE
pp. 604–607 Exs. 14–38 even, 40–42, 44–52 even, 55–62, Quiz 1 (with 11.4)

Extra Practice

• Student Edition, p. 695
• Chapter 11 Resource Book, pp. 27–28

Homework Check

To quickly check student understanding of key concepts, go over the following exercises:

Basic: 31, 33, 35, 43, 47
Average: 31, 37, 39, 44, 48
Advanced: 34, 38, 41, 45, 51

✗ Common Error

In Exercises 13–18, and again in Exercises 19–26, watch for students who confuse minor and major arcs. Remind them that major arcs have a measure greater than 180° and are named with three letters.

2. *Sample answer:*

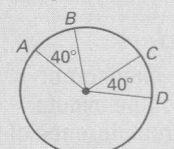

Guided Practice

Vocabulary Check

3. The measure of an arc is the degree measure of the related central angle (or 360° − the measure of the related central angle), while an arc length is a portion of the circumference of a circle.

Skill Check

1. In the diagram at the right, identify a *major arc*, a *minor arc*, and a *semicircle*.
major arc: $\overarc{BAD}$ or $\overarc{ADB}$; minor arc: $\overarc{AB}$ or $\overarc{BD}$; semicircle: $\overarc{ABD}$

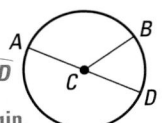

2. Draw a circle with a pair of congruent arcs. **See margin.**

3. What is the difference between *arc measure* and *arc length*?

Find the measure in ⊙T.

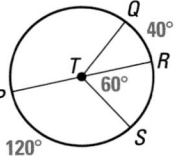

4. $m\overarc{RS}$ 60°

5. $m\overarc{RPS}$ 300°

6. $m\overarc{PQR}$ 180°

7. $m\overarc{QS}$ 100°

8. $m\overarc{QSP}$ 220°

9. $m\angle QTR$ 40°

Find the blue arc length. Round your answer to the nearest hundredth.

10. Length of $\overarc{AB}$
1.40 yd

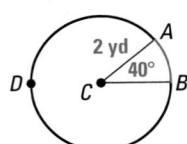

11. Length of $\overarc{DE}$
10.47 cm

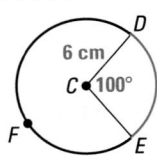

12. Length of $\overarc{FGH}$
19.20 m

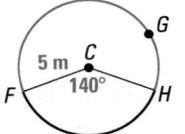

Practice and Applications

Extra Practice
See p. 695.

Naming Arcs Name the blue minor arc and find its measure.

13. $\overarc{PQ}$; 135°

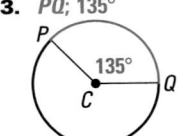

14. $\overarc{DE}$; 130°

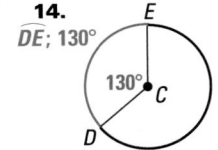

15. $\overarc{LN}$; 150°

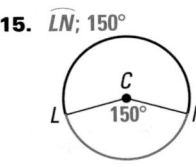

Naming Arcs Name the blue major arc and find its measure.

16. $\overarc{ADB}$; 285°

17. $\overarc{WXY}$; 200°

18. $\overarc{GFH}$; 330°

Homework Help

Example 1: Exs. 13–39
Example 2: Exs. 30–42
Example 3: Exs. 43–46
Example 4: Exs. 47–54

Types of Arcs Determine whether the arc is a *minor arc*, a *major arc*, or a *semicircle* of ⊙R. $\overline{PT}$ and $\overline{QU}$ are diameters.

19. $\overparen{PQ}$ minor arc

20. $\overparen{SU}$ minor arc

21. $\overparen{PQT}$ semicircle

22. $\overparen{QT}$ minor arc

23. $\overparen{TUQ}$ major arc

24. $\overparen{TUP}$ semicircle

25. $\overparen{QUT}$ major arc

26. $\overparen{PUQ}$ major arc

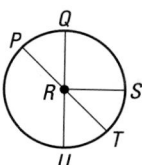

Finding the Central Angle Find the measure of ∠ACB.

27. 165°
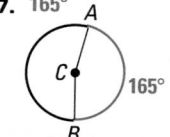

28. 90°
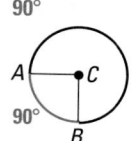

29. 180° 180°
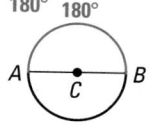

Student Help

VISUAL STRATEGY
In Exs. 30–39, copy the diagram and add information to it as you solve the exercises, as shown on p. 588.

Measuring Arcs and Central Angles $\overline{KN}$ and $\overline{JL}$ are diameters. Find the measure.

30. $m\overparen{KL}$ 60°

31. $m\overparen{MN}$ 55°

32. $m\overparen{LNK}$ 300°

33. $m\overparen{MKN}$ 305°

34. $m\overparen{NJK}$ 180°

35. $m\angle MQL$ 65°

36. $m\overparen{ML}$ 65°

37. $m\angle JQN$ 60°

38. $m\overparen{JM}$ 115°

39. $m\overparen{LN}$ 120°

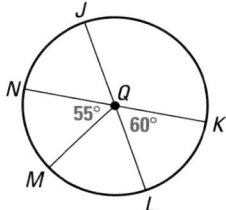

Time Zone Wheel **In Exercises 40–42, use the following information.**
To find the time in Tokyo when it is 4 P.M. in San Francisco, rotate the small wheel until 4 P.M. and San Francisco line up as shown. Then look at Tokyo to see that it is 9 A.M. there.

When it is 9 A.M. in Tokyo . . .

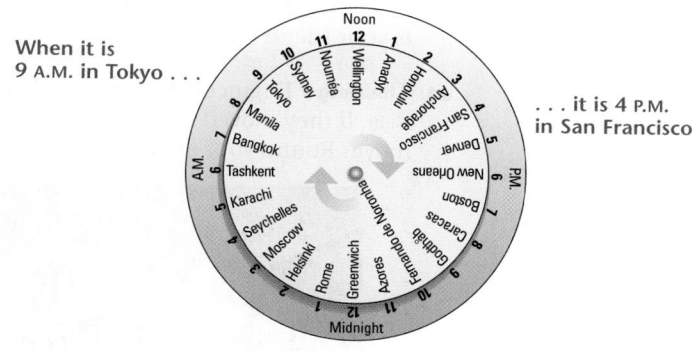

. . . it is 4 P.M. in San Francisco

40. What is the arc measure for each time zone on the wheel? 15°

41. What is the measure of the minor arc from the Tokyo zone to the Anchorage zone? 90°

42. If two cities differ by 180° on the wheel, then it is 3:00 P.M. in one city when it is __?__ in the other city. 3 A.M.

11.3 Arcs and Central Angles **605**

Visualize It!

Refer students to the Student Help note to the left of Exercises 30–39. Suggest that students make their copy of the circle larger than the one in the textbook so there will be plenty of room to add all the measures. Students can label angle measures with one color and arc measures with a different color.

Teaching Tip
When assigning Exercises 40–42, point out that the arcs between cities are congruent because the arcs representing the hours are all equal in length.

✗ Common Error

Exercise 53 points out a very common error that students make. Use this exercise to once again stress to students that to have the same arc length, the arcs in Exercise 53 must be on the *same* circle or *congruent* circles. Refer students back to parts a and b of Example 4 on page 603 which demonstrate that if the arcs are not on the same circle or on congruent circles, then their arc lengths are *not* equal.

Student Help
CLASSZONE.COM

HOMEWORK HELP
Extra help with problem solving in Exs. 47–52 is at classzone.com

43. No; the circles are not congruent.

44. Yes; ∠ACD ≅ ∠BCE since they are vertical angles; $m\widehat{AD} = m\widehat{BE}$, so $\widehat{AD} \cong \widehat{BE}$.

45. Yes; $\widehat{UW}$ and $\widehat{XZ}$ are arcs of congruent circles with the same measure.

46. No; F is not the center of the circle, so you cannot determine the measures of $\widehat{JK}$ and $\widehat{GH}$.

Naming Congruent Arcs Are the blue arcs congruent? Explain.

43.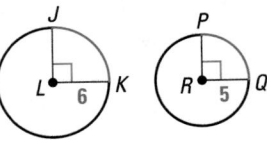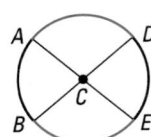

44.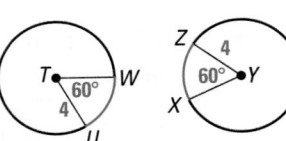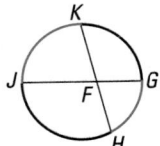

45.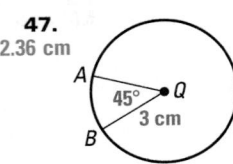

46.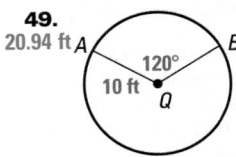

Finding Arc Length Find the length of $\widehat{AB}$. Round your answer to the nearest hundredth.

47. 2.36 cm

48. 7.33 in.

49. 20.94 ft

50. 2.09 cm

51. 15.71 m

52. 15.71 in.

53. **You be the Judge** A friend tells you two arcs from different circles have the same arc length if their central angles are equal. Is your friend correct? Explain your reasoning.
No; they have the same arc length only if the two circles are congruent circles.

54. **Challenge** Engineers reduced the lean of the Leaning Tower of Pisa. If they moved it back 0.46°, what was the arc length of the move? Round your answer to the nearest whole number.   45 cm

5588 cm

55. Multiple Choice What is the length of $\overarc{AC}$ in $\odot P$ shown below? C

ⓐ 5.6 ft   Ⓑ 16.8 ft

Ⓒ 19.5 ft   Ⓓ 25.1 ft

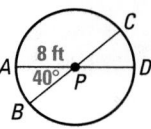

Mixed Review

Finding Leg Lengths Find the lengths of the legs of the triangle. Round your answers to the nearest tenth. *(Lesson 10.5)*

56.
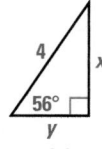

$x \approx 3.3; y \approx 2.2$

57.
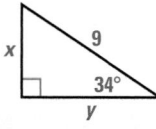

$x \approx 5.0; y \approx 7.5$

58.
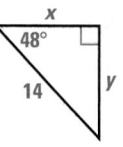

$x \approx 9.4; y \approx 10.4$

Algebra Skills

Simplifying Ratios Simplify the ratio. *(Skills Review, p. 660)*

59. $\dfrac{2 \text{ km}}{400 \text{ km}}$ $\dfrac{1}{200}$ **60.** $\dfrac{5 \text{ ft}}{72 \text{ in.}}$ $\dfrac{5}{6}$ **61.** $\dfrac{3 \text{ yards}}{27 \text{ ft}}$ $\dfrac{1}{3}$ **62.** $\dfrac{4 \text{ ounces}}{8 \text{ pounds}}$ $\dfrac{1}{32}$

Quiz 1

Tell whether the given line, segment, or point is best described as a *chord*, a *secant*, a *tangent*, a *diameter*, a *radius*, or a *point of tangency*. *(Lesson 11.1)*

1. $\overleftrightarrow{AB}$ tangent   **2.** $\overleftrightarrow{JH}$ secant

3. $\overline{GE}$ diameter   **4.** $\overline{JH}$ chord

5. $\overline{CE}$ radius   **6.** D point of tangency

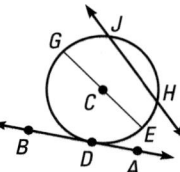

$\overline{PQ}$ and $\overline{PR}$ are tangent to $\odot C$. Find the value of x. *(Lesson 11.2)*

7. 15
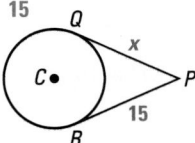

8. 6
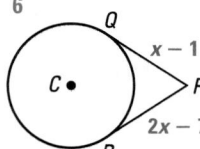

9. P 4
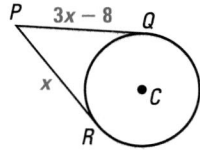

Find the length of $\overarc{AB}$. Round your answer to the nearest hundredth. *(Lesson 11.3)*

10.
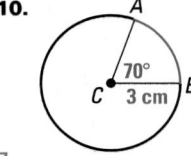

$\dfrac{7}{6}\pi \approx 3.67$ cm

11.
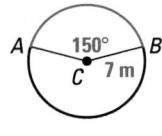

$\dfrac{35}{6}\pi \approx 18.33$ m

12.
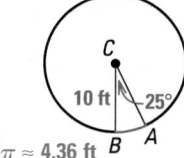

$\dfrac{25}{18}\pi \approx 4.36$ ft

11.3 *Arcs and Central Angles* **607**

④ Assess

Assessment Resources

The Mini-Quiz below is also available on blackline (*Chapter 11 Resource Book*, p. 36) and on transparency. For more assessment resources, see:
• Chapter 11 Resource Book
• Standardized Test Practice
• Test and Practice Generator

Mini-Quiz

Find the measure.

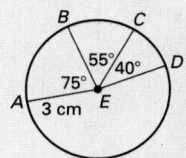

1. $m\overarc{BC}$ 55°
2. $m\overarc{AC}$ 130°
3. $m\angle BED$ 95°
4. Find the length of $\overarc{CD}$. Round your answer to the nearest hundredth. 2.09 cm

Plan

Pacing
Suggested Number of Days

Basic: 2 days
Average: 2 days
Advanced: 2 days
Block Schedule: 0.5 block with 11.3
0.5 block with 11.5

Teaching Resources

Blacklines
(See page 586B.)

Transparencies
• Warm-Up with Quiz
• Answers

Technology
• Electronic Teacher Tools
• Test and Practice Generator
• Online Lesson Planner
• Internet Support

Teach

Content and Teaching Strategies
For background information on geometric concepts and teaching strategies related to this lesson, see pages 586E and 586F in this Teacher's Edition.

Tips for New Teachers
As you discuss the theorems about chords of circles, remind students to use their prior knowledge about right angles and right triangles. See the Tips for New Teachers on pp. 1–2 of the *Chapter 11 Resource Book* for additional notes about Lesson 11.4.

Goal
Use properties of chords of circles.

Key Words
• congruent arcs p. 602
• perpendicular bisector p. 274

By finding the perpendicular bisectors of two chords, an archaeologist can recreate a whole plate from just one piece.

This approach relies on Theorem 11.5, and is shown in Example 2.

THEOREM 11.4

Words If a diameter of a circle is perpendicular to a chord, then the diameter bisects the chord and its arc.

Symbols If $\overline{BG} \perp \overline{FD}$, then $\overline{DE} \cong \overline{EF}$ and $\overarc{DG} \cong \overarc{GF}$.

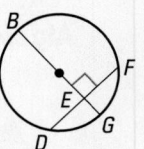

EXAMPLE 1 Find the Length of a Chord

In $\odot C$ the diameter $\overline{AF}$ is perpendicular to $\overline{BD}$. Use the diagram to find the length of $\overline{BD}$.

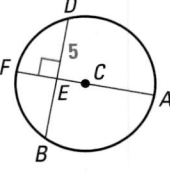

Solution

Because $\overline{AF}$ is a diameter that is perpendicular to $\overline{BD}$, you can use Theorem 11.4 to conclude that $\overline{AF}$ bisects $\overline{BD}$. So, $BE = ED = 5$.

$BD = BE + ED$	Segment Addition Postulate
$= 5 + 5$	Substitute 5 for BE and ED.
$= 10$	Simplify.

ANSWER ▶ The length of $\overline{BD}$ is 10.

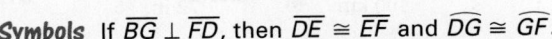

Checkpoint Find the Length of a Segment

1. Find the length of $\overline{JM}$.

12

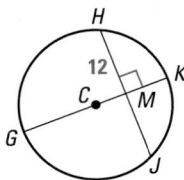

2. Find the length of $\overline{SR}$.

30

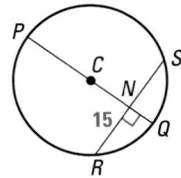

THEOREM 11.5

Words If one chord is a perpendicular bisector of another chord, then the first chord is a diameter.

Symbols If $\overline{JK} \perp \overline{ML}$ and $\overline{MP} \cong \overline{PL}$, then $\overline{JK}$ is a diameter.

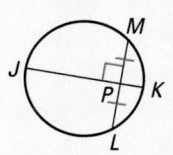

All diameters of a circle include the center of the circle. Therefore, the point where two diameters intersect is the center of the circle.

EXAMPLE 2 Find the Center of a Circle

Suppose an archaeologist finds part of a circular plate. Show how to reconstruct the original shape of the plate.

Solution

❶ Draw any two chords that are not parallel to each other.

❷ Draw the perpendicular bisector of each chord. These lines contain diameters.

❸ The diameters intersect at the circle's center. Use a compass to draw the rest of the plate.

Student Help

STUDY TIP
In the same circle or in congruent circles, if two central angles are congruent then their corresponding arcs are congruent.

THEOREM 11.6

Words In the same circle, or in congruent circles:
- If two chords are congruent, then their corresponding minor arcs are congruent.
- If two minor arcs are congruent, then their corresponding chords are congruent.

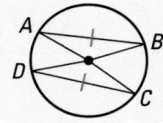

Symbols If $\overline{AB} \cong \overline{DC}$, then $\overarc{AB} \cong \overarc{DC}$.
If $\overarc{AB} \cong \overarc{DC}$, then $\overline{AB} \cong \overline{DC}$.

11.4 *Arcs and Chords* 609

Extra Example 1

In ⊙W, the diameter $\overline{RT}$ is perpendicular to $\overline{US}$. Use the diagram to find the length of $\overline{US}$.

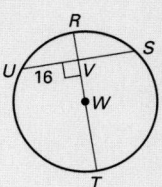

32

Extra Example 2

Explain how to find a radius of a circular object, given only one third of the object. **Draw two chords. Draw the perpendicular bisector of each chord. The perpendicular bisectors lie on diameters of the circular object and intersect at its center. Any segment from the center to the edge of the circular object is a radius.**

Teaching Tip

In Step 2 of Example 2, make sure students understand why the perpendicular bisector of a chord must lie on a diameter.

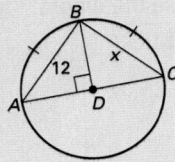
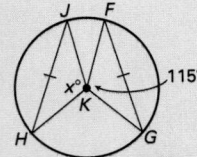

Find the value of x.

a.

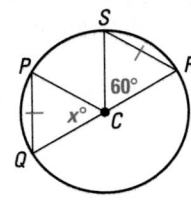

b.

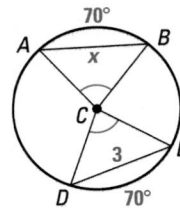

Solution

a. Because $\overline{QP} \cong \overline{RS}$, it follows that $\overparen{QP} \cong \overparen{RS}$. So, $m\overparen{QP} = m\overparen{RS} = 60°$, and $x = 60$.

b. Because $\overparen{AB} \cong \overparen{DE}$, it follows that $\overline{AB} \cong \overline{DE}$. So, $x = DE = 3$.

Checkpoint ✓ **Find Measures of Angles and Chords**

Find the value of x.

3. 4

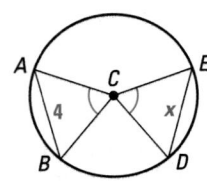

4. 3

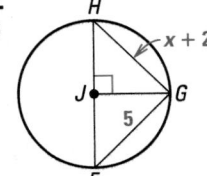

5. 30

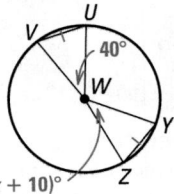

11.4 Exercises

Guided Practice

Vocabulary Check

1. Identify a diameter.
$\overline{BE}$

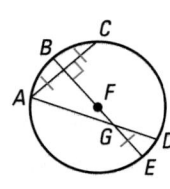

2. Identify a pair of congruent arcs.
$\overparen{PQ}$ and $\overparen{RS}$

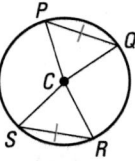

Skill Check

Find the value of x.

3. 7

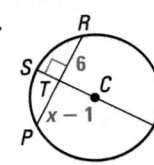

4. 45

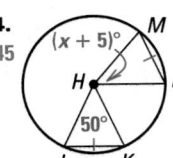

5. 8

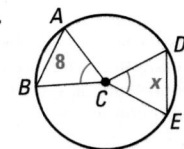

Practice and Applications

Extra Practice

See p. 695.

6. No; $\overline{AB}$ is not perpendicular to $\overline{CD}$, so $\overline{AB}$ is not a diameter of the circle.

7. No; $\overline{AB}$ does not bisect $\overline{CD}$, so $\overline{AB}$ is not a diameter of the circle.

8. Yes; $\overline{AB}$ is a perpendicular bisector of $\overline{CD}$, so $\overline{AB}$ is a diameter of the circle.

Identifying Diameters Determine whether $\overline{AB}$ is a diameter of the circle. Explain your reasoning.

6.
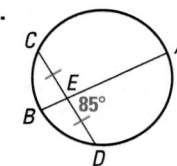

7.
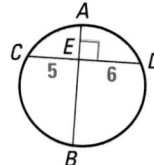

8.
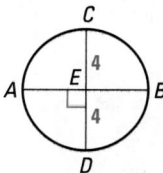

Finding Chords and Central Angles Find the value of x.

9.
7
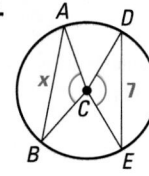

10.
29
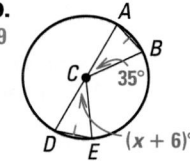

11.
3
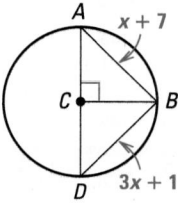

Logical Reasoning Name any congruent arcs, chords, or angles. State a postulate or theorem that justifies your answer.

12. $\overline{AB} \cong \overline{BC}$ (given) and $\overparen{AB} \cong \overparen{BC}$ (If two chords are congruent, then their corresponding minor arcs are congruent.)

13. $\overparen{AB} \cong \overparen{CD}$ (given) and $\overline{AB} \cong \overline{CD}$ (If two minor arcs are congruent, then their corresponding chords are congruent.)

14. $\overparen{AB} \cong \overparen{CD}$ (given) and $\overline{AB} \cong \overline{CD}$ (If two chords are congruent, then their corresponding minor arcs are congruent.) $\angle AQB \cong \angle CQD$ by the definition of the measure of a minor arc.

12.
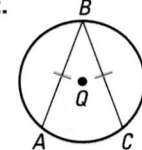

13.
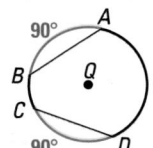

14.
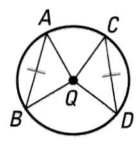

Finding Measures Find the measure of the red segment or arc.

15.
10
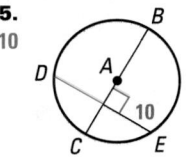

16.
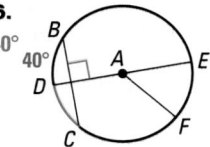

17.
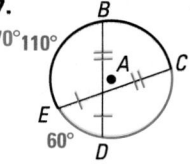

Homework Help

Example 1: Exs. 6–8, 15–20
Example 2: Exs. 21–22
Example 3: Exs. 9–20

⟨x/y⟩ Using Algebra Find the value of x.

18.
40
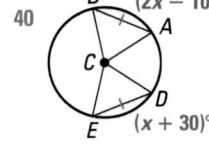

19.
7
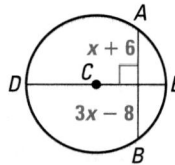

20.
15
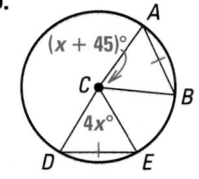

21. **Visualize It!** Draw a large circle and cut it out. Tear part of it off and ask another student to recreate your circle. **Check students' work, which should be based on the method shown in Example 2 on page 609.**

11.4 *Arcs and Chords* **611**

❸ Apply

Assignment Guide

BASIC
Day 1: pp. 611–612 Exs. 6–8, 15, 16, 19, 24–29
Day 2: pp. 611–612 Exs. 9–14, 17, 18, 20, 23, 30–32

AVERAGE
Day 1: pp. 611–612 Exs. 6–8, 15, 16, 19, 21, 24–29
Day 2: pp. 611–612 Exs. 9–14, 17, 18, 20, 23, 30–32

ADVANCED
Day 1: pp. 611–612 Exs. 15, 16, 19, 21, 22, 24–29
Day 2: pp. 611–612 Exs. 9–14, 17, 18, 20, 23, 30–32;
EC: TE p. 586D*, classzone.com

BLOCK SCHEDULE
pp. 611–612 Exs. 6–8, 15, 16, 19, 21, 24–29 (with 11.3)
pp. 611–612 Exs. 9–14, 17, 18, 20, 23, 30–32 (with 11.5)

Extra Practice
• Student Edition, p. 695
• Chapter 11 Resource Book, pp. 37–38

Homework Check
To quickly check student understanding of key concepts, go over the following exercises:

Basic: 6, 9, 12, 15, 18
Average: 10, 13, 16, 19, 21
Advanced: 11, 14, 17, 20, 21

✗ Common Error
In Exercises 18–20, watch for students who make errors in their calculation because they try to solve the algebraic equations mentally. Remind them to write the equation and each step used to solve it. Once students have found a value for x, remind them to substitute the value for x in the expressions to check that the two measures have the same value.

Assessment Resources

The Mini-Quiz below is also available on blackline (*Chapter 11 Resource Book*, p. 44) and on transparency. For more assessment resources, see:
- Chapter 11 Resource Book
- Standardized Test Practice
- Test and Practice Generator

Mini-Quiz

Find the value of *x*.

1.

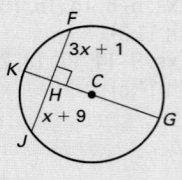

9

2.

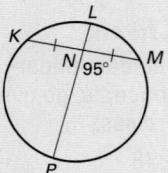

4

3. Determine whether $\overline{LP}$ is a diameter of the circle. Explain your reasoning.

No; $\overline{LP}$ is not perpendicular to $\overline{KM}$. Thus, $\overline{LP}$ is not a diameter.

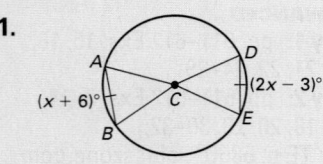
22. The searcher is constructing a chord of the beacon's circle and the perpendicular bisector of the chord, which is a diameter of the circle. By locating the midpoint of the diameter, the searcher locates the center of the circle, which is the location of the beacon.

Standardized Test Practice

23. a. In a circle, if two chords are congruent, then their corresponding minor arcs are congruent.

---

22. Avalanche Rescue Beacon An avalanche rescue beacon is a device used by backcountry skiers. It gives off a signal that is detectable within a circle of a certain radius. In a practice drill, a ski patrol uses the following steps to locate a beacon buried in the snow. Explain how it works. **See margin.**

❶ Walk in a straight line until the signal disappears. Turn around and walk back until the signal disappears again.

❷ Walk back to the halfway point, and walk away from the line at a 90° angle until the signal disappears.

❸ Turn around and walk in a straight line until the signal disappears again.

❹ Walk back to the halfway point. You will be near the center of the circle. The beacon is under you.

23. Multi-Step Problem Use the diagram below.

a. Explain why $\widehat{AD} \cong \widehat{BE}$. See margin.

b. Find the value of *x*. **10**

c. Find $m\widehat{AD}$ and $m\widehat{BE}$.
$m\widehat{AD} = m\widehat{BE} = 110°$

d. Find $m\widehat{BD}$. **100°**

Mixed Review

Measuring Arcs In the diagram below, $\overline{AD}$ and $\overline{BE}$ are diameters of $\odot F$. Find the measure. *(Lesson 11.3)*

24. $m\widehat{DE}$ **40°**

25. $m\widehat{BC}$ **75°**

26. $m\widehat{AE}$ **140°**

27. $m\widehat{BCD}$ **140°**

28. $m\widehat{ABC}$ **115°**

29. $m\widehat{ADE}$ **220°**

Algebra Skills

Comparing Numbers Compare the two numbers. Write the answer using <, >, or =. *(Skills Review, p. 662)*

30. -26 and -29
$-26 > -29$

31. $\frac{15}{20}$ and $-\frac{3}{4}$ $\frac{15}{20} > -\frac{3}{4}$

32. 0.2 and $\frac{1}{5}$ $0.2 = \frac{1}{5}$

Question

How are inscribed angles related to central angles?

Materials
- compass
- straightedge
- protractor

Explore

1 Use a compass to draw a circle. Label the center *P*. Use a straightedge to draw a central angle. Label it ∠*RPS*.

2 Locate three points on ⊙*P* in the exterior of ∠*RPS* and label them *T*, *U*, and *V*.

3 Draw ∠*RTS*, ∠*RUS*, and ∠*RVS*. These are called *inscribed angles*. Measure each angle.

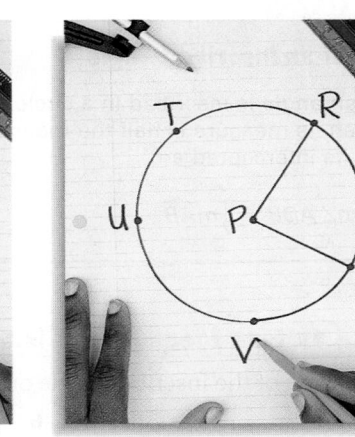

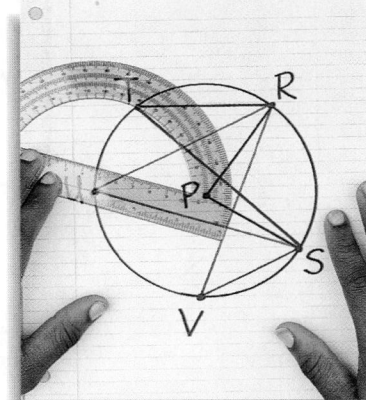

Step 3. Angle measures will vary, but the measures of all three angles will be equal.

Think About It

1. Make a table similar to the one below. Record the angle measures for ⊙*P* in the table.

Student Help

LOOK BACK
To review the measures of central angles, see p. 454.

	Central Angle	Inscribed Angle 1	Inscribed Angle 2	Inscribed Angle 3
Name	∠*RPS*	∠*RTS*	∠*RUS*	∠*RVS*
Measure	?	?	?	?

1, 2. Answers will vary, but in each case, *m*∠*RTS* = *m*∠*RUS* = *m*∠*RVS* = $\frac{1}{2}$*m*∠*RPS*.

3. The measure of an inscribed angle is equal to half the measure of the corresponding central angle.

2. Draw two more circles. Repeat Steps 1–3, using different central angles. Record the measures in a table similar to the one above.

3. Use the results in your table to make a conjecture about how the measure of an inscribed angle is related to the measure of the corresponding central angle.

Activity 613

1 *Planning the Activity*

Goal
Students measure angles to discover the relationship between the measure of an inscribed angle and the measure of the central angle that intercepts the same arc.

Materials
See the margin of the student page.

LINK TO LESSON ➡
In Example 1 on page 614, students find the measure of an inscribed angle. Ask students to give the measure of the central angle that intersects the same arc.

2 *Managing the Activity*

Tips for Success
Point out that ∠*RPS* does not have to be a right angle. Also point out that the points *T*, *U*, and *V* do not need to be evenly spaced along the major arc from *R* to *S*.

3 *Closing the Activity*

KEY DISCOVERY
The measure of an inscribed angle is half the measure of the corresponding central angle.

Activity Assessment
Use Exercise 3 to assess student understanding.

Pacing
Suggested Number of Days

Basic: 2 days
Average: 2 days
Advanced: 2 days
Block Schedule: 0.5 block with 11.4
0.5 block with 11.6

Teaching Resources

 **Blacklines**
(See page 586B.)

Transparencies
• Warm-Up with Quiz
• Answers

Technology
• Electronic Teacher Tools
• Test and Practice Generator
• Online Lesson Planner
• Internet Support

─**Content and Teaching Strategies**─
For background information on geometric concepts and teaching strategies related to this lesson, see pages 586E and 586F in this Teacher's Edition.

Tips for New Teachers
Activity 11.5 on page 613 leads nicely into the theorem for inscribed angles on this page. Confirm that students remember that the measure of a central angle is equal to the measure of its intercepted arc. See the Tips for New Teachers on pp. 1–2 of the *Chapter 11 Resource Book* for additional notes about Lesson 11.5.

 **11.5**

Inscribed Angles and Polygons

Goal
Use properties of inscribed angles.

Key Words
• inscribed angle
• intercepted arc
• inscribed
• circumscribed

An **inscribed angle** is an angle whose vertex is on a circle and whose sides contain chords of the circle.

The arc that lies in the interior of an inscribed angle and has endpoints on the angle is called the **intercepted arc** of the angle.

Activity 11.5 shows the relationship between an inscribed angle and its intercepted arc.

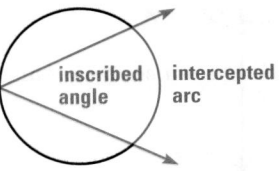

THEOREM 11.7

Measure of an Inscribed Angle

Words If an angle is inscribed in a circle, then its measure is half the measure of its intercepted arc.

Symbols $m\angle ADB = \frac{1}{2}m\widehat{AB}$

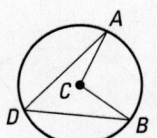

EXAMPLE 1 Find Measures of Inscribed Angles and Arcs

Find the measure of the inscribed angle or the intercepted arc.

a.

b.
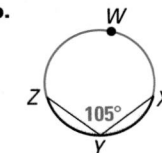

Student Help
CLASSZONE.COM

MORE EXAMPLES
More examples at classzone.com

Solution

a. $m\angle NMP = \frac{1}{2}m\widehat{NP}$ The measure of an inscribed angle is half the measure of its intercepted arc.

$= \frac{1}{2}(100°)$ Substitute 100° for $m\widehat{NP}$.

$= 50°$ Simplify.

b. $m\angle ZYX = \frac{1}{2}m\widehat{ZWX}$ The measure of an inscribed angle is half the measure of its intercepted arc.

$105° = \frac{1}{2}m\widehat{ZWX}$ Substitute 105° for $m\angle ZYX$.

$210° = m\widehat{ZWX}$ Multiply each side by 2.

Checkpoint ✓ *Find Measures of Inscribed Angles and Arcs*

Find the measure of the inscribed angle or the intercepted arc.

1.

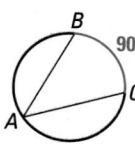

$m\angle BAC = 45°$

2.

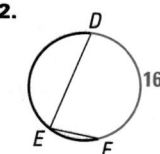

$m\angle DEF = 80°$

3.

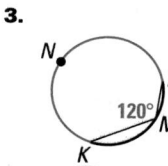

$m\widehat{KNP} = 240°$

Inscribed and Circumscribed If all the vertices of a polygon lie on a circle, the polygon is **inscribed** in the circle and the circle is **circumscribed** about the polygon. The polygon is an *inscribed polygon* and the circle is a *circumscribed circle*.

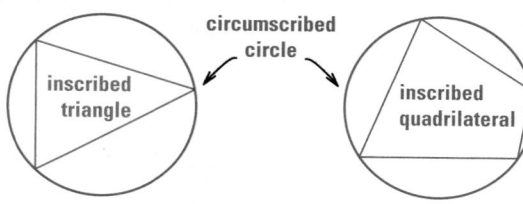

circumscribed circle

inscribed triangle

inscribed quadrilateral

THEOREM 11.8

Words If a triangle inscribed in a circle is a right triangle, then the hypotenuse is a diameter of the circle.

If a side of a triangle inscribed in a circle is a diameter of the circle, then the triangle is a right triangle.

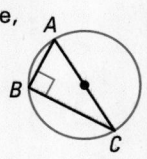

EXAMPLE 2 **Find Angle Measures**

Find the values of x and y.

Solution

Because $\triangle ABC$ is inscribed in a circle and $\overline{AB}$ is a diameter, it follows from Theorem 11.8 that $\triangle ABC$ is a right triangle with hypotenuse $\overline{AB}$.

Therefore, $x = 90$. Because $\angle A$ and $\angle B$ are acute angles of a right triangle, $y = 90 - 50 = 40$.

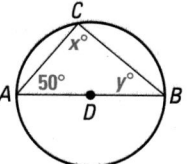

Student Help

LOOK BACK
To review the Corollary of the Triangle Sum Theorem, see p. 180.

11.5 Inscribed Angles and Polygons | **615**

Geometric Reasoning

While discussing Theorem 11.7, point out that $m\angle ADB$ is also equal to $\frac{1}{2}m\angle ACB$.

Extra Example 1

Find the measure of the inscribed angle or the intercepted arc.

a.

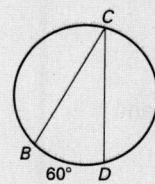

$m\angle BCD = 30°$

b.

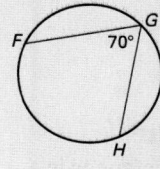

$m\widehat{FH} = 140°$

Study Skills

Vocabulary Invite students to suggest memory clues they can use to help them distinguish between the terms *inscribed* and *circumscribed*. For example, it will help students to think of *inscribed* figures as being *inside* another figure.

Extra Example 2

Find the values of x and y.

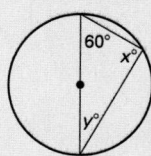

$x = 90$, $y = 30$

Extra Example 3

Find the values of *g* and *h*.

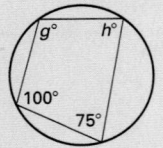

$g = 105$, $h = 80$

✓ Concept Check

Under what circumstances is a triangle inscribed in a circle a right triangle? **when the hypotenuse is a diameter of the circle**

🧩 Daily Puzzler

An inscribed angle has a measure of 90°. How are the endpoints of the intercepted arc related to the diameter of the circle? **If you join the endpoints of the arc with a segment, the segment is a diameter of the circle.**

Find the values of *x* and *y* in ⊙C.

4.
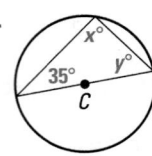

$x = 90$; $y = 55$

5.
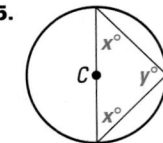

$x = 45$; $y = 90$

6.
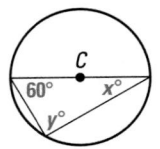

$x = 30$; $y = 90$

Visualize It!

∠*D* and ∠*F* are *opposite* angles. ∠*E* and ∠*G* are *opposite* angles.

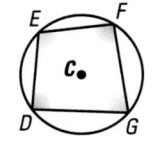

THEOREM 11.9

Words If a quadrilateral can be inscribed in a circle, then its opposite angles are supplementary.

If the opposite angles of a quadrilateral are supplementary, then the quadrilateral can be inscribed in a circle.

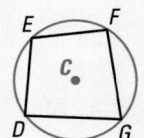

EXAMPLE 3 Find Angle Measures

Find the values of *y* and *z*.

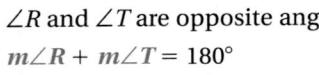

Solution

Because *RSTU* is inscribed in a circle, by Theorem 11.9 opposite angles must be supplementary.

∠*S* and ∠*U* are opposite angles.

$m\angle S + m\angle U = 180°$

$120° + y° = 180°$

$y = 60$

∠*R* and ∠*T* are opposite angles.

$m\angle R + m\angle T = 180°$

$z° + 80° = 180°$

$z = 100$

Checkpoint ✓ Find Angle Measures

Find the values of *x* and *y* in ⊙C.

7.

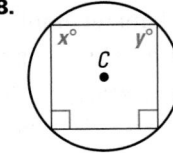

$x = 85$; $y = 80$

8.

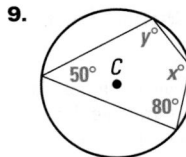

$x = 90$; $y = 90$

9.

$x = 130$; $y = 100$

11.5 Exercises

Guided Practice

Vocabulary Check

In Exercises 1 and 2, use the diagram at the right.

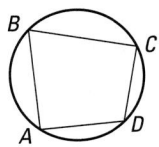

1. Name the *inscribed angles*. ∠A, ∠B, ∠C, ∠D

2. Identify the two pairs of *opposite angles* in the inscribed quadrilateral.
   ∠A and ∠C, ∠B and ∠D

Skill Check

Find the measure of the blue intercepted arc.

3.
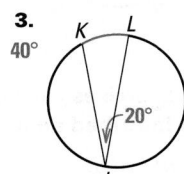
40° 20°

4.
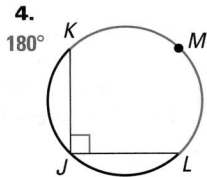
180°

5.
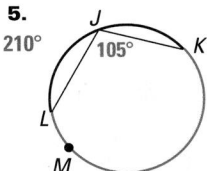
210° 105°

Find the value of each variable.

6.
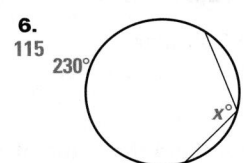
115 230° x°

7.
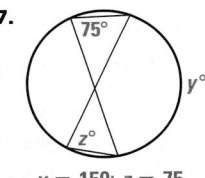
75° y° z°
y = 150; z = 75

8.
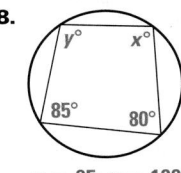
y° x° 85° 80°
x = 95; y = 100

Practice and Applications

Extra Practice

See p. 696.

Angle Measures Find the measure of the inscribed angle.

9.
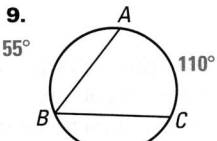
55° 110°

10.
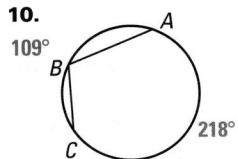
109° 218°

11.
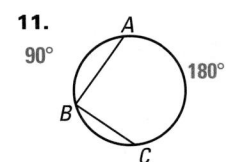
90° 180°

Homework Help

Example 1: Exs. 9–27
Example 2: Exs. 28–31
Example 3: Exs. 32–38

12.
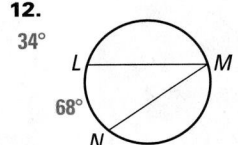
34° 68°

13.
67° 134°

14.
238° 119°

11.5 *Inscribed Angles and Polygons*   617

3 Apply

Assignment Guide

BASIC
Day 1: pp. 617–619 Exs. 9–20, 39, 50–55
Day 2: pp. 617–619 Exs. 28–30, 32–36, 40–49

AVERAGE
Day 1: pp. 617–619 Exs. 12–26, 39, 50–55
Day 2: pp. 617–619 Exs. 28–38, 40–49

ADVANCED
Day 1: pp. 617–619 Exs. 12–17, 21–27, 39, 50–55
Day 2: pp. 617–619 Exs. 28–38, 40–49; EC: TE p. 586D*, classzone.com

BLOCK SCHEDULE
pp. 617–619 Exs. 12–26, 39, 50–55 (with 11.4)
pp. 617–619 Exs. 28–38, 40–49 (with 11.6)

Extra Practice

• Student Edition, p. 696
• Chapter 11 Resource Book, pp. 45–46

Homework Check

To quickly check student understanding of key concepts, go over the following exercises:

Basic: 12, 15, 18, 28, 32
Average: 13, 18, 22, 29, 33
Advanced: 14, 17, 27, 30, 34

31. *Sample answer:* Position the vertex of the tool on the circle and mark the two points at which the sides intersect the circle; draw a segment to connect the two points, forming a diameter of the circle. Repeat these steps, placing the vertex at a different point on the circle. The center is the point at which the two diameters intersect.

Arc Measures Find the measure of the blue intercepted arc.

15.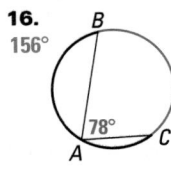

16.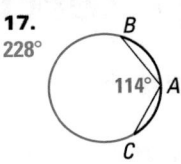

17.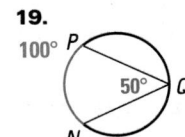

18.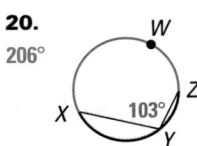

19.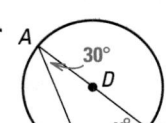

20.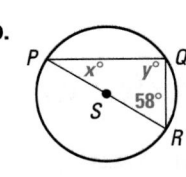

Arc and Angle Measures In Exercises 21–26, use the diagram below to find the intercepted arc or inscribed angle.

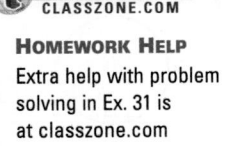

21. $m\widehat{BE}$ 94°

22. $m\angle BDE$ 47°

23. $m\angle AED$ 53°

24. $m\widehat{AD}$ 106°

25. $m\angle ABD$ 53°

26. $m\widehat{DE}$ 60°

27. Are $\triangle ABC$ and $\triangle DEC$ similar? Explain your reasoning. **Yes. Explanations may vary;** *Sample answer:* $m\angle BAC = 47° = m\angle CDE$ (from Ex. 22) and $m\angle DCE = m\angle ACB$ (vertical angles), so $\triangle ABC \sim \triangle DEC$ by the AA Similarity Postulate.

28. $\triangle ABC$ is an inscribed triangle and $\overline{AC}$ is a diameter, so $\triangle ABC$ is a right triangle with diameter $\overline{AC}$; $x = 90$; $y = 90 - 30 = 60$.

29. $\triangle KLM$ is an inscribed triangle and $\overline{KM}$ is a diameter, so $\triangle KLM$ is a right triangle with diameter $\overline{KM}$; $x = 90$; $y = 90 - 40 = 50$.

30. $\triangle PQR$ is an inscribed triangle and $\overline{PR}$ is a diameter, so $\triangle PQR$ is a right triangle with diameter $\overline{PR}$; $y = 90$; $x = 90 - 58 = 32$.

Inscribed Right Triangles Find the value of each variable. **Explain your reasoning.**

28.

29.

30.

31. **Carpenter's Square** A carpenter's square is an L-shaped tool used to draw right angles. Suppose you are making a toy truck. To make the wheels you trace a circle on a piece of wood. How could you use a carpenter's square to find the center of the circle? **See margin.**

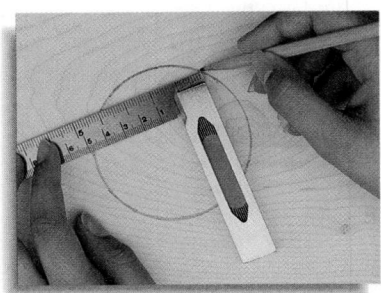

35. Yes; both pairs of opposite angles are right angles, which are supplementary angles.

36. Yes; both pairs of opposite angles of an isosceles trapezoid are supplementary.

37. No; if a rhombus is not a square, then the opposite angles are not supplementary.

38. Yes; both pairs of opposite angles are right angles, which are supplementary angles.

Inscribed Quadrilaterals Find the values of x and y.

32.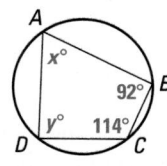

$x = 66; y = 88$

33.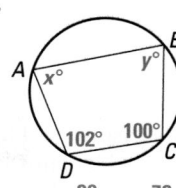

$x = 80; y = 78$

34.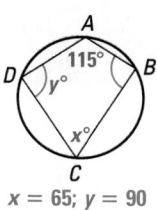

$x = 65; y = 90$

You be the Judge Can the quadrilateral always be inscribed in a circle? Explain your answer.

35. square

36. isosceles trapezoid

37. rhombus

38. rectangle

Standardized Test Practice

39. Multiple Choice In the diagram at the right, if $\angle ACB$ is a central angle and $m\angle ACB = 80°$, what is $m\angle ADB$? **B**

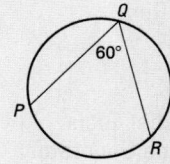

Ⓐ 20°   Ⓑ 40°

Ⓒ 80°   Ⓓ 160°

40. Multiple Choice In the diagram at the right, what are the values of x and y? **H**

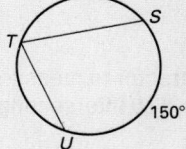

Ⓕ $x = 80, y = 95$ Ⓖ $x = 85, y = 100$

Ⓗ $x = 95, y = 80$ Ⓙ $x = 95, y = 85$

Mixed Review

Multiplying Radicals Multiply the radicals. Then simplify if possible. *(Lesson 10.1)*

41. $\sqrt{5} \cdot \sqrt{7}$ $\sqrt{35}$

42. $\sqrt{2} \cdot \sqrt{2}$ 2

43. $\sqrt{6} \cdot \sqrt{14}$ $2\sqrt{21}$

44. $(8\sqrt{2})^2$ 128

45. $(3\sqrt{3})^2$ 27

46. $2\sqrt{5} \cdot \sqrt{10}$ $10\sqrt{2}$

Solving Right Triangles Solve the right triangle. Round decimals to the nearest tenth. *(Lesson 10.6)*

47. $m\angle B = 46°$; $AC \approx 8.3$; $AB \approx 11.5$
48. $m\angle J = 55°$; $JK \approx 6.3$; $KL \approx 9.0$
49. $m\angle R = 40°$; $RP \approx 6.0$; $QR \approx 7.8$

47.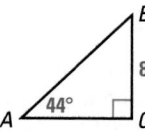

48.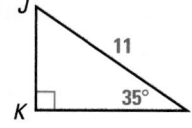

49.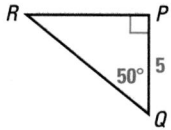

Algebra Skills

Evaluating Expressions Evaluate the expression when $x = 2$. *(Skills Review, p. 670)*

50. $3x + 5$ 11

51. $8x - 7$ 9

52. $x^2 + 9$ 13

53. $(x + 4)(x - 4)$ −12

54. $x^2 + 3x - 2$ 8

55. $x^3 + x^2$ 12

11.5 *Inscribed Angles and Polygons*   **619**

---

④ Assess

Assessment Resources

The Mini-Quiz below is also available on blackline (*Chapter 11 Resource Book*, p. 52) and on transparency. For more assessment resources, see:
• Chapter 11 Resource Book
• Standardized Test Practice
• Test and Practice Generator

Mini-Quiz

Find the measure of the intercepted arc or the inscribed angle.

1.

$m\overparen{PR} = 120°$

2.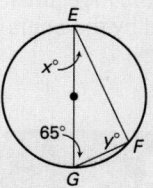

$m\angle STU = 75°$

Find the value of each variable.

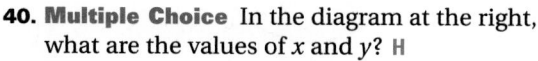

3. x 25

4. y 90

Pacing
Suggested Number of Days

Basic: 2 days
Average: 2 days
Advanced: 2 days
Block Schedule: 0.5 block with 11.5
0.5 block with 11.7

Teaching Resources

 **Blacklines**
(See page 586B.)

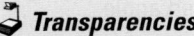 **Transparencies**
• Warm-Up with Quiz
• Answers

Technology
• Electronic Teacher Tools
• Test and Practice Generator
• Online Lesson Planner
• Internet Support

Geo-Activity

Goal Use a protractor to measure the angles formed by intersecting chords.

Key Discovery If two chords intersect in the interior of a circle, then the measure of each angle formed is one half the sum of the measures of the arcs intercepted by the angle and its vertical angle.

Goal
Use properties of chords in a circle.

Key Words
• chord p. 589

Step 2. $m\angle AEB = 50°$

Step 3. The measure of $\angle AEB$ is 50° for every student.

Step 4. Answers will vary, but the measure of $\angle AEB$ will always be $\frac{1}{2}(m\widehat{AB} + m\widehat{CD})$.

Step 5. The measure of an angle formed by intersecting chords of a circle is equal to half the sum of the measures of the intercepted arcs.

11.6 Properties of Chords

In Lessons 11.3 and 11.5, you saw how to find the measure of an angle formed by chords that intersect *at the center* of a circle or *on* a circle. The Geo-Activity below explores the angles formed by chords that intersect *inside* a circle.

Geo-Activity ▶ Properties of Angles Formed By Chords

❶ Draw a circle with two central angles measuring 70° and 30°. Label as shown.

❷ Draw chords $\overline{AC}$ and $\overline{BD}$. Label the intersection E, as shown. Find $m\angle AEB$.

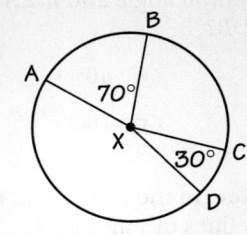

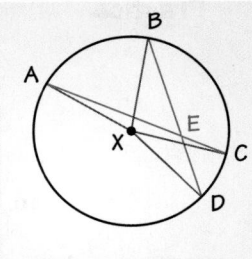

❸ Compare your angle measures with those of other students. What do you notice?

❹ Repeat Steps 1 and 2 for different central angles.

❺ What can you say about an angle formed by intersecting chords?

The result demonstrated by the Geo-Activity is summarized in the theorem below.

THEOREM 11.10

Words If two chords intersect inside a circle, then the measure of each angle formed is one half the *sum* of the measures of the arcs intercepted by the angle and its vertical angle.

Symbols $m\angle 1 = \frac{1}{2}(m\widehat{CD} + m\widehat{AB})$,

$m\angle 2 = \frac{1}{2}(m\widehat{BC} + m\widehat{AD})$

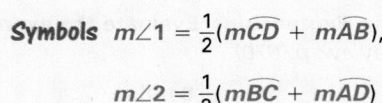

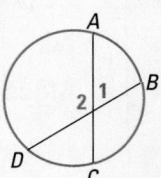

EXAMPLE 1 Find the Measure of an Angle

Find the value of x.

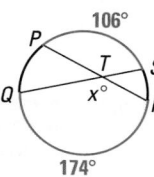

106°

174°

Solution

$x° = \frac{1}{2}(m\widehat{PS} + m\widehat{RQ})$ Use Theorem 11.10.

$x° = \frac{1}{2}(106° + 174°)$ Substitute 106° for $m\widehat{PS}$ and 174° for $m\widehat{RQ}$.

$x = \frac{1}{2}(280)$ Add.

$x = 140$ Multiply.

Student Help
CLASSZONE.COM

MORE EXAMPLES
More examples at classzone.com

EXAMPLE 2 Find the Measure of an Arc

Find the value of x.

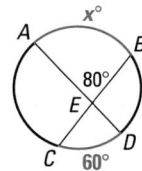

$x°$

80°

60°

Solution

$80° = \frac{1}{2}(m\widehat{AB} + m\widehat{CD})$ Use Theorem 11.10.

$80° = \frac{1}{2}(x° + 60°)$ Substitute $x°$ for $m\widehat{AB}$ and 60° for $m\widehat{CD}$.

$80 = \frac{1}{2}x + 30$ Use the distributive property.

$50 = \frac{1}{2}x$ Subtract 30 from each side.

$100 = x$ Multiply each side by 2.

Checkpoint ✓ Find the Measure of an Angle and an Arc

Find the value of x.

1. 130

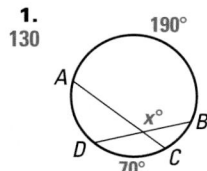

190°

70°

2. 68

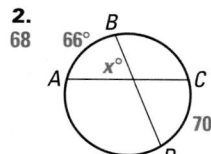

66°

70°

3. 45

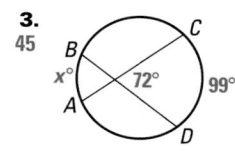

72°

99°

11.6 *Properties of Chords* **621**

Tips for New Teachers

The Geo-Activity on page 620 leads to the statement of Theorem 11.10 about the measures of the angles formed by intersecting chords. Have students recall information about vertical angles prior to discussing the theorem. See the Tips for New Teachers on pp. 1–2 of the *Chapter 11 Resource Book* for additional notes about Lesson 11.6.

Extra Example 1

Find the value of x. 115

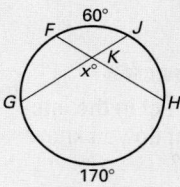

60°

170°

Extra Example 2

Find the value of x. 98

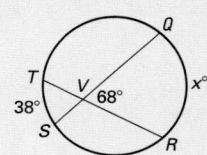

38°

68°

621

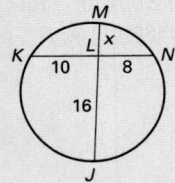
Intersecting Chords When two chords intersect in a circle, four segments are formed. The following theorem shows the relationship among these segments.

THEOREM 11.11

Words If two chords intersect inside a circle, then the product of the lengths of the segments of one chord is equal to the product of the lengths of the segments of the other chord.

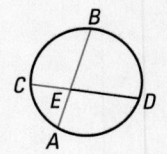

Symbols $EA \cdot EB = EC \cdot ED$

EXAMPLE 3 Find Segment Lengths

Find the value of x.

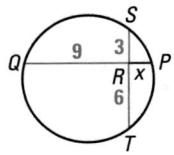

Solution

Notice that $\overline{ST}$ and $\overline{QP}$ are chords that intersect at R.

$RS \cdot RT = RQ \cdot RP$	Use Theorem 11.11.
$3 \cdot 6 = 9 \cdot x$	Substitute 3 for RS, 6 for RT, 9 for RQ, and x for RP.
$18 = 9x$	Simplify.
$2 = x$	Divide each side by 9.

Checkpoint ✓ Find Segment Lengths

Find the value of x.

4.
6

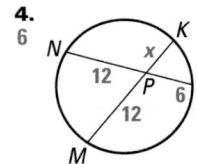

5.
4

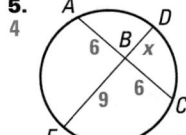

6.
3

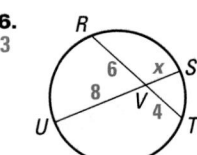

7.
5

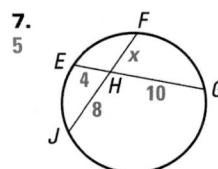

8.
5

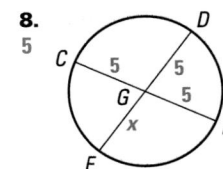

9.
32

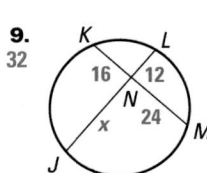

Guided Practice

Vocabulary Check

1. In the diagram, name the points inside the circle. *B and E*

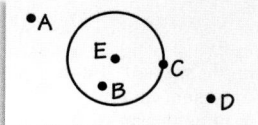

Skill Check

Find the measure of ∠1.

2. 60°
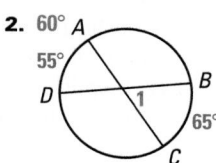

3. 88°
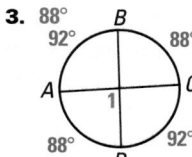

4. 139° 110°

Find the value of x.

5. 3
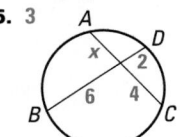

6. 16
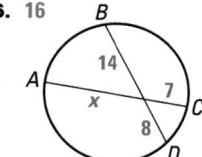

7. 12
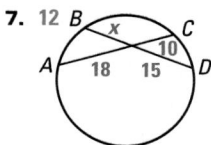

Practice and Applications

Extra Practice

See p. 696.

Matching Match each angle with the correct expression you can use to find its measure.

8. $m\angle 1$ C

9. $m\angle 2$ B

10. $m\angle 3$ D

11. $m\angle 4$ A

A. $\frac{1}{2}(m\widehat{BF} + m\widehat{DE})$

B. $\frac{1}{2}(m\widehat{AB} + m\widehat{CE})$

C. $\frac{1}{2}(m\widehat{AE} + m\widehat{BC})$

D. $\frac{1}{2}(m\widehat{BD} + m\widehat{FE})$

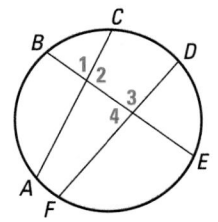

Finding Angle Measures Find the value of x.

12. 148
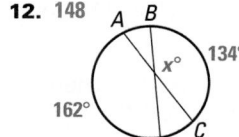

13. 50
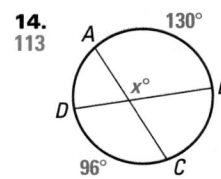

14. 130°
113

Homework Help

Example 1: Exs. 8–14
Example 2: Exs. 15–18,
Example 3: Exs. 19–26

11.6 *Properties of Chords*   **623**

③ Apply

Assignment Guide

BASIC
Day 1: pp. 623–625 Exs. 8–17, 31–38
Day 2: pp. 623–625 Exs. 20–25, 27–30, Quiz 2

AVERAGE
Day 1: pp. 623–625 Exs. 8–17, 31–38
Day 2: pp. 623–625 Exs. 19–30, Quiz 2

ADVANCED
Day 1: pp. 623–625 Exs. 12–18, 31–38
Day 2: pp. 623–625 Exs. 19–30, Quiz 2; EC: TE p. 586D*, classzone.com

BLOCK SCHEDULE
pp. 623–625 Exs. 8–17, 31–38 (with 11.5)
pp. 623–625 Exs. 19–30, Quiz 2 (with 11.7)

Extra Practice
• Student Edition, p. 696
• Chapter 11 Resource Book, pp. 56–57

Homework Check

To quickly check student understanding of key concepts, go over the following exercises:

Basic: 10, 12, 15, 20, 23
Average: 11, 13, 16, 21, 24
Advanced: 14, 17, 18, 22, 25

18. Yes; *Sample answer:* If two chords intersect and the measure of each angle formed is the same as the measure of the arc intercepted by the angle, then the circle is divided into two pairs of congruent arcs since vertical angles are congruent. Suppose one pair of arcs has measure $x°$ and an angle intercepting the arc is $\angle 1$.

Then $m\angle 1 = \frac{1}{2}(x° + x°) =$

$\frac{1}{2}(2x°) = x°$. The angle formed by the chords has the same measure as its intercepted arc, so it is a central angle.

Finding Arc Measures Find the value of x.

15. 21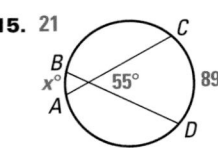

16. 48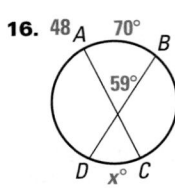

17. 186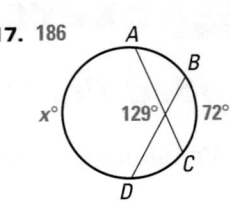

18. ⚖ **You be the Judge** A student claims if two chords intersect and the measure of each angle formed is the same as the measure of the arc intercepted by the angle, then each angle must be a central angle. Is he correct? Explain.

19. **Animation Design** You are designing an animated logo for a Web site. You want sparkles to leave point C and reach the circle at the same time. To find out how far each sparkle moves between frames, you need to know the distances from C to the circle. Three distances are shown below. Find CN. 18

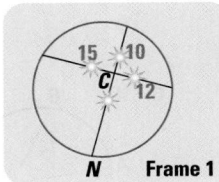
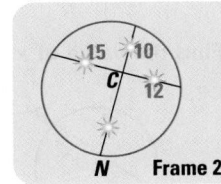
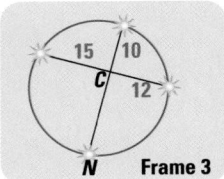

Chords in a Circle Find the value of x.

20. 20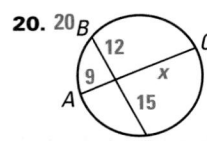

21. 12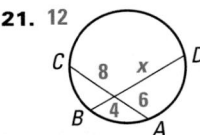

22. 7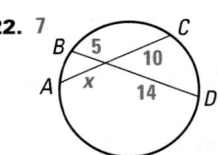

23. 103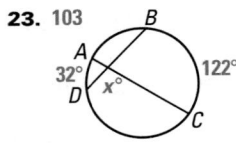

24. 66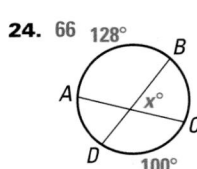

25. 60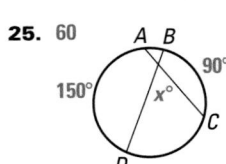

26. $EA \cdot EB = EC \cdot ED$; If two chords intersect inside a circle, then the product of the lengths of the segments of one chord is equal to the product of the lengths of the segments of the other chord.

26. **Technology** Use geometry drawing software.

❶ Draw a circle and label points A, B, C, and D as shown.

❷ Draw lines $\overleftrightarrow{AB}$ and $\overleftrightarrow{CD}$. Label the point of intersection E.

❸ Measure EA, EB, EC, and ED. Then calculate $EA \cdot EB$ and $EC \cdot ED$.

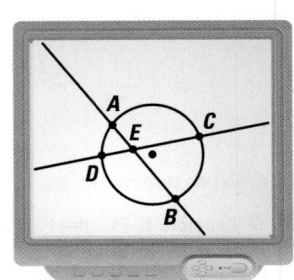

What do you notice? What theorem does this demonstrate?

 Chapter 11 Circles

d. Yes. Explanations may vary.
   Sample answer:
   $\angle ACB \cong \angle ECD$ (vertical
   angles); $\frac{AC}{EC} = \frac{12}{9} = \frac{4}{3} = \frac{BC}{CD}$;
   by the SAS Similarity Theorem,
   $\triangle ACB \sim \triangle ECD$.

27. Multi-Step Problem In the diagram, $AC = 12$, $CD = 3$, and
$EC = 9$.

a. Find BC. 4

b. What is the measure of $\angle ACB$? 90°

c. What is the measure of $\widehat{AE}$? 144°

d. Is $\triangle ACB$ similar to $\triangle ECD$? Explain
   your reasoning.

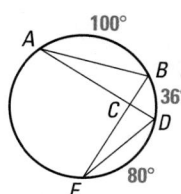

Mixed Review

Finding Side Lengths Find the unknown side length. Round your
answer to the nearest tenth if necessary. *(Lesson 4.4)*

28.
9.5
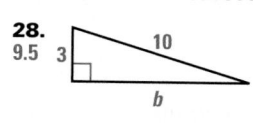

29.
34
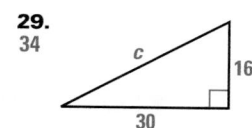

30.
13.7
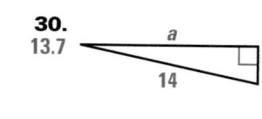

Algebra Skills

Absolute Values Evaluate. *(Skills Review, p. 662)*

31. $|-3|$ 3 **32.** $|1|$ 1 **33.** $|-19|$ 19 **34.** $|50|$ 50

35. $|2.7|$ 2.7 **36.** $|-8|$ 8 **37.** $|-10.01|$ 10.01 **38.** $|-100|$ 100

Quiz 2

Find the value of x in $\odot C$. *(Lesson 11.4)*

1.
5
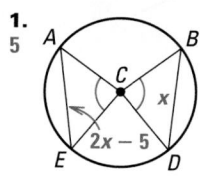

2.
3
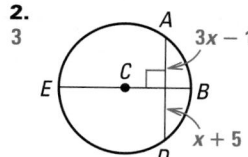

3.
31
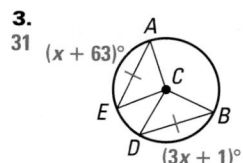

Find the value of each variable. Explain your reasoning. *(Lesson 11.5)*

4. 102; the measure of an
   inscribed angle is half the
   measure of its intercepted
   arc, so $51 = \frac{1}{2}x$; $x = 102$.

5. $x = 58$ and $y = 41$; $x° =$
   $m\angle B = \frac{1}{2}m\widehat{AD} = m\angle C = 58°$;
   $y° = m\angle D = \frac{1}{2}m\widehat{BC} =$
   $m\angle A = 41°$.

6. $x = 75$ and $y = 82$; the
   opposite angles of an
   inscribed quadrilateral
   are supplementary, so
   $x° = 180° - 105° = 75°$ and
   $y° = 180° - 98° = 82°$.

4.
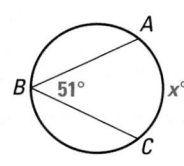

5.
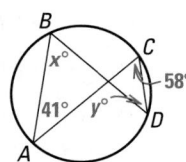

6.
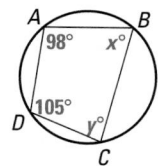

Find the value of x. *(Lesson 11.6)*

7.
62
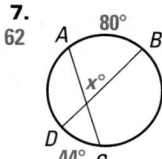

8.
163
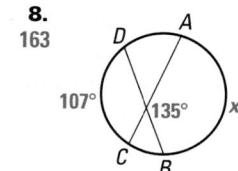

9.
6
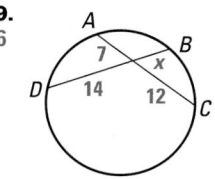

11.6 *Properties of Chords* **625**

Assessment Resources
The Mini-Quiz below is also avail-
able on blackline (*Chapter 11
Resource Book*, p. 64) and on trans-
parency. For more assessment
resources, see:
• Chapter 11 Resource Book
• Standardized Test Practice
• Test and Practice Generator

④ Assess

Mini-Quiz
Find the value of x.

1.

145

2.
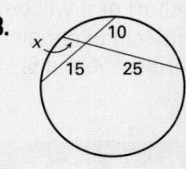
85

3.
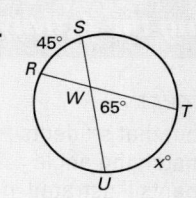
6

Planning the Activity

Goal

Students explore the relationship between the measures of segments of intersecting secants.

Materials

See the Technology Activity Keystrokes Masters in the *Chapter 11 Resource Book.*

LINK TO LESSON

In Lesson 11.6 students learned about the relationships between segments that intersect inside of circles. This activity presents relationships for segments that pass through circles and intersect outside of the circles.

Tips for Success

In Step 1, point out that students do not need to match the angle between the secants illustrated in the figure. Also, the segments should not be too short or it will be difficult for students to make their conjectures in Exercises 4 and 5.

KEY DISCOVERY

When two secants intersect outside a circle, the product of the length of the segment from the point of intersection to the near side of the circle and the length of the segment from the point of intersection to the far side of the circle is equal to a similar product for the other secant. Also, the measure of the angle formed by the secants is half the positive difference of the measures of the intercepted arcs.

Activity Assessment

Use Exercises 4 and 5 to assess student understanding.

626

Technology Activity 11.6 Intersecting Secants

Question

What is the relationship between segments of secants that intersect?

Explore

1 Construct a circle. Then construct two secants to the circle that intersect outside the circle.

2 Label the intersections as shown.

3 Measure *JH, JG, JK,* and *JL.*

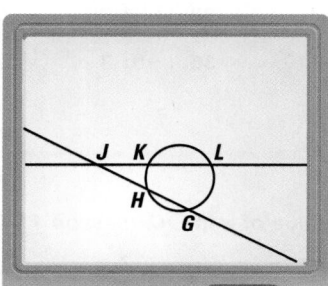

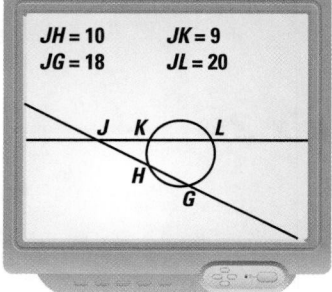

$JH = 10 \quad JK = 9$
$JG = 18 \quad JL = 20$

4. If $\overleftrightarrow{JL}$ and $\overleftrightarrow{JG}$ are secants of a circle intersecting at a point *J* outside the circle and points *K, L, G,* and *H* are points on the circle as shown in the diagram on page 626, then $JH \cdot JG = JK \cdot JL$.

5. Measures will vary, but
$m\angle KJH = \frac{1}{2}(m\widehat{LG} - m\widehat{KH})$
in each case. If 2 secants intersect outside a circle, then the measure of the angle formed is half the positive difference of the measures of the intercepted arcs.

Think About It

1. Calculate $JH \cdot JG$ and $JK \cdot JL$. What do you notice?
   Product will vary, but in each case, $JH \cdot JG = JK \cdot JL$.
2. Drag point *J.* What do you notice?
   The relationship $JH \cdot JG = JK \cdot JL$ is true.
3. Draw two more circles and repeat Steps 1 through 3. Are your results different? no

4. Make a conjecture about the segments formed by the intersection of two secants.

5. **Extension** Find the measures of $\angle KJH$, $\widehat{KH}$, and $\widehat{LG}$. Compare the sum of $m\widehat{KH}$ and $m\widehat{LG}$ with $m\angle KJH$. What do you notice? Make a conjecture about the angle formed by the intersection of two secants and the corresponding intercepted arcs.

11.7 Equations of Circles

Goal
Write and graph the equation of a circle.

Key Words
- standard equation of a circle

In the circle below, let point (x, y) represent any point on the circle whose center is at the origin. Let r represent the radius of the circle.

In the right triangle,

r = length of hypotenuse,

x = length of a leg,

y = length of a leg.

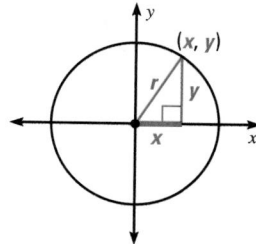

By the Pythagorean Theorem, you can write

$$x^2 + y^2 = r^2.$$

This is an equation of a circle with center at the origin.

EXAMPLE 1 **Write an Equation of a Circle**

Write an equation of the circle.

Solution

The radius is 4 and the center is at the origin.

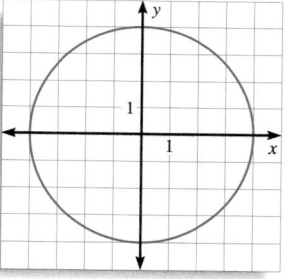

$x^2 + y^2 = r^2$ Write an equation of a circle with center at the origin.

$x^2 + y^2 = 4^2$ Substitute 4 for r.

$x^2 + y^2 = 16$ Simplify.

ANSWER ▶ An equation of the circle is $x^2 + y^2 = 16$.

Checkpoint ✓ **Write an Equation of a Circle**
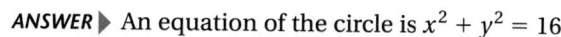

Write an equation of the circle.

1.

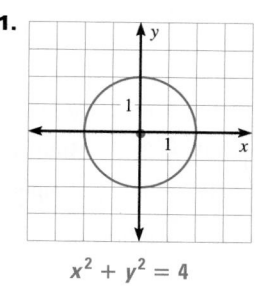

$x^2 + y^2 = 4$

2.

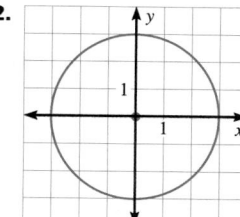

$x^2 + y^2 = 9$

11.7 *Equations of Circles*   **627**

Pacing
Suggested Number of Days

Basic: 1 day
Average: 1 day
Advanced: 1 day
Block Schedule: 0.5 block with 11.6

Teaching Resources

📄 **Blacklines**
(See page 586B.)

✍ **Transparencies**
- Warm-Up with Quiz
- Answers

⊞ **Technology**
- Electronic Teacher Tools
- Test and Practice Generator
- Online Lesson Planner
- Internet Support

2 Teach

Content and Teaching Strategies
For background information on geometric concepts and teaching strategies related to this lesson, see pages 586E and 586F in this Teacher's Edition.

Tips for New Teachers
Show students how to sketch a circle when they know the center and the radius. The technique of marking a few points in the vertical and horizontal directions from the center will work nicely whenever the students do not have a compass to use. This technique is very easy to use when a coordinate plane is involved. Sketch the circle lightly at first to keep it rounded. See the Tips for New Teachers on pp. 1–2 of the *Chapter 11 Resource Book* for additional notes about Lesson 11.7.

Extra Example 1
See next page.

Write an equation of the circle.

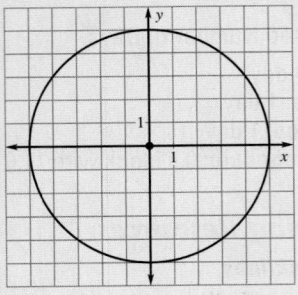

$$x^2 + y^2 = 25$$

Extra Example 2

Write the standard equation of the circle with center (3, −2) and radius 5. $(x − 3)^2 + (y + 2)^2 = 25$

Geometric Reasoning

As you discuss the standard equation of a circle, point out that when the circle is centered at the origin, h and k are both 0 and therefore the equation can be written without parentheses.

Student Help

SKILLS REVIEW
To review absolute value, see p. 662.

Standard Equation of a Circle If the center of a circle is not at the origin, you can use the Distance Formula to write an equation of the circle.

For example, the circle shown at the right has center (3, 5) and radius 4.

Let (x, y) represent any point on the circle. Use the Distance Formula to find the lengths of the legs.

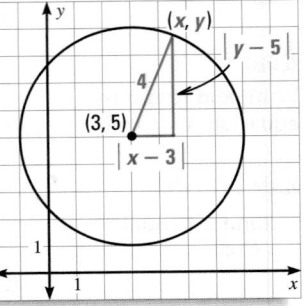

leg: $|x − 3|$

leg: $|y − 5|$

hypotenuse: 4

Use these expressions in the Pythagorean Theorem to find an equation of the circle.

$$(x − 3)^2 + (y − 5)^2 = 4^2$$

This is an example of the **standard equation of a circle**.

STANDARD EQUATION OF A CIRCLE

In the coordinate plane, the standard equation of a circle with center at (h, k) and radius r is

$$(x − h)^2 + (y − k)^2 = r^2.$$

x-coordinate of the center

y-coordinate of the center

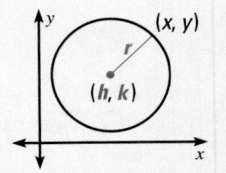

EXAMPLE 2 Write the Standard Equation of a Circle

Write the standard equation of the circle with center (2, −1) and radius 3.

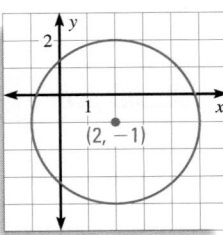

Solution

$(x − h)^2 + (y − k)^2 = r^2$	Write the standard equation of a circle.
$(x − 2)^2 + (y − (−1))^2 = 3^2$	Substitute 2 for h, −1 for k, and 3 for r.
$(x − 2)^2 + (y + 1)^2 = 9$	Simplify.

ANSWER ▶ The standard equation of the circle is $(x − 2)^2 + (y + 1)^2 = 9$.

EXAMPLE 3 Graph a Circle

Graph the given equation of the circle.

a. $(x - 1)^2 + (y - 2)^2 = 4$

b. $(x + 2)^2 + y^2 = 4$

Solution

a. Rewrite the equation of the circle as $(x - 1)^2 + (y - 2)^2 = 2^2$. The center is (1, 2) and the radius is 2.

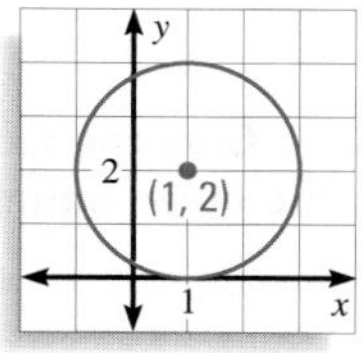

b. Rewrite the equation of the circle as $(x - (-2))^2 + (y - 0)^2 = 2^2$. The center is $(-2, 0)$ and the radius is 2.

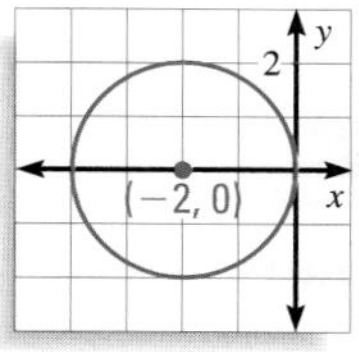

Checkpoint Circles Not Centered at the Origin

3. Write the standard equation of the circle with center (–4, –6) and radius 5. $(x + 4)^2 + (y + 6)^2 = 25$

Graph the given equation of the circle. 4, 5. See margin.

4. $(x - 1)^2 + y^2 = 25$

5. $(x + 2)^2 + (y - 4)^2 = 16$

11.7 Exercises

Guided Practice

Vocabulary Check

1. Which of the following is a *standard equation of a circle*? C

A. $(x+ 2)^2 = 16y$

B. $(x^2 - 5) + (y^2 - 8) = 16$

C. $(x - 4)^2 + (y - 3)^2 = 16$

D. $2x^2 + 3y - 5 = 16$

Skill Check

Give the radius and the coordinates of the center. Write the equation of the circle in standard form.

2.

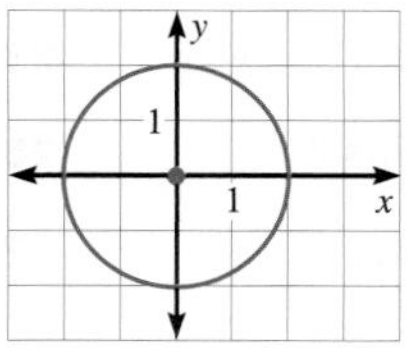

3.

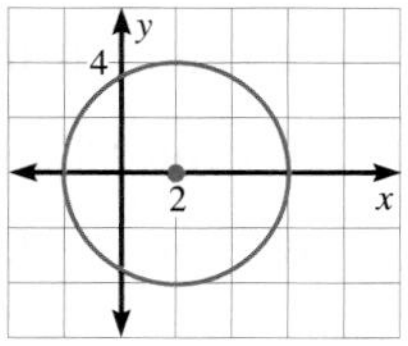

4.

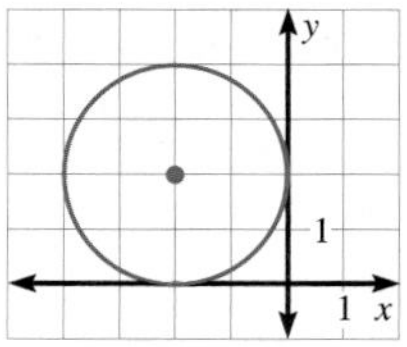

2. radius, 2; center, (0, 0); equation, $x^2 + y^2 = 4$

3. radius, 4; center, (2, 0); equation, $(x - 2)^2 + y^2 = 16$

4. radius, 2; center, $(-2, 2)$; equation, $(x + 2)^2 + (y - 2)^2 = 4$

Extra Example 3

Graph the given equation of the circle.

a. $(x - 2)^2 + (y + 3)^2 = 16$

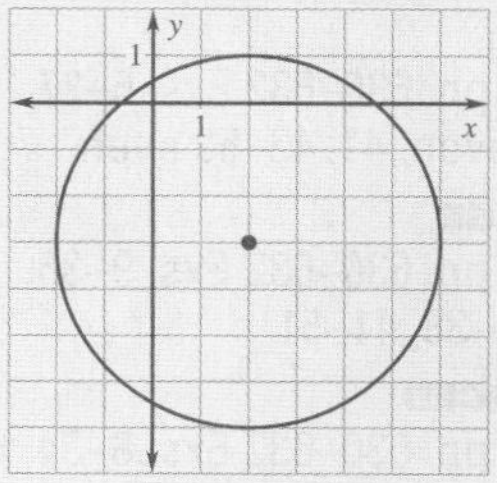

b. $x^2 + (y - 1)^2 = 9$

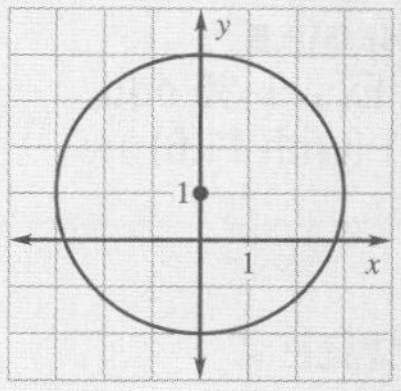

✓ Concept Check

Describe the graph of the equation $(x + 5)^2 + (y - 3)^2 = 4$. a circle of radius 2 centered at $(-5, 3)$

Daily Puzzler

Describe the circles that have the equation $(x - a)^2 + (y - a)^2 = a^2$, where a is a positive integer. The circles are in the first quadrant, tangent to both axes, with a center at (a, a), such as (1, 1), (2, 2), and (3, 3).

4.

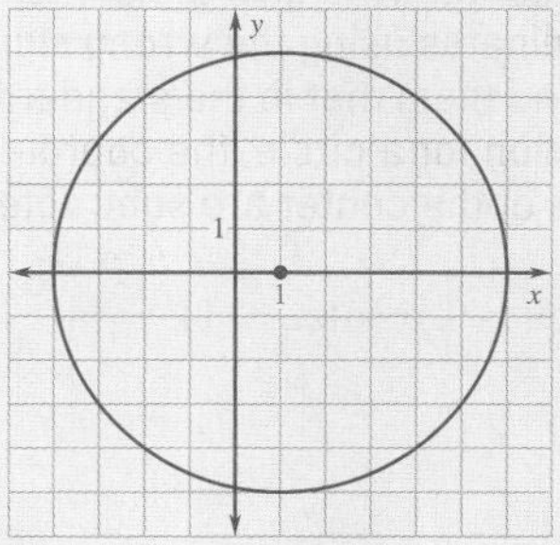

5.

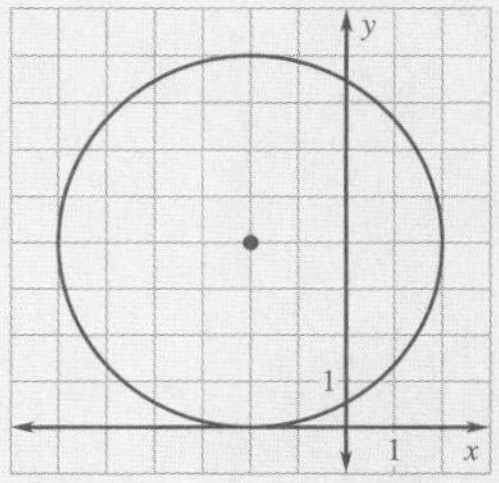

Practice and Applications

Extra Practice

• Student Edition, p. 696
• Chapter 11 Resource Book, pp. 67–68

Homework Check

To quickly check student understanding of key concepts, go over the following exercises:

Basic: 8, 12, 16, 22, 24
Average: 9, 13, 17, 23, 29
Advanced: 8, 14, 20, 26, 34

✗ Common Error

Throughout the exercises, watch for students who give the center coordinates using the wrong signs. Remind them that in the standard equation for a circle, the coordinates of the center are subtracted.

8. radius, 6; center, (0, 0);

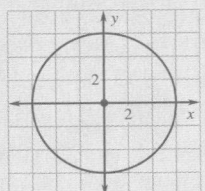

9. radius, 1; center, (0, 0);

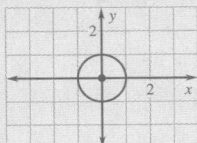

Extra Practice

See p. 696.

Matching Equations Match each graph with its equation.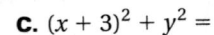

A. $x^2 + y^2 = 4$ **B.** $(x - 3)^2 + y^2 = 4$ **C.** $(x + 3)^2 + y^2 = 4$

5. B 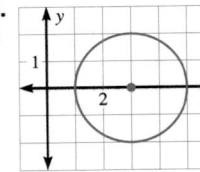   **6.** A 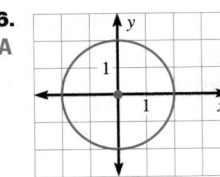   **7.** C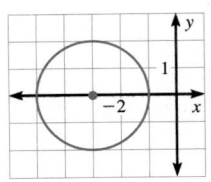

Using Standard Equations Give the radius and the coordinates of the center of the circle with the given equation. Then graph the circle.

8. $x^2 + y^2 = 36$ 8–15. See margin.   9. $x^2 + y^2 = 1$

10. $(x - 2)^2 + (y - 6)^2 = 49$ 11. $(x - 4)^2 + (y - 3)^2 = 16$

12. $(x - 5)^2 + (y - 1)^2 = 25$ 13. $(x + 2)^2 + (y - 3)^2 = 36$

14. $(x - 2)^2 + (y + 5)^2 = 4$ 15. $x^2 + (y - 5)^2 = 64$

Using Graphs Give the radius and the coordinates of the center of the circle. Then write the standard equation of the circle.

16. radius, 2; center, (−3, 2); equation, $(x + 3)^2 + (y - 2)^2 = 4$

17. radius, 2; center, (0, 1); equation, $x^2 + (y - 1)^2 = 4$

18. radius, 1; center, (3, 3); equation, $(x - 3)^2 + (y - 3)^2 = 1$

19. radius, 2.5; center, (0.5, 1.5); equation, $(x - 0.5)^2 + (y - 1.5)^2 = 6.25$

20. radius, 4; center, (2, 2); equation, $(x - 2)^2 + (y - 2)^2 = 16$

21. radius, 6; center, (0, 0); equation, $x^2 + y^2 = 36$

16. 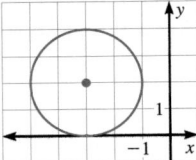   **17.** 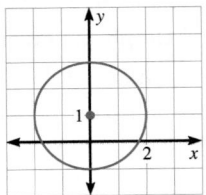   **18.**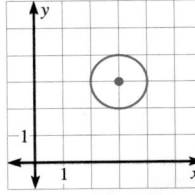

19. 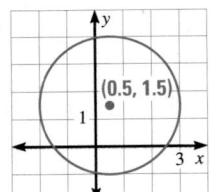   **20.** 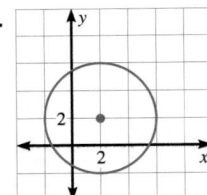   **21.**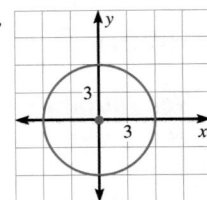

Writing Equations Write the standard equation of the circle with the given center and radius.

22. center (0, 0), radius 10
    $x^2 + y^2 = 100$

23. center (4, 0), radius 4
    $(x - 4)^2 + y^2 = 16$

24. center (3, −2), radius 2
    $(x - 3)^2 + (y + 2)^2 = 4$

25. center (−1, −3), radius 6
    $(x + 1)^2 + (y + 3)^2 = 36$

26. center (−3, 5), radius 3
    $(x + 3)^2 + (y - 5)^2 = 9$

27. center (1, 0), radius 7
    $(x - 1)^2 + y^2 = 49$

EXAMPLE Use the Equation of a Circle

The equation of a circle is $(x - 5)^2 + (y - 1)^2 = 9$. Without sketching the circle, tell whether the point is *on* the circle, *inside* the circle, or *outside* the circle.

a. (6, 0)　　　　　　　　　　**b.** (8, 2)

Solution

Substitute the coordinates of the point into the equation.

If the left side is *less than* the right side, the point is *inside* the circle.

If the left side is *greater than* the right side, the point is *outside* the circle.

Student Help

STUDY TIP
If the left side is *equal* to the right side, the point is *on* the circle.

a.
$$(x - 5)^2 + (y - 1)^2 = 9$$
$$(6 - 5)^2 + (0 - 1)^2 \overset{?}{=} 9$$
$$1^2 + (-1)^2 \overset{?}{=} 9$$
$$2 < 9$$

Because $2 < 9$, the point (6, 0) is *inside* the circle.

b.
$$(x - 5)^2 + (y - 1)^2 = 9$$
$$(8 - 5)^2 + (2 - 1)^2 \overset{?}{=} 9$$
$$3^2 + 1^2 \overset{?}{=} 9$$
$$10 > 9$$

Because $10 > 9$, the point (8, 2) is *outside* the circle.

Equation of a Circle The equation of a circle is $(x - 2)^2 + (y + 3)^2 = 4$. Tell whether the point is *on* the circle, *inside* the circle, or *outside* the circle. Use the example above as a model.

28. $R(0, 0)$ outside　　**29.** $A(2, -4)$ inside　　**30.** $X(0, -3)$ on　　**31.** $K(3, -1)$ outside

32. $M(1, -4)$ inside　　**33.** $T(2, -5)$ on　　**34.** $D(2, 0)$ outside　　**35.** $Z(2.5, -3)$ inside

Link to
Communications

Cell Phones In Exercises 36 and 37, use the following information.
A cellular phone network uses towers to transmit calls. Each tower transmits to a circular area. On a grid of a town, the coordinates of the towers and the circular areas covered by the towers are shown.

36. Write the equations that represent the transmission boundaries of the towers.

37. Tell which towers, if any, transmit to phones located at $J(1, 1)$, $K(4, 2)$, $L(3.5, 4.5)$, $M(2, 2.8)$, and $N(1, 6)$.

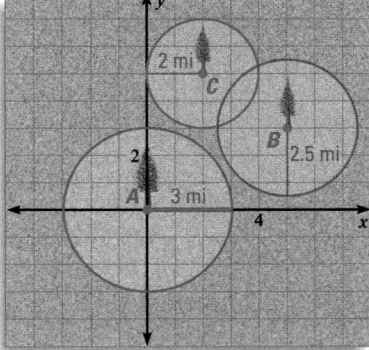

36. circle A: $x^2 + y^2 = 9$;
circle B: $(x - 5)^2 + (y - 3)^2 = 6.25$;
circle C: $(x - 2)^2 + (y - 5)^2 = 4$

37. Tower A transmits to J; tower B transmits to K; towers B and C transmit to L; no tower transmits to M; tower C transmits to N.

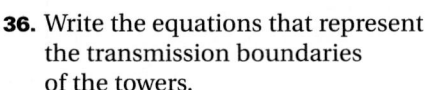

CELL PHONE towers are sometimes built to look like trees so that they blend in with their environment. Other cell phone towers have also been built to resemble farm silos and cactus plants.

10. radius, 7; center, (2, 6);
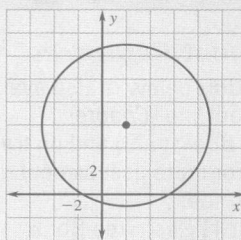

11. radius, 4; center, (4, 3);
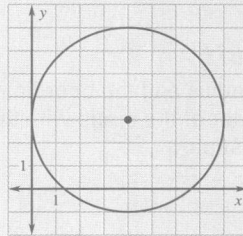

12. radius, 5; center, (5, 1);
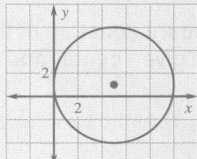

13. radius, 6; center, (-2, 3);
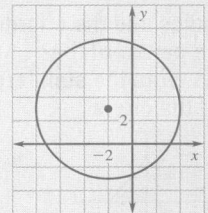

14. radius, 2; center, (2, -5);
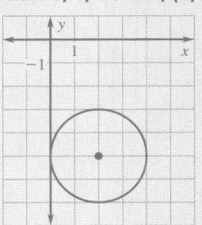

15. radius, 8; center, (0, 5);
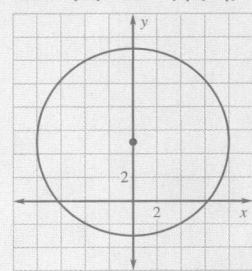

Assessment Resources

The Mini-Quiz below is also available on blackline (*Chapter 11 Resource Book*, p. 75) and on transparency. For more assessment resources, see:
• Chapter 11 Resource Book
• Standardized Test Practice
• Test and Practice Generator

Mini-Quiz

Give the radius and the coordinates of the center of the circle with the given equation.

1. $(x - 4)^2 + y^2 = 144$
   center (4, 0), radius 12

2. $(x + 7)^2 + (y - 3)^2 = 100$
   center (−7, 3), radius 10

Write the standard equation of the circle with the given center and radius.

3. center $(-1, 1)$, radius 2
   $(x + 1)^2 + (y - 1)^2 = 4$

4. center (0, 5), radius 8
   $x^2 + (y - 5)^2 = 64$

5. Give the radius and the coordinates of the center of the circle. Then write the standard equation of the circle.

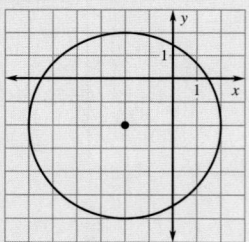

radius: 4; center: (−2, −2);
equation: $(x + 2)^2 + (y + 2)^2 = 16$

---

38. *Sample answer:* The student did not subtract the coordinates of the center from *x* and *y* in the equation and did not square the radius. The equation should be $(x + 1)^2 + (y - 2)^2 = 4$.

Standardized Test Practice

43. $P'(3, 3)$; $Q'(8, 1)$; $R'(6, -2)$; $S'(1, 0)$

44. $P'(-3, 4)$; $Q'(2, 2)$; $R'(0, -1)$; $S'(-5, 1)$

45. $P'(0, 2)$; $Q'(5, 0)$; $R'(3, -3)$; $S'(-2, -1)$

46. $P'(4, 9)$; $Q'(9, 7)$; $R'(7, 4)$; $S'(2, 6)$

Algebra Skills

---

38. **Error Analysis** A student was asked to write the standard equation of the circle below. Why is the equation incorrect?

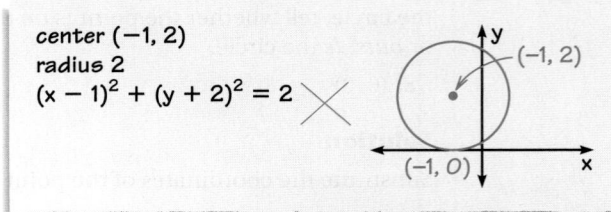

Challenge Use the given information to write the standard equation of the circle.

39. The center is (1, 2). A point on the circle is (4, 6). $(x - 1)^2 + (y - 2)^2 = 25$

40. The center is (3, 2). A point on the circle is (5, 2). $(x - 3)^2 + (y - 2)^2 = 4$

41. **Multiple Choice** What is the standard form of the equation of a circle with center $(-3, 1)$ and radius 2?   **D**

   (A) $(x - 3)^2 + (y - 1)^2 = 2$ (B) $(x + 3)^2 + (y - 1)^2 = 2$

   (C) $(x - 3)^2 + (y - 1)^2 = 4$ (D) $(x + 3)^2 + (y - 1)^2 = 4$

42. **Multiple Choice** The center of a circle is $(-3, 0)$ and its radius is 5. Which point does *not* lie on the circle?   **H**

   (F) (2, 0)   (G) (0, 4)   (H) $(-3, 0)$ (J) $(-3, -5)$

Finding an Image Find the coordinates of *P'*, *Q'*, *R'*, and *S'*, using the given translation. *(Lesson 3.7)*

43. $(x, y) \rightarrow (x + 2, y)$

44. $(x, y) \rightarrow (x - 4, y + 1)$

45. $(x, y) \rightarrow (x - 1, y - 1)$

46. $(x, y) \rightarrow (x + 3, y + 6)$

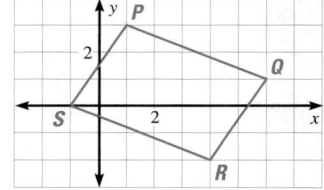

Identifying Dilations Tell whether the dilation is a *reduction* or an *enlargement*. Then find its scale factor. *(Lesson 7.6)*

47.

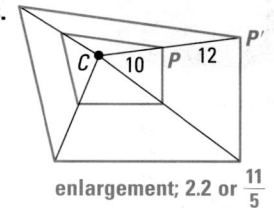

reduction; $\frac{3}{8}$

48.

enlargement; 2.2 or $\frac{11}{5}$

Solving Equations Solve the equation. *(Skills Review, p. 673)*

49. $14 = -3x - 7$ -7 50. $11 - x = -2$ 13   51. $20 = 5x - 12 - x$ 8

Goal

Identify rotations and rotational symmetry.

Key Words

- rotation
- center of rotation
- angle of rotation
- rotational symmetry

Step 4. 120°; yes, you can turn the paper clockwise or counterclockwise.

Step 5. rectangle: 180° clockwise or counterclockwise; square: 90° or 180° clockwise or counterclockwise

Geo-Activity · Rotating a Figure

1 Draw an equilateral triangle. Label as shown. Draw a line from the center to one of the vertices.

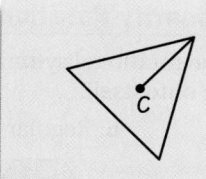

2 Copy the triangle onto a piece of tracing paper.

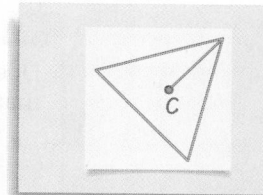

3 Place a pencil on the center point and turn the tracing paper over the original triangle until it matches up with itself.

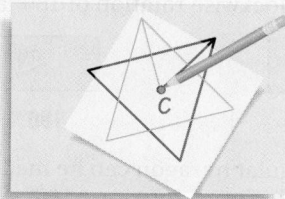

4 How many degrees did you turn the triangle? Is there more than one way to turn the triangle so that it matches up with itself?

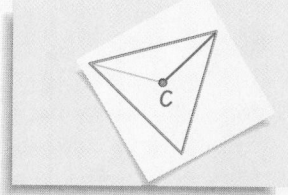

5 Draw a rectangle and a square. Repeat Steps 1 through 4. How many degrees did you turn each figure until it matched up with itself?

Visualize It!

Clockwise means to go in the direction of the hands on a clock.

Counterclockwise means to go in the opposite direction.

A **rotation** is a transformation in which a figure is turned about a fixed point. The fixed point is the **center of rotation**. In the Geo-Activity above, point *C* is the center of rotation. Rays drawn from the center of rotation to a point and its image form an angle called the **angle of rotation**. Rotations can be clockwise or counterclockwise.

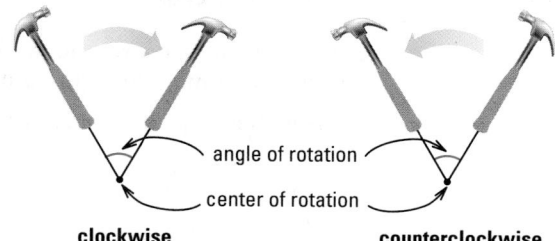

angle of rotation

center of rotation

clockwise　　　**counterclockwise**

11.8 *Rotations*　**633**

---

① Plan

Pacing

Suggested Number of Days

Basic: 2 days
Average: 2 days
Advanced: 2 days
Block Schedule: 1 block

Teaching Resources

📄 **Blacklines**
(See page 586B.)

🖇 **Transparencies**
- Warm-Up with Quiz
- Answers

🖥 **Technology**
- Electronic Teacher Tools
- Test and Practice Generator
- Online Lesson Planner
- Internet Support

Geo-Activity

Goal Use a protractor and compass to draw the rotation of a figure.

Key Discovery Certain figures, such as equilateral triangles, rectangles, and squares, can be rotated clockwise or counterclockwise 180° or less and match up with themselves.

② Teach

Content and Teaching Strategies

For background information on geometric concepts and teaching strategies related to this lesson, see pages 586E and 586F in this Teacher's Edition.

Rotational Symmetry A figure in a plane has **rotational symmetry** if the figure can be mapped onto itself by a rotation of 180° or less. For instance, the figure below has rotational symmetry because it maps onto itself by a rotation of 90°.

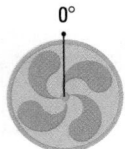 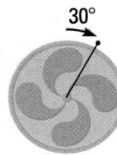 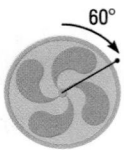 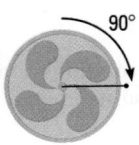

EXAMPLE 1 Identify Rotational Symmetry

Does the figure have rotational symmetry? If so, describe the rotations that map the figure onto itself.

a. Rectangle   **b.** Regular hexagon   **c.** Trapezoid

 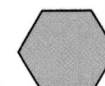 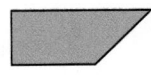

Solution

a. Yes. A rectangle can be mapped onto itself by a clockwise or counterclockwise rotation of 180° about its center.

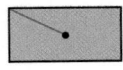 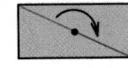

     0°   **180°**

b. Yes. A regular hexagon can be mapped onto itself by a clockwise or counterclockwise rotation of 60°, 120°, or 180° about its center.

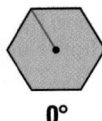 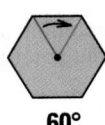 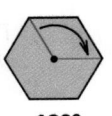 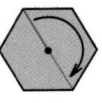

  0°   **60°**   **120°**   **180°**

c. No. A trapezoid does not have rotational symmetry.

 **Checkpoint ✓ Identify Rotational Symmetry**

2. yes; a clockwise or counterclockwise rotation of 180° about its center

3. yes; a clockwise or counterclockwise rotation of 45°, 90°, 135°, or 180° about its center

Does the figure have rotational symmetry? If so, describe the rotations that map the figure onto itself.

1. Isosceles trapezoid   **2.** Parallelogram   **3.** Regular octagon
  no

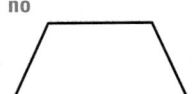  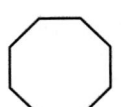

EXAMPLE 2 Rotations

Rotate △*FGH* 50° counterclockwise about point *C*.

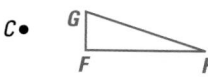

Solution

❶ To find the image of point *F*, draw $\overline{CF}$ and draw a 50° angle. Find *F'* so that *CF* = *CF'*.

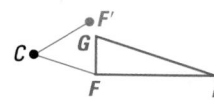

❷ To find the image of point *G*, draw $\overline{CG}$ and draw a 50° angle. Find *G'* so that *CG* = *CG'*.

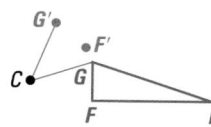

❸ To find the image of point *H*, draw $\overline{CH}$ and draw a 50° angle. Find *H'* so that *CH* = *CH'*. Draw △*F'G'H'*.

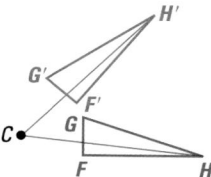

EXAMPLE 3 Rotations in a Coordinate Plane

Sketch the quadrilateral with vertices *A*(2, −2), *B*(4, 1), *C*(5, 1), and *D*(5, −1). Rotate it 90° counterclockwise about the origin and name the coordinates of the new vertices.

Solution

Plot the points, as shown in blue.

Use a protractor and a ruler to find the rotated vertices.

The coordinates of the vertices of the image are *A'*(2, 2), *B'*(−1, 4), *C'*(−1, 5), and *D'*(1, 5).

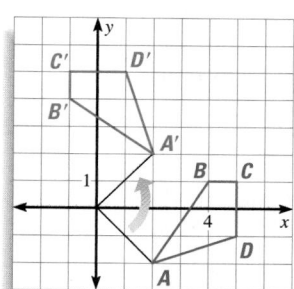

Checkpoint ✓ **Rotations in a Coordinate Plane**

4. Sketch the triangle with vertices *A*(0, 0), *B*(3, 0), and *C*(3, 4). Rotate △*ABC* 90° counterclockwise about the origin. Name the coordinates of the new vertices *A'*, *B'*, and *C'*. See margin.

11.8 *Rotations* **635**

Extra Example 2

Rotate △*ABC* 100° clockwise about point *X*.

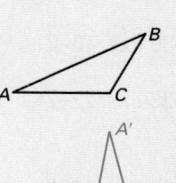

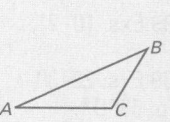

Extra Example 3

A quadrilateral has vertices at *F*(−5, −2), *G*(−1, −1), *H*(−2, −4), and *J*(−4, −5). Rotate *FGHJ* 90° clockwise about the origin and name the coordinates of the new vertices. *F'*(−2, 5), *G'*(−1, 1), *H'*(−4, 2), *J'*(−5, 4)

✓ **Concept Check**

How can you tell if a figure has rotational symmetry? It can be mapped onto itself by a rotation, either clockwise or counterclockwise, of 180° or less.

🧩 **Daily Puzzler**

A rectangle 4 units by 6 units is rotated 90° about its center. What fraction of the rectangle's original area is covered by the image of the rotated rectangle? $\frac{2}{3}$

4.

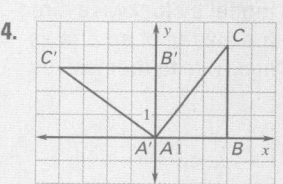

A'(0, 0), *B'*(0, 3), *C'*(−4, 3)

Extra Practice
• Student Edition, p. 696
• Chapter 11 Resource Book, pp. 78–79

Homework Check
To quickly check student understanding of key concepts, go over the following exercises:

Basic: 10, 13, 16, 22, 27
Average: 11, 14, 20, 24, 29
Advanced: 12, 15, 21, 26, 30

✗ Common Error
Throughout the exercises, watch for students who confuse clockwise and counterclockwise. Have them stand and model a clockwise direction using the sweep of an arm as they say *clockwise* to themselves.

11.8 Exercises

Guided Practice

Vocabulary Check

1. What is a *center of rotation*? the fixed point about which a figure is turned when it is rotated

2. Explain how you know if a figure has *rotational symmetry*. A figure in a plane has rotational symmetry if the figure looks the same after it is rotated 180° or less.

Does the figure have rotational symmetry? If so, describe the rotations that map the figure onto itself.

Skill Check

3. yes; a clockwise or counterclockwise rotation of 180° about its center

4. yes; a clockwise or counterclockwise rotation of 180° about its center

10. yes; a clockwise or counterclockwise rotation of 90° or 180° about its center

11. yes; a clockwise or counterclockwise rotation of 72° or 144° about its center

13. The wheel hub can be mapped onto itself by a clockwise or counterclockwise rotation of 72° or 144° about its center.

14. The wheel hub can be mapped onto itself by a clockwise or counterclockwise rotation of 36°, 72°, 108°, 144°, or 180° about its center.

3.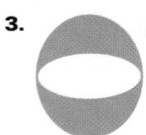

4.

5.  no

The diagonals of the regular hexagon shown form six equilateral triangles. Use the diagram to complete the statement.

6. A clockwise rotation of 60° about *P* maps *R* onto __?__. S

7. A counterclockwise rotation of 60° about __?__ maps *R* onto *Q*. P

8. A clockwise rotation of 120° about *Q* maps *R* onto __?__. W

9. A counterclockwise rotation of 180° about *P* maps *V* onto __?__. R

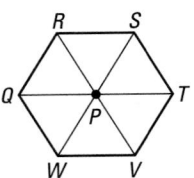

Practice and Applications

Extra Practice
See p. 696.

15. The wheel hub can be mapped onto itself by a clockwise or counterclockwise rotation of 45°, 90°, 135° or 180° about its center.

Rotational Symmetry Does the figure have rotational symmetry? If so, describe the rotations that map the figure onto itself.

10, 11. See margin.

10.

11.

12.  no

Wheel Hubs Describe the rotational symmetry of the wheel hub.
13–15. See margin.

13.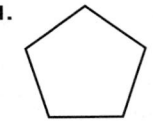

14.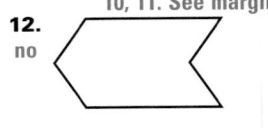

15.

Homework Help
Example 1: Exs. 10–15
Example 2: Exs. 16–26
Example 3: Exs. 27–30

636

Rotating a Figure Trace the polygon and point *P* on paper. Use a straightedge and protractor to rotate the polygon clockwise the given number of degrees about *P*. 16–21. See margin.

16. 150°

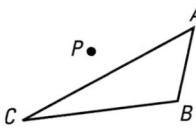

17. 135°

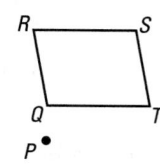

18. 60°

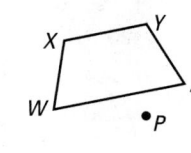

27. *J′*(1, 2); *K′*(4, 1); *L′*(4, −3); *M′*(1, −3); the coordinates of the image of the point (*x*, *y*) after a 90° clockwise rotation about the origin are (*y*, −*x*).

19. 40°

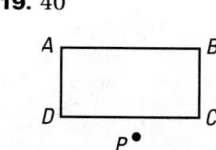

20. 100°

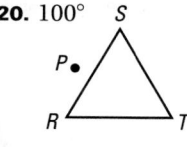

21. 120°

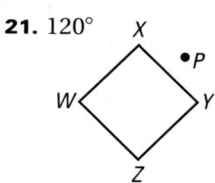

Visualize It!

Rotating a figure 180° clockwise is the same as rotating a figure 180° counterclockwise.

180° counterclockwise

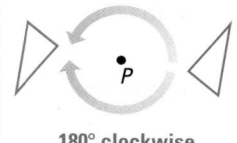

180° clockwise

Describing an Image State the segment or triangle that represents the image.

22. 90° clockwise rotation of $\overline{AB}$ about *P* $\overline{CD}$

23. 90° clockwise rotation of $\overline{KF}$ about *P* $\overline{LH}$

24. 180° rotation of $\triangle BCJ$ about *P* $\triangle FGL$

25. 180° rotation of $\triangle KEF$ about *P* $\triangle MAB$

26. 90° counterclockwise rotation of $\overline{CE}$ about *E* $\overline{GE}$

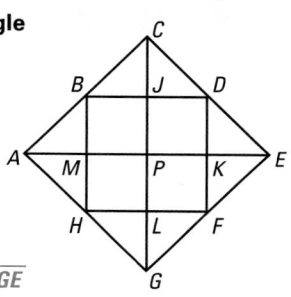

Finding a Pattern Use the given information to rotate the figure about the origin. Find the coordinates of the vertices of the image and compare them with the vertices of the original figure. Describe any patterns you see.   27–30. See margin.

27. 90° clockwise

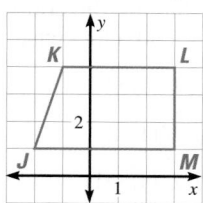

28. 90° counterclockwise

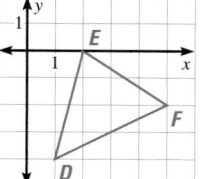

28. *D′*(4, 1); *E′*(0, 2); *F′*(2, 5); the coordinates of the image of the point (*x*, *y*) after a 90° counterclockwise rotation about the origin are (−*y*, *x*).

29. *A′*(−1, −1); *B′*(−4, 2); *C′*(−2, 5); the coordinates of the image of the point (*x*, *y*) after a 90° counterclockwise rotation about the origin are (−*y*, *x*).

30. *X′*(2, 3); *O′*(0, 0); *Z′*(−3, 4); the coordinates of the image of the point (*x*, *y*) after a 180° rotation about the origin are (−*x*, −*y*).

29. 90° counterclockwise

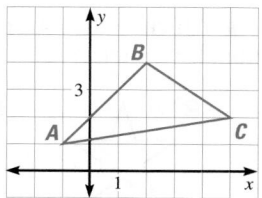

30. 180°

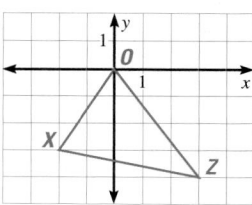

11.8 *Rotations*   **637**

16.

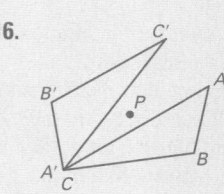

17.

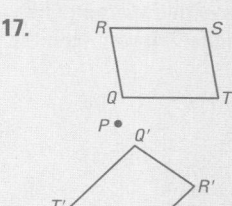

18.

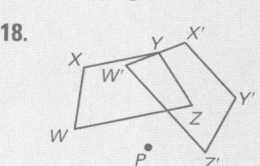

19.

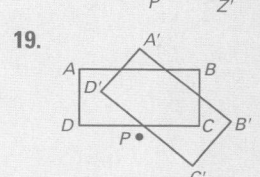

20.

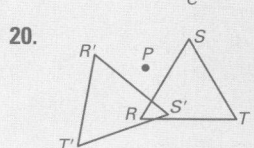

21.

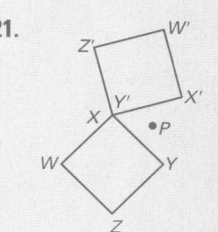

34. Yes; the answer would change to a clockwise or counterclockwise rotation of 90° or 180° about its center. If you disregard the colors of the figures, then, for example, the green fish in the middle map onto the orange fish and the orange fish map onto the green fish.

3.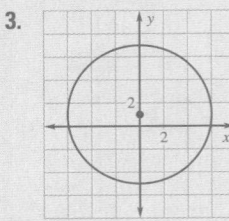

4.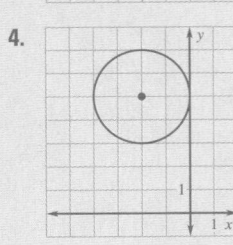

5.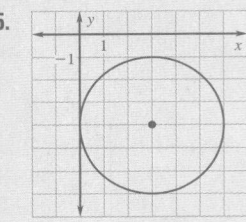

6.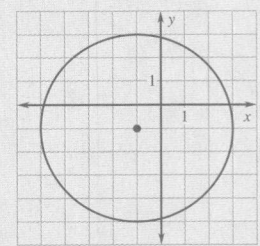

Link to Careers

GRAPHIC DESIGNERS may create symbols to represent a company or organization. These symbols often appear on packaging, stationery, and Web sites.

Career Links
CLASSZONE.COM

31. The design has rotational symmetry about its center; it can be mapped onto itself by a clockwise or counterclockwise rotation of 180°.

32. The design has rotational symmetry about its center; it can be mapped onto itself by a clockwise or counterclockwise rotation of 90° or 180°.

Standardized Test Practice

33. Yes. The image can be mapped onto itself by a clockwise or counterclockwise rotation of 180° about its center.

35. the center of the circle

36. Yes; this piece could be hung upside down because the image can be mapped onto itself by a clockwise or counterclockwise rotation of 180° about its center.

Graphic Design A music store, Ozone, is running a contest for a store logo. Two of the entries are shown. What do you notice about them?
31, 32. See margin.

31.

32.

Rotations in Art In Exercises 33–36, refer to the image below by M.C. Escher. The piece is called *Circle Limit III* and was completed in 1959. 33–36. See margin.

33. Does the piece have rotational symmetry? If so, describe the rotations that map the image onto itself.

34. Would your answer to Exercise 33 change if you disregard the color of the figures? Explain your reasoning.

35. Describe the center of rotation.

36. Is it possible that this piece could be hung upside down and have the same appearance? Explain.

37. **Multiple Choice** What are the coordinates of the vertices of the image of △*JKL* after a 90° clockwise rotation about the origin? **D**

   (A) $J'(1, 2)$, $K'(4, 2)$, $L'(1, 4)$

   (B) $J'(2, 1)$, $K'(4, 2)$, $L'(1, 4)$

   (C) $J'(4, 2)$, $K'(2, 1)$, $L'(4, -1)$

   (D) $J'(2, 4)$, $K'(1, 2)$, $L'(-1, 4)$

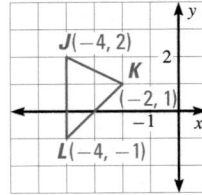

38. **Multiple Choice** Which of the four polygons shown below does not have rotational symmetry? **H**

   (F)    (G)    (H) 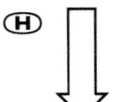   (J)

Mixed Review

Area of Polygons **Find the area of the polygon.** *(Lessons 8.3, 8.5, 8.6)*

39. rectangle *ABCD*   **40.** parallelogram *EFGH* **41.** trapezoid *JKMN*

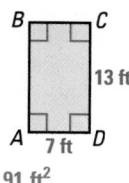

13 ft

7 ft

91 ft²

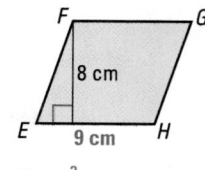

8 cm

9 cm

72 cm²

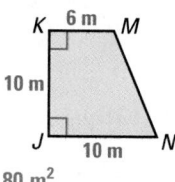

6 m

10 m

10 m

80 m²

Algebra Skills

Evaluating Radicals **Evaluate. Give the exact value if possible. If not, approximate to the nearest tenth.** *(Skills Review, p. 668)*

42. $\sqrt{42}$ **43.** $\sqrt{90}$ **44.** $\sqrt{256}$ **45.** $\sqrt{0}$
6.5   9.5   16   0

Quiz 3

1. What are the center and the radius of the circle whose equation is $(x + 1)^2 + (y - 6)^2 = 25$? *(Lesson 11.7)*   center, (−1, 6); radius, 5

2. Write the standard equation of the circle with center $(0, -4)$ and radius 3. *(Lesson 11.7)* $x^2 + (y + 4)^2 = 9$

Graph the equation. *(Lesson 11.7)* 3–6. See margin.

3. $x^2 + (y - 1)^2 = 36$ **4.** $(x + 2)^2 + (y - 5)^2 = 4$

5. $(x - 3)^2 + (y + 4)^2 = 9$ **6.** $(x + 1)^2 + (y + 1)^2 = 16$

7. yes; a clockwise or
counterclockwise rotation of
90° or 180° about its center

8. yes; a clockwise or
counterclockwise rotation
of 180° about its center

10. $A'(-2, -4)$; $B'(-1, -1)$;
$C'(1, -3)$; the coordinates
of the image of the point
(x, y) after a 180° rotation
about the origin are
$(-x, -y)$.

11. $A'(-4, 1)$; $B'(-4, 4)$;
$C'(-1, 4)$; $D'(-1, 2)$; the
coordinates of the image of
the point (x, y) after a 90°
counterclockwise rotation
about the origin are $(-y, x)$.

12. $A'(0, -2)$; $B'(-1, -4)$;
$C'(-3, -3)$; the coordinates
of the image of the point
(x, y) after a 90° clockwise
rotation about the origin are
$(y, -x)$.

Does the figure have rotational symmetry? If so, describe the rotations that map the figure onto itself. *(Lesson 11.8)*

7.

8.
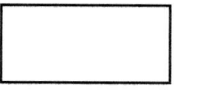

9.

no

Use the given information to rotate the figure about the origin. Find the coordinates of the vertices of the image and compare them with the vertices of the original figure. Describe any patterns you see. *(Lesson 11.8)*

10. 180°   **11.** 90° counterclockwise   **12.** 90° clockwise

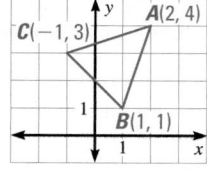
$C(-1, 3)$ $A(2, 4)$
$B(1, 1)$

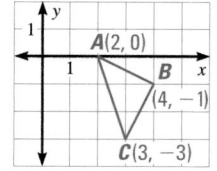
$A(1, 4)$ $B(4, 4)$
$D(2, 1)$ $C(4, 1)$

$A(2, 0)$
B
$(4, -1)$
$C(3, -3)$

11.8   Rotations   **639**

Planning the Activity

Goal
Students explore the relationship between reflections and rotations.

Materials
See the Technology Activity Keystrokes Masters in the *Chapter 11 Resource Book*.

LINK TO LESSON
In Lessons 5.7 and 11.8 students studied reflections and rotations, respectively. This activity shows how the two transformations are related.

Managing the Activity

Tips for Success
In Step 1, urge students not to draw the lines too close to the triangle or the angle measures in Step 3 will be more difficult to obtain. Stress that lines *k* and *m* should not intersect the triangle.

Closing the Activity

KEY DISCOVERY
The image of a figure after a second reflection over two intersecting lines is a rotation of the original figure, with the angle of rotation being twice the measure of the acute angle formed at the intersection of the lines.

Activity Assessment
Use Exercise 3 to assess student understanding.

Question

How are reflections related to rotations?

Explore

1 Draw △*ABC* and lines *k* and *m*. Label the intersection of *k* and *m* as point *P*.

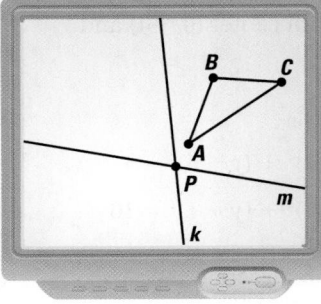

2 Reflect △*ABC* about line *k*.

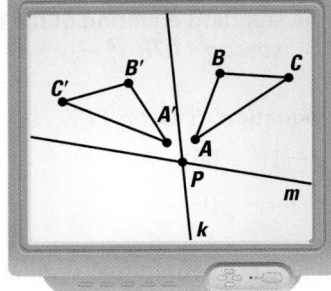

3 Reflect △*A'B'C'* about line *m*.

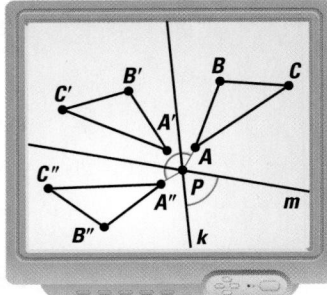

Student Help

LOOK BACK
To review reflections, see p. 282.

2. $AP = A''P$

3. The measure of ∠*APA''* is twice the measure of the acute angle formed at the intersection of lines *m* and *k*.

4. The measure of ∠*APA''* changes so it is still twice the measure of the acute angle formed at the intersection of lines *m* and *k*.

Think About It

1. How are △*ABC* and △*A''B''C''* related? *Sample answer: △A''B''C'' is a rotation of △ABC about point P.*

2. Draw $\overline{AP}$ and $\overline{A''P}$ and measure them. What do you notice?

3. Find the measure of ∠*APA''* and the measure of the acute angle formed at the intersection of the lines of reflection. What do you notice?

4. Change the positions of lines *k* and *m* so the angle formed by the lines changes. How does this affect the measure of ∠*APA''*?

5. Repeat Steps 1–3 and Exercises 1–4 for a different triangle. Do you get the same results? *yes*

Additional Resources

The following resources are available to help review the materials in this chapter.

📘 **Chapter 11 Resource Book**
• Chapter Review Games and Activities, p. 86
• Cumulative Review, Chs. 1–11

VOCABULARY

- **chord,** *p. 589*
- **secant,** *p. 589*
- **tangent,** *p. 589*
- **point of tangency,** *p. 589*
- **tangent segment,** *p. 597*
- **minor arc,** *p. 601*
- **major arc,** *p. 601*
- **measure of a minor arc,** *p. 601*

- **measure of a major arc,** *p. 601*
- **semicircle,** *p. 601*
- **congruent circles,** *p. 602*
- **congruent arcs,** *p. 602*
- **arc length,** *p. 603*
- **inscribed angle,** *p. 614*
- **intercepted arc,** *p. 614*
- **inscribed,** *p. 615*

- **circumscribed,** *p. 615*
- **standard equation of a circle,** *p. 628*
- **rotation,** *p. 633*
- **center of rotation,** *p. 633*
- **angle of rotation,** *p. 633*
- **rotational symmetry,** *p. 634*

VOCABULARY REVIEW

Fill in the blank.

1. A ___?___ is a line that intersects a circle in two points. secant

2. A polygon is ___?___ in a circle if all of its vertices lie on the circle. inscribed

3. A line in the plane of a circle that intersects the circle in exactly one point is called a ___?___. tangent

4. If the endpoints of an arc are the endpoints of a diameter, then the arc is a ___?___. semicircle

5. An ___?___ is an angle whose vertex is on a circle and whose sides contain chords of the circle. inscribed angle

6. A ___?___ is a segment whose endpoints are points on a circle. chord

7. A ___?___ is a transformation in which a figure is turned about a fixed point. rotation

---

11.1 PARTS OF A CIRCLE

Examples on pp. 589–590

> **EXAMPLE** Identify a chord, a secant, a tangent, a diameter, the center, and a point of tangency.
>
> $\overline{MP}$ is a chord. $\overleftrightarrow{MP}$ is a secant. $\overleftrightarrow{TK}$ is a tangent.
>
> $\overline{LT}$ is a diameter. R is the center. T is a point of tangency.

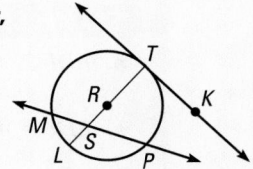

Tell whether the point, line, or segment is best described as a *chord*, a *secant*, a *tangent*, a *diameter*, a *radius*, the *center*, or a *point of tangency*.

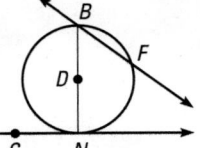

8. $\overline{BN}$ diameter **9.** N point of tangency **10.** $\overline{BF}$ chord

11. D center **12.** $\overleftrightarrow{CN}$ tangent **13.** $\overleftrightarrow{BF}$ secant

11.2 PROPERTIES OF TANGENTS

Examples on pp. 595–597

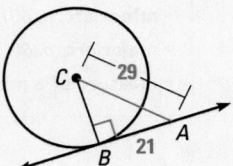

> **EXAMPLE** $\overleftrightarrow{AB}$ is tangent to ⊙*C*. Find *CB*.
>
> $(AC)^2 = (AB)^2 + (CB)^2$ Pythagorean Theorem
>
> $29^2 = 21^2 + (CB)^2$ Substitute 29 for *AC*, and 21 for *AB*.
>
> $841 = 441 + (CB)^2$ Multiply.
>
> $400 = (CB)^2$ Subtract 441 from each side.
>
> $20 = CB$ Find the positive square root.

$\overleftrightarrow{AB}$ is tangent to ⊙*C*. **Find the value of *r*.**

14.

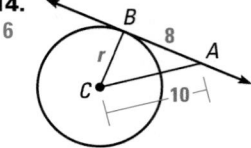

15.

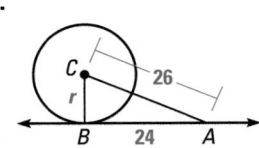

16.

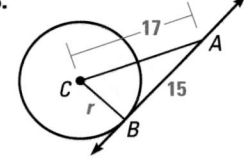

$\overline{AB}$ and $\overline{AD}$ are tangent to ⊙*C*. **Find the value of *x*.**

17.

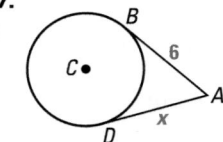

18.

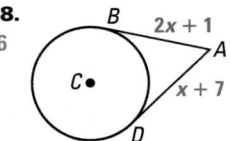

19.

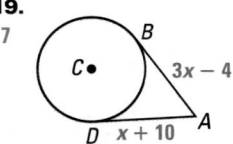

11.3 ARCS AND CENTRAL ANGLES

Examples on pp. 601–603

> **EXAMPLES** Find the measure of the arc.
>
> **a.** $m\widehat{DF}$ **b.** $m\widehat{DA}$ **c.** $m\widehat{ABD}$
>
>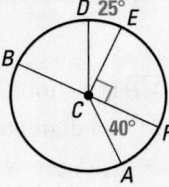
>
> **a.** $m\widehat{DF} = m\widehat{DE} + m\widehat{EF} = 25° + 90° = 115°$
>
> **b.** $m\widehat{DA} = m\widehat{DF} + m\widehat{AF} = 115° + 40° = 155°$
>
> **c.** $m\widehat{ABD} = 360° - m\widehat{DA} = 360° - 155° = 205°$

$\overline{AD}$ is a diameter and $m\widehat{CE} = 121°$. Find the measure of the arc.

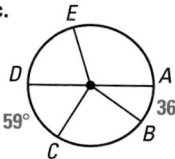

20. $\widehat{DE}$ 62°   **21.** $\widehat{AE}$ 118°   **22.** $\widehat{AEC}$ 239°

23. $\widehat{BC}$ 85°   **24.** $\widehat{BDC}$ 275°   **25.** $\widehat{BDA}$ 324°

Find the length of the red arc. Round your answer to the nearest hundredth.

26.
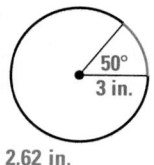
50°
3 in.

2.62 in.

27.
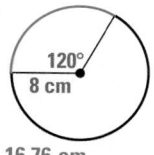
120°
8 cm

16.76 cm

28.
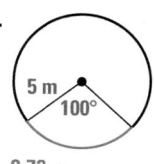
5 m
100°

8.73 m

11.4   ARCS AND CHORDS

Examples on pp. 608–610

> **EXAMPLE**   Find the value of *x*.
>
>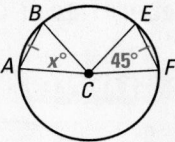
>
> Because $\overline{AB} \cong \overline{EF}$, it follows that $\widehat{AB} \cong \widehat{EF}$.
> So, $m\widehat{AB} = m\widehat{EF} = 45°$, and $x = 45$.

Find the value of *x*.

29.
3
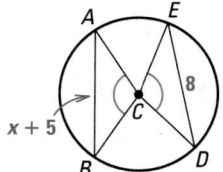
x + 5
8

30.
2
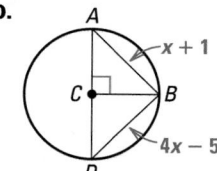
x + 1
4x − 5

31.
35
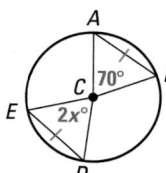
70°
2x°

32.
6
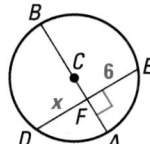
6
x

33.
60
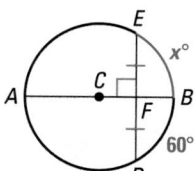
x°
60°

34.
115
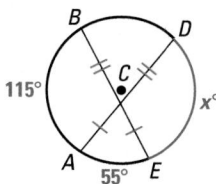
115°
x°
55°

11.5   INSCRIBED ANGLES AND POLYGONS

Examples on pp. 614–616

> **EXAMPLE**   Find the measure of the inscribed angle.
>
>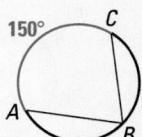
> 150°
>
> $m\angle ABC = \frac{1}{2}m\widehat{AC} = \frac{1}{2}(150°) = 75°$

Find the measure of the inscribed angle or intercepted arc.

35.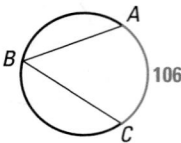
53°

36.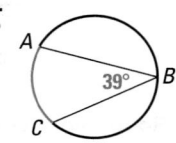
78°

37.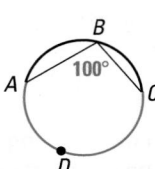
200°

Find the value of x and y.

38.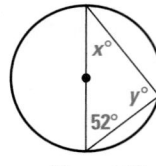

$x = 38;\ y = 90$

39.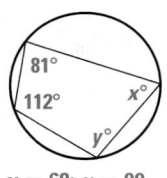

$x = 68;\ y = 99$

40.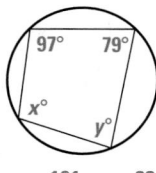

$x = 101;\ y = 83$

11.6   PROPERTIES OF CHORDS

Examples on pp. 620–622

EXAMPLES   **Find the value of x.**

a.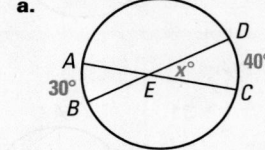

$$x° = \frac{1}{2}(m\widehat{AB} + m\widehat{DC})$$

$$= \frac{1}{2}(30° + 40°)$$

$$= 35°$$

b.

$$EC \cdot EA = ED \cdot EB$$

$$3 \cdot x = 4 \cdot 6$$

$$3x = 24$$

$$x = 8$$

Find the value of x.

41.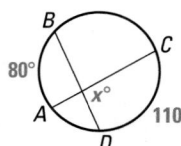
95

42.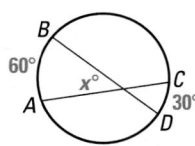
45

43.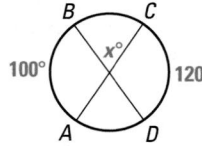
70

44.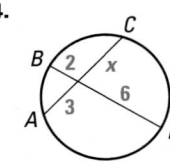
4

45.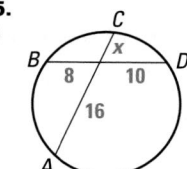
5

46.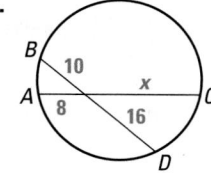
20

11.7 EQUATIONS OF CIRCLES

Examples on pp. 627–629

> **EXAMPLE** **Write the standard equation of the circle.**
>
> ⊙*C* has center (−3, −1) and radius 2. Its standard equation is
>
> $(x - (-3))^2 + (y - (-1))^2 = 2^2$
>
> $(x + 3)^2 + (y + 1)^2 = 4.$

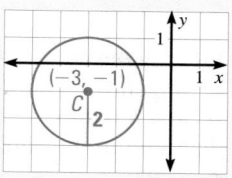

Write the standard equation of the circle with the given center and radius.

47. center (2, 5), radius 3
$(x - 2)^2 + (y - 5)^2 = 9$

48. center (−4, −1), radius 4
$(x + 4)^2 + (y + 1)^2 = 16$

49. center (5, −2), radius 7
$(x - 5)^2 + (y + 2)^2 = 49$

Give the radius and the coordinates of the center of the circle with the given equation. Then graph the circle. 50–52. See margin.

50. $(x + 4)^2 + (y - 1)^2 = 9$ **51.** $(x - 2)^2 + (y + 3)^2 = 16$ **52.** $x^2 + y^2 = 25$

11.8 ROTATIONS

Examples on pp. 633–635

> **EXAMPLE** **Rotate the triangle with vertices *F*(−4, 1), *G*(−3, 5), and *H*(−1, 2) 90° clockwise about the origin.**
>
>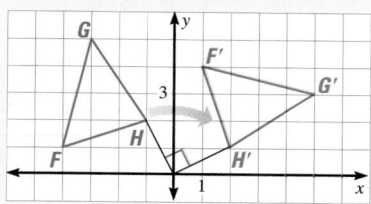

Trace the figure and point *P* on paper. Use a straightedge and protractor to rotate the figure clockwise the given number of degrees about *P*. 53–55. See margin.

53. 90° counterclockwise   **54.** 90° clockwise   **55.** 180°

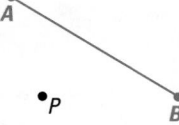

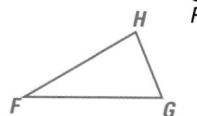

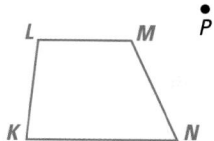

56. Does the figure shown at the right have rotational symmetry? If so, describe the rotations that map the figure onto itself.
yes; a clockwise or counterclockwise rotation of 90° or 180° about its center

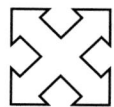

Chapter Summary and Review **645**

50. radius, 3; center, (−4, 1);

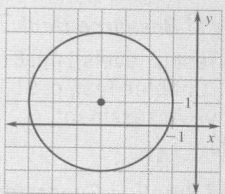

51. radius, 4; center, (2, −3);

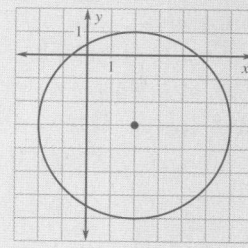

52. radius, 5; center, (0, 0);

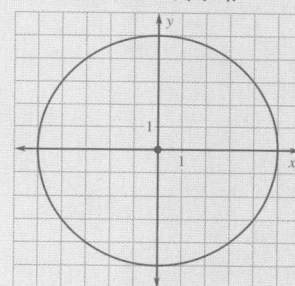

53.

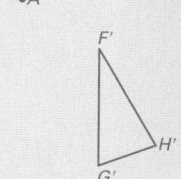

54.

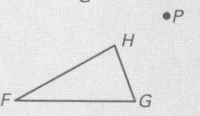

55.

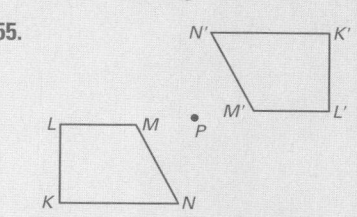

1. center, C; point of tangency, D; chord, $\overline{AB}$ or $\overline{BF}$; secant, $\overleftrightarrow{AB}$; radius, $\overline{FC}$ or $\overline{BC}$; diameter, $\overline{BF}$

14. $(x - 6)^2 + (y + 3)^2 = 25$

15.

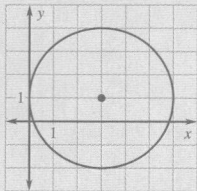

16.

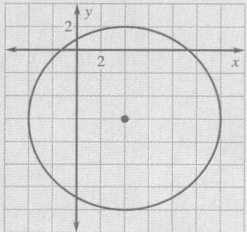

17.

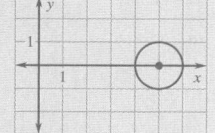

 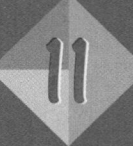
1. Identify the center, a point of tangency, a chord, a secant, a radius, and a diameter in the circle. **See margin.**

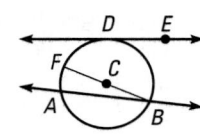

2. $\overline{CB}$ is tangent to $\odot A$. What is the length of $\overline{AB}$? **15**

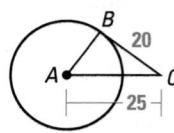

3. $\overline{RS}$ and $\overline{RV}$ are tangent to $\odot T$. What is the value of x? **4**

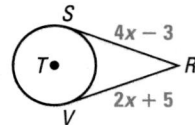

$\overline{AD}$ is a diameter. Find the measure.

4. $m\overset{\frown}{BC}$ **60°**

5. $m\overset{\frown}{AC}$ **110°**

6. $m\overset{\frown}{ABD}$ **180°**

7. $m\overset{\frown}{CAE}$ **250°**

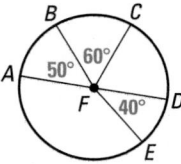

Find the length of the red arc. Round your answer to the nearest hundredth.

8.

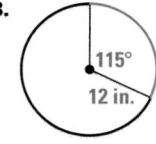

9.

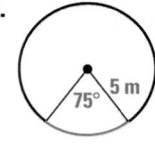

24.09 in.   6.54 m

Find the value of x.

10. **1**

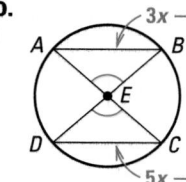

11. **12**

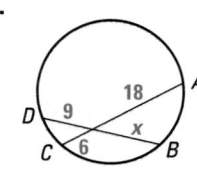

In Exercises 12 and 13, find the value of each variable.

12. **44**

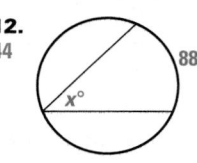

13.

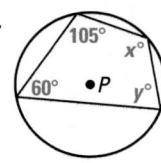

$x = 120$; $y = 75$

14. Write the standard equation of the circle with center $(6, -3)$ and radius 5. **See margin.**

15. Graph the circle with equation $(x - 3)^2 + (y - 1)^2 = 9$. **See margin.**

16. Graph the circle with equation $(x - 4)^2 + (y + 6)^2 = 64$. **See margin.**

17. Graph the circle with equation $(x - 5)^2 + y^2 = 1$. **See margin.**

18. Name the coordinates of the vertices of the image after rotating $\triangle ABC$ 90° counterclockwise about the origin. $A'(-3,0)$; $B'(-1, 3)$; $C'(1, -1)$

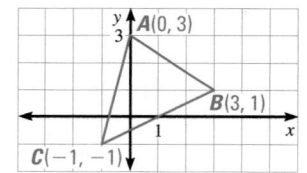

19. **Rock Circle** The rock circle shown below is in the Ténéré desert in the African country of Niger. The circle is about 60 feet in diameter. Suppose the center of the circle is at $(30, 30)$ on a grid measured in units of feet. Write an equation of the circle. $(x - 30)^2 + (y - 30)^2 = 900$

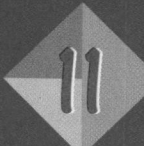
Additional Resources

📕 *Chapter 11 Resource Book*
• SAT/ACT Chapter Test, p. 91

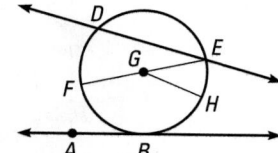 **Test Tip**

Write your work neatly and number it. If you have time to check your answers, it is faster to review this work rather than start over.

1. Which of the following is a secant in ⊙G? **D**

(A) $\overline{EF}$ **(B)** $\overline{GH}$

(C) $\overleftrightarrow{AB}$ **(D)** $\overleftrightarrow{DE}$

2. What is the approximate length of the blue arc? **G**

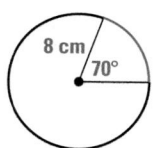

(F) 8.12 cm **(G)** 9.77 cm

(H) 10.85 cm **(J)** 11.27 cm

3. Which of the figures has rotational symmetry? **A**

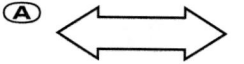 

4. Find the value of *x*. **H**

(F) 4

(G) 11

(H) 16

(J) 144

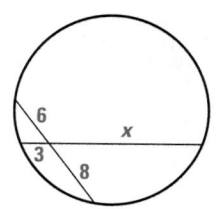

5. Find the value of *x*. **D**

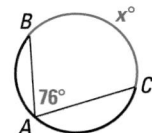

(A) 38 **(B)** 76

(C) 90 **(D)** 152

6. Find the measure of ∠1. **G**

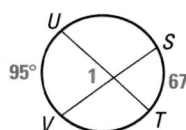

(F) 67 **(G)** 81

(H) 93 **(J)** 160

7. Which of the following is the equation of a circle with center (7, −1) and radius 8? **B**

(A) $(x + 7)^2 + (y - 1)^2 = 64$

(B) $(x - 7)^2 + (y + 1)^2 = 64$

(C) $(x + 7)^2 + (y - 1)^2 = 8$

(D) $(x - 7)^2 + (y - 1)^2 = 64$

8. Find the values of *x* and *y*. **F**

(F) *x* = 83, *y* = 98

(G) *x* = 98, *y* = 83

(H) *x* = 83, *y* = 88

(J) *x* = 98, *y* = 93

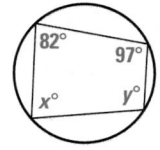

Chapter Standardized Test **647**

Teaching Tip

The Brain Games activity provides a motivating way to review selected content in the chapter. For a more comprehensive review, see the Chapter Summary and Review on pp. 641–645.

BraiN ·····▶GaMes·····

What Did I Describe? ◀·····

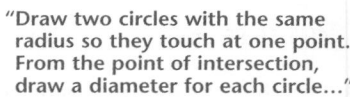

Materials
- circle cards
- pencil
- paper

Object of the Game   To correctly draw a picture from a description.

Step 1 ▶ One player picks a card and describes the picture to the other players.

Step 2 ▶ The other players try to draw the picture described. The team gets a point for each player who draws the picture correctly.

Step 3 ▶ Continue play until each player has a turn describing a picture. The team with the most points wins.

"Draw two circles with the same radius so they touch at one point. From the point of intersection, draw a diameter for each circle..."

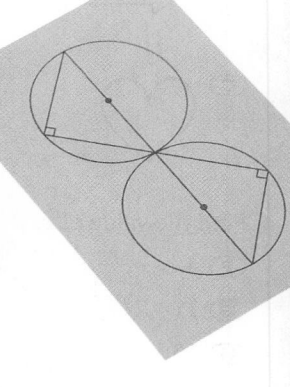

Another Way to Play   Draw your own picture of a circle with tangents, secants, chords, inscribed angles, and central angles. Play as shown above.

"Draw a circle and a tangent line. Draw a line through the circle's center and the point of tangency..."

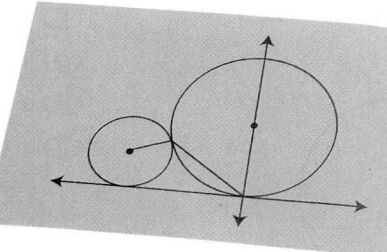

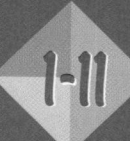
Additional Resources
A Cumulative Test covering
Chapters 1–11 is available in the
Chapter 11 Resource Book,
pp. 98–101.

16. In a plane, if two lines are perpendicular to the same line, then
they are parallel to each other.

Find the measure. (Lessons 1.5, 1.6)

1. $AC = \underline{\ \ ?\ \ }$ 15

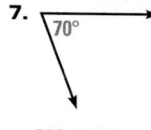

2. $ST = \underline{\ \ ?\ \ }$ 12

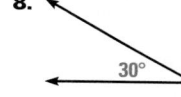

3. $m\angle STR = \underline{\ \ ?\ \ }$ 118°

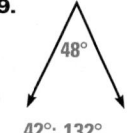

4. $m\angle JGH = \underline{\ \ ?\ \ }$ 45°

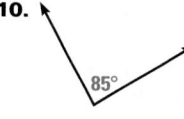

In Exercises 5 and 6, $\overline{ST}$ has endpoints $S(3, 9)$ and $T(-1, 5)$.

5. Find the coordinates of the midpoint of $\overline{ST}$. (Lesson 2.1) (1, 7)

6. Find the distance between the points. Round your answer
to the nearest tenth. (Lesson 4.4) 5.7

**Find the measure of a complement and a supplement of
the angle given.** (Lesson 2.3)

7.

70°

20°; 110°

8.

30°

60°; 150°

9.

48°

42°; 132°

10.

85°

5°; 95°

11. Find the value of x. Then use substitution
to find $m\angle APB$ and $m\angle BPC$. (Lesson 2.4)
$x = 7$; $m\angle APB = 28°$; $m\angle BPC = 152°$

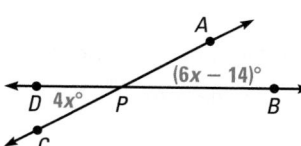

12. Show the conjecture is false by finding a counterexample: If two
lines do not intersect, then the lines are parallel. (Lessons 1.2, 3.1)
The two lines can be skew lines.

**Explain how you can show that $\ell \parallel m$. State any theorems or postulates
that you would use.** (Lessons 3.5, 3.6)

13.

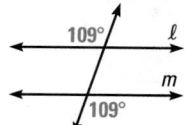

109°

ℓ

m

109°

Alternate Exterior
Angles Converse

14.

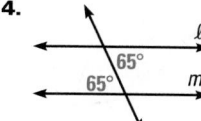

65°

65°

ℓ

m

Alternate Interior
Angles Converse

15.

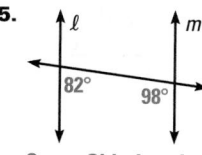

ℓ m

82°   98°

Same-Side Interior
Angles Converse

16.

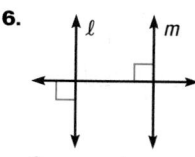

ℓ m

See margin.

17. An isosceles triangle has an angle of measure 124°. Find the
measures of the other two angles. Explain. (Lessons 4.2, 4.3) 28° and 28°; Since the triangle is isosceles, the
base angles are congruent. By the Triangle Sum Theorem, you can show that each base angle has measure 28°.

29. *Sample answer:* $\angle CBD \cong \angle CAE$ and $\angle BCD \cong \angle ACE$. Therefore, $\triangle BCD \sim \triangle ACE$ by the AA Similarity Postulate.

In Exercises 18 and 19, use $\triangle ABC$.

18. Classify the triangle as *acute, right,* or *obtuse.* (Lesson 4.5)
obtuse; $12^2 > 6^2 + 8^2$

19. Name the smallest and largest angles of $\triangle ABC$. (Lesson 4.7)
smallest, $\angle A$; largest, $\angle B$

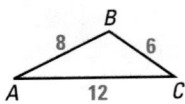

Does the diagram give enough information to show that the triangles are congruent? If so, state the postulate or theorem you would use. Explain your reasoning. (Lessons 5.2–5.4)

20.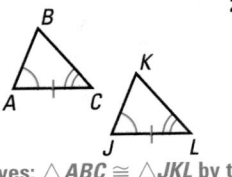
yes; $\triangle ABC \cong \triangle JKL$ by the ASA Congruence Postulate

21.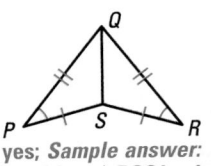
no

22.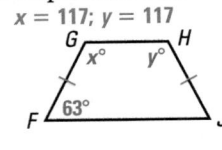
yes; *Sample answer:* $\triangle PQS \cong \triangle RQS$ by the SAS Congruence Postulate

23.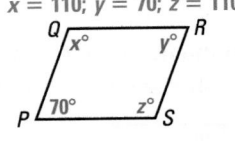
yes; $\triangle XYU \cong \triangle VWU$ by the HL Congruence Theorem

In Exercises 24–27, find the value of each variable. (Lessons 6.1, 6.2, 6.4, 6.5)

24. square *ABCD*
$x = 8; y = 90$

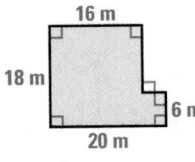

25. trapezoid *FGHJ*
$x = 117; y = 117$

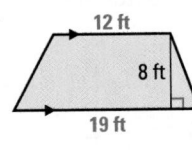

26. quadrilateral *KLMN*
67

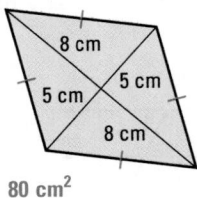

27. parallelogram *PQRS*
$x = 110; y = 70; z = 110$

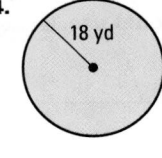

28. The perimeter of a rectangle is 60 centimeters. The ratio of the length to the width is 3 : 2. Sketch the rectangle. Then find the length and the width. (Lesson 7.1) Check sketches; 18 cm, 12 cm.

In Exercises 29 and 30, use the diagram at the right.

29. Show that the overlapping triangles are similar. Then write a similarity statement. (Lesson 7.4) See margin.

30. Find the value of *x*. (Lesson 7.5) 10

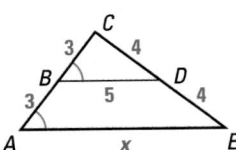

In Exercises 31–34, find the area of the figure. (Lessons 8.3, 8.5–8.7)

31.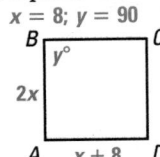
312 m²

32.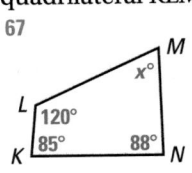
124 ft²

33.
80 cm²

34.
$324\pi \approx 1018$ yd²

35. A parallelogram has a base of 14 meters and an area of 84 square meters. Find the height. (Lesson 8.5) 6 m

Find the surface area and volume of the solid. If necessary, round your answer to the nearest whole number. (Lessons 9.2–9.6)

36.
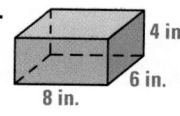
4 in.
6 in.
8 in.
208 in.²; 192 in.³

37.
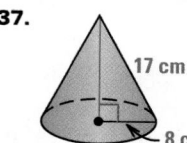
17 cm
8 cm
628 cm²; 1005 cm³

38.

6 m
10 m
603 m²; 1131 m³

39.

12 mm
1810 mm²; 7238 mm³

In Exercises 40 and 41, use the diagram of △RST. (Lessons 10.4–10.6)

40. Find sin R, cos R and tan R. Write your answers as decimals rounded to four decimal places. sin $R \approx 0.3846$; cos $R \approx 0.9231$; tan $R \approx 0.4167$

41. Find the measures of $\angle R$ and $\angle T$ to the nearest tenth of a degree. $m\angle R \approx 22.6°$; $m\angle T \approx 67.4°$

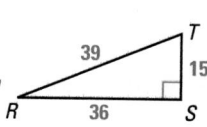
T
39
15
R
36
S

Solve the right triangle. Round decimals to the nearest tenth.
(Lessons 10.3–10.6)

42.
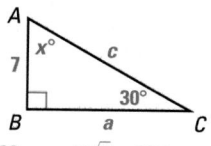
A
7
$x°$
c
30°
B
a
C
$x = 60$; $a = 7\sqrt{3} \approx 12.1$; $c = 14$

43.
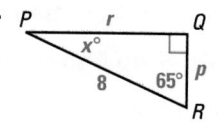
P
r
Q
$x°$
8
65°
p
R
$x = 25$; $p \approx 3.4$; $r \approx 7.3$

44.
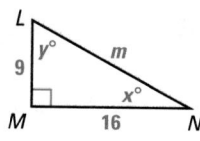
L
9
$y°$
m
$x°$
M
16
N
$m \approx 18.4$; $x \approx 29.4°$; $y \approx 60.6°$

45.
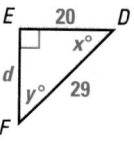
E
20
D
$x°$
d
29
$y°$
F
$d = 21$; $x \approx 46.4°$; $y \approx 43.6°$

In Exercises 46–48, use the diagram of ⊙O.

46. Find the area of the blue sector. (Lesson 8.7)
about 23.6 ft²

47. Find the measures of $\overset{\frown}{AB}$ and $\overset{\frown}{ACB}$. (Lesson 11.3)
See margin.

48. Find the length of $\overset{\frown}{AB}$. (Lesson 11.3)
about 7.9 ft

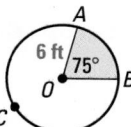
A
6 ft
75°
O
B
C

In Exercises 49–52, find the value of x. (Lessons 11.4–11.6)

49.
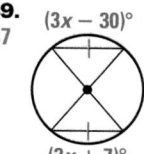
37
$(3x - 30)°$
$(2x + 7)°$

50.
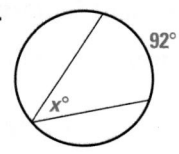
46
92°
$x°$

51.
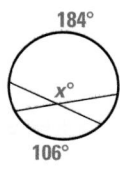
145
184°
$x°$
106°

52.
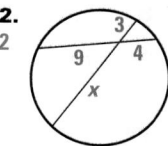
12
3
9
4
x

53. Write the standard equation of a circle with center $(0, -3)$ and radius 7. (Lesson 11.7) $x^2 + (y + 3)^2 = 49$

In Exercises 54–56, copy △ABC. 54, 55. See margin.

54. Draw the image of △ABC after the translation $(x, y) \rightarrow (x - 1, y + 1)$. (Lesson 3.7)

55. Draw the reflection of △ABC in the y-axis. (Lesson 5.7)

56. Rotate △ABC 180° about the origin. Find the coordinates of the vertices of △A′B′C′. (Lesson 11.8)
$A'(-2, -5)$; $B'(-4, -1)$; $C'(-1, -2)$

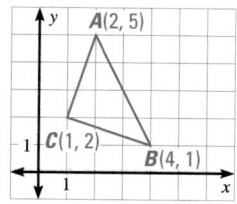
$A(2, 5)$
$C(1, 2)$
$B(4, 1)$

47. $m\overset{\frown}{AB} = 75°$; $m\overset{\frown}{ACB} = 285°$

54.
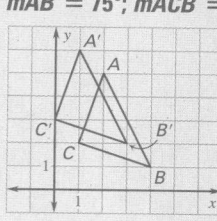
A'
A
C'
C
B'
B

55.
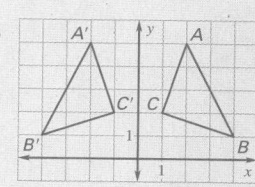
A'
A
C' C
B'
B

Contents of Student Resources

PROBLEM SOLVING

One of your primary goals in mathematics should be to become a good problem solver. It helps to approach a problem with a plan.

Step ❶ Understand the problem.
Read the problem carefully. Organize the given information and decide what you need to find. Check for unnecessary or missing information. Supply missing facts, if possible.

Step ❷ Make a plan to solve the problem.
Choose a problem-solving strategy. (See the next page for a list.) Choose the correct operations. Decide if you will use a tool such as a calculator, a graph, or a spreadsheet.

Step ❸ Carry out the plan to solve the problem.
Use the strategy and any tools you have chosen. Estimate before you calculate, if possible. Do any calculations that are needed. Answer the question that the problem asks.

Step ❹ Check to see if your answer is reasonable.
Reread the problem. See if your answer agrees with the given information and with any estimate you calculated.

EXAMPLE **Eight people can be seated evenly around a rectangular table, with one person on each end. How many people can be seated around three of these tables placed end-to-end?**

Solution

❶ You know each table is rectangular, seats eight people, and can fit one person on each end. You need to find the number of people that can be seated at three tables placed end-to-end.

❷ An appropriate strategy is to draw a diagram.

❸ Draw one table, with an X for each person seated. Notice that three people can be seated at each long side of the table. Then draw three tables placed end-to-end. Draw Xs and count them.

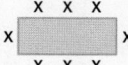 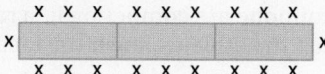

There are 20 Xs on the diagram, so 20 people can be seated.

❹ At three individual tables you can seat 24 people. Since seats are lost when tables are placed end-to-end, 20 is a reasonable answer.

In Step 2 of the problem-solving plan on the previous page you select a strategy. Here are some problem-solving strategies to consider.

Guess, check, and revise.	Use when you do not seem to have enough information.
Draw a diagram or a graph.	Use when words describe a visual representation.
Make an organized list or table.	Use when you have data or choices to organize.
Use an equation or a formula.	Use when you know a relationship between quantities.
Use a proportion.	Use when you know that two ratios are equal.
Look for a pattern.	Use when you can examine several cases.
Break into simpler parts.	Use when you have a multi-step problem.
Solve a simpler problem.	Use when smaller numbers help you understand the problem.
Work backward.	Use when you look for a fact leading to a known result.
Act out the situation.	Use when visualizing the problem is helpful.

Practice

1. During the month of May, Rosa made deposits of $128.50 and $165.19 into her checking account. She wrote checks for $25, $55.12, and $83.98. If her account balance at the end of May was $327.05, what was her balance at the beginning of May? **$197.46**

2. A rectangular room measures 9 feet by 15 feet. How many square yards of carpet are needed to cover the floor of this room? **15 yd^2**

3. You make 20 silk flower arrangements and plan to sell them at a craft show. Each arrangement costs $12 in materials, and your booth at the show costs $30. If you price the arrangements at $24 each, how many must you sell to make at least $100 profit? **at least 16 flower arrangements**

4. A store sells sweatshirts in small, medium, large, and extra large. The color choices are red, white, blue, gray, and black. How many different kinds of sweatshirts are sold at the store? **20 kinds**

5. If 4.26 pounds of ham cost $6.77, what would 3.75 pounds cost? **$5.96**

6. Roger bought some 34¢ stamps and some 20¢ stamps, and spent $2.90. How many of each type of stamp did Roger buy? **five 34¢ stamps and six 20¢ stamps**

7. Abigail, Bonnie, Carla, and Dominique are competing in a race. In how many different orders can the athletes finish the race? **24 different orders**

8. Five boys are standing in a line. Sam is before Mark and immediately after Alex. Eric is next to Charlie and Alex. List the order of the boys in line. **Charlie, Eric, Alex, Sam, Mark**

9. How many diagonals can be drawn on a stop sign? **20 diagonals**

654

Student Resources

654

DECIMALS

The steps for adding, subtracting, multiplying, and dividing with decimals are like those for computing with whole numbers.

EXAMPLES **Add or subtract.**

a. $15.2 - 8.65$

b. $3.8 + 0.19 + 7$

Solution Write in vertical form, lining up the decimal points. Use zeros as placeholders.

a.
```
  15.20   ← placeholder
-  8.65
  6.55
```

b.
```
   3.80   ← placeholder
   0.19
+  7.00   ← placeholders
  10.99
```

ANSWER ▸ $15.2 - 8.65 = 6.55$

ANSWER ▸ $3.8 + 0.19 + 7 = 10.99$

EXAMPLES **Multiply or divide.**

a. 6.75×4.9

b. $0.068 \div 0.4$

Solution

a. Write in vertical form.

The number of decimal places in the product is equal to the sum of the number of decimal places in the factors.

```
    6.75    ←2 decimal places
  ×  4.9    ←1 decimal place
    6 075
   27 000
   33.075   ←3 decimal places
```

b. Write in long division form: $0.4\overline{)0.068}$

Move the decimal points the same number of places so that the divisor is a whole number: $0.4\overline{)0.068}$ Then divide.

```
     0.17
  4)0.68       Line up the decimal point in
    0.40       the quotient with the decimal
     28        point in the dividend.
     28
      0
```

ANSWER ▸ $6.75 \times 4.9 = 33.075$

ANSWER ▸ $0.068 \div 0.4 = 0.17$

Practice

Evaluate.

1. $1.78 + 0.3$ **2.08**

2. $0.14 + 0.095$ **0.235**

3. $66 + 17.5 + 96.5$ **180**

4. $0.83 - 0.07$ **0.76**

5. $90 - 42.5$ **47.5**

6. $49.5 - 1.2 - 23.5$ **24.8**

7. 9×3.14 **28.26**

8. 0.15×6.2 **0.93**

9. 400×0.025 **10**

10. $124.2 \div 0.02$ **6210**

11. $0.8 \div 0.1$ **8**

12. $600 \div 0.15$ **4000**

FRACTION CONCEPTS

Multiply or divide the numerator and denominator of a fraction by the same nonzero number to write an **equivalent fraction**. To write a fraction in **simplest form**, divide the numerator and denominator by their greatest common factor. Two numbers are **reciprocals** if their product is 1.

EXAMPLE **Write two fractions equivalent to $\frac{4}{10}$.**

Solution

Divide the numerator and denominator by 2:

$$\frac{4}{10} = \frac{4 \div 2}{10 \div 2} = \frac{2}{5}$$

Multiply the numerator and denominator by 2:

$$\frac{4}{10} = \frac{4 \times 2}{10 \times 2} = \frac{8}{20}$$

EXAMPLE **Write the fraction $\frac{8}{12}$ in simplest form.**

Solution

The greatest common factor of 8 and 12 is 4.

Divide the numerator and denominator by 4: $\frac{8}{12} = \frac{8 \div 4}{12 \div 4} = \frac{2}{3}$.

EXAMPLES **Find the reciprocal of the number.** **a.** $\frac{4}{5}$ **b.** 10

Solution

a. Switch the numerator and the denominator. The reciprocal is $\frac{5}{4}$.

b. Write 10 as a fraction, $\frac{10}{1}$. The reciprocal is $\frac{1}{10}$.

Practice

Write two fractions equivalent to the given fraction. 1–5. Sample answers are given.

1. $\frac{6}{12}$ $\frac{1}{2}, \frac{2}{4}$ **2.** $\frac{18}{20}$ $\frac{9}{10}, \frac{36}{40}$ **3.** $\frac{1}{4}$ $\frac{2}{8}, \frac{3}{12}$ **4.** $\frac{4}{6}$ $\frac{2}{3}, \frac{8}{12}$ **5.** $\frac{3}{10}$ $\frac{6}{20}, \frac{15}{50}$

Write the fraction in simplest form.

6. $\frac{9}{12}$ $\frac{3}{4}$ **7.** $\frac{6}{9}$ $\frac{2}{3}$ **8.** $\frac{100}{200}$ $\frac{1}{2}$ **9.** $\frac{5}{15}$ $\frac{1}{3}$ **10.** $\frac{16}{64}$ $\frac{1}{4}$

Find the reciprocal of the number.

11. $\frac{2}{5}$ $\frac{5}{2}$ **12.** $\frac{1}{8}$ 8   **13.** $\frac{4}{3}$ $\frac{3}{4}$ **14.** 7 $\frac{1}{7}$ **15.** 1 1

FRACTIONS AND DECIMALS

Divide to write a fraction as a decimal. If the remainder is ever zero, the result is a **terminating decimal**. If the quotient has a digit or group of digits that repeats, the result is a **repeating decimal**.

EXAMPLES   **Write the fraction as a decimal.**

a. $\dfrac{5}{8}$

b. $\dfrac{2}{3}$

Solution   Divide the numerator by the denominator.

a. $\dfrac{0.625}{8)\overline{5.000}}$ terminating decimal

b. $\dfrac{0.666\ldots}{3)\overline{2.000}}$ repeating decimal

ANSWER ▸ $\dfrac{5}{8} = 0.625$

ANSWER ▸ $\dfrac{2}{3} = 0.\overline{6}$ A bar indicates repeating digits.

≈ 0.67 Round for an approximation.

EXAMPLES   **Write the decimal as a fraction in simplest form.**

a. 0.15

b. $0.8\overline{3}$

Solution

a. 0.15 is fifteen hundredths.

$$0.15 = \frac{15}{100} = \frac{3}{20}$$

Use simplest form.

ANSWER ▸ $0.15 = \dfrac{3}{20}$

b. $x = 0.8333\ldots$ Write an equation.

$10x = 8.3333\ldots$ Multiply each side by 10.

$9x = 7.5$ Find $10x - x$.

$x = \dfrac{7.5}{9} = \dfrac{75}{90} = \dfrac{5}{6}$ Solve and simplify.

ANSWER ▸ $0.8\overline{3} = \dfrac{5}{6}$

Practice

Write the fraction as a decimal. For repeating decimals, also round to the nearest hundredth for an approximation.

1. $\dfrac{7}{10}$ 0.7
2. $\dfrac{1}{4}$ 0.25
3. $\dfrac{1}{3}$ $0.\overline{3} \approx 0.33$
4. $\dfrac{7}{8}$ 0.875
5. $\dfrac{2}{9}$ $0.\overline{2} \approx 0.22$

6. $\dfrac{4}{5}$ 0.8
7. $\dfrac{1}{8}$ 0.125
8. $\dfrac{19}{100}$ 0.19
9. $\dfrac{1}{2}$ 0.5
10. $\dfrac{7}{15}$ $0.4\overline{6} \approx 0.47$

Write the decimal as a fraction in simplest form.

11. 0.5 $\dfrac{1}{2}$
12. 0.01 $\dfrac{1}{100}$
13. 0.2 $\dfrac{1}{5}$
14. 0.75 $\dfrac{3}{4}$
15. 0.375 $\dfrac{3}{8}$

16. 0.24 $\dfrac{6}{25}$
17. $0.\overline{6}$ $\dfrac{2}{3}$
18. $0.5\overline{3}$ $\dfrac{8}{15}$
19. $0.1\overline{3}$ $\dfrac{2}{15}$
20. $0.\overline{4}$ $\dfrac{4}{9}$

ADDING AND SUBTRACTING FRACTIONS

To add or subtract two fractions with the same denominator, add or subtract the numerators. Write the result in simplest form.

EXAMPLES Add or subtract.

a. $\dfrac{1}{8} + \dfrac{3}{8}$

b. $\dfrac{9}{10} - \dfrac{3}{10}$

Solution

a. $\dfrac{1}{8} + \dfrac{3}{8} = \dfrac{1+3}{8} = \dfrac{4}{8} = \dfrac{1}{2}$

b. $\dfrac{9}{10} - \dfrac{3}{10} = \dfrac{9-3}{10} = \dfrac{6}{10} = \dfrac{3}{5}$

To add or subtract two fractions with different denominators, write equivalent fractions with a common denominator. Then add or subtract and write the result in simplest form.

EXAMPLES Add or subtract.

a. $\dfrac{4}{15} + \dfrac{2}{5}$

b. $\dfrac{3}{4} - \dfrac{1}{6}$

Solution

a. Write $\dfrac{2}{5}$ as $\dfrac{6}{15}$.

$$\dfrac{4}{15} + \dfrac{2}{5} = \dfrac{4}{15} + \dfrac{6}{15} = \dfrac{10}{15} = \dfrac{2}{3}$$

b. Write $\dfrac{3}{4}$ as $\dfrac{9}{12}$ and $\dfrac{1}{6}$ as $\dfrac{2}{12}$.

$$\dfrac{3}{4} - \dfrac{1}{6} = \dfrac{9}{12} - \dfrac{2}{12} = \dfrac{7}{12}$$

Practice

Add or subtract. Write the answer in simplest form.

1. $\dfrac{2}{7} + \dfrac{4}{7}$ $\dfrac{6}{7}$

2. $\dfrac{7}{12} - \dfrac{1}{12}$ $\dfrac{1}{2}$

3. $\dfrac{2}{3} - \dfrac{1}{3}$ $\dfrac{1}{3}$

4. $\dfrac{1}{6} + \dfrac{1}{6}$ $\dfrac{1}{3}$

5. $\dfrac{3}{4} - \dfrac{1}{4}$ $\dfrac{1}{2}$

6. $\dfrac{4}{5} + \dfrac{1}{5}$ 1

7. $\dfrac{7}{20} + \dfrac{9}{20}$ $\dfrac{4}{5}$

8. $\dfrac{79}{100} - \dfrac{14}{100}$ $\dfrac{13}{20}$

9. $\dfrac{7}{8} - \dfrac{1}{4}$ $\dfrac{5}{8}$

10. $\dfrac{2}{3} + \dfrac{1}{6}$ $\dfrac{5}{6}$

11. $\dfrac{3}{4} + \dfrac{1}{2}$ $1\dfrac{1}{4}$

12. $\dfrac{5}{12} - \dfrac{1}{3}$ $\dfrac{1}{12}$

13. $\dfrac{5}{7} - \dfrac{2}{5}$ $\dfrac{11}{35}$

14. $\dfrac{3}{16} + \dfrac{5}{8}$ $\dfrac{13}{16}$

15. $\dfrac{3}{8} - \dfrac{1}{12}$ $\dfrac{7}{24}$

16. $\dfrac{2}{3} + \dfrac{1}{2}$ $1\dfrac{1}{6}$

17. $\dfrac{5}{8} + \dfrac{3}{8}$ 1

18. $\dfrac{24}{25} - \dfrac{3}{5}$ $\dfrac{9}{25}$

19. $\dfrac{5}{9} + \dfrac{14}{15}$ $1\dfrac{22}{45}$

20. $\dfrac{9}{10} - \dfrac{4}{5}$ $\dfrac{1}{10}$

21. $\dfrac{2}{3} + \dfrac{1}{5}$ $\dfrac{13}{15}$

22. $\dfrac{3}{8} - \dfrac{7}{40}$ $\dfrac{1}{5}$

23. $\dfrac{17}{20} - \dfrac{1}{5}$ $\dfrac{13}{20}$

24. $\dfrac{7}{11} + \dfrac{1}{3}$ $\dfrac{32}{33}$

MULTIPLYING AND DIVIDING FRACTIONS

To multiply two fractions, multiply the numerators and multiply the denominators. Then write the result in simplest form.

EXAMPLES **Multiply.**

a. $\frac{3}{4} \times \frac{5}{6}$

b. $20 \times \frac{4}{5}$

Solution

a. $\frac{3}{4} \times \frac{5}{6} = \frac{3 \times 5}{4 \times 6} = \frac{15}{24} = \frac{5}{8}$

b. $\frac{20}{1} \times \frac{4}{5} = \frac{20 \times 4}{1 \times 5} = \frac{80}{5} = 16$

To divide by a fraction, multiply by its reciprocal and write the product in simplest form.

EXAMPLES **Divide.**

a. $\frac{1}{5} \div \frac{5}{8}$

b. $9 \div \frac{7}{10}$

Solution

a. $\frac{1}{5} \div \frac{5}{8} = \frac{1}{5} \times \frac{8}{5} = \frac{1 \times 8}{5 \times 5} = \frac{8}{25}$

b. $9 \div \frac{7}{10} = \frac{9}{1} \times \frac{10}{7} = \frac{9 \times 10}{1 \times 7} = \frac{90}{7} = 12\frac{6}{7}$

Practice

Multiply or divide. Write the answer in simplest form.

1. $\frac{1}{2} \times \frac{3}{4}$ $\frac{3}{8}$

2. $\frac{2}{3} \times \frac{3}{11}$ $\frac{2}{11}$

3. $20 \times \frac{1}{8}$ $2\frac{1}{2}$

4. $65 \times \frac{2}{5}$ 26

5. $\frac{5}{12} \times \frac{4}{9}$ $\frac{5}{27}$

6. $\frac{7}{8} \times \frac{1}{4}$ $\frac{7}{32}$

7. $16 \times \frac{15}{16}$ 15

8. $10 \times \frac{1}{3}$ $3\frac{1}{3}$

9. $\frac{1}{3} \div \frac{1}{2}$ $\frac{2}{3}$

10. $\frac{3}{4} \div \frac{5}{6}$ $\frac{9}{10}$

11. $9 \div \frac{1}{7}$ 63

12. $14 \div \frac{1}{2}$ 28

13. $\frac{7}{8} \div \frac{7}{2}$ $\frac{1}{4}$

14. $\frac{4}{5} \div \frac{1}{10}$ 8

15. $100 \div \frac{7}{8}$ $114\frac{2}{7}$

16. $40 \div \frac{3}{4}$ $53\frac{1}{3}$

17. $\frac{2}{5} \div \frac{1}{5}$ 2

18. $\frac{1}{4} \times \frac{7}{10}$ $\frac{7}{40}$

19. $\frac{22}{7} \times 49$ 154

20. $\frac{8}{9} \div 4$ $\frac{2}{9}$

21. $\frac{8}{3} \times \frac{3}{8}$ 1

22. $\frac{1}{5} \div \frac{2}{5}$ $\frac{1}{2}$

23. $\frac{3}{11} \div \frac{3}{11}$ 1

24. $\frac{5}{7} \times \frac{2}{15}$ $\frac{2}{21}$

25. $12 \div \frac{3}{2}$ 8

26. $\frac{7}{2} \times \frac{1}{14}$ $\frac{1}{4}$

27. $\frac{6}{5} \div \frac{3}{10}$ 4

28. $4 \times \frac{3}{16}$ $\frac{3}{4}$

RATIO AND PROPORTION

The **ratio of a to b** is $\frac{a}{b}$. The ratio of a to b can also be written as "a to b"

or as $a : b$. Because a ratio is a quotient, its denominator cannot be zero.

EXAMPLES A geometry class consists of 16 female students, 12 male students, and 2 teachers. Write each ratio in simplest form.

a. male students : female students   **b.** students : teachers

Solution

a. $\dfrac{12}{16} = \dfrac{3}{4}$ **b.** $\dfrac{16 + 12}{2} = \dfrac{28}{2} = \dfrac{14}{1}$

EXAMPLES Simplify the ratio.

a. $\dfrac{12 \text{ cm}}{4 \text{ m}}$ **b.** $\dfrac{6 \text{ ft}}{18 \text{ in.}}$

Solution Express both quantities in the same units of measure so that the units divide out. Write the fraction in simplest form.

a. $\dfrac{12 \text{ cm}}{4 \text{ m}} = \dfrac{\overset{3}{\cancel{12} \text{ cm}}}{\underset{1}{\cancel{4} \cdot 100 \text{ cm}}} = \dfrac{3}{100}$

b. $\dfrac{6 \text{ ft}}{18 \text{ in.}} = \dfrac{\overset{1}{\cancel{6}} \cdot 12 \text{ in.}}{\underset{3}{\cancel{18} \text{ in.}}} = \dfrac{12}{3} = \dfrac{4}{1}$

A **proportion** is an equation showing that two ratios are equal.

If the ratio $\frac{a}{b}$ is equal to the ratio $\frac{c}{d}$, then the following

proportion can be written:

$\dfrac{a}{b} = \dfrac{c}{d}$ where a, b, c, and d are not equal to zero

The numbers a and d are the **extremes** of the proportion.
The numbers b and c are the **means** of the proportion.

Here are two properties that are useful when solving a proportion.

Cross Product Property   The product of the extremes equals the product of the means.

     If $\dfrac{a}{b} = \dfrac{c}{d}$, then $ad = bc$.

Reciprocal Property   If two ratios are equal, then their reciprocals are also equal.

     If $\dfrac{a}{b} = \dfrac{c}{d}$, then $\dfrac{b}{a} = \dfrac{d}{c}$.

Skills Review

EXAMPLES Solve the proportion.

a. $\dfrac{x}{6} = \dfrac{5}{9}$

b. $\dfrac{3}{x} = \dfrac{4}{7}$

Solution

a. $\dfrac{x}{6} = \dfrac{5}{9}$

$9x = 6(5)$ Use cross products.

$x = \dfrac{30}{9} = \dfrac{10}{3}$ or $3\dfrac{1}{3}$

b. $\dfrac{3}{x} = \dfrac{4}{7}$

$\dfrac{x}{3} = \dfrac{7}{4}$ Use reciprocals.

$x = 3 \cdot \dfrac{7}{4} = \dfrac{21}{4}$ or $5\dfrac{1}{4}$

Practice

An algebra class consists of 10 female students, 15 male students, and 2 teachers. Write the ratio in simplest form.

1. female students : teachers $\dfrac{5}{1}$

2. students : teachers $\dfrac{25}{2}$

3. female students : male students $\dfrac{2}{3}$

4. teachers : male students $\dfrac{2}{15}$

5. teachers : students $\dfrac{2}{25}$

6. male students : female students $\dfrac{3}{2}$

Write the ratio of length to width for the rectangle.

7. $\dfrac{12}{5}$

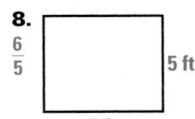

10 cm

24 cm

8. $\dfrac{6}{5}$

5 ft

6 ft

9. $\dfrac{1}{1}$

8 in.

8 in.

Simplify the ratio.

10. $\dfrac{3 \text{ yd}}{10 \text{ ft}}$ $\dfrac{9}{10}$

11. $\dfrac{4 \text{ lb}}{20 \text{ oz}}$ $\dfrac{16}{5}$

12. $\dfrac{40 \text{ cm}}{2 \text{ m}}$ $\dfrac{1}{5}$

13. $\dfrac{1 \text{ kg}}{450 \text{ g}}$ $\dfrac{20}{9}$

14. $\dfrac{18 \text{ in.}}{1 \text{ ft}}$ $\dfrac{3}{2}$

15. $\dfrac{6 \text{ mm}}{1 \text{ cm}}$ $\dfrac{3}{5}$

Solve the proportion.

16. $\dfrac{x}{4} = \dfrac{5}{12}$ $1\dfrac{2}{3}$

17. $\dfrac{5}{x} = \dfrac{2}{5}$ $12\dfrac{1}{2}$

18. $\dfrac{3}{x} = \dfrac{21}{49}$ 7

19. $\dfrac{x}{8} = \dfrac{1}{10}$ $\dfrac{4}{5}$

20. $\dfrac{x}{7} = \dfrac{11}{7}$ 11

21. $\dfrac{2}{x} = \dfrac{7}{9}$ $2\dfrac{4}{7}$

22. $\dfrac{2}{3} = \dfrac{x}{18}$ 12

23. $\dfrac{x}{10} = \dfrac{3}{100}$ 0.3 or $\dfrac{3}{10}$

24. $\dfrac{3}{4} = \dfrac{8}{x}$ $10\dfrac{2}{3}$

25. $\dfrac{2}{5} = \dfrac{10}{x}$ 25

26. $\dfrac{5}{4} = \dfrac{x}{14}$ $17\dfrac{1}{2}$

27. $\dfrac{x}{6} = \dfrac{7}{3}$ 14

28. $\dfrac{5}{8} = \dfrac{x}{20}$ $12\dfrac{1}{2}$

29. $\dfrac{x}{6} = \dfrac{6}{15}$ $2\dfrac{2}{5}$

30. $\dfrac{2}{1} = \dfrac{16}{x}$ 8

INEQUALITIES AND ABSOLUTE VALUE

When you compare two numbers a and b, a must be *less than, equal to,* or *greater than* b. You can compare two whole numbers or positive decimals by comparing the digits of the numbers from left to right. Find the first place in which the digits are different.

a is less than b.	$a < b$
a is equal to b.	$a = b$
a is greater than b.	$a > b$

> **EXAMPLES**   **Compare the two numbers.**
>
> **a.** 2.9 and 2.2   **b.** -4 and -1
>
> **Solution**   Use a number line. The numbers increase from left to right.
>
> **a.**
>
> **b.**
>
> 2.9 is to the right of 2.2, so $2.9 > 2.2$. -4 is to the left of -1, so $-4 < -1$.
> Also $9 > 2$, so $2.9 > 2.2$.

The absolute value of a number is its distance from zero on a number line. The symbol $|a|$ represents the absolute value of a.

> **EXAMPLES**   **Evaluate.**
>
> **a.** $|3|$ **b.** $|-2|$
>
> **Solution**   Use a number line.
>
> **a.**
>
> **b.**
>
> 3 is 3 units from 0, so $|3| = 3$. -2 is 2 units from 0, so $|-2| = 2$.

Practice

Compare the two numbers. Write the answer using <, >, or =.

1. 6 and -6 $6 > -6$ **2.** 4108 and 4117   **3.** -2 and -8 $-2 > -8$ **4.** -5 and 5 $-5 < 5$
                           $4108 < 4117$

5. -7.8 and -7.6 **6.** 16 and 16.5   **7.** 0.01 and 0.1   **8.** -3.14 and -3.141
   $-7.8 < -7.6$ $16 < 16.5$ $0.01 < 0.1$ $-3.14 > -3.141$

Write the numbers in order from least to greatest.

9. $-2, 0, 9, -8, 4, -6, 3$ **10.** $-0.5, 1, 1.5, -2.5, 0.05, -2, -1.5$
  $-8, -6, -2, 0, 3, 4, 9$ $-2.5, -2, -1.5, -0.5, 0.05, 1, 1.5$

11. 5124, 5421, 5214, 5142, 5412   **12.** $-0.39, -0.4, -0.26, -0.41, -0.32$
  5124, 5142, 5214, 5412, 5421 $-0.41, -0.4, -0.39, -0.32, -0.26$

Evaluate.

13. $|-6|$ 6   **14.** $|4|$ 4   **15.** $|0|$ 0   **16.** $|-20|$ 20   **17.** $|1.4|$ 1.4

INTEGERS

You can use a number line to add two integers. Move to the right to add a positive integer. Move to the left to add a negative integer.

EXAMPLES   **Add.**   **a.** $-4 + 3$ **b.** $-2 + (-3)$

Solution Use the number lines below.

a.

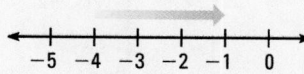

Start at -4. Go 3 units to the right.
End at -1. So, $-4 + 3 = -1$.

b.

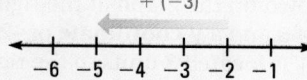

Start at -2. Go 3 units to the left.
End at -5. So, $-2 + (-3) = -5$.

EXAMPLES   **Subtract.**   **a.** $7 - 9$ **b.** $2 - (-6)$

Solution To subtract an integer, add its opposite.

a. $7 - 9 = 7 + (-9) = -2$ **b.** $2 - (-6) = 2 + 6 = 8$

When you multiply or divide integers, use these rules.

The product or quotient of two integers with the *same* sign is positive.

The product or quotient of two integers with *opposite* signs is negative.

EXAMPLES   **Multiply or divide.**

a. $(-3)(4)$ **b.** $-27 \div (-9)$

Solution

a. $(-3)(4) = -12$ ← negative product
      opposite signs

b. $-27 \div (-9) = 3$ ← positive quotient
      same signs

Practice

1. $-8 + (-2)$ -10 **2.** $10 + (-11)$ -1 **3.** $-4 + 6$ 2 **4.** $3 + (-9) + (-1)$ -7

5. $6 - (-5)$ 11 **6.** $-1 - 8$ -9 **7.** $-2 - (-2)$ 0 **8.** $4 - 12$ -8

9. $4(-4)$ -16 **10.** $(-7)(-1)$ 7 **11.** $(-3)(-20)$ 60 **12.** $(-10)(5)(2)$ -100

13. $80 \div (-2)$ -40 **14.** $-42 \div 7$ -6 **15.** $-16 \div (-8)$ 2 **16.** $-81 \div (-3)$ 27

17. $7 - (-9)$ 16 **18.** $20 \div (-4)$ -5 **19.** $-1 + (-5)$ -6 **20.** $(-9)(-2)(4)$ 72

21. $(-4)(25)$ -100 **22.** $-5 + 8$ 3 **23.** $-6 - 6$ -12 **24.** $-49 \div (-7)$ 7

25. $-13 - (-7)$ -6 **26.** $-45 \div 15$ -3 **27.** $(-3)(-3)$ 9 **28.** $9 + (-12) + 2$ -1

Skills Review Handbook   **663**

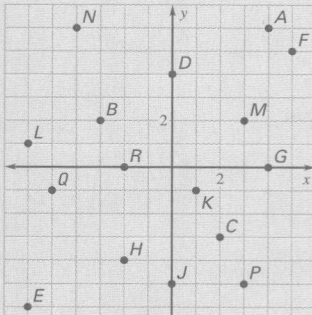

THE COORDINATE PLANE

A **coordinate plane** is formed by two number lines that intersect at the **origin**. The horizontal number line is the **x-axis**, and the vertical number line is the **y-axis**. Each point in a coordinate plane corresponds to an **ordered pair** of real numbers. The ordered pair for the origin is (0, 0).

Point $W(3, -2)$, shown on the graph at the right, has an **x-coordinate** of 3 and a **y-coordinate** of -2. From the origin, point W is located 3 units to the right and 2 units down.

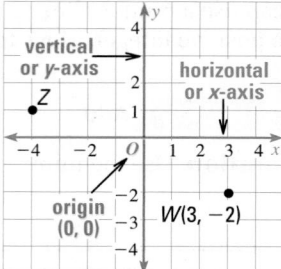

EXAMPLE   Use the graph above to name the coordinates of point Z.

Solution

From the origin, point Z is located 4 units to the left and 1 unit up. The coordinates of point Z are $(-4, 1)$.

EXAMPLES   Plot each point in a coordinate plane.

a. $P(-5, 2)$ **b.** $Q(3, 0)$

Solution

a. Start at the origin. Move 5 units left and 2 units up.

b. Start at the origin. Move 3 units right and 0 units up.

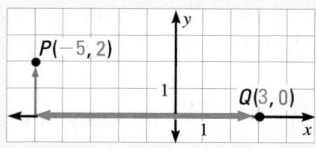

Practice

Name the coordinates of each point.

1. A $(-1, 0)$ **2.** B $(2, 3)$ **3.** C $(2, -2)$

4. D $(-5, -3)$ **5.** E $(-5, 3)$ **6.** F $(-1, -2)$

7. G $(5, -2)$ **8.** H $(4, 2)$ **9.** J $(-2, 5)$

10. K $(-4, 1)$ **11.** M $(1, 1)$ **12.** N $(-4, -1)$

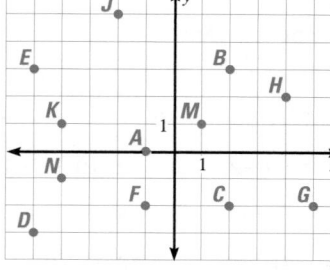

Plot the points in a coordinate plane. 13-28. See margin.

13. $A(4, 6)$ **14.** $B(-3, 2)$ **15.** $C(2, -3)$ **16.** $D(0, 4)$

17. $E(-6, -6)$ **18.** $F(5, 5)$ **19.** $G(4, 0)$ **20.** $H(-2, -4)$

21. $J(0, -5)$ **22.** $K(1, -1)$ **23.** $L(-6, 1)$ **24.** $M(3, 2)$

25. $N(-4, 6)$ **26.** $P(3, -5)$ **27.** $Q(-5, -1)$ **28.** $R(-2, 0)$

SLOPE OF A LINE

The **slope** of a line is the ratio of the vertical *rise* to the horizontal *run* between any two points on the line. You subtract coordinates to find the rise and the run. If a line passes through the points (x_1, y_1) and (x_2, y_2), then

$$\text{slope} = \frac{\text{rise}}{\text{run}} = \frac{y_2 - y_1}{x_2 - x_1}.$$

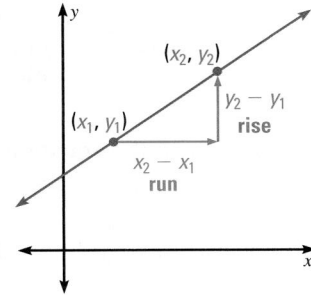

EXAMPLE Find the slope of the line that passes through the points $(-2, 1)$ and $(5, 4)$.

Solution Let $(x_1, y_1) = (-2, 1)$ and $(x_2, y_2) = (5, 4)$.

$$\text{slope} = \frac{y_2 - y_1}{x_2 - x_1}$$

$$= \frac{4 - 1}{5 - (-2)}$$

$$= \frac{3}{7}$$

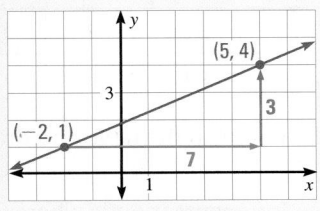

The slope of a line can be positive, negative, zero, or undefined.

Positive slope	Negative slope	Zero slope	Undefined slope
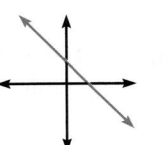	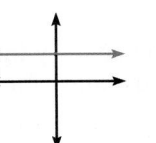	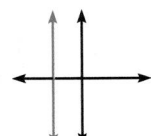	
rising line	falling line	horizontal line	vertical line

Practice

Find the slope of the line that passes through the points.

1. $(0, 3)$ and $(6, 1)$ $-\frac{1}{3}$

2. $(3, 2)$ and $(8, 4)$ $\frac{2}{5}$

3. $(1, 0)$ and $(3, 4)$ 2

4. $(1, 2)$ and $(5, 2)$ 0

5. $(1, 1)$ and $(-4, -4)$ 1

6. $(-3, 0)$ and $(2, -5)$ -1

7. $(4, -1)$ and $(-2, 3)$ $-\frac{2}{3}$

8. $(4, 2)$ and $(2, -6)$ 4

9. $(-4, -5)$ and $(0, -5)$ 0

Plot the points and draw the line that passes through them. Determine whether the slope is *positive, negative, zero,* or *undefined*. 10–15. Check graphs.

10. $(-3, 1)$ and $(-3, -5)$ undefined

11. $(1, 4)$ and $(4, -1)$ negative

12. $(-2, 2)$ and $(2, 4)$ positive

13. $(0, -1)$ and $(5, -1)$ zero

14. $(2, 0)$ and $(2, 2)$ undefined

15. $(-3, 1)$ and $(1, 2)$ positive

Skills Review Handbook **665**

1.

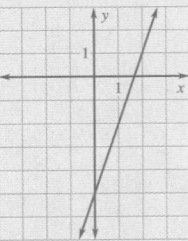

2.

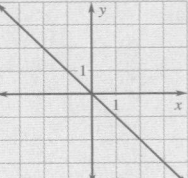

3.

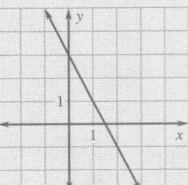

4.

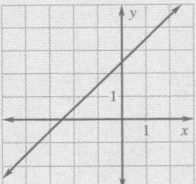

5.

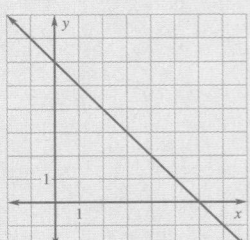

6.

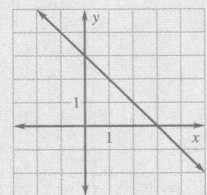

7.

8–12. See Additional Answers beginning on page AA1.

GRAPHING LINEAR EQUATIONS

Equations like $2x + 3y = -6$ and $y = 5x - 1$ are **linear equations**. A **solution** of a linear equation is an ordered pair (x, y) that makes the equation true. The graph of all solutions of a linear equation is a line.

> **EXAMPLE** Graph the equation $y - 4x = 2$.
>
> ### Solution
>
> ❶ Solve the equation for y: $y - 4x = 2$, so $y = 4x + 2$.
>
> ❷ Make a table of values to graph the equation.
>
> | x | $y = 4x + 2$ | (x, y) |
> |-----|--------------|----------|
> | -1 | $y = 4(-1) + 2 = -2$ | $(-1, -2)$ |
> | 0 | $y = 4(0) + 2 = 2$ | $(0, 2)$ |
> | 1 | $y = 4(1) + 2 = 6$ | $(1, 6)$ |
>
> ❸ Plot the points in the table and draw a line through them.

You can use **intercepts** (points where the graph crosses the x-axis and y-axis) to graph linear equations.

> **EXAMPLE** Graph the equation $3x - 4y = 12$.
>
> ### Solution
>
> ❶ To find the **x-intercept**, substitute 0 for y.
>
> $$3x - 4y = 12$$
> $$3x - 4(0) = 12$$
> $$x = 4$$
>
> So, $(4, 0)$ is a solution.
>
> ❷ To find the **y-intercept**, substitute 0 for x.
>
> $$3x - 4y = 12$$
> $$3(0) - 4y = 12$$
> $$y = -3$$
>
> So, $(0, -3)$ is a solution.
>
> ❸ Plot $(4, 0)$ and $(0, -3)$ and draw a line through them.

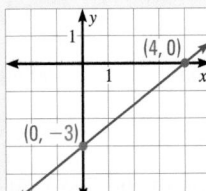

Practice

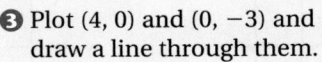

Graph the equation using a table of values or intercepts. See margin.

1. $y = x - 2$
2. $y = 3x - 5$
3. $y = -x$
4. $y = -2x + 3$

5. $y = 2.5 + x$
6. $y = 6 - x$
7. $x + y = 3$
8. $x - y = 4$

9. $2x + y = 6$
10. $2x - 3y = 12$
11. $-5x + 6y = 30$
12. $3.5x - 7y = -14$

SLOPE-INTERCEPT FORM

A linear equation $y = mx + b$ is written in **slope-intercept form**. The slope of the line is m and the y-intercept is b.

EXAMPLE Graph the equation $\frac{1}{2}x + 2y = 4$.

Solution

❶ Write the equation in slope-intercept form.

$$\frac{1}{2}x + 2y = 4$$

$$2y = -\frac{1}{2}x + 4$$

$$y = -\frac{1}{4}x + 2$$

The slope is $-\frac{1}{4}$ and the y-intercept is 2.

❷ Plot the point $(0, 2)$. Use the slope to locate other points on the line.

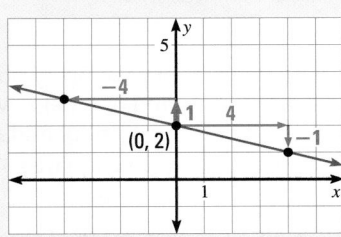

Graphs of linear equations of the form $y = b$ are horizontal lines with slope 0. Graphs of linear equations of the form $x = a$ are vertical lines with *undefined* slope.

EXAMPLES Graph the equation.

a. $y = -1$

b. $x = 3$

Solution

a.

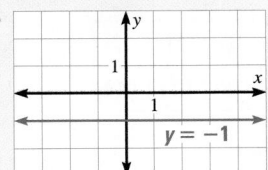

b.

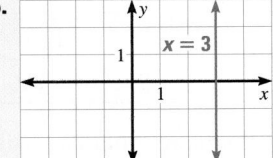

Practice

Graph the equation. See margin.

1. $y = x + 1$

2. $y = -2x + 5$

3. $y = 0.5x - 3$

4. $y - 3 = x$

5. $2y = 6x + 10$

6. $-2x + y = -6$

7. $y = -3$

8. $x = 5$

Lines with the same slope are parallel. Tell whether the graphs of the equations are *parallel* or *not parallel*.

9. $y = -3x + 2$
$y = -3x - 2$
parallel

10. $y = 4 - x$
$y = x - 4$
not parallel

11. $y - 2x = 4$
$2y - x = 4$
not parallel

12. $y = 4$
$y = 6$
parallel

Skills Review Handbook **667**

1.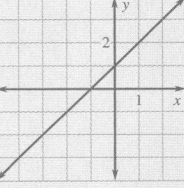

2.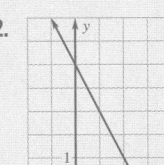

3.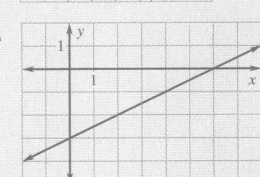

4.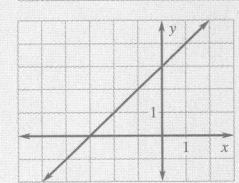

5.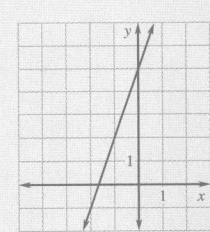

6.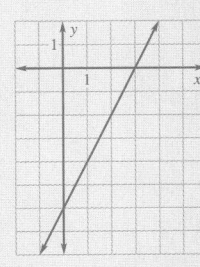

7.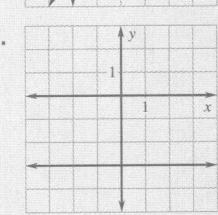

8.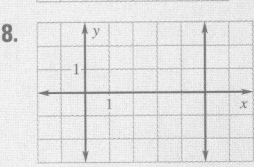

POWERS AND SQUARE ROOTS

An expression like 5^3 is called a **power.** The **exponent** 3 represents the number of times the **base** 5 is used as a factor: $5^3 = 5 \cdot 5 \cdot 5 = 125$.

EXAMPLES Evaluate.

a. 4^5

b. $(-10)^2$

Solution

a. $4^5 = 4 \cdot 4 \cdot 4 \cdot 4 \cdot 4 = 1024$

b. $(-10)^2 = (-10)(-10) = 100$

If $b^2 = a$, then b is a **square root** of a. Every positive number has two square roots, one positive and one negative. The two square roots of 16 are 4 and -4 because $4^2 = 16$ and $(-4)^2 = 16$. The radical symbol $\sqrt{}$ indicates the *nonnegative* square root, so $\sqrt{16} = 4$.

EXAMPLES Find all square roots of the number.

a. 25

b. -81

Solution

a. Since $5^2 = 25$ and $(-5)^2 = 25$, the square roots are 5 and -5.

b. Since -81 is negative, it has *no square roots*. There is no real number you can square to get -81.

The square of an integer is a **perfect square**, so the square root of a perfect square is an integer.

Integer, (n)	1	2	3	4	5	6	7	8	9	10	11	12
Perfect square, (n^2)	1	4	9	16	25	36	49	64	81	100	121	144

You can approximate the square root of a positive number that is *not* a perfect square by using a calculator and rounding.

EXAMPLES Evaluate. Give the exact value if possible. Otherwise, approximate to the nearest tenth.

a. $\sqrt{49}$

b. $\sqrt{5}$

Solution

a. Since 49 is a perfect square with $7^2 = 49$, $\sqrt{49} = 7$.

b. Since 5 is not a perfect square, use a calculator and round: $\sqrt{5} \approx 2.2$.

A number or expression inside a radical symbol is called a **radicand**. The **simplest form of a radical expression** is an expression that has no perfect square factors other than 1 in the radicand, no fractions in the radicand, and no radicals in the denominator of a fraction.

You can use the following properties to simplify radical expressions.

Product Property of Radicals $\sqrt{ab} = \sqrt{a} \cdot \sqrt{b}$ where $a \geq 0$ and $b \geq 0$

Quotient Property of Radicals $\sqrt{\dfrac{a}{b}} = \dfrac{\sqrt{a}}{\sqrt{b}}$ where $a \geq 0$ and $b > 0$

EXAMPLES   Simplify.

a. $\sqrt{18}$ **b.** $\sqrt{\dfrac{9}{4}}$ **c.** $\dfrac{1}{\sqrt{2}}$

Solution

a. $\sqrt{18} = \sqrt{9 \cdot 2} = \sqrt{9} \cdot \sqrt{2} = 3 \cdot \sqrt{2} = 3\sqrt{2}$ Factor using perfect square factor.

b. $\sqrt{\dfrac{9}{4}} = \dfrac{\sqrt{9}}{\sqrt{4}} = \dfrac{3}{2}$ Use the quotient property and simplify.

c. $\dfrac{1}{\sqrt{2}} = \dfrac{1}{\sqrt{2}} \cdot \dfrac{\sqrt{2}}{\sqrt{2}} = \dfrac{\sqrt{2}}{2}$ Write an equivalent fraction that has no radicals in the denominator.

Practice

Evaluate.

1. 8^2 64   **2.** $(-3)^2$ 9   **3.** $(-1)^3$ −1   **4.** 4^3 64

5. 2^5 32   **6.** 10^4 10,000   **7.** $(-9)^2$ 81   **8.** 6^3 216

Find all square roots of the number or write *no real square roots*.

9. 100 10, −10   **10.** −25 no real square roots   **11.** 1 1, −1   **12.** 49 7, −7

13. −9 no real square roots   **14.** 0 0   **15.** −36 no real square roots   **16.** 64 8, −8

Evaluate. Give the exact value if possible. If not, approximate to the nearest tenth.

17. $\sqrt{100}$ 10   **18.** $\sqrt{2}$ 1.4   **19.** $\sqrt{15}$ 3.9   **20.** $\sqrt{144}$ 12

21. $\sqrt{4}$ 2   **22.** $\sqrt{87}$ 9.3   **23.** $\sqrt{11}$ 3.3   **24.** $\sqrt{32}$ 5.7

25. $\sqrt{45}$ 6.7   **26.** $\sqrt{36}$ 6   **27.** $\sqrt{0}$ 0   **28.** $\sqrt{81}$ 9

Simplify.

29. $\sqrt{28}$ $2\sqrt{7}$ **30.** $\sqrt{27}$ $3\sqrt{3}$ **31.** $\sqrt{50}$ $5\sqrt{2}$ **32.** $\sqrt{48}$ $4\sqrt{3}$

33. $\sqrt{\dfrac{5}{16}}$ $\dfrac{\sqrt{5}}{4}$ **34.** $\sqrt{\dfrac{36}{49}}$ $\dfrac{6}{7}$ **35.** $\sqrt{\dfrac{1}{9}}$ $\dfrac{1}{3}$ **36.** $\sqrt{\dfrac{3}{25}}$ $\dfrac{\sqrt{3}}{5}$

37. $\dfrac{1}{\sqrt{3}}$ $\dfrac{\sqrt{3}}{3}$ **38.** $\dfrac{10}{\sqrt{2}}$ $5\sqrt{2}$ **39.** $\dfrac{5}{\sqrt{2}}$ $\dfrac{5\sqrt{2}}{2}$ **40.** $\dfrac{3}{\sqrt{3}}$ $\sqrt{3}$

EVALUATING EXPRESSIONS

To evaluate a **numerical expression** involving more than one operation, follow the **order of operations**.

❶ First do operations that occur within *grouping symbols*.

❷ Then evaluate *powers*.

❸ Then do *multiplications* and *divisions* from left to right.

❹ Finally, do *additions* and *subtractions* from left to right.

EXAMPLE   Evaluate the expression $2 - (4 - 7)^2 \div (-6)$.

Solution

$$2 - (4 - 7)^2 \div (-6) = 2 - (-3)^2 \div (-6) \qquad \text{Evaluate within parentheses.}$$

$$= 2 - 9 \div (-6) \qquad \text{Evaluate the power: } (-3)^2 = 9.$$

$$= 2 - (-1.5) \qquad \text{Do the division } 9 \div (-6).$$

$$= 3.5 \qquad \text{Do the subtraction } 2 - (-1.5).$$

To evaluate a **variable expression**, substitute a value for each variable and use the order of operations to simplify.

EXAMPLE   Evaluate the expression $x^2 + x - 18$ when $x = 5$.

Solution

$$x^2 + x - 18 = 5^2 + 5 - 18 \qquad \text{Substitute 5 for each } x.$$

$$= 25 + 5 - 18 \qquad \text{Evaluate the power: } 5^2 = 25.$$

$$= 12 \qquad \text{Add and subtract from left to right.}$$

Practice

Evaluate the expression.

1. $180 - (30 + 45)$ 105

2. $(8 - 2) \cdot 180$ 1080

3. $16 + 4 \cdot 2 - 3$ 21

4. $8^2 + (-6)^2$ 100

5. $-7 + 2^3 - 9$ −8

6. $9(7 - 2)^2$ 225

7. $\frac{1}{2}(100 - 74)$ 13

8. $\frac{5 + 7 \cdot 3}{6 + 7}$ 2

9. $\frac{3}{4} \cdot 24 + 4^2 - 1$ 33

Evaluate the expression when $n = 3$.

10. $-3n^2$ −27

11. $(-3n)^2$ 81

12. $n(n - 7)$ −12

13. $\frac{n + 2}{n - 2}$ 5

14. $n^2 - 7n + 6$ −6

15. $\frac{(n - 2) \cdot 180}{n}$ 60

16. $n^2 + 25$ 34

17. $-2n - 18$ −24

18. $(n + 2)(n - 2)$ 5

Student Resources

THE DISTRIBUTIVE PROPERTY

Here are four forms of the **distributive property**.

$$a(b + c) = ab + ac \qquad\qquad (b + c)a = ba + ca$$

$$a(b - c) = ab - ac \qquad\qquad (b - c)a = ba - ca$$

EXAMPLES   Use the distributive property to write the expression without parentheses.

a. $x(x + 4)$ **b.** $5 - (n - 2)$

Solution

a. $x(x + 4) = x(x) + x(4) = x^2 + 4x$

b. $5 - (n - 2) = 5 + (-1)(n - 2) = 5 + (-1)(n) + (-1)(-2) = 5 - n + 2 = 7 - n$

When an expression is written as a sum, the parts that are added are the
terms of the expression. **Like terms** are terms in an expression that have
the same variable raised to the same power. Numbers are also considered to
be like terms. You can use the distributive property to combine like terms.

EXAMPLES   Simplify the expression.

a. $-4x + 7x$ **b.** $2(x + y) - (4 - y)x$

Solution

a. $-4x + 7x = (-4 + 7)x = 3x$

b. $2(x + y) - (4 - y)x = 2x + 2y - 4x + xy$ Write without parentheses.

$\qquad\qquad\qquad\qquad\quad = (2 - 4)x + 2y + xy$ Group and combine like terms.

$\qquad\qquad\qquad\qquad\quad = -2x + 2y + xy$ Simplify within parentheses.

Practice

**Use the distributive property to write the expression without
parentheses.**

1. $2(a + 4)$ $2a + 8$ **2.** $(2k + 1)7$ $14k + 7$ **3.** $-(-3x + 2)$ $3x - 2$ **4.** $(7 - 2z)z$ $7z - 2z^2$

5. $y(y - 9)$ $y^2 - 9y$ **6.** $(j - 1)(-3)$ $-3j + 3$ **7.** $4b(b + 3)$ $4b^2 + 12b$ **8.** $-2(n - 6)$ $-2n + 12$

Simplify the expression.

9. $-m + 4 + 7m$ $6m + 4$ **10.** $6x - 9x + x$ $-2x$ **11.** $x^2 - 2x + 7x - 14$ $x^2 + 5x - 14$

12. $a^2 + 3a + 3a + 9$ **13.** $3 - (2x - 7)$ $10 - 2x$ **14.** $8 + 3(y - 4)$ $3y - 4$
$a^2 + 6a + 9$

15. $6h - 3h(h + 1)$ $3h - 3h^2$ **16.** $2(x + 4x) - 7$ $10x - 7$ **17.** $y(2y - 6) + y^2$ $3y^2 - 6y$

SOLVING ONE-STEP EQUATIONS

A **solution** of an equation is a value for the variable that makes a true statement. You can solve an equation by writing an *equivalent* equation (an equation with the same solution) that has the variable alone on one side. Here are four ways to solve a one-step equation.

• Add the same number to each side of the equation.

• Subtract the same number from each side of the equation.

• Multiply each side of the equation by the same nonzero number.

• Divide each side of the equation by the same nonzero number.

EXAMPLES　Solve the equation.

a. $y - 7 = 3$　　　　**b.** $x + 6 = -2$　　　　**c.** $\dfrac{n}{5} = 30$　　　　**d.** $12 = -4c$

Solution　Choose an operation to perform that will leave the variable alone on one side. Check your solution by substituting it back into the original equation.

a. Add 7 to each side.

$$y - 7 = 3$$
$$y - 7 + 7 = 3 + 7$$
$$y = 10$$

CHECK ✓ $10 - 7 = 3$

b. Subtract 6 from each side.

$$x + 6 = -2$$
$$x + 6 - 6 = -2 - 6$$
$$x = -8$$

CHECK ✓ $-8 + 6 = -2$

c. Multiply each side by 5.

$$\frac{n}{5} = 30$$
$$5 \cdot \frac{n}{5} = 5 \cdot 30$$
$$n = 150$$

CHECK ✓ $\dfrac{150}{5} = 30$

d. Divide each side by -4.

$$12 = -4c$$
$$\frac{12}{-4} = \frac{-4c}{-4}$$
$$-3 = c$$

CHECK ✓ $12 = -4(-3)$

Practice

Solve the equation.

1. $k - 6 = 0$　6　　　　**2.** $19 = r - (-9)$　10　　　　**3.** $w - 5 = -13$　-8　　　　**4.** $20 = y - 4$　24

5. $x + 12 = 25$　13　　　　**6.** $y + 7 = -16$　-23　　　　**7.** $n + (-4) = -1$　3　　　　**8.** $-6 = c + 4$　-10

9. $\dfrac{1}{2}x = -14$　-28　　　　**10.** $\dfrac{n}{3} = 6$　18　　　　**11.** $\dfrac{a}{5} = -1$　-5　　　　**12.** $-\dfrac{3}{4}d = 24$　-32

13. $-36 = -9a$　4　　　　**14.** $-32h = 4$　$-\dfrac{1}{8}$　　　　**15.** $12 = -12b$　-1　　　　**16.** $4z = 132$　33

17. $22 = x - 2$　24　　　　**18.** $-\dfrac{m}{6} = -12$　72　　　　**19.** $-15y = 75$　-5　　　　**20.** $w + 4 = 15$　11

Student Resources

SOLVING MULTI-STEP EQUATIONS

Sometimes solving an equation requires more than one step. Use the techniques for solving one-step equations given on the previous page. Simplify one or both sides of the equation first, if needed, by using the distributive property or combining like terms.

EXAMPLES   Solve the equation.

a. $2x - 3 = 5$ **b.** $6y + 1 = 5y - 9$

Solution

a. $2x - 3 = 5$ **b.** $6y + 1 = 5y - 9$

     $2x - 3 + 3 = 5 + 3$ Add 3. $6y - 5y + 1 = 5y - 5y - 9$ Subtract 5y.

            $2x = 8$ Simplify. $y + 1 = -9$ Simplify.

            $\dfrac{2x}{2} = \dfrac{8}{2}$ Divide by 2. $y + 1 - 1 = -9 - 1$ Subtract 1.

            $x = 4$ Simplify. $y = -10$ Simplify.

   CHECK $\checkmark 2(4) - 3 \overset{?}{=} 5$ **CHECK** $\checkmark 6(-10) + 1 \overset{?}{=} 5(-10) - 9$

             $8 - 3 = 5$ $-59 = -50 - 9$

Practice

Solve the equation.

1. $2x + 3 = 11$ 4   **2.** $4x - 2 = 10$ 3   **3.** $7 + 2x = 17$ 5

4. $3y - 4 = 20$ 8   **5.** $\dfrac{x}{7} + 2 = 1$ −7   **6.** $21 = -2z + 1$ −10

7. $10 - x = -16$ 26   **8.** $6 - \dfrac{3a}{2} = -6$ 8   **9.** $8n + 2 = -30$ −4

10. $\dfrac{3}{8}x - 6 = 18$ 64   **11.** $3 = 5 + \dfrac{1}{4}x$ −8   **12.** $-6m - 1 = 11$ −2

13. $3r - (2r + 1) = 21$ 22   **14.** $5(z + 3) = 30$ 3   **15.** $44 = 5g - 8 - g$ 13

16. $4(t - 7) + 6 = 30$ 13   **17.** $22d - (6 + 2d) = 4$ $\dfrac{1}{2}$ **18.** $85 = \dfrac{1}{2}(226 - x)$ 56

19. $12a - 5 = 7a$ 1   **20.** $75 + 7x = 2x$ −15   **21.** $5n - 9 = 3n - 1$ 4

22. $14r + 81 = -r$ $-\dfrac{27}{5}$ **23.** $4 - 6p = 2p - 12$ 2   **24.** $7y - 84 = 2y + 61$ 29

25. $1 + j = 2(2j + 1)$ $-\dfrac{1}{3}$ **26.** $3(x + 1) - 6 = 9$ 4   **27.** $5 - 2(r + 6) = 1$ −4

28. $\dfrac{1}{5}y + 7 = 3$ −20   **29.** $\dfrac{8 + x}{2} = 10$ 12   **30.** $\dfrac{n - 2}{2} = -6$ −10

31. $7(b - 3) = 8b + 2$ −23   **32.** $12 - 23c = 7(9 - c)$ $-\dfrac{51}{16}$ **33.** $4z + 2(z - 3) = 0$ 1

Using Formulas

A **formula** is an algebraic equation that relates two or more real-life quantities. Here are some formulas for the perimeter P, area A, and circumference C of some common figures.

Square

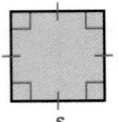

side length s

$P = 4s$

$A = s^2$

Rectangle

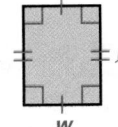

length ℓ and width w

$P = 2\ell + 2w$

$A = \ell w$

Triangle

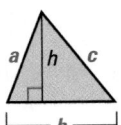

side lengths a, b, and c
base b and height h

$P = a + b + c$

$A = \frac{1}{2}bh$

Circle

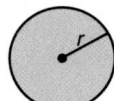

radius r

$C = 2\pi r$

$A = \pi r^2$

Pi (π) is the ratio of a circle's circumference to its diameter.

EXAMPLE **Find the length of a rectangle with perimeter 20 centimeters and width 4 centimeters.**

Solution

$P = 2\ell + 2w$	Write the appropriate formula.
$20 = 2\ell + 2(4)$	Substitute known values of the variables.
$20 = 2\ell + 8$	Simplify.
$12 = 2\ell$	Subtract 8 from each side.
$6 = \ell$	Divide each side by 2.

ANSWER ▸ The length of the rectangle is 6 centimeters.

Practice

1. The perimeter of a square is 24 meters. Find the side length. **6 m**

2. Find the area of a circle with radius 1.5 centimeters. (Use $\pi \approx 3.14$.) **$2.25\pi \approx 7.07$ cm²**

3. A triangle has a perimeter of 50 millimeters and two sides that measure 14 millimeters each. Find the length of the third side. **22 mm**

4. Find the width of a rectangle with area 32 square feet and length 8 feet. **4 ft**

5. The circumference of a circle is 8π inches. Find the radius. **4 in.**

6. Find the side length of a square with area 121 square centimeters. **11 cm**

7. Find the height of a triangle with area 18 square meters and base 4 meters. **9 m**

8. A square has an area of 49 square units. Find the perimeter. **28 units**

Chapter 1

Sketch the next figure you expect in the pattern. (Lesson 1.1) 1, 2. See margin.

1.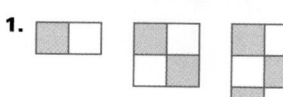

2.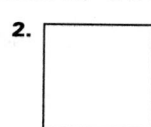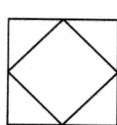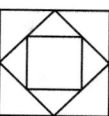

Describe a pattern in the numbers. Write the next number you expect in the pattern. (Lesson 1.1)

3. 1, 5, 25, 125, . . .
Each number is 5 times the previous number; 625.

4. 1, 2, 4, 7, 11, . . . Add 1, then 2, then 3, then 4, and so on; 16.

5. $-8, -5, -2, 1, . . .$ Each number is 3 more than the previous number; 4.

6. 405, 135, 45, 15, . . .
Each number is one third the previous number; 5.

7. $\frac{1}{1}, \frac{2}{3}, \frac{4}{9}, \frac{8}{27}, . . .$ Multiply the previous number by $\frac{2}{3}$; $\frac{16}{81}$.

8. 1, 1, 2, 6, 24, 120, . . .
Multiply by 1, then 2, then 3, and so on; 720.

9. Complete the conjecture based on the pattern you observe.
(Lesson 1.2)

> *Conjecture:* The square of a negative number is a(n) __?__ number positive

$$(-4)^2 = 16 \qquad (-1)^2 = 1 \qquad \left(-\frac{3}{4}\right)^2 = \frac{9}{16} \qquad (-0.5)^2 = 0.25$$

Show that the conjecture is false by finding a counterexample.
(Lesson 1.2)

10. Conjecture: All rectangles with an area of 12 ft^2 have the same perimeter. *Sample answer:* A rectangle that is 6 ft by 2 ft has an area of 12 ft^2 and a perimeter of 16 ft, but a rectangle that is 4 ft by 3 ft has an area of 12 ft^2 and a perimeter of 14 ft.

11. Conjecture: The absolute value of a number is always positive.
The absolute value of 0 is 0.

12. Conjecture: If the sum of two numbers is positive, then the numbers must be positive. *Sample answer:* $-3 + 8 = 5$, but -3 and 8 are not both positive numbers.

13. Conjecture: The quotient of two positive integers is a positive integer. *Sample answer:* $2 \div 6 = \frac{1}{3}$, which is not an integer.

Sketch the figure. (Lesson 1.3) 14–16. See margin for sample answers.

14. Draw three points, A, B, and C, that are not collinear. Then draw $\overleftrightarrow{AB}$ and $\overrightarrow{BC}$.

15. Draw two points, P and Q. Sketch $\overrightarrow{PQ}$. Then draw a point R that lies on $\overrightarrow{PQ}$ but is not on $\overline{PQ}$.

16. Draw two points J and K. Then sketch $\overleftrightarrow{JK}$. Add points L and M so that L is not on $\overleftrightarrow{JK}$ and M is between J and K. Draw $\overrightarrow{JL}$, $\overleftrightarrow{KL}$, and $\overline{LM}$.

1.

2.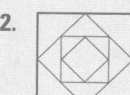

14–16. Sample answers are given.

14.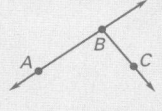

15. P Q R

16.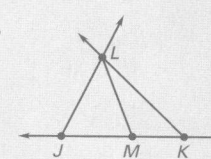

Extra Practice

22. Sample answer:

23. Sample answer:

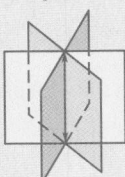

Use the diagram at the right. (Lessons 1.3, 1.4)

17. Name a point that is coplanar with *Q*, *N*, and *K*. *R*

18. Are points *L*, *M*, and *N* collinear? no

19. Name three segments that contain point *P*. $\overline{PQ}$, $\overline{PS}$, $\overline{PM}$

20. Name four lines that intersect $\overleftrightarrow{NK}$. $\overleftrightarrow{NQ}$, $\overleftrightarrow{NM}$, $\overleftrightarrow{KR}$, $\overleftrightarrow{KL}$

21. Describe the intersection of plane *KLM* and plane *QPM*. $\overleftrightarrow{NM}$

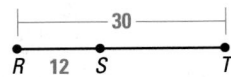

Sketch the figure. (Lesson 1.4) 22, 23. See margin.

22. Sketch a line and a plane that do not intersect.

23. Sketch three planes that intersect in a line.

Find the length. (Lesson 1.5)

24. Find *DF*. 14

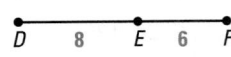

25. Find *ST*. 18

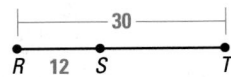

Plot the points in a coordinate plane. Decide whether $\overline{JK}$ and $\overline{MN}$ are congruent. (Lesson 1.5)

26. *J*(−2, −3), *K*(−2, −1), *M*(0, 4), *N*(−3, 4) no

27. *J*(5, 1), *K*(−1, 1), *M*(−2, 3), *N*(−2, −3) yes

28. *J*(3, −2), *K*(3, −5), *M*(3, 3), *N*(7, 3) no

Name the vertex and the sides of the angle. Then write two names for the angle. (Lesson 1.6)

29.

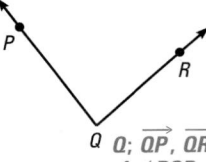

Q; $\overrightarrow{QP}$, $\overrightarrow{QR}$; any two of ∠*PQR*, ∠*RQP*, ∠*Q*

30.

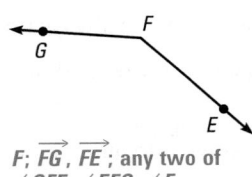

F; $\overrightarrow{FG}$, $\overrightarrow{FE}$; any two of ∠*GFE*, ∠*EFG*, ∠*F*

31.

B; $\overrightarrow{BA}$, $\overrightarrow{BC}$; ∠*ABC*, ∠*CBA*

Use the Angle Addition Postulate to find the measure of the specified angle. (Lesson 1.6)

32. *m*∠*STR* = __?__ 70°

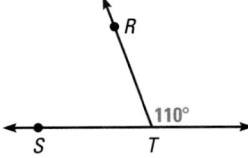

33. *m*∠*HJK* = __?__ 65°

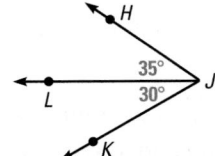

34. *m*∠*DEF* = __?__ 90°

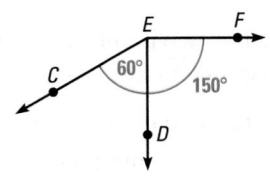

Extra Practice

Chapter 2

M is the midpoint of $\overline{AB}$. Find the indicated segment lengths or value of the variable. (Lesson 2.1)

1. Find *AB* and *AM*.   30; 15

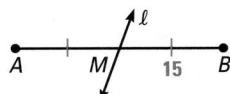

2. Find *AM* and *MB*.   1.8; 1.8

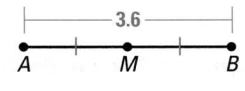

3. Find the value of *x*.   14

Find the coordinates of the midpoint of $\overline{PQ}$. (Lesson 2.1)

4. $P(-4, 2)$, $Q(8, -4)$ (2, −1)

5. $P(-1, 5)$, $Q(7, -5)$ (3, 0)

6. $P(-4, -6)$, $Q(-12, 4)$ (−8, −1)

7. $P(-3, 1)$, $Q(6, 5)$ (1.5, 3)

8. $P(0, 4)$, $Q(-7, -5)$ (−3.5, −0.5)

9. $P(-4, -4)$, $Q(1, 1)$ (−1.5, −1.5)

$\overrightarrow{BD}$ bisects $\angle ABC$. Find the indicated angle measure or value of the variable. (Lesson 2.2)

10. Find $m\angle ABD$.   42°

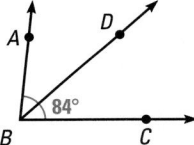

11. Find $m\angle ABC$.   50°

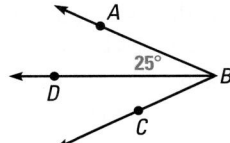

12. Find the value of *x*.   11

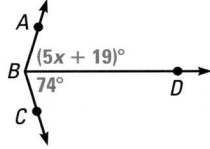

13. State whether the angles in the diagram are *complementary*, *supplementary*, or *neither*. Also state whether the angles are *adjacent* or *nonadjacent*. (Lesson 2.3)   neither; adjacent

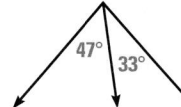

14. The measure of $\angle R$ is 12°. Find the measures of a complement and a supplement of $\angle R$. (Lesson 2.3)   78°; 168°

15. The measure of $\angle S$ is 53°. Find the measures of a complement and a supplement of $\angle S$. (Lesson 2.3)   37°; 127°

16. $\angle JKM$ and $\angle MKL$ are supplementary angles. Find the value of *y*. (Lesson 2.3)   36

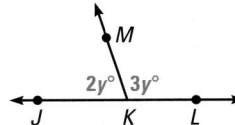

17. $\angle 1$ and $\angle 2$ are supplementary angles. Also $\angle 2$ and $\angle 3$ are supplementary angles. Name a pair of congruent angles. Explain your reasoning. (Lesson 2.3) $\angle 1$ and $\angle 3$; $\angle 1$ and $\angle 3$ are both supplementary to $\angle 2$, so $\angle 1 \cong \angle 3$ by the Congruent Supplements Theorem.

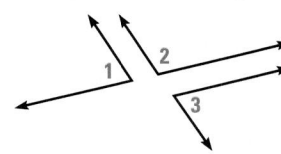

Find the measure of the numbered angle(s). (Lesson 2.4)

18.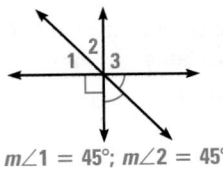

$m\angle 1 = 152°$

19.

$m\angle 1 = 95°$; $m\angle 2 = 85°$;
$m\angle 3 = 95°$

20.

$m\angle 1 = 45°$; $m\angle 2 = 45°$;
$m\angle 3 = 90°$

Rewrite the statement as an if-then statement. Then underline the hypothesis and circle the conclusion. (Lesson 2.5)

21. It must be true if you read it in a newspaper. **If you read it in a newspaper, then it must be true.**

22. An apple a day keeps the doctor away. **If you have an apple a day, then it will keep the doctor away.**

23. The square of an odd number is odd. **If a number is odd, then its square is odd.**

24. Use the Law of Detachment to reach a conclusion from the given true statements. (Lesson 2.5)

If E is between D and F, then $DF = DE + EF$.
E is between D and F. **DF = DE + EF**

25. Use the Law of Syllogism to write the if-then statement that follows from the pair of true statements. (Lesson 2.5)

If it is hot today, May will go to the beach.
If May goes to the beach, I will go too. **If it is hot today, then I will go to the beach.**

Name the property that the statement illustrates. (Lesson 2.6)

26. $\overline{MN} \cong \overline{MN}$ **Reflexive Property of Congruence**

27. If $ST = 4$, then $RS + ST = RS + 4$. **Addition Property of Equality**

28. If $\angle R \cong \angle Y$, then $\angle Y \cong \angle R$. **Symmetric Property of Congruence**

29. If $AB = BC$ and $BC = CD$, then $AB = CD$. **Transitive Property of Equality**

30. If $m\angle P = m\angle Q$, then $2 \cdot m\angle P = 2 \cdot m\angle Q$. **Multiplication Property of Equality**

31. If $AM = MB$ and $WX = XY$, then $AM - WX = MB - XY$. **Subtraction Property of Equality**

32. In the diagram, $\angle 3$ and $\angle 2$ are supplementary angles. Complete the argument to show that $m\angle 3 = m\angle 1$. (Lesson 2.6)

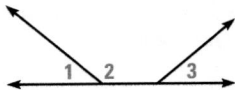

$\angle 3$ and $\angle 2$ are supplementary angles.	Given
$m\angle 3 + m\angle 2 = 180°$	_?_ Definition of supplementary angles
$m\angle 1 + m\angle 2 = 180°$	_?_ Postulate Linear Pair
$m\angle 3 + m\angle 2 = m\angle 1 + m\angle 2$	_?_ Substitution Property of Equality
$m\angle 3 = m\angle 1$	_?_ Subtraction Property of Equality

Chapter 3

Think of each segment in the diagram as part of a line. There may be more than one correct answer. (Lesson 3.1) 1–5. Sample answers are given.

1. Name a line that appears parallel to $\overleftrightarrow{AD}$. $\overleftrightarrow{BC}$, $\overleftrightarrow{GF}$, $\overleftrightarrow{HE}$

2. Name a line perpendicular to $\overleftrightarrow{AD}$. $\overleftrightarrow{GA}$, $\overleftrightarrow{FD}$

3. Name a line skew to $\overleftrightarrow{AD}$. $\overleftrightarrow{EC}$, $\overleftrightarrow{HB}$, $\overleftrightarrow{HG}$

4. Name a plane that appears parallel to plane GAB.
   FDC, DCE, CEF, EFD

5. Name a plane perpendicular to $\overleftrightarrow{AD}$. GAB, FDC

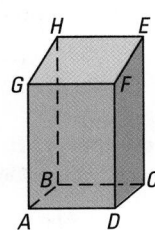

In the diagram, $r \perp s$ and $\angle 5 \cong \angle 6$. Determine whether enough information is given to conclude that the statement is true. (Lesson 3.2)

6. $m\angle 5 = 90°$ yes

7. $\angle 2 \cong \angle 3$ no

8. $m\angle 2 + m\angle 3 = 90°$ yes

9. $\angle 1 \cong \angle 5$ yes

10. $\angle 2 \cong \angle 4$ no

11. $r \perp t$ yes

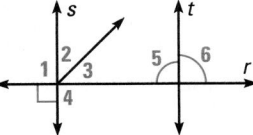

Find the value of x, given that $p \perp q$. (Lesson 3.2)

12. 50
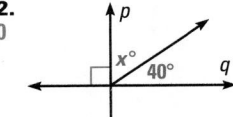

13. 15
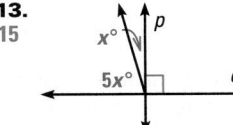

14. 48
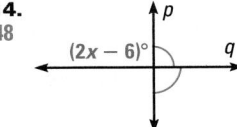

Use the diagram at the right to determine the relationship between the pair of angles. (Lesson 3.3)

15. $\angle 1$ and $\angle 8$
alternate exterior angles

16. $\angle 6$ and $\angle 2$
corresponding angles

17. $\angle 7$ and $\angle 2$
alternate interior angles

18. $\angle 1$ and $\angle 5$
corresponding angles

19. $\angle 6$ and $\angle 3$
alternate exterior angles

20. $\angle 5$ and $\angle 2$
same-side interior angles

Find $m\angle 1$ and $m\angle 2$. Explain your reasoning. (Lesson 3.4) 21–26. See margin for sample responses.

21.
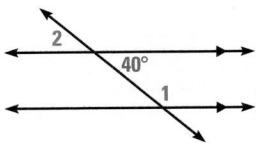

22.
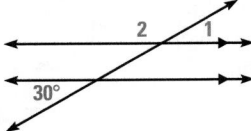

23.
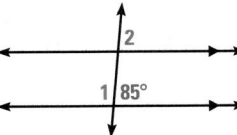

24.
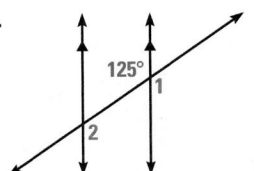

25.
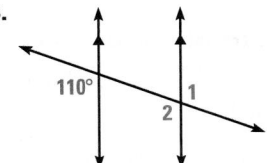

26.
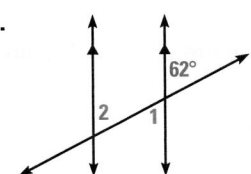

21. $m\angle 1 = 140°$, by the Same-Side Interior Angles Theorem; $m\angle 2 = 40°$, by the Vertical Angles Theorem

22. $m\angle 1 = 30°$, by the Alternate Exterior Angles Theorem; $m\angle 2 = 150°$, by the Linear Pair Postulate

23. $m\angle 1 = 95°$, by the Linear Pair Postulate; $m\angle 2 = 85°$, by the Corresponding Angles Postulate

24. $m\angle 1 = 125°$, by the Vertical Angles Theorem; $m\angle 2 = 125°$, by the Corresponding Angles Postulate

25. $m\angle 1 = 110°$, by the Alternate Exterior Angles Theorem; $m\angle 2 = 110°$, by the Corresponding Angles Postulate

26. $m\angle 1 = 62°$, by the Vertical Angles Theorem; $m\angle 2 = 62°$, by the Alternate Interior Angles Theorem

36.

Determine whether enough information is given to conclude that
$m \parallel n$. (Lesson 3.5)

27.
yes

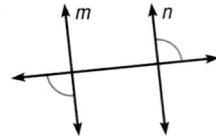

28.
no

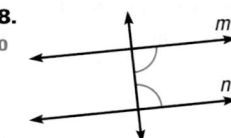

29.
yes

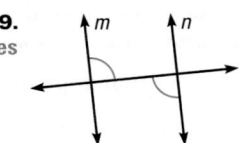

Explain how you would show that $a \parallel b$. State any theorems or
postulates that you would use. (Lesson 3.6)

30.

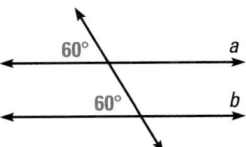

Corresponding Angles Converse

31.

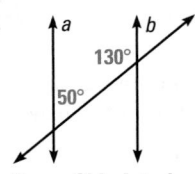

Same-Side Interior
Angles Converse

32.

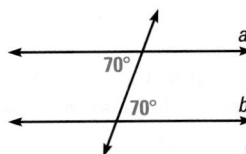

Alternate Interior
Angles Converse

Find a value of x so that $p \parallel q$. (Lesson 3.6)

33.
13

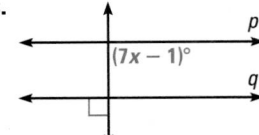

34.
19

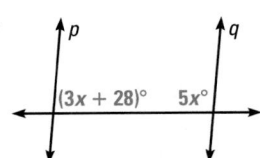

35.
5

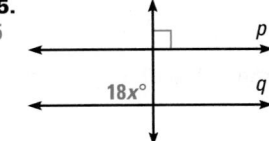

36. Draw a horizontal line ℓ and choose a point A above ℓ. Construct a
line m perpendicular to ℓ through point A. (Lesson 3.6) **See margin.**

Decide whether the red figure is a translation of the blue figure.
(Lesson 3.7)

37.
no

38.
no

39.
yes

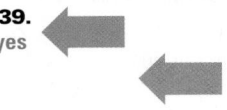

40. The red figure is a translation of the blue
figure. Describe the translation using words.
Then describe the translation using
coordinate notation. (Lesson 3.7)
Each point is moved 8 units to the left
and 1 unit down; $(x, y) \rightarrow (x - 8, y - 1)$.

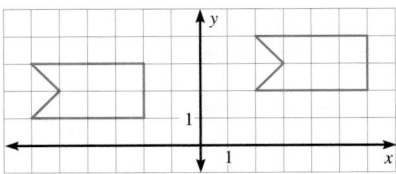

Find the image of each point using the translation $(x, y) \rightarrow (x - 3, y + 11)$.
(Lesson 3.7)

41. $(0, -6)$ $(-3, 5)$

42. $(5, 9)$ $(2, 20)$

43. $(-2, 0)$ $(-5, 11)$

44. $(4, -12)$ $(1, -1)$

45. $(-7, -9)$ $(-10, 2)$

46. $(3, -11)$ $(0, 0)$

Chapter 4

Classify the triangle by its sides. (Lesson 4.1)

1.
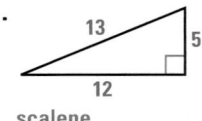
13
5
12
scalene

2.
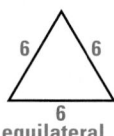
6   6
6
equilateral

3.
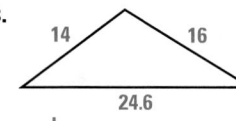
14   16
24.6
scalene

Classify the triangle by its angles. (Lesson 4.1)

4.

equiangular

5.
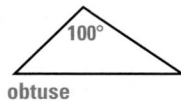
100°
obtuse

6.

70°
25°
acute

Identify which side is opposite each angle. (Lesson 4.1)

7.
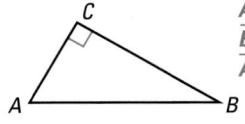
$\overline{AB}$ is opposite $\angle C$;
$\overline{BC}$ is opposite $\angle A$;
$\overline{AC}$ is opposite $\angle B$.

8.
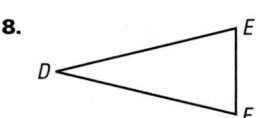
$\overline{DE}$ is opposite $\angle F$;
$\overline{EF}$ is opposite $\angle D$;
$\overline{DF}$ is opposite $\angle E$.

Find the measure of $\angle 1$. (Lesson 4.2)

9.

79°
73°
28°   1

10.
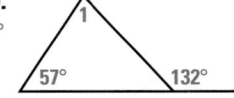
75°
1
57°   132°

11.
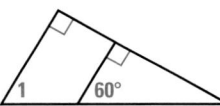
60°
1   60°

Find the value of x. (Lesson 4.3)

12.
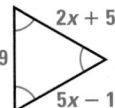
2
$2x + 5$
9
$5x - 1$

13.
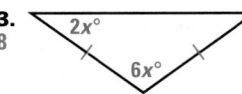
18
$2x°$
$6x°$

14.
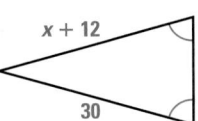
18
$x + 12$
30

15. Use the diagram to find the measures of $\angle 1$, $\angle 2$, and $\angle 3$.
(Lesson 4.3) 55°, 55°, 70°
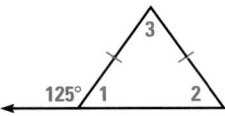
3
125° 1   2

Find the unknown side length. Round your answer to the nearest tenth, if necessary. (Lesson 4.4)

16.
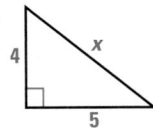
6.4
4   x
5

17.
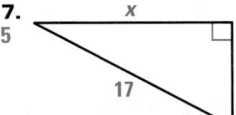
15
x
17   8

18.
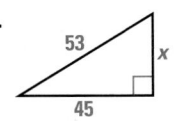
28
53
x
45

Find the distance between *A* and *B*. Round your answer to the nearest tenth, if necessary. (Lesson 4.4)

19. $A(-4, 0)$
$B(4, 6)$ 10

20. $A(5, 7)$
$B(0, -5)$ 13

21. $A(2, -1)$
$B(1, -4)$ 3.2

22. $A(-7, -2)$
$B(0, 7)$ 11.4

Classify the triangle as *acute*, *right*, or *obtuse*. (Lesson 4.5)

23.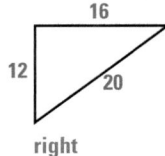

right

24.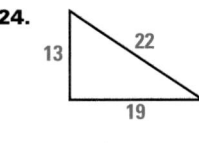

acute

25.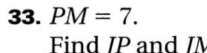

obtuse

Classify the triangle with the given side lengths as *acute*, *right*, or *obtuse*. (Lesson 4.5)

26. 10, 12, 14 acute

27. 65, 63, 16 right

28. 7, 9, 12 obtuse

29. $\sqrt{11}$, 5, 6 right

30. 5, 7, 9 obtuse

31. 8, 10, 12 acute

***P* is the centroid of $\triangle JKL$. Use the information to find the indicated lengths.** (Lesson 4.6)

32. $KN = 24$.
Find KP and PN.

33. $PM = 7$.
Find JP and JM.

34. $LP = 10$.
Find PR and LR.

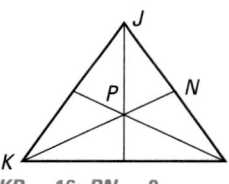

$KP = 16$; $PN = 8$

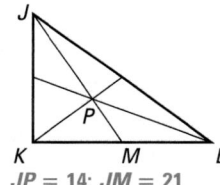

$JP = 14$; $JM = 21$

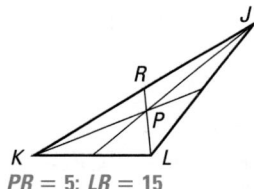

$PR = 5$; $LR = 15$

Name the angles from largest to smallest. (Lesson 4.7)

35.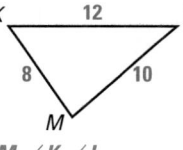

$\angle M, \angle K, \angle L$

36.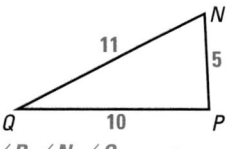

$\angle P, \angle N, \angle Q$

37.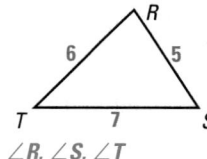

$\angle R, \angle S, \angle T$

Name the sides from longest to shortest. (Lesson 4.7)

38.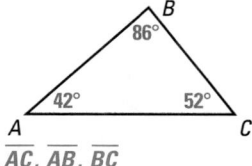

$\overline{AC}, \overline{AB}, \overline{BC}$

39.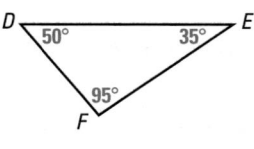

$\overline{DE}, \overline{EF}, \overline{DF}$

40.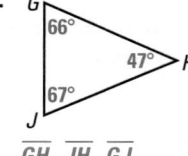

$\overline{GH}, \overline{JH}, \overline{GJ}$

Determine whether the side lengths of the triangle are possible. Explain your reasoning. (Lesson 4.7)

41. 7, 7, 7
Yes; $7 + 7 > 7$.

42. 5, 7, 12
No; $5 + 7 = 12$.

43. 1, 2, 4
No; $1 + 2 < 4$.

44. 10, 20, 29
Yes; $10 + 20 > 29$, $20 + 29 > 10$, and $29 + 10 > 20$.

Student Resources

Chapter 5

In Exercises 1–3, use the diagram at the right. (Lesson 5.1)

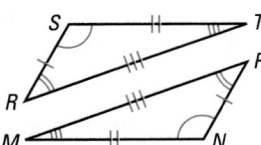

1. Name all pairs of corresponding congruent sides.
$\overline{RS} \cong \overline{PN}$; $\overline{ST} \cong \overline{NM}$; $\overline{TR} \cong \overline{MP}$

2. Name all pairs of corresponding congruent angles.
$\angle R \cong \angle P$, $\angle S \cong \angle N$, $\angle T \cong \angle M$

3. Write a congruence statement for the triangles.
$\triangle RST \cong \triangle PNM$

Determine whether the triangles are congruent. If so, write a congruence statement. (Lesson 5.1)

4.

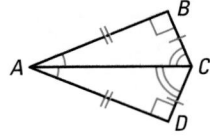

yes; $\triangle ABC \cong \triangle ADC$

5.

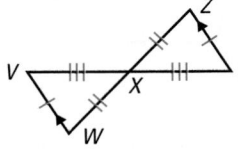

yes; $\triangle XVW \cong \triangle XYZ$

6.

no

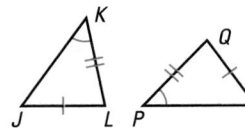

Decide whether enough information is given to show that the triangles are congruent. If so, state the congruence postulate you would use. (Lesson 5.2)

7.

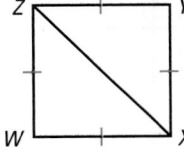

yes; SSS Congruence Postulate

8.

no

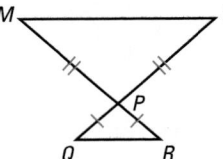

9.

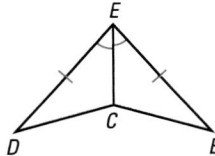

yes; SAS Congruence Postulate

Does the diagram give enough information to show that the triangles are congruent? If so, state the postulate or theorem you would use.
(Lesson 5.3)

10.

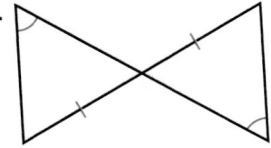

yes; AAS Congruence Theorem

11.

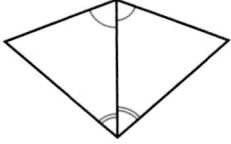

yes; ASA Congruence Postulate

12.

no

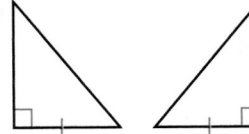

Decide whether enough information is given to show that the triangles are congruent. If so, state the postulate or theorem you would use. (Lesson 5.4)

13.

no

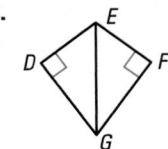

14.

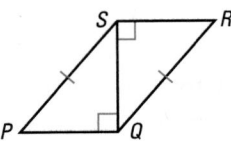

yes; HL Theorem

15.

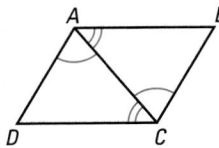

yes; ASA Congruence Postulate

Extra Practice

16.

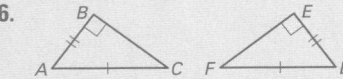

17. Statements (Reasons)

(1) $\overline{EG} \cong \overline{HG}$; $\overline{GF} \cong \overline{GJ}$ (Given)

(2) $\angle EGF \cong \angle HGJ$ (Vertical $\angle$ are $\cong$.)

(3) $\triangle GEF \cong \triangle GHJ$ (SAS Congruence Postulate)

(4) $\angle GEF \cong \angle GHJ$ (Corresp. parts of $\cong \triangle$ are $\cong$.)

18. Proofs may vary. An example is given.

Statements (Reasons)

(1) $\overline{AD} \parallel \overline{BC}$ (Given)

(2) $\angle ADE \cong \angle CBE$ (If 2 $\parallel$ lines are cut by a trans., alt. int. $\angle$ are $\cong$.)

(3) $\angle AED \cong \angle CEB$ (Vertical $\angle$ are $\cong$.)

(4) $\overline{DE} \cong \overline{BE}$ (Given)

(5) $\triangle AED \cong \triangle CEB$ (ASA Congruence Postulate)

(6) $\overline{AE} \cong \overline{CE}$ (Corresp. parts of $\cong \triangle$ are $\cong$.)

19. Statements (Reasons)

(1) $\overline{QT} \cong \overline{ST}$; $\overline{RQ} \cong \overline{RS}$ (Given)

(2) $\overline{RT} \cong \overline{RT}$ (Reflexive Property of Congruence)

(3) $\triangle RQT \cong \triangle RST$ (SSS Congruence Postulate)

(4) $\angle RQT \cong \angle RST$ (Corresp. parts of $\cong \triangle$ are $\cong$.)

16. Sketch the overlapping triangles separately. Mark all congruent angles and sides. Then tell what theorem or postulate you could use to show that $\triangle ABC \cong \triangle DEF$. (Lesson 5.5) See margin; HL Theorem

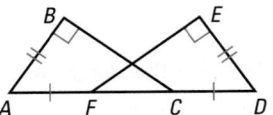

Use the information in the diagram to show that the given statement is true. (Lesson 5.5) 17–19. See margin.

17. $\angle GEF \cong \angle GHJ$

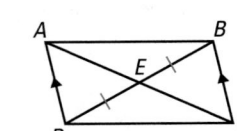

18. $\overline{AE} \cong \overline{CE}$

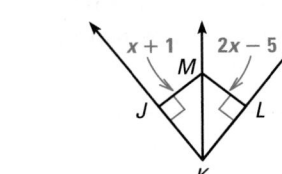

19. $\angle RQT \cong \angle RST$

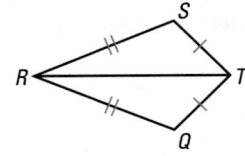

Use the diagram to find each missing measure. (Lesson 5.6)

20. Find CD. 5

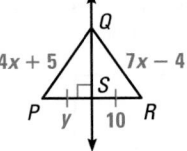

21. Find ML. 7

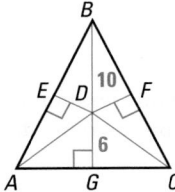

22. Find PS and PQ. $PS = 10$; $PQ = 17$

23. The perpendicular bisectors of $\triangle ABC$ meet at point D. Find AD. (Lesson 5.6) 10

Tell whether the grid shows a reflection in *the x-axis*, *the y-axis*, or *neither*. (Lesson 5.7)

24.

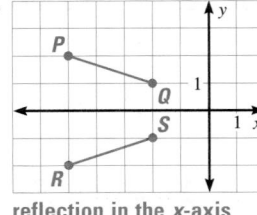

reflection in the *x*-axis

25.

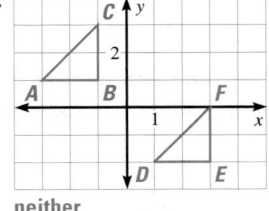

neither

26.

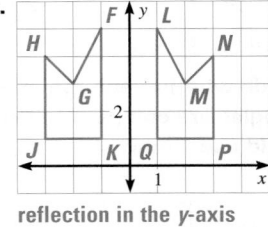

reflection in the *y*-axis

Determine the number of lines of symmetry in each figure. (Lesson 5.7)

27.

1

28.

6

29.

0

Chapter 6

Decide whether the figure is a polygon. If so, tell what type. If not, explain why. (Lesson 6.1)

1.
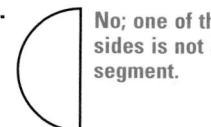
No; one of the sides is not a segment.

2.
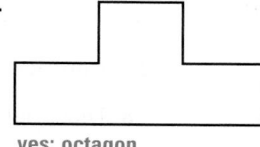
yes; octagon

3.
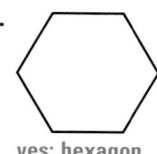
yes; hexagon

Find the value of x. (Lesson 6.1)

4.
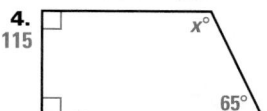

5.
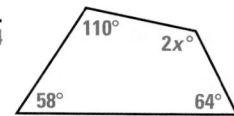

6.

Use the diagram of □VWXY at the right. Complete each statement and give a reason for your answer. (Lesson 6.2) 7–12. See margin for explanations.

7. $\overline{VW} \cong$ ___?___ $\overline{XY}$

8. $\angle VWX \cong$ ___?___ $\angle XYV$

9. $\overline{XW} \cong$ ___?___ $\overline{VY}$

10. $\overline{VT} \cong$ ___?___ $\overline{XT}$

11. $\overline{WX} \parallel$ ___?___ $\overline{VY}$

12. $\angle XYW \cong$ ___?___ $\angle VWY$

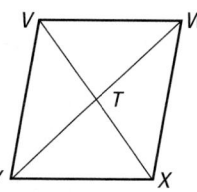

13. $\angle VYX$ is supplementary to ___?___ and ___?___.
$\angle YVW$ and $\angle YXW$; the consecutive angles of a □ are supplementary.

14. Point T is the midpoint of ___?___ and ___?___.
$\overline{VX}$ and $\overline{YW}$; the diagonals of a □ bisect each other.

ABCD is a parallelogram. Find the missing angle measures. (Lesson 6.2)

15.
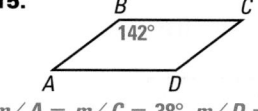
$m\angle A = m\angle C = 38°$, $m\angle D = 142°$

16.
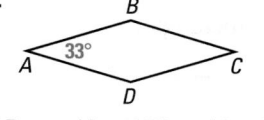
$m\angle B = m\angle D = 147°$, $m\angle C = 33°$

17.
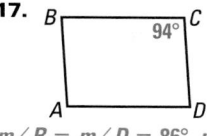
$m\angle B = m\angle D = 86°$, $m\angle A = 94°$

Tell whether the quadrilateral is a parallelogram. Explain your reasoning. (Lesson 6.3)

18.
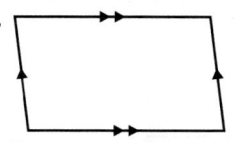
Yes; opposite sides are parallel.

19.
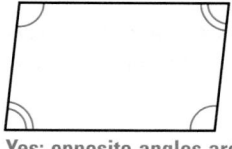
Yes; opposite angles are congruent.

20.
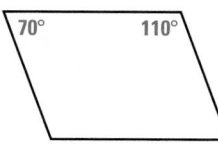
no

21.
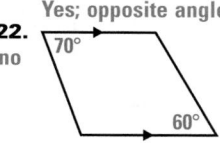
Yes; opposite sides are congruent.

22.
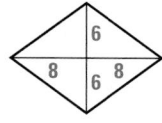
no

23.
Yes; the diagonals bisect each other.

24. In quadrilateral $EFGH$, $m\angle E = 12°$, $m\angle H = 12°$, and $m\angle F = 168°$. Is $EFGH$ a parallelogram? Explain your reasoning. (Lesson 6.3)
No; opposite angles must be congruent, but $\angle F$ and $\angle H$ are not congruent.

Extra Practice

7. Opposite sides of a parallelogram are congruent.

8. Opposite angles of a parallelogram are congruent.

9. Opposite sides of a parallelogram are congruent.

10. The diagonals of a parallelogram bisect each other.

11. By definition, a parallelogram is a quadrilateral with both pairs of opposite sides parallel.

12. $\overline{VW} \parallel \overline{XY}$ by the definition of a parallelogram; if two parallel lines are cut by a transversal, then alternate interior angles are congruent.

Extra Practice

List each quadrilateral for which the statement is true. (Lesson 6.4)

parallelogram

rectangle

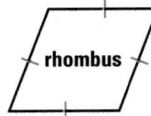
rhombus

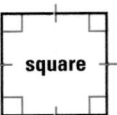
square

25. Consecutive angles are supplementary.
parallelogram, rectangle, rhombus, square

26. Consecutive angles are congruent.
rectangle, square

27. Consecutive sides are perpendicular.
rectangle, square

28. Diagonals are congruent.   rectangle, square

29. Consecutive sides are congruent.
rhombus, square

30. Opposite sides are parallel.
parallelogram, rectangle, rhombus, square

Find the missing angle measures in the trapezoid. (Lesson 6.5)

31.
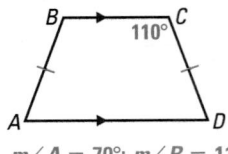
m∠A = 70°; m∠B = 110°;
m∠D = 70°

32.
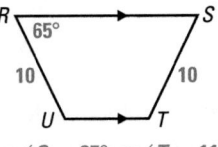
m∠S = 65°; m∠T = 115°;
m∠U = 115°

33.
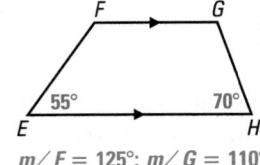
m∠F = 125°; m∠G = 110°

Find the value of *x* for the trapezoid. (Lesson 6.5)

34.
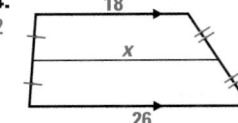

35.
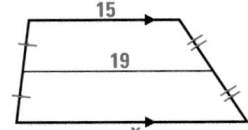

36.
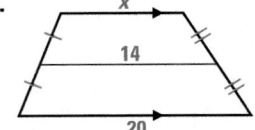

Are you given enough information to conclude that the figure is the given type of special quadrilateral? Explain your reasoning. (Lesson 6.6)

37. A parallelogram?

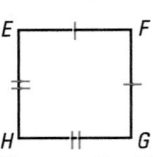

No; opposite sides are not congruent.

38. A rectangle?

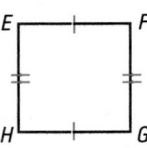

No; it is not given that the angles are right angles.

39. A rhombus?

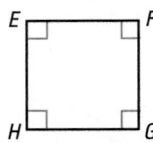

No; it is not given that the sides are all congruent.

Determine whether the quadrilateral is a *trapezoid*, an *isosceles trapezoid*, a *parallelogram*, a *rectangle*, a *rhombus*, or a *square*.
(Lesson 6.6)

40.
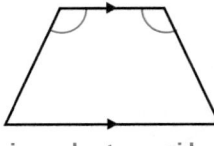
isosceles trapezoid

41.
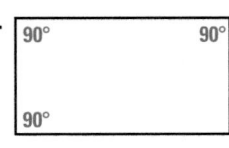
rectangle

42.
square

24. $\angle R \cong \angle R$ by the Reflexive Property of Congruence. It is given that $\angle RQS \cong \angle RPT$. The triangles are similar by the AA Similarity Postulate.

Chapter 7

Simplify the ratio. (Lesson 7.1)

1. $\dfrac{50\text{ m}}{250\text{ cm}}$ $\dfrac{20}{1}$

2. $\dfrac{15\text{ ft}}{4\text{ yd}}$ $\dfrac{5}{4}$

3. $\dfrac{15\text{ in.}}{2\text{ ft}}$ $\dfrac{5}{8}$

4. $\dfrac{10\text{ km}}{900\text{ m}}$ $\dfrac{100}{9}$

Sketch the rectangle described and use ratios to label its sides. Then find the length and the width of the rectangle. (Lesson 7.1)

5. The perimeter of a rectangle is 70 feet. The ratio of the length to the width is 4 : 3.   **Check sketches; 20 ft by 15 ft**

6. The perimeter of a rectangle is 128 meters. The ratio of the length to the width is 5 : 3.   **Check sketches; 40 m by 24 m**

Solve the proportion. (Lesson 7.1)

7. $\dfrac{a}{21} = \dfrac{1}{3}$ 7

8. $\dfrac{5}{b} = \dfrac{20}{7}$ $\dfrac{7}{4}$

9. $\dfrac{4}{9} = \dfrac{c}{72}$ 32

10. $\dfrac{4}{13} = \dfrac{16}{d}$ 52

11. $\dfrac{x+5}{24} = \dfrac{1}{2}$ 7

12. $\dfrac{r-2}{21} = \dfrac{2}{3}$ 16

13. $\dfrac{2y}{13} = \dfrac{12}{39}$ 2

14. $\dfrac{3}{2z+1} = \dfrac{1}{5}$ 7

In the diagram, _PQRS_ ~ _TVWX_. (Lesson 7.2)

15. List all pairs of congruent angles. $\angle P \cong \angle T$; $\angle Q \cong \angle V$; $\angle R \cong \angle W$; $\angle S \cong \angle X$

16. Find the scale factor of _PQRS_ to _TVWX_. $\dfrac{3}{2}$

17. Find the value of _u_.   9

18. Find the value of _y_.   4

19. Find the value of _z_.   10

20. Find the ratio of the perimeter of _PQRS_ to the perimeter of _TVWX_. $\dfrac{3}{2}$

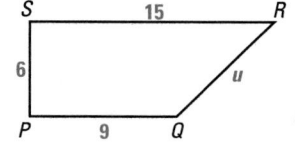
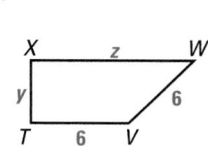

Determine whether the triangles are similar. If they are similar, write a similarity statement. (Lesson 7.3)

21.

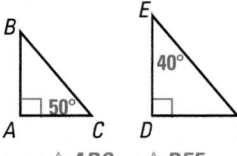

yes; $\triangle ABC \sim \triangle DEF$

22.

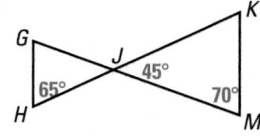

yes; $\triangle GHJ \sim \triangle MKJ$

23. no

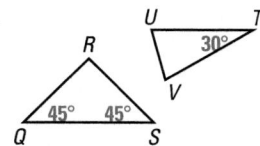

In Exercises 24 and 25, use the diagram. (Lesson 7.3)

24. Show that the triangles are similar.   See margin.

25. Write a similarity statement for the triangles.
  $\triangle RQS \sim \triangle RPT$

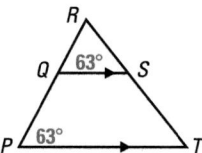

29. $\angle M \cong \angle M$ by the Reflexive Property of Congruence. Also, $\frac{15}{15+5} = \frac{3}{4}$ and $\frac{21}{21+7} = \frac{3}{4}$. The lengths of the sides that include $\angle M$ are proportional. So, by the SAS Similarity Theorem, the triangles are similar.

Determine whether the triangles are similar. If they are similar, state the similarity and the postulate or theorem that justifies your answer. (Lesson 7.4)

26.

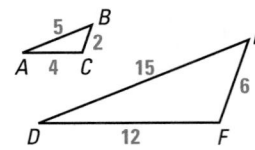

Yes; $\triangle ABC \sim \triangle DEF$ by the SSS Similarity Theorem.

27. no

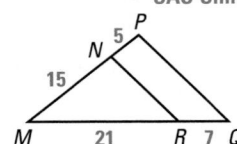

28.

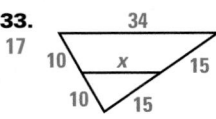

Yes; $\triangle PQR \sim \triangle STR$ by the SAS Similarity Theorem.

In Exercises 29 and 30, use the diagram. (Lesson 7.4)

29. Show that the overlapping triangles are similar. See margin.

30. Write a similarity statement for the triangles. $\triangle MNR \sim \triangle MPQ$

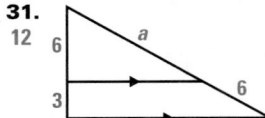

Find the value of the variable. (Lesson 7.5)

31.

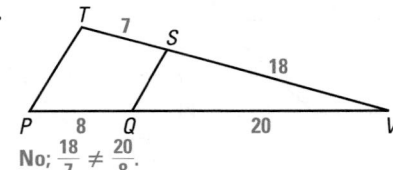

32.

33.

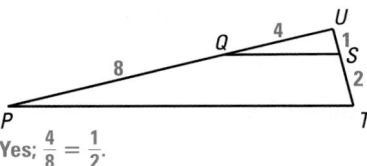

Given the diagram, determine whether $\overline{QS}$ is parallel to $\overline{PT}$. (Lesson 7.5)

34.

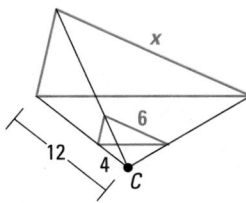

No; $\frac{18}{7} \neq \frac{20}{8}$.

35.

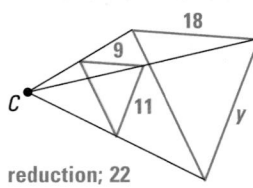

Yes; $\frac{4}{8} = \frac{1}{2}$.

The red figure is the image of the blue figure after a dilation. Tell whether the dilation is a reduction or an enlargement. Then find the value of the variable. (Lesson 7.6)

36.

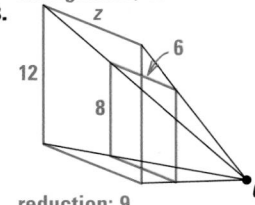

enlargement; 18

37.

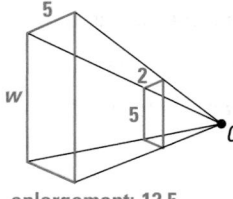

reduction; 22

38.

reduction; 9

39.

enlargement; 12.5

Chapter 8

Decide whether the polygon is *convex* or *concave*. (Lesson 8.1)

1.

concave

2.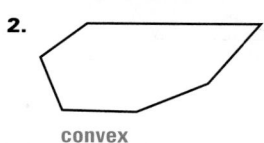

convex

3.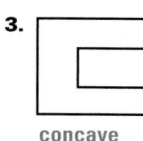

concave

Decide whether the polygon is regular. Explain your answer. (Lesson 8.1)

4. 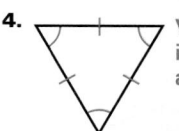 Yes; the polygon is equilateral and equiangular.

5. 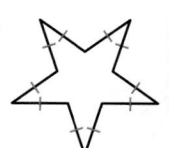 No; the polygon is equilateral, but it is not equiangular.

6. 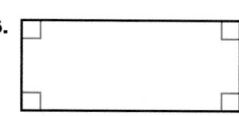 No; the polygon is equiangular, but it is not equilateral.

Find the sum of the measures of the interior angles of the convex polygon. (Lesson 8.2)

7. 12-gon 1800°

8. octagon 1080°

9. 18-gon 2880°

Find the value of *x*. (Lesson 8.2)

10. 140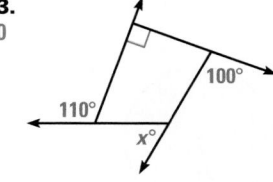

11. 108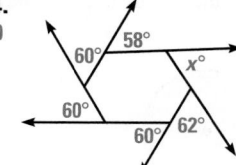

12. 120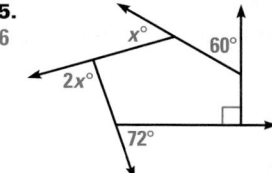

13. 60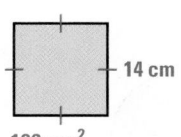

14. 60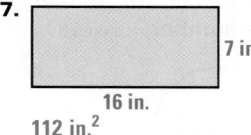

15. 46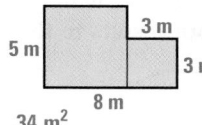

Find the area of the polygon. (Lesson 8.3)

16. 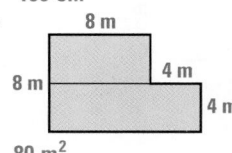 196 cm²

17. 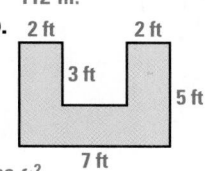 112 in.²

18. 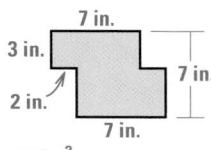 34 m²

19. 80 m²

20. 26 ft²

21. 49 in.²

Find the area of the polygon. (Lesson 8.4)

22.
12 ft²
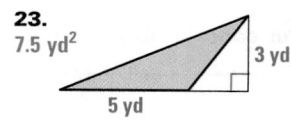
4 ft
6 ft

23.
7.5 yd²
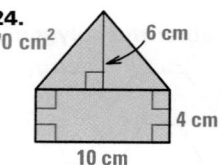
3 yd
5 yd

24.
70 cm²
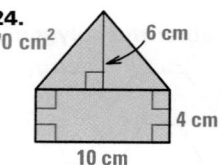
6 cm
4 cm
10 cm

25. A triangle has an area of 65 square feet and a base of 13 feet. Find the height. (Lesson 8.4)   10 ft

Find the area of the quadrilateral. (Lesson 8.5)

26.
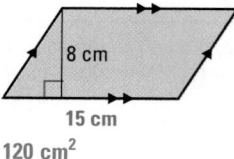
8 cm
15 cm
120 cm²

27.
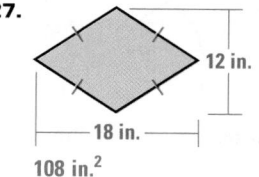
12 in.
18 in.
108 in.²

28.
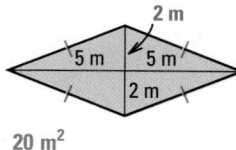
2 m
5 m   5 m
2 m
20 m²

Find the area of the trapezoid. (Lesson 8.6)

29.
9 ft
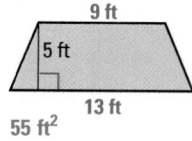
5 ft
13 ft
55 ft²

30.
16 cm
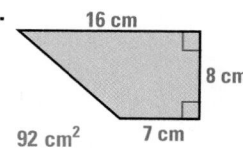
8 cm
92 cm²   7 cm

31.
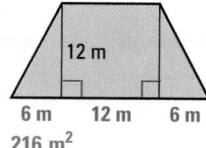
12 m
6 m   12 m   6 m
216 m²

32. A trapezoid has an area of 135 square units. The lengths of the bases are 12 units and 18 units. Find the height. (Lesson 8.6)   9 units

Find the circumference and area of the circle. Round your results to the nearest whole number. (Lesson 8.7)

33.
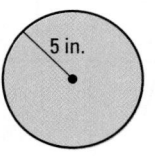
5 in.
31 in.; 79 in.²

34.
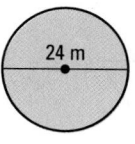
24 m
75 m; 452 m²

35.
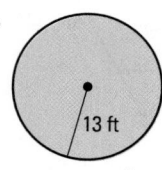
13 ft
82 ft; 531 ft²

Find the area of the blue sector. Start by finding the area of the circle. Round your results to the nearest whole number. (Lesson 8.7)

36.
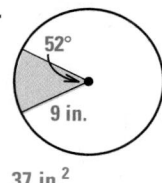
52°
9 in.
37 in.²

37.
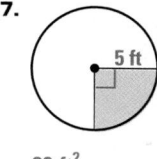
5 ft
20 ft²

38.
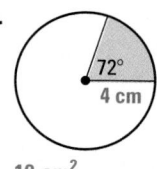
72°
4 cm
10 cm²

Chapter 9

Tell whether the solid is a polyhedron. If so, describe the shape of the base and then name the solid. (Lesson 9.1)

1.

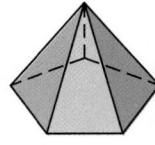

yes; pentagon; pentagonal pyramid

2.

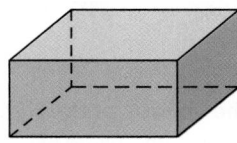

no

3.

yes; rectangle; rectangular prism

Name the polyhedron. Then count the number of faces and edges.
(Lesson 9.1)

4.

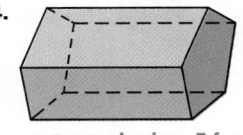

pentagonal prism; 7 faces; 15 edges

5.

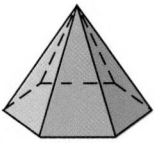

hexagonal pyramid; 7 faces; 12 edges

6.

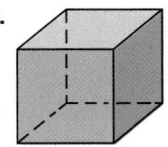

cube; 6 faces; 12 edges

Find the surface area of the prism. (Lesson 9.2)

7.

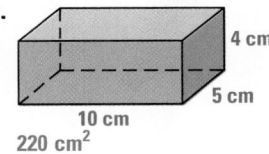

4 cm
5 cm
10 cm
220 cm^2

8.

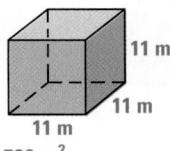

11 m
11 m
11 m
726 m^2

9.

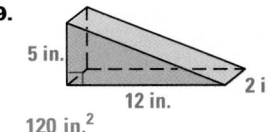

5 in.
12 in.
2 in.
120 in.^{2}

Find the surface area of the cylinder. Round your answer to the nearest whole number. (Lesson 9.2)

10.

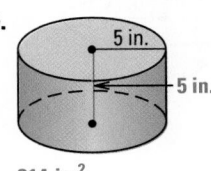

5 in.
5 in.
314 in.^{2}

11.

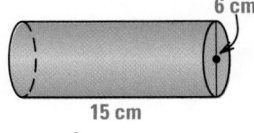

6 cm
15 cm
339 cm^2

12.

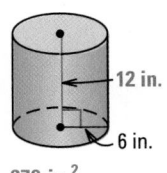

12 in.
6 in.
679 in.^{2}

Find the surface area of the pyramid or cone. If necessary, round your answer to the nearest whole number. (Lesson 9.3)

13.

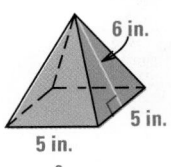

6 in.
5 in.
5 in.
85 in.^{2}

14.

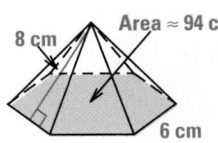

8 cm
Area ≈ 94 cm^2
6 cm
238 cm^2

15.

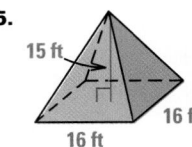

15 ft
16 ft
16 ft
800 ft^2

16.

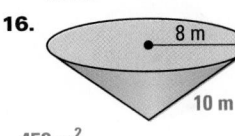

8 m
10 m
452 m^2

17.

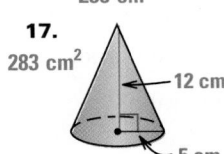

283 cm^2
12 cm
5 cm

18.

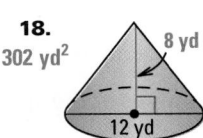

302 yd^2
8 yd
12 yd

Find the volume of the prism. (Lesson 9.4)

19.
200 cm³
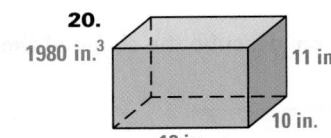
4 cm
5 cm
10 cm

20.
1980 in.³
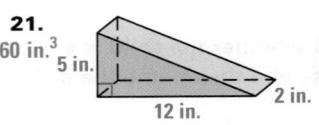
11 in.
10 in.
18 in.

21.
60 in.³
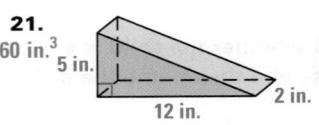
5 in.
12 in.
2 in.

Find the volume of the right cylinder. Round the result to the nearest whole number. (Lesson 9.4)

22.
393 in.³
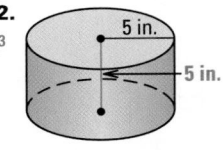
5 in.
5 in.

23.
424 cm³
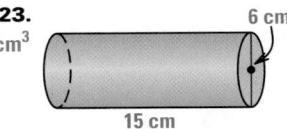
6 cm
15 cm

24.
1060 ft³

15 ft
6 ft

Find the volume of the pyramid. (Lesson 9.5)

25.
480 cm³
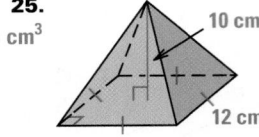
10 cm
12 cm

26.
103 in.³
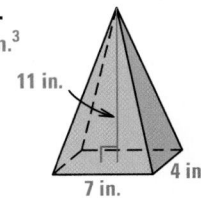
11 in.
7 in.
4 in.

27.
12 m³
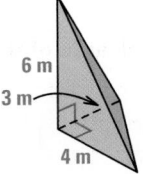
6 m
3 m
4 m

Find the volume of the cone. Round the result to the nearest whole number. (Lesson 9.5)

28.
513 in.³
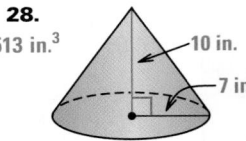
10 in.
7 in.

29.
871 ft³
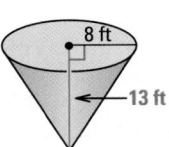
8 ft
13 ft

30.
402 m³
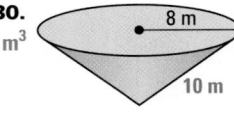
8 m
10 m

Find the surface area and volume of the sphere. Round the result to the nearest whole number. (Lesson 9.6)

31.
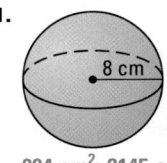
8 cm

804 cm²; 2145 cm³

32.
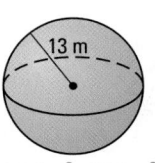
13 m

2124 m²; 9203 m³

33.
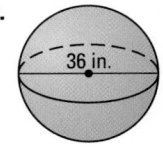
36 in.

4072 in.²; 24,429 in.³

In Exercises 34 and 35, use the diagram at the right. (Lesson 9.6)

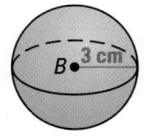
1 cm
B 3 cm
A

34. Find the surface areas of the spheres. What is their ratio?
4π cm², 36π cm²; 1 : 9

35. Find the volumes of the spheres. What is their ratio?
$\frac{4}{3}\pi$ cm³, 36π cm³; 1 : 27

Chapter 10

Find the square root. Round your answer to the nearest tenth.
(Lesson 10.1)

1. $\sqrt{58}$ 7.6 **2.** $\sqrt{7}$ 2.6 **3.** $\sqrt{41}$ 6.4 **4.** $\sqrt{134}$ 11.6 **5.** $\sqrt{106}$ 10.3

Multiply the radicals. Then simplify if possible. (Lesson 10.1)

6. $\sqrt{11} \cdot \sqrt{5}$ $\sqrt{55}$ **7.** $\sqrt{7} \cdot \sqrt{7}$ 7 **8.** $\sqrt{12} \cdot \sqrt{6}$ $6\sqrt{2}$ **9.** $(3\sqrt{2})^2$ 18 **10.** $\sqrt{48} \cdot 2\sqrt{3}$ 24

Simplify the radical expression. (Lesson 10.1)

11. $\sqrt{99}$ $3\sqrt{11}$ **12.** $\sqrt{28}$ $2\sqrt{7}$ **13.** $\sqrt{150}$ $5\sqrt{6}$ **14.** $\sqrt{32}$ $4\sqrt{2}$ **15.** $\sqrt{98}$ $7\sqrt{2}$

Show that the triangle is a 45°-45°-90° triangle. Then find the value of each variable. Write your answer in radical form. (Lesson 10.2) 16–19. See margin for explanations.

16.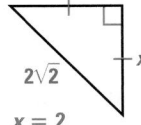
$x = 2$

17.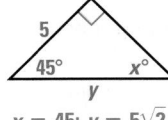
$x = 45; y = 5\sqrt{2}$

18.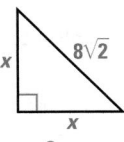
$x = 8$

19.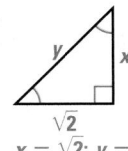
$x = \sqrt{2}; y = 2$

20. The hypotenuse of a 45°-45°-90° triangle has a length of 6. Sketch and label the triangle. Then find the length of each leg to the nearest tenth. (Lesson 10.2) Check sketches. The length of each leg is $\frac{6}{\sqrt{2}} \approx 4.2$.

Find the value of each variable. Write your answer in radical form.
(Lesson 10.3)

21.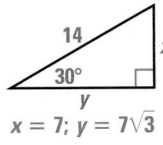
$x = 7; y = 7\sqrt{3}$

22.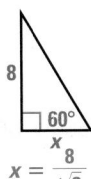
$x = \frac{8}{\sqrt{3}}$

23.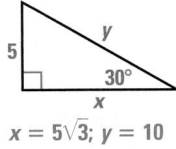
$x = 5\sqrt{3}; y = 10$

24.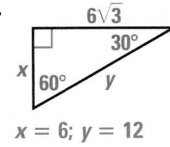
$x = 6; y = 12$

Find tan R and tan S. Write your answers as fractions and as decimals rounded to four decimal places. (Lesson 10.4)

25.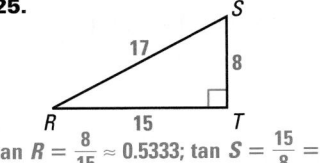
$\tan R = \frac{8}{15} \approx 0.5333$; $\tan S = \frac{15}{8} = 1.875$

26.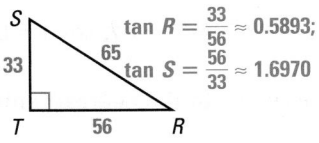
$\tan R = \frac{33}{56} \approx 0.5893$;
$\tan S = \frac{56}{33} \approx 1.6970$

27.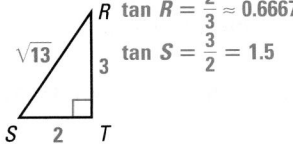
$\tan R = \frac{2}{3} \approx 0.6667$;
$\tan S = \frac{3}{2} = 1.5$

Use a calculator to approximate the value to four decimal places.
(Lesson 10.4)

28. $\tan 15°$ 0.2679 **29.** $\tan 72°$ 3.0777 **30.** $\tan 60°$ 1.7321 **31.** $\tan 9°$ 0.1584 **32.** $\tan 37°$ 0.7536

33. $\tan 10°$ 0.1763 **34.** $\tan 31°$ 0.6009 **35.** $\tan 76°$ 4.0108 **36.** $\tan 58°$ 1.6003 **37.** $\tan 49°$ 1.1504

Extra Practice **693**

16. Since the triangle has two congruent sides, the angles opposite those sides are congruent. Let each acute angle measure $d°$. By the Triangle Sum Theorem, $d° + d° + 90° = 180°$. So, $2d° = 90°$, and $d° = 45°$. Since the measure of each acute angle is 45°, the triangle is a 45°-45°-90° triangle.

17. By the Triangle Sum Theorem, $x° + 45° + 90° = 180°$. So $x° = 45°$. Since the measure of each acute angle is 45°, the triangle is a 45°-45°-90° triangle.

18. Since the triangle has two congruent sides, the angles opposite those sides are congruent. Let each acute angle measure $d°$. By the Triangle Sum Theorem, $d° + d° + 90° = 180°$. So, $2d° = 90°$, and $d° = 45°$. Since the measure of each acute angle is 45°, the triangle is a 45°-45°-90° triangle.

19. Let each acute angle measure $d°$. By the Triangle Sum Theorem, $d° + d° + 90° = 180°$. So, $2d° = 90°$, and $d° = 45°$. Since the measure of each acute angle is 45°, the triangle is a 45°-45°-90° triangle.

41. $\sin R = \dfrac{8}{17} \approx 0.4706;$

  $\cos R = \dfrac{15}{17} \approx 0.8824$

42. $\sin R = \dfrac{33}{65} \approx 0.5077;$

  $\cos R = \dfrac{56}{65} \approx 0.8615$

43. $\sin R = \dfrac{2}{\sqrt{13}} \approx 0.5547;$

  $\cos R = \dfrac{3}{\sqrt{13}} \approx 0.8321$

Extra Practice

Find the value of *x* to the nearest tenth. (Lesson 10.4)

38.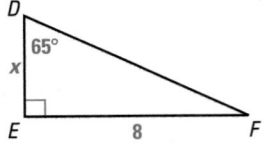

39.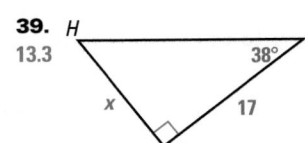

40.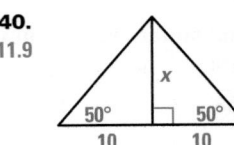

Find sin *R* and cos *R*. Write your answers as fractions and as decimals rounded to four decimal places. (Lesson 10.5) 41–43. See margin.

41.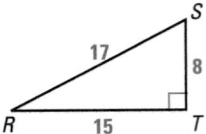

42.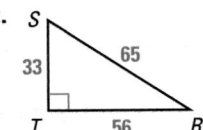

43.

Use a calculator to approximate the value to four decimal places.
(Lesson 10.5)

44. $\sin 33°$ 0.5446   **45.** $\cos 78°$ 0.2079   **46.** $\cos 8°$ 0.9903   **47.** $\sin 40°$ 0.6428   **48.** $\sin 57°$ 0.8387

49. $\sin 43°$ 0.6820   **50.** $\sin 80°$ 0.9848   **51.** $\cos 20°$ 0.9397   **52.** $\cos 89°$ 0.0175   **53.** $\sin 12°$ 0.2079

Find the lengths *x* and *y* of the legs of the triangle. Round your answers to the nearest tenth. (Lesson 10.5)

54.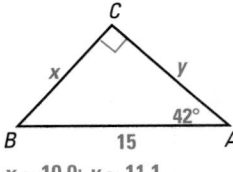

$x \approx 10.0;\ y \approx 11.1$

55.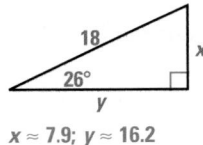

$x \approx 7.9;\ y \approx 16.2$

56.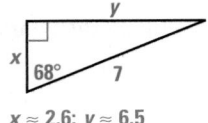

$x \approx 2.6;\ y \approx 6.5$

∠*A* is an acute angle. Use a calculator to approximate the measure of ∠*A* to the nearest tenth of a degree. (Lesson 10.6)

57. $\tan A = 0.8734$ 41.1° **58.** $\sin A = 0.8045$ 53.6° **59.** $\cos A = 0.2933$ 72.9° **60.** $\tan A = 1.6$ 58.0°

61. $\cos A = 0.8912$ 27.0° **62.** $\sin A = 0.2587$ 15.0° **63.** $\tan A = 3.123$ 72.2° **64.** $\cos A = 0.6789$ 47.2°

65. $\sin A = 0.3728$ 21.9° **66.** $\tan A = 0.5726$ 29.8° **67.** $\sin A = 0.9554$ 72.8° **68.** $\cos A = 0.0511$ 87.1°

Solve the right triangle. Round decimals to the nearest tenth.
(Lesson 10.6)

69.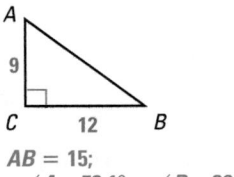

$AB = 15;$
$m\angle A \approx 53.1°;\ m\angle B \approx 36.9°$

70.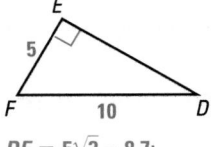

$DE = 5\sqrt{3} \approx 8.7;$
$m\angle F = 60°;\ m\angle D = 30°$

71.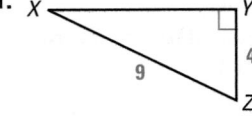

$XY = \sqrt{65} \approx 8.1;$
$m\angle X \approx 26.4°;\ m\angle Z \approx 63.6°$

Chapter 11

Use the diagram at the right. Name the term that best describes the given line segment, line, or point. (Lesson 11.1)

1. $\overline{BF}$ chord

2. $\overline{BD}$ diameter

3. $\overline{AC}$ radius

4. $\overleftrightarrow{BF}$ secant

5. $\overleftrightarrow{ED}$ tangent

6. D point of tangency

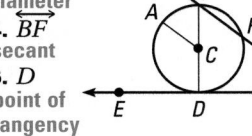

Use the diagram at the right. (Lesson 11.1)

7. Name the coordinates of the center of each circle.
   $\odot C$, (3, 1); $\odot D$, (7, 1)

8. Name the coordinates of the intersection of the two circles.
   (5, 1)

9. What is the length of the radius of $\odot C$? 2

10. What is the length of the diameter of $\odot D$? 4

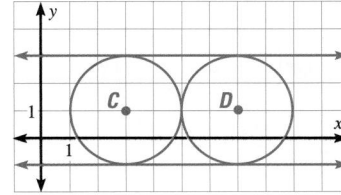

$\overleftrightarrow{AB}$, $\overline{PQ}$, and $\overline{PR}$ **are tangents to the circles. Find the value of the variable.** (Lesson 11.2)

11. 20

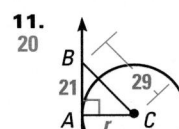

12. 2

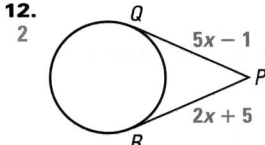

13. 6

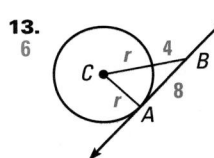

$\overline{AD}$ **and** $\overline{BE}$ **are diameters. Copy the diagram. Find the indicated measure.** (Lesson 11.3)

14. $m\widehat{AB}$ 35°

15. $m\angle DQC$ 55°

16. $m\widehat{BC}$ 90°

17. $m\widehat{DE}$ 35°

18. $m\angle CQE$ 90°

19. $m\angle AQE$ 145°

20. $m\widehat{AC}$ 125°

21. $m\widehat{BDC}$ 270°

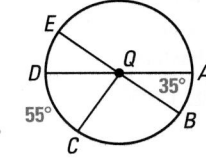

Find the length of $\widehat{AB}$. (Lesson 11.3)

22.

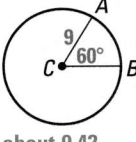

about 9.42

23.

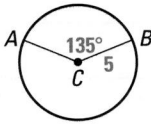

about 11.78

24.

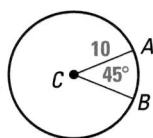

about 7.85

Find the value of x. (Lesson 11.4)

25. 22

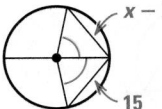

26. 6

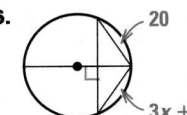

27. 14

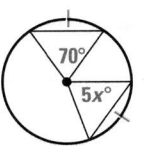

28. 4

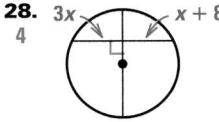

36.

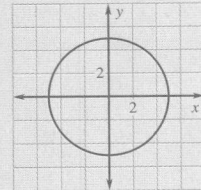

37.

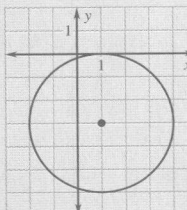

38.

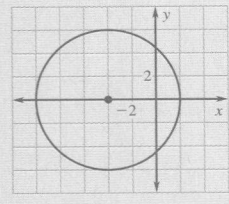

42. $J'(4, -4)$, $K'(1, -2)$, $M'(4, 0)$, $N'(7, -2)$; $(x, y) \rightarrow (-x, -y)$

43. $E'(5, -2)$, $F'(5, -5)$, $G'(2, -7)$, $H'(2, 0)$; $(x, y) \rightarrow (y, -x)$

44. $A'(2, -2)$, $B'(-1, -2)$, $C'(2, 3)$; $(x, y) \rightarrow (-y, x)$

Extra Practice

Find the value of each variable. (Lesson 11.5)

29. 65

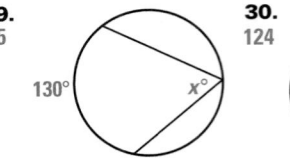

30. 124

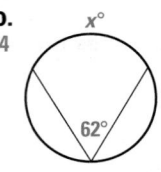

31.

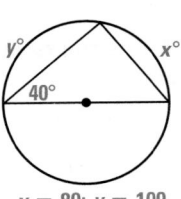

$x = 80$; $y = 100$

32.

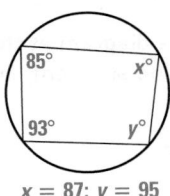

$x = 87$; $y = 95$

Find the value of x. (Lesson 11.6)

33. 80

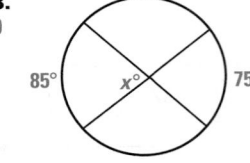

34. 4

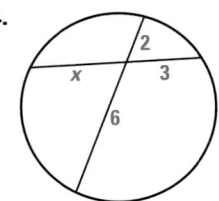

35. 107

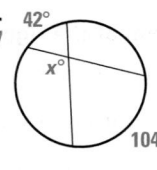

Give the radius and center of the circle with the given equation. Then graph the circle. (Lesson 11.7) 36–38. See margin for graphs.

36. $x^2 + y^2 = 25$
5; (0, 0)

37. $(x - 1)^2 + (y + 3)^2 = 9$
3; (1, −3)

38. $(x + 4)^2 + y^2 = 36$
6; (−4, 0)

Write the standard equation of the circle with the given center and radius. (Lesson 11.7)

39. center $(0, -1)$, radius 20
$x^2 + (y + 1)^2 = 400$

40. center $(-5, 7)$, radius 1
$(x + 5)^2 + (y - 7)^2 = 1$

41. center $(3, 4)$, radius 7
$(x - 3)^2 + (y - 4)^2 = 49$

Use the given information to rotate the figure about the origin. Find the coordinates of the vertices of the image and compare them with the vertices of the original figure. Describe any patterns you see. (Lesson 11.8) 42–44. See margin.

42. 180°

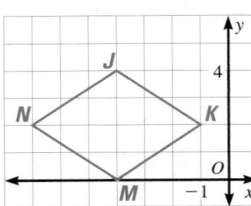

43. 90° clockwise

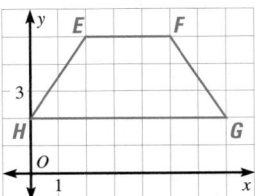

44. 90° counterclockwise

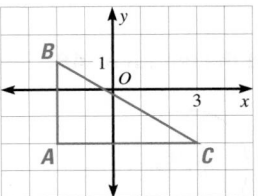

Does the figure have rotational symmetry? Is so, describe the rotations that map the figure onto itself. (Lesson 11.8)

45.

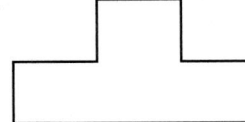

yes; 180°

46. no

47.

yes; 120° clockwise or counterclockwise

696   *Student Resources*

POINTS, LINES, PLANES, AND ANGLES

In Exercises 1–8, use the diagram at the right to name the following.

1. three collinear points F, G, H

2. two rays with endpoint F $\overrightarrow{FG}$, $\overrightarrow{FH}$

3. a segment bisector $\overrightarrow{EF}$ or $\overline{EF}$

4. a segment that is bisected $\overline{GH}$

5. a straight angle $\angle GFH$

6. an acute angle $\angle EFH$

7. a linear pair $\angle GFE$, $\angle EFH$

8. a point of intersection F

9. Sketch three possible intersections of a line and a plane. **See margin.**

10. Given $P(0, 7)$ and $Q(4, -9)$, find the midpoint of $\overline{PQ}$. (2, −1)

11. $\angle A$ is a complement of $\angle B$, and $m\angle A = 42°$. Find $m\angle B$. 48°

REASONING

12. Sketch the next figure you expect in the pattern at the right. **See margin.**

13. Show the conjecture is false by finding a counterexample.
    Conjecture: A quadrilateral is always convex. **See margin.**

 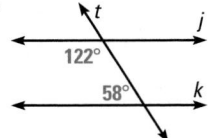 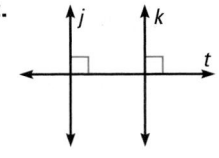

14. Write the statement as an if-then statement. Circle the conclusion.
    Statement: I will buy the video if it costs less than $16.00.
    If the video costs less than $16.00, then (I will buy it).

15. Name the property that the statement illustrates.
    If $\angle A \cong \angle B$ and $\angle B \cong \angle C$, then $\angle A \cong \angle C$. **Transitive Property of Congruence**

PARALLEL AND PERPENDICULAR LINES

In the diagram at the right, think of each segment as part of a line.
Fill in the blank with *parallel*, *perpendicular*, or *skew*.

16. $\overleftrightarrow{AB}$ and $\overleftrightarrow{CY}$ are __?__. skew

17. $\overleftrightarrow{XW}$ and $\overleftrightarrow{YZ}$ appear __?__. parallel

18. $\overleftrightarrow{CD}$ is __?__ to plane ADZ. perpendicular

19. Planes ABX and DCY appear __?__. parallel

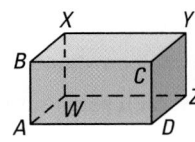

Is enough information given to conclude that $j \parallel k$? Explain. 20–22. See margin.

20.

21. 122° 58°

22.

---

9. **Sample answer:**

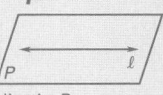

ℓ lies in P.

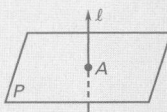

ℓ intersects P at point A.

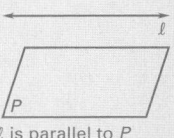

ℓ is parallel to P (does not intersect P).

12.

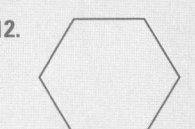

13. **Sample answer:**

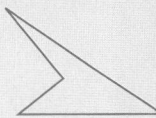

20. Yes; $j \parallel k$ by the Alternate Interior Angles Converse.

21. Yes; $122° + 58° = 180°$, so $j \parallel k$ by the Same-Side Interior Angles Converse.

22. Yes; in a plane, if two lines are perpendicular to the same line, then they are parallel to each other (or $j \parallel k$ by the Corresponding Angles Converse).

23–25. Sample answers are given.

23.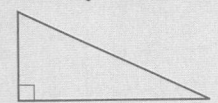

24.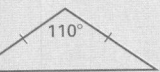

25.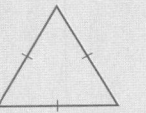

40.

TRIANGLES

Draw an example of the triangle. 23–25. See margin.

23. right scalene
24. obtuse isosceles
25. equilateral

In Exercises 26–28, find the value of x.

26.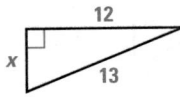

27. $7\sqrt{2} \approx 9.9$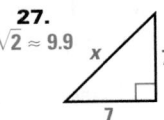

28.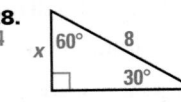

29. Complete the Triangle Sum Theorem.
The sum of the measures of the angles of a triangle is __?__. 180°

30. Find the distance between points $A(1, 4)$ and $B(9, -2)$. 10

31. Classify the triangle with side lengths 6, 9, and 12 as *acute*, *right*, or *obtuse*. obtuse

32. Can the side lengths 4, 4, and 9 form a triangle? Explain. No; 4 + 4 < 9.

CONGRUENCE

In Exercises 33–35, decide whether enough information is given to show that the triangles are congruent. If so, write a congruence statement and state the theorem or postulate you would use.

Yes; $\triangle QTP \cong \triangle STR$ by the SAS Congruence Postulate.

33.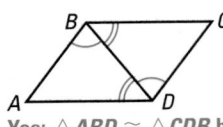
Yes; $\triangle ABD \cong \triangle CDB$ by the ASA Congruence Postulate.

34.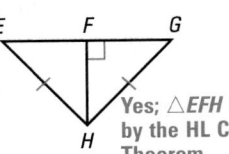
Yes; $\triangle EFH \cong \triangle GFH$ by the HL Congruence Theorem.

35.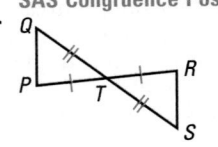

36. In the diagram at the right, $\overleftrightarrow{MZ}$ is the perpendicular bisector of $\overline{PQ}$. Complete the congruence statements.

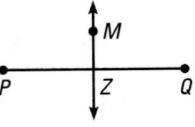

$\overline{PZ} \cong$ __?__ $\overline{QZ}$ $\overline{PM} \cong$ __?__ $\overline{QM}$

Use the coordinate plane at the right.

37. Which segment is a reflection of $\overline{AB}$ in the *x*-axis? $\overline{HG}$

38. Which segment is a reflection of $\overline{AB}$ in the *y*-axis? $\overline{DC}$

39. Which segment is a translation of $\overline{AB}$? Describe the translation using words. $\overline{FE}$; 5 units right and 4 units down

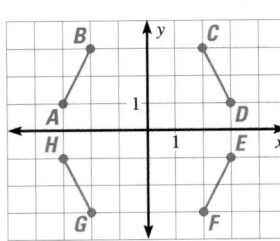

40. Determine the number of lines of symmetry in a rectangle. Draw a diagram to support your answer.
2 lines of symmetry; See margin.

QUADRILATERALS

Name the type of polygon with the given number of sides.

41. three triangle **42.** four quadrilateral **43.** five pentagon **44.** six hexagon **45.** eight octagon

Complete each statement with *parallelogram, rectangle, rhombus, square,* or *trapezoid*. Use each word once.

46. A __?__ is a quadrilateral with four congruent sides. rhombus

47. A __?__ is a quadrilateral with four right angles. rectangle

48. A __?__ is a quadrilateral with exactly one pair of parallel sides. trapezoid

49. A __?__ is both a rectangle and a rhombus. square

50. A __?__ is a quadrilateral with both pairs of opposite sides parallel. parallelogram

Tell whether the quadrilateral is a parallelogram. Explain your reasoning.

51.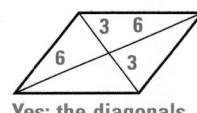

Yes; the diagonals bisect each other.

52.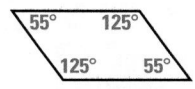

Yes; both pairs of opposite angles are congruent.

53.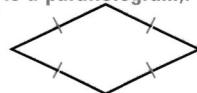

Yes; both pairs of opposite sides are congruent (or the quadrilateral is a rhombus, and every rhombus is a parallelogram).

SIMILARITY

Determine whether the polygons are similar. If they are similar, write a similarity statement. Explain your reasoning.

54.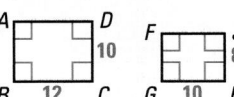

No; $\frac{10}{12} \neq \frac{8}{10}$.

55.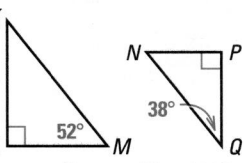

Yes; $m\angle K = 38°$ by the Triangle Sum Theorem, so $\triangle KLM \sim \triangle QPN$ by the AA Similarity Postulate.

56.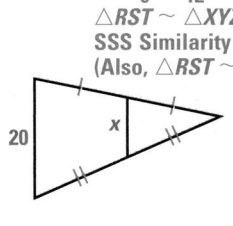

Yes; $\frac{6}{9} = \frac{8}{12} = \frac{8}{12}$, so $\triangle RST \sim \triangle XYZ$ by the SSS Similarity Theorem. (Also, $\triangle RST \sim \triangle YXZ$.)

In Exercises 57–59, find the value of *x*.

57.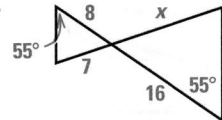

58.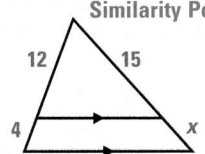

59.

60. What is the relationship between the original figure and its image after a dilation? They are similar figures.

81.

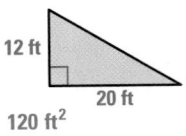

AREA AND VOLUME

Find the area of the figure. In Exercise 66, the polygon is made up of rectangles.

61.

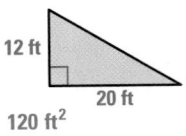

12 ft
20 ft
120 ft²

62.

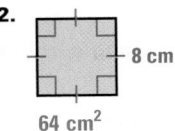

8 cm
64 cm²

63.

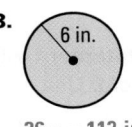

6 in.
36π ≈ 113 in.²

64.

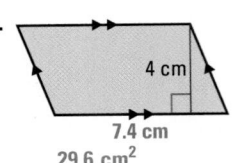

4 cm
7.4 cm
29.6 cm²

65.

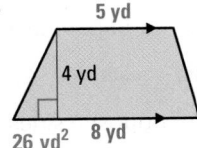

5 yd
4 yd
26 yd² 8 yd

66.

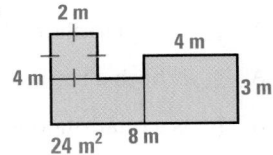

2 m
4 m
4 m
3 m
24 m² 8 m

67. Which figure in Exercises 61–66 is a regular polygon? Explain. The quadrilateral in Exercise 62 is a square, which is a regular polygon since it has 4 congruent sides and 4 congruent angles.

Name the solid. Then find its surface area and volume.

cone; 24π ≈ 75 cm²; 12π ≈ 38 cm³

68.

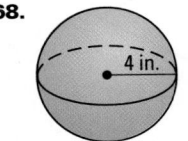

4 in.
sphere; 64π ≈ 201 in.²; $\frac{256\pi}{3}$ ≈ 268 in.³

69.

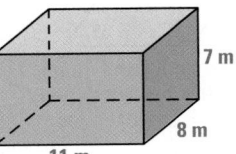

7 m
8 m
11 m
rectangular prism; 442 m²; 616 m³

70.

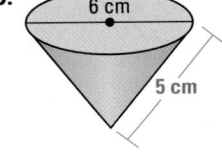

6 cm
5 cm

71.

7 ft
16 ft
cylinder; 322π ≈ 1012 ft²; 784π ≈ 2463 ft³

72.

4 in.
5 in.
7 in.
3 in. triangular prism; 96 in.²; 42 in.³

73.

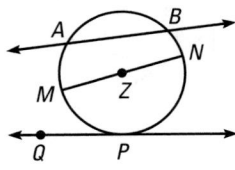

8 cm
12 cm
12 cm
square pyramid; 384 cm²; 384 cm³

CIRCLES

Tell whether the point, line, or segment is best described as a center, a chord, a secant, a tangent, a diameter, a radius, or a point of tangency.

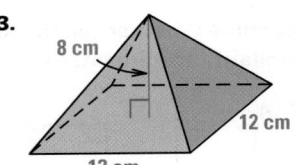

74. $\overleftrightarrow{QP}$ tangent

75. P point of tangency

76. $\overline{AB}$ chord

77. $\overleftrightarrow{AB}$ secant

78. $\overline{MZ}$ radius

79. $\overline{MN}$ diameter

80. What is the measure of a semicircle? 180°

81. Write the standard equation of a circle with center at the origin and radius 5. Graph the circle in a coordinate plane.
$x^2 + y^2 = 25$; see margin for graph.

Symbols

Symbol	Meaning	Page
$\overleftrightarrow{AB}$	line AB	16
$\overrightarrow{AB}$	ray AB	16
$\overline{AB}$	segment AB	16
AB	the length of $\overline{AB}$	28
$\parallel$	is parallel to	108
$\perp$	is perpendicular to	108
$\angle ABC$	angle ABC	35
$\triangle ABC$	triangle ABC	179
$\square ABCD$	parallelogram $ABCD$	310
n-gon	polygon with n sides	304
$\odot P$	circle P	589
$m\angle A$	measure of angle A	36
$m\overset{\frown}{AB}$	measure of minor arc AB	601
$m\overset{\frown}{ABC}$	measure of major arc ABC	601
$\cdot$	multiplication, times	9
$\circ$	degrees	36
A'	A prime	153
A''	A double prime	158
π	pi	452

Symbol	Meaning	Page
$=$	is equal to	9
$\cong$	is congruent to	30
$\sim$	is similar to	365
$\approx$	is approximately equal to	193
$\neq$	is not equal to	194
$\overset{?}{=}$	is it equal to?	200
$<$	is less than	201
$>$	is greater than	201
$\ldots$	and so on	3
$\lvert a \rvert$	absolute value of a	28
$\dfrac{a}{b}$, $a:b$	ratio of a to b	357
$\sqrt{a}$	square root of a	537
(x, y)	ordered pair	664
$\tan$	tangent	557
$\sin$	sine	563
$\cos$	cosine	563
$\tan^{-1}$	inverse tangent	569
$\sin^{-1}$	inverse sine	570
$\cos^{-1}$	inverse cosine	570

Properties

Properties of Equality and Congruence for Segments

	Equality	Congruence
Reflexive (p. 88)	$AB = AB$	$\overline{AB} \cong \overline{AB}$
Symmetric (p. 88)	If $AB = CD$, then $CD = AB$.	If $\overline{AB} \cong \overline{CD}$, then $\overline{CD} \cong \overline{AB}$.
Transitive (p. 88)	If $AB = CD$ and $CD = EF$, then $AB = EF$.	If $\overline{AB} \cong \overline{CD}$ and $\overline{CD} \cong \overline{EF}$, then $\overline{AB} \cong \overline{EF}$.

Properties of Equality and Congruence for Angles

	Equality	Congruence
Reflexive (p. 88)	$m\angle A = m\angle A$	$\angle A \cong \angle A$
Symmetric (p. 88)	If $m\angle A = m\angle B$, then $m\angle B = m\angle A$.	If $\angle A \cong \angle B$, then $\angle B \cong \angle A$.
Transitive (p. 88)	If $m\angle A = m\angle B$ and $m\angle B = m\angle C$, then $m\angle A = m\angle C$.	If $\angle A \cong \angle B$ and $\angle B \cong \angle C$, then $\angle A \cong \angle C$.

Algebraic Properties of Equality

	Property	Example
Addition (p. 90)	If $a = b$, then $a + c = b + c$.	If $x - 4 = 9$, then $x - 4 + 4 = 9 + 4$.
Subtraction (p. 90)	If $a = b$, then $a - c = b - c$.	If $y + 1 = 6$, then $y + 1 - 1 = 6 - 1$.
Multiplication (p. 90)	If $a = b$, then $ac = bc$.	If $z = 5$, then $z \cdot 3 = 5 \cdot 3$.
Division (p. 90)	If $a = b$ and $c \neq 0$, then $\dfrac{a}{c} = \dfrac{b}{c}$.	If $7x = 14$, then $\dfrac{7x}{7} = \dfrac{14}{7}$.
Cross Product (p. 359)	If $\dfrac{a}{b} = \dfrac{c}{d}$, then $ad = bc$.	If $\dfrac{3}{4} = \dfrac{x}{12}$, then $3 \cdot 12 = 4 \cdot x$.

Right Triangles

Pythagorean Theorem (p. 192)

$$a^2 + b^2 = c^2$$

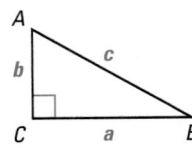

Trigonometric Ratios
(pp. 557, 563)

$$\sin A = \frac{a}{c} \qquad \cos A = \frac{b}{c} \qquad \tan A = \frac{a}{b}$$

Special Right Triangles (pp. 542, 549)

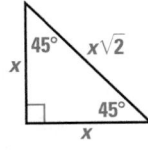

45°-45°-90° triangle

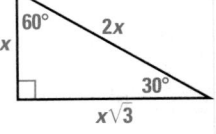

30°-60°-90° triangle

Formulas

Perimeter and Circumference

Square (p. 672)	$P = 4s$	where s = side length
Rectangle (p. 672)	$P = 2\ell + 2w$	where ℓ = length and w = width
Triangle (p. 672)	$P = a + b + c$	where a, b, c = lengths of the sides of a triangle
Circle (p. 452)	$C = \pi d$ or $C = 2\pi r$	where d = diameter and r = radius

Area

Square (p. 424)	$A = s^2$	where s = side length
Rectangle (p. 425)	$A = bh$	where b = base and h = height
Triangle (p. 431)	$A = \frac{1}{2}bh$	where b = base and h = height
Parallelogram (p. 439)	$A = bh$	where b = base and h = height
Rhombus (p. 441)	$A = \frac{1}{2}d_1 d_2$	where d_1, d_2 = diagonals
Trapezoid (p. 446)	$A = \frac{1}{2}h(b_1 + b_2)$	where h = height and b_1, b_2 = bases
Circle (p. 453)	$A = \pi r^2$	where r = radius

Surface Area

Prism (p. 484)	$S = 2B + Ph$	where B = area of base, P = perimeter, and h = height
Cylinder (p. 485)	$S = 2\pi r^2 + 2\pi rh$	where r = radius and h = height
Pyramid (p. 492)	$S = B + \frac{1}{2}P\ell$	where B = area of base, P = perimeter, and ℓ = slant height
Cone (p. 493)	$S = \pi r^2 + \pi r\ell$	where r = radius and ℓ = slant height
Sphere (p. 517)	$S = 4\pi r^2$	where r = radius

Volume

Prism (p. 500)	$V = Bh$	where B = area of base and h = height
Cylinder (p. 502)	$V = \pi r^2 h$	where r = radius and h = height
Pyramid (p. 510)	$V = \frac{1}{3}Bh$	where B = area of base and h = height
Cone (p. 511)	$V = \frac{1}{3}\pi r^2 h$	where r = radius and h = height
Sphere (p. 518)	$V = \frac{4}{3}\pi r^3$	where r = radius

Coordinate Geometry

Midpoint (p. 55)	The midpoint between $A(x_1, y_1)$ and $B(x_2, y_2)$ is $\left(\dfrac{x_1 + x_2}{2}, \dfrac{y_1 + y_2}{2} \right)$.
Distance (p. 194)	The distance between $A(x_1, y_1)$ and $B(x_2, y_2)$ is $\sqrt{(x_2 - x_1)^2 + (y_2 - y_1)^2}$.

Tables

Tables

Squares and Square Roots

No.	Square	Sq. Root	No.	Square	Sq. Root	No.	Square	Sq. Root
1	1	1.000	51	2,601	7.141	101	10,201	10.050
2	4	1.414	52	2,704	7.211	102	10,404	10.100
3	9	1.732	53	2,809	7.280	103	10,609	10.149
4	16	2.000	54	2,916	7.348	104	10,816	10.198
5	25	2.236	55	3,025	7.416	105	11,025	10.247
6	36	2.449	56	3,136	7.483	106	11,236	10.296
7	49	2.646	57	3,249	7.550	107	11,449	10.344
8	64	2.828	58	3,364	7.616	108	11,664	10.392
9	81	3.000	59	3,481	7.681	109	11,881	10.440
10	100	3.162	60	3,600	7.746	110	12,100	10.488
11	121	3.317	61	3,721	7.810	111	12,321	10.536
12	144	3.464	62	3,844	7.874	112	12,544	10.583
13	169	3.606	63	3,969	7.937	113	12,769	10.630
14	196	3.742	64	4,096	8.000	114	12,996	10.677
15	225	3.873	65	4,225	8.062	115	13,225	10.724
16	256	4.000	66	4,356	8.124	116	13,456	10.770
17	289	4.123	67	4,489	8.185	117	13,689	10.817
18	324	4.243	68	4,624	8.246	118	13,924	10.863
19	361	4.359	69	4,761	8.307	119	14,161	10.909
20	400	4.472	70	4,900	8.367	120	14,400	10.954
21	441	4.583	71	5,041	8.426	121	14,641	11.000
22	484	4.690	72	5,184	8.485	122	14,884	11.045
23	529	4.796	73	5,329	8.544	123	15,129	11.091
24	576	4.899	74	5,476	8.602	124	15,376	11.136
25	625	5.000	75	5,625	8.660	125	15,625	11.180
26	676	5.099	76	5,776	8.718	126	15,876	11.225
27	729	5.196	77	5,929	8.775	127	16,129	11.269
28	784	5.292	78	6,084	8.832	128	16,384	11.314
29	841	5.385	79	6,241	8.888	129	16,641	11.358
30	900	5.477	80	6,400	8.944	130	16,900	11.402
31	961	5.568	81	6,561	9.000	131	17,161	11.446
32	1,024	5.657	82	6,724	9.055	132	17,424	11.489
33	1,089	5.745	83	6,889	9.110	133	17,689	11.533
34	1,156	5.831	84	7,056	9.165	134	17,956	11.576
35	1,225	5.916	85	7,225	9.220	135	18,225	11.619
36	1,296	6.000	86	7,396	9.274	136	18,496	11.662
37	1,369	6.083	87	7,569	9.327	137	18,769	11.705
38	1,444	6.164	88	7,744	9.381	138	19,044	11.747
39	1,521	6.245	89	7,921	9.434	139	19,321	11.790
40	1,600	6.325	90	8,100	9.487	140	19,600	11.832
41	1,681	6.403	91	8,281	9.539	141	19,881	11.874
42	1,764	6.481	92	8,464	9.592	142	20,164	11.916
43	1,849	6.557	93	8,649	9.644	143	20,449	11.958
44	1,936	6.633	94	8,836	9.695	144	20,736	12.000
45	2,025	6.708	95	9,025	9.747	145	21,025	12.042
46	2,116	6.782	96	9,216	9.798	146	21,316	12.083
47	2,209	6.856	97	9,409	9.849	147	21,609	12.124
48	2,304	6.928	98	9,604	9.899	148	21,904	12.166
49	2,401	7.000	99	9,801	9.950	149	22,201	12.207
50	2,500	7.071	100	10,000	10.000	150	22,500	12.247

Tables

Trigonometric Ratios

Angle	Sine	Cosine	Tangent
1°	.0175	.9998	.0175
2°	.0349	.9994	.0349
3°	.0523	.9986	.0524
4°	.0698	.9976	.0699
5°	.0872	.9962	.0875
6°	.1045	.9945	.1051
7°	.1219	.9925	.1228
8°	.1392	.9903	.1405
9°	.1564	.9877	.1584
10°	.1736	.9848	.1763
11°	.1908	.9816	.1944
12°	.2079	.9781	.2126
13°	.2250	.9744	.2309
14°	.2419	.9703	.2493
15°	.2588	.9659	.2679
16°	.2756	.9613	.2867
17°	.2924	.9563	.3057
18°	.3090	.9511	.3249
19°	.3256	.9455	.3443
20°	.3420	.9397	.3640
21°	.3584	.9336	.3839
22°	.3746	.9272	.4040
23°	.3907	.9205	.4245
24°	.4067	.9135	.4452
25°	.4226	.9063	.4663
26°	.4384	.8988	.4877
27°	.4540	.8910	.5095
28°	.4695	.8829	.5317
29°	.4848	.8746	.5543
30°	.5000	.8660	.5774
31°	.5150	.8572	.6009
32°	.5299	.8480	.6249
33°	.5446	.8387	.6494
34°	.5592	.8290	.6745
35°	.5736	.8192	.7002
36°	.5878	.8090	.7265
37°	.6018	.7986	.7536
38°	.6157	.7880	.7813
39°	.6293	.7771	.8098
40°	.6428	.7660	.8391
41°	.6561	.7547	.8693
42°	.6691	.7431	.9004
43°	.6820	.7314	.9325
44°	.6947	.7193	.9657
45°	.7071	.7071	1.0000

Angle	Sine	Cosine	Tangent
46°	.7193	.6947	1.0355
47°	.7314	.6820	1.0724
48°	.7431	.6691	1.1106
49°	.7547	.6561	1.1504
50°	.7660	.6428	1.1918
51°	.7771	.6293	1.2349
52°	.7880	.6157	1.2799
53°	.7986	.6018	1.3270
54°	.8090	.5878	1.3764
55°	.8192	.5736	1.4281
56°	.8290	.5592	1.4826
57°	.8387	.5446	1.5399
58°	.8480	.5299	1.6003
59°	.8572	.5150	1.6643
60°	.8660	.5000	1.7321
61°	.8746	.4848	1.8040
62°	.8829	.4695	1.8807
63°	.8910	.4540	1.9626
64°	.8988	.4384	2.0503
65°	.9063	.4226	2.1445
66°	.9135	.4067	2.2460
67°	.9205	.3907	2.3559
68°	.9272	.3746	2.4751
69°	.9336	.3584	2.6051
70°	.9397	.3420	2.7475
71°	.9455	.3256	2.9042
72°	.9511	.3090	3.0777
73°	.9563	.2924	3.2709
74°	.9613	.2756	3.4874
75°	.9659	.2588	3.7321
76°	.9703	.2419	4.0108
77°	.9744	.2250	4.3315
78°	.9781	.2079	4.7046
79°	.9816	.1908	5.1446
80°	.9848	.1736	5.6713
81°	.9877	.1564	6.3138
82°	.9903	.1392	7.1154
83°	.9925	.1219	8.1443
84°	.9945	.1045	9.5144
85°	.9962	.0872	11.4301
86°	.9976	.0698	14.3007
87°	.9986	.0523	19.0811
88°	.9994	.0349	28.6363
89°	.9998	.0175	57.2900

Tables

Postulates

1 Two Points Determine a Line
Through any two points there is exactly one line. *(p. 14)*

2 Three Points Determine a Plane
Through any three points not on a line there is exactly one plane. *(p. 14)*

3 Intersection of Two Lines
If two lines intersect, then their intersection is a point. *(p. 22)*

4 Intersection of Two Planes
If two planes intersect, then their intersection is a line. *(p. 22)*

5 Segment Addition Postulate
If B is between A and C, then $AC = AB + BC$.
If $AC = AB + BC$, then B is between A and C.
(p. 29)

6 Angle Addition Postulate
If P is in the interior of $\angle RST$, then the measure of $\angle RST$ is the sum of the measures of $\angle RSP$ and $\angle PST$. *(p. 37)*

7 Linear Pair Postulate
If two angles form a linear pair, then they are supplementary. *(p. 75)*

8 Corresponding Angles Postulate
If two parallel lines are cut by a transversal, then corresponding angles are congruent. *(p. 128)*

9 Corresponding Angles Converse
If two lines are cut by a transversal so that corresponding angles are congruent, then the lines are parallel. *(p. 137)*

10 Parallel Postulate
If there is a line and a point not on the line, then there is exactly one line through the point parallel to the given line. *(p. 144)*

11 Perpendicular Postulate
If there is a line and a point not on the line, then there is exactly one line through the point perpendicular to the given line. *(p. 144)*

12 Side-Side-Side Congruence Postulate (SSS)
If three sides of one triangle are congruent to three sides of a second triangle, then the two triangles are congruent. *(p. 241)*

13 Side-Angle-Side Congruence Postulate (SAS)
If two sides and the included angle of one triangle are congruent to two sides and the included angle of a second triangle, then the two triangles are congruent. *(p. 242)*

14 Angle-Side-Angle Congruence Postulate (ASA)
If two angles and the included side of one triangle are congruent to two angles and the included side of a second triangle, then the two triangles are congruent. *(p. 250)*

15 Angle-Angle Similarity Postulate (AA)
If two angles of one triangle are congruent to two angles of another triangle, then the two triangles are similar. *(p. 372)*

16 Arc Addition Postulate
The measure of an arc formed by two adjacent arcs is the sum of the measures of the two arcs. *(p. 602)*

Theorems

2.1 Congruent Complements Theorem
If two angles are complementary to the same angle, then they are congruent. *(p. 69)*

2.2 Congruent Supplements Theorem
If two angles are supplementary to the same angle, then they are congruent. *(p. 69)*

2.3 Vertical Angles Theorem
Vertical angles are congruent. *(p. 76)*

3.1
All right angles are congruent. *(p. 114)*

3.2
If two lines are perpendicular, then they intersect to form four right angles. *(p. 114)*

3.3
If two lines intersect to form adjacent congruent angles, then the lines are perpendicular. *(p. 115)*

3.4
If two sides of adjacent acute angles are perpendicular, then the angles are complementary. *(p. 115)*

3.5 Alternate Interior Angles Theorem
If two parallel lines are cut by a transversal, then alternate interior angles are congruent. *(p. 129)*

3.6 Alternate Exterior Angles Theorem
If two parallel lines are cut by a transversal, then alternate exterior angles are congruent. *(p. 130)*

3.7 Same-Side Interior Angles Theorem
If two parallel lines are cut by a transversal, then same-side interior angles are supplementary. *(p. 131)*

3.8 Alternate Interior Angles Converse
If two lines are cut by a transversal so that alternate interior angles are congruent, then the lines are parallel. *(p. 138)*

3.9 Alternate Exterior Angles Converse
If two lines are cut by a transversal so that alternate exterior angles are congruent, then the lines are parallel. *(p. 138)*

3.10 Same-Side Interior Angles Converse
If two lines are cut by a transversal so that same-side interior angles are supplementary, then the lines are parallel. *(p. 138)*

3.11
If two lines are parallel to the same line, then they are parallel to each other. *(p. 145)*

3.12
In a plane, if two lines are perpendicular to the same line, then they are parallel to each other. *(p. 145)*

4.1 Triangle Sum Theorem
The sum of the measures of the angles of a triangle is 180°. *(p. 179)*

Corollary to the Triangle Sum Theorem
The acute angles of a right triangle are complementary. *(p. 180)*

4.2 Exterior Angle Theorem
The measure of an exterior angle of a triangle is equal to the sum of the measures of the two nonadjacent interior angles. *(p. 181)*

4.3 Base Angles Theorem
If two sides of a triangle are congruent, then the angles opposite them are congruent. *(p. 185)*

4.4 Converse of the Base Angles Theorem
If two angles of a triangle are congruent, then the sides opposite them are congruent. *(p. 186)*

4.5 Equilateral Theorem
If a triangle is equilateral, then it is equiangular. *(p. 187)*

4.6 Equiangular Theorem
If a triangle is equiangular, then it is equilateral. *(p. 187)*

4.7 Pythagorean Theorem
In a right triangle, the square of the length of the hypotenuse is equal to the sum of the squares of the lengths of the legs. *(p. 192)*

4.8 Converse of the Pythagorean Theorem
If the square of the length of the longest side of a triangle is equal to the sum of the squares of the lengths of the other two sides, then the triangle is a right triangle. *(p. 200)*

4.9 Intersection of Medians of a Triangle
The medians of a triangle intersect at the centroid, a point that is two thirds of the distance from each vertex to the midpoint of the opposite side. *(p. 208)*

4.10
If one side of a triangle is longer than another side, then the angle opposite the longer side is larger than the angle opposite the shorter side. *(p. 212)*

4.11
If one angle of a triangle is larger than another angle, then the side opposite the larger angle is longer than the side opposite the smaller angle. *(p. 212)*

4.12 Triangle Inequality
The sum of the lengths of any two sides of a triangle is greater than the length of the third side. *(p. 213)*

5.1 Angle-Angle-Side Congruence Theorem (AAS)
If two angles and a non-included side of one triangle are congruent to two angles and the corresponding non-included side of a second triangle, then the two triangles are congruent. *(p. 251)*

5.2 Hypotenuse-Leg Congruence Theorem (HL)
If the hypotenuse and a leg of a right triangle are congruent to the hypotenuse and a leg of a second right triangle, then the two triangles are congruent. *(p. 257)*

5.3 Angle Bisector Theorem
If a point is on the bisector of an angle, then it is equidistant from the two sides of the angle. *(p. 273)*

5.4 Perpendicular Bisector Theorem
If a point is on the perpendicular bisector of a segment, then it is equidistant from the endpoints of the segment. *(p. 274)*

6.1 Quadrilateral Interior Angles Theorem
The sum of the measures of the interior angles of a quadrilateral is 360°. *(p. 305)*

6.2
If a quadrilateral is a parallelogram, then its opposite sides are congruent. *(p. 310)*

6.3
If a quadrilateral is a parallelogram, then its opposite angles are congruent. *(p. 311)*

6.4
If a quadrilateral is a parallelogram, then its consecutive angles are supplementary. *(p. 311)*

6.5
If a quadrilateral is a parallelogram, then its diagonals bisect each other. *(p. 312)*

Theorems

6.6
If both pairs of opposite sides of a quadrilateral are congruent, then the quadrilateral is a parallelogram. *(p. 316)*

6.7
If both pairs of opposite angles of a quadrilateral are congruent, then the quadrilateral is a parallelogram. *(p. 316)*

6.8
If an angle of a quadrilateral is supplementary to both of its consecutive angles, then the quadrilateral is a parallelogram. *(p. 317)*

6.9
If the diagonals of a quadrilateral bisect each other, then the quadrilateral is a parallelogram. *(p. 318)*

Rhombus Corollary
If a quadrilateral has four congruent sides, then it is a rhombus. *(p. 326)*

Rectangle Corollary
If a quadrilateral has four right angles, then it is a rectangle. *(p. 326)*

Square Corollary
If a quadrilateral has four congruent sides and four right angles, then it is a square. *(p. 326)*

6.10
The diagonals of a rhombus are perpendicular. *(p. 327)*

6.11
The diagonals of a rectangle are congruent. *(p. 327)*

6.12
If a trapezoid is isosceles, then each pair of base angles is congruent. *(p. 332)*

6.13
If a trapezoid has a pair of congruent base angles, then it is isosceles. *(p. 332)*

7.1 Perimeters of Similar Polygons
If two polygons are similar, then the ratio of their perimeters is equal to the ratio of their corresponding side lengths. *(p. 368)*

7.2 Side-Side-Side Similarity Theorem (SSS)
If the corresponding sides of two triangles are proportional, then the triangles are similar. *(p. 379)*

7.3 Side-Angle-Side Similarity Theorem (SAS)
If an angle of one triangle is congruent to an angle of a second triangle and the lengths of the sides that include these angles are proportional, then the triangles are similar. *(p. 380)*

7.4 Triangle Proportionality Theorem
If a line parallel to one side of a triangle intersects the other two sides, then it divides the two sides proportionally. *(p. 386)*

7.5 Converse of the Triangle Proportionality Theorem
If a line divides two sides of a triangle proportionally, then it is parallel to the third side. *(p. 388)*

7.6 Midsegment Theorem
The segment connecting the midpoints of two sides of a triangle is parallel to the third side and is half as long. *(p. 389)*

8.1 Polygon Interior Angles Theorem
The sum of the measures of the interior angles of a convex polygon with n sides is $(n - 2) \cdot 180°$. *(p. 417)*

Theorems

8.2 Polygon Exterior Angles Theorem
The sum of the measures of the exterior angles of a convex polygon, one angle at each vertex, is 360°. *(p. 419)*

8.3 Areas of Similar Polygons
If two polygons are similar with a scale factor of $\frac{a}{b}$, then the ratio of their areas is $\frac{a^2}{b^2}$. *(p. 433)*

10.1 45°-45°-90° Triangle Theorem
In a 45°-45°-90° triangle, the length of the hypotenuse is the length of a leg times $\sqrt{2}$. *(p. 542)*

10.2 30°-60°-90° Triangle Theorem
In a 30°-60°-90° triangle, the hypotenuse is twice as long as the shorter leg, and the longer leg is the length of the shorter leg times $\sqrt{3}$. *(p. 549)*

11.1
If a line is tangent to a circle, then it is perpendicular to the radius drawn at the point of tangency. *(p. 595)*

11.2
In a plane, if a line is perpendicular to a radius of a circle at its endpoint on the circle, then the line is tangent to the circle. *(p. 595)*

11.3
If two segments from the same point outside a circle are tangent to the circle, then they are congruent. *(p. 597)*

11.4
If a diameter of a circle is perpendicular to a chord, then the diameter bisects the chord and its arc. *(p. 608)*

11.5
If one chord is a perpendicular bisector of another chord, then the first chord is a diameter. *(p. 609)*

11.6
In the same circle, or in congruent circles:

If two chords are congruent, then their corresponding minor arcs are congruent.

If two minor arcs are congruent, then their corresponding chords are congruent. *(p. 609)*

11.7 Measure of an Inscribed Angle
If an angle is inscribed in a circle, then its measure is half the measure of its intercepted arc. *(p. 614)*

11.8
If a triangle inscribed in a circle is a right triangle, then the hypotenuse is a diameter of the circle.

If a side of a triangle inscribed in a circle is a diameter of the circle, then the triangle is a right triangle. *(p. 615)*

11.9
If a quadrilateral can be inscribed in a circle, then its opposite angles are supplementary.

If the opposite angles of a quadrilateral are supplementary, then the quadrilateral can be inscribed in a circle. *(p. 616)*

11.10
If two chords intersect inside a circle, then the measure of each angle formed is one half the sum of the measures of the arcs intercepted by the angle and its vertical angle. *(p. 620)*

11.11
If two chords intersect inside a circle, then the product of the lengths of the segments of one chord is equal to the product of the lengths of the segments of the other chord. *(p. 622)*

Glossary

A

acute angle (p. 36) An angle with measure between 0° and 90°.

acute triangle (p. 174) A triangle with three acute angles.

adjacent angles (p. 68) Two angles with a common vertex and side but no common interior points.

alternate exterior angles (p. 121) Two angles that are formed by two lines and a transversal, and lie outside the two lines on the opposite sides of the transversal. In the diagram below, ∠1 and ∠8 are alternate exterior angles.

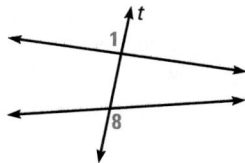

alternate interior angles (p. 121) Two angles that are formed by two lines and a transversal, and lie between the two lines on the opposite sides of the transversal. In the diagram below, ∠3 and ∠6 are alternate interior angles.

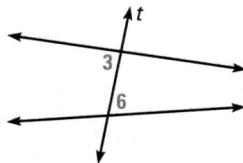

angle (p. 35) Consists of two rays with the same endpoint. The rays are the *sides* of the angle, and the endpoint is the *vertex* of the angle.

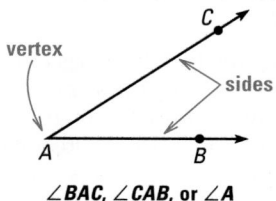

∠BAC, ∠CAB, or ∠A

angle bisector (p. 61) A ray that divides an angle into two angles that are congruent.

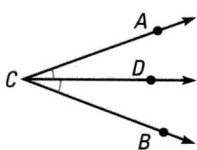

$\overrightarrow{CD}$ bisects ∠ACB.
m∠ACD = m∠BCD.

angle of rotation (p. 633) *See* rotation.

arc length (p. 603) A portion of the circumference of a circle.

area (p. 424) The amount of surface covered by a figure.

B

base of a cone (p. 493) *See* cone.

bases of a cylinder (p. 485) *See* cylinder.

base of an isosceles triangle (p. 185) *See* legs of an isosceles triangle.

bases of a parallelogram (p. 439) Either pair of parallel sides of a parallelogram are called the bases. *See also* height of a parallelogram.

bases of a prism (p. 473) *See* prism.

base of a pyramid (p. 491) *See* pyramid.

bases of a trapezoid (p. 332) *See* trapezoid.

base of a triangle (p. 431) *See* height of a triangle.

base angles of an isosceles triangle (p. 185) The two angles at the base of an isosceles triangle. *See also* legs of an isosceles triangle.

base angles of a trapezoid (p. 332) If trapezoid ABCD has bases $\overline{AB}$ and $\overline{CD}$, then there are two pairs of base angles: ∠A and ∠B, and ∠C and ∠D.

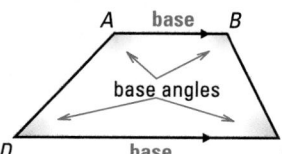

between (p. 29) When three points lie on a line, one of them is *between* the other two.

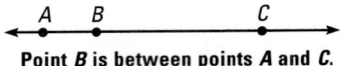

Point B is between points A and C.

bisect (pp. 53, 61) To divide into two congruent parts.

center of a circle (p. 452) *See* circle.

center of dilation (p. 393) *See* dilation.

center of rotation (p. 633) *See* rotation.

center of a sphere (p. 517) *See* sphere.

central angle (p. 454) An angle whose vertex is the center of a circle.

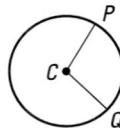

∠PCQ is a central angle.

centroid of a triangle (p. 208) The point at which the three medians of a triangle intersect.

chord (p. 589) A segment whose endpoints are points on a circle.

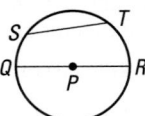

Chords: $\overline{QR}$, $\overline{ST}$

circle (pp. 452, 589) The set of all points in a plane that are the same distance from a given point, called the *center* of the circle.

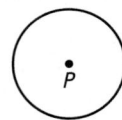

Circle with center P, or ⊙P

circumference (p. 452) The distance around a circle.

circumscribed circle (p. 615) *See* inscribed polygon.

collinear points (p. 15) Points that lie on the same line.

compass (p. 143) A construction tool used to draw arcs and circles.

complement (p. 67) The sum of the measures of an angle and its *complement* is 90°.

complementary angles (p. 67) Two angles whose measures have a sum of 90°.

concave polygon (p. 411) *See* convex polygon.

conclusion (p. 82) The "then" part of an if-then statement. In the statement "If it is cold, then I will wear my coat," the conclusion is "I will wear my coat."

cone (p. 493) A solid with a circular *base* and a vertex that is not in the same plane as the base. The *height* of a cone is the perpendicular distance between the vertex and the base. The *radius* of a cone is the radius of the base. The *slant height* of a cone is the distance between the vertex and a point on the base edge.

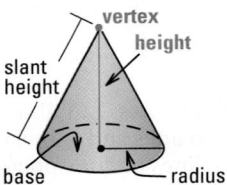

congruent angles (p. 36) Angles that have the same measure.

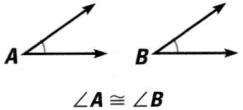

∠A ≅ ∠B

congruent arcs (p. 602) Two arcs of the same circle or of congruent circles that have the same measure.

congruent circles (p. 602) Two circles that have the same radius.

congruent figures (p. 233) Two geometric figures that have exactly the same size and shape. When two figures are congruent, all pairs of corresponding angles and corresponding sides are congruent.

congruent segments (p. 30) Segments that have the same length.

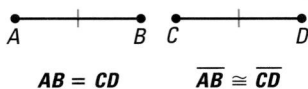

$$AB = CD \qquad \overline{AB} \cong \overline{CD}$$

conjecture (p. 8) An unproven statement that is based on a pattern or observations.

construction (p. 143) A geometric drawing that uses a limited set of tools, usually a compass and a straightedge.

converse (p. 136) The statement formed by switching the hypothesis and the conclusion of an if-then statement.

convex polygon (p. 411) A polygon is convex if no line that contains a side of the polygon passes through the interior of the polygon. A polygon that is not convex is called *concave*.

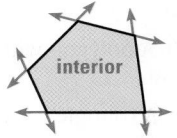 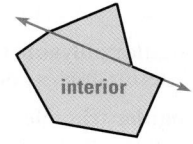

Convex polygon   Concave polygon

coordinate (p. 28) The real number that corresponds to a point on a line.

coplanar lines (p. 15) Lines that lie on the same plane.

coplanar points (p. 15) Points that lie on the same plane.

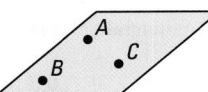

corollary to a theorem (p. 180) A statement that can be proved easily using the theorem.

corresponding angles (p. 121) Two angles that are formed by two lines and a transversal, and occupy corresponding positions. In the diagram below, ∠1 and ∠5 are corresponding angles.

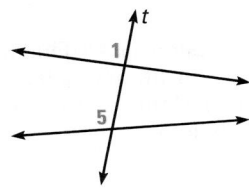

corresponding parts of congruent figures (p. 233) The corresponding sides and angles in congruent figures.

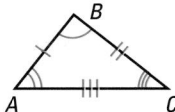 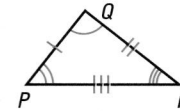

∠**B** and ∠**Q** are corresponding angles.
$\overline{BC}$ and $\overline{QR}$ are corresponding sides.

cosine (p. 563) A trigonometric ratio, abbreviated as *cos* and computed as the ratio of the length of the leg adjacent to the angle to the length of the hypotenuse.

$$\cos A = \frac{\text{leg adjacent to } \angle A}{\text{hypotenuse}} = \frac{b}{c}$$

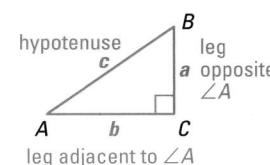

counterexample (p. 10) An example that shows that a conjecture is false.

cross product property (pp. 349, 359) If $\frac{a}{b} = \frac{c}{d}$, then $ad = bc$.

cylinder (p. 485) A solid with two congruent circular *bases* that lie in parallel planes. The *height* of a cylinder is the perpendicular distance between the bases. The *radius* of the cylinder is the radius of a base.

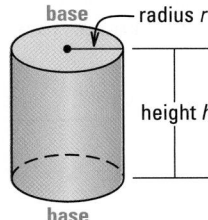

deductive reasoning (p. 83) Using facts, definitions, accepted properties, and the laws of logic to make a logical argument.

Glossary

degrees (°) (p. 36) *See* measure of an angle.

diagonal of a polygon (p. 303) A segment that joins two nonconsecutive vertices of a polygon.

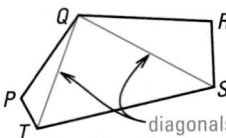

diameter (pp. 452, 589) The distance across the circle, through the center. The *diameter* is twice the *radius*. A chord that passes through the center of the circle is also called a *diameter*.

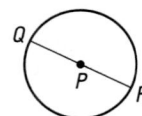

Diameter: $\overline{QR}$ or QR

dilation (p. 393) A transformation with center C and scale factor k that maps each point P to an image point P' so that P' lies on $\overrightarrow{CP}$ and $CP' = k \cdot CP$.

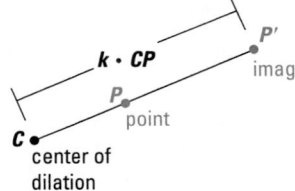

distance between two points on a line (p. 28) The distance AB is the absolute value of the difference of the coordinates of A and B. AB is also called the *length* of $\overline{AB}$.

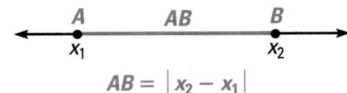

Distance Formula (p. 194) If $A(x_1, y_1)$ and $B(x_2, y_2)$ are points in a coordinate plane, then the distance between A and B is

$$\sqrt{(x_2 - x_1)^2 + (y_2 - y_1)^2}.$$

distance from a point to a line (p. 273) The length of the perpendicular segment from the point to the line.

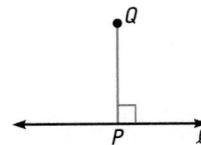

The distance from Q to line ℓ is QP.

edge of a polyhedron (p. 474) *See* polyhedron.

endpoint of a ray (p. 16) *See* ray.

endpoints of a segment (p. 16) *See* segment.

enlargement (p. 394) A dilation in which the image is larger than the original figure. The scale factor of an enlargement is greater than 1.

equiangular polygon (p. 412) A polygon with all of its interior angles congruent.

equiangular triangle (p. 174) A triangle with three congruent angles.

equidistant (p. 273) The same distance.

equilateral polygon (p. 412) A polygon with all of its sides congruent.

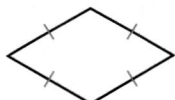

Equilateral polygon

equilateral triangle (p. 173) A triangle with three congruent sides.

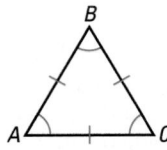

Equilateral triangle

evaluate an expression (p. 49) Find the value of an expression by substituting values for the variables, and then simplifying the result using the order of operations.

exponent (p. 467) *See* power.

exterior angles of a triangle (p. 181) When the sides of a triangle are extended, the angles that are adjacent to the *interior angles* of the triangle.

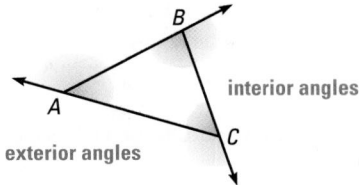

extremes of a proportion (p. 359) The extremes of the proportion $\frac{a}{b} = \frac{c}{d}$ are a and d. *See also* means of a proportion.

face (p. 474) *See* polyhedron.

height of a cone (p. 493) *See* cone.

height of a cylinder (p. 485) *See* cylinder.

height of a parallelogram (p. 439) The shortest distance between the bases. The segment that represents the height is perpendicular to the bases.

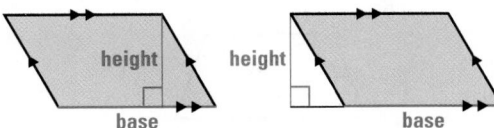

A height can be inside the parallelogram.

A height can be outside the parallelogram.

height of a prism (p. 484) *See* prism.

height of a pyramid (p. 491) The perpendicular distance between the vertex and base.

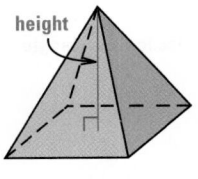

Pyramid

height of a trapezoid (p. 446) The shortest distance between the bases. The segment that represents the height is perpendicular to the bases.

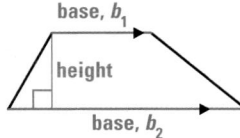

base, b_1

height

base, b_2

height of a triangle (p. 431) The perpendicular segment from a vertex to the line containing the opposite side, called the *base* of the triangle. The term *height* is also used to represent the length of the segment.

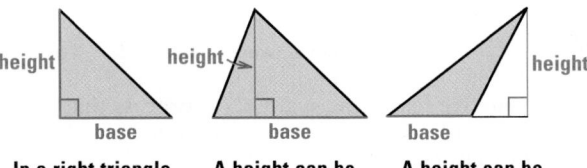

height | height | height
base | base | base

In a right triangle, a leg is a height.

A height can be inside a triangle.

A height can be outside a triangle.

hemisphere (p. 517) A geometric plane passing through the center of a sphere divides it into two *hemispheres*.

heptagon (p. 304) A polygon with seven sides.

hexagon (p. 304) A polygon with six sides.

hypotenuse (pp. 192, 257) In a right triangle, the side opposite the right angle. The hypotenuse is the longest side of a right triangle. *See also* legs of a right triangle.

hypothesis (p. 82) The "if" part of an if-then statement. In the statement "If it is cold, then I will wear my coat," the hypothesis is "it is cold."

if-then statement (p. 82) A statement with two parts: an "if" part that contains the hypothesis and a "then" part that contains the conclusion.

image (pp. 152, 282, 393, 633) The new figure that results from the transformation of a figure in a plane.

included angle (p. 242) An angle of a triangle whose vertex is the shared point of two sides of the triangle.

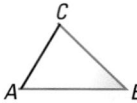

∠*B* is included between $\overline{AB}$ and $\overline{BC}$.

included side (p. 250) A side of a triangle whose endpoints are the vertices of two angles of the triangle.

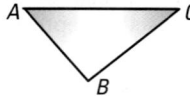

$\overline{AC}$ is included between ∠*A* and ∠*C*.

inductive reasoning (pp. 8, 83) A process that includes looking for patterns and making conjectures.

inscribed angle (p. 614) An angle whose vertex is on a circle and whose sides contain chords of the circle. The arc that lies in the interior of an inscribed angle and has endpoints on the angle is the *intercepted arc*.

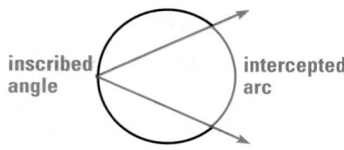

inscribed polygon (p. 615) A polygon whose vertices all lie on a circle. The circle is *circumscribed about* the polygon and is called a *circumscribed circle*.

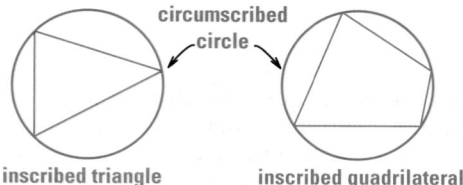

inscribed triangle   inscribed quadrilateral

intercepted arc (p. 614) The arc that lies in the interior of an inscribed angle and has endpoints on the angle. *See also* inscribed angle.

interior of an angle (p. 37) A point is in the *interior* of an angle if it is between points that lie on each side of the angle.

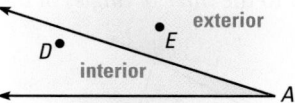

interior angles of a triangle (p. 181) *See* exterior angles of a triangle.

intersect (p. 22) Figures intersect if they have any points in common.

intersection (p. 22) The intersection of two or more figures is the point or points that the figures have in common.

inverse cosine (pp. 570, 571) A function, available on a scientific calculator as $\cos^{-1}x$, which can be used to find the measure of an angle when you know the cosine of the angle.

inverse sine (pp. 570, 571) A function, available on a scientific calculator as $\sin^{-1}x$, which can be used to find the measure of an angle when you know the sine of the angle.

inverse tangent (p. 569) A function, available on a scientific calculator as $\tan^{-1}x$, which can be used to find the measure of an angle when you know the tangent of the angle.

isosceles trapezoid (p. 332) A trapezoid with congruent legs.

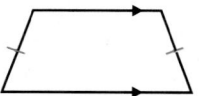

Isosceles trapezoid

isosceles triangle (p. 173) A triangle with at least two congruent sides. *See also* legs of an isosceles triangle.

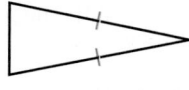

Isosceles triangle

lateral area of a cylinder (p. 485) The area of the curved surface of the cylinder.

lateral area of a prism (p. 484) The sum of the areas of the lateral faces.

lateral faces of a prism (p. 484) *See* prism.

lateral faces of a pyramid (p. 491) *See* pyramid.

leg adjacent to an angle (p. 557) *See* tangent of an angle.

leg opposite an angle (p. 557) *See* tangent of an angle.

legs of an isosceles triangle (p. 185) The congruent sides of an isosceles triangle. The third side is the *base*.

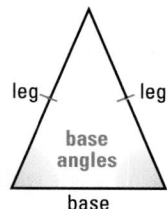

legs of a right triangle (p. 192) The sides that form the right angle.

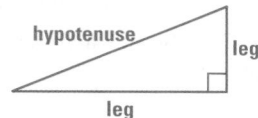

Right triangle

legs of a trapezoid (p. 332) *See* trapezoid.

length of a segment (p. 28) The distance between the endpoints of a segment. *See also* distance between two points on a line.

like terms (p. 101) Terms in an expression that have the same variable raised to the same power. Constant terms, such as 2 and −5, are also considered like terms.

line (p. 14) A line has one dimension and extends without end in two directions. It is represented by a line with two arrowheads. *See also* undefined term.

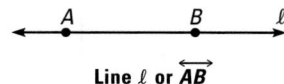

Line ℓ or $\overleftrightarrow{AB}$

line of reflection (p. 282) *See* reflection.

line of symmetry (p. 284) A figure in the plane has a line of symmetry if the figure can be reflected onto itself by a reflection in the line.

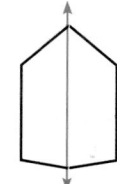 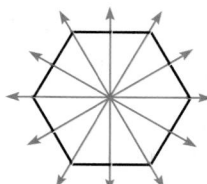

Hexagon with one line of symmetry   **Hexagon with six lines of symmetry**

line perpendicular to a plane (p. 109) A line that intersects a plane in a point and is perpendicular to every line in the plane that intersects it.

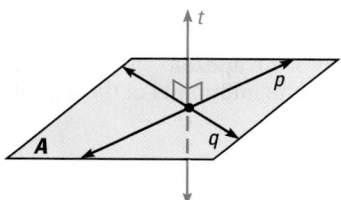

Line *t* is perpendicular to plane *A*.

linear pair (p. 75) Two adjacent angles whose noncommon sides are on the same line.

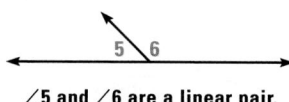

∠5 and ∠6 are a linear pair.

major arc (p. 601) *See* minor arc.

means of a proportion (p. 359) The means of the proportion $\frac{a}{b} = \frac{c}{d}$ are *b* and *c*. *See also* extremes of a proportion.

measure of an angle (p. 36) The size of an angle, written in units called *degrees* (°). The measure of ∠*A* is denoted by *m*∠*A*.

measure of a major arc (p. 601) The difference of 360° and the measure of the related minor arc.

measure of a minor arc (p. 601) The measure of its central angle.

median of a triangle (p. 207) A segment from a vertex to the midpoint of the opposite side.

midpoint (p. 53) The point on a segment that divides it into two congruent segments.

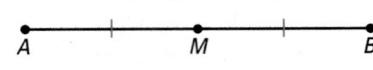

M is the midpoint of $\overline{AB}$.

Midpoint Formula (p. 55) The midpoint of the segment joining $A(x_1, y_1)$ and $B(x_2, y_2)$ is

$$M\left(\frac{x_1 + x_2}{2}, \frac{y_1 + y_2}{2}\right).$$

midsegment of a trapezoid (p. 333) The segment that connects the midpoints of the legs of a trapezoid.

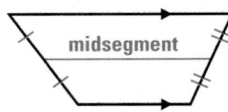

midsegment of a triangle (p. 389) A segment that connects the midpoints of two sides of a triangle.

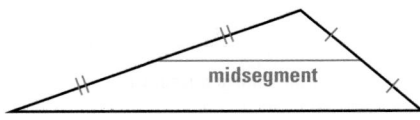

minor arc (p. 601) Points *A* and *B* on a circle *C* determine a minor arc and a major arc.

If the measure of $\angle ACB$ is less than 180°, then *A*, *B*, and all the points on circle *C* that lie in the interior of $\angle ACB$ form a *minor arc*.

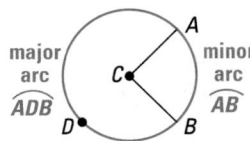

Points *A*, *B*, and all the points of circle *C* that do not lie on $\overline{AB}$ form a *major arc*.

net (p. 483) A flat representation of all the faces of a polyhedron.

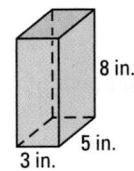
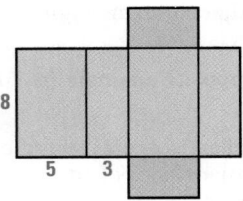

Net

n-gon (p. 304) A polygon with *n* sides.

obtuse angle (p. 36) An angle with measure between 90° and 180°.

obtuse triangle (p. 174) A triangle with one obtuse angle.

octagon (p. 304) A polygon with eight sides.

opposite side (p. 175) The side across from an angle of a triangle.

parallel lines (p. 108) Two lines that lie in the same plane and do not intersect. The symbol for "is parallel to" is ∥.

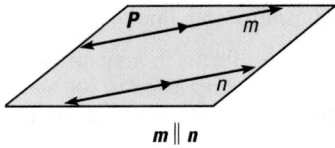

m ∥ *n*

parallel planes (p. 109) Two planes that do not intersect.

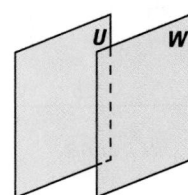

Plane *U* is parallel to plane *W*.

parallelogram (p. 310) A quadrilateral with both pairs of opposite sides parallel. The symbol for parallelogram *PQRS* is □ *PQRS*.

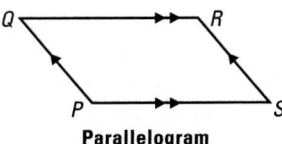

Parallelogram

pentagon (p. 304) A polygon with five sides.

perpendicular bisector (p. 274) A line that is perpendicular to a segment at its midpoint.

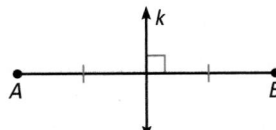

Line *k* is the perpendicular bisector of $\overline{AB}$.

perpendicular lines (p. 108) Two lines that intersect to form a right angle. The symbol for "is perpendicular to" is ⊥. The red angle mark shown below indicates a right angle.

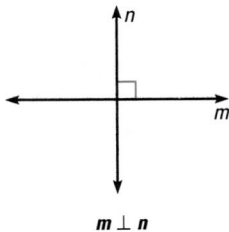

m* ⊥ *n

pi (p. 452) The ratio of the circumference of a circle to its diameter. Pi is an irrational number denoted by π and is approximately equal to 3.14.

plane (p. 14) A plane has two dimensions. It is represented by a shape that looks like a floor or wall. You have to imagine that it extends without end, even though the drawing of a plane appears to have edges. *See also* undefined term.

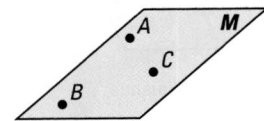

Plane *M* or plane *ABC*

point (p. 14) A point has no dimension. It is represented by a small dot. *See also* undefined term.

point of tangency (p. 589) *See* tangent.

polygon (p. 303) A plane figure that is formed by three or more segments called *sides*. Each side intersects exactly two other sides at each of its endpoints. Each endpoint is a *vertex* of the polygon.

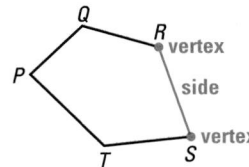

polyhedron (pp. 473, 474) A solid that is formed by polygons. The plane surfaces are called *faces* and the segments joining the vertices are called *edges*.

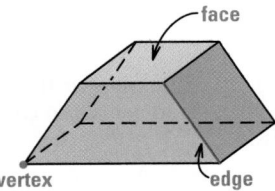

postulate (p. 14) A statement that is accepted without further justification.

power (p. 467) An expression like 7^3. The *exponent* 3 represents the number of times the *base* 7 is used as a factor: $7^3 = 7 \cdot 7 \cdot 7$.

prism (pp. 473, 475, 483) A polyhedron with two congruent faces, called *bases*, that lie in parallel planes. The other faces are called *lateral faces*. The *height* is the perpendicular distance between the bases.

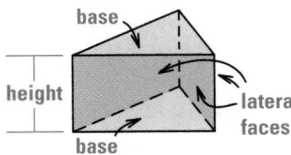

proof (p. 243) A convincing argument that shows why a statement is true.

Glossary

Glossary **719**

719

proportion (pp. 349, 359) An equation that states that two ratios are equal. *Example:* $\frac{a}{b} = \frac{c}{d}$

pyramid (pp. 473, 475, 491) A polyhedron in which the *base* is a polygon and the *lateral faces* are triangles with a common vertex. *See also* height of a pyramid, slant height of a pyramid.

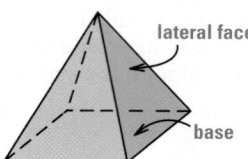

Pythagorean triple (p. 196) Three positive integers a, b, and c that satisfy the equation $c^2 = a^2 + b^2$.

Q

quadrilateral (p. 304) A polygon with four sides.

R

radical (p. 537) An expression written with a radical symbol $\sqrt{\ }$. A radical is also called a *radical expression*.

radical symbol (p. 537) The square root symbol, $\sqrt{\ }$, which indicates the nonnegative square root of a number. For example, $\sqrt{25} = 5$.

radicand (p. 537) The number or expression written inside a radical symbol. In the radical $\sqrt{25}$, the radicand is 25.

radius (pp. 452, 589) The distance from the center to a point on the circle. A segment whose endpoints are the center of the circle and a point on the circle is also called a *radius*. The plural of radius is *radii*.

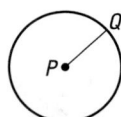

Radius: *PQ* or $\overline{PQ}$

radius of a cone (p. 493) *See* cone.

radius of a cylinder (p. 485) *See* cylinder.

radius of a sphere (p. 517) *See* sphere.

rate of *a* to *b* (p. 227) The quotient $\frac{a}{b}$ if a and b are two quantities that have different kinds of units of measure.

ratio of *a* to *b* (pp. 227, 357) A comparison of a number a and a nonzero number b using division. The ratio of a to b can be written as the fraction $\frac{a}{b}$, as $a : b$, or as "a to b."

ray (p. 16) The ray $\overrightarrow{AB}$ consists of the *endpoint A* and all points on $\overleftrightarrow{AB}$ that lie on the same side of A as B.

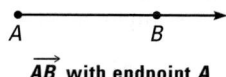

$\overrightarrow{AB}$ with endpoint *A*

rectangle (p. 325) A parallelogram with four right angles.

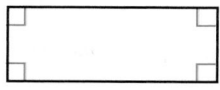

Rectangle

reduction (p. 394) A dilation in which the image is smaller than the original figure. The scale factor of a reduction is less than 1.

reflection (p. 282) A transformation that creates a mirror image. The original figure is reflected in a *line of reflection*.

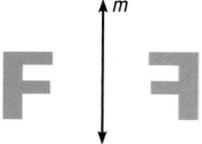

Line *m* is a line of reflection.

regular polygon (p. 412) A polygon that is both equilateral and equiangular.

rhombus (p. 325) A parallelogram with four congruent sides.

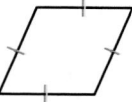

Rhombus

right angle (p. 36) An angle with measure 90°.

right triangle (p. 174) A triangle with one right angle. *See also* legs of a right triangle.

rotation (p. 633) A transformation in which a figure is turned about a fixed point, called the *center of rotation*. Rays drawn from the center of rotation to a point and its image form an angle called the *angle of rotation*.

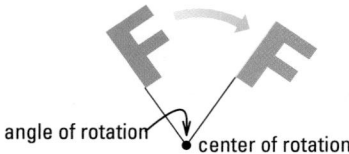

rotational symmetry (p. 634) A figure in the plane has rotational symmetry if the figure can be mapped onto itself by a rotation of 180° or less.

same-side interior angles (p. 121) Two angles that are formed by two lines and a transversal, and lie between the two lines on the same side of the transversal. In the diagram below, ∠3 and ∠5 are same-side interior angles.

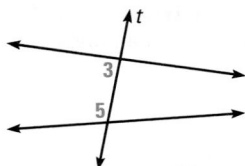

scale factor (p. 366) The ratio of the lengths of two corresponding sides of two similar polygons.

scale factor of a dilation (p. 394) The value of k where $k = \dfrac{CP'}{CP}$. *See also* dilation.

scalene triangle (p. 173) A triangle with no congruent sides.

secant (p. 589) A line that intersects a circle in two points.

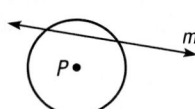

sector (p. 454) A region of a circle determined by two radii and a part of the circle.

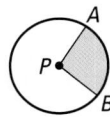

segment (p. 16) Part of a line that consists of two points, called *endpoints*, and all points on the line that are between the endpoints.

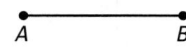

$\overline{AB}$ with endpoints *A* and *B*

segment bisector (p. 53) A segment, ray, line, or plane that intersects a segment at its midpoint.

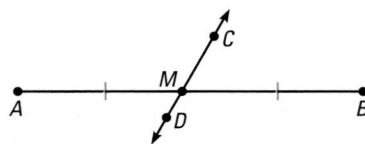

$\overleftrightarrow{CD}$ is a bisector of $\overline{AB}$.

semicircle (p. 601) An arc whose central angle measures 180°.

side of an angle (p. 35) *See* angle.

side of a polygon (p. 303) *See* polygon.

similar polygons (p. 365) Two polygons are similar polygons if corresponding angles are congruent and corresponding side lengths are proportional. The symbol for "is similar to" is ~.

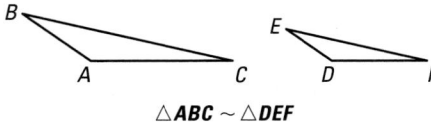

△*ABC* ~ △*DEF*

similarity statement (p. 365) A statement, such as △*ABC* ~ △*DEF*, that indicates that two polygons are similar.

sine (p. 563) A trigonometric ratio, abbreviated as *sin* and computed as the ratio of the length of the leg opposite the angle to the length of the hypotenuse.

$$\sin A = \frac{\text{leg opposite } \angle A}{\text{hypotenuse}} = \frac{a}{c}$$

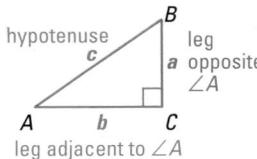

skew lines (p. 108) Two lines that do not lie in the same plane and do not intersect.

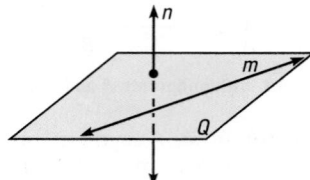

Lines *m* and *n* are skew lines.

slant height of a cone (p. 493) *See* cone.

slant height of a pyramid (p. 491) The height of a lateral face of a pyramid. The letter ℓ is used to represent the slant height.

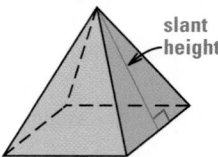

slant height

slope (pp. 150, 299, 665) The ratio of the vertical change (rise) to the horizontal change (run) between any two points on a line.

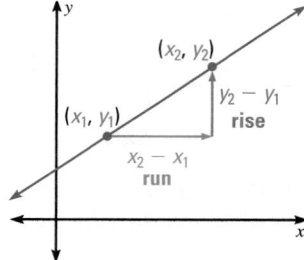

solid (p. 473) A three-dimensional shape.

solve a right triangle (p. 569) Find the measures of both acute angles and the lengths of all three sides.

sphere (p. 517) The set of all points in space that are the same distance from a point, called the *center* of the sphere. The *radius* of a sphere is the length of a segment from the center to a point on the sphere.

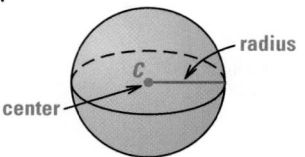

radius

center

square (p. 325) A parallelogram with four congruent sides and four right angles.

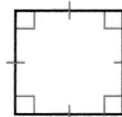

square root (p. 167) If $b^2 = a$, then b is a square root of a.

standard equation of a circle (p. 628) In the coordinate plane, the standard equation of a circle with center at (h, k) and radius r is
$$(x - h)^2 + (y - k)^2 = r^2.$$

straight angle (p. 36) An angle with measure 180°.

straightedge (p. 143) A construction tool used to draw segments. A ruler without marks.

supplement (p. 67) The sum of the measures of an angle and its *supplement* is 180°.

supplementary angles (p. 67) Two angles whose measures have a sum of 180°.

surface area of a polyhedron (p. 483) The sum of the areas of the faces of a polyhedron.

T

tangent (p. 589) A line in the plane of a circle that intersects the circle in exactly one point, called a *point of tangency.*

Line *n* is a tangent.
R is the point of tangency.

tangent of an angle (p. 557) A trigonometric ratio, abbreviated as *tan* and computed as the ratio of the length of the leg opposite the angle to the length of the leg adjacent to (contained in) the angle.

$$\tan A = \frac{\text{leg opposite } \angle A}{\text{leg adjacent to } \angle A} = \frac{a}{b}$$

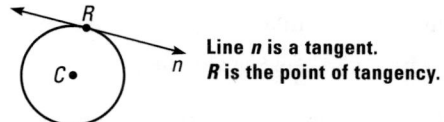

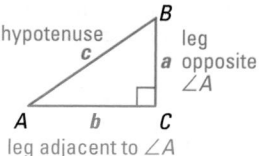

hypotenuse
c
leg
a opposite
$\angle A$

A b C
leg adjacent to $\angle A$

tangent segment (p. 597) A segment that touches a circle at one of the segment's endpoints and lies in the line that is tangent to the circle at that point.

theorem (p. 69) A true statement that follows from other true statements.

transformation (p. 152) An operation that *maps*, or moves, a figure onto an image. *See also* dilation, reflection, rotation, translation.

translation (p. 152) A transformation that slides each point of a figure the same distance in the same direction.

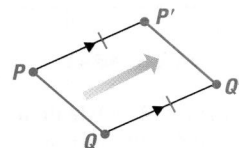

transversal (p. 121) A line that intersects two or more coplanar lines at different points.

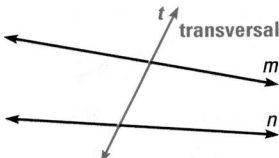

trapezoid (p. 332) A quadrilateral with exactly one pair of parallel sides, called *bases*. The nonparallel sides are the *legs*.

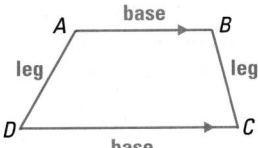

triangle (p. 173) A figure formed by three segments joining three noncollinear points, called *vertices*. The triangle symbol is △.

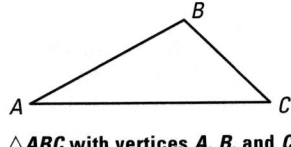

△*ABC* with vertices *A*, *B*, and *C*

trigonometric ratio (p. 557) A ratio of the lengths of two sides of a right triangle. *See also* cosine, sine, tangent of an angle.

undefined term (p. 14) A word, such as *point*, *line*, or *plane*, that is not mathematically defined using other known words, although there is a common understanding of what the word means.

vertex of an angle (p. 35) *See* angle.

vertex of a polygon (p. 303) *See* polygon.

vertex of a triangle (p. 175) A point that joins two sides of a triangle. The plural is *vertices*. *See also* triangle.

vertical angles (p. 75) Two angles that are not adjacent and whose sides are formed by two intersecting lines.

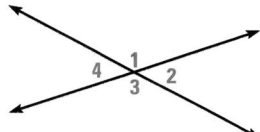

∠**1 and** ∠**3 are vertical angles.**
∠**2 and** ∠**4 are vertical angles.**

volume of a solid (p. 500) The number of cubic units contained in the interior of the solid.

Glossary

A

acute angle (p. 36) **ángulo agudo** Ángulo que mide entre 0° and 90°.

acute triangle (p. 174) **triángulo agudo** Triángulo con tres ángulos agudos.

adjacent angles (p. 68) **ángulos adyacentes** Dos ángulos con un vértice y un lado común pero sin puntos internos comunes.

alternate exterior angles (p. 121) **ángulos exteriores alternos** Dos ángulos formados por dos rectas y una transversal y situados fuera de las dos rectas en lados opuestos de la transversal. En el siguiente diagrama ∠1 y ∠8 son ángulos exteriores alternos.

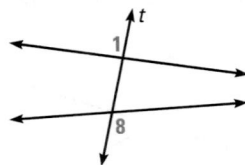

alternate interior angles (p. 121) **ángulos interiores alternos** Dos ángulos formados por dos rectas y una transversal y situados entre las dos rectas en lados opuestos de la transversal. En el siguiente diagrama ∠3 y ∠6 son ángulos interiores alternos.

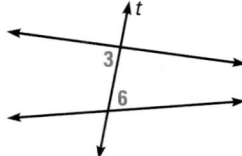

angle (p. 35) **ángulo** Figura formada por dos rayos que comienzan en el mismo extremo. Los rayos son los *lados* del ángulo y el extremo es el *vértice* del ángulo.

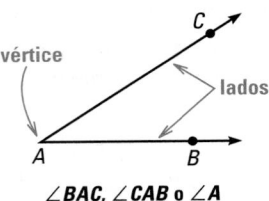

∠*BAC,* ∠*CAB* o ∠*A*

angle bisector (p. 61) **bisectriz de un ángulo** Rayo que divide un ángulo en dos ángulos congruentes.

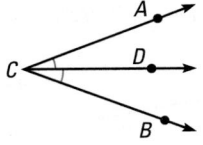

$\overrightarrow{CD}$ biseca a ∠*ACB.*

$m∠ACD = m∠BCD$

angle of rotation (p. 633) **ángulo de rotación** *Ver* rotation/rotación.

arc length (p. 603) **longitud de arco** Parte de la circunferencia de un círculo.

area (p. 424) **área** Medida de la superficie cubierta por una figura.

B

base of a cone (p. 493) **base de un cono** *Ver* cone/cono.

bases of a cylinder (p. 485) **bases de un cilindro** *Ver* cylinder/cilindro.

base of an isosceles triangle (p. 185) **base de un triángulo isósceles** *Ver* legs of an isosceles triangle/catetos de un triángulo isósceles.

bases of a parallelogram (p. 439) **bases de un paralelogramo** Cualquier par de lados paralelos de un paralelogramo se llama *bases. Ver también* height of a parallelogram/altura de un paralelogramo.

bases of a prism (p. 473) **bases de un prisma** *Ver* prism/prisma.

base of a pyramid (p. 491) **base de una pirámide** *Ver* pyramid/pirámide.

bases of a trapezoid (p. 332) **bases de un trapezoide** *Ver* trapezoid/trapezoide.

base of a triangle (p. 431) **base de un triángulo** *Ver* height of a triangle/altura de un triángulo.

base angles of an isosceles triangle (p. 185) **ángulos base de un triángulo isósceles** Los dos ángulos en la base de un triángulo isósceles. *Ver también* legs of an isosceles triangle/catetos de un triángulo isósceles.

base angles of a trapezoid (p. 332) **ángulos base de un trapezoide** Si un trapezoide *ABCD* tiene $\overline{AB}$ y $\overline{CD}$ como bases, entonces hay dos pares de ángulos base: $\angle A$ y $\angle B$, y $\angle C$ y $\angle D$.

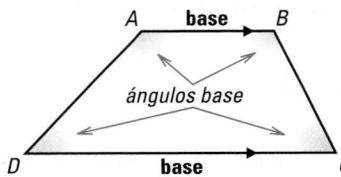

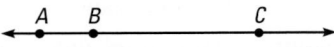

between (p. 29) **entre** Cuando tres puntos se encuentran en una recta, uno de ellos está *entre* los otros dos.

El punto *B* está entre los puntos *A* y *C*.

bisect (pp. 53, 61) **bisecar** Dividir en dos partes congruentes.

center of a circle (p. 452) **centro de un círculo** *Ver* circle/círculo.

center of dilation (p. 393) **centro de dilatación** *Ver* dilation/dilatación.

center of rotation (p. 633) **centro de rotación** *Ver* rotation/rotación.

center of a sphere (p. 517) **centro de una esfera** *Ver* sphere/esfera.

central angle (p. 454) **ángulo central** Ángulo cuyo vértice es el centro de un círculo.

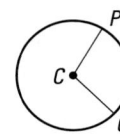

$\angle PCQ$ es un ángulo central.

centroid of a triangle (p. 208) **centroide de un triángulo** Punto en el que se intersecan las tres medianas de un triángulo.

chord (p. 589) **cuerda** Segmento cuyos extremos son puntos de un círculo.

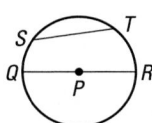

Cuerdas $\overline{QR}$, $\overline{ST}$

circle (pp. 452, 589) **círculo** Conjunto de todos los puntos de un plano que está a la misma distancia de un punto dado llamado *centro* del círculo.

Círculo con centro *P*, o ⊙*P*

circumference (p. 452) **circunferencia** Distancia en torno a un círculo.

circumscribed circle (p. 615) **círculo circunscrito** *Ver* inscribed polygon/polígono inscrito.

collinear points (p. 15) **puntos colineales** Puntos ubicados en la misma recta.

compass (p. 143) **compás** Instrumento de construcción utilizado para dibujar arcos y círculos.

complement (p. 67) **complemento** La suma de las medidas de un ángulo y su *complemento* es 90°.

complementary angles (p. 67) **ángulos complementarios** Dos ángulos cuyas medidas suman 90°.

concave polygon (p. 411) **polígono cóncavo** Ver convex polygon/polígono convexo.

conclusion (p. 82) **conclusión** La parte que comienza con "entonces" de un enunciado de tipo "si..., entonces...". En el enunciado "Si hace frío, entonces me pondré un abrigo", la conclusión es "me pondré un abrigo".

cone (p. 493) **cono** Sólido que tiene una *base* circular y un vértice que no pertenece al mismo plano que la base. La *altura* de un cono es la distancia perpendicular entre el vértice y la base. El *radio* de un cono es el radio de la base. La *altura inclinada* de un cono es la distancia entre el vértice y un punto del borde de la base.

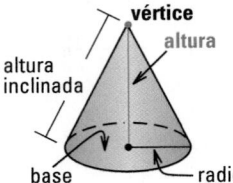

congruent angles (p. 36) **ángulos congruentes** Ángulos que tienen la misma medida.

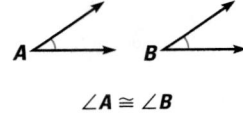

$$\angle A \cong \angle B$$

congruent arcs (p. 602) **arcos congruentes** Dos arcos del mismo círculo o de círculos congruentes que tienen la misma medida.

congruent circles (p. 602) **círculos congruentes** Dos círculos que tienen el mismo radio.

congruent figures (p. 233) **figuras congruentes** Dos figuras geométricas que tienen exactamente el mismo tamaño y la misma forma. Cuando dos figuras son congruentes, todos los pares de ángulos y lados correspondientes son congruentes.

congruent segments (p. 30) **segmentos congruentes** Segmentos que tienen la misma longitud.

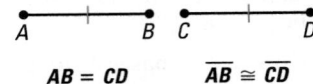

$$AB = CD \qquad \overline{AB} \cong \overline{CD}$$

conjecture (p. 8) **conjetura** Enunciado no comprobado que se basa en un patrón o en observaciones.

construction (p. 143) **construcción** Dibujo geométrico realizado con varios instrumentos, generalmente un compás y una regla.

converse (p. 136) **inverso** Enunciado formado al cambiar la hipótesis y la conclusión de un enunciado de tipo "si..., entonces...".

convex polygon (p. 411) **polígono convexo** Un polígono es convexo si ninguna recta que contiene un lado del polígono pasa por el interior del polígono. El polígono que no es convexo se llama *cóncavo*.

Polígono convexo   **Polígono cóncavo**

coordinate (p. 28) **coordenada** Número real que corresponde a un punto de una recta.

coplanar lines (p. 15) **rectas coplanarias** Rectas que pertenecen a un mismo plano.

coplanar points (p. 15) **puntos coplanarios** Puntos que pertenecen a un mismo plano.

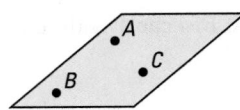

corollary to a theorem (p. 180) **corolario de un teorema** Enunciado que se puede probar fácilmente usando el teorema.

corresponding angles (p. 121) **ángulos correspondientes** Ángulos que están en la misma posición relativa cuando una transversal cruza dos rectas. En el siguiente diagrama, $\angle 1$ y $\angle 5$ son ángulos correspondientes.

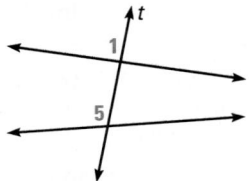

corresponding parts of congruent figures (p. 233) **partes correspondientes de figuras congruentes** Lados y ángulos correspondientes de figuras congruentes.

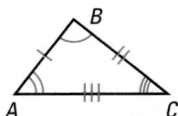 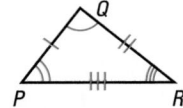

$\angle B$ y $\angle Q$ son ángulos correspondientes.

$\overline{BC}$ y $\overline{QR}$ son lados correspondientes.

cosine (p. 563) **coseno** Razón trigonométrica que se abrevia *cos* y se calcula como la razón entre la longitud del cateto adyacente al ángulo y la longitud de la hipotenusa.

$$\cos A = \frac{\text{cateto adyacente a } \angle A}{\text{hipotenusa}} = \frac{b}{c}$$

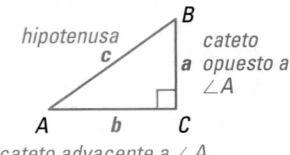

counterexample (p. 10) **contraejemplo** Ejemplo que muestra que la conjetura es falsa.

cross product property (pp. 349, 359) **propiedad de los productos cruzados** Si $\frac{a}{b} = \frac{c}{d}$, entonces $ad = bc$.

cylinder (p. 485) **cilindro** Figura sólida con dos *bases* circulares congruentes que yace sobre planos paralelos. La *altura* de un cilindro es la distancia perpendicular entre las bases. El *radio* de un cilindro es el radio de una de las bases.

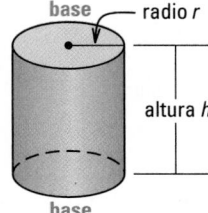

deductive reasoning (p. 83) **razonamiento deductivo** El uso de hechos, definiciones, propiedades aceptadas y las reglas de la lógica para establecer un argumento lógico.

degrees (°) (p. 36) **grados** (°) *Ver* measure of an angle/medida de un ángulo.

diagonal of a polygon (p. 303) **diagonal de un polígono** Segmento que conecta dos vértices no consecutivos de un polígono.

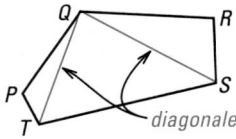

diameter (pp. 452, 589) **diámetro** Distancia a través de un círculo que pasa por su centro. La longitud del *diámetro* es el doble de la longitud del *radio*. Una cuerda que pasa por el centro del círculo también se denomina *diámetro*.

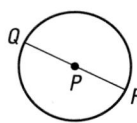

Diámetro: $\overline{QR}$ o QR

dilation (p. 393) **dilatación** Transformación con un centro *C* y un factor de escala *k* que refleja cada punto *P* en un punto de imagen *P′* de manera que *P′* se encuentra en $\overrightarrow{CP}$ y $CP′ = k \cdot CP$.

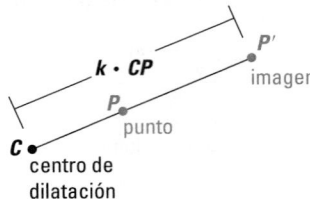

centro de dilatación

distance between two points on a line (p. 28) **distancia entre dos puntos en una recta** La distancia *AB* es el valor absoluto de la diferencia de las coordenadas de *A* y *B*. $\overline{AB}$ también se conoce como la longitud de $\overline{AB}$.

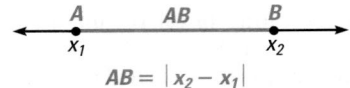

$$AB = |x_2 - x_1|$$

Distance Formula (p. 194) **Fórmula de distancia** Si $A(x_1, y_1)$ y $B(x_2, y_2)$ son puntos de un plano de coordenadas, la distancia entre *A* y *B* es

$$\sqrt{(x_2 - x_1)^2 + (y_2 - y_1)^2}.$$

distance from a point to a line (p. 273) **distancia de un punto a una recta** La longitud del segmento perpendicular de un punto a una recta.

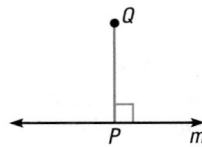

La distancia de *Q* a la recta *m* es *QP*.

edge of a polyhedron (p. 474) **arista de un poliedro** *Ver* polyhedron/poliedro.

endpoint of a ray (p. 16) **extremo de un rayo** *Ver* ray/rayo.

endpoints of a segment (p. 16) **extremos de un segmento** *Ver* segment/segmento.

enlargement (p. 394) **ampliación** Dilatación en la cual la imagen es más grande que la figura original. El factor de escala de una ampliación es mayor que 1.

equiangular polygon (p. 412) **polígono equiangular** Polígono cuyos ángulos interiores son congruentes.

equiangular triangle (p. 174) **triángulo equiangular** Triángulo en el que los tres ángulos tienen la misma medida.

equidistant (p. 273) **equidistante** La misma distancia.

equilateral polygon (p. 412) **polígono equilátero** Polígono en el que todos los lados son congruentes.

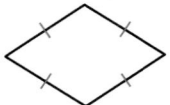

Polígono equilátero

equilateral triangle (p. 173) **triángulo equilátero** Triángulo que tiene tres lados congruentes.

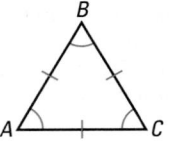

Triángulo equilátero

evaluate an expression (p. 49) **evaluar una expresión** Hallar el valor de una expresión sustituyendo valores por variables y simplificando el resultado realizando operaciones en orden.

exponent (p. 467) **exponente** *Ver* power/potencia.

exterior angles of a triangle (p. 181) **ángulos externos de un triángulo** Los ángulos adyacentes a los *ángulos internos* del triángulo cuando se amplía los lados de un triángulo.

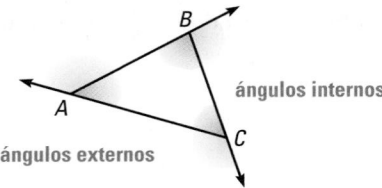

ángulos internos

ángulos externos

extremes of a proportion (p. 359) extremos de una proporción Los extremos de una proporción $\frac{a}{b} = \frac{c}{d}$ son *a* y *d*. *Ver también* means of a proportion/medias de una proporción.

face (p. 474) cara *Ver* polyhedron/poliedro.

height of a cone (p. 493) altura de un cono *Ver* cone/cono.

height of a cylinder (p. 485) altura de un cilindro *Ver* cylinder/cilindro.

height of a parallelogram (p. 439) altura de un paralelogramo La distancia más corta entre las bases. El segmento que representa la altura es perpendicular a las bases.

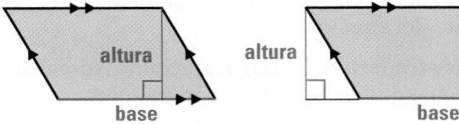

La altura puede medirse dentro del paralelogramo.

La altura puede medirse fuera del paralelogramo.

height of a prism (p. 484) altura de un prisma *Ver* prism/prisma.

height of a pyramid (p. 491) altura de una pirámide La distancia perpendicular entre el vértice y la base.

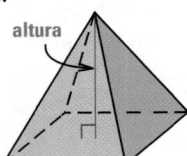

Pirámide

height of a trapezoid (p. 446) altura de un trapezoide La distancia más corta entre las bases. El segmento que representa la altura es perpendicular a las bases.

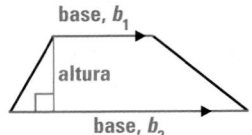

height of a triangle (p. 431) altura de un triángulo El segmento perpendicular que va de un vértice a la recta que contiene el lado opuesto, llamado *base* del triángulo. El término *altura* también se utiliza para representar la longitud del segmento.

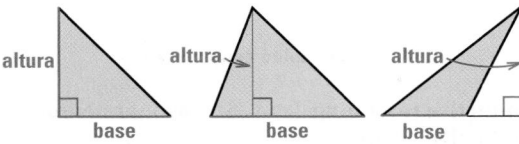

hemisphere (p. 517) hemisferio Un plano geométrico que pasa por el centro de una esfera la divide en dos *hemisferios*.

heptagon (p. 304) heptágono Polígono de siete lados.

hexagon (p. 304) hexágono Polígono de seis lados.

hypotenuse (pp. 192, 257) hipotenusa Lado de un triángulo recto que está opuesto al ángulo recto. Es el lado más largo de un triángulo recto. *Ver también* legs of a right triangle/catetos de un triángulo recto.

hypothesis (p. 82) hipótesis La parte "si" de un enunciado de tipo "si…, entonces…". En el enunciado "Si hace frío, entonces me pondré un abrigo", la hipótesis es "hace frío".

if-then statement (p. 82) enunciado de tipo "si…, entonces…" Enunciado con dos partes: una parte (si…) que expresa una hipótesis y una parte (entonces…) que expresa una conclusión.

image (pp. 152, 282, 393, 633) imagen Nueva figura formada por la transformación de una figura en un plano.

included angle (p. 242) ángulo incluido Ángulo de un triángulo cuyo vértice es el punto común de dos lados del triángulo.

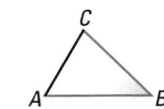

$\angle B$ está incluido entre $\overline{AB}$ y $\overline{BC}$.

included side (p. 250) **lado incluido** Lado de un triángulo cuyos extremos son los vértices de dos ángulos del triángulo.

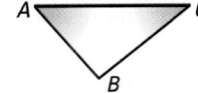

$\overline{AC}$ está incluido entre ∠*A* y ∠*C*.

inductive reasoning (pp. 8, 83) **razonamiento inductivo** Proceso que consiste en buscar patrones para realizar conjeturas.

inscribed angle (p. 614) **ángulo inscrito** Ángulo cuyo vértice pertenece a un círculo y cuyos lados tienen cuerdas del círculo. El arco en el interior de un ángulo inscrito que tiene extremos en el ángulo es el *arco interceptado*.

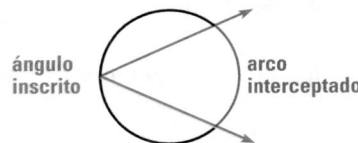

inscribed polygon (p. 615) **polígono inscrito** Polígono cuyos vértices pertenecen a un círculo. El círculo *es circunscrito al* polígono y se denomina *círculo circunscrito*.

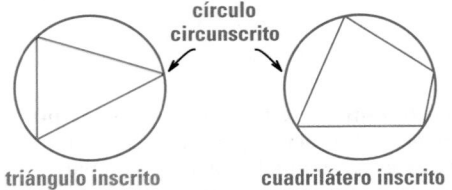

intercepted arc (p. 614) **arco interceptado** Arco situado en el interior de un ángulo inscrito y que tiene extremos en el ángulo. *Ver también* inscribed angle/ángulo inscrito.

interior of an angle (p. 37) **interior de un ángulo** Un punto está en el interior de un ángulo si está situado entre puntos de cada lado del ángulo.

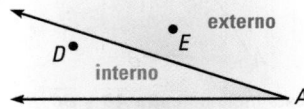

interior angles of a triangle (p. 181) **ángulos internos de un triángulo** *Ver* exterior angles of a triangle/ángulos externos de un triángulo.

intersect (p. 22) **intersecar** Las figuras se intersecan si tienen puntos en común.

intersection (p. 22) **intersección** La intersección de dos o más figuras es el punto o los puntos que dichas figuras tienen en común.

inverse cosine (pp. 570, 571) **coseno inverso** Función, representada en una calculadora científica por $\cos^{-1} x$, que se puede usar para hallar la medida de un ángulo si se sabe el coseno del ángulo.

inverse sine (pp. 570, 571) **seno inverso** Función, representada en una calculadora científica por $\sin^{-1} x$, que se puede usar para hallar la medida de un ángulo si se sabe el seno del ángulo.

inverse tangent (p. 569) **tangente inversa** Función, representada en una calculadora científica por $\tan^{-1} x$, que se puede usar para hallar la medida de un ángulo si se sabe la tangente del ángulo.

isosceles trapezoid (p. 332) **trapezoide isósceles** Trapezoide que tiene catetos congruentes.

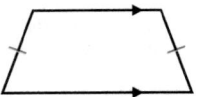

Trapezoide isósceles

isosceles triangle (p. 173) **triángulo isósceles** Triángulo que tiene al menos dos lados congruentes. *Ver también* legs of an isosceles triangle/catetos de un triángulo isósceles.

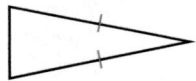

Triángulo isósceles

lateral area of a cylinder (p. 485) **área lateral de un cilindro** Área de la superficie curva de un cilindro.

lateral area of a prism (p. 484) **área lateral de un prisma** Suma de las áreas de las caras laterales.

lateral faces of a prism (p. 484) **caras laterales de un prisma** *Ver* prism/prisma.

lateral faces of a pyramid (p. 491) **caras laterales de una pirámide** *Ver* pyramid/pirámide.

leg adjacent to an angle (p. 557) **cateto adyacente a un ángulo** *Ver* tangent of an angle/tangente de un ángulo.

leg opposite an angle (p. 557) **cateto opuesto al ángulo** *Ver* tangent of an angle/tangente de un ángulo.

legs of an isosceles triangle (p. 185) **catetos de un triángulo isósceles** Lados congruentes de un triángulo isósceles. El tercer lado es la *base*.

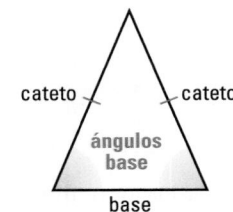

legs of a right triangle (p. 192) **catetos de un triángulo recto** Lados que forman el ángulo recto.

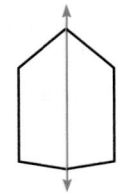

Triángulo recto

legs of a trapezoid (p. 332) **catetos de un trapezoide** *Ver* trapezoid/trapezoide.

length of a segment (p. 28) **longitud de un segmento** Distancia entre los extremos de un segmento. *Ver también* distance between two points on a line/distancia entre dos puntos en una recta.

like terms (p. 101) **términos semejantes** Términos de una expresión que tienen las mismas variables elevadas a las mismas potencias. Los términos constantes, como 2 y −5, también son considerados términos semejantes.

line (p. 14) **línea recta** o **recta** Una recta tiene una dimensión y se extiende de manera infinita en dos direcciones. Su representación es una línea recta con dos flechas. *Ver también* undefined term/término indefinido.

Recta ℓ o $\overleftrightarrow{AB}$

line of reflection (p. 282) **eje de reflexión** *Ver* reflection/reflexión.

line of symmetry (p. 284) **eje de simetría** Una figura en un plano tiene eje de simetría si se puede reflejar en sí misma a través de una reflexión en la recta.

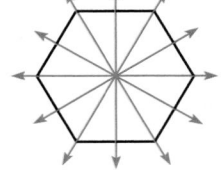

Hexágono con eje de simetría   **Hexágono con seis ejes de simetría**

line perpendicular to a plane (p. 109) **recta perpendicular a un plano** Recta que interseca un plano en un punto y que es perpendicular a cada recta que la interseca en el plano.

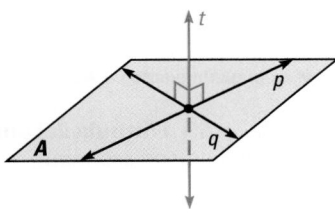

La recta *t* es perpendicular al plano *A*.

linear pair (p. 75) **par lineal** Dos ángulos adyacentes cuyos lados no comunes están en la misma recta.

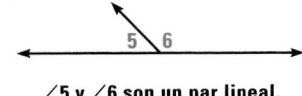

∠5 y ∠6 son un par lineal.

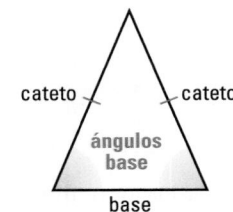

English-to-Spanish Glossary

English-to-Spanish Glossary

major arc (p. 601) **arco mayor** *Ver* minor arc/arco menor.

means of a proportion (p. 359) **medias de una proporción** Las medias de una proporción $\frac{a}{b} = \frac{c}{d}$ son *b* y *c*. *Ver también* extremes of a proportion/extremos de una proporción.

measure of an angle (p. 36) **medida de un ángulo** Tamaño de un ángulo que se expresa en unidades llamadas *grados* (°). $m\angle A$ denota la medida de $\angle A$.

measure of a major arc (p. 601) **medida de un arco mayor** La diferencia de 360° y la medida del arco menor relacionado.

measure of a minor arc (p. 601) **medida de un arco menor** Medida de su ángulo central.

median of a triangle (p. 207) **mediana de un triángulo** Segmento de un vértice al punto medio del lado opuesto.

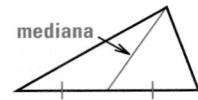

midpoint (p. 53) **punto medio** Punto de un segmento que lo divide en dos segmentos congruentes.

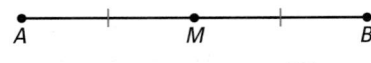

M es el punto medio de $\overline{AB}$.

Midpoint Formula (p. 55) **Fórmula del punto medio** El punto medio de un segmento que une $A(x_1, y_1)$ y $B(x_2, y_2)$ es

$$M\left(\frac{x_1 + x_2}{2}, \frac{y_1 + y_2}{2}\right).$$

midsegment of a trapezoid (p. 333) **segmento medio de un trapezoide** Segmento que une los puntos medios de los catetos de un trapezoide.

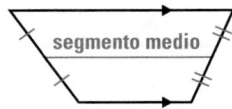

midsegment of a triangle (p. 389) **segmento medio de un triángulo** Segmento que une los puntos medios de dos lados de un triángulo.

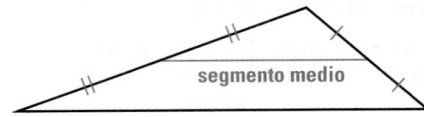

minor arc (p. 601) **arco menor** Los puntos *A* y *B* de un círculo *C* determinan un arco menor y un arco mayor.

Si la medida de $\angle ACB$ es menor que 180°, entonces *A*, *B* y todos los puntos del círculo *C* situados en el interior de $\angle ACB$ forman un *arco menor*.

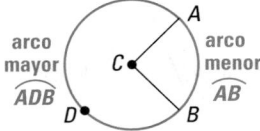

Los puntos *A*, *B* y todos los puntos del círculo *C* que no están situados en $\overarc{AB}$ forman un *arco mayor*.

net (p. 483) **red** Representación plana de todas las caras de un poliedro.

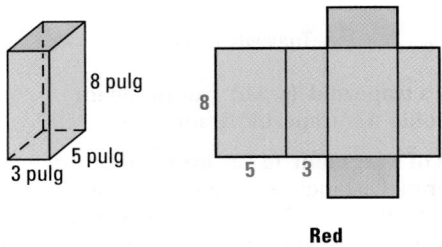

Red

n-gon (p. 304) **n-gono** Polígono con *n* número de lados.

obtuse angle (p. 36) **ángulo obtuso** Ángulo que mide entre 90° y 180°.

obtuse triangle (p. 174) **triángulo obtuso** Triángulo que tiene un ángulo obtuso.

octagon (p. 304) **octágono** Polígono de ocho lados.

opposite side (p. 175) **lado opuesto** Lado situado frente a un ángulo en un triángulo.

parallel lines (p. 108) **rectas paralelas** Rectas del mismo plano que no se intersecan. El símbolo usado para representar "es paralelo a" es ∥.

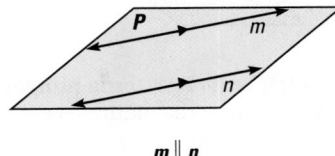

$m \parallel n$

parallel planes (p. 109) **planos paralelos** Dos planos que no se intersecan.

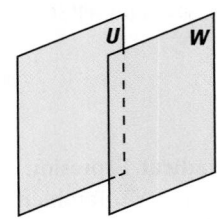

El plano **U** es paralelo al plano **W**.

parallelogram (p. 310) **paralelogramo** Cuadrilátero cuyos lados opuestos son paralelos. El símbolo ▱ *PQRS* se usa para representar el paralelogramo *PQRS*.

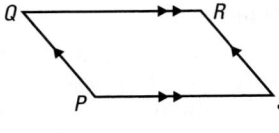

Paralelogramo

pentagon (p. 304) **pentágono** Polígono de cinco lados.

perpendicular bisector (p. 274) **bisectriz perpendicular** Recta perpendicular que pasa por el punto medio de un segmento.

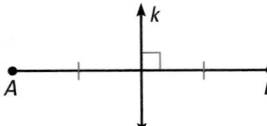

La recta **k** es la bisectriz perpendicular de $\overline{AB}$.

perpendicular lines (p. 108) **rectas perpendiculares** Dos rectas que se intersecan para formar un ángulo recto. El símbolo ⊥ representa la expresión "es perpendicular a". La marca roja del ángulo que se muestra abajo indica un ángulo recto.

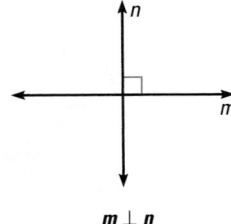

$m \perp n$

pi (p. 452) **pi** Número que representa la relación entre la circunferencia de un círculo y su diámetro. Pi es un número irracional que se denota con π y es aproximadamente igual a 3.14.

plane (p. 14) **plano** Un plano tiene dos dimensiones y se representa con una forma parecida a una pared o al piso. Es necesario imaginar que se extiende de manera infinita aunque el dibujo de un plano parezca tener bordes. *Ver también* undefined term/término indefinido.

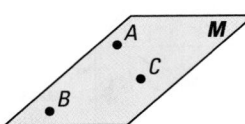

Plano **M** o plano **ABC**

point (p. 14) **punto** Un punto no tiene dimensiones. Se representa por un punto pequeño. *Ver también* undefined term/término indefinido.

point of tangency (p. 589) **punto de tangencia** *Ver* tangent/tangente.

polygon (p. 303) **polígono** Figura plana formada por tres o más segmentos llamados *lados*. Cada lado interseca otros dos lados exactamente en sus extremos. Cada extremo es un *vértice* del polígono.

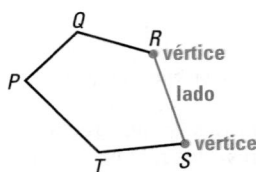

polyhedron (pp. 473, 474) **poliedro** Sólido formado por polígonos. Las superficies planas de un poliedro se llaman *caras* y los segmentos que unen los vértices se llaman *aristas*.

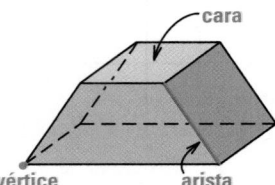

postulate (p. 14) **postulado** Enunciado que se admite sin ser justificado.

power (p. 467) **potencia** Expresión tal como 7^3. El *exponente* 3 representa el número de veces que se usa la *base* 7 como factor: $7^3 = 7 \cdot 7 \cdot 7$.

prism (p. 483) **prisma** Poliedro que tiene dos caras congruentes, llamadas *bases,* situadas en planos paralelos. Las otras caras se denominan *caras laterales*. La *altura* es la distancia perpendicular entre las bases.

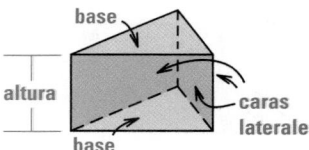

proof (p. 243) **prueba** Argumento convincente que muestra que un enunciado es cierto.

proportion (pp. 349, 359) **proporción** Ecuación que muestra que dos razones son iguales. Por ejemplo: $\frac{a}{b} = \frac{c}{d}$.

pyramid (pp. 475, 491) **pirámide** Poliedro cuya *base* es un polígono y cuyas *caras laterales* son triángulos que tienen un vértice común. *Ver también* height of a pyramid, slant height of a pyramid/altura de una pirámide, altura inclinada de una pirámide.

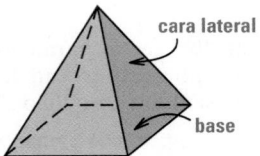

Pythagorean triple (p. 196) **terna pitagórica** Conjunto de tres números naturales *a*, *b* y *c* que cumple con la ecuación $c^2 = a^2 + b^2$.

quadrilateral (p. 304) **cuadrilátero** Polígono de cuatro lados.

radical (p. 537) **radical** Expresión escrita con un símbolo de radical $\sqrt{}$. También se llama *expresión radical*.

radical symbol (p. 537) **símbolo de radical** Símbolo de la raíz cuadrada, $\sqrt{}$, que indica la raíz cuadrada positiva de un número. Por ejemplo: $\sqrt{25} = 5$.

radicand (p. 537) **radicando** Número o expresión escrita dentro de un símbolo de radical. En el radical $\sqrt{25}$, el radicando es 25.

radius (pp. 452, 589) **radio** Distancia del centro a un punto de un círculo. Todo segmento cuyos extremos son el centro de un círculo y un punto del círculo también se llama *radio*.

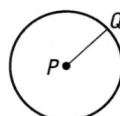

Radio: *PQ* o $\overline{PQ}$

radius of a cone (p. 493) **radio de un cono** *Ver* cone/cono.

radius of a cylinder (p. 485) **radio de un cilindro** *Ver* cylinder/cilindro.

radius of a sphere (p. 517) **radio de una esfera** *Ver* sphere/esfera.

rate of *a* to *b* (p. 227) **tasa de *a* a *b*** El cociente $\frac{a}{b}$ si *a* y *b* son dos cantidades que se miden en diferentes unidades.

ratio of *a* to *b* (pp. 227, 357) **razón de *a* a *b*** Comparación de un número *a* y un número *b* que no es cero usando la división. La razón de *a* a *b* se puede escribir como la fracción $\frac{a}{b}$, como *a* : *b* o como "*a* a *b*".

ray (p. 16) **rayo** El rayo $\overrightarrow{AB}$ incluye el *extremo A* y todos los puntos de $\overleftrightarrow{AB}$ situados en el mismo lado de *A* como *B*.

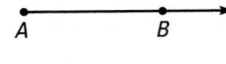

$\overrightarrow{AB}$ con extremo *A*

rectangle (p. 325) **rectángulo** Paralelogramo que tiene cuatro ángulos rectos.

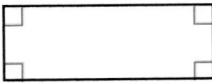

Rectángulo

reduction (p. 394) **reducción** Dilatación en la cual la imagen es más pequeña que la figura original. El factor de escala de una reducción es menor que 1.

reflection (p. 282) **reflexión** Transformación que crea una imagen especular. La figura original se refleja en el *eje de reflexión*.

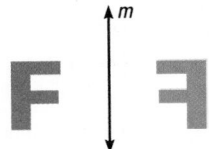

La recta *m* es un eje de reflexión.

regular polygon (p. 412) **polígono regular** Polígono equilátero y equiangular.

rhombus (p. 325) **rombo** Paralelogramo que tiene cuatro lados congruentes.

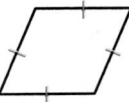

Rombo

right angle (p. 36) **ángulo recto** Ángulo que mide 90°.

right triangle (p. 174) **triángulo recto** Triángulo que tiene un ángulo recto. *Ver también* legs of a right triangle/catetos de un triángulo recto.

rotation (p. 633) **rotación** Transformación que hace girar una figura en torno a un punto fijo llamado *centro de rotación*. Los rayos que se extienden del centro de rotación a un punto y su imagen forman un ángulo llamado *ángulo de rotación*.

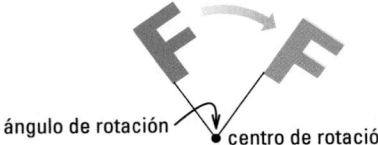

ángulo de rotación   centro de rotación

rotational symmetry (p. 634) **simetría rotacional** Una figura en el plano tiene simetría rotacional si coincide con sí misma luego de rotar 180° o menos en torno a un punto.

same-side interior angles (p. 121) **ángulos internos del mismo lado** Dos ángulos formados por dos rectas y una transversal y situados entre las dos rectas en el mismo lado de la transversal. En el diagrama de abajo, ∠3 y ∠5 son ángulos internos del mismo lado.

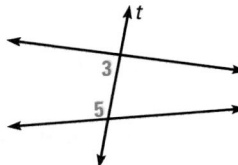

scale factor (p. 366) **factor de escala** Razón de las longitudes de dos lados correspondientes de dos polígonos semejantes.

scale factor of a dilation (pp. 392, 394) **factor de escala de dilatación** El valor de k donde $k = \frac{CP'}{CP}$. *Ver tambien* dilation/dilatación.

scalene triangle (p. 173) **triángulo escaleno** Triángulo que no tiene lados congruentes.

secant (p. 589) **secante** Recta que interseca un círculo en dos puntos.

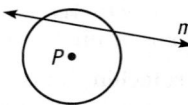

sector (p. 454) **sector** Región de un círculo determinada por dos radios y una parte del círculo.

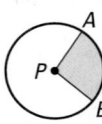

segment (p. 16) **segmento** Parte de una recta que consta de dos puntos, llamados *extremos,* y todos los puntos en la recta situados entre los extremos.

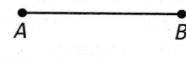

$\overline{AB}$ con extremos **A** y **B**

segment bisector (p. 53) **bisectriz de un segmento** Segmento, rayo, recta o plano que interseca un segmento en su punto medio.

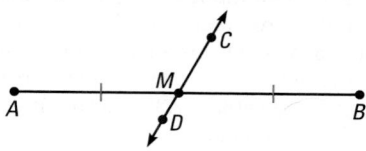

$\overleftrightarrow{CD}$ es una bisectriz de $\overline{AB}$.

semicircle (p. 601) **semicírculo** Arco cuyo ángulo central mide 180°.

side of an angle (p. 35) **lado de un ángulo** *Ver* angle/ángulo.

side of a polygon (p. 303) **lado de un polígono** *Ver* polygon/polígono.

similar polygons (p. 365) **polígonos semejantes** Dos polígonos son semejantes si sus ángulos correspondientes son congruentes y las longitudes de sus lados correspondientes son proporcionales. El símbolo $\sim$ significa "es semejante a".

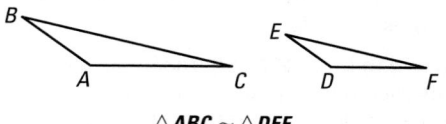

$\triangle \textbf{ABC} \sim \triangle \textbf{DEF}$

similarity statement (p. 365) **enunciado de semejanza** Enunciado tal como $\triangle ABC \sim \triangle DEF$, que indica que dos polígonos son semejantes.

sine (p. 563) **seno** Razón trigonométrica que se abrevia *sen* y que se calcula como la razón entre la longitud del cateto opuesto al ángulo y la longitud de la hipotenusa.

$$\operatorname{sen} A = \frac{\text{cateto opuesto a } \angle A}{\text{hipotenusa}} = \frac{a}{c}$$

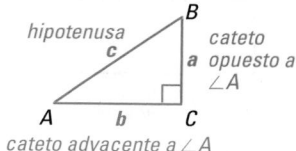

skew lines (p. 108) **rectas oblicuas** Dos rectas que no están situadas en el mismo plano y no se intersecan.

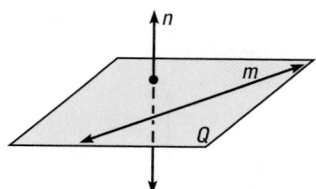

Las rectas **m** y **n** son oblicuas.

slant height of a cone (p. 493) **altura inclinada de un cono** *Ver* cone/cono.

slant height of a pyramid (p. 491) **altura inclinada de una pirámide** Altura de la cara lateral de una pirámide. La letra ℓ se usa para representar la altura inclinada.

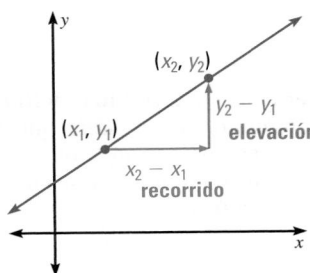

slope (pp. 150, 299, 665) **inclinación** Razón de la diferencia entre las coordenadas del eje vertical (elevación) y la diferencia entre las coordenadas del eje horizontal (recorrido) de dos puntos cualesquiera de una recta.

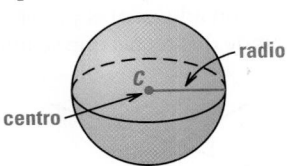

solid (p. 473) **sólido** Figura tridimensional.

solve a right triangle (p. 569) **resolver un triángulo recto** Hallar las medidas de los dos ángulos agudos y las longitudes de los tres lados.

sphere (p. 517) **esfera** Conjunto de puntos en el espacio que están a la misma distancia de un punto, llamado *centro* de la esfera. El *radio* de una esfera es la longitud de un segmento del centro a un punto de la esfera.

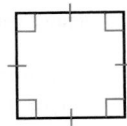

square (p. 325) **cuadrado** Paralelogramo que tiene cuatro lados congruentes y cuatro ángulos rectos.

square root (p. 167) **raíz cuadrada** Si $b^2 = a$, entonces b es la raíz cuadrada de a.

standard equation of a circle (p. 628) **ecuación estándar de un círculo** En el plano de coordenadas, la ecuación estándar de un círculo con centro en (h, k) y radio r es

$$(x - h)^2 + (y - k)^2 = r^2.$$

straight angle (p. 36) **ángulo llano** Ángulo que mide 180°.

straightedge (p. 143) **regla** Instrumento de construcción que se usa para dibujar segmentos; una regla sin marcas.

supplement (p. 67) **suplemento** La suma de las medidas de un ángulo y su *suplemento* es 90°.

supplementary angles (p. 67) **ángulos suplementarios** Dos ángulos cuyas medidas suman 180°.

surface area of a polyhedron (p. 483) **área de la superficie de un poliedro** Suma de las áreas de las caras de un poliedro.

tangent (p. 589) **tangente** Recta en el plano de un círculo que interseca el círculo exactamente en un punto, llamado *punto de tangencia*.

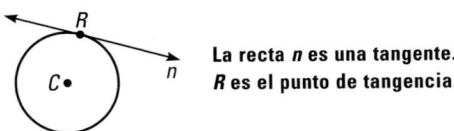

La recta *n* es una tangente. *R* es el punto de tangencia.

tangent of an angle (p. 557) **tangente de un ángulo** Razón trigonométrica que se abrevia *tan* y se calcula como la razón entre la longitud del cateto opuesto al ángulo y la longitud del cateto adyacente al ángulo.

$$\tan A = \frac{\text{cateto opuesto a } \angle A}{\text{cateto adyacente a } \angle A} = \frac{a}{b}$$

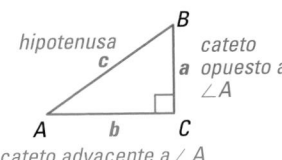

tangent segment (p. 597) **segmento tangente** Segmento que toca un círculo en uno de los extremos del segmento y que está ubicado en la recta tangente al círculo en ese punto.

theorem (p. 69) **teorema** Enunciado verdadero derivado de otros enunciados verdaderos.

transformation (p. 152) **transformación** Operación que *desplaza,* o mueve, una figura a una imagen. *Ver también* dilation, reflection, rotation, translation/dilatación, reflexión, rotación, traslación.

translation (p. 152) **traslación** Transformación que desplaza cada punto de una figura la misma distancia en la misma dirección.

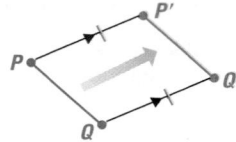

transversal (p. 121) **transversal** Recta que interseca dos o más rectas coplanarias en diferentes puntos.

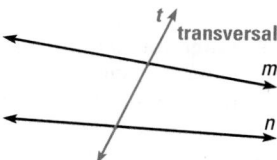

trapezoid (p. 332) **trapezoide** Cuadrilátero que tiene exactamente un par de lados paralelos, llamados *bases.* Los lados que no son paralelos se llaman *catetos.*

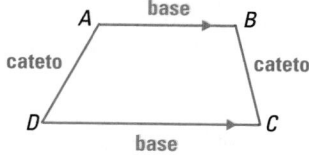

triangle (p. 173) **triángulo** Figura formada por tres segmentos que se unen a tres puntos no colineales, llamados *vértices.* El símbolo del triángulo es △.

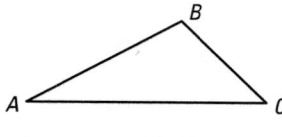

△**ABC con vértices A, B y C**

trigonometric ratio (p. 557) **razón trigonométrica** Razón de las longitudes de dos lados de un triángulo recto. *Ver también* cosine, sine, tangent of an angle/coseno, seno, tangente de un ángulo.

undefined term (p. 14) **término indefinido** Palabra, como *punto, recta* o *plano,* que no se define matemáticamente usando otras palabras conocidas, aunque comúnmente se entiende lo que la palabra significa.

vertex of an angle (p. 35) **vértice de un ángulo** *Ver* angle/ángulo.

vertex of a polygon (p. 303) **vértice de un polígono** *Ver* polygon/polígono.

vertex of a triangle (p. 175) **vértice de un triángulo** Punto en el que se unen dos puntos de un triángulo. *Ver también* triangle/triángulo.

vertical angles (p. 75) **ángulos verticales** Dos ángulos que no son adyacentes y cuyos lados se forman a partir de dos rectas que se intersecan.

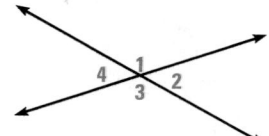

∠**1 y** ∠**3 son ángulos verticales.**

∠**2 y** ∠**4 son ángulos verticales.**

volume of a solid (p. 500) **volumen de un sólido** Número de unidades cúbicas que contiene el interior del sólido.

Credits

Credits

Pre-Course Practice

Decimals and Fractions (p. xxii) **1.** 1.715 **3.** 13.73
5. 0.37 **7.** 13.08 **9.** 20.74 **11.** 8.82 **13.** 3350
15. 10 **17–21.** Sample answers are given.
17. $\frac{4}{6}, \frac{20}{30}$ **19.** $\frac{1}{2}, \frac{5}{10}$ **21.** $\frac{8}{14}, \frac{20}{35}$ **23.** $\frac{1}{9}$ **25.** $\frac{2}{5}$
27. 1 **29.** $\frac{1}{3}$ **31.** $\frac{7}{15}$ **33.** $\frac{1}{8}$ **35.** $\frac{3}{16}$ **37.** $5\frac{5}{8}$ **39.** $\frac{1}{2}$
41. 30 **43.** 0.5 **45.** $0.\overline{4}$; 0.44 **47.** 0.75 **49.** $\frac{1}{20}$
51. $\frac{3}{10}$

Ratio and Proportion (p. xxiii) **1.** $\frac{5}{4}$ **3.** $\frac{4}{9}$ **5.** $\frac{3}{1}$
7. $\frac{17}{20}$ **9.** $2\frac{6}{7}$ **11.** $1\frac{1}{4}$ **13.** 25 **15.** $1\frac{4}{5}$

Inequalities and Absolute Value (p. xxiii)
1. $-8 > -10$ **3.** $0 > -5$ **5.** $\frac{1}{4} = 0.25$ **7.** $8.65 < 8.74$
9. $-9, -6, 0, 3, 4$ **11.** 8256, 8265, 8526, 8562, 8652
13. 9 **15.** 1 **17.** 2.5

Integers (p. xxiii) **1.** -12 **3.** -4 **5.** -11 **7.** 5
9. -56 **11.** 32 **13.** -12 **15.** -3

The Coordinate Plane (p. xxiv) **1.** (1, 3) **3.** (2, 0)
5. $(-4, 2)$ **7.** (3, 3) **9.** $(-3, -1)$

Slope of a Line (p. xxiv) **1.** $\frac{1}{3}$ **3.** 0 **5.** $\frac{1}{2}$ **7.** zero
9. undefined **11.** zero

Powers and Square Roots (p. xxiv) **1.** 125 **3.** 16
5. 1 **7.** 4, -4 **9.** no real square roots **11.** 5
13. 3.2 **15.** 7 **17.** $10\sqrt{6}$ **19.** $3\sqrt{7}$ **21.** $\frac{2}{9}$
23. $\frac{\sqrt{2}}{2}$ **25.** $2\sqrt{5}$

Evaluating Expressions (p. xxv) **1.** 0 **3.** 100
5. 23 **7.** 7 **9.** 32 **11.** -8 **13.** 42 **15.** 64

The Distributive Property (p. xxv) **1.** $3y - 15$
3. $-6x - 6$ **5.** $2x + 2y$ **7.** $x^2 - 4x$ **9.** $-4x + 8$
11. $2z - 1$ **13.** $2x + 3$ **15.** $9 + x$

Solving Equations (p. xxv) **1.** 22 **3.** -7 **5.** -6
7. 1 **9.** -7 **11.** $\frac{5}{2}$ **13.** 35 **15.** -88

Using Formulas (p. xxv) **1.** 28 cm **3.** 7 m

Chapter 1

Study Guide (p. 2) **1.** A **2.** J **3.** A **4.** J

1.1 Guided Practice (p. 5)
1.  **3.**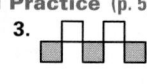

5. Each number is 8 more than the previous
number; 35; 43. **7.** Each number is 0.5 more
than the previous number; 9.0; 9.5.
9. Each number is $\frac{1}{4}$ the previous number; 1; $\frac{1}{4}$.

1.1 Practice and Applications (pp. 5–7)
11. **13.** **15.**

17. Each number is 5 more than the previous
number; 24. **19.** Each number is $\frac{1}{4}$ the previous
number; $\frac{5}{4}$. **21.** Begin with 1 and add 2, then 3,
then 4, and so on; 21. **23.** Each point has a
y-coordinate of 3. *Sample answer:* (5, 3)
25. For each point, the y-coordinate is 2 less than
half the opposite of the x-coordinate. *Sample
answer:* (2, -3) **27.** Each distance after the first
is 4 more than the one before it. **29.** 40
31.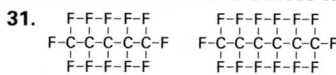

33. Each of the second ten letters has the same
pattern of dots as the letter above it, with an
additional dot inserted in the first column of the
third row.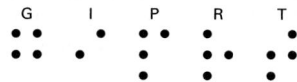

35. 28 blocks **39.** *Sample answer:* 3 square feet,
8 square feet, 15 square feet, 24 square feet,
35 square feet **41.** 11.3 **43.** 1.04 **45.** 45.24
46–53.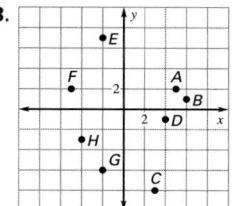

1.2 Guided Practice (p. 11)
3. even   **5.** *Sample answer:* 6 is divisible by 2, but it is not divisible by 4.
7. *Sample answer:*

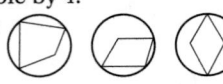

A circle cannot be drawn around any of the four-sided shapes shown.

1.2 Practice and Applications (pp. 11–13)
9. 15; 21   **11.** 121; 12,321; 1,234,321; The square of the number with *n* digits, all of which are 1, is the number obtained by writing the digits from 1 to *n* in increasing order, then the digits from $n - 1$ to 1 in decreasing order. (This pattern does not continue forever.)   **13.** even   **15.** *Conjecture:* The next two shapes will have 14 diagonals and 20 diagonals.   **17.** *Sample answer:* $(-4)(-5) = 20$
19. *Sample answer:*

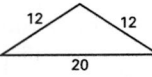

The triangle above shows that the conjecture is false.
25.

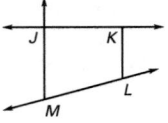

27. 7   **29.** -10 **31.** 14   **33.** 4   **35.** 2   **37.** 11   **39.** -3 **41.** 54

1.3 Guided Practice (p. 17)   **3.** false   **5.** true
7. true   **9.** true   **11.** false   **13.** false

1.3 Practice and Applications (pp. 17–20)   **15.** any four of: A, B, C, D, E   **17.** plane S or plane ABC
19. false   **21.** true   **23.** true   **25.** false   **27.** K
29. M   **31.** L   **33.** J   **35.** *Sample answers:* N, P, and R; N, Q, and R; P, Q, and R   **37.** *Sample answers:* W, A, and Z; W, X, and Y   **39.** G   **41.** H
43. E   **45.** H   **47.** K, N, Q, and R   **49.** P, Q, N, and M   **51.** L, M, P, and S   **53.** M, N, R, and S
55. *Sample answers:* A, Q, and B; A, B, and C
57. $\overleftrightarrow{BE}$, $\overleftrightarrow{CF}$, $\overleftrightarrow{RU}$, and $\overleftrightarrow{SP}$ **59.** either three of the points J, K, N, and Q, or three of the points K, L, M, and N   **61.** *Sample answers:* plane LKN and plane QNM
63. *Sample answer:*

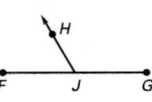

65. *Sample answer:*

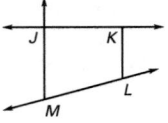

67. yes   **69.** no   **71.** 50   **73.** 2500   **75.** 0.5   **77.** 0.6
79. $0.\overline{6}$; 0.67   **81.** $0.\overline{7}$; 0.78

Quiz 1 (p. 20)
1. 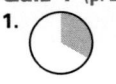   **2.**   **3.** *Sample answer:* 30 is divisible by 10, but not by 20.
4. *Sample answer:*

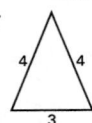

5. *Sample answer:* $2 + 0 = 2$, and 2 is not greater than 2.   **6.** If you fold the square along a line segment that joins two opposite corners and cut along the fold, you will create two triangles, not two rectangles.
7. *Sample answer:*

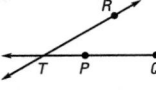

8. *Sample answer:*

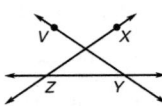

1.4 Guided Practice (p. 25)   **3.** false   **5.** true

1.4 Practice and Applications (pp. 25–27)
7. intersecting lines   **9.** intersecting planes
11. They do not appear to intersect.   **13.** point T
15. $\overleftrightarrow{RS}$, $\overleftrightarrow{ST}$, or $\overleftrightarrow{RT}$ **17.** line ℓ **19.** line k
21. no intersection   **23.** point A   **25.** $\overleftrightarrow{DC}$ **27.** $\overleftrightarrow{AD}$
29. *Sample answer:*

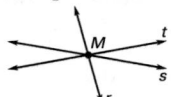

31. *Sample answer:*

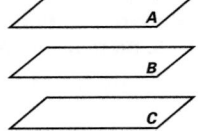

35. Each point has a *y*-coordinate of -3. *Sample answer:* $(4, -3)$ **37.** For each point, the *y*-coordinate is twice the opposite of the *x*-coordinate. *Sample answer:* $(2, -4)$
39. *Sample answers:* B, A, and C; B, A, and D; B, C, and D   **41.** 11.4   **43.** 3.7   **45.** 8.57   **47.** 7   **49.** 0
51. 2   **53.** 2

1.5 Guided Practice (p. 31)   **3.** 13   **5.** $\overline{CD}$ and $\overline{EF}$

1.5 Practice and Applications (pp. 31–33)
7. 31 mm   **9.** 25 mm   **11.** 18 mm
13. $DE + EF = DF$ **15.** $NM + MP = NP$
17. true   **19.** true   **21.** false   **23.** 21   **25.** 6
27. $\overline{CD}$, $\overline{EF}$, and $\overline{GH}$

29. yes

31. $8x - 1$

37. *Sample answer:* A rectangle with a length of 5 inches and a width of 3.5 inches has a perimeter of 17 inches.

39. *Sample answer:*

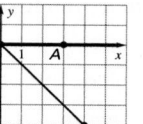

41. 26   **43.** 8   **45.** 14

1.6 Guided Practice (p. 38)   **5.** E; $\overrightarrow{ED}$, $\overrightarrow{EF}$; about 35°   **7.** J; $\overrightarrow{JH}$, $\overrightarrow{JK}$; about 75°   **9.** straight   **11.** obtuse   **13.** yes; $m\angle DEF = m\angle FEG$

1.6 Practice and Applications (pp. 38–41)

15. X; $\overrightarrow{XF}$, $\overrightarrow{XT}$ **17.** Q; $\overrightarrow{QR}$, $\overrightarrow{QS}$ **19.** any two of $\angle C$, $\angle BCD$, $\angle DCB$ **21.** 55°   **23.** 140°   **25.** right; about 90°   **27.** 105°   **29.** 140°

31.

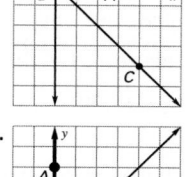

acute angle

33.

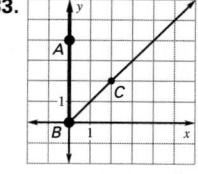

acute angle

35. 82°   **37.** 117°   **39.** 150°   **41.** 150°
43. 60°   **45. a.** $\angle AOB$, $\angle BOC$, $\angle COD$, $\angle DOE$, $\angle EOF$, $\angle FOG$, $\angle GOH$, $\angle HOA$ **b.** $\angle AOC$, $\angle BOD$, $\angle COE$, $\angle DOF$, $\angle EOG$, $\angle FOH$, $\angle GOA$, $\angle HOB$ **c.** $\angle AOD$, $\angle BOE$, $\angle COF$, $\angle DOG$, $\angle EOH$, $\angle FOA$, $\angle GOB$, $\angle HOC$ **47.** $\overrightarrow{NM}$ or $\overrightarrow{NQ}$ **49.** $XY + YZ = XZ$
51. $AB + BC = AC$ **53.** 12   **55.** 16   **57.** 12   **59.** 25

Quiz 2 (p. 41)   **1.** *Sample answers:* $\overleftrightarrow{AB}$ and $\overleftrightarrow{EF}$, $\overleftrightarrow{AB}$ and $\overleftrightarrow{DE}$, $\overleftrightarrow{BC}$ and $\overleftrightarrow{DE}$, $\overleftrightarrow{BC}$ and $\overleftrightarrow{EF}$
2. two of the following: $\overleftrightarrow{AB}$, $\overleftrightarrow{BC}$, $\overleftrightarrow{BE}$
3. two of the following: $\overleftrightarrow{BE}$, $\overleftrightarrow{DE}$, $\overleftrightarrow{EF}$

4. yes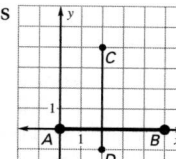

5. no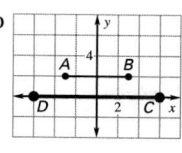

6. right   **7.** obtuse   **8.** straight   **9.** acute
10. acute   **11.** acute   **12.** 80°   **13.** 75°   **14.** 60°

Chapter Summary and Review (pp. 42–45)
1. coplanar lines   **3.** conjecture   **5.** intersect
7. collinear   **9.** congruent   **11.** obtuse   **13.** Each number is 9 more than the previous number; 41; 50.   **15.** Each number is 10 less than the previous number; 60; 50.

17.

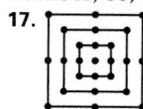

19. *Sample answer:* $(-2)^3 = -8$ and $-8 < -2$
21. false

23.

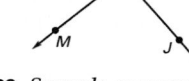

25. $\overleftrightarrow{QS}$ **27.** $\overleftrightarrow{UV}$

29. *Sample answer:*

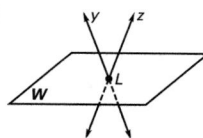

31. 26   **33.** yes

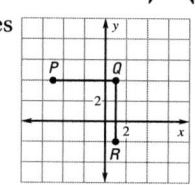

35. about 90°; right   **37.** 105°   **39.** 70°

Algebra Review (p. 49)   **1.** 34   **3.** 19   **5.** 0   **7.** -11
9. 15   **11.** 6   **13.** 5   **15.** 4   **17.** -9 **19.** -7

Chapter 2

Study Guide (p. 52)   **1.** C   **2.** G   **3.** C

2.1 Guided Practice (p. 56)   **3.** 25; 25   **5.** 20; 40
7. 3   **9.** $\left(-\frac{1}{2}, -1\right)$

2.1 Practice and Applications (pp. 56–59)
11. No; M does not divide $\overline{AB}$ into two congruent segments.   **13.** No; M does not lie on $\overline{AB}$.

15. *Sample answer:*

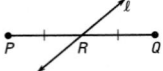

17. 41; 41   **19.** 1.35; 1.35   **21.** 15; 30   **23.** 3.6; 7.2
25. Caitlin will bike 0.95 mi and Laurie will bike
2.25 mi.   **27.** 12   **29.** -1 **31.** (5, 7)   **33.** $\left(-3, \frac{1}{2}\right)$
35. $(-2, -1)$ **37.** 54   **39.** (35.75°N, 118.3°W)
41. (36.1°N, 121°W)   **47.** true   **49.** true   **51.** true
53. K; $\overrightarrow{KJ}$, $\overrightarrow{KL}$; obtuse   **55.** 70   **57.** 170   **59.** 19
61. 12   **63.** -33

2.2 Guided Practice (p. 64)   **3.** 41°   **5.** 60°   **7.** 78°

2.2 Practice and Applications (pp. 64–66)
9. 54°; 54°   **11.** 33.5°; 33.5°   **13.** 45.5°; 45.5°
15. 90°; 90°   **17.** 22°; 44°; acute   **19.** 38°; 76°; acute
21. 17°; 34°; acute   **23.** $\angle JRL$, $\angle HAM$, $\angle FAN$,
$\angle DAP$ **25.** 67°   **27.** Yes; *Sample answer:*
$m\angle PKR = m\angle QKR = 90°$ and $m\angle JKR = m\angle LKR$,
so $m\angle PKR - m\angle JKR = m\angle QKR - m\angle LKR$.
Therefore, $m\angle JKP = m\angle LKQ$.   **29.** 7   **31.** no; yes
35. Each number is 5 more than the previous
number; 7.   **37.** Each number is half the previous
number; 3.75.   **39.** obtuse   **41.** straight   **43.** 12
45. -7 **47.** 20

2.3 Guided Practice (p. 70)   **3.** complementary;
adjacent   **5.** supplementary; nonadjacent   **7.** 71°

2.3 Practice and Applications (pp. 70–73)
9. supplementary; adjacent   **11.** supplementary
13. complementary   **15.** 49°   **17.** 66°   **19.** 81°
21. 166°   **23.** 84°   **25.** 51°; 141°   **27.** 36°; 126°
29. $\angle QPT$ **31.** $\angle QPS$, $\angle SPT$ **33.** 64   **35.** 6   **37.** 16
39. Congruent Supplements Theorem   **41.** 5°
47. 12   **49.** $(-2, 2)$ **51.** $(-2, 6)$ **53.** $(3, -2)$
55. 3.21   **57.** 0.2318   **59.** 42.12

Quiz 1 (p. 73)   **1.** 17; 34   **2.** (4, 1)   **3.** (1, 1)
4. $(-1, 0)$ **5.** 41°   **6.** 22°   **7.** 116°   **8.** 79°
9. 7°; 97°

2.4 Guided Practice (p. 78)   **3.** 161°
5. $m\angle 1 = 108°$; $m\angle 2 = 72°$; $m\angle 3 = 108°$ **7.** 4

2.4 Practice and Applications (pp. 78–81)
9. neither   **11.** vertical angles   **13.** linear pair
15. 143°   **17.** 44°   **19.** 56°   **21.** 160°
23. false   **25.** true   **27.** false
29. $m\angle 1 = 98°$; $m\angle 2 = 82°$; $m\angle 3 = 98°$
31. $m\angle 1 = 125°$; $m\angle 2 = 55°$; $m\angle 3 = 125°$
33. $m\angle 1 = 40°$; $m\angle 2 = 140°$; $m\angle 3 = 40°$
35. $m\angle 1 = 90°$; $m\angle 2 = 90°$; $m\angle 3 = 90°$
37. $m\angle 1 = 95°$; $m\angle 2 = 85°$; $m\angle 3 = 95°$
39. $m\angle 1 = 30°$; $m\angle 2 = 44°$; $m\angle 3 = 106°$;
$m\angle 4 = 30°$ **41.** $\angle FGE$ **43.** $\angle DGF$ **45.** 22

47. 72   **49.** 158   **51.** 79   **53.** 58   **55.** 23; 157°
59. a. $6x = 4x + 8$ **b.** 4   **c.** 24°   **d.** 156°
61. Each number is 5 times the previous number;
1875.   **63.** Each number is 10 more than the
previous number; 444.   **65.** $\overline{LM} \cong \overline{NP} \cong \overline{QR}$
67. $7 - a$ **69.** $5b^2 + 6b$ **71.** $2w^2 + 4$

2.5 Guided Practice (p. 85)   **3.** If two angles are
adjacent angles, then they share a common side.
5. The midpoint of $\overline{AB}$ is at (2, 0).

2.5 Practice and Applications (pp. 85–87)
7. hypothesis: the car is running; conclusion: the
key is in the ignition   **9.** If <u>a number is divisible
by 6</u>, then ⌐it is also divisible by 3 and 2⌐.

11. If <u>a shape is a square</u>, then ⌐it has four sides⌐.
13. The intersection of the planes is a line.
15. The school will celebrate.   **17.** If the sun is
shining, then we will have a picnic.   **19.** If Chris
goes to the movies, then Gabriela will go to the
movies.   **21.** Answers will vary.   **23.** is

25. If <u>you want to do things</u>, then ⌐you first have⌐
⌐to expect things of yourself⌐.

27. If <u>a person is happy</u>, then ⌐he or she will⌐
⌐make others happy too⌐.   **29.** acute   **31.** right

33. straight   **35.** $(-1, 3)$ **37.** $(3, -1)$ **39.** $-\frac{5}{2}$

41. 2   **43.** 1

2.6 Guided Practice (p. 91)   **7.** Reflexive Property
of Congruence   **9.** Symmetric Property of
Congruence

2.6 Practice and Applications (pp. 91–94)
11. $m\angle Q = m\angle P$ **13.** $\angle GHJ$ **15.** $\overline{IJ} \cong \overline{PQ}$
17. Substitution Property of Equality
19. Substitution; Subtraction
21. $PQ = RS$ — Given
$PQ + QR = RS + QR$ — Addition Property of Equality
$PQ + QR = PR$ — Segment Addition Postulate
$RS + QR = QS$ — Segment Addition Postulate
$PR = QS$ — Substitution Property of Equality

23.

$m\angle 1 + m\angle 2 = 90°$	Definition of complementary angles
$m\angle 3 + m\angle 2 = 90°$	Definition of complementary angles
$m\angle 1 + m\angle 2 =$ $m\angle 3 + m\angle 2$	Substitution Property of Equality
$m\angle 1 = m\angle 3$	Subtraction Property of Equality
$\angle 1 \cong \angle 3$	Definition of congruent angles

25.

$AB = BC$	Given
$BC = CD$	Given
$AB = CD$	Transitive Property of Equality
$AB = 3t + 1$	Given
$CD = 7$	Given
$3t + 1 = 7$	Substitution Property of Equality
$3t = 6$	Subtraction Property of Equality
$t = 2$	Division Property of Equality

31. D **33.** F

35. *Sample answer:*

36–43.

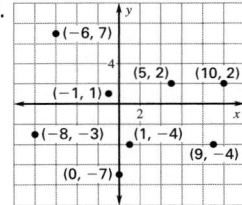

Quiz 2 (p. 94) **1.** $m\angle 1 = 126°$; $m\angle 2 = 54°$; $m\angle 3 = 126°$ **2.** $m\angle 4 = 40°$; $m\angle 5 = 140°$; $m\angle 6 = 40°$ **3.** $m\angle 7 = 49°$; $m\angle 8 = 90°$; $m\angle 9 = 49°$; $m\angle 10 = 41°$ **4.** If a figure is a square, then it has four sides. **5.** If $x = 5$, then the value of x^2 is 25. **6.** If we charter a boat, then we will be gone all day. **7.** bisector; bisector; Transitive

Chapter Summary and Review (pp. 95–97)
1. supplementary **3.** segment bisector

5. adjacent **7.** 5.5; 11 **9.** $\left(2, \dfrac{3}{2}\right)$ **11.** $50°$ **13.** $45°$

15. $m\angle 1 = 135°$; $m\angle 2 = 45°$; $m\angle 3 = 135°$
17. $m\angle 1 = 99°$; $m\angle 2 = 81°$; $m\angle 3 = 99°$

19. If <u>a computer is in the store</u>, then the computer is on sale. **21.** If Mike goes to the concert, then Jeannine will go to the concert.
23. A **25.** C

Algebra Review (p. 101) **1.** $22p$ **3.** $7q$ **5.** $3x + 3$
7. $19y + 40z$ **9.** 10 **11.** 6 **13.** -14 **15.** 11 **17.** 2

Chapter 3

Study Guide (p. 106) **1.** B **2.** H **3.** D

3.1 Guided Practice (p. 110) **3.** $\parallel$ **5.** $\parallel$ **7.** C

3.1 Practice and Applications (pp. 110–113)
9. neither **11.** parallel **13.** Yes; m and n do not intersect in space. **15.** parallel **17.** skew
19. parallel **21.** $\overleftrightarrow{UV}$, $\overleftrightarrow{TS}$, or $\overleftrightarrow{XW}$
23. $\overleftrightarrow{SW}$, $\overleftrightarrow{TX}$, $\overleftrightarrow{VW}$, or $\overleftrightarrow{UX}$
25. *Sample answer:*

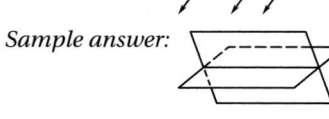

27. *Sample answer:*

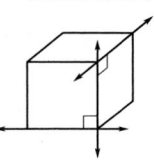

29. *Sample answer:*

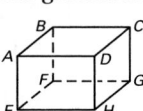

31. parallel **33.** three **35.** Yes; plane A remains parallel to the ground level for the entire time that the step is moving from ground level to the top.
37. *Sample answer:*

$\overline{AE}$ and $\overline{DH}$ appear parallel.
45. hypothesis: $m\angle 5 = 120°$; conclusion: $\angle 5$ is obtuse **47.** hypothesis: we can get tickets; conclusion: we'll go to the movies **49.** $\angle 2 \cong \angle 1$

51. $\dfrac{1}{26}$ **53.** $\dfrac{1}{10}$ **55.** 15 **57.** 11 **59.** 26 **61.** 80

3.2 Guided Practice (p. 117) **3.** If two lines intersect to form adjacent congruent angles, then the lines are perpendicular. **5.** $x = 90$
7. $70 + x = 90$

3.2 Practice and Applications (pp. 117–120) **9.** no
11. Yes; perpendicular lines intersect to form 4 right angles; all right angles are congruent.

13. *Sample answers:* ∠1 and ∠2 are right angles; ∠1 ≅ ∠2.   **15.** The equation should be $(x + 4)° + 56° = 90°$.   **17.** 90   **19.** 35   **21.** 15   **23.** 10; 70°   **25.** 35; 35°   **27.** No. Since ∠1 and ∠3 are congruent and complementary, each has measure 45°. There is no information given about the measure of ∠2, so you cannot conclude that $\overrightarrow{BA} \perp \overrightarrow{BC}$.   **29.** 50°   **31.** Yes; ∠DJE and ∠EJF are adjacent acute angles and $\overline{BF} \perp \overline{HD}$, so ∠DJE and ∠EJF are complementary.   **33.** No; there is no information given about ∠AJG or the lines that form ∠AJG.   **37.** acute; about 80°   **39.** 53°   **41.** 15   **43.** 3   **45.** 11.79   **47.** 8.8   **49.** 333

Quiz 1 (p. 120) **1–4.** Sample answers are given.
1. $\overleftrightarrow{BF}, \overleftrightarrow{CG}$ **2.** $\overleftrightarrow{AD}$ **3.** $\overleftrightarrow{BC}$ **4.** $\overleftrightarrow{BC}$ **5.** 67   **6.** 11   **7.** 34

3.3 Guided Practice (p. 123) **3.** A   **5.** one of: ∠1 and ∠5, ∠2 and ∠6, ∠3 and ∠7, or ∠4 and ∠8   **7.** ∠1 and ∠7, or ∠2 and ∠8

3.3 Practice and Applications (pp. 123–125)
9. same-side interior angles   **11.** corresponding angles   **13.** *Sample answer:* ∠2 and ∠3   **15.** *Sample answer:* ∠2 and ∠4   **17.** alternate exterior   **19.** corresponding   **21.** alternate interior   **23.** alternate exterior angles   **25.** alternate exterior angles   **27.** corresponding angles   **29.** ∠5 and ∠12, ∠7 and ∠10   **31.** ∠6 and ∠11, ∠8 and ∠9   **37.** $\overrightarrow{AB}, \overrightarrow{GH},$ or $\overleftrightarrow{EF}$ **39.** ABC   **41.** 24   **43.** $\frac{2}{5}$ **45.** 1   **47.** 52   **49.** 15

3.4 Guided Practice (p. 132) **9.** Alternate Exterior Angles Theorem   **11.** Same-Side Interior Angles Theorem

3.4 Practice and Applications (pp. 132–135)
15.
17. 90°   **19.** 37°   **21.** 94°   **23.** 94°

25. $m∠1 = 42°$ by the Alternate Interior Angles Theorem   **27.** $m∠1 = 82°$ by the Alternate Interior Angles Theorem; $m∠2 = 82°$ by the Vertical Angles Theorem.   **29.** 131°   **31.** 76°   **33.** 23   **35.** 23   **37.** 25
39.
43. parallel   **45.** skew   **47.** ∠1 and ∠5, ∠2 and ∠6, ∠3 and ∠7, ∠4 and ∠8   **49.** ∠1 and ∠8, ∠3 and ∠6   **51.** 8   **53.** −15   **55.** 1

Quiz 2 (p. 135) **1.** alternate exterior angles   **2.** same-side interior angles   **3.** corresponding angles   **4.** alternate exterior angles   **5.** alternate interior angles   **6.** alternate interior angles   **7.** $m∠1 = 104°$; $m∠2 = 76°$ **8.** $m∠1 = 78°$; $m∠2 = 102°$ **9.** $m∠1 = 107°$; $m∠2 = 73°$ **10.** 21   **11.** 26   **12.** 11

3.5 Guided Practice (p. 139) **3.** C   **5.** B

3.5 Practice and Applications (pp. 140–142)
7. If $m∠1 + m∠2 = 180°$, then ∠1 and ∠2 are supplementary; true.   **9.** If ∠B is obtuse, then ∠B measures 123°; false.   **11.** yes, by the Corresponding Angles Converse   **13.** yes, by the Alternate Interior Angles Converse   **15.** No; the angles have no special relationship.
17. *Sample answer:*

19. Yes; *Sample explanation:* $m∠ABE = 143°$ by the Vertical Angles Theorem; $143° + 37° = 180°$, so $\overrightarrow{AC} \parallel \overrightarrow{DF}$ by the Same-Side Interior Angles Converse.   **21.** Yes; *Sample explanation:* $m∠CBE = 115°$ by the Vertical Angles Theorem; $115° + 65° = 180°$, so $\overrightarrow{AC} \parallel \overrightarrow{DF}$ by the Same-Side Interior Angles Converse.   **23.** 20   **25.** 21   **27.** 11   **29.** 10   **31.** $\overleftrightarrow{HJ} \parallel \overleftrightarrow{AB}$ by the Alternate Exterior Angles Converse.   **33.** 32°   **37.** $m∠1 = m∠2 = 90°$ **39.** ∠1 and ∠2 are right angles.
41. corresponding angles   **43.** 108°   **45.** $\frac{7}{11}$ **47.** $\frac{2}{3}$

3.6 Guided Practice (p. 147) **3.** In a plane, if two lines are perpendicular to the same line, then they are parallel to each other.

3.6 Practice and Applications (pp. 147–149)
5. In a plane, if two lines are perpendicular to the same line, then they are parallel to each other.   **7.** Corresponding Angles Converse   **9.** Alternate Interior Angles Converse   **11.** In a plane, if two lines are perpendicular to the same line, then they are parallel to each other.   **13.** $p \parallel q$ by the Corresponding Angles Converse; $q \parallel r$ by the Same-Side Interior Angles Converse. Then, since $p \parallel q$ and $q \parallel r$, $p \parallel r$ (if two lines are parallel to the same line, then they are parallel to each other).   **15.** a and b are each perpendicular to d, so $a \parallel b$ (in a plane, if two lines are perpendicular to the same line, then they are parallel to each other); c and d are each perpendicular to a, so $c \parallel d$ (in a plane, if two lines are perpendicular to the same line, then they are parallel to each other).

17. The 8th fret is parallel to the 9th fret, and the 9th fret is parallel to the 10th fret. Therefore, the 8th fret is parallel to the 10th fret (if two lines are parallel to the same line, then they are parallel to each other). **19.** 11 **21.** 6

23. *Sample answer:*

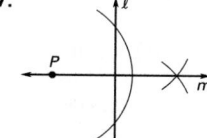

25. No; if the wind is constant, then the boats' paths will be parallel by the Corresponding Angles Converse. They will never cross.
29. true **31.** true

33–36. 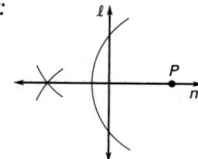 **37.** −32 **39.** 2 **41.** −33

3.7 Guided Practice (p. 155) **3.** yes **5.** yes
7. false; shift 3 units to the left and 2 units down

3.7 Practice and Applications (pp. 155–159) **9.** no
11. no **13.** no **15.** C **17.** B **19.** Each point is moved 2 units to the right and 3 units up.
21. Each point is moved 2 units to the left and 3 units up. **23.** $(x, y) \rightarrow (x - 4, y + 1)$ **25.** (1, 4)
27. (8, −9) **29.** (0, 0) **31.** (3, −4) **33.** $P'(-4, 3)$, $Q'(-1, 6)$, $R'(3, 5)$, $S'(-1, 1)$ **35.** $P'(-1, -2)$, $Q'(2, 1)$, $R'(6, 0)$, $S'(2, -4)$ **37.** Move 2 units to the right and 1 unit down.

39. 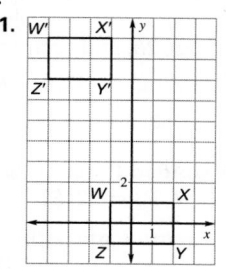 **41.**

43. (−7, 5) **45.** Yes; the image of (2, −2) after the translation $(x, y) \rightarrow (x - 6, y + 4)$ is (−4, 2). This is the point that is labeled C'. **47.** 180°
51. acute; about 50° **53.** acute; about 80°
55. 24 orders **57.** −4, −1.5, −1.2, 0, 0.7, 1.1, 3.4
59. −6.8, −6.12, −6.1, 6, 6.09, 6.3

Quiz 3 (p. 159) **1.** yes, by the Alternate Exterior Angles Converse **2.** No; there is no information about line m. **3.** yes, by the Alternate Interior Angles Converse

4. In a plane, if two lines are perpendicular to the same line, then they are parallel to each other.
5. If two lines are parallel to the same line, then they are parallel to each other. **6.** Use the Corresponding Angles Converse or Theorem 3.12 (in a plane, if two lines are perpendicular to the same line, then they are parallel to each other).
7. 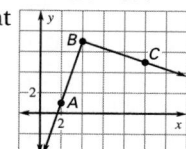 **8.** $(x, y) \rightarrow (x + 3, y - 3)$
9. $(x, y) \rightarrow (x - 2, y + 3)$

Chapter Summary and Review (pp. 160–163)
1. parallel **3.** perpendicular **5.** construction
7. parallel **9.** skew **11.** Yes; since $f \perp g$, $\angle 11$ is a right angle. **13.** Yes; h and j intersect to form adjacent congruent angles, so the lines are perpendicular. **15.** corresponding
17. same-side interior **19.** 99° **21.** 71°
23. 20 **25.** If two lines are parallel to the same line, then they are parallel to each other.
27. Corresponding Angles Converse **29.** yes
31. Each point is moved 5 units to the right and 3 units down.

Algebra Review (p. 167) **1.** $x > 15$ **3.** $z \leq -4$
5. $p \leq -3$ **7.** 16 **9.** 9 **11.** 5 **13.** 9

Cumulative Practice (pp. 168–169) **1.** You begin with 10, and add 2, then 3, then 4, and so on; 30.
3. B **5.** B **7.** 16 **9.** 7
11. right 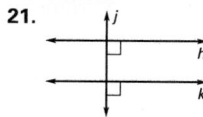 **13.** 17 **15.** 55

17. If two lines intersect, then the lines are coplanar. **19.** Transitive Property of Equality
21. 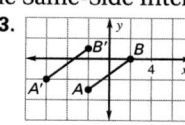 **23.** $\angle 1$ and $\angle 7$, $\angle 4$ and $\angle 9$, $\angle 2$ and $\angle 4$, $\angle 5$ and $\angle 10$ **25.** 55°

27. yes, by the Alternate Interior Angles Converse
29. *Sample answer:* yes, by the Same-Side Interior Angles Converse **31.** $\angle 1$ **33.**

Chapter 4

Study Guide (p. 172) **1.** D **2.** H **3.** B

4.1 Guided Practice (p. 175) **3.** $\overline{PR}$ **5.** isosceles
7. scalene **9.** right

4.1 Practice and Applications (pp. 176–178)
11. scalene **13.** equilateral **15.** scalene
17. obtuse **19.** right **21.** right **23.** An acute
triangle has three acute angles, so the triangle is
not an acute triangle. An obtuse triangle has one
obtuse angle and two acute angles. **25.** right
isosceles triangle **27.** right scalene triangle
29. acute scalene triangle **31.** E **33.** D **35.** C
37. acute **39.** acute **41.** *Sample answer:* B, D,
and E **43.** $\overline{DE}$ is opposite $\angle F$; $\overline{EF}$ is opposite $\angle D$;
$\overline{DF}$ is opposite $\angle E$. **45.** $\overline{KL}$ is opposite $\angle M$;
$\overline{LM}$ is opposite $\angle K$; $\overline{KM}$ is opposite $\angle L$.
47. $\overline{RS}$ is opposite $\angle T$; $\overline{ST}$ is opposite $\angle R$; $\overline{RT}$ is
opposite $\angle S$. **49–53.** Sample answers are given.

49. **51.**

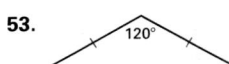

53.

57. 11 **59.** (0, 9) **61.** (−3, 6) **63.** (−6, 2)
65. (−8, 8) **67.** 39 **69.** 20 **71.** −175

4.2 Guided Practice (p. 182) **3.** 61 **5.** 35

4.2 Practice and Applications (pp. 182–184)
7. 110° **9.** 60° **11.** 45° **13.** 139° **15.** $\overline{ML}$
17. 45° to 60° **19.** 20 **21.** $x = 40$; $y = 50$
23. 77° **27.** Corresponding Angles Converse
29. Same-Side Interior Angles Converse
31. 3.5 > 3.06 **33.** 1.75 > 1.57 **35.** 2.055 < 2.1

Quiz 1 (p. 184) **1.** obtuse isosceles triangle
2. acute scalene triangle **3.** right scalene triangle
4. 30° **5.** 78° **6.** 65°

4.3 Guided Practice (p. 188) **3.** $\angle R \cong \angle T$;
$\overline{RS} \cong \overline{TS}$ **5.** 50; Base Angles Theorem

4.3 Practice and Applications (pp. 188–190)
7. 55; Base Angles Theorem **9.** 45; Corollary to
the Triangle Sum Theorem and the Base Angles
Theorem **11.** 2 **13.** 2 **15.** 18 **17.** 120° **19.** 90°
21. 5 **23.** 5 **25.** 3 **27.** no

29.

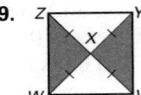

31. $\triangle WXZ$, $\triangle VXW$, $\triangle YXV$, $\triangle YXZ$
35. $m\angle DBC = 42°$; $m\angle ABC = 84°$
37. $m\angle DBC = 150°$; $m\angle ABC =$
75° **39.** 50 **41.** 7 **43.** 1

4.4 Guided Practice (p. 195) **3.** 2.2 **5.** 5.8 **7.** 6.3

4.4 Practice and Applications (pp. 195–198)
9. 41 **11.** 26 **13.** 17 **15.** 80 **17.** 4; yes
19. $\sqrt{170} \approx 13.0$; no **21.** $\sqrt{495} \approx 22.2$; no
23. yes; **25.** no **27.** 5
29. 6.1 **31.** yes
33. *A* to *B*: 115 yd;
B to *C*: 80 yd;
C to *A*: 65 yd
41. 1.05 **43.** 0.02 **45.** $x \geq 2$ **47.** $x \leq 0$ **49.** $\frac{2}{25}$
51. $\frac{3}{25}$ **53.** $\frac{173}{1000}$ **55.** $\frac{1}{9}$

Quiz 2 (p. 198) **1.** 6 **2.** 27 **3.** 7 **4.** $\sqrt{34} \approx 5.8$
5. $\sqrt{17} \approx 4.1$ **6.** $\sqrt{13} \approx 3.6$ **7.** $\sqrt{11} \approx 3.3$ ft

4.5 Guided Practice (p. 203)
3. obtuse **5.** C **7.** D

4.5 Practice and Applications (pp. 203–205)
9. $15^2 + 20^2 = 25^2$ **11.** $1^2 + 4^2 = (\sqrt{17})^2$
13. $4^2 + 6^2 > 7^2$ **15.** $2^2 + 6^2 < 7^2$ **17.** $13^2 + 16^2 <$
22^2 **19.** obtuse **21.** obtuse **23.** acute **25.** right
27. obtuse **29.** acute **31.** right **33.** acute
35. acute **37.** No; $714^2 < 599^2 + 403^2$, so the
triangle is not a right triangle. **39.** No; the
doubled side lengths will also form a right
triangle. *Sample answer:* Let *a*, *b*, and *c* be the
lengths of the sides of a right triangle with
$a^2 + b^2 = c^2$. Then $(2a)^2 + (2b)^2 = 4a^2 + 4b^2 =$
$4(a^2 + b^2) = 4c^2 = (2c)^2$. Since $(2a)^2 + (2b)^2 =$
$(2c)^2$, 2*a*, 2*b*, and 2*c* are also the side lengths of a
right triangle. **43.** 67 **45.** 15 **47.** $\frac{9}{32}$ **49.** $\frac{1}{3}$
51. 14 **53.** $3\frac{1}{2}$

4.6 Guided Practice (p. 209)
5. 11 **7.** $PT = 22$; $ST = 11$

4.6 Practice and Applications (pp. 210–211)
9. **11.** $PN = 6$; $QP = 3$
13. $PN = 20$; $QP = 10$
15. $CD = 22$; $CE = 33$

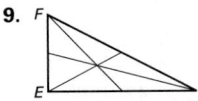

17. The equation should be $AD = \frac{2}{3}AE$, so $AD =$
$\frac{2}{3}(18) = 12$, and $DE = 18 - 12 = 6$. **19.** $Q(5, 0)$;
$R(2, 2)$; $S(8, 4)$ **21.** (5, 2) **25.** 60° **27.** 125°
29. 147° **31–41.** Sample answers are given.

31. $\frac{2}{10}, \frac{10}{50}$ **33.** $\frac{4}{7}, \frac{80}{140}$ **35.** $\frac{1}{10}, \frac{5}{50}$ **37.** $\frac{2}{9}, \frac{4}{18}$
39. $\frac{6}{8}, \frac{3}{4}$ **41.** $\frac{8}{11}, \frac{48}{66}$

4.7 Guided Practice (p. 214) **3.** $\overline{AC}$ **5.** shortest, $\overline{EF}$; longest, $\overline{DE}$ **7.** Yes; $6 + 10 > 15$, $6 + 15 > 10$, and $10 + 15 > 6$. **9.** Yes; $7 + 8 > 13$, $7 + 13 > 8$, and $8 + 13 > 7$. **11.** No; $5 + 5 = 10$.

4.7 Practice and Applications (pp. 215–218)
13. smallest, $\angle R$; largest, $\angle Q$ **15.** shortest, $\overline{RT}$; longest, $\overline{ST}$ **17.** shortest, $\overline{JK}$; longest, $\overline{JH}$
19. $\angle P$; $\angle Q$; $\angle N$ **21.** $\angle C$; $\angle A$; $\angle B$ **23.** $\angle D$; $\angle E$; $\angle F$ **25.** No; $3 + 5 < 9$, so the side lengths do not satisfy the Triangle Inequality. **27.** $\overline{EF}$; $\overline{DE}$; $\overline{DF}$
29. $\overline{AC}$; $\overline{BC}$; $\overline{AB}$ **31.** $\overline{FG}$; $\overline{GH}$; $\overline{FH}$ **33.** $3 + 10 < 14$, so the side lengths do not satisfy the Triangle Inequality. **35.** no
37.

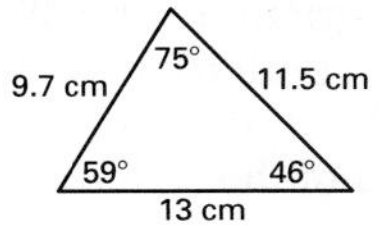

39. The diagonal and the sidewalks along Pine St. and Union St. form a triangle. The walk along the diagonal is shorter than staying on the sidewalks by the Triangle Inequality. **41.** raised **43.** Yes; when the boom is lowered and AB is greater than 100, $AB > BC$ and so $\angle ACB$ will be larger than $\angle BAC$. **45.** $\overline{RT}$ **47.** $\overline{RS}, \overline{ST}$ **49.** $58°$ **51.** $34°$
53. 1 **55.** 7 **57.** 70

Quiz 3 (p. 218) **1.** obtuse **2.** acute **3.** right
4. $KN = 4$; $MN = 2$ **5.** $KN = 26$; $MN = 13$
6. $KN = 40$; $MN = 20$ **7.** $\overline{QM}$; $\overline{LM}$; $\overline{LQ}$
8. $\overline{PQ}$; $\overline{MP}$; $\overline{MQ}$ **9.** $\overline{MN}$; $\overline{NP}$; $\overline{MP}$

Chapter Summary and Review (pp. 219–223)
1. triangle **3.** corollary **5.** vertex **7.** centroid
9. equilateral **11.** right **13.** isosceles **15.** $142°$
17. $82°$, $82°$ **19.** 5 **21.** 4 **23.** 52 **25.** 50 **27.** 14.8
29. 9.8 **31.** 8.1 **33.** 12.2 **35.** 5.4 **37.** obtuse
39. acute **41.** obtuse **43.** $KP = 28$; $PM = 14$
45. $CE = 12$; $DE = 4$ **47.** $CE = 42$; $DE = 14$
49. $\angle U$; $\angle T$; $\angle S$ **51.** $\overline{AC}$; $\overline{BC}$; $\overline{AB}$ **53.** $\overline{GH}$; $\overline{GJ}$; $\overline{HJ}$
55. Yes; $21 + 23 > 25$, $21 + 25 > 23$, and $23 + 25 > 21$. **57.** No; $6 + 6 = 12$. **59.** Yes; $2 + 3 > 4$, $2 + 4 > 3$, and $3 + 4 > 2$. **61.** Yes; $11 + 11 > 20$ and $11 + 20 > 11$.

Algebra Review (p. 227)
1. $\frac{3}{8}$ **3.** $\frac{1}{6}$ **5.** 7 hours/day **7.** $9.50/hour

Chapter 5

Study Guide (p. 232) **1.** D **2.** G **3.** D

5.1 Guided Practice (p. 236) **5.** $\angle F$ **7.** $\overline{ED}$
9. $105°$ **11.** 11.6 **13.** yes; *Sample answer:* $\triangle EFG \cong \triangle KLJ$

5.1 Practice and Applications (pp. 236–239)
15. neither **17.** corresponding sides
19. corresponding angles **21.** $\angle N$ **23.** $\angle NLM$
25. $\triangle QRP$ **27.** yes **29.** yes **31.** no **33.** $\overline{QR} \cong \overline{TU}$; $\overline{RP} \cong \overline{US}$; $\overline{PQ} \cong \overline{ST}$; $\angle Q \cong \angle T$; $\angle R \cong \angle U$; $\angle P \cong \angle S$; *Sample answer:* $\triangle QRP \cong \triangle TUS$
35. *Sample answer:*

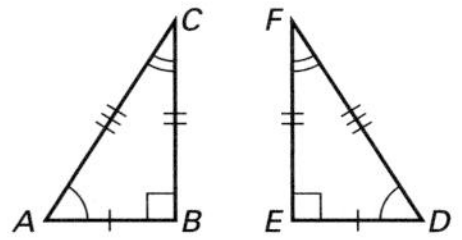

37. 5; $100°$ **39.** 6; $50°$ **41.** C **43.** yes; *Sample answer:* $\triangle JKL \cong \triangle PNM$ **45.** yes; *Sample answer:* $\triangle ABC \cong \triangle EDC$ **47.** no **49.** Base Angles Theorem **51.** Reflexive Property of Congruence
55. Reflexive Property of Congruence
57. Transitive Property of Equality **59.** acute
61. 11.95 **63.** 19.33 **65.** 17.076

5.2 Guided Practice (p. 245) **7.** no

5.2 Practice and Applications (pp. 245–249)
9. $\angle ABD$ **11.** $\angle C$ **13.** $\angle BDC$ **15.** Yes; all three pairs of corresponding sides of the triangles are congruent. **17.** No; there is no information about $\overline{PS}$ and $\overline{RS}$. **19.** No; the congruent angles are not included between the congruent sides.
21. no **23.** yes; SSS Congruence Postulate
25. yes; SAS Congruence Postulate or SSS Congruence Postulate **27.** $\overline{AB} \cong \overline{CD}$
29. $\overline{BC} \cong \overline{EF}$ **31.** $\overline{PQ} \cong \overline{RQ}$ or $\angle PSQ \cong \angle RSQ$
33. *Sample answer:* Since $\overline{AB} \parallel \overline{CD}$, $\angle ABC \cong \angle DCB$ by the Alternate Interior Angles Theorem. $\overline{AB} \cong \overline{DC}$ is given and $\overline{BC} \cong \overline{BC}$ by the Reflexive Property. So, the triangles are congruent by the SAS Congruence Postulate.

35. Statements	Reasons
1. $\overline{SP} \cong \overline{TP}$	1. Given
2. $\overline{PQ}$ bisects $\angle SPT$.	2. <u>Given</u>
3. $\angle SPQ \cong \angle TPQ$	3. <u>Definition of angle bisector</u>
4. $\overline{PQ} \cong \overline{PQ}$	4. Reflexive Prop. of Cong.
5. $\triangle SPQ \cong \triangle TPQ$	5. <u>SAS Congruence Postulate</u>

37. The congruent angles are not included between the congruent sides.   **43.** same-side interior angles   **45.** alternate interior angles   **47.** same-side interior angles   **49.** yes   **51.** 1.7   **53.** 6.3   **55.** 3.8   **57.** 1.1

Quiz 1 (p. 249) **1.** $\angle C$ **2.** $\overline{QP}$ **3.** $\triangle PQR$ **4.** $\triangle CBA$ **5.** *Sample answer:* $\triangle EFG \cong \triangle YXZ$ **6.** No; the congruent angles are not included between the congruent sides.   **7.** Yes; *Sample answer:* the three sides of $\triangle DGF$ are congruent to the three sides of $\triangle EGF$, so $\triangle DGF \cong \triangle EGF$ by the SSS Congruence Postulate.   **8.** Yes; two sides and the included angle of $\triangle JKF$ are congruent to two sides and the included angle of $\triangle NMF$, so the triangles are congruent by the SAS Congruence Postulate.

5.3 Guided Practice (p. 253) **5.** $\overline{AB} \cong \overline{DE}$ **7.** yes; ASA Congruence Postulate   **9.** no

5.3 Practice and Applications (pp. 254–256)
11. included   **13.** not included   **15.** AAS Congruence Theorem; two angles and a non-included side of $\triangle JLM$ are congruent to two angles and the corresponding non-included side of $\triangle PNM$.   **17.** Yes; SAS Congruence Postulate; two sides and the included angle of $\triangle PQR$ are congruent to two sides and the included angle of $\triangle TSM$.   **19.** Yes; ASA Congruence Postulate; vertical angles are congruent, so $\angle RVS \cong \angle UVT$; two angles and the included side of $\triangle RVS$ are congruent to two angles and the included side of $\triangle UVT$.   **21.** Yes; SSS Congruence Postulate; $\overline{XY} \cong \overline{XY}$; all three sides of $\triangle WXY$ are congruent to the three sides of $\triangle ZXY$.   **23.** AAS Congruence Theorem   **25.** $\angle K \cong \angle Q$ **27.** $\angle C \cong \angle D$
29. $\overline{AC} \cong \overline{AD}$
31. *Sample answer:*

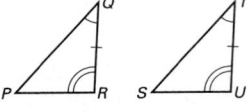

33. *Sample answer:*

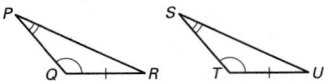

35.

Statements	Reasons
1. $\overline{BC} \cong \overline{EC}$	1. Given
2. $\overline{AB} \perp \overline{AD}$	2. Given
3. $\overline{DE} \perp \overline{AD}$	3. Given
4. $\angle A$ and $\angle D$ are right angles.	4. Perpendicular lines form right angles.
5. $\angle A \cong \angle D$	5. Right angles are congruent.
6. $\angle ACB \cong \angle DCE$	6. Vertical Angles Theorem
7. $\triangle ABC \cong \triangle DEC$	7. AAS Congruence Theorem

37. 15.0   **39.** 21.2   **41.** $\overline{ED}$ **43.** $\overline{BC}$ **45.** $\overline{FE}$
47. $\frac{1}{5}$ **49.** 2   **51.** $\frac{5}{8}$ **53.** $\frac{1}{33}$

5.4 Guided Practice (p. 260) **7.** Yes; all three sides of $\triangle EDG$ are congruent to the three sides of $\triangle GFE$ so the triangles are congruent by the SSS Congruence Postulate.   **9.** No; both triangles are equilateral and equiangular, but there is no information about the lengths of the sides of the triangles.

5.4 Practice and Applications (pp. 260–263)
11. Yes; $\overline{JL} \cong \overline{JL}$, so the hypotenuse and a leg of right $\triangle JKL$ are congruent to the hypotenuse and a leg of right $\triangle JML$.   **13.** You need to know that the wires have the same length.   **15.** Yes; SAS Congruence Postulate; two sides and the included angle of $\triangle ABC$ are congruent to two sides and the included angle of $\triangle DEF$.   **17.** No; the congruent angles are not included between the congruent sides, so there is not enough information to prove the triangles congruent.   **19.** Yes; AAS Congruence Theorem; perpendicular lines form right angles and all right angles are congruent, so $\angle UTS \cong \angle UTV$; also, $\overline{UT} \cong \overline{UT}$; two angles and a non-included side of $\triangle STU$ are congruent to two angles and the corresponding non-included side of $\triangle VTU$.   **21.** Yes; *Sample answer:* HL Congruence Theorem; the hypotenuse and a leg $(\overline{DB})$ of right $\triangle ABD$ are congruent to the hypotenuse and a leg $(\overline{DB})$ of right $\triangle CBD$.   **23.** Yes; *Sample answer:* SAS Congruence Postulate since $\overline{DF} \cong \overline{DF}$; two sides and the included angle of $\triangle CDF$ are congruent to two sides and the included angle of $\triangle EDF$.   **25.** *Sample answer:*

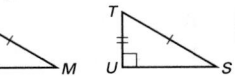

27. *Sample answer:*

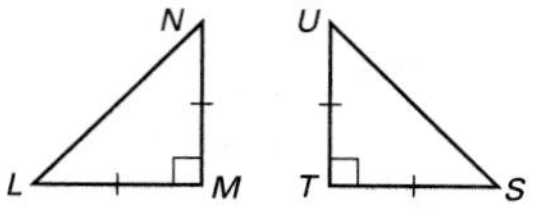

29. $\angle F \cong \angle J$, ASA Congruence Postulate; or $\angle G \cong \angle L$, AAS Congruence Theorem; or $\overline{GH} \cong \overline{LK}$, SAS Congruence Postulate

31. $\overline{WX} \cong \overline{ZX}$, ASA Congruence Postulate; or $\overline{VW} \cong \overline{YZ}$, AAS Congruence Theorem; or $\overline{VX} \cong \overline{YX}$, AAS Congruence Theorem

33. a. *Sample answer:* $\overline{BD} \cong \overline{BD}$; the hypotenuse and a leg of right $\triangle ABD$ are congruent to the hypotenuse and a leg of right $\triangle CBD$, so $\triangle ABD \cong \triangle CBD$ by the HL Congruence Theorem. **b.** Yes; *Sample answer:* It is given that $\overline{AB} \cong \overline{BC}$. By the Base Angles Theorem, $\angle BAD \cong \angle BCD$. Perpendicular lines form right angles and all right angles are congruent, so $\angle BDA \cong \angle BDC$. Therefore, $\triangle ABD \cong \triangle CBD$ by the AAS Congruence Theorem. **c.** Yes; *Sample answer:* The hypotenuse and a leg of $\triangle ABD$ are congruent to the hypotenuse and a leg of $\triangle CBD$, $\triangle CEG$, and $\triangle FEG$ by the HL Congruence Theorem.

35. $m\angle 1 = 82°$ by the Corresponding Angles Postulate; $m\angle 2 = 82°$ by the Alternate Exterior Angles Theorem (or by the Vertical Angles Theorem). **37.** Yes; AAS Congruence Theorem; two angles and a non-included side of $\triangle ABC$ are congruent to two angles and the corresponding non-included side of $\triangle DEF$. **39.** Yes; AAS Congruence Theorem; vertical angles are congruent, so $\angle JLK \cong \angle NLM$; two angles and a non-included side of $\triangle JKL$ are congruent to two angles and the corresponding non-included side of $\triangle NML$. **41.** 0 **43.** 32 **45.** 5

Quiz 2 (p. 263) **1.** yes **2.** yes **3.** no **4.** no **5.** No; only two pairs of sides are congruent. **6.** Yes; AAS Congruence Theorem; two angles and a non-included side of $\triangle JKL$ are congruent to two angles and the corresponding non-included side of $\triangle QRP$. **7.** Yes; ASA Congruence Postulate; two angles and the included side of $\triangle STU$ are congruent to two angles and the included side of $\triangle VUT$. **8.** Yes; HL Congruence Theorem; the hypotenuse and a leg of right $\triangle HEF$ are congruent to the hypotenuse and a leg of right $\triangle FGH$.

9. Yes; ASA Congruence Postulate or AAS Congruence Theorem; when two parallel lines are cut by a transversal, alternate interior angles are congruent, so $\angle R \cong \angle U$ and $\angle S \cong \angle V$; to use AAS, use congruent vertical angles $\angle RTS$ and $\angle UTV$ along with one pair of alternate interior angles. **10.** Yes; ASA Congruence Postulate or AAS Congruence Theorem; $\overline{JL} \cong \overline{JL}$; select the angles that include $\overline{JL}$ to use ASA or a pair of angles that do not include $\overline{JL}$ to use AAS.

5.5 Guided Practice (p. 268) **3.** When two parallel lines are cut by a transversal, alternate interior angles are congruent, so $\angle VSU \cong \angle TUS$ and $\angle TSU \cong \angle VUS$; also $\overline{SU} \cong \overline{SU}$; $\triangle STU \cong \triangle UVS$ by the ASA Congruence Postulate; $\angle STU \cong \angle UVS$ since corresponding parts of congruent triangles are congruent. **5.** $\overline{LN} \cong \overline{LK}$, $\angle LNM \cong \angle LKJ$, and $\angle L \cong \angle L$; $\triangle JKL \cong \triangle MNL$ by the ASA Congruence Postulate; $\angle J \cong \angle M$ since corresponding parts of congruent triangles are congruent.

5.5 Practice and Applications (pp. 268–271) **7.** $\triangle ABC \cong \triangle DBC$ **9.** $\triangle ABC \cong \triangle EDF$ **11.** Check diagrams; AAS Congruence Theorem. **13.** ASA Congruence Postulate; $\angle GJH \cong \angle KMH$, and $\overline{MH} \cong \overline{JH}$; $\angle H \cong \angle H$ by the Reflexive Property of Congruence.

15.

Statements	Reasons
1. $\overline{AD} \cong \overline{CD}$	1. Given
2. $\angle ABD$ and $\angle CBD$ are right angles.	2. Given
3. $\triangle ABD$ and $\triangle CBD$ are right triangles.	3. Def. of right triangle
4. $\overline{BD} \cong \overline{BD}$	4. Reflexive Property of Congruence
5. $\triangle ABD \cong \triangle CBD$	5. HL Congruence Theorem
6. $\angle A \cong \angle C$	6. Corresp. parts of $\cong$ triangles are $\cong$.

17. *Sample answer:*

Statements	Reasons
1. $\overline{RQ} \cong \overline{TS}$	1. Given
2. $\angle RTQ \cong \angle TRS$	2. Given
3. $\angle Q$ and $\angle S$ are right angles.	3. Given
4. $\angle Q \cong \angle S$	4. Right angles are congruent.
5. $\triangle RTQ \cong \triangle TRS$	5. AAS Congruence Theorem
6. $\overline{QT} \cong \overline{SR}$	6. Corresp. parts of $\cong$ triangles are $\cong$.

19.

Statements	Reasons
1. $\overline{BD}$ and $\overline{AE}$ bisect each other at C.	1. Given
2. $\overline{BC} \cong \overline{DC}$	2. Def. of segment bisector
3. $\overline{AC} \cong \overline{EC}$	3. Def. of segment bisector
4. $\angle BCA \cong \angle DCE$	4. Vertical Angles Theorem
5. $\triangle ABC \cong \triangle EDC$	5. SAS Congruence Postulate
6. $\angle A \cong \angle E$	6. Corresp. parts of $\cong$ triangles are $\cong$.

25. 18   **27.** 42   **29.** 11   **31.** -9 **33.** 7   **35.** 9

5.6 Guided Practice (p. 276) **3.** 16   **5.** 12

5.6 Practice and Applications (pp. 276–280)

7.

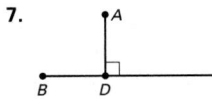

9. Paige cannot assume that $\overline{PC} \perp \overleftrightarrow{AC}$.   **11.** 6
13. $\overline{BC} \cong \overline{BD}$ by the Angle Bisector Theorem
15. 18   **17.** 3   **19.** $AD = 12$; $BC = 16$ **21.** ℓ is the perpendicular bisector of the goal line.
23. $CG = AG = 2$ **25.** $\overline{FJ} \cong \overline{JG}$; $\overline{FK} \cong \overline{KH}$; $\overline{GL} \cong \overline{LH}$; $\overline{JM} \cong \overline{KM}$; $\overline{MG} \cong \overline{MH} \cong \overline{FM}$

26–28.

29. The fire station at A; X is in the red region, closest to station A.

31. Yes; by the Perpendicular Bisector Theorem, if D lies on the perpendicular bisector of $\overline{AB}$, then $\overline{DA}$ and $\overline{DB}$ will always be congruent segments.
37. $(8, -5)$ **39.** $(-1, -10)$ **41.** $(9, -4)$ **43.** $(13, 6)$
45. SAS Congruence Postulate; $\overline{DB} \cong \overline{DB}$ and all right angles are congruent, so two sides and the included angle of $\triangle ADB$ are congruent to two sides and the included angle of $\triangle CDB$.
47. ASA Congruence Postulate; vertical angles are congruent, so two angles and the included side of $\triangle JNK$ are congruent to two angles and the included side of $\triangle LNM$.   **49.** $-1.25, -0.75,$ $-0.25, 0.25, 1, 4$ **51.** $-3.9, -3.3, -3, 3.1, 3.5, 3.8$
53. $-2.5, 1, 2.1, 3.2, 3.25, 5$ **55.** -1 **57.** 3
59. -8 **61.** 3

5.7 Guided Practice (p. 286) **3.** yes   **5.** 3   **7.** 5

5.7 Practice and Applications (pp. 286–290)
9. Yes; all three properties of a reflection are met.
11. the y-axis   **13.** the x-axis   **15.** $\overline{CD}$; C; D
17.   **19.**

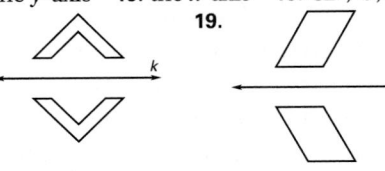

21. no   **23.** yes   **25.** 4   **27.** all shown
29. not all shown;

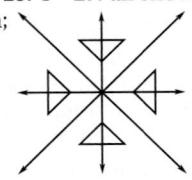

31. b, d, p, q   **33.** yes; $\leftarrow$ HOOK $\rightarrow$
35. yes; $\rightarrow$ BIB $\rightarrow$ **37.** $45°$ **39.** $60°$

41. $AB = 2 = DE$; $AC = 3 = DF$; $BC = \sqrt{13} = EF$; by the SSS Congruence Postulate, $\triangle ABC \cong \triangle DEF$.
45. 72   **47.** $67°$ **49.** $46°$ **51.** $>$ **53.** $>$ **55.** $<$

Quiz 3 (p. 290) **1.** AAS Congruence Theorem   **2.** 2
3. $ML = 9$; $JK = 25$ **4.** 13   **5.** 1   **6.** 2   **7.** 3

Chapter Summary and Review (pp. 291–295)
1. congruent   **3.** perpendicular bisector
5. reflection   **7.** corresponding sides   **9.** neither
11. corresponding angles   **13.** yes; SAS Congruence Postulate   **15.** $\angle S \cong \angle X$
17. $\angle JKM \cong \angle LMK$ **19.** HL Congruence Theorem
21. AAS Congruence Theorem
23.

Statements	Reasons
1. $\angle JMN$ and $\angle JKL$ are right angles.	1. Given
2. $\angle JMN \cong \angle JKL$	2. All right angles are congruent.
3. $\overline{MN} \cong \overline{KL}$	3. Given
4. $\angle J \cong \angle J$	4. Reflexive Prop. of Congruence
5. $\triangle JMN \cong \triangle JKL$	5. AAS Congruence Theorem
6. $\overline{JM} \cong \overline{JK}$	6. Corresp. parts of $\cong$ triangles are $\cong$.

25. 3   **27.** 7   **29.** no; 1
31.

Algebra Review (p. 299) **1.** yes   **3.** no   **5.** no
7. $\dfrac{6}{5}$ **9.** -1 **11.** 0

Chapter 6

Study Guide (p. 302) **1.** C **2.** G **3.** C

6.1 Guided Practice (p. 306) **3.** Yes; the figure is a polygon formed by five straight lines. **5.** No; two of the sides intersect only one other side. **7.** 67°

6.1 Practice and Applications (pp. 306–308) **9.** no; not formed by segments **11.** 3; in order for each side to intersect exactly two other sides at each of its endpoints, a polygon must have at least three sides; triangle.
13. *Sample answer:*

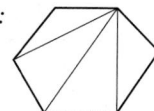

15. 71° **17.** 75° **19.** 20 **21.** 8; octagon
23. $\overline{MP}, \overline{MQ}, \overline{MR}, \overline{MS}, \overline{MT}$ **25.** 8; octagon
27. 17; 17-gon **31.** parallel **33.** neither **35.** 72°
37. $4x + 12$ **39.** $-2x + 14$ **41.** $-15x + 6$

6.2 Guided Practice (p. 313) **3.** yes **5.** $\angle KJM$; opposite $\measuredangle$ of a parallelogram are $\cong$. **7.** $\overline{NL}$; diagonals of a parallelogram bisect each other.
9. $\overline{NK}$; diagonals of a parallelogram bisect each other. **11.** 9

6.2 Practice and Applications (pp. 313–315) **13.** B; diagonals of a parallelogram bisect each other.
15. D; diagonals of a parallelogram bisect each other. **17.** E; vertical $\measuredangle$ are $\cong$. **19.** G; if 2 ∥ lines are cut by a transversal, then alternate interior $\measuredangle$ are $\cong$. **21.** $\overline{HG}$; if you sketch parallelogram $EFGH$, you see that $\overline{EF}$ intersects $\overline{GF}$ and $\overline{EF}$ is parallel to $\overline{HG}$. **23.** $EF = 25$; $FG = 17$
25. $m\angle K = 129°$; $m\angle L = 51°$; $m\angle M = 129°$
27. $m\angle J = 90°$; $m\angle K = 90°$; $m\angle L = 90°$ **29.** 13
31. $x = 12$; $y = 3$ **33.** $x = 2$; $y = 7$ **35.** $m\angle B$ increases. **37.** The height increases.
39. $\angle 5$ and $\angle 8$ **45.** No; same-side interior angles are not supplementary. **47.** 31 **49.** 5 **51.** 3
53. $-\dfrac{1}{3}$ **55.** $\dfrac{3}{5}$

6.3 Guided Practice (p. 320) **5.** No; you can tell that one pair of sides must be parallel, but not both. **7.** yes, by the definition of a parallelogram

6.3 Practice and Applications (pp. 320–323)
9. Yes; both pairs of opposite sides are congruent.
11. Yes; both pairs of opposite angles are congruent. **13.** No; opposite angles are not congruent (or consecutive angles are not supplementary).

15. No; consecutive angles are not supplementary.
17. Yes; the diagonals bisect each other.
19. No; the diagonals do not bisect each other.
21. The parallel sides may not be congruent so the congruent sides may not be parallel. So the quadrilateral may not be a parallelogram.
23. The diagonals of the resulting quadrilateral were drawn to bisect each other. Therefore, the resulting quadrilateral is a parallelogram.
25. The slope of $\overline{FG}$ = slope of $\overline{JH}$ = -3, and slope of $\overline{FJ}$ = slope of $\overline{GH}$ = 0, so the opposite sides are parallel and $FGHJ$ is a parallelogram by definition. **29.** 86° **31.** 71° **33.** 14 **35.** 17
37. 22 **39.** 17 **41.** -11 **43.** 32

Quiz 1 (p. 323) **1.** yes; pentagon **2.** yes; hexagon
3. no; not formed by segments **4.** $x = 16$; $y = 15$
5. $x = 58$; $y = 122$; $z = 58$ **6.** $x = 12$; $y = 9$
7. Yes; one angle is supplementary to both of its consecutive angles. **8.** No; you can tell that one pair of sides must be parallel, but not both.
9. Yes; both pairs of opposite sides are parallel, so the quadrilateral is a parallelogram by the definition of a parallelogram.

6.4 Guided Practice (p. 328)
3. B, C, D **5.** A, B, C, D

6.4 Practice and Applications (pp. 328–330)
7. 4; 4; 4 **9.** 90; 3; 3 **11.** rhombus **13.** No, a quadrilateral with congruent diagonals does not have to be a rectangle; *Sample answer:* You need to know that the figure is a parallelogram, or that its diagonals bisect each other, or that all of its angles are right. **15.** parallelogram, rectangle, rhombus, square **17.** rhombus, square **19.** 18
21. The consecutive angles of a parallelogram are supplementary, so the two angles consecutive to $\angle J$ must also be right angles. Also, opposite angles of a parallelogram are congruent, so the fourth angle must also be a right angle. By definition, a parallelogram with four right angles is a rectangle. **29.** 90° **31.** 145° **33.** $\dfrac{2}{1}$

6.5 Guided Practice (p. 334)
3. isosceles trapezoid **5.** trapezoid **7.** 5

6.5 Practice and Applications (pp. 334–336)
9. D **11.** B **13.** A **15.** $m\angle K = 60°$; $m\angle J = 120°$; $m\angle M = 120°$ **17.** $m\angle R = 102°$; $m\angle T = 48°$
19. $m\angle Q = 90°$; $m\angle S = 30°$ **21.** 15 **23.** 6 **25.** 19

27.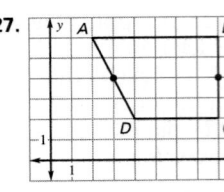

29. $\overleftrightarrow{AD}$ and $\overleftrightarrow{BC}$
31. Corresponding Angles Postulate
37. No; you can tell that one pair of sides must be parallel, but not both.

39. 10   **41.** 17   **43.** $\frac{2}{7}$ **45.** $\frac{5}{18}$

6.6 Practice and Applications (pp. 339–341)
9. trapezoid   **11.** rectangle   **13.** Yes; since $140° + 40° = 180°$, the top and bottom sides are parallel by the Same-Side Interior Angles Converse. Since $60° + 40° \neq 180°$, the other two sides are not parallel. The figure has exactly one pair of parallel sides, so it is a trapezoid.   **15.** Yes; the four angles are congruent, so the four angles are right angles by the Quadrilateral Interior Angles Theorem. A quadrilateral with four right angles is a rectangle.   **17.** Yes; both pairs of opposite sides are congruent, so the figure is a parallelogram.   **19.** isosceles trapezoid   **21.** true   **23.** true   **25.** isosceles trapezoid   **27.** 1   **29.** 16   **31.** 10   **33.** 16   **35.** 0.2   **37.** $0.8333\ldots \approx 0.83$

Quiz 2 (p. 341) **1.** 35   **2.** 6   **3.** 45   **4.** trapezoid
5. isosceles trapezoid   **6.** square, rectangle, rhombus, and parallelogram

Chapter Summary and Review (pp. 342–345)
1. parallelogram   **3.** rhombus   **5.** square
7. vertex   **9.** legs; bases   **11.** yes; hexagon
13. *Sample answer:*

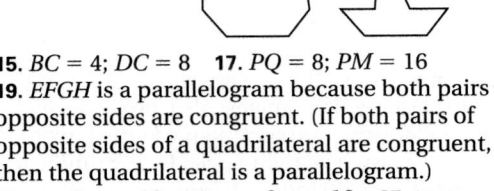

15. $BC = 4$; $DC = 8$ **17.** $PQ = 8$; $PM = 16$
19. *EFGH* is a parallelogram because both pairs of opposite sides are congruent. (If both pairs of opposite sides of a quadrilateral are congruent, then the quadrilateral is a parallelogram.)
21. $x = 8$; $y = 13$ **23.** $x = 6$; $y = 10$ **25.** true
27. true   **29.** 71°   **31.** No; both pairs of opposite angles are congruent, so the figure is a parallelogram. There is no information about the sides, so you cannot conclude that the figure is a rhombus.   **33.** Yes; *Sample answer:* The diagonals bisect each other, so the figure is a parallelogram. One angle of the figure is given as a right angle. Consecutive angles of a parallelogram are supplementary, so the other three angles can be shown to be right angles. Because the figure is a parallelogram with four right angles, it is a rectangle by definition.

Algebra Review (p. 349) **1.** 30   **3.** 15   **5.** 2
7. $\frac{22}{7} = 3\frac{1}{7}$ **9.** $1\frac{1}{6}$ **11.** 74

Cumulative Practice (pp. 350–351) **1.** Each number is 9 more than the previous number; 38, 47.
3. $(-1, -1)$ **5.** $(1, 1)$ **7.** 9; obtuse   **9.** 62°
11. same-side interior   **13.** alternate interior
15. 105   **17.** 75   **19.** 58   **21.** 48   **23.** 5   **25.** 10
27. no; $3 + 6 < 12$ **29.** no   **31.** yes; SAS Congruence Postulate
33. *Sample answer:*

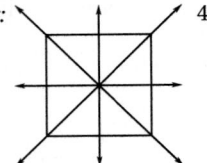

35. $x = 18$; $y = \frac{5}{2}$

37. A trapezoid; by the Quadrilateral Interior Angles Theorem, $m\angle H = 113°$; since $\angle E$ and $\angle F$ are supplementary, $\overline{EH} \parallel \overline{FG}$, but the other two sides are not parallel; a quadrilateral with exactly one pair of parallel sides is a trapezoid.   **39.** Yes; $\overleftrightarrow{AB}$ and $\overleftrightarrow{DE}$ are both perpendicular to $\overleftrightarrow{AD}$ and the three lines are coplanar, so $\overleftrightarrow{AB} \parallel \overleftrightarrow{DE}$.

Chapter 7

Study Guide (p. 356) **1.** C   **2.** G   **3.** B

7.1 Guided Practice (p. 361)
3. $\frac{6}{5}$ **5.** $\frac{3}{1}$ **7.** $FG = 32$; $GH = 24$ **9.** 1   **11.** 15

7.1 Practice and Applications (pp. 361–363)
13. $\frac{4}{1}$ **15.** $\frac{1}{4}$ **17.** $\frac{8}{1}$ **19.** $\frac{35}{12}$ **21.** $\frac{3}{1}$ **23.** $\frac{3}{5}$
25. $\frac{20\text{ mm}}{16\text{ mm}}$; $\frac{5}{4}$ **27.** $\frac{2\text{ ft}}{12\text{ in.}}$ or $\frac{24\text{ in.}}{12\text{ in.}}$; $\frac{2}{1}$ **29.** $\frac{1}{1}$ **31.** $\frac{7}{2}$
33. $JK = 30$; $KL = 12$ **35.** length, 44 inches; width, 11 inches   **37.** length, 48 meters; width, 18 meters
39. 15   **41.** 56   **43.** 4   **45.** 5   **47.** 1440 inches
49. about 1.0 inches   **51.** about 300 miles
53. about 60 miles   **57.** $\angle P$ **59.** yes; pentagon
61. yes; octagon   **63.** 18.34   **65.** 19.6   **67.** 12

7.2 Guided Practice (p. 368) **3.** $\angle A \cong \angle L$;
$\angle B \cong \angle M$; $\angle C \cong \angle N$ **5.** $\frac{3}{4}$ **7.** no; $\frac{7}{15} \neq \frac{4}{10}$

7.2 Practice and Applications (pp. 369–371)
9. $\angle A \cong \angle Q$; $\angle B \cong \angle R$; $\angle C \cong \angle S$; $\angle D \cong \angle T$;
$\angle E \cong \angle U$; *Sample answer:* $\frac{AB}{QR} = \frac{BC}{RS} = \frac{CD}{ST} = \frac{DE}{TU} = \frac{EA}{UQ}$ **11.** yes; $\triangle GHJ \sim \triangle DEF$; $\frac{3}{4}$ **13.** yes; $JKLM \sim EFGH$; $\frac{5}{4}$ **15.** no $\left(\frac{3}{4} \neq \frac{4}{6}\right)$

17. $x = 30$; $y = 14$ **19.** $x = 8$; $y = 20$ **21.** 196 in. by 73.5 in.; 16 ft 4 in. by 6 ft $1\frac{1}{2}$ in. **23.** 80° **25.** 15

27. 6 : 7 **29.** No; the ratio $\frac{length}{length} = \frac{4}{16} = \frac{1}{4}$ and the ratio $\frac{width}{width} = \frac{3}{9} = \frac{1}{3}$; $\frac{1}{4} \neq \frac{1}{3}$. **31.** sometimes

33. sometimes **35.** yes; ASA Congruence Postulate **37.** yes; HL Congruence Theorem or AAS Congruence Theorem **39.** *Sample answer:* $\frac{4}{10}, \frac{20}{50}$ **41.** *Sample answer:* $\frac{18}{8}, \frac{54}{24}$

7.3 Guided Practice (p. 375) **3.** Yes; $\angle D \cong \angle G$ and $\angle F \cong \angle H$, so $\triangle DEF \sim \triangle GJH$ by the AA Similarity Postulate. **5.** Yes; $\angle Q \cong \angle Q$ and $\angle QRT \cong \angle QSU$, so $\triangle QRT \sim \triangle QSU$ by the AA Similarity Postulate.

7.3 Practice and Applications (pp. 375–378) **7.** no **9.** yes; $\triangle ABC \sim \triangle DEF$ **11.** yes; $\triangle XYZ \sim \triangle GFH$ **13.** Yes; $\angle U \cong \angle U$ and $\angle UVW \cong \angle S$, so $\triangle UVW \sim \triangle UST$ by the AA Similarity Postulate.

15. Yes; *Sample answer:* $\overline{DE} \parallel \overline{BA}$, so $\angle CED \cong \angle A$ and $\angle CDE \cong \angle B$; $\triangle CDE \sim \triangle CBA$ by the AA Similarity Postulate. **17.** MN **19.** 16 **21.** 14 **23.** 24 **25.** 35 **27.** true **29.** true **37.** Meredith is right; $m\angle ACB = 55°$, so $\angle ACB \cong \angle ADE$; $\angle A \cong \angle A$, so $\triangle ABC \sim \triangle AED$ by the AA Similarity Postulate. **39.** 34 **41.** $\overline{ST}$ **43.** 17 **45.** 15

46–53.

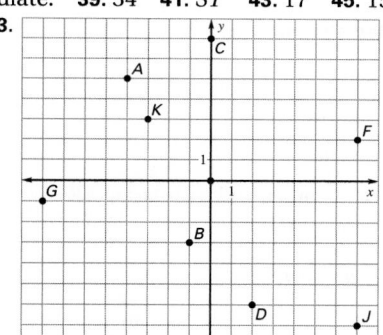

Quiz 1 (p. 378) **1.** 12 **2.** 40 **3.** 30 **4.** 10 **5.** 5 **6.** 20 **7.** Yes; $\angle A \cong \angle F$ and $\angle B \cong \angle G$, so $\triangle ABC \sim \triangle FGH$ by the AA Similarity Post. **8.** Yes; $\angle J \cong \angle R$ and $\angle K \cong \angle P$, so $\triangle JKL \sim \triangle RPQ$ by the AA Similarity Postulate. **9.** Yes; vertical angles are congruent, so $\angle SUT \cong \angle WUV$; also $\angle T \cong \angle V$, so $\triangle STU \sim \triangle WVU$ by the AA Similarity Postulate.

7.4 Guided Practice (p. 382) **3.** No; $\angle D \cong \angle R$ but $\frac{8}{6} \neq \frac{10}{8}$.

7.4 Practice and Applications (pp. 382–385) **5.** yes, $\triangle CDE \sim \triangle FGH$; $\frac{2}{3}$ **7.** no

9. yes, $\triangle UVW \sim \triangle JGH$; $\frac{3}{2}$ **11.** $\triangle RST$ is not similar to $\triangle ABC$ because the ratios of corresponding sides are not equal. $\triangle ABC \sim \triangle XYZ$ by the SSS Similarity Theorem because the ratios of corresponding sides are all equal to $\frac{3}{2}$. **13.** $\triangle RST$ is not similar to $\triangle ABC$ because the ratios of corresponding sides are not equal. $\triangle ABC \sim \triangle XYZ$ by the SSS Similarity Theorem because the ratios of corresponding sides are all equal to $\frac{3}{4}$.

15. yes; $\triangle ABC \sim \triangle DFE$ **17.** no **19.** $\angle A \cong \angle A$, $\frac{AD}{AB} = \frac{9}{30} = \frac{3}{10}$, and $\frac{AE}{AC} = \frac{6}{20} = \frac{3}{10}$; the lengths of the sides that include $\angle A$ are proportional, so $\triangle ADE \sim \triangle ABC$ by the SAS Similarity Theorem. **21.** Yes, $\triangle JKL \sim \triangle XYZ$ (or $\triangle YXZ$) by the SSS Similarity Theorem. **23.** no **25.** Yes; $\triangle PQR \sim \triangle DEF$ by the SAS Similarity Theorem.

27. $\frac{AE}{AC} = \frac{AD}{AB}$ or $\frac{AE}{AC} = \frac{DE}{BC}$ **29.** *Sample answer:* Jon is correct. $\frac{6}{27} = \frac{4}{18}$, so the triangles are similar when $x = 6$. **31.** SAS Similarity Theorem

35. $\angle VTS$ **37.** Yes; *Sample answer:* $\overrightarrow{TV}$ bisects $\angle STU$, so $\angle STV \cong \angle UTV$; $\angle TSV \cong \angle TUV$ and $\overline{TV} \cong \overline{TV}$, so $\triangle STV \cong \triangle UTV$ by the AAS Congruence Theorem; corresponding parts of congruent triangles are congruent, so $\angle TVS \cong \angle TVU$. **39.** 54 **41.** 24 **43.** $\frac{1}{4}$ **45.** $\frac{22}{25}$ **47.** $\frac{11}{20}$ **49.** $\frac{17}{50}$

7.5 Guided Practice (p. 390) **3.** AE **5.** AB **7.** 12 **9.** 8

7.5 Practice and Applications (pp. 390–392) **11.** $\frac{5}{6}$ **13.** 36 **15.** 18 **17.** 8 **19.** 15 **21.** yes **23.** yes **25.** 22 **27.** 10 **29.** 13.5 **33.** $\overline{BC}$ **35.** 7.5 **37.** 9.5 **39.** proportional **41.** yes; ~~OHIO~~ **43.** yes; ~~BOOK~~

45. 1 **47.** $-\frac{2}{3}$ **49.** $\frac{5}{2}$ **51.** 3

7.6 Guided Practice (p. 396) **3.** Enlargement; the red image is larger than the blue original figure. **5.** enlargement

7.6 Practice and Applications (pp. 396–398) **7.** enlargement; $\frac{8}{3}$ **9.** enlargement; $\frac{9}{2}$ **11.** enlargement; 21 **13.** reduction; 3 **15.** 4 **19.** 47° **21.** 3 **23.** 51 **25.** 3 **27.** 16 **29.** 26

Quiz 2 (p. 398) **1.** yes (by the SAS Similarity Theorem); $\triangle ABC \sim \triangle FGH$ **2.** yes (by the SSS Similarity Theorem); $\triangle JKL \sim \triangle PQR$ **3.** 8 **4.** 18 **5.** 44 **6.** enlargement; 2 **7.** reduction; $\frac{1}{3}$

Chapter Summary and Review (pp. 400–403)
1. g; h; f; j **3.** scale factor **5.** midsegment
7. enlargement **9.** 42 **11.** 20 **13.** 20
15. $\angle P \cong \angle P$ and $\angle STP \cong \angle QRP$, so $\triangle PST \sim \triangle PQR$ by the AA Similarity Postulate.
17. $\angle B \cong \angle F$ and $\angle A \cong \angle D$, so $\triangle ABC \sim \triangle DFE$ by the AA Similarity Postulate.
19. $\angle QSR \cong \angle UST$, $\frac{QS}{US} = \frac{35}{28} = \frac{5}{4}$, and $\frac{RS}{TS} = \frac{30}{24} = \frac{5}{4}$; the lengths of the sides that include the congruent angles are proportional, so $\triangle QRS \sim \triangle UTS$ by the SAS Similarity Theorem.
21. 3 **23.** 10 **25.** enlargement; $\frac{5}{2}$

Algebra Review (p. 407) **1.** $2\sqrt{3}$ **3.** $2\sqrt{11}$ **5.** $5\sqrt{6}$
7. $4\sqrt{3}$ **9.** $h = \frac{A}{b}$ **11.** $P = \frac{S - 2B}{h}$

Chapter 8

Study Guide (p. 410) **1.** A **2.** H

8.1 Guided Practice (p. 413)
5. concave **7.** C **9.** B

8.1 Practice and Applications (pp. 413–415)
11. concave **13.** equiangular **15.** neither
17. regular (both equilateral and equiangular)
19. regular (both equilateral and equiangular)
21. not regular (equiangular but not equilateral)
23. concave **25.** C; A, B **27.** Yes. The sum of the measures of the angles of a triangle is 180°, or 3(60°). So, the missing angle measure is 60°. All three angles are congruent, so the triangle is equiangular.
29. Yes. *Sample answer:*

31. Yes. *Sample answer:*

33. 15 in. **35.** 48 ft **37.** 6 **41.** yes; pentagon
43. yes; octagon **45.** 16 **47.** 32 **49.** 1 **51.** 2.2

8.2 Guided Practice (p. 420)
3. 1260° **5.** $m\angle 1 + 66° + 130° + 100° = 360°$
7. $m\angle 1 + 85° + 59° + 57° + 117° = 360°$

8.2 Practice and Applications (pp. 421–423)
9. 1080° **11.** 1440° **13.** 3240° **15.** 9000° **17.** 99°

19. 127° **21.** 120° **23.** 60° **25.** 32 **27.** 540°
29. 108° **31.** always **33.** sometimes **41.** $BE = 4$, $ED = 2$ **43.** $BE = 34\frac{2}{3}$, $ED = 17\frac{1}{3}$ **45.** 11 **47.** 0
49. 2 **51.** 1.11

Quiz 1 (p. 423) **1.** not regular (equilateral but not equiangular) **2.** regular (both equilateral and equiangular) **3.** not regular (not equilateral and not equiangular) **4.** 140° **5.** 120° **6.** 109°
7. 101° **8.** 110°

8.3 Guided Practice (p. 427) **3.** A **5.** True; the area of the polygon is Area A + Area B + Area C.
7. True; the height is 3 + 2 = 5 units.

8.3 Practice and Applications (pp. 427–429)
9. 144 m^2 **11.** 10 cm^2 **13.** 60.5 m^2 **15.** $A = 44$ m^2
17. 196 m^2 **19.** 81 m^2 **21.** $b = 3$ cm **23.** Yes. Since the perimeter is 28 feet, each side measures $28 \div 4 = 7$ feet. Area $= 7^2 = 49$ square feet.
25. $b = 3$ ft, $h = 6$ ft **27.** 95 m^2 **29.** 116 yd^2
31. 207,500 ft^2 **33.** about $433 **35.** C; Diagonals of a parallelogram bisect each other. **37.** D; If two parallel lines are cut by a transversal, then alternate interior angles are congruent.
39. Yes; $\triangle SUT \sim \triangle PRQ$ by the SAS Similarity Theorem. **41.** $8 > -18$ **43.** $-10 < 0$
45. $2.44 > 2.044$

8.4 Guided Practice (p. 434)
3.

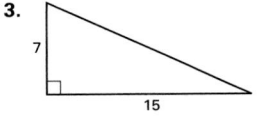

8.4 Practice and Applications (pp. 434–437)
5. 42 cm^2 **7.** 7.5 yd^2 **9.** 14 yd^2 **11.** 72 in.^{2}
13. 63 ft^2
15. *Sample answer:*

17. $h = 9$ cm **19.** 12 in. **21.** 4 in.^{2} **23.** 42 m^2
25. 68 cm^2 **27.** $\frac{3}{4}$; $\frac{9}{16}$ **29.** 8 triangles
31. 210.6 in.^{2} **33.** 120 square units **37.** 12
39. $(-3, 3)$ **41.** $(6, 2)$ **43.** $(3, -1)$

Quiz 2 (p. 437) **1.** 144 cm^2 **2.** 21 in.^{2} **3.** 46 m^2
4. 154 ft^2 **5.** 56 mm^2 **6.** 96 yd^2 **7.** $b = 8$ in.
8. $h = 12$ m **9.** $b = 14$ cm

8.5 Guided Practice (p. 442) **3.** A **5.** 8 ft, 11 ft
7. 8 cm, 10 cm

8.5 Practice and Applications (pp. 442–445)
9. 66 square units **11.** 8.4 square units
13. 264 square units **15.** 192 cm^2
17. The formula is $A = bh$, not $A = \frac{1}{2}bh$. So, the
area is $(14)(11)$, or 154 square units. **19.** $h = 6$ in.
21. 8 m **23.** 104 square units **25.** 40 square units
27. 234 square units **29.** 89.49 square units
31. 4029.18 square units **33.** 175.5 cm^2
35. 9 square units **37.** 8 square units **43.** 1 **45.** 8
47. concave **49.** $2\sqrt{3}$ **51.** $2\sqrt{13}$ **53.** $10\sqrt{3}$
55. $6\sqrt{5}$

8.6 Guided Practice (p. 448)
3. $h = 12, b_1 = 9, b_2 = 19$ **5.** C **7.** B

8.6 Practice and Applications (pp. 448–450)
9. 64 square units **11.** 224 square units
13. 330 square units **15.** The areas are equal.
17. Answers will vary; they are the same.
19. $h = 11$ in. **21.** 4 units **23.** about 3897 ft^2
25. about 4182 ft^2 **27.** 1824 in.^{2}
29. 1280 square units **31.** 517.5 square units
35. E **37.** B **39.** 1 **41.** $\frac{5}{6}$

8.7 Guided Practice (p. 455)
3. 7 in. **5.** 7 m **7.** 8 **9.** 7

8.7 Practice and Applications (pp. 456–459)
11. 31 m **13.** 82 yd **15.** 50 ft^2 **17.** 113 in.^{2}
19. 201 ft^2 **21.** about 95 m^2 **23.** *Sample answer:*
Disagree. The area of the circle with radius 5 is
about 79. The area of the circle with radius 10 is
about 314. Since $314 \div 79 \approx 4$, the area is
quadrupled, not doubled. **25.** 13 m **27.** about
4 units **29.** about 6 units **31.** $\frac{x}{12} = \frac{60°}{360°}$
33. 61 m^2 **35.** 25 cm^2 **37.** 182 ft^2 **39.** 7 m^2
41. trapezoid **43.** rhombus **45.** 88 ft^2 **47.** 3
49. 44 **51.** -11 **53.** -16 **55.** 31

Quiz 3 (p. 459) **1.** 63 ft^2 **2.** 31.72 cm^2 **3.** 24 m^2
4. 18 yd^2 **5.** 28 mm^2 **6.** 13 in.^{2} **7.** $h = 13$ cm
8. $b_2 = 12$ ft **9.** $r \approx 8$ in.

Chapter Summary and Review (pp. 460–463)
1. convex **3.** regular **5.** height **7.** circle
9. not regular (not equilateral and not
equiangular) **11.** not regular (equiangular but
not equilateral) **13.** 91° **15.** 50° **17.** 72 ft^2
19. 44 m^2 **21.** 58.5 cm^2 **23.** 26 ft **25.** 56 ft^2
27. 22.5 in.^{2} **29.** 11 m **31.** 36 ft^2 **33.** 7 m^2
35. $C \approx 82$ in.; $A \approx 531$ in.^{2} **37.** $C \approx 25$ ft; $A \approx 50$ ft^2
39. about 8 ft

Algebra Review (p. 467) **1.** 16 **3.** -243 **5.** 1000
7. 51 **9.** 45%

Chapter 9

Study Guide (p. 472) **1.** A **2.** G **3.** C

9.1 Guided Practice (p. 476) **11.** no; cylinder
13. hexagonal prism; 8 faces and 18 edges;
congruent faces: $ABVU \cong BCWV \cong CDXW \cong$
$DEYX \cong EFZY \cong FAUZ$ and $ABCDEF \cong UVWXYZ$;
congruent edges: $\overline{AB} \cong \overline{BC} \cong \overline{CD} \cong \overline{DE} \cong \overline{EF} \cong$
$\overline{FA} \cong \overline{UV} \cong \overline{VW} \cong \overline{WX} \cong \overline{XY} \cong \overline{YZ} \cong \overline{ZU}$ and
$\overline{AU} \cong \overline{BV} \cong \overline{CW} \cong \overline{DX} \cong \overline{EY} \cong \overline{FZ}$ **15.** cube or
square prism; 6 faces and 12 edges; congruent
faces: $JKLM \cong TUVW \cong JKUT \cong KLVU \cong$
$LMWV \cong MJTW$

9.1 Practice and Applications (pp. 477–480)
17. no; cone **19.** yes; rectangular; rectangular
pyramid **21.** true **23.** false **25.** F **27.** A **29.** B
31. triangular prism **33.** yes; rectangular;
rectangular pyramid **35.** no **37.** triangular
prism; 5 faces and 9 edges; congruent faces:
$\triangle FGH \cong \triangle JKL$; congruent edges: $\overline{FG} \cong \overline{JK}$,
$\overline{GH} \cong \overline{KL}, \overline{FH} \cong \overline{JL}, \overline{FJ} \cong \overline{GK} \cong \overline{HL}$ **39.** True;
they are all three-dimensional shapes. **41.** False;
if a prism is not rectangular, then the bases are
the congruent faces that are not rectangular.
43. 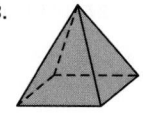 **45.**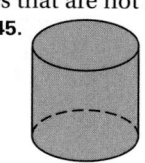

47. *Sample answer:*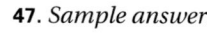

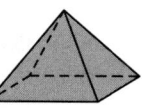

49. *Sample answer:*

51. *Sample answer:*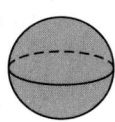

53. 6 vertices **55.** 12 vertices **57.** 7 faces
63. 28 m^2 **65.** 13 ft **67.** 31 cm; 79 cm^2 **69.** 41
71. 28 **73.** 4 **75.** 30 **77.** 10π **79.** 9π

9.2 Guided Practice (p. 487) **5.** 248 in.^{2} **7.** 36 m^2

9.2 Practice and Applications (pp. 487–490)
9. 6 in.^{2} **11.** 5 m **13.** about 25 m **15.** 24 ft^2
17. 408 ft^2 **19.** 216 m^2 **21.** rectangular prism

23. triangular prism   **25.** 302 m^2 **27.** 377 in.^{2}
29. 152 in.^{2} **31.** 1180 cm^2

33. 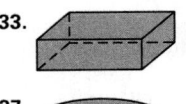   **35.**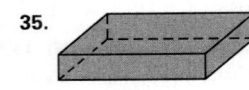

37. 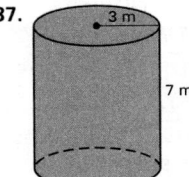   about 188 m^2

39. 112 m^2 **41.** 120 ft^2 **43.** 25 in.^{2} **45.** 106,240 m^2
49. 69 ft^2 **51.** 69 in.^{2} **53.** 15   **55.** 21   **57.** 39

9.3 Guided Practice (p. 495)
5. 13 in.   **7.** about 44 in.^{2}

9.3 Practice and Applications (pp. 495–499)
9. height   **11.** slant height   **13.** 15 mm
15. 105 m^2 **17.** about 1498 cm^2 **19.** 2304 mm^2
21. The slant height is the hypotenuse of a right triangle formed by the height of the pyramid and half the length of the base. Since the hypotenuse of a right triangle is always the longest side of the triangle, the slant height is always greater than the height.   **23.** 877 m^2 **25.** 704 m^2 **27.** 1414 yd^2
29. 196 cm^2 **31.** about 176 in.^{2}
33.   184 m^2

13 ft, 8 ft, 8 ft, 8 ft

35.   14 in.   about 855 in.^{2}

10 in.

37. right cone; about 50 cm^2 **39.** about 47 in.^{2}
41. 23 cm   **47.** 108   **49.** 15   **51.** 60   **53.** 50 m^2
55. 86 cm^2 **57.** 4   **59.** $4x - 4$ **61.** $13c - 5$

Quiz 1 (p. 499)   **1.** triangular; triangular prism; yes; 5 faces   **2.** circular; cone; no   **3.** rectangular; rectangular pyramid; yes; 5 faces   **4.** 138 ft^2
5. 242 in.^{2} **6.** 302 m^2 **7.** 300 in.^{2} **8.** 280 m^2
9. 377 cm^2

9.4 Guided Practice (p. 503)
5. about 763.2 cm^3 **7.** about 280.8 cm^3

9.4 Practice and Applications (pp. 503–507)
9. 48 unit cubes; 16 unit cubes per layer and 3 layers in all   **11.** 100 in.^{3} **13.** 216 m^3 **15.** 343 ft^3
17. 112 m^3 **19.** 78 ft^3 **21.** 350 m^3 **23.** 4 boxes
25. 1,408,000 ft^3 **27.** 452 in.^{3} **29.** 6786 m^3

31. 1810 ft^3 **33.** the pool in Exercise 32
35. The solid on the right; the solid with the vertical line of rotation has a volume that is almost twice the volume of the solid with the horizontal line of rotation.   **37.** about 216,773 gal
39. about 785 cm^3; about 6283 cm^3 **41.** 92 in.^{3}
43. 60 m^3 **45.** $7x^2$ **53.** 4.1   **55.** 188 ft^2
57. 267 yd^2 **59.** -11 **61.** 32   **63.** 17

9.5 Guided Practice (p. 513)   **5.** 35 m^3 **7.** 400 ft^3

9.5 Practice and Applications (pp. 513–516)
9. 30 ft^2 **11.** about 201 cm^2 **13.** 224 in.^{3}
15. 80 ft^3 **17.** 94 yd^3 **19.** 679 m^3 **21.** 209 cm^3
23. The volume is doubled.   **25.** The student used the slant height instead of the height of the pyramid. The volume should be $\frac{1}{3}(10 \cdot 10)(12) =$
400 in.^{3} **27.** 4 in.   **29.** 6 m   **31.** 1280 m^3
33. 302 cm^3 **35.** 48 ft^3 **37.** 3   **39.** about 173.4 in.^{3}
41. Yes; the feeder holds about 173.4 in.^{3} and only 72 in.^{3} are needed for five days.
43. about 0.04 mi^3 **47.** $C \approx 88$ in.; $A \approx 616$ in.^{2}
49. 262 in.^{2} **51.** 679 m^2 **53.** positive
55. positive   **57.** negative

9.6 Guided Practice (p. 519)
3. 113 ft^2 **5.** 113 ft^3 **7.** 2094 yd^3

9.6 Practice and Applications (pp. 520–523)
9. 3217 in.^{2} **11.** Bob wrote V rather than S for surface area, used the wrong formula, used the diameter rather than the radius, and wrote the answer in cubic units rather than square units.
$S = 4\pi r^2 = 4\pi(5^2) = 100\pi \approx 314$ mm^2.
13. 137 cm^2 **15.** 9 in.^{2} **17.** 290 cm^2 **19.** about 197,000,000 mi^2 **21.** The surface area of Earth is about 13.4 times the surface area of the moon.
23. 2145 m^3 **25.** 4189 cm^3 **27.** 1437 ft^3
29. 4 times greater; 9 times greater; 16 times greater   **31.** Answers may vary. The volume is multiplied by 8 when the radius is doubled, and the volume is multiplied by 27 when the radius is tripled.   **33.** about 23,779 ft^2 **35.** 857,375 ft^3
37. 718 cm^3 **39.** 1527 in.^{3} **41.** 905 ft^3 **43.** about 33,917 ft^3 **45.** about 3619 ft^2 **47.** 145 m^2
49. 1018 cm^2 **51.** 4.2   **53.** 20   **55.** 9.9   **57.** 11.0
59. 40 in.   **61.** 11 yd

Quiz 2 (p. 523)   **1.** 144 in.^{3} **2.** 210 ft^3 **3.** 550 m^3
4. 400 ft^3 **5.** 2145 m^3 **6.** 38 in.^{3} **7.** about 201 in.^{3}
8. about 1018 cm^2

Chapter Summary and Review (pp. 524–527)
1. base(s)   **3.** prism   **5.** volume   **7.** no   **9.** 78 ft^2
11. 264 m^2 **13.** 478 in.^{2} **15.** 302 cm^2 **17.** 283 ft^2

19. 30 m^3 **21.** 2036 ft^3 **23.** 4398 in.^{3} **25.** 314 in.^{3}
27. 531 m^2; 1150 m^3 **29.** 186 ft^2; 239 ft^3

Algebra Review (p. 531) **1.** 0.4   **3.** 0.9   **5.** 3.8
7. 2.5   **9.** 3.6   **11.** $\frac{5}{7}$ **13.** $\frac{9}{11}$ **15.** $\frac{7}{4}$ **17.** $\frac{\sqrt{2}}{3}$
19. $\frac{7\sqrt{3}}{20}$

Cumulative Practice (pp. 532–533) **1.** acute
3. $m\angle 1 = m\angle 2 = 90°$ **5.** $m\angle 5 = 99°$, $m\angle 6 = 81°$,
$m\angle 7 = 99°$ **7.** acute   **9.** SSS Congruence
Postulate; $\overline{AB} \cong \overline{DC}$ (Given); $\overline{BC} \cong \overline{BC}$ (Reflexive
Property of Congruence); $\overline{AC} \cong \overline{DB}$ (Given)
11. HL Congruence Theorem; $\overline{PR} \cong \overline{ST}$ (Given);
$\overline{QR} \cong \overline{RT}$ (Given)   **13.** always   **15.** sometimes
17. $x = 3$, $y = 70$, $z = 110$ **19.** AA Similarity
Postulate   **21.** 3   **23.** Convex; *Sample answer:*
none of the extended sides pass through the
interior.   **25.** 135°   **27.** 196 m^2 **29.** 112 cm^2
31. 304 m^2, 320 m^3 **33.** 226 cm^2, 254 cm^3
35. 3217 in.^{2}

Chapter 10

Study Guide (p. 536) **1.** B   **2.** G   **3.** B

10.1 Guided Practice (p. 539)
7. 4.5   **9.** $2\sqrt{7}$ **11.** $3\sqrt{6}$

10.1 Practice and Applications (pp. 539–541)
13. 2.4; $\sqrt{4} < \sqrt{6} < \sqrt{9}$, so $2 < 2.4 < 3$.   **15.** 5.8;
$\sqrt{25} < \sqrt{34} < \sqrt{36}$, so $5 < 5.8 < 6$.   **17.** 12.2;
$\sqrt{144} < \sqrt{148} < \sqrt{169}$, so $12 < 12.2 < 13$.   **19.** 13.6;
$\sqrt{169} < \sqrt{186} < \sqrt{196}$, so $13 < 13.6 < 14$.   **21.** 7
23. 4.1   **25.** 2.8   **27.** 5   **29.** $2\sqrt{35}$ **31.** $11\sqrt{2}$
33. 75   **35.** $3\sqrt{2}$ **37.** $4\sqrt{3}$ **39.** $2\sqrt{14}$ **41.** $10\sqrt{2}$
43. $2\sqrt{11}$ **45.** no   **47.** 62.0   **49.** 48   **51.** 15.9
57. 45°   **59.** 68°   **61.** 16   **63.** $x^2 + 5x$
65. $3x^2 + 4x$ **67.** $-3 + 3x$

10.2 Guided Practice (p. 545) **3.** $6\sqrt{2}$ **5.** $2\sqrt{3}$

10.2 Practice and Applications (pp. 545–547)
7. $7\sqrt{2}$ **9.** $\sqrt{10}$ **11.** $2\sqrt{5}$ **13.** 4   **15.** 1   **17.** 14
19. No; you cannot determine the measures of the
other two angles.   **21.** No; the triangle is
isosceles, but there is no information about the
angle measures.   **23.** The triangle has congruent
acute angles. By Example 3, the triangle is a
45°-45°-90° triangle. $x = \frac{9}{\sqrt{2}} \approx 6.4$.   **25.** By
Example 3, $x = 45$ and the triangle is a 45°-45°-90°
triangle. $y = \frac{8}{\sqrt{2}} \approx 5.7$.   **27.** By the Triangle Sum
Theorem, the third angle measures 45°. So the
triangle is a 45°-45°-90° triangle. $x = \frac{35}{\sqrt{2}} \approx 24.7$.

29. Lengths may vary, but AC, CB, and CD are all
equal. Since $\triangle ACD$ and $\triangle BCD$ are 45°-45°-90°
triangles, they are isosceles triangles by the
Converse of the Base Angles Theorem. Therefore,
$AC = CD$ and $CD = BC$.   **31.** If the hypotenuse
has length $5\sqrt{2}$, the legs should each have length
5. If the legs have length $\sqrt{5}$, then the hypotenuse
has length $\sqrt{5} \cdot \sqrt{2}$, or $\sqrt{10}$.   **33.** the right triangle
with legs of length 1 and hypotenuse of length
$r = \sqrt{2}$ **35.** right   **37.** right   **39.** $3\sqrt{7}$ **41.** 8
43. 14   **45.** $3\sqrt{13}$ **47.** 0.6   **49.** 0.33   **51.** 0.15
53. 0.1666... $\approx$ 0.17

10.3 Guided Practice (p. 552) **3.** false   **5.** true
7. true   **9.** 4

10.3 Practice and Applications (pp. 552–555)
11. 6   **13.** 22   **15.** 2   **17.** $4\sqrt{3}$ **19.** $13\sqrt{3}$
21. $16\sqrt{3}$ **23.** 2.3   **25.** 10.4   **27.** 14   **29.** 30 feet
31. $x = 4$; $y = 4\sqrt{3}$ **33.** $x = \frac{9}{2}$; $y = \frac{9}{2}\sqrt{3}$ **35.** $x = \frac{15}{2}$;
$y = \frac{15}{2}\sqrt{3}$ **37.** about 390 ft^2 **39.** about 21 cm^2
41. hypotenuse, 30 cm; longer leg, $15\sqrt{3}$ cm
45. $\frac{5}{3}$ **47.** $\frac{5}{8}$ **49.** yes; $\triangle ABC \sim \triangle FHG$ **51.** -16
53. 9   **55.** 37

Quiz 1 (p. 555) **1.** $2\sqrt{6}$ **2.** $\sqrt{30}$ **3.** 12   **4.** 20
5. $3\sqrt{3}$ **6.** $4\sqrt{11}$ **7.** $2\sqrt{13}$ **8.** $6\sqrt{5}$ **9.** $\frac{2}{\sqrt{3}}$ **10.** $\sqrt{6}$
11. $x = 14$; $y = 14\sqrt{3}$

10.4 Guided Practice (p. 560)
3. $\tan A = \frac{4}{3} \approx 1.3333$ **5.** $\tan A = \frac{1}{\sqrt{3}} \approx 0.5774$
7. 1.8807   **9.** 0.9325

10.4 Practice and Applications (pp. 560–562)
11. $\sqrt{3}$ **13.** $\tan P = \frac{7}{24} \approx 0.2917$; $\tan R = \frac{24}{7} \approx$
3.4286   **15.** $\tan P = \frac{4}{3} \approx 1.3333$; $\tan R = \frac{3}{4} = 0.75$
17. 1.3764   **19.** 57.2900   **21.** 0.8391   **23.** 8.1443
25. $\tan 39° = \frac{33}{x}$, $\tan 51° = \frac{x}{33}$; 40.8   **27.** 59.8
29. 15.1   **31.** 13.4 m   **33.** 12.2   **35.** 29.4   **37.** 11.9
41. 1018 m^3 **43.** 513 in.^{3} **45.** 5   **47.** -3 **49.** 13

10.5 Guided Practice (p. 566)
5. $\frac{4}{5} = 0.8$ **7.** $\frac{4}{3} \approx 1.3333$ **9.** $\frac{4}{5} = 0.8$

10.5 Practice and Applications (pp. 566–568)
11. $\sin A = \frac{11}{61}$; $\cos A = \frac{60}{61}$ **13.** $\sin A = \frac{12}{13}$;
$\cos A = \frac{5}{13}$ **15.** $\sin P = \frac{6\sqrt{2}}{11} \approx 0.7714$;
$\cos P = \frac{7}{11} \approx 0.6364$ **17.** 0.6428   **19.** 0.9848
21. 0.8572   **23.** 1   **25.** $x \approx 5.8$; $y \approx 5.6$ **27.** $x \approx 5.3$;
$y \approx 9.6$ **29.** $x \approx 9.0$; $y \approx 12.0$ **31.** 14 ft

33. Answers will vary. **35.** $(\sin A)^2 + (\cos A)^2 = 1$
37. Both students will get the correct answer.
41. $x = 7\sqrt{3}; y = 14$ **43.** 0.6249 **45.** 1.4826
47. 0.3057 **49.** $-18, -8, -1.8, -0.8, 0, 0.08, 1.8$
51. $-0.61, -0.6, -0.56, -0.5, -0.47$

10.6 Guided Practice (p. 572) **3.** false **5.** 6.4
7. 79.6° **9.** 67.4° **11.** $d \approx 7.2; m\angle D \approx 31.0°$;
$m\angle E \approx 59.0°$

10.6 Practice and Applications (pp. 573–575)
13. 45° **15.** 13.0° **17.** 76.4° **19.** *Sample answer:*
Use the inverse tangent function; 48.9°.
21. $x = 7\sqrt{2} \approx 9.9; m\angle A = 45°$ **23.** $x = 29$;
$m\angle A \approx 43.6°$ **25.** 81.3° **27.** 63.3° **29.** 39.5°
31. *Sample answer:* 28 ft (Any length greater than
or equal to 27.6 ft will meet the standards.)
33. 14.9° **35.** 36.9° **37.** $LM \approx 5.7; m\angle L \approx 70.3°$;
$m\angle K \approx 19.7°$ **39.** $ST \approx 13.7; m\angle R \approx 66.4°$;
$m\angle S \approx 23.6°$ **43.** 50 cm; 201 cm² **45.** 107 yd;
908 yd² **47.** 11,494 cm³ **49.** 0.554 **51.** 32.55
53. 2000

Quiz 2 (p. 575) **1.** 6.7 **2.** 3.9 **3.** $x \approx 9.2; y \approx 7.7$
4. $x \approx 6.8; y \approx 17.7$ **5.** 7.8 **6.** $x \approx 4.6; y \approx 6.6$
7. 3.0777 **8.** 0.7880 **9.** 0.8090 **10.** $m\angle N = 50°$;
$m \approx 13.4; q \approx 20.9$ **11.** $m\angle P \approx 61.0°; m\angle Q \approx 29.0°$;
$q \approx 3.9$ **12.** $m\angle K \approx 76.0°; m\angle L \approx 14.0°; k \approx 12.0$

Chapter Summary and Review (pp. 576–579)
1. radical **3.** trigonometric ratio **5.** solve
7. tangent **9.** sine **11.** 12 **13.** 112 **15.** 5
17. 99 **19.** $6\sqrt{2}$ **21.** $2\sqrt{17}$ **23.** $4\sqrt{5}$ **25.** $13\sqrt{3}$
27. 50.9 **29.** $15\sqrt{2}$ **31.** $\sqrt{14}$ **33.** 3 **35.** $x = 50$;
$y = 25\sqrt{3} \approx 43.3$ **37.** $x = 94; y = 188$
39. $\tan A = \dfrac{8}{15} \approx 0.5333; \tan B = \dfrac{15}{8} = 1.875$
41. $\tan A = \dfrac{3}{2} = 1.5; \tan B = \dfrac{2}{3} \approx 0.6667$
43. 6.3138 **45.** 0.4452 **47.** $\sin A = \dfrac{5}{9} \approx 0.5556$;
$\cos A = \dfrac{2\sqrt{14}}{9} \approx 0.8315$ **49.** 0.8387 **51.** 0.8572
53. $x \approx 5.2; y \approx 8.6$ **55.** 5.1° **57.** 50.4°
59. $q \approx 26.9; m\angle P \approx 58.7°; m\angle R \approx 31.3°$

Algebra Review (p. 583) **1.** $(-2, 3)$ **3.** $(-5, 18)$
5. $(1, 2)$ **7.** $(1, 1)$ **9.** $(3, -4)$

Chapter 11

Study Guide (p. 588) **1.** D **2.** F **3.** D

11.1 Guided Practice (p. 591) **3.** E **5.** A **7.** F
9. $(3, 0)$ and $(3, 6)$ **11.** $(0, 3)$ and $(3, 0)$

11.1 Practice and Applications (pp. 591–593)
13. 7.5 cm **15.** 1.5 ft **17.** 52 in. **19.** 17.4 m
21. chord **23.** diameter **25.** point of tangency

27. chord **29.** diameter **31.** tangent **33.** $\overline{EG}$ is a
chord $\left(\text{as is } \overline{EF}\right); \overleftrightarrow{EG}$ is a secant; $\overline{EF}$ is a diameter;
$\overline{CE}$ is a radius $\left(\text{as are } \overline{CF} \text{ and } \overline{CG}\right); D$ is a point of
tangency. **35.** $\overline{LM}$ is a chord $\left(\text{as is } \overline{PN}\right); \overleftrightarrow{LM}$ is a
secant; $\overline{PN}$ is a diameter; $\overline{QR}$ is a radius $\big($as are
$\overline{QP}$ and $\overline{QN}\big); K$ is a point of tangency. **37.** any
two of $\overline{GD}, \overline{HC}, \overline{FA},$ and $\overline{EB}$ **39.** Yes; $\overleftrightarrow{JK}$ is a
tangent through J, G, and K. **41.** 2; 2 **43.** $\odot A$:
$(3, 2); \odot B$: $(3, 3)$; intersection: $(3, 0)$; tangent line:
x-axis **45.** 2; 4
47.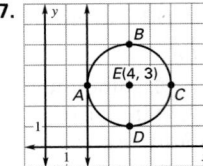

51. SSS Congruence
Postulate; three sides of
$\triangle JKL$ are congruent to
three sides of $\triangle PQR$,
so $\triangle JKL \cong \triangle PQR$ by
the SSS Congruence
Postulate.

53. yes **55.** 5.7 **57.** 6.3 **59.** 9.9 **61.** 15.8 **63.** 7
65. -1 **67.** -9

11.2 Guided Practice (p. 598) **3.** No; $5^2 + 5^2 \neq 7^2$,
so $\triangle ABD$ is not a right triangle and $\overline{AB}$ is not
perpendicular to $\overline{BD}$. Therefore, $\overline{BD}$ is not
tangent to $\odot C$. **5.** 2

11.2 Practice and Applications (pp. 598–600) **7.** 3
9. 15 **11.** 15 **13.** $x^2 + 4x + 4$ **15.** $x^2 + 14x + 49$
17. 3 **19.** 20 **21.** No; $5^2 + 14^2 \neq 15^2$, so $\triangle ABC$ is
not a right triangle and $\overline{AB}$ is not perpendicular
to $\overline{AC}$. Therefore, $\overline{AB}$ is not tangent to $\odot C$.
23. $\angle ABC \cong \angle ADC; \angle DAC \cong \angle BAC$;
$\angle BCA \cong \angle DCA$.
25.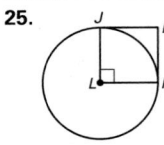

27. $BA = BC \approx 15{,}977$ miles
31. Yes; $5 + 11 > 14$,
$5 + 14 > 11$, and $11 + 14 > 5$.
33. Yes; $15 + 3 > 13$, $15 + 13 > 3$,
and $3 + 13 > 15$.
35. No; $3 + 7 = 10$.

37. 10 **39.** 19 **41.** $-\dfrac{2}{3}$ **43.** $\dfrac{5}{2}$ **45.** $\dfrac{1}{3}$

11.3 Guided Practice (p. 604) **5.** 300° **7.** 100°
9. 40° **11.** 10.47 cm

11.3 Practice and Applications (pp. 604–607)
13. $\overparen{PQ}$; 135° **15.** $\overparen{LN}$; 150° **17.** $\overparen{WXY}$; 200°
19. minor arc **21.** semicircle **23.** major arc
25. major arc **27.** 165° **29.** 180° **31.** 55°
33. 305° **35.** 65° **37.** 60° **39.** 120° **41.** 90°
43. No; the circles are not congruent. **45.** Yes;
$\overparen{UW}$ and $\overparen{XZ}$ are arcs of congruent circles with the
same measure. **47.** 2.36 cm **49.** 20.94 ft

51. 15.71 m   **53.** No; they have the same arc length only if the two circles are congruent circles.   **57.** $x \approx 5.0$; $y \approx 7.5$ **59.** $\frac{1}{200}$ **61.** $\frac{1}{3}$

Quiz 1 (p. 607)   **1.** tangent   **2.** secant   **3.** diameter   **4.** chord   **5.** radius   **6.** point of tangency   **7.** 15   **8.** 6   **9.** 4   **10.** $\frac{7}{6}\pi \approx 3.67$ cm   **11.** $\frac{35}{6}\pi \approx 18.33$ m   **12.** $\frac{25}{18}\pi \approx 4.36$ ft

11.4 Guided Practice (p. 610)   **3.** 7   **5.** 8

11.4 Practice and Applications (pp. 611–612)   **7.** No; $\overline{AB}$ does not bisect $\overline{CD}$, so $\overline{AB}$ is not a diameter of the circle.   **9.** 7   **11.** 3   **13.** $\overarc{AB} \cong \overarc{CD}$ (given) and $\overline{AB} \cong \overline{CD}$ (If two minor arcs are congruent, then their corresponding chords are congruent.)   **15.** 10   **17.** 170°   **19.** 7   **23. a.** In a circle, if two chords are congruent, then their corresponding minor arcs are congruent.   **b.** 10   **c.** $m\overarc{AD} = m\overarc{BE} = 110°$ **d.** 100°   **25.** 75°   **27.** 140°   **29.** 220°   **31.** $\frac{15}{20} > -\frac{3}{4}$

11.5 Guided Practice (p. 617)   **3.** 40°   **5.** 210°   **7.** $y = 150$; $z = 75$

11.5 Practice and Applications (pp. 617–619)   **9.** 55°   **11.** 90°   **13.** 67°   **15.** 64°   **17.** 228°   **19.** 100°   **21.** 94°   **23.** 53°   **25.** 53°   **27.** Yes. Explanations may vary; *Sample answer:* $m\angle BAC = 47° = m\angle CDE$ (from Ex. 22) and $m\angle DCE = m\angle ACB$ (vertical angles), so $\triangle ABC \sim \triangle DEC$ by the AA Similarity Postulate.

29. $\triangle KLM$ is an inscribed triangle and $\overline{KM}$ is a diameter, so $\triangle KLM$ is a right triangle with hypotenuse $\overline{KM}$; $x = 90$; $y = 90 - 40 = 50$.

31. *Sample answer:* Position the vertex of the tool on the circle and mark the two points at which the sides intersect the circle; draw a segment to connect the two points, forming a diameter of the circle. Repeat these steps, placing the vertex at a different point on the circle. The center is the point at which the two diameters intersect.

33. $x = 80$; $y = 78$ **35.** Yes; both pairs of opposite angles are right angles, which are supplementary angles.   **37.** No; if a rhombus is not a square, then the opposite angles are not supplementary.

41. $\sqrt{35}$ **43.** $2\sqrt{21}$ **45.** 27   **47.** $m\angle B = 46°$; $AC \approx 8.3$; $AB \approx 11.5$ **49.** $m\angle R = 40°$; $RP \approx 6.0$; $QR \approx 7.8$ **51.** 9   **53.** -12 **55.** 12

11.6 Guided Practice (p. 623)   **3.** 88°   **5.** 3   **7.** 12

11.6 Practice and Applications (pp. 623–625)   **9.** B   **11.** A   **13.** 50   **15.** 21   **17.** 186   **19.** 18   **21.** 12   **23.** 103   **25.** 60   **27. a.** 4   **b.** 90°   **c.** 144°   **d.** Yes. Explanations may vary. *Sample answer:* $\angle ACB \cong \angle ECD$ (vertical angles); $\frac{AC}{EC} = \frac{12}{9} = \frac{4}{3} = \frac{BC}{CD}$; by the SAS Similarity Theorem, $\triangle ACB \sim \triangle ECD$.   **29.** 34   **31.** 3   **33.** 19   **35.** 2.7   **37.** 10.01

Quiz 2 (p. 625)   **1.** 5   **2.** 3   **3.** 31   **4.** 102; the measure of an inscribed angle is half the measure of its intercepted arc, so $51 = \frac{1}{2}x$; $x = 102$.

5. $x = 58$ and $y = 41$; $x° = m\angle B = \frac{1}{2}m\overarc{AD} = m\angle C = 58°$; $y° = m\angle D = \frac{1}{2}m\overarc{BC} = m\angle A = 41°$.

6. $x = 75$ and $y = 82$; the opposite angles of an inscribed quadrilateral are supplementary, so $x° = 180° - 105° = 75°$ and $y° = 180° - 98° = 82°$.   **7.** 62   **8.** 163   **9.** 6

11.7 Guided Practice (p. 629)   **3.** radius, 4; center, (2, 0); equation, $(x - 2)^2 + y^2 = 16$

11.7 Practice and Applications (pp. 630–632)   **5.** B   **7.** C

9. radius, 1; center, (0, 0);

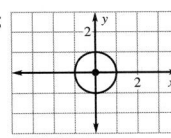

11. radius, 4; center, (4, 3);

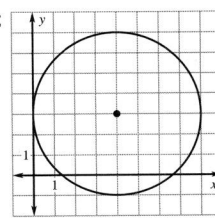

13. radius, 6; center, (−2, 3);

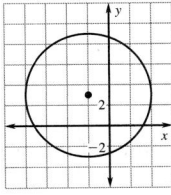

15. radius, 8; center, (0, 5);

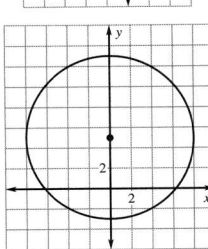

17. radius, 2; center, (0, 1); equation, $x^2 + (y - 1)^2 = 4$ **19.** radius, 2.5; center, (0.5, 1.5); equation, $(x - 0.5)^2 + (y - 1.5)^2 = 6.25$ **21.** radius, 6; center, (0, 0); equation, $x^2 + y^2 = 36$ **23.** $(x - 4)^2 + y^2 = 16$ **25.** $(x + 1)^2 + (y + 3)^2 = 36$ **27.** $(x - 1)^2 + y^2 = 49$ **29.** inside **31.** outside **33.** on **35.** inside **37.** Tower A transmits to J; tower B transmits to K; towers B and C transmit to L; no tower transmits to M; tower C transmits to N. **43.** $P'(3, 3); Q'(8, 1); R'(6, -2); S'(1, 0)$ **45.** $P'(0, 2); Q'(5, 0); R'(3, -3); S'(-2, -1)$ **47.** reduction; $\frac{3}{8}$ **49.** -7 **51.** 8

11.8 Guided Practice (p. 636) **3.** yes; a clockwise or counterclockwise rotation of 180° about its center **5.** no **7.** P **9.** R

11.8 Practice and Applications (pp. 636–639) **11.** yes; a clockwise or counterclockwise rotation of 72° or 144° about its center **13.** The wheel hub can be mapped onto itself by a clockwise or counterclockwise rotation of 72° or 144° about its center. **15.** The wheel hub can be mapped onto itself by a clockwise or counterclockwise rotation of 45°, 90°, 135°, or 180° about its center.

17. 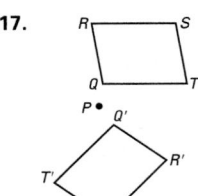 **19.**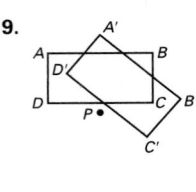

21. 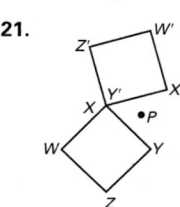 **23.** $\overline{LH}$ **25.** $\triangle MAB$ **27.** $J'(1, 2); K'(4, 1); L'(4, -3); M'(1, -3);$ the coordinates of the image of the point (x, y) after a 90° clockwise rotation about the origin are $(y, -x)$.

29. $A'(-1, -1); B'(-4, 2); C'(-2, 5);$ the coordinates of the image of the point (x, y) after a 90° counterclockwise rotation about the origin are $(-y, x)$. **31.** The design has rotational symmetry about its center; it can be mapped onto itself by a clockwise or counterclockwise rotation of 180°. **33.** Yes. The image can be mapped onto itself by a clockwise or counterclockwise rotation of 180° about its center. **35.** the center of the circle **39.** 91 ft^2 **41.** 80 m^2 **43.** 9.5 **45.** 0

Quiz 3 (p. 639) **1.** center, $(-1, 6)$; radius, 5 **2.** $x^2 + (y + 4)^2 = 9$

3. 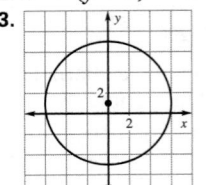 **4.**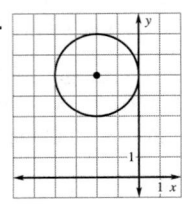

5. 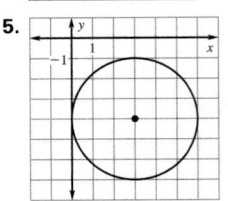 **6.**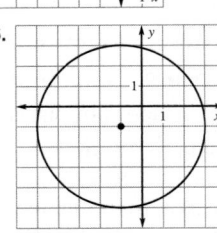

7. yes; a clockwise or counterclockwise rotation of 90° or 180° about its center **8.** yes; a clockwise or counterclockwise rotation of 180° about its center **9.** no **10.** $A'(-2, -4); B'(-1, -1); C'(1, -3);$ the coordinates of the image of the point (x, y) after a 180° rotation about the origin are $(-x, -y)$. **11.** $A'(-4, 1); B'(-4, 4); C'(-1, 4); D'(-1, 2);$ the coordinates of the image of the point (x, y) after a 90° counterclockwise rotation about the origin are $(-y, x)$. **12.** $A'(0, -2); B'(-1, -4); C'(-3, -3);$ the coordinates of the image of the point (x, y) after a 90° clockwise rotation about the origin are $(y, -x)$.

Chapter Summary and Review (pp. 641–645) **1.** secant **3.** tangent **5.** inscribed angle **7.** rotation **9.** point of tangency **11.** center **13.** secant **15.** 10 **17.** 6 **19.** 7 **21.** 118° **23.** 85° **25.** 324° **27.** 16.76 cm **29.** 3 **31.** 35 **33.** 60 **35.** 53° **37.** 200° **39.** $x = 68; y = 99$ **41.** 95 **43.** 70 **45.** 5 **47.** $(x - 2)^2 + (y - 5)^2 = 9$ **49.** $(x - 5)^2 + (y + 2)^2 = 49$ **51.** radius, 4; center, $(2, -3)$;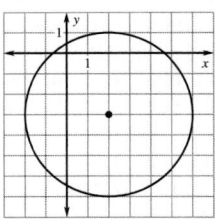

53.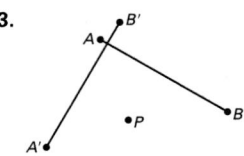

55.

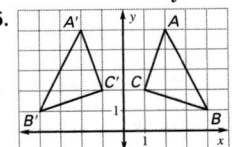

Cumulative Practice (pp. 649–651) **1.** 15 **3.** 118°
5. (1, 7) **7.** 20°; 110° **9.** 42°; 132° **11.** $x = 7$;
$m\angle APB = 28°$; $m\angle BPC = 152°$ **13.** Alternate
Exterior Angles Converse **15.** Same-Side Interior
Angles Converse **17.** 28° and 28°; Since the
triangle is isosceles, the base angles are
congruent. By the Triangle Sum Theorem, you
can show that each base angle has measure 28°.
19. smallest, $\angle A$; largest, $\angle B$ **21.** no **23.** Yes;
$\triangle XYU \cong \triangle VWU$ by the HL Congruence Theorem.
25. $x = 117$; $y = 117$ **27.** $x = 110$; $y = 70$; $z = 110$
29. *Sample answer:* $\angle CBD \cong \angle CAE$ and $\angle BCD \cong$
$\angle ACE$. Therefore, $\triangle BCD \sim \triangle ACE$ by the AA
Similarity Postulate. **31.** 312 m² **33.** 80 cm²
35. 6 m **37.** 628 cm²; 1005 cm³ **39.** 1810 mm²;
7238 mm³ **41.** $m\angle R \approx 22.6°$; $m\angle T \approx 67.4°$
43. $x = 25$; $p \approx 3.4$; $r \approx 7.3$ **45.** $d = 21$; $x \approx 46.4°$;
$y \approx 43.6°$ **47.** $m\widehat{AB} = 75°$; $m\widehat{ACB} = 285°$ **49.** 37
51. 145 **53.** $x^2 + (y + 3)^2 = 49$
55.

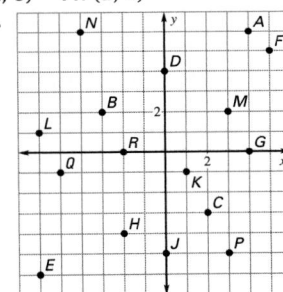

Skills Review Handbook

Problem Solving (p. 654) **1.** $197.46 **3.** at least
16 flower arrangements **5.** $5.96 **7.** 24 different
orders **9.** 20 diagonals

Decimals (p. 655)
1. 2.08 **3.** 180 **5.** 47.5 **7.** 28.26 **9.** 10 **11.** 8

Fraction Concepts (p. 656)
1–5. Sample answers are given. **1.** $\frac{1}{2}, \frac{2}{4}$ **3.** $\frac{2}{8}, \frac{3}{12}$
5. $\frac{6}{20}, \frac{15}{50}$ **7.** $\frac{2}{3}$ **9.** $\frac{1}{3}$ **11.** $\frac{5}{2}$ **13.** $\frac{3}{4}$ **15.** 1

Fractions and Decimals (p. 657)
1. 0.7 **3.** $0.\overline{3} \approx 0.33$ **5.** $0.\overline{2} \approx 0.22$ **7.** 0.125 **9.** 0.5
11. $\frac{1}{2}$ **13.** $\frac{1}{5}$ **15.** $\frac{3}{8}$ **17.** $\frac{2}{3}$ **19.** $\frac{2}{15}$

Adding and Subtracting Fractions (p. 658)
1. $\frac{6}{7}$ **3.** $\frac{1}{3}$ **5.** $\frac{1}{2}$ **7.** $\frac{4}{5}$ **9.** $\frac{5}{8}$ **11.** $1\frac{1}{4}$ **13.** $\frac{11}{35}$
15. $\frac{7}{24}$ **17.** 1 **19.** $1\frac{22}{45}$ **21.** $\frac{13}{15}$ **23.** $\frac{13}{20}$

Multiplying and Dividing Fractions (p. 659)
1. $\frac{3}{8}$ **3.** $2\frac{1}{2}$ **5.** $\frac{5}{27}$ **7.** 15 **9.** $\frac{2}{3}$ **11.** 63 **13.** $\frac{1}{4}$
15. $114\frac{2}{7}$ **17.** 2 **19.** 154 **21.** 1 **23.** 1 **25.** 8 **27.** 4

Ratio and Proportion (p. 661) **1.** $\frac{5}{1}$ **3.** $\frac{2}{3}$ **5.** $\frac{2}{25}$
7. $\frac{12}{5}$ **9.** $\frac{1}{1}$ **11.** $\frac{16}{5}$ **13.** $\frac{20}{9}$ **15.** $\frac{3}{5}$ **17.** $12\frac{1}{2}$
19. $\frac{4}{5}$ **21.** $2\frac{4}{7}$ **23.** 0.3 or $\frac{3}{10}$ **25.** 25 **27.** 14
29. $2\frac{2}{5}$

Inequalities and Absolute Value (p. 662)
1. $6 > -6$ **3.** $-2 > -8$ **5.** $-7.8 < -7.6$
7. $0.01 < 0.1$ **9.** $-8, -6, -2, 0, 3, 4, 9$ **11.** 5124,
5142, 5214, 5412, 5421 **13.** 6 **15.** 0 **17.** 1.4

Integers (p. 663) **1.** -10 **3.** 2 **5.** 11 **7.** 0 **9.** -16
11. 60 **13.** -40 **15.** 2 **17.** 16 **19.** -6 **21.** -100
23. -12 **25.** -6 **27.** 9

The Coordinate Plane (p. 664)
1. $(-1, 0)$ **3.** $(2, -2)$ **5.** $(-5, 3)$ **7.** $(5, -2)$
9. $(-2, 5)$ **11.** $(1, 1)$
13–28.

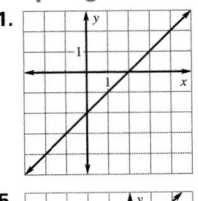

Slope of a Line (p. 665) **1.** $-\frac{1}{3}$ **3.** 2 **5.** 1 **7.** $-\frac{2}{3}$
9. 0 **11.** negative **13.** zero **15.** positive

Graphing Linear Equations (p. 666)
1.

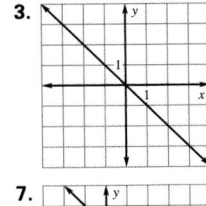

3.

5.

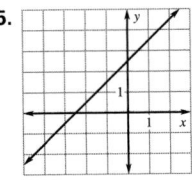

7.

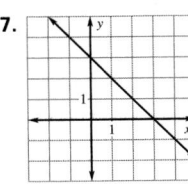

9. 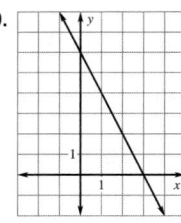   **11.**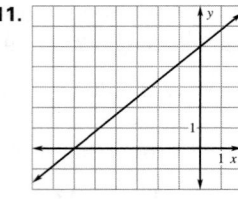

Slope-Intercept Form (p. 667)

1. 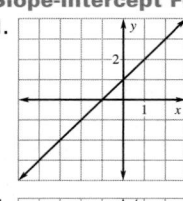   **3.**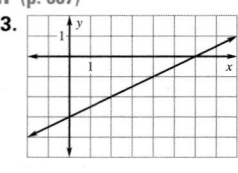

5. 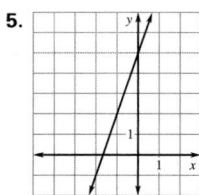   **7.**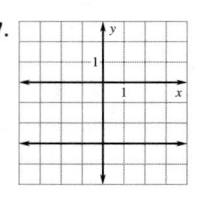

9. parallel   **11.** not parallel

Powers and Square Roots (p. 669) **1.** 64 **3.** −1
5. 32 **7.** 81 **9.** 10, −10 **11.** 1, −1 **13.** no real
square roots **15.** no real square roots **17.** 10
19. 3.9 **21.** 2 **23.** 3.3 **25.** 6.7 **27.** 0 **29.** $2\sqrt{7}$
31. $5\sqrt{2}$ **33.** $\frac{\sqrt{5}}{4}$ **35.** $\frac{1}{3}$ **37.** $\frac{\sqrt{3}}{3}$ **39.** $\frac{5\sqrt{2}}{2}$

Evaluating Expressions (p. 670) **1.** 105 **3.** 21
5. −8 **7.** 13 **9.** 33 **11.** 81 **13.** 5 **15.** 60
17. −24

The Distributive Property (p. 671) **1.** $2a + 8$
3. $3x − 2$ **5.** $y^2 − 9y$ **7.** $4b^2 + 12b$ **9.** $6m + 4$
11. $x^2 + 5x − 14$ **13.** $10 − 2x$ **15.** $3h − 3h^2$
17. $3y^2 − 6y$

Solving One-Step Equations (p. 672)
1. 6 **3.** −8 **5.** 13 **7.** 3 **9.** −28 **11.** −5 **13.** 4
15. −1 **17.** 24 **19.** −5

Solving Multi-Step Equations (p. 673) **1.** 4 **3.** 5
5. −7 **7.** 26 **9.** −4 **11.** −8 **13.** 22 **15.** 13
17. $\frac{1}{2}$ **19.** 1 **21.** 4 **23.** 2 **25.** $-\frac{1}{3}$ **27.** −4
29. 12 **31.** −23 **33.** 1

Using Formulas (p. 674) **1.** 6 m **3.** 22 mm
5. 4 in. **7.** 9 m

Extra Practice

Chapter 1 (pp. 675–676)

1.   **3.** Each number is 5 times the previous number; 625. **5.** Each number is 3 more than the previous number; 4.

7. Multiply the previous number by $\frac{2}{3}$; $\frac{16}{81}$.

9. positive **11.** The absolute value of 0 is 0.

13. *Sample answer:* $2 \div 6 = \frac{1}{3}$, which is not an

integer. **15.** *Sample answer:* $\overset{\bullet\quad\bullet\quad\bullet}{\underset{P\quad Q\quad R}{\longrightarrow}}$

17. R **19.** $\overrightarrow{PQ}, \overrightarrow{PS}, \overrightarrow{PM}$ **21.** $\overleftrightarrow{NM}$

23. *Sample answer:*

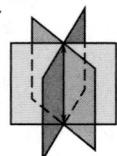

25. 18 **27.** yes **29.** Q; $\overrightarrow{QP}, \overrightarrow{QR}$; any two of $\angle PQR$, $\angle RQP, \angle Q$ **31.** B; $\overrightarrow{BA}, \overrightarrow{BC}$; $\angle ABC, \angle CBA$ **33.** 65°

Chapter 2 (pp. 677–678) **1.** 30; 15 **3.** 14 **5.** (3, 0)
7. (1.5, 3) **9.** (−1.5, −1.5) **11.** 50° **13.** neither;
adjacent **15.** 37°; 127° **17.** $\angle 1$ and $\angle 3$; $\angle 1$ and
$\angle 3$ are both supplementary to $\angle 2$, so $\angle 1 \cong \angle 3$
by the Congruent Supplements Theorem.
19. $m\angle 1 = 95°$; $m\angle 2 = 85°$; $m\angle 3 = 95°$ **21.** If you

read it in a newspaper, then it must be true.

23. If a number is odd, then its square is odd.

25. If it is hot today, then I will go to the beach.
27. Addition Property of Equality **29.** Transitive
Property of Equality **31.** Subtraction Property
of Equality

Chapter 3 (pp. 679–680) **1–5.** Sample answers are
given. **1.** $\overleftrightarrow{BC}, \overleftrightarrow{GF}, \overleftrightarrow{HE}$ **3.** $\overleftrightarrow{EC}, \overleftrightarrow{HB}, \overleftrightarrow{HG}$
5. *GAB, FDC* **7.** no **9.** yes **11.** yes **13.** 15
15. alternate exterior angles **17.** alternate
interior angles **19.** alternate exterior angles
21. $m\angle 1 = 140°$, by the Same-Side Interior Angles
Theorem; $m\angle 2 = 40°$, by the Vertical Angles
Theorem **23.** $m\angle 1 = 95°$, by the Linear Pair
Postulate; $m\angle 2 = 85°$, by the Corresponding
Angles Postulate **25.** $m\angle 1 = 110°$, by the
Alternate Exterior Angles Theorem; $m\angle 2 = 110°$,
by the Corresponding Angles Postulate **27.** yes
29. yes **31.** Same-Side Interior Angles Converse
33. 13 **35.** 5 **37.** no **39.** yes **41.** (−3, 5)
43. (−5, 11) **45.** (−10, 2)

Chapter 4 (pp. 681–682) **1.** scalene **3.** scalene
5. obtuse **7.** $\overline{AB}$ is opposite $\angle C$; $\overline{BC}$ is opposite
$\angle A$; $\overline{AC}$ is opposite $\angle B$. **9.** 79° **11.** 60° **13.** 18
15. 55°, 55°, 70° **17.** 15 **19.** 10 **21.** 3.2 **23.** right
25. obtuse **27.** right **29.** right **31.** acute
33. $JP = 14$; $JM = 21$ **35.** $\angle M$, $\angle K$, $\angle L$ **37.** $\angle R$,
$\angle S$, $\angle T$ **39.** $\overline{DE}$, $\overline{EF}$, $\overline{DF}$ **41.** Yes; $7 + 7 > 7$.
43. No; $1 + 2 < 4$.

Chapter 5 (pp. 683–684) **1.** $\overline{RS} \cong \overline{PN}$, $\overline{ST} \cong \overline{NM}$,
$\overline{TR} \cong \overline{MP}$ **3.** $\triangle RST \cong \triangle PNM$ **5.** yes;
$\triangle XVW \cong \triangle XYZ$ **7.** yes; SSS Congruence
Postulate **9.** yes; SAS Congruence Postulate
11. yes; ASA Congruence Postulate **13.** no
15. yes; ASA Congruence Postulate
17.

Statements	Reasons
1. $\overline{EG} \cong \overline{HG}$; $\overline{GF} \cong \overline{GJ}$	1. Given
2. $\angle EGF \cong \angle HGJ$	2. Vertical $\angle$ are $\cong$.
3. $\triangle GEF \cong \triangle GHJ$	3. SAS Congruence Postulate
4. $\angle GEF \cong \angle GHJ$	4. Corresp. parts of $\cong \triangle$ are $\cong$.

19.

Statements	Reasons
1. $\overline{QT} \cong \overline{ST}$; $\overline{RQ} \cong \overline{RS}$	1. Given
2. $\overline{RT} \cong \overline{RT}$	2. Reflexive Property of Congruence
3. $\triangle RQT \cong \triangle RST$	3. SSS Congruence Postulate
4. $\angle RQT \cong \angle RST$	4. Corresp. parts of $\cong \triangle$ are $\cong$.

21. 7 **23.** 10 **25.** neither **27.** 1 **29.** 0

Chapter 6 (pp. 685–686) **1.** No; one of the sides is
not a segment. **3.** yes; hexagon **5.** 64 **7.** $\overline{XY}$;
opposite sides of a parallelogram are congruent.
9. $\overline{VY}$; opposite sides of a parallelogram are
congruent. **11.** $\overline{VY}$; by definition, a
parallelogram is a quadrilateral with both pairs
of opposite sides parallel. **13.** $\angle YVW$ and $\angle YXW$;
the consecutive angles of a parallelogram are
supplementary. **15.** $m\angle A = m\angle C = 38°$,
$m\angle D = 142°$ **17.** $m\angle B = m\angle D = 86°$, $m\angle A = 94°$
19. Yes; opposite angles are congruent. **21.** Yes;
opposite sides are congruent. **23.** Yes; the
diagonals bisect each other. **25.** parallelogram,
rectangle, rhombus, square **27.** rectangle, square
29. rhombus, square **31.** $m\angle A = 70°$; $m\angle B = 110°$;
$m\angle D = 70°$ **33.** $m\angle F = 125°$; $m\angle G = 110°$

35. 23 **37.** No; opposite sides are not congruent.
39. No; it is not given that the sides are all
congruent. **41.** rectangle

Chapter 7 (pp. 687–688) **1.** $\frac{20}{1}$ **3.** $\frac{5}{8}$ **5.** 20 ft by 15 ft
7. 7 **9.** 32 **11.** 7 **13.** 2 **15.** $\angle P \cong \angle T$; $\angle Q \cong \angle V$;
$\angle R \cong \angle W$; $\angle S \cong \angle X$ **17.** 9 **19.** 10 **21.** yes;
$\triangle ABC \sim \triangle DEF$ **23.** no **25.** $\triangle RQS \sim \triangle RPT$
27. no **29.** $\angle M \cong \angle M$ by the Reflexive Property of
Congruence. Also, $\frac{15}{15 + 5} = \frac{3}{4}$ and $\frac{21}{21 + 7} = \frac{3}{4}$.
The lengths of the sides that include $\angle M$ are
proportional. So, by the SAS Similarity Theorem,
the triangles are similar. **31.** 12 **33.** 17 **35.** Yes;
$\frac{4}{8} = \frac{1}{2}$. **37.** reduction; 22 **39.** enlargement; 12.5

Chapter 8 (pp. 689–690) **1.** concave **3.** concave
5. No; the polygon is equilateral, but it is not
equiangular. **7.** 1800° **9.** 2880° **11.** 108 **13.** 60
15. 46 **17.** 112 in.^{2} **19.** 80 m^2 **21.** 49 in.^{2}
23. 7.5 yd^2 **25.** 10 ft **27.** 108 in.^{2} **29.** 55 ft^2
31. 216 m^2 **33.** 31 in.; 79 in.^{2} **35.** 82 ft; 531 ft^2
37. 20 ft^2

Chapter 9 (pp. 691–692) **1.** yes; pentagon;
pentagonal pyramid **3.** yes; rectangle;
rectangular prism **5.** hexagonal pyramid; 7 faces;
12 edges **7.** 220 cm^2 **9.** 120 in.^{2} **11.** 339 cm^2
13. 85 in.^{2} **15.** 800 ft^2 **17.** 283 cm^2 **19.** 200 cm^3
21. 60 in.^{3} **23.** 424 cm^3 **25.** 480 cm^3 **27.** 12 m^3
29. 871 ft^3 **31.** 804 cm^2; 2145 cm^3 **33.** 4072 in.^{2};
24,429 in.^{3} **35.** $\frac{4}{3}\pi$ cm^3, 36π cm^3; $1:27$

Chapter 10 (pp. 693–694) **1.** 7.6 **3.** 6.4 **5.** 10.3
7. 7 **9.** 18 **11.** $3\sqrt{11}$ **13.** $5\sqrt{6}$ **15.** $7\sqrt{2}$
17. $x = 45$; $y = 5\sqrt{2}$; By the Triangle Sum Theorem,
$x° + 45° + 90° = 180°$. So $x° = 45°$. Since the
measure of each acute angle is 45°, the triangle is
a 45°-45°-90° triangle. **19.** $x = \sqrt{2}$; $y = 2$; Let each
acute angle measure $d°$. By the Triangle Sum
Theorem, $d° + d° + 90° = 180°$. So, $2d° = 90°$, and
$d° = 45°$. Since the measure of each acute angle
is 45°, the triangle is a 45°-45°-90° triangle.
21. $x = 7$; $y = 7\sqrt{3}$ **23.** $x = 5\sqrt{3}$; $y = 10$
25. $\tan R = \frac{8}{15} \approx 0.5333$; $\tan S = \frac{15}{8} = 1.875$
27. $\tan R = \frac{2}{3} \approx 0.6667$; $\tan S = \frac{3}{2} = 1.5$
29. 3.0777 **31.** 0.1584 **33.** 0.1763 **35.** 4.0108
37. 1.1504 **39.** 13.3 **41.** $\sin R = \frac{8}{17} \approx 0.4706$;
$\cos R = \frac{15}{17} \approx 0.8824$ **43.** $\sin R = \frac{2}{\sqrt{13}} \approx 0.5547$;
$\cos R = \frac{3}{\sqrt{13}} \approx 0.8321$ **45.** 0.2079 **47.** 0.6428

49. 0.6820   **51.** 0.9397   **53.** 0.2079   **55.** $x \approx 7.9$;
$y \approx 16.2$ **57.** 41.1°   **59.** 72.9°   **61.** 27.0°   **63.** 72.2°
65. 21.9°   **67.** 72.8°   **69.** $AB = 15$; $m\angle A \approx 53.1°$;
$m\angle B \approx 36.9°$ **71.** $XY = \sqrt{65} \approx 8.1$; $m\angle X \approx 26.4°$;
$m\angle Z \approx 63.6°$

Chapter 11 (pp. 695–696) **1.** chord   **3.** radius
5. tangent   **7.** $\odot C$, (3, 1); $\odot D$, (7, 1)   **9.** 2   **11.** 20
13. 6   **15.** 55°   **17.** 35°   **19.** 145°   **21.** 270°
23. about 11.78   **25.** 22   **27.** 14   **29.** 65
31. $x = 80$; $y = 100$ **33.** 80   **35.** 107
37. 3; (1, −3);

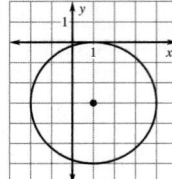

39. $x^2 + (y + 1)^2 = 400$ **41.** $(x - 3)^2 + (y - 4)^2 = 49$ **43.** $E'(5, -2)$, $F'(5, -5)$, $G'(2, -7)$, $H'(2, 0)$;
$(x, y) \rightarrow (y, -x)$ **45.** yes; 180°   **47.** yes; 120°
clockwise or counterclockwise

Teacher's Edition Index

Paper folding

Height
of a cone, 493
of a cylinder, 485
of a parallelogram, 439
of a pyramid, 491
of a rectangle, 425
slant
of a cone, 493
of a pyramid, 491
of a solid figure, 474
of a trapezoid, 446
of a triangle, 431
Hemisphere, 517
volume of, 522
Heptagon, 304
Herbst, Brett, 429
Hexagon, 304
Hexagonal pyramid, 474
Homework Check, *Occurs at the beginning of each exercise set*
Hypotenuse, 192, 257, *See also* Right triangle(s)
Hypotenuse-Leg (HL) Congruence Theorem, 257–262, 293
Hypothesis, 82

Icosahedron, 479
If-then statement, 22, 82–87, 97, 678
converse of, 136
law of detachment and, 83–86
law of syllogism and, 83–86
Image
dilation, 393
orientation, 282
transformation, 152, 393
Included angle, 242
Included side, 250–251
Indirect measurement
using similar triangles, 584–585
using trigonometric ratios, 557, 585
Inductive reasoning, 8–13, 43, 83
Inequality (Inequalities)
review of, 167, 662
triangle, 212–217, 223
Inscribed angle
definition of, 614
intercepted arc and, 613–619, 643
theorems about, 614, 615, 616

Inscribed polygon, 615
theorems about, 615, 616
Integer, operations, 663
Integration, *See* Algebra; Trigonometric ratios
Intercept, 666
Intercepted arc, 614–619
Interior, of an angle, 37
Interior angle(s), 121–135, 138–142
alternate, 121–125
of a polygon, 416–423, 461
of a quadrilateral, 305
same side, 121–135, 138–142
of a triangle, 181
Internet Links
Applications, 7, 12, 40, 65, 133, 189, 196, 247, 255, 321, 377, 391, 436, 521, 561, 599
Careers, 6, 39, 71, 80, 86, 134, 183, 275, 277, 288, 329, 340, 360, 370, 414, 516, 553, 559, 574, 609, 612, 638
Homework Help, 12, 18, 32, 57, 72, 79, 118, 133, 148, 189, 197, 211, 238, 247, 261, 270, 278, 307, 315, 336, 363, 376, 383, 397, 421, 449, 457, 497, 504, 506, 514, 515, 522, 540, 562, 567, 599, 606, 630
More Examples, 4, 10, 24, 63, 84, 90, 137, 153, 174, 180, 202, 208, 213, 244, 258, 267, 283, 317, 326, 338, 366, 387, 412, 425, 441, 486, 494, 518, 551, 570, 590, 614, 635
Intersecting bisectors, of a triangle, 275
Intersection
of chords, 620–625
of lines, 22–27, 44
of medians, 206, 208, 228–229
of planes, 21–27, 44
Intersection of Two Lines Postulate, 22
Intersection of Two Planes Postulate, 22
Inverse cosine function, 570–574, 705
Inverse sine function, 570–574, 705
Inverse tangent function, 569–570, 573, 705
Irrational number, 452, *See also* Radical expression; Square roots
pi, 452
Isosceles trapezoid, 332–333
identifying, 338–340
Isosceles triangle(s), 173
properties of, 185–190, 220–221
right, 542–547, 577

Labeling
images, 153
shapes, 154
Lateral area, 484
of a cone, 493
of a cylinder, 485–486
of a prism, 484
of a pyramid, 492
Lateral face(s)
of a prism, 484
of a pyramid, 492
Latitude, 58
Law of detachment, 83–84
Law of syllogism, 83–84
Leg(s)
adjacent, 557
of an isosceles triangle, 185
opposite, 557
of a right triangle, 192, 257
of a trapezoid, 332
Length
arc, 603–606
chord, 608, 610, 620, 622
of a line segment, 28
of the midsegment of a trapezoid, 333
Like terms, 101, 671
Line(s), 14–20, 44
coplanar, 15
graphing, 666–667
intersecting, 22–27
parallel, 108–113, 667, 680, *See also* Parallel lines
perpendicular, 108–113, *See also* Perpendicular lines
of reflection, 282
secant, 589, 592
skew, 108–113
slope of, 150–151, 299, 665, 667
of symmetry, 284–289
tangent, 589
transversals, 121–135, 161, 162
Linear equation(s)
graphing, 666–667
systems of, 583
Linear pair, 75–76
Linear Pair Postulate, 75
Line of reflection, 282
Line segment, *See* Segment(s)
Line symmetry, 284–289, 295

Teacher's Edition Index (side tab)

Additional Answers

Chapter 1

1.1 Checkpoint (p. 4)

5. 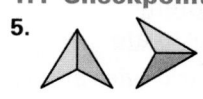 **6.**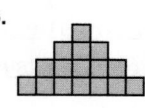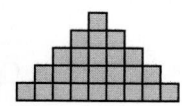

1.1 Practice and Applications (pp. 5–7)

1.  **2.** 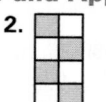 **3.** 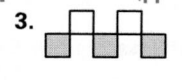 **4.**

11. 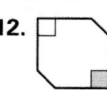 **12.** **13.** **14.**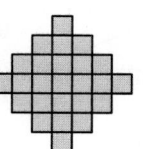

15. 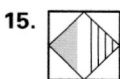 **16.**

1.2 Practice and Applications (pp. 11–13) **7.** *Sample answer:*

 A circle cannot be drawn around any of the four-sided shapes shown.

10–12. Sample answers are given. **10.** The product of 101 and a two-digit number is the four-digit number formed by writing the two digits in order twice. **11.** The square of the number with *n* digits, all of which are 1, is the number obtained by writing the digits from 1 to *n* in increasing order, then the digits from *n* − 1 to 1 in decreasing order. (This pattern does not continue forever.) **12.** Form two numbers, each with *n* digits; the first consists of *n* 3's, and the second consists of *n* − 1 3's and the last digit a 4; the product of these two numbers is the number with 2*n* digits of which the first *n* digits are 1's and the last *n* digits are 2's.

16.

	July–Aug.	Aug.–Sept.	Sept.–Oct.	Oct.–Nov.	Nov.–Dec.
Number of Days	29	30	29	30	29

1.3 Checkpoint (p. 16)

7. 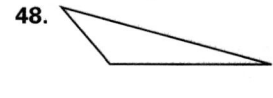 **8.**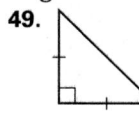

9. 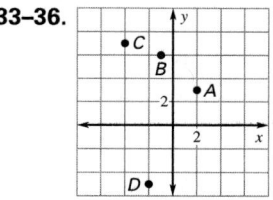 **10.**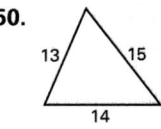

Chapter Test (p. 46)

6.

Figure Number	1	2	3	4	5
Distance (units)	8	10	12	14	16

Chapter 2

2.5 Practice and Applications (pp. 85–87) **11.** If <u>a shape is a square</u>, then [it has four sides]. **12.** If <u>two angles form a linear pair</u>, then [the angles are supplementary].

Chapter 3

3.6 Practice and Applications (pp. 147–149)
26. *Sample answer:*

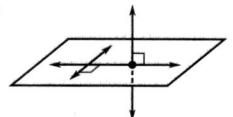

33–36.

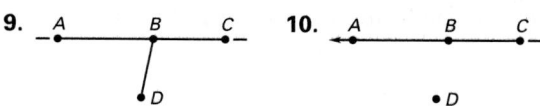

Chapter 4

4.1 Practice and Applications (p. 178)
48–53. Sample answers are given.

48. **49.** **50.**

51. **52.** **53.**

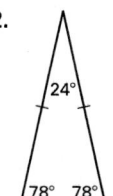 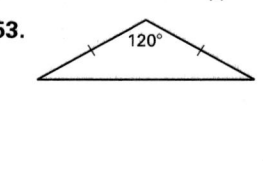

4.4 Activity (p. 191) **3.** *Sample answer:* Divide the square drawn from the hypotenuse as shown below.

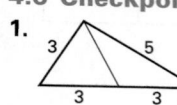

The area of the entire square is equal to the sum of the area of the small square and the areas of the four congruent right triangles.

That is, $A = 3^2 + 4\left(\frac{1}{2} \cdot 4 \cdot 1\right) = 9 + 4(2) = 17$.

4.6 Checkpoint (p. 207) **1–3.** Sample answers are given.

1.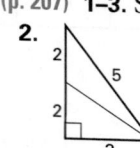

2.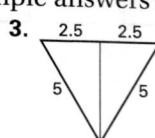

3.

Chapter 5

5.4 Practice and Applications (p. 261)
25–28. Sample answers are given.

25.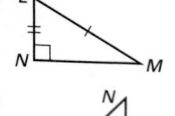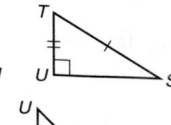

26.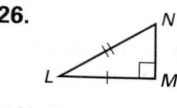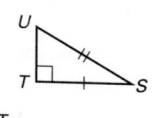

27.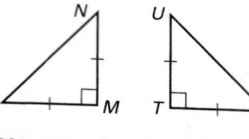

28.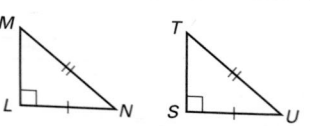

5.5 Checkpoint (p. 267) **2–3.** Check students' drawings.

2.

Statements	Reasons
1. $\overline{KJ} \cong \overline{KL}$	1. Given
2. $\angle J \cong \angle L$	2. Given
3. $\angle K \cong \angle K$	3. Reflexive Prop. of Congruence
4. $\triangle KJN \cong \triangle KLM$	4. ASA Congruence Postulate
5. $\overline{NJ} \cong \overline{ML}$	5. Corresp. parts of $\cong$ triangles are $\cong$.

3.

Statements	Reasons
1. $\angle SPR \cong \angle QRP$	1. Given
2. $\angle Q \cong \angle S$	2. Given
3. $\overline{PR} \cong \overline{RP}$	3. Reflexive Prop. of Congruence
4. $\triangle PQR \cong \triangle RSP$	4. AAS Congruence Theorem

Chapter Test (p. 296)

8.

Statements	Reasons
1. $\overline{BC} \cong \overline{DC}$	1. Given
2. $\angle B$ and $\angle D$ are right angles.	2. Given
3. $\angle B \cong \angle D$	3. Right angles are congruent.
4. $\angle ACB \cong \angle ECD$	4. Vertical Angles Theorem
5. $\triangle ABC \cong \triangle EDC$	5. ASA Congruence Postulate
6. $\overline{AB} \cong \overline{ED}$	6. Corresp. parts of $\cong$ triangles are $\cong$.

Chapter 10

10.5 Practice and Applications (pp. 566–568) **36.** *Sample answer:*

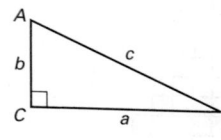

In $\triangle ABC$ shown, $\sin A = \frac{a}{c}$, $\cos A = \frac{b}{c}$, and $\tan A = \frac{a}{b}$. Then $\frac{\sin A}{\cos A} = \frac{a}{c} \div \frac{b}{c} = \frac{a}{c} \cdot \frac{c}{b} = \frac{ac}{bc} = \frac{a}{b} = \tan A$.

Skills Review Handbook

Graphing Linear Equations (p. 666)

8.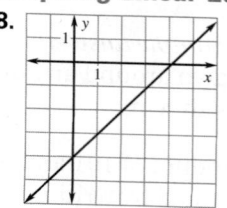

9.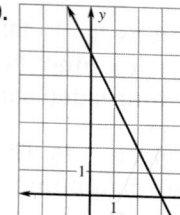

10.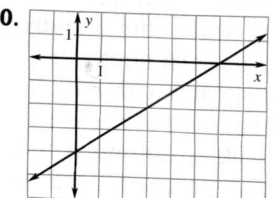

11.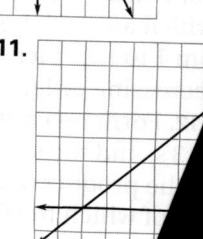

12.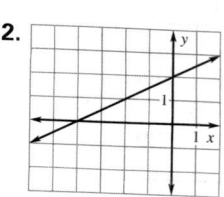